Electrostatics and Current Electricity

A Concept-Focused Text for JEE Main and Advanced, NEET, Olympiads, and Other Engineering & Medical Entrance Examinations

Each Problem is Solved with a Dedicated Problem-Solving Approach and Numerical Questions Range from Basic to JEE Advanced and Olympiad Level

Sanjay Kumar

M.Tech, PhD

Managing Director
Quanta Classes, Lucknow

Former Senior Faculty of Physics
Jain Classes, Jhansi
Bansal Classes, Madhya Pradesh
TATA Aarambh (Engineering & Medical Simplified), Lucknow

K 423 A Sector K Ashiyana Colony Lucknow

SK Education
Quanta Classes: K-423 A, Sector K, Ashiyana Colony, Lucknow (U.P.)
JEE (Main & Advanced) / NEET / Physics Olympiads / Foundation
📞 +91-9453763058
✉ spphysicsworld@gmail.com

Concepts and Problems in Physics
Title: Electrostatics and Current Electricity

Second Edition: 2025
ISBN: 9798882825088

© 2025, Sanjay Kumar
All rights reserved.

No part of this book may be reproduced, stored in a retrieval system, or transmitted in any form or by any means—electronic, mechanical, photocopying, recording, or otherwise—without the prior written permission of the author.

To the aspirants of JEE (Main & Advanced), NEET, and Physics Olympiads

With the hope that this work will inspire
a deeper interest in physics and serve as a reliable guide to its understanding.

This page intentionally left blank

Preface

This physics book is the product of more than twenty years of teaching and innovation in preparing students for JEE Main and Advanced, NEET, Physics Olympiads, and other competitive examinations. Our primary goals in writing this book are:

- To present the fundamental concepts and principles of physics that are essential for success in JEE Main, JEE Advanced, NEET, Physics Olympiads, and other engineering and medical entrance examinations.
- To maintain a balanced emphasis on both quantitative reasoning and conceptual understanding, with particular focus on areas where students commonly face difficulties.
- To systematically develop students' problem-solving skills and enhance their confidence.
- To engage students by incorporating real-world examples that connect physics to everyday experiences.

What's New?

This edition incorporates substantial improvements in structure, content, and teaching methodology. Many elements are newly developed and presented with a strong emphasis on learner-focused explanations. Below are five key highlights:

1. Each concept is explained in accessible, student-friendly language, followed by various types of solved problems. Every solution is structured around a problem-solving approach and includes explanatory discussion.
2. Checkpoint questions are strategically placed within sections, encouraging students to pause and test their understanding before moving forward. Thorough, fully worked-out solutions with a clear problem-solving approach to these checkpoint questions are provided at the end of each chapter, supporting effective self-assessment and independent learning.
3. Special care has been taken with tricky and conceptually dense topics to ensure they can be mastered with clarity and enjoyment.
4. After the theoretical discussions, each chapter concludes with a collection of miscellaneous solved problems that integrate multiple concepts.
5. To further enhance students' understanding, a comprehensive set of conceptual questions, practice problems, and multiple-choice questions—including problems from previous years' JEE Main and Advanced examinations—is provided at the end of each chapter. These exercises span various difficulty levels, denoted by the number of dots: straightforward (basic-level) problems by a single dot (•), intermediate problems by two dots (••), and challenging (advanced-level) problems by three dots (• • •). Complete, step-by-step solutions for all end-of-chapter exercises, including hints where required and following a clear problem-solving approach, are provided at the end of each chapter to support independent practice and effective self-assessment.

I hope this book serves as a rigorous and insightful guide for students preparing for JEE Main, JEE Advanced, NEET, Physics Olympiads, and other equivalent engineering and medical entrance examinations. Your feedback, suggestions, and corrections are most welcome and will be gratefully acknowledged in future editions.

Sanjay Kumar
Lucknow, July 2025

Online Physics Classes
by
Dr. Sanjay Kumar

JEE (Main & Advanced) / NEET / Physics Olympiads / Foundation (IX - XII)
Quanta Classes: K 423 A Sector K Ashiyana Colony Kanpur Road Lucknow
Mo. +919453763058
Email: spphysicsworld@gmail.com

Contents

Preface   vii

1 Electric Charge and Field   1
1.1 Electric Charge- a Property of Matter 1
    1.1.1 Electric Fluid Model 1
    1.1.2 Atomic Model of Charge 1
1.2 Properties of Electric Charge 2
    1.2.1 Conservation of Charge 2
    1.2.2 Quantization of Charge 3
        1.2.2.1 Number of Electrons in 1C of Charge 3
    1.2.3 Charge is Always Associated with Mass 4
        1.2.3.1 Effect of Charge on the Mass of a Given Object 4
    1.2.4 Charge is Transferable 4
    1.2.5 An Accelerated Charge Always Radiates Energy 4
1.3 Quarks 5
1.4 Grounding a Conductor 5
    1.4.1 Check Point 1 5
1.5 Classification of Solids 6
    1.5.1 Conductors 6
    1.5.2 Insulators 6
    1.5.3 Semiconductors 6
    1.5.4 Superconductors 6
1.6 Methods of Charging 6
    1.6.1 Charging by Friction 6
        1.6.1.1 Friction is Used to Charge Insulators 6
        1.6.1.2 Charging Conductors by Friction is Very Difficult 6
        1.6.1.3 Triboelectric Series 6
    1.6.2 Charging by Direct Contact 7
    1.6.3 Charging by Induction 7
    1.6.4 Charging by Field Emission 8
    1.6.5 Charging by Thermionic Emission 8
    1.6.6 Charging by Photoelectric Effect 8
1.7 Coulomb's Law 8
        1.7.0.1 Value of k in CGS Units 9
    1.7.1 Experimental Verification of Coulomb's Law 9
    1.7.2 Vector Form of Coulomb's Law 10
1.8 Superposition Principle 11
1.9 Coulomb's Law in a Dielectric Medium 11
        1.9.0.1 Key Points Regarding Coulomb's Law 12
1.10 Equilibrium of Charged Particles 17
    1.10.1 Equilibrium of Three Point Charges 17
    1.10.2 Equilibrium of Symmetric Geometrical Point Charged Systems 17
    1.10.3 Equilibrium of Suspended Point Charge System 18
1.11 Check Point 2 24
1.12 The Electric Field 26
1.13 The Electric Field of Point Charges 27
    1.13.1 The Electric Field Due to a Single Point Charge 27
    1.13.2 Electric Field Due To a Group of Point Charges (Superposition of Electrostatic Fields) 27
1.14 Electric Dipole 32
    1.14.1 Dipole Moment 32
    1.14.2 Dipole Moment of a System of Discrete Charges 33
    1.14.3 Induced Dipole Moment 33
    1.14.4 Electric Field at any Point due to an Electric Dipole 33
        1.14.4.1 Electric field at an Axial Point, i.e., in End on Position 33
        1.14.4.2 At a Point on its Perpendicular Bisector, i.e. in Broadside on Position 33
        1.14.4.3 At any General Point 34
1.15 Check Point 3 35

- 1.16 Electric Field of a Continuous Charge Distribution 35
  - 1.16.1 Electric Field Due to a Uniformly Charged Rod of Finite Length at a Perpendicular Distance y 37
    - 1.16.1.1 Electric Field Due to an Infinite Line Charge 38
    - 1.16.1.2 Electric Field of a Finite Line Charge at a Point on its Extension 38
    - 1.16.1.3 Approximating Equations 1.62 and 1.63 on the Symmetry Plane 39
  - 1.16.2 Electric Field at the Centre of a Charged Circular Arc 41
  - 1.16.3 Electric Field at a Point on the Axis of a Thin Charged Ring 42
    - 1.16.3.1 Motion of a Negatively Charged Particle Close to the Centre of the Positively Charged Ring on it's Axis 43
  - 1.16.4 Electric Field on the Axis of a Charged Disc 44
- 1.17 The Shape of Lightning Rods 44
- 1.18 Check Point 4 47
- 1.19 Electric Field Lines 48
- 1.20 Theorem on Circulation of Vector $\vec{E}$ 48
  - 1.20.1 Deduction of Pattern of Field Lines 49
  - 1.20.2 Properties of Electric Field Lines 50
- 1.21 Check Point 5 50
- 1.22 Action of the Electric Field on Charges 51
  - 1.22.1 Motion of a charged Particle in a Uniform Electric Field 51
    - 1.22.1.1 Motion of a Charged Particle Along an Electric Field 52
    - 1.22.1.2 Motion of a Charged Particle Perpendicular to an Electric Field 52
- 1.23 Electric Dipole in an External Electric Field 53
  - 1.23.1 Electric force and torque on electric dipole in a uniform external electric field 53
  - 1.23.2 The Angular Acceleration and Time period of a Dipole in an External Uniform Electric Field 54
  - 1.23.3 A Dipole in a Non-uniform Field 54
- 1.24 Earnshaw's Theorem 57
- 1.25 Check Point 6 58
- 1.26 Conductors in Electrostatic Equilibrium 58
- 1.27 Limitations of Coulomb's Law for a Relativistically Moving Charge 59
  - 1.27.1 Retarded Field 60
- 1.28 Check Point 7 60
- 1.29 Solid Angle 61
- 1.30 Electric Flux 63
    - 1.30.0.1 The Electric Flux Through a Closed Surface 64
- 1.31 Check Point 8 65
- 1.32 Gauss's Law 66
  - 1.32.1 General Integral Form of Gauss's Law 67
  - 1.32.2 Differential Form of Gauss's Law (Optional) 70
- 1.33 Applications of Gauss's law 71
  - 1.33.1 Gauss's Law: Problem Solving Approach 71
  - 1.33.2 Cylindrical Symmetry 73
  - 1.33.3 Spherical Symmetry 76
- 1.34 Check Point 9 79
- 1.35 Conductors in Electrostatic Equilibrium 81
  - 1.35.1 Field in a Substance 81
    - 1.35.1.1 Micro- and Macroscopic Fields 81
    - 1.35.1.2 The Influence of a Substance on a Field 82
  - 1.35.2 Fields Inside and Outside a Conductor 82
    - 1.35.2.1 Inside a Conductor $E = 0$ 82
    - 1.35.2.2 The Field Near a Conductor Surface 83
  - 1.35.3 Mechanical Pressure (or Surface Density of Force) on the Surface of a Charged Conductor 84
  - 1.35.4 Faraday Cage Protection 85
- 1.36 Check Point 10 89
- 1.37 Questions and Exercises 89
  - 1.37.1 Conceptual Questions 89
  - 1.37.2 Problems 90
  - 1.37.3 Multiple Choice Questions 93
    - 1.37.3.1 Level 1 93
    - 1.37.3.2 Level 2 97
    - 1.37.3.3 Level 3 98
    - 1.37.3.4 Level 4 (Previous Years JEE Main & Advanced Questions) 99
- 1.38 Answer Keys and Solutions 109
  - 1.38.1 Check Point 1 109
  - 1.38.2 Check Point 2 110

    1.38.3 Check Point 3 114
    1.38.4 Check Point 4 115
    1.38.5 Check Point 5 116
    1.38.6 Check Point 6 117
    1.38.7 Check Point 7 117
    1.38.8 Check Point 8 120
    1.38.9 Check Point 9 121
    1.38.10 Check Point 10 124
    1.38.11 Conceptual Questions 125
    1.38.12 Problems 126
    1.38.13 Multiple Choice Assignments 140
        1.38.13.1 Level 1 140
        1.38.13.2 Level 2 141
        1.38.13.3 Level 3 141
        1.38.13.4 Level 4 141

2 Electric Potential   143
2.1 Electric Potential Energy and Potential Difference in Fields 143
    2.1.1 Potential Difference in a Uniform Field 146
    2.1.2 Electrostatic Potential Energy Difference and Potential Energy of a System of Two Point Charges 147
2.2 Check Point 1 149
2.3 Electric Potential Energy of a System of Point Charges: Self Electric Energy 149
    2.3.1 Potential Energy of a System of Two Charges in an External Field 152
    2.3.2 Potential Energy of a Dipole in a Uniform Electric Field 152
    2.3.3 The Electric Potential Due to a Point Charge 153
2.4 Equipotential Surfaces 154
    2.4.1 Various Equipotential Surfaces 154
        2.4.1.1 Equipotential Surfaces For a Point Charge 154
        2.4.1.2 Equipotential Surfaces Due to an Electric Dipole 154
        2.4.1.3 Equipotential Surfaces Due to Two Identical Positive Charges: 155
        2.4.1.4 Equipotential Surfaces of a Charged Wire of Infinite Length 155
        2.4.1.5 Uniformly Charged Plane Surface of Infinite Dimensions 155
    2.4.2 Properties of Equipotential surface 156
2.5 Calculating the Field From the Potential: Potential Gradient 156
2.6 Electric Potential For a System of Charges 158
    2.6.1 Electric Potential at a Point $P(r, \theta)$ Due to an Electric Dipole 161
2.7 Check Point 2 162
    2.7.1 Electric Potential due to a Line charge of Finite Length 164
    2.7.2 Potential Due to an Infinite Charged Wire 164
    2.7.3 Electric Potential due to a Charged Thin Plane Sheet 164
    2.7.4 Electric Potential Due to a Charged Ring 165
    2.7.5 Electric Potential Due to a Charged Disc at a Point on it's Geometric Axis 166
    2.7.6 Electric Potential Due to a Charged Conducting Shell or Sphere 168
    2.7.7 Electric Potential Due to a Non-Conducting Charged Sphere 168
    2.7.8 Potential on the Edge of a Uniformly Charged Disc: 169
2.8 Equipotentials and Conductors 170
2.9 Connected Conducting Spheres 173
2.10 Earthing 174
2.11 Corona Discharge 174
    2.11.1 Electric Potential Energy for Continuous Charge System 175
        2.11.1.1 Self (or Internal) Energy of a Uniformly Charged Conducting Spherical Shell or Conducting Sphere 175
        2.11.1.2 Self Energy of a Uniformly Charged Non Conducting Solid Sphere 175
2.12 The Van de Graaff Generator 176
2.13 The Millikan Oil-Drop Experiment 177
2.14 Check Point 3 178
2.15 Questions and Exercises 178
    2.15.1 Conceptual Questions 178
    2.15.2 Problems 181
    2.15.3 Multiple Choice Questions 186
        2.15.3.1 Level 2 188
        2.15.3.2 Level 3 189
        2.15.3.3 Level 4 (Previous Years JEE Main and Advanced Questions) 190
2.16 Answer Keys and Solutions 197
    2.16.1 Check Point 1 197

- 2.16.2 Check Point 2 198
- 2.16.3 Check Point 3 200
- 2.16.4 Conceptual Questions 201
- 2.16.5 Problems 205
- 2.16.6 Multiple Choice Assignments 212
  - 2.16.6.1 Level 1 212
  - 2.16.6.2 Level 2 212
  - 2.16.6.3 Level 3 212
  - 2.16.6.4 Level 4 212

3 Capacitance and Dielectrics — 215
- 3.1 Capacitor 215
- 3.2 Capacitance of an Isolated Spherical Conductor 215
  - 3.2.1 q Versus V Graph 216
- 3.3 Capacitance of System of Two Conductors 216
- 3.4 Parallel-plate Capacitor 217
  - 3.4.1 Parallel Plate Capacitor with With Unequal Plate Charges 218
- 3.5 Electric Force Between the Plates of a Parallel Plate Capacitor .. 220
  - 3.5.1 Spherical Capacitor 220
  - 3.5.2 The Cylindrical Capacitor 220
    - 3.5.2.1 Capacitance of two long parallel wires 221
    - 3.5.2.2 Circuit Symbols of Capacitors 222
- 3.6 Check Point 1 222
- 3.7 Capacitors in Series and in Parallel 223
  - 3.7.1 Capacitors in Series 223
  - 3.7.2 Capacitors in Parallel 224
- 3.8 An Infinite Network 229
- 3.9 Wheatstone's Balanced Capacitance Bridge 230
- 3.10 Energy Stored in the Capacitor's Electric Field 234
  - 3.10.1 Electric Field Energy Density 236
    - 3.10.1.1 Alternative Method 236
  - 3.10.2 Redistribution of Charges 237
    - 3.10.2.1 Energy Loss in Redistribution of Charges 238
- 3.11 Dielectrics 241
  - 3.11.1 Molecular View of a Dielectric 242
    - 3.11.1.1 Mathematical Description of Dielectric Constant 243
  - 3.11.2 Magnitude of the Bound or Induced Charge 244
  - 3.11.3 Electric Displacement Vector 244
  - 3.11.4 Dielectric Breakdown and Dielectric Strength 244
  - 3.11.5 Change in Various Factors when a Dielectric Slab is Introduced Between the Plates of a Capacitor 245
    - 3.11.5.1 When the Battery is Disconnected 245
    - 3.11.5.2 When the Battery Remains Connected 246
- 3.12 Guass's Law in Dielectrics 247
  - 3.12.1 Capacitor with Partially Filled Dielectric Medium 248
  - 3.12.2 Capacitor Filled with a Variable Dielectric Medium 250
- 3.13 Effects of Nonuniform Electric Fields on Dielectrics 254
  - 3.13.1 Energy Method for Calculating the Force Between Dielectrics and Conductors 254
    - 3.13.1.1 Calculation of Force on a Dielectric Slab in a Fringing Field of a Capacitor 255
- 3.14 Check Point 2 258
- 3.15 Types of Commercial Capacitors 259
- 3.16 Solved Examples 259
- 3.17 Questions and Exercises 263
  - 3.17.1 Conceptual Questions 263
- 3.18 Problems 264
  - 3.18.1 Multiple Choice Questions 268
    - 3.18.1.1 Level 1 268
    - 3.18.1.2 Level 2 273
    - 3.18.1.3 Level 3 275
    - 3.18.1.4 Level 4 (Previous Years JEE Main & Advanced Questions) .. 276
- 3.19 Answer Keys and Solutions 286
  - 3.19.1 Check Point 1 286
  - 3.19.2 Check Point 2 287
  - 3.19.3 Conceptual Questions 289
  - 3.19.4 Problems 291

- 3.19.5 Multiple Choice Assignments 297
  - 3.19.5.1 Level 1 297
  - 3.19.5.2 Level 2 297
  - 3.19.5.3 Level 3 297
  - 3.19.5.4 Level 4 298

4 Current Electricity and DC Circuits   299

- 4.1 Electric Current 299
  - 4.1.1 Calculating Charge From Electric Current 299
  - 4.1.2 Rate of Flow of Electrons Through a Given Cross-Section of Conductor 300
  - 4.1.3 Current due to Circular Motion of a Point Charge 300
  - 4.1.4 The Direction of Electric Current 301
  - 4.1.5 Classification of Materials According to Conductivity 301
    - 4.1.5.1 Conductors and Insulators 301
    - 4.1.5.2 Semiconductor 301
  - 4.1.6 Check Point 1 301
- 4.2 Microscopic Model of Current 301
  - 4.2.1 In Absence of External Electric Field 301
  - 4.2.2 In Presence of External Electric Field 302
  - 4.2.3 Electric Current in Terms of Drift Velocity 303
    - 4.2.3.1 Current Remains Same at Each Point in a Single Branched Closed Loop 304
  - 4.2.4 Current Density 304
  - 4.2.5 Calculation of drift velocity 306
  - 4.2.6 Check Point 2 306
- 4.3 Mobility, Conductivity and Resistivity 307
  - 4.3.1 Mobility 307
  - 4.3.2 Conductivity 307
  - 4.3.3 Resistivity (or Specific Resistance) 309
  - 4.3.4 Thermistor 310
    - 4.3.4.1 Key Points 310
- 4.4 Electrical Resistance, Conductance and Ohm's Law 311
  - 4.4.1 Electrical Resistance and Conductance 311
  - 4.4.2 Dependence of Resistance on Temperature 312
  - 4.4.3 Effects of Change in Geometry on Resistance of the Conductor 315
  - 4.4.4 Effect of Percentage Change in Length of Wire 315
  - 4.4.5 The Radial Resistance of a Hollow Spherical Conductor 316
  - 4.4.6 The Radial Resistance of a Coaxial Cable 316
  - 4.4.7 Linear Resistance of a Hollow Cylinder 317
  - 4.4.8 Resistance of a Truncated Cone 317
  - 4.4.9 Ohm's Law 318
- 4.5 Charge at a Junction 319
- 4.6 Colour Code for Carbon Resistors 319
- 4.7 Specific Use of Conducting Materials 320
  - 4.7.1 Check Point 3 320
- 4.8 Power in Electric Circuits and Joule's law 321
  - 4.8.1 Check Point 4 322
- 4.9 Electric Cells and Batteries 322
- 4.10 Symbols for Circuit Components 323
- 4.11 Resistors in Series and Parallel 323
  - 4.11.1 Resistors in Series 323
  - 4.11.2 Resistors in Parallel 324
- 4.12 Voltage and Power Rating 325
  - 4.12.1 Check Point 5 326
- 4.13 Electromotive Force (EMF) 327
  - 4.13.1 Internal Resistance (r) 328
  - 4.13.2 Potential Changes Around a Circuit 329
  - 4.13.3 Power Supplied by an Ideal emf Source 329
- 4.14 Kirchhoff's Rules for Resistive Circuits 330
  - 4.14.1 Kirchhoff's Rules for Circuits having Resistors and Battery 330
- 4.15 Applications of Kirchhoff's Rules 331
  - 4.15.1 Resistors in Series 333
  - 4.15.2 Resistors in Parallel 333
- 4.16 Terminal Potential Difference 336
  - 4.16.1 When Cell Is Discharging 336

- 4.16.1.1 When Cell is in Open circuit 337
- 4.16.1.2 When Cell is Short Circuited 337
- 4.16.2 When Cell is Charging 337
- 4.17 Maximum Power Transfer Theorem 337
- 4.18 Division of Current in Resistors Joined in Parallel 338
- 4.19 Division of Voltage in Resistors Joined in Series 338
- 4.20 Relative Potential 338
- 4.21 Grouping of Cells 339
  - 4.21.1 Series Grouping 339
  - 4.21.2 Parallel Grouping 340
  - 4.21.3 Mixed Grouping 342
- 4.22 The Wheatstone Bridge 342
- 4.23 The meter Bridge 346
  - 4.23.1 Check Point 6 347
- 4.24 Electrical Meters 349
  - 4.24.1 The Galvanometer 349
  - 4.24.2 Ammeter Design 349
    - 4.24.2.1 Conversion of Galvanometer to ammeter 350
    - 4.24.2.2 Maximum Current Reading of an Ammeter 350
    - 4.24.2.3 Modification of Ammeter to Obtain Other Range 350
  - 4.24.3 Voltmeter Design 352
  - 4.24.4 Check Point 7 354
- 4.25 Stretched-Wire Potentiometer 355
  - 4.25.1 Comparison of Emf's of Two Batteries 356
  - 4.25.2 Measurement of Internal Resistance of a Battery 357
  - 4.25.3 Check Point 8 358
- 4.26 An Infinite Network 358
- 4.27 Symmetrical Electric Circuits 360
  - 4.27.1 Check Point 10 364
- 4.28 Star and Delta Networks 365
  - 4.28.1 Delta to Star Conversion 365
  - 4.28.2 Star to Delta Conversion 365
  - 4.28.3 Check Point 11 369
- 4.29 Kirchhoff's Rules for Capacitive Circuits 369
- 4.30 RC Circuits 370
  - 4.30.1 Charging 370
  - 4.30.2 $i(t)$ for a Charging Capacitor 371
    - 4.30.2.1 Time constant 371
  - 4.30.3 Discharging 371
    - 4.30.3.1 Discharging Current 372
  - 4.30.4 Check Point 12 378
- 4.31 Solved Problems 378
- 4.32 Solved Objective Questions 388
- 4.33 Questions and Problems 398
  - 4.33.1 Conceptual Questions 398
  - 4.33.2 Problems 398
- 4.34 Multiple Choice Assignments 401
  - 4.34.1 Level 1 401
  - 4.34.2 Level 2 405
  - 4.34.3 Level 3 408
  - 4.34.4 Level 4 (Previous Years JEE Main & Advanced Questions) 411
- 4.35 Answer Keys and Solutions 433
  - 4.35.1 Check Point 1 433
  - 4.35.2 Check Point 2 434
  - 4.35.3 Check Point 3 434
  - 4.35.4 Check Point 4 435
  - 4.35.5 Check Point 5 436
  - 4.35.6 Check Point 6 438
  - 4.35.7 Check Point 7 442
  - 4.35.8 Check Point 8 443
  - 4.35.9 Check Point 9 444
  - 4.35.10 Check Point 10 444
  - 4.35.11 Check Point 11 444
  - 4.35.12 Check Point 12 445

- 4.35.13 Conceptual Questions 445
- 4.35.14 Problems 447
- 4.35.15 Multiple Choice Assignments 453
- 4.35.16 Level 1 453
- 4.35.17 Level 2 453
- 4.35.18 Level 3 453
- 4.35.19 Level 4 453

This page intentionally left blank

Chapter 1
Electric Charge and Field

Electromagnetism governs virtually all that we see of the physical world. Electromagnetic forces control the structure of atoms and all materials. In this chapter, we introduce electric charge, a property of atomic constituents, and we discuss the fundamental law of the interaction of two charges at rest, Coulomb's law. This force law is as fundamental as the universal law of gravitation. The interaction between charges has the same space dependence as gravitation. To discuss the action at distance, we further introduce the concept of an electric field in terms of Faraday's electric field lines. Gauss' law, which relates electric fields to charges, is developed in terms of these electric field lines. Finally, we apply Gauss' law to a variety of physical situations.

1.1 Electric Charge- a Property of Matter

You have probably seen that on combing hair, on a dry day, hair is attracted to the comb. When we rub objects together such as wool against amber or silk against glass, we find that after rubbing, the objects acquire a property due to which they attract to each other; a silk cloth is attracted to a glass rod that it was rubbed against. The, rubbed objects can also attract other objects. For example, small bits of paper are attracted to a comb that had been rubbed through hair.

When objects behave in this way, they are said to be electrified (from the Greek word elektron meaning amber) or electrically charged.

You can easily electrify your body by vigorously rubbing your shoes on a wool carpet. Evidence of the electric charge on your body can be detected by touching a metal doorknob. Under the right conditions, you will feel a shock when you touch the metal doorknob. (Experiments such as these work best on a dry day because an excessive amount of moisture in the air can cause any charge you build up to "leak" from your body to the Earth.) There are two types of electric charge, as the following simple experiments show. Suppose a hard rubber rod that has been vigorously rubbed on fur is suspended by a string as shown in Figure 1.1. When a glass rod that has been rubbed on silk is brought near the rubber rod, the two attract each other (Fig. 1.1 a). On the other hand, if two charged rubber rods (or two charged glass rods) are brought near each other as shown in Figure 1.1b, the two repel each other. This observation shows that the rubber and glass have two different types of charge on them. Each type of charge repels the same type but attracts the opposite type. That is: **unlike charges attract; like charges repel**.

On the basis of above discussion, we can define electric charge as *a fundamental property of matter that gives rise to electric (more general term "electromagnetic") interactions*. Like mass, it is also a scalar quantity.

1.1.1 Electric Fluid Model

According to Benjamin Franklin all objects are full of an electric fluid. When you bring two objects (for example – a glass rod and a silk cloth) in contact, some electric fluid may transfer from one object to the other. As a result of which, one object gets a surplus of electric fluid and the other has a deficit of electric fluid. Franklin called the object with a surplus of electric fluid as positive or plus, and the object with a deficit of electric fluid as

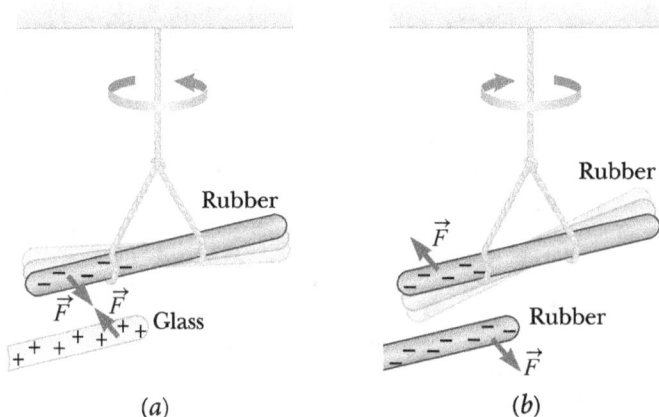

Figure 1.1: The electric force between (a) oppositely charged objects and (b) like-charged objects.

negative or minus. According to Franklin, initially, both of them (the glass rod and the silk cloth) have some amount of electric fluid, and after they are rubbed together, some of the electric fluid is transferred from the silk to the glass. The silk then has a deficit of electric fluid, so Franklin said the silk is negative. The glass has a surplus of electric fluid, so he said it is positive.

Franklin's choice was arbitrary in this model; he assumed the electric fluid was transferred from the silk to the glass. There was no way for Franklin to know whether that was true. Because there was no experimental evidence to determine which way the electric fluid flowed, Franklin could have imagined that electric fluid was transferred from the glass to the silk. Subsequent scientists have kept his arbitrary choice in their model of electric charge.

It was only long afterward, with the discovery of electrons and protons, that electrons were found to be attracted to a positively charged glass rod while protons were repelled. Thus by convention electrons have a negative charge and protons have a positive charge. The magnitudes of charge on both the particles are equal.

1.1.2 Atomic Model of Charge

Today we have experimental evidence supporting the model that all objects are made up of atoms, and atoms are made up of three types of particles—neutrons, protons, and electrons. Neutrons are neutral, protons are positively charged and electrons are negatively charged particles. The neutrons and protons are tightly packed into the central region of the atom known as the nucleus. The electrons move rapidly outside the nucleus, forming a cloud. According to this model, when two objects are placed in contact, the electrons can move from atoms of the object in which these are loosely bound to atoms of the object in which electrons are relatively tightly bound. The protons and neutrons are tightly bound in the nucleus, and so they are not transferred.

Therefore, when glass is rubbed against silk, electrons are transferred from the glass to the silk. As a result of which, the glass has a surplus of protons and the silk has a deficit of protons (Fig.1.2). In terms of transferred electrons, we can also say that, the glass has a deficit of electrons and the silk has a surplus of electrons.

As electrons and protons are particles of matter. Their motion is governed by Newton's laws. Electrons can move from one object to another when the objects are in contact, but neither electrons nor protons can jump through the air from one object to another. So, an object can not be charged by simply bringing it closer to a charged object.

Charge is represented by the symbol q (or sometimes Q). A macroscopic object, such as a plastic rod, has charge -
$$q = N_p e - N_e e = (N_p - N_e)e \quad (1.1)$$
where N_p and N_e are the number of protons and electrons contained in the object. An object with an equal number of protons and electrons has no net charge (i.e., $q = 0$) and is said to be electrically neutral.

Thus, neutral does not mean "no charges" but, instead, no net charge.

If $N_e < N_p$, then from Eq.1.1, $q > 0$, i.e., an object is positively charged. Similarly, if $N_e > N_p$, then from Eq.1.1, $q < 0$, i.e., an object is negatively charged.

In practice, objects acquire a positive charge not by gaining protons, but by losing electrons. Protons are extremely tightly bound within the nucleus and cannot be added to or removed from atoms. Electrons, on the other hand, are bound rather loosely and can be removed without great difficulty. The process of removing an electron from the electron cloud of an atom is called ionization. An atom that is missing an electron is called a positive ion. Its net charge is $q = +e$.

Some atoms can accommodate an extra electron and thus become a negative ion with net charge $q = -e$. A saltwater solution is a good example. When table salt (the chemical sodium chloride, NaCl) dissolves, it separates into positive sodium ions Na$^+$ and negative chlorine ions Cl$^-$.

In ground state, Na atom has 11 protons and 11 electrons, i.e., $N_p = 11$ and $N_e = 11$. So, net charge on Na atom in ground state-
$$q_{Na} = (N_p - N_e)e = (11 - 11)e = 0$$
Similarly, in ground state of Cl atom, $N_p = 17$, $N_e = 17$, therefore net charge on it in ground state-
$$q_{Cl} = (N_p - N_e)e = (17 - 17)e = 0$$
When 1 electron is transferred from Na to Cl, Na ionizes to Na$^+$ and Cl ionizes to Cl$^-$.

In Na$^+$, we have $N_p = 11$, $N_e = 10$, therefore net charge on Na$^+$:
$$q_{Na^+} = (N_p - N_e)e = (11 - 10)e = +e$$
In Cl$^-$, we have $N_p = 17$, $N_e = 18$, therefore net charge on Cl$^-$:
$$q_{Cl^-} = (N_p - N_e)e = (17 - 18)e = -e$$

☞ Particles that exert electric forces are said to have an electric charge; particles that do not exert electric forces are said to have no electric charge. Thus, charge of a material body or particle is the property (acquired or natural) due to which it produces and experiences electric force. Some of naturally occurring charged particles are electrons, protons, α-particles etc.

So, Electric charge is thought of as the source of electric force, just as mass is the source of gravitational force.

Unit of Charge: Charge is a derived physical quantity and is measured in "coulomb (C)" in SI unit. 1 C is a very large amount of charge, in practice, we use smaller units milli coulomb (mC) (1 mC = 10^{-3}C), micro coulomb (μC) [1 μC = 10^{-6}C], nano coulomb (nC) [1 nC = 10^{-9}C] and pico coulomb (pC) [1 pC = 10^{-12}C] etc.

CGS unit of charge is "electrostatic unit (esu)."
1C = 2997924579.9996 esu $\approx 3 \times 10^9$ esu of charge.
Dimensional formula of charge = $[M^0 L^0 T^1 A^1]$
Magnitude of the smallest known charge is $e = 1.6 \times 10^{-19} C$ (magnitude of charge of one electron or proton).

Note: Practically, mass of a body is always positive * whereas a charge can either be positive or negative. Particles are the sub-

---
*Theoretically, there is also a concept of $-ve$ mass. If we bring a negative mass near a positive mass, there is a gravitational repulsion between them.

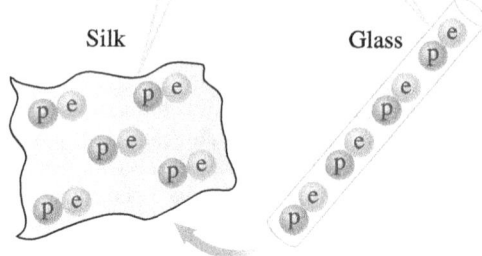

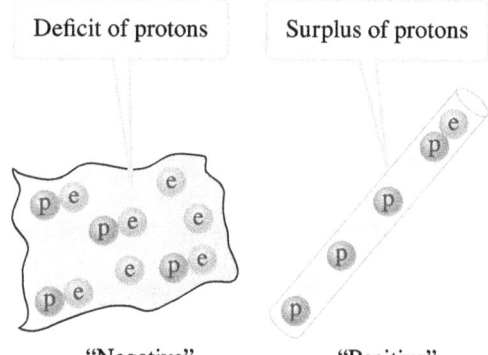

Figure 1.2: Electric charge model

stance and charge happens to be one of their intrinsic properties, just as mass is.

1.2 Properties of Electric Charge

Charging a body implies transfer of charge (electrons) from one body to another. Positively charged body means loss of electrons i.e. deficiency of electrons. Negatively charged body means excess of electrons. Since each electron has mass $m_e \approx 9.1 \times 10^{-31}$ kg. So, excess of electron means excess of mass. Therefore, the mass of a negatively charged body will always be greater than the original mass of the neutral body. A positively charged body means loss of electrons and hence loss of mass.

If n electrons are transferred from a body A to body B, then the body A losses its mass by an amount of $\Delta m = n m_e$, whereas body B gains the same amount of mass. Here m_e is the mass of electron.

1.2.1 Conservation of Charge

Experiments show that- "*In an isolated system, total charge (sum of positive and negative charges) remains constant whatever change takes place in that system*". This law is called **law of conservation of charge**. This law was first put forward by Benjamin Franklin which holds both for large-scale charged bodies and for atoms, nuclei, and elementary particles. No exceptions have ever been found. Thus, law of conservation of charge is a fundamental physical law like- law of conservation of energy, law of conservation of momentum (both linear and angular momentum).

Example of charge conservation occurs when an electron e^- (charge = $-e$) and its antiparticle, the positron e^+ (charge = $+e$), undergo an annihilation process, transforming into two gamma rays (high-energy light):
$$e^- + e^+ \to \gamma + \gamma \quad \text{(annihilation)} \quad (1.2)$$
In pair production, the converse of annihilation, charge is also

conserved. In this process a gamma ray transforms into an electron and a positron:

$$2\gamma \rightarrow e^- + e^+ \quad \text{(pair production)} \quad (1.3)$$

In above Equations 1.2 and 1.3, we use two gamma photons in pair production and annihilation to ensure conservation of both energy and momentum. A single photon cannot satisfy both laws simultaneously in these processes.

Note: In above subsection we have seen that charge is conserved for an isolated system. So, unlike mass, the charge on a body does not change, whatever be its speed, however specific charge (q/m) depends on speed as mass depends on speed:

$$m = \frac{m_0}{\sqrt{1 - \frac{v^2}{c^2}}}$$

here m is called the dynamic mass and m_0 the rest mass.

EXAMPLE 1. Sixty four conducting drops each of radius 0.02 m and each carrying a charge of 5 μC are combined to form a bigger drop at constant temperature. Find the ratio of surface charge density of bigger drop to the smaller drop.

APPROACH Since smaller drops are combined at a constant temperature, there will not be any thermal expansion or contraction, and hence the principle of conservation of volume is applicable. By applying it, we can determine a relation between the radii of a smaller drop and the final bigger drop. Charge is also a conserved quantity; therefore, by applying the charge conservation principle, we can find a relation between the charge on a smaller drop and the final bigger drop. Now, the surface charge densities of smaller and bigger drops can be calculated by applying the relation:

$$\sigma = \frac{\text{total surface charge}}{\text{surface area}}$$

After finding the surface charge densities, we can calculate their ratio.

SOLUTION Let R denotes the radius of the combined drop, and r denotes the radius of the smaller drop. Conservation of volume gives:

$$\frac{4}{3}\pi R^3 = 64 \times \frac{4}{3}\pi r^3 \implies R = 4r$$

If q is the charge on each smaller drop, and Q is the charge on newly formed bigger drop, then by principle of conservation of charge, we can write:

$$Q = 64q;$$

If σ_{bigger} and σ_{smaller} denote the surface charge densities of bigger and smaller drops, then ratio of surface charge densities on the bigger and smaller drops

$$\frac{\sigma_{\text{bigger}}}{\sigma_{\text{smaller}}} = \frac{\frac{Q}{4\pi R^2}}{\frac{q}{4\pi r^2}} = \frac{Q}{q} \cdot \frac{r^2}{R^2} = 64 \frac{r^2}{16r^2} = 4$$

$$\implies \sigma_{\text{bigger}} : \sigma_{\text{smaller}} = 4 : 1$$

1.2.2 Quantization of Charge

First, let's talk about what quantization means. Quantization is like putting things into specific groups or amounts instead of letting them be in between. It's when something can only be a certain amount, not anything in the middle. Imagine it's like having stairs instead of a smooth ramp.

Suppose, we have a quantity Q and let it is specified in terms of fundamental unit q, then according to quantization rule,

$$Q = Nq \quad (1.4)$$

where N is some integer which shows excess or deficiency of q. For example, if n is the number of excess of q, then $N = +n$ and if it is the deficiency of q, then $N = -n$. So, in terms of n Eq. 1.4 can also be written as:

$$Q = \pm q \quad (1.5)$$

Here, +ve sign shows the excess and −ve sign shows the deficiency of q.

Now, come to the charge. Quantization of charge refers to the observation that electric charge exists only in discrete or quantized amounts. In other words, charge is always found in specific multiples of a fundamental unit of charge. The fundamental unit of charge is the elementary charge, denoted as "e", which is approximately equal to 1.602×10^{-19} coulombs.

According to experimental observations, all charged particles in nature, such as electrons and protons, have charges that are integer multiples of the elementary charge (e). For example, an electron has a charge of "$-1e$", while a proton has a charge of "$+1e$". There are no particles found with charges that are fractions or arbitrary values of the elementary charge. This elementary charge can also be called "the quantum of charge".

Other quantized quantities are energy and angular momentum. The quantum of energy is $h\nu$ (i.e., photon) and that of angular momentum is $h/2\pi$. The quantum of mass not known till date.

The quantization of charge was first suggested by the experimental laws of electrolysis discovered by English experimentalist Faraday. In 1912, it was experimentally demonstrated by physicist Robert Millikan in his oil drop experiment. The experiment demonstrated that the charge on an object could only change in discrete steps, corresponding to the presence or absence of whole numbers of elementary charges. For example, if Q is the charge on object, then

$$Q = Ne \quad (1.6)$$

Here, N is the integer which represents excess or deficiency of fundamental unit of charge e.

If there is an excess of n protons on an object, then $N = n$. Therefore from Eq.1.4, the excess charge on object becomes:

$$Q = +ne \quad (1.7)$$

and for deficiency of n protons $N = -n$, therefore, Eq.1.4 takes form:

$$Q = -ne \quad (1.8)$$

According to atomic model, these are electrons not protons which move from one atom to other. So, as for as charge is concerned, the excess of electrons is equivalent to deficiency of protons and deficiency of electrons is equivalent to excess of protons. Therefore, above results can be generalized as:

$$Q = \pm ne \quad (1.9)$$

in which e, the elementary charge, has the approximate value $1.602 \times 10^{-19} C$ and $n = 0, 1, 2, ...$

+ve sign is used for deficiency and negative sign is used for excess of electrons.

☞ No unit of charge smaller than e has been detected on a free particle; current theories, however, propose the existence of particles called quarks (the constituent particles of protons and neutrons) having charges $\pm e/3$ and $\pm 2e/3$. Although there is considerable experimental evidence for such particles inside nuclear matter, free quarks have never been detected. For this reasons, we do not take their charges to be the elementary charge.

1.2.2.1 Number of Electrons in 1C of Charge

Here, we will find the number of deficient electrons in 1 C of positive charge. By quantization of charge (Eq.1.6), we have

$$Q = Ne \quad \text{or} \quad N = \frac{Q}{e}$$

Here, $Q = 1$ C, $e \approx 1.6 \times 10^{-19}$C, $N = ?$

Substituting these values in above equation and simplifying for N:

$$N = \frac{Q}{e} = \frac{1C}{1.6 \times 10^{-19}C} = 6.25 \times 10^{18}$$

Therefore, the number deficient electrons in $1C$ charge is

6.25×10^{18}, which is a huge number. Thus, the step size is very small as compared to the charges usually found on many cases. At macroscopic level, we deal with charges that are enormous compared to the magnitude of minimum charge i.e. e ($1.6 \times 10^{-19}C$). In this case, the increase or decrease in units of e is not very different from saying that charges are continuous. So, just for simplicity, at macroscopic level we can ignore the quantization of electric charge.

☞ As $1C$ charge contains 6.25×10^{18} electrons, which is a huge number. So "1 coulomb" is a huge amount of charge.

☞ We sometimes speak of the fundamental unit of charge e as the charge of an electron but e is a positive quantity. The charge of an electron, properly, is $-e$.

1.2.3 Charge is Always Associated with Mass

Every charged particle, such as electrons and protons, has mass. Mass is a property of matter that tells us how heavy or bulky something is. When particles have a charge, they also have a mass. charge cannot exist without mass, though mass can exist without charge. Particles such as neutrino or photon have no rest mass, so they cannot have any charge.

Reason: Particles interact with a special field called the Higgs field*, which is present throughout space. This interaction with the Higgs field gives particles their mass. The strength of this interaction depends on the particle's charge. So, in simple terms, charged particles have mass because they interact with the Higgs field, which gives them their weight or "heaviness".

1.2.3.1 Effect of Charge on the Mass of a Given Object

Let M_0 be the mass of a neutral object, m be the mass of an electron and $-e$ be the charge on an electron. If the object is given a positive charge Q by taking away $N = Q/e$ electrons from it, then the mass of the positively charged object will be-

$$M_{pos} = M_0 - mN = M_0 - mQ/e$$

Now, if another similar object is given a negative charge $-Q$ by putting $N = Q/e$ electrons on it, then, the mass of the negatively charged object-

$$M_{neg} = M_0 + mN = M_0 + mQ/e$$

☞ Here, you may wonder why the mass of an electron changes with velocity according to relativistic equation $m = m_0/\sqrt{1 - \frac{v^2}{c^2}}$, then why not it's charge?

The answer is that, the charge is a fundamental property of matter which is a quantum mechanical effect whereas the mass is a relativistic effect. These two effects are independent of each other and they do not effect each other.

1.2.4 Charge is Transferable

Charge is transferable from one charged body to another body which may be charged or uncharged, if they are put in contact. The process of charge transfer is called conduction. Whole of the charge cannot be transferred by conduction from one body to another except in case when a charged body is enclosed by a conducting body and connected to it (it will be discussed later in this chapter).

1.2.5 An Accelerated Charge Always Radiates Energy

Accelerated charge always radiates energy, which is a fundamental principle in classical electrodynamics. According to Maxwell's equations, when an electric charge undergoes acceleration, it emits electromagnetic radiation. This phenomenon is known as bremsstrahlung, which is a German term meaning "braking radiation".

When a charged particle accelerates or changes its velocity, it creates a changing electric field and a changing magnetic field around it. These changing fields propagate as electromagnetic waves, carrying energy away from the accelerating charge. This energy loss manifests as radiation.

The emitted radiation can have various forms, depending on the nature of the acceleration and the properties of the charged particle. For example, when an electron accelerates or decelerates in the vicinity of an atomic nucleus, it emits X-rays or gamma rays. In contrast, when a charged particle moves in a circular path, such as in a particle accelerator, it radiates synchrotron radiation.

EXAMPLE 2. Is it possible for a body to have an electric charge of $2.0 \times 10^{-19}C$? $3.2 \times 10^{-19}C$?
(A) Yes; yes (B) Yes; no
(C) No; yes   (D) No; no

APPROACH According to quantization of charge, the net charge on any charged object is given by-
$$Q = \pm ne$$
Here, n is an integer value. Substitute given values of Q and $e = 1.6 \times 10^{-19}C$ and solve for n. Acceptable values of n are only the integer one, i.e., $n = 0, 1, 2, \ldots$

SOLUTION (C) In first case, $2.0 \times 10^{-19}C$, therefore-
$$n = \frac{2.0 \times 10^{-19}C}{1.6 \times 10^{-19}C} = \frac{2}{3}$$
Since, $n = 2/3$, is a fraction which is not acceptable.
In second case, $Q = 3.2 \times 10^{-19}C$, therefore-
$$n = \frac{Q}{e} = \frac{3.2 \times 10^{-19}C}{1.6 \times 10^{-19}C} = 2$$
Since, $n = 2$ is an integer value so it is acceptable.
Thus the first charge given is impossible, while the second charge given, twice the fundamental charge, is possible.

EXAMPLE 3. How much negative charge and how much positive charge are there on the electrons and the protons in a cup of water (0.25 kg)?

APPROACH To find total negative and positive charge in 250 g of water, we first calculate total number of electrons and protons in it. Then by using the expression, $Q = -ne$ we can find the total negative charge in it. $Q = +ne$ will give total positive charge in it.

SOLUTION The "molecular mass" of water is 18 g; therefore, number of moles in 250 g of water $= 250/18$ moles.
Since, each mole has N_A, i.e., 6.02×10^{23} molecules, therefore, number of molecules in 250/18 moles of water
$$= (250/18) \times 6.0 \times 10^{23}$$
Each molecule consists of two hydrogen atoms (one electron for each one) and one oxygen atom (eight electrons). Thus, there are 10 electrons in each molecule. Therefore, total number of electrons in 250 g of water-
$$n = (250/18) \times 6.0 \times 10^{23} \times 10$$
Total negative charge on 250 g of water,
$$Q = -ne = -(250/18) \times 6.02 \times 10^{23} \times 10 \, e$$
$$= -(250/18) \left(6.02 \times 10^{23}\right) \times 10 \times \left(1.60 \times 10^{-19}C\right)$$
$$= -1.3 \times 10^7 C$$

The positive charge on the protons is the opposite of this. You can also calculate it by using $Q = +ne$.

EXAMPLE 4. We have an iron piece of mass 56 milligram (Atomic number = 26, molar mass = 56 gram) if $10^{-6}\%$ times of

---

*The Higgs field is a special field that fills all of space. It's like an invisible substance that exists everywhere. When particles interact with this field, they gain mass. It's similar to how a ball moving through molasses (thick liquid) feels heavier and harder to push. The Higgs field slows down particles, giving them their mass or "heaviness". The discovery of the Higgs boson particle in 2012 confirmed the existence of this field and how it gives mass to other particles.

1.3. QUARKS

total electron are removed from this body then calculate charge appear on this body.

SOLUTION Number of moles
$$= \frac{\text{mass of substance}}{\text{molar mass}} = \frac{56 \times 10^{-3}}{56} = 10^{-3} \text{ moles}$$
Number of atom = No. of moles × Avogadro No.
$$= 10^{-3} \times 6.022 \times 10^{23} \approx 6 \times 10^{20}$$

Therefore, number of electrons in 56 milligram of iron,
= Atomic number × No. of atoms
$$= 26 \times \left(6 \times 10^{20}\right)$$

So, number of electron removed $n = 26 \times 6 \times 10^{20} \times 10^{-8}$
Charge on body, $Q = ne$
$$= 156 \times 10^{12} \times 1.6 \times 10^{-19} \, C = 25 \, \mu C$$
Since, electrons are removed, therefore the nature of charge on the iron piece will be positive.

1.3 Quarks

Quarks are truly elementary particles that carry electric charges which are fractions of the elementary charge $\left(+\frac{2}{3}e \text{ or } -\frac{1}{3}e\right)$.

There are six types of quarks (referred to as six flavours of quarks) and these are: (i) up (u), (ii) charmed (c), (iii) top or truth (t), all having charge $+(2/3)e$, and (iv) down (d), (v) sideways or strange (s) (vi) bottom or beauty (b), all having charge $-(1/3)e$ (see Table 1.1).

The up, charmed, and top quarks have electric charges of +2/3 that of the proton, whereas the down, strange, and bottom quarks have charges of −1/3 that of the proton.

Table 1.1: Quarks and their charges

Quark	Symbol	Charge
Up	u	$+(2/3)e$
Down	d	$-(1/3)e$
Strange	s	$-(1/3)e$
Charmed	c	$+(2/3)e$
Bottom	b	$-(1/3)e$
Top	t	$+(2/3)e$

Two kinds (or flavors) of quarks, up (u) and down (d), are sufficient to describe normal matter. The electric charges are $+\frac{2}{3}e$ and $-\frac{1}{3}e$ for the u and d quarks, respectively. As shown in Fig.1.3, the proton consists of two u and one d valence quarks, and the neutron of one u and two d valence quarks:
$$p = +\frac{2e}{3} + \frac{2e}{3} - \frac{e}{3} = +e \quad \text{i.e., uud}$$

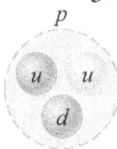
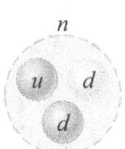

(a) Proton (uud), $q = +e$ (b) Neutron (udd), $q = 0$
Figure 1.3: Quark composition of proton and neutron

$$n = +\frac{2e}{3} - \frac{e}{3} - \frac{e}{3} = 0 \quad \text{i.e., udd}$$

There is firm experimental evidence of the existence of all six quarks and their six antiquarks within the nucleus, but free quarks have not been detected yet. Current theory implies that direct detection of quarks may, in principle, be impossible.

1.4 Grounding a Conductor

Earth is a conductor because of the presence of ions and moisture in it. It is also large enough that for many purposes it can be considered as a limitless reservoir of charge (i.e., the addition or subtraction of electrons has a negligible effect on it). So, the ground remains essentially neutral at all times. Grounding a conductor means to provide a conducting path between it and the Earth. The copper rod in Figure1.6 is connected to the ground through your body. When something is connected to the ground by a conductor, we say that it is grounded. Every building (with a contemporary electrical system) has a wire connected between the electrical system and a copper pipe. The copper pipe is connected through other copper pipes to the Earth. This ensures that the third connection on electrical sockets is grounded. The connection to the ground keeps electrical appliances from building up excess charge.

1.4.1 Check Point 1

1. •If you charge a pocket comb by rubbing it with a silk scarf, how can you determine if the comb is positively or negatively charged?
2. •Why does a shirt or blouse taken from a clothes dryer sometimes cling to your body?
3. •Explain why fog or rain droplets tend to form around ions or electrons in the air.
4. •A positively charged rod is brought close to a neutral piece of paper, which it attracts. Draw a diagram showing the separation of charge in the paper, and explain why attraction occurs.
5. •Why does a plastic ruler that has been rubbed with a cloth have the ability to pick up small pieces of paper? Why is this difficult to do on a humid day?
6. •Contrast the net charge on a conductor to the "free charges" in the conductor.
7. •Figures 1.4 (a) and (b) show how a charged rod placed near an uncharged metal object can attract (or repel) electrons. There are a great many electrons in the metal, yet only some of them move as shown. Why not all of them?

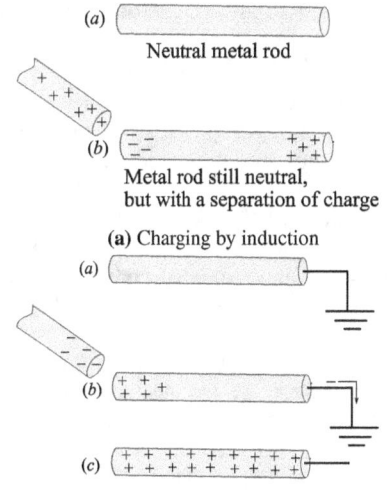

Figure 1.4: Comparison of Ben Franklin's model and our contemporary model of electric charge.

8. •The fact that the electron has a negative charge and the proton has a positive charge is due to a convention established by Benjamin Franklin. Would there have been any significant consequences if Franklin had chosen the opposite convention? Is there any advantage to naming charges plus and minus as opposed to, say, A and B?
9. •Small bits of paper are attracted to an electrically charged comb, but as soon as they touch the comb they are strongly

repelled. Explain this behaviour.
10. • Find the total electric charge of 2.5 kg of (a) electrons and (b) protons. Given: mass of an electron $\approx 9.1 \times 10^{-31}$ kg, mass of a proton $\approx 1.67 \times 10^{-27}$ kg.
11. •A container holds a gas consisting of 2.85 moles of oxygen molecules. One in a million of these molecules has lost a single electron. What is the net charge of the gas?
12. •A copper penny ($Z = 29$) has a mass of 3.10 grams. What is the total charge of all the electrons in the penny? Given that molar mass of Cu is 63.5 gram.

1.5 Classification of Solids
Solids can be classified in either of the following four categories:
(i) Conductors (ii) Insulators
(iii) Semiconductors (iv) Superconductors

1.5.1 Conductors
These are the materials that allow flow of charge through them. Examples include metals (such as copper in common lamp wire), tap water, and the human body (which is similar to tap water). This category generally comprises of metal but may sometimes contain non-metals too. (Ex. Carbon in form of graphite).

1.5.2 Insulators
These are the materials which do not allow movement of charge through them. Examples include rubber (such as the insulation on common lamp wire), plastic, glass, and chemically pure water.

1.5.3 Semiconductors
These are materials that are intermediate between conductors and insulators; Examples include silicon and germanium in computer chips.

1.5.4 Superconductors
These are materials that are perfect conductors, allowing charge to move without any hindrance.
In this chapter we discuss only conductors and insulators.

1.6 Methods of Charging
1.6.1 Charging by Friction
Charging by friction is the oldest form of charging. It was found that when an amber rod is rubbed with fur, the rod became negatively charged. The two bodies acquire opposite signs of electricity; one gets positively charged, while the other becomes negatively charged. When two bodies are charged by friction, they acquire the same magnitude of charge. Furthermore, the bodies retain these excess charges even when they are separated from each other.

Rubbing causes a transfer of electrons from one object to another. If some electrons are transferred from an object A to another object B, then A becomes positively charged and B negatively charged.

Note: Electrons are matter particles. Mass of electron is $m_e = 9.1 \times 10^{-31}$ kg. So, charging involves transformation of mass.

1.6.1.1 Friction is Used to Charge Insulators
The electrons in some materials—such as rubber—do not move freely. Such materials are known as insulators. When a surplus of charged particles (positive or negative) builds up on some part of an insulator, the excess remains there. So, if you hold one end of a rubber rod while the far end is being rubbed with silk (Fig.1.5a), the far end of the rod will acquire a surplus of electrons and those electrons will remain at that end, never flowing into your hand (Fig. 1.5b). So, insulators can be charged by friction.

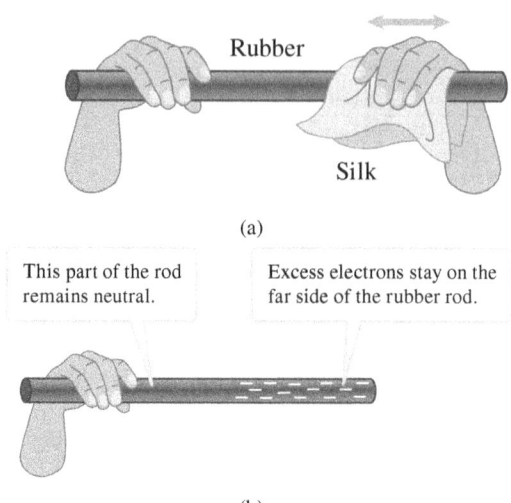

Figure 1.5: (a) When a rubber rod is rubbed with a silk cloth, (b) the excess charge stays on the part of the rod that was in contact with the cloth.

1.6.1.2 Charging Conductors by Friction is Very Difficult
If you try to charge a copper rod in the same way, you will find that you cannot build up charge on the copper. Copper is an example of a conductor. A conductor is a material in which the charged particles (usually electrons) can flow freely. When you hold one end of the copper rod and rub the other end with silk, electrons are transferred from the silk to the copper rod, and those excess electrons are free to flow. Because like charges repel, the electrons move away from one another, which means they travel through the rod into your hand, as shown in Fig.1.6. The human body is also a conductor, so charged particles move freely through your body toward the Earth. If there are no insulators between you and the ground—such as when you are barefoot—the charge will continue to flow into the Earth. As a result, you and the copper rod remain neutral despite the rod being rubbed with silk.
• The human body consists largely of salt water. Pure water is not a very good conductor, but salt water, with its Na^+ and Cl^- ions, is. Consequently, humans are reasonably good conductors.

Now, considering a case, when two conductive materials are rubbed together. In this case, electrons can be transferred from one material to the other. This transfer occurs due to the different affinity for electrons that each material possesses. The material with a higher affinity for electrons tends to gain electrons and becomes negatively charged, while the other material loses electrons and becomes positively charged. This electron transfer leads to a charge separation between the two materials.

However, due to the conductive nature of the materials involved, the excess charges rapidly spread and distribute themselves evenly over the surface of the conductors. This redistribution occurs because the excess charges repel each other and move to the outer surface of the conductor, where they can be evenly distributed.

As a result, any localized charge imbalance that occurs during the charging process quickly disappears, and the conductor as a whole remains electrically neutral. The excess charges are effectively cancelled out by their mutual repulsion and redistribution. From above discussion, we can say that charging by friction can be applied only if at least one material is insulator.

1.6.1.3 Triboelectric Series
In general, when two materials are rubbed together, the magnitude and sign of the charge that each material acquires depends on how strongly it holds onto its electrons. For example, if silk is rubbed against glass, the glass acquires a positive charge. It means electrons have moved from the glass to the silk, giving the

1.6. METHODS OF CHARGING

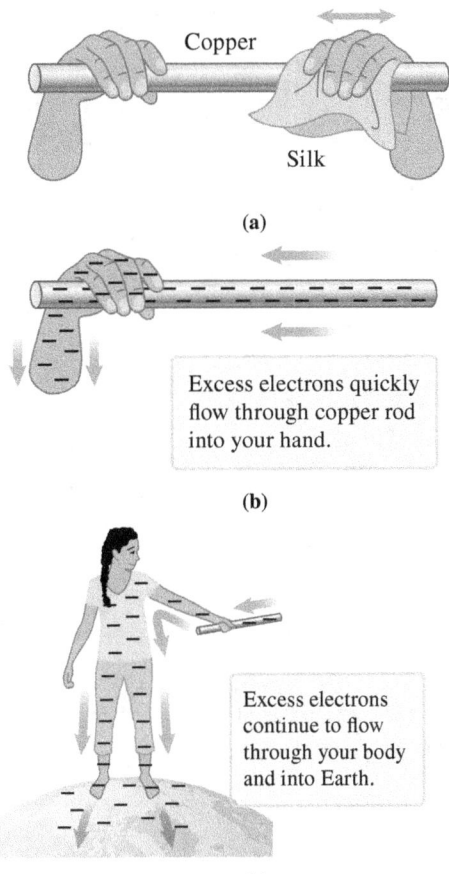

Figure 1.6: (a) When you rub a copper rod with a cloth, (b) the excess charge moves all over the rod and (c) through your body into the Earth.

silk a negative charge. If silk is rubbed against amber, however, the silk becomes positively charged, as electrons in this case pass from the silk to the amber.

Table 1.2: Triboelectric series

Material	Relative charging with rubbing
Air	Most positive
Asbestos	↑
Rabbit fur	
Glass	
Human hair, oily skin	
mica	
Nylon	
Wool (no charge)	
Lead	
Cat Fur	
Silk	
Aluminum	
Paper	Least positive
Cotton	Least negative
Wood	↑
Amber	
Rubber	
Vinyl (PVC)	
Teflon	
Ebonite	Most negative

Table 1.2, shows the relative charging due to rubbing for a variety of materials. This relative charging is known as triboelectric charging. The more plus signs associated with a material, the more readily it gives up electrons and becomes positively charged. Similarly, the more minus signs for a material, the more readily it acquires electrons. For example, we know that amber becomes negatively charged when rubbed against fur, but a greater negative charge is obtained if rubber, PVC, or Teflon is rubbed with fur instead. In general, when two materials in Table 1.2 are rubbed together, the one higher in the list becomes positively charged, and the one lower in the list becomes negatively charged. The greater the separation on the list, the greater the magnitude of the charge.

1.6.2 Charging by Direct Contact

Suppose, you have a conducting sphere as shown in Fig.1.7a. You can charge it by touching it with a charged object such as a negatively charged plastic rod (Fig.1.7b). When the rod is in contact with the sphere, some of the electrons are transferred from the rod to the sphere. Remember that charge cannot flow through the rod because it is an insulator, so you may need to roll the rod around the surface of the sphere in order to transfer charge from many parts of the rod. If we bring a negatively charged conducting rod in contact of an uncharged conducting sphere, then, off course, rolling is not required. In this case, electrons can easily be transferred from charged rod to uncharged sphere. Because electrons move freely throughout the conductor and because like charges repel, the electrons quickly redistribute themselves in the conductor, moving as far apart as possible. For a spherical conductor, *"as far apart as possible"* means that the electrons are uniformly distributed on the outside surface of the sphere. If you try to charge a conductor of another shape, the charge is again distributed on the outside surface, although for nonspherical shapes the charge distribution is not uniform.

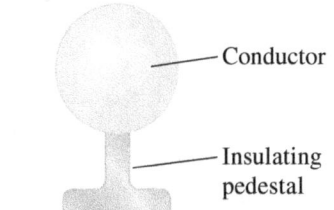

(a) A conductor rests on top of an insulating pedestal.

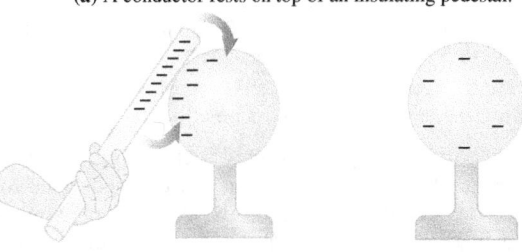

(b) When the conductor is touched by a charged rod, some charge is transferred from the rod to the conductor.

(c) The excess charge quickly distributes over the surface of the conductor.

Figure 1.7

- The uncharged body acquires the same sign of charge as the charged body. The total charge is distributed between the two bodies.

1.6.3 Charging by Induction

There is another way to build up charge on a conductor. Suppose you use the same equipment as in Figure1.7b. This time, however, you hold the negatively charged rod near but not touching the sphere as shown in Fig.1.8a. Because the electrons in the sphere are free to move, they flow to the side opposite the rod due to repulsion of negatively charged rod. If you ground the side of the conducting sphere where there is a surplus of electrons, those electrons will flow to the ground (Fig.1.8b). When you remove the connection to the ground, keeping the charged rod near the sphere, the net charge of the sphere is positive and the ground remains essentially neutral (Fig.1.8c). This excess positive charge on sphere is called called the induced positive charge. Now, if you remove the rod too, this charge get uniformly distributed over the surface of sphere due to repulsion.

Note that, no electrons are lost from the rod; it has the same charge throughout the process. In fact, you could reuse the rod to charge another conductor by same method without recharging the rod again.

"This process of charging, in which a charged plastic rod gives another body (metal sphere) a charge of opposite sign without losing any of its own charge, is called charging by induction. In this process the charged rod never touches the sphere."

Note: You can also charge two bodies by induction as follows-

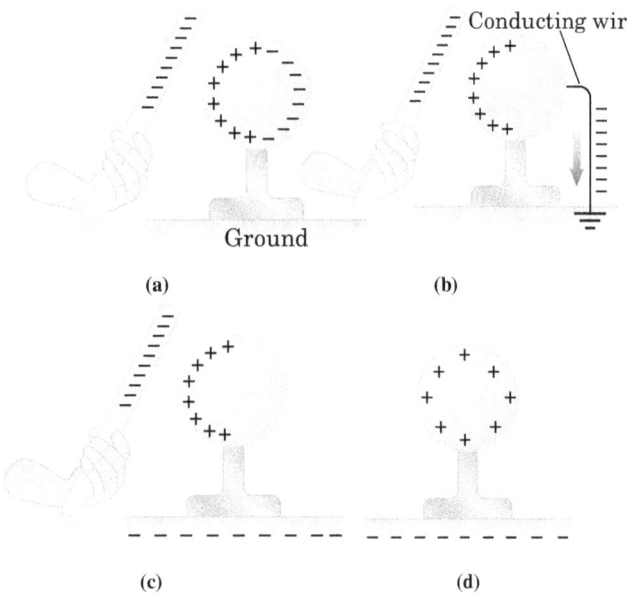

Figure 1.8: (a) When a charged rod is held near a neutral conductor, electrons in conductor move away from negatively charged rod and the conductor becomes polarized. (b) The far side of the conductor is connected to the ground, and some electrons flow out of the conductor into the Earth. (c) The connection is removed. (d) When the rod is removed, the electrons redistribute, leaving the surplus positive charge uniformly distributed over the surface of the conductor.

Take an isolated neutral conductor and then bring a charged rod near it. In Fig.1.9 the the rod is negatively charged. Due to the charged rod, charges will induce on the conductor. Now, connect another neutral conductor with it. Due to repulsion of the negatively charged rod, some free electrons get transferred from the left conductor to the right conductor and due to deficiency of electrons positive charge appears on the left conductor and on the right conductor, there is an excess of electrons due to transfer from left conductor. Now disconnect the connecting wire and remove the rod. The first conductor will get negatively charged and the second conductor will get positively charged. ☞ Note that only conductors can be charged by induction.

1.6.4 Charging by Field Emission

When a large magnitude electric field $\left(\vec{E}\right)$ is applied near the metal surface, some electrons come out from the metal surface due to the electric force applied by the external electric field, and

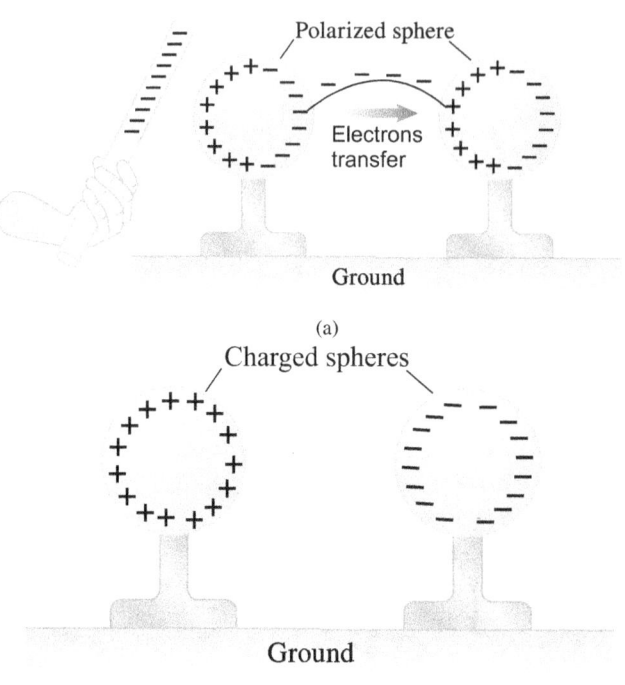

Figure 1.9: Charging by induction

hence the metal becomes positively charged, as shown in Fig. 1.10.

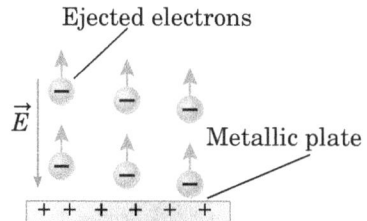

Figure 1.10: Charging by field emission

1.6.5 Charging by Thermionic Emission

When the metal is heated at a high temperature then some electrons of metals are ejected and the metal becomes positively charged.

1.6.6 Charging by Photoelectric Effect

When light photons, having energy greater than work function* of material of a given metal sheet, are incident on it's surface then some electrons gain energy from these incident photons and come out of the metal surface. As a result, there becomes a deficiency of electrons in the metal surface. So, remaining metal becomes positively charged. These ejected electrons are called photo electrons. It is shown in Fig.1.11. In this figure, $h\nu$ is the energy of a photon having frequency ν. If ϕ represents work function of metal then for ejection of photoelectrons, the required condition is $h\nu \geq \phi$.

☞ Neutral does not mean "chargeless". A neutral body always has equal amount of positive and negative charge, i.e., there is a microscopic balance of −ve and +ve charge.

1.7 Coulomb's Law

Coulomb's law is a fundamental law in physics that describes the electrostatic interaction between point charges[†] quantitatively.

---

*The amount of energy that must be supplied to break the bond between a metal and one of it's electrons, is called the work function (ϕ). Each metal has its own characteristic work function.

[†]A point charge is a point like object with a non zero electric charge and non zero mass. Note that a point like object is small enough that its internal structure

1.7. COULOMB'S LAW

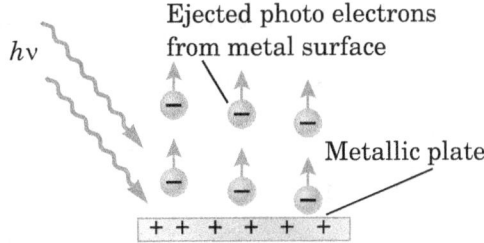

Figure 1.11: Charging by photoelectric effect

It is an empirical law, derived from observations rather than theoretical derivation.

It states that- the electrostatic force acting between two point charges is directly proportional to the of product of magnitude of charges and inversely proportional to the square of the distance between them.

Mathematically, the electrostatic force F that a particle of charge q_1 exerts on a particle of charge q_2 at a distance r, can be written as-

$$F \propto \frac{|q_1||q_2|}{r^2}$$

or
$$\boxed{F = k\frac{|q_1||q_2|}{r^2}} \qquad (1.10)$$

Here, k is a proportionality constant, its numerical value depends on the system of units used. its value is $8.99 \times 10^9 \text{N.m}^2/\text{C}^2$
In one significant figure, $k = 9 \times 10^9 \text{ N.m}^2/\text{C}^2$

Eq.1.10 holds only for charged objects whose sizes are much

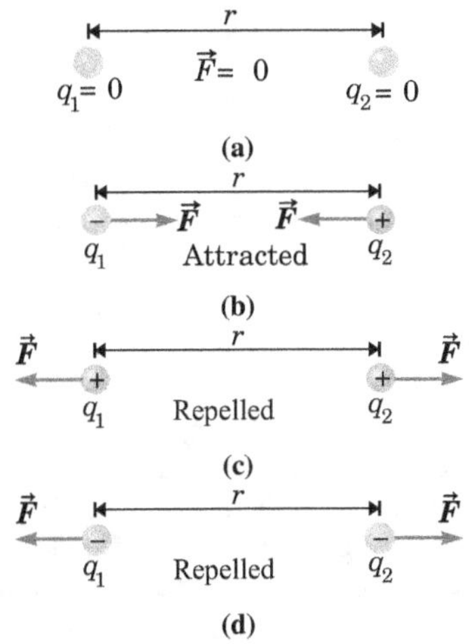

(a) $q_1 = 0$, $\vec{F} = 0$, $q_2 = 0$ at distance r

(b) q_1 — $\vec{F}$ → ← $\vec{F}$ — q_2, Attracted

(c) ← $\vec{F}$ — q_1 ... q_2 — $\vec{F}$ → , Repelled

(d) ← $\vec{F}$ — q_1 ... q_2 — $\vec{F}$ → , Repelled

Figure 1.12: (a) There is no electrostatic force between neutral objects. (b) The electrostatic force between two oppositely charged objects is attractive. (c) The electrostatic force between two positively charged objects is repulsive. (d) The electrostatic force between two negatively charged objects is repulsive.

smaller than the distance between them. We often say that it holds only for point charges.

The directions of the forces the two charged particles exert on each other are always along the line joining them. If any one or both the charges are zero, then the electrostatic force between the particles will be zero (Fig.1.12a). When the charges q_1 and q_2 have the same sign, either both positive or both negative, the forces are repulsive (see Fig.1.12c and 1.12d); when the charges have opposite signs, the forces are attractive (Fig.1.12b).

In the SI system, the constant k is expressed in the following form:

$$k = \frac{1}{4\pi\epsilon_0} \qquad (1.11)$$

Although the choice of this form for the constant k appears to make Coulomb's law needlessly complex, it ultimately results in a simplification of formulas of electromagnetism that are used more often than Coulomb's law.

Here, ϵ_0 ("epsilon nought") is called the electric permittivity* of free space. It's value, determined by the adopted value of the speed of light, is $\epsilon_0 = 8.854 \times 10^{-12} \text{C}^2/\text{N} \cdot \text{m}^2$
The constant k has the corresponding value (to three significant figures)

$$\boxed{k = \frac{1}{4\pi\epsilon_0} = 8.99 \times 10^9 \text{ N} \cdot \text{m}^2/\text{C}^2}$$

For simplicity, we'll often use the approximate value

$$\boxed{k = \frac{1}{4\pi\epsilon_0} = 9.0 \times 10^9 \text{ N} \cdot \text{m}^2/\text{C}^2}$$

With this choice of the constant k, Coulomb's law can be written as-

$$\boxed{F = \frac{1}{4\pi\epsilon_o}\frac{|q_1||q_2|}{r^2}} \qquad (1.12)$$

1.7.0.1 Value of k in CGS Units

In SI units, $k \approx 9.0 \times 10^9 \text{ N.m}^2/\text{C}^2$
Here, we use the method of unit conversion. If n_1, u_1 are numeric value and unit respectively in SI, whereas n_2, u_2 in CGS, then
$n_1 u_1 = n_2 u_2$, gives

$$n_2 = n_1 \frac{u_1}{u_2} = 9.0 \times 10^9 \frac{\text{N.m}^2/\text{C}^2}{\text{dyne.cm}^2/\text{esu}^2}$$

$$= 9.0 \times 10^9 \frac{10^5 \text{dyne}.10^4 (\text{cm})^2/(3 \times 10^9 \text{esu})^2}{\text{dyne}.(\text{cm})^2/(\text{esu})^2} = 1$$

Therefore, in CGS, $k = 1$ dyne $(\text{cm})^2/(\text{esu})^2$

1.7.1 Experimental Verification of Coulomb's Law

The experimental setup designed by Charles Augustin Coulomb (1736 - 1806) to measured electrical attractions and repulsions quantitatively, is shown in Fig.1.13. One small metal sphere is charged and fixed in place. A pair of small metal spheres is attached to a lightweight insulated rod. The rod is suspended from a torsion spring of known torsion spring constant.

Imagine that the pair of spheres is initially uncharged. The rod is twisted, bringing one of those spheres in contact with the charged fixed sphere and thus transferring charge. If the spheres are identical, half of the charge on the fixed sphere will be transferred to the movable sphere. So these spheres repel each other, causing the rod to twist. The amount of twist is measured. Because Coulomb knew the torsion spring constant, he could calculate the force exerted between the spheres. In many trials, Coulomb varied the amount of charge on the spheres as well as the torsion spring constant. From these trials, he found that-
"The electrostatic force acting between two charged spheres is

---

is of no importance. The electron can be treated as a point charge, since there is no experimental evidence for any internal structure. The proton does have internal structure—it contains three particles called quarks bound together—but, since its size is only about 10^{-15}m, it too can be treated as a point charge for most purposes. Any small charged bodies, provided that the sizes of these bodies are much smaller than the distances between them can also be treated like point charges. For example, a charged metal sphere of radius 10 cm can be treated as a point charge if it interacts with another such sphere 100 m away, but not if the two spheres are only a few centimetres apart.

*Electric permittivity is the measure of resistance that is encountered when forming an electric field (Section 1.12) in a particular medium. More electric permittivity means more resistance is offered by the medium in forming electric field inside the medium, i.e., the medium is more insulating.

Inside a conductor the electric field is always zero, therefore, its electric permittivity is infinite.

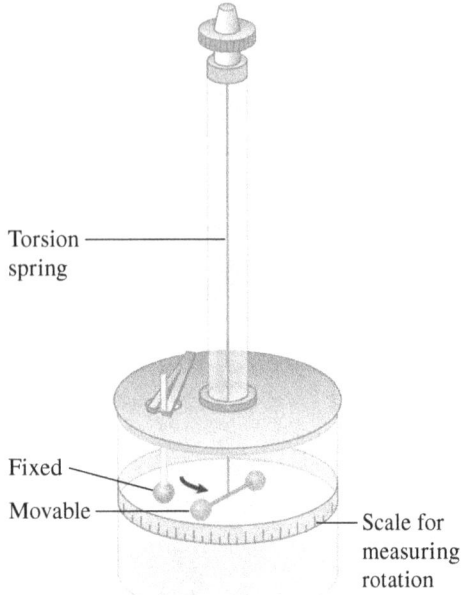

Figure 1.13: Torsion balance designed by Coulomb. Two (blue) spheres are attached to a rod suspended from a torsion spring. The fixed (red) sphere is positively charged, and then one of the movable spheres is brought in contact with the charged, fixed one. Because they are both conductors, charge is shared between the spheres, and the movable (blue) sphere becomes positive. Both spheres have charge of the same sign, so they are repelled, which causes the rod to rotate. The rod's rotation is measured using a scale (shown in yellow).

directly proportional to the of product of magnitude of charges and inversely proportional to the square of the distance between them."

☞ For given charges, if we plot a graph of the measured force versus the inverse square of the distance $(1/r^2)$, the graph should be a straight line and the slope of the line will give you the product of the charges divided by the electrostatic constant, $(q_1 q_2/k)$.

EXAMPLE 5. Find the electric force between two point charges of $1C$ each separated by 1 m apart in free space

SOLUTION By Coulomb's law (Eq.1.10), we have

$$F = k\frac{|q_1||q_2|}{r^2} \quad \cdots (1)$$

Substituting, the values of k, q_1, q_2 and r in Eq.(1), we get-

$$F = (9.0 \times 10^9 \text{N.m}^2/\text{C}^2)\frac{|1C||1C|}{(1m)^2} = 9.0 \times 10^9 \text{N}$$

which is a very large value and keeping these charges at 1 m apart, is practically very difficult. This is the reason, why we consider systems having smaller charges, such as- micro coulomb ($1\,\mu C = 10^{-6}C$), nano coulomb ($1\,nC = 10^{-9}C$) and pico coulomb ($1\,pC = 10^{-12}C$) etc.

1.7.2 Vector Form of Coulomb's Law

So far we have considered only the magnitude of the force between two point charges determined according to Coulomb's law. Force, being a vector, has directional properties as well. In the case of Coulomb's law, the direction of the force is determined by the relative sign of the two electric charges.

As illustrated in Fig.1.14, suppose we have two like point charges q_1 and q_2 at positions $\vec{r}_1$ and $\vec{r}_2$ respectively.

The position vector of charge q_1 with respect to q_2 is given by-
$$\vec{r}_{12} = \vec{r}_1 - \vec{r}_2$$

The unit vector in the direction of q_1 from q_2 is -
$$\hat{r}_{12} = \frac{\vec{r}_{12}}{r_{12}} = \frac{\vec{r}_1 - \vec{r}_2}{|\vec{r}_1 - \vec{r}_2|}$$

Here, r_{12} represents the magnitude of the vector $\vec{r}_{12}$, i.e., the separation between q_1 and q_2.

The repulsive electric force exerted by charge q_2 on q_1 is given by-

$$\boxed{\vec{F}_{12} = \frac{1}{4\pi\epsilon_0}\frac{q_1 q_2}{r_{12}^2}\hat{r}_{12}} \quad (1.13)$$

Similarly, electric force on charge q_2 due to charge q_1 is given by-

$$F_{21} = \frac{1}{4\pi\epsilon_0}\frac{q_1 q_2}{r_{21}^2}$$

In vector form, above equation can be written as

$$\boxed{\vec{F}_{21} = \frac{1}{4\pi\epsilon_0}\frac{q_1 q_2}{r_{21}^2}\hat{r}_{21}} \quad (1.14)$$

Here $\hat{r}_{21}$ is a unit vector that points from charge q_1 to charge q_2; that is, it would be the unit vector in the direction of point charge q_2 if the origin of coordinates were at the location of charge q_1. Since, $\hat{r}_{21} = -\hat{r}_{12}$, therefore-

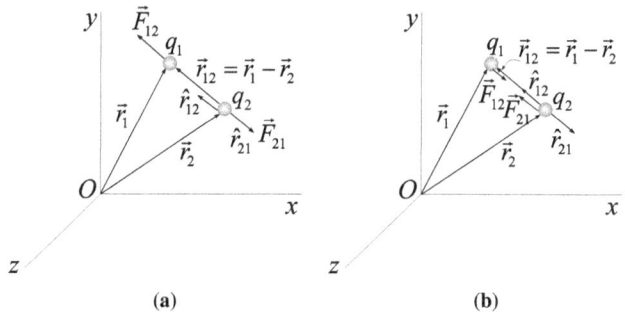

(a) (b)

Figure 1.14: (a) Two point charges q_1 and q_2 of the same sign exert equal and opposite repulsive forces on one another. The vector $\vec{r}_{12}$ locates q_1 relative to q_2, and the unit vector $\hat{r}_{12}$ points in the direction of $\vec{r}_{12}$. Note that $\vec{F}_{12}$ is parallel to $\vec{r}_{12}$. (b) The two charges now have opposite signs, and the force is attractive. Note that $\vec{F}_{12}$ is antiparallel to $\vec{r}_{12}$

$$\boxed{\vec{F}_{21} = -\vec{F}_{12}} \quad (1.15)$$

Thus, electric forces form an action-reaction pair.

The significance of Coulomb's law goes far beyond the description of the forces acting between charged particles. This law, when incorporated into the structure of quantum physics, correctly describes-

1. the electrical forces that bind the electrons of an atom to its nucleus.
2. the forces that bind atoms together to form molecules, and
3. the forces that bind atoms and molecules together to form solids or liquids.

Thus most of the forces of our daily experience that are not gravitational in nature are electrical. Moreover, unlike Newton's law of gravitation, which can be considered a useful everyday approximation of the more basic general theory of relativity, Coulomb's law is an exact result for stationary charges and not an approximation from some higher law. It holds not only for ordinary objects, but also for the most fundamental "point" particles such as electrons and quarks. Coulomb's law remains valid in the quantum limit (for example, in calculating the electrostatic force between the proton and the electron in an atom of hydrogen). When charged particles move at speeds close to the speed of light, such as in a high-energy accelerator, Coulomb's law does not give a complete description of their electromagnetic interactions; instead, a more complete analysis based on Maxwell's equations* must be done.

1.8 Superposition Principle

The vector form of Coulomb's law is useful because it carries within it the directional information about $\vec{F}$ and whether the force is attractive or repulsive. Using the vector form is of critical importance when we consider the forces acting on an assembly of more than two charges. In this case, Eq.1.15 would hold for every pair of charges, and the total force on any one charge would be found by taking the vector sum of the forces due to each of the other charges. For example, the force on point charge q_1 in an assembly would be-

$$\boxed{\vec{F}_1 = \vec{F}_{12} + \vec{F}_{13} + \vec{F}_{14} + \cdots} \quad (1.16)$$

where $\vec{F}_{12}$ is the force on point charge q_1 due to point charge q_2, $\vec{F}_{13}$ is the force on point charge q_1 due to point charge q_3, and so on. Equation 1.16 is the mathematical representation of the **principle of superposition** applied to electric forces. It permits us to calculate the force due to any pair of charges as if the other charges were not present. For instance, the force $\vec{F}_{13}$ that point charge q_3 exerts on q_1 is completely unaffected by the presence of point charge q_2.

Note that - if the electromagnetic force were proportional to the square of the total source charge, then, the principle of superposition would not hold, since $(q_1 + q_2)^2 \neq q_1^2 + q_2^2$ (there would be an extra term "$2q_1q_1$" to consider). Superposition is not a logical necessity, but an experimental fact.

The electric force on test charge q_1, due to the point source charge q_i (say) ($i \neq 1$), not only depends on the position of q_i, it also depends on both velocities and on the acceleration.

Since, an electromagnetic signal or effect produced by the source charge q_i travels at the speed of light, so to find the effect of q_i on q_1, we need the position, velocity, and acceleration of q_i at some earlier time, when the signal left.

Hence, we can say that the principle of superposition does not hold in many situations when the source charge is moving. In this chapter, we will be considering only the stationary source charges, therefore, the principle of superposition is valid.

1.9 Coulomb's Law in a Dielectric Medium

A non conducting material (for example, air, glass, paper, or wood) is called a dielectric. If two point charges q_1 and q_2 are placed at separation r in a liquid dielectric medium of electric permittivity ϵ, then the electric force acting between them is given by-

$$\boxed{F = \frac{1}{4\pi\epsilon} \frac{|q_1|\,|q_2|}{r^2}} \quad (1.17)$$

If ϵ_r or κ denotes relative permittivity or dielectric constant of the medium, then-

$$\epsilon_r \quad \text{or} \quad \kappa = \frac{\epsilon}{\epsilon_0}$$

or

$$\boxed{\epsilon = \epsilon_0 \kappa}$$

Substituting this value of ϵ in Eq.1.17, we get-

$$\boxed{F = \frac{1}{4\pi\epsilon_0 \kappa} \frac{q_1 q_2}{r^2}}^\dagger \quad (1.18)$$

Eq.1.18 shows that effective electrostatic force between two point charges decreases to $1/\kappa$ times due to presence of dielectric medium.

☞ A dielectric medium consists of atoms which are neutral but not chargeless. When a dielectric or any other substance is present in vicinity of a charge, the positive and negative charges of atoms (nuclei and electrons) experience electric force, which in turn leads to a partial separation of these charges. These partially separated charges present on atoms apply additional electric force on charges which in combination with the force of interaction between q_1 and q_2 gives the resultant force. If we know the force of interaction between q_1 and q_2 in absence of dielectric medium and the additional electric force due to induced charges of the medium, we can forget about the presence of the substance itself while calculating the resultant force, since the role of the substance has already been taken into account with the help of induced charges. The effective force acting between charges q_1 and q_2 in presence of a dielectric is the result of superposition of actual force in absence of medium and the force applied by induced charges of medium due to it's presence.

Thus, although the electrostatic force of q_1 on q_2 or vice versa remains unchanged, the effective force between the charges in a dielectric medium decreases to $1/\kappa$ times of it's value in free space.

For free space or air, $\epsilon = \epsilon_0$, therefore, $\kappa = \epsilon_0/\epsilon_0 = 1$
Now, Eq.1.18 can also be written as

$$F = \frac{1}{4\pi\epsilon_0} \frac{q_1 q_2}{(r\sqrt{\kappa})^2}$$

which is similar to an expression for electric force between two charged particles placed at a separation $r\sqrt{\kappa}$ in vacuum. So, the separation r of charged particles placed in a dielectric medium produces same electric force if these charged particles are placed at separation $r\sqrt{\kappa}$ in vacuum.

Now, we calculate the electric force between two point charges when the vacuum separation between the charges is partially filled with a dielectric medium of dielectric constant κ (Fig.1.15).

Suppose two charges q_1 and q_2 are separated by distance r in

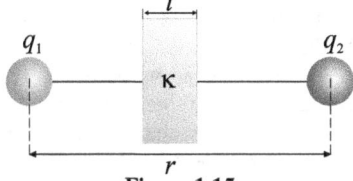

Figure 1.15

vacuum. If, we insert a liquid dielectric of dielectric constant κ and thickness $t(< r)$ between the charges, then, the width of vacuum part between the charges will become $(r-t)$ and effective vacuum width corresponding to the thickness t of dielectric slab is $t\sqrt{\kappa}$. Therefore, effective vacuum separation between the charges can be written as-

$$r_{\text{vacuum}} = (r - t + t\sqrt{\kappa})$$

So, in this case, Coulombs law can be written as -

$$\boxed{F = \frac{1}{4\pi\epsilon_0} \frac{q_1 q_2}{(r - t + t\sqrt{\kappa})^2}} \quad (1.19)$$

EXAMPLE 6. If two charges q_1 and q_2 are separated with distance 'r' and placed in a medium of dielectric constant κ. What

---

*Maxwell's equations will be discussed in chapter "Electromagnetic Waves"

†Eq.1.18 is applicable only for space filled with a liquid dielectric. According to Eq.1.18, the force depends on the fact that κ is a constant, which is only approximately true for most real materials. Finally, the microscopic electric field created by dielectric also applies electric force on the charged particles or conductors in it (see subsection "3.13.1.1 Calculation of Force on a Dielectric Slab in a Fringing Field of a Capacitor" at page 255 of chapter "3. Capacitance and Dielectrics").

will be the equivalent distance between charges in air for the same electrostatic force?

SOLUTION The electric force (F) between charges q_1 and q_2 separated by a distance r in a medium of dielectric constant κ is given by
$$F = \frac{1}{(4\pi\epsilon_0)} \frac{q_1 q_2}{\kappa r^2}$$
And electric force in air (F_{air}) between charges q_1 and q_2 separated by a distance r', is given by
$$F_{\text{air}} = \frac{1}{4\pi\epsilon_0} \frac{q_1 q_2}{r'^2}$$
If $F = F_{\text{air}}$, then
$$\frac{q_1 q_2}{4\pi\epsilon_0 \kappa r^2} = \frac{q_1 q_2}{4\pi\epsilon_0 r'^2} \implies r' = r\sqrt{\kappa}$$

1.9.0.1 Key Points Regarding Coulomb's Law

1. Electric forces always form action-reaction pairs, i.e., the force which first charge exerts on the second is equal and opposite to the force which the second charge exerts on the first.
2. The direction of force is always along the line joining the two charges.
3. Electrostatic force between two point charges is independent of presence or absence of other charges.
4. The electrostatic force is conservative, i.e., the work done in moving a point charge once round a closed path under the action of Coulomb's force, is zero.
5. The net Coulomb's force on two charged particles q_1 and q_2 separated by distance r in free space and in a uniform liquid dielectric medium, are-
   (i) $F = \frac{1}{4\pi\epsilon_0} \frac{q_1 q_2}{r^2}$ (in vacuum or free space)
   (ii) $F' = \frac{1}{4\pi\epsilon_0 \kappa} \frac{q_1 q_2}{r^2}$ (in the dielectric medium)
   Therefore, $\frac{F}{F'} = \frac{\epsilon}{\epsilon_0} = \kappa$

   So, the dielectric constant (κ) of a liquid dielectric medium, is numerically equal to the ratio of the force on two point charges in free space to that in the fully filled dielectric medium.
6. The law expresses the force between two point charges at rest. In applying it to the case of extended bodies of finite size care should be taken in assuming the whole charge of a body to be concentrated at its 'centre' as this is true only for spherical charged body, that too for external point.
   Although net electric force on both the charged particles changes in the presence of dielectric but force due to one charged particle on another charge particle does not depend on the medium between them.
7. Coulomb's law resembles Newton's inverse square law of gravitation, $F = Gm_1 m_2/r^2$. Both are inverse square laws, and the charge q plays the same role in Coulomb's law that the mass m plays in Newton's law of gravitation. Main differences between the two forces are-
   - Practically, (not theoretically*) gravitational forces are always attractive, while electrostatic forces can be repulsive or attractive, depending on whether the two charges have the same or opposite signs.
   - Electric force depends on the nature of medium between the charges while gravitational force does not.
   - There is another important difference between the two laws. In using the law of gravitation, we were able to define mass from Newton's second law, $F = ma$, and then by applying the law of gravitation to known masses we could determine

---
*Theoretically, there is a concept of negative mass. It is assumed that there is a gravitational repulsion between a positive and a negative mass.

the constant G. In using Coulomb's law, we take the reverse approach—we define the constant k to have a particular value, and we then use Coulomb's law to determine the basic unit of electric charge as the quantity of charge that produces a standard unit of force.

For example, consider the force between two equal charges of magnitude q. We could adjust q until the force has a particular value, say 1 N for a separation of $r = 1$ m, and define the resulting q as the basic unit of charge. *It is, however, more precise to measure the magnetic force between two wires carrying equal currents, and therefore the fundamental SI electrical unit is the unit of current, from which the unit of charge is derived.*

EXAMPLE 7. If a charged body is placed near a neutral conductor, will it attract the conductor or repel it?

ANSWER The charged body will attract the conductor.

EXPLANATION Suppose, a charged body A having charge $+q$ is brought near a neutral conductor (Fig.1.16) B, the charged body induces a redistribution of charges within the conductor. As here, the charged body is positively charged, it will attract the negative charges in the neutral conductor B towards it, causing an accumulation of the negative charge ($-q'$) on the side of conductor facing the charged body and an equal positive charge ($+q'$) accumulated on the side of the conductor facing away from the charged body A. Remember that the magnitude of induced charge produced on each side is always less than the magnitude of the source charge, i.e., $|q'| < |q|$ (See EXAMPLE 68). These induced charges can be assumed to be concentrated at their centre of charges.

Clearly, the separation between the centre of charge of positively charged body A from the centre of negative induced charge in conductor is smaller as compared to the separation between the centre of charge of positively charged body A and the centre of positive induced charge in neutral conductor B. Therefore, from $F \propto 1/r^2$, we can say that force of attraction between positively charged body A and negative induced charge of neutral body B is greater than the repulsion between positively charged body A and positive induced charge on the neutral conductor B. So, the resultant force is attractive. Thus, a positively charged body attracts a neutral body.

Similarly, we can show that a negatively charged body also attracts a neutral body.

From the above example, we can conclude that- "A charged body

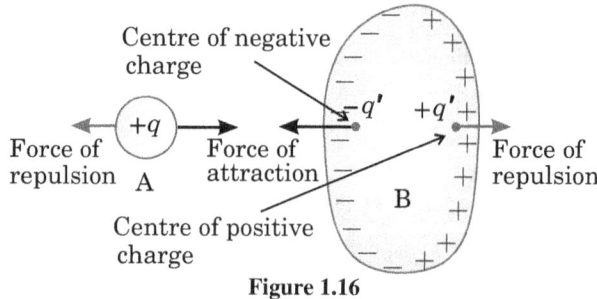

Figure 1.16

can attract a neutral body."

If there is attraction between two bodies then one of them may be neutral. But if there is repulsion between two bodies, both must be likely charged.

So, "repulsion is the sure test of electrification" whereas attraction is not.

EXAMPLE 8. Compare the magnitudes of the gravitational force of attraction and of the electric force of attraction between the electron and the proton in a hydrogen atom. According to Newtonian mechanics, what is the acceleration of the electron?

1.9. COULOMB'S LAW IN A DIELECTRIC MEDIUM

Assume that the distance between these particles in a hydrogen atom is 5.3×10^{-11} m.

APPROACH Gravitational force between two masses m_1 and m_2 separated by distance r, is given by-
$$F_g = G \frac{m_1 m_2}{r^2} \quad (1)$$
with $G = 6.67 \times 10^{-11}$ N.m^2/kg^2

Electrostatic force between two point charges q_1 and q_2 separated by distance r, is given by-
$$F_e = k \frac{q_1 q_2}{r^2} \quad (2)$$
with $k \approx 9.0 \times 10^9$ N.m^2/C^2

For electron-proton system- $m_1 = m_e = 9.11 \times 10^{-31}$kg, $q_1 = -e = -1.6 \times 10^{-19}$C, $m_2 = m_p = 1.67 \times 10^{-27}$kg and $q_2 = +e = +1.6 \times 10^{-16}$C

Substitute these values in Eq.(1) and (2) and then compare F_g and F_e.

SOLUTION From Eq.1, we get-
$$F_g = G \frac{m_e m_p}{r^2}$$
$$= \left(6.67 \times 10^{-11} \text{ N} \cdot \text{m}^2/\text{kg}^2\right)$$
$$\times \frac{(9.11 \times 10^{-31} \text{ kg})(1.67 \times 10^{-27} \text{ kg})}{(5.3 \times 10^{-11} \text{ m})^2}$$
$$= 3.6 \times 10^{-47} \text{ N}$$

and from Eq. 2, the magnitude of the electric force is-
$$F_e = \frac{1}{4\pi\epsilon_0} \frac{e \times e}{r^2}$$
$$= \left(8.99 \times 10^9 \text{ N} \cdot \text{m}^2/\text{C}^2\right) \times \frac{(1.60 \times 10^{-19}\text{C})^2}{(5.3 \times 10^{-11} \text{ m})^2}$$
$$= 8.2 \times 10^{-8} \text{ N}$$

The ratio of these forces is-
$$\frac{F_e}{F_g} = \frac{(8.2 \times 10^{-8} \text{ N})}{(3.6 \times 10^{-47} \text{ N})} = 2.3 \times 10^{39}$$

Thus the electric force is very large as compared to the gravitational force. Since the gravitational force is insignificant compared with the electric force, it can be neglected. The acceleration of the electron is then
$$a = \frac{F}{m} = \frac{8.2 \times 10^{-8} \text{ N}}{9.11 \times 10^{-31} \text{ kg}} = 9.0 \times 10^{22} \text{ m/s}^2$$

This is a very large acceleration. If it occurred along the electron's motion instead of centripetally, such an acceleration could boost the electron's velocity close to one-third of the speed of light in only a femtosecond $(10^{-15}$ s$)$.

Comment: For the ratio of the electric force and the gravitational force between the proton and electron, we would obtain the same immense value 2.3×10^{39} whatever the separation between the two particles, since both are inverse square forces. Also notice that for the given atomic-scale distance, the electric force has a measurable value, the same as weighing an 8 μg mass, whereas the gravitational force is far below the current limits of detection (the highest sensitivity attained by a measurement of force is near 10^{-20} N).

EXAMPLE 9. In EXAMPLE 3, what is the magnitude of the attractive force exerted by the electrons in a cup of water(0.25 kg water) on the protons in a second cup of water at a distance of 10 m?

APPROACH Substitute the values of charges, obtained in EXAMPLE 3, in Coulomb's force formula $F = k \frac{q_1 q_2}{r^2}$ and simplify for F_e.

SOLUTION According to the preceding example, the charge on the electrons in the cup is -1.3×10^7C and the charge on the protons is $+1.3 \times 10^7$C. If we treat both of these charges as point charges, the force on the protons is
$$F = \frac{1}{4\pi\epsilon_0} \frac{q_1 q_2}{r^2}$$
$$= \left(8.99 \times 10^9 \text{ N} \cdot \text{m}^2/\text{C}^2\right) \frac{(-1.3 \times 10^7\text{C})(1.3 \times 10^7\text{C})}{(10 \text{ m})^2}$$
$$= -1.5 \times 10^{22} \text{ N}$$

This is approximately the weight of a billion billion tons! This enormous attractive force on the protons is precisely canceled by an equally large repulsive force exerted by the protons in one cup on the protons in the other cup. Thus, the cups exert no net forces on each other.

EXAMPLE 10. The force of electrostatic repulsion between two fixed point charges separated by distance of 1 m is F. Now if we replace the two point charges by two metallic charged spheres each of radius 25 cm having the same charges as that of given point charges. then compare the force of repulsion in two cases.

APPROACH The electrostatic force acting between two point charges separated by distance r is given by
$$F = k \frac{q_1 q_2}{r^2}$$
So, if magnitudes of charges q_1 and q_2 remains fixed, then $F \propto 1/r^2$.

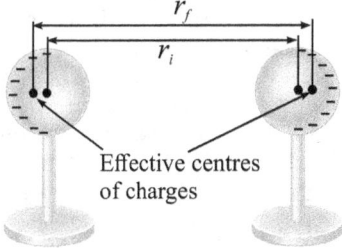

(a) Isolated charged metal sphere on an insulated stand.
(b) Centres of charges shift towards each other

(c) Centres of charges shift away from each other

Figure 1.17: Shifting of effective centers of charges when two charged metal spheres placed close to each other

In case of charged conducting spheres, due to mutual interaction between the charges, the effective distance (r) between centre of charges get altered, so the electrostatic force also get altered. If the charges of spherical conductors have opposite nature, then due to electrostatic attraction, the charge of each sphere moves towards the other and hence the effective distance between the centres of charge of the two spheres, decreases (Fig.1.17b). So, force of electrostatic attraction between the charges, increases. But, if the spheres have identical charges, then due to mutual repulsion between the charge of one sphere and the the other, the charge of of each sphere moves away from other sphere. So, in this case, the effective distance between the centres of charges of the two sphere, increases (Fig.1.17c) and hence force of electrostatic repulsion between the charged conducting sphere decreases.

SOLUTION Since, radii of both spheres (25 cm) cannot be neglected as compared to separation between their centres (1 m), therefore, given spheres cannot be cosidered as point objects. Now, in 2nd case due to mutual repulsion, the effective distance between their centre of charges will be increased $(r_f > r_i)$ so force of repulsion decreases (as $F \propto \frac{1}{r^2}$)

EXAMPLE 11. Five Styrofoam balls A, B, C, D and E are used in an experiment. Several experiments are performed on the balls and the following observations are made:
(i) Ball A repels C and attracts B.
(ii) Ball D attracts B and has no effect on E.
(iii) A negatively charged rod attracts both A and E.
For your information, an electrically neutral Styrofoam ball is very sensitive to charge induction and gets attracted considerably, if placed nearby a charged body. What are the charges, if any, on each ball?

APPROACH Balls can be considered like extended bodies unlike just point particle. Also keep in mind that a neutral body is attracted towards both positive as well as negative charges. Now observe the given observations and decide the sign of the balls.

SOLUTION From (i), as A repels C, so both A and C must be likely charged. Either both are +ve or both are −ve. As A also attract B, so charge on B should be opposite of A or B may be uncharged conductor.

From (ii) as D has no effect on E, so both D and E should be uncharged and as B attracts uncharged D, so B must be charged and D must be an uncharged conductor.

From (iii), a negatively charged rod attracts the charged ball A, so A must be +ve and from experiment (i) C must also be +ve and B must be −ve.

ANSWER: The nature of charge on each ball is shown in following table:

A	B	C	D	E
+	−	+	0	0

EXAMPLE 12. A particle of mass m carrying charge '$+q_1$' is revolving around a fixed charge '$-q_2$' in a circular path of radius r. Calculate the period of revolution.

APPROACH To revolve the charged particle q_1 of mass m around the charge $-q_2$, with a constant angular speed ω in a circular orbit of radius r (Fig.1.18), the required centripetal force $(F_c = mr\omega^2)$ is provided by the electrostatic force of attraction of charge $-q_2$ on charge q_1 $\left(F_e = \frac{1}{4\pi\epsilon_0}\frac{q_1q_2}{r^2}\right)$. Here, $\omega = 2\pi/T$. T is the time period of q_1 around the charge $-q_2$.

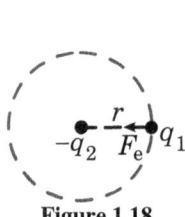

Figure 1.18

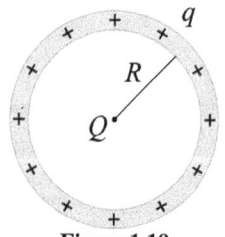

Figure 1.19

So, the condition for the charged particle q_1 to revolve around $-q_2$ in a circular orbit of radius r, is $F_e = F_c$, i.e.,

$$\frac{1}{4\pi\epsilon_0}\frac{q_1q_2}{r^2} = mr\omega^2 \qquad (1.20)$$

SOLUTION Substituting angular speed $\omega = \frac{2\pi}{T}$, in Eq.3.73, we get

$$\frac{1}{4\pi\epsilon_0}\frac{q_1q_2}{r^2} = mr\left(\frac{2\pi}{T}\right)^2$$

or $\quad \frac{1}{4\pi\epsilon_0}\frac{q_1q_2}{r^2} = \frac{4\pi^2 mr}{T^2}$

or $\quad T^2 = \frac{(4\pi\epsilon_0)r^2(4\pi^2 mr)}{q_1q_2}$ or $\quad T = 4\pi r\sqrt{\frac{\pi\epsilon_0 mr}{q_1q_2}}$

here the vector $\vec{r}$ is drawn from source charge $(-q_2)$ to charge q_1.

EXAMPLE 13. A ring of radius R has uniformly distributed charge q. A point charge Q is placed at the centre of the ring (Fig.1.19).
(a) Find the increase in tension in the ring after the point charge is placed at its centre.
(b) Find the increase in force between the two semicircular parts of the ring after the point charge is placed at the centre.
(c) Using the result found in part (b) find the force that the point charge exerts on one half of the ring.

APPROACH If there is no charge at the centre, there is some tension in the ring due to repulsion of charge present on the ring.

(a) When Q is placed at the centre the tension increases by ΔT. The direction of ΔT is tangential to the ring at each point on it. Now, consider any infinitesimally small portion ACB(say) of the ring and resolve components of ΔT, at both ends, into two mutually perpendicular directions - one component towards centre of the ring whereas other perpendicular to it. In equilibrium, the magnitude of net component of ΔT towards centre of the ring will be equal to the magnitude of outward electrostatic force on the segment ACB.

SOLUTION Let the infinitesimally small segment ACB of the ring subtends a very small angle $d\theta$ at the centre of the ring (see Fig.1.20a). From Fig.1.20a, it is clear that horizontal compo-

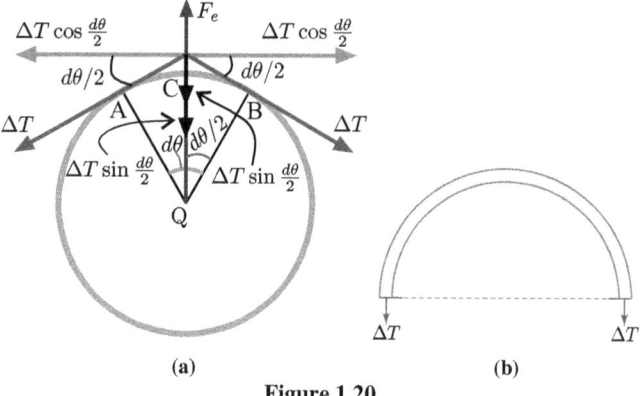

Figure 1.20

nents of ΔT at each end A and B is $\Delta T \cos\frac{d\theta}{2}$. These components are opposite in directions so get canceled.

Vertical components of ΔT at each end A and B is $\Delta T \sin\frac{d\theta}{2}$. These two components are directed towards the centre of the ring. Therefore, net inward force on the segment ACB due to tension of the ring is $2\Delta T \sin\frac{d\theta}{2}$. In mechanical equilibrium, this force will be equal to electrostatic repulsion F_e of Q at centre on the segment of the ring under consideration. Therefore,-

$$2\Delta T \sin\left(\frac{d\theta}{2}\right) = F_e \qquad (i)$$

If λ is the linear charge density of the ring, then charge on the segment ACB will be -

$$dq = \text{length} ACB \times \lambda = (Rd\theta)\lambda$$

Therefore, $F_e = \frac{kQdq}{R^2} = \frac{kQRd\theta}{R^2}$, here R is the radius of the ring.

Since, $d\theta/2$ is a very small quantity, therefore, $\sin(d\theta/2) = d\theta/2$.

On substituting these values in Eq.(i), we get -

1.9. COULOMB'S LAW IN A DIELECTRIC MEDIUM

or
$$2\Delta T \cdot \frac{d\theta}{2} = \frac{kQ\lambda R d\theta}{R^2}$$
$$\Delta T = \frac{kQ\lambda}{R} = \frac{kQq}{2\pi R^2}$$

(b) From Fig.1.20b it is clear that the answer is $2\Delta T = \frac{kQq}{\pi R^2}$

(c) Answer to (b) must be the answer to (c) also.

EXAMPLE 14. Three charges $q_1 = 1\ \mu C$, $q_2 = -2\ \mu C$ and $q_3 = 3\ \mu C$ are placed at the vertices of an equilateral triangle of side 1.0 m [Fig.1.21]. Find the net electric force acting on charge q_1.

APPROACH Charge q_2 will attract charge q_1 (along the line

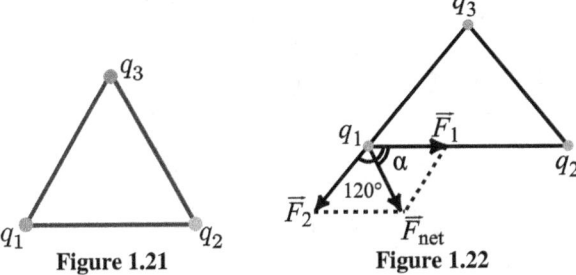

Figure 1.21   **Figure 1.22**

joining them) and charge q_3 will repel charge q_1 [Fig.1.22]. Therefore, two forces will act on q_1, one due to q_2 and another due to q_3. Since, the force is a vector quantity both of these forces (say $\vec{F}_1$ and $\vec{F}_2$) will be added by vector method. Now you can apply any one of the following methods of vector addition.

SOLUTION Method 1. In the Figure 1.22, the magnitude of force of q_2 on q_1

$$|\vec{F}_1| = F_1 = \frac{1}{4\pi\epsilon_0} \cdot \frac{q_1 q_2}{r^2}$$
$$= \frac{(9.0 \times 10^9)(1.0 \times 10^{-6})(2.0 \times 10^{-6})}{(1.0)^2} = 1.8 \times 10^{-2}\ \text{N}$$

Similarly, the magnitude of force of q_3 on q_1 -

$$|\vec{F}_2| = F_2 = \frac{1}{4\pi\epsilon_0} \cdot \frac{q_1 q_3}{r^2}$$
$$= \frac{(9.0 \times 10^9)(1.0 \times 10^{-6})(3.0 \times 10^{-6})}{(1.0)^2} = 2.7 \times 10^{-2}\ \text{N}$$

Now, $|\vec{F}_{net}| = \sqrt{F_1^2 + F_2^2 + 2F_1F_2 \cos 120°}$

$$= \left(\sqrt{(1.8)^2 + (2.7)^2 + 2(1.8)(2.7)\left(-\frac{1}{2}\right)}\right) \times 10^{-2}\ \text{N}$$

$$= 2.38 \times 10^{-2}\ \text{N}$$

Angle made by F_{net} from the direction of F_1, is given by
$$\tan\alpha = \frac{F_2 \sin 120°}{F_1 + F_2 \cos 120°}$$
$$= \frac{(2.7 \times 10^{-2})(0.87)}{(1.8 \times 10^{-2}) + (2.7 \times 10^{-2})\left(-\frac{1}{2}\right)}$$

or $\alpha = 79.2°$

Thus, the net force on charge q_1 is 2.38×10^{-2} N at an angle $\alpha = 79.2°$ with a line joining q_1 and q_2 as shown in the Fig. 1.22.

Method 2. In this method we assume a coordinate axes with q_1 at origin as shown in Fig.1.23. The coordinates of q_1, q_2 and q_3 in this coordinate system are $(0,0,0)$, $(1\ m, 0, 0)$ and $(0.5\ m, 0.87\ m, 0)$ respectively.

Now, force on charge q_1 due to charge q_2 is-
$$\vec{F}_1 = \frac{1}{4\pi\epsilon_0} \cdot \frac{q_1 q_2}{|\vec{r}_1 - \vec{r}_2|^3}(\vec{r}_1 - \vec{r}_2)$$
$$= \frac{(9.0 \times 10^9)(1.0 \times 10^{-6})(-2.0 \times 10^{-6})}{(1.0)^3}$$

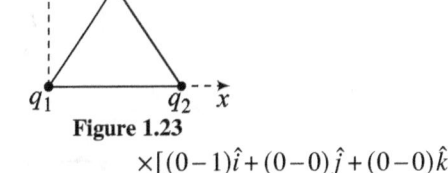

Figure 1.23

$$\times [(0-1)\hat{i} + (0-0)\hat{j} + (0-0)\hat{k}]$$
$$\Rightarrow \vec{F} = (1.8 \times 10^{-2}\hat{i})\ \text{N}$$

and force on q_1 due to charge q_3 is given by-
$$\vec{F}_2 = \frac{1}{4\pi\epsilon_0} \cdot \frac{q_1 q_3}{|\vec{r}_1 - \vec{r}_3|^3}(\vec{r}_1 - \vec{r}_3)$$
$$= \frac{(9.0 \times 10^9)(1.0 \times 10^{-6})(3.0 \times 10^{-6})}{(1.0)^3} \times$$
$$[(0-0.5)\hat{i} + (0-0.87)\hat{j} + (0-0)\hat{k}]$$
$$\Rightarrow \vec{F} = (-1.35\hat{i} - 2.349\hat{j}) \times 10^{-2}\ \text{N}$$

Therefore, net force on q_1 is-
$$\vec{F} = \vec{F}_1 + \vec{F}_2$$
$$= (0.45\hat{i} - 2.349\hat{j}) \times 10^{-2}\ \text{N}$$

Note: Once you write a vector in terms of $\hat{i}$, $\hat{j}$ and $\hat{k}$, there is no need of writing the magnitude and direction of vector separately.

EXAMPLE 15. Four identical particles, each having charge $+q$, are fixed at the corners of a square of side L. A fifth point charge $-Q$ lies a distance z along the line perpendicular to the plane of the square and passing through the centre of the square (Fig.1.24). (a) Show that the force exerted by the other four charges on $-Q$ is

$$\vec{F} = -\frac{4kqQz}{\left[z^2 + (L^2/2)\right]^{3/2}}\hat{k}$$

Note that this force is directed toward the centre of the square whether z is positive ($-Q$ above the square) or negative ($-Q$ below the square). (b) If z is small compared with L, the above expression reduces to $\vec{F} \approx -(\text{constant})z\hat{k}$. Why does this imply that the motion of the charge $-Q$ is simple harmonic, and what is the period of this motion if the mass of $-Q$ is m?

APPROACH 1 In first approach we break the forces applied

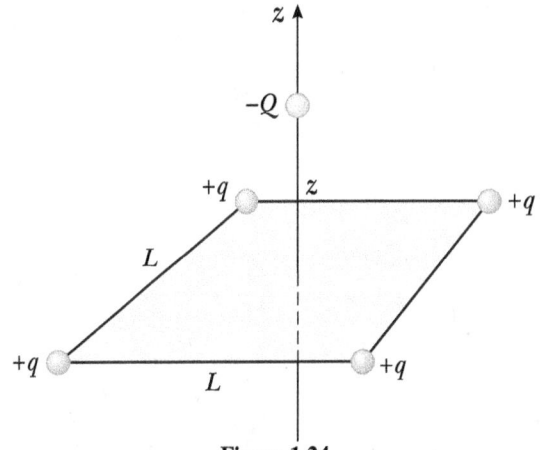

Figure 1.24

by all positive charges $+q$ on $-Q$ into horizontal and vertical components and then add these components to find the net force on Q.

SOLUTION (a) The distance from each corner to the centre of the square of side length L, is $L/\sqrt{2}$

From Fig.1.25, the distance from each positive charge to $-Q$ is-

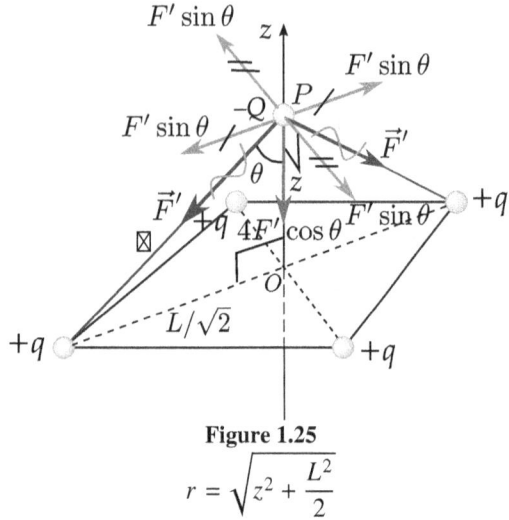

Figure 1.25

$$r = \sqrt{z^2 + \frac{L^2}{2}}$$

Each positive charge exerts a force directed along the line joining $+q$ and $-Q$, of magnitude

$$F' = k\frac{Qq}{r^2} = \frac{kQq}{z^2 + L^2/2}$$

The line of force makes an angle θ with the z-axis. From Fig.1.25

$$\cos\theta = \frac{z}{\sqrt{z^2 + L^2/2}}$$

The four charges together exert forces whose horizontal components $F'\sin\theta$ add to zero, while the vertical z-components $F'\cos\theta$ add to

$$\vec{F} = 4F'\cos\theta(-\hat{k})$$

$$\Rightarrow \quad \vec{F} = -\frac{4kQqz}{(z^2 + L^2/2)^{3/2}}\hat{k}$$

If m is the mass and $\frac{d^2z}{dt^2}$ is the acceleration of the charged particle corresponding to displacement z from position P, then above expression can also be written as

$$m\frac{d^2z}{dt^2} = -\frac{4kQqz}{(z^2 + L^2/2)^{3/2}}$$

$$\Rightarrow \quad \frac{d^2z}{dt^2} = -\frac{4kQqz}{m(z^2 + L^2/2)^{3/2}}$$

For $z \ll L$, above expression changes to-

$$\frac{d^2z}{dt^2} = -\left(\frac{8\sqrt{2}kQq}{mL^3}\right)z$$

Which is of the form of SHM, $\frac{d^2z}{dt^2} = -\omega^2 z$.

Here, $\omega = \sqrt{\left(\frac{8\sqrt{2}kQq}{mL^3}\right)}$ is the angular frequency of the SHM.

Time period of SHM, $T = \frac{2\pi}{\omega} = 2\pi\sqrt{\frac{mL^3}{8\sqrt{2}kQq}}$.

APPROACH 2 In second approach [Fig.1.26], we use Coulomb's law in vector form and then apply the superposition principle of electric forces

$$\vec{F} = \vec{F}_1 + \vec{F}_2 + \vec{F}_3 + \vec{F}_4$$

$$= \frac{q_1 q_0 (\vec{r}_p - \vec{r}_1)}{4\pi\epsilon_0 |\vec{r}_p - \vec{r}_1|^3} + \frac{q_2 q_0 (\vec{r}_p - \vec{r}_2)}{4\pi\epsilon_0 |\vec{r}_p - \vec{r}_2|^3} + \frac{q_3 q_0 (\vec{r}_p - \vec{r}_3)}{4\pi\epsilon_0 |\vec{r}_p - \vec{r}_3|^3}$$

$$+ \frac{q_4 q_0 (\vec{r}_p - \vec{r}_4)}{4\pi\epsilon_0 |\vec{r}_p - \vec{r}_4|^3}$$

SOLUTION The arrangement of charges is shown in Fig.1.26.

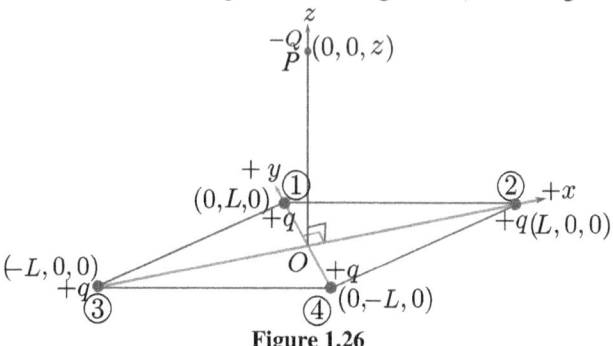

Figure 1.26

According' to given problem,
$q_1 = q_2 = q_3 = q_4 = +q$, $\vec{r}_1 = L\hat{j}$, $\vec{r}_2 = L\hat{i}$, $\vec{r}_3 = -L\hat{i}$; $\vec{r}_4 = -L\hat{j}$ and $\vec{r}_p = z\hat{k}$

$\therefore \quad \vec{r}_p - \vec{r}_1 = z\hat{k} - L\hat{i}$,

$$|\vec{r}_p - \vec{r}_1| = \sqrt{z^2 + L^2} = (z^2 + L^2)^{1/2}$$

$$\vec{r}_p - \vec{r}_2 = z\hat{k} - L\hat{j}, |\vec{r}_p - \vec{r}_2| = (z^2 + L^2)^{1/2}$$

$$\vec{r}_p - \vec{r}_3 = z\hat{k} + L\hat{i}, |\vec{r}_p - \vec{r}_3| = (z^2 + L^2)^{1/2}$$

and $\vec{r}_p - \vec{r}_4 = z\hat{k} + L\hat{j}, |\vec{r}_p - \vec{r}_4| = (z^2 + L^2)^{1/2}$

$$\vec{F} = \vec{F}_1 + \vec{F}_2 + \vec{F}_3 + \vec{F}_4$$

$$\Rightarrow \vec{F} = \frac{q_1 q_0 (\vec{r}_p - \vec{r}_1)}{4\pi\epsilon_0 |\vec{r}_p - \vec{r}_1|^3} + \frac{q_2 q_0 (\vec{r}_p - \vec{r}_2)}{4\pi\epsilon_0 |\vec{r}_p - \vec{r}_2|^3}$$

$$+ \frac{q_3 q_0 (\vec{r}_p - \vec{r}_3)}{4\pi\epsilon_0 |\vec{r}_p - \vec{r}_3|^3} + \frac{q_4 q_0 (\vec{r}_p - \vec{r}_4)}{4\pi\epsilon_0 |\vec{r}_p - \vec{r}_4|^3}$$

Putting the values, we get

$$\vec{F} = -\frac{qQ}{4\pi\epsilon_0 (z^2 + L^2)^{3/2}}\left[(z\hat{k} - L\hat{i}) + (z\hat{k} - L\hat{j}) + (z\hat{k} + L\hat{i}) + (z\hat{k} + L\hat{j})\right]$$

$$= -\frac{qQ}{4\pi\epsilon_0 (z^2 + L^2)^{3/2}}(4z\hat{k})$$

Since, $\frac{1}{4\pi\epsilon_0} = k$, therefore, above expression can also be written as-

$$\vec{F} = -\frac{4kqQz}{(z^2 + L^2)^{3/2}}(\hat{k})$$

EXAMPLE 16. Three charges $q/2$, q and $q/2$ are placed at the corners A, B and C of a square of side 'a' as shown in Fig.1.27. What is the magnitude of electric field ($\vec{E}$) at the corner D of the square?

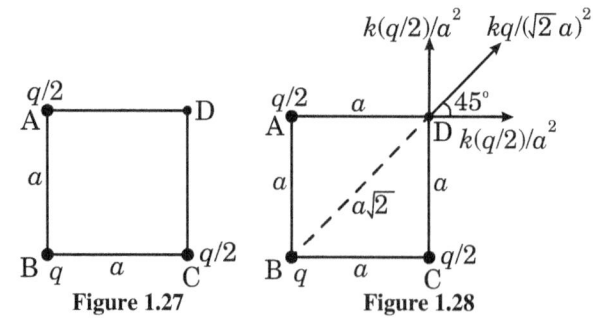

Figure 1.27   Figure 1.28

SOLUTION From Fig.1.28, the net electric field at D

$$(E_{net})_D = k\frac{q}{\left(a\sqrt{2}\right)^2} + k\frac{q/2}{a^2}\cos 45° + k\frac{q/2}{a^2}\cos 45°$$

$$= \frac{kq}{2a^2} + \frac{\sqrt{2}kq}{2a^2} = \frac{q}{4\pi\epsilon_0 a^2}\left(\frac{1}{2} + \frac{1}{\sqrt{2}}\right)$$

1.10 Equilibrium of Charged Particles

If in a system of charged particles, the net electric force on every particle is zero, then they are said to be in equilibrium. In 1842, British mathematician Samuel Earnshaw proved that a collection of point charges cannot be maintained in a stable stationary equilibrium configuration solely by the electrostatic interaction of the charges. Discussion of this statement is given in "Earnshaw's theorem" in section 1.24.

1.10.1 Equilibrium of Three Point Charges

Suppose, three positive point charges q_1, q_2 and q are placed along a line as shown in Fig.1.29. The separation between charges q_1 and q_2 is r. Charge q is placed at distance x from charge q_1. For the equilibrium of all the three charges, the net electrostatic

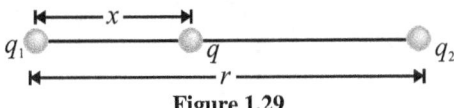

Figure 1.29

force on each of the above charges should be zero.
Now, for the equilibrium of charge q, the net electrostatic force on it should be zero, i.e., $F_q = 0$

$$\text{or} \quad k\frac{q_1 q}{x^2} - k\frac{q_2 q}{(r-x)^2} = 0 \quad (1.21)$$

Similarly, for the equilibrium of charge q_1, we have
$F_{q_1} = 0$, i.e., $-k\frac{q_1 q_2}{r^2} - k\frac{q_1 q}{x^2} = 0$

$$\text{or} \quad k\frac{q_1 q_2}{r^2} + k\frac{q_1 q}{x^2} = 0 \quad (1.22)$$

From Eq. (1.21), we have

$$\frac{q_1}{x^2} = \frac{q_2}{(r-x)^2} \Rightarrow \left(\frac{r-x}{x}\right)^2 = \frac{q_2}{q_1}$$

$$\Rightarrow \left(\frac{r-x}{x}\right) = \pm\sqrt{\frac{q_2}{q_1}} \Rightarrow \left(\frac{r}{x} - 1\right) = \pm\sqrt{\frac{q_2}{q_1}}$$

$$\Rightarrow \frac{r}{x} = 1 \pm \sqrt{\frac{q_2}{q_1}}$$

Solving for x, we get-

$$x = \frac{r}{1 \pm \sqrt{\frac{q_2}{q_1}}} \quad (1.23)$$

From Eq. (1.23), it is clear if q_1 and q_2 are unlike charges, then x will be imaginary. So, for a real value of x, both q_1 and q_2 must be like charges, i.e., either both of them must be positive or both of them must be negative.
Now, from Eq.(1.22), we have

$$q = -\frac{q_1 q_2}{r^2}\frac{x^2}{q_1} = -\frac{q_2}{r^2}x^2$$

on substituting the value of x, we get

$$q = -\frac{q_2}{r^2}\left(\frac{r}{1 \pm \sqrt{\frac{q_2}{q_1}}}\right)^2 \quad (1.24)$$

Negative sign in above Eq.(1.24) shows that, q and q_2 will be opposite in nature. But q_2 and q_1 are like charges, therefore, the nature of q is also opposite to q_1.
So, the result is that the nature of q_1 and q_2 will be similar whereas that of q is opposite to q_1 and q_2
Eq. (1.23) and (1.24) can also be written as -

$$x = \frac{\sqrt{q_1}}{\sqrt{q_1} \pm \sqrt{q_2}}r \quad \text{and} \quad q = \frac{-q_1 q_2}{\left(\sqrt{q_1} \pm \sqrt{q_2}\right)^2}$$

respectively.

EXAMPLE 17. Two particles A and B having charges 8.0×10^{-6}C and -2.0×10^{-6}C respectively are held fixed with a separation of 20 cm. Where should a third charged particle be placed so that it does not experience a net electric force?
APPROACH Since, the net electric force on charge C should be equal to zero, the force due to A and B must be equal and opposite in direction. Hence, the particle should be placed on the line AB. As A and B have charges of opposite signs, C cannot be between A and B. Also, A has larger magnitude of charge than B. So, we expect the charge C to be closer to B, the charge that has smaller magnitude, than A. The situation is shown in Figure 1.30

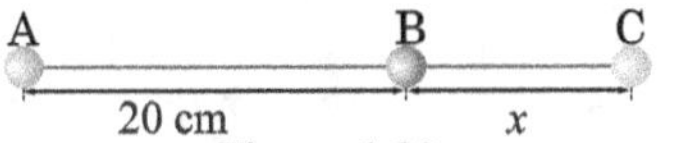

Figure 1.30

SOLUTION Suppose $BC = x$ and the charge on C is Q.

The force due to $A = \dfrac{(8 \cdot 0 \times 10^{-6}\text{C})Q}{4\pi\epsilon_0(20\,\text{cm} + x)^2}$.

The force due to $B = \dfrac{(2 \cdot 0 \times 10^{-6}\text{C})Q}{4\pi\epsilon_0 x^2}$.

They are oppositely directed and to have a zero resultant, they should be equal in magnitude. Thus,

$$\frac{8}{(20\,\text{cm} + x)^2} = \frac{2}{x^2} \Rightarrow x = 20\,\text{cm}$$

$$\Rightarrow \left(\frac{20\,\text{cm} + x}{x}\right)^2 = \frac{8}{2} = 4 \Rightarrow \frac{20\,\text{cm} + x}{x} = \pm 2$$

$$\Rightarrow \frac{20\,\text{cm}}{x} = -1 \pm 2 \Rightarrow x = \frac{20}{-1 \pm 2}\,\text{cm}$$

From here, we get- $x = 20\,\text{cm}, -\dfrac{20}{3}\,\text{cm}$.

Since, x cannot be negative, therefore we take $x = 20\,\text{cm}$.

EXAMPLE 18. A point charge $q_1 = 4q_0$ is placed at origin. Another point charge $q_2 = -q_0$ is placed at $x = 12$ cm. Charge of proton is q_0. The proton is placed on x-axis so that the electrostatic force on the proton in zero. In this situation, find the position of the proton from the origin.

SOLUTION Since the given charges have opposite natures, the zero electric field position will be on the line joining the given charges, outside the charges, and closer to the charge with the smaller magnitude.
This situation is shown in Fig.1.31. P is the position of the zero electric field point, which is located at a distance x_0 from point A, where a charge of $-q_0$ is placed. Now, if a proton is placed at P, it experiences zero electric force.
Since, net electric field is zero at point P, therefore-

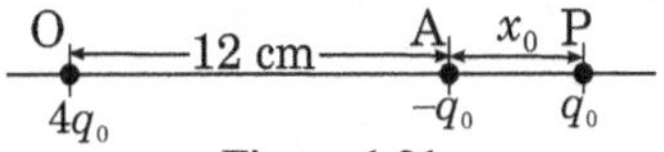

Figure 1.31

$$\frac{q_0}{x_0^2} = \frac{4q_0}{(x_0 + 12)^2} \Rightarrow x_0 + 12 = 2x_0 \Rightarrow x_0 = 12\,\text{cm}$$

Therefore, the position of proton from origin O is,
$$x = x_0 + 12 = 24\,\text{cm}.$$

1.10.2 Equilibrium of Symmetric Geometrical Point Charged Systems

Figure 1.32 shows an equilateral triangle and a square of side length a. A charged particle of equal magnitude q is placed at each corner and a point charge Q at their circumcentre. If the system is in electrostatic equilibrium, then net electrostatic force on each point charge should be.

In equilibrium, to determine the value of charge Q placed at

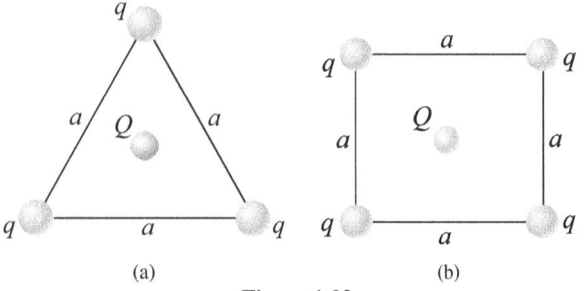

Figure 1.32

the centre, you have to apply the condition of equilibrium (i.e., $\Sigma \vec{F} = 0$) at any corner charge only not on the charge Q placed at circumcentre becase in that case each term get cancelled out and you wont find Q.

(i) Calculation of Q Placed at the Center of Triangle:
This situation with free body diagram of charge q at vertex A, is shown in Fig.1.33. From Fig.1.33, the distance $OA = (a/2)\sec 30° = a/\sqrt{3}$.

The electrostatic forces acting on charge q at A are-

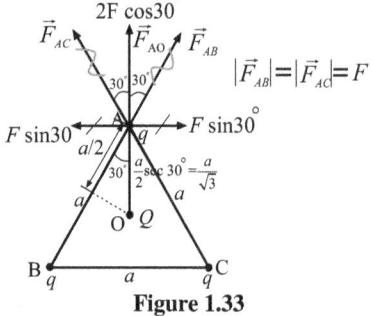

Figure 1.33

1. The force $\vec{F}_{AB}$ (along BA) due to charge q at B.
2. The force $\vec{F}_{AC}$ (along CA) due to charge q at C.
3. The force $\vec{F}_{AO}$ (along OA) due to charge Q at centre O.

Note that, $|\vec{F}_{AB}| = |\vec{F}_{AC}| = k\frac{q^2}{a^2} = F$ (say)

The horizontal components of each of these two forces are $F\sin 30°$ and are oppositely directed, therefore they get cancelled with each other. However, the vertical components due to each force is $F\cos 30°$ and directed along OA, therefore they get added. So, the resultant of these forces will be $2F\cos 30°$ (along OA)

The electrostatic force applied by the charge Q at centre O on charge q at A is -

and $|\vec{F}_{AO}| = k\frac{Qq}{(a/\sqrt{3})^2} = k\frac{3Qq}{a^2}$

Forces F_{AO} and $2F\cos 30°$ both are directed along OA, therefore, net force on A-

$\Sigma F = F_{AO} + 2F\cos 30° = k\frac{3Qq}{a^2} + k\frac{q^2\sqrt{3}}{a^2}$

In equilibrium, $\Sigma F = 0$

$$\Rightarrow \boxed{Q = -\frac{q}{\sqrt{3}}} \quad (1.25)$$

Here, negative sign indicates that Q will be opposite in nature to q.

(ii) Calculation of Q Placed at the Center of Square: This situation with free body diagram of a charge q at C is shown in Fig.1.34. Considering AB along the positive direction of x axis and AD along the positive direction of y axis.
The electrostatic forces acting on charge q at C are-
1. Force $\vec{F}_{CA}$, due to charge q at A
2. Force $\vec{F}_{CB}$, due to charge q at B
2. Force $\vec{F}_{CD}$, due to charge q at D
2. Force $\vec{F}_{CO}$, due to charge Q at O
Net electrostatic force on charge q at C, is-

$$\Sigma\vec{F} = \vec{F}_{CA} + \vec{F}_{CB} + \vec{F}_{CD} + \vec{F}_{CO} \quad (1.26)$$

Here, $\vec{F}_{CA} = k\frac{q^2}{(a\sqrt{2})^2}\left(\frac{\hat{i}+\hat{j}}{\sqrt{2}}\right)$, $\vec{F}_{CB} = k\frac{q^2}{a^2}\hat{j}$, $\vec{F}_{CD} = k\frac{q^2}{a^2}\hat{i}$,

and $\vec{F}_{CO} = k\frac{qQ}{(a/\sqrt{2})^2}\left(\frac{\hat{i}+\hat{j}}{\sqrt{2}}\right) = k\frac{2qQ}{a^2}\left(\frac{\hat{i}+\hat{j}}{\sqrt{2}}\right)$

Substituting these values in Eq.1.26, we get-

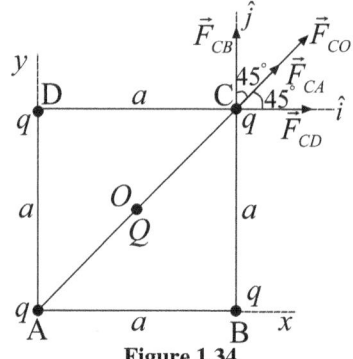

Figure 1.34

$$\Sigma\vec{F} = k\frac{q}{a^2}\left[\left(\frac{q}{2\sqrt{2}} + q + Q\sqrt{2}\right)\hat{i} + \left(\frac{q}{2\sqrt{2}} + q + Q\sqrt{2}\right)\hat{j}\right]$$

In equilibrium, $\Sigma\vec{F} \Rightarrow \frac{q}{2\sqrt{2}} + q + Q\sqrt{2} = 0$

$$\Rightarrow \boxed{Q = \frac{-q\left(2\sqrt{2}+1\right)}{4}} \quad (1.27)$$

Here, again negative sign shows that Q will be opposite in nature to q.

Comment: Students are advised to keep results 1.25 and 1.27 in their memory for simplifying such types of problems in a relatively quicker way.

1.10.3 Equilibrium of Suspended Point Charge System

If two charged point objects are suspended by threads in a given medium, then in equilibrium state, net force on each object must be zero. To solve such kind of problems, make FBD of of any one object (if necessary make FBDs of both objects). Resolve all forces acting on any one object in to two mutually perpendicular components along x and y direction (say). If ΣF_x and ΣF_y are the net forces along x and y directions, then, in equilibrium, $\Sigma F_x = 0$ and $\Sigma F_y = 0$. In other words, net force acting along positive direction of x axis must get balanced with net force acting in negative x direction and also net force acting along positive direction of y axis must get balanced with the net force acting in negative y direction. Don't forget to consider upthrust applied by medium if it's density is not negligible as compared to water. If V is the immersed volume of object in the medium, g is gravitational acceleration and ρ is the density of medium, then upthrust applied by medium on the object is $V\rho g$. The upthrust force of air can be neglected because average density of air at 1 atm and 20° is $1.204 kg/m^3$ which is very small as compared to density of water which is $10^3 kg/m^3$ at 1 atm and 20°.

To understand the working see following solved examples:

EXAMPLE 19. A simple electroscope for the detection and

1.10. EQUILIBRIUM OF CHARGED PARTICLES

measurement of electric charge consists of two small foil-covered cork balls of 1.5×10^{-4} kg each suspended by threads 10 cm long (see Fig.1.35). When equal electric charges are placed on the balls, the electric repulsive force pushes them apart, and the angle between the threads indicates the magnitude of the electric charge. If the equilibrium angle between the threads is 60°, what is the magnitude of the charge?

APPROACH Make FBD of any one sphere. Resolve all forces

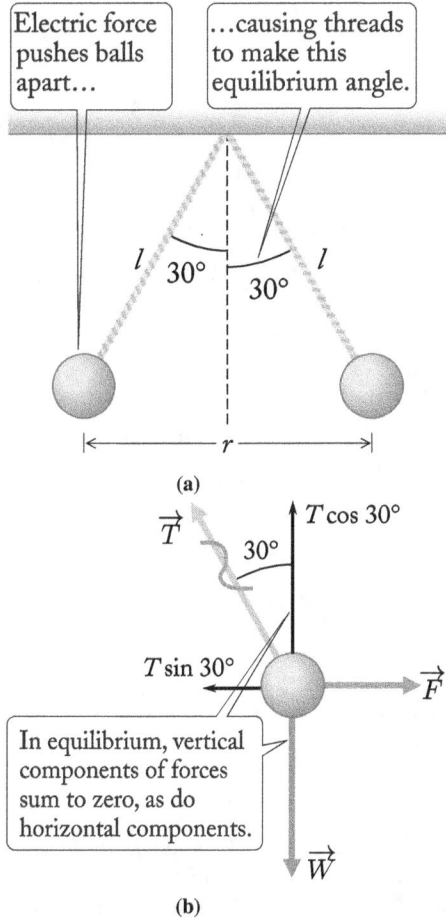

Figure 1.35: (a) Two equal charged balls suspended by threads. (b) "Free-body" diagram for the right ball.

acting on it in to two mutually perpendicular components along x and y direction (say). If ΣF_x and ΣF_y are the net forces along x and y directions, then, in equilibrium, $\Sigma F_x = 0$ and $\Sigma F_y = 0$.

SOLUTION Figure 1.35b shows a "free-body" diagram for right ball. The forces acting on the right ball are-
1. Electrostatic force F_e (along positive x direction)
2. Gravitational force mg (along negative y direction)
3. tension applied by string T (away from sphere along the string) x and y components of tension T are $T \sin 30°$ and $T \cos 30°$ respectively.
Net force on the ball, along x axis is-
$$\Sigma F_x = F - T \sin 30°$$
Net force on the sphere, along y axis is-
$$\Sigma F_y = T \cos 30° - W = T \cos 30° - mg$$
In equilibrium, $\sum F_x = 0 \Rightarrow F = T \sin 30°$ ... (1)
and $\sum F_y = 0 \Rightarrow mg = T \cos 30°$ ... (2)
We can eliminate the tension from above equations by taking the ratio of these equations, yielding
$$F = mg \tan 30° \qquad \cdots (3)$$
From Fig.1.35a, we see that the distance between the balls is $r = 2l \sin 30°$, so, Coulomb's Law gives

$$F = \frac{1}{4\pi\epsilon_0} \frac{q^2}{(2l \sin 30°)^2} \qquad \cdots (4)$$

Equating Eq.(3) and (4) for F, we get
$$mg \tan 30° = \frac{1}{4\pi\epsilon_0} \frac{q^2}{(2l \sin 30°)^2}$$
and $q = \sqrt{4\pi\epsilon_0 mg \tan 30°} \times 2l \sin 30°$
$= [(4\pi) (8.85 \times 10^{-12} \text{C}^2/\text{N} \cdot \text{m}^2) (1.5 \times 10^{-4} \text{ kg})$
$\times (9.81 \text{ m/s}^2) (\tan 30°)]^{1/2} \times (2)(0.10 \text{ m})(\sin 30°)$
$= 3.1 \times 10^{-8}$ C

EXAMPLE 20. Two identically charged spheres are suspended by strings of equal length L. The charge on each sphere is q and mass is m. Find the separation between them in equilibrium- (a) if the the system is located in gravitational field of earth, (b) if the system is located in gravity free space.

APPROACH Make FBD of either sphere. Resolve all forces

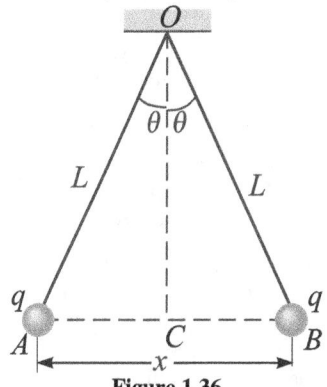

Figure 1.36

acting on it in to two mutually perpendicular components one along x and other along y direction (say). If ΣF_x and ΣF_y are the net forces along x and y directions, then, in equilibrium, $\Sigma F_x = 0$ and $\Sigma F_y = 0$.

If whole setup is placed in gravity free space, then there will be no downward gravitational force on any sphere. In this case, the angle between the strings becomes 180°.

SOLUTION FBD of left sphere is shown in Fig.1.37. The forces acting on the left sphere are-
1. Electrostatic force F_e (along negative x direction)
2. Gravitational force mg (along negative y direction)
3. tension applied by string T (away from sphere along the string) x and y components of tension T are $T \sin \theta$ and $T \cos \theta$ respectively. The separation between the spheres in equilibrium is x.
Net force on the sphere, along x axis is-
$$\Sigma F_x = T \sin \theta - F_e$$
Net force on the sphere, along y axis is-
$$\Sigma F_y = T \cos \theta - mg$$
In equilibrium,
$$\Sigma F_x = 0 \Rightarrow T \sin \theta = F_e \qquad (i)$$
and $\Sigma F_y = 0 \Rightarrow T \cos \theta = mg \qquad (ii)$
On dividing Eq. (i) by (ii), we get-
$$\tan \theta = \frac{F_e}{mg}$$
Here, electrostatic force, $F_e = k \frac{q^2}{x^2}$, therefore,
$$\tan \theta = \frac{kq^2}{mgx^2}$$
If θ is small then-
$$\tan \theta \approx \sin \theta = \frac{x}{2L} \Rightarrow \frac{x}{2L} = \frac{kq^2}{x^2 mg}$$

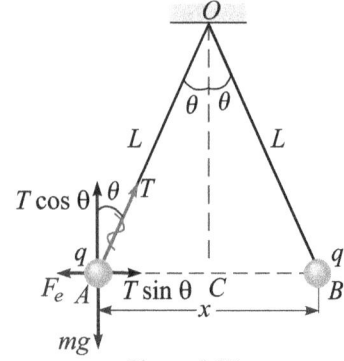

Figure 1.37

$$\Rightarrow x^3 = \frac{2kq^2 L}{mg} \Rightarrow x = \left[\frac{q^2 L}{2\pi \epsilon_0 \, mg}\right]^{\frac{1}{3}}$$

(b) If whole set up is taken into a gravity free space or in an artificial satellite ($g_{eff} \simeq 0$), then from FBD shown in Fig.1.38

$$T = F_e = \frac{kq^2}{4L^2}$$

EXAMPLE 21. Two identical balls each having a density

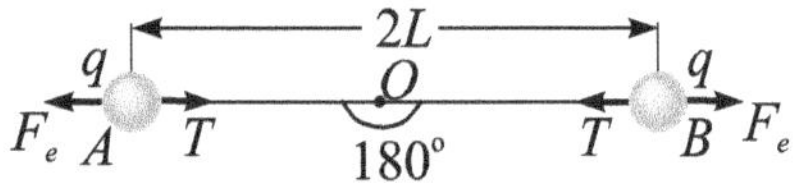

Figure 1.38

ρ are suspended from a common point by two insulating strings of equal length. Both the balls have equal mass and charge. In equilibrium each string makes an angle θ with vertical. Now, both the balls are immersed in a liquid. As a result the angle θ does not change. The density of the liquid is σ. Find the dielectric constant of the liquid.

APPROACH Make FBD of either ball in air as well as in liquid. Write the force balance equations for equilibrium in air and in liquid and then solve for dielectric constant.

SOLUTION Each ball is in equilibrium under the following three forces:
(i) tension,
(ii) electric force and
(iii) weight, so, Lami's theorem* can be applied.

The weight of ball in air, $W = mg = V\rho g$
The upthrust applied by liquid is, $F_{th} = V\sigma g$
Therefore, the apparent weight of the ball in liquid,
$$W' = W - F_{th} = V\rho g - V\sigma g = V(\rho - \sigma)g$$
The electrostatic force between the balls in liquid-
$$F'_e = \frac{F_e}{\kappa}$$
where, F_e is the electrostatic force between the balls in air and κ = dielectric constant of liquid.

Applying Lami's theorem in vacuum (Fig.1.39a)
$$\frac{W}{\sin(90° + \theta)} = \frac{F_e}{\sin(180° - \theta)}$$
$$\Rightarrow \frac{W}{\cos\theta} = \frac{F_e}{\sin\theta} \qquad (1.28)$$
Similarly in liquid (Fig.1.39b),

---

*Lami's Theorem: If $\vec{P}, \vec{Q}$ and $\vec{R}$ are three concurrent forces such that the angle between $\vec{Q}$ and $\vec{R}$ is α, between $\vec{R}$ and $\vec{P}$ it is β and between $\vec{P}$ and $\vec{Q}$ the angle is γ (Fig.1.49a), then in equilibrium-

$$\boxed{\frac{P}{\sin\alpha} = \frac{Q}{\sin\beta} = \frac{R}{\sin\gamma}}$$

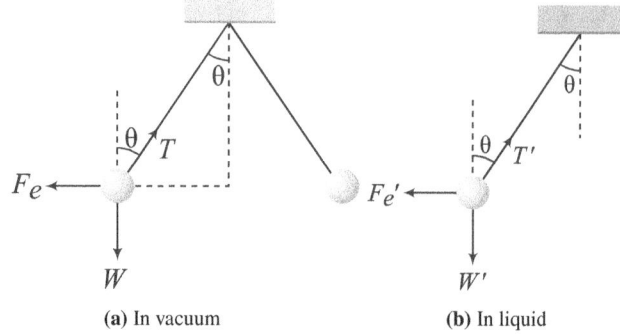

(a) In vacuum (b) In liquid

Figure 1.39

$$\frac{W'}{\cos\theta} = \frac{F'_e}{\sin\theta} \qquad (1.29)$$

Dividing Eq.(1.28) by Eq. (1.29), we get-
$$\frac{F}{F'} = \frac{W}{W'} = \frac{V\rho g}{V(\rho - \sigma)g} \quad (V = \text{volume of ball})$$
or
$$\kappa = \frac{\rho}{\rho - \sigma}$$

Note: In the liquid F_e and W have changed. Therefore, T will also change.

EXAMPLE 22. For the system shown in Fig.1.40a, find Q for which resultant force on q is zero.

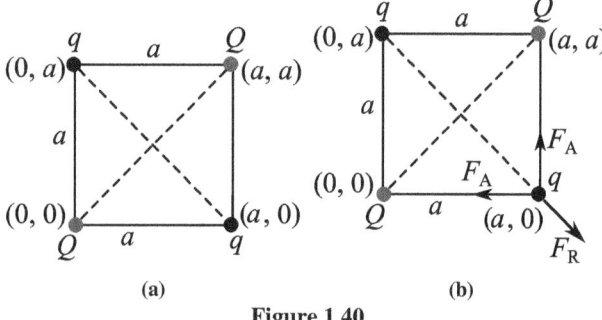

Figure 1.40

APPROACH Make FBD of charge q. Write force balance equations and simplify for required value of Q.

SOLUTION FBD of charge q at $(a, 0)$ is shown in Fig.1.40b. For force on q at $(a, 0)$ to be zero, charges q and Q must be of opposite nature.

Net force of attraction of both Q on q at $(a, 0)$ = force of repulsion of q at $(0, a)$ on q at $(a, 0)$.

$$\sqrt{F_A^2 + F_B^2 + 2F_A F_B \cos 90} = F_R$$

Here, F_A is the force between each Q and q whereas F_R is the repulsion between each q and q.

Simplifying above equation, we get
$$\sqrt{2}F_A = F_R \Rightarrow \sqrt{2}\frac{kQq}{a^2} = \frac{kq^2}{(\sqrt{2}a)^2}$$
$$\Rightarrow q = 2\sqrt{2}Q.$$

Therefore, the required charge = $-2\sqrt{2}\, Q$

EXAMPLE 23. Two identically charged spheres are suspended by strings of equal length. The strings make an angle of 30° with each other in air (Fig.1.41). When suspended in a liquid of density 0.8 g/cc the angle remains same. What is the dielectric constant of liquid. Density of sphere = 1.6 g/cc.

APPROACH Make FBD of either ball in air and in liquid. Write translational equilibrium conditions in both case. Now simplify for dielectric constant κ of liquid.

SOLUTION In air: This situation is shown in Fig.1.42a. The forces acting on the left charged sphere are-

1.10. EQUILIBRIUM OF CHARGED PARTICLES

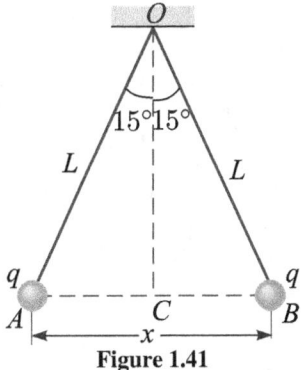

Figure 1.41

1. Gravitational force mg, acting in downward direction,
2. Electrostatic repulsive force F_e, acting towards left,
3. Tension in the string T, acting upward along the string.

In equilibrium, all forces and their components acting on left charged sphere, are shown in Fig.1.42
In horizontal direction we have-

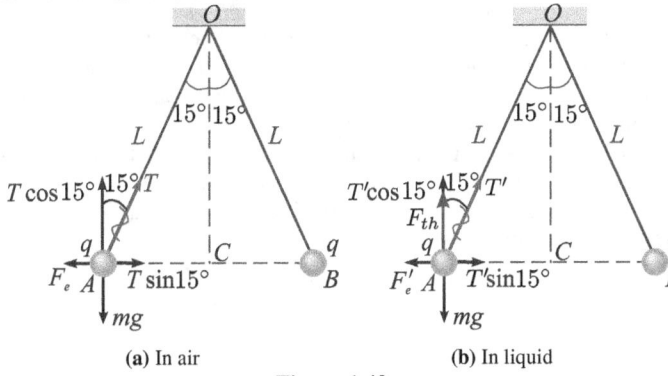

(a) In air (b) In liquid

Figure 1.42

$$T \sin 15° = F_e \qquad (i)$$

In vertical direction we have-
$$T \cos 15° = mg \qquad (ii)$$

On dividing Eq. (i) by Eq. (ii), we get-

$$\boxed{\tan 15° = \frac{F_e}{mg}} \qquad (iii)$$

In liquid: This situation is shown in Fig.1.42a. When set up is immersed in the liquid medium as shown in Figure 1.42b, the forces on the sphere are -
1. Gravitational force mg, acting in downward direction,
2. Electrostatic repulsive force F'_e, acting towards left,
3. Tension in the string T', acting upward along the string.
4. Upthrust of liquid F_{th}, acting in upward direction

In equilibrium, all forces and their components acting on left charged sphere, are shown in Fig.1.42b
In horizontal direction we have -
$$T' \sin 15° = F'_e \qquad (iv)$$
In vertical direction we have-
$$T' \cos 15° = mg - F_{th} \qquad (v)$$
On dividing Eq. (iv) by Eq. (v), we get-

$$\boxed{\tan 15° = \frac{F'_e}{mg - F_{th}}} \qquad (vi)$$

From Eq. (iii) and (vi), we can write-
$$\frac{F_e}{mg} = \frac{F'_e}{mg - F_{th}}$$

$$\Rightarrow \quad \frac{F_e}{mg} = \frac{F_e/\kappa}{mg - F_{th}} \qquad [\because F'_e = F_e/\kappa]$$

$$\Rightarrow \quad \kappa = \frac{mg}{mg - F_{th}}$$

$$\Rightarrow \quad \kappa = \frac{mg}{mg - V\rho_l g} \qquad [\because F_{th} = V\rho_l g]$$

$$\Rightarrow \quad \kappa = \frac{1}{1 - \frac{V\rho_l}{m}} = \frac{1}{1 - \frac{V\rho_l}{V\rho_s}} \qquad [\because m = V\rho_s]$$

$$\Rightarrow \quad \kappa = \frac{1}{1 - \frac{0.8}{1.6}} = 2$$

EXAMPLE 24. Two small equally charged spheres, each of mass m, are suspended from the same point by silk threads of length l. In equilibrium, the distance between the spheres is x ($x \ll l$). Find the rate dq/dt with which the charge leaks off each sphere if their approach velocity varies as $v = a/\sqrt{x}$, where a is a constant.

APPROACH Initially, the charged spheres were in equilibrium when they were separated by distance x. Now consider only left sphere in equilibrium. The forces acting on this sphere are-

1. Gravitational force $m\vec{g}$ - acting downwards.
2. Tension applied by thread (T) - acting away along the thread.
3. Electrostatic repulsion F_e applied by second sphere- acting leftwards.

If ΣF_x and ΣF_y are the net electrostatic forces acting on the left sphere along x and y direction respectively, then, in equilibrium -
(i) $\Sigma F_x = T \sin\theta - F_e = 0 \quad \Rightarrow \quad T \sin\theta = F_e$, and
(ii) $\Sigma F_y = T \cos\theta - mg = 0 \quad \Rightarrow \quad T \cos\theta = mg$

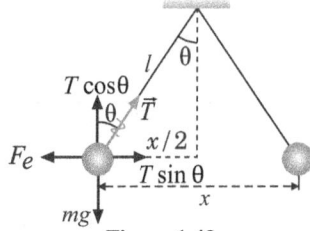

Figure 1.43

Dividing condition (i) by (ii), we get -
$$\tan\theta = \frac{F_e}{mg} \qquad (i)$$

But from the Figure 1.43,
$$\tan\theta = \frac{x}{2\sqrt{l^2 - \left(\frac{x}{2}\right)^2}} \cong \frac{x}{2l} \text{ as } x \ll l \qquad (ii)$$

and $\quad F_e = \frac{q^2}{4\pi\epsilon_0 x^2} \qquad (iii)$

Now, substitute the values obtained in Eq.(ii) and (iii) in Eq.i and simplify for q. From here you will get a relation between q and x, i.e., $q = f(x)$ in equilibrium.

Differentiation of this relation with respect to time 't', will give you dq/dt, i.e., the rate of leakage of charge.

SOLUTION On substituting the values $\tan\theta$ and F_e from Eq.ii and iii in Eq. i, we get-
$$\frac{x}{2l} = \frac{q^2}{4\pi\epsilon_0 x^2 mg}$$

or $\quad q^2 = \frac{2\pi\epsilon_0 mg x^3}{l} \qquad (iii)$

On differentiating Eq.(iii) with respect to time 't', we get-
$$2q \frac{dq}{dt} = \frac{2\pi\epsilon_0 mg}{l} 3x^2 \frac{dx}{dt}$$

According to the problem $\frac{dx}{dt} = v = \frac{a}{\sqrt{x}}$ (approach velocity is dx/dt) so,

$$\left(\frac{2\pi\epsilon_0 mg}{l}x^3\right)^{1/2} \frac{dq}{dt} = \frac{3\pi\epsilon_0 mg}{l}x^2 \frac{a}{\sqrt{x}}$$

Hence, $\frac{dq}{dt} = \frac{3}{2}a\sqrt{\frac{2\pi\epsilon_0 mg}{l}}$

EXAMPLE 25. Three identical charged balls each of charge 2 C are suspended from a common point P by silk threads of 2 m each (as shown in Fig.1.44). They form an equilateral triangle of side 1 m. Find the ratio of net electric force on a charged ball to the force between any two charged balls.

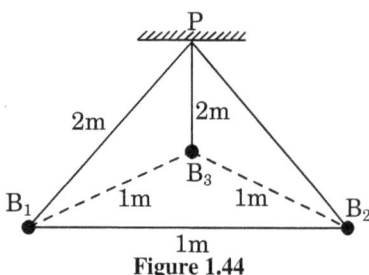

Figure 1.44

SOLUTION Fig.1.45 shows the FBD of ball B_2. Each ball has equal charge. Since, charged balls form equilateral triangle, therefore, the angle between forces applied by charged balls B_1 and B_3 on B_2 is $60°$.
If F is the magnitude of force applied by each charged particle

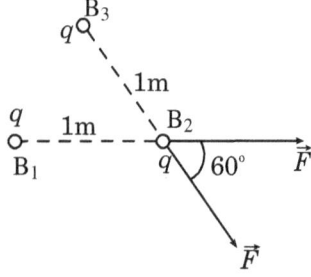

Figure 1.45

B_1 and B_3 on the particle B_2, then
$$F = k\frac{q_1 q_2}{r^2} = \frac{k(2)(2)}{(1)^2}$$

Therefore, net electric force on charged particle B_2 is
$$F_{B_2} = \sqrt{F^2 + F^2 + 2FF\cos 60°} = F\sqrt{3}$$

Therefore, $\frac{F_{net}}{F} = \frac{F\sqrt{3}}{F} = \sqrt{3}$

EXAMPLE 26. Given a cube with point charges q on each of its vertices. Calculate the force exerted on any of the charges due to rest of the 7 charges.

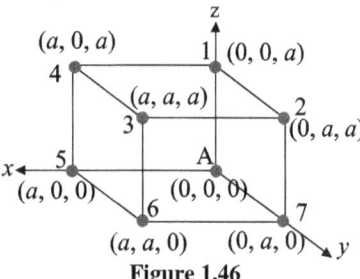

Figure 1.46

APPROACH All charges (1 to 7) are applying electrostatic forces simultaneously on charge at A. Therefore by applying superposition principle, you can find the net force on the charge at A.

SOLUTION The net force on particle A can be given by vector sum of force experienced by this particle due to all the other charges on vertices of the cube. For this we use vector form of coulomb's law $\vec{F} = \frac{kq_1q_2}{|\vec{r}_1 - \vec{r}_2|^3}(\vec{r}_1 - \vec{r}_2)$

From the Fig.1.46 the forces acting on A due to charges at 1, 2, 3, ..., 7, are given as -

$\vec{F}_{A_1} = \frac{kq^2(-a\hat{k})}{a^3}$, $\vec{F}_{A_2} = \frac{kq^2(-a\hat{j}-a\hat{k})}{(\sqrt{2}a)^3}$,

$\vec{F}_{A_3} = \frac{kq^2(-a\hat{i}-a\hat{j}-a\hat{k})}{(\sqrt{3}a)^3}$, $\vec{F}_{A_4} = \frac{kq^2(-a\hat{i}-a\hat{k})}{(\sqrt{2}a)^3}$

$\vec{F}_{A_5} = \frac{kq^2(-a\hat{i})}{a^3}$, $\vec{F}_{A_6} = \frac{kq^2(-a\hat{i}-a\hat{j})}{(\sqrt{2}a)^3}$, $\vec{F}_{A_7} = \frac{kq^2(-a\hat{j})}{a^3}$

respectively

By superposition principle, the net force experienced by A can be given as -

$\vec{F}_{net} = \vec{F}_{A_1} + \vec{F}_{A_2} + \vec{F}_{A_3} + \vec{F}_{A_4} + \vec{F}_{A_5} + \vec{F}_{A_6} + \vec{F}_{A_7}$

$= \frac{-kq^2}{a^2}\left[\left(\frac{1}{3\sqrt{3}} + \frac{1}{\sqrt{2}} + 1\right)(\hat{i}+\hat{j}+\hat{k})\right]$

EXAMPLE 27. Five point charges, each of value $+q$ are placed at five vertices of a regular hexagon of side L m. What is the magnitude of the force on a point charge of value $-q$ coulomb placed at the centre of the hexagon?

APPROACH If there had been a sixth charge $+q$ at the remaining vertex of hexagon, the net force due to all the six charges on $-q$ at O would be zero (as the forces due to individual charges will balance each other).

Now if $\vec{f}$ is the force due to sixth charge and $\vec{F}$ due to remaining

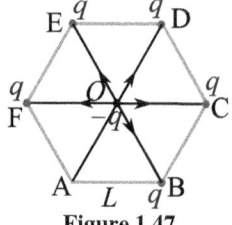

Figure 1.47

five charges. $\vec{F} + \vec{f} = 0 \Rightarrow \vec{F} = -\vec{f} \Rightarrow F = f$
Now, by using Coulomb's law, find f

SOLUTION By Coulomb's law, we have
$$f = k\frac{q_1 q_2}{r^2} = \frac{1}{4\pi\epsilon_0}\frac{q^2}{L^2}$$
$$= \frac{1}{4\pi\epsilon_0}\frac{q \times q}{L^2} = \frac{q^2}{4\pi\epsilon_0 L^2}$$

EXAMPLE 28. A particle A having a charge of 5.0×10^{-7}C is fixed in a vertical wall. A second particle B of mass 100 g and having equal charge is suspended by a silk thread of length 30 cm from the wall. The point of suspension is 30 cm above the particle A. Find the angle of the thread with the vertical when it stays in equilibrium.

APPROACH The situation is shown in Fig.1.48. Suppose the point of suspension is O and let θ be the angle between the thread and the vertical. Forces on the particle B are

(i) weight mg downwards
(ii) tension T along the thread and
(iii) electric force of repulsion F along AB.

For equilibrium, these forces should add to zero.
We will be solving the problem by using Lami's theorem. The distance between A and B can be obtained by sine rule.

Sine Rule: In any $\triangle ABC$ (Fig.1.49b), we have-

1.10. EQUILIBRIUM OF CHARGED PARTICLES

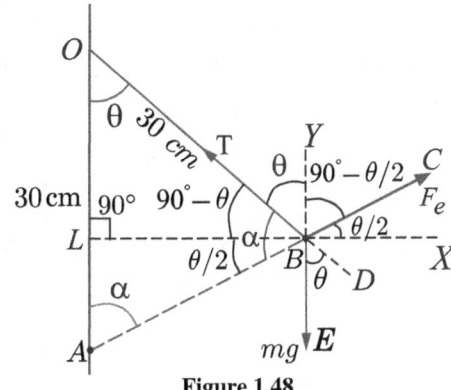

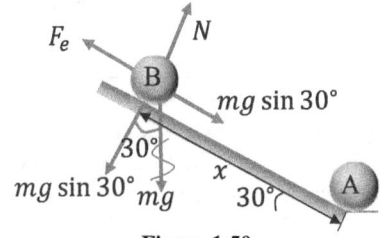

(a) Lami's Theorem (b) Sine Rule
Figure 1.49

$$\frac{a}{\sin A} = \frac{b}{\sin B} = \frac{c}{\sin C}$$

here a, b and c are sides against angles A, B and C respectively.

SOLUTION The FBD of problem is shown in Fig.1.48. Let B is the equilibrium position of ball and $\angle AOB = \theta$. Let us draw a perpendicular BL on side OA.
Note that $OA = OB$, so ΔOAB is an isosceles triangle.
Therefore, $\angle OAB = \angle OBA = \alpha$ (say)
So, in ΔOAB, $\angle AOB + \angle OAB + \angle OBA = 180°$
$\Longrightarrow \theta + \alpha + \alpha = 180°$
$\Longrightarrow 2\alpha = 180° - \theta \Longrightarrow \alpha = 90° - \theta/2$
Again in right angle ΔBLA, $\alpha + 90° + \angle LBA = 180°$
$\Longrightarrow \angle LBA = 90° - \alpha = 90° - \left(90° - \frac{\theta}{2}\right)$
$= \frac{\theta}{2} = \angle CBX$ (opposite angles)
All angles between forces on the particle are shown in Fig.1.48.
By Lami's theorem at B, we have-

$$\frac{F_e}{\sin \angle OBE} = \frac{mg}{\sin \angle OBC}$$

$\Longrightarrow \dfrac{F_e}{\sin(180 - \theta)} = \dfrac{mg}{\sin(\theta + 90° - \frac{\theta}{2})}$

$\Longrightarrow \dfrac{F_e}{\sin \theta} = \dfrac{mg}{\sin(90° + \frac{\theta}{2})} \Longrightarrow \dfrac{F_e}{\sin \theta} = \dfrac{mg}{\cos \frac{\theta}{2}}$

$\Longrightarrow \dfrac{F_e}{2 \sin \frac{\theta}{2} \cos \frac{\theta}{2}} = \dfrac{mg}{\cos \frac{\theta}{2}}$

$\Longrightarrow F_e \cos \frac{\theta}{2} - 2mg \sin \frac{\theta}{2} \cos \frac{\theta}{2} = 0$

$\Longrightarrow \cos \frac{\theta}{2} \left[F_e - 2mg \sin \frac{\theta}{2}\right] = 0$

If $\cos \frac{\theta}{2} = 0$, then $\frac{\theta}{2} = \frac{\pi}{2}$ or $\theta = \pi$, which is not possible.
Now, $F_e - 2mg \sin \frac{\theta}{2} = 0$
$\Longrightarrow \sin \frac{\theta}{2} = \frac{F_e}{2mg}$ ...(i)

From sine rule, we have-

$$\frac{AB}{\sin \theta} = \frac{OA}{\sin \alpha} \Longrightarrow AB = \frac{OA \sin \theta}{\sin \alpha}$$

$\Longrightarrow AB = \dfrac{OA(2 \sin \frac{\theta}{2} \cos \frac{\theta}{2})}{\sin(90° - \frac{\theta}{2})} = \dfrac{2 OA \sin \frac{\theta}{2} \cos \frac{\theta}{2}}{\cos \frac{\theta}{2}}$

$\Longrightarrow AB = 2 OA \sin \frac{\theta}{2}$

Electrostatic force between A and B is,

$$F_e = k\frac{q^2}{(AB)^2} = k\frac{q^2}{(2 OA \sin \frac{\theta}{2})^2} = k\frac{q^2}{4 OA^2 \sin^2 \frac{\theta}{2}}$$

On substituting this value in Eq.(i), we get -

$$\sin \frac{\theta}{2} = k\frac{q^2}{4(OA)^2 \sin^2 \frac{\theta}{2}(2mg)}$$

$\Longrightarrow \sin^3 \frac{\theta}{2} = k\frac{q^2}{8(OA)^2 mg}$

On substituting given values in above expression, we get-

$$\sin^3 \frac{\theta}{2} = \frac{9 \times 10^9 \times 25 \times 10^{-14} \text{ N}}{4 \times 9 \times 10^{-2} \times 2 \times (100 \times 10^{-3} \text{ kg}) \times 9 \cdot 8 \text{ m s}^{-2}}$$
$= 0 \cdot 0032$

$\Longrightarrow \sin \frac{\theta}{2} = 0 \cdot 15 \Longrightarrow \theta = 17°$

EXAMPLE 29. A particle 'A' having a charge of 2×10^{-6}C and a mass of 100 g is fixed at the bottom of a smooth inclined plane of inclination 30°. Where should another particle B, having same charge and mass be placed on the incline so that it may remain in equilibrium?

Figure 1.50

APPROACH Make FBD for B and apply the condition of translational static equilibrium along and perpendicular to the inclined plane of Fig.1.50

SOLUTION Suppose, the equilibrium distance of particle B from A is x. FBD of particle B, is shown in Fig.1.50.
  The forces acting on the particle B are-
1. Gravitational force: mg (downwards)
2. Normal reaction of the plane: N (perpendicular to the plane in upward sense)
3. Electrostatic repulsion of A: F_e (along the plane in upward sense)

We resolve the gravitational force mg into components— one ($mg \sin 30°$) along the plane and other ($mg \cos 30°$) perpendicular to it.
In static equilibrium, we have -

Along the plane: $mg \sin 30° = F_e = \dfrac{kq^2}{x^2}$

$\Longrightarrow \dfrac{mg}{2} = \dfrac{kq^2}{x^2} \Longrightarrow x = \sqrt{\dfrac{2kq^2}{mg}}$

On substituting the given values- $q = 2 \times 10^{-6}$C, $m = 0.1$ kg and $k = 9.0 \times 10^9$ N.m^2/C^2, we get -
$x = 0.27$ m $= 27$ cm

EXAMPLE 30. A rigid insulated wire frame in the form of a right angled triangle ABC, is set in a vertical plane as shown in the figure. Two beads of equal masses m each and carrying charges q_1 and q_2 are connected by a cord of length l and can slide without friction on the wires. Considering the case when

the beads are stationary, determine.
(a) (i) The angle α. (ii) The tension in the cord. (iii) The normal

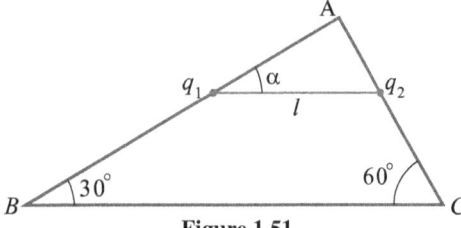

Figure 1.51

reaction on the beads.
(b) If the cord is now cut what are the values of the charges for which the beads continue to remain stationary?
APPROACH Make FBDs of both the beads. Write translational equilibrium conditions along and perpendicular to their respective planes and then simplify them for required variables.
SOLUTION The forces acting on the charge q_1 are its weight mg, normal reaction N_1, string tension T and Coulomb's force $F_e = q_1q_2/(4\pi\epsilon_0 l^2)$. Similarly, forces on the charge q_2 are its weight mg, normal reaction N_2, string tension T and Coulomb's force $F_e = q_1q_2/(4\pi\epsilon_0 l^2)$ The forces are shown in the Fig.1.52.

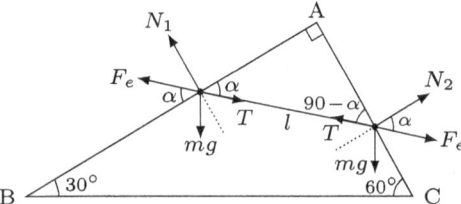

Figure 1.52

In triangle ABC, $\angle A = 90°$. Resolve the forces on q_1 and q_2 in the directions parallel and perpendicular to the sides of the frame. Balancing the forces on q_1, we get-

$$N_1 + F_e \sin\alpha = mg\cos 30° + T\sin\alpha \quad (1.30)$$

$$F_e \cos\alpha + mg\sin 30° = T\cos\alpha \quad (1.31)$$

and on q_2 to get

$$N_2 + F_e \cos\alpha = mg\cos 60° + T\cos\alpha, \quad (1.32)$$

$$F_e \sin\alpha + mg\sin 60° = T\sin\alpha \quad (1.33)$$

On simplifying equations (1.31) and (1.33), we get $\alpha = 60°$ and $T = mg + F_e$
Now, substitution of these values in equations (1.30) and (1.32) gives $N_1 = \sqrt{3}mg$ and $N_2 = mg$.
The tension becomes zero when the cord is cut. Substituting $T = 0$ in $T = mg + F_e$ gives $F_e = -mg$ i.e., $q_1 q_2 = -4\pi\epsilon_0 l^2 mg$
Ans. (a) (i) 60° (ii) $\frac{1}{4\pi\epsilon_0}\frac{q_1q_2}{l^2} + mg$ (iii) $\sqrt{3}mg, mg$ (b) $q_1q_2 = -4\pi\epsilon_0 mgl^2$

EXAMPLE 31. Three particles, each of mass 1 g and carrying a charge q, are suspended from a common point by insulated massless strings, each 100 cm long. If the particles are in equilibrium and are located at the corners of an equilateral triangle of side length 3 cm, calculate the charge q on each particle. [Take $g = 10$ m/s^2.]
APPROACH Make FBD of either particle and apply the condition of translational equilibrium
SOLUTION Let the three charged particles, each of charge q, be located at the three corners of an equilateral triangle of side $a = 3$ cm in the horizontal plane. The distance of the charges from the centroid O is r. On applying sine rule in triangle OBC, we get -

$$\frac{r}{\sin\angle OBC} = \frac{a}{\sin\angle BOC}$$

$$\Rightarrow \frac{r}{\sin 30°} = \frac{a}{\sin 120°} \Rightarrow r = a/\sqrt{3} = \sqrt{3} \text{ cm}.$$

Consider the electrostatic force on the charge at A due to the

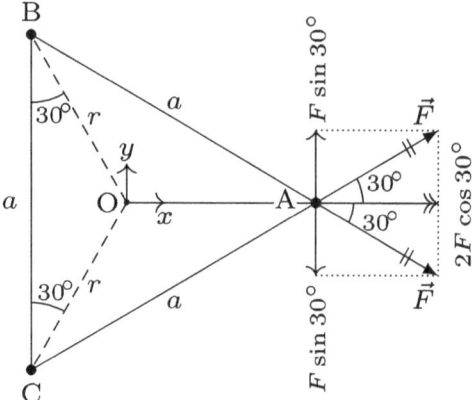

Figure 1.53

charges at B and C. The magnitude of forces due to each charge is

$$|\vec{F}| = \frac{q^2}{4\pi\epsilon_0 a^2}$$

and their directions are as shown in the Fig.1.53. Resolve $\vec{F}$ along and perpendicular to AO. The resultant electrostatic force on charge at A is given by

$$\vec{F}_e = 2F\cos 30°\hat{i} = \frac{\sqrt{3}q^2}{4\pi\epsilon_0 a^2}\hat{i}$$

The charges are connected with strings of length $l = 100$ cm. Let

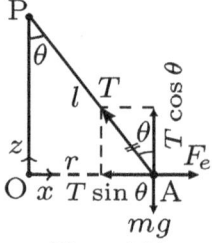

Figure 1.54

the other end of each string is hanging from the point P, which is vertically above O (Fig.1.54). In triangle OAP,

$$\tan\theta = \frac{r}{\sqrt{l^2 - r^2}} = \frac{\sqrt{3}}{\sqrt{100^2 - 3}} = 0.017$$

The forces on the charge at A are F_e, tension T, and weight mg. Resolve T in the horizontal and the vertical directions. The equilibrium condition on charge at A gives

$$T\sin\theta = F_e = \frac{\sqrt{3}q^2}{4\pi\epsilon_0 a^2} \quad (1)$$

$$T\cos\theta = mg \quad (2)$$

On dividing equation (1) by (2), we get

$$q = \left[\frac{4\pi\epsilon_0 a^2 mg \tan\theta}{\sqrt{3}}\right]^{1/2}$$

$$= \left[\frac{(0.03)^2 (10^{-3})(10)(0.017)}{9\times 10^9 (1.73)}\right]^{1/2} = 3.17\times 10^{-9}\text{C}$$

1.11 Check Point 2

1. •Suppose that the electric force between two point charges is attractive. What can you conclude about the signs of these point charges?
2. •Can two similarly charged bodies attract each other?
3. •Does in charging the mass of a body change?
4. •Why a third hole in a socket provided for grounding?

5. •Two balls, separated by some distance, carry equal electric charges and exert a repulsive electric force on each other. If we transfer a fraction of the electric charge of one ball to the other, will the electric force increase or decrease?
6. ••What would be the interaction force between two copper spheres, each of mass 1g, separated by the distance 1 m, if the total electronic charge in them differed from the total charge of the nuclei by one percent? [Given: molar mass of copper = 63.54 gram, atomic number of copper = 29]
7. ••Six charges are kept at the vertices of a regular hexagon as shown in the Fig.1.55. If magnitude of force applied by $+Q$ on $+q$ charge is F, then net electric force on the $+Q$ is nF. Find the value of n.

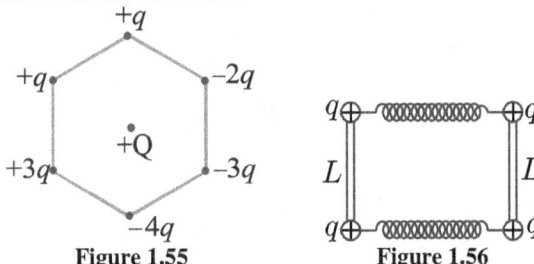

Figure 1.55   Figure 1.56

8. ••Two stiff non conducting rods have length L each and have small balls connected to their ends. The rods are placed parallel to each other and the balls are connected by two identical springs as shown in Fig.1.56. When each ball is given a charge q, the system stays in equilibrium when it is in the shape of a square. If natural relaxed length of each spring is $L/2$ find the force constant (k) for them.
9. ••A charge Q is to be divided on two objects. What should be the values of the charges on the objects so that the force between the objects can be maximum?
10. ••Will a point charge have a position of stable equilibrium when it is at the mid-point between two other equal point charges, the signs of which are either the same or opposite to that of the first charge?
11. ••Three equal charges q of the same sign are located at the vertices of an equilateral triangle. What charge Q of opposite sign must be placed at the centre of the triangle in order for the resultant of the forces acting on each charge to be zero?

Multiple Choice Questions

12. ••Two point charges A and B, halving charges $+Q$ and $-Q$ respectively, are placed at certain distance apart and force acting between them is F. If 25% charge of A is transferred to B, then force between the charges becomes -
    (A) $\frac{4F}{3}$ (B) F (C) $\frac{9F}{16}$ (D) $\frac{16F}{9}$
13. ••Suppose the charge of a proton and an electron differ slightly. One of them is $-e$, the other is $(e + \Delta e)$. If the net of electrostatic force and gravitational force between two hydrogen atoms placed at a distance d (much greater than atomic size) apart is zero, then Δe is of the order of [Given: mass of hydrogen $m_h = 1.67 \times 10^{-27}$ kg]
    (A) 10^{-23}C (B) 10^{-37}C (C) 10^{-47}C (D) 10^{-20}C
14. ••Two identical charged spheres suspended from a common point by two massless strings of lengths l, are initially at a distance $d (d \ll l)$ apart because of their mutual repulsion. The charges begin to leak from both the spheres at a constant rate. As a result, the spheres approach each other with a velocity v. Then v varies as a function of the distance x between the spheres, as
    (A) $v \propto x^{-1/2}$ (B) $v \propto x^{-1}$ (C) $v \propto x^{1/2}$ (D) $v \propto x$
15. ••Three identically charged, small spheres each of mass m are suspended from a common point by insulated light strings each of length l. The spheres are always on vertices of an equilateral triangle of length of the sides $x (\ll l)$. The rate dq/dt with which charge on each sphere increases, if length of the sides of the equilateral triangle increases slowly according to law $\frac{dx}{dt} = \frac{a}{\sqrt{x}}$, is -
    (A) $\sqrt{\frac{3\pi\epsilon_0 mga^2}{l}}$ (B) $\sqrt{\frac{\pi\epsilon_0 mga^2}{l}}$
    (C) $\sqrt{\frac{\pi\epsilon_0 mga^2}{3}}$ (D) $\sqrt{\frac{2\pi\epsilon_0 mga^2}{l}}$
16. ••Two pith* balls carrying equal charges are suspended from a common point by strings of equal length, the equilibrium separation between them is r. Now the strings are rigidly clamped at half the height. The equilibrium separation between the balls now become [Fig.1.57]
    (A) $\left(\frac{2r}{\sqrt{3}}\right)$ (B) $\left(\frac{2r}{3}\right)$ (C) $\left(\frac{1}{\sqrt{2}}\right)^2$ (D) $\left(\frac{r}{\sqrt[3]{2}}\right)$

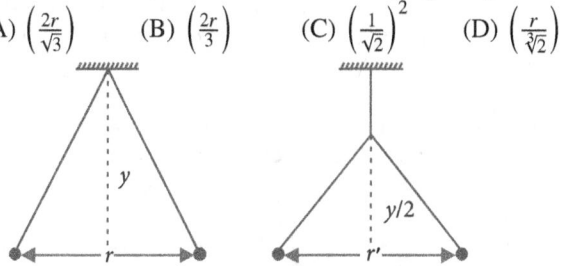

Figure 1.57

17. ••Two positive ions, each carrying a charge q, are separated by a distance d. If F is the force of repulsion between the ions, the number of electrons missing from each ion will be (e being the charge on an electron)
    (A) $\frac{4\pi\epsilon_0 Fd^2}{e^2}$ (B) $\sqrt{\frac{4\pi\epsilon_0 Fe^2}{d^2}}$
    (C) $\sqrt{\frac{4\pi\epsilon_0 Fd^2}{e^2}}$ (D) $\frac{4\pi\epsilon_0 Fd^2}{q^2}$
18. ••When air is replaced by a liquid dielectric medium of constant κ, the maximum force of attraction between two charges separated by a fixed distance
    (A) increases to κ times (B) remains unchanged
    (C) decreases to κ times (D) decreases to κ^{-1} times.
19. ••A charge q is placed at the centre of the line joining two equal charges Q. The system of the three charges will be in equilibrium if q is equal to
    (A) $-Q/4$ (B) $Q/4$ (C) $-Q/2$ (D) $Q/2$
20. ••Point charges $+4q$, $-q$ and $+4q$ are kept on the X-axis at point $x = 0, x = a$ and $x = 2a$ respectively. Then
    (A) only $-q$ is in stable equilibrium
    (B) all the charges are in stable equilibrium
    (C) all of the charges are in unstable equilibrium
    (D) none of the charges is in equilibrium.
21. ••A positively charged body 'A' attracts a body 'B' then charge on body 'B' may be:
    (A) positive (B) negative (C) zero (D) can't say
22. ••Three identical point charges are at the vertices of an equilateral triangle. A fourth, identical point charge is placed at the midpoint of one side of the triangle. As a result of the three electric force contributions from the vertex charges, the fourth charge is
    (A) in equilibrium and remains at rest
    (B) pushed toward the centre of the triangle
    (C) pushed outside the triangle
    (D) in unstable equilibrium

---
*A pith ball is a very small, lightweight styrofoam (or similar material) ball coated with conductive paint that picks up electric charge quite well.

1.12 The Electric Field

The concept of a field* was developed by Michael Faraday. According to him, an electric field is said to exist in the region of space around a charge, if another charge experiences an electric force (F_e) within it. Thus, the interaction between two charges is a two-step process—a charge produces an electric field around itself; this field then exerts an electric force on the other charge.

The electric field is a vector quantity generally denoted by $\vec{E}$.

In analogy with the gravitational field, we define the electric field intensity or strength associated with a certain collection of charges in terms of the force exerted on a positive test charge q_0 at a particular point, as

$$\boxed{E = \frac{F}{q_0}} \tag{1.34}$$

i.e., *the electric force per unit test charge is called the intensity of electric field at the position of test charge.* The direction of the vector $\vec{E}$ is the same as the direction of $\vec{F}$, because q_0 is a positive scalar.

The SI unit of electric field intensity is newton/coulomb (N/C), although it is more often given in the equivalent unit of volt/meter (V/m). The term volt will be discussed in next chapter, "Electric Potential".

Note the similarity with the gravitational field strength 'g', which can also be expressed as the force per unit mass in units of newton/kilogram. Both the gravitational and electric field strengths can be expressed as a force divided by a property (mass or charge) of the test body.

In Fig.1.58, the charge Q, which is producing the electric field, is

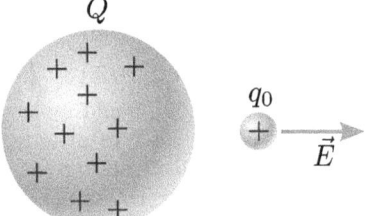

Figure 1.58

called a source charge and the charge q_0, which tests the effect of a source charge, is called a test charge. Note that the source charge Q must remain at its original location. However, if a charge q_0 is brought at any point around Q, Q itself is bound to experience an electrical force due to q_0 and will tend to move. To sort out this difficulty, take q_0 negligibly small. The force $\vec{F}$ is then negligibly small but the ratio $\vec{E} = \lim_{q_0 \to 0} \dfrac{\vec{F}}{q_0}$ is finite and defines the electric field.

This limit in actuality cannot be taken to 0 because the test charge can never be smaller than the elementary charge e. If we are calculating (rather than measuring) the electric field due to a specified collection of charges at fixed positions, neither the magnitude nor the sign of q_0 affects the result.

Also, note that an electric field is a property of its source, the presence of the test charge is not necessary for the field to exist. The test charge is only used to detect the electric field.

When using $E = F/q_0$ we must assume that the test charge q_0 is small enough that it does not disturb the charge distribution responsible for the electric field. If vanishingly small test charge q_0 is placed near a uniformly charged metallic sphere, as in Fig. 1.59 (a) the charge on the metallic sphere, which produces the electric field, remains uniformly distributed. Now, if we consider a new test charge q'_0 which is great enough ($q'_0 \gg q_0$), as in Fig. 1.59b the charge on the metallic sphere is redistributed and the ratio of the force to the test charge is different. If in this case, the electric force on the test charge q'_0 is F', then-

$$\frac{F'}{q'_0} \neq \frac{F}{q_0}$$

Because of this redistribution of charge on the metallic sphere, the electric field it sets up is different from the field it sets up in the presence of the much smaller test charge q_0.

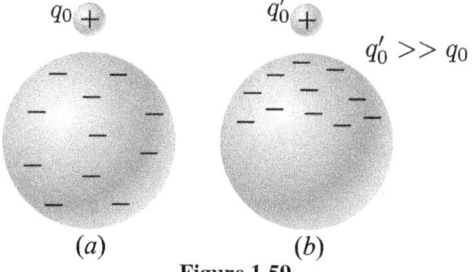

Figure 1.59

Important Points Regarding Electric Field

1. The static electric field produced by any point charge is always conservative. A conservative force or field must be a function of only spatial coordinates. The work done by such a force or field depends only on start and ending location in the field, and not on the actual path taken between those points. Now, suppose we place a point charge q near the surface of a conductor. It produces an induced charge over the surface of the conductor and this surface induced charge also produces an electric field in space which is also conservative. But if we move this point charge q, continuously, near the surface, the amount and distribution both of the induced surface charge changes continuously with time. So electric field produced by it also changes with time. Therefore, this induced electric field is time dependent and hence it is non-conservative.

2. The electric field of induced charge can be either conservative or non-conservative. In electrostatic situations (i.e., when charge remains in equilibrium under electric forces) without changing magnetic fields, it is conservative. In dynamic situations (i.e, when charge moves) or when there are changing magnetic fields, it can be non-conservative.

EXAMPLE 32. Suppose someone discovers that blue and yellow objects attract each other, that two blue objects repel each other, and that two yellow objects repel each other. The strength of this "chromatic interaction" is found to depend on color depth: The deeper the color, the greater the magnitude of the interaction. How would you define the magnitude and direction of the "chromatic field" of an object?

SOLUTION The gravitational field is defined as the gravitational force per unit of mass, with the field direction the same as the direction of the force. The electric field is defined as the electric force per unit of charge, with the field direction parallel to that of the force exerted on a positively charged particle. Therefore, the chromatic field can be defined as the chromatic force per unit of color, with the field direction parallel to that of the force exerted on a particle carrying some chosen color.

EXAMPLE 33. What happens when a high energy X-ray beam falls on a small metal ball suspended in a uniform electric field with the help of an insulated thread?

SOLUTION Since, the high energy X-rays cause ejection of the

---

*Every position in a field is assigned a particular value of some quantity. For example, a temperature field is a scalar field and every position is associated with a temperature. If the field is a **vector field** (**for example gravitational or electric fields**), each position in the field is associated with a particular magnitude and direction.

1.13 The Electric Field of Point Charges

In this section we consider the electric field of point charges, first a single charge and then an assembly of individual charges. Later we generalize to continuous distributions of charge.

1.13.1 The Electric Field Due to a Single Point Charge

Let a positive test charge q_0 be placed a distance r from a point charge q. The magnitude of the force acting on q_0 is given by Coulomb's law.
$$\vec{F} = k\frac{qq_0}{r^2}\hat{r}$$
The magnitude of the electric field at the position of the test charge is, from Eq.1.34
$$\vec{E} = \frac{\vec{F}}{q_0} = k\frac{q}{r^2}\hat{r} = \frac{1}{4\pi\epsilon_0}\frac{q}{r^2}\hat{r} \tag{1.35}$$
where $\hat{r}$ is a unit vector directed from q towards q_0. The direction of $\vec{E}$ is the same as the direction of $\vec{F}$, along a radial line from q, pointing outward(i.e., along $\hat{r}$), if q is positive and inward (i.e., opposite to $\hat{r}$) if q is negative.

If q is negative, as in Fig.1.60c, the force on the test charge

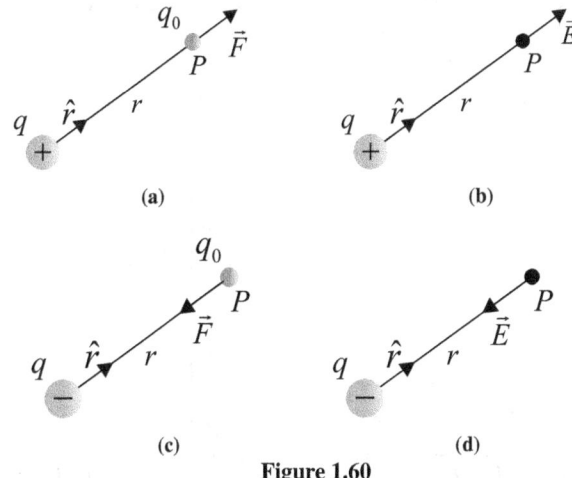

Figure 1.60

is toward the source charge, so the electric field at P is directed toward the source charge, as in Fig.1.60d.

Fig.1.60 shows the magnitude and direction of the electric field $\vec{E}$ at a point P due to positive and negative point charges.

EXAMPLE 34. (a) Does an electrically neutral particle that has mass interact with an electric field? (b) Does a charged particle interact with a gravitational field?

SOLUTION (a) No. A particle has no extent and therefore cannot be polarized. So, an uncharged particles* don't interact with electric fields.
(b) Yes, because any particle which is charged would certainly have mass. so it will be, interacted with a gravitational field.

EXAMPLE 35. A water droplet of mass 3.00×10^{-12} kg is located in the air near the ground during a stormy day. An atmospheric electric field of magnitude 6.00×10^3 N/C points vertically downward in the vicinity of the water droplet. The droplet remains suspended at rest in the air. What is the electric

---
*An uncharged body is a different case. It is a collection of particles. It has equal amount of positive and negative charge. So it is always attracted towards a charged body or charged particle.

charge on the droplet?

APPROACH Since, the droplet is hovering at rest in the air in a downward atmospheric electric field and gravitational field of earth, so there will be a downward gravitational force of earth on the droplet and to keep it in equilibrium, there must be an equal upward force on it. This upward force is provided by downward electric field $\vec{E}$.

SOLUTION Suppose, the mass of droplet is m and it's charge

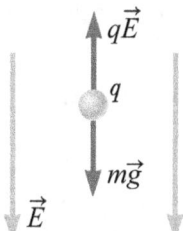

Figure 1.61: FBD of water droplet

is q. If the downward electric field is $\vec{E}$, then The forces acting on the droplet are-
1. Downward gravitational force $m\vec{g}$
2. Upward electrostatic force $q\vec{E}$

The free body diagram (FBD) of the droplet is shown in Fig.1.61. In equilibrium, we have
$$qE = mg \quad \Rightarrow \quad q = \frac{mg}{E}$$
On substituting the given values, we get
$$q = \frac{(3.00 \times 10^{-12} \text{ kg})(9.80 \text{ m/s}^2)}{6.00 \times 10^3 \text{ N/C}} = 4.90 \times 10^{-15} \text{C}$$
Since, the direction of electrostatic force is opposite to the electric field $\vec{E}$, therefore the nature of charge on the drop is negative. So, the charge on the droplet, is -4.90×10^{-15}C

EXAMPLE 36. A vertical electric field of magnitude 4.9×10^5 N/C just prevents a water droplet of a mass 0.1 g from falling. Find the magnitude of charge on the droplet. (Given $g = 9.8$ m/s²)

SOLUTION Let m is the mass of droplet and q is the charge on it. In equilibrium under gravitational field $\vec{g}$ and electric field $\vec{E}$, the forces acting on the charged droplet, are-
1. Downward Gravitational force $m\vec{g}$
2. Upward electric force $q\vec{E}$.
If q is positive, then electric force on q will be in the direction of $\vec{E}$ and if q is negative then it will be opposite to $\vec{E}$.
Since, the droplet is in equilibrium under these two forces only, therefore, net force on it must be zero.
Taking upward direction positive, we have
$$qE - mg = 0 \quad \Longrightarrow \quad q = \frac{mg}{E}$$
Clearly, if E is positive, i.e., in upward direction, then q will be positive and if E is negative, i.e., in downward direction, then q will also be negative.
Substituting the given values in above expression, we get
$$q = \frac{(0.1 \times 10^{-3}) \times 9.8}{4.9 \times 10^5} = 2 \times 10^{-9} \text{ C}$$

1.13.2 Electric Field Due To a Group of Point Charges (Superposition of Electrostatic Fields)

To find $\vec{E}$ for a group of N point charges, the procedure is as follows: (1) Calculate E_i due to each charge i at the given point as if it were the only charge present. (2) Add these separately calculated fields vectorially to find the resultant field E at the

point. In equation form,
$$\vec{E} = \vec{E}_1 + \vec{E}_2 + \vec{E}_3 + \cdots$$
$$= \sum \vec{E}_i \quad (i = 1, 2, 3, \ldots, N) \quad (1.36)$$

The sum is a vector sum, taken over all the charges. *Eq. 1.36 is an example of the application of the principle of superposition, which states that at a given point the electric fields due to separate charge distributions simply add up (vectorially) or superimpose independently.* This principle may fail when the magnitudes of the fields are extremely large*, but it will be valid in all situations we discuss in this text.

Above equation can also be written as-
$$\vec{E} = k \sum_i \frac{q_i}{r_i^2} \hat{r}_i \quad (1.37)$$

where r_i is the distance of point P from i^{th} source charge q_i. The unit vector $\hat{r}$ is directed from q_i toward P.

If some more charges are added, more terms are added to the summation. However, there is no change to the terms that were already there, provided that the original charges do not move.

EXAMPLE 37. Two point charges are separated by a distance of 10.0 cm. One has a charge of $-25\mu C$ and the other $+50\mu C$. (a) Determine the direction and magnitude of the electric field at a point P between the two charges that is 2.0 cm from the negative charge (Fig.(1.62)a). (b) If an electron (mass = 9.11×10^{-31} kg) is placed at rest at P and then released, what will be its initial acceleration (direction and magnitude)?

APPROACH (a) The electric field at P will be the vector sum of the fields created separately by Q_1 and Q_2. The field due to the negative charge Q_1 points toward Q_1, and the field due to the positive charge Q_2 points away from Q_2. Thus both fields point to the left as shown in Fig. 1.62b, and we can add the magnitudes of the two fields together algebraically, ignoring the signs of the charges.

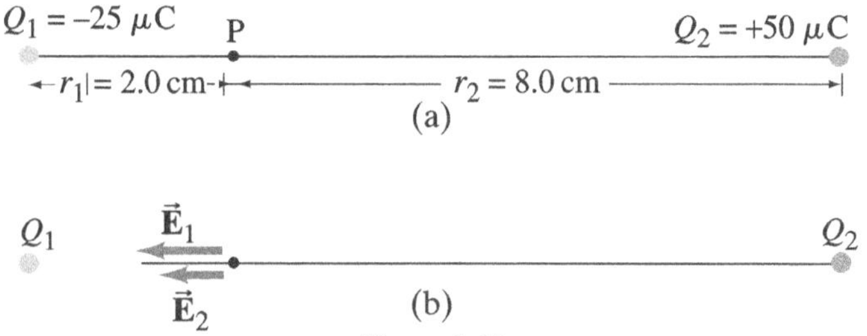

(b) In this part, we use Newton's second law $\left(\sum \vec{F} = m\vec{a}\right)$ to find the acceleration, where $\sum \vec{F} = q\sum \vec{E}$.

SOLUTION (a) Each field is due to a point charge as given by Eq. 1.35, $E = kq/r^2$. The total field points to the left and has magnitude

$E = k\frac{Q_1}{r_1^2} + k\frac{Q_2}{r_2^2} = k\left(\frac{Q_1}{r_1^2} + \frac{Q_2}{r_2^2}\right)$

$= (9.0 \times 10^9 \text{ N} \cdot \text{m}^2/\text{C}^2) \left(\frac{25\times 10^{-6}\text{C}}{(2.0\times 10^{-2} \text{ m})^2} + \frac{50\times 10^{-6}\text{C}}{(8.0\times 10^{-2} \text{ m})^2}\right)$

$= 6.3 \times 10^8$ N/C

(b) The electric field points to the left, so the electron will feel a force to the right since it is negatively charged. Therefore the acceleration $a = F/m$ (Newton's second law) will be to the right. The force on a charge q in an electric field E is $F = qE$. Hence the magnitude of the electron's initial acceleration is

$a = \frac{F}{m} = \frac{qE}{m} = \frac{(1.60\times 10^{-19}\text{C})(6.3\times 10^8 \text{ N/C})}{9.11\times 10^{-31} \text{ kg}} = 1.1 \times 10^{20}$ m/s²

---
*Large electric field can cause the material to become polarized, which can interact with the external electric field.

EXAMPLE 38. Two point charges A and B of magnitude $+8\times 10^{-6}$ C and -8×10^{-6} C respectively are placed at a distance d apart. The electric field at the middle point O between the charges is 6.4×10^4 NC^{-1}. Calculate the distance 'd' between the point charges A and B.

APPROACH Electric fields at point O, due to both charges, are directed along the same direction (here it is directed from O to B). Therefore, by superposition principle, the magnitude of net electric field at point O will be equal to the sum of their magnitudes, i.e.

SOLUTION $\quad E_O = 2 \times \frac{kq}{(d/2)^2} \implies E_0 = 8\frac{kq}{d^2}$

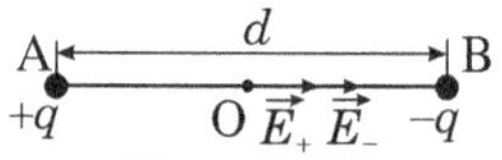

Figure 1.63

$\implies d^2 = \frac{8 \times 9 \times 10^9 \times 8 \times 10^{-6}}{6.4 \times 10^4} \implies d = 3$ m

EXAMPLE 39. Charges q_1 and q_2 are located on the x axis, at distances a and b, respectively, from the origin as shown in Fig.1.64.

(a) Find the components of the net electric field at the point P, which is at position $(0, y)$.

(b) Evaluate the electric field at point P in the special case that $|q_1| = |q_2|$ and $a = b$.

(c) Find the electric field, in part (b), when point P is at a distance $y \gg a$ from the origin.

$|q_1| = |q_2|$ and $a = b$.

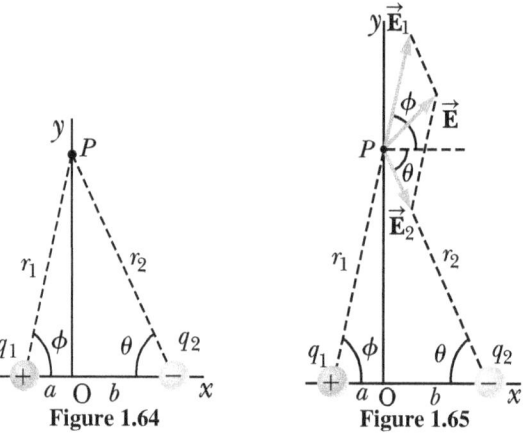

Figure 1.64   Figure 1.65

(a) APPROACH First of all calculate the intensities of electric fields at point P due to charges q_1 and q_2, then add these electric field vectors to find the net electric field at point P.

SOLUTION Find the magnitude of the electric field at P due to charge q_1: $E_1 = k\frac{|q_1|}{r_1^2} = k\frac{|q_1|}{a^2+y^2}$

Find the magnitude of the electric field at P due to charge q_2: $E_2 = k\frac{|q_2|}{r_2^2} = k\frac{|q_2|}{b^2+y^2}$

Write the electric field vector for each charge in unit-vector form:
$\vec{E}_1 = k_e \frac{|q_1|}{a^2+y^2} \cos\phi \hat{i} + k_e \frac{|q_1|}{a^2+y^2} \sin\phi \hat{j}$,
$\vec{E}_2 = k_e \frac{|q_2|}{b^2+y^2} \cos\theta \hat{i} - k_e \frac{|q_2|}{b^2+y^2} \sin\theta \hat{j}$

The X and Y components of the net electric field vector:
$E_x = E_{1x} + E_{2x} = k_e \frac{|q_1|}{a^2+y^2} \cos\phi + k_e \frac{|q_2|}{b^2+y^2} \cos\theta$,
$E_y = E_{1y} + E_{2y} = k_e \frac{|q_1|}{a^2+y^2} \sin\phi - k_e \frac{|q_2|}{b^2+y^2} \sin\theta$

(b) APPROACH Fig.1.66 shows the situation when $|q_1| = |q_2|$ and $a = b$. Such symmetrical charge distribution is called an electric dipole.

Because Fig.1.66 is a special case of the general case shown in

1.13. THE ELECTRIC FIELD OF POINT CHARGES

Fig.1.65, so, we can take the result of part (a) and substitute the appropriate values of the variables.
SOLUTION Evaluate the values of E_x and E_y obtained in from

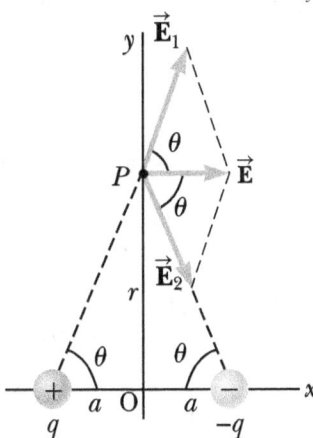

Figure 1.66: When the charges in Fig 1.65 are of equal magnitude and equidistant from the origin, the situation becomes symmetric as shown here.

part (a) with $a = b$, $|q_1| = |q_2| = q$, and $\phi = \theta$:
$$E_x = k\frac{q}{a^2+y^2}\cos\theta + k\frac{q}{a^2+y^2}\cos\theta = 2k\frac{q}{a^2+y^2}\cos\theta$$
$$E_y = k\frac{q}{a^2+y^2}\sin\theta - k\frac{q}{a^2+y^2}\sin\theta = 0$$
From the geometry in Fig.1.66, evaluate $\cos\theta$:
$$\cos\theta = \frac{a}{r} = \frac{a}{(a^2+y^2)^{1/2}}$$
Now substitute this value of $\cos\theta$ in the above expression for E_x:
$$E_x = 2k\frac{q}{a^2+y^2}\left[\frac{a}{(a^2+y^2)^{1/2}}\right] = k\frac{2aq}{(a^2+y^2)^{3/2}}$$
(c) In the solution to part (b), because $y \gg a$, neglect a^2 compared with y^2 and write the expression for E in this case:
$$E \approx k\frac{2aq}{y^3}$$

EXAMPLE 40. Two identical positive point charges, each of magnitude q, are placed on the axis at $x = -a$ and $x = +a$, as shown in Fig.1.67.
(a) Find an expression of net electric field on x axis and plot the variation of electric field E along the x-axis.
(b) Find an expression of net electric field on y axis and plot the variation of E along the y-axis.

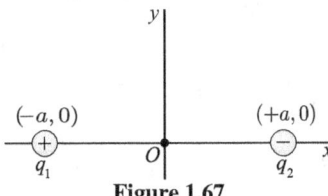

Figure 1.67

(a) APPROACH Let $\vec{E}_1$ and $\vec{E}_2$ be the electric fields due to q_1 and $q_{2'}$ respectively. Because q_1 is positive, $\vec{E}_1$ points away from q_1 everywhere, and because q_2 is also positive, so, $\vec{E}_2$ also points away from q_2 everywhere (here, $q_1 = q_2 = q$).
By principle of superposition, the net electric field at any position is: $\vec{E} = \vec{E}_1 + \vec{E}_2$.
SOLUTION (a) 1. Draw the charge configuration and place the field point A on the x axis at the appropriate place. Draw vectors representing the electric field at A due to each point charge. Repeat this procedure for field point B (Fig.1.68):

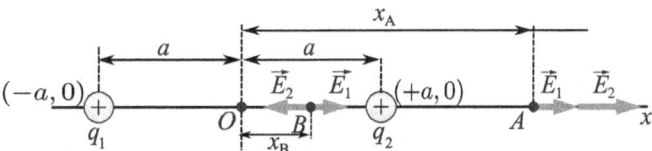

Figure 1.68: Because q_1 is a positive charge, $\vec{E}_1$ points away from q_1, at both point A and point B. Because q_2 is a positive charge, $\vec{E}_2$ points away from q_2 at both point A and point B.

2. Calculate $\vec{E}$ at point A using, $r_{1A} = |x_A + a|$ and $r_{2A} = |x_A - a|$:
$$\vec{E} = \vec{E}_1 + \vec{E}_2 = \frac{kq_1}{r_{1A}^2}\hat{r}_{1A} + \frac{kq_2}{r_{2A}^2}\hat{r}_{2A}$$
$$= \frac{kq_1}{(x_A+a)^2}\hat{i} + \frac{kq_2}{(x_A-a)^2}\hat{i}$$
$$= kq\left[\frac{1}{(x_A+a)^2}\hat{i} + \frac{1}{(x_A-a)^2}\hat{i}\right] \quad (\because q_1 = q_2 = q)$$
Special Positions: (i) When $x \to a$, then $\vec{E} = +\infty\,\hat{i}$
(ii) When $x \to +\infty$, then $\vec{E} = 0\,\hat{i}$
(b) Calculate $\vec{E}$ at point B, where $r_{1B} = |x_B + a|$ and $r_{2B} = |x_B - a|$:
$$\vec{E} = \vec{E}_1 + \vec{E}_2 = \frac{kq_1}{r_{1B}^2}\hat{r}_{1B} + \frac{kq_2}{r_{2B}^2}\hat{r}_{2B}$$
$$= \frac{kq_1}{(x_B+a)^2}\hat{i} + \frac{kq_2}{(x_B-a)^2}(-\hat{i})$$
$$= kq\left[\frac{1}{(x_B+a)^2}\hat{i} + \frac{1}{(x_B-a)^2}(-\hat{i})\right] \quad (\because q_1 = q_2 = q)$$
Special Positions: (i) When $x \to a$, then $\vec{E} = -\infty\,\hat{i}$
(ii) When $x = 0$, then $\vec{E} = 0\,\hat{i}$
(iii) When $x \to -a$, then $\vec{E} = +\infty\,\hat{i}$
(c) If we consider another point C in the left of charge q_1 (not shown in the diagram,), the calculation will be similar to solution step 2(a) except that the net field is directed towards the negative direction of x-axis (students are advised to think this situation and draw diagram themselves). In this case-
$$\vec{E} = \vec{E}_1 + \vec{E}_2 = \frac{kq_1}{r_{1C}^2}\hat{r}_{1C} + \frac{kq_2}{r_{2C}^2}\hat{r}_{2C}$$
$$= \frac{kq_1}{(x_C-a)^2}(-\hat{i}) + \frac{kq_2}{(x_C+a)^2}(-\hat{i})$$
$$= kq\left[\frac{1}{(x_C-a)^2} + \frac{1}{(x_C+a)^2}\right](-\hat{i}) \quad (\because q_1 = q_2 = q)$$
Special Positions: (i) When $x \to a$, then, $\vec{E} = -\infty\hat{i}$
(ii) When $x \to -\infty$, then, $\vec{E} = 0\,\hat{i}$
E **vs** x **graph:** The resultant electric field at source points close to q_1 is dominated by the field $\vec{E}_1$ due to q_1. There is one point between q_1 and q_2 where the resultant electric field is zero. Since, $q_1 = q_2 = q$, therefore it is the origin, where net electric field due to both the charges is zero. A test charge placed at this point would experience no electric force.
A sketch of E versus x for this charge configuration is shown in Fig.1.69.
(b)APPROACH We'll assume that q is positive when drawing pictures, but the solution should allow for the possibility that q is negative. The question does not ask about any specific point, so we will be looking for a symbolic expression in terms of the unspecified position y. Fig.1.70 shows the charges, the coordinate system, and the two electric field vectors $\vec{E}_1$ and $\vec{E}_2$. Each of these field points away from its source charge because of the assumption that q is positive. We need to find the vector sum

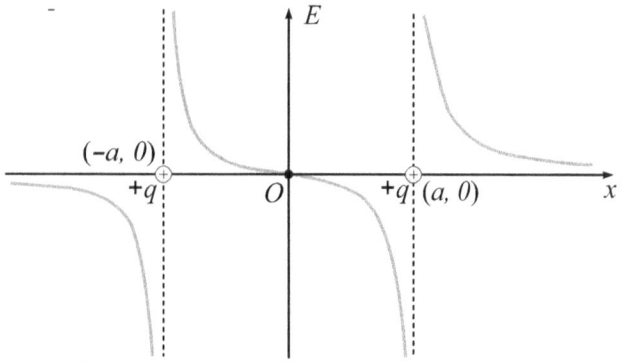

Figure 1.69: The variation of electric field E along the x axis. Electric field directed along $+ve$ x-axis is taken as poisitive
$$\vec{E}_{net} = \vec{E}_1 + \vec{E}_2$$
Before rushing into a calculation, we can make our task much easier by first thinking qualitatively about the situation. For example, the fields $\vec{E}_1$ and $\vec{E}_2$ both lie in the xy-plane, hence we can conclude without any calculations that $(E_{net})_z = 0$. Next, look at the x-components of the fields. The fields $\vec{E}_1$ and $\vec{E}_2$ have equal magnitudes and are tilted away from the y-axis by the same angle θ. Consequently, the x-components of $\vec{E}_1$ and $\vec{E}_2$ will cancel when added. The only component we need to calculate is $(E_{net})_y$.

SOLUTION The y-component of the field is

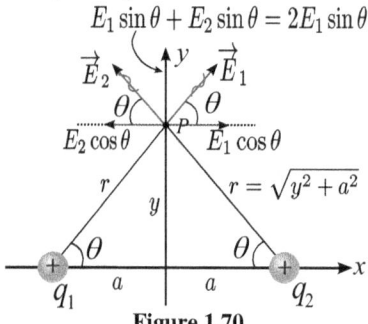

Figure 1.70

$$(E_{net})_y = (E_1)_y + (E_2)_y = 2(E_1)_y$$

where we used the fact that fields $\vec{E}_1$ and $\vec{E}_2$ have equal y-components.
Vector $\vec{E}_1$ is at angle θ from the x-axis, so its y-component is
$$(E_1)_y = E_1 \sin\theta = k\frac{q_1}{r^2}\sin\theta$$
where r is the distance from q_1. This expression for $(E_1)_y$ is correct, but it is not yet sufficient. Both the distance r and the angle θ vary with the position y and need to be expressed as functions of y. From the Pythagorean theorem, $r = (y^2 + a^2)^{1/2}$. Thus,
$$\sin\theta = \frac{y}{r} = \frac{y}{(y^2+a^2)^{1/2}}$$
By combining these pieces, we see that $(E_1)_y$ is
$$(E_1)_y = k\frac{q}{y^2+a^2}\frac{y}{(y^2+a^2)^{1/2}}$$
$$\Rightarrow \quad (E_1)_y = k\frac{qy}{(y^2+a^2)^{3/2}}$$
$$\therefore \quad (E_{net})_y = 2(E_1)_y = k\frac{2qy}{(y^2+a^2)^{3/2}}$$
The other two components of $\vec{E}_{net}$ are zero, hence the electric field of the two charges at a point on the y-axis is
$$\vec{E}_{net} = k\frac{2qy}{(y^2+a^2)^{3/2}}\hat{j} = \frac{1}{4\pi\epsilon_0}\frac{2qy}{(y^2+a^2)^{3/2}}\hat{j}$$
$$\left[\because \quad k = \frac{1}{4\pi\epsilon_0}\right]$$
This is the electric field only at points on the y-axis. Furthermore, this expression is valid only for $y > 0$. The electric field to the downward of the charges points in the opposite direction. Our expression is valid for both positive and negative q. A negative value of q makes $(E_{net})_y$ negative, which would be an electric field pointing in negative direction of y-axis.

Tracing E vs y Graph

1. **Maximum and Minimum values of Electric Field:** For maximum value of E, on y axis, we have
$$\frac{dE}{dy} = 0$$
We have already obtained the expression for $\vec{E}$, on y axis as -
$$\vec{E}_{net} = k\frac{2qy}{(y^2+a^2)^{3/2}}\hat{j}$$
Therefore,
$$\frac{dE}{dy} = 2kq\frac{d}{dy}\frac{y}{(y^2+a^2)^{3/2}}$$
$$= 2kq\left[\frac{(y^2+a^2)^{3/2}1 - y\frac{3}{2}(y^2+a^2)^{1/2}2y}{(y^2+a^2)^3}\right]$$
$$= 2kq(y^2+a^2)^{1/2}\left[\frac{a^2-2y^2}{(y^2+a^2)^3}\right]$$
Therefore, $\frac{dE}{dy} = 0 \Rightarrow 2kq(y^2+a^2)^{1/2}\frac{a^2-2y^2}{(y^2+a^2)^3} = 0$
or $\quad y = \pm\frac{a}{\sqrt{2}}$

So, the value of electric field will be maximum at $y = \pm\frac{a}{\sqrt{2}}$
Substituting this value of y in the expression for E, we get
$$\vec{E}_{max} = \pm 2kq\frac{a/\sqrt{2}}{\left(\frac{a^2}{2}+a^2\right)^{3/2}}\hat{j} = \pm\sqrt{\frac{16}{27}}\frac{kq}{a^2}\hat{j}$$

☞ If a point charge is placed at a distance $y \pm a/\sqrt{2}$, it experiences the maximum Coulomb's force

2. From the equation for $\vec{E}$, at $y = 0$, the electric field $\vec{E} = 0$.
3. When $y \to \pm\infty$, then, again $\vec{E} \to 0$

Keeping these points in mind, the variation of E with y is given in Fig.1.71.

EXAMPLE 41. A charge $+q$ is at $x = a$ and a second charge

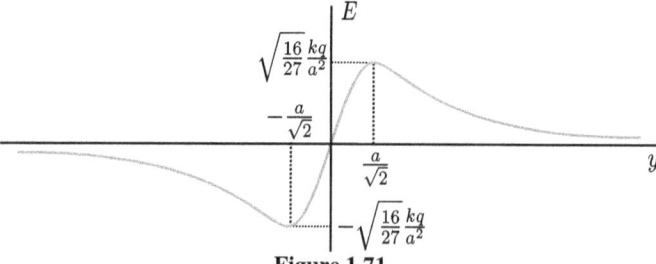

Figure 1.71

$-q$ is at $x = -a$ (Figure 1.73). (a) Find the electric field on the

1.13. THE ELECTRIC FIELD OF POINT CHARGES

x axis at an arbitrary point $x > a$. (b) Find the limiting form of the electric field for $x \gg a$ (c) Plot the variation of E along the x-axis.

APPROACH We calculate the electric field at point P using

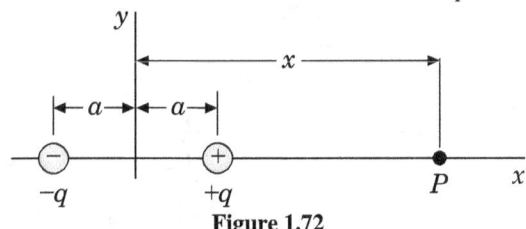

Figure 1.72

the principle of superposition, $\vec{E}_P = \vec{E}_{1P} + \vec{E}_{2P}$. For $x > a$, the electric field $\vec{E}_+$ due to the positive charge is in the $+x$ direction and the electric field $\vec{E}_-$ due to the negative charge is in the $-x$ direction. The distances are $x - a$ to the positive charge and $x - (-a) = x + a$ to the negative charge.

SOLUTION (a) 1. Draw the charge configuration on a coordinate axis and label the distances from each charge to the field point (Fig.1.73):

2. Calculate $\vec{E}$ due to the two charges for $x > a$: (Note that the

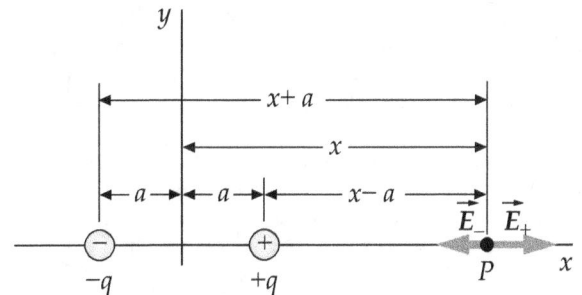

Figure 1.73

equation given below holds only for $x > a$.)

$$\vec{E} = \vec{E}_+ + \vec{E}_- = \frac{kq}{[x-a]^2}\hat{i} + \frac{kq}{[x-(-a)]^2}(-\hat{i})$$

$$= kq\left[\frac{1}{(x-a)^2} - \frac{1}{(x+a)^2}\right]\hat{i}$$

On simplifying, we get-

$$\vec{E} = kq\left[\frac{(x+a)^2 - (x-a)^2}{(x+a)^2(x-a)^2}\right]\hat{i} = kq\frac{4ax}{(x^2-a^2)^2}\hat{i} \quad x > a$$

(b) In the limit $x \gg a$, we can neglect a^2 compared with x^2 in the denominator:

$$\vec{E} = kq\frac{4ax}{(x^2-a^2)^2}\hat{i} \approx kq\frac{4ax}{x^4}\hat{i} = \frac{4kqa}{x^3}\hat{i} \quad x \gg a$$

The answer approaches zero as x approaches infinity, which is as expected.

(c) Fig.1.74 shows E_x versus x for all x, for $q = 1.0$ nC and $a = 1.0$ m. For $|x| \gg a$ (far from the charges), the field is given by

$$\vec{E} = \frac{4kqa}{|x|^3}\hat{i} \quad |x| \gg a$$

Between the charges, the contribution from each charge is in the negative direction. An expression for $\vec{E}$ at distance $x(< a)$, is

$$\vec{E} = -\frac{kq}{(x-a)^2}\hat{i} - \frac{kq}{(x+a)^2}\hat{i} \quad -a < x < a$$

In the range, $-a \leq x \leq +a$: corresponding to $x \to \pm a$, above equation gives-

$$\vec{E} \to -\infty\,\hat{i}$$

At origin, i.e., at $x = 0$, above equation gives-

$$\vec{E} = -\frac{kq}{(-a)^2}\hat{i} - \frac{kq}{(a)^2}\hat{i} = -\frac{2kq}{a^2}\hat{i}$$

In right side of charge $+q$, if the point is very close to it, i.e, $x \to a$, then point (3) of part (a), gives

$$\vec{E} = kq\lim_{x\to+a}\frac{4ax}{(x^2-a^2)^2}\hat{i} = +\infty\,\hat{i}$$

In right side of charge $+q$, if the point is very far from it, i.e, $x \to \infty$, then point (3) of part (a), gives

$$\vec{E} = kq\lim_{x\to\infty}\frac{4ax}{(x^2-a^2)^2}\hat{i} = 0\,\hat{i}$$

In left of charge $-q$, the net electric field at distance x from the origin is given by

$$\vec{E} = \vec{E}_+ + \vec{E}_- = \frac{kq}{[|x|+a]^2}\hat{i} + \frac{kq}{[|x|-a]^2}\hat{i}$$

$$= kq\left[\frac{1}{(|x|+a)^2} + \frac{1}{(|x|-a)^2}\right]\hat{i}$$

or $\quad \vec{E} = kq\frac{4a|x|}{(|x|^2-a^2)^2}\hat{i}$

when, $x \to a$, then

or $\quad \vec{E} = kq\lim_{x\to -a}\frac{4a|x|}{(|x|^2-a^2)^2}\hat{i} = +\infty\hat{i}$

and when $x \to -\infty$, then

$$\vec{E} = kq\lim_{x\to -\infty}\frac{4a|x|}{(|x|^2-a^2)^2}\hat{i} = 0\hat{i}$$

Keeping these points in mind, the variation of E with x is given in Fig.1.74

EXAMPLE 42. Three point charges lie along the x axis

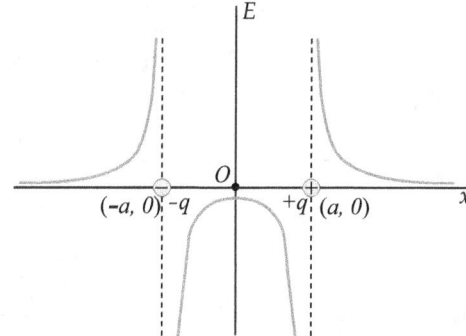

Figure 1.74: A plot of E versus x on the x axis for the given charge distribution

as shown in Fig.1.75. The positive charge $q_1 = 24.0\ \mu C$ is at $x = 3.00$ m, the positive charge $q_2 = 6.00\ \mu C$ is at the origin, and the resultant force acting on q_3 is zero. What is the x coordinate of q_3?

APPROACH Since the charges q_1 and q_2 are of same nature

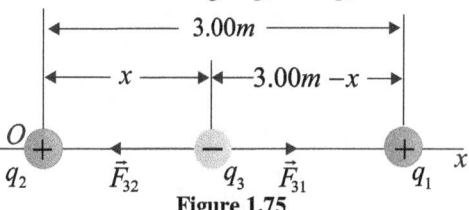

Figure 1.75

(here both are positive), therefore the null point lies between them near the smaller charge.

If x is the distance of null point from smaller charge q_2, then net electric force on any charge q_3, placed at this point will be zero,

i.e.,
$$|\vec{F}_{32}| = |\vec{F}_{31}| \quad \text{(i)}$$
here $\vec{F}_{32}$ is the electrostatic force of charge q_2 on the charge q_3 and $\vec{F}_{31}$ is the electrostatic force of charge q_1 on the charge q_3. Now, find the forces $\vec{F}_{32}$ and $\vec{F}_{31}$ by using Coulomb's law and substitute these values in Eq.(i) to get the required value of x.

☞ Be careful that Eq.(i) is the relation between magnitudes only. So, for each value of x obtained from Eq.(i), the magnitudes of forces on charge q_3 will be equal but it does not tell anything about directions of forces on q_3. Since the position of null point should be between q_1 and q_2, therefore, only $+ve$ values of x will be acceptable.

SOLUTION $\quad k\dfrac{|q_2||q_3|}{x^2} = k\dfrac{|q_1||q_3|}{(3.00-x)^2}$
$\Rightarrow \quad (3.00-x)^2 |q_2| = x^2 |q_1|$

On simplifying this quadratic equation for x, we find that the positive root is $x = 1$ m. There is also a second root, $x = -3$ m. This is another location at which the magnitudes of the forces on q_3 are equal, but both forces are in the same direction at this location.

EXAMPLE 43. Point charges q and $-q$ are located at the vertices of a square with diagonals $2l$ as shown in Fig.1.76. Find the magnitude of the electric field strength at a point located symmetrically with respect to the vertices of the square at a distance x from its centre.

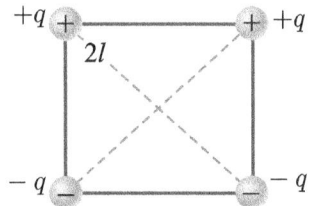

Figure 1.76

APPROACH Apply superposition principle of electric field:
$$\vec{E} = \vec{E}_1 + \vec{E}_2 + \vec{E}_3 + \vec{E}_4$$
$$= \dfrac{q_1(\vec{r}_p - \vec{r}_1)}{4\pi\epsilon_0 |\vec{r}_p - \vec{r}_1|^3} + \dfrac{q_2(\vec{r}_p - \vec{r}_2)}{4\pi\epsilon_0 |\vec{r}_p - \vec{r}_2|^3} + \dfrac{q_3(\vec{r}_p - \vec{r}_3)}{4\pi\epsilon_0 |\vec{r}_p - \vec{r}_3|^3}$$
$$+ \dfrac{q_4(\vec{r}_p - \vec{r}_4)}{4\pi\epsilon_0 |\vec{r}_p - \vec{r}_4|^3}$$

SOLUTION The arrangement of charges is shown in Fig.1.77.

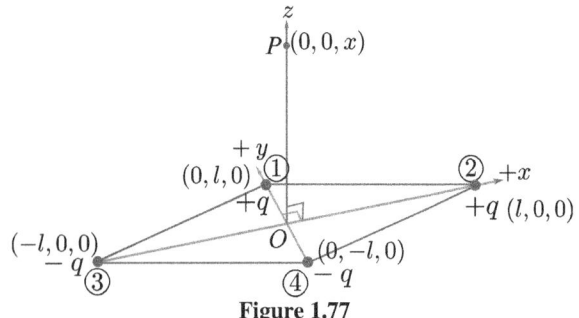

Figure 1.77

According' to given problem,
$q_1 = +q, \vec{r}_1 = l\hat{j}, q_2 = +q, \vec{r}_2 = l\hat{i}$
$q_3 = -q, \vec{r}_3 = -l\hat{i}; q_4 = -q, \vec{r}_4 = -l\hat{j}$
$\vec{r}_p = x\hat{k}$
$\therefore \quad \vec{r}_p - \vec{r}_1 = x\hat{k} - l\hat{j},$
$|\vec{r}_p - \vec{r}_1| = \sqrt{x^2 + l^2} = (x^2 + l^2)^{1/2}$

$\vec{r}_p - \vec{r}_2 = x\hat{k} - l\hat{i}, \quad |\vec{r}_p - \vec{r}_2| = (x^2 + l^2)^{1/2}$
$\vec{r}_p - \vec{r}_3 = x\hat{k} + l\hat{i}, \quad |\vec{r}_p - \vec{r}_3| = (x^2 + l^2)^{1/2}$
and $\vec{r}_p - \vec{r}_4 = x\hat{k} + l\hat{j}, |\vec{r}_p - \vec{r}_4| = (x^2 + l^2)^{1/2}$
$\vec{E} = \vec{E}_1 + \vec{E}_2 + \vec{E}_3 + \vec{E}_4$
$\Rightarrow \vec{E} = \dfrac{q_1(\vec{r}_p - \vec{r}_1)}{4\pi\epsilon_0 |\vec{r}_p - \vec{r}_1|^3} + \dfrac{q_2(\vec{r}_p - \vec{r}_2)}{4\pi\epsilon_0 |\vec{r}_p - \vec{r}_2|^3}$
$$+ \dfrac{q_3(\vec{r}_p - \vec{r}_3)}{4\pi\epsilon_0 |\vec{r}_p - r_3|^3} + \dfrac{q_4(\vec{r}_p - \vec{r}_4)}{|\vec{r}_p - \vec{r}_4|^3}$$

Putting the values, we get
$$\vec{E} = \dfrac{q}{4\pi\epsilon_0 (x^2+l^2)^{3/2}}[(x\hat{k}-l\hat{j})+(x\hat{k}-l\hat{i})-(x\hat{k}+l\hat{i})$$
$$-(x\hat{k}+l\hat{j})]$$
$$= \dfrac{q}{4\pi\epsilon_0(x^2+l^2)^{3/2}}(-2l\hat{j}-2l\hat{i})$$
$\therefore \quad E = \dfrac{q}{4\pi\epsilon_0(x^2+l^2)^{3/2}}\sqrt{(-2l)^2+(-2l)^2}$
$\Rightarrow \quad E = \dfrac{2ql\sqrt{2}}{4\pi\epsilon_0(x^2+l^2)^{3/2}} = \dfrac{ql}{\sqrt{2}\pi\epsilon_0(x^2+l^2)^{3/2}}$

EXAMPLE 44. A point charge of 10 C is placed at the origin. At what location on the X-axis should a point charge of 40 μC be placed so that the net electric field is zero at $x = 2$ cm on the X-axis?

SOLUTION Suppose point A is located at $x = x_0$ from the origin O, where the second charge is placed, which is 40 μC. As the given charges have the same nature, the zero-field point will be between the charges and closer to the smaller charge, i.e., 10 μC. In Fig.1.78, P is the zero field point at distance 2 cm from origin O where 10 μC charge is placed.

Now, $\quad E_P = \dfrac{k \times 10}{2^2} - \dfrac{k \times 40}{(x_0-2)^2} = 0 \implies \dfrac{1}{2} = \dfrac{2}{x_0-2}$

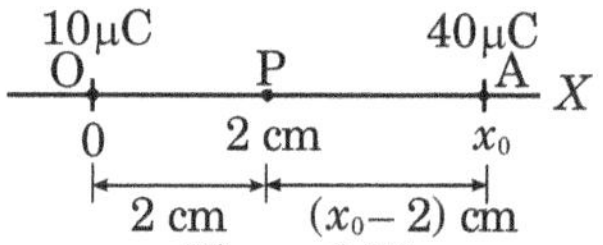

```
   10μC            40μC
    O       P       A
    |───────|───────|─── X
    0     2 cm      x₀
    |─2 cm─|(x₀−2) cm|
```
Figure 1.78

$\implies \quad x_0 = 6$ cm

1.14 Electric Dipole

Two equal and opposite point charges $+q$ and $-q$ placed a short distance ($2l$, say) apart, form an electric dipole. The particles are attracted to each other, but their separation is maintained so that the distance between them remains constant as shown in Fig.1.79.

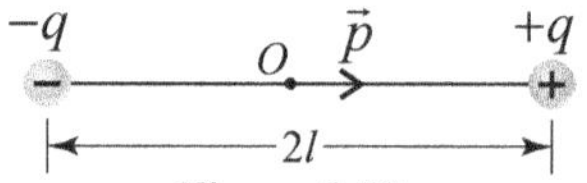

```
  −q         O   p⃗    +q
   ●─────────·──→────⊕
   |←────── 2l ──────→|
```
Figure 1.79

1.14.1 Dipole Moment

The product of the magnitude of either charge and the distance between the charges is called the electric dipole moment. The electric dipole moment–also called the dipole moment–is a vector quantity whose direction is along the axis of dipole pointing from negative towards the positive charge in a dipole. It is denoted by $\vec{p}$.

Magnitude of dipole moment is given as-
$$\boxed{p = q.(2l) = 2ql} \quad (1.38)$$

1.14. ELECTRIC DIPOLE

The S.I. unit of dipole moment is "coulomb-metre".
The practical unit of dipole moment is debye.
1 debye (D) = 3.3×10^{-30} coulomb-meter.
The direction of $\vec{p}$ identifies the orientation of the dipole, and the dipole-moment magnitude $p = 2ql$ determines the electric field strength.

1.14.2 Dipole Moment of a System of Discrete Charges

Since dipole moment is a vector quantity, hence in case of two or more than two dipoles, the resultant dipole moment will be the vector sum of the dipole moments of individual dipoles, i.e.,
$$\vec{p} = \vec{p}_1 + \vec{p}_2 + \vec{p}_3 + \ldots$$
The real dipoles are the chemical dipoles such as H_2O and CO_2. H_2O has some dipole moment but CO_2 being linear has zero dipole moment.

1.14.3 Induced Dipole Moment

Many molecules, such as water (H_2O), have *permanent* electric dipole moments. In other molecules like: CO_2 and CH_4 (called *nonpolar molecules*) and in every isolated atom, the centres of the positive and negative charges coincide (Fig.1.80a) and thus no dipole moment is set up. However, if we place an atom or a nonpolar molecule in an external electric field, the field distorts the electron orbits and separates the centres of positive and negative charge (Fig.1.80b). Because the electrons are negatively charged, they tend to be shifted in a direction opposite the field. This shift sets up a dipole moment $\vec{p}$ that points in the direction of the field. This dipole moment is said to be *induced* by the field, and the atom or molecule is then said to be *polarized* by the field (that is, it has a positive side and a negative side). When the field is removed, the induced dipole moment and the polarization disappear.

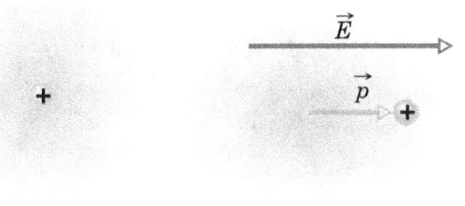

(a) (b)
Figure 1.80: The electric field shifts the positive and negative charges, creating a dipole.

☞ In non-polar molecules, centre of the positive charge coincides with centre of negative charge, i.e., separation between the centres of positive and negative charges is zero, therefore net dipole moment of non-polar molecules is zero.

☞ When non polar material is placed in an external field, centre of negative charge shifts opposite to applied electric field. So, in presence of external field, the centres of opposite charges do not coincide. Therefore, a net dipole moment induces in it from centre of negative charge to centre of positive charge, i.e., in the direction of applied electric field.

1.14.4 Electric Field at any Point due to an Electric Dipole

1.14.4.1 Electric field at an Axial Point, i.e., in End on Position

Let us consider that there is no external field, and we want to determine the electric field produced by the dipole at a point P on the axis of the dipole AB [Fig.1.81]. If O is the centre of the dipole, then $OP = r > l$.
Let E_A and E_B are the magnitudes of electric fields at the axial

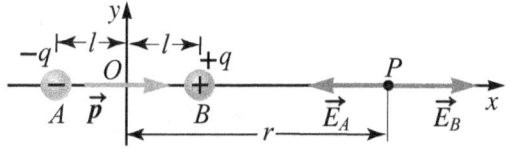

Figure 1.81

position P due to charges $-Q$ at A and $+Q$ at B respectively, then

$$\vec{E}_A = k\frac{q}{(r+l)^2} \text{ (Along } \overrightarrow{PO}) \qquad (1.39)$$

$$\vec{E}_B = k\frac{q}{(r-l)^2} \text{ (Along } \overrightarrow{OP}) \qquad (1.40)$$

Net intensity of electric field at P
$$\vec{E} = E_B - E_A \text{ (Along } \overrightarrow{OP})$$
$$= kq\left(\frac{1}{(r-l)^2} - \frac{1}{(r+l)^2}\right) \text{ (Along } \overrightarrow{OP})$$
$$= kq\left(\frac{(r+l)^2 - (r-l)^2}{(r^2-l^2)^2}\right) \text{ (Along } \overrightarrow{OP})$$
$$= kq\left(\frac{4rl}{(r^2-l^2)^2}\right) \text{ (Along } \overrightarrow{OP})$$
$$= k\frac{2(2ql)r}{(r^2-l^2)^2} \text{ (Along } \overrightarrow{OP})$$
$$= k\frac{2pr}{(r^2-l^2)^2} \text{ (Along } \overrightarrow{OP})$$

$$\Rightarrow \boxed{E = k\frac{2pr}{(r^2-l^2)^2} \text{ (Along } \overrightarrow{OP})} \qquad (1.41)$$

In vector form,
$$\Rightarrow \boxed{\vec{E} = k\frac{2\vec{p}r}{(r^2-l^2)^2} \text{ (Along } \overrightarrow{OP})} \qquad (1.42)$$

As $r \gg l$, therefore
$$\boxed{E = k\frac{2p}{r^3} \text{ (Along } \overrightarrow{OP})} \qquad (1.43)$$

In vector form
$$\boxed{\vec{E} = k\frac{2\vec{p}}{r^3} \text{ (Along } \overrightarrow{OP})} \qquad (1.44)$$

In this case, $\vec{E}$ is in the direction of $\vec{p}$.
Note: E vs r graph in axial position will be similar to graph shown in Fig.1.74. The only difference is that letter a is used for l and x is used for r.

1.14.4.2 At a Point on its Perpendicular Bisector, i.e. in Broadside on Position

Let us consider mid point O of dipole as the origin. Let P be any point at distance $OP = r$ on the perpendicular bisector of the electric dipole. $\vec{E}_1$ and $\vec{E}_2$ are the intensities at P due to charges $+q$ and $-q$ respectively.

$$\vec{E}_1 = k\frac{q}{r^2+l^2} \quad \text{(along } \overrightarrow{BP} \text{ direction)} \qquad (1.45)$$

$$\vec{E}_2 = k\frac{q}{r^2+l^2} \quad \text{(along } \overrightarrow{PA} \text{ direction)} \qquad (1.46)$$

From Eq.1.45 and 1.46, it is clear that $E_1 = E_2$
Now to get the resultant of $\vec{E}_1$ and $\vec{E}_2$ we resolve $\vec{E}_1$ and $\vec{E}_2$ into components:
(i) Vertical components $E_1\sin\theta$, and $E_2\sin\theta$ are equal in magnitude and oppositely directed to each other, therefore they get cancelled.
(ii) Horizontal components $E_1\cos\theta$ and $E_2\cos\theta$ are directed along same direction, so they get added

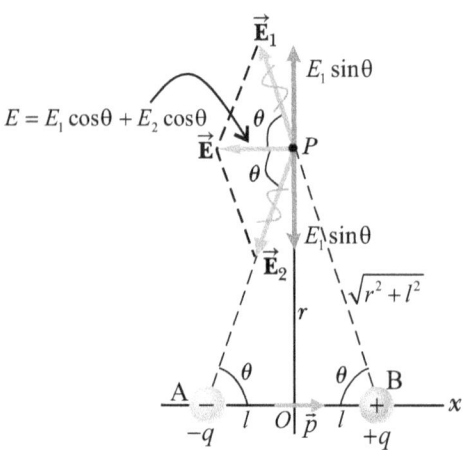

Figure 1.82

Therefore, net electric field
$$E = E_1\cos\theta + E_2\cos\theta = 2E_1\cos\theta$$
Now, from Fig.1.82,
$$\cos\theta = \frac{l}{\sqrt{r^2+l^2}}$$
Therefore, above equation gives
$$E = 2k\left(\frac{q}{r^2+l^2}\right)\left(\frac{l}{\sqrt{r^2+l^2}}\right)$$
$$\Rightarrow \quad E = k\frac{2ql}{(r^2+l^2)^{3/2}} \quad (1.47)$$

Since, $2ql = p$ (dipole moment), therefore Eq.(1.47) gives-
$$\boxed{E = k\frac{p}{(r^2+l^2)^{3/2}}} \quad (1.48)$$

In vector form-
$$\boxed{\vec{E} = -k\frac{\vec{p}}{(r^2+l^2)^{3/2}}} \quad (1.49)$$

In this case, the direction of $\vec{E}$ is opposite to that of $\vec{p}$. If $r^2 \gg l^2$, then,
$$\boxed{E = k\frac{p}{r^3}} \quad (1.50)$$

In vector form-
$$\boxed{\vec{E} = -k\frac{\vec{p}}{r^3}} \quad (1.51)$$

Thus, we see that for $r \gg l$, the intensity of electric field in axial position of dipole is double in magnitude to that of intensity at the same distance on perpendicular bisector.

Special Cases: (i) At $r = 0$, the electric field $E = k\frac{p}{l^3}$. It is the maximum value of electric field in broad side on position
(ii) At $r \to \pm\infty$, the electric field $E \to 0$.
Keeping these point in mind, the variation of E with r is plotted in Fig.1.83.

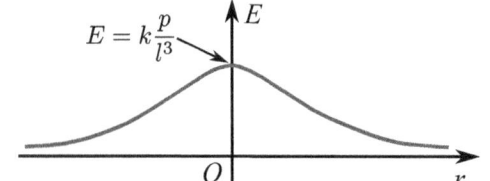

Figure 1.83: Variation of E with r in broad side on position

1.14.4.3 At any General Point

Let P (r, θ) be any general point where intensity of electric field due to electric dipole is to be determined. The direction of dipole moment is along axis A to B.
Now, resolving vector $\vec{p}$ into two components
(i) $p_1 = p\cos\theta$ (along the direction $\overrightarrow{OP}$),
(ii) $p_2 = p\sin\theta$ (perpendicular to $\overrightarrow{OP}$)
Point P is now at end on position of a dipole of dipole moment

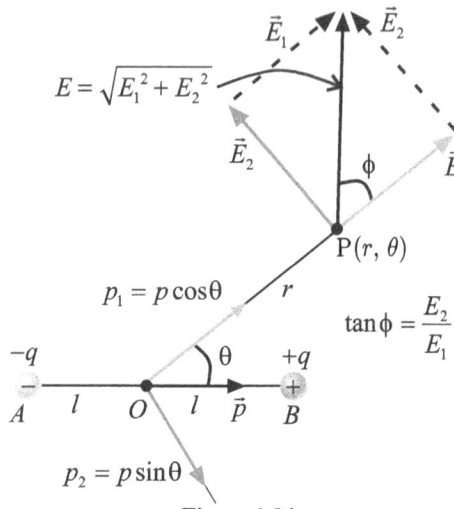

Figure 1.84

p_1 and it is on perpendicular bisector position of dipole of dipole moment p_2.
Electric field $\vec{E}_1$ at P due to dipole of dipole moment p_1 will be along $\overrightarrow{OP}$ vector. From Eq.1.43, the magnitude of $\vec{E}_1$ is given by
$$E_1 = k\frac{2p_1}{r^3} = k\frac{2p\cos\theta}{r^3}$$
Electric field $\vec{E}_2$ at P due to dipole of dipole moment p_2 will be perpendicular to $\overrightarrow{OP}$ and opposite to the direction of $\vec{p}_2$ vector. From Eq.1.50, the magnitude of $\vec{E}_2$ is given by
$$E_2 = k\frac{2p_2}{r^3} = k\frac{p\sin\theta}{r^3}$$
Resultant electric field at P,
$$E = \sqrt{E_1^2 + E_2^2} = k\frac{p}{r^3}\sqrt{4\cos^2\theta + \sin^2\theta}$$
$$\Rightarrow \quad E = k\frac{p}{r^3}\sqrt{4\cos^2\theta + (1-\cos^2\theta)}$$
$$\Rightarrow \quad \boxed{E = k\frac{p}{r^3}\sqrt{3\cos^2\theta + 1}} \quad (1.52)$$

The angle between the resultant field and the radial direction $\overrightarrow{OP}$ (O is the centre point of the dipole axis and P is that at which electric field is being calculated) is ϕ. From Fig.1.84, we can write
$$\tan\phi = \frac{E_2}{E_1}$$
$$\Rightarrow \quad \boxed{\phi = \tan^{-1}\left(\frac{\tan\theta}{2}\right)} \quad (1.53)$$

Special cases:
1. $\theta = 0°$: In this case P is on the axis of the dipole. This position is called an end-on position.
In this case,
$$E_1 = k\frac{2p\cos 0°}{r^3} = k\frac{2p}{r^3},$$
$$E_2 = k\frac{p\sin 0°}{r^3} = 0$$
and $E = k\frac{p}{r^3}\sqrt{3\cos^2 0 + 1} = k\frac{2p}{r^3}$

$$\phi = \tan^{-1}\left(\frac{\tan\theta}{2}\right) = \tan^{-1} 0 = 0$$

2. $\theta = 90°$: In this case P is on the perpendicular bisector of the dipole axis.

$$E_1 = k\frac{2p\cos 90°}{r^3} = 0,$$
$$E_2 = k\frac{p\sin 90°}{r^3} = k\frac{p}{r^3}$$

and $E = k\frac{p}{r^3}\sqrt{3\cos^2 90° + 1} = k\frac{p}{r^3}$,

$$\phi = \tan^{-1}\left(\frac{\tan 90}{2}\right) = \tan^{-1}\infty = \frac{\pi}{2}$$

1.15 Check Point 3

1. ••Shown in Figure 1.85 is an array of six charges. The top charges are all positive and of the same magnitude, $+Q$, and there is a distance a between the charges. At a distance $2a$ below this line of charges are three other charges, with the outer two being $+Q$ and the central one $-Q$. What is the net electric field at a point P at the centre of the charge array?

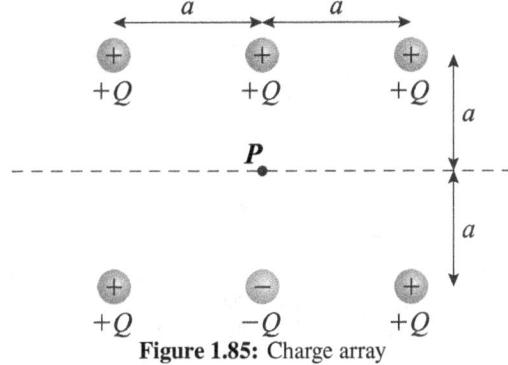

Figure 1.85: Charge array

2. ••A charge $q = 1\mu C$ is placed at point $(1\text{ m}, 2\text{ m}, 4\text{ m})$. Find the electric field at point $P(0, -4\text{m}, 3\text{ m})$.

3. ••A positive point charge $50\ \mu C$ is located in the plane xy at the point with radius vector $\vec{r}_0 = 2\hat{i} + 3\hat{j}$, where $\hat{i}$ and $\hat{j}$ are the unit vector of the x and y axis. Find the vector of the electric field strength $\vec{E}$ and its magnitude at the point with radius vector $\hat{r} = 8\hat{i} - 5\hat{j}$. Here r_0 and r are expressed in metres.

4. ••Four identical positive charges (each of magnitude q) are fixed at the corners of a square of side l. Find electric field at point P which is at a distance $z(\ll l)$ lying on the line perpendicular to the plane of the square passing through the centre of square. Now, if we place a test charge q_0 at point P, find the net electric force on it.

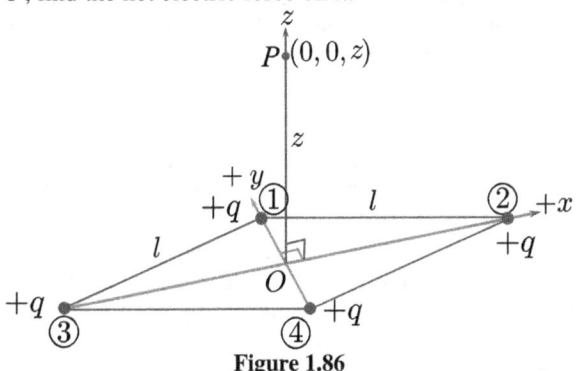

Figure 1.86

5. ••Charges Q_1 and Q_2 are at point A and B of a right angle triangle OAB [Fig.1.87]. The resultant electric field at point O is perpendicular to the hypotenuse, then find the value of Q_1/Q_2.

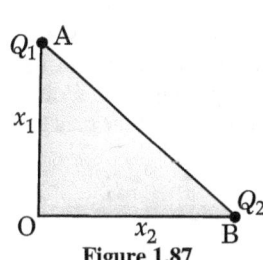

Figure 1.87

6. ••A positively charged ball hangs from a long silk thread. We wish to measure E at a point in the same horizontal plane as that of the hanging charge. To do so, we put a positive test charge q_0 at the point and measure F/q_0 [Fig.1.88]. Will F/q_0 be less than, equal to, or greater than E at the point in the question?

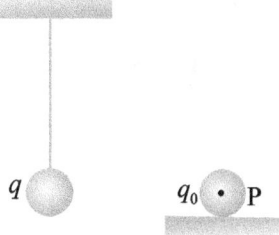

Figure 1.88

7. ••In Fig.1.89, find distance from $4Q$, where total electric field is zero.

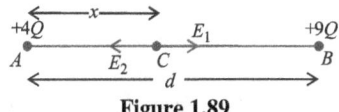

Figure 1.89

8. ••Infinite charges of equal magnitude $20\mu c$ each are placed at $x = 1$ cm, 2 cm, 4 cm, 8 cm, 16 cm, $\ldots\infty$. Calculate intensity of electric field at origin.

9. ••Infinite charges are placed on x-axis such that $+q$ at $x = a, -q$ at $x = \sqrt{2}a, +q$ at $x = \sqrt{3}a, -q$ at $x = 2a, +q$ at $x = \sqrt{5}a$ etc. Calculate electric field intensity at origin.

Multiple Choice Questions

10. ••A small electric dipole is placed at origin with its dipole moment directed along positive x-axis. The direction of electric field at point $(2, 2\sqrt{2}, 0)$ is along
    (A) z-axis (B) y-axis
    (C) negative y-axis (D) negative z-axis

11. ••What is the angle between electric dipole moment and the electric field strength due to dipole on the broadside on position.
    (A) $0°$ (B) $90°$
    (C) $180°$ (D) none of these

1.16 Electric Field of a Continuous Charge Distribution

So far we have dealt with only charged particles because Coulomb's law applies only to charged particles. However, most charged objects— like charged comb, charged sphere, charged sheet, line charge etc. — are not particles. Instead, they are extended bodies. Although every macroscopic object consists of very large numbers of charged particles-protons and electrons-it is not practical to calculate the individual field of each of these particles and then add them vectorially. Instead, we shall treat any macroscopic charged object as having a continuous charge distribution (Fig.1.90). The field set up by a continuous charge distribution can be computed by dividing the distribution into infinitesimal elements dq. Each element of charge establishes a field $d\vec{E}$ at a point P, and the resultant field at P is then found from the superposition principle by adding (that is, integrating) the field contributions due to all the charge elements, i.e.,

$$\vec{E} = \int d\vec{E} \qquad (1.54)$$

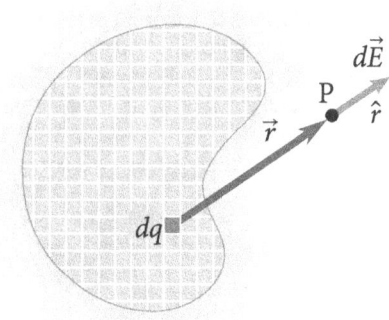

Figure 1.90

In Fig.1.90, the electric field produced by small charge element dq at point P is,

$$d\vec{E} = k\frac{dq}{r^2}\hat{r} \qquad (1.55)$$

Therefore, net electric field at point P, will be given as-

$$\vec{E} = \int d\vec{E} = k \int \frac{dq}{r^2}\hat{r} \qquad (1.56)$$

In order to evaluate this integral, we must express dq, $1/r^2$, and $\hat{r}$ in terms of the same coordinate(s). To do so, it is necessary to express the charge on the object in terms of a charge density. There are three types of charge densities:

1. Volume charge density ρ (lowercase Greek letter rho) is the amount of charge per unit volume (Fig. 1.91). For a uniform charge distribution,

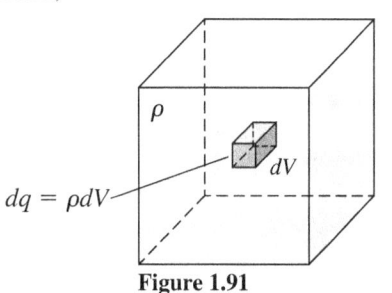

Figure 1.91

$$\rho \equiv \frac{Q}{V} \qquad (1.57)$$

Volume charge density has the dimensions charge per volume, with SI units C/m³. The amount of charge dq in a small volume dV can be written in terms of the volume charge density:

$$\begin{pmatrix} \text{charge contained} \\ \text{in small volume} \end{pmatrix} = \begin{pmatrix} \text{charge per} \\ \text{unit volume} \end{pmatrix} \times (\text{volume})$$

$$\boxed{dq = \rho\, dV} \qquad (1.58)$$

2. Surface charge density σ (lowercase Greek letter sigma) is the amount of charge per unit area (Fig. 1.92). For a uniform charge distribution,

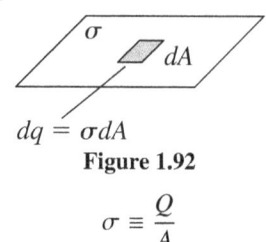

Figure 1.92

$$\sigma \equiv \frac{Q}{A}$$

The dimensions of σ are charge per area, with SI units C/m². The amount of charge dq in a small area dA can be written in terms of the surface charge density:

$$\begin{pmatrix} \text{charge contained} \\ \text{in small area} \end{pmatrix} = \begin{pmatrix} \text{charge per} \\ \text{unit area} \end{pmatrix} \times (\text{area})$$

$$\boxed{dq = \sigma\, dA} \qquad (1.59)$$

3. Linear charge density λ (lowercase Greek letter lambda) is the amount of charge per unit length (Fig.1.93). For a uniform charge distribution,

$$\lambda \qquad dL$$

$$dq = \lambda dL$$

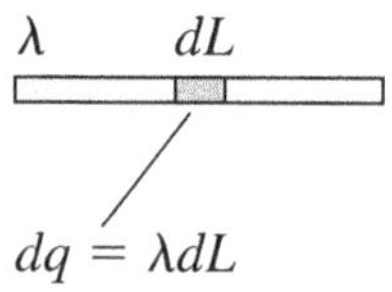

Figure 1.93

$$\lambda \equiv \frac{Q}{L} \qquad (1.60)$$

The dimensions of λ are charge per length, with SI units C/m. The amount of charge dq in a small length element dL can be written in terms of the linear charge density:

$$\begin{pmatrix} \text{charge contained in} \\ \text{small length element} \end{pmatrix} = \begin{pmatrix} \text{charge per} \\ \text{unit length} \end{pmatrix} \times (\text{length})$$

$$\boxed{dq = \lambda\, dL} \qquad (1.61)$$

Problem solving tactics for calculating the electric field from continuous charge distributions:

1. Identify the type of charge distribution and compute the charge density λ, σ or ρ.
2. Divide the charge distribution into infinitesimal charges dq, each of which will act as a tiny point charge.
3. The amount of charge dq, i.e., within a small element dl, dA or dV is
   $dq = \lambda dl$ (charge distributed along length)
   $dq = \sigma dA$ (charge distributed over a surface)
   $dq = \rho dV$ (charge distributed throughout a volume)
4. Draw at point P the dE vector produced by the charge dq. The magnitude of dE is

$$dE = \frac{1}{4\pi\epsilon_0}\frac{dq}{r^2}$$

Vector $d\vec{E}$ is along radial line joining dq to P, $d\vec{E}$ is directed away for positive charge dq while directed towards dq for negative dq.

5. Resolve the $d\vec{E}$ vector into its components. Identify any special symmetry features to show whether any component(s) of the field that are not get cancelled by other components.
6. Write the distance r and any trigonometric factors in terms of given coordinates and parameters.
7. The electric field is obtained by summing over all the infinitesimal contributions.

$$\vec{E} = \int d\vec{E} = \int \frac{dq}{4\pi\epsilon_0 r^2}$$

1.16.1 Electric Field Due to a Uniformly Charged Rod of Finite Length at a Perpendicular Distance y

Let us consider a rod of length L as shown in Fig. 1.94. suppose a positive charge Q is uniformly distributed over the surface of the rod along it's length. P is an arbitrarily positioned point, where the electric field $\vec{E}$ is to be calculated.

The x axis is along the rod, and the y axis passes through the field point P. Let y be the perpendicular distance of P from the x axis. To calculate the electric field $\vec{E}$ at P, we separately calculate E_x and E_y. Using Equations 1.55 and 1.56, first find the electric field increment $d\vec{E}$ at P due to an arbitrary increment dq of the charge distribution. Then integrate each component of $d\vec{E}$ over the entire charge distribution. (Because Q is distributed uniformly, the linear charge density λ equals Q/L.)

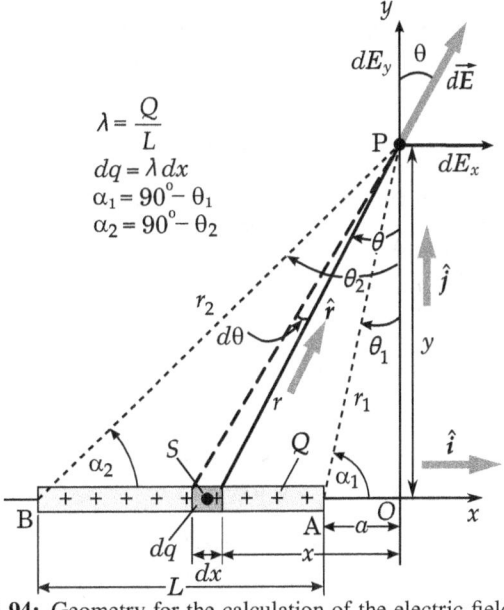

Figure 1.94: Geometry for the calculation of the electric field at field point P due to a uniformly charged rod.

Stepwise Calculation

1. The situation is shown in Fig. 1.94. In addition, figures also shows an arbitrary increment of the line charge at point S (at $OS = x$) that has a length dx and a charge dq, and the electric field $d\vec{E}$ at P due to dq. The electric field vector $d\vec{E}$ is drawn by assuming if dq is positive.

2. $\vec{E} = E_x\hat{i} + E_y\hat{j}$. Now, we find the expressions for dE_x and dE_y in terms of dE_r and θ, where dE_r, is the component of $d\vec{E}$ in the direction away from S toward P:
$$d\vec{E} = dE_r \hat{r}$$
so $\quad dE_x = dE_r \sin\theta \quad$ and $\quad dE_y = dE_r \cos\theta$

3. First we solve for E_x. Express dE_r using Equation 1.55, where r is the distance from the source point S to the field point P. From Fig. 1.94 $\cos\theta = y/r$. In addition, use $dq = \lambda dx$:
$$dE_r = \frac{k\, dq}{r^2} \quad \text{and} \quad \sin\theta = \frac{x}{r}$$
so $\quad dE_x = \frac{k\, dq}{r^2}\sin\theta = \frac{k\sin\theta\, \lambda dx}{r^2}$

4. Integrate the step-3 result:
$$dE_x = \int_a^{a+L} \frac{k\sin\theta\, \lambda dx}{r^2} = k\lambda \int_a^{a+L} \frac{\sin\theta\, dx}{r^2}$$

5. Now, change the integration variable from x to θ. From Figure 1.94, find the relation between x and θ and between r and θ.
$$\tan\theta = \frac{x}{y}, \quad \text{so} \quad x = \frac{y}{\tan\theta}$$
$$\cos\theta = \frac{y}{r}, \quad \text{so} \quad r = \frac{y}{\cos\theta} = y\sec\theta$$

6. Differentiate the step 5 result to obtain an expression for dx (the field point P remains fixed, so y is constant):
$$dx = y\sec^2\theta\, d\theta$$

7. Substitute $y\sec^2\theta\, d\theta$ for dx and $y\sec\theta$ for r in the integral in step 4 and simplify:
$$\int_a^{a+L} \frac{\sin\theta\, dx}{r^2} = \int_{\theta_1}^{\theta_2} \frac{\sin\theta\, y\sec^2\theta\, d\theta}{y^2\sec^2\theta}$$
$$= \frac{1}{y}\int_{\theta_1}^{\theta_2} \sin\theta\, d\theta \quad (y \neq 0)$$

8. Evaluate the integral and solve for E_x:
$$E_x = k\lambda \frac{1}{y}\int_{\theta_1}^{\theta_2} \sin\theta\, d\theta = -\frac{k\lambda}{y}(\cos\theta_2 - \cos\theta_1)$$
Again from Fig. 1.94, $\theta_1 = 90° - \alpha_1$ and $\theta_2 = 90° - \alpha_2$, therefore in terms of angles $\alpha_1 =$ and α_2, above expression can also be written as
$$E_x = -\frac{k\lambda}{y}(\sin\alpha_2 - \sin\alpha_1)$$
Now, substituting $\cos\theta_2 = y/r_2$ and $\cos\theta_1 = y/r_1$ in first expression, we get
$$E_x = -\frac{k\lambda}{y}\left(\frac{y}{r_2} - \frac{y}{r_1}\right)$$
$$= k\lambda\left(\frac{1}{r_1} - \frac{1}{r_2}\right) \quad (r_1 > 0 \text{ and } r_2 > 0)$$
From this expression, it is clear that $E_x > 0$ at all points on the x axis in the region $x > a$. Above expression is not defined at end points of the rod, because at end point A, $r_1 = 0$ and at end point B, $r_2 = 0$.

9. Similar to E_x, use steps steps 3-7 to get E_y:
$$E_y = k\lambda \frac{1}{y}\int_{\theta_1}^{\theta_2} \cos\theta\, d\theta = \frac{k\lambda}{y}(\sin\theta_2 - \sin\theta_1)$$
In terms of α_1 and α_2,
$$E_y = \frac{k\lambda}{y}(\cos\alpha_2 - \cos\alpha_1)$$
From step 5, $\cos\theta_2 = y/r_2$ or $y = r_2\cos\theta_2$, and $\cos\theta_1 = y/r_1$ or $y = r_1\cos\theta_1$ therefore,
$$E_y = k\lambda\left(\frac{\sin\theta_2}{y} - \frac{\sin\theta_1}{y}\right) =$$
$$k\lambda\left(\frac{\sin\theta_2}{r_2\cos\theta_2} - \frac{\sin\theta_1}{r_1\cos\theta_1}\right)$$
or $\quad E_y = k\lambda\left(\dfrac{\tan\theta_2}{r_2} - \dfrac{\tan\theta_1}{r_1}\right) \quad (y \neq 0)$
And in terms of α_1 and α_2 above expresion can be written as
$$E_y = k\lambda\left(\frac{\cot\alpha_2}{r_2} - \frac{\cot\alpha_1}{r_1}\right) \quad (y \neq 0)$$
Note that if point P lies on the x-axis, then $y = 0$. In this case, you should not write, $\theta_1 = \theta_2 = 180°$, or $\alpha_1 = \alpha_2 = 0°$. It is because the first expression for E_y:

$E_y = \dfrac{k\lambda}{y}(\sin\theta_2 - \sin\theta_1)$ is valid everywhere in the xy plane but not on the x axis where ($y = 0$). The two tangent functions in the expression for E_y are given by
$$\tan\theta_1 = \dfrac{a}{y} \text{ and } \tan\theta_2 = \dfrac{a+L}{y}$$
and neither of these functions is defined on the x axis (where $y = 0$). Therefore, to get E_y on x axis, we have to use Eq.1.55. By recognizing that on the x axis $\hat{r} = \pm\hat{i}$, we can see that Equation 1.55 tells us that $d\vec{E} = \pm dE\hat{i}$, which implies $E_y = 0$.

10. Combine steps 8 and 9 to obtain an expression for the electric field at P:
$$\vec{E} = E_x\hat{i} + E_y\hat{j}$$

Making Sense of the Result: Consider the plane that is perpendicular to and bisecting the rod. At points on this plane, symmetry dictates that $\vec{E}$ points directly away from the centre of the rod. That is, we expect that $E_x = 0$ throughout this plane. At all points on this plane $r_1 = r_2$. The step- 8 result gives $E_x = 0$ if $r_1 = r_2$, as expected.

Generalization of Result

Now, consider Fig.1.95. The electric field at point P due to a thin uniformly charged rod located on the z axis is given by $\vec{E} = E_z\hat{k} + E_R\hat{R}$, where

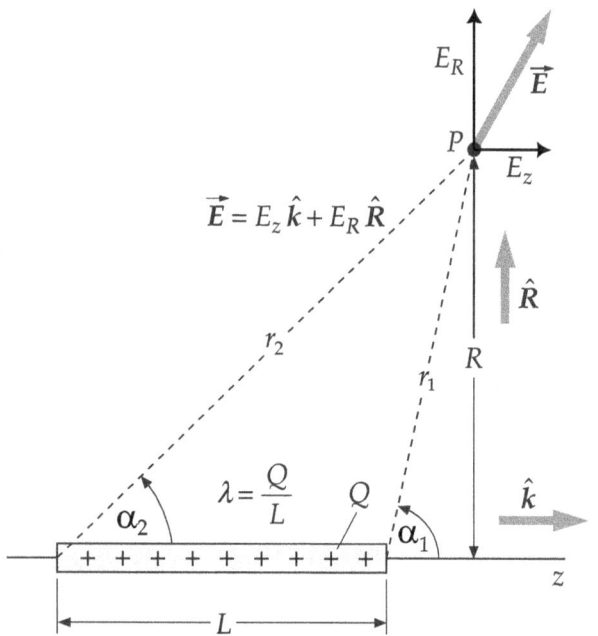

Figure 1.95: The electric field due to a uniformly charged thin rod.

$$E_z = -\dfrac{k\lambda}{R}(\sin\alpha_2 - \sin\alpha_1)$$
$$\text{or } E_z = k\lambda\left(\dfrac{1}{r_1} - \dfrac{1}{r_2}\right) \quad (r_1 \neq 0) \text{ and } (r_2 \neq 0) \tag{1.62}$$
$$\text{and } E_R = \dfrac{k\lambda}{R}(\cos\alpha_2 - \cos\alpha_1)$$
$$\text{or } E_R = k\lambda\left(\dfrac{\cot\alpha_2}{r_2} - \dfrac{\cot\alpha_1}{r_1}\right) \quad (R \neq 0) \tag{1.63}$$

The expressions for E_z (Equation (1.62)) are undefined at the end points of the thin charged rod and the expressions for E_R (Equation (1.63) are undefined at all points on the z axis (where $R = 0$). However, $E_R = 0$ at all points where $R = 0$.

Here, it is to be noted that α_1 and α_2 should to be used with proper sign.

1.16.1.1 Electric Field Due to an Infinite Line Charge

If $L \to \infty$, then from the Figure 1.95, we see that $\alpha_1 \to \pi$ and $\alpha_2 \to 0$. Now, put these values in Equations (1.62) and (1.63) and solve for E_z and E_R. The resultant electric field at point P can be obtained by using the expression-
$$\vec{E} = E_z\hat{k} + E_R\hat{R}$$

Step Wise Calculations

1. Choose the first expression for the electric field in each of Equations (1.62) and (1.63):
$$E_z = -\dfrac{k\lambda}{R}(\sin\alpha_2 - \sin\alpha_1)$$
$$E_R = \dfrac{k\lambda}{R}(\cos\alpha_2 - \cos\alpha_1)$$

2. Take the limit as both $\alpha_1 \to \pi$ and as $\alpha_2 \to 0$
$$E_z = -\dfrac{k\lambda}{R}(\sin 0 - \sin\pi) = -\dfrac{k\lambda}{R}(0 - 0) = 0$$
$$E_R = \dfrac{k\lambda}{R}(\cos 0 - \cos\pi) = \dfrac{k\lambda}{R}(1 - (-1)) = 2\dfrac{k\lambda}{R}$$

3. Express the electric field in vector form:
$$\vec{E} = E_z\hat{k} + E_R\hat{R} = 0\hat{k} + \dfrac{2k\lambda}{R}\hat{R} = \dfrac{2k\lambda}{R}\hat{R}$$
$$\implies \boxed{\vec{E} = \dfrac{2k\lambda}{R}\hat{R}} \tag{1.64}$$

Thus, the magnitude of electric field decreases inversely with the radial distance from the line charge.

Special Results: Semi Infinite Wire

For a semi-infinite wire, as shown in Fig.1.96, we have-
$$\alpha_1 = \pi, \quad \alpha_2 = \pi/2$$

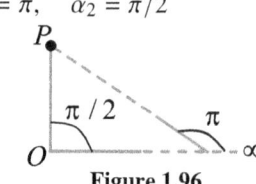

Figure 1.96

Therefore, $E_z = -\dfrac{k\lambda}{R}(\sin\alpha_2 - \sin\alpha_1) = -\dfrac{k\lambda}{R}\left(\sin\dfrac{\pi}{2} - \sin\pi\right)$
$$= -\dfrac{k\lambda}{R}(1 - 0) = -\dfrac{k\lambda}{R}$$
and $E_R = \dfrac{k\lambda}{R}(\cos\alpha_2 - \cos\alpha_1) = \dfrac{k\lambda}{R}\left(\cos\dfrac{\pi}{2} - \cos\pi\right)$
$$= \dfrac{k\lambda}{R}(0 - (-1)) = \dfrac{k\lambda}{R}$$

In this case, net electric field,
$$E = \sqrt{E_z^2 + E_R^2} = \sqrt{(-k\lambda/R)^2 + (k\lambda/R)^2} = \dfrac{k\lambda}{R}\sqrt{2}$$

The angle between net electric field vector $\vec{E}$ and length of the rod, i.e., z-axis, is given by-
$$\theta = \tan^{-1}\dfrac{E_R}{E_z} = \tan^{-1} 1 = \dfrac{\pi}{4}$$

1.16.1.2 Electric Field of a Finite Line Charge at a Point on its Extension

EXAMPLE 45. A charge Q is uniformly distributed along the z axis, from $z = -\tfrac{1}{2}L$ to $z = +\tfrac{1}{2}L$. Show that for large values of z the expression for the electric field of the line charge on the z axis approaches the expression for the electric field of a point charge Q at the origin.

APPROACH To find $\vec{E}$, at an axial point P, use Eq. (1.62).

Step wise Calculation

1. The electric field on the z axis has only a z component, given by Equation (1.62):

1.16. ELECTRIC FIELD OF A CONTINUOUS CHARGE DISTRIBUTION

$$E_z = k\lambda \left(\frac{1}{r_2} - \frac{1}{r_1} \right)$$

2. The line charge along the z axis, the field point P, r_1 and r_2 are shown in Figure 1.97:

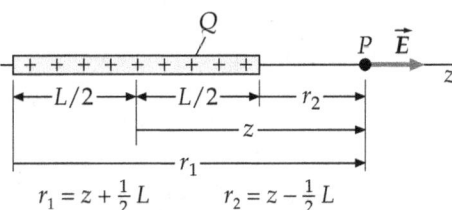

$$r_1 = z + \tfrac{1}{2}L \qquad r_2 = z - \tfrac{1}{2}L$$

Figure 1.97: Geometry for the calculation of the electric field on the axis of a uniformly charged rod.

3. Substitute with $r_1 = z - \tfrac{1}{2}L$ and $r_2 = z + \tfrac{1}{2}L$ into the step 1 result and simplify:

$$\begin{aligned} E_z &= k\lambda \left(\frac{1}{z - \tfrac{1}{2}L} - \frac{1}{z + \tfrac{1}{2}L} \right) \\ &= \frac{kQ}{L} \frac{L}{z^2 - \left(\tfrac{1}{2}L\right)^2} \\ &= \frac{kQ}{z^2 - \left(\tfrac{1}{2}L\right)^2} \quad \left(z > \tfrac{1}{2}L\right) \end{aligned}$$

This is the required expression for electric field at an axial position of a uniformly charged rod.
Note that the result is valid for the region $L/2 < z < \infty$ only. The result is not valid in the region $-L/2 < z < +L/2$?
For $z \gg L$, the above result gives-

$$E_z \approx \frac{kQ}{z^2} \quad (z \gg L)$$

This expression is the same as the expression for the electric field of a point charge Q located at the origin.

EXAMPLE 46. What is the electric field at any point on the axis of a uniformly charged rod of length 'L' having linear charge density 'λ'? The point is separated from the nearer end by distance 'a' [Fig.1.98].

Note: Although this type of problem has been already discussed

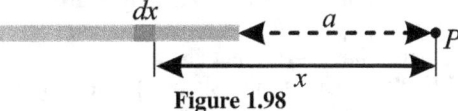

Figure 1.98

in above article, here we solve it by using the method of integration.
SOLUTION Consider an elementary length dx of the rod at distance x, from point P, where electric field E is to be determined. The elemental charge, $dq = \lambda dx$
The small electric field at P due to this element,

$$dE = k \cdot \frac{\lambda dx}{x^2}$$

or $\quad E = k\lambda \int_a^{a+L} \frac{1}{x^2} dx = k\lambda \left[-\frac{1}{x} \right]_a^{a+L} = k\lambda \left[\frac{-1}{a+L} + \frac{1}{a} \right]$

Thus, $\quad E = \frac{\lambda}{4\pi\epsilon_o} \left[\frac{1}{a} - \frac{1}{L+a} \right] \quad \left[\because k = \frac{1}{4\pi\epsilon_0} \right]$

1.16.1.3 Approximating Equations 1.62 and 1.63 on the Symmetry Plane

EXAMPLE 47. A charge Q is uniformly distributed along the z axis, from $z = -\tfrac{1}{2}L$ to $z = +\tfrac{1}{2}L$. (a) Find an expression for the electric field on the $z = 0$ plane as a function of R, the radial distance of the field point from the z axis. (b) Show that for $R \gg L$, the expression found in Part (a) approaches that of a point charge at the origin of charge Q. (c) Show that for $R \ll L$, the expression found in Part (a) approaches that of an infinitely long line charge on the z axis with a uniform linear charge density $\lambda = Q/L$.

APPROACH The charge distribution is uniform and the linear charge density is $\lambda = Q/L$. The line charge is along the z axis and the field point is in the $z = 0$ plane. We use Equations (1.62) and (1.63) to find the electric field expression for part (a). The electric field due to a point charge decreases inversely with the square of the distance from the charge. Examine the part (a) result to see how it approaches that of a point charge at the origin for $R \gg L$. The electric field due to a uniform line charge of infinite length decreases inversely with the radial distance from the line (Eq.1.64). Examine the Part (a) result to see how it approaches the expression for the electric field of a line charge of infinite length for $R \ll L$.

SOLUTION (a) Step wise solution is given below-
1. From Eq.(1.62) and (1.62), we have-

$$E_z = -\frac{k\lambda}{R}(\sin\alpha_2 - \sin\alpha_1), \; E_R = \frac{k\lambda}{R}(\cos\alpha_2 - \cos\alpha_1)$$

2. The charge configuration with the line charge on the z axis from $z = -\tfrac{1}{2}L$ to $z = +\tfrac{1}{2}L$ is shown in Fig.1.99. The field point P is in the $z = 0$ plane at a distance R from the origin.

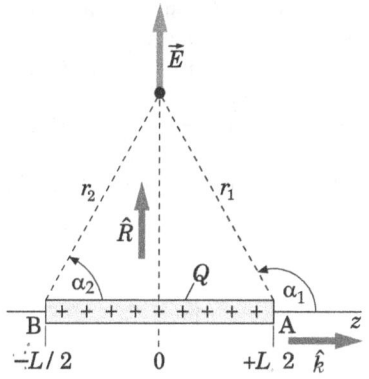

Figure 1.99

3. From the Figure 1.99, $r_1 = r_2$, therefore, the interior angle at the right end B of the rod will also be α_2. Therefore, at point B, we can write $\alpha_1 + \alpha_2 = \pi$, so $\sin\alpha_1 = \sin(\pi - \alpha_2) = \sin\alpha_2$ and $\cos\alpha_1 = \cos(\pi - \alpha_2) = -\cos\alpha_2$. Substitution of these values in equations of step 1 gives:

$$E_z = -\frac{k\lambda}{R}(\sin\alpha_2 - \sin\alpha_2) = 0,$$
$$E_R = \frac{k\lambda}{R}(\cos\alpha_2 + \cos\alpha_2) = \frac{2k\lambda}{R}\cos\alpha_2$$

4. Express $\cos\alpha_2$ in terms of R and L and substitute into the step-3 result:

$$\cos\alpha_2 = \frac{\tfrac{1}{2}L}{\sqrt{R^2 + \left(\tfrac{1}{2}L\right)^2}}$$

so, $E_R = \dfrac{2k\lambda}{R} \dfrac{\tfrac{1}{2}L}{\sqrt{R^2 + \left(\tfrac{1}{2}L\right)^2}} = \dfrac{k\lambda L}{R\sqrt{R^2 + \left(\tfrac{1}{2}L\right)^2}}$

5. Express the electric field in vector form, and substitute Q for λL:

$$\vec{E} = E_z\hat{k} + E_R\hat{R} = 0\hat{k} + E_R\hat{R}$$
$$\text{so } \vec{E} = E_R\hat{R} = \frac{kQ}{R\sqrt{R^2 + \left(\frac{1}{2}L\right)^2}}\hat{R}$$

(b) The step wise solution is given below-
1. Examine the step-5 result of part (a). If $R \gg L$ then $R^2 + \left(\frac{1}{2}L\right)^2 \approx R^2$. Substitute R^2 for $R^2 + \left(\frac{1}{2}L\right)^2$:
$$\vec{E} \approx \frac{kQ}{R\sqrt{R^2}}\hat{R} = \frac{kQ}{R^2}\hat{R} \quad (R \gg L)$$
2. This (approximate) expression for the electric field decreases inversely with the square of the distance from the origin, just as it would for a point charge Q at the origin.
$$\vec{E} \approx \frac{kQ}{R^2}\hat{R} \quad (R \gg L)$$

(c) Examine the Part (a), step-5 result. If $R \ll L$ then $R^2 + \left(\frac{1}{2}L\right)^2 \approx \left(\frac{1}{2}L\right)^2$. Substitute $\left(\frac{1}{2}L\right)^2$ for $R^2 + \left(\frac{1}{2}L\right)^2$. This (approximate) expression for the electric field falls off inversely with the radial distance from the line charge, just as the exact expression for an infinite line charge (1.64) would.
$$\vec{E} \approx \frac{k\lambda L}{R\sqrt{\left(\frac{1}{2}L\right)^2}}\hat{R} = \frac{2k\lambda}{R}\hat{R} \quad (R \ll L)$$

EXAMPLE 48. Figure 1.100 shows a non-conducting rod that has a uniform positive charge density $+\lambda$ and a total charge Q along its right half, and a uniform negative charge density $-\lambda$ and a total charge $-Q$ along its left half. What is the direction and magnitude of the net electric field at point P that shown in Fig. 1.100?

SOLUTION When we consider a segment dx at distance x from

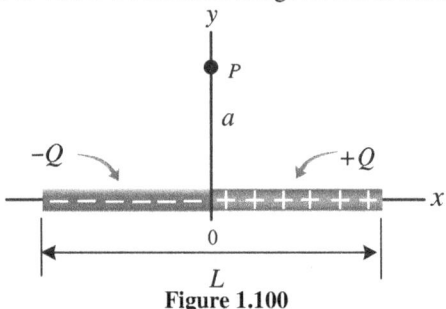

Figure 1.100

O on the right side of the rod, the charge on this segment will be $dq = \lambda dx$, as shown in Fig. 1.101.

The electric field $d\vec{E}_+$ at P due to this segment is directed outwards and away from the positive charge dq and has a magnitude:
$$dE_+ = k\frac{dq}{r^2} = k\frac{\lambda dx}{r^2}$$
A symmetric segment on the opposite side of the rod, but with a negative charge, creates an electric field $d\vec{E}_-$ that is directed inwards and toward this segment and has the same magnitude as $d\vec{E}_+$, i.e. $dE_+ = dE_-$. The resultant electric field $d\vec{E}$ from both symmetric segments will be a vector to the left (Fig. 1.101), and its magnitude will be given by:
$$dE = dE_+ \cos\theta + dE_+ \cos\theta = 2dE_+ \cos\theta$$
From Fig.1.101, $\cos\theta = x/r$ and $r = \sqrt{x^2 + a^2}$, therefore
$$dE = 2k\frac{\lambda dx}{r^2}\frac{x}{r} = k\lambda(x^2 + a^2)^{-3/2}(2x)dx$$
The total electric field at P due to all segments of the rod is found by integrating dE from $x = 0$ to only $x = L/2$, since the negative charge of the rod is considered in evaluating dE. Thus:

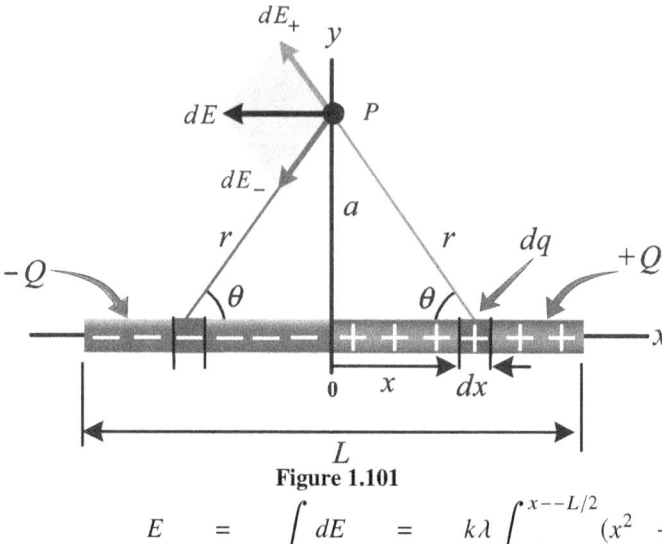

Figure 1.101

$$E = \int dE = k\lambda \int_{x=0}^{x=L/2}(x^2 + a^2)^{-3/2}(2xdx) \quad \cdots (1)$$

We can simplify this integration by using substitution method.
Let $x^2 + a^2 = u$, then $2xdx = du$
When $x = 0$, then $u = x^2 + a^2$ gives $u = 0^2 + a^2 = a^2$ and when $x = L/2$, then $u = (L/2)^2 + a^2$.
Substituting these values and limits in Eq.(1), we get
$$E = k\lambda \int_{u=0}^{(L/2)^2+a^2} u^{-3/2}du = k\lambda\left[\frac{-2}{\sqrt{u}}\right]$$
$$= k\lambda\left[\frac{-2}{\sqrt{(L/2)^2 + a^2}} - \frac{-2}{a}\right]$$
$$= 2k\lambda\left[\frac{1}{a} - \frac{1}{\sqrt{(L/2)^2 + a^2}}\right]$$

When we use the fact that the magnitude of the charge Q is given by $Q = \lambda L/2$, we get:
$$E = \frac{4kQ}{L}\left[\frac{1}{a} - \frac{1}{\sqrt{(L/2)^2 + a^2}}\right]$$

When P is very far away from the rod, i.e. $a \gg L$, we can neglect $(L/2)^2$ in the denominator of this equation and hence get $E \approx 0$. In this situation, the two oppositely charged halves of the rod would appear to point P as if they were two coinciding point charges and hence have a zero net charge.

EXAMPLE 49. An infinite sheet of charge is lying on the xy-plane as shown in Fig. 1.102. A positive charge is distributed uniformly over the plane of the sheet with a charge per unit area σ. Calculate the electric field at a point P located a distance a from the plane.

SOLUTION Let us divide the plane into narrow strips parallel to the y-axis. A strip of width dx can be considered as an infinitely long wire of charge per unit length λ given as:
$$\lambda = \lim_{L\to\infty}\frac{\text{charge on the length } L \text{ of strip of thickness } dx}{L}$$
$$= \lim_{L\to\infty}\frac{\sigma(L(dx))}{L} = \sigma dx$$

From Eq. (1.64), at point P, the strip sets up an electric field $d\vec{E}$ lying in the xz-plane of magnitude:
$$dE = 2k\frac{\lambda}{r} = 2k\frac{\sigma dx}{r}$$

This electric field vector can be resolved into two components $d\vec{E}_x$ and $d\vec{E}_z$. When we consider the entire sheet of charge, we find a symmetrically opposite component corresponding to each com-

1.16. ELECTRIC FIELD OF A CONTINUOUS CHARGE DISTRIBUTION

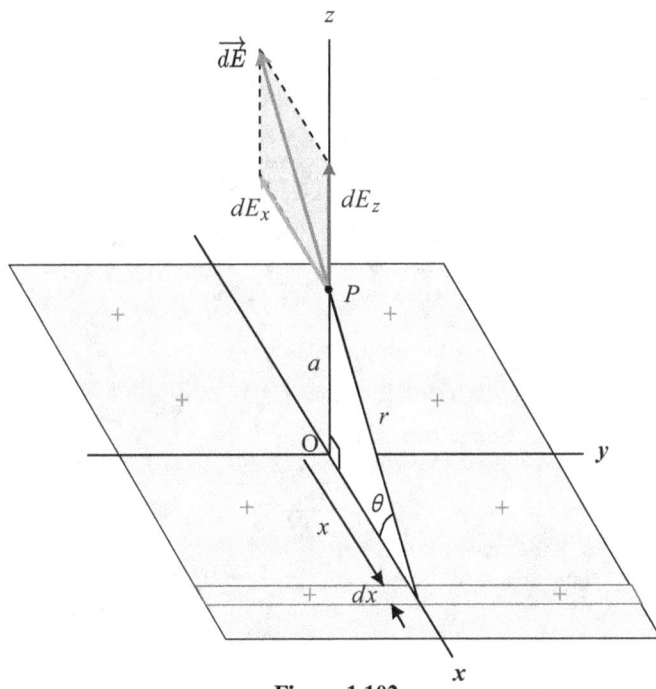

Figure 1.102

ponent of $d\vec{E}_x$. So, it will sum to zero. Therefore, the resultant electric field at point P will be in the z-direction, perpendicular to the sheet. From Fig. 1.102, we find the following:

$$dE_z = dE \sin\theta$$

and hence:

$$E = \int dE_z = 2k\sigma \int_{-\infty}^{+\infty} \frac{\sin\theta \, dx}{r}$$

To perform the integration of this expression, we must first relate the variables $\theta, x,$ and r. One approach is to express θ and r in terms of x. From the geometry of Fig.1.102, we have: $r = \sqrt{x^2 + a^2}$ and $\sin\theta = \dfrac{a}{r} = \dfrac{a}{\sqrt{x^2 + a^2}}$

Therefore, $E = 2k\sigma a \displaystyle\int_{-\infty}^{+\infty} \frac{dx}{x^2 + a^2} = 2k\sigma a \left[\frac{1}{a}\tan^{-1}\frac{x}{a}\right]_{-\infty}^{+\infty}$

$= 2k\sigma \left[\tan^{-1}(\infty) - \tan^{-1}(-\infty)\right]$

$= 2k\sigma \left[\dfrac{\pi}{2} + \dfrac{\pi}{2}\right] = 2\pi k\sigma = 2\pi \dfrac{1}{4\pi\epsilon_0}\sigma = \dfrac{\sigma}{2\epsilon_0}$

i.e., $\boxed{E = \dfrac{\sigma}{2\epsilon_o}}$ \hfill (1.65)

We note that the distance a from the plane to the point P does not appear in the final result of E. This means that the electric field set up at any point by an infinite plane sheet of charge is independent of how far the point is from the plane. In other words, the electric field is *uniform* and *normal* to the plane.

Also, the same result is obtained if the point P lies below the xy-plane. That is, the field below the plane has the same magnitude as that above the plane but as a vector it points in the opposite direction.

EXAMPLE 50. Two infinite plane sheets with uniform surface charge densities $+\sigma$ and $-\sigma$ are placed parallel to each other with separation d (Fig.1.103). Find the electric field between the sheets, above the upper sheet, and below the lower sheet.

APPROACH Equation (1.65) gives the electric field due to a single infinite plane sheet of charge. To find the field due to two such sheets, we combine the fields by using the principle of superposition (Fig.1.103).

SOLUTION From Eq.(1.65), both $\vec{E}_1$ and $\vec{E}_2$ have the same magnitude at all points, independent of distance from either sheet:

$$E_1 = E_2 = \frac{\sigma}{2\epsilon_0}$$

$\vec{E}_1$ is everywhere directed away from sheet 1, and $\vec{E}_2$ is everywhere directed toward sheet 2. Between the sheets, $\vec{E}_1$ and $\vec{E}_2$ reinforce each other; above the upper sheet and below the lower sheet, they cancel each other. Thus the total field is

$$\vec{E} = \vec{E}_1 + \vec{E}_2 = \begin{cases} 0 & \text{above the upper sheet} \\ \dfrac{\sigma}{\epsilon_0}\hat{j} & \text{between the sheets} \\ 0 & \text{below the lower sheet} \end{cases} \quad (1.66)$$

Discussion: Because we considered the sheets to be infinite, our result does not depend on the separation d. Our result shows that the field between oppositely charged plates is essentially uniform if the plate separation is much smaller than the dimensions of the plates.

important Point Electric fields are not some kind of physical substance that "flows". So, it is not correct to say that the field $\vec{E}_1$ of sheet 1 would be unable to "penetrate" sheet 2, and that field $\vec{E}_2$ caused by sheet 2 would be unable to "penetrate" sheet 1. Infact, the electric fields $\vec{E}_1$ and $\vec{E}_2$ depend on only the individual charge distributions that create them. The total field at every point is just the vector sum of $\vec{E}_1$ and $\vec{E}_2$.

EXAMPLE 51. A force of 10 N acts on a charged particle placed between two thin large charged metal plates. Now, one plate is removed, find the force acting on that particle.

SOLUTION If σ is the magnitude of surface charge density on the metal plates, then from Eq.1.66, the electric field between the plates:

$$E = \frac{\sigma}{\epsilon_0}$$

Now, according to given problem, 10 N force is acting on a charged particle placed between the plates. So, if Q is the charge on charged particle placed between the plates, then

$$F = QE = \frac{Q\sigma}{\epsilon_0} \implies \frac{Q\sigma}{\epsilon_0} = 10 \text{ N} \quad \cdots (1)$$

If one plate is removed then from Eq.1.65, electric field produced due to single plate will be

$$E' = \frac{\sigma}{2\epsilon_0}$$

Therefore, the force on the charged particle due to single plate

$$F' = QE' = \frac{Q\sigma}{2\epsilon_0} = \frac{1}{2}\left(\frac{Q\sigma}{\epsilon_0}\right)$$

From Eq.(1), $Q\sigma/\epsilon_0 = 10$ N, therefore above equation gives

$$F' = \frac{1}{2}(10 \text{ N}) = 5 \text{ N}$$

1.16.2 Electric Field at the Centre of a Charged Circular Arc

Let us consider a thin circular arc of radius R. Suppose, total positive charge Q is uniformly distributed over it. This arc forms central angle of α rad (Fig.1.104a). To find the electric field at the centre P of this arc, we place coordinate axes such that the axis of symmetry of the arc lies along the x axis and the origin is at the arc's center. If λ represent the linear charge density of this arc which has a length $R\alpha$, then:

$$\lambda = \frac{Q}{R\alpha}$$

For an arc element ds subtending an angle $d\theta$ at P, we have:

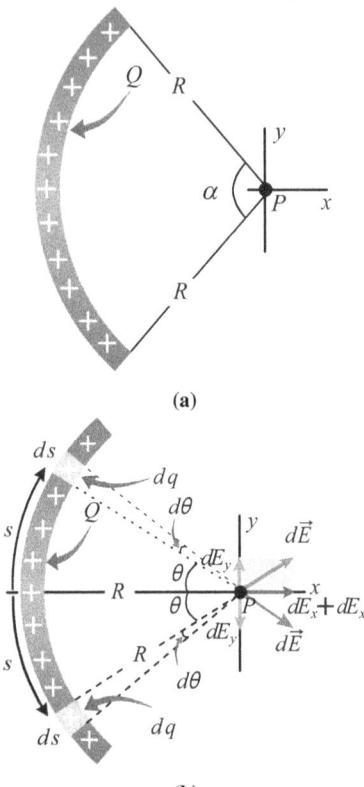

Figure 1.104: (a) A circular arc of radius R, central angle α, and centre P has a uniformly distributed positive charge Q. (b) The figure shows the electric fields $d\vec{E}$ at P due to lower and upper arc elements (each of length ds and charge dq). From symmetry, the vertical components of all elements cancel out and the total field is along the x-axis

$ds = Rd\theta$

Therefore, the charge dq on this arc element will be given by:

$$dq = \lambda ds = \frac{Q}{R\alpha}Rd\theta = \frac{Q}{\alpha}d\theta$$

To find the electric field at point P, we first calculate the magnitude of the electric field dE at P due to this element of charge dq, [Fig.1.104b], as follows:

$$dE = k\frac{dq}{R^2} = \frac{kQ}{R^2\alpha}d\theta \qquad (1.67)$$

This field has a vertical component $dE_y = dE\sin\theta$ along the positive y-axis and a horizontal component $dE_x = dE\cos\theta$ along the positive x-axis, as shown in Fig.1.104b. The y-component created at P by any charge element dq is canceled by a symmetric charge element on the opposite side of the arc. Thus, the perpendicular components of all of the charge elements sum to zero. The horizontal component will take the form:

$$dE_x = dE\cos\theta = \frac{kQ}{R^2\alpha}\cos\theta d\theta \qquad (1.68)$$

Consequently, the total electric field at P due to all elements of the arc is given by the integration of the x-component as follows:

$$E = \int dE_x = \frac{kQ}{R^2\alpha}\int_{-\alpha/2}^{+\alpha/2}\cos\theta d\theta = \frac{kQ}{R^2\alpha}[\sin\theta]_{-\alpha/2}^{+\alpha/2}$$

$$= \frac{kQ}{R^2\alpha}\left[\sin\frac{\alpha}{2} - \sin\left(-\frac{\alpha}{2}\right)\right]$$

$$\implies E = \frac{kQ}{R^2\alpha}\left[\sin\frac{\alpha}{2} + \sin\frac{\alpha}{2}\right]$$

or $\qquad \boxed{E = \frac{kQ}{R^2}\frac{\sin\alpha/2}{\alpha/2}} \qquad (1.69)$

So, the total electric field at P will be along the x-axis and it's magnitude given by Eq.(1.69).

There are three special cases to Eq. (1.69):

1. $\alpha = 0$ (Point charge)
   When we apply the limiting case
   $\lim_{\alpha\to 0}[\sin(\alpha/2)/(\alpha/2)] = 1$, we get: $E = \frac{kQ}{R^2}$

2. $\alpha = \pi$ (A Semicircular arc)
   When we substitute with $\sin(\pi/2)/(\pi/2) = 2/\pi$, we get:
   $$E = \frac{2kQ}{\pi R^2}$$

3. $\alpha = 2\pi$ (A ring of radius R)
   When we substitute with $\sin(2\pi/2) = 0$, we get: $E = 0$
   This is an expected result, since we shall see that Eq. (1.76) gives $E = 0$ when P is at the centre of the ring, i.e. when $a = 0$.

1.16.3 Electric Field at a Point on the Axis of a Thin Charged Ring

Fig.1.105 shows a ring-shaped conductor with radius a carries a total charge Q uniformly distributed around it. Axis of the ring is along x axis. P is an arbitrary point that lies on the axis of the ring at a distance x from its centre. Our target variable is the electric field at such a point as a function of the coordinate x.

☞This is a clear case of the superposition of electric fields. Note that the charge is distributed continuously around the ring rather than in a number of point charges.

Suppose, the ring divided into infinitesimal segments of length ds, Each segment has charge dQ and acts as a point-charge source of electric field as shown in Fig.1.105. Let $d\vec{E}$ be the electric field from one such segment; the net electric field at P is then the sum of all contributions $d\vec{E}$ from all the segments that make up the ring.

Now, Consider two segments at the top and bottom of the ring: The contributions $d\vec{E}$ to the field at P from these segments have the same x-component but opposite y-components. Hence the total y-component of field due to this pair of segments is zero. When we add up the contributions from all such pairs of segments, the total field $\vec{E}$ will have only a component along the ring's symmetry axis (the x-axis), with no component perpendicular to that axis (that is, no component in yz-plane). So the field at P is described completely by its x-component E_x.

CALCULATION OF $\vec{E}$ AT POINT P: If r be the distance of point P from the upper charge segment dQ, then from Fig.1.105, we have

$$r = \sqrt{x^2 + a^2} \qquad (1.70)$$

and, $\qquad \cos\alpha = \frac{x}{r} \qquad (1.71)$

The magnitude of the electric field at P due to the upper segment ds of charge dQ is

$$dE = k\frac{dQ}{r^2} \qquad (1.72)$$

The x-component of this field-

$$dE_x = dE\cos\alpha \qquad (1.73)$$

1.16. ELECTRIC FIELD OF A CONTINUOUS CHARGE DISTRIBUTION

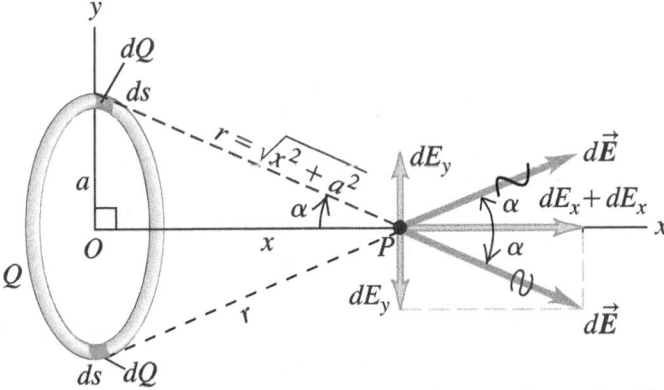

Figure 1.105: The total electric field at P is along the x axis. The perpendicular component of the field at P due to upper charge element segment dQ is cancelled by the perpendicular component due to lower charge element segment dQ.

Substituting, the values of dE, $\cos\alpha$ and r from Eq.1.72, 1.71 and 1.70 respectively in Eq.1.73, we get

$$dE_x = \left(k\frac{dQ}{r^2}\right)\frac{x}{r} = \frac{kx}{(x^2+a^2)^{3/2}}dQ \quad (1.74)$$

All elements of the ring make the same contribution to the field at P because they are all equidistant from this point. Thus, we can integrate to obtain the total field at P

$$E_x = \int \frac{kx}{(x^2+a^2)^{3/2}}dQ$$

Since x does not vary as we move from point to point around the ring, all the factors on the right side except dQ are constant and can be taken outside the integral. The integral of dQ is just the total charge Q, and we finally get

$$E_x = \frac{kx}{(x^2+a^2)^{3/2}}\int dQ$$

or

$$\boxed{E_x = \frac{kQx}{(x^2+a^2)^{3/2}}} \quad (1.75)$$

As Q is positive, therefore, field is directed away from the centre of the ring. In vector form Eq.(1.75) can be written as-

$$\boxed{\vec{E} = E_x\hat{\imath} = k\frac{Qx}{(x^2+a^2)^{3/2}}\hat{\imath}} \quad (1.76)$$

Here, $k = 1/4\pi\epsilon_0$

Special Cases

1. When $x = 0$ i.e., the field point P is at the centre of the ring, the Eq.1.76 gives: $\vec{E} = 0$.
2. When $x \gg a$, i.e., the field point P is much farther from the ring than its size, the denominator in Eq.1.76 becomes approximately equal to x^3, and the expression becomes approximately

$$\vec{E} = \frac{1}{4\pi\epsilon_0}\frac{Q}{x^2}\hat{\imath}$$

In other words, when we are so far from the ring that its size a is negligible in comparison to the distance x, its field is the same as that of a point charge.

Tracing E vs x Graph

1. **Maximum and Minimum values of Electric Field:** For maximum value of E, on x axis, we have

$$\frac{dE}{dx} = 0$$

From Eq.1.76, the expression for E, on x axis as-

$$E = k\frac{Qx}{(x^2+a^2)^{3/2}}$$

Therefore,

$$\frac{dE}{dx} = kQ\frac{d}{dx}\frac{x}{(x^2+a^2)^{3/2}}$$

$$= kQ\left[\frac{(x^2+a^2)^{3/2}1 - x\frac{3}{2}(x^2+a^2)^{1/2}2x}{(x^2+a^2)^3}\right]$$

$$= kQ(x^2+a^2)^{1/2}\left[\frac{a^2-2x^2}{(x^2+a^2)^3}\right]$$

Therefore, $\frac{dE}{dx} = 0 \Rightarrow kQ(x^2+a^2)^{1/2}\frac{a^2-2x^2}{(x^2+a^2)^3} = 0$

or $\quad x = \pm\frac{a}{\sqrt{2}}$

So, the value of electric field will be maximum at $x = \pm\frac{a}{\sqrt{2}}$.
Substituting this value of x in the expression for E, we get

$$E_{\max} = \pm kQ\frac{a/\sqrt{2}}{\left(\frac{a^2}{2}+a^2\right)^{3/2}} = \pm\frac{2}{\sqrt{27}}\frac{kQ}{a^2}$$

2. We have already shown that, at $x = 0$, the electric field $E = 0$.
3. When $x \to \pm\infty$, then, again $\vec{E} \to 0$

Keeping these points in mind, the variation of E with x is given in Fig.1.106

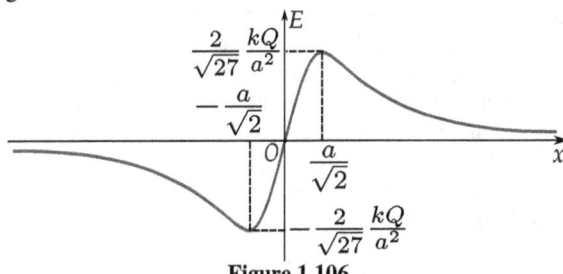

Figure 1.106

1.16.3.1 Motion of a Negatively Charged Particle Close to the Centre of the Positively Charged Ring on it's Axis

In the expression for the field (Eq.1.76) due to a ring of charge Q, we let $x \ll a$, which results in

$$E_x = \frac{kQ}{a^3}x$$

Thus, from $F_x = Q_0 E_x$ the force on a charge $-Q_0$ placed near the centre of the ring is

$$F_x = -k\frac{QQ_0}{a^3}x$$

Negative sign shows that the force is restoring in nature. Because this force has the form of SHM, the motion will be simple harmonic.
If mass of the charged particle is m and it's acceleration towards the centre of the ring is d^2x/dt^2, then above equation can also be written as-

$$m\frac{d^2x}{dt^2} = -k\frac{QQ_0}{a^3}x \implies \frac{d^2x}{dt^2} = -\frac{kQQ_0}{ma^3}x$$

$$\implies \frac{d^2x}{dt^2} = -\omega^2 x$$

here, $\omega = \sqrt{\frac{kQQ_0}{ma^3}}$ is the angular frequency of SHM.

Time period of oscillation of the particle about the centre of the ring-

$$T = \frac{2\pi}{\omega} = 2\pi\sqrt{\frac{ma^3}{kQQ_0}}$$

Note: Always look for the condition of SHM as $F \propto -x$

1.16.4 Electric Field on the Axis of a Charged Disc

Let us consider a point P on the axis of a uniformly charged disc of radius R at a distance x from the centre of it. Suppose, the surface charge density of it is σ.

We have already calculated the electric field at an axial point of a

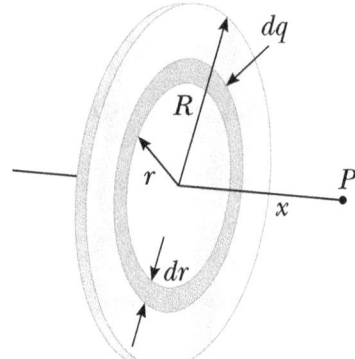

Figure 1.107: A uniformly charged disc of radius R. The electric field at an axial point P is directed along the central axis, perpendicular to the plane of the disc.

uniformly charged ring. Now, the disc can be considered as made of number of concentric very thin continuous rings with continuously varying radii from $r = 0$ to $r = R$. To find the electric field on the axis of the disc, we first consider an elementary concentric ring of radius r and thickness dr as shown in Fig.1.107. The surface area of this ring is $2\pi r dr$. The charge dq on this ring is $2\pi r dr \sigma$. We can easily find the electric field at point P due to this elementary charged ring and then integration from $r = 0$ to $r = R$ gives the electric field at P due to the complete uniformly charged disc of radius R.

Now, from Eq.1.75, the electric field at point P due to this elementary charged ring

$$dE_x = \frac{kx}{(x^2 + r^2)^{3/2}}(2\pi\sigma r dr)$$

To obtain the total field at P, we integrate this expression over the limits $r = 0$ to $r = R$. Here, x is a constant.

$$E_x = \pi k\sigma x \int_0^R \frac{2r dr}{(x^2 + r^2)^{3/2}} = 2\pi k\sigma \left(1 - \frac{x}{(x^2 + R^2)^{1/2}}\right)$$

Since, $k = 1/4\pi\epsilon_0$, therefore

$$\boxed{E_x = \frac{\sigma}{2\epsilon_0}\left(1 - \frac{x}{(x^2 + R^2)^{1/2}}\right)} \quad (1.77)$$

This result is valid for all values of $x > 0$ and $x < 0$, but for $x = 0$ the answer is not valid as there is discontinuity at $x = 0$. We can calculate the field close to the disc along the axis by assuming that $R >> x$; thus, the expression in bracket reduces to unity to give us the near-field approximation.

In this case, just right side of the centre of disc, the electric field is directed towards the positive direction of x-axis and is given by-

$$E_x = 2\pi k\sigma = \frac{\sigma}{2\epsilon_0} \quad (\text{if} \quad x \to 0^+)$$

Just left side of the centre of disc, the electric field is directed towards the negative direction of x-axis and is given by-

$$E_x = -2\pi k\sigma = -\frac{\sigma}{2\epsilon_0} \quad (\text{if} \quad x \to 0^-)$$

In this case, the disc can be regarded as a plane sheet of charge. If Q is the the total charge on the disc, then $\sigma = Q/\pi R^2$. Substi-

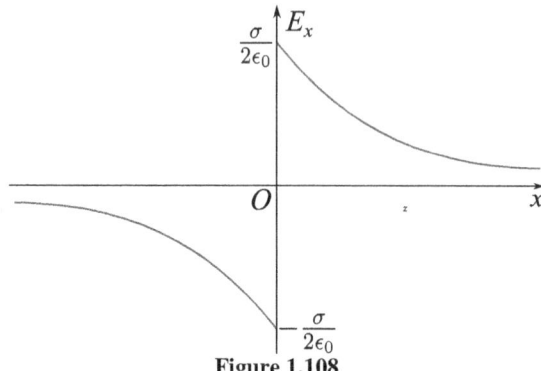

Figure 1.108

tuting this value of σ in Eq.1.77, we get

$$\boxed{E_x = \frac{Q}{2\pi\epsilon_0 R^2}\left(1 - \frac{x}{(x^2 + R^2)^{1/2}}\right)} \quad (1.78)$$

This is an expression of electric field at any point on the axis of the charged disc in terms of it's charge Q.

1.17 The Shape of Lightning Rods

According to Benjamin Franklin, the lightning was a giant electrical spark. To support his theory, he suggested that a metal rod be placed on top of a tall structure to capture "electric fluid"—what we call charged particles.

After conducting his experiments that supported his theory of lightning as a giant spark, Franklin came up with a plan to avoid lightning strikes. He knew that objects could be charged by rubbing. He also knew that you cannot charge a metal object—a conductor—by rubbing it if you connect the conductor to ground. Today we know why: The Earth acts as a source and sink of electrons. If you build up a few excess electrons on a grounded conductor, these electrons quickly travel through the conductor and along the pathway to ground (Fig.1.109).

Franklin reasoned that during a storm, the atmosphere builds up excess charge much as a glass rod builds up excess charge when it is rubbed with silk. He knew from his experience with charged objects that a spark can make its way through the air. His indoor experiments demonstrated that if he used a pointed object (like a knitting needle) to draw charge from an object through the air, only a small spark occurred compared to the larger spark drawn to a blunt object (like his thumb). Franklin reasoned that when a small charge builds up in the atmosphere, a pointed lightning rod will draw the small charge through the air and into the Earth continuously (Fig. 1.109b). If only a small charge travels through the air, it will not be visible and there will be no giant spark of lightning.

Without a lightning rod, charge still builds up in the atmosphere. Because buildings are connected to the Earth, they attract charge of the opposite sign (Fig. 1.109c). When sufficient charge builds up, the air acts as a conductor and a large charge is transferred in a giant lightning spark. Such a violent spark causes great damage to buildings and can be very dangerous. Franklin published the following recommendations for lightning rods:

1. Just outside each building an iron rod **should be planted 3 feet to 4 feet into the moist ground**.
2. The rod should extend 6 feet to 8 feet **above the tallest part of the structure**.
3. On top of the rod should be a foot of brass wire **sharpened to a fine point**.

Benjamin Wilson—a contemporary of Franklin's—believed that lightning rods should be blunt. He argued that a pointed rod would draw down lightning that might have just passed harm-

1.17. THE SHAPE OF LIGHTNING RODS

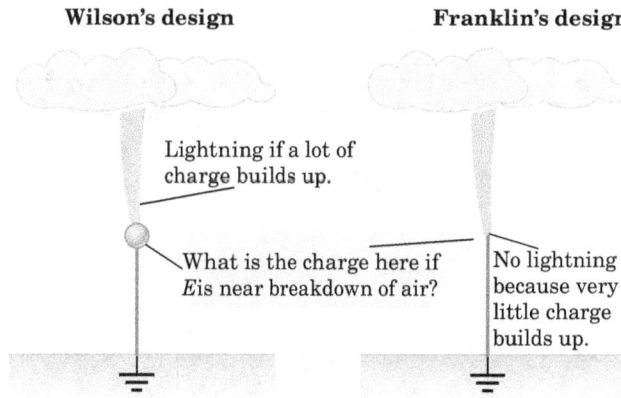

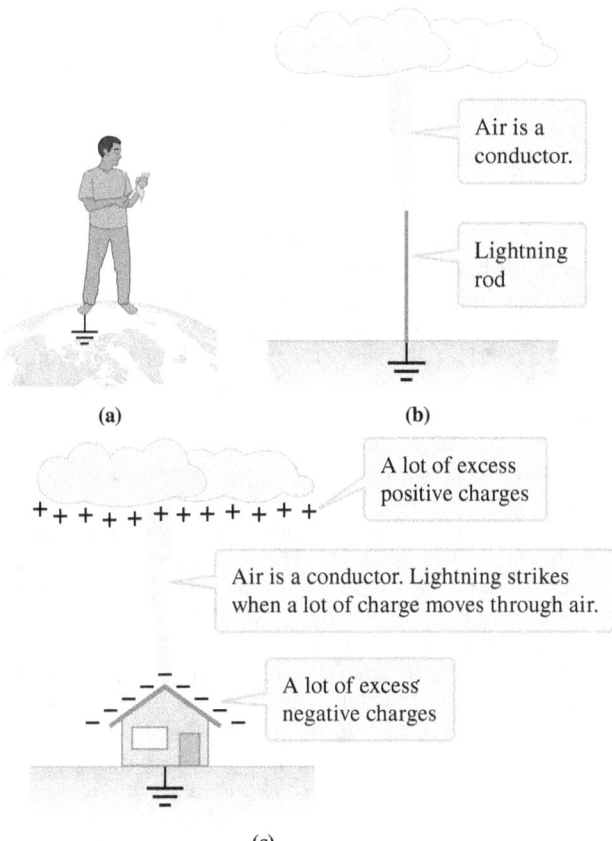

Figure 1.109: (a) Franklin knew that a grounded person cannot charge a conducting rod by rubbing. (b) According to Franklin, lightning strikes could be prevented if a town put up a lot of lightning rods. Then the atmosphere would be connected to ground, so it could not build up charge. (c) Without a lightning rod, Franklin reasoned, the atmosphere discharges in a violent strike.

lessly overhead. Wilson argued that pointed lightning rods were more dangerous than having no rods at all.

In this section, we explore both types of rods (Fig.1.110) in following two solved examples. In order for air to breakdown* and become a conductor, the electric field in the air must be 3×10^6 N/C. Let's assume that in order for a lightning rod to work, the electric field at its surface must equal to breakdown electric field. We will calculate the amount of charge on the surface of each conductor. The one with the least amount of charge is the better design because a smaller amount of charge on the surface of the conductor means a smaller amount of charge travels through the air. Calculating the charge on each rod gives us a way to compare their effectiveness. A rod that does not require much charge to have a strong electric field on its surface will not generate a big spark to discharge the atmosphere. Because both designs use conductors, we will assume the excess charge is uniformly spread over the surface of the lightning rod.

EXAMPLE 52. At the end of Wilson's rod was a cannonball (Fig.1.111) with a radius of $R = 0.1m$. If the electric field on the surface of the ball is $E = 3 \times 10^6$ N/C (breakdown field strength for air), what is the charge on the ball?

APPROACH Because the ball is so much larger than the thickness of the supporting rod, we will ignore that rod and model Wilson's device as a suspended sphere connected to ground. The

---

*If the magnitude of an electric field in air is so great, that the air becomes ionized and begins to conduct electricity, then the electric field is called the breakdown field for air and the phenomenon is called **dielectric breakdown** of air

ball is a sphere, so we can use the magnitude of $\vec{E}(r) = (kQ/r^2)\hat{r}$ to find the charge on the ball.

SOLUTION For the charge Q on the surface of the sphere, where $r = R$, we have-

$$E(R) = k\frac{Q}{R^2}$$
$$\Rightarrow \quad Q = \frac{ER^2}{k} = \frac{(3 \times 10^6 \text{N/C})(0.1 \text{ m})^2}{(8.99 \times 10^9 \text{N} \cdot \text{m}^2/\text{C}^2)}$$
$$\Rightarrow \quad Q = 3.3 \times 10^{-6} C = 3.3 \mu C$$

Result This is not very much charge, especially compared to the amount of charge involved in a lightning strike, which is of the order of hundreds of coulombs.

EXAMPLE 53. Let's model the end of a Franklin rod as a tiny ball of radius $R = 2$ mm - something like the end of a knitting needle. If the electric field on the surface of the ball is $E = 3 \times 10^6$ N/C, what is the charge on the ball?

APPROACH Repeat the calculation from above example for the Wilson rod, this time for a much smaller ball. Solve for the charge q on the surface of the sphere, where $r = R$.

SOLUTION $E(R) = k\frac{q}{R^2}$

$$\Rightarrow \quad q = \frac{ER^2}{k} = \frac{(3 \times 10^6 \text{N/C})(0.002 \text{ m})^2}{(8.99 \times 10^9 \text{N} \cdot \text{m}^2/\text{C}^2)}$$
$$\Rightarrow \quad q = 1.3 \times 10^{-9} C = 1.3 \text{ nC}$$

So, the charge on Franklin's rod is about 2500 times lower than on Wilson's.

From above two examples, we can say that an ideal Franklin rod (one that is infinitely thin) would be better than the blunt Wilson rod.

EXAMPLE 54. Find an expression for the electric field at the tip of Franklin's pointed lightning rod. Evaluate Franklin's design by considering a point very close to the end of the rod.

APPROACH Fig. 1.111 shows a vertical rod of length L and P is a point at a distance d from the upper end of the rod. For a limiting case, when P is just above the upper tip of the rod, we have $d \rightarrow 0$. As the charge will get distributed uniformly on the metal rod, therefore, to find electric field at point P, we have to consider a differential charge element $dq(= \lambda \, dl)$ of length dl of the rod at distance l from the point P. Here λ is the linear charge density of the rod. Since, each elementary piece of the rod produces electric field, at point P, in the same direction, therefore, the integration from $l = d$ to $l = d + L$ will give the net electric field at P due to complete length of the charged rod.

Electric field at P due to elementary charge dq, is given by-

$$d\vec{E} = k\frac{dq}{r^2}\hat{j}, \text{ with } r = l$$

To derive an expression for $d\vec{E}$, integrate this expression from $l = d$ to $l = d + L$

For the electric field just above the tip of the rod, we have $d \to 0$.

SOLUTION Electric field at P, due to the infinitesimal charge

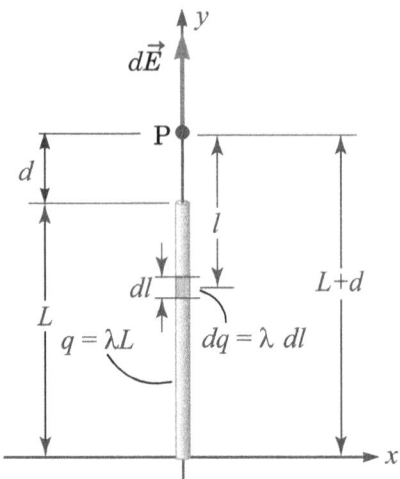

Figure 1.111

$dq = \lambda dl$, is-
$$d\vec{E} = k\frac{dq}{l^2}\hat{j} \quad \Rightarrow \quad d\vec{E} = k\frac{\lambda dl}{l^2}\hat{j}$$

Now, on integrating over the entire length of the rod from $l = d$ to $l = d + L$, we get

$$\int_0^E d\vec{E} = \int_0^l k\frac{\lambda dl}{l^2}\hat{j}$$

$$\Rightarrow \quad \vec{E} = k\lambda\left[-\frac{1}{l}\right]_d^{d+L}\hat{j} = -k\lambda\left[\frac{1}{d+L} - \frac{1}{d}\right]\hat{j}$$

$$\Rightarrow \quad \vec{E} = -k\lambda\left[\frac{-L}{d(d+L)}\right]\hat{j} = k\left[\frac{\lambda L}{d(d+L)}\right]\hat{j}$$

The total charge on the rod is $q = \lambda L$. Therefore, on substituting $\lambda L = q$ in above expression, we get-

$$\vec{E} = \frac{kq}{d(d+L)}\hat{j} \quad (1.79)$$

If point P is far from the rod $(d \gg L)$, then $d + L \approx d$. In this case, the electric field at P is given by

$$\vec{E} = \frac{kq}{d^2}\hat{j}$$

So, $\vec{E}$ approaches the electric field due to a charged particle. For the point, just above the rod, $d \to 0$, therefore

$$\vec{E} = \lim_{d \to 0}\left[\frac{kq}{d(d+L)}\right] = \infty$$

So, it is clear that, as point P gets closer to the end of the rod, the electric field increases. At the tip of the rod, the electric field approaches infinity. So, in principle, even a very tiny charge on the rod would lead to a very high electric field at its end. Franklin's rod would continually draw a small amount of charge from the atmosphere and prevent a large lightning strike.

EXAMPLE 55. A thin non-conducting ring of radius R has a linear charge $\lambda = \lambda_0 \cos\theta$, where λ_0 is the value of λ at $\theta = 0$ [see Fig.1.112]. Find the net electric dipole moment for this charge distribution.

APPROACH The charge density of the given ring is $\lambda = \lambda_0 \cos\theta$. The value of $\cos\theta$ is positive for $-\pi/2 < \theta < \pi/2$ and negative for $\pi/2 < \theta < 3\pi/2$. So, the charge of the ring is positive in first and fourth quadrants whereas, it is negative in second and third quadrants. Since, $|\cos\theta| = |\cos(2\pi \pm \theta)| = |\cos(\pi \pm \theta)|$. Therefore, corresponding to each positive elementary charge in

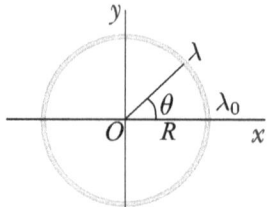

Figure 1.112

first and fourth quadrant, there will be an equal and opposite charge in third and fourth quadrant. So, each elementary charge is able to form an electric dipole.

Let us consider a small part of length dl of the ring as shown in Fig.1.113a. If this part subtends angle $d\theta$ at the centre of the ring, then
$$dl = R d\theta$$
The small charge on this part of ring is given as-
$$dq = \lambda\, dl = (\lambda_0 \cos\theta)(R d\theta)$$
The symmetrically opposite part to this part has charge $-dq$. So, these parts form a dipole. The location of charge $-dq$ may be considered in quadrant fourth or in quadrant second (Fig. 1.113). For, Fig.1.113a, dipole length is $2R$, whereas for dipole shown in Fig.1.113b, its lemgth is $2R\cos\theta$.

Now, for the dipole considered in Fig.1.113(a), the magnitude of dipole moment of this elementary dipole is given by-
$$dp = 2R dq = 2R\lambda_0 \cos\theta\, Rd\theta$$
The direction of this dipole moment is from $-dq$ to $+dq$.
x and y components of this dipole moment are,
$$dp_x = 2R\lambda_0 \cos\theta R d\theta \cos\theta \quad (1.80a)$$
$$dp_y = 2R\lambda_0 \cos\theta R d\theta \sin\theta \quad (1.80b)$$

To get, total electric dipole moment of the ring, we have to integrate these expressions from $\theta = -\pi/2$ to $+\pi/2$. For the

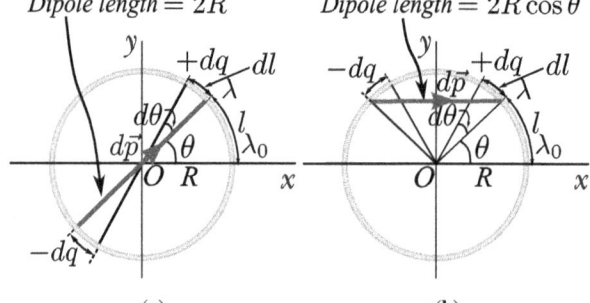

Figure 1.113: Charged ring

dipole shown in Fig.1.113b, the dipole length is $2R\cos\theta$. It's dipole moment is given by-
$$dp = 2R\cos\theta\, dq = 2R\cos\theta\, (\lambda_0 \cos\theta)\, Rd\theta$$
$$\Rightarrow \quad dp = 2R^2\lambda_0 \cos^2\theta\, d\theta \quad (1.81)$$

The direction of this dipole moment is from $-dq$ to $+dq$ (Fig1.113b), i.e., along $+ve$ x- direction. In this case, y- component of dipole moment is zero.

To get net dipole moment of the ring, we have to integrate it from $\theta = -\pi/2$ to $+\pi/2$.

Now, we find the net dipole moment of the ring by using both above methods, i.e., by considering both the dipoles of Fig.1.113.

SOLUTION Method 1. Considering, the dipole shown in Fig.1.113, From, Eq.(1.80) the x y components of dipole moment are -
$$dp_x = 2R^2\lambda_0 \cos^2\theta d\theta$$
and
$$dp_y = 2R\lambda_0 \cos\theta \sin\theta R\, d\theta$$
respectively.

On integrating both of the above expressions, from $\theta = -\pi/2$ to

$+\pi/2$, we get-
$$p_x = \int_{-\pi/2}^{+\pi/2} 2R^2 \lambda_0 \cos^2\theta \, d\theta$$
and $p_y = \int_{-\pi/2}^{+\pi/2} 2R^2 \lambda_0 \cos\theta \sin\theta \, d\theta$
$$\implies p_x = \int_{-\pi/2}^{+\pi/2} 2R^2 \lambda_0 \cos^2\theta \, d\theta$$
$$= R^2 \lambda_0 \int_{-\pi/2}^{+\pi/2} (1 + \cos 2\theta) \, d\theta$$
$$= R^2 \lambda_0 \left[[\theta]_{-\pi/2}^{+\pi/2} + \left[\frac{\sin 2\theta}{2}\right]_{-\pi/2}^{+\pi/2} \right] = \pi R^2 \lambda_0$$
$$\implies p_x = \pi R^2 \lambda_0$$
& $p_y = R^2 \lambda_0 \int_{-\pi/2}^{+\pi/2} \sin 2\theta \, d\theta = R^2 \lambda_0 \left[-\frac{\cos 2\theta}{2}\right]_{-\pi/2}^{+\pi/2} = 0$

Method 2. In this method, we integrate Eq.(1.81) from limit $\theta = -\pi/2$ to $\theta = +\pi/2$
Integrating, we get-
$$p = 2R^2 \lambda_0 \int_{-\pi/2}^{\pi/2} \cos^2\theta R d\theta$$
Simplifying as like we have done for p_x, we get
$$p = \pi R^2 \lambda_0$$
Here, $\vec{p}$ is directed along the $+ve$ direction of x axis.

1.18 Check Point 4

1. ●●In Fig.1.114, a thin glass rod forms a semicircle of radius $R = 5.00$ cm. Charge is uniformly distributed along the rod, with $+q = 4.50$ pC in the upper half and $-q = -4.50$ pC in the lower half. What are the (a) magnitude and (b) direction (relative to the positive direction of the x axis) of the electric field $\vec{E}$ at the centre (P) of the semicircle?

2. ●●In Fig.1.115, two curved plastic rods, one of charge $+q$ and the other of charge $-q$, form a circle of radius $R = 8.50$ cm in an xy plane. The x axis passes through both of the connecting points, and the charge is distributed uniformly on both rods. If $q = 15.0$ pC, what are the (a) magnitude and (b) direction (relative to the positive direction of the x axis) of the electric field $\vec{E}$ produced at P, the centre of the circle?

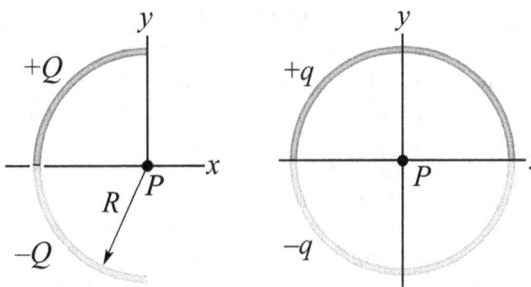

Figure 1.114   Figure 1.115

3. ●●Find the electric dipole moment of a non-conducting ring of radius a, made of two semicircular rings having linear charge densities $-\lambda$ and $+\lambda$ as shown in Fig. 1.116.

4. ●●Find the electric dipole moment of a non-conducting ring of radius a, having linear charge densities $+\lambda$ and $-\lambda$ and arranged as shown in Fig.1.117.

5. ●●Fig.1.118 shows two discs and a flat ring, each with the same uniform charge Q. Rank the objects according to the magnitude of the electric field they create at points P (which are at the same vertical heights) in decreasing order.

6. ●●A positive point charge, Q, is located at a distance h directly above the centre of a charged thin non-conducting circular plate of radius R. The plate carries a total positive charge, Q, spread uniformly over its surface area. What will be the electrical force on the point charge?

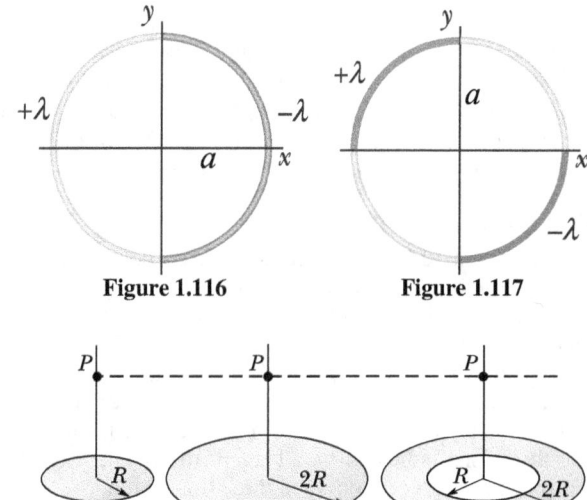

Figure 1.116   Figure 1.117

Figure 1.118

7. ●●●A thin nonconducting ring of radius R has a linear charge density $\lambda = \lambda_0 \cos\phi$, where λ_0 is a constant, and azimuthal angle ϕ is shown in Fig.1.119. Find the magnitude of the electric field strength
(a) at the centre of the ring;

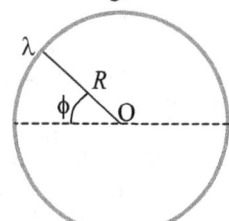

Figure 1.119

(b) on the axis of the ring as a function of the distance z from its centre. Investigate the obtained function at $z \gg R$.

8. ●●A thread carrying a uniform charge λ per unit length has the configurations shown in Fig. 1.120a and 1.120b. Assuming a curvature radius R to be considerably less than the length of the thread, find the magnitude of the electric field strength at the point O.

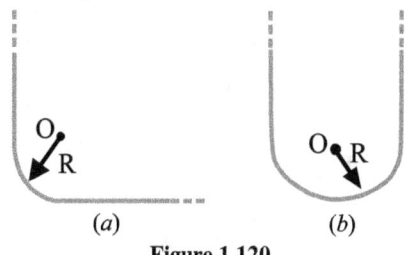

Figure 1.120

Multiple Choice Questions

9. ●●Supposing that the earth has a surface charge density of 1 electron/m^2; calculate electric field just outside earth's surface. The electronic charge is -1.6×10^{-19}C and earth's radius is 6.4×10^6m. $\left(\epsilon_0 = 8.9 \times 10^{12} C^2/N.m^2\right)$

(A) 1.8×10^{-8} N/C   (B) $+1.8 \times 10^{-8}$N/C
(C) 1.8×10^{-9} N/C   (D) $+1.8 \times 10^{-9}$N/C

10. ●●A circular wire of radius R carries a total charge Q distributed uniformly over its circumference. A small length of the wire subtending angle θ at the centre is cut off. Find the electric field at the centre due to the remaining portion.
   (A) $Q/(4\pi^2\epsilon_0 R^2)\sin\theta$ (B) $Q/(4\pi^2\epsilon_0 R^2)\sin\theta/2$
   (C) $Q/(4\pi^2\epsilon_0 R^2)\sin\theta$ (D) $Q/(4\pi^2\epsilon_0 R^2)\sin\theta/8$

1.19   Electric Field Lines

The picture of field lines (sometimes called lines of force) was invented by Faraday to develop an intuitive non-mathematical way of visualizing electric fields around charged configurations. An electric field line is, in general, a curve drawn in such a way that the tangent to it at each point is in the direction of the net field at that point (Fig.1.121). An arrow on the curve is obviously necessary to specify the direction of electric field from the two possible directions indicated by a tangent to the curve. A field line is a space curve, i.e., a curve in three dimensions.

To understand the dependence of the field lines on the area, or

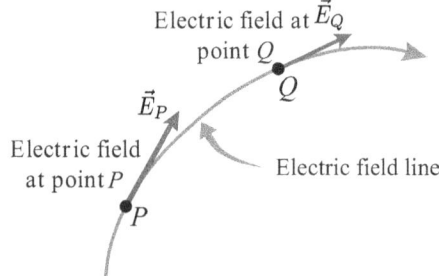

Figure 1.121: The direction of the electric field at any point is the tangent to the electric field line at this point

rather the solid angle subtended by an area element, let us try to relate the area with the solid angle, a generalization of angle to three dimensions. Recall how a (plane) angle is defined in two-dimensions. Let a small transverse line element Δl be placed at a distance r from a point O. Then the angle subtended by Δl at O can be approximated as $\Delta\theta = \Delta l/r$. Similarly, in three-dimensions the solid angle subtended by a small perpendicular plane area ΔS, at a distance r, can be written as $\Delta\Omega = \Delta S/r^2$. In Fig.1.122, for two points P_1 and P_2 at distances r_1 and r_2 from the charge, the element of area subtending the solid angle $\Delta\Omega$ is $r_1^2\Delta\Omega$ at P_1 and an element of area $r_2^2\Delta\Omega$ at P_2, respectively. The number of lines (say n) cutting these area elements are the same. The number of field lines, cutting unit area element is therefore $n/(r_1^2\Delta\Omega)$ at P_1 and $n/(r_2^2\Delta\Omega)$ at P_2, respectively. Since n and $\Delta\Omega$ are common, the strength of the field clearly has a $1/r^2$ dependence.

The electric field is strong where field lines are close together

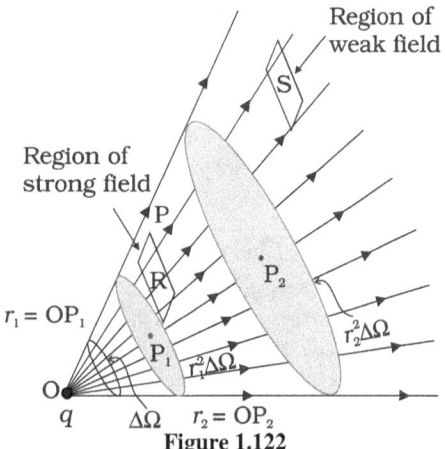

Figure 1.122

and weak where they are far apart (Fig.1.123). (More specifically, if you imagine a small surface perpendicular to the field lines, the magnitude of the field is proportional to the number of lines that cross the surface divided by the area.)

☞ Electric field lines represent the net electric field at any point.

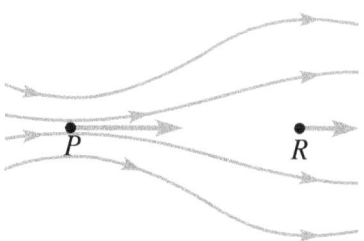

Figure 1.123: The magnitude of the electric field at point P is larger than the magnitude at R.

This means that they take into account the contribution of all the charges in the region.

Rules for Drawing Electric Field Lines: The rules for drawing electric field lines are as follows:

- The lines must begin on a positive charge and terminate on a negative charge. In the case of an excess of one type of charge, some lines will begin or end infinitely far away.
- The number of field lines emerging from a positive charge or ending at a negative charge is proportional to the magnitude of the charge.
- At large distances from a system of charges that has a non-zero net charge, the field lines are equally spaced and radial, as if they emanated from (or terminated on) a single point charge equal to the total charge of the system.
- Electric field lines cannot cross each other. The electric field at any point has a unique direction; if two field lines crossed, the field would have two directions at the same point (Fig.1.124)

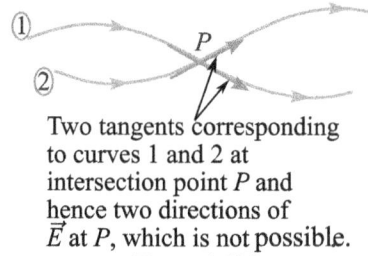

Two tangents corresponding to curves 1 and 2 at intersection point P and hence two directions of $\vec{E}$ at P, which is not possible.
Figure 1.124

We choose the number of field lines starting from any positively charged object to be Cq and the number of lines ending on any negatively charged object to be $C|q|$, where C is an arbitrary proportionality constant and q is charge on the object. Once C is chosen, the number of lines is fixed. For example, if object 1 has charge Q_1 and object 2 has charge Q_2, then the ratio of number of lines is $N_2/N_1 = Q_2/Q_1$. The electric field lines for two isolated point charges of magnitude $+q$ and $|-2q|$ are shown in Fig.1.125. Because the magnitudes of charges are in the ratio of 1 : 2, the number of field lines that originated from $+q$ and terminated on $-2q$ are also in the ratio of 1 : 2. For example, 8 field line are originated from $+q$ and 16 lines are terminated on $-2q$, so $N_1 : N_2 = 8 : 16$, i.e., 1 : 2.

1.20   Theorem on Circulation of Vector $\vec{E}$

In mechanics we have studied that any stationary field of central forces is conservative, i.e. the work done by the forces of this field is independent of the path and depends only on the position of the initial and final points. This property is inherent in the electrostatic field, viz. the field created by a system of fixed charges. If we take a unit positive charge for the test charge and

1.20. THEOREM ON CIRCULATION OF VECTOR $\vec{E}$

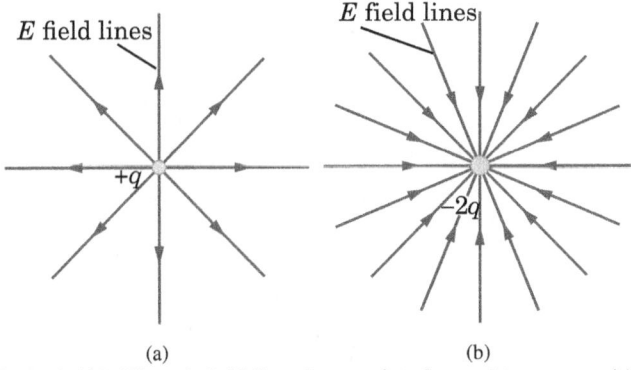

(a) (b)

Figure 1.125: Electric field lines for a point charge (a) near a positive charge the field lines point radially away from the charge. The lines start on the positive charge and end at infinity. (b) near a negative charge the field lines point radially inward. They start at infinity and end on a negative charge and are more dense where the field is more intense. notice that the number of lines drawn for part (b) is twice the number drawn for part (a), a reflection of the relative magnitudes of the charges carry it from initial position point i of a given field $\vec{E}$ to final position point f, the elementary work of the forces of the field done over the distance $d\vec{l}$ is-

$$dW = \vec{F_e} \cdot d\vec{l} = \vec{E} \cdot d\vec{l}$$

[because, for unit positive test charge, $\vec{F_e} = q_0\vec{E} = 1\vec{E} = \vec{E}$] and the total work of the field forces over the distance between points i and f is defined as

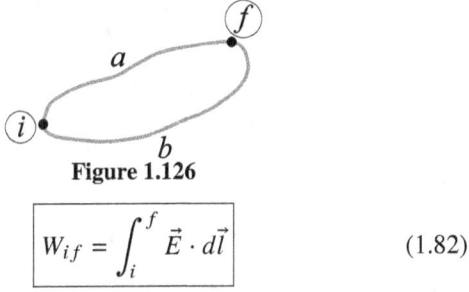

Figure 1.126

$$\boxed{W_{if} = \int_i^f \vec{E} \cdot d\vec{l}} \quad (1.82)$$

This integral is taken along a certain line (path) and is therefore called the line integral.

We shall now show that from the independence of line integral (1.82) of the path between two points it follows that when taken along an arbitrary closed path, this integral is equal to zero. Integral (1.82) along a closed path is called the circulation of vector $\vec{E}$ and is denoted by $\oint$.

Thus, we state that circulation of vector $\vec{E}$ in any electrostatic field is equal to zero, i.e.,

$$\boxed{\oint \vec{E} \cdot d\vec{l} = 0} \quad (1.83)$$

This statement is called the *theorem on circulation of vector $\vec{E}$*. In order to prove this theorem, we break an arbitrary closed path into two parts iaf and fbi (Fig.1.126). Since line integral (1.82), we denote it by $\left(\int_{12}\right)$, does not depend on the path between points i and f, we have $\int_{if}^{(a)} = \int_{if}^{(b)}$. On the other hand, it is clear that $\int_{if}^{(b)} = -\int_{fi}^{(b)}$, where $\int_{fi}^{(b)}$ is the integral over the same segment b but taken in the opposite direction. Therefore,

$$\oint \vec{E} \cdot d\vec{l} = \int_{if}^{(a)} \vec{E} \cdot d\vec{l} + \int_{fi}^{(b)} \vec{E} \cdot d\vec{l}$$
$$= \int_{if}^{(a)} \vec{E} \cdot d\vec{l} - \int_{if}^{(b)} \vec{E} \cdot d\vec{l} = 0,$$

A field having property (1.83) is called the potential field. Hence, any electrostatic field is a potential field. The term 'potential' will be discussed in next chapter.

> ☞ In next chapter, we will see that the term potential is defined only for conservative force field, therefore, the field for which Eq.1.83 holds, is also known as a conservative field.

The theorem on circulation of vector $\vec{E}$ makes it possible to draw a number of important conclusions without resorting to calculations. Let us consider two examples.

EXAMPLE 56. The field lines of an electrostatic field $\vec{E}$ cannot be closed. Why?

SOLUTION Suppose, the opposite is true and some lines of field $\vec{E}$ are closed like Fig.1.127. The arrows on the contour indicate the direction of circumvention as well as the direction of electric field.

Taking the circulation of vector $\vec{E}$ along the line shown in this figure, we get-

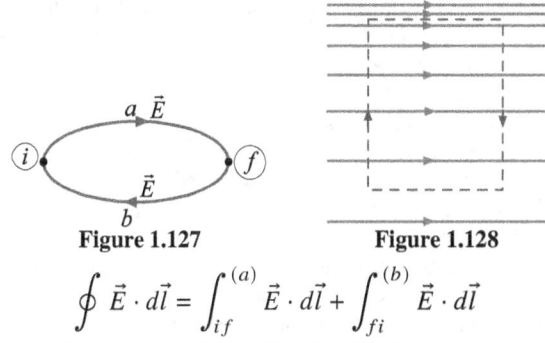

Figure 1.127 **Figure 1.128**

$$\oint \vec{E} \cdot d\vec{l} = \int_{if}^{(a)} \vec{E} \cdot d\vec{l} + \int_{fi}^{(b)} \vec{E} \cdot d\vec{l}$$

For both the terms of right hand side of above expression, $\vec{E}$ and $d\vec{l}$ are directed in the same direction. Therefore, the scalar product $\vec{E} \cdot d\vec{l} > 0$. So,

$$\oint \vec{E} \cdot d\vec{l} > 0$$

It contradicts the theorem 1.20. This means that electric field lines cannot form any closed curve in an electrostatic field: *the lines emerge from positive charges and terminate on negative ones (or go to infinity)*.

EXAMPLE 57. Is the configuration of an electrostatic field shown in Fig.1.128 possible?

SOLUTION No, it is not. It can be easily shown by applying the theorem on circulation of vector $\vec{E}$ to the closed contour shown in the Fig.1.128 by the dashed line. The arrows on the contour indicate the direction of circumvention. With such a special choice of the contour, the contribution to the circulation from its vertical parts is equal to zero, since in this case $\vec{E} \perp d\vec{l}$ and $\vec{E} \cdot d\vec{l} = 0$. It remains for us to consider the two horizontal segments of equal lengths. The Fig.1.128 shows that the contributions to the circulation from these regions are opposite in sign, and unequal in magnitude (the contribution from the upper segment is larger since the field lines are denser, and hence the value of $\vec{E}$ is larger). Therefore, the circulation of $\vec{E}$ differs from zero, which contradicts to (1.83).

1.20.1 Deduction of Pattern of Field Lines

Fig.1.129 shows the sketch of field lines for two point charges $2q$ and $-q$.

The pattern of field lines can be deduced by considering the

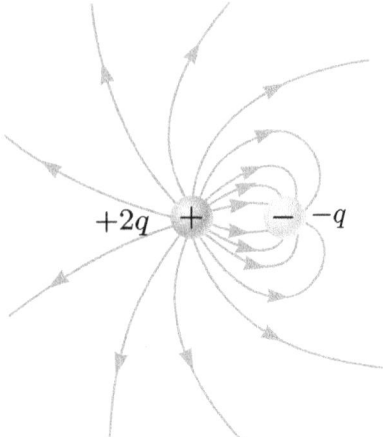

Figure 1.129: Two field lines leave $+2q$ for every one that terminates on $-q$.

following points:
1. **Symmetry:** For every point above the line joining the two charges there is an equivalent point below it. Therefore, the pattern must be symmetrical about the line joining the two charges.
2. **Near field:** Very close to a charge, its own field predominates. Therefore, the lines are radial and spherically symmetric. The high density of lines near the charges indicates a region of strong electric field.
3. **Far field:** Far from the system of charges, the pattern should look like that of a single point charge of value $(2Q - Q) = +Q$, i.e., the lines should be radially outward.
4. **Null point:** The point at which $E = 0$, is called null point. No lines should pass through this point.
5. **Number of lines:** Number of lines is directly proportional to magnitude of charge. So, number of field lines that leave $+2Q$ is twice that of terminating on charge $-Q$.

The field lines around a dipole [Fig.1.130a], show clearly a vivid pictorial description of the mutual attraction between the two charges. The total charge of the dipole is zero, but because the charges are separated, the electric field does not vanish. Instead, the field lines start from positive charge and terminate at negative charge.

The field lines around a system of two positive charges (q, q) [Fig.1.130b] give a vivid pictorial description of their mutual repulsion.

Fig.1.131 shows the electric field lines for a configuration of two positive and two negative charges, all of equal magnitude.

1.20.2 Properties of Electric Field Lines

The field lines follow some important general properties:
1. Field lines start from positive charges and end at negative charges. If there is a single charge, they may start or end at infinity.
2. In a charge-free region, electric field lines can be taken to be continuous curves without any breaks.
3. Tangent to the lines of force at any point gives the direction of the electric field.
4. Two field lines can never cross each other. (If they did, the field at the point of intersection will not have a unique direction, which is absurd.)
5. Electrostatic field lines do not form any closed loops. This follows from the conservative nature of electric field.
6. Lines of force originate or terminate perpendicular to the metal surface. Reason is given in section 1.26
7. The density of the lines at any point (the number of lines per unit area through a surface element normal to the lines) is

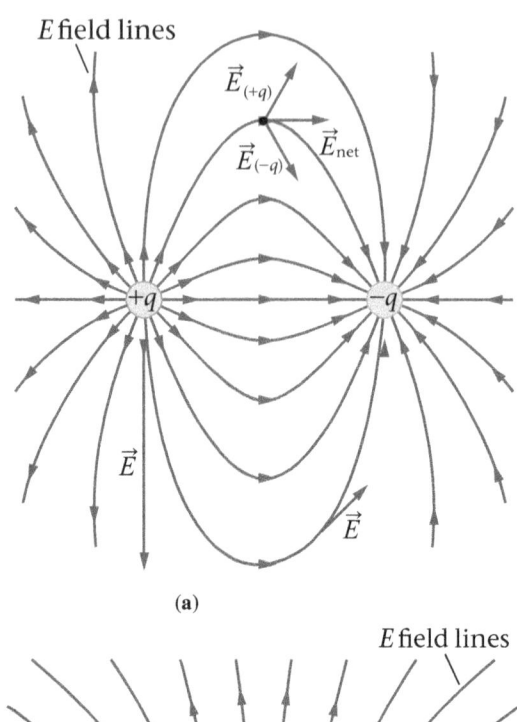

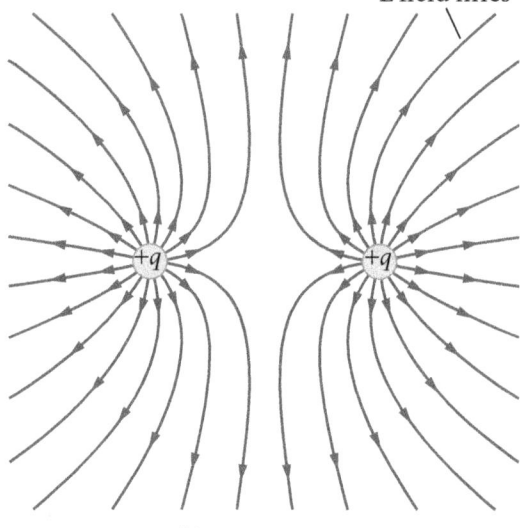

Figure 1.130: Electric field lines for systems of charges (a) The electric field lines for a dipole form closed loops that become more widely spaced with distance from the charges. Note that at each point in space, the electric field vector $\vec{E}$ is tangent to the field lines. (b) All of the field lines in a system with charges of the same sign extend to infinity

proportional to the magnitude of the field there.

1.21 Check Point 5

1. •The electric field lines for a system of two charges are shown in Figure 1.132 (a) What is the sign of charge 1? (b) What is the sign of charge 2? (c) Is the magnitude of charge 1 greater than, less than, or equal to the magnitude of charge 2? Explain.
2. •Rank the magnitudes of the electric field at points A, B, and C shown in Figure1.133 (greatest magnitude first).
3. •For the drawing shown in Figures 1.134, write the ratio of charges q_1 and q_2.
4. •The electric field lines surrounding three charges are shown in Figure1.135 (a) Which of the charges q_A, q_B, and q_C are positively charged, and which are negatively charged? (b) Rank the charges in order of increasing magnitude.
5. •Sketch the electric field lines for the system of charges

1.22. ACTION OF THE ELECTRIC FIELD ON CHARGES 51

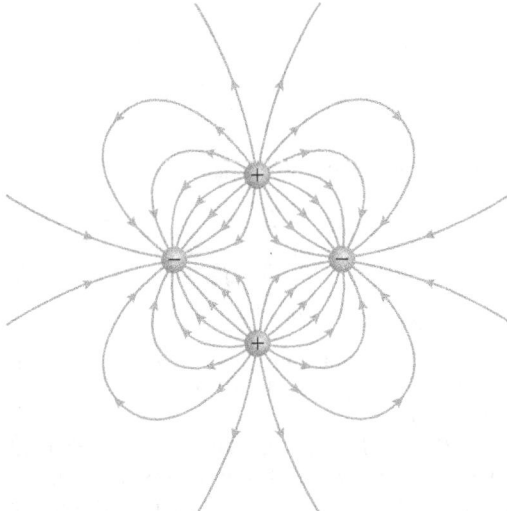

Figure 1.131: Electric field lines for a configuration of two positive and two negative charges, all of equal magnitude. Note that the field lines always start on a positive charge and end on a negative charge when there are equal numbers of both charges.

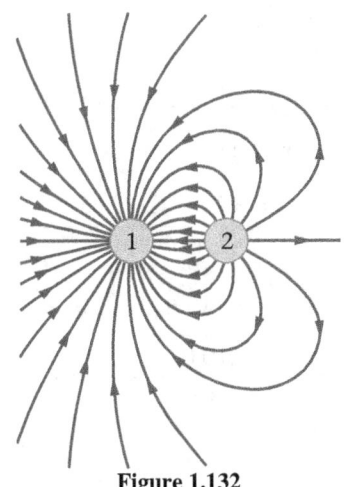

Figure 1.132

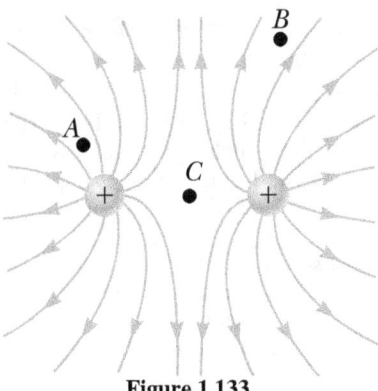

Figure 1.133

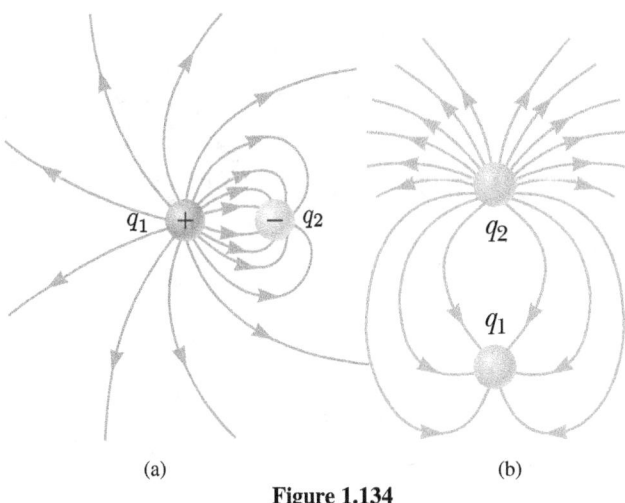

Figure 1.134

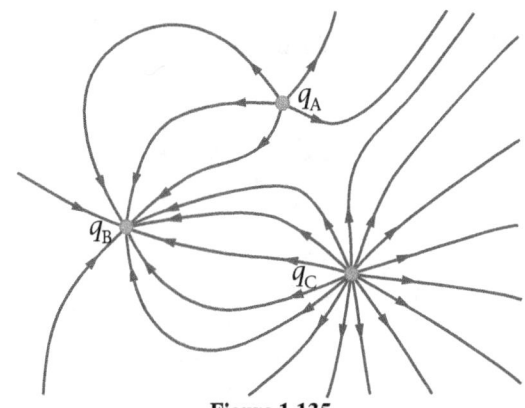

Figure 1.135

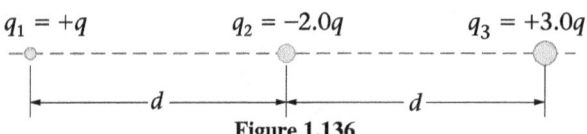

Figure 1.136

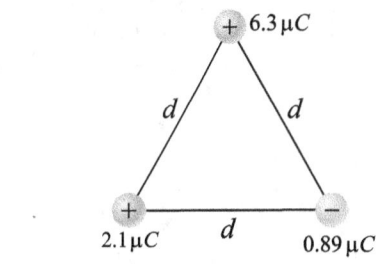

Figure 1.137

(A) 2, 3, 4, 1   (B) 2, 1, 3, 4
(C) 1, 4, 3, 2   (D) 4, 3, 1, 2

1.22 Action of the Electric Field on Charges

A uniform electric field can exert a force on a single charged particle and can exert both a torque and a net force on an electric dipole.

1.22.1 Motion of a charged Particle in a Uniform Electric Field

When a particle of charge q and mass m is in an external electric field of strength $\vec{E}$, a force $q\vec{E}$ will be exerted on this particle. If $q\vec{E}$ is the only acting force on the particle, then according to

shown in Fig.1.136.
6. •Sketch the electric field lines for the system of charges described in Figure 1.137.
Multiple Choice Questions
7. •Which lines in Figure1.138 cannot represent an electric field? Explain.
8. •In the Figure1.139, rank points 1 – 4 in order of increasing field strength.

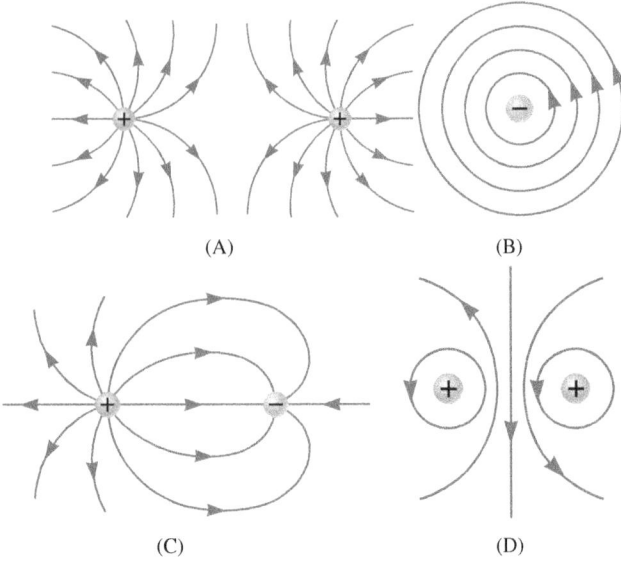

Figure 1.138

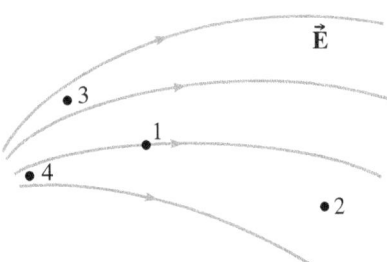

Figure 1.139

Newton's second law, $\Sigma\vec{F} = m\vec{a}$, the acceleration of the particle will be given by:
$$\vec{a} = q\vec{E}/m \qquad (1.84)$$

1.22.1.1 Motion of a Charged Particle Along an Electric Field

Consider a particle of positive charge q and mass m in a uniform horizontal electric field $\vec{E}$ produced by two charged plates that are separated by a distance d as shown in Fig.1.140.

If the particle is released from rest at the positive plate and $q\vec{E}$ is

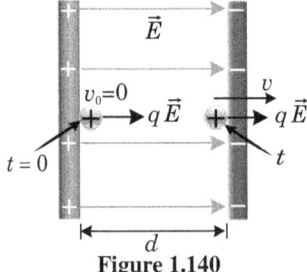

Figure 1.140

the only force that acts on the particle, then the particle will move horizontally along the x - axis with an acceleration $\vec{a} = q\vec{E}/m$. In such a case, we can apply the kinematics equations on the initial and final motion as follows:

- The particle's time of flight t:
$$x = v_\circ t + \frac{1}{2}at^2 \implies d = 0 + \frac{1}{2}\frac{qE}{m}t^2$$
$$\implies t = \sqrt{\frac{2md}{qE}} \qquad (1.85)$$

- The speed of the particle v:

$$v = v_\circ + at \implies v = 0 + \frac{qE}{m}\sqrt{\frac{2md}{qE}}$$
$$\implies v = \sqrt{\frac{2qEd}{m}} \qquad (1.86)$$

- The kinetic energy of the particle K:
$$K = \frac{1}{2}mv^2$$
$$\implies K = \frac{1}{2}m\left(\sqrt{\frac{2qEd}{m}}\right)^2 = qEd \qquad (1.87)$$

The last result can also be obtained from the application of the work-energy theorem $W = \Delta K$ because $W = (qE)d$ and $\Delta K = K_\text{f} - K_\text{i} = K$.

EXAMPLE 58. In Fig.1.140, assume that the charged particle is a proton of charge $q = +e$. The proton is released from rest at the positive plate. In this case, each of the two oppositely charged plates which are $d = 2$ cm apart has a charge per unit area of $\sigma = 5\mu C/m^2$. (a) What is the magnitude of the electric field between the two plates? (b) What is the speed of the proton as it strikes the second plate?

SOLUTION (a) Since, the electric field arises from two infinite thin plates, therefore,
$$E = \frac{\sigma}{2\epsilon_\circ} + \frac{\sigma}{2\epsilon_\circ} = \frac{\sigma}{\epsilon_\circ}$$
$$= \frac{5 \times 10^{-6} C/m^2}{8.85 \times 10^{-12} C/N \cdot m^2} = 5.65 \times 10^5 \text{ N/C}$$

(b) We first find the proton's acceleration from Newton's second law:
$$a = \frac{F}{m} = \frac{eE}{m} = \frac{(1.6 \times 10^{-19} C)(5.65 \times 10^5 \text{ N/C})}{1.67 \times 10^{-27} \text{ kg}}$$
$$= 5.41 \times 10^{13} \text{ m/s}^2$$

Then, using $x = v_\circ t + \frac{1}{2}at^2$, we find that $d = \frac{1}{2}at^2$. Thus:
$$t = \sqrt{\frac{2d}{a}} = \sqrt{\frac{2(0.02 \text{ m})}{5.41 \times 10^{13} \text{ m/s}^2}} = 2.72 \times 10^{-8} \text{ s}$$

Finally, we use $v = v_\circ + at$ to find the speed of the proton as follows:
$$v = at = \left(5.41 \times 10^{13} \text{ m/s}^2\right)\left(2.72 \times 10^{-8} \text{ s}\right)$$
$$= 1.47 \times 10^6 m/s$$

1.22.1.2 Motion of a Charged Particle Perpendicular to an Electric Field

Let a uniform electric field $\vec{E}$ is created between two parallel, charged plates as shown in Fig.1.141. A charged particle of charge q enters at O (0, 0) in the field symmetrically between the plates with an initial speed v_0. The length of each plate is l.

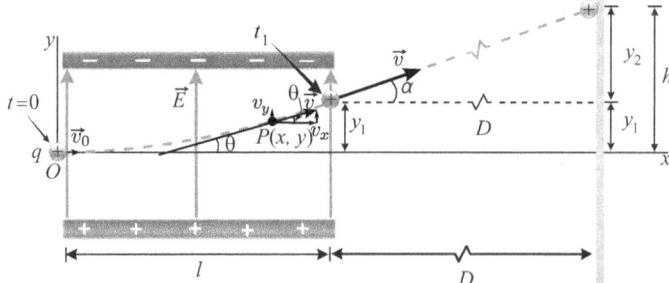

Figure 1.141: The effect of an upward force $q\vec{E}$ exerted on charged particle projected horizontally with speed v_0 into an upward uniform electric field $\vec{E}$

The acceleration of the charged particle is $a = \frac{qE}{m}$ in upward direction. The horizontal velocity remains v_0 as there is no acceleration in horizontal direction. Thus, the time taken by charged particle in reaching to a general point $P(x,y)$ in electric field $\vec{E}$ is given by-

$$t = \frac{x}{v_0} \tag{1.88}$$

The upward component of the velocity of the charged particle at point $P(x, y)$ in the field region is

$$v_y = v_{0y} + a_y t = 0 + \frac{qEx}{mv_0}$$

$$\text{or} \quad v_y = \frac{qEx}{mv_0} \tag{1.89}$$

At point P, the horizontal component of velocity remains $v_x = v_0$. The angle θ made by the resultant velocity with the original direction is given by

$$\tan\theta = \frac{v_y}{v_x} = \frac{qEx}{mv_0^2}$$

Thus, the charged particle deviates by an angle,

$$\theta = \tan^{-1}\frac{qEx}{mv_0^2} \tag{1.90}$$

Vertical displacement y can be given by-

$$y = v_{0y}t + \frac{1}{2}a_y t^2 = 0 + \frac{1}{2}\frac{qEx^2}{mv_0^2}$$

$$\Rightarrow \quad \boxed{y = \frac{1}{2}\frac{qEx^2}{mv_0^2}} \tag{1.91}$$

This is an expression for path of a charged particle in a uniform electric field. Since Eq.(1.91) is of the form of a parabola, therefore the path of charged particle will be parabolic in uniform electric field.
When, charged particle emerges from the field, $x = l$, from Eq.1.91, the vertical displacement can be written as-

$$y_1 = \frac{1}{2}\frac{qEl^2}{mv_0^2} \tag{1.92}$$

The deviation of charged particle at $x = l$, is given by-

$$\alpha = \tan^{-1}\frac{qEl}{mv_0^2} \tag{1.93}$$

The extra vertical distance y_2 that the charged particle will move before hitting the screen, which is located at a horizontal distance D from the plates, is given by:

$$y_2 = D\tan\alpha = D\frac{qEl}{mv_0^2}$$

Finally, the total vertical distance h that the charged particle will move is:

$$h = y_1 + y_2 = \frac{qEl}{mv_0^2}\left(\frac{l}{2} + D\right)$$

EXAMPLE 59. An electron enters the region of a uniform electric field as shown in Fig.1.141, with $v_0 = 3.00\times 10^6$ m/s and $E = 200$ N/C. The horizontal length of the plates is $l = 0.100$ m.
(A) Find the acceleration of the electron while it is in the electric field.
(B) If the electron enters the field at time $t = 0$, find the time at which it leaves the field.
(C) If the vertical position of the electron as it enters the field is $y_0 = 0$, what is its vertical position when it leaves the field?
SOLUTION (A) The charge on the electron has an absolute value of $q = 1.60 \times 10^{-19}$C, and $m_e = 9.11 \times 10^{-31}$ kg. Therefore,

acceleration of electron-

$$\vec{a} = -\frac{eE}{m_e}\hat{j} = -\frac{(1.60 \times 10^{-19}\text{C})(200 \text{ N/C})}{9.11 \times 10^{-31} \text{ kg}}\hat{j}$$

$$= -3.51 \times 10^{13}\hat{j}\text{m/s}^2$$

(B) The horizontal distance across the field is $l = 0.100$ m. Using Equation (1.88) with $x = l$, we find that the time at which the electron exits the electric field is

$$t = \frac{l}{v_0} = \frac{0.100 \text{ m}}{3.00 \times 10^6 \text{ m/s}} = 3.33 \times 10^{-8} \text{ s}$$

(C) Vertical displacement of electron is given by-

$$y = 0 + \frac{1}{2}at^2 = -\frac{1}{2}\left(3.51 \times 10^{13} \text{ m/s}^2\right)\left(3.33 \times 10^{-8} \text{ s}\right)^2$$

$$= -0.0195 \text{ m} = -1.95 \text{ cm}$$

If the electron enters just below the negative plate in Figure 1.141 and the separation between the plates is less than the value we have just calculated, the electron will strike the positive plate.

EXAMPLE 60. A block of mass m having a charge q is placed on a smooth horizontal table and is connected to a wall through an unstressed spring of spring constant k as shown in Fig.1.142. A horizontal electric field E parallel to the spring is switched on. Find the extension in the spring. Now, if we turn off the field, find the amplitude of the resulting motion of the block.

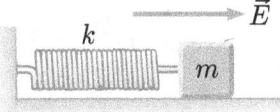

Figure 1.142

SOLUTION Suppose, the initial position of the block is at $x = 0$. At this instant, the spring has it's natural length. Now, if we turn on the horizontal electric field $\vec{E}$ as shown in Fig.1.143, it applies an electric force $F_e = qE$ on the block in the direction of electric field that causes the spring to stretch to an equilibrium length $x = x_0$. In equilibrium, the forces acting on the block are-
1. Gravitational force $m\vec{g}$: acting downwards.

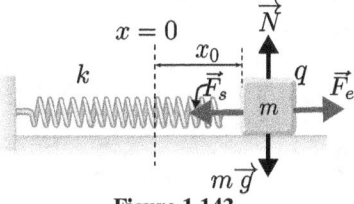

Figure 1.143

2. Normal reaction $\vec{N}$: acting upwards.
3. Electric force $\vec{F_e}$ ($F_e = qE$): acting towards right
4. Spring force $\vec{F_s}$ ($F_s = kx$): acting towards left.
For vertical equilibrium of the block, $N = mg$.
For horizontal equilibrium, we have-

$$F_s = F_e \quad \Rightarrow \quad kx_0 = qE \quad \Rightarrow \quad x_0 = \frac{qE}{k}$$

Now, if we turn-off the electric field, the block begins to oscillate about mean position $x = 0$ with amplitude of $x_0 = qE/k$.

1.23 Electric Dipole in an External Electric Field

1.23.1 Electric force and torque on electric dipole in a uniform external electric field

Fig.1.144 shows an electric dipole AB of length $2l$ in a uniform external electric field $\vec{E}$. The charges at A and B are $-q$ and $+q$ respectively. The direction of dipole moment vector $\vec{p}$, is from $-q$

to +q i.e., from point A to B. The angle between dipole moment vector $\vec{p}$ (which is along $\overrightarrow{AB}$) and electric field vector $\vec{E}$ is θ. The electric forces on $-q$ and $+q$ have same magnitude ($=qE$)

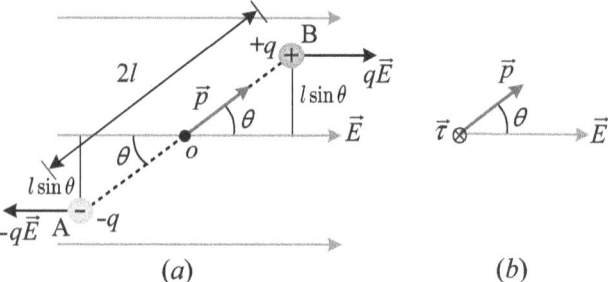

(a)   (b)

Figure 1.144: (a) An electric dipole has an electric dipole moment $\vec{p}$ in an external uniform electric field $\vec{E}$. The angle between $\vec{p}$ and $\vec{E}$ is θ. The line connecting the two charges represents their rigid connection and their centre of mass is assumed to be midway between them. (b) Representing the electric dipole by a vector $\vec{p}$ in the external electric field $\vec{E}$ and showing the direction of the torque $\vec{\tau}$ into the page by the symbol ⊗

but their directions are opposite to each other (Fig.1.144a). The electric force on $-q$ is opposite to $\vec{E}$ and on $+q$, it is parallel to $\vec{E}$.

The net force on the dipole is $F_{net} = -qE + qE = 0$ Thus, net force on an electric dipole in a uniform external electric field is zero, so the dipole will have no translational acceleration.

Since, the lines of action of these forces, are not same, so they produce torque about the centre of mass of the dipole. This torque tends to rotate the dipole to bring $\vec{p}$ into alignment with $\vec{E}$.

If $\vec{\tau}_1$ and $\vec{\tau}_2$ are the respective torques produced by forces on $+q$ and $-q$ about the centre of mass of the dipole, then from Fig.1.144, the magnitude of net torque on the dipole-

$$\tau = \tau_1 + \tau_2 = qE(l\sin\theta) + qE(l\sin\theta)$$
$$= 2qlE\sin\theta = pE\sin\theta$$
$$\implies \tau = pE\sin\theta \tag{1.94}$$

In vector form, torque on dipole due to external uniform electric field $\vec{E}$ can be written as-

$$\vec{\tau} = \vec{p} \times \vec{E} \tag{1.95}$$

which is consistent with the directional relationships for the cross product, as shown by the three vectors in 1.144b

Unit: Unit of torque is "newton-meter".

Maximum and Minimum Values of Torque:

(i) Maximum torque applied by electric field on the dipole is given by-

$$\tau_{max} = pE(\sin\theta)_{max}$$

Since, $(\sin\theta)_{max} = 1$ and corresponding to it $\theta = \pi/2$, therefore-

$$\tau_{max} = pE$$

(ii) Minimum torque applied by electric field on the dipole is given by-

$$\tau_{max} = pE|(\sin\theta)_{min}|$$

Since, $|(\sin\theta)_{min}| = 0$ and corresponding to it $\theta = 0$, therefore-

$$\tau_{min} = 0$$

The torque is greatest when $\vec{p}$ is perpendicular to $\vec{E}$, zero when $\vec{p}$ is aligned with or opposite to $\vec{E}$.

EXAMPLE 61. Two charges each of magnitude 0.01 C and separated by a distance of 0.4 mm constitute an electric dipole. If the dipole is placed in a uniform electric field $|\vec{E}|$ of 10 dyne/C making $30°$ angle with $\vec{E}$, find the magnitude of torque acting on it.

APPROACH To determine the magnitude of torque of electric field on the dipole, apply Eq.1.94:

$$\tau = pE\sin\theta \tag{1}$$

here, $p = q(2l)$ (see Eq.1.38) is the dipole moment.

SOLUTION Given: $q = 0.01$ C, $2l = 0.4$ mm $= 4.0 \times 10^{-4}$ m and $E = 10$ dyne/C $= 10^{-4}$ N/C

Dipole moment of dipole,

$$p = q(2l) = 0.01 \times 4.0 \times 10^{-4} = 4.0 \times 10^{-6} \text{ Cm}$$

Substituting given values and the value of dipole moment in Eq.(1), we get

$$\tau = pE\sin\theta = 4 \times 10^{-6} \times 10 \times 10^{-5} \times \sin 30°$$
$$= 4 \times 10^{-6-5+1} \times \frac{1}{2} = 2.0 \times 10^{-10} \text{ Nm}$$

1.23.2 The Angular Acceleration and Time period of a Dipole in an External Uniform Electric Field

Since, the torque applied by electric field on the dipole is restoring, therefore we can write equation (1.94) as follows-

$$\tau = -pE\sin\theta$$

If I is the moment of inertia of dipole about an axis passing through its centre and perpendicular to the plane of the Fig.1.144a, then torque can also be written as $\tau = I\frac{d^2\theta}{dt^2}$, here $d^2\theta/dt^2$ is the angular acceleration of the dipole. For small angular displacement, $\sin\theta \approx \theta$, therefore, we can write above equation as-

$$I\frac{d^2\theta}{dt^2} = -pE\theta$$
$$\implies \frac{d^2\theta}{dt^2} = -\frac{pE}{I}\theta \tag{1.96}$$

which is of the form of angular SHM: $\frac{d^2\theta}{dt^2} = -\omega^2\theta$, with $\omega = \sqrt{pE/I}$

Thus, if in uniform electric field the dipole is rotated by a small angular displacement θ and released, the electric dipole performs angular SHM with time period-

$$\boxed{T = \frac{2\pi}{\omega} = 2\pi\sqrt{\frac{I}{pE}}} \tag{1.97}$$

1.23.3 A Dipole in a Non-uniform Field

Suppose that a dipole is placed in a non-uniform electric field in which the electric field is changing along x-axis. Let E and $E + dE$ are magnitudes of electric fields at positions A and B respectively. The first response of the dipole is to rotate until it is aligned with the field, with the dipole's positive end pointing in the same direction as the field. Now, however, there is a slight difference between the forces acting on the two ends of the dipole. This difference occurs because the electric field is not same at positions A, and B. The net force exerted by external non-uniform electric field on the dipole is given by-

$$F = \underbrace{q(E+dE)}_{\text{Electric force on charge }+q\text{ at B}} - \underbrace{qE}_{\text{Electric force on charge }-q\text{ at A}}$$

$$\implies F = q\frac{dE}{dx}dx = qdx\frac{dE}{dx} \tag{1.98}$$

From Fig.1.145, $dx = AC = 2l\cos\theta$, therefore, on substituting this value of dx in above equation, we get-

$$\boxed{F = q(2l\cos\theta)\frac{dE}{dx} = p\cos\theta\frac{dE}{dx}} \tag{1.99}$$

Eq.(1.99) gives the electric force on a dipole inclined at an

1.23. ELECTRIC DIPOLE IN AN EXTERNAL ELECTRIC FIELD

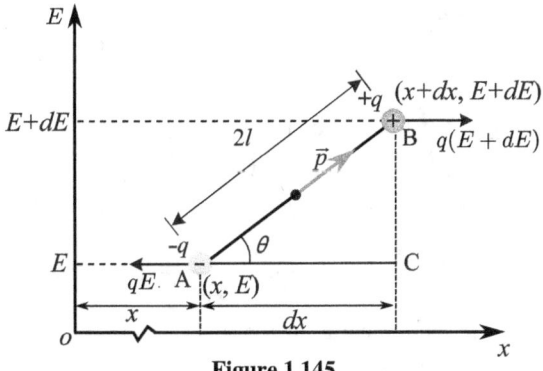

Figure 1.145

angle θ with the direction of non uniform electric field $\vec{E}$.

When the dipole is aligned in the direction of field, $\theta = 0$, therefore, $\cos\theta = 1$. In this case, Eq.1.98 becomes-

$$\boxed{F = (2ql)\frac{dE}{dx} = p\frac{dE}{dx}} \quad (1.100)$$

So, once the dipole is aligned, the rightward force on its positive end is slightly stronger than the leftward force on its positive end. This causes a net force towards the stronger region of the field. For any non-uniform electric field (except at maximum or minimum field points*), the net force on a dipole is toward the direction of the strongest field. Because any finite-size charged object, such as a charged rod or a charged disc, has a field strength that increases as you get closer to the object, we can conclude that a dipole will experience a net force toward any charged object.

EXAMPLE 62. A dipole consists of charges $+2\,\mu C$ and $-2\,\mu C$ separated by a distance of $4\,cm$. It is placed in a non-uniform electric field given by

$$E(x) = 200\,x^2 \text{ N/C},$$

where x is in metres. The axis of the dipole makes an angle of $60°$ with the x-axis.

Calculate the magnitude of the net force acting on the dipole when its centre is at $x = 0.50\,m$.

SOLUTION Given: $q = 2 \times 10^{-6}$ C, $2l = 0.04$ m, $\theta = 60°$

The electric field is

$$E(x) = 200\,x^2 \quad \Rightarrow \quad \frac{dE}{dx} = 400x.$$

At $x = 0.50$ m:

$$\frac{dE}{dx} = 400 \times 0.50 = 200 \text{ N/C·m}.$$

The force on the dipole is

$$F = q\,(2l\cos\theta)\frac{dE}{dx}.$$

Substituting the value in above expression, we get

$$F = (2 \times 10^{-6})\,(0.04 \times \cos 60°)\,(200)$$
$$= (2 \times 10^{-6})(0.04 \times 0.5)(200) = 8.0 \times 10^{-6} \text{ N}$$
$$\Rightarrow F = 8.0 \times 10^{-6} \text{ N} = 8.0\,\mu N$$

EXAMPLE 63. A wheel having mass m, charge $+q$ and $-q$ on diametrically opposite points, is in equilibrium on a rough inclined plane in the presence of uniform vertical electric field $\vec{E}$ as shown in Fig.1.146. Find the magnitude of $\vec{E}$.

APPROACH A wheel can be considered as a rigid body. The system containing wheel, attached with charges $+q$ and $-q$ at it's

---
*For maximum or minimum value of electric field, $\frac{dE}{dx} = 0$, therefore from Eq.1.100, $F = 0$.
So, if the dipole is kept at the point where the electric field is at its maximum or minimum, then the force on it will be zero.

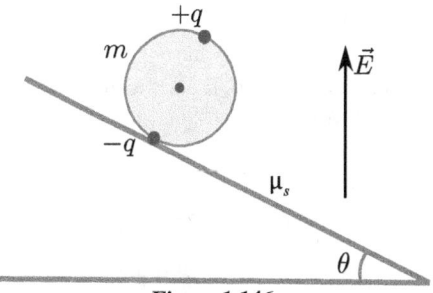

Figure 1.146

diametrically opposite ends, is shown in Fig.1.146.
The forces acting on the wheel are-

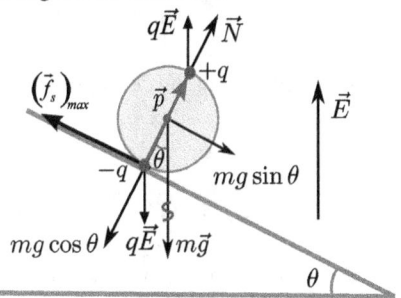

Figure 1.147

1. Gravitational force $m\vec{g}$: acting in downward direction
2. Normal reaction $\vec{N}$: acting perpendicular to the inclined plane
3. Force of static friction $\vec{f_s}$: acting along the plane in upward sense.
4. Electrostatic force of upward electric field $\vec{E}$: On charge $-q$, it is $-q\vec{E}$ and directed opposite to $\vec{E}$ i.e., in downward direction and on charge $+q$, it is $+q\vec{E}$ in the direction of $\vec{E}$, i.e., in upward direction (Fig.1.147). These two forces are equal in magnitude (qE) but opposite in direction. So, the net electrostatic force on the wheel due to dipole attached to it, is zero. However, the lines of action of these electrostatic forces are not passing through the centre of mass (CM) of the wheel, so it will definitely provide a mechanical torque about CM.

Conditions for Static Equilibrium for the Wheel-Dipole System

Let positive direction of x axis is along the inclined plane in downward direction and positive direction of y axis perpendicular to it in upward sense. In static equilibrium:
1. Conditions for Translational Equilibrium: For translational equilibrium along x and y direction-
$$\Sigma F_x = 0,\; \Sigma F_y = 0$$
2. Condition for Rotational Equilibrium: In rotational equilibrium, the net torque on the dipole - wheel system about centre of mass (CM),
$$\Sigma \tau_{CM} = 0$$
Now, apply these conditions and simplify for minimum required value of electric field.

SOLUTION If μ is the coefficient of static friction between the wheel and the inclined plane, then for translational equilibrium of the wheel, we have-
1. $\Sigma F_y = 0 \quad \Rightarrow \quad N - mg\cos\theta = 0$
$$\Rightarrow N = mg\cos\theta \quad (i)$$
2. $\Sigma F_x = 0 \quad \Rightarrow \quad (f_s)_{max} - mg\sin\theta = 0$
$$\Rightarrow (f_s)_{max} = mg\sin\theta \quad (ii)$$
In limiting translational equilibrium, we have-
$$(f_s)_{max} = \mu_s N$$
Substituting this value of $(f_s)_{max}$ in Eq.(ii), we get-
$$\mu_s N = mg\sin\theta \quad (iii)$$

From Eq.(i) and (iii), we get-
$$\mu_s = \tan\theta \quad \text{(iv)}$$
And for rotational equilibrium, the net torque on the wheel about it's centre of mass should be zero, i.e.,
$$\Sigma\tau = \tau_{\text{electric field}} + \tau_{mg} + \tau_{\text{friction}} = 0 \quad \text{(v)}$$
Considering, perpendicularly outward to the plane of figure as a positive direction of torque, the torque of electric field about the centre of mass of the wheel
$$\tau_{\text{electric field}} = pE\sin\theta = +2qrE\sin\theta \quad \text{(vi)}$$
The direction of this torque is normally outward to the plane of the figure.
Torque of weight mg about the centre of mass of the wheel is zero, because it passes through the centre of mass of the wheel.
$$\tau_{mg} = 0 \quad \text{(vii)}$$
Torque of force of friction about the centre of mass of the wheel-
$$\tau_{\text{friction}} = -r \times \mu_s N \quad \text{(viii)}$$
The direction of this torque is perpendicularly inward to the plane of the Fig.1.147
Substituting the value of $\mu_s N$, from Eq. (iii) in Eq.(viii), we get-
$$\tau_{\text{friction}} = -rmg\sin\theta \quad \text{(ix)}$$
Using, Eq.(vi), (vii) and (ix) in Eq. (v), we get -
$$2qrE\sin\theta - rmg\sin\theta = 0$$
$$\implies E = \frac{mg}{2q} \quad \text{(x)}$$
This is the required electric field needed for equilibrium of the wheel.

EXAMPLE 64. A system consists of a thin uniformly charged wire ring of radius R and a very long uniformly charged thin rod oriented along the axis of the ring, with one of its ends coinciding with the centre of the ring. The total charge of the ring is equal to Q. The linear charge density of the rod is λ. Find the interaction force between the ring and the rod.

Note: To determine the total electric force between the rod and ring, we can use following two approaches-

APPROACH 1. Fig.1.148 shows a thin charged wire ring of radius R carries a total charge Q uniformly distributed around it. A very long uniformly charged wire having uniform charge density λ is placed along it's axis in such a way that it's one end coincides with the centre of the ring.

Now, consider an elementary length dx of the rod at distance

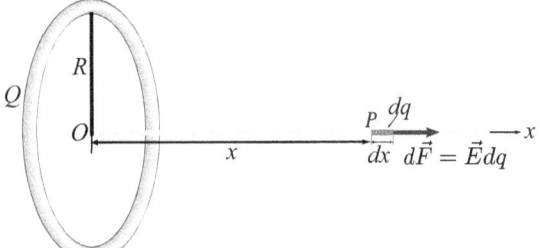

Figure 1.148

x from the centre of the ring and find the electric field E of the ring at the position of the elementary charge of rod. If the elementary charge of the rod is $dq(=\lambda dx)$ then electric force on this elementary charge due to electric field of ring is-
$$dF = Edq \quad \implies \quad dF = E\lambda dx$$
The integration of this expression from $x = 0$ to $x = \infty$, will give the net force F on the rod exerted by charged ring, i.e.,
$$F = \int_0^\infty E\lambda dx \quad (1.101)$$

SOLUTION The electric field at an axial position of charged ring is given by-
$$E = \frac{kQx}{(R^2+x^2)^{3/2}} \quad (1.102)$$
Substituting this value in Eq.(1.101), we get-
$$F = \int_0^\infty \left(\frac{kQ\lambda x}{(R^2+x^2)^{3/2}}\right)\lambda dx$$
or $$F = kQ\lambda \int_0^\infty \left(\frac{x}{(R^2+x^2)^{3/2}}\right)\lambda dx \quad (1.103)$$
To solve right hand integration, we let-
$$R^2 + x^2 = t \quad \implies \quad xdx = \frac{1}{2}dt$$
when $x \to 0$, then $t \to R^2$
and when $x \to \infty$, then $t \to \infty$
Now, substituting these values in Eq.(1.103), we get
$$F = \frac{kQ\lambda}{2}\int_{R^2}^\infty \frac{dt}{t^{3/2}}$$
$$= \frac{kQ\lambda}{2}\int_{R^2}^\infty t^{-3/2}dt = kQ\lambda\left[-\frac{1}{\sqrt{t}}\right]_{R^2}^\infty = \frac{kQ\lambda}{R}$$
So, the force on the rod is-
$$\boxed{F = \frac{kQ\lambda}{R}} \quad (1.104)$$
By, Newton's third law, a force of equal in magnitude but opposite in direction, will act on the ring.

APPROACH 2. In second approach, we first calculate the electric field of the rod at any point on the circumference of ring. Since the ring is symmetrical about the rod, therefore, the magnitude of electric field of the rod will be equal for each position on the circumference of the ring.

Now, calculate the force on an elementary segment of ring due to this field. Break it in component form and use symmetrically opposite segment to know which component will get cancelled and which will sum to provide the net force on the ring. Now integration of this component for the total circumference gives us the net force on ring due to charged rod.

SOLUTION Electric field of the charged thin rod, at any point

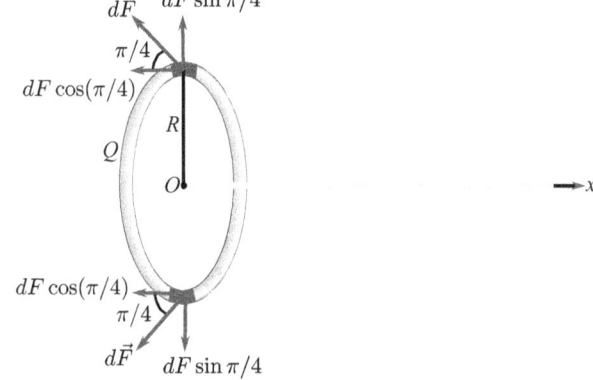

Figure 1.149

on the circumference of the ring is given by -
$$E_{\text{rod}} = \frac{k\lambda\sqrt{2}}{R}$$
This field is directed at angle $\pi/4$ from the negative direction of x-axis (Fig.1.149)
Now consider an elementary part of ring having charge dQ. The

electric force on this part of the ring is given by-
$$dF = E_{\text{rod}} dQ = \frac{k\lambda\sqrt{2}}{R} dQ$$
This force has two components-
1. $dF_x = dF\cos(\pi/4)$: acting along the negative direction of the axis of the rod.
2. $dF_y = dF\sin(\pi/4)$: acting perpendicular to the axis of the ring, i.e., acting along the outward radius vector of the ring.

Now, Consider two segments at the top and bottom of the ring: The perpendicular force component on the top segment $dF\sin(\pi/4)$ get cancelled by equal and opposite perpendicular force component of bottom segment. Hence the total perpendicular component of force due to this pair of segments is zero. When we add up the components of all such pairs of segments, the total force $\vec{F}$ will have only a component along the negative direction of ring's symmetry axis (the $-x$-axis), with no component perpendicular to that axis (that is, no component in yz-plane).

So, net force on the ring is given by-
$$F = \int dF_x = \int dF\sin\frac{\pi}{4} = \int_0^Q \frac{k\lambda\sqrt{2}}{R} dQ \sin\frac{\pi}{4} = \int_0^Q \frac{k\lambda}{R} dQ$$
or
$$\boxed{F = \frac{kQ\lambda}{R}} \qquad (1.105)$$

As expected, Eq.1.104 and 1.105 are same. If the charge density on the ring is also same as that of rod, then-
$$Q = \lambda(2\pi R)$$
Substituting this value and $k = 1/4\pi\epsilon_0$ in Eq.1.104 or 1.105 gives us the result-
$$\boxed{F = \frac{\lambda^2}{2\epsilon_0}} \qquad (1.106)$$

1.24 Earnshaw's Theorem

According to this theorem, *A charged particle cannot be held in a stable equilibrium by electrostatic forces alone*. As an example, imagine three negative charges at the corners of an equilateral triangle in a horizontal plane. Now, put a positive charge at the centre of the triangle and ignore gravity for the moment (although including it would not change the results). The net electrostatic force on the positive charge is zero at the center. So the charged particle is in equilibrium. Now, suppose we displace it from its equilibrium position and observe the direction of net force at new position, we find that it is not restoring, i.e., not towards the centre of the triangle, so the equilibrium will be unstable.

This theorem also holds for a complicated arrangement of charges held together in fixed relative positions—with rods. For example, consider two equal charges fixed on a rod. This combination too cannot be in a stable equilibrium in some electrostatic field, because, the total force on the rod cannot be restoring for displacements in every direction.

The extension of the argument shows that no rigid combination of any number of charges can have a position of stable equilibrium in an electrostatic field in free space.

If want to make equilibrium stable, we have to use pivots or other mechanical constraints. For example, consider a hollow tube in which a charge can move back and forth freely, but not sideways. Now, produce inward electric fields at both the ends by fixing two identical positive charges at the ends of tube. Near the centre of the tube, electric fields also directed towards the sidewall of the tube[Fig.1.150]. This time, the equilibrium is a stable one. The sidewall restricts its lateral motion by applying non-electrical normal force to it whereas inward electric fields provide restoring force to it. A theoretical proof of the theorem, by applying Gauss's law, is given in 'EXAMPLE 82'

EXAMPLE 65. Earnshaw's theorem says that no particle can

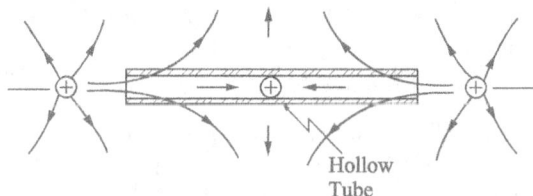

Figure 1.150: A charge can be in equilibrium if there are mechanical constraints.

be in stable equilibrium under the action of electrostatic forces alone. Consider, however, point P at the centre of a square of four equal positive charges, as in Figure1.151. If you put a positive test charge there it might seem to be in stable equilibrium. Every one of the four external charges pushes it toward P. Yet Earnshaw's theorem holds. Can you explain how?

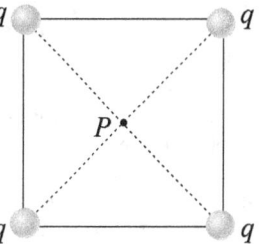

Figure 1.151

SOLUTION The equilibrium of positive test charge at the point P is unstable. Because, on shifting it from mean position (including along diagonal) there is no net restoring force on it towards the position P. So, the equilibrium is unstable. For a stable equilibrium, the total force on the test charge should always be restoring for displacements in every direction.

EXAMPLE 66. Earnshaw's theorem states that a point charge cannot be in stable equilibrium while purely electrostatic forces act on the point charge.

Consider a ring that is uniformly positively charged, with a positive charge at the centre. It appears that the centre charge suffers an identical repulsive force from every direction. How can the theorem be true?

SOLUTION If we shift the positive charge along the symmetric axis of the ring, there will be a repulsion of ring charge on it. As a result of it, the charge moves away from the center. So, the equilibrium at the centre of the ring is unstable. For a stable equilibrium, there must be a restoring force on the charge.

EXAMPLE 67. If a dielectric in the form of a sphere is introduced into a homogeneous electric field. A, B and C are three points as shown in Fig.1.152, then find the points where the intensity of electric field increases and where it decreases.

ANSWER The intensity of electric field at points A, C will in-

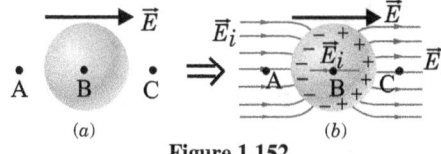

Figure 1.152

crease and at B it will decrease.

Reason: The dielectric polarises when placed in an external electric field. $-ve$ charge produces towards A and an equal positive charge towards B (Fig.1.152). These induced charges produce their own electric fields. In Fig.**??**b induced fields are represented by $\vec{E}_i$. Clearly at points A the induced field $\vec{E}_i$ due to negative charge is towards the sphere which enhances $\vec{E}$. Similarly, the induced field $\vec{E}_i$ at point C due to positive induced charge is away from the sphere which also enhances external field $\vec{E}$. So, at A and C, the net electric field will be greater than E. At point B, the induced electric field is from right to left which opposes the external electric field, so net electric field at B will be less than E.

EXAMPLE 68. Show that the charges of each sign induced on a conductor B, when it is approached by a point charge charge A of magnitude $+q$ (Fig.1.153), is always less than q.

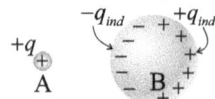

Figure 1.153

SOLUTION Since the sum of the charges in the system as a whole is not zero, and the conductor B does not enclose the charges $+q$, part of the lines of forces from $+q$ must go to infinity (or terminate on other conductors), and only a fraction of the lines of force from $+q$ terminate in the induced charge of B. The induced charge (q_{ind}) is thus less than q.

EXAMPLE 69. If a positive point charge is brought in an electric field, how the field near the point charge get effected.

Answer Suppose the electric field's direction is to the right at a given observation point, and a positive point charge is placed to the left of this point. The additional electric field produced by the given point charge at the observation point is directed away from the positive point charge, that is, towards the right. In this case, both electric fields are directed in the same direction at the observation point, and hence the net electric field increases.

Now, let's consider a scenario where the positive point charge is placed to the right of the observation point. In this case, at the observation point, the electric field produced by the positive point charge will be directed to the left, i.e., opposite to the given electric field. As a result, the field at the observation point decreases.

1.25 Check Point 6

Multiple Choice Questions

1. •A charged drop of mass m floats in air in electric field E. The magnitude of charge on the drop is-
   (A) mg/E (B) E/mg
   (C) m/Eg (D) none of these

2. ••A point charge q moves from point P to S along the path PQRS in a uniform electric field $\vec{E}$ pointing parallel to the positive direction of the x-axis. The coordinates of the point P, Q, R and S are $(a, b, 0)$, $(2a, 0, 0)$, $(a, -b, 0)$ and $(0, 0, 0)$ respectively. The work done by the field in the above process is given by the expression-
   (A) qaE (B) $-qaE$
   (C) $q\sqrt{(a^2+b^2)}E$ (D) $3qE\sqrt{(a^2+b^2)}$

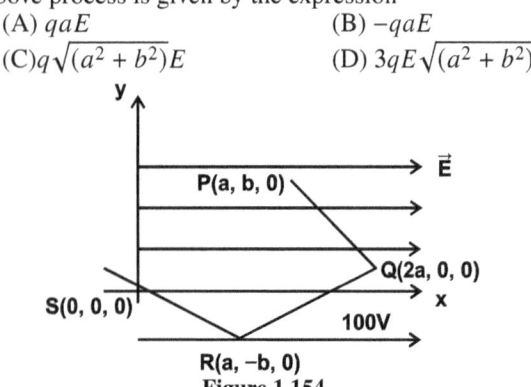

Figure 1.154

3. ••A simple pendulum consists of a small sphere of mass m suspended by a thread of length l. The sphere carries a positive charge q. The pendulum is placed in a uniform electric field of strength E directed vertically upwards. With what period will the pendulum oscillate if the electrostatic force acting on the sphere is less that the gravitational force? Assume the oscillations to be small
   (A) $T = 2\pi \left(\frac{l}{g}\right)^{1/2}$ (B) $T = 2\pi \left(\frac{ml}{qE}\right)^{1/2}$
   (C) $T = 2\pi \left[\frac{l}{\left(g-\frac{qE}{m}\right)}\right]^{1/2}$ (D) $T = 2\pi \left[\frac{l}{\left(g+\frac{qE}{m}\right)}\right]^{1/2}$

4. •••A simple pendulum of mass m and length l carries a charge q. It's time period when it is suspended in a uniform electric field region as shown in Fig.1.155, is-
   (A) $4\pi \sqrt{\frac{l}{\sqrt{g^2+(Eq/m)^2}}}$ (B) $2\pi \sqrt{\frac{l}{\sqrt{g^2+(Eq/m)^2}}}$
   (C) $2\pi^2 \sqrt{\frac{l}{\sqrt{g^2+(Eq/m)^2}}}$ (D) $2\pi \frac{l^2}{\sqrt{g^2+(Eq/m)^2}}$

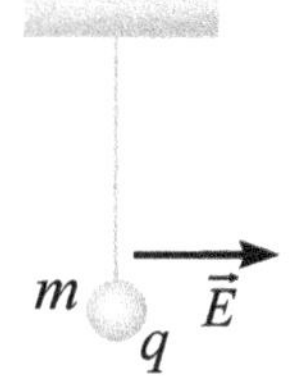

Figure 1.155

5. ••An electric dipole, made up of a positive and a negative charge, each of magnitude 1 μC and placed at a distance 2 cm apart, is placed in an electric field 10^5 N/C. Compute the maximum torque which the field can exert on the dipole.
   (A) 6×10^{-3} N.m (B) 3×10^{-3} N.m
   (C) 4×10^{-3} N.m (D) 2×10^{-3} N.m

6. ••An electric dipole of dipole moment p is placed in a uniform electric field E in stable equilibrium position. Its moment of inertia about the centroidal axis is I. If it is rotated slightly from its mean position find the period of small oscillation.
   (A) $T = 2\pi \sqrt{\frac{PE}{I}}$ (B) $T = 2\pi \sqrt{\frac{I}{PE}}$
   (C) $T = 4\pi \sqrt{\frac{I}{PE}}$ (D) $T = 4\pi^2 \sqrt{\frac{I}{PE}}$

1.26 Conductors in Electrostatic Equilibrium

A good electric conductor like copper, although electrically neutral, contains charges (electrons) that aren't bound to any atom and are free to move about within the material. When no net motion of charge occurs within a conductor, the conductor is said to be in **electrostatic equilibrium**. An isolated conductor (one that is insulated from ground) has the following properties:

1. *In electrostatic equilibrium, the electric field is zero everywhere inside the conducting material. In terms of field lines, we can say that, there are no field lines within the conducting material*[Fig.1.156].

   Reason: If the electric field within a conducting material is nonzero, it exerts a force on each of the mobile charges (usually electrons) and makes them move preferentially in a certain direction. With mobile charge in motion, the conductor cannot be in electrostatic equilibrium.

2. *Any excess charge on an isolated conductor resides entirely on its surface.*

   Reason: This Property is a direct result of the $1/r^2$ repulsion between like charges described by Coulomb's law. If by some means an excess of charge is placed inside a conductor, the

1.27. LIMITATIONS OF COULOMB'S LAW FOR A RELATIVISTICALLY MOVING CHARGE

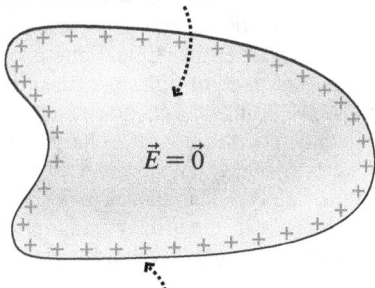

The electric field inside the conductor is zero.

All excess charge is on the surface.

Figure 1.156: The electric field inside e a charged conductor.

repulsive forces between the like charges push them as far apart as possible, causing them to quickly migrate to the surface. So, the electric field is zero within the conducting material, but is not necessarily zero outside.

3. *The electric field just outside a charged conductor is perpendicular to the conductor's surface (Fig.1.157). In terms of field lines, we can say that- the field lines that start or stop on the surface of a conductor are perpendicular to the surface where they intersect it.*

Proof: Let us start with a contradictory statement, that elec-

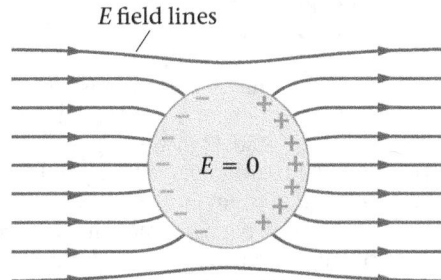

Figure 1.157: Electric field lines meet a conducting surface at right angles and the electric field $\vec{E}$ vanishes inside the conductor

tric field lines are not perpendicular to the conductor surface and the angle between $\vec{E}$ and conductor surface is $\theta \, (\neq \pi/2)$. The component of field along the surface (Fig.1.158)
$$E_x = E \cos \theta$$
The electric force applied by this field on the electronic charge $q(=-e)$ on the surface
$$F_x = qE_x = qE \cos \theta$$
As in electrostatic equilibrium, charges on the surface remains at rest, i.e., no net force acting along the surface. Therefore,
$$F_x = 0 \quad \Rightarrow \quad qE \cos \theta = 0$$
$$\Rightarrow \quad \cos \theta = 0 \quad \Rightarrow \quad \theta = \pi/2$$
which contradicts our initial assumption that $\vec{E}$ is not perpendicular to the surface. Therefore, it is proved that just outside a conductor, the electric field $\vec{E}$ is perpendicular to it's surface.

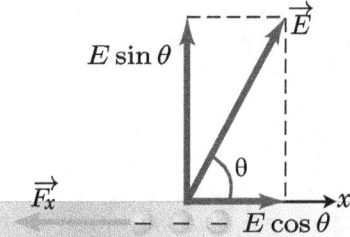

Figure 1.158: This situation is impossible if the conductor is in electrostatic equilibrium. If the electric field $\vec{E}$ had component parallel to the surface, an electric force would be exerted on the charges along the surface and they would move to the left.

4. *On an irregularly shaped conductor, the charge accumulates at sharp points, where the radius of curvature of the surface is smallest. In terms of field lines this can be stated that the field lines are much denser near the more sharply curved surface than the less sharply curved surface.*[Fig.1.159b]

Reason: Think of the charges as being constrained to move

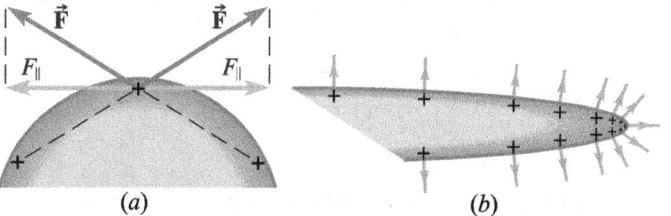

Figure 1.159: (a) Repulsive forces on a charge constrained to move along a curved surface due to two of its neighbors. The parallel components of the forces $(F_{\parallel})$ determine the spacing between the charges.(b) For a conductor in electrostatic equilibrium, the surface charge density is largest where the radius of curvature of the surface is smallest and the electric field just outside the surface is strongest there.

along the surfaces of the conductor. On flat surfaces, repulsive forces between neighboring charges push parallel to the surface, making the charges spread apart evenly. On a curved surface, only the components of the repulsive forces parallel to the surface, $F_{\parallel}$, are effective at making the charges spread apart (Fig.1.159a). If charges were spread evenly over an irregular surface, the parallel components of the repulsive forces would be smaller for charges on the more sharply curved regions (since, at more sharply curved surface, the angle between $\vec{E}$ and $F_{\parallel}$ is larger that that of less sharply curved surface) and charge would tend to move toward these regions.

The electric field lines just outside a conductor are densely packed at sharp points because each line starts or ends on a surface charge. Since the density of field lines reflects the magnitude of the electric field, the electric field outside the conductor is largest near the sharpest points of the conducting surface. The electric field near very sharp points may be strong enough to ionize the air around it.

EXAMPLE 70. Fig.1.160 shows a positively charged metal sphere above a conducting plate with a negative charge. Sketch the electric field lines.

APPROACH Field lines start on positive charges and end on

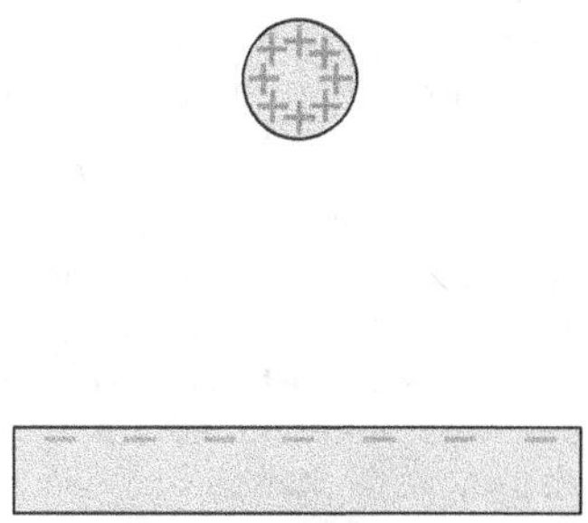

Figure 1.160

negative charges. Thus we draw the field lines from the positive sphere to the negative plate, perpendicular to both surfaces, as shown in Fig.1.161. The single field line that goes upward tells us that there is a field above the sphere, but that it is weak.

1.27 Limitations of Coulomb's Law for a Relativistically Moving Charge

Coulomb's law:
$$\vec{F} = \frac{1}{4\pi\varepsilon_0} \frac{q_1 q_2}{r^2} \hat{r}$$

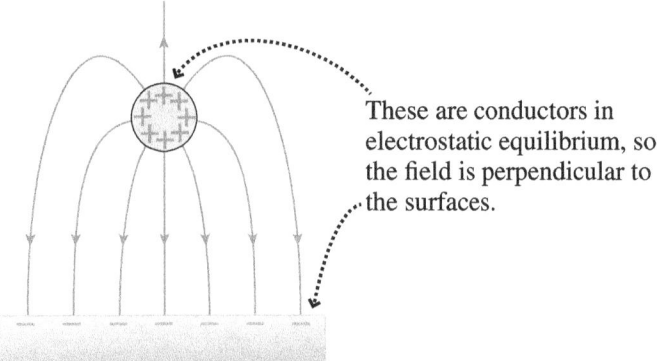

Figure 1.161: Drawing field lines from sphere to plate.

gives the electrostatic force between two stationary charges. In practice, it is also applied when charges move very slowly compared to the speed of light, so that the finite time taken for interaction can be ignored. The law assumes that the force depends only on the instantaneous separation between the charges, as if any change in position were transmitted instantly.

In reality, the force on a charge is determined by the electric field at its location:
$$\vec{F} = q\vec{E}$$
The electric field itself is generated by other charges, but any change in their state does not influence the field everywhere instantly. Instead, information about the change propagates outward at the finite speed of light c.

This fact becomes significant in two distinct cases:

- **Charges at very large separations:** If the charges are at rest and have always been at rest, Coulomb's law still holds, even if they are separated by vast distances (e.g., light-years). The electric field is static, so there is no new information that needs to travel. However, if one charge suddenly moves or changes, the other will continue to experience the field due to its *previous* state for a time interval equal to the signal travel time, r/c. In such cases, the assumption of instantaneous interaction breaks down.
- **Charges moving at relativistic speeds:** Even if the distance is not large, if a charge moves at a speed comparable to the speed of light, it generates not just an electric field but also a magnetic field. The force it exerts on another charge depends not only on their separation but also on velocity, acceleration, and the geometry of motion. Coulomb's law, which accounts only for static electric interactions, is therefore invalid in such scenarios.

Because of these limitations, applying Coulomb's law blindly to moving charges or across astronomical distances leads to incorrect predictions. The correct description requires the concept of *retarded fields* (see subsection **1.27.1** "**Retarded Field**," page "60"), which take into account both the finite speed of propagation and the fact that moving charges influence the field in more complicated ways. However, a detailed treatment of retarded fields belongs to advanced electrodynamics and is **beyond the scope of JEE, NEET, or Olympiad syllabi**. Here, we will only introduce the idea and definition for awareness.

> **Note for Exams**
>
> In JEE, NEET, and Olympiads, Coulomb's law is always applied in the *static or quasi-static limit*. You will not be required to calculate using retarded fields. Only the basic idea and definition are included here for completeness.

1.27.1 Retarded Field

A **retarded field** is the electric (or magnetic) field observed at a point in space, not based on the charge's current position, but on where the charge was at an earlier time. This is because changes in the field travel at the speed of light, not instantly.

Example: Imagine a point charge placed exactly 1 light-year away from you in vacuum. Suppose the charge suddenly changes its position or starts moving. Will you feel the change in the electric force immediately? The answer is **no**. Electromagnetic effects (such as changes in electric or magnetic fields) do not propagate instantaneously–they travel at the finite speed of light, $c = 3 \times 10^8$ m/s.

The distance between you and the charge is:
$$\text{Distance} = 1 \text{ light-year} = c \times 1 \text{ year}$$
So the time taken for any change in the charge's motion or position to affect you is:
$$\text{Time taken} = \frac{\text{Distance}}{\text{Speed}} = \frac{1 \text{ light-year}}{c} = \frac{c \times 1 \text{ year}}{c} = 1 \text{ year}$$
Therefore, for the next 1 year, you will continue to experience the electric field *as if the charge were still at its old position*.

This delayed influence is due to the finite speed at which electromagnetic information travels and is known as the **retarded field**. It reflects the state of the charge (position, velocity, etc.) from 1 year ago—not the present moment.

1.28 Check Point 7

1. ●●A simple pendulum has a bob of mass m which carries a charge q on it. Length of the pendulum is L. There is a uniform electric field E in the region. Calculate the time period of small oscillations for the pendulum about its equilibrium position in following cases:
   (a) E is vertically down having magnitude $E = mg/q$
   (b) E is vertically up having magnitude $E = 2mg/q$
   (c) E is horizontal having magnitude $E = mg/q$
   (d) $\vec{E}$ has magnitude of $E = \sqrt{2}mg/q$ and is directed upward making an angle of 45° with the horizontal.

2. ●●In a region of space an electric field line is in the shape of a semicircle of radius R. Magnitude of the field at all point is E. A particle of mass m having charge q is constrained to move along this field line. The particle is released from rest at A [Fig.1.162].

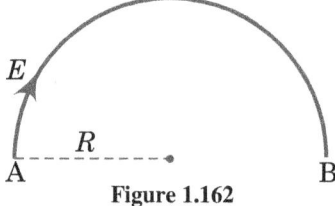

Figure 1.162

   (a) Find its kinetic energy when it reaches point B.
   (b) Find the acceleration of the particle when it is at midpoint of the path from A to B.

3. ●●Electric field $E = -bx + a$ exists in a region parallel to the x direction (a and b are positive constants). A charge particle having charge q and mass m is released from the origin $x = 0$. Find the acceleration of the particle at the instant its speed becomes zero for the first time after release.

4. ●●Two identical blocks resting on a frictionless, horizontal surface are connected by a light spring having a spring constant $k = 100$ N/m and an unstretched length $L_i = 0.400$ m as shown in Figure 1.163a. A charge Q is slowly placed on each block, causing the spring to stretch to an equilibrium length $L = 0.500$ m as shown in Figure 1.163b. Determine

the value of Q, modelling the blocks as charged particles.

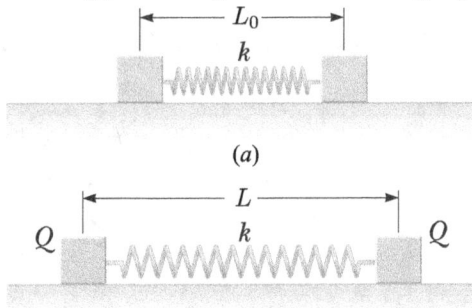

Figure 1.163

Multiple Choice Questions

5. ••In absence of gravity, an electron falls from rest through a vertical distance h in a uniform and vertically upward directed electric field E. The direction of electric field is now reversed, keeping its magnitude the same. A proton is allowed to fall from rest in it through the same vertical distance h. The time of fall of the electron, in comparison to the time of fall of the proton is
   (A) smaller   (B) 5 times greater
   (C) 10 times greater   (D) equal

6. ••A toy car with charge q moves on a frictionless horizontal plane surface under the influence of a uniform electric field $\vec{E}$. Due to the force $q\vec{E}$, its velocity increases from 0 to 6 m s^{-1} in one second duration. At that instant the direction of the field is reversed. The car continues to move for two more seconds under the influence of this field. The average velocity and the average speed of the toy car between 0 to 3 seconds are respectively-
   (A) 2 m s^{-1}, 4 m s^{-1} (B) 1 m s^{-1}, 3 m s^{-1}
   (C) 1 m s^{-1}, 3.5 m s^{-1} (D) 1.5 m s^{-1}, 3 m s^{-1}

7. ••A particle of mass m and charge q is placed at rest in a uniform electric field E and then released. The kinetic energy attained by the particle after moving a distance y is
   (A) qEy (B) qE^2y
   (C) qEy^2 (D) q^2Ey

8. ••A charged particle (mass m and charge q) moves along X axis with velocity v_0. When it passes through the origin it enters a region having uniform electric field $\vec{E} = -E\hat{j}$ which extends upto $x = d$ [Fig.1.164]. Equation of path of electron in the region $x > d$ is:
   (A) $y = \frac{qEd}{m v_0^2} x$ (B) $y = \frac{qEd}{m v_0^2}(x-d)$
   (B) $y = \frac{qEd}{m v_0^2}\left(\frac{d}{2}-x\right)$ (D) $y = \frac{qEd^2}{m v_0^2} x$

9. ••A small point mass carrying some positive charge on it, is released from the edge of a table. There is a uniform electric field in this region in the horizontal direction [Fig.1.165]. Which of the following options then correctly describe the trajectory of the mass [Fig.1.166]? (Curves are drawn schematically and are not to scale)

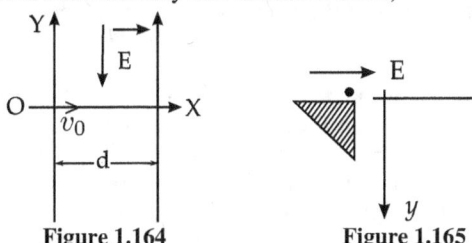

Figure 1.164   Figure 1.165

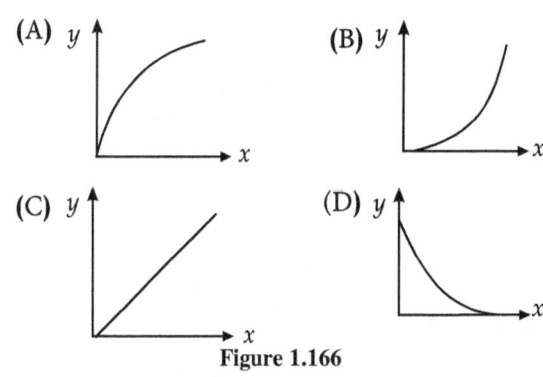

Figure 1.166

10. ••A ring has charge Q and radius R. If a charge q is placed at its centre then the increase in tension in the ring is [Fig.1.167]
    (A) $\frac{Qq}{4\pi\epsilon_0 R^2}$ (B) zero
    (C) $\frac{Qq}{4\pi^2\epsilon_0 R^2}$ (D) $\frac{Qq}{8\pi^2\epsilon_0 R^2}$

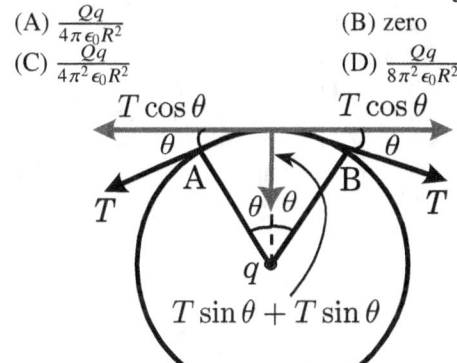

Figure 1.167

11. •A point Charge q is placed inside the cavity of a metallic shell. Which one of the diagram [Fig.1.168], correctly represents the electric lines of force.

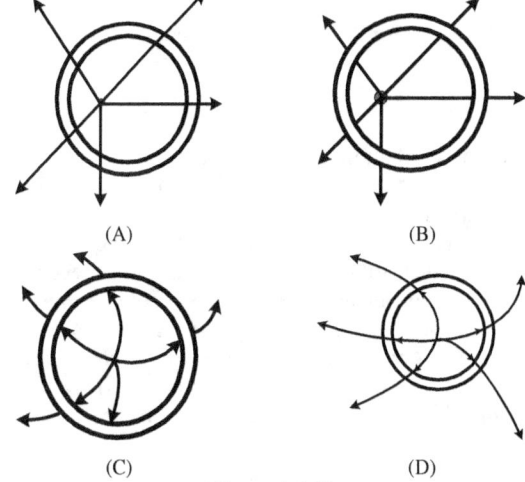

Figure 1.168

1.29 Solid Angle

The solid angle is the extension of the concept of angle from two to three dimension. So let's start with a two dimensional geometry- a circle (Fig.1.169) and pick two rays OA and OC starting from the centre O. They will divide the circumference in two parts ABC and ADC, called arcs. The length of each arc divided by the length of the radius will be the measure of the angle subtended by the arc itself, i.e.,

$$\text{angle} = \frac{\text{arc}}{\text{radius}}$$

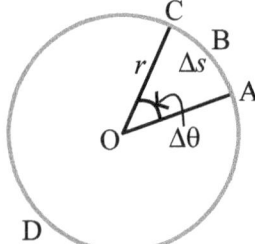

Figure 1.169: Plane angle

From Fig.1.169, we have

$$\Delta\theta = \frac{\text{arcABC}}{OA} = \frac{\Delta s}{r} \quad (1.107)$$

Here, Δs is the arc length and r is the radius of the circle. The SI unit of plane angle is "radian (rad)".

Now, extend this idea to three dimensions. Instead of a circle take a sphere, and instead of picking two rays pick a cone centered in the centre of the sphere [Fig.1.170]. The cone will cross the surface of the sphere: and now to define the *solid angle* measure the area (ΔA) of the surface bounded by the cone, divided by the square of the length of the radius (R^2) (so that we have an area divided by an area).

i.e., solid angle,

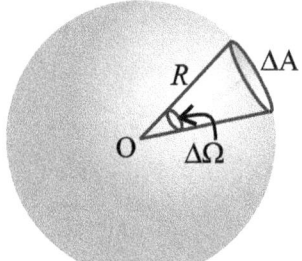

Figure 1.170: A solid angle measured on a sphere.

$$\Delta\Omega = \frac{\Delta A}{R^2} \quad (1.108)$$

The SI unit of solid angle is steradian (sr).

Here, it is important to note that the line joining centre of sphere O to ΔA is normal to ΔA.

Since, plane angles and the solid angles are the ratios of same physical quantities, therefore they are dimensionless quantities. Note that a small surface area as seen from a short distance can cover the same solid angle as a large area as seen from a long distance. For example, in Fig.1.171 different areas A and $A'(A' > A)$, at positions R, R' covers same solid angle at O.

Definition: *The solid angle is a three dimensional angle subtended by an object (two dimensional or three dimensional) at a certain point in the space. It merely depends on the relative distance of the object and its configuration with respect to the given point in the space. Solid angle subtended by a straight line or a point is always zero.*

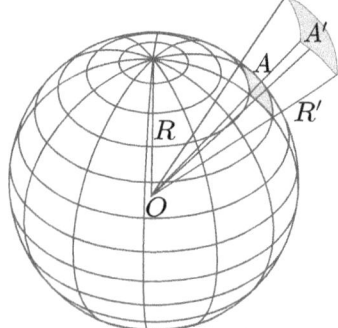

Figure 1.171

In Fig.1.172, we see that the projection of the area element

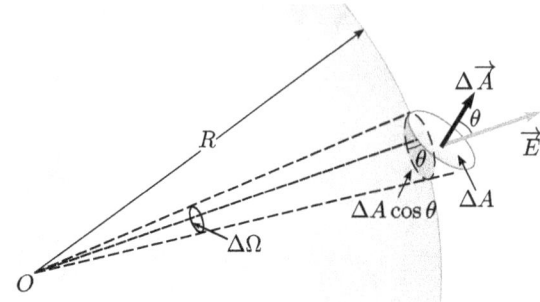

Figure 1.172

perpendicular to the radius vector is $\Delta A \cos\theta$. Thus, the quantity $\Delta A \cos\theta / R^2$ is equal to the solid angle $\Delta\Omega$ that the surface element ΔA subtends at the *origin O*. We also see that $\Delta\Omega$ is equal to the solid angle subtended by the area element of a spherical surface of radius R.

☞ If the line joining O to ΔA makes an angle θ with the normal to ΔA (Fig.1.172), we can write:

$$\Delta\Omega = \frac{\Delta A \cos\theta}{R^2}$$

A complete circle subtends an angle -

$$\theta = \frac{\Delta s}{r} = \frac{2\pi r}{r} = 2\pi \text{ rad}$$

at the centre. In fact, any closed curve subtends an angle 2π at any of the internal points. Similarly, a complete sphere subtends a solid angle-

$$\Omega = \frac{A}{R^2} = \frac{4\pi R^2}{R^2} = 4\pi \text{ sr}$$

at the centre. Also, any closed surface subtends a solid angle 4π sr at any internal point.

So for, we have defined the solid angles at an internal point. Now, we find the plane and solid angles at an external point.

From Fig.1.173, it is clear that on gradually closing the curve,

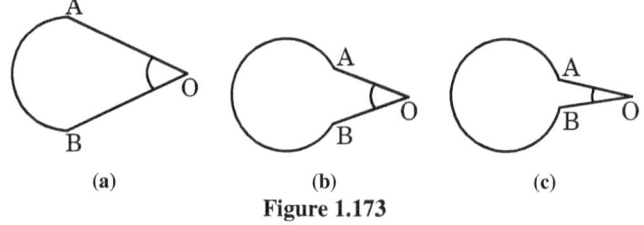

Figure 1.173

the angle subtended by arc AB at an external point O, finally diminishes to zero. So, we can say that a closed curve subtends zero angle at an external point. Similarly, a closed surface also subtends zero solid angle at an external point.

Calculation of Solid Angle at the Centre of a Sphere:

Mathematically, the solid angle can be defined as-

$$\Omega = \frac{A}{R^2}$$

The definition of a solid angle 'Ω,' is analogous to the definition of a plane angle. Just as the arc length s is everywhere perpendicular to the radius r, the area A must be everywhere perpendicular to the radius.

Fig.1.174 shows a sphere of radius R. Let us consider a conical section AOF having vertex at O. Suppose semi vertex angle of this cone is $\angle AOC = \theta$ and radius of its base is r. Now, consider a strip $ABEF$ of radius r on this sphere. If $\angle AOB = d\theta$, then thickness of the strip $AB = Rd\theta$ and radius of strip, $r = R\sin\theta$. Curved area of this strip is given by -

1.30. ELECTRIC FLUX

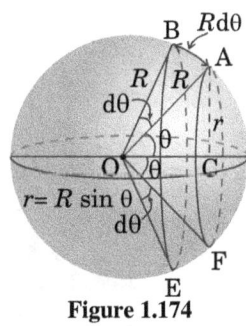

Figure 1.174

$$dA = 2\pi r \times (\text{thickness } AB)$$
$$\because \quad r = R\sin\theta, \text{ therefore -}$$
$$dA = 2\pi R \sin\theta \, R \, d\theta$$
$$\Longrightarrow \quad A = 2\pi R^2 \int_0^\theta \sin\theta \, d\theta \quad \Longrightarrow \quad A = 2\pi R^2 \, (1 - \cos\theta)$$
$$\therefore \quad \Omega = \frac{A}{R^2} = \frac{2\pi R^2 \, (1 - \cos\theta)}{R^2} = 2\pi \, (1 - \cos\theta)$$
$$\Longrightarrow \quad \boxed{\Omega = 2\pi \, (1 - \cos\theta) = 4\pi \sin^2 \frac{\theta}{2}} \quad (1.109)$$

Eq.(1.109) gives the relation between semi vertex plane angle and the solid angle formed at the vertex of cone having its vertex at the centre of the sphere of radius R.
At $\theta = 0°$, $\Omega = 2\pi \, (1 - \cos 0°) = 0$ sr
For a hemispherical surface, at it's centre we have-
$\theta = 90°$, $\Omega = 2\pi \, (1 - \cos 90°) = 2\pi$ sr
For a complete spherical surface, at it's centre, we have-
$\theta = 180°$, $\Omega = 2\pi \, (1 - \cos 180°) = 4\pi$ sr
Not only a spherical surface, all closed surfaces subtend a solid angle of $= 4\pi$ sr at their centres.

1.30 Electric Flux

Latin: flux = "to flow"

Analogy With Flow of Water and Concept of Flux: Imagine holding a ring with inside area A in a stream of water flowing with velocity $\vec{v}$, as shown in Fig.1.175. The area vector, $\vec{A}$, of the ring is defined as a vector with magnitude A pointing in a direction perpendicular to the plane of the ring. Of course, for a ring, disc or other "open" surface that is not part of a three-dimensional volume, there are two possible directions perpendicular to the surface. Choose the direction of $\vec{A}$ so that it makes the smallest possible angle with respect to the velocity vector. In Fig.1.175a, the area vector of the ring is parallel to the flow velocity, and the flow velocity is perpendicular to the plane of the ring. The product Av gives the amount of water passing through the ring per unit time, where v is the magnitude of the flow velocity. If the plane of the ring is tilted with respect to the direction of the flowing water (Fig.1.175b), the amount of water flowing through the ring per second, is given by $Av \cos\theta$, where θ is the angle between the area vector of the ring and the direction of the velocity of the flowing water. The amount of water flowing through the ring is called the flux, $\Phi = Av \cos\theta = \vec{A} \cdot \vec{v}$. Since flux is a measure of volume per unit time, its units are cubic meters per second (m^3/s). For the case of the electric field, we define an analogous quantity and call it *electric flux*. We should however note that there is no *flow* of a physically observable quantity unlike the case of water flow.

Consider a uniform electric field of magnitude E passing through a given area A (Fig.1.176). Again, the area vector is $\vec{A}$, with a direction normal to the surface of the area and a magnitude A. The angle θ is the angle between the vector electric field and the area vector, as shown in Fig.1.176. The electric field passing

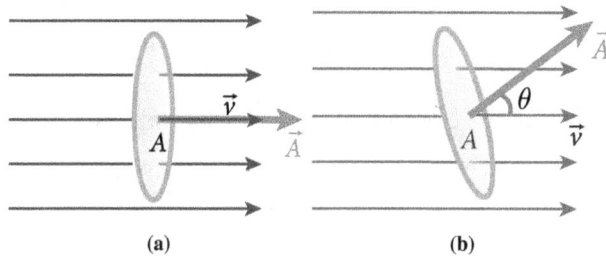

Figure 1.175: Water flowing with velocity of magnitude v through a ring of area A. (a) The area vector is parallel to the flow velocity. (b) The area vector is at an angle θ to the flow velocity.

through a given area A is called the electric flux and is given by-
$$\Phi_E = \vec{E} \cdot \vec{A} = EA \cos\theta \quad (1.110)$$
In simple words, the electric flux is proportional to the number of electric field lines passing through the area.
The flux through a surface of area A has a maximum value EA when the surface is perpendicular to the field (i.e. when $\theta = 0°$), and is zero when the surface is parallel to the field (i.e. when $\theta = 90°$).
When we apply Eq.(1.110), it is often best to sketch $\vec{E}$ and $\vec{A}$ with their tails touching and θ is the angle between the two vectors From Eq.(1.110), we can also say that the electric flux Φ_E is the

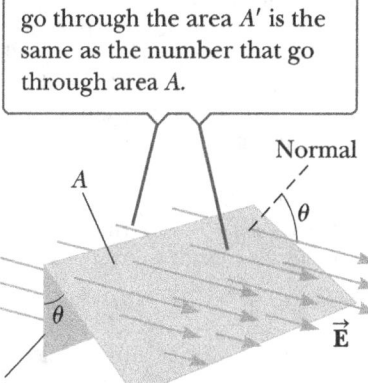

$A' = A\cos\theta$
Figure 1.176: Field lines for a uniform electric field through an area A that is at an angle of $(90° - \theta)$ to the field.

magnitude of the electric field $|\vec{E}|$ times the magnitude of the component of the area vector $\vec{A}'$ that is parallel to the electric field (Fig.1.176):
$$\Phi_E = EA' \quad (1.111)$$
where $A' = A\cos\theta$. The unit of electric flux is $NC^{-1}m^2$.
If you were to shine a light parallel to the electric field, the shadow cast by a surface of area A would have an area equal to the magnitude of $\vec{A'}$ (Fig.1.176).

A More General Expression for Electric Flux

In Eq.1.110 we have assumed that the electric field $\vec{E}$ was constant over the surface. What if a surface is curved and/or the field varies with position? In such a case, we divide the surface into many infinitesimally small pieces, each piece is so small that we can consider it as essentially flat and the field is uniform over it (Fig.1.177). If $d\vec{A}$ is the area vector of one such a small piece and the electric field at this piece is $\vec{E}$, then the flux through one such small piece is $d\Phi_E = \vec{E} \cdot d\vec{A}$, and the flux through the entire surface comes from integrating the fluxes through each small pieces, i.e.,

$$\Phi_E = \int_{\text{surface}} \vec{E} \cdot d\vec{A} \quad (1.112)$$

Equation 1.177 is a surface integral, which means it must be

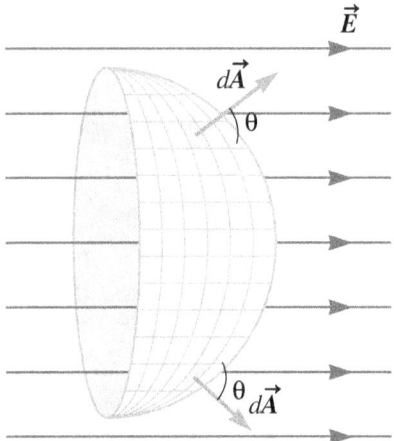

Figure 1.177

evaluated over the surface in question. Generally, the value of Φ_E depends both on the field pattern and on the surface.

1.30.0.1 The Electric Flux Through a Closed Surface

Our final step, to calculate the electric flux through a closed surface such as a box, a cylinder, or a sphere, requires nothing new. We've already learned how to calculate the electric flux through flat and curved surfaces, and a closed surface is nothing more than a surface that happens to be closed.

However, the mathematical notation for the surface integral over a closed surface differs slightly from what we've been using. It is customary to use a little circle on the integral sign to indicate that the surface integral is to be performed over a closed surface. With this notation, the electric flux through a closed surface is

$$\Phi_E = \oint \vec{E} \cdot d\vec{A} = \oint E_n dA \quad (1.113)$$

where E_n represents the component of the electric field normal to the surface.

Note that only the notation has changed. The electric flux is still the summation of the fluxes through a vast number of tiny pieces, that now cover a closed surface.

Direction of Area Vector: A closed surface has a distinct inside and outside. The direction of the area vector $d\vec{A}$ is chosen so that the vector points outward from the surface. If the area element is not part of a closed surface, the direction of the area vector is chosen so that the angle between the area vector and the electric field vector is less than or equal to 90°.

In case of a closed surface, the sign of the flux depends on the angle between $\vec{E}$ and $d\vec{A}$ as follows:

1. If $\theta < 90°$, then $\vec{E}$ crosses the surface from the inside to the outside and hence $d\Phi_E = \vec{E} \cdot d\vec{A}$ is positive.
2. If $\theta = 90°$, then $\vec{E}$ grazes the surface and hence $d\Phi_E = \vec{E} \cdot d\vec{A}$ is zero.
3. If $90° < \theta < 180°$, then $\vec{E}$ crosses the surface from the outside to the inside and hence $d\Phi_E = \vec{E} \cdot d\vec{A}$ is negative.

Consider the closed surface in Fig.1.178. The vectors $d\vec{A}_i$ point in different directions for the various surface elements, but for each element they are normal to the surface and point outward. At the element labeled ①, the field lines are crossing the surface from the inside to the outside and $\theta < 90°$; hence, the flux $d\Phi_1 = \vec{E} \cdot d\vec{A}_1$ through this element is positive. For element ②, the field

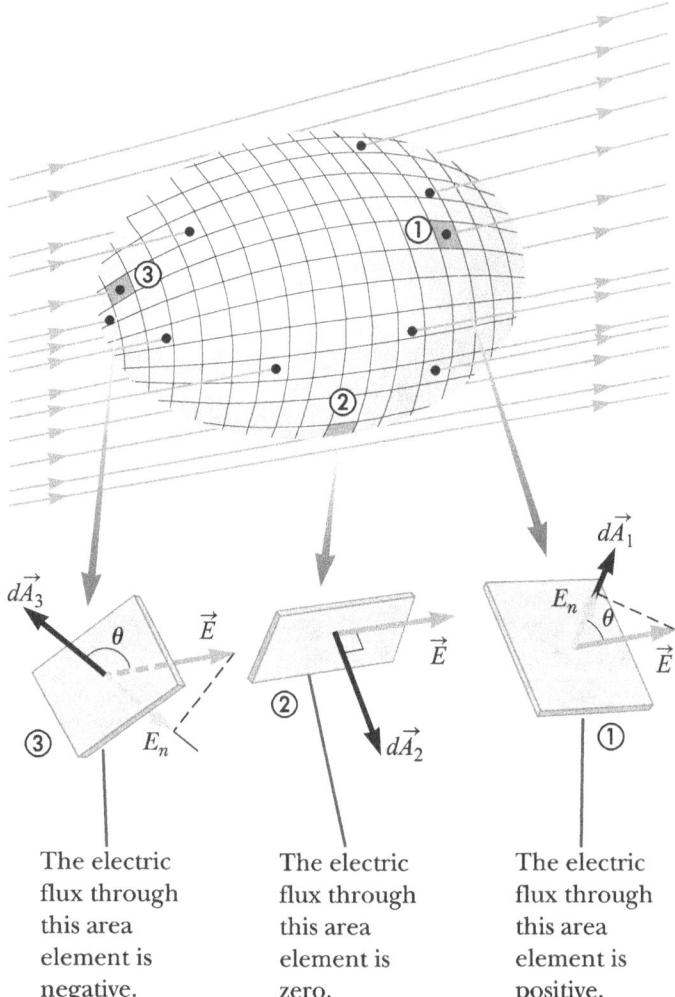

The electric flux through this area element is negative.

The electric flux through this area element is zero.

The electric flux through this area element is positive.

Figure 1.178: A closed surface in an electric field. The area vectors are, by convention, pointed normally outward to the surface.

lines graze the surface (perpendicular to $d\vec{A}_2$); therefore, $\theta = 90°$ and the flux is zero. For elements such as ③, where the field lines are crossing the surface from outside to inside, $90° < \theta < 180°$ and the flux is negative because $\cos\theta$ is negative. The *net* flux through the surface is proportional to the net number of lines leaving the surface, where the net number means the number of lines leaving the surface minus the number of lines entering the surface. If more lines are leaving than entering, the net flux is positive. If more lines are entering than leaving, the net flux is negative.

Now we're ready to calculate the flux through a closed surface.

EXAMPLE 71. Fig.1.179 shows a cube that has edge length l in a uniform electric field, $\vec{E}$, that is directed along the positive x-axis and perpendicular to the plane of one face of the cube. What is the net electric flux passing though the cube?

APPROACH To find the net flux through the cube, determine the flux passing through each surface and add them algebraically.

SOLUTION In Fig.1.179, the electric field lines pass through two faces perpendicularly and are parallel to four other faces of the cube. So, the flux through four of the faces (③, ④, and the unnumbered faces) is zero because $\vec{E}$ is parallel to the four faces and therefore perpendicular to $d\vec{A}$ on these faces.

The net flux through faces ① and ②:

$$\Phi_{12} = \Phi_1 + \Phi_2 = \int_① \vec{E} \cdot d\vec{A} + \int_② \vec{E} \cdot d\vec{A}$$

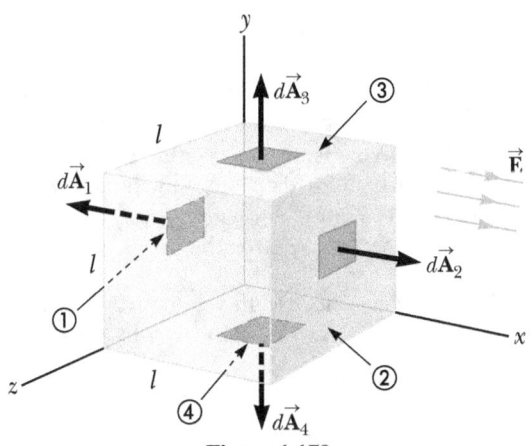

Figure 1.179

For face ①, $\vec{E}$ is constant and directed inward but $d\vec{A}_1$ is directed outward ($\theta = 180°$). So, the flux through surface ①,

$$\Phi_1 = \int_① \vec{E} \cdot d\vec{A} = \int_① E(\cos 180°)\, dA$$

$$= -E\int_① dA = -EA = -El^2 \quad (\because \cos 180° = -1)$$

For face ②, $\vec{E}$ is constant and outward and in the same direction as $d\vec{A}_2$ ($\theta = 0°$). Therefore, flux through this face:

$$\Phi_2 = \int_② \vec{E} \cdot d\vec{A} = \int_② E(\cos 0°)\, dA$$

$$= E\int_② dA = +EA = El^2 \quad (\because \cos 0° = 1)$$

Therefore, net flux through the cube

$$\Phi_E = \Phi_{12} + \Phi_{3456}$$
$$= -El^2 + El^2 + 0 + 0 + 0 + 0 = 0$$

EXAMPLE 72. Expression for an electric field is given by $\vec{E} = 4000x^2\hat{i}$ V/m. Find the electric flux through the cube of side 20 cm when placed in electric field (as shown in the Fig.1.180).

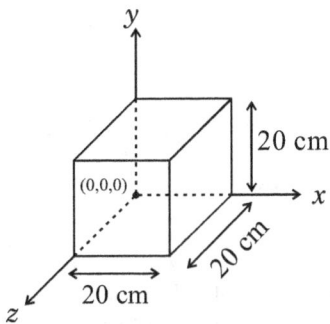

Figure 1.180

SOLUTION Electric flux (ϕ) is defined as $\phi = \vec{E} \cdot \vec{A}$. Clearly, flux will be zero if area vector is perpendicular to $\vec{E}$.

It is given that $\vec{E} = 4000x^2\hat{i}$ V/m. It is directed along positive direction of X-axis. So, electric flux is zero corresponding to planes $y = 0$, $y = 0.2$, $z = 0$ and $z = 0.2$ m.

Now, electric flux corresponding to plane $x = 0$

$$\phi_1 = \vec{E} \cdot \vec{A} = \left[4000x^2\hat{i}\ \text{V/m} \cdot (0.20\ \text{m})^2(-\hat{i})\right]_{x=0} = 0$$

Electric flux corresponding to plane $x = 0.20$ m$\hat{i}$

$$\phi_1 = \vec{E} \cdot \vec{A} = \left[4000x^2\hat{i}\ \text{V/m} \cdot (0.20\ \text{m})^2\hat{i}\right]_{x=0.20}$$
$$= 4000 \times 16 \times 10^{-4}\ \text{Vm} = 640\ \text{Vcm}$$

Therefore, net electric flux linked to the given cube:

$$\phi = \phi_1 + \phi_2 + \underbrace{\phi_3}_{\text{flux passing through remaining four surfaces}}$$

$$= 0 + 640\ \text{V}\cdot\text{cm} + 0 = 640\ \text{V}\cdot\text{cm}$$

1.31 Check Point 8

1. •A person is placed in a large, hollow, metallic sphere that is insulated from ground. If a large charge is placed on the sphere, will the person be harmed upon touching the inside of the sphere?

2. •Why must hospital personnel wear special conducting shoes while working around oxygen in an operating room? What might happen if the personnel wore shoes with rubber soles?

3. ••Consider a closed triangular box resting within a horizontal electric field of magnitude $E = 7.80 \times 10^4$ N/C as shown in Fig.1.181. Calculate the electric flux through (a) the vertical rectangular surface, (b) the slanted surface, and (c) the entire surface of the box.

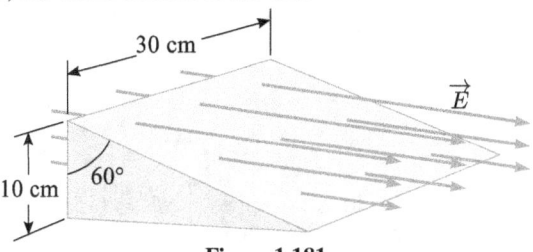

Figure 1.181

4. ••A uniform electric field $a\hat{i} + b\hat{j}$ intersects a surface of area A. What is the flux through this area if the surface lies (a) in the yz plane? (b) in the xz plane? (c) in the xy plane?

5. ••A point charge q is located at the centre of a uniform ring having linear charge density λ and radius a, as shown in Fig.1.182. Determine the total electric flux through a sphere centered at the point charge and having radius R where $R < a$.

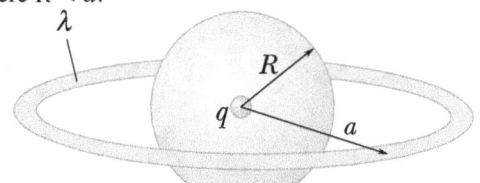

Figure 1.182

6. ••A pyramid with horizontal square base, 6.00 m on each side, and a height of 4.00 m is placed in a vertical electric field of 52.0 N/C [Fig.1.183]. Calculate the total electric flux through the pyramid's four slanted surfaces.

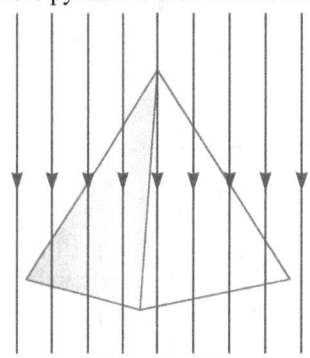

Figure 1.183

7. ••A cone with base radius R and height h is located on a

horizontal table. A horizontal uniform field E penetrates the cone, as shown in Figure 1.184. Determine the electric flux that enters the left-hand side of the cone.

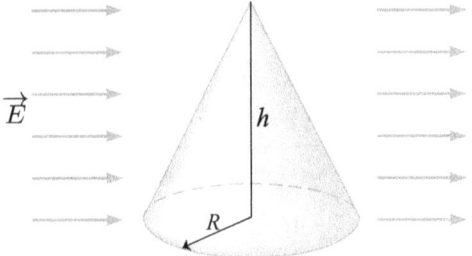

Figure 1.184

8. ●●A hemispherical surface of radius R is kept in a uniform electric field $\vec{E}$ such that $\vec{E}$ is parallel to the axis of hemi-sphere [Fig.1.185]. Find the net flux linked with hemispherical surface-

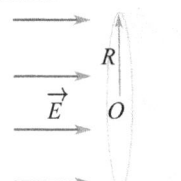

Figure 1.185

9. ●●An electric field is described by $\vec{E} = (15\hat{i} + 25\hat{j})$N/C. Find the electric flux through a surface whose area vector is $\vec{A} = (0.65\hat{i} + 0.35\hat{j})m^2$

10. ●●Calculate the electric flux through the surfaces shown in Fig.1.186

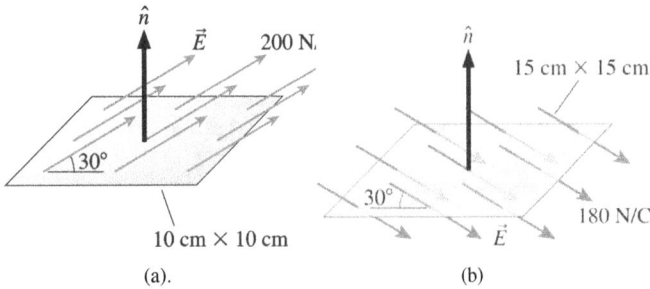

Figure 1.186

11. ●●The electric flux through the surface shown in Fig.1.187 is 25 N m^2/C. What is the electric field strength?

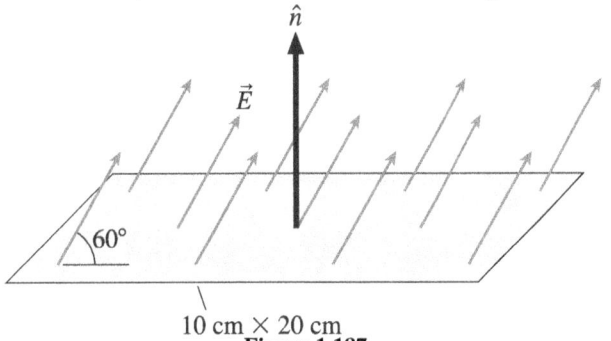

Figure 1.187

12. ●●A 2.0 cm × 3.0 cm rectangle lies in the xy-plane. What is the electric flux through the rectangle if
    (a) $\vec{E} = (100\hat{i} + 50\hat{k})$N/C?
    (b) $\vec{E} = (100\hat{i} + 50\hat{j})$N/C?

13. ●●A 2.0 cm × 3.0 cm rectangle lies in the xz-plane. What is the electric flux through the rectangle if

    (a) $\vec{E} = (100\hat{i} + 50\hat{k})$ N/C?
    (b) $\vec{E} = (100\hat{i} + 50\hat{j})$ N/C?

14. ●●A 3.0 cm diameter circle lies in the xz-plane in a region where the electric field is $\vec{E} = (1500\hat{i} + 1500\hat{j} - 1500\hat{k})$ N/C. What is the electric flux through the circle?

15. ●●A 1.0 cm×1.0 cm×1.0 cm box with its edges aligned with the xyz-axes, is in the electric field $\vec{E} = (350x + 150)\hat{i}$N/C, where x is in meters. What is the net electric flux through the box?

16. ●●What is the net electric flux through the two cylinders shown in Fig.1.188? Give your answer in terms of R and E.

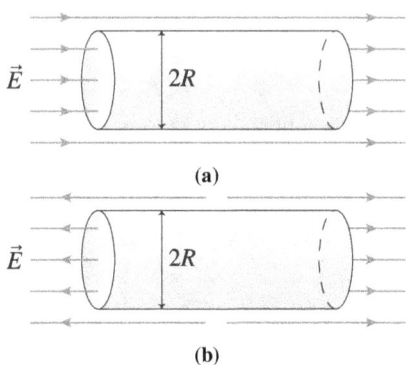

Figure 1.188

Multiple Choice Questions

17. ●●A square surface of side L meter in the plane of the paper is placed in a uniform electric field E (volt/m) acting along the same plane at an angle θ with the horizontal side of the square as shown in Fig.1.189. The electric flux linked to the surface, in units of volt m is-

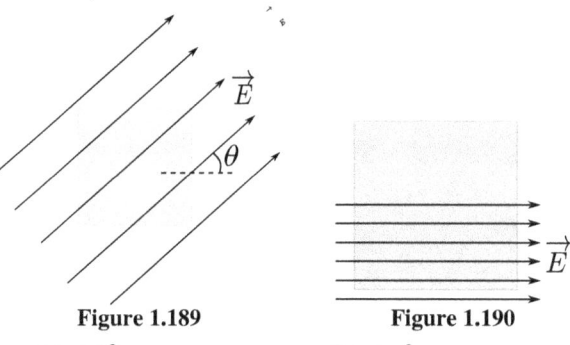

Figure 1.189   **Figure 1.190**

(A) EL^2 (B) $EL^2 \cos\theta$
(C) $EL^2 \sin\theta$ (D) zero

18. ●●A square surface of side L metres is in the plane of the paper. A uniform electric field $\vec{E}$(volt /m), also in the plane of the paper is limited only to the lower half of the square surface (Fig.1.190). The electric flux in SI units associated with the surface is
    (A) EL^2 (B) $EL^2/2\epsilon_0$
    (C) $EL^2/2$ (D) zero

1.32 Gauss's Law

Gauss's law is equivalent to Coulomb's law for static charges, although Gauss's law will look very different. It provides a different way to express the relationship between electric charge and electric field.

The purpose of learning Gauss's law is twofold:

- Gauss's law allows the electric fields of some continuous distributions of charge to be found much more easily than does Coulomb's law.

1.32. GAUSS'S LAW

- Gauss's law is more general in that it also covers the case of a rapidly moving charge. For such charges the electric lines of force become compressed in a plane at right angles to the direction of motion, thus losing their spherical symmetry. Coulomb's law is not valid for moving charges.

Thus Gauss's law is ultimately a more fundamental statement about electric fields than is Coulomb's law.

Statement of Gauss's Law: Gauss's Law states that the net flux Φ_E of electric field passing through any closed surface is proportional to the net charge Q_{encl} that is enclosed by the same surface. The proportionally constant is $1/\epsilon_o$, i.e.

$$\Phi_E = \oint \vec{E} \cdot d\vec{A} = \frac{Q_{encl}}{\epsilon_0}$$

Proof: Let us consider a positive point charge q located at the centre of a sphere of radius r, as shown in Fig.1.191. The electric field created by this charge at any point the surface of the Gaussian sphere is directed outwards and be normal to the surface. It's magnitude is also constant and given by-

$$E = k\frac{q}{r^2} = \frac{1}{4\pi\epsilon_0}\frac{q}{r^2} \quad (1.114)$$

Therefore:

$$\vec{E} \cdot d\vec{A} = E\,dA\cos 0° = E\,dA = \frac{1}{4\pi\epsilon_0}\frac{q}{r^2}dA$$

Hence, the total flux Φ_E through the entire Gaussian surface is

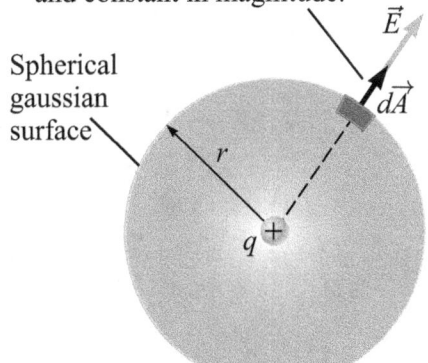

Figure 1.191

$$\Phi_E = \oint \vec{E} \cdot d\vec{A} = \frac{q}{4\pi\epsilon_0 r^2} \oint dA$$
$$= \frac{q}{4\pi\epsilon_0 r^2} 4\pi r^2 = \frac{q}{\epsilon_o}$$

or $$\boxed{\Phi_E = \oint \vec{E} \cdot d\vec{A} = \frac{q}{\epsilon_o}} \quad (1.115)$$

From Eq.(1.115), it is clear that the flux through a spherical surface of radius r is equal to the charge q inside the sphere divided by the permittivity of free space ϵ_0.

Since, entire Gaussian surface of any shape forms a solid angle of magnitude of 4π sr, therefore, this flux is linked with solid angle 4π sr. Now consider several closed Gaussian surfaces surrounding the charge as shown in Fig.1.192(a). The number of electric field lines passing through the spherical surface S_1 is the same as the number of lines passing through the non-spherical surfaces S_2 and S_3. Therefore, we conclude that the electric flux through any closed surface is independent of the shape of the surface that encloses the charge q, and it also does not depend upon the particular location of q inside the surface. The magnitude of the flux through any closed Gaussian surface surrounding the point charge q is q/ϵ_o.

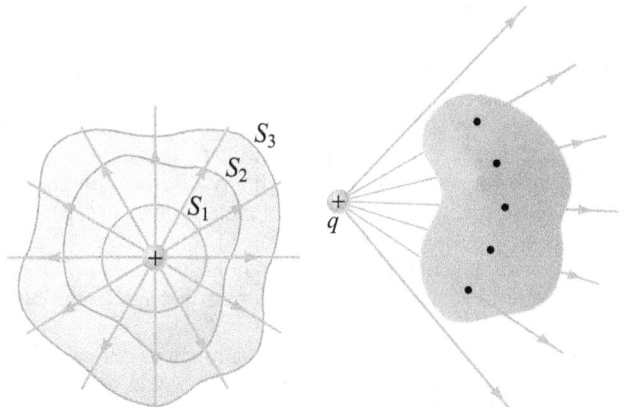

(a) The net electric flux is the same through all surfaces.

(b) The number of field lines entering the surface equals the number leaving the surface.

Figure 1.192: (a) Closed surfaces of various shapes surrounding a positive charge. (b) A point charge located outside a closed surface.

Now consider a point charge located outside a closed surface of arbitrary shape as shown in Fig.1.192(b). In such a case, any electric field line entering the surface leaves the surface at another point. The number of electric field lines entering the surface equals the number leaving the surface. Therefore, the net electric flux through a closed surface that surrounds no charge is zero.

1.32.1 General Integral Form of Gauss's Law

Now, suppose the surface encloses not just one point charge q but several charges $q_1, q_2, q_3, \cdots$. The total (resultant) electric field $\vec{E}$ at any point is the vector sum of the $\vec{E}$ fields of the individual charges. Let Q_{encl} be the total charge enclosed by the surface: $Q_{encl} = q_1 + q_2 + q_3 + \cdots$. Also let $\vec{E}$ be the total field at the position of the surface area element $d\vec{A}$, then we can write an equation like Eq.(1.115) for each charge and its corresponding field and add the results. When we do, we obtain the general statement of Gauss's law:

$$\boxed{\Phi_E = \oint \vec{E} \cdot d\vec{A} = \frac{Q_{encl}}{\epsilon_0}} \quad \text{(Gauss's law)} \quad (1.116)$$

The net electric flux through any closed surface is equal to the net charge inside the surface divided by the permittivity of free space ϵ_0.

In terms of solid angle, we can say that the complete flux linked with solid angle 4π sr is Q_{encl}/ϵ_0

Note: If we have a closed surface of irregular geometry, then during the integration the value of E may be different at various locations on the surface, and the angle between $\vec{E}$ and dA may also vary as we sum the various contributions over the surface. But, interestingly, regardless of the shape of the surface the net flux through the closed surface is always Q_{encl}/ϵ_0

Key Points

1. Gauss's law is true for any closed surface, no matter what its shape or size.
2. The term Q_{encl} on the right side of Gauss's law includes the sum of all charges enclosed by the surface. The charges may be located anywhere inside the surface.
3. In the situation when the surface is so chosen that there are some charges inside and some outside the Gaussian surface, the electric field [whose flux appears on the left side of Gauss's law] is caused partly by charges inside the surface and partly

by charges outside. But as Fig.1.192(b) shows, the outside charges do not contribute to the total (net) flux through the surface.

4. The closed surface that we choose for the application of Gauss's law is called the Gaussian surface. It is an imaginary surface. There need not be any material object at the position of the surface. You may choose any Gaussian surface and apply Gauss's law. However, take care not to let the Gaussian surface pass through any discrete charge. This is because electric field due to a system of discrete charges is not well defined at the location of any charge. (As you go close to the charge, the electric field increases to infinity.) However, the Gaussian surface can pass through a continuous charge distribution.

5. Gauss's law is often useful towards a much easier calculation of the electrostatic field when the system has some symmetry. This is facilitated by the choice of a suitable Gaussian surface.

6. Finally, Gauss's law is based on the inverse square dependence on distance contained in the Coulomb's law. Any violation of Gauss's law will indicate departure from the inverse square law.

EXAMPLE 73. Fig. 1.193 shows two Gaussian surfaces in uniform electric fields. In Fig.1.193(a) the Gaussian surface is a square box, and in Fig.1.193(b) the Gaussian surface is a closed cylinder. In each case, integrate $\Phi_E = \oint \vec{E} \cdot d\vec{A} = q_{\text{encl}}/\epsilon_0$ to find the amount of charge enclosed in each Gaussian surface.

SOLUTION (a) Note: We can easily determine whether the

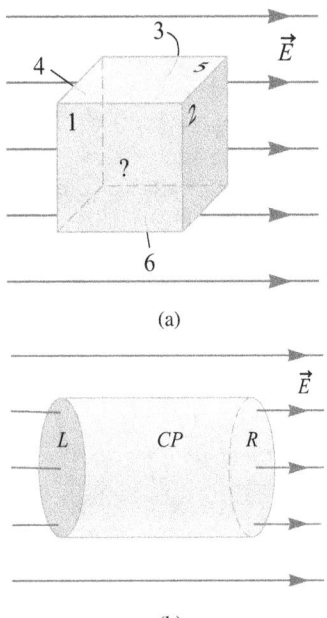

Figure 1.193

charge is positive, negative, or zero by looking at the electric field lines (Fig. 1.193). Because the electric field lines pass completely through the Gaussian surfaces, the net charge inside each surface is zero.

By Using Gauss's Law Although we already know the results, we are asked to use integration in order to practice using Gauss's law with different Gaussian surfaces.

A closed box is made up of six sides. According to Gauss's law, we must integrate over the entire closed box, so we break up the integral into six pieces, one for each side of the box. The six subscripts on the integrals correspond to the six sides of the box (Fig. 1.193(a)).

$$\Phi_E = \oint \vec{E} \cdot d\vec{A}$$
$$= \int_1 \vec{E} \cdot d\vec{A} + \int_2 \vec{E} \cdot d\vec{A} + \int_3 \vec{E} \cdot d\vec{A}$$
$$+ \int_4 \vec{E} \cdot d\vec{A} + \int_5 \vec{E} \cdot d\vec{A} + \int_6 \vec{E} \cdot d\vec{A}$$

To calculate the six dot products, we need to know the angle θ between $\vec{E}$ and $d\vec{A}$ for each side. A separate drawing of each side is helpful (Fig. 1.194). The area vector $d\vec{A}$ for each side is perpendicular to that side and points outward.

The dot product is zero whenever the angle θ between $\vec{E}$ and

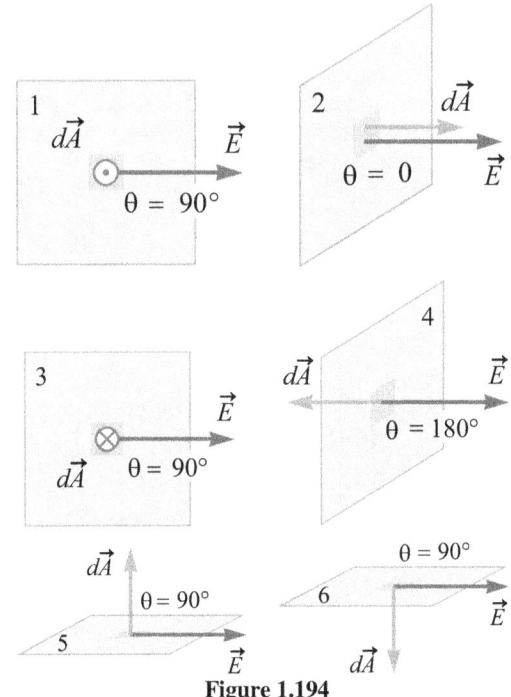

Figure 1.194

$d\vec{A}$ is 90°, so the dot product is zero for sides 1, 3, 5, and 6. That means these four integrals are zero, and we are left with just two integrals.

$$\Phi_E = \oint \vec{E} \cdot d\vec{A}$$
$$= 0 + \int_2 \vec{E} \cdot d\vec{A} + 0 + \int_4 \vec{E} \cdot d\vec{A} + 0 + 0$$
$$= \int_2 \vec{E} \cdot d\vec{A} + \int_4 \vec{E} \cdot d\vec{A}$$
$$\Rightarrow \quad \Phi_E = \int_2 \vec{E} \cdot d\vec{A} + \int_4 \vec{E} \cdot d\vec{A}$$

Write each dot product in terms of the magnitude of the vectors and the angle between them.

$$\Phi_E = \int_2 E\, dA \cos\theta + \int_4 E\, dA \cos\theta$$
$$\Rightarrow \quad \Phi_E = \int_2 E\, dA \cos 0 + \int_4 E\, dA \cos 180°$$
$$\Rightarrow \quad \Phi_E = \int_2 E\, dA - \int_4 E\, dA$$

The electric field is uniform, so as we integrate over dA, E is a constant that we can pull outside the integrals.

$$\Phi_E = E \int_2 dA - E \int_4 dA$$

The integrals are identical: Each equals the area A of one square side.

$$\Phi_E = EA - EA = 0$$

1.32. GAUSS'S LAW

The net electric flux is zero, and according to Gauss's law, that means the charge enclosed by the Gaussian surface is zero.

$$\Phi_E = \frac{Q_{encl}}{\epsilon_0} = 0 \quad \Rightarrow \quad Q_{encl} = 0$$

Note: This is exactly what we predicted because the number of electric field lines entering the box is equal to the number of electric field lines leaving the box.

(b) The closed cylinder is made up of three surfaces - the left cap, the right cap, and the curved part. These are labeled L, R, and CP (Fig. 1.195). Break the integral up into three pieces, one for each surface in Fig.1.195.

$$\Phi_E = \oint \vec{E} \cdot d\vec{A}$$

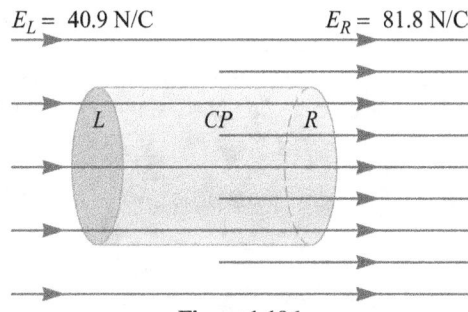

Figure 1.195

$$= \int_L \vec{E} \cdot d\vec{A} + \int_{CP} \vec{E} \cdot d\vec{A} + \int_R \vec{E} \cdot d\vec{A}$$

Figure 1.195 shows the angle θ between $\vec{E}$ and $d\vec{A}$ for each surface.

The area vector points outward for all small pieces of the curved part, while the electric field points to the right. Therefore, the angle $\theta = 90°$ for the curved part and the corresponding dot product and integral are zero, leaving two integrals.

$$\Phi_E = \int_L \vec{E} \cdot d\vec{A} + \int_R \vec{E} \cdot d\vec{A}$$

Write each dot product in terms of the magnitude of the vectors and the angle between the vectors.

$$\Phi_E = \int_L E\, dA \cos 180° + \int_R E\, dA \cos 0$$

$$= -\int_L E\, dA + \int_R E\, dA$$

As for the box, the electric field is constant and the integrals are identical. This time, each is the area of the end cap.

$$\Phi_E = -EA + EA = 0$$

The electric flux is zero, so the charge inside the Gaussian surface is zero.

$$\Phi_E = \frac{q_{encl}}{\epsilon_0} = 0 \quad \Rightarrow \quad q_{encl} = 0$$

☞ We find the same result whether we use a Gaussian cylinder or a box. The process is similar, but there is slightly less work when we use the cylinder because the Gaussian cylinder is made up of just three surfaces instead of the six surfaces of a box. Because we are free to choose a convenient Gaussian surface, it may be helpful to choose a closed cylinder instead of a box when that is possible.

EXAMPLE 74. The Gaussian surface in Figure 1.196 is a closed cylinder. An electric field points to the right throughout, made up of two uniform components with magnitude $E_L = 40.9$ N/C on the left side of the cylinder and $E_R = 81.8$ N/C on the right side. Integrate $\Phi = \oint \vec{E} \cdot d\vec{A} = q_{encl}/\epsilon_0$ to find the amount of charge enclosed in the Gaussian surface. Express your result in terms of the surface charge density (charge per unit area) σ. (Assume σ is uniform.)

SOLUTION This is similar to last problem (Fig.1.193 (b)), ex-

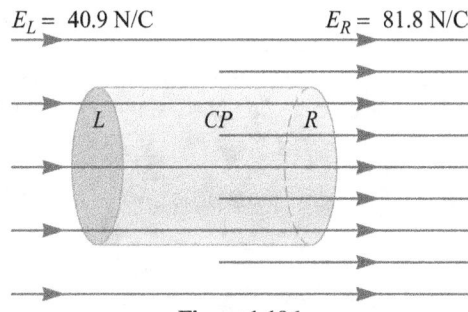

Figure 1.196

cept the electric field is not uniform. Because more electric field lines emerge from the Gaussian surface than enter it, we expect the net charge inside to be positive.

As like last problem, the net flux associated with the cylinder is-

$$\Phi_E = -E_L \int_L dA + E_R \int_R dA$$

The electric field is constant over each end cap, so it can be pulled outside the integrals. The magnitude of the electric field depends on the position, and the subscripts refer to the magnitudes in Fig.1.195.

$$\Phi_E = -E_L \int_L dA + E_R \int_R dA$$

The integrals are identical, equaling the area A of each end cap.

$$\Phi_E = -E_L A + E_R A = A(E_R - E_L)$$

According to Gauss's law, the electric flux is proportional to the charge inside.

$$\Phi_E = A(E_R - E_L) = \frac{Q_{encl}}{\epsilon_0}$$

The surface charge density σ is charge per unit area.

$$\sigma = \frac{Q_{encl}}{A} = \epsilon_0 (E_R - E_L)$$

$$\sigma = \left(8.85 \times 10^{-12} C^2/N \cdot m^2\right)(81.8 \text{ N/C} - 40.9 \text{ N/C})$$

$$= 3.62 \times 10^{-10} C/m^2$$

Note: As expected, the charge inside the Gaussian surface is positive.

EXAMPLE 75. A charge is placed at the centre of a cylindrical surface. Find total flux passing through lateral curved surface.

SOLUTION Total flux linked can be divided into two parts

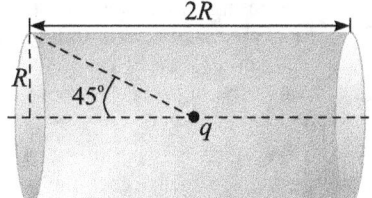

Figure 1.197

(i) Flux through longitudinal cylindrical curved surface ϕ_L

(ii) Flux through end caps ϕ_A
If end cap subtends a solid angle Ω at the centre, then
$$\Phi_L + \Phi_A = \frac{q}{\epsilon_\circ} \qquad (1.117)$$
The solid angle corresponding to semi vertical angle θ at the vertex of a cone [Fig.1.198] is given by-

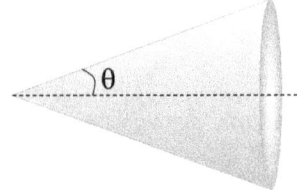

Figure 1.198

$$\Omega = 2\pi(1 - \cos\theta) \qquad (1.118)$$

Since, the electric flux corresponding to solid angle 4π sr is q/ϵ_0, therefore, the electric flux corresponding to solid angle Ω is $\frac{\Omega}{4\pi}\frac{q}{\epsilon_0}$, i.e., $\frac{2\pi(1-\cos\theta)}{4\pi}\frac{q}{\epsilon_0}$ or $\frac{(1-\cos\theta)}{2}\frac{q}{\epsilon_0}$.

The electric flux passing through left end cap of Fig.1.197,
$$\Phi_1 = \frac{(1-\cos 45°)}{2}\frac{q}{\epsilon_0} = \frac{1}{2}\left(1 - \frac{1}{\sqrt{2}}\right)\frac{q}{\epsilon_0}$$
Similarly, the electric flux passing through the right cap is-
$$\Phi_2 = \frac{1}{2}\left(1 - \frac{1}{\sqrt{2}}\right)\frac{q}{\epsilon_0}$$
Therefore, the electric flux passing through both end caps-
$$\Phi_A = \Phi_1 + \Phi_2 = \left(1 - \frac{1}{\sqrt{2}}\right)\frac{q}{\epsilon_0}$$
Substituting this value in Eq.(1.117), we get-
$$\Phi_L = \frac{q}{\epsilon_\circ} - \left(1 - \frac{1}{\sqrt{2}}\right)\frac{q}{\epsilon_0} = \frac{q}{\sqrt{2}\epsilon_0}$$
So, the flux passing through curved cylindrical surface is $q/\sqrt{2}\epsilon_\circ$

EXAMPLE 76. A cone of base radius R and height h is located in a uniform electric field $\vec{E}$ parallel to its base [Fig.1.199]. Find the electric flux entering the cone.

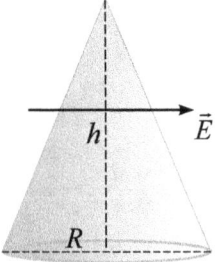

Figure 1.199

APPROACH You require to take only the projection of area of cone on a plane normal to electric field and it is a triangle of height h and base length $2R$ for the cone. To determine the required flux, just multiply this projection with electric field E.

SOLUTION Area of the triangle $= \frac{1}{2} \times 2R \times h = Rh$. Hence flux $= ERh$. (since, incoming flux is positive)

EXAMPLE 77. If a charge q is placed at the centre of a closed hemispherical non-conducting surface as shown in Fig.1.200, the total flux passing through the flat surface would be:
(A) 0   (B) $\frac{q}{2\epsilon_0}$ (C) $\frac{q}{4\epsilon_0}$ (D) $\frac{q}{2\pi\epsilon_0}$

SOLUTION As the flat surface is passing through the charge, therefore field lines emitted from charge wont be able to cross the flat surface. So flux passing through the flat surface will be zero. Therefore option A is correct.

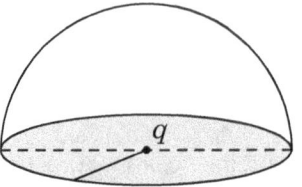

Figure 1.200

EXAMPLE 78. A charge q is surrounded by a closed surface consisting of an inverted cone of height h and base radius R, and a hemisphere of radius R as shown in the Fig.1.201. The electric flux through the conical surface is $\frac{nq}{6\epsilon_0}$ (in SI units). Find the value of n.

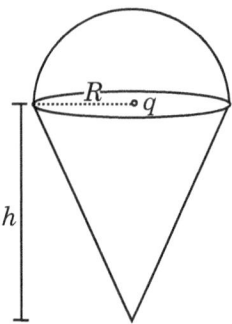

Figure 1.201

SOLUTION From Gauss law,
$$\phi_{\text{hemisphere}} + \phi_{\text{cone}} = \frac{q}{\epsilon_0} \qquad \cdots (1)$$
From Eq.1.109, the solid angle corresponding to plane angle 2θ, as shown in Fig.1.174, is
$$\Omega = 2\pi(1-\cos\theta)$$
Since, total electric flux corresponding to 4π sr solid angle is $Q_{\text{encl}}/\epsilon_0$, therefore flux corresponding to solid angle Ω is-
$$\phi = \frac{Q_{\text{encl}}\Omega}{4\pi\epsilon_0} = \frac{2\pi(1-\cos\theta)Q_{\text{encl}}}{4\pi\epsilon_0} = \frac{Q_{\text{encl}}}{2\epsilon_0}(1-\cos\theta)$$
For hemisphere, $2\theta = \pi$, i.e., $\theta = \pi/2$ and given that $Q_{\text{encl}} = q$, therefore
$$\phi_{\text{hemisphere}} = \frac{q}{2\epsilon_0}\left(1 - \cos\frac{\pi}{2}\right) = \frac{q}{2\epsilon_0}$$
Substituting this value in Eq.(1), we get
$$\frac{q}{2\epsilon_0} + \phi_{\text{cone}} = \frac{q}{\epsilon_0} \implies \phi_{\text{cone}} = \frac{q}{2\epsilon_0}$$
It is also given that, $\phi_{\text{cone}} = \frac{nq}{6\epsilon_0}$, therefore
$$\frac{nq}{6\epsilon_0} = \frac{q}{2\epsilon_0} \implies n = 3$$

Alternatively, $\phi \propto$ no of electric field lines passing through the given surface. As the given point charge q, is equally distributed between the hemispherical and conical parts, therefore half of electric field lines will pass through hemisphere & other half will pass through conical surface.

1.32.2 Differential Form of Gauss's Law (Optional)

In contrast to Eq.1.116 which is called the integral form we shall seek the differential form of the Gauss theorem, which establishes the relation between the volume charge density ρ and the changes in the field intensity E in the vicinity of a given point in space. For this purpose, we first represent the charge q in the volume V enveloped by a closed surface S in the form $Q_{\text{encl}} = \langle\rho\rangle V$, where $\langle\rho\rangle$ is the volume charge density, averaged over the volume V. Then we substitute this expression into Eq. (1.116) and divide

both its sides by V, which gives

$$\frac{1}{V}\oint \vec{E}\cdot d\vec{A} = \langle\rho\rangle/\epsilon_0 \quad (1.119)$$

We now make the volume V tend to zero by contracting it to the point we are interested in. In this case, $\langle\rho\rangle$ will obviously tend to the value of ρ at the given point of the field, and hence the ratio on the left-hand side of Eq. 1.119 will tend to ρ/ϵ_0.

The quantity which is the limit of the ratio of $\oint \vec{E}\cdot d\vec{A}$ to V as $V \to 0$ is called the divergence of the field $\vec{E}$ and is denoted by div $\vec{E}$. Thus, by definition,

$$\text{div}\,\vec{E} = \lim_{V\to 0}\frac{1}{V}\oint \vec{E}d\vec{A} \quad (1.120)$$

The divergence of any other vector field is determined in a similar way. It follows from definition (1.120) that divergence is a scalar function of coordinates.

In order to obtain the expression for the divergence of the field $\vec{E}$, we must, in accordance with (1.120), take an infinitely small volume V, determine the flux of $\vec{E}$ through the closed surface enveloping this volume, and find the ratio of this flux to the volume. The expression obtained for the divergence will depend on the choice of the coordinate system (in different systems of coordinates it turns out to be different). For example, in Cartesian coordinates it is given by

$$\text{div}\,\vec{E} = \frac{\partial E_x}{\partial x} + \frac{\partial E_y}{\partial y} + \frac{\partial E_z}{\partial z} \quad (1.121)$$

Thus, we have found that as $V \to 0$ in Eq.1.119, its right-hand side tends to ρ/ϵ_0, while the left-hand side tends to div $\vec{E}$. Consequently, the divergence of the field $\vec{E}$ is related to the charge density at the same point through the equation

$$\boxed{\text{div}\,\vec{E} = \frac{\rho}{\epsilon_0}} \quad (1.122a)$$

or

$$\boxed{\frac{\partial E_x}{\partial x} + \frac{\partial E_y}{\partial y} + \frac{\partial E_z}{\partial z} = \frac{\rho}{\epsilon_0}} \quad (1.122b)$$

Equations (1.122a) and (1.122b) represent the Gauss theorem in the differential form. The form of many expressions and their applications can be considerably simplified if we introduce the vector differential operator $\vec{\nabla}$. In Cartesian coordinates, the operator $\vec{\nabla}$ has the form

$$\vec{\nabla} = \hat{i}\frac{\partial}{\partial x} + \hat{j}\frac{\partial}{\partial y} + \hat{k}\frac{\partial}{\partial z}$$

where $\hat{i}$, $\hat{j}$, and $\hat{k}$ are the unit vectors along the X-, Y-, and Z-axes respectively. The operator $\vec{\nabla}$ itself does not have any meaning. It becomes meaningful only in combination with a scalar or vector function by which it is symbolically multiplied. For example, if we form the scalar product of vector $\vec{\nabla}$ and vector $\vec{E} = E_x\hat{i} + E_y\hat{j} + E_z\hat{k}$, we obtain-

$$\vec{\nabla}\cdot\vec{E} = \frac{\partial}{\partial x}E_x + \frac{\partial}{\partial y}E_y + \frac{\partial}{\partial z}E_z$$

It follows from Eq.1.121 that this is just the divergence of $\vec{E}$. Thus, the divergence of the field $\vec{E}$ can be written as div $\vec{E}$ or $\vec{\nabla}\cdot\vec{E}$ (in both cases it is read as "the divergence of $\vec{E}$"). So, the Gauss theorem (Eq.1.122) in terms of $\vec{\nabla}$ operator can be written as-

$$\boxed{\vec{\nabla}\cdot\vec{E} = \frac{\rho}{\epsilon_0}} \quad (1.123)$$

The Gauss theorem in the differential form is a local theorem: the divergence of the field $\vec{E}$ at a given point depends only on the electric charge density ρ at this point. This is one of the remarkable properties of electric field. For example, the field $\vec{E}$ of a point charge is different at different points. Generally, this refers to the spatial derivatives $\partial E_x/\partial x$, $\partial E_y/\partial y$, and $\partial E_z/\partial z$ as well. However, the Gauss theorem states that the sum of these derivatives, which determines the divergence of $\vec{E}$, turns out to be equal to zero at all points of the field (outside the charge itself). At the points of the field where the divergence of $\vec{E}$ is positive, we have the sources of the field (positive charges), while at the points where it is negative, we have sinks (negative charges). The field lines emerge from the field sources and terminate at the sinks. Eq.1.123 can also be written as-

$$\vec{\nabla}\cdot\left(\epsilon_0\vec{E}\right) = \rho$$

or

$$\boxed{\vec{\nabla}\cdot\vec{D} = \rho} \quad (1.124)$$

In Eq.1.124, $\vec{D} = \epsilon_0\vec{E}$ is called electric displacement vector. Detailed discussion of displacement vector is out of scope of this book.

1.33 Applications of Gauss's law

Selecting a Gaussian Surface: While applying Gauss's law we are interested in evaluating the integral

$$\Phi_E = \oint \vec{E}\cdot d\vec{A} = \oint EdA\cos\theta \quad (1.125)$$

Now note that, this integration may be complicated if product of E and $\cos\theta$ is not constant. Therefore we always select a Gaussian surface in such a way that product of E and $\cos\theta$ remains constant. That is we have to select some symmetrical surfaces. The types of symmetry are illustrated in Fig.1.202 and summarized in Table 1.3. If the object does not have any of these three types of symmetry, Gauss's law, even though still true, is unlikely to be helpful in calculating the electric field strength.

Table 1.3: Type of symmetry and the corresponding Gaussian surface that should be used

Symmetry	Type of Gaussian Surface	Objects
spherical	sphere concentric with given charged object	point charge, charged sphere, spherical shell
cylindrical	cylinder coaxial with given charged object	line of charge, charged cylinder, coaxial cylindrical shells
planar	cylinder or box perpendicular to given plane charged object	charged parallel plane(s), large flat object

1.33.1 Gauss's Law: Problem Solving Approach

To apply Gauss's law, we simply follow the following steps-
1. Sketch the charge distribution, as well as the position(s) where the electric field is to be calculated.
2. Determine the symmetry of its electric field.
3. Draw the appropriate closed Gaussian surface (see Table 1.3) based on the symmetry, choosing the size of the surface according to where the electric field is to be determined. For

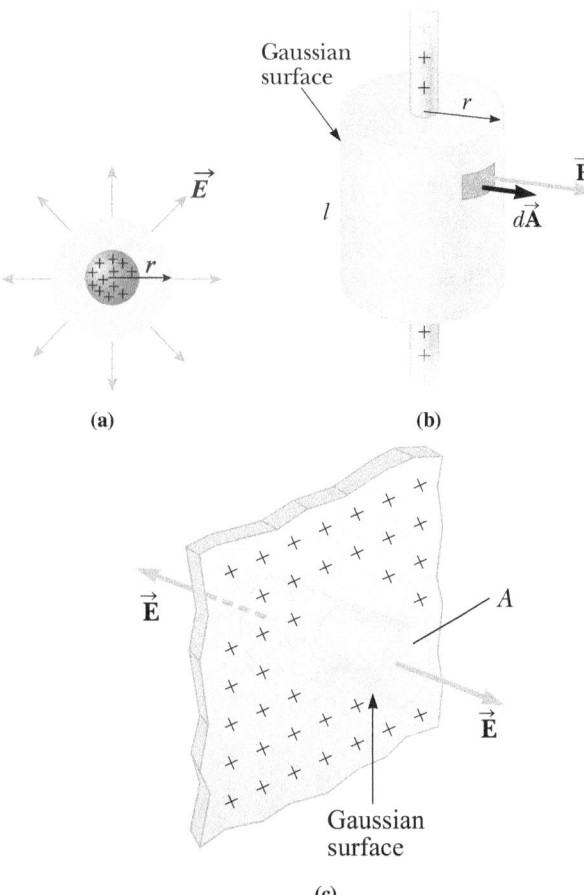

(a)

(b)

(c)

Figure 1.202: In spherical symmetry (a), the points on a sphere have constant electric field strength. For cylindrical symmetry (b), points at a common distance from a line have constant electric field strength. In planar symmetry (c), points on a plane parallel to the charged plane have constant electric field strength.

example, in the case of a charged sphere, a Gaussian sphere is drawn concentric with the centre of the object and is of radius equal to the radial distance where the electric field is to be computed. Sketch in the directions of $\vec{E}$ and $d\vec{A}$ on your diagram. You need not enclose all the charge within the Gaussian surface.

4. Be sure every part of the Gaussian surface satisfies one or more of the following conditions:
   (a) The value of the electric field can be argued by symmetry to be constant over the portion of the surface.
   (b) Gaussian surface is either tangent to or perpendicular to the electric field.
      i. If the electric field is everywhere tangent to a surface, the electric flux through the surface is $\Phi_E = \vec{E}.\vec{A} = 0$
      ii. If the electric field is everywhere perpendicular to a surface and has the same magnitude E at every point, the electric flux through the surface is $\Phi_E = \vec{E}.\vec{A} = EA$
   (c) The electric field is zero over the portion of the surface. Different portions of the Gaussian surface can satisfy different conditions as long as every portion satisfies at least one condition.
5. *If necessary, divide the closed surface into parts.* In some cases, the electric field is not constant over the entire surface but is known over each part. In this case, divide the Gaussian surface into different parts, ensuring that the entire closed surface is included. For example, we earlier divided the cylinder into two end caps and the curved side (see Ex.75).
6. For each part, *determine the orientation of $\vec{E}$ relative to $d\vec{A}$* and use that to find the dot product for that part.
7. Finally *integrate over the Gaussian surface to compute the electric flux.* However, if you have a surface with uniform electric field strength (and constant direction relative to the surface normal) you can take E out of the integral and simply multiply the constant E by the surface area multiplied by the $\cos\theta$ value.
8. *Calculate the net charge enclosed by the Gaussian surface.* Remember that only the charge inside the surface needs to be considered, and when there are positive and negative charges inside the surface it is the net charge that is used.
9. *Use mathematical expression of Gauss's law (with the expression for the electric flux (from step 6) and the enclosed charge (step 7)) to solve for the electric field strength-*
$$\Phi_E = \oint \vec{E} \cdot d\vec{A} = \frac{Q_{encl}}{\epsilon_0}$$
10. Finally you can check that your answer makes sense with limiting cases (i.e., does your relationship make sense just outside a surface, or at infinite distance).

EXAMPLE 79. Using Gauss's law, find the electric field at a distance r from a positive point charge q, and compare it with Coulomb's law.

APPROACH By definition, a point is a sphere of negligible radius, i.e., a sphere of zero radius, therefore we can assume a point charge as spherical and draw a spherical Gaussian surface of radius r around it (Fig.1.203). From the symmetry of the Gaussian surface, we know that at each point on it, the electric field $\vec{E}$ is perpendicular to the surface and directed outwards from the spherical center. Thus, $\vec{E}//d\vec{A}$ and $\vec{E}.d\vec{A} = EdA$, with $Q_{encl} = q$. Now, apply Gauss's law-

$$\Phi_E = \oint \vec{E} \cdot d\vec{A} = \frac{Q_{encl}}{\epsilon_0} \quad (1.126)$$

and solve for E

SOLUTION Substituting, $\vec{E} \cdot d\vec{A} = EdA$ and $Q_{encl} = q$ in Eq.

When the charge is at the center of the sphere, the electric field is everywhere normal to the surface and constant in magnitude.

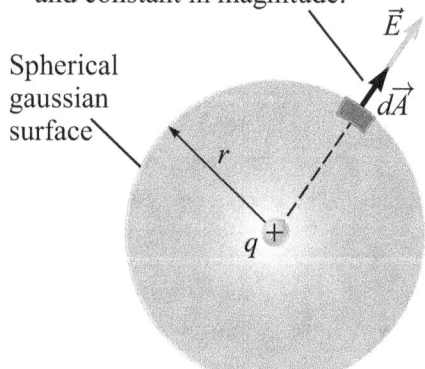

Figure 1.203

(1.126), we get
$$\oint E dA = \frac{q}{\epsilon_0}$$
$$\Rightarrow E \oint dA = \frac{q}{\epsilon_0} \quad \Rightarrow \quad 4\pi r^2 E = \frac{q}{\epsilon_0}$$

$$\implies \boxed{E = \frac{1}{4\pi\epsilon_\circ}\frac{q}{r^2} = k\frac{q}{r^2}}$$

which is simply Coulomb's law. This proves that Gauss's law and Coulomb's law are equivalent.

EXAMPLE 80. A single positive point charge, $+Q$, is placed at the exact centre of a cube of side length L (see Fig.1.204).
(a) Is it possible to calculate the electric flux through one face of

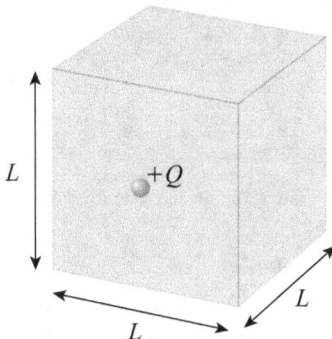

Figure 1.204: A single point charge, $+Q$, is at the exact centre of a cube. The length of each side of the cube is L.

the cube? If so, what is the value?
(b) Can Gauss's law be used to find the electric field on the face of the cube without considering the symmetry due to all faces? If so, what is the value?
SOLUTION (a) Yes. The enclosed charge is $+Q$, so by Gauss's law the total electric flux through all six faces of the cube is $\frac{Q}{\epsilon_0}$. By symmetry, all the faces are equal, so the electric flux through any one face is just $1/6$ of this value, i.e., $Q/6\epsilon_0$.
(b) No! Since, we don't know the distribution of electric field strength across the faces, so we cannot take E out of the integral.

EXAMPLE 81. Find electric flux through square of side a, due to charge placed at distance $a/2$ from centre of a square[Fig.1.205].
SOLUTION Let us enclose the charge q by a cubical gaussian surface with q at its centre[Fig.1.206].

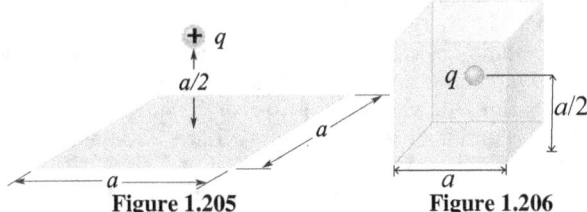

Figure 1.205   Figure 1.206

Net flux, coming out from the cubical Gaussian surface-
$$\phi = \frac{1}{\epsilon_0}Q_{\text{encl}} = \frac{1}{\epsilon_0}q$$
By symmetry all six faces have equal flux. Therefore, flux coming out from a single face, $\phi = q/6\epsilon_0$

EXAMPLE 82. By applying Gauss's law, show that the stable equilibrium of a charge under the effect of an electric field only, is impossible(Earnshaw's Theorem 1.24).
SOLUTION Suppose, in vacuum, we have a system of fixed

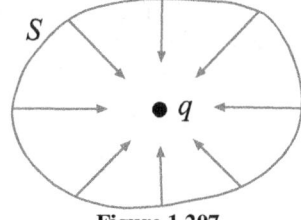

Figure 1.207

point charges in equilibrium. Now, consider, one of these charges, e.g. a charge q. Let us envelop the charge q by a small closed surface S (Fig.1.207). For the sake of definiteness, we assume that $q > 0$. For the equilibrium of this charge to be stable, it is necessary that the electric field $\vec{E}$ created by all the remaining charges of the system at all the points of the surface S be directed towards the charge q. Only in this case any small displacement of the charge q from the equilibrium position will give rise to a restoring force, and the equilibrium state will actually be stable. But such a configuration of the field $\vec{E}$ around the charge q is in contradiction to the Gauss theorem: the flux of $\vec{E}$ through the surface S is negative, while in accordance with the Gauss theorem it must be equal to zero since it is created by charges lying outside the surface S. On the other hand, the fact that E is equal to zero indicates that at some points of the surface S vector $\vec{E}$ is directed inside it and at some other points it is directed outside.
Hence it follows that in any electrostatic field a charge cannot be in stable equilibrium.

1.33.2 Cylindrical Symmetry

The following examples illustrate situations where cylindrical symmetry can be applied.

EXAMPLE 83. **Field of an Infinite Line Charge:** An infinitely long, straight line of uniform positive charge has a charge per unit length of λ (in units of charge per length). Find the electric field $\vec{E}$ at an arbitrary distance r from the line.
APPROACH From each incremental charge along the line, an electric field emanates equally in all directions. However, by symmetry, the superposition of the fields from all of the incremental charges results in a cancellation of fields parallel to the line of charge*. The result is a net field directed radially outward from the line. At all points at a given distance r from the line (in any direction), the field has the same magnitude.
Therefore, we match this symmetry with a Gaussian surface in the form of a cylinder of radius r and length L whose axis is the line of charge (Fig.1.208). At every point on the cylindrical curved side S_1 of the cylinder, $\vec{E}$ is parallel to the area elements dA, and it has the same magnitude everywhere. On the flat end caps S_1 and S_2 of the cylinder, $\vec{E}$ is perpendicular to $d\vec{A}$ everywhere. The net charge Q_{encl} inside the cylinder is λL. Applying Gauss's law, we obtain-
$$\Phi_E = \oint \vec{E}\cdot d\vec{A} = \frac{Q_{\text{encl}}}{\epsilon_0}$$
$$\implies \int_{S_1}\vec{E}\cdot d\vec{A} + \int_{S_2}\vec{E}\cdot d\vec{A} + \int_{S_3}\vec{E}\cdot d\vec{A} = \frac{Q_{\text{encl}}}{\epsilon_0} \quad (1.127)$$
From Fig.1.208a, we see that, $\vec{E}\perp \overrightarrow{dA}$ at end caps S_1 and S_2, therefore-
$$\int_{S_2}\vec{E}\cdot d\vec{A} = \int_{S_3}\vec{E}\cdot d\vec{A} = 0$$
and $\vec{E}\,||\,\vec{A}$ at whole curved surface S_3, so -
$$\int_{S_3}\vec{E}\cdot d\vec{A} = E\,dA$$
Substituting these values and $Q_{\text{encl}} = \lambda L$ in Eq.(??), we get-
$$0 + 0 + \int_{S_3} E\,dA = \frac{\lambda L}{\epsilon_0} \implies E(2\pi rL) = \frac{\lambda L}{\epsilon_0}$$
$$\implies \boxed{E = \frac{\lambda}{2\pi\epsilon_0 r} = k\frac{2\lambda}{r} \quad \text{(radially outward)}} \quad (1.128)$$

---
*We have already seen this in case of applications of Coulomb's law for "electric field due to an infinite line charge"

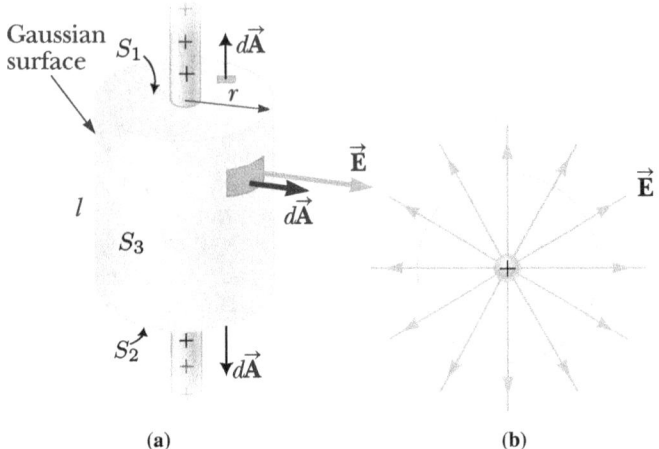

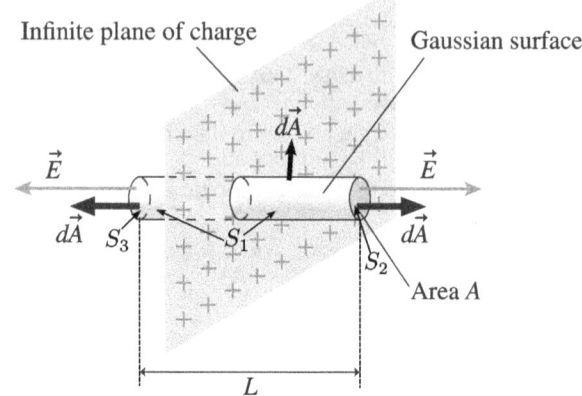

Figure 1.208: (a) An infinite line of charge surrounded by a cylindrical Gaussian surface concentric with the line. (b) An end view shows that the electric field at the cylindrical surface is constant in magnitude and perpendicular to the surface.

here, $k = 1/4\pi\epsilon_0$.

Eq.1.128 gives the electric field strength, E, at a distance r from an infinite line with charge density λ. The direction of the electric field is radially away from the line of charge.

EXAMPLE 84. What if the line segment in Example 83 were not infinitely long?

SOLUTION If the line charge in Example 83 were of finite length, the electric field would not be given by Eq.1.128. A finite line charge does not possess sufficient symmetry to make use of Gauss's law because the magnitude of the electric field is no longer constant over the surface of the Gaussian cylinder: the field near the ends of the line would be different from that far from the ends. Therefore, condition (4a) of problem solving approach of Gauss's law "Subsection 1.33.1", would not be satisfied in this situation. Furthermore, $\vec{E}$ is not perpendicular to the cylindrical surface at all points: the field vectors near the ends would have a component parallel to the line. Therefore, condition (4b(ii)) of problem solving approach of Gauss's law "Subsection 1.33.1", would not be satisfied. For points close to a finite line charge and far from the ends, Equation 1.128 gives a good approximation of the value of the field.

EXAMPLE 85. The Electric Field of an Infinite Plane Sheet of Charge: Use Gauss's law to find the electric field of an infinite plane of charge with surface charge density σ (in C/m^2).

APPROACH From symmetry we can say that, as long as we are not near an edge, the electric field must extend perpendicularly away from the plane on both sides. (There is no asymmetry that would cause the field lines to bend to one side or the other as they extend away from the positive charges.) We match the symmetry of this field by considering a Gaussian surface in the form of a cylinder or pill box, of cross-sectional area A, whose axis is perpendicular to the plane and whose ends are equidistant from the plane (Fig.1.209. The net charge enclosed by the surface is $Q_{encl} = \sigma A$. By symmetry, the field emerges uniformly and perpendicularly from each end and is tangent to the curved side of the cylinder. Now, apply Gauss's law, and solve for $\vec{E}$.

SOLUTION By, Gauss's law, we have

$$\oint \vec{E} \cdot d\vec{A} = \frac{Q_{encl}}{\epsilon_0} = \frac{A\sigma}{\epsilon_0} \quad (1.129)$$

Now, $\oint \vec{E} \cdot d\vec{A} = \int_{S_1} \vec{E} \cdot d\vec{A} + \int_{S_2} \vec{E} \cdot d\vec{A} + \int_{S_3} \vec{E} \cdot d\vec{A}$

here, S_1 represents curved surface whereas S_2, S_3 represent two planer surfaces.

Figure 1.209: The Gaussian surface extends to both sides of a plane of charge.

From Fig.1.209, we have -

$\int_{S_1} \vec{E} \cdot d\vec{A} = 0 \quad \because \vec{E} \perp d\vec{A}$ over whole surface S_1, and

$\int_{S_2} \vec{E} \cdot d\vec{A} + \int_{S_3} \vec{E} \cdot d\vec{A} = 2EA \quad (\because \vec{E} \parallel d\vec{A}, \vec{E} \parallel d\vec{A})$

Note that, for S_2, both $\vec{E}$ and $d\vec{A}$ pointed towards right side, while for S_3, both $\vec{E}$ and $d\vec{A}$ pointed towards left (Fig.1.209).

Substituting these values in Eq.1.129, we get -

$$\therefore \quad 2EA = \frac{A\sigma}{\epsilon_0}$$

$$\implies \boxed{E = \frac{\sigma}{2\epsilon_0}} \quad (1.130)$$

This electric field is away from the plane, in both left and right side of it.

Because the distance from the surface does not appear in the Eq.(1.130), we conclude that the field has the same constant value for all distances on either side of the plane of charge.

EXAMPLE 86. Explain why Gauss's law cannot be used to calculate the electric field near an electric dipole, a charged disc, or a triangle with a point charge at each corner.

SOLUTION The charge distributions of all these configurations do not have sufficient symmetry to make the use of Gauss's law practical. We cannot find a closed surface surrounding any of these distributions for which all portions of the surface satisfy one or more of conditions listed in "Gauss's Law: Problem solving Approach (Subsection 1.33.1)."

EXAMPLE 87. Electric Field Near a Charged Conducting Surface: Using Gauss's law, find the electric field just outside the surface of a conductor carrying a positive surface charge density σ.

APPROACH Near the conducting surface, the surface appears to be a flat (Fig.1.210, close-up), infinite plane, just as the surface of the ocean seems to be flat from our perspective standing on a beach. So, the surface charge density σ over a small part of the conductor is uniform*.

Notice that In the static case, no electric field $\vec{E}$ can exist within a conductor (because it would make the conduction charges move). So field lines extend away from the conductor, perpendicular to the surface, Fig.1.210. From symmetry considerations, we choose a cylindrical Gaussian surface as in the previous example, but in this case, we can draw two types of cylindrical Gaussian surfaces-

1. Gaussian cylinder crosses the conductor

---

*In general, the surface charge density σ is not constant on the surface of a conductor but depends on the shape of the conductor.

2. One end of Gaussian cylinder is inside the conductor, whereas the other end out side the surface of the conductor.
Apply Gauss's law in both cases and solve for electric field $\vec{E}$.
SOLUTION Case 1. Let Gaussian cylinder crosses the conductor completely. 1, 2 are planer surfaces and 3 is curved surface of Gaussian cylinder

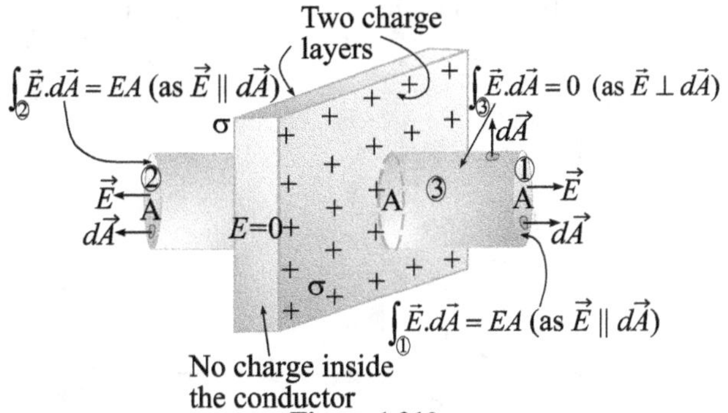

Figure 1.210

Net flux linked with the Gaussian cylinder

$$\Phi_E = \oint E.dA = \frac{Q_{encl}}{\epsilon_0}$$

$$\Rightarrow \underbrace{\int_1 \vec{E}.d\vec{A}}_{\vec{E}\|d\vec{A}} + \underbrace{\int_2 \vec{E}.d\vec{A}}_{\vec{E}\|d\vec{A}} + \underbrace{\int_3 \vec{E}.d\vec{A}}_{\vec{E}\perp d\vec{A}} = \frac{2\sigma A}{\epsilon_0}$$

$$\Rightarrow EA + EA + 0 = \frac{2\sigma A}{\epsilon_0}$$

$$\Rightarrow \boxed{E = \frac{\sigma}{\epsilon_0}} \quad (1.131)$$

Thus electric field near the conducting surface is $E = \sigma/\epsilon_0$
This field is twice the value that we found for thin sheet of charge in the previous example, and it has a constant value for all distances from the infinite plane conductor.

Case 2. Let Gaussian surface does not cross the conductor i.e., one end is inside the conductor whereas other end is outside it [Fig.1.211]
Net flux linked with the Gaussian cylinder

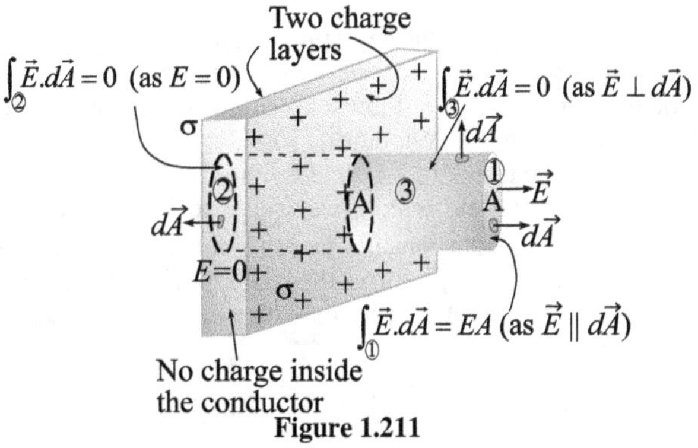

Figure 1.211

$$\Phi_E = \oint E.dA = \frac{Q_{encl}}{\epsilon_0}$$

$$\Rightarrow \underbrace{\int_1 \vec{E}.d\vec{A}}_{\vec{E}\|d\vec{A}} + \underbrace{\int_2 \vec{E}.d\vec{A}}_{\vec{E}=0} + \underbrace{\int_3 \vec{E}.d\vec{A}}_{\vec{E}\perp d\vec{A}} = \frac{\sigma A}{\epsilon_0} \Rightarrow EA = \frac{\sigma A}{\epsilon_0}$$

$$\Rightarrow \boxed{E = \frac{\sigma}{\epsilon_0}} \quad (1.132)$$

Thus electric field near the conducting surface is $E = \sigma/\epsilon_0$
Explanation: A solid thick conductor can be assumed as a combination of two thin plane sheets of charges having same charge densities as that of conductor.
The directions of the fields to the left, between, and right of

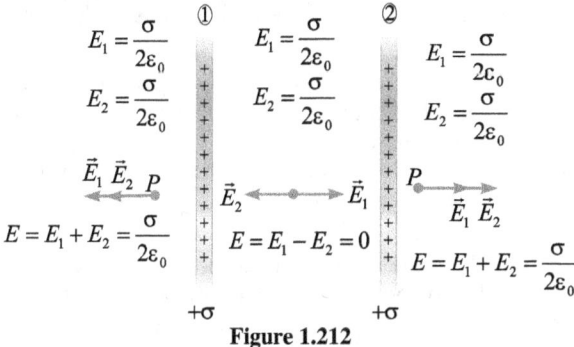

Figure 1.212

the sheets are shown in Fig.1.212. The resultant field depends on the direction and values of E_1 and E_2 due to sheet ① and ② respectively.

Electric field inside the conductor: Inside the conductor, $\vec{E}_1$ and $\vec{E}_2$ are oppositely directed, therefore the resultant electric field-

$$E = \frac{\sigma}{2\epsilon_0} - \frac{\sigma}{2\epsilon_0} = 0$$

here, σ is the surface charge density of both sheets.
Electric field outside the conductor: Outside left or right of the conductor, $\vec{E}_1$ and $\vec{E}_2$ both are directed in same direction, therefore the resultant electric field-

$$E = \frac{\sigma}{2\epsilon_0} + \frac{\sigma}{2\epsilon_0} = \frac{\sigma}{\epsilon_0}$$

EXAMPLE 88. A non-conducting very long cylinder of radius R has a positive uniform volume charge density ρ throughout. Derive expressions for the electric field both (a) inside ($r < R$) and (b) outside ($r > R$) the cylinder.
SOLUTION In this case, we have cylindrical symmetry, so the appropriate closed surface is a coaxial Gaussian cylinder of radius r [Fig.1.213]. The length of the Gaussian cylinder, L, is arbitrary and will cancel out in the final answer.

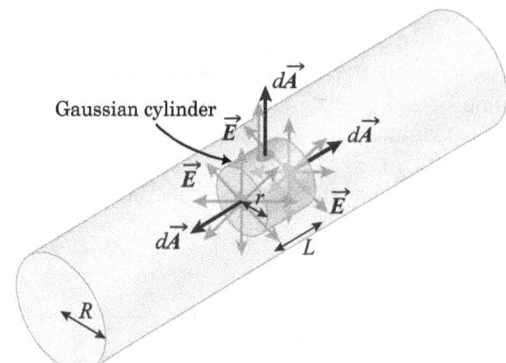

Figure 1.213: The very long charged cylinder (yellow) has a uniform positive charge density. In grey is drawn the smaller Gaussian cylinder for finding the electric field inside the charged cylinder.

(a) Inside the cylinder ($r < R$) : For this part, we draw the Gaussian cylinder with radius r less than the radius, R, of the charged cylinder (i.e., the Gaussian surface is inside the physical cylinder, but coaxial with it). The situation is illustrated in Figure 1.213. From the symmetry of the situation, the positively charged cylinder results in an electric field that points radially outward and has constant electric field strength at any particular radial distance. This is a situation in which we need to divide the closed Gaussian surface into three parts, the two end caps and the curved side of the cylinder. In all cases, $d\vec{A}$ points outward from the closed Gaussian surface. This is pictured for the Gaussian surface in Figure 1.213. For the two end caps, $d\vec{A}$ is parallel to the axis of the cylinder, pointing in opposite directions at the two ends. For the curved surface of the cylinder, $d\vec{A}$ is everywhere radially outward (we only show $\vec{E}$ and $d\vec{A}$ at one point on the curved Gaussian surface, but both are similarly radially outward for all other points on the curved surface).

We find the electric flux by combining the contributions from the three parts of the surface:

$$\Phi_{E\,net} = \Phi_{E\,left\,cap} + \Phi_{E\,curved\,surface} + \Phi_{E\,right\,cap}$$

Everywhere along the curved surface, $\vec{E}$ and $d\vec{A}$ are parallel to each other, while on both end caps they are perpendicular to each other (Figure 1.213). This means that the vector dot product of $\vec{E}$ with $d\vec{A}$ yields

left end cap: $\vec{E} \cdot d\vec{A} = E\, dA \cos 90° = E\, dA(0) = 0$
curved wall: $\vec{E} \cdot d\vec{A} = E\, dA \cos 0° = E\, dA(1.00) = E\,dA$
right end cap: $\vec{E} \cdot d\vec{A} = E\, dA \cos 90° = E\,dA(0) = 0$

Here E represents the magnitude of the electric field, and dA is the differential area element without a vector direction. Therefore, substituting into the relationship for electric flux, we have

$$\Phi_{E\,left\,cap} = \oint_{left\,cap} \vec{E} \cdot d\vec{A} = 0,$$

$$\Phi_{E\,curved\,wall} = \oint_{curved\,wall} \vec{E} \cdot d\vec{A} = \oint_{curved\,wall} E\, dA$$

$$= E \oint_{curved\,wall} dA = E A_{curved\,wall},$$

$$\Phi_{E\,right\,cap} = \oint_{right\,cap} \vec{E} \cdot d\vec{A} = 0$$

The area of the curved wall can be determined if we imagine it has been unwrapped into a corresponding rectangle. One side of the rectangle is L and the other is the circumference of the surface, $2\pi r$:

$$\Phi_{E\,curved\,wall} = E A_{curved\,wall} = EL(2\pi r) = 2\pi r L E$$

Since the two end caps have zero electric flux, this is also the net electric flux:

$$\Phi_{net} = 2\pi r L E$$

Now we need to find the net enclosed charge. We multiply the volume charge density, ρ, by the volume of the Gaussian cylinder. The volume of a cylinder is the cross-sectional area times the length of the cylinder:

$$Q_{encl} = \pi r^2 L \rho$$

Finally, we invoke Gauss's law and solve for the electric field strength:

$$\Phi_E = \frac{Q_{encl}}{\epsilon_0} \implies 2\pi r L E = \frac{\pi r^2 L \rho}{\epsilon_0}$$

$$\implies \boxed{E = \frac{r\rho}{2\epsilon_0}} \quad (1.133)$$

The direction of the electric field points radially outward from the axis of the cylinder.

(b) Outside the charged cylinder ($r > R$): In this case, we make a Gaussian cylinder of radius larger than that of the charged cylinder, but otherwise the solution is similar. The electric flux process is identical to that given in part (a), with the same result:

$$\Phi_{net} = 2\pi r L E$$

When computing the enclosed charge, we now use the volume of the charged cylinder, rather than the larger Gaussian cylinder:

$$Q_{encl} = \pi R^2 L \rho$$

We invoke Gauss's law for the electric field strength.

$$\Phi_E = \frac{q_{encl}}{\epsilon_0} \implies 2\pi r L E = \frac{\pi R^2 L \rho}{\epsilon_0}$$

$$\implies \boxed{E = \frac{R^2 \rho}{2r\epsilon_0}} \quad (1.134)$$

Again, the direction of the electric field is radially outward from the axis of the cylinder. So, within the cylinder, the field is directly

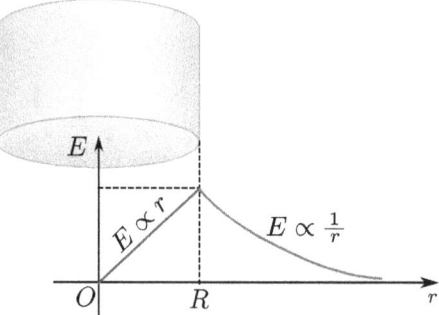

Figure 1.214: A graph of the electric field E vs. r for a uniform volume charge density throughout an infinitely long cylinder of radius R. In the graph, E is positive when $\vec{E}$ is directed outward.

proportional to the distance from the axis. Outside the cylinder, the field is the same as for the charged wire of infinite length. Fig.1.214 shows a plot of E vs. r.

Making sense of the result: Inside the charged cylinder, the electric field increases linearly with distance from the axis. At the exact centre of the cylinder, the electric field is zero. Outside the charged cylinder, the electric field is proportional to $1/r$, approaching zero at infinite distance. When $r = R$, we require that the results from the two parts be consistent, which they are. Note that this solution is for the case of a non-conducting cylinder with the charge uniformly distributed.

1.33.3 Spherical Symmetry

The following examples are based on the situations in which spherical symmetry exists.

EXAMPLE 89. Field of a Charged Conducting Sphere: We place positive charge q on a solid conducting sphere (or spherical shell) with radius R (Fig.1.215). Find $\vec{E}$ at any point inside or outside the sphere.

APPROACH For a conducting sphere, all the charge lies on its surface. So, the system has spherical symmetry. To take advantage of the symmetry, we draw a spherical Gaussian surface of radius r centred on the conductor. To calculate the field outside the conductor, we take r to be greater than the conductor's radius R; to calculate the field inside, we take r to be less than R. In either case, the point where we want to calculate $\vec{E}$ lies on the Gaussian surface. Now, we use the mathematical expression of Gauss's law (Eq.1.135) to calculate the electric field E at the required position-

$$\Phi_E = \oint \vec{E} \cdot d\vec{A} = \frac{Q_{encl}}{\epsilon_0} \quad \text{(Gauss's Law)} \quad (1.135)$$

SOLUTION (i) For a point outside the conducting sphere ($r \geq R$): In this case, the entire conductor is within the Gaussian surface, so the enclosed charge $Q_{encl} = q$

By symmetry, we conclude that the field $\vec{E}$ can only be radially outward, outside the sphere. Furthermore, for a given value of r, $\vec{E}$ has the same magnitude everywhere. Therefore, Eq.(1.135) can be written as

$$\oint E\, dA = \frac{q}{\epsilon_0}$$

Since, the field at each point on Gaussian surface has same magnitude, so E can be taken out of integral, so-

1.33. APPLICATIONS OF GAUSS'S LAW

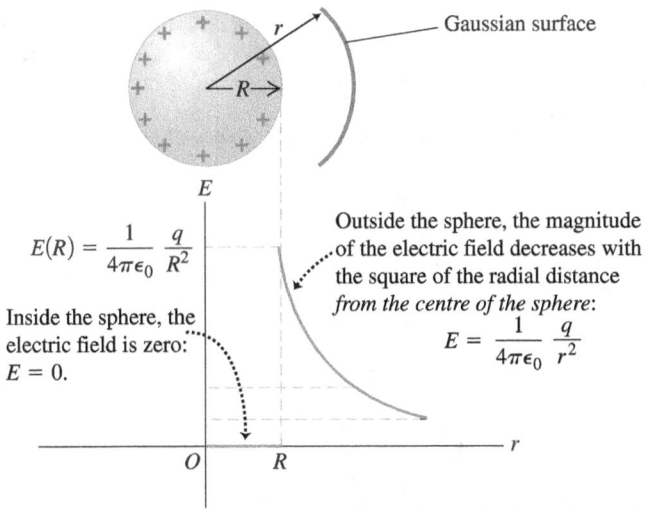

Figure 1.215: Calculating the electric field of a conducting sphere with positive charge q. Outside the sphere, the field is the same as if all of the charge were concentrated at the centre of the sphere.

$$E \oint dA = \frac{q}{\epsilon_0} \quad \Rightarrow \quad E(4\pi r^2) = \frac{q}{\epsilon_0}$$

$$\Rightarrow \boxed{E = \frac{1}{4\pi\epsilon_0}\frac{q}{r^2} = k\frac{q}{r^2}} \quad \text{(For } r > R\text{)} \quad (1.136)$$

This is just the inverse-square-law field for a point charge q concentrated at the centre of the sphere.
Just outside the surface of the sphere, where $r = R$, Eq. (1.136) takes the form-

$$\boxed{E = \frac{1}{4\pi\epsilon_0}\frac{q}{R^2} = k\frac{q}{R^2}} \quad (1.137)$$

Note: *A point charge q can be considered to be the limiting case of a small spherical conductor whose radius tends to zero and electric field to infinity.*

Comment: Flux can be positive or negative: Note that we have chosen the charge q to be positive. If the charge is negative, the electric field is radially inward instead of radially outward, and the electric flux through the Gaussian surface is negative. The electric field magnitudes outside and at the surface of the sphere are given by the same expressions as above, except that q denotes the magnitude (absolute value) of the charge.

(ii) For a point inside the conducting sphere ($r < R$): In this case Gaussian surface does not include any charge, i.e. $Q_{encl} = 0$, therefore Eq.(1.135) gives-

$$\oint E\, dA = 0 \quad \Rightarrow \quad E \oint dA = 0 \quad \Rightarrow \quad E(4\pi r^2) = 0$$

$$\Rightarrow \boxed{E = 0} \quad (1.138)$$

So, the electric field inside the conductor is zero.

Comment: We already knew that $\vec{E} = 0$ inside the conductor, as it must be inside any solid conductor when the charges are at rest. Figure 1.215 shows E as a function of the distance r from the centre of the sphere. Note that in the limit as $R \to 0$, the sphere becomes a point charge; there is then only an 'outside', and the field is everywhere given by $E = q/4\pi\epsilon_0 r^2$. Thus we have again deduced Coulomb's law from Gauss's law.

EXAMPLE 90. A thin spherical shell of radius R has a total positive charge Q distributed uniformly over its surface. Find the electric field inside and outside the shell.

APPROACH Here, we use the method described in the last problem. By symmetry, if any field exists inside the shell, it must be radial. For any point outside or on the surface of the conducting shell, it behaves like a solid conducting sphere. So, construct spherical Gaussian surfaces for both cases and apply Gauss's law–

$$\Phi_E = \oint \vec{E} \cdot d\vec{A} = \frac{Q_{encl}}{\epsilon_0} \quad (1.139)$$

and solve above equation for E in each case.

SOLUTION (i) Inside the shell (r<R): Let us construct a spherical Gaussian surface of radius $r < R$ concentric with the shell (Fig.1.216). Since, there is no enclosed charge within the Gaussian surface, i.e., $Q_{encl} = 0$, therefore, Eq.1.139 gives-

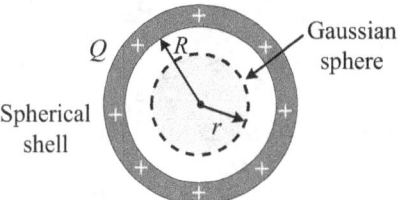

Figure 1.216: Cross sectional view of shell with Gaussian surface inside it.

$$\oint \vec{E} \cdot d\vec{A} = E\left(4\pi r^2\right) = 0$$

$$\Rightarrow \boxed{E = 0} \quad (1.140)$$

So, we conclude that there is no electric field inside a uniformly charged spherical shell.

(ii) Outside the shell ($r \geq R$): Outside the shell, we construct a spherical Gaussian surface of radius $r > R$ concentric with the charged shell as shown in Fig.1.217. Symmetry suggests that $E = $ constant on that surface and $\vec{E}$ is parallel to $d\vec{A}$, i.e. $\oint \vec{E} \cdot d\vec{A} = E\left(4\pi r^2\right)$. Since the net charge Q_{encl} inside the Gaussian surface is equal to the total charge Q on the shell, the shell is equivalent to a point charge located at the centre. Therefore, Eq.1.139 gives-

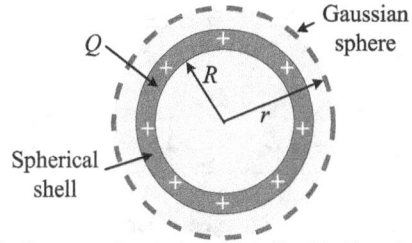

Figure 1.217: Cross sectional view of shell with Gaussian surface outside it.

$$\boxed{E = \frac{1}{4\pi\epsilon_o}\frac{Q}{r^2} = k\frac{Q}{r^2}} \quad (r > R) \quad (1.141)$$

(iii) At the surface of the shell ($r = R$): In this case, Eq.(1.141) takes the form,

$$\boxed{E = \frac{1}{4\pi\epsilon_o}\frac{Q}{r^2} = k\frac{Q}{R^2}} \quad (1.142)$$

Electric field E vs r graph will be similar to the graph obtained in the case of solid conducting sphere (Fig.1.215).

EXAMPLE 91. A solid sphere of radius R has a uniform volume charge density ρ and carries a total positive charge Q. Find and sketch the electric field at any distance r away from the sphere's center.

APPROACH The whole space can be divided into two parts:(i) $0 \leq r \leq R$ and (ii) $r \geq R$. Observe the symmetry in each case and find corresponding Q_{encl}. Apply Gauss's law and simplify for electric field E.

SOLUTION (1) For $0 \leq r \leq R$: When dealing with a spherically symmetric charge distribution, we chose a spherical Gaussian surface of radius $r < R$ concentric with the charged sphere as shown in Fig.1.218.

By symmetry, the magnitude of the electric field is constant

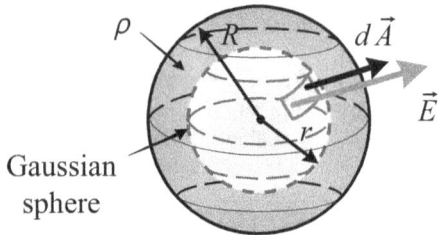

Figure 1.218

everywhere on the spherical Gaussian surface and normal to the surface at any point, i.e. $\vec{E} // d\vec{A}$. Thus:

$$\oint \vec{E} \cdot d\vec{A} = \oint E dA = E \oint dA = E\left(4\pi r^2\right)$$

It is important to note that the volume, say V', of the Gaussian sphere encloses a net charge $Q_{encl} = \rho V'$; that is:

$$Q_{encl} = \rho V' = \rho \left(\frac{4}{3}\pi r^3\right)$$

We can now use Gauss's law to find electric field as follows:

$$\Phi_E = \oint \vec{E} \cdot d\vec{A} = \frac{Q_{encl}}{\epsilon_o} \Rightarrow E\left(4\pi r^2\right) = \frac{\rho\left(\frac{4}{3}\pi r^3\right)}{\epsilon_o}$$

Then: $\boxed{E = \frac{\rho}{3\epsilon_o} r \quad (0 \le r \le R)}$ (1.143)

In vector form,

$$\boxed{\vec{E} = \frac{\rho}{3\epsilon_o} \vec{r} \quad (0 \le r \le R)}$$ (1.144)

Using the definition $\rho = Q / \left(\frac{4}{3}\pi R^3\right)$ and $k = 1/(4\pi\epsilon_o)$, we get:

$$E = \frac{Q}{\frac{4}{3}\pi R^3 (3\epsilon_0)} = \frac{Q}{4\pi\epsilon_o R^3} r$$

$$\Rightarrow \boxed{E = k\frac{Q}{R^3} r \quad (0 \le r \le R)}$$ (1.145)

(2) For $r \ge R$ Again, because the charge distribution is spherically symmetric, we can construct a Gaussian sphere of radius $r > R$ concentric with the charged sphere, as shown in Fig.1.219.

Just as when $r < R$, $\oint \vec{E} \cdot d\vec{A} = E\left(4\pi r^2\right)$, but in this case,

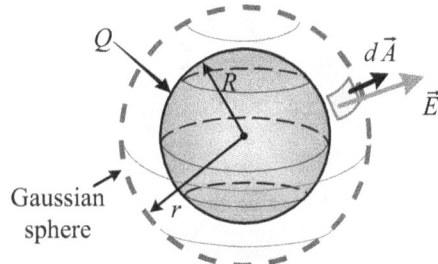

Figure 1.219

$Q_{encl} = Q$. Therefore, by Gauss's law:

$$\Phi_E = \oint \vec{E} \cdot d\vec{A} = \frac{Q_{encl}}{\epsilon_o} \Rightarrow E\left(4\pi r^2\right) = \frac{Q}{\epsilon_o}$$

i.e., $\boxed{E = \frac{1}{4\pi\epsilon_o}\frac{Q}{r^2} = k\frac{Q}{r^2} \quad (r \ge R)}$ (1.146)

Note that this is identical to the result obtained for a point charge. Therefore, we conclude that the electric field outside any uniformly charged sphere is equivalent to that of a point charge located at the centre of the sphere. At $r = R$, the two cases give identical results $E = kQ/R^2$. A plot of E versus r is shown in Fig.1.220. This figure shows the continuation of E and it is maximum at $r = R$.

EXAMPLE 92. A Uniformly charged solid non-conducting

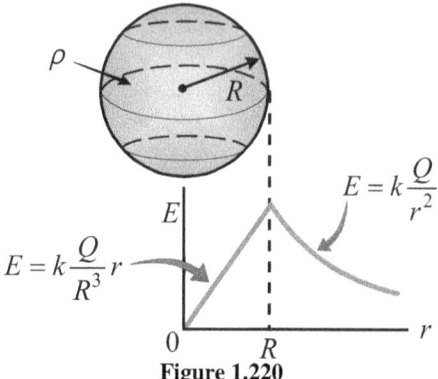

Figure 1.220

sphere of uniform volume charge density ρ and radius R has a concentric spherical cavity of radius r (Fig.1.221). Find out electric field intensity at
(i) Point A, (ii) Point B, (iii) Point C, (iv) Center of the sphere.

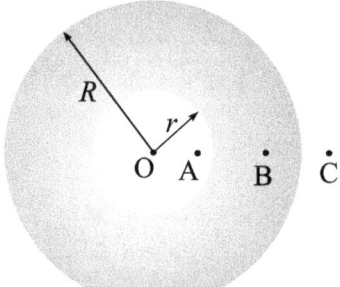

Figure 1.221

SOLUTION Method I:
(i) For point A: Since point A lies inside the cavity, therefore for concentric Gaussian surface through A, $Q_{encl} = 0$, so electric field intensity due to all shells will be zero.

$$\vec{E_A} = 0$$

(ii) For point B: For concentric Gaussian surface through B, $Q_{encl} = \frac{4}{3}\pi\left((OB)^3 - r^3\right)\rho$, therefore by Gauss's theorem

$$\oint \vec{E} \cdot d\vec{A} = \frac{Q_{encl}}{\epsilon_0}$$

$$\Rightarrow E(4\pi(OB)^2) = \frac{\frac{4}{3}\pi\left((OB)^3 - r^3\right)\rho}{\epsilon_0}$$

$$\Rightarrow E = \frac{1}{4\pi\epsilon_0}\left[\frac{\frac{4}{3}\pi((OB)^3 - r^3)\rho}{(OB)^2}\right] = k\left[\frac{\frac{4}{3}\pi((OB)^3 - r^3)\rho}{(OB)^2}\right]$$

This electric field is radially away for $+ve$ charge density, therefore in vector form it can be written as

$$\vec{E} = k\left[\frac{\frac{4}{3}\pi\left((OB)^3 - r^3\right)\rho}{(OB)^2}\right]\widehat{(OB)}$$

or $\quad \vec{E} = k\left[\frac{\frac{4}{3}\pi\left((OB)^3 - r^3\right)\rho}{(OB)^2}\right]\frac{\overrightarrow{OB}}{OB}$

So, $\vec{E_B} = \frac{k\frac{4}{3}\pi(OB^3 - r^3)\rho}{OB^3}\overrightarrow{OB} = \frac{\rho}{3\epsilon_0}\frac{[OB^3 - r^3]}{OB^3}\overrightarrow{OB}$

(iii) For point C: In this case, $Q_{encl} = \frac{4}{3}\pi\left[R^3 - r^3\right]\rho$

Now, simplifying like part (ii), we get

$$\vec{E}_C = \frac{k\frac{4}{3}\pi(R^3 - r^3)\rho}{OC^3}\overrightarrow{OC} = \frac{\rho}{3\epsilon_0}\frac{[R^3 - r^3]}{OC^3}\overrightarrow{OC}$$

Method: II We can consider that the spherical cavity is filled with charge density ρ and also $-\rho$, thereby making net charge density zero after combining. We can consider two concentric solid spheres: One of radius R and charge density ρ and other of radius r and charge density $-\rho$ [Fig.1.222].

From Eq1.144, at A, $\vec{E}_\rho = \frac{\rho(\overrightarrow{OA})}{3\epsilon_0}$ and $\vec{E}_{-\rho} = \frac{-\rho(\overrightarrow{OA})}{3\epsilon_0}$

Applying superposition principle:

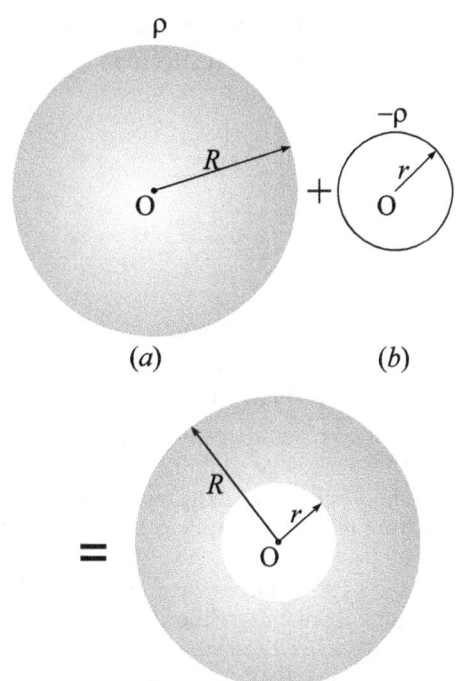

Figure 1.222

(i) $\vec{E}_A = \vec{E}_\rho + \vec{E}_{-\rho} = \frac{\rho(\overrightarrow{OA})}{3\epsilon_0} + \frac{[-\rho(\overrightarrow{OA})]}{3\epsilon_0} = 0$

(ii) Point B is inside the sphere of density ρ and outside the sphere of density $-\rho$. $\vec{E}_B = \vec{E}_\rho + \vec{E}_{-\rho}$

here, $\vec{E}_\rho = \frac{\rho(\overrightarrow{OB})}{3\epsilon_0}$ and $\vec{E}_{-\rho} = \frac{kQ_{encl}}{(OB)^2}\overrightarrow{OB}$

Therefore, $\vec{E}_B = \frac{\rho(\overrightarrow{OB})}{3\epsilon_0} + \frac{k\left[\frac{4}{3}\pi r^3(-\rho)\right]}{(OB)^3}\overrightarrow{OB}$

$= \left[\frac{\rho}{3\epsilon_0} - \frac{r^3\rho}{3\epsilon_0(OB)^3}\right]\overrightarrow{OB} = \frac{\rho}{3\epsilon_0}\left[1 - \frac{r^3}{OB^3}\right]\overrightarrow{OB}$

(iii) $\vec{E}_C = \vec{E}_\rho + \vec{E}_{-\rho} = \frac{K\left(\frac{4}{3}\pi R^3\rho\right)}{OC^3}\overrightarrow{OC} + \frac{K\left(\frac{4}{3}\pi r^3(-\rho)\right)}{OC^3}\overrightarrow{OC}$

$= \frac{\rho}{3\epsilon_0(OC)^3}[R^3 - r^3]\overrightarrow{OC}$

(iv) $\vec{E}_O = \vec{E}_\rho + \vec{E}_{-\rho} = 0 + 0 = 0$

EXAMPLE 93. Solve Example 92, if cavity is not concentric and centered at point P [Fig.1.223].

SOLUTION Again assume ρ as the charge densities of solid sphere and $(-\rho)$ as that in cavity (similar to the previous example):

(i) $\vec{E}_A = \vec{E}_\rho + \vec{E}_{-\rho}$

$\Rightarrow \vec{E}_A = \frac{\rho[\overrightarrow{OA}]}{3\epsilon_0} + \frac{-\rho[\overrightarrow{PA}]}{3\epsilon_0}$

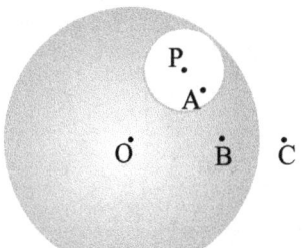

Figure 1.223

$= \frac{\rho}{3\epsilon_0}\left[\overrightarrow{OA} - \overrightarrow{PA}\right] = \frac{\rho}{3\epsilon_0}\left[\overrightarrow{OP}\right]$

Note: From above it is clear that the electric field intensity at point A is independent of position of point A inside the cavity. Also the electric field is along the line joining the centres of the sphere and the spherical cavity.

(ii) $\vec{E}_B = \vec{E}_\rho + \vec{E}_{-\rho} = \frac{\rho(\overrightarrow{OB})}{3\epsilon_0} + \frac{k\left[\frac{4}{3}\pi r^3(-\rho)\right]}{(PB)^3}\overrightarrow{PB}$

(iii) $\vec{E}_C = \vec{E}_\rho + \vec{E}_{-\rho} = \frac{k\left(\frac{4}{3}\pi R^3\rho\right)}{OC^3}\overrightarrow{OC} + \frac{k\left(\frac{4}{3}\pi r^3(-\rho)\right)}{PC^3}\overrightarrow{PC}$

(iv) $\vec{E}_O = \vec{E}_\rho + \vec{E}_{-\rho} = 0 + \frac{k\left[\frac{4}{3}\pi r^3(-\rho)\right]}{[PO]^3}\overrightarrow{PO}$

EXAMPLE 94. A non-conducting solid sphere has volume charge density that varies as $\rho = \rho_0 r$, where ρ_0 is a constant and r is distance from centre. Find out electric field intensities at following positions.
(i) $r < R$ (ii) $r \geq R$

SOLUTION Method I: (i) For $r<R$: The sphere can be considered to be made of large number of spherical shells [Fig.1.224]. Each shell has uniform charge density. So, the previous results of the spherical shell can be used. Consider a shell of radius x and thickness dx as an element. Charge on shell $dq = (4\pi x^2 dx)\rho_0 x$. Therefore, the electric field intensity at point P due to shell,

$$dE = \frac{kdq}{x^2}$$

Since all the shells will have electric field in same direction

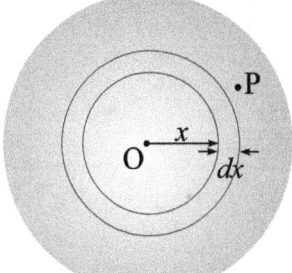

Figure 1.224

therefore, $E = \int_0^E dE = \int_0^r dE + \int_r^R dE$

Due to shells which lie between region $r < x \leq R$, electric field at point P will be zero.

$\vec{E} = \int_0^r \frac{kdq}{r^2} + 0 = \int_0^r \frac{k.4\pi x^2 dx \rho_0 x}{r^2} = \frac{4\pi k\rho_0}{r^2}\left[\frac{x^4}{4}\right]_0^r = \frac{\rho_0 r^2}{4\epsilon_0}\hat{r}$

(ii) $r \geq R$

$E = \int_0^R dE = \int_0^R \frac{k.4\pi x^2 dx\rho_0 x}{r^2} = \frac{\rho_0 R^4}{4\epsilon_0 r^2}\hat{r}$

1.34 Check Point 9

1. ●●A point charge Q is located just above the centre of the flat face of a hemisphere of radius R as shown in Figure 1.225.

What is the electric flux (a) through the curved surface and (b) through the flat face?

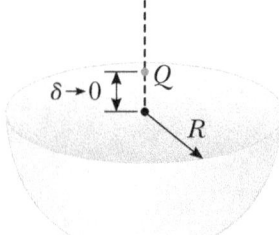

Figure 1.225

2. ••Fig.1.226(a) shows three charges. Draw these charges on your paper four times. Then draw two-dimensional cross sections of three-dimensional closed surfaces through which the electric flux is (a) $2q/\epsilon_0$, (b) q/ϵ_0, (c) 0, and (d) $5q/\epsilon_0$.

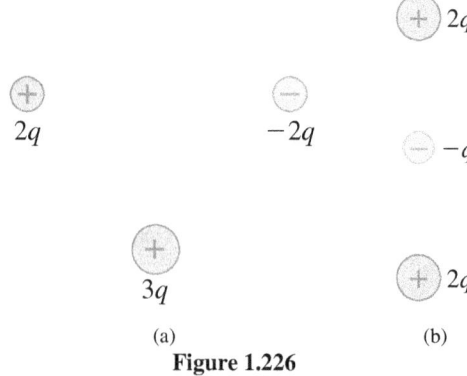

Figure 1.226

3. ••Fig.1.226(b) shows three charges. Draw these charges on your paper four times. Then draw two-dimensional cross sections of three-dimensional closed surfaces through which the electric flux is (a) $-q/\epsilon_0$, (b) q/ϵ_0, (c) $3q/\epsilon_0$, and (d) $4q/\epsilon_0$.

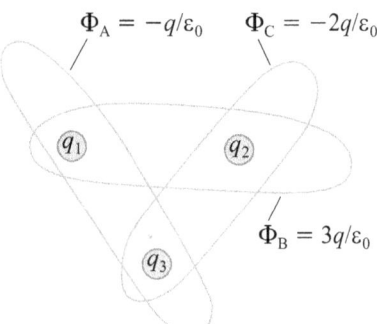

Figure 1.227

4. ••Fig.1.227 shows three Gaussian surfaces and the electric flux through each. What are the three charges q_1, q_2, and q_3?
5. ••As shown in Fig.1.228 a closed surface intersects a spherical conductor. What will be the nature of the electric flux coming out of the closed surface, if we place a negative charge at point P?
6. ••Find the electric field due to an infinitely long cylindrical charge distribution of radius R and having linear charge density λ at a distance half of the radius from its axis.
7. ••Five charges q_1, q_2, q_3, q_4, and q_5 are fixed at their positions as shown in Figure 1.229. S is Gaussian surface. The Gauss's law is given by $\int \vec{E} \cdot \vec{ds} = \frac{q}{\epsilon_0}$. Which of the following statements is correct?

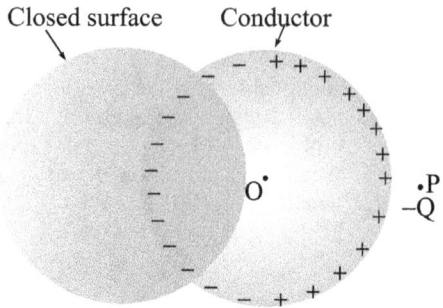

Figure 1.228

(A) $\vec{E}$ on the LHS of the above equation will have a contri-

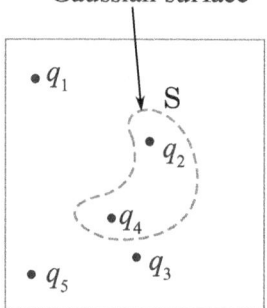

Figure 1.229

bution from q_1, q_5, and q_3 while q on the RHS will have a contribution from q_2 and q_4 only.
(B) $\vec{E}$ on the LHS of the above equation will have a contribution from all charges while q on the RHS will have a contribution from q_2 and q_4 only.
(C) $\vec{E}$ on the LHS of the above equation will have a contribution from all charges while q on the RHS will have a contribution from q_1, q_3, and q_5 only.
(D) Both $\vec{E}$ on the LHS and q on the RHS will have a contribution from q_2 and q_4 only.

8. ••What is the net electric flux through the torus (i.e., doughnut shape) of Fig.1.230

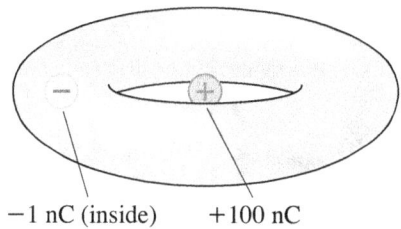

Figure 1.230

9. ••What is the net electric flux through the cylinder of Figure 1.231?

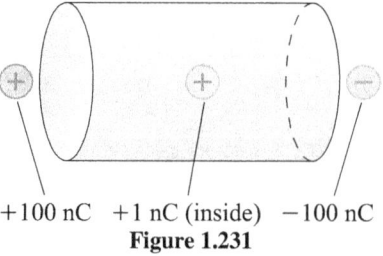

Figure 1.231

10. ••The net electric flux through an octahedron is $-1000 \text{ Nm}^2/\text{C}$. How much charge is enclosed within the octahedron?
11. ••55.3 million excess electrons are inside a closed surface.

What is the net electric flux through the surface?

12. ●●What is the electric flux through each face of a cube of side a (Fig.1.232) if a point charge of magnitude q is at one of its corner?

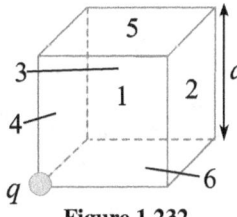

Figure 1.232

Multiple Choice Questions

13. ●●What is the flux through a cube of side a if a point charge of q is at one of its corner?
    (A) $2q/\epsilon_0$ (B) $q/8\epsilon_0$
    (C) q/ϵ_0 (D) $q/2\epsilon_0 6a^2$

14. ●A charge Q is enclosed by a Gaussian spherical surface of radius R. If the radius is doubled, then the outward electric flux will -
    (A) increase four times (B) be reduced to half
    (C) remain the same (D) be doubled

15. ●●A hollow cylinder has a charge q coulomb within it[Fig.1.233]. If ϕ is the electric flux in units of volt-meter associated with the curved surface B, the flux linked with the plane surface A in units of V.m will be-
    (A) $\frac{q}{2\epsilon_0}$ (B) $\frac{\Phi}{3}$
    (C) $\frac{q}{\epsilon_0} - \phi$ (D) $\frac{1}{2}\left(\frac{q}{\epsilon_0} - \phi\right)$

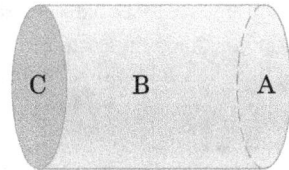

Figure 1.233

16. ●A charge q is located at the centre of a cube. The electric flux through any face is
    (A) $\frac{2\pi q}{6(4\pi\epsilon_0)}$ (B) $\frac{4\pi q}{6(4\pi\epsilon_0)}$
    (C) $\frac{\pi q}{6(4\pi\epsilon_0)}$ (D) $\frac{q}{6(4\pi\epsilon_0)}$

17. ●A charge Q μC is placed at the centre of a cube, the flux coming out from each face will be
    (A) $\frac{Q}{6\epsilon_0} \times 10^{-6}$ (B) $\frac{Q}{6\epsilon_0} \times 10^{-3}$
    (C) $\frac{Q}{24\epsilon_0}$ (D) $\frac{Q}{8\epsilon_0}$

18. ●A point charge $+q$ is placed at the centre of a cube of side l. The electric flux emerging from the cube is
    (A) $6ql^2/\epsilon_0$ (B) $q/6l^2\epsilon_0$
    (C) zero (D) q/ϵ_0

19. ●●A hollow metal sphere of radius R is uniformly charged. The electric field due to the sphere at a distance r from the centre
    (A) decreases as r increases for $r < R$ and for $r > R$
    (B) increases as r increases for $r < R$ and for $r > R$
    (C) zero as r increases for $r < R$, decreases as r increases for $r > R$
    (D) zero as r increases for $r < R$, increases as r increases for $r > R$

20. ●●Two parallel infinite line charges with linear charge densities $+\lambda$C/m and $-\lambda$C/m are placed at a distance of $2R$ in free space. The electric field midway between the two line charges?

    (A) $\frac{\lambda}{2\pi\epsilon_0 R}$ N/C (B) zero
    (C) $\frac{2\lambda}{\pi\epsilon_0 R}$ N/C (D) $\frac{\lambda}{\pi\epsilon_0 R}$ N/C

21. ●●The electric field at a distance $3R/2$ from the centre of a charged conducting spherical shell of radius R is E. The electric field at a distance $R/2$ from the centre of the sphere is
    (A) 0 (B) E (C) $E/2$ (D) $E/3$

22. ●●A hollow insulated conducting sphere is given a positive charge of 10μC. What will be the electric field at the centre of the sphere if its radius is 2 meter?
    (A) 20 μCm^{-2} (B) 5 μCm^{-2}
    (C) zero (D) 8 mCm^{-2}

23. ●●Figure1.234 shows a uniformly charged hemisphere of radius R. It has volume charge density ρ. If the electric field at a point A at $2R$ distance above its centre is E then the electric field at point B which is $2R$ below its centre, is?
    (A) $\frac{\rho R}{6\epsilon_0} + E$ (B) $\frac{\rho R}{12\epsilon_0} - E$
    (C) $\frac{-\rho R}{6\epsilon_0} + E$ (D) $\frac{\rho R}{24\epsilon_0} + E$

Figure 1.234

24. ●●For a spherically symmetrical charge distribution, electric field at a distance r from the centre of sphere is $\vec{E} = kr^7\hat{r}$, where k is a constant. What will be the volume charge density at a distance r from the centre of sphere?
    (A) $\rho = 9k\epsilon_0 r^6$ (B) $\rho = 5k\epsilon_0 r^3$
    (C) $\rho = 3k\epsilon_0 r^4$ (D) $\rho = 9k\epsilon_0 r^0$

25. ●●A long string with a charge of λ per unit length passes through an imaginary cube of edge a. The maximum possible flux of the electric field through the cube may be-
    (A) $\lambda a/\epsilon_o$ (B) $\sqrt{2}\lambda a/\epsilon_o$
    (C) $6\lambda a^2/\epsilon_o$ (D) $\sqrt{3}\lambda a/\epsilon_o$

26. ●●The total electric flux, leaving spherical surface of radius 1 cm, and surrounding an electric dipole is-
    (A) q/ϵ_0 (B) zero
    (C) $2q/\epsilon_0$ (D) $8\pi r^2 q/\epsilon_0$

27. ●●An electric dipole is placed inside a conducting shell. Mark the correct statement(s)
    (A) the flux of the electric field through the shell is zero
    (B) the electric field is zero at every point on the shell
    (C) the electric field is not zero anywhere on the shell
    (D) the electric field is zero on a circle on the shell.

1.35 Conductors in Electrostatic Equilibrium

1.35.1 Field in a Substance

1.35.1.1 Micro- and Macroscopic Fields

The real electric field in any substance (which is called the microscopic field) varies abruptly both in space and in time. It is different at different points of atoms and in the interstices. In order to find the intensity $\vec{E}$ of a real field at a certain point at a given instant, we should vectorially sum up the intensities of the fields of all individual charged particles of the substance, viz. electrons and nuclei. The solution of this problem is obviously not feasible. In any case, the result would be so complicated that it would be impossible to use it. Moreover, the knowledge of this field is not required for the solution of macroscopic problems. In many cases it is sufficient to have a simpler and rougher description which we shall be using henceforth.

Under the electric field $\vec{E}$ in a substance (which is called the macroscopic field) we shall understand the microscopic field averaged over space (in this case time averaging is worthless). This averaging is performed over what is called a *physically infinitesimal volume*, viz. the volume containing a large number of atoms and having the dimensions that are many times smaller than the distances over which the macroscopic field noticeably changes. The averaging over such volumes smoothens all irregular and rapidly varying fluctuations of the microscopic field over the distances of the order of atomic ones, but retains smooth variations of the macroscopic field over macroscopic distances.

Thus, the field in the substance is

$$E = E_{macro} = \langle E_{micro} \rangle \qquad (1.147)$$

1.35.1.2 The Influence of a Substance on a Field

If any substance is introduced into an electric field, the positive and negative charges (nuclei and electrons)* are displaced, which in turn leads to a partial separation of these charges. In certain regions of the substance, uncompensated charges of different signs appear. This phenomenon is called the electrostatic induction, while the charges appearing as a result of separation are called *induced charges*.

Induced charges create an additional electric field which in combination with the initial (external) field forms the resultant field. Knowing the external field and the distribution of induced charges, we can forget about the presence of the substance itself while calculating the resultant field, since the role of the substance has already been taken into account with the help of induced charges. Thus, the resultant field in the presence of a substance is determined simply as the superposition of the external field and the field of induced charges.

1.35.2 Fields Inside and Outside a Conductor

1.35.2.1 Inside a Conductor $E = 0$

Let us place a metallic conductor into an external electrostatic field or impart a certain charge to it. In both cases, the electric field will act on all the charges of the conductor, and as a result all the negative charges (free electrons) will be displaced in the direction against the field. This displacement (current) will continue until (this practically takes a small fraction of a second) a certain charge distribution sets in, at which the electric field at all the points inside the conductor vanishes. Thus, in the static case the electric field inside a conductor is absent ($E = 0$).

Further, since $E = 0$ everywhere in the conductor, the density of excess (uncompensated) charges inside the conductor is also equal to zero at all points ($\rho = 0$). This can be easily explained with the help of the Gauss theorem. Indeed, since inside the conductor $E = 0$, the flux of E through any closed surface inside the conductor is also equal to zero. And this means that there are no excess charges inside the conductor.

Excess charges appear only on the conductor surface with a certain density σ which is generally different for different points of the surface. *It should be noted that the excess surface charge is located in a very thin surface layer (whose thickness amounts to one or two interatomic distances).* When there is no net motion of electrons within the conductor, the conductor is said to be in electrostatic equilibrium. In electrostatic equilibrium, all conductors show following four properties:

1. The electric field is zero everywhere inside the conductor.
2. If an isolated conductor carries a charge, the charge resides on its surface.
3. The electric field just outside a charged conductor is perpendicular to the surface of the conductor and has a magnitude σ/ϵ_0, where σ is the surface charge density at that point.
4. On an irregularly shaped conductor, the surface charge density is greatest at locations where the radius of curvature of the surface is smallest.

Explanation of Above Properties on the Basis of Gauss's Law: Here, we verify only the first three properties for the conductors. The fourth property is presented here so that we have a complete list of properties for conductors in electrostatic equilibrium, but cannot be verified until the next Chapter "Electric Potential".

Property 1. We can elaborate more about the first property by considering the conducting slab in electrostatic equilibrium on the left of Fig.1.235, where the free electrons are uniformly distributed throughout the slab, i.e. $\vec{E}_{int} = 0$. When we place the slab in an external electric field $\vec{E}_{ext}$ as in the right part of Fig.1.235, the free electrons respond to the electric force $-e\vec{E}$ by moving-in the direction, opposite the field. In Fig.1.235, the direction of applied external electric field is towards right side, therefore electrons move to the left. In time, more negative and positive charges accumulate on the left and right surfaces, respectively. The resulting charge separation gives rise to an increasing electric field $\vec{E}_{int}$ within the conductor that's opposite to the applied field. As more charge moves, this internal field becomes stronger and after awhile, $\vec{E}_{int}$ will compensate $\vec{E}_{ext}$ resulting in a zero net electric field inside the conductor, i.e. $\vec{E}_{net} = \vec{E}_{ext} - \vec{E}_{int} = 0$. The time to reach this new electrostatic equilibrium is of the order 10^{-6} s.

Therefore, *The electric field is zero inside a conductor in elec-*

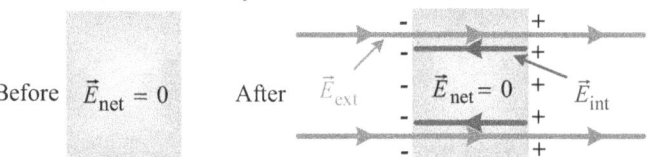

Figure 1.235: An external electric field $\vec{E}_{ext}$ creates an internal electric field $\vec{E}_{int}$ in the conductor such that the net electric field $\vec{E}_{net}$ is zero

trostatic equilibrium.

It could not be otherwise: Since a conductor contains free charges, the presence of any internal electric field would result in bulk charge motion, and we wouldn't have equilibrium. This result doesn't depend on the size or shape of the conductor, the magnitude or direction of the applied field, or even the nature of the material as long as it's a conductor. This is a macroscopic view; it considers only average fields within the material. At the atomic and molecular level, there are still strong electric fields near individual electrons and positive ions. But the average field, taken over larger distances, is zero inside a conductor in electrostatic equilibrium.

Explanation of Property 2. Figure 1.236a shows a cross-sectional view of a conductor with a Gaussian surface drawn just below the material surface. In equilibrium there's no electric field inside the conductor, and thus the field is zero everywhere on the Gaussian surface. Therefore, from Gauss's law, we have-

$$\phi_E = \oint \vec{E} \cdot d\vec{A} = \frac{Q_{encl}}{\epsilon_0}$$

---

*In conductors, only free electron can move not nuclei. So there will be a relative displacement between nuclei and electrons due to motion of free electrons only.

1.35. CONDUCTORS IN ELECTROSTATIC EQUILIBRIUM

$$\implies \oint 0\, d\vec{A} = \frac{Q_{encl}}{\epsilon_0}$$

$$\implies 0 = \frac{Q_{encl}}{\epsilon_0} \implies Q_{encl} = 0$$

Thus, net electric flux ϕ, through the Gaussian surface and hence net charge enclosed within the Gaussian surface is also zero. This is true no matter where the Gaussian surface is as long as it's inside the conductor. We can move it arbitrarily close to the conductor surface and it still encloses no net charge. If there is a net charge q_C (say) on the conductor, it lies outside the Gaussian surface, and therefore we conclude: If a conductor in electrostatic equilibrium carries a net charge, that charge must reside on the conductor surface.

A Cavity Inside the Conductor: Fig.1.236(b) shows a cross-

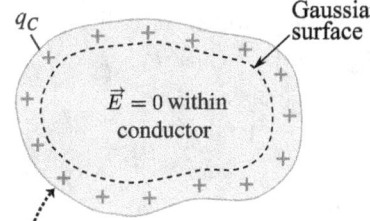

The charge q_C resides entirely on the surface of the conductor.

(a)

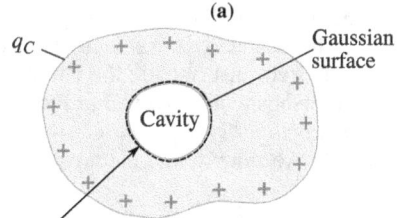

Because $\vec{E} = 0$ at all points within the conductor, the electric field at all points on the Gaussian surface must be zero.

(b)

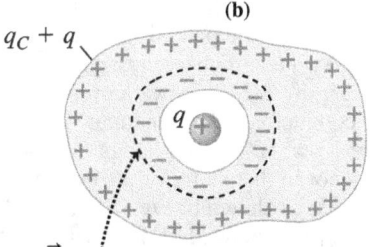

For $\vec{E}$ to be zero at all points on the Gaussian surface, the surface of the cavity must have a total charge $-q$.

(c)

Figure 1.236: (a) A cross sectional view of a solid conductor of arbitrary shape. The broken line represents a gaussian surface that can be as close to the surface of the conductor as we wish (b) The same conductor with an internal cavity (c) An isolated charge q placed in the cavity

sectional view of same charged conductor with a cavity that is totally within the conductor. Can there be charge on this interior surface? To find out, we place a Gaussian surface around the cavity, infinitesimally close but entirely within the conductor. Because $\vec{E} = 0$ inside the conductor, there can be no flux through this new Gaussian surface. Therefore, from Gauss' law, we have-

$$\phi_E = \oint \vec{E} \cdot d\vec{A} = \frac{Q_{encl}}{\epsilon_0}$$

$$\implies \oint 0\, d\vec{A} = \frac{Q_{encl}}{\epsilon_0} \implies 0 = \frac{Q_{encl}}{\epsilon_0}$$

$$\implies Q_{encl} = 0$$

So, we conclude that there is no net charge on the cavity walls; all the excess charge remains on the outer surface of the conductor, as in Fig.1.236b.

Suppose that, by some process, the excess charges could be "frozen" into position on the conductor surface of Fig.1.236b, perhaps by embedding them in a thin plastic coating, and suppose that the conductor could then be removed completely. This is equivalent to enlarging the cavity of Fig.1.236b until it consumes the entire conductor, leaving only the charges. The electric field pattern would not change at all; it would remain zero inside the thin shell of charge and would remain unchanged for all external points. The electric field is set up by the charges and not by the conductor. The conductor simply provides a pathway so that the charges can change their positions.

Now, Suppose we place a small body with a charge q inside a cavity within a conductor (Fig.1.236(c)). The conductor is uncharged and is insulated from the charge q. Again $\vec{E} = 0$ everywhere on the Gaussian surface. If we assume that the charge on the cavity wall is x, then by Gausses law. we have-

$$\phi_E = \oint \vec{E} \cdot d\vec{A} = \frac{Q_{encl}}{\epsilon_0} \implies \oint 0\, d\vec{A} = \frac{Q_{encl}}{\epsilon_0}$$

$$\implies 0 = \frac{Q_{encl}}{\epsilon_0} \implies Q_{encl} = 0 \implies q + x = 0 \implies x = -q$$

So according to Gauss's law there must be a charge $-q$ distributed on the surface of the cavity, drawn there by the charge q inside the cavity. Let charge y appears on the outer surface of conductor, then by charge conservation principle, we have

$$x + y = 0 \implies y = -x \implies y = -(-q) = q$$

Therefore, we conclude that the charge $+q$ must appear on the outer surface of the conductor.

If the conductor originally had a charge q_C, then charge conservation principle gives-

$$x + y = q_C \implies y = q_C - x = q_C - (-q) = q_C + q$$

i.e., the total charge on the outer surface must be $q_C + q$ after the charge q is inserted into the cavity.

Explanation of Property 3. There can't be an electric field within a conductor in electrostatic equilibrium, but there may be a field right at the conductor surface (Fig.1.237). Such a field must be perpendicular to the surface; otherwise, charge would move along the surface in response to the field's parallel component, and we wouldn't have equilibrium.

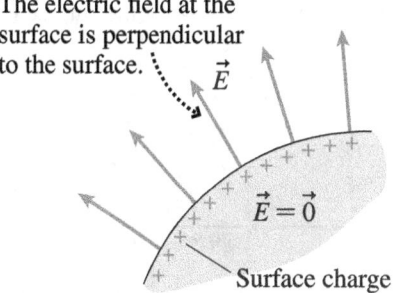

Figure 1.237: The electric field at the surface of a charged conductor is perpendicular to the conductor surface.

1.35.2.2 The Field Near a Conductor Surface

Here we will show that the electric field intensity in the immediate vicinity of the surface of a conductor is connected with the local charge density at the conductor surface through a simple relation. This relation can be established with the help of the Gauss theorem.

Suppose that the region of the conductor surface we are interested in borders on a vacuum. The field lines are normal to the conductor surface. Hence for a closed surface we shall take a small cylinder and arrange it as is shown in Fig.1.238. Then

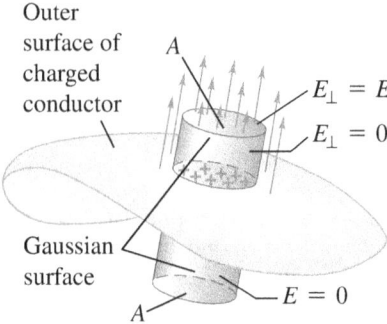

Figure 1.238: The field just outside a charged conductor is perpendicular to the surface, and its perpendicular component $E_\perp$ is equal to σ/ϵ_0.

the flux of $\vec{E}$ through this surface will be equal only to the flux through the "outer" endface of the cylinder. (the fluxes through the lateral surface and the inner endface are equal to zero). Thus we obtain $\vec{E} \cdot \vec{A} = E_\perp A = Q_{encl}/\epsilon_0 = \sigma \vec{A}/\epsilon_0$, where $E_\perp$ is the projection of vector $\vec{E}$ onto the outward normal (with respect to the conductor), A is the cross-sectional area of the cylinder and σ is the local surface charge density of the conductor. Cancelling both sides of this expression by A, we get

$$E_\perp = \frac{\sigma}{\epsilon_0} \quad (1.148)$$

If $\sigma > 0$, then $E_\perp > 0$, i.e., vector $\vec{E}$ is directed from the conductor surface (coincides in direction with the outward normal). If $\sigma < 0$, then $E_\perp < 0$, and vector $\vec{E}$ is directed towards the conductor surface.

Relation (1.148) may lead to the erroneous conclusion that the field $\vec{E}$ in the vicinity of a conductor depends only on the local charge density σ. This is not so. The intensity $\vec{E}$ is determined by all the charges of the system under consideration as well as the value of σ itself.

1.35.3 Mechanical Pressure (or Surface Density of Force) on the Surface of a Charged Conductor

Let us consider the case when a charged region of the surface of a conductor borders, is in a vacuum. It's surface charge density is σ. A small patch of charge on a conducting surface, is shown in Figure 1.239.

Total electric field anywhere outside the surface is -

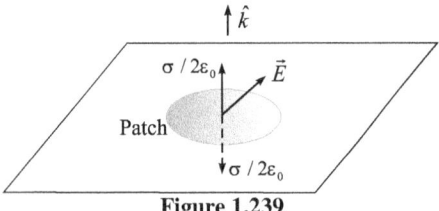

Figure 1.239

$$\vec{E} = \vec{E}_{patch} + \vec{E}' \quad (1.149)$$

where $\vec{E}_{patch}$ is the electric field due to charge on the patch, and $\vec{E}'$ is the electric field due all other charges. Since by Newton's third law, the patch cannot exert a force on itself, the force on the patch must come solely from $\vec{E}'$. Assuming the patch to be a flat surface, from Gauss's law, the electric field due to the patch is

$$\vec{E} = \begin{cases} \frac{+\sigma}{2\epsilon_0}\hat{k} & z > 0 \\ -\frac{\sigma}{2\epsilon_0}\hat{k} & z < 0 \end{cases} \quad (1.150)$$

By superposition principle, the electric field above the conducting surface is

$$\vec{E}_{above} = \left(\frac{\sigma}{2\epsilon_0}\right)\hat{k} + \vec{E}' \quad (1.151)$$

Similarly, below the conducting surface, the electric field is

$$\vec{E}_{below} = -\left(\frac{\sigma}{2\epsilon_0}\right)\hat{k} + \vec{E}' \quad (1.152)$$

Notice that $\vec{E}'$ is continuous across the boundary. This is due to the fact that if the patch were removed, the field in the remaining "hole" exhibits no discontinuity. Using the two equations above, we find

$$\vec{E}' = \frac{1}{2}\left(\vec{E}_{above} + \vec{E}_{below}\right) = \vec{E}_{avg} \quad (1.153)$$

But, for a thick conductor which is parallel to xy-plane,

$$\vec{E}_{above} = \left(\frac{\sigma}{\epsilon_0}\right)\hat{k} \text{ and } \vec{E}_{below} = 0,$$

therefore, above equation gives-

$$\vec{E}_{avg} = \frac{1}{2}\left(\frac{\sigma}{\epsilon_0}\hat{k} + 0\right) = \frac{\sigma}{2\epsilon_0}\hat{k} \quad (1.154)$$

Thus, the force acting on the patch is

$$\vec{F} = q\vec{E}_{avg} = (\sigma A)\frac{\sigma}{2\epsilon_0}\hat{k} = \frac{\sigma^2 A}{2\epsilon_0}\hat{k} \quad (1.155)$$

where A is the area of the patch. This is precisely the force needed to drive the charges on the surface of a conductor to an equilibrium state where the electric field just outside the conductor takes on the value σ/ϵ_0 and vanishes inside. *Note that irrespective of the sign of σ, the force tends to pull the patch into the field.*

Using the above result, we may define electrostatic pressure on the patch as

$$P = \frac{F}{A} = \frac{\sigma^2}{2\epsilon_0} = \frac{1}{2}\epsilon_0\left(\frac{\sigma}{\epsilon_0}\right)^2 = \frac{1}{2}\epsilon_0 E^2 \quad (1.156)$$

where E is the magnitude of the field just above the patch. The pressure is being transmitted via the electric field.

This electrostatic pressure on the patch is also called the surface density of force on the surface of charged conductor.

EXAMPLE 95. Figure 1.240 shows a cross-sectional view of a thick spherical conductor. The conductor is neutral, and a small charged sphere ($q = +29.5\ \mu C$) hangs from an insulating thread. The sphere is not in the centre of the conductor; instead, it is closer to the left side as shown in Fig. 1.240.

(a) Find the charge q_{wall} on the wall of the cavity and the charge

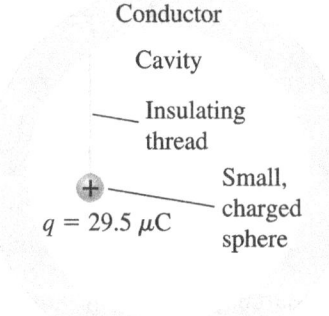

Figure 1.240: Cross-sectional view

q_{out} on the outer surface of the conductor. Start by sketching the electric field for all regions-inside the cavity, inside the body of the conductor, and outside the conductor.

(b) If the radius of the conductor is $R = 14.4$ cm, what is the magnitude of the electric field just outside the conductor?

1.35. CONDUCTORS IN ELECTROSTATIC EQUILIBRIUM

(a) APPROACH The positive charge on the small sphere attracts electrons in the conductor. These electrons move close to the walls of the cavity. If the positively charged sphere were in the centre of the cavity, the electrons would be uniformly distributed on the cavity wall. However, the electrons are more concentrated on the left side of the cavity because the positive sphere is closer to the left side (Fig.1.241). The electric field in the body of any conductor in electrostatic equilibrium is zero. By choosing a Gaussian sphere that is concentric with the conductor and embedded in it, we can determine the amount of charge on the walls of the cavity.

SOLUTION The electric flux through the Gaussian surface

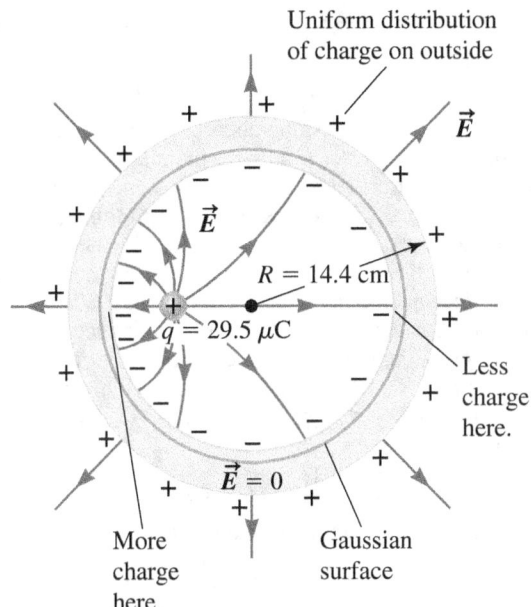

Figure 1.241

(Fig. 1.241) is zero because the electric field in the body of the conductor is zero. Therefore, the net charge inside the Gaussian sphere is zero, so the total charge on the inside wall of the cavity is negative, equal in magnitude to the charge on the small sphere inside.

$$\Phi_E = \oint \vec{E} \cdot d\vec{A} = 0 = \frac{Q_{encl}}{\epsilon_0}$$
$$\Rightarrow \quad Q_{encl} = 0 = q + q_{wall}$$
$$\Rightarrow \quad q_{wall} = -q = -29.5 \, \mu C$$

Because the conductor is neutral, therefore by charge conservation, the charge on its surface must be positive and equal to the charge on the cavity wall.

$$q_{out} = -q_{wall} = -(-29.5 \mu C) = +29.5 \mu C$$

The positive charge q_{out} on the outer surface of the conductor is uniformly distributed, as excess charge always is on the surface of a spherical conductor. To see why this is so, imagine that free electrons move toward the cavity wall, leaving positively charged ions in place.

(b) APPROACH The electric field just outside any conductor depends only on the surface charge density, so we need to find the surface charge density so that we can find the electric field.

SOLUTION The charge is uniformly distributed on a sphere of radius R. To find surface charge density, we divide the charge q_{out} by the surface area of the sphere.

$$\sigma = \frac{q_{out}}{A} = \frac{q_{out}}{4\pi R^2} = \frac{29.5 \times 10^{-6} C}{4\pi (14.4 \times 10^{-2} \, m)^2}$$
$$\Rightarrow \quad \sigma = 1.13 \times 10^{-4} C/m^2$$

The magnitude of the electric field just outside a conductor is given by-

$$E = \frac{\sigma}{\epsilon_0} = \frac{1.13 \times 10^{-4} C/m^2}{8.85 \times 10^{-12} C^2/N \cdot m^2} = 1.28 \times 10^7 \, N/C$$

EXAMPLE 96. The Fig.1.242 shows a cross section of a spherical metal shell of inner radius R. A point charge of $-5.0 \, \mu C$ is located at a distance $R/2$ from the centre of the shell. If the shell is electrically neutral, what are the (induced) charges on its inner and outer surfaces?

ANSWER Total induced charge on inner surface = $5.0 \, \mu C$. It's

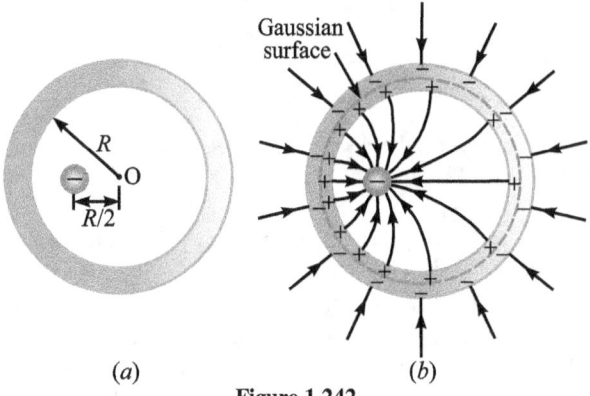

Figure 1.242

distribution is non-uniform. The total induce charge on outer surface = $-5.0 \, \mu C$. This charge will be distributed uniformly.

EXAMPLE 97. An uncharged spherical conductor centered at the origin has a cavity of some weird shape carved out of it (Fig.1.243). Somewhere within the cavity is a charge q. What is the field outside the sphere?

SOLUTION The answer is -

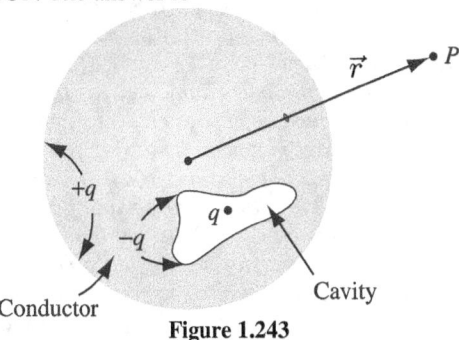

Figure 1.243

$$\vec{E} = \frac{1}{4\pi\epsilon_0} \frac{q}{r^2} \hat{r}$$

The conductor conceals from us all information concerning the nature of the cavity, revealing only the total charge it contains. How can this be? The charge $+q$ induces an opposite charge $-q$ on the wall of the cavity, which distributes itself in such a way that its field cancels that of q, for all points exterior to the cavity. Since the conductor carries no net charge, this leaves $+q$ to distribute itself uniformly over the surface of the sphere. (It is *uniform* because the asymmetrical influence of the point charge $+q$ is negated by that of the induced charge $-q$ on the inner surface.) For points outside the sphere, then, the only thing that survives is the field of the leftover $+q$ uniformly distributed over the outer surface.

1.35.4 Faraday Cage Protection

We have already seen that the electric field is zero inside any cavity within a conductor unless there is a charge in the cavity. This conclusion has an important practical application. For example, suppose we need to exclude the electric field from the region in Fig.1.244(a) enclosed within dashed lines. We can do so by surrounding this region with the neutral conducting box of Fig.1.244(b).

This region of space is now a cavity inside a conductor, thus

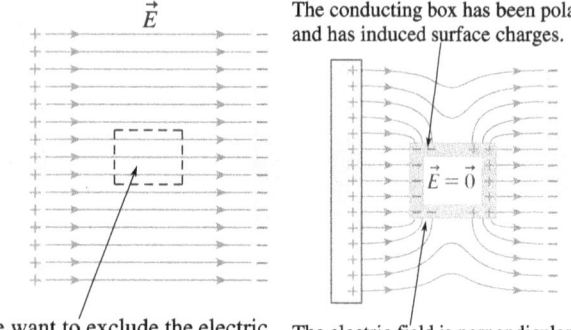

(a)   (b)

Figure 1.244: The electric field can be excluded from a region of space by surrounding it with a conducting box.

the interior electric field is zero. The use of a conducting box to exclude electric fields from a region of space is called ***screening***. Solid metal walls are ideal, but in practice wire screen or wire mesh provides sufficient screening for all the sensitive equipment. In such case, the exterior field is now very complicated. Such shielding enclosures are called Faraday cages, after Michael Faraday. In 1816, Faraday used them to demonstrate that no charge was present inside a conducting shell.

Applications: *We can use these ideas to explain how antistatic bags work and why your cell phone doesn't work in a closed elevator. The antistatic bag is coated with a conductor, and sensitive equipment is sealed inside. If the bag comes in contact with a charged object, the outside of the bag becomes charged but the inside remains uncharged, and the electric field inside the bag is zero. The antistatic bag is a Faraday cage. In order for your cell phone to work, it must be able to receive signals that consist of changing electric and magnetic fields. If your cell phone is inside a closed elevator, the elevator acts as a Faraday cage. The electric field inside the elevator is zero, so your cell phone cannot receive signals. The same physics tells you that one of the safest places to be in a lightning storm is inside a car; if the car is struck by lightning, the charge tends to remain on the metal skin of the vehicle, and little or no electric field is produced inside the passenger compartment.*

EXAMPLE 98. A conducting sphere of radius R carries a net positive charge $2Q$. A conducting spherical shell of inner radius R_1 ($R_1 > R$) and outer radius R_2 carries a net negative charge $-Q$. This shell is concentric with the conducting sphere. Find the magnitude of the electric field at a distance r away from the common centre when: (a) $r < R$, (b) $R < r < R_1$, (c) $R_1 < r < R_2$, and (d) $r > R_2$.

SOLUTION The charge distributions under consideration are characterized by being spherically symmetrical around the common centre c. This suggests that a spherical Gaussian surface of radius r is to be constructed in each case such as S_1, S_2, S_3, and S_4 that are displayed in Fig.1.245. In addition, we use the fact that the electric field inside a conductor is zero and all the excess charge will lie entirely on the outer surface of the isolated conductor.

(a) In this region the Gaussian sphere S_1 of Fig. 1.245 satisfies the condition $r < R$. Because there is no charge inside the conductor in this region, i.e. $Q_{encl} = 0$; then, $E_1 = 0$.

(b) In this region the Gaussian sphere S_2 of Fig. 1.245 satisfies the condition $R < r < R_1$. Because $Q_{encl} = 2Q$ inside this surface

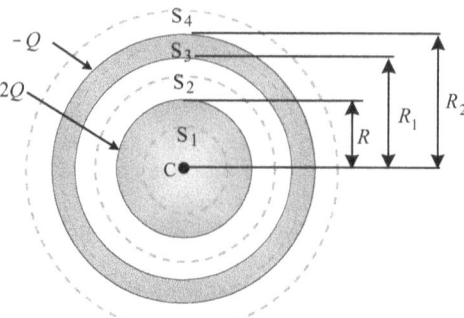

Figure 1.245

and because $\oint \vec{E}_2 \cdot d\vec{A} = E_2(4\pi r^2)$, we can use Gauss's law to find:

$$E_2 = \frac{1}{4\pi\epsilon_0}\frac{2Q}{r^2} = k\frac{2Q}{r^2} \quad (R < r < R_1)$$

(c) In this region, the Gaussian sphere S_3 of Fig. 1.245 satisfies the condition $R_1 < r < R_2$. Because the electric field inside an equilibrium conductor is zero, i.e. $E_3 = 0$; then, based on Gauss's law, the net charge Q_{encl} must be zero. From this argument, we find that an induced charge $-2Q$ must be established on the inner surface of the shell to cancel the charge $+2Q$ on the solid sphere. In addition, because the net charge on the whole shell is $-Q$, we conclude that its outer surface must carry an induced charge $+Q$.

(d) In this region, the Gaussian sphere S_4 of Fig. 1.245 satisfies the condition $r > R_2$. Because $Q_{encl} = 2Q - Q = Q$ inside this surface and because $\oint \vec{E}_4 \cdot d\vec{A} = E_4(4\pi r^2)$, we can use Gauss's law to find:

$$E_4 = \frac{1}{4\pi\epsilon_0}\frac{Q}{r^2} = k\frac{Q}{r^2} \quad (r > R_2)$$

Figure 1.246 shows a graphical representation of the variation of

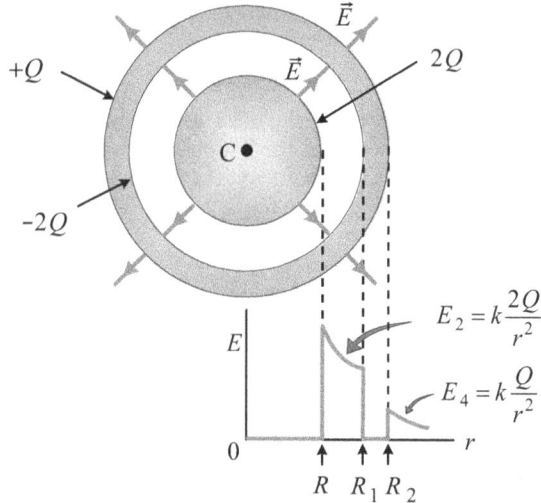

Figure 1.246

the electric field E with r. In addition, the Fig.1.246 shows the final distribution of the charge on the two conductors.

EXAMPLE 99. **Electric Field Due to Plane Plus Point Charge:** A very large (and thin) non-conducting plane slab carries a uniform surface charge density σ (assume positively charged). A single positive point charge, $+Q$, is located a distance d above the plane. What is the electric field at a point between the point charge and the plane and at a distance b from the plane?

APPROACH The situation is drawn in Fig.1.247. We will use the superposition principle to solve this problem.

SOLUTION The electric field at point P due to the charged plane is-

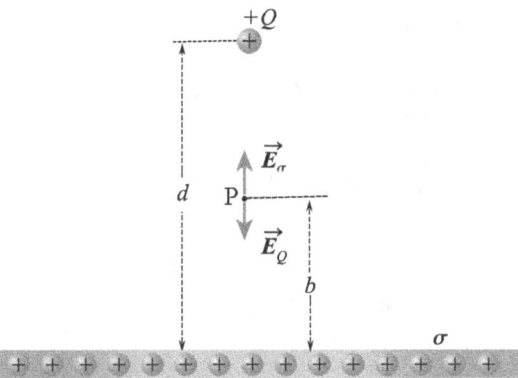

Figure 1.247: Situation of EXAMPLE 99 with the electric field at point P shown separately due to the non-conducting plane of charge $\vec{E}_\sigma$ and the point charge $\vec{E}_Q$

$$E_\sigma = \frac{\sigma}{2\epsilon_0}$$

The direction of this electric field is perpendicular to the plane and away from it (due to the positive charge), as shown in Figure 1.247

The electric field at point P due to the point charge Q is-

$$E_Q = k\frac{Q}{(d-b)^2}$$

The direction of this electric field is away from positive point charge Q and directed perpendicularly towards the charged plane. The net electrical field at P is given by-

$$\vec{E}_{net} = \vec{E}_\sigma + \vec{E}_Q = \left(\frac{\sigma}{2\epsilon_0} - k\frac{Q}{(d-b)^2}\right)\hat{k}$$

The net electric field could be either upward or downward, depending on the relative magnitudes of the point charge and the surface charge density.

Note: While the superposition technique can be powerful, you have to be careful not to use it when it is not valid. If you have a conductor, the presence of another charged object will change the distribution of charges on the conductor, so we could not assume that the plane is still uniformly charged.

EXAMPLE 100. Why is it normally not possible to use the technique of superposition along with Gauss's law when one of the objects is a conductor?
(A) Conductors have zero interior electric fields.
(B) Charges on conductors are all on the outer surface.
(C) The electric fields on conductors (just outside) are perpendicular to the surface.
(D) Charges in conductors are free to move, which will often destroy the symmetry needed for Gauss's law.
ANSWER: (D) While the first three statements are true, that is not why we can't use the Gauss's law plus superposition technique. When we have two charged objects, and one of them is a conductor, the presence of the other object will distort the symmetry of the charge distribution and we can no longer make assumptions such as a constant-magnitude of electric field on a surface.

EXAMPLE 101. Two conducting plates A and B are placed parallel to each other. A is given a charge Q_1 and B a charge Q_2. Find the distribution of charges on the four surfaces.
SOLUTION Consider a Gaussian surface as shown in Fig.1.248. Two faces of this closed surface lie completely inside the conductor where the electric field is zero. The flux through these faces is, therefore, zero. The other parts of the closed surface which are outside the conductor are parallel to the electric field and hence the flux on these parts is also zero. The total flux of the electric field through the closed surface is, therefore, zero. So, from Gauss's law, the total charge inside this closed surface should be zero. The charge on the inner surface of A should be equal and opposite to the inner surface of B.

The distribution should be like the one shown in figure1.249. To

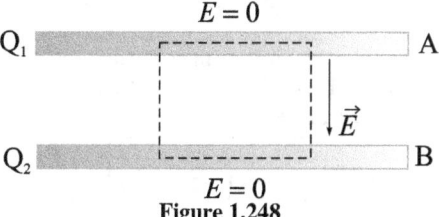

Figure 1.248

find the value of q, consider the field at a point P inside the plate A. Suppose, the surface area of each plate is A. Let σ_1, σ_2, are respective surface charge densities on the outer, inner surfaces of the plate A and σ_3, σ_4 are respectively the surface charge densities on inner and outer surface of the plate B.

Now, the electric field at point P inside the plate A-

$$\begin{array}{c}Q_1 - q \quad P\\ \hline Q_1 \quad\quad\quad\quad\quad\bullet\quad\quad A\\ +q\end{array}$$

$$\begin{array}{c} -q\\ \hline Q_2 \quad\quad\quad\quad\quad\quad\quad B\\ Q_2 + q\end{array}$$

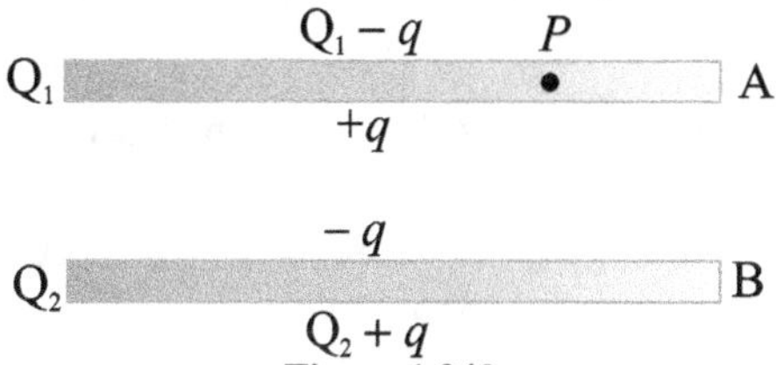

Figure 1.249

(i) due to the charge $Q_1 - q$ on the outer surface of plate A, is given by-

$$E_1 = \frac{\sigma_1}{2\epsilon_0}$$

Since, $\sigma_1 = (Q_1 - q)/A$, therefore,

$$E_1 = \frac{Q_1 - q}{2A\epsilon_0} \quad \text{(downwards)},$$

(ii) due to the charge $+q$ on the inner surface of plate A,

$$E_2 = \frac{\sigma_2}{2\epsilon_0}$$

Since, $\sigma_2 = q/A$, therefore,

$$E_2 = \frac{q}{2A\epsilon_0} \quad \text{(upwards)},$$

(iii) Due to the charge $-q$ on the inner surface of plate B,

$$E_3 = \frac{\sigma_3}{2\epsilon_0}$$

Since, $\sigma_3 = q/A$, therefore,

$$E_3 = \frac{q}{2A\epsilon_0} \quad \text{(downwards)},$$

(iv) due to the charge $Q_2 + q$ on the outer surface of the plate B,

$$E_4 = \frac{\sigma_4}{2\epsilon_0}$$

Since, $\sigma_3 = (Q_2 + q)/A$, therefore,

$$E_4 = \frac{Q_2 + q}{2A\epsilon_0} \quad \text{(upward)},$$

If we consider, upward direction as positive direction of electric field, then, net electric field at point P due to all the four charged surfaces is (in the downward direction)

$$E = -E_1 + E_2 - E_3 + E_4$$
$$= -\frac{Q_1 - q}{2A\epsilon_0} + \frac{q}{2A\epsilon_0} - \frac{q}{2A\epsilon_0} + \frac{Q_2 + q}{2A\epsilon_0}$$
$$= -\frac{Q_1}{2A\epsilon_0} + \frac{2q}{2A\epsilon_0} + \frac{Q_2}{2A\epsilon_0}$$
or $$E = \frac{Q_2 - Q_1}{2A\epsilon_0} + \frac{2q}{2A\epsilon_0}$$

As the point P is inside the conductor, this field should be zero. Hence,

$$E = \frac{Q_2 - Q_1}{2A\epsilon_0} + \frac{2q}{2A\epsilon_0} = 0$$

$$\Rightarrow \boxed{q = \frac{Q_1 - Q_2}{2}} \quad (1.157)$$

It is the electric charge on the inner surface of plate A (which has the net electric charge Q_1). The electric charge on the inner surface of the plate B, is $-q$, i.e.,

$$\boxed{-q = -\left(\frac{Q_1 - Q_2}{2}\right)} \quad (1.158)$$

Thus, the charge on outer surface of the plate A-

$$\boxed{Q_1 - q = \frac{Q_1 + Q_2}{2}} \quad (1.159)$$

and the charge on the outer surface of the plate B-

$$\boxed{Q_2 + q = \frac{Q_1 + Q_2}{2}} \quad (1.160)$$

From Eq.1.159 and 1.160, we see that the charge distribution on the outer surfaces of plates A and B remains identical. Final

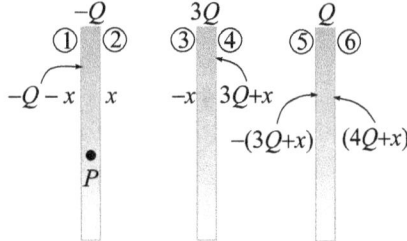

Figure 1.250

charge distribution is shown in Fig.1.250

EXAMPLE 102. Figure 1.251 shows three large metallic plates with charges $-Q$, $3Q$ and Q respectively. Determine the final charges on all the surfaces.

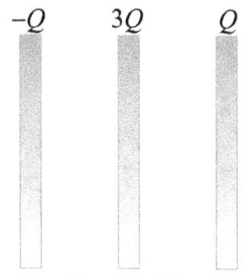

Figure 1.251

APPROACH In Fig.1.252 we have numbered each surface of the given plates. Now, assume that the charge on surface 2 is x. Following conservation of charge, we see that surfaces 1 has charge $(-Q - x)$ i.e., $-(Q + x)$. Similarly, we can also find the charge distribution on other plates. Since, in electrostatic equilibrium, the net electric field inside a metal plate is always zero, therefore, the net electric field at P should be zero, i.e., $E_P = 0$.

SOLUTION Resultant electric field at point P inside the left most plate -

$$\frac{Q+x}{2A\epsilon_0} + \frac{x}{2A\epsilon_0} + \frac{3Q}{2A\epsilon_0} + \frac{Q}{2A\epsilon_0} = 0$$

(Since electric fields due to all surface charges are directed in same direction (left))

$$\Rightarrow \quad 2x + 5Q = 0 \quad \Rightarrow x = -5Q/2$$

Therefore, charge on surface ①

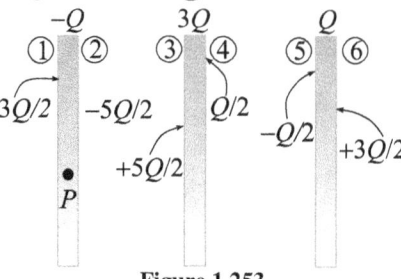

Figure 1.253

$$= -(Q + x) = -\left(Q - \frac{5Q}{2}\right) = +\frac{3Q}{2}$$

Since, the front surfaces of two parallel metal plates always acquire equal and opposite electric charges, therefore, electric charge on surface ③ will be $-x$, i.e., $+5Q/2$.

By charge conservation, the electric charge on surface ④

$$= 3Q - \frac{5Q}{2} = \frac{Q}{2}$$

Therefore, the electric charge on the opposite front surface ⑤ will be $-Q/2$.

Again by charge conservation, the electric charge on the surface ⑥-

$$= Q - \left(-\frac{Q}{2}\right) = \frac{3Q}{2}$$

Final charge distribution is shown in Fig.1.253.

EXAMPLE 103. An isolated conducting sheet of area A and carrying a charge Q is placed in a uniform electric field $\vec{E}$, such that electric field is perpendicular to sheet and covers all the sheet[Fig.1.254]. Find out charges appearing on its two surfaces.

SOLUTION Let the electric charge on the left surface of the

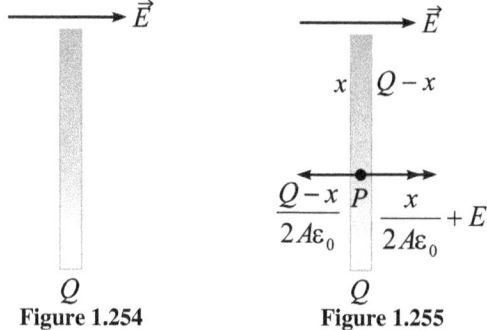

Figure 1.254   **Figure 1.255**

plate is x, then by charge conservation, the electric charge on the right surface of the plate will be $Q - x$ [see Fig.1.255].

Now, consider a point P inside the metal plate. Find the net electric field at P and equate it with zero because in electrostatic static equilibrium the net electric field inside the plate should always be zero, i.e., $E_P = 0$

From Fig.1.255, the net electric field at P–

$$E_P = \frac{x}{2A\epsilon_0} - \frac{Q-x}{2A\epsilon_0} + E = 0$$

$$\Rightarrow \quad 2x + 2A\epsilon_0 E - Q = 0 \quad \Rightarrow \quad x = \frac{Q}{2} - A\epsilon_0 E$$

So charge on one side is $\frac{Q}{2} - A\epsilon_0 E$ and other side

$$= Q - x = Q - \left(\frac{Q}{2} - A\epsilon_0 E\right) = \frac{Q}{2} + A\epsilon_0 E$$

Remark Solve this question for $Q = 0$ without using the above answers and match that answers with the answers that you will get by putting $Q = 0$ in the above answers.

1.36 Check Point 10

1. ••The electric field strength just above one face of a copper penny is 2000 N/C. What is the surface charge density on this face of the penny?
2. ••A spark occurs at the tip of a metal needle if the electric field strength exceeds 3.0×10^6 N/C, the field strength at which air breaks down. What is the minimum surface charge density for producing a - spark?
3. ••The conducting box in Fig.1.256 has been given an excess negative charge. The surface density of excess electrons at the centre of the top surface is 5.0×10^{10} electrons/m². What are the electric field strengths E_1 to E_3 at points 1 to 3?

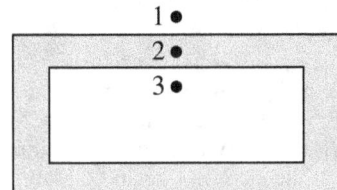

Figure 1.256

4. ••A thin, horizontal, 10 cm diameter copper plate is charged to 3.5 nC. If the electrons are uniformly distributed on the surface, what are the strength and direction of the electric field
   (a) 0.1 mm above the centre of the top surface of the plate?
   (b) at the plate's centre of mass?
   (c) 0.1 mm below the centre of the bottom surface of the plate?
5. ••Figure 1.257 shows a hollow cavity within a neutral conductor. A point charge Q is inside the cavity. What is the net electric flux through the closed surface that surrounds the conductor?

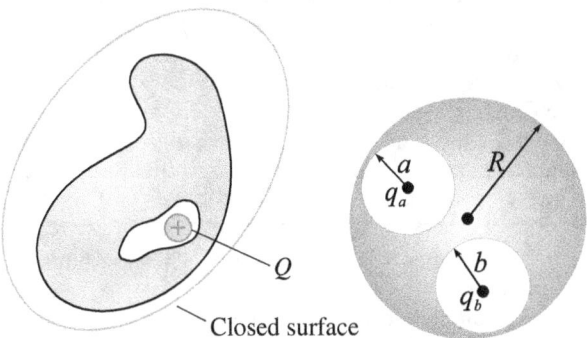

Figure 1.257   **Figure 1.258**

6. ••Two spherical cavities, of radii a and b, are hollowed out from the interior of a (neutral) conducting sphere of radius R (Fig.1.258). At the centre of each cavity a point charge is placed-call these charges q_a and q_b.
   (a) Find the surface charge densities σ_a, σ_b, and σ_R.
   (b) What is the field outside the conductor?
   (c) What is the field within each cavity?
   (d) What is the force on q_a and q_b?
   (e) Which of these answers would change if a third charge, q_c, were brought near the conductor?
7. ••Electronic charge e may be determined by Millikan's oil drop method. Oil drops of radius r acquire a terminal speed v_1 with downward electric field E and a speed v_2 with the upward electric field. Find the value of e in terms of E, v_1, v_2, r and η, the viscosity of oil in air.

Multiple Choice Questions

8. ••A soap bubble has radius R, charge Q, surface tension T. Find the excess pressure in it.
   (A) $\dfrac{32\pi^2 R^2 \epsilon_0 T - q^2}{32\pi^2 R^4 \epsilon_0}$
   (B) $\dfrac{64\pi^2 R^3 \epsilon_0 T - q^2}{32\pi^2 R^4 \epsilon_0}$
   (C) $\dfrac{128\pi^2 R^3 \epsilon_0 T - q^2}{32\pi^2 R^4 \epsilon_0}$
   (D) none of these

1.37 Questions and Exercises

1.37.1 Conceptual Questions

1. Why does a plastic ruler that has been rubbed with a cloth have the ability to pick up small pieces of paper? Why is this difficult to do on a humid day?
2. A popular classroom demonstration consists of rubbing a plastic rod with fur to give the rod charge, and then placing the rod near an empty soda can that is on its side (Figure 1.259). Explain why the can will roll toward the rod.

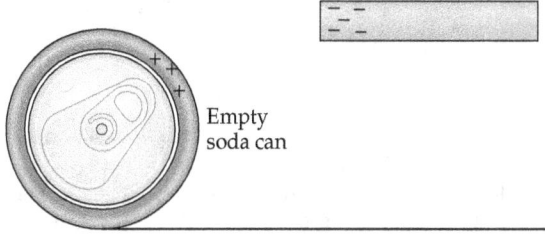

Figure 1.259

3. A metal sphere is positively charged. Is it possible for the sphere to electrically attract another positively charged ball? Explain your answer
4. We are not normally aware of the gravitational or electric force between two ordinary objects. What is the reason in each case? Give an example where we are aware of each one and why.
5. What are the two major reasons for introducing the field concept?
6. Explain why the test charges we use when measuring electric fields must be small.
7. When determining an electric field, must we use a *positive* test charge, or would a negative one do as well? Explain.
8. Why do electric field lines point away from positive charges and toward negative charges?
9. An electron initially moving horizontally near Earth's surface enters a uniform electric field and is deflected upward. What can you say about the direction of the electric field (assuming no other interaction)? What can you say about the direction of the electric field if the electron is deflected downward?
10. Consider the electric field at the three points indicated by the letters A, B, and C in Fig. 1.260. First draw an arrow at each point indicating the direction of the net force that a positive test charge would experience if placed at that point, then list the letters in order of *decreasing* field strength (strongest first). Explain.
11. Given two point charges, Q and $2Q$, a distance l apart, is there a point along the straight line that passes through them where $E = 0$ when their signs are (a) opposite, (b) the same? If yes, state roughly where this point will be.
12. Consider a small positive test charge located on an electric field line at some point, such as point P in Fig.1.261. Is the direction of the velocity and/or acceleration of the test charge along this line? Discuss.

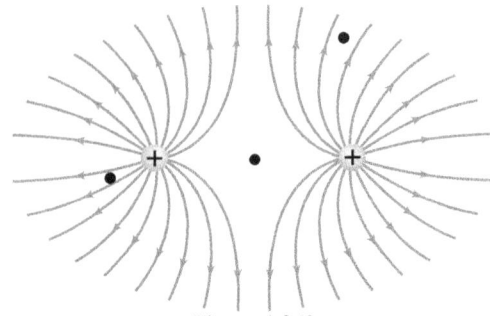

Figure 1.260

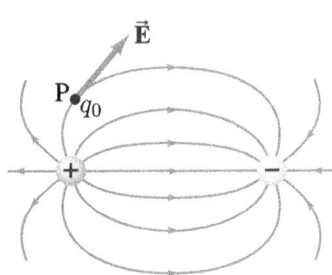

Figure 1.261

13. Is it possible for a dipole that is initially moving in a straight line to be deflected by a uniform electric field?
14. An electron is travelling in circular motion at constant speed due to the effect of an electric field. Is it possible for the electric field to be uniform?
15. A point charge is surrounded by a spherical Gaussian surface of radius r. If the sphere is replaced by a cube of side r, will Φ_E be larger, smaller, or the same? Explain.
16. Eight electrons are the only charged particles inside an isolated balloon, producing a field line flux through its surface. (a) If eight additional electrons are set outside the balloon to form the eight corners of a cube, how does the field line flux through the balloon change? (b) How does the flux change if the eight additional electrons are all placed at one location outside the balloon instead of being distributed at the corners of a cube?
17. Under what conditions is the electric field magnitude nonzero inside the cavity enclosed by a hollow conducting sphere?
18. What is the distinction between field line flux and electric flux? Why is electric flux the preferred variable for analyzing electric fields?
19. Objects are composed of atoms which are composed of charged particles (protons and electrons); however, we rarely observe the effects of the electrostatic force. Explain why we do not observe these effects.
20. Occasionally, people who gain static charge by shuffling their feet on the carpet will have their hair stand on end. Why does this happen?

1.37.2 Problems

Discrete Charge Distributions

1. •How many electrons are required to yield a total charge of 1.00 C?
2. •A plastic rod is rubbed against a wool shirt, thereby acquiring a charge of $-0.80\ \mu C$. How many electrons are transferred from the wool shirt to the plastic rod?
3. •What is the total charge of all of the protons in 1.00 kg of carbon (C)? (Given: atomic number of C is 6 and it's molar mass is 12 gram)
4. ••The faraday is a unit of charge frequently encountered in electrochemical applications and named for the British physicist and chemist Michael Faraday. It consists of exactly 1 mole of elementary charges (i.e., 1 mol of electrons). Calculate the number of coulombs in 1 faraday (1F).
5. ••Suppose a cube of aluminum which is 1.00 cm on a side accumulates a net charge of $+2.50$ pC. (a) What percentage of the electrons originally in the cube was removed? (b) By what percentage has the mass of the cube decreased because of this removal?

Electric Force

6. ••What is the magnitude of the electric force between two protons separated by 2.00×10^{-15} m?
7. •A point charge $q_1 = 4.0\ \mu C$ is at the origin and a point charge $q_2 = 6.0\ \mu C$ is on the x axis at $x = 3.0$ m. (a) Find the electric force on charge q_2. (b) Find the electric force on q_1. (c) How would your answers for Parts (a) and (b) differ if q_2 were $-6.0\ \mu C$?
8. •Three point charges are on the x-axis: $q_1 = -6.0\mu C$ is at $x = -3.0$ m, $q_2 = 4.0\ \mu C$ is at the origin, and $q_3 = -6.0\ \mu C$ is at $x = 3.0$ m. Find the electric force on q_1.
9. ••A 2.0 μC point charge and a 4.0 μC point charge are a distance L apart. Where should a third point charge be placed so that the electric force on that third charge is zero?
10. ••A $-2.0\ \mu C$ point charge and a 4.0 μC point charge are a distance L apart. Where should a third point charge be placed so that the electric force on that third charge is zero?
11. ••A point particle that has a charge of $-2.5\ \mu C$ is located at the origin. A second point particle that has a charge of 6.0 μC is at $x = 1.0$ m, $y = 0.50$ m. A third point particle, an electron, is at a point with coordinates (x, y). Find the values of x and y such that the electron is in equilibrium.
12. •••Five identical point charges, each having charge Q, are equally spaced on a semicircle of radius R as shown in Figure 1.262. Find the force (in terms of k, Q, and R) on a charge q located equidistant from the five other charges.
13. •••The structure of the NH_3 molecule is approximately that of an equilateral tetrahedron, with three H^+ ions forming the base and an N^{3-} ion at the apex of the tetrahedron. The length of each side is 1.64×10^{-10} m. Calculate the electric force that acts on each ion.

The Electric Field

14. •A point charge of 4.0 μC is at the origin. What are the magnitude and direction of the electric field on the x axis at (a) $x = 6.0$ m and (b) $x = -10$ m ? (c) Sketch the function E_x versus x for both positive and negative values of x. (Remember that E_x is negative when $\vec{E}$ points in the $-x$ direction.)
15. ••Two point charges, each $+4.0\mu C$, are on the x axis; one point charge is at the origin and the other is at $x = 8.0$ m. Find the electric field on the x axis at (a) $x = -2.0$ m, (b) $x = 2.0$ m, (c) $x = 6.0$ m, and (d) $x = 10$m. (e) At what point on the x axis is the electric field zero? (f) Sketch E_x versus x for -3.0 m $< x < 11$ m.
16. ••A point charge of $+5.0\ \mu C$ is located on the x axis at $x = -3.0$ cm, and a second point charge of $-8.0\ \mu C$ is located on the x axis at $x = +4.0$ cm. Where should a third charge of $+6.0\ \mu C$ be placed so that the electric field at the origin is zero?
17. ••A $-5.0\ \mu C$ point charge is located at $x = 4.0$ m, $y = -2.0$ m and a 12 μC point charge is located at $x = 1.0$ m, $y = 2.0$ m. (a) Find the magnitude and direction of the electric field at $x = -1.0$ m, $y = 0$. (b) Calculate the magnitude and direction of the electric force on an electron that is placed at $x = -1.0$ m, $y = 0$.
18. ••Two point particles, each having a charge q, sit on the base

of an equilateral triangle that has sides of length L as shown in Figure 1.263. A third point particle that has a charge equal to 2q sits at the apex of the triangle. Where must a fourth point particle that has a charge equal to q be placed in order that the electric field at the centre of the triangle be zero? (The centre is in the plane of the triangle and equidistant from the three vertices.)

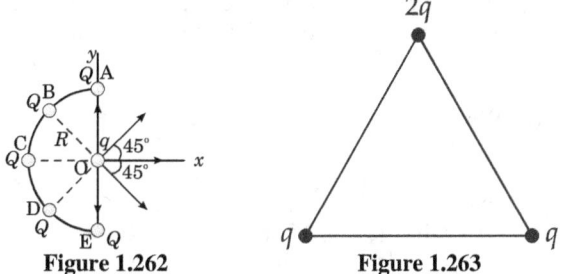

Figure 1.262 **Figure 1.263**

19. • • •Two equal positive point charges $+q$ are on the y axis; one is at $y = +a$ and the other is at $y = -a$. The electric field at the origin is zero. A test charge q_0 placed at the origin will therefore be in equilibrium. (a) Discuss the stability of the equilibrium for a positive test charge by considering small displacements from equilibrium along the x axis and small displacements along the y axis. (b) Repeat Part (a) for a negative test charge. (c) Find the magnitude and sign of a charge q_0 that when placed at the origin results in a net force of zero on each of the three charges.

20. ••Two equal positive charges q are on the y axis; one point charge is at $y = +a$ and the other is at $y = -a$. (a) Show that on the x axis the x component of the electric field is given by $E_x = 2kqx/(x^2 + a^2)^{3/2}$. (b) Show that near the origin, where x is much smaller than a, $E_x \approx 2kqx/a^3$. (c) Show that for values of x much larger than a, $E_x \approx 2kq/x^2$. Explain why a person might expect this result even without deriving it by taking the appropriate limit.

21. ••Two positive point charges $+q$ are on the y axis at $y = +a$ and $y = -a$. A bead of mass m and charge $-q$ slides without friction along a taut thread that runs along the x axis. Let x be the position of the bead. (a) Show that for $x \ll a$, the bead experiences a linear restoring force (a force that is proportional to x and directed toward the equilibrium position at $x = 0$) and therefore undergoes simple harmonic motion. (b) Find the period of the motion.

22. •The acceleration of a particle in an electric field depends on the charge-to-mass ratio of the particle. (a) Compute q/m for a proton, and find its acceleration in a uniform electric field that has a magnitude of 100 N/C. (b) Find the time it takes for a proton initially at rest in such a field to reach a speed of $0.01c$* (where c is the speed of light). Given: mass of proton = 1.673×10^{-27} kg and charge of a proton = 1.602×10^{-19} C.

23. ••An electron is released from rest in a weak electric field $\vec{E} = (-1.50 \times 10^{-10} \text{ N/C})\hat{j}$. After the electron has travelled a vertical distance of $1.0\,\mu$m, what is its speed? (Do not neglect the gravitational force on the electron.)

24. ••A 2.00 g charged particle is released from rest in a region that has a uniform electric field $\vec{E} = (300 \text{ N/C})\hat{i}$. After travelling a distance of 0.500 m in this region, the particle has a kinetic energy of 0.120 J. Determine the charge of the particle.

---

*When the speed of an electron approaches the speed of light c, relativistic kinematics must be used to calculate its motion, but at speeds of $0.01c$ or less, non-relativistic kinematics is sufficiently accurate for most purposes.

25. ••A charged particle leaves the origin with a speed of 3.00×10^6 m/s at an angle of 35° above the x axis. A uniform electric field given by $\vec{E} = -E_0\hat{j}$ exists throughout the region. Find E_0 such that the particle will cross the x axis at $x = 1.50$ cm if the particle is (a) an electron, and (b) a proton.

Dipoles

26. ••Two point charges, $q_1 = 2.0$ pC and $q_2 = -2.0$ pC, are separated by 4.0 μm. (a) What is the magnitude of the dipole moment of this pair of charges? (b) Sketch the pair and show the direction of the dipole moment.

27. ••An electron (charge $-e$, mass m) and a positron (charge $+e$, mass m) revolve around their common centre of mass under the influence of their attractive coulomb force. Find the speed v of each particle in terms of e, m, k, and their separation distance L.

28. ••A point particle of mass m and charge q is constrained to move vertically inside a narrow, friction less cylinder (Figure 1.264). At the bottom of the cylinder is a point charge Q having the same sign as q. (a) Show that the particle whose mass is m will be in equilibrium at a height $y_0 = (kqQ/mg)^{1/2}$ (b) Show that if the particle is displaced from its equilibrium position by a small amount and released, it will exhibit simple harmonic motion with angular frequency $\omega = (2g/y_0)^{1/2}$

29. ••During the Millikan experiment used to determine the charge on the electron, a charged polystyrene microsphere is released in still air in a known vertical electric field. The charged microsphere will accelerate in the direction of the net force until it reaches terminal speed. The charge on the microsphere is determined by measuring the terminal speed. During one such experiment, the microsphere has radius of $r = 5.50 \times 10^7$m, and the field has a magnitude $E = 6.00 \times 10^4$N/C. The magnitude of the drag force on the sphere is given by $F_D = 6\pi\eta rv$, where v is the speed of the sphere and η is the viscosity of air ($\eta = 1.8 \times 10^{-5}$N·s/m²). Polystyrene has density 1.05×10^3kg/m³. (a) If the electric field is pointing down and the polystyrene microsphere is rising with a terminal speed of 1.16×10^{-4}m/s, what is the charge on the sphere? (b) How many excess electrons are on the sphere? (c) If the direction of the electric field is reversed but its magnitude remains the same, what is the new terminal speed?

30. • • •In problem 29, there is a description of the Millikan experiment used to determine the charge on the electron. During the experiment, a switch is used to reverse the direction of the electric field without changing its magnitude, so that one can measure the terminal speed of the microsphere both as it is moving upward and as it is moving downward. Let v_u represent the terminal speed when the particle is moving up, and v_d the terminal speed when moving down. (a) If we let $u = v_u + v_d$, show that $q = 3\pi\eta ru/E$, where q is the microsphere's net charge. For the purpose of determining q, what advantage does measuring both v_u and v_d have over measuring only one terminal speed? (b) Because charge is quantized, u can only change by steps of magnitude N, where N is an integer. Using the data from Problem 29, calculate Δu.

Continuous Charge Distribution

31. ••Two infinite nonconducting sheets of charge are parallel to each other, with sheet A in the $x = -2.0$ m plane and sheet B in the $x = +2.0$ m plane. Find the electric field in the region $x < -2.0$ m, in the region $x > +2.0$ m, and between

the sheets for the following situations. (a) When each sheet has a uniform surface charge density equal to +3.0 μC/m² and (b) when sheet A has a uniform surface charge density equal to +3.0 μC/m² and sheet B has a uniform surface charge density equal to −3.0 μC/m². (c) Sketch the electric field line pattern for each case.

32. •A non-conducting disc of radius R lies in the $z = 0$ plane with its centre at the origin. The disc has a uniform surface charge density σ. Find the value of z for which $E_z = \sigma/(4\epsilon_0)$. Note that at this distance, the magnitude of the electric field strength is half the electric-field strength at points on the x axis that are very close to the disc.

33. •A ring that has radius a lies in the $z = 0$ plane with its centre at the origin. The ring is uniformly charged and has a total charge Q. Find E_z on the z axis at (a) $z = 0.2a$, (b) $z = 0.5a$, (c) $z = 0.7a$, (d) $z = a$, and (e) $z = 2a$. (f) Use your results to plot E_z versus z for both positive and negative values of z. (Assume that these distances are exact.)

34. •A line charge that has a uniform linear charge density λ lies along the x axis from $x = x_1$ to $x = x_2$ where $x_1 < x_2$. Show that the x component of the electric field at a point on the y-axis is given by $E_x = \frac{k\lambda}{y}(\cos\theta_2 - \cos\theta_1)$, where $\theta_1 = \tan^{-1}(x_1/y)$, $\theta_2 = \tan^{-1}(x_2/y)$ and $y \neq 0$.

35. ••A thin hemispherical shell of radius R has a uniform surface charge σ. Find the electric field at the centre of the base of the hemispherical shell.

Gauss's Law

36. ••An imaginary right circular cone (Figure 1.265) that has a base angle θ and base radius R is in a charge free region that has a uniform electric field $\vec{E}$ (field lines are vertical and parallel to the cone's axis). What is the ratio of the number of field lines per unit area penetrating the base to the number of field lines per unit area penetrating the conical surface of the cone? Use Gauss's law in your answer. (The field lines in the figure are only a representative sample.)

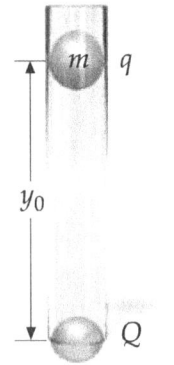
Figure 1.264

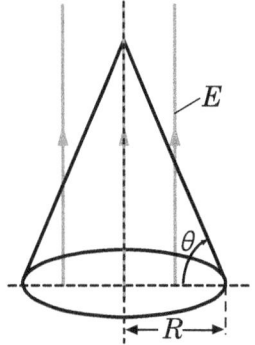
Figure 1.265

37. ••A non-conducting solid sphere of radius R has a volume charge density that is proportional to the distance from the centre. That is, $\rho = Ar$ for $r \leq R$, where A is a constant. (a) Find the total charge on the sphere. (b) Find the expressions for the electric field inside the sphere ($r < R$) and outside the sphere ($r > R$) (c) Sketch the magnitude of the electric field as a function of the distance r from the sphere's centre.

38. •••A non-conducting spherical shell of inner radius R_1 and outer radius R_2 has a uniform volume charge density ρ. (a) Find the total charge on the shell. (b) Find expressions for the electric field everywhere.

Gauss's Law Applications in Cylindrical Symmetry Situations

39. ••An infinitely long non-conducting solid cylinder of radius a has a uniform volume charge density of ρ_0. Show that the electric field is given by the following expressions: $E_a = \rho_0 a/(2\epsilon_0)$ for $0 \leq r < a$ and $E_a = \rho_0 a^2/(2\epsilon_0 r)$ for $r > a$, where r is the distance from the long axis of the cylinder.

40. ••Consider two infinitely long, coaxial thin cylindrical shells. The inner shell has a radius a_1 and has a uniform surface charge density of σ_1, and the outer shell has a radius a_2 and has a uniform surface charge density of σ_2. (a) Use Gauss's law to find expressions for the electric field in the three regions: $0 \leq r < a_1$, $a_1 < r < a_2$, and $r > a_2$, where r is the distance from the axis.
(b) What is the ratio of the surface charge densities σ_2/σ_1 and their relative signs if the electric field is to be zero everywhere outside the largest cylinder? (c) For the case in Part (b), what would be the electric field between the shells?

Electric Charge and Field at Conductor Surfaces

41. •An uncharged penny is in a region that has a uniform electric field of magnitude 1.60 kN/C directed perpendicular to its faces. (a) Find the charge density on each face of the penny, assuming the faces are planes. (b) If the radius of the penny is 1.00 cm, find the total charge on one face.

42. •A thin metal slab has a net charge of zero and has square faces that have 12 cm long sides. It is in a region that has a uniform electric field that is perpendicular to its faces. The total charge induced on one of the faces is 1.2 nC. What is the magnitude of the electric field?

43. •••If the magnitude of an electric field in air is as great as 3.0×10^6 N/C, the air becomes ionized and begins to conduct electricity. This phenomenon is called **dielectric breakdown**. A charge of 18 μC is to be placed on a conducting sphere. What is the minimum radius of a sphere that can hold this charge without breakdown?

44. •••A quantum-mechanical treatment of the hydrogen atom shows that the electron in the atom can be treated as a smeared-out distribution of negative charge of the form $\rho(r) = -\rho_0 e^{-2r/a}$. Here r represents the distance from the centre of the nucleus and a represents the first *Bohr radius* which has a numerical value of 0.0529 nm. Recall that the nucleus of a hydrogen atom consists of just one proton and treat this proton as a positive point charge. (a) Calculate ρ_0, using the fact that the atom is neutral. (b) Calculate the electric field at any distance r from the nucleus.

45. •••A thin, non-conducting, uniformly charged spherical shell of radius R (Figure 1.266a) has a total positive charge of Q. A small circular plug is removed from the surface. (a) What are the magnitude and direction of the electric field at the centre of the hole? (b) The plug is now put back in the hole (Figure 1.266b). Using the result of Part (a), find the electric force acting on the plug. (c) Using the magnitude of the force, calculate the "electrostatic pressure" (force/unit area) that tends to expand the sphere.

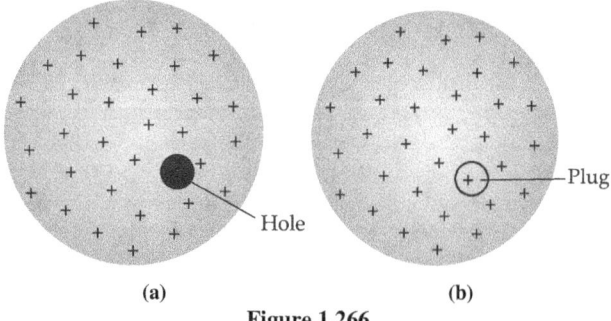
Figure 1.266

46. ••A uniformly charged, infinitely long line of negative

charge has a linear charge density of $-\lambda$ and is located on the z axis. A small positively charged particle that has a mass m and a charge q is in a circular orbit of radius R in the xy plane centred on the line of charge. (a) Derive an expression for the speed of the particle. (b) Obtain an expression for the period of the particle's orbit.

47. ••A uniformly charged non-conducting solid sphere of radius R has its centre at the origin and has a volume charge density of ρ. (a) Show that at a point within the sphere a distance r from the centre $\vec{E} = \frac{\rho}{3\epsilon_0} r \hat{r}$. (b) Material is removed from the sphere leaving a spherical cavity that has a radius $b = R/2$ and its centre at $x = b$ on the x axis (Figure 1.267). Calculate the electric field at points 1 and 2 shown in Figure 1.267.

Hint: Model the sphere-with-cavity as two uniform spheres of equal positive and negative charge densities.

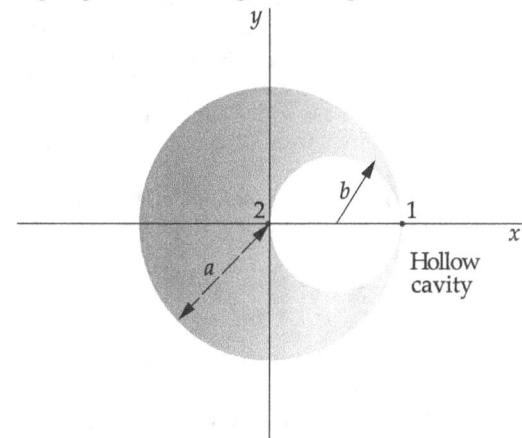

Figure 1.267

48. ••Show that the electric field throughout the cavity of Problem 47(b) is uniform and is given by $\vec{E} = \frac{\rho}{3\epsilon_0} b \hat{i}$

49. ••The cavity in Problem 47(b) is now filled with a uniformly charged non-conducting material with a total charge of Q. Calculate the new values of the electric field at points 1 and 2 shown in Figure 1.267

50. ••An electric dipole that has a dipole moment of $\vec{p}$ is located at a perpendicular distance R from an infinitely long line charge that has a uniform linear charge density λ. Assume that the dipole moment is in the same direction as the field of the line of charge. Determine an expression for the electric force on the dipole.

1.37.3 Multiple Choice Questions
1.37.3.1 Level 1
Charge & its Properties

1. The ratio of electric force to gravitational force acting between two electrons will be
   (A) 1×10^{36} (B) 2×10^{39}
   (C) 6×10^{45} (D) 4×10^{42}

2. F_g and F_e represent the gravitational and electrostatic force respectively between two electrons situated at some distance. The ratio of F_g to F_e is of the order of -
   (A) 10^{36} (B) 10^1 (C) $10°$ (D) 10^{-43}

3. One quantum of charge should be at least be equal to the charge in coulomb:
   (A) 1.6×10^{-17} C. (B) 1.6×10^{-19} C.
   (C) 1.6×10^{-10} C (D). 4.8×10^{-10} C.

4. The unit of charge is coulomb in SI system and esu of charge (or stat coulomb) in C.G.S. system 1 coulomb equals

   (A) 3×10^9 esu (B) $(1/3 \times 10^9)$ esu
   (C) $(1/3 \times 10^8)$ esu (D) (9×10^9) esu

5. The relative strengths of gravitational, electromagnetic and strong nuclear forces are-
   (A) $1 : 10^{39} : 10^{36}$ (B) $1 : 10^{36} : 10^{39}$
   (C) $1 : 10^{-26} : 10^{-39}$ (D) $1 : 10^{-39} : 10^{-36}$

6. An electron at rest has a charge of 1.6×10^{-19}C. It starts moving with a velocity $v = c/2$, where c is the speed of light, then the new charge on it is -
   (A) 1.6×10^{-19} coulomb
   (B) $1.6 \times 10^{-19} \sqrt{1 - \left(\frac{1}{2}\right)^2}$ coulomb
   (C) $1.6 \times 10^{-19} \sqrt{\left(\frac{2}{1}\right)^2 - 1}$ coulomb
   (D) $\frac{1.6 \times 10^{-19}}{\sqrt{1 - \left(\frac{1}{2}\right)^2}}$ coulomb

7. If a glass rod is rubbed with silk, it acquires a positive charge because-
   (A) Protons are added to it.
   (B) Protons are removed from it.
   (C) Electrons are added to it.
   (D) Electrons are removed from it.

8. Which one of the following is the unit of electric charge?
   (A) Coulomb (B) Newton
   (C) Volt (D) Coulomb/Volt

9. An accelerated or decelerated charge produces-
   (A) Electric field only
   (B) Magnetic field only
   (C) Localized electric and magnetic fields
   (D) Electric and magnetic fields that are radiated

10. Which one of the following statement regarding electrostatics is wrong ?
    (A) Charge is quantized
    (B) Charge is conserved
    (C) There is an electric field near an isolated charge at rest
    (D) A stationary charge produces both electric and magnetic fields

11. The dielectric constant for water is-
    (A) 1 (B) 40 (C) 81 (D) 0.3

12. In M.K.S. System, $1/4\pi\epsilon_0$ equals-
    (A) 9×10^9 N.m^2/C^2
    (B) 1 N.m^2/C^2
    (C) 1 dyne .cm^2/ stat C^2
    (D) 9×10^9 dyne ×cm^2/ stat C^2

13. A stationary electric charge produces-
    (A) Only electric fields
    (B) Only magnetic field
    (C) Both electric and magnetic field
    (D) Neither electric nor magnetic field

14. Charges reside on the-
    (A) Outer surface of the charged conductor
    (B) Inner surface of the charged conductor
    (C) Inner as well as outer surface of the charged conductor
    (D) None of the above

15. An isolated solid metallic sphere is charged with +Q charge .The distribution of their +Q charge on the sphere will be-
    (A) uniform but on the surface alone
    (B) non uniform but on the surface alone
    (C) uniform inside the volume
    (D) non uniform inside the volume

Coulomb's Law

16. Two similar charge of +Q , as shown in Figure1.268 are placed at A and B. ˘q charge is placed at point C midway between A and B. ˘q charge will oscillate if
    (A) It is moved towards A.
    (B) It is moved towards B.
    (C) It is moved upwards AB.
    (D) Distance between A and B is reduced.

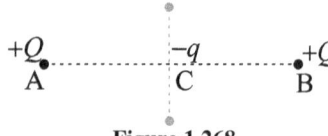

Figure 1.268

17. When the distance between two charged particle is halved, the force between them becomes-
    (A) One fourth (B) One half
    (C) Double (D) Four times
18. The force between two point charges in vacuum is 15N, if a brass plate is introduced between the two charges, then force between them will-
    (A) Become zero (B) Remain the same
    (C) Become 30 N (D) Become 60 N
19. The force between an α particle and an electron separated by a distance of 1 Å is-
    (A) 2.3×10^{-8} N attractive
    (B) 2.3×10^{-8} N Repulsive
    (C) 4.6×10^{-8} N attractive
    (D) 4.6×10^{-8} repulsive
20. Two charges are at distance (d) apart in air. Coulomb force between them is F. If a dielectric material of dielectric constant (κ) is placed between them, the coulomb force now becomes.
    (A) F/κ (B) $F\kappa$ (C) F/κ^2 (D) $\kappa^2 F$
21. Two point charges in air at a distance of 20 cm. from each other interact with a certain force. At what distance from each other should these charges be placed in oil of relative permittivity 5 to obtain the same force of interaction –
    (A) 8.94×10^{-2} m (B) 0.894×10^{-2} m
    (C) 89.4×10^{-2} m (D) 8.94×10^{2} m
22. A certain charge Q is divided at first into two parts, (q) and $(Q - q)$. Later on the charges are placed at a certain distance. If the force of interaction between the two charges is maximum then-
    (A) $Q/q = 4/1$ (B) $Q/q = 2/1$
    (C) $Q/q = 3/1$ (D) $Q/q = 5/1$
23. A unit charge is one which when placed in vacuum one cm from an equal charge of the same kind will repel it with a force of-
    (A) 1 N (B) 1 dyne (C) 2 dyne (D) 4 dyne
24. The permittivity ϵ_0 of vacuum is 8.86×10^{-12} C^2/N – m^2 and the dielectric constant of water is 81 . The permittivity of water in C^2/Nm^2 is-
    (A) $81 \times 8.86 \times 10^{-12}$ (B) 8.86×10^{-12}
    (C) $(8.86 \times 10^{-12})/81$ (D) $81/(8.86 \times 10^{-12})$
25. The force between two point charges placed in vacuum at distance 1 mm is 18 N. If a glass plate of thickness 1 mm and dielectric constant 6, be kept between the charges then new force between them would be-
    (A) 18 N (B) 10^8 N
    (C) 3 N (D) 3×10^{-6} N
26. Two similar and equal charges repel each other with force of 1.6 N, when placed 3 m apart. Strength of each charge is -
    (A) 40 μC (B) 20 μC
    (C) 4 μC (D) 2 μC

27. There are two charges +1 micro-coulomb and +5 micro-coulomb, the ratio of force on them will be -
    (A) 10^{43} (B) 1 : 1 (C) $10°$ (D) 10^{-43}

Superposition Principle

28. A charge Q is divided in two parts Q_1 and Q_2 and these charges are placed at distance R. there will be maximum repulsion between them, when-
    (A) $Q_2 = (Q/R), Q_1 = Q - (Q/R)$
    (B) $Q_2 = (Q/3), Q_1 = (2Q/3)$
    (C) $Q_2 = (Q/4), Q_1 = (3Q/4)$
    (D) $Q_1 = Q_2 = Q/2$
29. The three charges each of 5×10^{-6} coloumb are placed at vertex of an equilateral triangle of side 10 cm. The force exerted on the charge of 1μC placed at centre of triangle in newton will be-
    (A) 13.5 (B) zero (C) 4.5 (D) 6.75
30. A point charge q_1 exerts a force F upon another charge q_2. If one other charge q_3 be placed quite near to charge q_2, then the froce that charge q_1 exerts on the charge q_2 will be
    (A) F (B) > F (C) (C) < F (D) zero
31. A mass particle (mass = m and charge = q) is placed bewteen two point charges of charge q separtion between these two charge is 2 L. The frequency of oscillation of mass particle, if it is displaced for a small distance along the line joining the charges-
    (A) $\dfrac{q}{2\pi}\sqrt{\dfrac{1}{m\pi\epsilon_0 L^3}}$ (B) $\dfrac{q}{2\pi}\sqrt{\dfrac{4}{m\pi\epsilon_0 L^3}}$
    (C) $\dfrac{q}{2\pi}\sqrt{\dfrac{1}{4m\pi\epsilon_0 L^3}}$ (D) $\dfrac{q}{2\pi}\sqrt{\dfrac{1}{16\pi\epsilon_0 m L^3}}$
32. Two small balls having equal positive charge Q (Coulomb) on each are suspended by two insulating strings of equal length 'L' metre, from a hook fixed to a stand. The whole set up is taken in a satellite in to space where there is no gravity (state of weightlessness) Then the angle (θ) between the two strings is-
    (A) 0° (B) 90°
    (C) 180° (D) $0° < q < 180°$
33. *ABC* is a right angle triangle $AB = 3$ cm, BC = 4 cm charges +15, +12, −12 esu are placed at *A*, *B* and C respectively. The magnitude of the force experienced by the charge at *B* in dyne is-
    (A) 125 (B) 35 (C) 22 (D) 0
34. Equal charges of each 2μC are placed at a point $x = 0, 2, 4$, and 8 cm on the x-axis. The force experienced by the charge at $x = 2$ cm is equal to -
    (A) 5 Newton (B) 10 Newton
    (C) 0 Newton (D) 15 Newton
35. Three equal charges (q) are placed at corners of a equilateral triangle. The force on any charge is-
    (A) Zero (B) $\sqrt{3}\dfrac{Kq^2}{a^2}$
    (C) $\dfrac{Kq^2}{\sqrt{3}a^2}$ (D) $3\sqrt{3}\dfrac{Kq^2}{a^2}$
36. Two identical charges of magnitude 'q' are placed at $(-a, 0)$ and $(a, 0)$. Same nature charged particle is placed at origin. It executes S.H.M. If it is displaced-
    (A) in x-direction
    (B) in y-direction
    (C) at an angle of 45° from the x-axis
    (D) along perpendicular to the plane.
37. Two equal negative charge $(-q)$ are fixed at the points $(0, a)$ and $(0, -a)$ on the y-axis. A positive charge (Q) is released from rest at the point $(2a, 0)$ on the x-axis. The charge Q

will -
(A) execute simple harmonic motion about the origin.
(B) move to the origin and remains at rest
(C) move to infinity
(D) execute oscillatory but not simple harmonic motion

38. Five point charges, each of value +q coulomb, are placed on five vertices of a regular hexagon of side L meter. The magnitude of the force on a point charge of value $-q$ coulomb placed at the centre of the hexagon is-
(A) $\dfrac{kq^2}{L^2}$ (B) $\sqrt{5}\dfrac{kq^2}{L^2}$
(C) $\sqrt{3}\dfrac{kq^2}{L^2}$ (D) Zero

Electric Field

39. A pendulum bob of mass 80 mg and carrying a charge of 2×10^{-8} C. is at rest in a horizontal uniform electric field of 20,000 V m^{-1}. Find the tension in the thread of pendulum-
(A) 8.8×10^{-2} N (B) 8.8×10^{-3} N
(C) 8.8×10^{-4} N (D) 8.8×10^{-5} N

40. Two charges $4q$ and q are placed 30 cm. apart. At what point the value of electric field will be zero
(A) 10 cm away from q and between the charge
(B) 20 cm away from q and between the charge
(C) 10 cm away from q and out side the line joining the charge.
(D) 10 cm away from $4q$ and out side the line joining them.

41. One unit of electric field intensity is newtons/coulomb. The other unit of this can be
(A) Vm (B) Vm^2
(C) V/m (D) V/m^2

42. If $Q = 2$ C coulomb and force on it is $F = 100$ N. Then the value of field intensity will be-
(A) 100 N/C (B) 50 N/C
(C) 200 N/C (D) 10 N/C

43. Four equal but like charge are placed at four corners of a square. The electric field intensity at the centre of the square due to any one charge is E then the resultant electric field intensity at centre of square will be
(A) Zero (B) 4E (C) E (D) 1/2 E

44. Two charges $9e$ and $3e$ are placed at a distance r. The distance of the point where the electric field intensity will be zero is
(A) $\left(\dfrac{r}{1+\sqrt{3}}\right)$ from $9e$ charge
(B) $\left(\dfrac{r}{1+\sqrt{1/3}}\right)$ from $9e$ charge
(C) $\left(\dfrac{r}{1-\sqrt{3}}\right)$ from $3e$ charge
(D) $\left(\dfrac{r}{1+\sqrt{1/3}}\right)$ from $3e$ charge.

45. An electric field can deflect-
(A) X-rays (B) Neutrons
(C) α-particles (D) γ - rays

46. Which one of the following relations is correct-
(A) 1 N/C = 10^8 Volt /m
(B) 1 N/C = 10^{-6} V/m
(C) 1 N/C = 1 V/m
(D) 1 N/C = 10^{-8} V/m

47. If mass of the electron = 9.1×10^{-31}kg. Charge on the electron = 1.6×10^{-19} coulomb and g = 9.8 m/s^2. Then the intensity of the electric field required to balance the weight of an electron is-
(A) 5.6×10^{-9} N/C (B) 5.6×10^{-11} N/C
(C) 5.6×10^{-8} N/C (D) 5.6×10^{-7} N/C

48. Six charges +Q each are placed at the corners of a regular hexagon of side (a), the electric field at the centre of hexagon is-
(A) Zero (B) $\dfrac{1}{4\pi\epsilon_0} \cdot \dfrac{6Q^2}{a^2}$
(C) $\dfrac{1}{4\pi\epsilon_0} \cdot \dfrac{Q^2}{a^2}$ (D) $\dfrac{1}{4\pi\epsilon_0} \cdot \dfrac{6Q^2}{a\sqrt{2}}$

49. Two charged spheres A and B are charged with the charges of +10 and +20 coulomb respectively and separated by a distance of 80 cm. The electric field at a point on the line joining the centers of the two spheres will be zero at a distance ... from sphere A.
(A) 20 cm (B) 33 cm (C) 55 cm (D) 60 cm

50. Four charges $+q, +q, -q$ and $-q$ are placed respectively at the corners A, B, C and D of a square of side (a), arranged in the given order. Calculate the intensity at (O) the centre of the square
(A) $\dfrac{4\pi\epsilon_0 \cdot a^2}{4\sqrt{2}q}$ (B) $\dfrac{4\sqrt{2}q}{4\pi\epsilon_0 \cdot a^2}$
(C) $\dfrac{\pi\epsilon_0 \cdot a^2}{4\sqrt{2}q}$ (D) $\dfrac{4\sqrt{2}q}{\pi\epsilon_0 \cdot a^2}$

51. A ring of radius (R) carries a uniformly distributed charge $+Q$. A point charge $-q$ is placed on the axis of the ring at a distance 2R from the centre of the ring and released from rest. The particle-
(A) Becomes in rest condition immediately.
(B) Executes simple harmonic motion
(C) Motion is not SHM
(D) Come at the centre of ring immediately.

52. A small circular ring has a uniform charge distribution. On a far-off axial point distance x from the centre of the ring, the electric field is proportional to-
(A) x^{-1} (B) $x^{-3/2}$ (C) x^{-2} (D) $x^{5/4}$

Electric flux and Gauss Laws

53. The tangent drawn at a point on a line of electric force shows the-
(A) intensity of gravity field
(B) intensity of magnetic field
(C) intensity of electric field
(D) direction of electric field

54. Which of the following statements concerning the electrostatics is correct-
(A) electric line of force never intersect each other
(B) electric lines of force start from positive charge and end at the negative charge
(C) electric lines of force start or ends perpendicular to the surface of a charged metal.
(D) all of the above

55. When no charge is confined with in the Gauss's surface, it implies that-
(A) $E = 0$
(B) $\vec{E}$ and $d\vec{A}$ are parallel
(C) $\vec{E}$ and $d\vec{A}$ are mutually perpendicular
(D) $\vec{E}$ and $d\vec{A}$ are inclined at some angle

56. If electric field flux coming out of a closed surface is zero, the electric field at the surface will be-

(A) zero
(B) same at all places
(C) dependent upon the location of points
(D) infinite

57. If three electric di-poles are placed in some closed surface, then the electric flux emitting from the surface will be-
(A) zero (B) positive
(C) negative (D) None

58. For which of the following fields, Gauss's law is valid-
(A) fields following inverse square law
(B) uniform field
(C) all types of field
(D) this law has no concern with the field

59. A charge of Q coloumb is located at the centre of a cube. If the corner of the cube is taken as the origin, then the flux coming out from the faces of the cube in the direction of X - axis will be-
(A) $4\pi Q$ (B) $Q/6\epsilon_0$
(C) $Q/3\epsilon_0$ (D) $Q/4\epsilon_0$

60. A rectangular surface of 2 metre width and 4 metre length, is placed in an electric field of intensity 20 newton/C, there is an angle of 60° between the perpendicular to surface and electrical field intensity. Then total flux emitted from the surface will be- (In volt-metre)
(A) 80 (B) 40 (C) 20 (D) 160

61. A charge q is inside a closed surface and charge $-q$ is outside. The out going electric flux is-
(A) $-q/\epsilon_0$ (B) zero (C) q/ϵ_0 (D) $2q/\epsilon_0$

62. If the electric field is uniform, then the electric lines of forces are-
(A) Divergent (B) Convergent
(C) Circular (D) Parallel

63. Electric lines of forces-
(A) Exist everywhere
(B) Are imaginary
(C) Exist only in the immediate vicinity of electric charges
(D) None of the above

64. Which one of the following diagrams Fig.1.269 shows the correct lines of force ?

65. In Fig.1.270 shown the electric lines of force emerging from a charged body. If the electric fields at A and B are E_A and E_B are respectively, If the distance between A and B is r then
(A) $E_A > E_B$ (B) $E_A < E_B$
(C) $E_A = E_B$ (D) $E_A = (E_B)/r^2$

Application of Gauss law

66. Three charges $q_1 = 1 \mu C, q_2 = 2 \mu C$ and $q_3 = -3 \mu C$ and four surfaces S_1, S_2, S_3 and S_4 are shown in Fig.1.271. The flux emerging through surface S_2 in N.m²/C is-
(A) $36\pi \times 10^3$ (B) $-36\pi \times 10^3$
(C) $36\pi \times 10^9$ (D) $-36\pi \times 10^9$

67. A surface enclosed an electric dipole, the flux through the surface is -
(A) Infinite (B) Positive
(C) Negative (D) Zero

68. Total flux coming out of some closed surface is-
(A) q/ϵ_0 (B) ϵ_0/q (C) $q\epsilon_0$ (D) $\sqrt{\frac{q}{\epsilon_0}}$

69. A square of side 20 cm. is enclosed by a concentric spherical surface of radius 80 cm. Four charges $+2 \times 10^{-6}$ C, -5×10^{-6} C, -3×10^{-6} C, $+6 \times 10^{-6}$ C, are located at the four corners of a square, then out going total flux from spherical surface in N m²/C, will be-
(A) zero (B) $(16\pi) \times 10^{-6}$
(C) $(8\pi) \times 10^{-6}$ (D) $(36\pi) \times 10^{-6}$

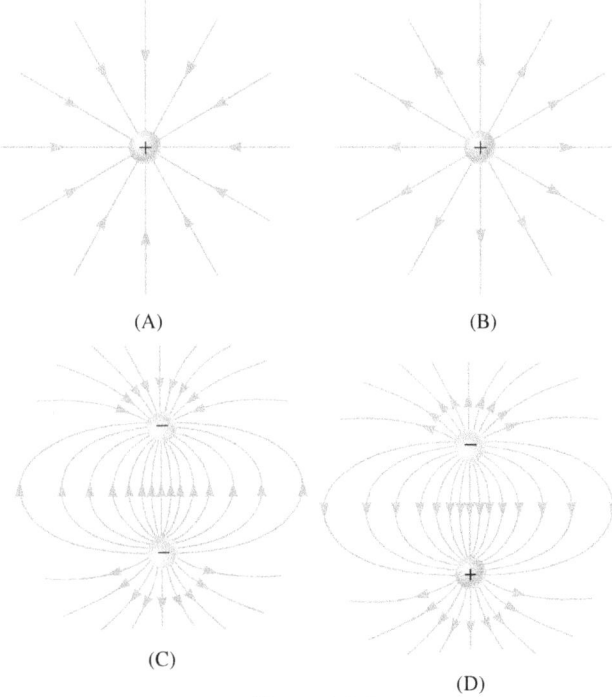

(A) (B)

(C) (D)

Figure 1.269

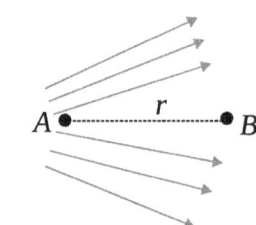

Figure 1.270

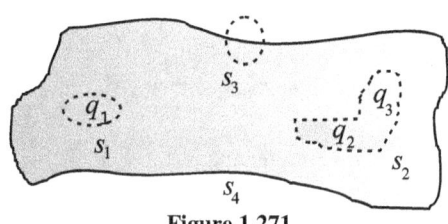

Figure 1.271

70. The flux emerging out from any one face of the cube will be -
(A) $\frac{q}{6\epsilon_0}$ (B) $\frac{q}{3\epsilon_0}$ (C) $\frac{q}{\epsilon_0}$ (D) $\frac{q}{4\epsilon_0}$

71. The electric field inside a spherical shell of uniform surface charge density is -
(A) Zero
(B) Constant, different from zero
(C) Proportional to the distance from the centre
(D) None of the above

72. A cubical box of side 1 m is immersed a uniform electric field of strength 10^4 N/C. The flux through the cube is-
(A) 10^4 (B) 6×10^4 (C) 2×10^4 (D) Zero

73. A charge (q) is located at the centre of a cube. The electric flux through any face of the cube is-
(A) q/ϵ_0 (B) $q/2\epsilon_0$ (C) $q/4\epsilon_0$ (D) $q/6\epsilon_0$

74. A large isolated metal sphere of radius (R) carries a fixed charge. A small charge is placed at a distance (r) from its surface experiences a force which is -

(A) Proportional to R
(B) Independent of R and
(C) Inversely proportional to $(R + r)^2$
(D) inversely proportional to r^2

75. A hollow sphere of charge does not produce an electric field at any-
    (A) Interior point   (B) Outer point
    (C) Surface point   (D) None of the above

76. A spherical conductor of radius 50 cm has a surface charge density of 8.85×10^{-6} C/m^2. The electric field near the surface in N/C is-
    (A) 8.85×10^{-6} (B) 8.85×10^6
    (C) 1×10^6 (D) Zero

77. The electric field intensity at a point located at distance $r (r < R)$ from the centre of a spherical conductor (radius R) of charge Q will be-
    (A) kQR/r^3 (B) kQr/R^3
    (C) kQ/r^2 (D) zero.

Electric Dipole

78. If an electric dipole is kept in a uniform electric field, then it will experience
    (A) a force
    (B) a couple and moves
    (C) a couple and rotates
    (D) a force and moves.

79. An electric dipole consists of two opposite charges each of magnitude 1×10^{-6} C separated by a distance 2 cm. The dipole is placed in an external field of 10×10^5 N/C. The maximum torque on the dipole is -
    (A) 0.2×10^{-3} N.m   (B) 1.0×10^{-3} N.m
    (C) 2×10^{-3} N.m   (D) 4×10^{-3} N.m

80. The ratio of the electric field due to an electric dipole on its axis and on the perpendicular bisector of the dipole is -
    (A) 1 : 2   (B) 2 : 1   (C) 1 : 4   (D) 4 : 1

81. The region surrounding a stationary electric dipole has-
    (A) electric field only
    (B) magnetic field only
    (C) both electric and magnetic fields
    (D) neither electric non magnetic field

1.37.3.2 Level 2

1. 5×10^5 lines of electric flux are entering in a closed surface and 4×10^5 lines come out of the surface the charge enclosed by the surface is -
   (A) 0.885×10^{-6} C   (B) 8.85×10^{-6} C
   (C) -8.85×10^{-7} C   (D) 8.85×10^{-8} C

2. A cylinder of radius (R) and length (L) is placed in a uniform electrical field (E) parallel to the axis of the cylinder the total flux for the surface of the cylinder is given by-
   (A) $2\pi R^2 E$ (B) $\pi R^2 E$
   (C) $\dfrac{\pi R^2 + \pi R^2}{E}$ (D) zero

3. A hemisphere (radius R) is placed in electric field as shown in Fig.1.272 Total outgoing flux is -
   (A) $\pi R^2 E$ (B) $2\pi R^2 E$
   (C) $4\pi R^2 E$ (D) $(\pi R^2 E)/2$

4. Three identical charges each of $1\ \mu C$ are kept on the circumference of a circle of radius 1 metre forming equilateral triangle. The electric intensity at the centre of the circle in N/C is-
   (A) 9×10^3 (B) 13.5×10^3 (C) 27×10^3 (D) Zero

5. The number of electrons, falling on spherical conductor

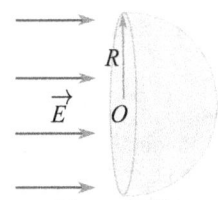

Figure 1.272

(radius = 0.1 m) to produce 0.036 N/C electric field at the surface of conductor, is-
   (A) 2.7×10^5 (B) 2.5×10^2
   (C) 2.6×10^5 (D) 2.4×10^5

6. A particle of mass 6 μg carrying a charge of 10^{-9} C, is placed in the electric field of strength $E = 6 \times 10^5$ V/m, the acceleration acquired by the particle is-
   (A) 10^2 m/s^2 (B) 10^3 m/s^2
   (C) 10^5 m/s^2 (D) 10^{20} m/s^2

7. Two small balls having equal positive charge Q on each are suspended by two insulating strings at equal length L meter, from a hook fixed to a stand. The whole set up is taken in a satellite into space where there is no gravity. Then the angle θ between two strings and tension in each string is-
   (A) $0, \dfrac{kq^2}{L^2}$ (B) $\pi, \dfrac{kq^2}{2L^2}$
   (C) $\pi, \dfrac{kq^2}{4L^2}$ (D) $\dfrac{\pi}{2}, \dfrac{kq^2}{2L^2}$

8. The magnitude of the electric field strength (E) such that an electron placed in the field would experience an electrical force equal to its weight is [assume g = 10 m/sec^2]
   (A) 5.68×10^{-11} N/C vertically up.
   (B) 5.68×10^{-11} N/C vertically down.
   (C) 5.68×10^{-10} N/C vertically up.
   (D) 5.68×10^{-10} N/C vertically down.

9. The electric field at the surface of a charged spherical conductor is 10 kV/m. The electric field at a distance equal to the diameter from its centre will be -
   (A) 2.5 V/m   (B) 2.5 kV/m
   (C) 5.0 kV/m   (D) 5.0 V/m

10. An Electron is situated 3×10^{-9} m from one α - particle and 4×10^{-9} m from another α - particle. The magnitude of force on the electron, when two α - particles are 5×10^{-9} m apart is -
    (A) 5.64×10^{-11} N   (B) 56.4×10^{-11} N
    (C) 0.564×10^{-11} N   (D) 564×10^{-11} N.

11. Two large metal plates, each of area A, carry charges $+q$ and $-q$ and face each other. the plates are separated by a small distance d the electric field between the plates would be-
    (A) $\dfrac{2q}{\epsilon_0 A}$ (B) $\dfrac{qA}{\epsilon_0 A}$ (C) $\dfrac{q}{\epsilon_0 A}$ (D) $\dfrac{A}{q\epsilon_0}$

12. Two parallel plates of infinite dimensions are uniformly charged [Fig.1.273]. The surface charge density on one plate is σ_A and on the other is σ_B, field intensity at point C will be-
    (A) $\propto (\sigma_A - \sigma_B)$ (B) $\propto (\sigma_A + \sigma_B)$
    (C) zero   (D) $2\sigma_A$

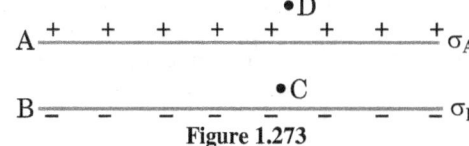

Figure 1.273

13. A charged particle of mass m is kept in equilibrium in the electric field between the plates of Millikan oil drop experiment. If the direction of the electric field between the

plate is reversed. then acceleration of the charged particle will be-
(A) Zero (B) $g/2$ (C) g (D) $2g$

14. Two identical small balls, each of mass, are suspended by two light inelastic conducting threads each of length l from the same fixed point support. If the distance d between two balls is very small as compared to l, then d is equal to-
(A) $\left(\dfrac{2kq^2}{mg}\right)^{1/3}$ (B) $\left(\dfrac{2klq^2}{mg}\right)^{2/3}$
(C) $\left(\dfrac{kq^2}{2mg}\right)^{2/3}$ (D) none of these

15. A metal sphere A of radius R has a charge of Q on it The field at a point B outside the sphere is E. Now another sphere of radius R having a charge $-3Q$ is placed at point B. The total field at a point midway between A and B due to both sphere is-
(A) $4E$ (B) $8E$ (C) $12E$ (D) $16E$

16. A uniformly charged rod with charge per unit length λ is bent in to the shape of a semicircle of radius R. The electric field at the centre is -
(A) $2k\lambda/R$ (B) $k\lambda/2R$ (C) Zero (D) None

17. A thin stationary ring of radius 1 m has a positive charge 10 μC uniformly distributed over it. A particle of mass 0.9 gram and having a negative charge of 1μC is placed on the axis at a distance of 1 cm from the centre of the ring and released then time period of oscillation of particle will be-
(A) 0.6 s (B) 0.2 s (C) 0.3 s (D) 0.4 s

18. Three charges $+3q$, $+q$ and Q are placed on a straight line with equal separation. In order to make the net force on q to be zero, the value of Q will be-
(A) $+3q$ (B) $+2q$ (C) $-3q$ (D) $-4q$

19. As per diagram 1.274 point charge $+q$ is placed at the origin O. Work done in taking another point charge $-Q$ from the point A [coordinates $(0, a)$] to another point B coordinates $(a, 0)$] along the straight path AB is-
(A) Zero
(B) $\left(\dfrac{-qQ}{4\pi\epsilon_0}\dfrac{1}{a^2}\right)\sqrt{2}a$
(C) $\left(\dfrac{qQ}{4\pi\epsilon_0}\dfrac{1}{a^2}\right)\dfrac{a}{\sqrt{2}}$
(D) $\left(\dfrac{qQ}{4\pi\epsilon_0}\dfrac{1}{a^2}\right)\sqrt{2}a$

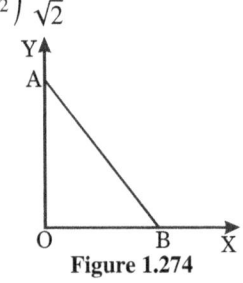
Figure 1.274

20. The electric field due to an electric dipole at a distance r from its centre in axial position is E. if the dipole is rotated through an angle of 90° about its perpendicular axis, the electric field at the same point will be-
(A) E (B) $E/4$ (C) $E/2$ (D) $2E$

21. Two electric dipoles of moment p and $64p$ are placed in opposite direction on a line at a distance of 25 cm. The electric field will be zero at point between the dipoles whose distance from the dipole of moment p is-
(A) 5 cm (B) $\frac{25}{9}$ cm (C) 10 cm (D) $\frac{4}{13}$ cm

22. When an electric dipole of dipole moment $\vec{p}$, is placed in a uniform electric field $\vec{E}$ then at what angle between $\vec{p}$ and $\vec{E}$ the value of torque will be maximum -
(A) 90° (B) 0° (C) 180° (D) 45°

23. An electric dipole has the magnitude of its charge as q and its dipole moment is p. It is placed in a uniform electric field E. If its dipole moment is along the direction of the field, the force on it is-
(A) $2qE$ (B) qE (C) 0 (D) $-qE$

24. Two opposite and equal charges 4×10^{-8} coulomb when placed 2×10^{-2} m away, form a dipole. If this dipole is placed in an external electric field 4×10^3 newton/coulomb, the value of maximum torque and the work done in rotating it through 180° will be-
(A) 64×10^{-4}Nm and 64×10^{-4} J
(B) 32×10^{-4}Nm and 32×10^{-4} J
(C) 64×10^{-4}Nm and 32×10^{-4} J
(D) 32×10^{-4}Nm and 64×10^{-4} J

1.37.3.3 Level 3

1. If an electron enters into a space between the plates of a parallel plate capacitor at an angle α with the plates and leaves at an angle β with the plates. The ratio of its kinetic energy while entering the capacitor to that while leaving will be -
(A) $\left(\dfrac{\sin\beta}{\sin\alpha}\right)^2$ (B) $\left(\dfrac{\cos\beta}{\cos\alpha}\right)^2$
(C) $\left(\dfrac{\cos\alpha}{\cos\beta}\right)^2$ (D) $\left(\dfrac{\sin\alpha}{\sin\beta}\right)^2$

2. Force between two identical charges placed at a distance of r in vacuum is F. Now a liquid of dielectric constant $\kappa = 4$ is inserted between these two charges. The thickness of the liquid is $r/2$. The force between the charges will now become -
(A) $F/4$ (B) $F/2$ (C) $3F/5$ (D) $4F/9$

3. In a certain region of surface there exists a uniform electric field of $2 \times 10^3 \hat{k}$ V/m. A rectangular coil of dimensions 10 cm $\times$ 20 cm is placed in xy plane. The electric flux through the coil is-
(A) Zero (B) 30 V.m (C) 40 V.m (D) 50 V.m

4. The electric flux from a cube of edge l is ϕ. What will be its value if edge of cube is made $2\,l$ and charge enclosed is halved -
(A) $\phi/2$ (B) 2ϕ (C) 4ϕ (D) ϕ

5. Each of the two point charges are doubled and their distance is halved. Force of interaction becomes n times, where n is -
(A) 4 (B) 1 (C) 1/16 (D) 16

6. Two point charges repel each other with a force of 100 N. One of the charges is increased by 10% and other is reduced by 10%. The new force of repulsion at the same distance would be -
(A) 100 N (B) 121 N
(C) 99 N (D) None of these

7. A spherical charged conductor has σ as the surface density of charge. The electric field on its surface is E. If the radius of the sphere is doubled keeping the surface density of charge unchanged, what will be the electric field on the surface of the new sphere -
(A) $E/4$ (B) $E/2$ (C) E (D) $2E$

8. Three equal and similar charges are placed at $(-a, 0, 0)$, $(0, 0, 0)$ and $(+a, 0, 0)$. What is the nature of equilibrium of the charge at the origin-
(A) Stable when moved along the Y-axis
(B) Stable when moved along Z-axis
(C) Stable when moved along X-axis
(D) Unstable in all of the above cases

9. The electric field strength due to a ring of radius R at a distance x from its centre on the axis of ring carrying charge

Q is given by
$$E = \frac{1}{4\pi\epsilon_0} \frac{Qx}{(R^2+x^2)^{3/2}}$$
The value of x corresponding to maximum electric field, is-
(A) R (B) $R/2$ (C) $R/\sqrt{2}$ (D) $\sqrt{2}R$

10. Two point charges Q and $-3Q$ are placed certain distance apart. If the electric field at the location of Q be $\vec{E}$, then at the location of $-3Q$ will be-
(A) $3\vec{E}$ (B) $-3\vec{E}$ (C) $\vec{E}/3$ (D) $-\vec{E}/3$

11. A and B are two points on the axis and the perpendicular bisector respectively of an electric dipole. A and B are far away from the dipole and at equal distances from it. The fields at A and B are $\vec{E}_A$ and $\vec{E}_B$ respectively such that-
(A) $\vec{E}_A = \vec{E}_B$ (B) $\vec{E}_A = 2\vec{E}_B$
(C) $\vec{E}_A = -2\vec{E}_B$ (D) $\vec{E}_A = \vec{E}_B/2$

12. A long string with a charge of λ per unit length passes through an imaginary cube of edge l. The maximum possible flux of the electric field through the cube will be
(A) $\lambda l/\epsilon_0$ (B) $\sqrt{2}\lambda l/\epsilon_0$ (C) $6\lambda l^2/\epsilon_0$ (D) $\sqrt{3}\lambda l/\epsilon_0$

13. A charge Q is placed at each of two opposite corners of a square and a charge q is placed at other two opposite corners of the square. If the resultant electric field at the position of Q is zero, then -
(A) $Q = -\frac{q}{2\sqrt{2}}$ (B) $Q = -2\sqrt{2}q$
(C) $Q = -2q$ (D) $Q = 2\sqrt{2}q$

14. The electric field outside a charged long straight wire is given by $E = -\frac{5000}{r}$Vm^{-1}. It is radially inward. The value of $V_B - V_A$ is -
[Given $r_B = 60$ cm and $r_A = 30$ cm]
(A) $5000\log_e 2$ V (B) 0 V (C) 2 V (D) 2500 V

15. An electron moves with velocity $\vec{v}$ in x - direction. An electric field acts on it in y - direction. The force on the electron acts in -
(A) +ve direction of Y - axis
(B) −ve direction of Y - axis
(C) +ve direction of Z - axis
(D) −ve direction of Z - axis

16. Two identical simple pendulums A and B, are suspended from the same point. The bobs are given positive charges, with A having more charge than B. They diverge and reach equilibrium, with A and B making angles θ_1 and θ_2 with the vertical respectively. T_A and T_B are tensions in A and B respectively. Which of the following is correct-
(A) $\theta_1 > \theta_2$ (B) $\theta_1 < \theta_2$
(C) $\theta_1 = \theta_2$ (D) $T_A > T_B$

Statements Type Question:
Each of the questions given below consist of Statement - I and Statement- II. Use the following Key to choose the appropriate answer.
(A) If both Statement-I and Statement-II are true, and Statement-II is the correct explanation of Statement-I.
(B) If both Statement -I and Statement -II are true but Statement- II is not the correct explanation of Statement-I.
(C) If Statement -I is true but Statement -II is false.
(D) If Statement -I is false but Statement -II is true.

17. Statement I: Force between two charges decreases when air separating the charges is replaced by water.
Statement II : Medium intervening the charges has no effect on force.

18. Statement I: The number of lines of force emanating from 1μC charge in vacuum is 1.13×10^5*
Statement II: This follow from Gauss's theorem in electrostatics.

1.37.3.4 Level 4 (Previous Years JEE Main & Advanced Questions)

Section A: JEE Main

1. Two points are separated by a certain distance. A charge Q is placed at each of these two points. Find the third charge which is placed at the mid point of the line joining the charges so that the system is in equilibrium - [2002]
(A) $-Q/4$ (B) $-Q/2$ (C) $-Q/3$ (D) $-Q$

2. A charged particle q is placed at the centre O of cube of length L (ABCDEFGH). Another same charge q is placed at a distance L from O. Then the electric flux through BCFG is[Fig.1.275]- [2002]
(A) L (B) zero (C) $3L$ (D) $\frac{q}{3\pi\epsilon_0 L}$

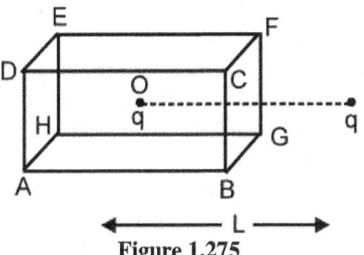

Figure 1.275

3. Three charges $-q_1$, $+q_2$ and $-q_3$ are placed as shown in Fig.1.276. The x - component of the force on $-q_1$ is proportional to – [2003]
(A) $\frac{q_2}{b^2} + \frac{q_3}{a^2}\sin\theta$ (B) $\frac{q_2}{b^2} + \frac{q_3}{a^2}\cos\theta$
(C) $\frac{q_2}{b^2} + \frac{q_3}{a^2}\cos\theta$ (D) $\frac{q_2}{b^2} - \frac{q_3}{a^2}\cos\theta$

4. If the electric flux entering and leaving an enclosed surface respectively is ϕ_1 and ϕ_2, the electric charge inside the surface will be – [2003]
(A) $(\phi_1 + \phi_2)/\epsilon_0$ (B) $(\phi_2 - \phi_1)/\epsilon_0$
(C) $(\phi_1 + \phi_2)\epsilon_0$ (D) $(\phi_2 - \phi_1)\epsilon_0$

5. Two spherical conductors B and C having equal radii and carrying equal charges on them repel each other with a force F when kept apart at some distance. A third spherical conductor having same radius as that of B but uncharged is brought in contact with B, then brought in contact with C and finally removed away from both. The new force of repulsion between B and C is -[2004]
(A) $F/4$ (B) $3F/4$ (C) $F/8$ (D) $3F/8$

6. Four charges equal to $-Q$ are placed at the four corners of a square and a charge q is at its centre. If the system is in equilibrium the value of q is - [2004]
(A) $-\frac{Q}{4}(1+2\sqrt{2})$ (B) $\frac{Q}{4}(1+2\sqrt{2})$
(C) $-\frac{Q}{2}(1+2\sqrt{2})$ (D) $\frac{Q}{2}(1+2\sqrt{2})$

7. A charged oil drop is suspended in a uniform field of 3×10^4 V/m so that it neither falls nor rises. The charge on the drop will be (Take the mass of the charge = 9.9×10^{-15} kg and $g = 10$ m/s^2) - [2004]
(A) 3.3×10^{-18}C (B) 3.2×10^{-18}C
(C) 1.6×10^{-18}C (D) 4.8×10^{-18}C

8. A charged ball B hangs from a silk thread S which makes an angle θ with a large charged conducting sheet P, as shown in the Figure 1.277. The surface charge density σ of the sheet is proportional to- [2005]
(A) $\cos\theta$ (B) $\cot\theta$ (C) $\sin\theta$ (D) $\tan\theta$

---

*The number of field lines emitted from a positive charge (or terminated to a negative charge) is not a well defined quantity.

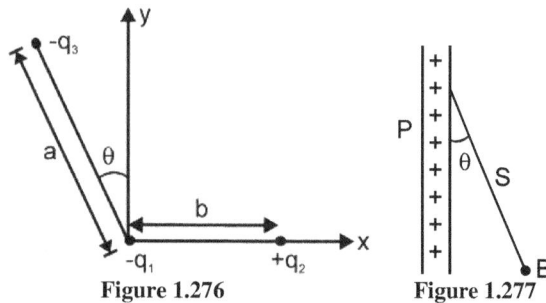

Figure 1.276   **Figure 1.277**

9. Two point charges $+8q$ and $-2q$ are located at $x = 0$ and $x = L$ respectively. The location of a point on the x axis at which the net electric field due to these two point charges is zero is - [2005]
   (A) $2L$ (B) $L/4$ (C) $8L$ (D) $4L$

10. An electric dipole is placed at an angle of 30° to a non-uniform electric field. the dipole will experience - [2006]
    (A) a torque as well as a translational force
    (B) a torque only
    (C) a translational force only in the direction of the field
    (D) a translational force only in a directin normal to the direction of the field

11. If g_E and g_M are the accelerations due to gravity on the surfaces of the earth and the moon respectively and if Millikan's oil drop experiment could be performed on the two surfaces, one will find the ratio (electronic charge on the moon/ electronic charge on the earth) to be- [2007]
    (A) 1   (B) 0   (C) g_E/g_M (D) g_M/g_E

12. A thin spherical shell of radius R has charge Q spread uniformly over its surface. Which of the following graphs [Fig.1.278] most closely represents the electric field $E(r)$ produced by the shell in the range $0 \le r < \infty$, where r is the distance from the centre of the shell? [2008]

Questions 13 and 14 consist of Statement-I and Statement-

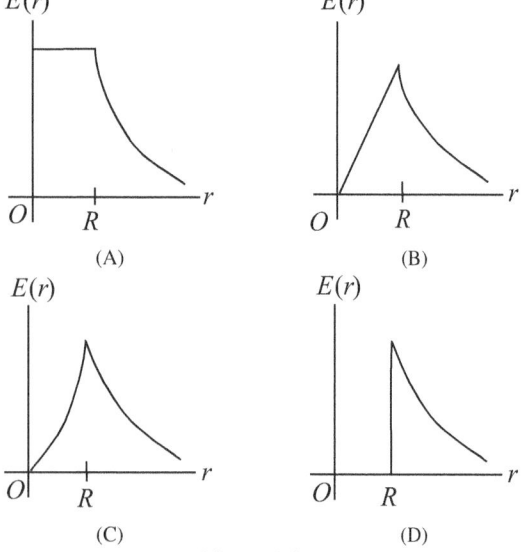

Figure 1.278

II. Use the following Key to choose the appropriate answer.
(**A**) If both Statement-I and Statement-II are true, and Statement - II is the correct explanation of Statement- I.
(**B**) If both Statement -I and Statement -II are true but Statement - II is not the correct explanation of Statement -I.
(**C**) If Statement -I is true but Statement -II is false.
(**D**) If Statement -I is false but Statement -II is true.

13. **Statement-1:** For a mass M kept at the centre of a cube of side 'a', the flux of gravitational field passing through its sides is $4\pi GM$.
    Statement-2: If the direction of a field due to a point source is radial and its dependence on the distance 'r' from the source is given as $\frac{1}{r^2}$, its flux through a closed surface depends only on the strength of the source enclosed by the surface and not on the size or shape of the surface. [2008]

14. **Statement-1:** For a charged particle moving from point P to point Q, the net work done by an electrostatic field on the particle is independent of the path connecting point P to point Q.
    Statement-2: The net work done by a conservative force on an object moving along a closed loop is zero. [2009]

15. Let $P(r) = \frac{Q}{\pi R^4} r$ be the charge density distribution for a solid sphere of radius R and total charge Q. For point 'P' inside the sphere at distance r_1 from the centre of the sphere, the magnitude of electric field is - [2009]
    (A) 0   (B) $\frac{Q}{4\pi\epsilon_0 r_1^2}$ (C) $\frac{Qr_1^2}{4\pi\epsilon_0 R^4}$ (D) $\frac{Qr_1^2}{3\pi\epsilon_0 R^4}$

16. A charge Q is placed at each of the opposite corners of a square. A charge q is placed at each of the other two corners. If the net electrical force on Q is zero, then Q/q equals - [2009]
    (A) $-2\sqrt{2}$ (B) -1 (C) 1 (D) $-\frac{1}{\sqrt{2}}$

17. A thin semi-circular ring of radius r has a positive charge q distributed uniformly over it[Fig.1.279]. The net field $\vec{E}$ at the centre O is [2010]
    (A) $\frac{q}{4\pi^2\epsilon_0 r^2} j$ (B) $-\frac{q}{4\pi^2\epsilon_0 r^2} j$
    (C) $-\frac{q}{2\pi^2\epsilon_0 r^2} j$ (D) $\frac{q}{2\pi^2\epsilon_0 r^2} j$

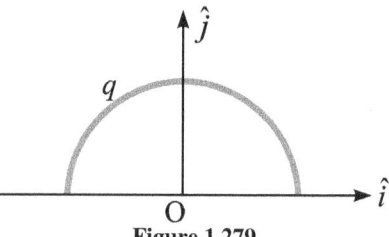

Figure 1.279

18. Let there be a spherically symmetric charge distribution with charge density varying as $\rho(r) = \rho_0 \left(\frac{5}{4} - \frac{r}{R}\right)$ upto $r = R$, and $r(r) = 0$ for $r > R$, where r is the distance from the origin. The electric field at a distance $r (r < R)$ from the origin is given by [2010]
    (A) $\frac{4\pi\rho_0 r}{3\epsilon_0}\left(\frac{5}{3} - \frac{r}{R}\right)$ (B) $\frac{\rho_0 r}{4\epsilon_0}\left(\frac{5}{3} - \frac{r}{R}\right)$
    (C) $\frac{4\rho_0 r}{3\epsilon_0}\left(\frac{5}{4} - \frac{r}{R}\right)$ (D) $\frac{\rho_0 r}{3\epsilon_0}\left(\frac{5}{4} - \frac{r}{R}\right)$

19. Two identical charged spheres suspended from a common point by two massless strings of length l are initially a distance $d(d << 1)$ apart because of their mutual repulsion. The charge begins to leak from both the spheres at a constant rate. As a result the charges approach each other with a velocity v. Then v as a function of distance x between them - [2011]
    (A) $v \propto x^{-1}$ (B) $v \propto x^{1/2}$ (C) $v \propto x$ (D) $v \propto x^{-1/2}$

20. In a uniformly charged sphere of total charge Q and radius R, the electric field E is plotted as a function of distance from the centre. The graph which would correspond to the above will be[Fig.1.280]- [2012]

21. Let $[\epsilon_0]$ denote the dimensional formula of the permittivity of vacuum. If M = mass, L = length, T = time and A = electric current, then $[\epsilon_0]$ is: [2013]

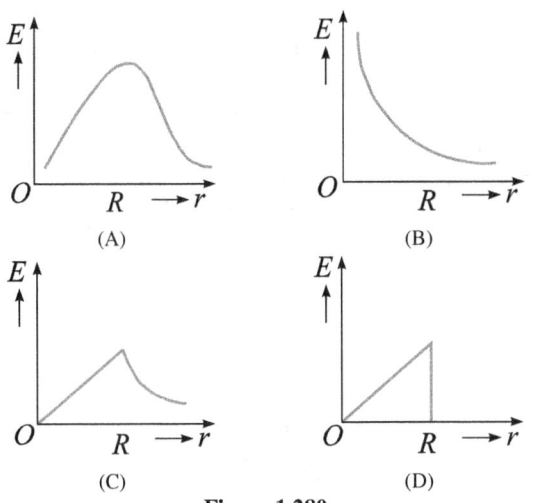

Figure 1.280

(A) $[M^{-1}L^{-3}T^2A]$ (B) $[M^{-1}L^{-3}T^4A^2]$
(C) $[M^{-1}L^2T^{-1}A^{-2}]$ (D) $[M^{-1}L^2T^{-1}A]$

22. Two charges, each equal to q, are kept at $x = -a$ and $x = a$ on the x-axis. A particle of mass m and charge $q_0 = q/2$ is placed at the origin. If charge q_0 is given a small displacement ($y \ll a$) along the y-axis, the net force acting on the particle is proportional to: [2013]
   (A) y (B) $-y$ (C) $1/y$ (D) $-1/y$

23. The region between two concentric spheres of radii 'a' and 'b', respectively (Figure 1.281), has volume charge density $\rho = A/r$, where A is a constant and r is the distance from the centre. At the centre of the spheres is a point charge Q. The value of A such that the electric field in the region between the spheres will be constant, is: [2016]
   (A) $\dfrac{2Q}{\pi a^2}$ (B) $\dfrac{e}{2\pi a^2}$
   (C) $\dfrac{Q}{2\pi(b^2-a^2)}$ (D) $\dfrac{2Q}{\pi(a^2-b^2)}$

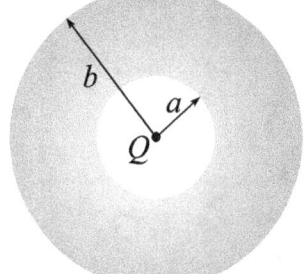

Figure 1.281

24. An electric dipole has a fixed dipole moment $\vec{p}$, which makes angle θ with respect to x-axis. When subjected to an electric field $\vec{E}_1 = E\hat{\imath}$, it experiences a torque $\vec{\tau}_1 = \tau\hat{k}$. When subjected to another electric field $\vec{E}_2 = E_1\hat{\imath}$ it experiences a torque $\vec{\tau}_2 = -\vec{\tau}_1$. The angle θ is: [2017]
   (A) 90°   (B) 30°   (C) 45°   (D) 60°

25. Two identical conducting spheres A and B, carry equal charge. They are separated by a distance much larger than their diameter, and the force between them is F. A third identical conducting sphere, C, is uncharged. Sphere C is first touched to A, then to B, and then removed. As a result, the force between A and B would be equal to [9 Jan 2018]
   (A) $3F/4$ (B) $F/2$ (C) F (D) $3F/8$

26. A solid ball of radius R has a charge density ρ given by $\rho = \rho_0\left(1 - \dfrac{r}{R}\right)$ for $0 \le r \le R$. The electric field outside the ball is: [15 April 2018]
   (A) $\dfrac{\rho_0 R^3}{\epsilon_0 r^2}$ (B) $\dfrac{4\rho_0 R^3}{3\epsilon_0 r^2}$
   (C) $\dfrac{3\rho_0 R^3}{4\epsilon_0 r^2}$ (D) $\dfrac{\rho_0 R^3}{12\epsilon_0 r^2}$

27. A charge Q is placed at a distance $a/2$ above the centre of the square surface of edge a as shown in the figure 1.282. The electric flux through the square surface is: [15 April 2018]

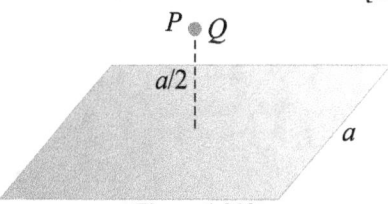

Figure 1.282

   (A) $Q/3\epsilon_0$ (B) $Q/6\epsilon_0$ (C) $Q/2\epsilon_0$ (D) Q/ϵ_0

28. Three charges $+Q$, q, $+Q$ are placed respectively, at distance, $d/2$ and d from the origin, on the x-axis. If the net force experienced by $+Q$, placed at $x = 0$, is zero, then value of q is: [2019]
   (A) $-Q/4$ (B) $+Q/2$ (C) $+Q/4$ (D) $-Q/2$

29. Charge is distributed within a sphere of radius R with a volume charge density $p(r) = \dfrac{A}{r^2}e^{-2r/a}$ where A and a are constants. If Q is the total charge of this charge distribution, the radius R is: [2019]
   (A) $a\log\left(1 - \dfrac{Q}{2\pi aA}\right)$ (B) $\dfrac{a}{2}\log\left(\dfrac{1}{1-\dfrac{Q}{2\pi aA}}\right)$
   (C) $a\log\left(\dfrac{1}{1-\dfrac{Q}{2\pi aA}}\right)$ (D) $\dfrac{a}{2}\log\left(1 - \dfrac{Q}{2\pi aA}\right)$

30. Four point charges $-q$, $+q$, $+q$ and $-q$ are placed on y-axis at $y = -2d$, $y = -d$, $y = +d$ and $y = +2d$, respectively. The magnitude of the electric field E at a point on the x-axis at $x = D$, with $D \gg d$, will behave as: [9 April 2019, (II)]
   (A) $E \propto 1/D^3$ (B) $E \propto 1/D$
   (C) $E \propto 1/D^4$ (D) $E \propto 1/D^2$

31. The bob of a simple pendulum has mass 2g and a charge of $5.0^{1/4}$C. It is at rest in a uniform horizontal electric field of intensity 2000 V/m. At equilibrium, the angle that the pendulum makes with the vertical is: (take $g = 10$ m/s^2) [8 April 2019(I)]
   (A) $\tan^{-1}(2.0)$ (B) $\tan^{-1}(0.2)$
   (C) $\tan^{-1}(5.0)$ (D) $\tan^{-1}(0.5)$

32. For a uniformly charged ring of radius R, the electric field on its axis has the largest magnitude at a distance h from its centre. Then value of h is: [9 Jan. 2019 (I)]
   (A) $R/\sqrt{5}$ (B) $R/\sqrt{2}$ (C) R (D) $R\sqrt{2}$

33. Shown in the Fig.1.283, is a shell made of a conductor. It has inner radius a and outer radius b, and carries charge Q. At its centre is a dipole having dipole moment $\vec{p}$ as shown. In this case: [12 April 2019(I)]
   (A) surface change density on the inner surface is uniform and equal to $\dfrac{Q/2}{4\pi a^2}$
   (B) electric field outside the shell is the same as that of a point charge at the centre of the shell.
   (C) surface charge density on the outer surface depends on $|\vec{p}|$
   (D) surface charge density on the inner surface of the shell is zero everywhere.

34. An electric dipole is formed by two equal and opposite charges q with separation d. The charges have same mass

m. It is kept in a uniform electric field E. If it is slightly rotated from its equilibrium orientation, then its angular frequency ω is: [8 April 2019, (II)]
(A) $\sqrt{\frac{qE}{md}}$ (B) $\sqrt{\frac{2qE}{md}}$ (C) $2\sqrt{\frac{qE}{md}}$ (D) $\sqrt{\frac{qE}{2md}}$

35. Charges Q_1 and Q_2 are at points A and B of a right angle triangle OAB (Fig.1.284). The resultant electric field at point O is perpendicular to the hypotenuse, then Q_1/Q_2 is proportional to: [06 Sep 2020 (I)]
(A) x_1^3/x_2^3 (B) x_2/x_1 (C) x_1/x_2 (D) x_2^2/x_1^2

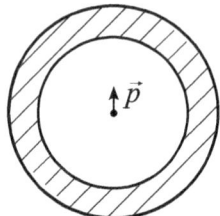
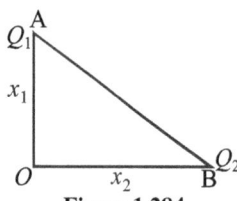

Figure 1.283   Figure 1.284

36. Consider the force F on a charge 'q' due to a uniformly charged spherical shell of radius R carrying charge Q distributed uniformly over it. Which one of the following statements is true for F, if 'q' is placed at distance r from the centre of the shell? [06 Sep. 2020 (II)]
(A) $F = \frac{1}{4\pi\epsilon_0}\frac{Qq}{R^2}$ for $r < R$
(B) $\frac{1}{4\pi\epsilon_0}\frac{Qq}{R^2} > F > 0$ for $r < R$
(C) $F = \frac{1}{4\pi\epsilon_0}\frac{Qq}{R^2}$ for $r > R$
(D) $F = \frac{1}{4\pi\epsilon_0}\frac{Qq}{R^2}$ for all r

37. Two charged thin infinite plane sheets of uniform surface charge density σ_+ and σ_-, where $|\sigma_+| > |\sigma_-|$, intersect at right angle. Which of the diagram of Fig.1.285 best represents the electric field lines for this system? [04 Sep. 2020 (I)]

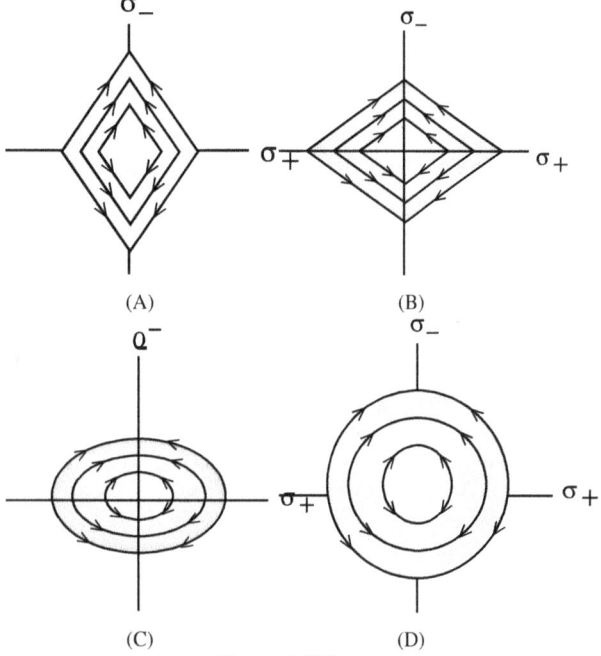

Figure 1.285

38. A particle of charge q and mass m is subjected to an electric field $E = E_0\left(1 - ax^2\right)$ in the x-direction, where a and E_0 are constants. Initially the particle was at rest at $x = 0$. Other than the initial position the kinetic energy of the particle becomes zero when the distance of the particle from the origin is: [04 Sep 2020 (II)]
(A) a (B) $\sqrt{2/a}$ (C) $\sqrt{3/a}$ (D) $\sqrt{1/a}$

39. A small point mass carrying some positive charge on it, is released from the edge of a table [Fig.1.286]. There is a uniform electric field in this region in the horizontal direction. Which of the given options then correctly describe the trajectory of the mass in Fig.1.287? (Curves are drawn schematically and are not to scale). [02 Sep 2020 (II)]

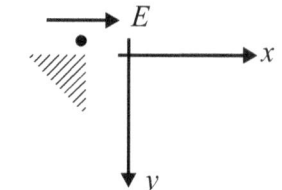

Figure 1.286

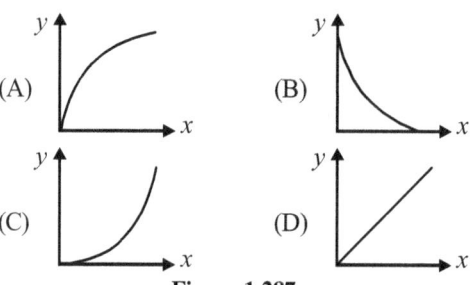

Figure 1.287

40. Consider a sphere of radius R which carries a uniform charge density ρ. If a sphere of radius $R/2$ is carved out of it as shown in Fig.1.288, then the ratio $|\vec{E}_A|/|\vec{E}_B|$ of magnitude of electric field $\vec{E}_A$ and $\vec{E}_B$, respectively, at points A and B due to the remaining portion is: [9 Jan. 2020 (I)]
(A) $\frac{21}{34}$ (B) $\frac{18}{34}$ (C) $\frac{17}{54}$ (D) $\frac{18}{54}$

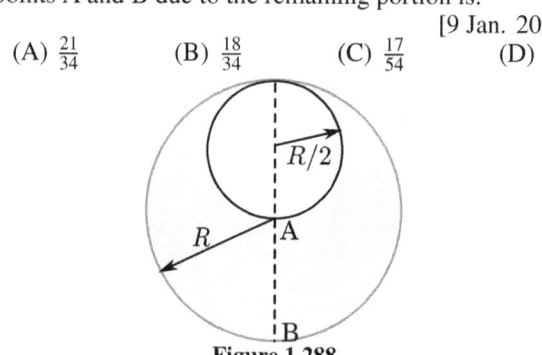

Figure 1.288

41. An electric dipole of moment $\vec{p} = (\hat{i} - 3\hat{j} + 2\hat{k}) \times 10^{-29}$ C.m is at the origin $(0,0,0)$. The electric field due to this dipole at $\vec{r} = +\hat{i} + 3\hat{j} + 5\hat{k}$ (note that $\vec{r}\cdot\vec{p} = 0$) is parallel to: [9 Jan. 2020 (I)]
(A) $(+\hat{i} - 3\hat{j} - 2\hat{k})$ (B) $(-\hat{i} + 3\hat{j} - 2\hat{k})$
(C) $(+\hat{i} + 3\hat{j} - 2\hat{k})$ (D) $(-\hat{i} - 3\hat{j} + 2\hat{k})$

42. A particle of mass m and charge q is released from rest in a uniform electric field. If there is no other force on the particle, the dependence of its speed v on the distance x travelled by it is correctly given by[Fig.1.289] (graphs are schematic and not drawn to scale) [8 Jan. 2020 (II)]

43. Two infinite planes each with uniform surface charge density $+\sigma$ are kept in such a way that the angle between them is $30°$. The electric field in the region shown between them is given by: [7 Jan. 2020, (I)]

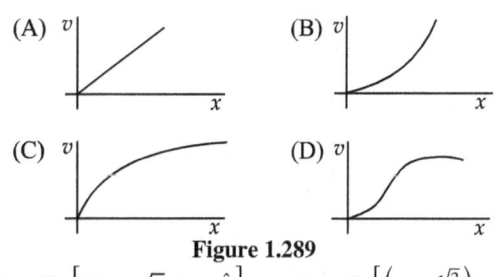

Figure 1.289

(A) $\frac{\sigma}{2\epsilon_0}\left[(1+\sqrt{3})\hat{y} - \frac{\hat{x}}{2}\right]$ (B) $\frac{\sigma}{\epsilon_0}\left[\left(1+\frac{\sqrt{3}}{2}\right)\hat{y} + \frac{\hat{x}}{2}\right]$

(C) $\frac{\sigma}{2\epsilon_0}\left[(1+\sqrt{3})\hat{y} + \frac{\hat{x}}{2}\right]$ (D) $\frac{\sigma}{2\epsilon_0}\left[\left(1-\frac{\sqrt{3}}{2}\right)\hat{y} - \frac{\hat{x}}{2}\right]$

44. A particle of mass m and charge q has an initial velocity $\vec{v} = v_0\hat{j}$. If an electric field $\vec{E} = E_0\hat{i}$ and magnetic field $\vec{B} = B_0\hat{i}$ act on the particle, its speed will double after a time: [7 Jan 2020, (II)]

(A) $\frac{2mv_0}{qE_0}$ (B) $\frac{3mv_0}{qE_0}$ (C) $\frac{\sqrt{3}mv_0}{qE_0}$ (D) $\frac{\sqrt{2}mv_0}{qE_0}$

45. Two identical electric point dipoles have dipole moments $\vec{p}_1 = p\hat{i}$ and $\vec{p}_2 = -p\hat{i}$ and are held on the x axis at distance 'a' from each other. When released, they move along x axis with the direction of their dipole moments remaining unchanged. If the mass of each dipole is 'm', their speed when they are infinitely far apart is: [06 Sep. 2020 (II)]

(A) $\frac{p}{a}\sqrt{\frac{1}{\pi\epsilon_0 ma}}$ (B) $\frac{p}{a}\sqrt{\frac{1}{2\pi\epsilon_0 ma}}$

(C) $\frac{p}{a}\sqrt{\frac{2}{\pi\epsilon_0 ma}}$ (D) $\frac{p}{a}\sqrt{\frac{2}{2\pi\epsilon_0 ma}}$

46. In finding the electric field using Gauss law the formula $|\vec{E}| = \frac{q_{encl}}{\epsilon_0|A|}$ is applicable. In the formula ϵ_0 is permittivity of free space, A is the area of Gaussian surface and q_{enc} is charge enclosed by the Gaussian surface. This equation can be used in which of the following situation? [8 Jan 2020, (I)]

(A) Only when the Gaussian surface is an equipotential surface.
(B) Only when the Gaussian surface is an equipotential surface and $|\vec{E}|$ is constant on the surface.
(C) Only when $|\vec{E}|$ = constant on the surface.
(D) For any choice of Gaussian surface.

47. An electric field $\vec{E} = 4x\hat{i} - (y^2 + 1)\hat{j}$ N/C passes through the box shown in Fig.1.290. The flux of the electric field through surfaces $ABCD$ and $BCGF$ are marked as ϕ_1 and ϕ_{11} respectively. The difference $(\phi_1 - \phi_{11})$ is ____ Nm²/C. [9 Jan 2020 (II)]

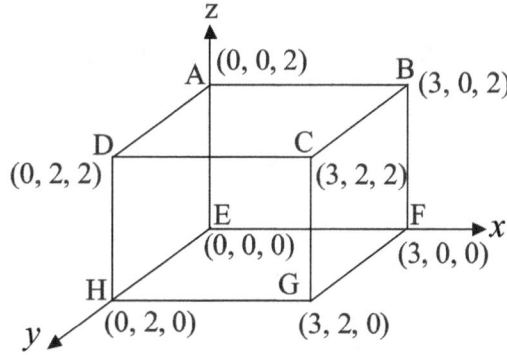

Figure 1.290

48. A uniformly charged disc of radius R having surface charge density σ is placed in the xy plane with its centre at the origin. Find the electric field intensity along the z-axis at a distance Z from origin: [27 Aug.2021 (I)]

(A) $E = \frac{\sigma}{2\epsilon_0}\left(\frac{1}{Z^2+R^2}\right) + \frac{1}{Z^2}$

(B) $E = \frac{2\epsilon_0}{2\sigma}\left(\frac{1}{(Z^2+R^2)^{1/2}} + Z\right)$

(C) $E = \frac{\sigma}{2\epsilon_0}\left(1 + \frac{Z}{(Z^2+R^2)^{1/2}}\right)$

(D) $E = \frac{\sigma}{2\epsilon_0}\left(1 - \frac{Z}{(Z^2+R^2)^{1/2}}\right)$

49. Figure 1.291 shows a rod AB, which is bent in a 120° circular arc of radius R. A charge $(-Q)$ is uniformly distributed over rod AB. What is the electric field $\vec{E}$ at the centre of curvature O? [27-Aug-2021 (I)]

(A) $\frac{3\sqrt{3}Q}{8\pi^2\epsilon_0 R^2}(\hat{i})$ (B) $\frac{3\sqrt{3}Q}{8\pi\epsilon_0 R^2}(\hat{i})$

(C) $\frac{3\sqrt{3}Q}{8\pi^2\epsilon_0 R^2}(-\hat{i})$ (D) $\frac{3\sqrt{3}Q}{16\pi^2\epsilon_0 R^2}(\hat{i})$

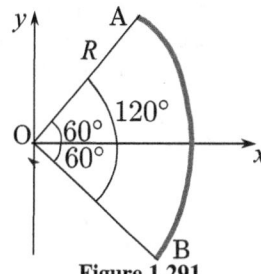

Figure 1.291

50. A vertical electric field of magnitude 4.9×10^5 N/C just prevents a water droplet of a mass 0.1 g from falling. The value of charge on the droplet will be: (Given $g = 9.8$ m/s²) [24 June 22 Shift I]

(A) 1.6×10^{-9} C (B) 2.0×10^{-9} C
(C) 3.2×10^{-9} C (D) 0.5×10^{-9} C

51. In the Fig.1.292, a very large plane sheet of positive charge is shown. P_1 and P_2 are two points at distance l and $2l$ from the charge distribution. If σ is the surface charge density, then the magnitude of electric fields E_1 and E_2 at P_1 and P_2 respectively are: [25 June 22 Shift I]

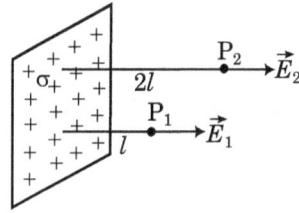

Figure 1.292

(A) $E_1 = \sigma/\epsilon_0$, $E_2 = \sigma/2\epsilon_0$
(B) $E_1 = 2\sigma/\epsilon_0$, $E_2 = \sigma/\epsilon_0$
(C) $E_1 = E_2 = \sigma/2\epsilon_0$
(D) $E_1 = E_2 = \sigma/\epsilon_0$

52. **Assertion (A)**: Non-polar materials do not have any permanent dipole moment.

Reason (R): When an non-polar material is placed in an electric field. the centre of the positive charge distribution of it's individual atom or molecule coincides with the centre of the negative charge distribution.

In the light of above statements, choose the most appropriate answer from the options given below. [26 June 22 Shift I]

(A) Both (A) and (R) are correct and (R) is the correct explanation of (A).
(B) Both (A) and (R) are correct and (R) is not the correct explanation of (A).
(C) (A) is correct but (R) is not correct.
(D) (A) is not correct but (R) is correct.

53. Sixty four conducting drops each of radius 0.02 m and each carrying a charge of 5 μC are combined to form a bigger drop at constant temperature. The ratio of surface density of bigger drop to the smaller drop will be: [26 June 22 Shift II]
(A) 1 : 4 (B) 4 : 1 (C) 1 : 8 (D) 8 : 1

54. A force of 10 N acts on a charged particle placed between two plates of a charged capacitor. If one plate of capacitor is removed, then the force acting on that particle will be: [27 June 22 Shift I]
(A) 5 N (B) 10 N (C) 20 N (D) Zero

55. If a charge q is placed at the centre of a closed hemispherical non-conducting surface as shown in Fig.1.293, the total flux passing through the flat surface would be: [27 June 22 Shift II]

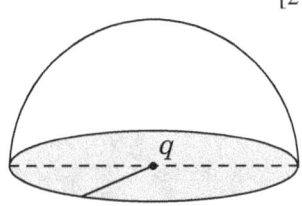

Figure 1.293

(A) 0 (B) $\frac{q}{2\epsilon_0}$ (C) $\frac{q}{4\epsilon_0}$ (D) $\frac{q}{2\pi\epsilon_0}$

56. Three identical charged balls each of charge 2 C are suspended from a common point P by silk threads of 2 m each (as shown in Fig.1.294). They form an equilateral triangle of side 1 m. The ratio of net electric force on a charged ball to the force between any two charged balls will be: [27 June 22 Shift II]
(A) 1 : 1 (B) 1 : 4 (C) $\sqrt{3}$: 2 (D) $\sqrt{3}$: 1

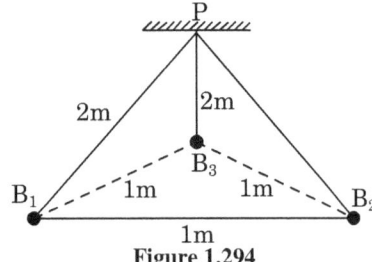

Figure 1.294

57. Given below are two statements: [28 June 22 Shift I]
Statement-I: A point charge is brought in an electric field. The value of electric field at a point near to the charge may increase if the charge is positive.
Statement-II: An electric dipole is placed in a non-uniform electric field. The net electric force on the dipole will not be zero.
Choose the correct answer from the options given below:
(A) Both statement-I and statement-II are true.
(B) Both statement-I and statement-I are false.
(C) Statement-I is true but statement-II is false.
(D) Statement-I is false but statement-II is true.

58. The three charges $q/2$, q and $q/2$ are placed at the corners A, B and C of a square of side 'a' as shown in Fig.1.295. The magnitude of electric field ($\vec{E}$) at the corner D of the square, is: [28 June 22 Shift I]

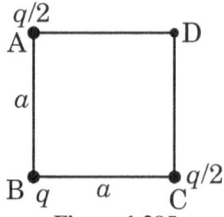

Figure 1.295

(A) $\frac{q}{4\pi\epsilon_0 a^2}\left(\frac{1}{\sqrt{2}} + \frac{1}{2}\right)$ (B) $\frac{q}{4\pi\epsilon_0 a^2}\left(1 + \frac{1}{\sqrt{2}}\right)$
(C) $\frac{q}{4\pi\epsilon_0 a^2}\left(1 - \frac{1}{\sqrt{2}}\right)$ (D) $\frac{q}{4\pi\epsilon_0 a^2}\left(\frac{1}{\sqrt{2}} - \frac{1}{2}\right)$

59. Two point charges A and B of magnitude $+8 \times 10^{-6}$ C and -8×10^{-6} C respectively are placed at a distance d apart. The electric field at the middle point O between the charges is 6.4×10^4 NC^{-1}. The distance 'd' between the point charges A and B is: [28 June 22 Shift II]
(A) 2.0 m (B) 3.0 m (C) 1.0 m (D) 4.0 m

60. Two point charges Q each are placed at a distance d apart. A third point charge q is placed at a distance x from mid-point on the perpendicular bisector. The value of x at which charge q will experience the maximum Coulomb's force is: [29 June 23 Shift II]
(A) d (B) $d/2$ (C) $d/\sqrt{2}$ (D) $\dfrac{d}{2\sqrt{2}}$

61. If two charges q_1 and q_2 are separated with distance 'd' and placed in a medium of dielectric constant κ. What will be the equivalent distance between charges in air for the same electrostatic force? [24 Jan 2023 Shift I]
(A) $d\sqrt{k}$ (B) $k\sqrt{d}$ (C) $1.5d\sqrt{k}$ (D) $2d\sqrt{k}$

62. A point charge of 10 C is placed at the origin. At what location on the X–axis should a point charge of 40 μC be placed so that the net electric field is zero at $x = 2$ cm on the X–axis? [25 Jan 2023 Shift II]
(A) $x = 6$ cm (B) $x = 4$ cm
(C) $x = 8$ cm (D) $x = -4$ cm

63. A point charge $q_1 = 4q_0$ is placed at origin. Another point charge $q_2 = -q_0$ is placed at $x = 12$ cm. Charge of proton is q_0. The proton is placed on x-axis so that the electrostatic force on the proton in zero. In this situation, the position of the proton from the origin is ____ cm. [29 Jan 2023 Shift I]

64. A point charge 2×10^{-2} C is moved from P to S in a uniform electric field of 30 NC^{-1} directed along positive x-axis as shown in Fig.1.296. If coordinates of P and S are (1 m, 2 m, 0 m) and (0 m, 0 m, 0 m) respectively, the work done by electric field will be [29 Jan 2023 Shift II]
(A) 1200 mJ (B) 600 mJ
(C) –600 mJ (D) –1200 mJ

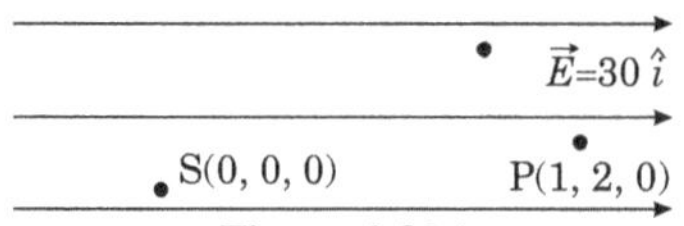

Figure 1.296

65. Electric field in a certain region is given by $\vec{E} = \left(\dfrac{A}{x^2}\hat{i} + \dfrac{B}{y^3}\hat{j}\right)$. The SI unit of A and B are: [30 Jan 2023 Shift I]
(A) Nm^{3}C^{-1}; Nm^{2}C^{-1} (B) Nm^{2}C^{-1}; Nm^{3}C^{-1}
(C) Nm^{3}C; Nm^{2}C (D) Nm^{2}C; Nm^{3}C

66. As shown in the Fig.1.297, a point charge Q is placed at the centre of conducting spherical shell of inner radius a and outer radius b. The electric field due to charge Q in three different regions I, II and III is given by : ($I : r < a$,

$II : a < r < b, III : r > b$) [30 Jan 2023 Shift II]
(A) $E_I = 0, E_{II} = 0, E_{III} \neq 0$
(B) $E_I \neq 0, E_{II} = 0, E_{II} \neq 0$
(C) $E_I \neq 0, E_{II} = 0, E_{II} = 0$
(D) $E_I = 0, E_{II} = 0, E_{II} = 0$

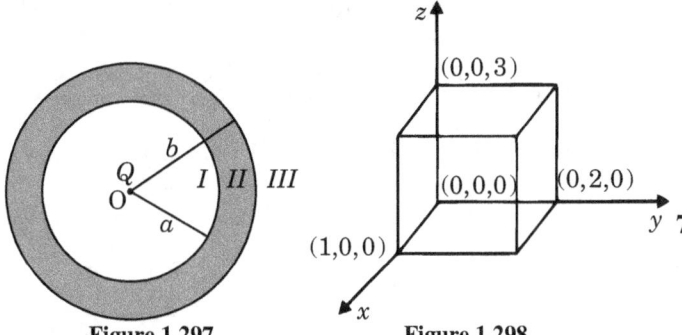

Figure 1.297 Figure 1.298

67. As shown in Fig.1.298, a cuboid lies in a region with electric field $\vec{E} = 2x^2\hat{i} - 4y\hat{j} + 6\hat{k}$ N/C. The magnitude of charge within the cuboid is $n\epsilon_0$ C. The value of n is ____ (if dimension of cuboid is $1 \times 2 \times 3$ m^3) [30 Jan 2023 Shift II]

68. Expression for an electric field is given by $\vec{E} = 4000x^2\hat{i}$ V/m. The electric flux through the cube of side 20 cm when placed in electric field (as shown in the Fig.1.299) is ____ V·cm. [31 Jan 2023 Shift I]

69. Let σ be the uniform surface charge density of two infinite thin plane sheets shown in Fig.1.300. Then the electric fields in three different region E_I, E_{II} and E_{III} are: [01 Feb 23 Shift I]

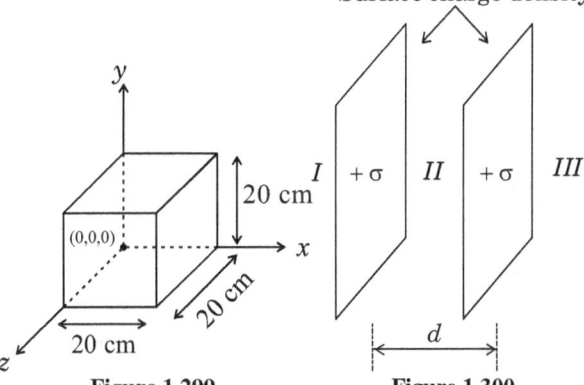

Figure 1.299 Figure 1.300

(A) $\vec{E}_I = \frac{2\sigma}{\epsilon_0}\hat{n}, \vec{E}_{II} = 0, \vec{E}_{III} = \frac{2\sigma}{\epsilon_0}\hat{n}$
(B) $\vec{E}_I = 0, \vec{E}_{II} = \frac{\sigma}{\epsilon_0}\hat{n}, \vec{E}_{III} = 0$
(C) $\vec{E}_I = \frac{\sigma}{2\epsilon_0}\hat{n}, \vec{E}_{II} = 0, \vec{E}_{III} = \frac{\sigma}{2\epsilon_0}\hat{n}$
(D) $\vec{E}_I = -\frac{\sigma}{\epsilon_0}\hat{n}, \vec{E}_{II} = 0, \vec{E}_{III} = \frac{\sigma}{\epsilon_0}\hat{n}$

70. In Fig.1.301, two equal positive point charges are separated by a distance $2a$. The distance of a point from the centre of the line joining two charges on the equatorial line (perpendicular bisector) at which force experienced by a test charge q_0 becomes maximum is $a/\sqrt{x}$. The value of x is ____. [01 Feb 23 Shift I]

71. A cubical volume is bounded by the surfaces $x = 0$, $x = a$, $y = 0$, $y = a$, $z = 0$, $z = a$. The electric field in the region is given by $\vec{E} = E_0 x \hat{i}$. Where $E_0 = 4 \times 10^4$ NC^{-1}m^{-1}. If $a = 2$ cm, the charge contained in the cubical volume is $Q \times 10^{-14}$ C. The value of Q is ____. (Take $\epsilon_0 = 9 \times 10^{-12}$ C^2/Nm^2) [01 Feb 23 Shift II]

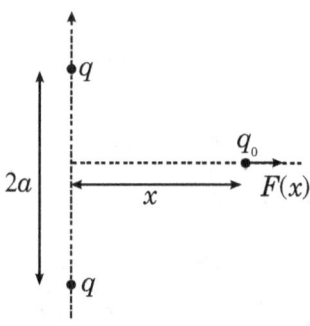

Figure 1.301

72. A dipole comprises of two charged particles of identical magnitude q and opposite in nature. The mass 'm' of the positive charged particle is half of the mass of the negative charged particle. The two charges are separated by a distance 'l'. If the dipole is placed in a uniform electric field '$\vec{E}$'; such a way that dipole axis makes a very small angle with the electric field, '$\vec{E}$'. The angular frequency of the oscillations of the dipole when released is given by: [06 April 23 Shift II]

(A) $\sqrt{\frac{3qE}{2ml}}$ (B) $\sqrt{\frac{4qE}{ml}}$ (C) $\sqrt{\frac{4qE}{3ml}}$ (D) $\sqrt{\frac{8qE}{ml}}$

73. Which of the following graphs, shown in Fig.1.303, represents the graphical variation of the electric field due to a uniformly charged insulating solid sphere of radius R at a distance r from the centre O, as depicted in Fig.1.302? [08 April 23 Shift I]

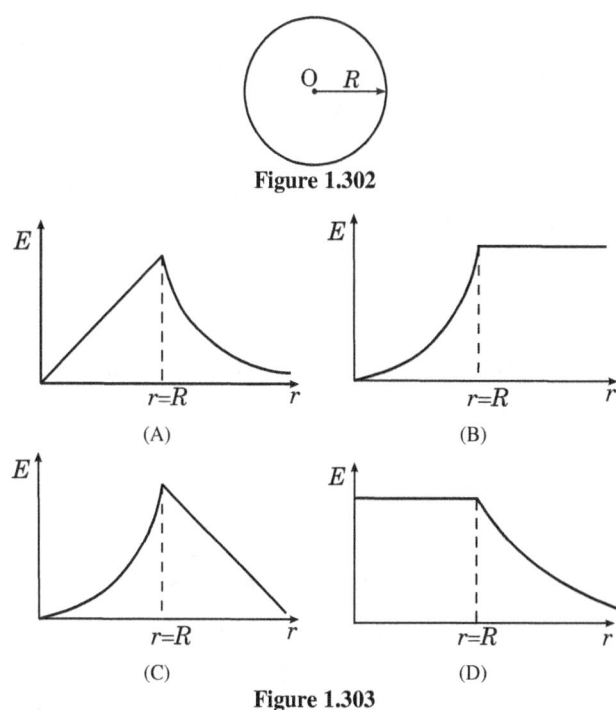

Figure 1.302

Figure 1.303

74. Given below are two statement: one is labelled as Assertion A and the other is labelled as Reason R. [12 April 23 Shift I]

Assertion A: If an electric dipole of dipole moment 30×10^{-5} cm is enclosed by a closed surface, the net flux coming out of the surface will be zero.

Reason R: Electric dipole consists of two equal and opposite charges.

In the light of above, statements, choose the correct answer

from the options given below:
(A) Both A and R are true and R is the correct explanation of A.
(B) A is true but R is false.
(C) Both A and R true but R is not the correct explanation of A.
(D) A is false but R is true.

75. A thin infinite sheet charge and an infinite line charge of respective charge densities $+\sigma$ and $+\lambda$ are placed parallel at 5 m distance from each other. Points 'A' and 'B' are at $(3/\pi)$ m and $(4/\pi)$ m perpendicular distance from line charge towards sheet charge, respectively. 'E_A' and 'E_B' are the magnitudes of resultant electric field intensities at point 'A' and 'B', respectively. If $E_A/E_B = 4/a$ for $2|\sigma| = |\lambda|$. Then the value of a is [13 April 23 Shift I]

76. Two charges each of magnitude 0.01 C and separated by a distance of 0.4 mm constitute an electric dipole. If the dipole is placed in a uniform electric field $|\vec{E}|$ of 10 dyne/C making $30°$ angle with $\vec{E}$, the magnitude of torque acting on dipole is- [13 April 23 Shift I]
(A) 4.0×10^{-10} Nm   (B) 2.0×10^{-10} Nm
(C) 1.0×10^{-8} Nm   (D) 15×10^{-9} Nm

77. A 10 μC charge is divided into two parts and placed at 1 cm distance so that the repulsive force between them is maximum. The charges of the two parts are:
[13 April 23 Shift II]
(A) 9 μC, 1 μC   (B) 5 μC, 5 μC
(C) 7 μC, 3 μC   (D) 8 μC, 2 μC

78. The electric field due to a short electric dipole at a large distance (r) from centre of dipole on the equatorial plane varies with distance as: [15 April 23 Shift I]
(A) r (B) $1/r$ (C) $1/r^3$ (D) $1/r^2$

Section B: JEE Advanced

1. Five point charges, each of value $+q$, are placed on five vertices of a regular hexagon of side L. The magnitude of the force on a point charge of value $-q$ placed at the centre of the hexagon is [1992]
(A) $\kappa q^2/L^2$ (B) $\kappa q^2/4L^2$ (C) $\kappa q^2/2L^2$ (D) $\kappa q^2/8L^2$

2. Two point charges $+q$ and $-q$ are held fixed at $(-d, 0)$ and $(d, 0)$ respectively of a (x, y) coordinate system, then
[1995]
(A) The electric field $\vec{E}$ at all points on the x-axis has the same direction
(B) $\vec{E}$ at all points on the Y - axis is along $\hat{i}$
(C) Work has to be done in bringing a test charge from infinity to the origin
(D) The dipole moment is $2qd$ directed along $\hat{i}$

3. A metallic solid sphere is placed in a uniform electric field. The lines of force follow the path(s) shown in Fig.1.304 as [1996]
(A) 1   (B) 2   (C) 3   (D) 4

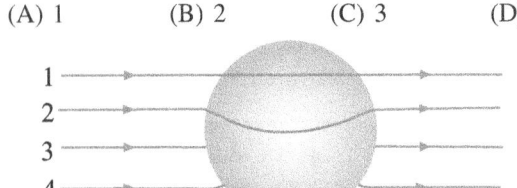

Figure 1.304

4. An electron of mass m_e, initially at rest, moves through a certain distance in a uniform electric field in time t_1. A proton of mass m_p, also, initially at rest, takes time t_2 to move through an equal distance in this uniform electric field. Neglecting the effect of gravity, the ratio t_2/t_1 is nearly equal to [1997]
(A) 1   (B) $(m_p/m_e)^{1/2}$ (C) $(m_e/m_p)^{1/2}$ (D) 1836

5. A non-conducting solid sphere of radius R is uniformly charged. The magnitude of the electric field due to the sphere at a distance r from its centre [1998]
(A) increases as r increases, for $r < R$
(B) decreases as r increases, for $0 < r < \infty$
(C) increases as r increases, for $R < r < \infty$
(D) is discontinuous at $r = R$

6. Three positive charges of equal value q are placed at the vertices of an equilateral triangle. The resulting lines of force should be sketched as in Fig.1.305 [2001]

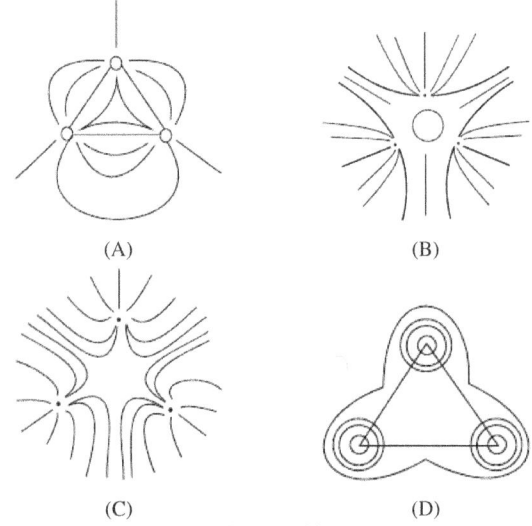

Figure 1.305

7. A point charge 'q' is placed at a point inside a hollow conducting sphere. Which of the following electric lines of force pattern is correct[Fig.1.306]? [2003]

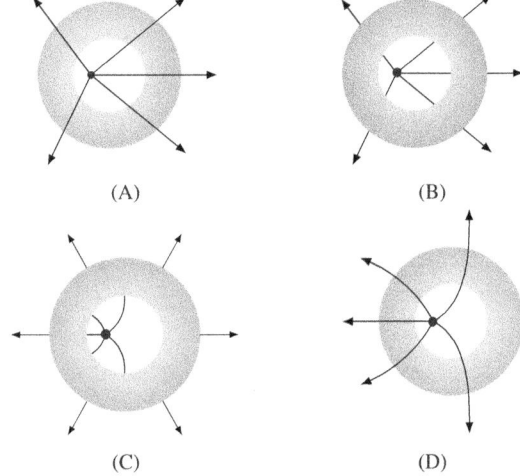

Figure 1.306

8. In Fig.1.307, charges q_1 and $-q_1$ are inside a Gaussian surface. Where as charge q_2 is outside the surface. Electric field on the Gaussian surface will be [2004]
(A) only due to q_2
(B) zero on the Gaussian surface
(C) uniform on the Gaussian surface
(D) due to all

9. Six charges of equal magnitude are placed at six corners of a regular hexagon [Fig.1.308]. Find arrangement the charges in order PQRSTU which produce double electric field at

centre as compared to electric field produce by single charges $+q$ at R- [2004]

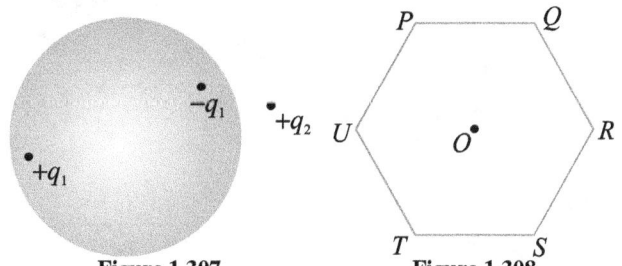

Figure 1.307   Figure 1.308

(A) + + + – – –   (B) + – + + –
(C) – + + + –   (D) – + + – –

10. Three large charged sheets having surface charge density as shown in the Figure1.309. The sheets are placed parallel to XY plane. Then electric field at point P- [2005]
(A) $\frac{-4\sigma}{\epsilon_0}k$ (B) $\frac{4\sigma}{\epsilon_0}k$ (C) $\frac{2\sigma}{\epsilon_0}k$ (D) $-\frac{2\sigma}{\epsilon_0}k$

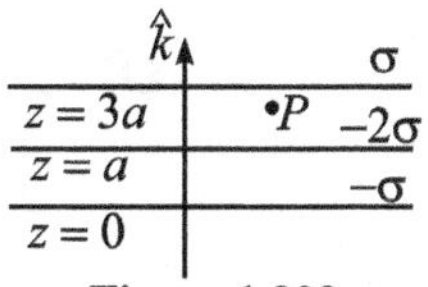

Figure 1.309

11. Consider a neutral conducting sphere. A positive point charge is placed outside the sphere. The net charge on the sphere is then, [2007]
(A) negative and distributed uniformly over the surface of the sphere
(B) negative and appears only at the point on the sphere closest to the point charge
(C) negative and distributed non-uniformly over the entire surface of the sphere
(D) zero

12. A disc of radius $a/4$ having a uniformly distributed charge 6C is placed in the x-y plane with its centre at $(-a/2, 0, 0)$. A rod of length 'a' carrying a uniformly distributed charge 8C is placed on the x-axis from $x = a/4$ to $x = 5a/4$ [Fig.1.310]. Two point charges $-7C$ and $3C$ are placed at $(a/4, -a/4, 0)$ and $(-3a/4, 3a/4, 0)$ respectively. Consider a cubical surface formed by six surfaces $x = \pm a/2$, $y = \pm a/2$, $z = \pm a/2$. The electric flux this cubical surface is- [2009]
(A) $-2C/\epsilon_0$ (B) $2C/\epsilon_0$ (C) $10C/\epsilon_0$ (D) $12C/\epsilon_0$

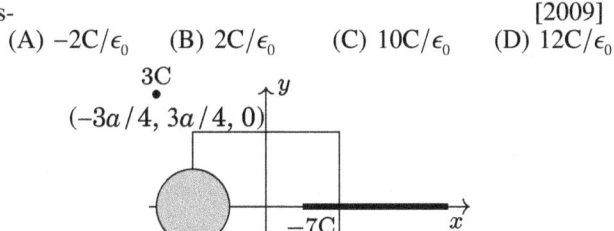

Figure 1.310

13. Under the influence of the Coulomb field of charge $+Q$, a charge $-q$ is moving around it in an elliptical orbital. Find out the correct statement(s) - [2009]
(A) The angular momentum of the charge $-q$ is constant
(B) The linear momentum of the charge $-q$ is constant
(C) The angular velocity of the charge $-q$ is constant
(D) The linear speed of the charge $-q$ is constant

14. A few electric field lines for a system of two charges Q_1 and Q_2 fixed at two different points on the x-axis are shown in the Fig.1.311. These lines suggest that [2010]
(A) $|Q_1| > |Q_2|$
(B) $|Q_1| < |Q_2|$
(C) at a finite distance to the left of Q_1 the electric field is zero
(D) at a finite distance to the right of Q_2 the electric field is zero

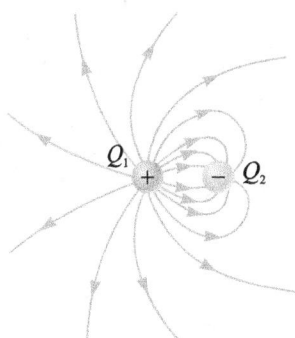

Figure 1.311

15. A uniformly charged thin spherical shell of radius R carries uniform surface charge density of σ per unit area. It is made of two hemispherical shells, held together by pressing them with force F (Figure 1.312). F is proportional to [2010]
(A) $\frac{1}{\epsilon_0}\sigma^2 R^2$ (B) $\frac{1}{\epsilon_0}\sigma^2 R$ (C) $\frac{1}{\epsilon_0}\frac{\sigma^2}{R}$ (D) $\frac{1}{\epsilon_0}\frac{\sigma^2}{R^2}$

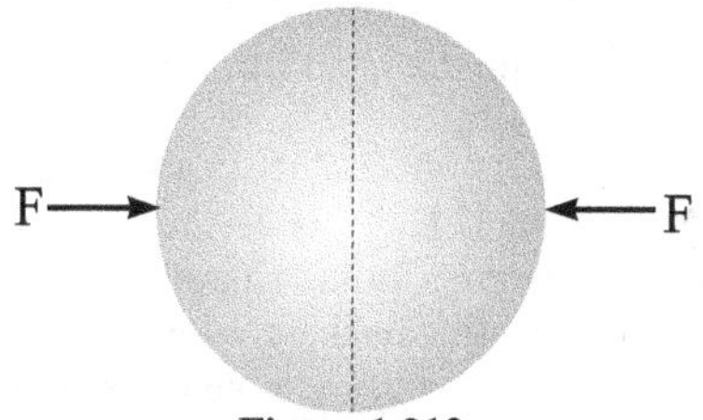

Figure 1.312

16. A tiny spherical oil drop carrying a net charge q is balanced in still air with a vertical uniform electric field of strength $\frac{81\pi}{7} \times 10^5 \text{Vm}^{-1}$. When the field is switched off, the drop is observed to fall with terminal velocity 2×10^{-3} ms^{-1} Given g = 9.8 ms^{-2}, viscosity of the air = 1.8×10^{-5} Nsm^{-2} and the density of oil = 900 kg m^{-3}, the magnitude of q is [2010]
(A) 1.6×10^{-19}C (B) 3.2×10^{-19}C
(C) 4.8×10^{-19}C (D) 8.0×10^{-19}C

17. Consider an electric field $E = E_0 x$, where E_0 is a constant. The flux through the shaded area (as shown in the Fig.1.313) due to this field is [2011]
(A) $2E_0 a^2$ (B) $\sqrt{2}E_0 a^2$ (C) $E_0 a^2$ (D) $\frac{E_0 a^2}{\sqrt{2}}$

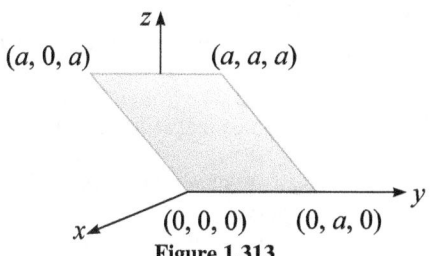

Figure 1.313

18. A cubical region of side a has its centre at the origin. It encloses three fixed point charges [Fig. 1.314], $-q$ at $(0, -a/4, 0)$, $+3q$ at $(0, 0, 0)$ and $-q$ at $(0, +a/4, 0)$. Choose the correct options(s) [2012]

(A) The net electric flux crossing the plane $x = +a/2$ is equal to the net electric flux crossing the plane $x = -a/2$
(B) The net electric flux crossing the plane $y = +a/2$ is more than the net electric flux crossing the plane $y = -a/2$.
(C) The net electric flux crossing the entire region is q/ϵ_0.
(D) The net electric flux crossing the plane $z = +a/2$ is equal to the net electric flux crossing the plane $x = +a/2$.

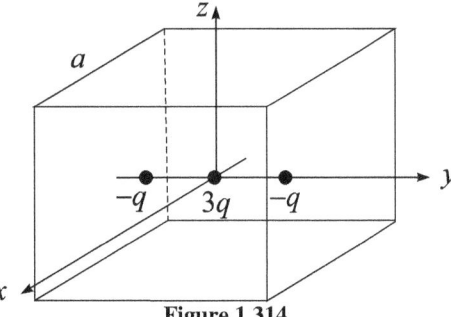

Figure 1.314

19. Two non-conducting solid spheres of radii R and $2R$, having uniform volume charge densities ρ_1 and ρ_2 respectively, touch each other. The net electric field at a distance $2R$ from the centre of the smaller sphere, along the line joining the centres of the spheres, is zero. The ratio (ρ_1/ρ_2) can be [2013]
    (A) -4 (B) $-(32/25)$ (C) $(32/25)$ (D) 4

20. Let $E_1(r)$, $E_2(r)$ and $E_3(r)$ be the respective electric fields at a distance r from a point charge Q, an infinitely long wire with constant linear charge density λ, and an infinite plane with uniform surface charge density σ. If $E_1(r_0) = E_2(r_0) = E_3(r_0)$ at a given distance r_0, then [2014]
    (A) $Q = 4\sigma\pi r_0^2$
    (B) $r_0 = \frac{\lambda}{2\pi\sigma}$
    (C) $E_1\left(\frac{r_0}{2}\right) = 2E_2\left(\frac{r_0}{2}\right)$
    (D) $E_2\left(\frac{r_0}{2}\right) = 4E_3\left(\frac{r_0}{2}\right)$

21. Charges Q, $2Q$ and $4Q$ are uniformly distributed in three dielectric solid spheres 1, 2 and 3 of radii $R/2$, R and $2R$ respectively, as shown in Fig.1.315. If magnitudes of the electric fields at point P at a distance R from the centre of spheres 1, 2 and 3 are E_1, E_2 and E_3 respectively, then [2014]
    (A) $E_1 > E_2 > E_3$
    (B) $E_3 > E_1 > E_2$
    (C) $E_2 > E_1 > E_3$
    (D) $E_3 > E_2 > E_1$

22. A long cylindrical shell carries positive surface charge σ in the upper half and negative surface charge $-\sigma$ in the lower half. The electric field lines around the cylinder will look like Fig.1.316 given in : (figures are schematic and not drawn to scale) [2015]

23. A solid sphere of radius R has a charge Q distributed in its volume with a charge density $\rho = \kappa r^a$, where κ and a are constants and r is the distance from its centre. If the electric field at $r = \frac{R}{2}$ is $1/8$ times that at $r = R$, find the value of a. [2009]

24. Four point charges, each of $+q$, are rigidly fixed at the four corners of a square planar soap film of side 'a'. The surface tension of the soap film is y. The system of charges and planar film are in equilibrium, and $a = k\left[\frac{q^2}{\gamma}\right]^{1/N}$, where '$k$' is a constant. Then N is [2011]

25. An infinitely long solid cylinder of radius R has a uniform volume charge density ρ. It has a spherical cavity of radius $R/2$ with its centre on the axis of the cylinder, as shown in the Figure 1.317. The magnitude of the electric field at

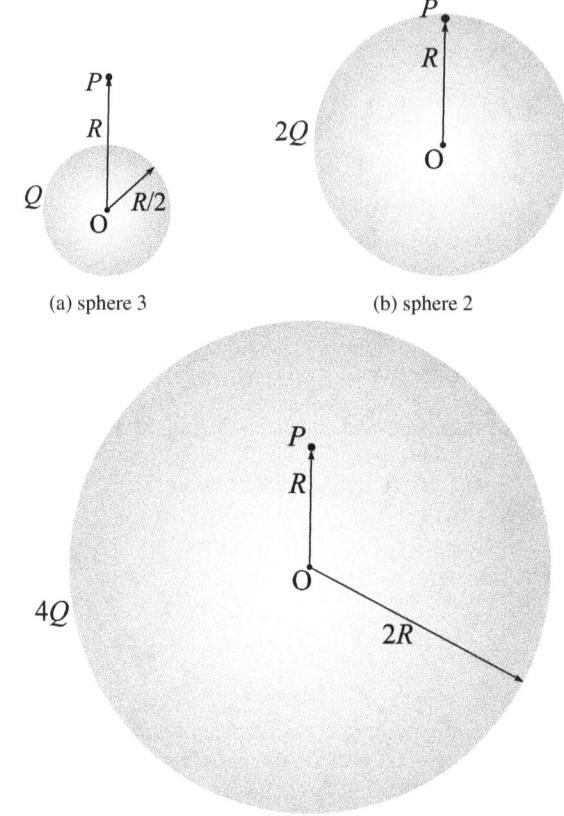

(a) sphere 3

(b) sphere 2

(c) sphere 3

Figure 1.315

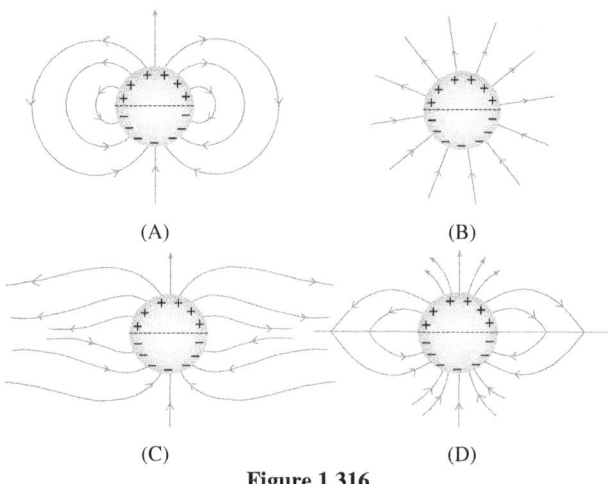

Figure 1.316

the point P, which is at a distance $2R$ from the axis of the cylinder, is given by the expression $\frac{23\rho R}{16 k\epsilon_0}$. The value of k is [2013]

26. An infinitely long thin non-conducting wire is parallel to the z-axis and carries a uniform line charge density λ. It pierces a thin non-conducting spherical shell of radius R in such a way that the arc PQ subtends an angle 120° at the centre O of the spherical shell, as shown in the Figure 1.318. The permittivity of free space is ϵ_0. Which of the following statements is (are) true? [2018]

(A) The electric flux through the shell is $\sqrt{3}R/\lambda\epsilon_0$
(B) The z-component of the electric field is zero at all the points on the surface of the shell
(C) The electric flux through the shell is $\sqrt{2}R/\lambda\epsilon_0$
(D) The electric field is normal to the surface of the shell at all points

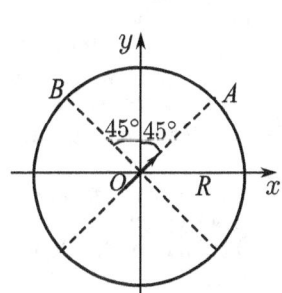

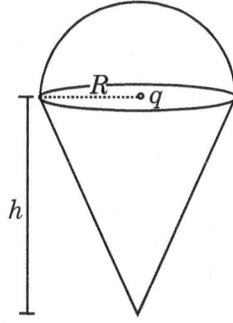

Figure 1.319 **Figure 1.320**

(A) The magnitude of total electric field on any two points of the circle will be same
(B) Total electric field at point B is $\vec{E}_B = 0$
(C) Total electric field at point A is $\vec{E}_A = \sqrt{2}E_0(\hat{i}+\hat{j})$
(D) $R = \left(\dfrac{p_0}{4\pi\epsilon_0 E_0}\right)^{1/3}$

29. A charged shell of radius R carries a total charge Q. Given ϕ as the flux of electric field through a closed cylindrical surface of height h, radius r and with its centre same as that of the shell. Here, centre of the cylinder is a point on the axis of the cylinder which is equidistant from its top and bottom surfaces. Which of the following option(s) is/are correct? [ϵ_0 is the permittivity of free space] [2019]
    (A) If $h > 2R$ and $r = 3R/5$ then $\phi = Q/5\epsilon_0$
    (B) If $h > 2R$ and $r > R$ then $\phi = Q/\epsilon_0$
    (C) If $h < 8R/5$ and $r = 3R/5$ then $\phi = 0$
    (D) If $h > 2R$ and $r > 4R/5$ then $\phi = Q/5\epsilon_0$

30. Two identical non-conducting solid spheres of same mass and charge are suspended in air from a common point by two non-conducting, massless strings of same length. At equilibrium, the angle between the strings is α. The spheres are now immersed in a dielectric liquid of density 800 kg.m^{-3} and dielectric constant 21. If the angle between the strings remains the same after the immersion, then [2020]
    (A) electric force between the spheres remains unchanged
    (B) electric force between the spheres reduces
    (C) mass density of the spheres is 840 kgm^{-3}
    (D) the tension in the strings holding the spheres remains unchanged

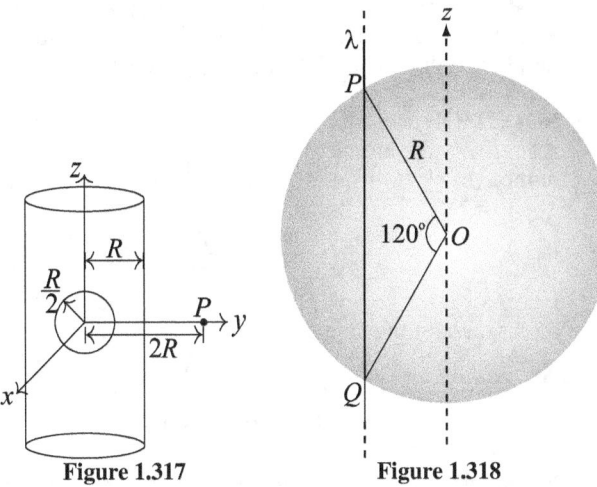

Figure 1.317 **Figure 1.318**

27. The electric field E is measured at a point P $(0,0,d)$ generated due to various charge distributions and the dependence of E on d is found to be different for different charge distributions. List-I contains different relations between E and d. List-II describes different electric charge distributions, along with their locations. Match the functions in List-I with the related charge distributions in List-II. [2018]

List I	List II
P. E is independent of d	1. A point charge Q at the origin
Q. $E \propto 1/d$	2. A small dipole with point charges Q at $(0,0,1)$ and $-Q$ at $(0,0,-1)$. Take $2 < d$
R. $E \propto 1/d^2$	3. An infinite line charge coincident with the x-axis, with uniform linear charge density λ.
S. $E \propto 1/d^x$	4. Two infinite wires carrying uniform linear Charge density parallel to the x-axis. The one along $(y = 0, z = 1)$ has a charge density $+\lambda$ and the one along $(y = 0, z = -1)$ has a charge density $-\lambda$. Take $2\lambda << d$
	5. Infinite plane charge coincident with the xy-plane with uniform surface charge density

(A) $P \to 5; Q \to 3,4; R \to 1; S \to 2$
(B) $P \to 5; Q \to 3; R \to 1,4; S \to 2$
(C) $P \to 5; Q \to 3; R \to 1,2; S \to 4$
(D) $P \to 4; Q \to 2,3; R \to 1; S \to 5$

28. An electric dipole with moment $\dfrac{p_0}{\sqrt{2}}(\hat{i}+\hat{j})$ is held fixed at the origin O in the presence of a uniform electric field of magnitude E_0. If the potential is constant on a circle of radius R centered at the origin as shown in Fig.1.319, then the correct statement(s) is/are: (ϵ_0 is permittivity of free space. $R \gg$ dipole size) [2019]

31. A circular disc of radius R carries surface charge density $\sigma(r) = \sigma_0\left(1 - \dfrac{r}{R}\right)$, where σ_0 is a constant and r is the distance from the centre of the disc. Electric flux through a large spherical surface that encloses the charged disc completely is ϕ_0. Electric flux through another spherical surface of radius $R/4$ and concentric with the disc is ϕ. Then the ratio ϕ_0/ϕ is ____. [2020]

32. A charge q is surrounded by a closed surface consisting of an inverted cone of height h and base radius R, and a hemisphere of radius R as shown in the Fig.1.320. The electric flux through the conical surface is $\dfrac{nq}{6\epsilon_0}$ (in SI units). The value of n is ____. [2022]

1.38 Answer Keys and Solutions
1.38.1 Check Point 1
1. Rub a glass rod with silk and use it to charge an electroscope. The electroscope will end up with a net positive charge. Bring the pocket comb close to the electroscope. If the electroscope leaves move farther apart, then the charge on the comb is positive, the same as the charge on the electroscope.

If the leaves move close to each other, then the charge on the comb is negative, opposite the charge on the electroscope.

2. The shirt or blouse becomes charged as a result of rubbing against the dryer sides and other clothes. When you put on the charged object (shirt), it causes charge separation within the molecules of your skin, which results in attraction between the shirt and your skin.
3. Fog or rain droplets tend to form around ions because water is a polar molecule, with a positive region and a negative region. The charge centers on the water molecule will be attracted to the ions (positive to negative).
4. The negatively charged electrons in the paper are attracted to the positively charged rod and move towards it within their molecules. The attraction occurs because the negative charges in the paper are closer to the positive rod than are the positive charges in the paper, and therefore the attraction between the unlike charges is greater than the repulsion between the like charges.

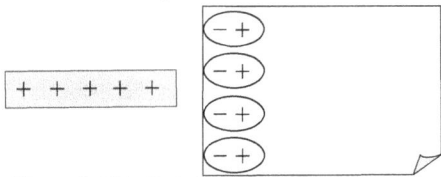

Figure 1.321: Polarization in the paper atoms.

5. A plastic ruler that has been rubbed with a cloth is charged. When brought near small pieces of paper, it will cause separation of charge in the bits of paper, which will cause the paper to be attracted to the ruler. On a humid day, polar water molecules will be attracted to the ruler and to the separated charge on the bits of paper, neutralizing the charges and thus eliminating the attraction.
6. The net charge on a conductor is the difference between the total positive charge and the total negative charge in the conductor. The "free charges" in a conductor are the electrons that can move about freely within the material because they are only loosely bound to their atoms. The "free electrons" are also referred to as "conduction electrons." A conductor may have a zero net charge but still have substantial free charges.
7. Most of the electrons are strongly bound to nuclei in the metal ions. Only a few electrons per atom (usually one or two) are free to move about throughout the metal. These are called the "conduction electrons". The rest are bound more tightly to the nucleus and are not free to move. Furthermore, in the cases shown in Figures 1.4a and 1.4b, not all of the conduction electrons will move. In Figure 1.4a, electrons will move until the attractive force on the remaining conduction electrons due to the incoming charged rod is balanced by the repulsive force from electrons that have already gathered at the left end of the neutral rod. In Figure 1.4a, conduction electrons will be repelled by the incoming rod and will leave the stationary rod through the ground connection until the repulsive force on the remaining conduction electrons due to the incoming charged rod is balanced by the attractive force from the net positive charge on the stationary rod.
8. No, the basic physics of electric charges would not have been affected at all by an opposite assignment of positive and negative labels. The use of + and – signs, as opposed to labels such as A and B, has the distinct advantage that it gives zero net charge to an object that contains equal amounts of positive and negative charge.
9. Initially, the bits of paper are uncharged and are attracted to the comb by polarization effects. When one of the bits of paper comes into contact with the comb, it acquires charge from the comb. Now the piece of paper and the comb have charge of the same sign, and hence there is a repulsive force between them.
10. (a) -4.4×10^{11} C (b) $2.4 \times 10^8 C$
11. 0.275 C
12. **APPROACH** Since, total charge on a body having N electrons is given by, $Q = Ne$. So, first of all, find the total number of electrons (N) in 3.10 g of copper and then use above relation.

    SOLUTION As, the number of electrons in a copper (Cu) atom is 29 (the atomic number of copper). Therefore, total number of electrons in 3.10 gram of copper is given by-
    N = No. of Cu atoms in 3.10 gram of Cu × 29
    $$i.e., \quad N = \frac{\text{Given mass of copper}}{\text{molar mass of copper}} \times N_A \times 29$$
    Here, N_A is avogadro's number and it's value is 6.02×10^{23}, molar mass of the copper is 63.5 g, therefore,
    $$N = \frac{3.10 \ g}{63.5 \ g} \times 6.02 \times 10^{23} \times 29 = 8.53 \times 10^{23} \text{electrons}$$
    So, the total charge on the copper penny,
    $Q = Ne = 8.53 \times 10^{23} \times 1.6 \times 10^{-19} \text{C} = 1.37 \times 10^5$ C
    Since, electrons are negatively charged particles, so the nature of this charge will be negative, i.e., $Q = -1.37 \times 10^5$ C.

1.38.2 Check Point 2

1. For point charges, the two charges must have opposite signs.
2. Yes, when the charge on one body (q_1) is much greater than that on the other (q_2) and they are close enough to each other so that force of attraction between q_1 and induced charge on the other exceeds the force of repulsion between q_1 and q_2 plus similar induced charge. However, two similar point charges can never attract each other because no induction will take place here.
3. Yes, as charging a body means addition or removal of electrons and electron has a mass.
4. All electric appliances may end with some charge due to faulty connections. In such a situation charge will be accumulated on the appliance. When the user touches the appliance, he may get a shock. By providing the third hole for grounding all accumulated charge is discharged to the ground and the appliance is safe.
5. Consider any two equal charges q, with an electric force $F = kq^2/r^2$. If we transfer a charge δ from one to the other, the charges become $(q+\delta)$ and $(q-\delta)$, so the force becomes $F' = k(q+\delta)(q-\delta)/r^2$. But $(q+\delta)(q-\delta) = q^2 - \delta^2$, which is less than q^2, so the force decreases.
6. Total number of atoms in the sphere of mass 1 gram
    $$N = \frac{\text{mass of copper sphere}}{\text{molar mass of copper}} \times \text{Avogadro number}$$
    $$= \frac{1}{63.54} \times 6.023 \times 10^{23}$$
    So the total nuclear charge
    Q_1 = number of atoms × atomic number
    i.e., $Q_1 = N \times 29$
    $$= \frac{6.023 \times 10^{23}}{63.54} \times 1.6 \times 10^{-19} \times 29$$
    Now the charge on the sphere = total nuclear charge – total electronic charge
    $$= \frac{6.023 \times 10^{23}}{63.54} \times 1.6 \times 10^{-19} \times \frac{29 \times 1}{100}$$
    $$= 4.298 \times 10^2 C$$

Hence, force of interaction between these two spheres,
$$F = k \cdot \frac{[4.398 \times 10^2]^2}{1^2} \text{ N}$$
$$= 9 \times 10^9 \times 10^4 \times 19.348 \text{ N} = 1.74 \times 10^{15} \text{ N}$$

7. Given situation is shown in Fig.1.322.
Since it is given that the magnitude of force between $+q$ and $+Q$ is F, therefore, magnitude of force between $2q$ and Q is $2F$; between $-3q$ and Q is $3F$ and between $-4q$ and Q is $4F$.

Charge $+q$ at A applies repulsion F on charge $+Q$ at O. It

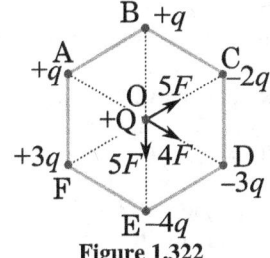

Figure 1.322

is directed from O to D.
Charge $-q$ at D applies attraction $3F$ on charge $+Q$ at O. It is also directed from O to D.
So, resultant of these two forces is $4F$ and it is along OD.
Similarly, Charge $+q$ at B applies repulsion of F on charge $+Q$ at O. It is directed from O to E.
Charge $-4q$ at E applies attraction of $4F$ on charge $+Q$ at O. It is also directed from O to E.
So, resultant of these two forces is $5F$ and it is along OE.
Charge $+3q$ at F applies repulsion of $3F$ on charge $+Q$ at O. It is directed from O to C.
Charge $-2q$ at C applies attraction of $2F$ on charge $+Q$ at O. It is also directed from O to C.
So, resultant of these two forces is $5F$ and it is along OC.
These opposite pair resultants are shown in Fig.1.322.
Now, if F_1 is the resultant of $5F$ acting along OC and $5F$ along OE, then
$$F_1 = \sqrt{(5F)^2 + (5F)^2 + 2(5F)(5F)\cos 120°} = 5F$$
$$[\because \angle COE = 120°]$$
This resultant of two equal force bisects the angle F_1 bisects the angle $\angle COE$ and hence it will be along OD, i.e., along third force of magnitude $4F$. The resultant of these two forces acting along the same direction is
$$F_{\text{net}} = F_1 + 4F = 5F + 4F = 9F$$
Therefore, $n = 9$

8. Consider the equilibrium of charge at A,
F = force due to other charge at distance $L = \dfrac{kq^2}{L^2}$

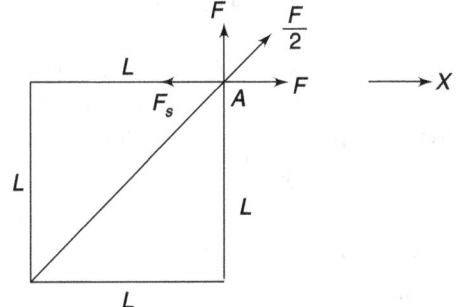

Figure 1.323

Force due to diagonally opposite charge,
$$\frac{kq^2}{(\sqrt{2}L)^2} = \frac{F}{2}$$
The resultant force in X direction,
$$F_x = F + \frac{F}{2}\cos 45° = F\left(1 + \frac{1}{2\sqrt{2}}\right)$$
This gets balanced by the spring force.
$$k \cdot \frac{L}{2} = F\left(1 + \frac{1}{2\sqrt{2}}\right)$$
$$kL = \frac{q^2}{4\pi \epsilon_0 L^2}\left(\frac{2\sqrt{2} + 1}{\sqrt{2}}\right)$$
$$\therefore \quad k = \frac{q^2}{4\pi \epsilon_0 L^3}\left(\frac{2\sqrt{2} + 1}{\sqrt{2}}\right)$$

9. **APPROACH 1.** Suppose the charge on one object is q and on the other is $Q - q$. Since, the electric force between these charges depends on the magnitude of q, therefore, for maximum force between the charges, we have-
$$\frac{dF}{dq} = 0$$
So, first find the expression for electric force F and then use above expression and solve for q.
SOLUTION The force between the objects is
$$F = k\frac{q(Q-q)}{r^2} = k\frac{(qQ - q^2)}{r^2}$$
where r is the separation between them.
Now, $\dfrac{dF}{dq} = \dfrac{d}{dq}\left(k\dfrac{(qQ - q^2)}{r^2}\right) = \left(k\dfrac{(Q - 2q)}{r^2}\right)$
For maximum force, $\dfrac{dF}{dq} = 0$, i.e.,
$$\left(k\frac{(Q - 2q)}{r^2}\right) = 0 \implies q = Q/2$$
Thus, the charge should be divided equally on the two objects.

10. The equilibrium will be unstable.
Explanation We take the case when the central charge is of opposite sign to the two other charges. If this charge is moved slightly along the line joining the three charges, the attraction due to the nearer charge increases, whilst that due to the more distant charge diminishes, with the result that the charge moves still further from the equilibrium position. Its equilibrium is therefore unstable.
If the central charge is of the same sign as the other two, and it moves slightly along the line joining the charges, forces will arise that tend to return it to its equilibrium position. However, if it moves at right angles to the line joining the charges, the resultant of the repulsions will no longer be zero and will act in the direction in which it has moved. As a result the charge will tend to move further from its equilibrium position. The equilibrium is thus unstable.
This result, which we have obtained for an elementary case, is always valid. If only Coulomb forces of interaction are present in a system of free electric charges, the equilibrium is always unstable.

11. $Q = q/\sqrt{3}$ (the equilibrium is unstable).

12. (C) : In case I:
$$F = -\frac{1}{4\pi\varepsilon_0}\frac{Q^2}{r^2} \qquad (1)$$

In Case II : $Q_A = Q - \frac{Q}{4}, Q_B = -Q + \frac{Q}{4}$

$\therefore \quad F' = \frac{1}{4\pi\varepsilon_0} \frac{\left(Q - \frac{Q}{4}\right)\left(-Q + \frac{Q}{4}\right)}{r^2}$

$\Rightarrow F' = \frac{1}{4\pi\varepsilon_0} \frac{\left(\frac{3}{4}Q\right)\left(\frac{-3}{4}Q\right)}{r^2} = -\frac{1}{4\pi\varepsilon_0}\frac{9}{16}\frac{Q^2}{r^2}$ (2)

From equations (1) and (2), $F' = \frac{9}{16}F$

13. **(B):** A hydrogen atom consists of an electron and a proton.
    $\therefore$ Charge on one hydrogen atom
    $= q_e + q_p = -e + (e + \Delta e) = \Delta e$
    Since a hydrogen atom carries a net charge Δe

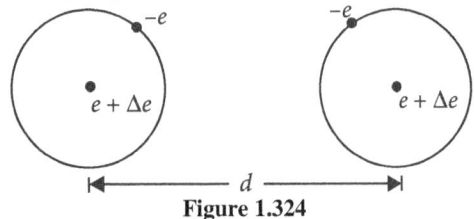

Figure 1.324

$\therefore$ Electrostatic force,
$$F_e = \frac{1}{4\pi\varepsilon_o}\frac{(\Delta e)^2}{d^2} \quad (1)$$

will act between two hydrogen atoms. The gravitational force between two hydrogen atoms is given as
$$F_g = \frac{Gm_h m_h}{d^2} \quad (2)$$

Since, the net force on the system is zero, therefore
$$F_e = F_g$$

Using Eq. 1 and 2, we get
$$\frac{(\Delta e)^2}{4\pi\varepsilon_o d^2} = \frac{Gm_h^2}{d^2}$$

or $(\Delta e)^2 = 4\pi\varepsilon_0 Gm_h^2$
$= 6.67\times 10^{-11} \times (1.67\times 10^{-27})^2 \left[1/(9\times 10^9)\right]$

or $\Delta e \approx 10^{-37}$ C

14. **(B)**
$$T\cos\theta = mg \quad (1)$$
$$T\sin\theta = \frac{kq^2}{x^2} \quad (2)$$

From Equations (1) and (2), we have

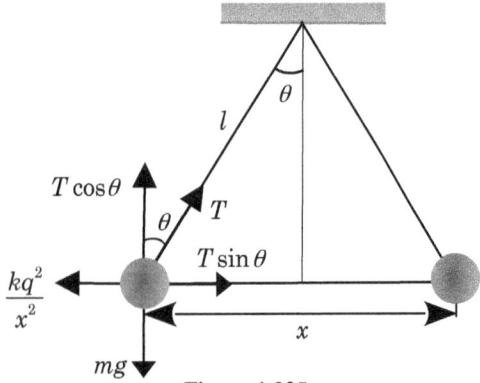

Figure 1.325

$\tan\theta = \frac{kq^2}{x^2 mg}$

Since θ is small, therefore, $\tan\theta = \sin\theta = \frac{x}{2l}$

so, $\frac{x}{2l} = \frac{kq^2}{x^2 mg} \Rightarrow q^2 = x^3\frac{mg}{2lk}$

or $q \propto x^{3/2}$

$\Rightarrow \frac{dq}{dt} \propto \frac{3}{2}\sqrt{x}\frac{dx}{dt} = \frac{3}{2}\left(\sqrt{x}\right)v$

Since, $\frac{dq}{dt}$ = constant, therefore, $v \propto \frac{1}{\sqrt{x}}$

15. **(A)** The given situation is drawn in Fig.1.326. A, B and C are three small charged spheres each has charge q and mass m. They are suspended from three strings AS, BS and CS respectively to the common point of suspension S. Length of each string is l, i.e., $AS = BS = CS = l$ and the angle of these strings from vertical downward direction is θ. The separation between charges is same and at any instant it is x. These charges are in a same horizontal plane and the triangle ABC is an equilateral triangle of side x. $\angle CAB = \angle ABC = \angle BCA = 60°$. O is the centroid of the triangle, therefore, $OA = OB = OC$. SO is perpendicular to the plane of triangle ABC. If we extrapolate median BO behind O, it perpendicularly bisects side CA at point D. So, $DA = x/2$.

Since, lines joining the vertices of an equilateral triangle to it's centroid bisects the angle, therefore $\angle DOA = 60°/2 = 30°$. Also in right angle $\triangle ADO$, we have
$$\cos 30° = \frac{DA}{OA}$$
$$\Rightarrow OA = \frac{DA}{\cos 30°} = \frac{x/2}{\sqrt{3}/2} = \frac{x}{\sqrt{3}} \quad (1)$$

In Fig.1.326, the positive directions of X, Y and Z axes are

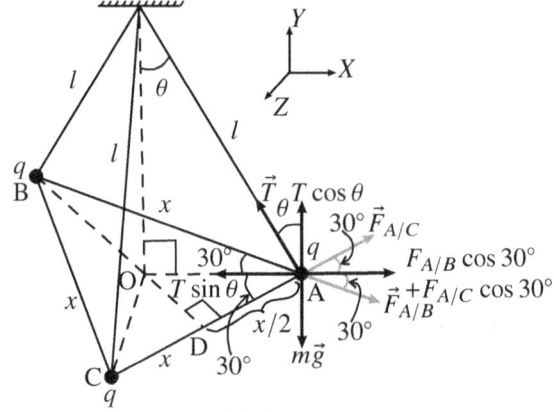

Figure 1.326

also shown in separate diagram.

Now, consider the sphere A. The forces acting on it are:

(a) **Gravitational force $m\vec{g}$:** It is acting vertically downwards along negative direction of Y axis:
$$m\vec{g} = mg\hat{j}$$

(b) **Tension $\vec{T}$:** It is acting along $\overrightarrow{AS}$ in XOY plane
$$\vec{T} = -T\sin\theta\hat{i} + T\sin\theta\hat{j}$$

(c) **Electric force $\vec{F}_{A/B}$:** It is applied by sphere B along $\overrightarrow{BA}$ in XOZ plane:
$$\vec{F}_{A/B} = F_{A/B}\cos 30°\hat{i} + F_{A/B}\sin 30°\hat{k}$$

(d) **Electric force $\vec{F}_{A/C}$:** It is applied by sphere B along $\overrightarrow{CA}$ in XOZ plane:
$$\vec{F}_{A/C} = F_{A/C}\cos 30°\hat{i} - F_{A/C}\sin 30°\hat{k}$$

All forces and their components are shown in Fig.1.326 except Z components of $\vec{F}_{A/B}$ and $\vec{F}_{A/C}$ which will be

outward and inward respectively to the plane pf paper.
Since, charges and distances of B and C from A are identical, therefore they apply same electric force on A. It is given as:
$$F_{A/B} = F_{A/C} = k\frac{q^2}{x^2} \qquad (2)$$
Net force on the charged sphere A,
$$\vec{F} = m\vec{g} + \vec{T} + \vec{F}_{A/B} + \vec{F}_{A/C}$$
$$= -mg\hat{j} + (-T\sin\theta\hat{i} + T\sin\theta\hat{j})$$
$$+ \left(F_{A/B}\cos 30°\hat{i} + F_{A/B}\sin 30°\hat{k}\right)$$
$$+ \left(F_{A/C}\cos 30°\hat{i} - F_{A/C}\sin 30°\hat{k}\right)$$
$$\Rightarrow \vec{F} = (2F_{A/B}\cos 30° - T\sin\theta)\hat{i} + (T\cos\theta - mg)\hat{j} \qquad (3)$$
$$[\because F_{A/B} = F_{AC}]$$
Now, according to problem, the length of side of triangle $\triangle ABC$ is slowly increasing, i.e., the acceleration of each sphere is zero. So, the sphere A is in dynamic equilibrium and hence net force on it must be zero. Therefore from Eq.(3), we get-
$$\vec{F} = (2F_{A/B}\cos 30° - T\sin\theta)\hat{i} + (T\cos\theta - mg)\hat{j} = 0$$
On comparing both sides, we get
$$2F_{A/B}\cos 30° - T\sin\theta = 0 \quad \text{and} \quad T\cos\theta - mg = 0$$
or $\qquad T\sin\theta = 2F_{A/B}\cos 30° = F_{A/B}\sqrt{3} \qquad (4)$
and $\qquad T\cos\theta = mg \qquad (5)$
Dividing Eq.(4) by Eq.(5), we get
$$\tan\theta = \frac{F_{A/B}\sqrt{3}}{mg} \qquad (6)$$
Also in right angle $\triangle SOA$, we have
$$\tan\theta = \frac{OA}{OS} = \frac{OA}{\sqrt{[(AS)^2 - (OA)^2]}} = \frac{x/\sqrt{3}}{\sqrt{[l^2 - (x/2)^2]}} \qquad (7)$$
Since, $x \ll l$, therefore $l^2 - (x/2)^2 \approx l^2$.
Substituting it in Eq.(7), we get-
$$\tan\theta = \frac{x}{l\sqrt{3}} \qquad (8)$$
Comparing the values of $\tan\theta$ obtained from Eq.(6) and (8), we get
$$\frac{F_{A/B}\sqrt{3}}{mg} = \frac{x}{l\sqrt{3}}$$
$$F_{A/B} = \frac{xmg}{3l} \qquad \ldots (9)$$
Putting the value of $F_{A/B}$ from Eq.(2) in (9), we get
$$k\frac{q^2}{x^2} = \frac{xmg}{3l} \implies q = \sqrt{\left[\frac{mg}{3lk}\right]}x^{3/2}$$
Differentiating both sides with respect to time 't', we get
$$\frac{dq}{dt} = \sqrt{\left[\frac{mg}{3lk}\right]}\frac{3}{2}x^{1/2}\frac{dx}{dt}$$
It is also given that, $\frac{dx}{dt} = \frac{a}{\sqrt{x}}$, therefore, above equation gives
$$\frac{dq}{dt} = \sqrt{\left[\frac{mg}{3lk}\right]}\frac{3}{2}x^{1/2}\frac{a}{\sqrt{x}} = \sqrt{\frac{3mga^2}{4lk}} \qquad (10)$$
Since, $k = 1/4\pi\epsilon_0$, therefore above equation can also be written as
$$\frac{dq}{dt} = \sqrt{\frac{3\pi\epsilon_0 mga^2}{l}}$$

16. (D) : Let m be mass of each ball and q be charge on each ball. Force of repulsion,
$$F = \frac{1}{4\pi\epsilon_0}\frac{q^2}{r^2}$$

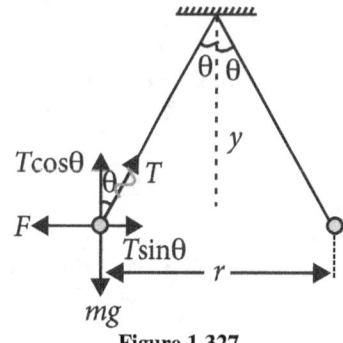

Figure 1.327

In equilibrium, we have
$$T\cos\theta = mg \qquad (1)$$
$$T\sin\theta = F \qquad (2)$$
Dividing Eq.(2) by (1), we get,
$$\tan\theta = \frac{F}{mg} = \frac{\frac{1}{4\pi\epsilon_0}\frac{q}{r^2}}{mg}$$
From Fig.1.327,
$$\frac{r/2}{y} = \frac{\frac{1}{4\pi\epsilon_0}\frac{q}{r^2}}{mg} \qquad (3)$$
From Fig.1.328, we have

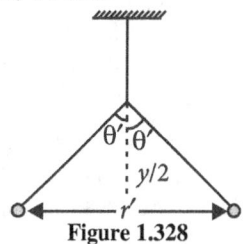

Figure 1.328

$$\tan\theta' = \frac{\frac{1}{4\pi\epsilon_0}\frac{q^2}{r'^2}}{mg} \implies \frac{r'/2}{y/2} = \frac{\frac{1}{4\pi\epsilon_0}\frac{q^2}{r'^2}}{mg} \qquad (4)$$
Divide Eq.(4) by (3), we get
$$\frac{2r'}{r} = \frac{r^2}{r'^2} \implies r^3 = \frac{r^3}{2} \implies r' = \frac{r}{\sqrt[3]{2}}$$

17. (C) According to Coulomb's law, the force of repulsion between the two positive ions each of charge q, separated by a distance d is given by -
$$F = \frac{1}{4\pi\epsilon_0}\frac{(q)(q)}{d^2} = \frac{q^2}{4\pi\epsilon_0 d^2}$$
$$\implies \qquad q^2 = 4\pi\epsilon_0 F d^2$$
$$\text{or} \quad q = \sqrt{4\pi\epsilon_0 F d^2} \qquad (1)$$
Since, $q = ne$ where, n = number of electrons missing from each ion, e = magnitude of charge on electron
$$\therefore \qquad n = \frac{q}{e} = \sqrt{\frac{4\pi\epsilon_0 F d^2}{e^2}}$$

18. (D): $F_m = F_0/\kappa$, i.e., decreases to κ^{-1} times.
Note that it is the net force on either charged particles due to polarisation effect of medium. The individual force acting between the charges remains unchanged.

19. (A) The situation is as shown in the Fig.1.329.

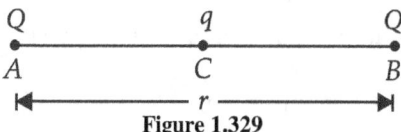

Figure 1.329

Let two equal charges Q each placed at points A and B at

a distance r apart. C is the centre of AB where charge q is placed.

For equilibrium, net force on charge $Q = 0$

$\therefore \quad \dfrac{1}{4\pi\varepsilon_0}\dfrac{QQ}{r^2} + \dfrac{1}{4\pi\varepsilon_0}\dfrac{Qq}{(r/2)^2} = 0$

or $\quad \dfrac{1}{4\pi\varepsilon_0}\dfrac{Q^2}{r^2} = -\dfrac{1}{4\pi\varepsilon_0}\dfrac{4Qq}{r^2}$

or $\quad Q = -4q \quad$ or $\quad q = -Q/4$

20. (C) Net force on each of the charge due to the other charges is zero. However, disturbance in any direction other than along the line on which the charges lie, will not make the charges return.

21. A, B, C

22. (C) Is pushed outside the triangle. Like charges repel. The forces from the two charges at vertices adjacent to the fourth charge are equal but opposite and cancel. The net force is thus the same as the contribution from the charge at the opposite vertex; that repulsion will push the fourth charge away, out of the triangle.

1.38.3 Check Point 3

1. **Approach** We can use symmetry to significantly reduce the number of actual calculations. If we place a small test charge at point P, the forces from the top left and the bottom right will be of equal magnitude but opposite direction and will cancel. The same is true for the charges at the top right and the bottom left. Therefore, we only need to consider the two central charges. Since the top one is positive and the bottom one is negative, a positive test charge placed at P would experience downward forces from both charges. We simply calculate the equal electric fields using the point charge relationship and then use the superposition principle to combine.

SOLUTION $E = k\dfrac{Q}{a^2} + k\dfrac{Q}{a^2} = 2k\dfrac{Q}{a^2}$ (downwards)

2. If $\vec{r}_0$ is the position vector of position O (1 m, 2 m, 4 m) of charge and $\vec{r}_P$ is the position vector of position P (0, −4 m, 3 m) with respect to origin (0, 0, 0), then

$\vec{r}_0 = (\hat{i} + 2\hat{j} + 4\hat{k})$ m and $\vec{r}_P = (-4\hat{j} + 3\hat{k})$ m

Therefore, $\vec{r}_{OP} = \vec{r}_P - \vec{r}_O = (-\hat{i} - 6\hat{j} - \hat{k})$ m

and $\vec{r}_{OP} = \sqrt{(-1)^2 + (-6)^2 + (-1)^2} = \sqrt{38}$ m

So, $E = k\dfrac{Q}{|\vec{r}_{OP}|^2} = \dfrac{9.0 \times 10^9 \times 10^{-6}}{38}$

$\approx 0.24 \times 10^3 \text{N/C} = 240 \text{ N/C}$

3. Here, $r - r_0 = (8\hat{i} - 5\hat{j}) - (2\hat{i} + 3\hat{j}) = 6\hat{i} - 8\hat{j}$

Therefore, $|r - r_0| = \sqrt{6\hat{i} - 8\hat{j}} = 10\ m$

So, Required field strength,

$|\vec{E}| = \dfrac{1}{4\pi\varepsilon_0}\dfrac{q}{|\vec{r} - \vec{r}_0|^2} = 9.0 \times 10^9 \dfrac{50 \times 10^{-6}}{10^2}$

$= 4.5 \times 10^3$ V/m $= 4.5$ kV/m

4. **APPROACH** follow the approach of Example 43. If $\vec{E}$ is the electric field at point P, then, the electric force on test charge placed at point P, is given by $\vec{F} = q_0\vec{E}$.

Answer $\vec{E} = \dfrac{qz.2\sqrt{2}}{ql^3}\hat{k}$, $\vec{F} = q_0\vec{E} = \dfrac{qq_0z.2\sqrt{2}}{ql^3}\hat{k}$

5. Given: Charges Q_1 and Q_2 are located on vertices of a right angled triangle OAB,
   The resultant electric field at O due to Q_1 and Q_2 is perpendicular to side AB.
   To find $\dfrac{Q_1}{Q_2}$: From the Fig.1.330,

   $\tan\theta = \dfrac{x_1}{x_2} = \dfrac{F_2}{F_1}$

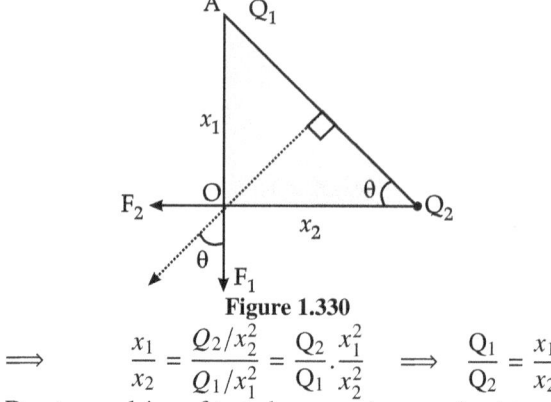

Figure 1.330

$\Rightarrow \dfrac{x_1}{x_2} = \dfrac{Q_2/x_2^2}{Q_1/x_1^2} = \dfrac{Q_2}{Q_1}\cdot\dfrac{x_1^2}{x_2^2} \quad \Rightarrow \quad \dfrac{Q_1}{Q_2} = \dfrac{x_1}{x_2}$

6. Due to repulsion of test charge q_0 (assumed not too small), placed at point P, the suspended charged ball repel away from q_0, therefore the separation between the two charges increases. As a result of it, the measured value of electric field $E_{\text{measured}} = \dfrac{F}{q_0}$ will be less than the actual value E_{act}.

7. $k\dfrac{4Q}{x^2} = k\dfrac{9Q}{(d-x)^2}$

$\Rightarrow \dfrac{4}{x^2} = \dfrac{9}{(d-x)^2}$

$\Rightarrow \dfrac{2}{x} = \pm\dfrac{3}{(d-x)}$

$\Rightarrow 2d - 2x = \pm 3x$

On simplifying above expression, we get

$x = \dfrac{2d}{5}$ and $x = -2d$

The only acceptable answer is, $x = 2d/5$

8. Net intensity of electric field at origin is given by-

$E_O = \dfrac{kq}{r_1^2} + \dfrac{kq}{r_2^2} + \dfrac{kq}{r_3^2} + \ldots \infty$

$= kq \times 10^4 \left[\dfrac{1}{1} + \dfrac{1}{4} + \dfrac{1}{16} + \dfrac{1}{64} + \ldots \infty\right]$

$= 9 \times 10^9 \times 20 \times 10^{-6} \times 10^4 \times \left(\dfrac{1}{1 - 1/4}\right)$

$= 24 \times 10^8$ N/C

9. **APPROACH** Apply the principle of superposition of electric fields and the expansion-

$\ln(1+x) = x - \dfrac{x^2}{2} + \dfrac{x^3}{3} - \dfrac{x^4}{4} + \ldots$

For, $x = 1$, above expression gives-

$\ln 2 = 1 - \dfrac{1}{2} + \dfrac{1}{3} - \dfrac{1}{4} + \ldots$

SOLUTION Net intensity of electric field at origin is given by-

$E_O = \dfrac{kq}{a^2} - \dfrac{kq}{2a^2} + \dfrac{kq}{3a^2} - \dfrac{kq}{3a^2} - \dfrac{kq}{4a^2} + \dfrac{kq}{5a^2} - \ldots$

$= \dfrac{kq}{a^2}\left[1 - \dfrac{1}{2} + \dfrac{1}{3} - \dfrac{1}{4} + \ldots\right]$

$= \dfrac{kq}{a^2}[\ln 2]$

10. (B) From Fig.1.331, in right angle $\triangle OLP$ we have-

$\tan\theta = \dfrac{LP}{OL} = \dfrac{2\sqrt{2}}{2} = \sqrt{2} \qquad \cdots (1)$

Now, from Eq.1.53, we have

$\tan\phi = \dfrac{\tan\theta}{2} = \dfrac{\sqrt{2}}{2} = \dfrac{1}{\sqrt{2}} = \dfrac{1}{\tan\theta} = \cot\theta$

[since from Eq.(1), $\tan\theta = \sqrt{2}$]

$\Rightarrow \tan\phi = \tan(90° - \theta)$

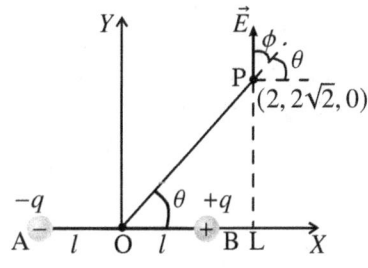

Figure 1.331

$$\Rightarrow \quad \theta + \phi = 90°$$

i.e., $\vec{E}$ is along positive y-axis.

11. (C) Direction of dipole moment will be opposite to the electric field

1.38.4 Check Point 4

1. **APPROACH** From Eq.1.69, the total electric field at the centre of the uniformly charged circular arc of radius R, subtending an angle α at the centre, is given by-

$$E = \frac{kQ}{R^2} \frac{\sin \alpha/2}{\alpha/2} \quad (1)$$

Here, Q is the charge on circular arc. For positive charge, the direction of field is away from the circular arc along the symmetry axis i.e., net electric field vector passing through the mid point of the arc at at angle $\alpha/2$ at the center. Similarly, for negative charge, the direction of $\vec{E}$ is towards the circular arc at angle $\alpha/2$ from the line dividing the circular arc.

In this problem, each charged quarter-circle produces a field of magnitude

$$E = \frac{kQ}{R^2} \frac{\sin \pi/4}{\pi/4} = \frac{2\sqrt{2}kQ}{\pi R^2}$$

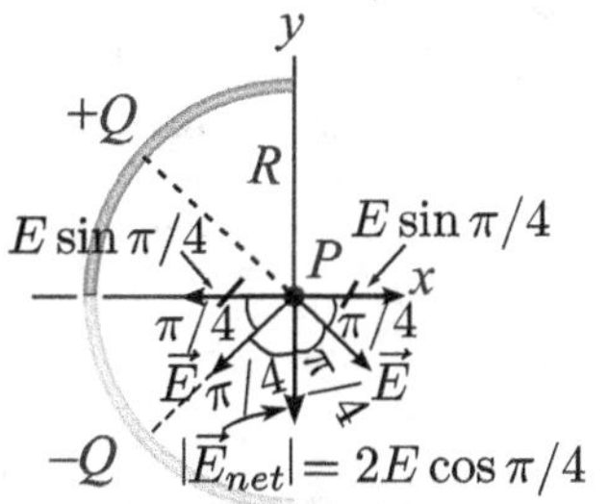

Figure 1.332

From Fig.1.332, it is clear that the x components of these electric fields are opposite to each other and hence get cancelled with each other. The components along negative direction of y-axis add up to each other and give the net electric field $\vec{E}_{net}$.

SOLUTION Net electric field at point P,

$$E_{net} = 2E \cos \pi/4 = 2\left(\frac{2\sqrt{2}kQ}{\pi R^2}\right) \cos \pi/4 = \frac{4kQ}{\pi R^2}$$

On substituting the given values in above expression, we get-

$$E = \frac{(9.0 \times 10^9 \text{ N} \cdot \text{m}^2/\text{C}^2) 4 (4.50 \times 10^{-12}\text{C})}{\pi (5.00 \times 10^{-2} \text{ m})^2}$$

$$= 20.6 \text{ N/C}$$

(b) By symmetry, the net field points vertically downward in the $-\hat{j}$ direction, or $-90°$ counter-clockwise from the $+x$ axis.

2. **APPROACH** From symmetry, we see that the net field at P is twice the field caused by the upper semicircular charge $+q$ (and that it points downward).

The electric field at centre P, due to each halves, can be obtained by using expression-

$$\vec{E} = -\frac{kq}{R^2} \frac{\sin \alpha/2}{\alpha/2} \hat{j} \quad \text{(with } \alpha = \pi\text{)}$$

Therefore, the net electric field at point P due to both halves can be given as-

$$\vec{E}_{net} = 2\vec{E} = -\frac{kq}{R^2} \frac{\sin \alpha/2}{\alpha/2} \hat{j}$$

Now, substitute the values in above expression and solve for $\vec{E}_{net}$.

SOLUTION (a) With $R = 8.50 \times 10^{-2}$m, $\alpha = \pi$ and $q = 1.50 \times 10^{-8}$C, $|\vec{E}_{net}| = 23.8$N/C.

(b) The net electric field $\vec{E}_{net}$ points in the $-\hat{j}$ direction, or $-90°$ clockwise from the $+x$ axis.

3. $4\lambda a^2$

   Hint: Follow the approach used in solved Ex. 55.

4. $2\sqrt{2}\lambda a^2$

   Hint: Follow the approach used in solved Ex.55

5. From Eq.1.78, we have

$$E_{(a)} = \frac{Q}{2\pi\epsilon_0 R^2}\left(1 - \frac{x}{(x^2+R^2)^{1/2}}\right),$$

$$E_{(b)} = \frac{Q}{2\pi\epsilon_0 (2R)^2}\left(1 - \frac{x}{(x^2+(2R)^2)^{1/2}}\right),$$

and $$E_{(c)} = \frac{Q}{2\pi\epsilon_0 (2R)^2}\left(1 - \frac{x}{(x^2+(2R)^2)^{1/2}}\right)$$

$$-\frac{Q}{2\pi\epsilon_0 R^2}\left(1 - \frac{x}{(x^2+R^2)^{1/2}}\right)$$

Clearly, $E_{(a)} > E_{(b)} > E_{(c)}$. So, the correct order is a > b > c.

6. From Eq.1.78,

$$E_x = \frac{Q}{2\pi\epsilon_0 R^2}\left(1 - \frac{x}{(x^2+R^2)^{1/2}}\right)$$

Since, charge Q is placed at distance h from the centre of charged disc, therefore putting $x = h$ in above expression, we get

$$E_h = \frac{Q}{2\pi\epsilon_0 R^2}\left(1 - \frac{h}{(h^2+R^2)^{1/2}}\right)$$

Therefore, $F = QE_h = \frac{Q^2}{2\pi\epsilon_0 R^2}\left(1 - \frac{h}{(h^2+R^2)^{1/2}}\right)$

The direction of this force is directly away from the charged plate.

7. The angle swept by radius vector is called azimuthal angle. Since, λ is cosine function of ϕ, so, ring is not uniformly charged. A part of ring is positively charged and remaining part of ring is negatively charged. The distribution of charge on ring is shown in Fig. 1.333.

Since, $\lambda = \lambda_0 \cos \phi$, therefore, first and fourth quadrants are positively charged (because from fourth to first $-\pi/2 < \phi < +\pi/2$) but second and third quadrants are negatively charged (because from second to third $\pi/2 < \phi < +3\pi/2$).

In Fig.1.333, we have considered two symmetrical elementary charges dq in fourth and first quadrants. Each of these charges produces an electric field $d\vec{E}$ at centre of the ring. Vertical components of each of these fields are oppositely directed and each has equal magnitude of $dE \sin \phi$. So, they get cancelled. The horizontal components are directed in same direction and each has magnitude $dE \cos \phi$. So, they

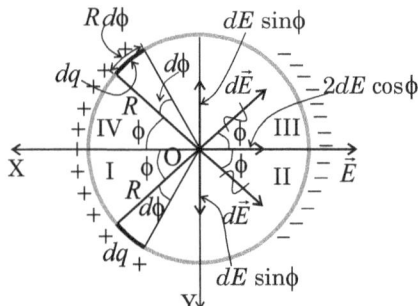

Figure 1.333

get added and their resultant is $2dE \cos\phi$. It is along the negative direction of X-axis.

Note that $dq = \lambda R d\phi = \lambda_0 \cos\phi R d\phi$.

Now, if we consider similar elementary $-ve$ charges in second and third quadrants, they also produce the field along $-ve$ X-axis. Their resultant will also be $2dE \cos\phi$. So, resultant of above field is $4dE \cos\phi$. It is directed along negative direction of X-axis.

Therefore, net field at the centre of ring can be obtained by integrating above resultant from limits $\phi = 0$ to $\phi = \pi/2$.

$$E_{\text{net}} = \int_{\phi=0}^{\pi/2} 4 \, dE \cos\phi$$

Here, $dE = k\dfrac{dq}{R^2}$ and $dq = \lambda_0 \cos\phi \, R d\phi$, therefore,

$$dE = k \frac{\lambda_0 \cos\phi \, R d\phi}{R^2}$$

So, $E_{\text{net}} = 4 \displaystyle\int_{\phi=0}^{\pi/2} k \frac{\lambda_0 \cos^2\phi \, R d\phi}{R^2}$

$$= \frac{4k\lambda_0}{R} \int_{\phi=0}^{\pi/2} \cos^2\phi \, R d\phi$$

Since, $\cos 2\phi = 2\cos^2\phi - 1 \implies \cos^2\phi = \tfrac{1}{2}(\cos 2\phi + 1)$

Therefore, $\displaystyle\int_{\phi=0}^{\pi/2} \cos^2\phi \, R d\phi = \int_0^{\pi/2} \frac{1}{2}(\cos 2\phi + 1)$

$$= \frac{1}{2} \int_0^{\pi/2} (\cos 2\phi + 1) \, d\phi$$

$$= \frac{1}{2}\left[\frac{\sin 2\phi}{2}\right]_0^{\pi/2} + \frac{1}{2}[\phi]_0^{\pi/2} = \frac{\pi}{4}$$

Therefore, $E_{\text{net}} = \dfrac{4k\lambda_0}{R} \dfrac{\pi}{4}$

Since, $k = 1/4\pi\epsilon_0$, therefore, $E_{\text{net}} = \dfrac{\lambda_0}{4\epsilon_0 R}$

(b) Let us take differential length element of the ring at an azimuthal angle ϕ from the x-axis, the element subtends an angle $d\phi$ at the centre, and carries charge $dq = \lambda R d\phi = (\lambda_0 \cos\phi) R \, d\phi$.

Taking the plane of ring as $x-y$ plane and centre of the ring as origin O, locations of field point $\vec{r}$, of charge element $\vec{R}$ and of field point relative to charge element $\vec{r}$ are shown in the Fig.1.334.

Electric field strength at the field point due to considered charge element

$$d\vec{E} = \frac{1}{4\pi\epsilon_0} \frac{dq}{r_1^3} \vec{r_1}$$

Using $\vec{r_1} = \vec{r} - \vec{R} = z\hat{k} - (R\cos\phi \, \hat{i} + R\sin\phi \, \hat{j})$

and $r_1^3 = (R^2 + z^2)^{3/2}$

$d\vec{E} = \dfrac{1}{4\pi\epsilon_0} \dfrac{(\lambda_0 \cos\phi)R \, d\phi}{(R^2+z^2)^{3/2}} \{z \, \hat{k} - (R\cos\phi \, \hat{i} + R\sin\phi \, \hat{j})\}$

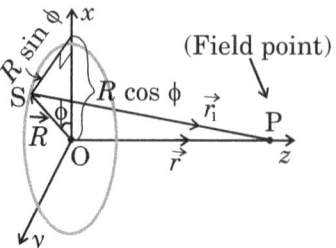

Figure 1.334

So, the net electric field strength

$$\vec{E} = \int d\vec{E} = \frac{\lambda_0 R}{4\pi\epsilon_0 (R^2 + z^2)^{3/2}} \left[z\hat{k} \int_0^{2\pi} \cos\phi \, d\phi \right.$$

$$\left. - R\hat{i} \int_0^{2\pi} \cos^2\phi \, d\phi - R\hat{j} \int_0^{2\pi} \sin 2\phi \, d\phi \right]$$

Taking into account

$$\int_0^{2\pi} \cos\phi \, d\phi = 0, \quad \int_0^{2\pi} \sin 2\phi \, d\phi = 0$$

and $\displaystyle\int_0^{2\pi} \cos^2\phi \, d\phi = \pi$

We get $\vec{E} = \dfrac{1}{4\pi\epsilon_0} \dfrac{\pi \lambda_0 R^2 (-\hat{i})}{(R^2 + z^2)^{3/2}}$

Hence, $E = \dfrac{\lambda_0 R^2}{4\epsilon_0 (R^2 + z^2)^{3/2}}$

For $z \gg R$, $E_x = \dfrac{p}{4\pi\epsilon_0 z^3}$, where $p = \lambda_0 \pi R^2$

8. (a) $E = \dfrac{\lambda\sqrt{2}}{4\pi\epsilon_0 R}$; (b) $E = 0$

9. (A) The electric field E just outside the earth's surface is same as if the entire charge q were concentrated at it's centre. Thus,

$$E = \frac{1}{4\pi\epsilon_0} \frac{q}{R^2} = \frac{1}{4\pi\epsilon_0} \frac{4\pi R^2 \sigma}{R^2} = \frac{\sigma}{\epsilon_0}$$

Substituting the given value:

$$E = \frac{(-1.6 \times 10^{-19}) \text{ C/m}^2}{8.9 \times 10^{-12} \text{C}^2/\text{N.m}^2} = -1.8108 \text{ N/C}$$

The minus sign indicates that E is radially inward.

10. Electric field due to an arc at its centre is $(k\lambda/R)2\sin\theta/2$, where $k = 1/4\pi\epsilon_0$

θ = angle subtended by the wire at the centre,

λ = Linear density of charge.

Let E be the electric field due to remaining portion. Since intensity at the centre due to the circular wire is zero. Applying principle of superposition.

$$\frac{k\lambda}{R} 2 \sin\frac{\theta}{2} - E = 0$$

$\Rightarrow \quad E = \dfrac{1}{4\pi\epsilon_0 R} \dfrac{Q}{2\pi R} 2\sin\dfrac{\theta}{2} = \dfrac{Q}{4\pi^2 \epsilon_0 R^2} \sin\dfrac{\theta}{2}$

1.38.5 Check Point 5

1. (a) Charge ① is negative. (b) Charge ② is positive. (c) Since, the number of field lines terminating on charge ① is greater than that of emitting from ②, therefore the magnitude of charge ① is greater than the magnitude of charge 2.

2. $A > B > C$.

3. (a) 2 : 1 (b) 1 : 3

4. (a) charges A and C are positive and charge B is negative
   (b) $|q_A| : |q_B| : |q_C| = 5 : 10 : 15 = 1 : 2 : 3$
   i.e., $|q_A| < |q_B| < |q_C|$
5. See Fig.1.335

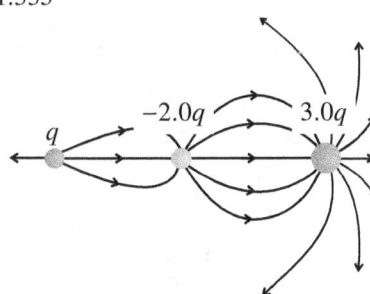

Figure 1.335

6. See Fig.1.336

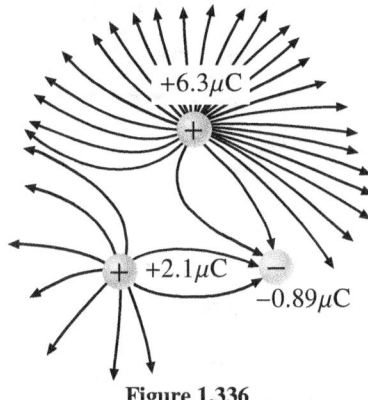

Figure 1.336

7. (B), (D)
8. (B) The electric field depends on the density of electric field lines. The field will be greater where the density of field lines is greater.

1.38.6 Check Point 6

1. (A) FBD of the charged particle is shown in Fig.1.337. As the charged particle is floating in air (neglecting the buoyant force due to air) we obtain-
$$mg = qE \implies q = \frac{mg}{E}$$

2. (B) Since, the electric field is a conservative field, therefore, the work done by electric field is independent of the path followed and is equal to-
$$W = (q\vec{E}) \cdot \vec{r}$$
where $\vec{r}$ = displacement from P to S.
Here, $\vec{r} = -a\hat{i} - b\hat{j}$, while $\vec{E} = E\hat{i}$
Therefore, work = $-(qE\hat{i}) \cdot (a\hat{i} + b\hat{j}) = -qaE$

3. (C) Let x be the small displacement given to the pendulum such that the angle θ is small. The forces acting at A are [Fig.1.338]-
   (i) tension $\vec{T}_0$ along the thread

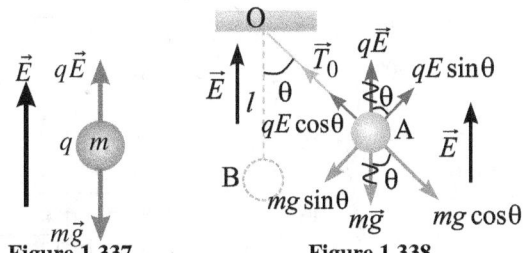

Figure 1.337   **Figure 1.338**

   (ii) weight $m\vec{g}$ acting vertically downwards.

   (iii) electrical force qE vertically upwards.
   The resultant force, vertically downwards, is $(mg - qE)$.
   If g_{eff} is the effective acceleration of bob, then
   $$mg_{\text{eff}} = mg - qE \implies g_{\text{eff}} = g - \frac{qE}{m}$$
   So, time period, $T = 2\pi\sqrt{\frac{l}{g_{\text{eff}}}} = 2\pi\left[\frac{l}{g - \frac{qE}{m}}\right]^{1/2}$

4. (B) FBD of given situation is shown in Fig.1.339
   From Fig.1.339, the resultant of vertical downward force

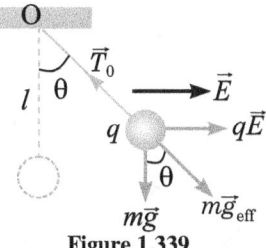

Figure 1.339

$m\vec{g}$ and horizontal force $q\vec{E}$ is $\sqrt{(mg)^2 + (qE)^2}$. If g_{eff} is the effective acceleration, then
$$mg_{\text{eff}} = \sqrt{(mg)^2 + (qE)^2} \implies g_{\text{eff}} = \sqrt{g^2 + \left(\frac{qE}{m}\right)^2}$$
Therefore, $T = 2\pi\sqrt{\frac{l}{g_{\text{eff}}}} = 2\pi\sqrt{\frac{l}{\sqrt{g^2 + \left(\frac{qE}{m}\right)^2}}}$.

5. (D) The torque exerted by an electric field E on a dipole of moment p is given by
$$\tau = pE\sin\theta$$
where θ is the angle which the dipole moment is making with the electric-field.
Corresponding to maximum torque, $\theta = 90°$, therefore-
$$\tau_{\max} = pE$$
Here, $p = q(2l) = 1 \times 10^{-6} \times 0.02$ C/m and $E = 10^5$ N/C
Therefore, $\tau_{\max} = 1 \times 10^{-6} \times 0.02 \times 10^5 = 2 \times 10^{-3}$ N.m

6. (B) When the dipole is given a small angular displacement θ from its mean position, the magnitude of restoring torque on it, is
$$\tau = -pE\sin\theta$$
For small angular displacement $\sin\theta \approx \theta$, therefore-
$$\tau = -pE\theta$$
If α is the angular acceleration of dipole, then $\tau = I\alpha$
Therefore, $I\alpha = -pE\theta \implies \alpha = -\frac{pE\theta}{I} = -\omega^2\theta$
where $\omega^2 = \frac{pE}{I}$
Therefore, $T = 2\pi\sqrt{\frac{I}{pE}}$

1.38.7 Check Point 7

1. **APPROACH** Time period of a simple pendulum is given by-
$$T = 2\pi\sqrt{\frac{L}{|g_{\text{eff}}|}} \qquad (1)$$
here, g_{eff} is the effective gravitational acceleration. Now, find the value of g_{eff} in given situations and by using Eq.1, calculate the value of time period T.
   (a) In this case, from FBD [Fig.1.340a], the effective value of g can be given by
$$mg_{\text{eff}} = mg + qE = mg + mg \quad \left(\because \quad E = \frac{mg}{q}\right)$$

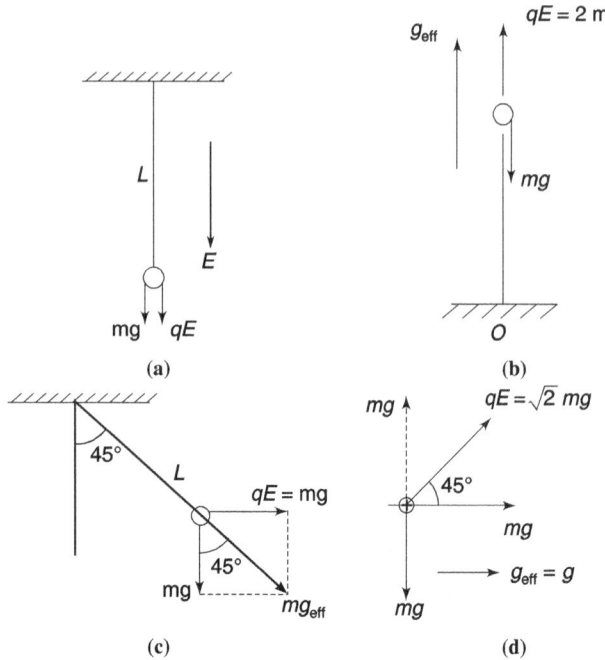

Figure 1.340: Different situations of pendulum

$$\implies g_{\text{eff}} = 2g$$

Substituting this value of g_{eff} in Eq.1, we get

$$T = 2\pi\sqrt{\frac{L}{2g}}$$

(b) In this case, from FBD (Fig.1.340b), the effective acceleration is given by-

$$mg_{\text{eff}} = mg - qE = mg - 2mg \quad \left(\because E = \frac{2mg}{q}\right)$$
$$\implies g_{\text{eff}} = -g$$

Here, $-ve$ sign just says that the effective gravitational acceleration is directed in upward direction. So, the pendulum get inverted and oscillates about the fixed point as shown in Fig.1.340b. On substituting this value of g_{eff} in Eq.1, we get

$$T = 2\pi\sqrt{\frac{L}{g}}$$

(c) Equilibrium position is shown in Fig.1.340c
In this case, the effective gravitational acceleration is given by-

$$mg_{\text{eff}} = \sqrt{(mg)^2 + qE^2} = mg\sqrt{2} \quad \left(\because E = \frac{mg}{q}\right)$$
$$\implies g_{\text{eff}} = g\sqrt{2}$$

On substituting this value of g_{eff} in Eq.1, we get

$$T = 2\pi\sqrt{\frac{L}{g\sqrt{2}}}$$

(d) In this case, the equilibrium is shown in Fig.1.340d (thread is horizontal). From, this figure, we have-

$$mg_{\text{eff}} = \sqrt{(mg - qE\sin 45°)^2 + (qE\cos 45°)^2}$$
$$= \sqrt{\left(mg - \frac{qE}{\sqrt{2}}\right)^2 + \left(\frac{qE}{\sqrt{2}}\right)^2}$$
$$= \sqrt{\left(mg - \frac{qE}{\sqrt{2}}\right)^2 + \left(\frac{qE}{\sqrt{2}}\right)^2} = mg$$
$$\left[\because E = \frac{\sqrt{2}mg}{q}\right]$$

$$\implies g_{\text{eff}} = g$$

Substituting this value of g_{eff} in Eq.1, we get

$$T = 2\pi\sqrt{\frac{L}{g}}$$

2. (a) At every point on the path, the electric force on the particle and it's displacement always remains tangential to the path, i.e., both remains in the same direction. So, work done by the electric force on the particle-

$$W = \int_A^B q\vec{E} \cdot d\vec{l} = qE \cdot \pi R$$

Now, by Work-Energy theorem, we have-
$$W = K_B - K_A$$
Therefore, $K_B - K_A = qE\pi R$
$$\implies K_B = \pi qER \qquad (\because K_A = 0)$$

(b) Suppose, at mid point the speed of the particle is v [Fig.1.341], then by work-energy theorem, we have

$$\frac{1}{2}mv^2 - 0 = qE\frac{\pi R}{2} \implies \frac{v^2}{R} = \frac{\pi qE}{m}$$

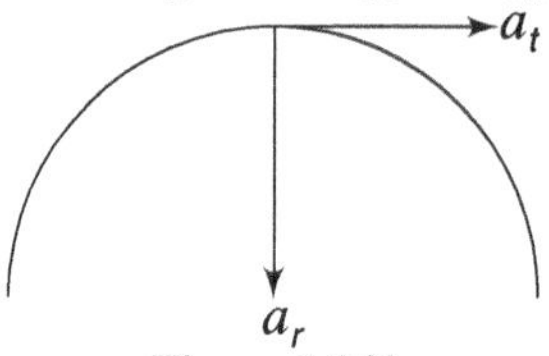

Figure 1.341

Therefore, radial acceleration of the particle,

$$a_r = \frac{v^2}{R} = \frac{\pi qE}{m}$$

This acceleration arises due to constraining forces.

Tangential acceleration, $a_t = \frac{F_t}{m} = \frac{qE}{m}$

∴ Resultant acceleration,
$$a = \sqrt{a_r^2 + a_t^2} = \frac{qE}{m}\sqrt{1 + \pi^2}$$

3. Electrostatic force,
$$F = qE \implies ma_0 = q(-bx + a) \quad [a_0 = \text{accleration}]$$
$$\implies a_0 = v\frac{dv}{dx} = -\frac{bq}{m}x + \frac{aq}{m} \qquad \cdots (1)$$
$$\int_0^v v\, dv = -\frac{bq}{m}\int_0^x x\, dx + \frac{aq}{m}\int_0^x dx$$
Now, $\frac{v^2}{2} = -\frac{bqx^2}{2m} + \frac{aqx}{m}$

$v = 0$ when $\frac{bqx^2}{2m} = \frac{aqx}{m} \implies x = 2a/b$

From Eq.(1), we have: $a_0 = \frac{q}{m}(-bx + a)$

at $x = 2a/b$, acceleration: $a_0 = \frac{q}{m}(-2a + a) = -\frac{qa}{m}$

4. Charge Q resides on each of the blocks, which repel as point charges. In equilibrium, we have-

$$F = \frac{1}{4\pi\epsilon_0}\frac{Q^2}{L^2} = k(L - L_i)$$

here, $k \to$ spring constant, $\frac{1}{4\pi\varepsilon_0} = 9.0 \times 10^9$ N.m^2/C^2.

Solving for Q, we get

$$Q = L\sqrt{\frac{k(L - L_i)}{1/4\pi\epsilon_0}}$$

Substituting the given values in above expression, we get

$$Q = (0.500 \text{ m})\sqrt{\frac{(100 \text{ N/m})(0.500 \text{ m} - 0.400 \text{ m})}{9.0 \times 10^9 \text{ N} \cdot \text{m}^2/\text{C}^2}}$$

= 1.67×10^{-5} C

5. **(A):** Force experienced by a charged particle in an electric field, $F = qE$

   As $F = ma$, therefore, $ma = qE \implies a = \dfrac{qE}{m}$ $\quad\cdots$ (1)

   As electron and proton both fall from same height at rest. Therefore, for both of them, initial velocity = 0.

   Now, from Newton's second equation of motion $y = v_0 t + \frac{1}{2}at^2$, with initial velocity $v_0 = 0$, we get
   $$h = \frac{1}{2}at^2$$
   Substituting the value of a from Eq.(1) in above expression, we get
   $$h = \frac{1}{2}\frac{qE}{m}t^2 \implies t = \sqrt{\frac{2hm}{qE}} \implies t \propto \sqrt{m}$$
   Since, 'q' is same for electron and proton and electron has smaller mass than proton, so electron will take smaller time.

6. **(B)** Given situation is shown in Fig.1.342.
   Acceleration, $a = \dfrac{6-0}{1} = 6$ ms^{-2}

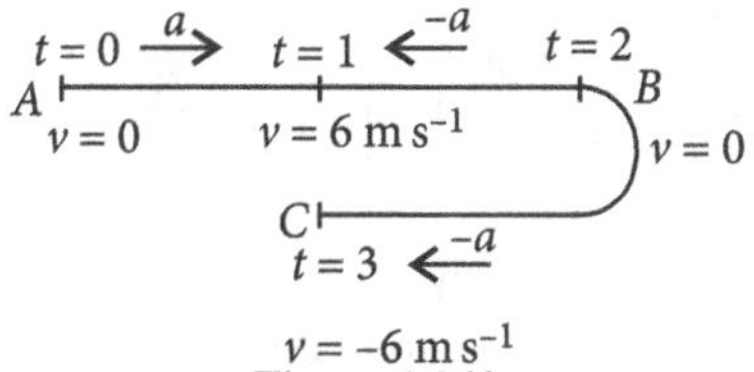

   Figure 1.342

   For $t = 0$ to $t = 1$ s, $x_1 = \frac{1}{2} \times 6(1)^2 = 3$ m
   For $t = 1$ s to $t = 2$ s, $x_2 = 6 \times 1 - \frac{1}{2} \times 6(1)^2 = 3$ m
   For $t = 2$ s to $t = 3$ s, $x_3 = 0 - \frac{1}{2} \times 6(1)^2 = -3$ m
   Total displacement $x = x_1 + x_2 + x_3 = 3$ m
   Average velocity = $3/3 = 1$ m s^{-1}
   Total distance travelled = 9 m
   Average speed = $9/3 = 3$ ms^{-1}

7. **(A)** Since, $v^2 = 0^2 + 2ay = 2(F/m)y = 2\left(\dfrac{qE}{m}\right)y$,

   Therefore-
   $$KE = \frac{1}{2}mv^2 = \frac{1}{2}m\left[2\frac{(qE)}{m}y\right] \implies KE = qEy$$

8. **(C) APPROACH** Equation of path always represents a relation between space coordinates, i.e., between x and y coordinates in two dimensions.

   Given situation can be divided in two parts as shown in Fig.1.343:
   1. Region *I* for which $0 \le x \le d$ and
   2. the region *II* for $x > d$.

   Charged particle enters in region first at O with velocity v_0 directed along positive X–direction. In first region ($0 \le x \le d$), electric field is directed along negative Y–direction, so in this region, the path of charged particle is parabolic. When it emerges out from this region at P(x_0, y_0), it has X and Y components of velocities v_x and v_y respectively. Now, with these velocities it enters in second region ($x > d$), where electric field is absent. So, in second region it will move in straight line path with constant velocities v_x and v_y obtained at boundary point P of region first.

   To find equation of path of charged particle in region *II*, find the position and velocity components of particle at P. Now, consider a position Q(x, y) of the particle at any instant t after emerging from P and apply relevant equation of motion to obtain a relation between x and y.

   SOLUTION Given that the mass of the charged particle is m, charge on the particle is q, initial velocity of particle before entering the region of electric field is v_0 and it is directed along positive direction of X-axis, electric field E in the region $x = 0$ to $x = d$ is directed along negative Y– axis.

   To find : The equation of the particle for the region $x > d$.

   Let after travelling through the electric field in parabolic

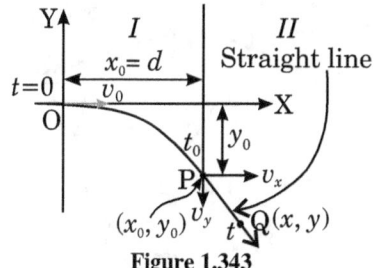

Figure 1.343

   path, the charged particle emerges out the field at point P (x_0, y_0). The components of velocity of the particle at point P will change as shown in Fig.1.343.

   Motion of charged particle from O to P

   Along X axis Given: $x_0 = d$, $v_{0x} = v_0$, $a_x = 0$, time = t_0.
   If at point P, X–component of velocity is v_x, then
   $$v_x = v_0 + a_x t = v_0 + 0(t) = v_0$$
   and $\quad x_0 = v_0 t_0 \implies t_0 = x_0/v_0 = d/v_0 \quad \cdots$ (1)

   Along Y axis Given: $v_{0y} = 0$, $a_y = -\dfrac{qE}{m} \quad \cdots$ (2)
   If at point P, Y–component of velocity is v_y, then
   $$v_y = 0 + a_y t_0 = -\frac{qEd}{mv_0} \quad \cdots (3)$$
   and displacement along Y–axis,
   $$y_0 = v_{0y}t_0 + \frac{1}{2}a_y t_0^2 = -\frac{1}{2}\frac{qEd^2}{mv_0^2} \quad \cdots (4)$$
   [since from Eq.(1), $t_0 = d/v_0$ and from (2), $a_y = -qE/m$]

   In the region $x > d$, $a_x = a_y = 0$. Therefore, the equation of motion of the charged particle will be a straight line. Let after emerging through electric field at point P, the charged particle reaches to point Q (x, y) in time t.

   Motion of charged particle from P to Q

   Along X axis: $(x - d) = v_0 t \implies t = (x - d)/v_0 \quad \cdots$ (5)
   Along Y axis: $y - y_0 = v_y t \implies y = y_0 + v_y t \quad \cdots$ (6)

   Substituting the values of y_0, v_y and t in Eq.(6), we get
   $$y = -\frac{1}{2}\frac{qEd^2}{mv_0^2} - \frac{qEd}{mv_0}\frac{(x-d)}{v_0} = -\frac{qEd}{mv_0^2}\left(\frac{d}{2} + x - d\right)$$
   $$\implies y = \frac{qEd}{mv_0^2}\left(\frac{d}{2} - x\right)$$

9. **(C)** Given : Charge on a point mass released from the edge of a table is $+q$, initial velocity of the point mass is $v_0 = 0$.

   To find: The trajectory of the point mass in presence of horizontal electrical and vertical gravitational field [Fig.1.344].

   Components of acceleration of the point mass, m :

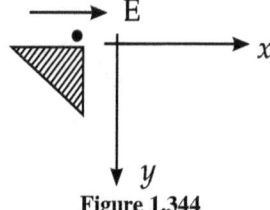

Figure 1.344

   $$a_x = \frac{qE}{m}, a_y = g$$
   Resultant acceleration of the point mass:
   $$a = \sqrt{\left(\frac{qE}{m}\right)^2 + g^2} = \text{constant}$$

As initial velocity of the point mass was $v_0 = 0$ and the acceleration is constant its trajectory will be straight line.

10. (D) Consider an elementary segment AB, subtending a very small angle 2θ at the centre of the given ring of radius R [Fig.1.345].

From Fig.1.345, we have $AB = R(2\theta)$

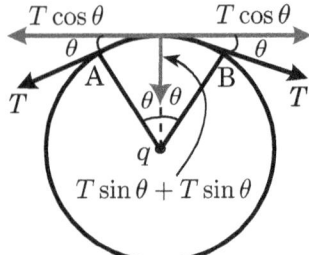

Figure 1.345

Charge on AB is, $dQ = \dfrac{Q}{2\pi R}(2R\theta) = \dfrac{Q\theta}{\pi}$

$2T\sin\theta = \dfrac{dQ \cdot q}{4\pi\epsilon_0 R^2} = \dfrac{Qq\theta}{4\pi^2\epsilon_0 R^2}$

Since, for a very small value of θ, $\sin\theta \approx \theta$, therefore

$2T\theta = \dfrac{Qq\theta}{4\pi^2\epsilon_0 R^2}$ or $T = \dfrac{Qq}{8\pi^2\epsilon_0 R^2}$

11. (C) Because electric field inside the conductor is zero and electric field lines are perpendicular to Gaussian surface.

1.38.8 Check Point 8

1. The answer depends on whether the person is initially (a) uncharged or (b) charged.
   (a) No. If the person is uncharged, the electric field inside the sphere is zero. The interior wall of the shell carries no charge. The person is not harmed by touching this wall.
   (b) If the person carries a (small) charge q, the electric field inside the sphere is no longer zero. Charge $-q$ is induced on the inner wall of the sphere. The person will get a (small) shock when touching the sphere, as all the charge on his body jumps to the metal.

2. Conducting shoes are worn to avoid the build up of a static charge on them as the wearer walks. Rubber-soled shoes acquire a charge by friction with the floor and could discharge with a spark, possibly causing an explosive burning situation, where the burning is enhanced by the oxygen.

3. (a) From Fig.1.346, we have
   $A' = (10.0 \text{ cm})(30.0 \text{ cm}) = 300 \text{ cm}^2 = 0.0300 \text{ m}^2$
   $\Phi_{E,A'} = EA'\cos\theta = (7.80\times 10^4)(0.0300)\cos 180°$
   $= -2.34 \text{ kN}\cdot\text{m}^2/\text{C}$

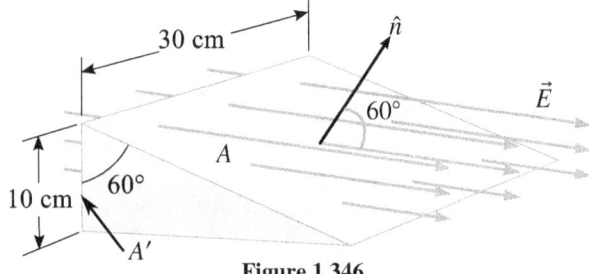

Figure 1.346

(b) The electric flux through the inclined plane is given by-
$\Phi_{E,A} = EA\cos\theta = (7.80\times 10^4)(A)\cos 60.0°$
$A = (30.0 \text{ cm})(w) = (30.0 \text{ cm})\left(\dfrac{10.0 \text{ cm}}{\cos 60.0°}\right)$
$= 600 \text{ cm}^2 = 0.0600 \text{ m}^2$
$\Rightarrow \Phi_{E,A} = (7.80\times 10^4)(0.0600)\cos 60.0°$
$= +2.34 \text{ kNm}^2/\text{C}$

(c) The bottom and the two triangular sides all lie parallel to $\vec{E}$, so $\Phi_E = 0$ for each of these. Thus,
$\Phi_{E,\text{total}} = -2.34 \text{ kN}\cdot\text{m}^2/\text{C} + 2.34 \text{ kN}\cdot\text{m}^2/\text{C} + 0 + 0 + 0$
$= 0$

4. (a) $\Phi_E = \vec{E}\cdot\vec{A} = (a\hat{i}+b\hat{j})\cdot A\hat{i} = aA$
   (b) $\Phi_E = (a\hat{i}+b\hat{j})\cdot A\hat{j} = bA$
   (c) $\Phi_E = (a\hat{i}+b\hat{j})\cdot A\hat{k} = 0$

5. Only the charge inside radius R contributes to the total flux.
$\Phi_E = q/\epsilon_0$

6. Electric flux passing through the base is given by [Fig 1.347]-
$\Phi_E = EA\cos\theta$

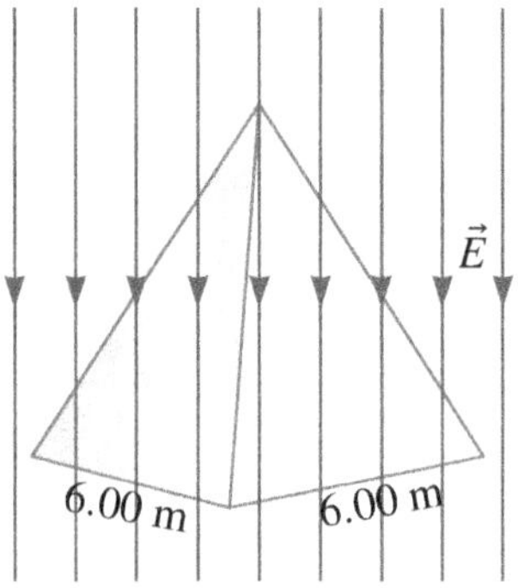

Figure 1.347

$= (52.0)(36.0)\cos 180° = -1.87 \text{ kNm}^2/\text{C}$

Note the same number of electric field lines go through the base as go through the pyramid's surface (not counting the base). For the slanting surfaces,
$\Phi_E = +1.87 \text{ kNm}^2/\text{C}$

7. The flux entering the closed surface equals the flux exiting the surface. The flux entering the left side of the cone is $\Phi_E = \int \vec{E}\cdot d\vec{A} = ERh$. This is the same as the flux that exits the right side of the cone. Note that for a uniform field only the cross sectional area matters, not shape.

8. The hemisphere seems like a much more complicated calculation; you may be tempted to break up the surface into many small pieces and integrate using Equation $\Phi_E = \oint \vec{E}\cdot d\vec{A}$. This would work, but there is a much simpler way. Imagine shining a light parallel to the electric field so that the shadow cast by the hemisphere is a disk of radius R. now by applying equation, $\Phi_E = \vec{E}\cdot\vec{A}$, we can find the electric flux from the area of the shadow, A.

   SOLUTION Net electric flux through the hemispherical surface-
   $\Phi_E = \oint \vec{E}\cdot d\vec{A}$
   $= E(\text{Area of surface perpendicular to }\vec{E})$
   $= E\cdot\pi R^2$

9. Electric flux, $\Phi_E = \vec{E}\cdot\vec{A} = (15\hat{i}+25\hat{j})\cdot(0.65\hat{i}+0.35\hat{j})$
   $= (9.75 + 8.75) \text{ m}^2/\text{C} = 18.5 \text{ m}^2/\text{C}$

10. In Fig.1.186(a), the electric field is uniform over the entire surface. The electric field vectors make an angle of $30°$ with the planar surface. Because the normal $\hat{n}$ to the planar surface is at an angle of $90°$ with the surface, the angle between $\hat{n}$ and $\vec{E}$ is $\theta = 60°$.
    The electric flux is
    $\Phi_E = \vec{E}\cdot\vec{A} = EA\cos\theta$
    $= (200 \text{ N/C})(1.0\times 10^{-2} \text{ m}^2)\cos 60°$

$$= 1.0 \text{ N m}^2/\text{C}$$

In Fig.1.186(b) the electric field vectors make an angle of 30° below the surface. Because the normal $\hat{n}$ to the planar surface is at an angle of 90° relative to the surface, the angle between $\hat{n}$ and $\vec{E}$ is $\theta = 120°$.
The electric flux is
$$\Phi_E = \vec{E} \cdot \vec{A} = EA \cos\theta$$
$$= (180 \text{ N/C})\left(15 \times 10^{-2} \text{ m}\right)^2 \cos 120°$$
$$= -2.3 \text{ N m}^2/\text{C}$$

11. **APPROACH** The electric field is uniform over the entire surface. Please refer to Fig.1.348. The electric field vectors make an angle of 60° above the surface. Because the normal $\hat{n}$ to the planar surface is at an angle of 90° relative to the surface, the angle between $\hat{n}$ and $\vec{E}$ is $\theta = 30°$.
SOLUTION The electric flux is
$$\Phi_E = \vec{E} \cdot \vec{A} = EA \cos\theta$$
$$E = \frac{\Phi_E}{A \cos\theta} = \frac{25 \text{ N m}^2/\text{C}}{(10 \times 10^{-2} \text{ m})(20 \times 10^{-2} \text{ m}) \cos 30°}$$
$$= 1.4 \times 10^3 \text{ N/C}$$

12. The electric field is uniform over the rectangle in the xy plane. (a) The area vector is perpendicular to the xy plane and points in the $\hat{k}$ direction. Thus
$$\vec{A} = (2.0 \text{ cm} \times 3.0 \text{ cm})\hat{k} = \left(6.0 \times 10^{-4} \text{ m}^2\right)\hat{k}$$
The electric flux through the rectangle is
$$\Phi_E = \vec{E} \cdot \vec{A} = (100\hat{i} + 50\hat{k}) \cdot \left(6.0 \times 10^{-4} \hat{k}\right) \text{Nm}^2/\text{C}$$
$$= 3.0 \times 10^{-2} \text{ N m}^2/\text{C}$$
(b) The electric flux is
$$\Phi_E = \vec{E} \cdot \vec{A} = (100\hat{i} + 50\hat{j}) \cdot \left(6.0 \times 10^{-4} \hat{k}\right) \text{Nm}^2/\text{C}$$
$$= 0.0 \text{ N m}^2/\text{C}$$
In (b), $\vec{E}$ is in the plane of the rectangle, which is why the flux is zero.

13. The electric field over the rectangle in the xz plane is uniform.
(a) The area vector is perpendicular to the xz plane and points in the $\hat{j}$ direction. Thus
$$\vec{A} = (2.0 \text{ cm} \times 3.0 \text{ cm})\hat{j} = \left(6.0 \times 10^{-4} \text{ m}^2\right)\hat{j}$$
The electric flux through the rectangle is
$$\Phi_E = \vec{E} \cdot \vec{A} = (100\hat{i} + 50\hat{k}) \text{N/C} \cdot \left(6.0 \times 10^{-4} \text{ m}^2\right)\hat{j}$$
$$= \left(600 \times 10^{-4} \text{ N m}^2/\text{C}\right)(\hat{i} \cdot \hat{j})$$
$$+ \left(300 \times 10^{-4} \text{ N m}^2/\text{C}\right)(\hat{k} \cdot \hat{j}) = 0.0 \text{ N m}^2/\text{C}$$
(b) The flux is
$$\Phi_E = \vec{E} \cdot \vec{A} = (100\hat{i} + 50\hat{j}) \text{ N/C} \cdot \left(6.0 \times 10^{-4} \text{ m}^2\right)\hat{j}$$
$$= \left(600 \times 10^{-4} \text{ N m}^2/\text{C}\right)(\hat{i} \cdot \hat{j})$$
$$+ \left(300 \times 10^{-4} \text{ N m}^2/\text{C}\right)\hat{j} \cdot \hat{j}$$
$$= 0 \text{ N m}^2/\text{C} + \left(3.0 \times 10^{-2} \text{ N m}^2/\text{C}\right)$$
$$= 3.0 \times 10^{-2} \text{ N m}^2/\text{C}$$

14. The area vector of the circle is
$$\vec{A} = \pi r^2 \hat{j} = \pi (0.015 \text{ m})^2 \hat{j} = \left(7.07 \times 10^{-4} \text{ m}^2\right)\hat{j}$$
Thus, the flux through the area of the circle is
$$\Phi_E = \vec{E} \cdot \vec{A} = (1500\hat{i} + 1500\hat{j} - 1500\hat{k}) \text{ N/C}$$
$$\cdot \left(7.07 \times 10^{-4} \text{ m}^2\right)\hat{j}$$
Using $\hat{i} \cdot \hat{j} = \hat{j} \cdot \hat{k} = 0$ and $\hat{j} \cdot \hat{j} = 1$, we find

$$\Phi_E = (1500 \text{ N/C})\left(7.07 \times 10^{-4} \text{ m}^2\right) = 1.1 \text{ Nm}^2/\text{C}$$

15. The electric field is uniform, and we take the area vectors to point outward from the box. In the Figure 1.348, the box is positioned with its edges aligned with the xyz axes, and the electric field is evaluated at the input face and the exit face.
The area vectors of the six box faces are-

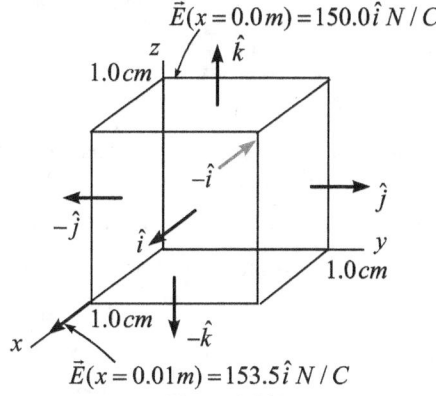

Figure 1.348

$\vec{A}_1 = (1.0 \times 10^{-2} \text{ m})^2 \hat{i} = (1.0 \times 10^{-4} \text{ m}^2)\hat{i}$,
$\vec{A}_2 = -(1.0 \times 10^{-4} \text{ m}^2)\hat{i}$, $\vec{A}_3 = (1.0 \times 10^{-4} \text{ m}^2)\hat{j}$,
$\vec{A}_4 = -(1.0 \times 10^{-4} \text{ m}^2)\hat{j}$, $\vec{A}_5 = (1.0 \times 10^{-4} \text{ m}^2)\hat{k}$,
and $\vec{A}_6 = -(1.0 \times 10^{-4} \text{ m}^2)\hat{k}$.

$$\Phi_E = \sum_{i=1}^{6} \vec{E} \cdot \vec{A}_i$$
$$= \vec{E}_{(x=0.01 \text{ m})} \cdot \vec{A}_1 + \vec{E}_{(x=0.0 \text{ m})} \cdot \vec{A}_2$$
$$= (350(0.01) + 150) \hat{i} \text{N/C} \cdot \left(1.0 \times 10^{-4}\right)\hat{i} \text{m}^2$$
$$+ (350(0.0) + 150) \hat{i} \text{N/C} \cdot \left(1.0 \times 10^{-4}\right)(-\hat{i}) \text{ m}^2$$
$$= (153.5 \text{ N/C})\left(1.0 \times 10^{-4} \text{ m}^2\right)$$
$$-(150 \text{ N/C})\left(1.0 \times 10^{-4} \text{ m}^2\right)$$
$$= 3.5 \times 10^{-4} \text{ N m}^2/\text{C}$$

16. The electric field through the two cylinders is uniform. Let $A = \pi R^2$ be the area of the end of the cylinder and let the area vector point outward from the cylinder ends. E is the electric field strength.
(a) There's no flux through the side walls of the cylinder because $\vec{E}$ is parallel to the wall. On the right end, where $\vec{E}$ points outward, $\Phi_{\text{right}} = EA \cos(0°) = \pi R^2 E$. The field points inward on the left, so $\Phi_{\text{left}} = EA \cos(180°) = -\pi R^2 E$. Altogether, the net flux is $\Phi_E = 0 \text{ N m}^2/\text{C}^2$
(b) The only difference from part (a) is that $\vec{E}$ points outward on the left end, making $\Phi_{\text{left}} = EA \cos(0°) = \pi R^2 E$. Thus the net flux through the cylinder is $\Phi_e = 2\pi R^2 E \text{ N m}^2/\text{C}^2$.

17. (D): Electric flux,
$$\Phi_E = \int \vec{E} \cdot d\vec{A} = \int E dA \cos\theta = \int E dA \cos 90° = 0$$
(because field lines are parallel to the surface.)

18. (D)

1.38.9 Check Point 9

1. (a) With δ very small, all points on the hemisphere are nearly at a distance R from the charge [Fig.1.349], so the field everywhere on the curved surface is kQ/R^2 radially outward (normal to the surface). Therefore, the flux is this field strength times the area of half a sphere:

$$\Phi_{\text{curved}} = \int \vec{E} \cdot d\vec{A} = E_{\text{local}} A_{\text{hemisphere}}$$

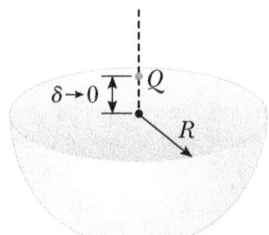

Figure 1.349

$$= \left(k\frac{Q}{R^2}\right)\left(\frac{1}{2}4\pi R^2\right) = \frac{1}{4\pi\epsilon_0}Q(2\pi) = \frac{+Q}{2\epsilon_0}$$

(b) The closed surface encloses zero charge so Gauss's law gives-

$$\Phi_{\text{curved}} + \Phi_{\text{flat}} = 0 \quad \text{or} \quad \Phi_{\text{flat}} = -\Phi_{\text{curved}} = \frac{-Q}{2\epsilon_0}$$

2. Required Gaussian surfaces are shown in Fig.1.350 For any

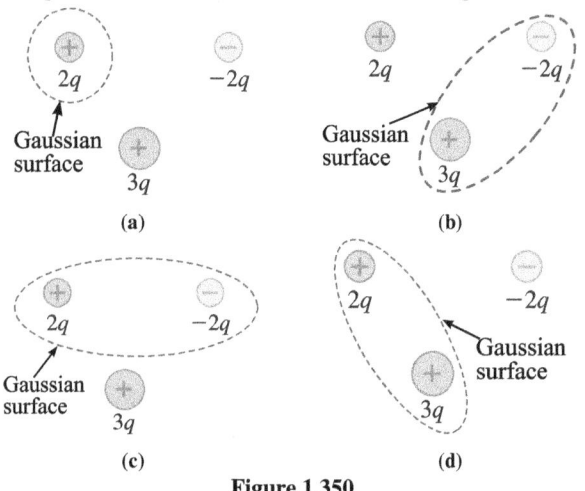

Figure 1.350

closed surface that encloses a total charge Q_{encl}, the net electric flux through the closed surface is $\Phi_e = Q_{\text{encl}}/\epsilon_0$.

3. The required Gaussian surfaces are shown in Fig.1.351 For any closed surface that encloses a total charge Q_{encl}, the net electric flux through the closed surface is $\Phi_E = Q_{\text{encl}}/\epsilon_0$.

4. Please refer to Figure 1.227. For any closed surface that encloses a total charge Q_{encl}, the net electric flux through the surface is $\Phi_E = Q_{\text{encl}}/\epsilon_0$. We can write three equations from the three closed surfaces in the Fig.1.227:

$$\Phi_A = -\frac{q}{\epsilon_0} = \frac{q_1 + q_3}{\epsilon_0} \implies q_1 + q_3 = -q$$

$$\Phi_B = \frac{3q}{\epsilon_0} = \frac{q_1 + q_2}{\epsilon_0} \implies q_1 + q_2 = 3q$$

$$\Phi_C = \frac{-2q}{\epsilon_0} = \frac{q_2 + q_3}{\epsilon_0} \implies q_2 + q_3 = -2q$$

Subtracting third equation from the first gives
$$q_1 - q_2 = +q$$
Adding second equation to this equation,
$$2q_1 = +4q \quad q_1 = 2q$$
That is, $q_1 = +2q, q_2 = +q$, and $q_3 = -3q$.

5. Point charge Q induces charge on conductor as shown in Figure 1.228.
Net charge enclosed by closed surface is negative so flux is negative.

6. From Eq.1.133, electric field inside a cylindrical geometry is given by-
$$E = \frac{r\rho}{2\varepsilon_0}$$
Since, $\rho = \frac{\lambda}{\pi R^2}$, therefore $E = \frac{\lambda}{2\pi\epsilon_0}\frac{r}{R^2}$

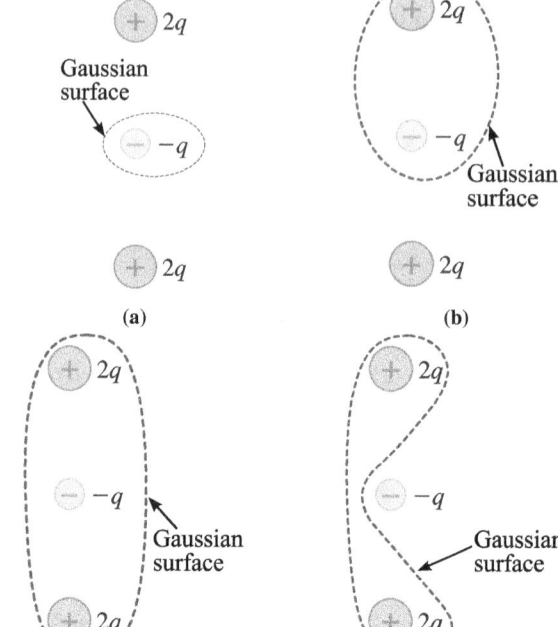

Figure 1.351

For $r = R/2$, we have-
$$E = \frac{\lambda}{2\pi\epsilon_0}\frac{R/2}{R^2} = \frac{\lambda}{4\pi\epsilon_0 R}$$

7. **(B)** In Gauss's law, $\oint \vec{E}\cdot d\vec{A} = \frac{Q_{\text{encl}}}{\varepsilon_0}$,
$\vec{E}$ on the LHS, the electric field vector, $\vec{E}$, will have a contribution from all charges while Q_{encl} on the RHS will have a contribution from enclosed charges q_2 and q_4 only.

8. Please refer to Figure 1.230. For any closed surface that encloses a total charge Q_{encl}, the net electric flux through the closed surface is $\Phi_E = Q_{\text{encl}}/\epsilon_0$. For the closed surface of the torus, Q_{encl} includes only the -1nC charge. Thus, the net flux through the torus is due only to this charge:
$$\Phi_E = \frac{-1\times 10^{-9}\text{C}}{8.85\times 10^{-12}\text{C}^2/\text{Nm}^2} = -0.11 \text{ kNm}^2/\text{C}$$
Here, negative sign indicates that the flux is inward.

9. Please refer to Figure 1.231. For any closed surface that encloses a total charge Q_{encl}, the net electric flux through the closed surface is $\Phi_E = Q_{\text{encl}}/\epsilon_0$. The cylinder encloses the $+1$ nC charge only as both the $+100$ nC and the -100 nC charges are outside the cylinder. Thus,
$$\Phi_E = \frac{1\times 10^{-9}\text{C}}{8.85\times 10^{-12}\text{ C}^2/\text{Nm}^2} = 0.11 \text{ kNm}^2/\text{C}$$
This is outward flux.

10. For any closed surface enclosing a total charge Q_{encl}, the net electric flux through the surface is
$$\Phi_E = \frac{Q_{\text{encl}}}{\epsilon_0}$$
$$\implies Q_{\text{encl}} = \epsilon_0\Phi_E$$
$$= (8.85\times 10^{-12}\text{C}^2/\text{Nm}^2)(-1000\text{Nm}^2/\text{C})$$
$$= -8.85 \text{ nC}$$

11. For any closed surface that encloses a total charge Q_{encl}, the net electric flux through the surface is
$$\Phi_E = \frac{Q_{\text{encl}}}{\epsilon_0} = \frac{(55.3\times 10^6)(-1.60\times 10^{-19}\text{C})}{8.85\times 10^{-12}\text{C}^2/\text{Nm}^2}$$
$$= -1.00 \text{ N m}^2/\text{C}$$

12. **APPROACH** our given information is that the cube side length is a and the cube has a point charge q at one corner (Fig.1.232). Our task here is to use Gauss's law, $\oint \vec{E} \cdot d\vec{A} = \frac{Q_{encl}}{\epsilon_0}$, to solve the problem of how much electric flux passes through each face of the cube. We know how to calculate the electric field surrounding the point charge, and we know the locations and orientations of the faces. We also know that a cube possesses symmetries that we can exploit to simplify the problem.

SOLUTION Because in Fig.1.352a, the point charge q

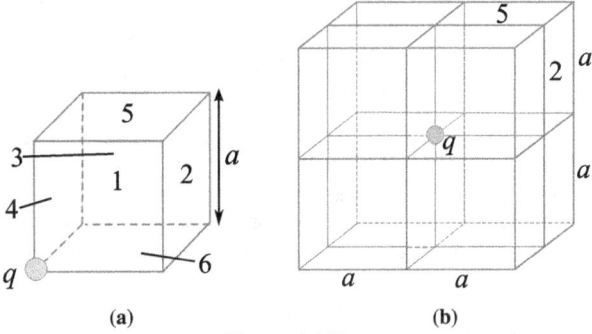

Figure 1.352

is at the corner of the cube, the cube does not enclose it completely and so cannot be our Gaussian surface. However, we can enclose the point charge q by attaching seven additional cubes, each of side length a, to our original cube, creating a cube of side length $2a$ and having q at its geometric centre as shown in Fig.1.352b. This large cube is our Gaussian surface. We can apply Gauss's law to get the electric flux through any face of this surface and then use symmetry to get the flux through any face of our original cube of side length a.

From Gauss's law, we know that the flux through the large cube enclosing a point charge q is q/ϵ_0 Because the *point charge q* is at the geometric center, symmetry tells us that the electric flux is the same through all six faces of the large cube. As Figure 1.352b shows, each face of the large cube is composed of four squares, each of side length a: One face of the large cube contains four faces of our original cube. Thus, if we apply Gauss's law and symmetry, the flux through each of these smaller faces must be-

$$\Phi_E = \frac{1}{(6)(4)} \frac{q}{\epsilon_0} = \frac{q}{24\epsilon_0}.$$

This result gives us the flux passing through each of the faces 2, 3 and 5 of our original cube shown in Fig.1.352a. Since, the electric field lines cannot cross through the faces 1, 4, and 6 of the original cube because they pass through the source charge q. Therefore the electric flux passing through each of the faces 1, 4, and 6 is still zero.

13. (B) As explained in previous answer total eight identical cubes are required so that the given charge q appears at the centre of the bigger cube. Thus, the electric flux passing through the given cube is-

$$\Phi_E = \frac{1}{8} \left(\frac{q}{\epsilon_0} \right) = \frac{q}{8\epsilon_0}$$

14. (C) : According to Gauss's law, $\phi_E = \frac{Q_{encl}}{\epsilon_0}$

From the Gauss's law given above, it is evident that the electric flux depends solely on the charge enclosed by the Gaussian surface and not on the radius of the Gaussian surface. Therefore, when the radius of the Gaussian surface is doubled, the electric flux will remain the same.

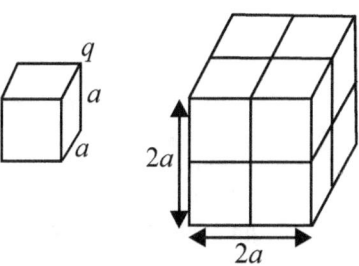

Figure 1.353

15. (D): Let Φ_A, Φ_B and ϕ_C are the electric flux linked with surfaces A, B and C respectively. According to Gauss theorem,

$$\Phi_A + \Phi_B + \Phi_C = \frac{q}{\epsilon_0}$$

Since, $\Phi_A = \Phi_C$, therefore $2\Phi_A + \Phi_B = \frac{q}{\epsilon_0}$

or $2\Phi_A = \frac{q}{\epsilon_0} - \Phi_B$ or, $2\Phi_A = \frac{q}{\epsilon_0} - \Phi$ (Given: $\Phi_B = \Phi$)

Therefore, $\Phi_A = \frac{1}{2} \left(\frac{q}{\epsilon_0} - \Phi \right)$

16. (B): The total flux through the cube $\Phi_{total} = \frac{q}{\epsilon_0}$, therefore, the electric flux through any face

$$\Phi_{face} = \frac{q}{6\epsilon_0} = \frac{4\pi q}{6(4\pi\epsilon_0)}$$

17. (A): For complete cube, $\Phi_E = \frac{Q_{encl}}{\epsilon_0} = \frac{Q}{\epsilon_0} \times 10^{-6}$.

Therefore, corresponding to each face, $\Phi_E = \frac{1}{6} \frac{Q}{\epsilon_0} \times 10^{-6}$

18. (D): Electric flux emerging from the cube does not depend on size of cube. Total flux $= q/\epsilon_0$

19. (C): In a uniformly charged hollow conducting sphere,
    (i) For $r < R, \vec{E} = 0$
    (ii) For $r > R, \vec{E} = \frac{1}{4\pi\epsilon_0} \frac{Q}{|\vec{r}^2|} \hat{r}; \vec{E}$ decreases

20. (D): Electric field due to an infinite line charge, [Fig.1.354]

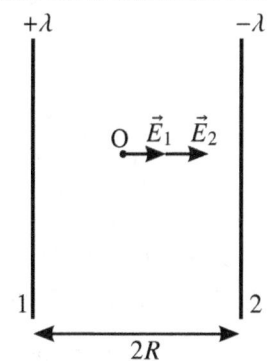

Figure 1.354

$$E = \frac{\lambda}{2\pi\epsilon_0 r}$$

Net electric field at mid-point O,
$$\vec{E}_0 = \vec{E}_1 + \vec{E}_2$$
As, $E_1 = E_2 = \frac{\lambda}{2\pi\epsilon_0 R}$

Therefore, $E_0 = 2E_1 = \frac{\lambda}{\pi\epsilon_0 R}$ NC^{-1}

21. (A): Electric field inside the charged spherical shell is zero as there is no charge inside it.

22. (C): Field inside a conducting sphere = 0.

23. **APPROACH** Apply the principle of superposition of electric fields.

    (B) **SOLUTION** Electric field due to a uniformly charged

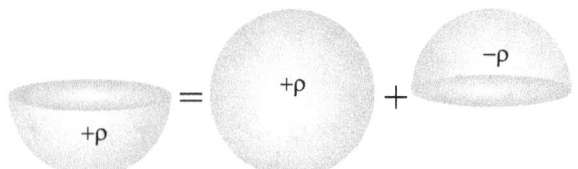

Figure 1.355

sphere, at any external point at distance r from it's center is given by-

$$E_1 = \frac{1}{4\pi\epsilon_0}\frac{Q}{r^2} = \frac{1}{4\pi\varepsilon_0}\frac{\rho(4\pi R^3)/3}{r^2}$$

Since, $r = 2R$, therefore, $E_1 = \frac{\rho R}{12\epsilon_0}$

The given electric field at a distance $2R$, above the hemisphere is-

$$E_2 = E$$

i.e., a hemispherical part creates an electric field E at distance $2R$ above it's centre.

Now, for a point at distance $r = 2R$, below the hemispherical section, we model the hemisphere as a super position of a sphere of charge density $+\sigma$ and an upper hemisphere of charge density $-\sigma$.

So, the net electric field at lower point at distance $2R$, from the centre of the hemisphere will be equal to the vector sum of field created by complete sphere and a hemisphere of negative charge density as shown in Fig.1.355, i.e.,

$$\vec{E}_{net} = \vec{E}_1 + \vec{E}_2 \implies E_{net} = E_1 - E_2 = \frac{\rho R}{12\varepsilon_0} - E$$

24. (A) By using Gauss law [Fig.1.356] -

$$\oint \vec{E}\cdot d\vec{A} = \frac{Q_{encl}}{\epsilon_0} \implies (E)\left(4\pi r^2\right) = \frac{\int \rho\left(4\pi r^2 dr\right)}{\epsilon_0}$$

$$\left(kr^7\right)\left(4\pi r^2\right) = \frac{\int_0^r \rho\left(4\pi r^2 dr\right)}{\epsilon_0} \implies k\varepsilon_0 r^9 = \int_0^r \rho r^2 dr$$

$$\implies k\epsilon_0 r^9 = \rho\frac{r^3}{3} \implies \rho \propto r^6$$

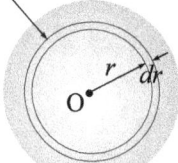

Gaussian surface

Figure 1.356

Remark : Students can also check dimensionally that $\rho \propto r^6$.

25. (D) The maximum length of the string which can fit into the cube is $a\sqrt{3}$, equal to its body diagonal. The total charge inside the cube is $\lambda a\sqrt{3}$, and hence the total flux through the cube is-

$$\Phi = \frac{Q_{encl}}{\varepsilon_0} = \frac{\lambda a\sqrt{3}}{\varepsilon_o}.$$

26. (C) Outgoing electric flux due to $+q$ charge of dipole $= +q/\epsilon_0$

Incoming electric flux due to $-q$ charge of dipole $= -q/\epsilon_0$

Therefore, net electric flux leaving the spherical surface due to dipole:

$$\Phi_{net} = \frac{q}{\epsilon_0} - \frac{q}{\epsilon_0} = 0$$

Alternatively, we can directly find the answer by applying Gauss's theorem:

$$\Phi_E = \frac{Q_{encl}}{\epsilon_0} = \frac{0}{\epsilon_0} = 0$$

(because for a dipole inside a sphere, $Q_{encl} = +q - q = 0$)

27. (A) Inside the surface the total charge is zero. So flux must be zero.

1.38.10 Check Point 10

1. **APPROACH** A copper penny is a conductor. Assume the penny to be a flat disc of radius much, much greater than the distance from the surface at which we are measuring the electric field.

   SOLUTION The excess charge on a conductor resides on the surface. The electric field just outside the surface of a charged conductor is

   $$\vec{E} = \left(\frac{\sigma}{\varepsilon_0}, \text{ perpendicular to surface}\right)$$

   $$\implies \sigma = \varepsilon_0 E = \left(8.85 \times 10^{-12} C^2/Nm^2\right)(2000\ N/C)$$
   $$= 17.7 \times 10^{-9} C/m^2$$

   ☞ Because the actual surface of a penny is not flat, the surface charge density will not be uniform. The result above gives the average surface charge density, far from the edges of the coin.

2. In electrostatic equilibrium of a conductor, the excess charge always resides on the outer surface. The electric field just above the surface of a charged conductor is

   $$\vec{E} = \left(\frac{\sigma}{\epsilon_0}, \text{ perpendicular to surface}\right)$$

   Therefore,

   $$\sigma = \epsilon_0 E = \left(8.85 \times 10^{-12} C^2/Nm^2\right)\left(3.0 \times 10^6\ N/C\right)$$
   $$= 2.7 \times 10^{-5} C/m^2$$

   ☞ It is the air molecules just above the surface that "break down" when the E-field becomes strong enough to accelerate stray charges to approximately $15\ eV$ between collisions, thus causing collisional ionization. It does not make any difference whether E points toward or away from the surface.

3. The excess charge on a conductor resides on the outer surface. In Fig.1.256, point 1 is just above the surface of a charged conductor, hence

   $$\vec{E} = \left(\frac{\sigma}{\epsilon_0}, \text{ perpendicular to surface}\right)$$

   $$\implies E = \frac{(5.0 \times 10^{10})(1.60 \times 10^{-19} C/m^2)}{8.85 \times 10^{-12} C^2/Nm^2} = 0.90\ kN/C$$

   At point 2 the electric field strength is zero because this point lies inside the conductor. The electric field strength at point 3 is zero because there is no excess charge on the interior surface of the box. This can be quickly seen by considering a Gaussian surface just inside the interior surface of the box as shown in Figure1.256.

4. The copper plate is a conductor. The excess charge resides on the surface of the plate. Ignore the charge that resides on the edge of the plate because the plate's thickness is much, much less than the radius.

   (a) One-half of the charge is located on the top surface and one-half on the bottom surface of the copper plate, so the surface charge density is

   $$\sigma = \frac{q}{A} = \frac{(3.5/2 nC)}{\pi(0.10/2\ m)^2} = 2.23 \times 10^{-7} C/m^2$$

   Thus, the electric field just above the surface of the plate is

   $$E = \frac{\sigma}{\epsilon_0} = \frac{2.23 \times 10^{-7} C/m^2}{8.85 \times 10^{-12} C^2/Nm^2}$$
   $$= 25.2 \times 10^3\ N/C \approx 25\ kN$$

   Because the charge on the plate is positive, the direction of the electric field is away from the plate. Thus the electric

field is $\vec{E}$ = 25 kN/C upward from the plate.
(b) The centre of mass of the plate is in the interior of the plate, so $E = 0.0$ N/C because the electric field within a conductor is zero.
(c) The electric field $E = 25.2 \times 10^3$ N/C, away from the plate, which is downward. Thus $\vec{E} = 25.2$ kN/C downward from the plate.

5. For any closed surface that encloses a total charge Q_{encl}, the net electric flux through the closed surface is $\Phi_E = Q_{encl}/\epsilon_0$. In the present case (Figure 1.257), the conductor is neutral and there is a point charge Q inside the cavity. Thus $Q_{encl} = Q$ and the flux is, $\Phi_E = Q/\epsilon_0$

6. (a) $\sigma_a = -\dfrac{q_a}{4\pi a^2}$; $\sigma_b = -\dfrac{q_b}{4\pi b^2}$; $\sigma_R = \dfrac{q_a+q_b}{4\pi R^2}$
   (b) $E_{out} = \dfrac{1}{4\pi\epsilon_0}\dfrac{(q_a+q_b)}{r^2}\hat{r}$, where $\hat{r}$ = vector from centre of large sphere.
   (c) $E_a = \dfrac{1}{4\pi\epsilon_0}\dfrac{q_a}{r_a^2}\hat{r}_a$, $E_b = \dfrac{1}{4\pi\epsilon_0}\dfrac{q_b}{r_b^2}\hat{r}_b$, where $\hat{r}_a$ and $\hat{r}_b$ are outward unit vectors from centres of cavities of radii a and b respectively.
   (d) Zero
   (e) σ_R changes (but not σ_a or σ_b); $E_{outside}$ changes (but not E_a or E_b); force on q_a and q_b still zero.

7. Equations of motion are-
$$- \underbrace{eE}_{\substack{\text{upward}\\\text{electric force}}} + \underbrace{mg}_{\substack{\text{downward}\\\text{weight}}} - \underbrace{6\pi\eta r v_1}_{\substack{\text{upward}\\\text{viscous force}}} = 0 \quad \text{(downward field)}$$

$$\underbrace{eE}_{\substack{\text{downward}\\\text{electric force}}} + \underbrace{mg}_{\substack{\text{downward}\\\text{weight}}} - \underbrace{6\pi\eta r v_2}_{\substack{\text{upward}\\\text{viscous force}}} = 0 \quad \text{(upward field)}$$

On subtracting first equation from second, we get
$$2eE - 6\pi\eta r (v_2 - v_1)$$
On simplifying above equation for e, we get
$$e = \dfrac{3\pi\eta r}{E}(v_2 - v_1)$$

8. (C) If T is the surface tension of the liquid of soap bubble and σ is the surface charge density, then, in electrostatic equilibrium of soab bubble, we have
$$P_{gas} + P_{el} = P_0 + \dfrac{4T}{R} \implies P_{gas} - P_0 = \dfrac{4T}{R} - P_{el}$$
Here, P_{gas} is the outward gas pressure, P_0 is the inward atmospheric pressure and P_{el} is the outward electric pressure on the surface of soap bubble.
Since, due to surface charge density σ on the soap bubble, the outward electric pressure $P_{el} = \sigma^2/2\epsilon_0$, therefore the excess pressure on the soap bubble is given by,
$$P_{excess} = P_{gas} - P_0 = \dfrac{4T}{R} - \dfrac{\sigma^2}{2\epsilon_0}$$
Since, $\sigma = \dfrac{Q}{4\pi R^2}$ therefore-
$$P_{excess} = \dfrac{4T}{R} - \dfrac{Q^2}{32\pi^2 R^4 \epsilon_0}$$

1.38.11 Conceptual Questions

1. A plastic ruler that has been rubbed with a cloth is charged. When brought near small pieces of paper, it will cause separation of charge (polarization) in the bits of paper, which will cause the paper to be attracted to the ruler. A small amount of charge is able to create enough electric force to be stronger than gravity. Thus the paper can be lifted. On a humid day this is more difficult because the water molecules in the air are polar. Those polar water molecules will be attracted to the ruler and to the separated charge on the bits of paper, neutralizing the charges and thus reducing the attraction.

2. Because the can is grounded, the presence of the negatively charged plastic rod induces a positive charge on it. The positive charges induced on the can are attracted, via the Coulomb interaction, to the negative charges on the plastic rod. Unlike charges attract, so the can will roll toward the rod.

3. It is possible only if one of them has very large positive charge as compared to other. Because a metal sphere is a conductor, the proximity of a positively charged ball (not necessarily a conductor), will induce a redistribution of charges on the metal sphere with the surface nearer the positively charged ball becoming negatively charged. Because the negative charges on the metal sphere are closer to the positively charged ball than are the positive charges on the metal sphere, the net force will be attractive.

4. For the gravitational force, we don't notice it because the force is very weak, due to the very small value of G, the gravitational constant, and the relatively small value of ordinary masses. For the electric force, we don't notice it because ordinary objects are electrically neutral to a very high degree. We notice our weight (the force of gravity) due to the huge mass of the Earth, making a significant gravity force. We notice the electric force when objects have a net static charge (like static cling from the clothes dryer), creating a detectable electric force.

5. It is impossible to deal with the interactions of moving charged particles without the field concept, and it is often easier to deal with fields than with distributions of charge.

6. The test charge creates its own electric field, The measured electric field is the sum of the original electric field plus the field of the test charge. If the test charge is small, then the field that it causes is small. Therefore, the actual measured electric field is not much different than the original field.

7. A negative test charge could be used. For the purposes of defining directions, the electric field might then be defined as the opposite of the force on the test charge, divided by the test charge. Equation $E = F/q_0$ might be changed to $\vec{E} = -\vec{F}/q_0, q_0 < 0$

8. The direction of the field is defined to be the direction of the force on a positively charged test particle. Positive charges always move away from other +ve charges and towards −ve charges.

9. A negatively charged particle placed in a uniform electric field is accelerated in the direction opposite the direction of $\vec{E}$, and so the upward acceleration of the electron tells you that $\vec{E}$ must have a component that is directed vertically downward. (There might also be a component parallel to the initial motion of the electron. This would change the electron's speed but not its direction.) If the electron is accelerated downward, $\vec{E}$ must have a component that is directed vertically upward.

10. See Fig.1.357. At point A, the direction of the net force on a positive test charge would be down and to the left, parallel to the nearby electric field lines. At point B, the direction of the net force on a positive test charge would be up and to the right, parallel to the nearby electric field lines. At point C, the net force on a positive test charge would be 0. In order of decreasing field strength, the points would be ordered A, B, C.

11. The two charges are located as shown in Fig.1.358.
    (a) If the signs of the charges are opposite, then the point on the line where $E = 0$ will lie to the left of Q. In that region

the electric fields from the two charges will point in opposite directions, and the point will be closer to the smaller charge.
(b) If the two charges have the same sign, then the point on

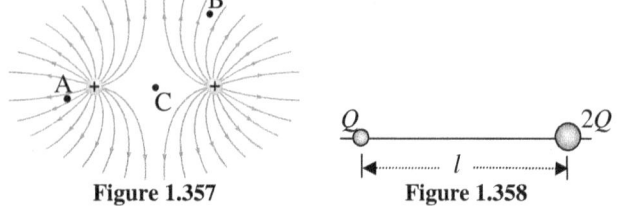

Figure 1.357   **Figure 1.358**

the line where $E = 0$ will lie between the two charges, closer to the smaller charge. In this region, the electric fields from the two charges will point in opposite directions.

12. We assume that there are no other forces (like gravity) acting on the test charge. The direction of the electric field line shows the direction of the force on the test charge. The acceleration is always parallel to the force by Newton's second law, so the acceleration lies along the field line. If the particle is at rest initially and then is released, the initial velocity will also point along the field line, and the particle will start to move along the field line. However, once the particle has a velocity, it will not follow the field line unless the line is straight. The field line gives the direction of the acceleration, or the direction of the change in velocity.

13. No, because the forces exerted on the two charged regions of the dipole are equal in magnitude but opposite in direction, making the vector sum of forces zero.

14. No. The acceleration is not constant (it is centripetal and always changing direction). Therefore the electric force exerted on the electron must be nonconstant, implying the electric field is not uniform.

15. The electric flux depends only on the charge enclosed by the Gaussian surface $\left(\Phi = \frac{Q_{encl}}{\epsilon_0}\right)$, not on the shape of the surface. Φ_E will be the same for the cube as for the sphere.

16. (a) Because the amount of charge enclosed by the balloon does not change, the field line flux through its surface does not change. (b) Again the field line flux does not change because the enclosed charge does not change. (However, in either case the field line pattern changes when the electrons are placed outside the balloon.)

17. If there are no charged particles inside the cavity, then the electric field is zero inside the cavity, no matter what the charge distribution is outside the sphere. The electric field inside the cavity can be non-zero when there are charged particles inside the cavity. In either case, the electric field is zero within the conducting material of the sphere itself.

18. The two are proportional to each other. The field line flux has an arbitrary value, depending on how many lines are drawn to represent a certain amount of charge. The electric flux is uniquely defined by $\Phi = \oint \vec{E} \cdot d\vec{A} = Q_{encl}/\epsilon_0$

19. The net charge on large objects is always very close to zero. Hence the most obvious force is the gravitational force.

20. The accumulation of static charge gives the individual hairs a charge. Since like charges repel and because the electrostatic force is inversely proportional to the charges separation distance squared, the hairs arrange themselves in a manner in which they are as far away from each other as possible. In this case that configuration is when the hairs are standing on end.

1.38.12 Problems

1. **APPROACH** Since, total charge is given therefore, to determine the number of electrons, apply the principle of quantization of charge: $q = ne$
SOLUTION Here, charge of each electron is $e = 1.602 \times 10^{-19}$ C, $q = 1$ C, therefore-
$$n = \frac{q}{e} = \frac{1\,C}{1.602 \times 10^{-19}C} = 6.18 \times 10^{18} \text{electrons}$$

2. The charge acquired by the plastic rod is an integral number of electronic charges, that is, $q = n(-e)$. Relate the charge acquired by the plastic rod to the number of electrons transferred from the wool shirt:
$$q = n(-e) \implies n = \frac{q}{-e}$$
Substitute numerical values and evaluate n:
$$n = \frac{-0.80\mu C}{-1.602 \times 10^{-19}C} = 5.0 \times 10^{12} \text{ electrons}$$

3. Number of moles in 1 kg., i.e., 10^3 gram of carbon:
$$n = \frac{w}{M} = \frac{10^3}{12}$$
Therefore, number of C atoms,
$$N = nN_A = \frac{10^3}{12} \times 6.022 \times 10^{23}$$
Since, each carbon atom has 6 protons in it's nucleus, therefore, number of protons in 10^3 gram carbon
$$N_p = 6N = 3.011 \times 10^{26}$$
And the total charge of protons
$$Q_p = N_p e = 6Ne = 3.011 \times 10^{26} \times 1.6 \times 10^{-19}$$
$$= 4.82 \times 10^7 C$$

4. **APPROACH** 1 mol of electrons = 6.02×10^{23} electrons, i.e., N_A(Avogadro's number) electrons.
Now, apply the principle of quantization of charge $q = ne$, with $n = N_A = 6.02 \times 10^{23}$ electrons.
SOLUTION 1 faraday = $N_A e$
$$= \left(6.02 \times 10^{23}\right) \times (1.602 \times 10^{-19}C) = 96470\,C$$

5. **APPROACH** (a) The percentage of the electrons originally in the cube that was removed can be found from the ratio of the number of electrons removed to the number of electrons originally in the cube. (b) The percentage decrease in the mass of the cube can be found from the ratio of the mass of the electrons removed to the mass of the cube.
SOLUTION (a) Express the ratio of the electrons removed to the number of electrons originally in the cube:
$$\frac{N_{rem}}{N_{ini}} = \frac{\frac{Q_{accumulated}}{e}}{N_{\substack{electrons \\ per\,atom}} N_{atoms}} \quad (1.161)$$
The number of atoms in the cube is the ratio of the mass of the cube to the mass of an aluminum atom:
$$N_{atoms} = \frac{m_{cube}}{m_{Al\,atom}} = \frac{\rho_{Al} V_{cube}}{m_{Al\,atom}}$$
The mass of an aluminum atom is its molar mass divided by Avogadro's number:
$$m_{Al\,atom} = \frac{M_{Al}}{N_A}$$
Substituting and simplifying yields:
$$N_{atoms} = \frac{\rho_{Al} V_{cube}}{\frac{M_{Al}}{N_A}} = \frac{\rho_{Al} V_{cube} N_A}{M_{Al}}$$
Substitute for N_{atoms} in equation (1.161) and simplify to obtain:
$$\frac{N_{rem}}{N_{ini}} = \frac{\frac{Q_{accumulated}}{e}}{N_{\substack{electrons \\ per\,atom}}\left(\frac{\rho_{Al} V_{cube} N_A}{M_{Al}}\right)}$$
$$= \frac{Q_{accumulated} M_{Al}}{N_{\substack{electrons \\ per\,atom}} \rho_{Al} V_{cube} e N_{per\,atom}}$$
Substitute numerical values and evaluate $\frac{N_{rem}}{N_{ini}}$, you get-

1.38. ANSWER KEYS AND SOLUTIONS

$$\frac{N_{\text{rem}}}{N_{\text{ini}}} \approx 1.99 \times 10^{-15}\%$$

(b) Express the ratio of the mass of the electrons removed to the mass of the cube:

$$\frac{m_{\text{rem}}}{m_{\text{cube}}} = \frac{N_{\text{rem}} \, m_{\text{electron}}}{\rho_{Al} V_{\text{cube}}}$$

From (a), the number of electrons removed is given by:

$$N_{\text{rem}} = \frac{Q_{\text{accumulated}}}{e}$$

Substituting and simplifying yields:

$$\frac{m_{\text{rem}}}{m_{\text{cube}}} = \frac{\frac{Q_{\text{accumulated}}}{e} m_{\text{electron}}}{\rho_{Al} V_{\text{cube}}}$$

$$= \frac{Q_{\text{accumulated}} \, m_{\text{electron}}}{e \rho_{Al} V_{\text{cube}}}$$

Substituting numerical values and evaluating for $\frac{m_{\text{rem}}}{m_{\text{cube}}} \times 100$, yields

$$\frac{m_{\text{rem}}}{m_{\text{cube}}} \times 100 = \frac{(2.50 \text{ pC})(9.109 \times 10^{-31} \text{kg})}{(1.602 \times 10^{-19} \text{C})\left(2.70 \frac{\text{g}}{\text{cm}^3}\right)(1.00 \text{ cm}^3)}$$

$$\approx 5.26 \times 10^{-19}\%$$

6. APPROACH The force on one proton is $F = \frac{kq_1 q_2}{r^2}$ away from the other proton.

SOLUTION The magnitude of above force,

$$\left(8.99 \times 10^9 \text{ N} \cdot \text{m/C}^2\right) \left(\frac{1.60 \times 10^{-19} \text{C}}{2 \times 10^{-15} \text{ m}}\right)^2 = 57.5 \text{ N}$$

7. APPROACH We can find the electric forces the two charges exert on each by applying Coulomb's law and Newton's 3rd law. Note that $\hat{r}_{1,2} = \hat{i}$ because the vector pointing from q_1 to q_2 is in the positive x direction. Figure 1.359 shows the situation for Parts (a) and (b).

SOLUTION (a) Use Coulomb's law to express the force

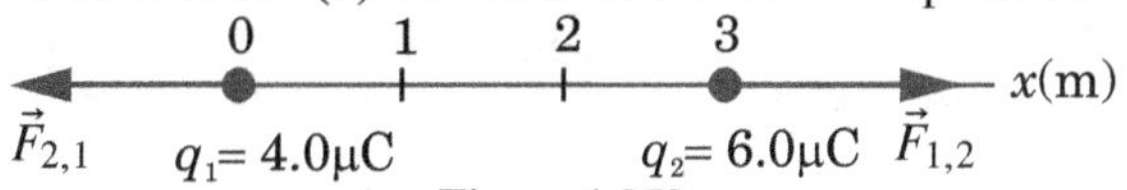

Figure 1.359

that q_1 exerts on q_2:

$$\vec{F}_{1,2} = \frac{kq_1 q_2}{r_{1,2}^2} \hat{r}_{1,2}$$

Substitute numerical values and evaluate $\vec{F}_{2,1}$:

$$\vec{F}_{1,2} = \frac{\left(8.988 \times 10^9 N \cdot m^2/C^2\right)(4.0\mu C)(6.0\mu C)}{(3.0m)^2} \hat{i}$$

$$= (24 mN)\hat{i}$$

(b) Because these are action and reaction forces, we can apply Newton's 3rd law to obtain:

$$\vec{F}_{2,1} = -\vec{F}_{1,2} = -(24 mN)\hat{i}$$

(c) If q_2 is $-6.0\mu C$, the force between q_1 and q_2 is attractive and both force vectors are reversed:

$$\vec{F}_{1,2} = \frac{(8.988 \times | 10^9 N \cdot m^2/C^2)(4.0 \, \mu C)(-6.0\mu C)}{(3.0 \text{ m})^2} \hat{i}$$

$$= -(24 \text{ mN})\hat{i}$$

and $\vec{F}_{2,1} = -\vec{F}_{1,2} = (24 \text{ mN})\hat{i}$

8. APPROACH q_2 exerts an attractive electric force $\vec{F}_{2,1}$ on point charge q_1 and q_3 exerts a repulsive electric force $\vec{F}_{3,1}$ on point charge q_1 [Fig. 1.360]. We can find the net electric force on q_1 by adding these forces (that is, by using the superposition principle).

SOLUTION Express the net force acting on q_1:

$$\vec{F}_1 = \vec{F}_{2,1} + \vec{F}_{3,1} \quad \cdots (1)$$

Express the force that q_2 exerts on q_1

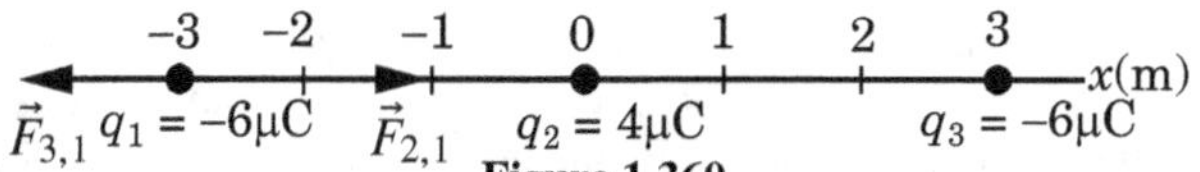

Figure 1.360

$$\vec{F}_{2,1} = \frac{k|q_1||q_2|}{r_{2,1}^2} \hat{i}$$

Express the force that q_3 exerts on q_1:

$$\vec{F}_{3,1} = \frac{k|q_1||q_3|}{r_{3,1}^2} (-\hat{i})$$

Substituting these values of $\vec{F}_{2,1}$ and $\vec{F}_{3,1}$ in Eq.(1), we get:

$$\vec{F}_1 = \frac{k|q_1||q_2|}{r_{2,1}^2} \hat{i} - \frac{k|q_1||q_3|}{r_{3,1}^2} \hat{i}$$

$$= k|q_1| \left(\frac{|q_2|}{r_{2,1}^2} - \frac{|q_3|}{r_{3,1}^2}\right) \hat{i}$$

Now, substituting numerical values in above equation and solving for $\vec{F}_1$, we get:

$$\vec{F}_1 = \left(8.988 \times 10^9 \text{ N} \cdot \text{m}^2/\text{C}^2\right) (6.0\mu C)$$

$$\times \left(\frac{4.0\mu C}{(3.0 \text{ m})^2} - \frac{6.0\mu C}{(6.0 \text{ m})^2}\right) \hat{i}$$

$$= (1.5 \times 10^{-2} \text{ N}) \hat{i}$$

9. APPROACH The third point charge should be placed at the location at which the forces on the third point charge due to each of the other two point charges cancel. There can be no such place except on the line between the two point charges. Denote the $2.0 \, \mu C$ and $4.0 \, \mu C$ point charges by the numerals 2 and 4, respectively, and the third point charge by the numeral 3. Assume that the $2.0\mu C$ point charge is to the left of the $4.0 \, \mu C$ point charge, let the $+x$ direction be to the right. Then the $4.0 \, \mu C$ point charge is located at $x = L$.

SOLUTION Apply the condition for translational equilibrium to the third point charge:

$$\vec{F}_{4,3} + \vec{F}_{2,3} = 0 \implies F_{4,3} = F_{2,3} \quad \cdots (1)$$

Letting the distance from the third point charge to the $4.0\mu C$ point charge be x, express the force that the $4.0\mu C$ point charge exerts on the third point charge:

$$F_{4,3} = \frac{kq_3 q_4}{(L-x)^2}$$

The force that the $2.0\mu C$ point charge exerts on the third charge is given by:

$$F_{2,3} = \frac{kq_3 q_2}{x^2}$$

Substitute in equation (1) to obtain:

$$\frac{kq_3 q_4}{(L-x)^2} = \frac{kq_3 q_2}{x^2} \implies \frac{q_4}{(L-x)^2} = \frac{q_2}{x^2}$$

Substituting $q_4 = 2q_2$ in above equation, we get

$$\frac{2}{(L-x)^2} = \frac{1}{x^2} \implies \left(\frac{x}{L-x}\right)^2 = \frac{1}{2} \implies \frac{x}{L-x} = \pm \frac{1}{\sqrt{2}}$$

Taking +ve sign, we get

$$\frac{x}{L-x} = \frac{1}{\sqrt{2}} \implies \sqrt{2}x = L - x$$

$$\implies x = \frac{L}{\sqrt{2}+1} = \left(\sqrt{2}-1\right)L = 0.41L$$

Similarly, taking $-$ve sign, we get

$$\frac{x}{L-x} = -\frac{1}{\sqrt{2}} \implies \sqrt{2}x = -(L-x)$$

$$\implies x = -\frac{L}{\sqrt{2}-1} = -\left(\sqrt{2}+1\right)L = 2.41L$$

The root corresponding to the negative sign is extraneous because it corresponds to a position to the left of the $2.0 \, \mu C$

point charge and is, therefore, not a physically meaningful root. Hence the third point charge should be placed between the point charges and a distance equal to $0.41\,L$ away from the $2.0\,\mu C$ charge.

10. APPROACH The third point charge should be placed at the location at which the forces on the third point charge due to each of the other two point charges cancel. There can be no such place between the two point charges. Beyond the $4.0\,\mu C$ point charge, and on the line containing the two point charges, the force due to the $4.0\,\mu C$ point charge overwhelms the force due to the $-2.0\,\mu C$ point charge. Beyond the $-2.0\,\mu C$ point charge, and on the line containing the two point charges, however, we can find a place where these forces cancel because they are equal in magnitude and oppositely directed. Denote the $-2.0\,\mu C$ and $4.0\,\mu C$ point charges by the numerals 2 and 4, respectively, and the third point charge by the numeral 3. Let the $+x$ direction be to the right with the origin at the position of the $-2.0\,\mu C$ point charge and the $4.0\,\mu C$ point charge be located at $x = L$.

SOLUTION Apply the condition for translational equilibrium to the third point charge:

$$\vec{F}_{4,3} + \vec{F}_{2,3} = 0 \implies F_{4,3} = F_{2,3} \quad (1)$$

Letting the distance from the third point charge to the $2.0\,\mu C$ point charge be x, express the force that the $4.0\,\mu C$ point charge exerts on the third point charge:

$$F_{4,3} = \frac{kq_3q_4}{(L+x)^2}$$

The force that the $-2.0\mu C$ point charge exerts on the third point charge is given by:

$$F_{2,3} = \frac{kq_3q_2}{x^2}$$

Substitute for $F_{4,3}$ and $F_{2,3}$ in Eq.(1) to obtain:

$$\frac{kq_3q_4}{(L+x)^2} = \frac{kq_3q_2}{x^2} \implies \frac{q_4}{(L+x)^2} = \frac{q_2}{x^2}$$

Substituting $q_4 = 2q_2$ and rewriting this equation explicitly as a quadratic equations yields:

$$x^2 - 2Lx - L^2 = 0$$

On simplifying above equation, we get-

$$x = \frac{2L \pm \sqrt{4L^2 + 4L^2}}{2} = L \pm \sqrt{2}L$$

The root corresponding to the positive sign between the terms is extraneous because it corresponds to a position to the right of the $2.0\,\mu C$ point charge and is, therefore, not a physically meaningful root. Hence the third point charge should be placed a distance equal to $0.41L$ from the $-2.0\,\mu C$ charge on the side away from the $4.0\,\mu C$ charge.

11. APPROACH The positions of the point particles are shown in Fig.1.361. It is apparent that the electron must be located along the line joining two point particles. Moreover, because it is negatively charged, it must be closer to the particle with a charge which is smaller in magnitude i.e., $-2.5\,\mu C$ than to the particle with a charge of $6.0\,\mu C$, as is indicated in the Fig.1.361. We can find the x and y coordinates of the electron's position by equating the two electrostatic forces acting on it and solving for its distance from the origin. We can use similar triangles to express this radial distance in terms of the x and y coordinates of the electron.

SOLUTION Express the condition that must be satisfied if the electron is to be in equilibrium:

$$F_{1,e} = F_{2,e}$$

here, $F_{1,e}$ is the force applied by q_1 on charge e whereas

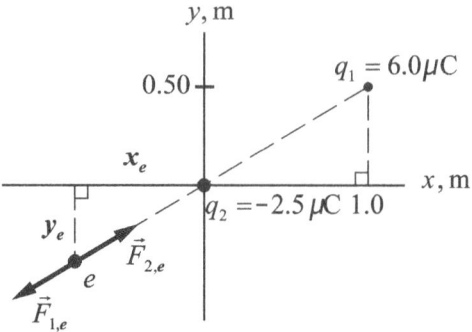

Figure 1.361

$F_{2,e}$ is the force applied by q_2 on e.
Let r represents the distance from the origin to the electron, express the magnitude of the force that the particle whose charge is q_1 exerts on the electron:

$$F_{1,e} = \frac{kq_1e}{(r+\sqrt{1.25})^2}$$

Express the magnitude of the force that the particle whose charge is q_2 exerts on the electron:

$$F_{2,e} = \frac{k|q_2|e}{r^2}$$

Substitute and simplify to obtain:

$$\frac{q_1}{(r+\sqrt{1.25})^2} = \frac{|q_2|}{r^2}$$

Substitute for q_1 and q_2 and simplifying for r, we get:

$$\frac{6}{(r+\sqrt{1.25})^2} = \frac{2.5}{r^2}$$

Solving for r yields:
$r = 2.036$ m and $r = -0.4386$ m.
Because $r < 0$ is unphysical, we'll consider only the positive root.
Use the similar triangles in the diagram to establish the proportion involving the y coordinate of the electron:

$$\frac{y_e}{0.50} = \frac{2.036}{1.12} \implies y_e = 0.909 \text{ m}$$

Use the similar triangles in the diagram to establish the proportion involving the x coordinate of the electron:

$$\frac{x_e}{1.0} = \frac{2.036}{1.12} \implies x_e = 1.82 \text{m}$$

Therefore, the coordinates of the electron's position are:
$$(x_e, y_e) = (-1.8 \text{ m}, -0.91 \text{ m})$$

Substituting for q_1 and q_2 and simplifying for r we get-

$$\frac{6}{(r+\sqrt{1.25})^2} = \frac{2.5}{r^2}$$

Solving for r yields: $r = 2.036$ m and $r = -0.4386$ m.
Because $r < 0$ is unphysical, we'll consider only the positive root. Use the similar triangles in the diagram to establish the proportion involving the y coordinate of the electron:

$$\frac{y_e}{0.50} = \frac{2.036}{1.12} \implies y_e = 0.909 \text{ m}$$

Use the similar triangles in the diagram to establish the proportion involving the x coordinate of the electron:

$$\frac{x_e}{1.0} = \frac{2.036}{1.12} \implies x_e = 1.82 \text{ m}$$

The coordinates of the electron's position are :
$(x_e, y_e) = (-1.8 \text{ m}, -0.91 \text{ m})$

12. APPROACH Apply Coulomb's law and the principle of superposition of forces to find the net force acting on the central charge q.

SOLUTION The given situation is shown in Fig.1.362. A, B, C, D and E are five equally spaced points on the semicircle of radius R. At each of these points a charge Q is placed. O os the centre of this semicircle where another point charge q is placed. Since each charge is at same distance from central charge at O, therefore magnitude of electric force applied by each point charge on the charge at O will be same and if it is F, then

$$F = k\frac{Qq}{R^2} \quad \cdots (1)$$

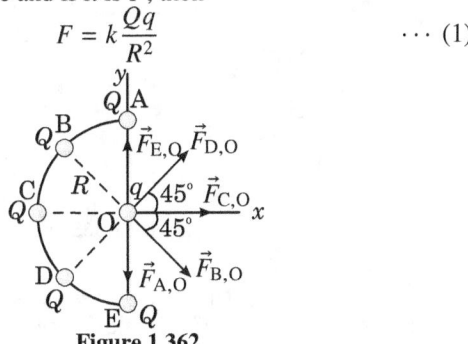

Figure 1.362

If $\vec{F}_{A,O}$, $\vec{F}_{B,O}$, $\vec{F}_{C,O}$, $\vec{F}_{D,O}$, and $\vec{F}_{E,O}$ are electric forces applied by charge Q at A, B, C, D and E respectively on the charge q at O, then
$\vec{F}_{AO} = -F\hat{j}$, $\vec{F}_{BO} = F\cos 45°\hat{i} - F\sin 45°\hat{j}$, $\vec{F}_{CO} = F\hat{i}$, $\vec{F}_{DO} = F\cos 45°\hat{i} + F\sin 45°\hat{j}$, $\vec{F}_{EO} = F\hat{j}$.
Net force on the charge q at O:
$$\vec{F}_{net} = \vec{F}_{AO} + \vec{F}_{BO} + \vec{F}_{CO} + \vec{F}_{DO} + \vec{F}_{EO}$$
Substituting the values of all forces, we get
$$\vec{F}_{net} = -F\hat{j} + (F\cos 45°\hat{i} - F\sin 45°\hat{j}) + F\hat{i}$$
$$+ (F\cos 45°\hat{i} + F\sin 45°\hat{j}) + F\hat{j}$$
$$\vec{F}_{net} = F(2\cos 45° + 1)\hat{i} = F(\sqrt{2} + 1)\hat{i} = \quad \cdots (2)$$
Substituting the value of F from Eq.(1) in Eq.(2), we get
$$\vec{F}_{net} = k\frac{Qq}{R^2}(\sqrt{2} + 1)\hat{i}$$

13. **APPROACH** Let the H$^+$ ions be in the x-y plane with one H$^+$ at $(0,0,0)$, one H$^+$ at $(a,0,0)$, and one H$^+$ at $\left(\frac{a}{2}, \frac{a\sqrt{3}}{2}, 0\right)$. These charges are denoted by q_1, q_2, and q_3 respectively in Fig.1.363. The N^{-3} ion, with charge q_4 in our notation, is then at $\left(\frac{a}{2}, \frac{a}{2\sqrt{3}}, a\sqrt{\frac{2}{3}}\right)$ where $a = 1.64 \times 10^{-10}$ m. To simplify our calculations we'll set $ke^2/a^2 = C = 8.56 \times 10^{-9}$ N. We can apply Coulomb's law and the principle of superposition of forces to find the net force acting on each ion.

SOLUTION Express the net force acting on point charge q_1:
$$\vec{F}_1 = \vec{F}_{2,1} + \vec{F}_{3,1} + \vec{F}_{4,1}$$
Find $\vec{F}_{2,1}$: $\vec{F}_{2,1} = \frac{kq_1q_2}{r_{2,1}^2}\hat{r}_{2,1} = C(-\hat{i}) = -C\hat{i}$

Find $\vec{F}_{3,1}$: $\vec{F}_{3,1} = \frac{kq_3q_1}{r_{3,1}^2}\hat{r}_{3,1} = C\frac{\left(0 - \frac{a}{2}\right)\hat{i} + \left(0 - \frac{a\sqrt{3}}{2}\right)\hat{j}}{a}$
$$= -C\left(\frac{1}{2}\hat{i} + \frac{\sqrt{3}}{2}\hat{j}\right)$$

Noting that the magnitude of point charge q_4 is three times that of the other point charges and that it is negative, express $\vec{F}_{4,1}$:
$$\vec{F}_{4,1} = 3C\hat{r}_{4,1}$$

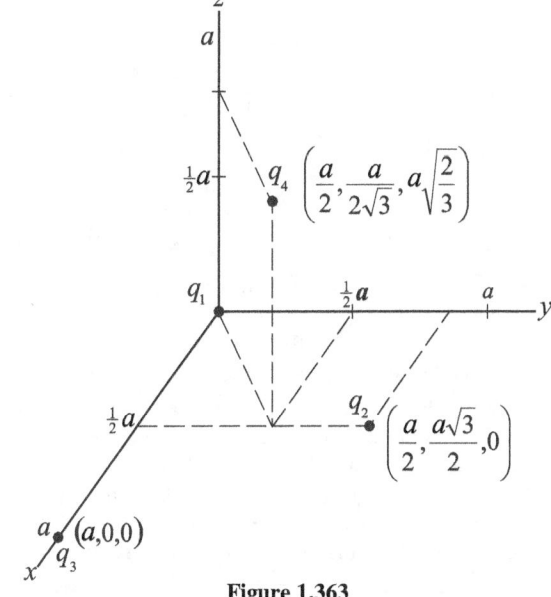

Figure 1.363

$$= -3C\frac{\left(0 - \frac{a}{2}\right)\hat{i} + \left(0 - \frac{a}{2\sqrt{3}}\right)\hat{j} + \left(0 - \frac{a\sqrt{2}}{\sqrt{3}}\right)\hat{k}}{\sqrt{\left(\frac{a}{2}\right)^2 + \left(\frac{a}{2\sqrt{3}}\right)^2 + \left(\frac{a\sqrt{2}}{\sqrt{3}}\right)^2}}$$

$$= 3C\frac{\left(\frac{a}{2}\right)\hat{i} + \left(\frac{a}{2\sqrt{3}}\right)\hat{j} + \left(\frac{a\sqrt{2}}{\sqrt{3}}\right)\hat{k}}{a}$$

$$= 3C\left[\left(\frac{1}{2}\right)\hat{i} + \left(\frac{1}{2\sqrt{3}}\right)\hat{j} + \left(\sqrt{\frac{2}{3}}\right)\hat{k}\right]$$

Substitute in the expression for $\vec{F}_1$ to obtain:
$$\vec{F}_1 = -C\hat{i} - C\left(\frac{1}{2}\hat{i} + \frac{\sqrt{3}}{2}\hat{j}\right)$$
$$+ 3C\left[\left(\frac{1}{2}\right)\hat{i} + \left(\frac{1}{2\sqrt{3}}\right)\hat{j} + \left(\sqrt{\frac{2}{3}}\right)\hat{k}\right] = C\sqrt{6}\hat{k}$$

From symmetry considerations:
$$\vec{F}_2 = \vec{F}_3 = \vec{F}_1 = C\sqrt{6}\hat{k}$$
Express the condition that the molecule is in equilibrium:
$$\vec{F}_1 + \vec{F}_2 + \vec{F}_3 + \vec{F}_4 = 0$$
Solve for and evaluate $\vec{F}_4$:
$$\vec{F}_4 = -\left(\vec{F}_1 + \vec{F}_2 + \vec{F}_3\right) = -3\vec{F}_1 = -3C\sqrt{6}\hat{k}$$

14. **APPROACH** Let q represent the point charge at the origin and use Coulomb's law for $\vec{E}$ due to a point charge to find the electric field at $x = 6.0$ m and -10 m.

SOLUTION (a) Express the electric field at a point P located a distance x from a point charge q:
$$\vec{E}(x) = \frac{kq}{x^2}\hat{r}_{P,0}$$
Evaluate this expression for $x = 6.0$ m:
$$\vec{E}(6.0 \text{ m}) = \frac{\left(8.988 \times 10^9 \frac{\text{N·m}^2}{\text{C}^2}\right)(4.0\mu\text{C})}{(6.0 \text{ m})^2}\hat{i} = (1.0\text{kN/C})\hat{i}$$

(b) Evaluate $\vec{E}$ at $x = -10$ m:
$$\vec{E}(-10 \text{ m}) = \frac{\left(8.988 \times 10^9 \frac{\text{N·m}^2}{\text{C}^2}\right)(4.0\mu\text{C})}{(10 \text{ m})^2}(-\hat{i})$$
$$= (-0.36 \text{ kN/C})\hat{i}$$

(c) E_x vs x graph is shown in Fig.1.364.

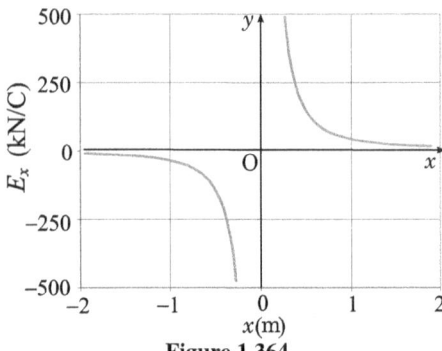

Figure 1.364

15. APPROACH Let q represent the point charges of $+4.0\ \mu C$ and use Coulomb's law for $\vec{E}$ due to a point charge and the principle of superposition for fields to find the electric field at the locations specified.

Noting that $q_1 = q_2$, use Coulomb's law and the principle of superposition to express the electric field due to the given charges at point P a distance x from the origin:

$$\vec{E}(x) = \vec{E}_{q_1}(x) + \vec{E}_{q_2}(x)$$
$$= \frac{kq_1}{x^2}\hat{r}_{q_1,P} + \frac{kq_2}{(8.0\text{ m}-x)^2}\hat{r}_{q_2,P}$$
$$= kq_1\left(\frac{1}{x^2}\hat{r}_{q_1,P} + \frac{1}{(8.0\text{ m}-x)^2}\hat{r}_{q_2,P}\right)$$
$$= \left(36\text{kN}\cdot\text{m}^2/\text{C}\right)\left(\frac{1}{x^2}\hat{r}_{q_1,P} + \frac{1}{(8.0\text{ m}-x)^2}\hat{r}_{q_2,P}\right)$$

(a) Apply this equation to the point at $x = -2.0$ m:
$$\vec{E}(-2.0\text{ m}) = \left(36\text{kN}\cdot\text{m}^2/\text{C}\right)\left[\frac{1}{(2.0\text{ m})^2}(-\hat{i}) + \frac{1}{(10\text{ m})^2}(-\hat{i})\right]$$
$$= (-9.4\text{kN/C})\hat{i}$$

(b) Evaluate $\vec{E}$ at $x = 2.0$ m:
$$\vec{E}(2.0\text{ m}) = \left(36\text{ kN}\cdot\text{m}^2/\text{C}\right)\left[\frac{1}{(2.0\text{ m})^2}(\hat{i}) + \frac{1}{(6.0\text{ m})^2}(-\hat{i})\right]$$
$$= (8.0\text{kN/C})\hat{i}$$

(c) Evaluate $\vec{E}$ at $x = 6.0$ m:
$$\vec{E}(6.0\text{ m}) = \left(36\text{ kN}\cdot\text{m}^2/\text{C}\right)\left[\frac{1}{(6.0\text{ m})^2}(\hat{i}) + \frac{1}{(2.0\text{ m})^2}(-\hat{i})\right]$$
$$= (-8.0\text{kN/C})\hat{i}$$

(d) Evaluate $\vec{E}$ at $x = 10$ m:
$$\vec{E}(10\text{ m}) = \left(36\text{kN}\cdot\text{m}^2/\text{C}\right)\left[\frac{1}{(10\text{ m})^2}(\hat{i}) + \frac{1}{(2.0\text{ m})^2}(\hat{i})\right]$$
$$= (9.4\text{ kN/C})\hat{i}$$

(e) From symmetry considerations:
$$\vec{E}(4.0\text{ m}) = 0$$

(f) E_x vs x graph is shown in Fig.1.365

16. APPROACH If the electric field at $x = 0$ is zero, both its x and y components must be zero. The only way this condition can be satisfied with point charges of $+5.0\ \mu C$ and $-8.0\ \mu C$ on the x axis is if the point charge $+6.0\ \mu C$ is also on the x axis. Let the subscripts 5, -8, and 6 identify the point charges and their fields. We can use Coulomb's law for $\vec{E}$ due to a point charge and the principle of superposition for fields to determine where the $+6.0\ \mu C$ point charge should be located so that the electric field at $x = 0$ is zero.

SOLUTION Express the electric field at $x = 0$ in terms of the fields due to the point charges of $+5.0\ \mu C$, $-8.0\ \mu C$, and

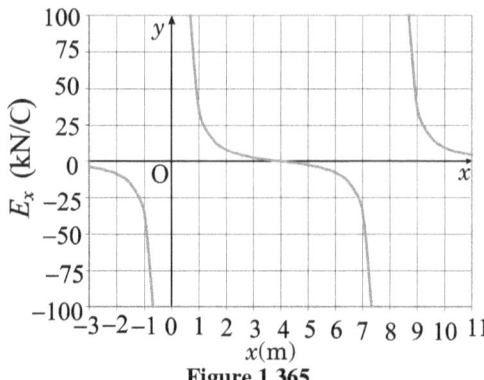

Figure 1.365

$+6.0\ \mu C$:
$$\vec{E}(0) = \vec{E}_{5\mu C} + \vec{E}_{-8\mu C} + \vec{E}_{6\mu C} = 0$$

Substitute for each of the fields to obtain:
$$\frac{kq_5}{r_5^2}\hat{r}_5 + \frac{kq_6}{r_6^2}\hat{r}_6 + \frac{kq_{-8}}{r_{-8}^2}\hat{r}_{-8} = 0$$

or
$$\frac{kq_5}{r_5^2}\hat{i} + \frac{kq_6}{r_6^2}(-\hat{i}) + \frac{kq_{-8}}{r_{-8}^2}(-\hat{i}) = 0$$

Divide out the unit vector $\hat{i}$ to obtain:
$$\frac{q_5}{r_5^2} - \frac{q_6}{r_6^2} - \frac{q_{-8}}{r_{-8}^2} = 0$$

Substitute numerical values to obtain r_6:
$$\frac{5}{(3.0\text{cm})^2} - \frac{6}{r_6^2} - \frac{-8}{(4.0\text{cm})^2} = 0 \implies r_6 = 2.4\text{ cm}$$

17. APPROACH Fig. 1.366 shows the electric field vectors at the point of interest P due to the two point charges. We can use Coulomb's law for $\vec{E}$ due to point charges and the superposition principle for electric fields to find $\vec{E}_P$. We can apply $\vec{F} = q\vec{E}$ to find the force on an electron at $(-1.0\text{m}, 0)$.

SOLUTION (a) Express the electric field at $(-1.0\text{ m}, 0)$

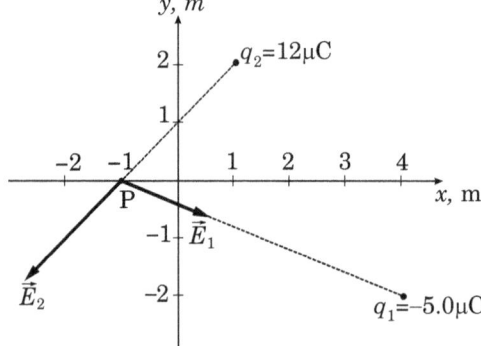

Figure 1.366

due to the point charges q_1 and q_2: $\vec{E}_P = \vec{E}_1 + \vec{E}_2$
Substitute numerical values and evaluate $\vec{E}_1$
$$\vec{E}_1 = \frac{kq_1}{r_{1,P}^2}\hat{r}_{1,P} = \frac{(8.988 \times 10^9\text{ N}\cdot\text{m}^2/\text{C}^2)(-5.0\mu C)}{(5.0\text{ m})^2 + (2.0\text{ m})^2}$$
$$\left(\frac{(-5.0\text{ m})\hat{i} + (2.0\text{ m})\hat{j}}{\sqrt{(5.0\text{ m})^2 + (2.0\text{ m})^2}}\right)$$
$$= (-1.55 \times 10^3\text{ N/C})(-0.928\hat{i} + 0.371\hat{j})$$
$$= (1.44\text{ kN/C})\hat{i} + (-0.575\text{ kN/C})\hat{j}$$

Substitute numerical values and evaluate $\vec{E}_2$:
$$\vec{E}_2 = \frac{kq_2}{r_{2,P}^2}\hat{r}_{2,P}$$
$$= \frac{(8.988 \times 10^9\text{ N}\cdot\text{m}^2/\text{C}^2)(12\mu C)}{(2.0\text{ m})^2 + (2.0\text{ m})^2}$$

$$\left(\frac{(-2.0\text{ m})\hat{i} + (-2.0\text{ m})\hat{j}}{\sqrt{(2.0\text{ m})^2 + (2.0\text{ m})^2}}\right)$$
$$= \left(13.5 \times 10^3 \text{ N/C}\right)(-0.707\hat{i} - 0.707\hat{j})$$
$$= (-9.54\text{kN/C})\hat{i} + (-9.54\text{kN/C})\hat{j} \text{ Substitute for}$$
$\vec{E}_1$ and $\vec{E}_2$ and simplify to find $\vec{E}_p$:
$$\vec{E}_P = (1.44\text{ kN/C})\hat{i} + (-0.575\text{ kN/C})\hat{j} + (-9.54\text{ kN/C})\hat{i}$$
$$+ (-9.54\text{ kN/C})\hat{j}$$
$$= (-8.10\text{ kN/C})\hat{i} + (-10.1\text{ kN/C})\hat{j}$$
The magnitude of $\vec{E}_P$ is:
$$E_P = \sqrt{(-8.10\text{ kN/C})^2 + (-10.1\text{ kN/C})^2} = 13\text{ kN/C}$$
The direction of $\vec{E}_p$ is:
$$\theta_E = \tan^{-1}\left(\frac{-10.1\text{ kN/C}}{-8.10\text{ kN/C}}\right) = 230°$$
Note that the angle returned by your calculator for $\tan^{-1}\left(\frac{-10.1\text{ kN/C}}{-8.10\text{ kN/C}}\right)$ is the reference angle and must be increased by 180° to yield θ_E.

(b) Express and evaluate the force on an electron at point P:
$$\vec{F} = q\vec{E}_P = \left(-1.602 \times 10^{-19}\text{C}\right)[(-8.10\text{kN/C})\hat{i}$$
$$+ (-10.1\text{kN/C})\hat{j}]$$
$$= \left(1.30 \times 10^{-15}\text{ N}\right)\hat{i} + \left(1.62 \times 10^{-15}\text{ N}\right)\hat{j}$$
Find the magnitude of $\vec{F}$:
$$F = \sqrt{\left(1.30 \times 10^{-15}\text{ N}\right)^2 + \left(1.62 \times 10^{-15}\text{ N}\right)^2}$$
$$= 2.1 \times 10^{-15}\text{ N Find the direction of } \vec{F}:$$
$$\theta_F = \tan^{-1}\left(\frac{1.62 \times 10^{-15}\text{ N}}{1.3 \times 10^{-15}\text{ N}}\right) = 51°$$

18. APPROACH The electric field of the 4th charged point particle must cancel the sum of the electric fields due to the other three charged point particles[Fig.1.367]. By symmetry, the position of the 4th charged point particle must lie on the vertical center line of the triangle. Using trigonometry, one can show that the center of an equilateral triangle is a distance $L/\sqrt{3}$ from each vertex, where L is the length of the side of the triangle. Note that the x components of the fields due to the base charged particles cancel each other, so we only need concern ourselves with the y components of the fields due to the charged point particles at the vertices of the triangle. Choose a coordinate system in which the origin is at the midpoint of the base of the triangle, the $+x$ direction is to the right, and the $+y$ direction is upward.

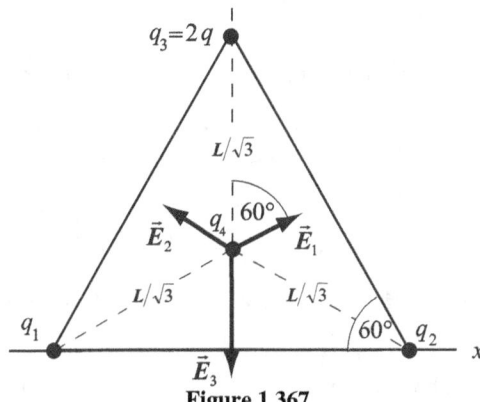

Figure 1.367

SOLUTION Express the condition that must be satisfied if the electric field at the center of the triangle is to be zero:
$$\sum_{i=1\text{to}4} \vec{E}_i = 0$$
Substituting for $\vec{E}_1, \vec{E}_2, \vec{E}_3$, and $\vec{E}_4$ yields:
$$\frac{k(q)}{\left(\frac{L}{\sqrt{3}}\right)^2}\cos 60°\hat{j} + \frac{k(q)}{\left(\frac{L}{\sqrt{3}}\right)^2}\cos 60°\hat{j}$$
$$-\frac{k(2q)}{\left(\frac{L}{\sqrt{3}}\right)^2}\hat{j} + \frac{kq}{y^2}\hat{j} = 0$$
On simplifying for y, we get-
$$y = \pm L/\sqrt{3}$$
The positive solution corresponds to the 4th point particle being a distance $L/\sqrt{3}$ above the base of the triangle, where it produces the same strength and same direction electric field caused by the three charges at the corners of the triangle. So the charged point particle must be placed a distance $L/\sqrt{3}$ below the midpoint of the triangle.

19. APPROACH We can determine the stability of the equilibrium in Part (a) and Part (b) by considering the forces the equal point charges q at $y = +a$ and $y = -a$ exert on the test charge when it is given a small displacement along either the x or y axis. (c) The application of Coulomb's law in Part (c) will lead to the magnitude and sign of the charge that must be placed at the origin in order that a net force of zero is experienced by each of the three point charges.

SOLUTION (a) Because E_x is in the x direction, a positive test charge that is displaced from $(0,0)$ in either the $+x$ direction or the $-x$ direction will experience a force pointing away from the origin and accelerate in the direction of the force. Consequently, the equilibrium at $(0, 0)$ is unstable for a small displacement along the x axis.

If the positive test charge is displaced in the direction of increasing y (the $+y$ direction), the charge at $y = +a$ will exert a greater force than the charge at $y = -a$, and the net force is then in the $-y$ direction; i.e., it is a restoring force. Similarly, if the positive test charge is displaced in the direction of decreasing y (the $-y$ direction), the charge at $y = -a$ will exert a greater force than the charge at $y = -a$, and the net force is then in the $-y$ direction; i.e., it is a restoring force. Consequently, the equilibrium at $(0,0)$ is stable for a small displacement along the y axis*.

(b) Following the same arguments as in Part (a) , one finds that, for a negative test charge, the equilibrium is stable at $(0, 0)$ for displacements along the x axis and unstable for displacements along the y axis.

(c) Express the net force acting on the charge at $y = +a$:
$$\sum F_{q\text{ at }y=+a} = \frac{kqq_0}{a^2} + \frac{kq^2}{(2a)^2} = 0$$
Solve for q_0 to obtain: $q_0 = -q/4$

☞ In Part (c), we could just as well have expressed the net force acting on the charge at $y = -a$. Due to the symmetric distribution of the charges at $y = -a$ and $y = +a$, summing the forces acting on q_0 at the origin does not lead to a relationship between q_0 and q.

20. APPROACH The diagram 1.368 shows the locations of point charges q_1 and q_2 and the point on the x axis at which we are to find $\vec{E}$. From symmetry considerations we can conclude that the y component of $\vec{E}$ at any point on the x

---

*It is important to note that the equilibrium is stable only along y axis not in all directions. So, Earnshaw's theorm (section 1.24) is not violated

axis is zero. We can use Coulomb's law for the electric field due to point charges and the principle of superposition of fields to find the field at any point on the x axis. We can establish the results called for in parts (b) and (c) by factoring the radicand and using the approximation $1 + \alpha \approx 1$ whenever $\alpha \ll 1$.

SOLUTION (a) Express the x-component of the electric field due to the point charges at $y = a$ and $y = -a$ as a function of the distance r from either charge to point P:

$$\vec{E}_x = 2\frac{kq}{r^2}\cos\theta\hat{i}$$

Substitute for $\cos\theta$ and r to obtain:

$$\vec{E}_x = 2\frac{kq}{r^2}\frac{x}{r}\hat{i} = \frac{2kqx}{r^3}\hat{i} = \frac{2kqx}{(x^2+a^2)^{3/2}}\hat{i} = \frac{2kqx}{(x^2+a^2)^{3/2}}\hat{i}$$

The magnitude of $\vec{E}_x$ is:

$$E_x = \frac{2kqx}{(x^2+a^2)^{3/2}}$$

(b) For $|x| \ll a$, $x^2 + a^2 \approx a^2$, so:

$$E_x \approx \frac{2kqx}{(a^2)^{3/2}} = \frac{2kqx}{a^3}$$

(c) For $x \gg a$, the charges separated by a would appear to be a single charge of magnitude $2q$. Its field would be given by $E_x = 2kq/x^2$.
Factorize the radicand to obtain:

$$E_x = 2kqx\left[x^2\left(1+\frac{a^2}{x^2}\right)\right]^{-3/2}$$

For $a \ll x$, $1 + \frac{a^2}{x^2} \approx 1$ and $E_x = 2kqx\left[x^2\right]^{-3/2} = \frac{2kq}{x^2}$

21. APPROACH From problem 20 The electric field on the x axis, due to equal positive point charges located at $(0,a)$ and $(0,-a)$, is given by $E_x = 2kqx(x^2+a^2)^{-3/2}$ We can use $T = 2\pi\sqrt{m/k'}$ to express the period of the motion of the bead in terms of the restoring constant k'.

(a) Express the force acting on the bead when its displacement from the origin is x:

$$F_x = -qE_x = -\frac{2kq^2x}{(x^2+a^2)^{3/2}} = -\frac{2kq^2x}{a^3\left(\frac{x^2}{a^2}+1\right)^{3/2}}$$

For $x \ll a$: $F_x = -\frac{2kq^2}{a^3}x$

That is, the bead experiences a linear restoring force.

(b) Express the period of a simple harmonic oscillator:

$$T = 2\pi\sqrt{\frac{m}{k'}}$$

Obtain k' from our result in Part (a): $k' = \frac{2kq^2}{a^3}$

Substitute for k' and simplify to obtain:

$$T = 2\pi\sqrt{\frac{m}{\frac{2kq^2}{a^3}}} = 2\pi\sqrt{\frac{ma^3}{2kq^2}}$$

22. APPROACH We can use Newton's second law of motion to find the acceleration of the proton in the uniform electric field and constant-acceleration equations to find the time required for it to reach a speed of $0.01c$ and the distance it travels while acquiring this speed.

SOLUTION (a) Compute e/m for a proton:

$$\frac{e}{m_p} = \frac{1.602 \times 10^{-19}\text{C}}{1.673 \times 10^{-27}\text{ kg}} = 9.58 \times 10^{7}\text{C/kg}$$

Apply Newton's second law to relate the acceleration of the proton to the electric field:

$$a = \frac{F_{\text{net}}}{m_p} = \frac{eE}{m_p}$$

Substitute numerical values and evaluate a:

$$a = \frac{(1.602 \times 10^{-19}\text{C})(100\text{ N/C})}{1.673 \times 10^{-27}\text{ kg}} = 9.576 \times 10^{9}\text{ m/s}^2$$
$$= 9.58 \times 10^{9}\text{ m/s}^2$$

The direction of the acceleration of a proton is in the direction of the electric field.

(b) Using the definition of acceleration, relate the time required for a proton to reach $0.01c$ to its acceleration:

$$\Delta t = \frac{v}{a} = \frac{0.01c}{a}$$

Substitute numerical values and evaluate Δt:

$$\Delta t = \frac{0.01(2.998 \times 10^{8}\text{ m/s})}{9.576 \times 10^{9}\text{ m/s}^2} = 0.3\text{ ms}$$

23. APPROACH Because the electric field is in the $-y$ direction, the force it exerts on the electron is in the $+y$ direction. Applying Newton's second law to the electron will yield an expression for the acceleration of the electron in the y direction. We can then use a constant-acceleration equation to relate its speed to its acceleration and the distance it has travelled.

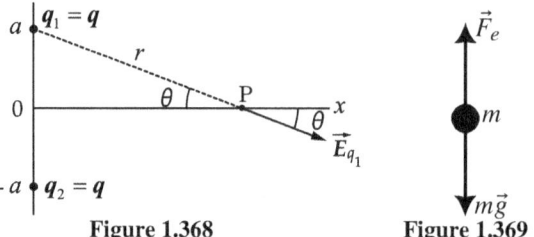

Figure 1.368 **Figure 1.369**

SOLUTION Applying Newton's second law: $\sum F_y = ma_y$ to the FBD of electron given in Fig.1.369, we get

$$F_e - F_g = ma_y$$

since, $F_e = eE$ and $F_g = mg$, therefore above equation gives:

$$eE - mg = ma_y \implies a_y = \frac{eE}{m} - g$$

Use a constant-acceleration equation to relate the speed of the electron to its acceleration and the distance it travels:

$$v_y^2 = v_0^2 + 2a_y\Delta y$$

Since, the electron starts at rest, i.e., $v_0 = 0$, therefore, above equation gives

$$v_y^2 = 2a_y\Delta y \implies v_y = \sqrt{2a_y\Delta y}$$

Substituting the value of for a_y in above expression for v_y, we get

$$v_y = \sqrt{2\left(\frac{eE}{m} - g\right)\Delta y}$$

Now, substituting numerical values, we get in above expression, we get:

$$v_y = \left[2\left\{\frac{(1.602 \times 10^{-19}\text{C})(1.50 \times 10^{-10}\text{ N/C})}{9.109 \times 10^{-31}\text{ kg}}\right.\right.$$
$$\left.\left. -9.81\text{ m/s}^2\right\}(1.0 \times 10^{-6}\text{ m})\right]^{1/2} = 5.8\text{ mm/s}$$

24. APPROACH We can apply the work-kinetic energy theorem to relate the change in the object's kinetic energy to the net force acting on it. We can express the net force acting on the charged body in terms of its charge and the electric field.

SOLUTION Using the work-kinetic energy theorem, express the kinetic energy of the object in terms of the net force acting on it and its displacement:

$$W = \Delta K = F_{net} \Delta x$$

Relate the net force acting on the charged particle to the electric field:

$$F_{net} = qE$$

Substitute for F_{net} to obtain:

$$\Delta K = K_f - K_i = qE\Delta x$$

Since, $K_i = 0$, therefore, $K_f = qE\Delta x \implies q = \dfrac{K_f}{E\Delta x}$

Substitute numerical values and evaluate q:

$$q = \dfrac{0.120 \text{ J}}{(300 \text{ N/C})(0.500 \text{ m})} = 800\mu C$$

25. APPROACH We can use constant-acceleration equations to express the x and y coordinates of the particle in terms of the parameter t and Newton's second law to express the constant acceleration in terms of the electric field. Eliminating t will yield an equation for y as a function of x, q, and m that we can solve for E_y.

SOLUTION Express the x and y coordinates of the particle as functions of time:

$$x = (v\cos\theta)t \qquad \cdots (1)$$

and

$$y = (v\sin\theta)t - \dfrac{1}{2}a_y t^2 \qquad \cdots (2)$$

Apply Newton's second law to relate the acceleration of the particle to the net force acting on it:

$$a_y = \dfrac{F_{net,y}}{m} = \dfrac{qE_0}{m} \qquad \cdots (3)$$

Substitute it in Eq.(2), to get y:

$$y = (v\sin\theta)t - \dfrac{qE_0}{2m}t^2 \qquad \cdots (4)$$

Eliminate the parameter t between the two equations (1) and (4) to obtain:

$$y = (\tan\theta)x - \dfrac{qE_0}{2mv^2\cos^2\theta}x^2$$

Set $y = 0$ and solve for E_0 to obtain:

$$E_0 = \dfrac{mv^2\sin 2\theta}{qx}$$

Substitute the non-particle-specific data to obtain:

$$E_0 = \dfrac{m(3.00\times 10^6 \text{ m/s})^2 \sin 70°}{q(0.0150 \text{ m})}$$

$$= (5.638\times 10^{14} \text{ m/s}^2)\dfrac{m}{q}$$

(a) Substitute for the mass and charge of an electron and evaluate E_0:

$$E_0 = \left(5.638\times 10^{14} \text{ m/s}^2\right)\dfrac{9.109\times 10^{-31} \text{ kg}}{-1.602\times 10^{-19}\text{C}}$$

$$= -3.2 \text{ kN/C}$$

(b) Substitute for the mass and charge of a proton and evaluate E_0:

$$E_0 = \left(5.64\times 10^{14} \text{ m/s}^2\right)\dfrac{1.673\times 10^{-27} \text{ kg}}{1.602\times 10^{-19}\text{C}} = 5.9 \text{ MN/C}$$

26. APPROACH We can use the definition of dipole moment to find the dipole moment of given pair of charges.

SOLUTION (a) Apply the definition of electric dipole moment to obtain:

$$\vec{p} = q\left(2\vec{l}\right)$$

and $\qquad p = (2.0 \text{ pC})(4.0 \; \mu m) = 8.0\times 10^{-18}$ C.m

(b) If the dipole is oriented to the right[Fig.1.370], then $\vec{p}$ is to the right; pointing from the negative charge toward the positive charge.

27. APPROACH The forces the electron and the proton exert on each other constitute an action-and-reaction pair. Because the magnitudes of their charges are equal and their masses are the same, we find the speed of each particle by

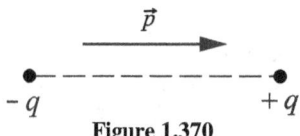

Figure 1.370

finding the speed of either one. We'll apply Coulomb's force law for point charges and Newton's second law to relate v to e, m, k, and their separation distance L.

SOLUTION Apply Newton's second law to the positron to obtain:

$$\dfrac{ke^2}{L^2} = m\dfrac{v^2}{\frac{1}{2}L} \implies \dfrac{ke^2}{L} = 2mv^2$$

Solving for v gives: $v = \sqrt{\dfrac{ke^2}{2mL}}$

28. APPROACH We can use Coulomb's force law for point particles and the condition for translational equilibrium to express the equilibrium position as a function of k, q, Q, m, and g. In Part (b) we'll need to show that the displaced point charge experiences a linear restoring force and, hence, will exhibit simple harmonic motion.

SOLUTION (a) Apply the condition for translational equilibrium to the particle:

$$\dfrac{kqQ}{y_0^2} - mg = 0 \implies y_0 = \sqrt{\dfrac{kqQ}{mg}}$$

(b) Express the restoring force that acts on the particle when it is displaced a distance Δy from its equilibrium position:

$$F = \dfrac{kqQ}{(y_0+\Delta y)^2} - \dfrac{kqQ}{y_0^2}$$

Since, $\Delta y \ll y_0$, therefore

$$F \approx \dfrac{kqQ}{y_0^2 + 2y_0\Delta y} - \dfrac{kqQ}{y_0^2}$$

Simplify this expression further by writing it with a common denominator:

$$F = -\dfrac{2y_0\Delta y kqQ}{y_0^4 + 2y_0^3\Delta y} = -\dfrac{2y_0\Delta y kqQ}{y_0^4\left(1+2\frac{\Delta y}{y_0}\right)} \approx -\dfrac{2\Delta y kqQ}{y_0^3}$$

(Since, $\Delta y \ll y_0$.)

From the part (a) of our solution:

$$\dfrac{kqQ}{y_0^2} = mg$$

Substitute for $\dfrac{kqQ}{y_0^2}$ and simplify to obtain:

$$F = -\dfrac{2mg}{y_0}\Delta y$$

Apply Newton's second law to the displaced particle to obtain:

$$m\dfrac{d^2\Delta y}{dt^2} = -\dfrac{2mg}{y_0}\Delta y$$

or $\qquad \dfrac{d^2\Delta y}{dt^2} + \dfrac{2g}{y_0}\Delta y = 0$

or $\qquad \dfrac{d^2\Delta y}{dt^2} + \omega^2\Delta y = 0$

which is the differential equation of simple harmonic motion with angular frequency

$$\omega = \sqrt{2g/y_0}$$

29. APPROACH The free body diagram 1.371 shows the forces acting on the microsphere of mass m and having an excess charge of $q = Ne$ when the electric field is downward. Under terminal-speed conditions the sphere is in dynamic equilibrium under the influence of the electric force $\vec{F}_e$, its weight $m\vec{g}$, and the drag force $\vec{F}_d$. We can apply Newton's second law, under terminal speed conditions, to relate the number of excess charges N on the sphere to its mass and, using Stokes' law, find its terminal speed.

(a) Apply Newton's second law to the microsphere to

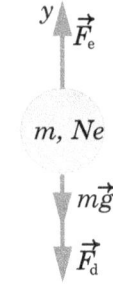

Figure 1.371

obtain:
$$F_e - mg - F_d = ma_y$$
since, $a_y = 0$, therefore
$$F_e - mg - F_{d,\,terminal} = 0$$
Substitute for F_e, m, and $F_{d,terminal}$ to obtain:
$$qE - \rho Vg - 6\pi\eta r v_t = 0$$
since, $q = Ne$, therefore
$$NeE - \frac{4}{3}\pi r^3 \rho g - 6\pi\eta r v_t = 0$$
Solve for Ne to obtain:
$$Ne = \frac{\frac{4}{3}\pi r^3 \rho g + 6\pi\eta r v_t}{E} \quad \cdots (1)$$
Substitute numerical values and evaluate $\frac{4}{3}\pi r^3 \rho g$:
$$\frac{4}{3}\pi r^3 \rho g = \frac{4}{3}\pi \left(5.50 \times 10^{-7}\text{ m}\right)^3 \left(1.05 \times 10^3 \text{ kg/m}^3\right)$$
$$\left(9.81 \text{ m/s}^2\right) = 7.18 \times 10^{-15}\text{ N}$$
Substitute numerical values and evaluate $6\pi\eta r v_t$:
$$6\pi\eta r v_t = 6\pi \left(1.8 \times 10^{-5}\text{ Pa}\cdot\text{s}\right)\left(5.50 \times 10^{-7}\text{ m}\right)$$
$$\left(1.16 \times 10^{-4}\text{ m/s}\right) = 2.16 \times 10^{-14}\text{ N}$$
Substitute numerical values in equation (1) and evaluate Ne:
$$Ne = \frac{7.18 \times 10^{-15}\text{ N} + 2.16 \times 10^{-14}\text{ N}}{6.00 \times 10^4 \text{ N/C}}$$
$$= 4.8 \times 10^{-19}\text{C}$$

(b) Divide the result in (a) by e to obtain:
$$N = \frac{4.80 \times 10^{-19}\text{C}}{1.602 \times 10^{-19}\text{C}} = 3$$

(c) With the field pointing upward, the electric force is downward and the application of $\sum F_y = ma_y$ to the bead yields:
$$F_{d,\,terminal} - F_e - mg = 0$$
or
$$6\pi\eta r v_t - NeE - \frac{4}{3}\pi r^3 \rho g = 0$$
Solve for v_t to obtain:
$$v_t = \frac{NeE + \frac{4}{3}\pi r^3 \rho g}{6\pi\eta r}$$
Substitute numerical values and evaluate v_t:
$$v_t = \left[3\left(1.602 \times 10^{-19}\text{C}\right)\left(6.00 \times 10^4 \frac{\text{N}}{\text{C}}\right)\right.$$
$$\left. + \frac{4}{3}\pi\left(5.50 \times 10^{-7}\text{ m}\right)^3\left(1.05 \times 10^3 \frac{\text{kg}}{\text{m}^3}\right)\left(9.81 \frac{\text{m}}{\text{s}^2}\right)\right]$$
$$\bigg/\left[6\pi\left(1.8 \times 10^{-5}\text{ Pa.s}\right)\left(5.50 \times 10^{-7}\text{ m}\right)\right] = 0.19 \text{ mm/s}$$

30. APPROACH The free body diagram Fig.1.371 shows the forces acting on the microsphere of mass m and having an excess charge of $q = Ne$ when the electric field is downward. Under terminal-speed conditions the sphere is in equilibrium under the influence of the electric force $\vec{F}_e$, its weight $m\vec{g}$, and the drag force $\vec{F}_d$. We can apply Newton's second law, under terminal speed conditions(i.e., dynamic equilibrium), to relate the number of excess charges N on the sphere to its mass and, using Stokes' law, to its terminal speed.

SOLUTION (a) Apply Newton's second law to the microsphere when the electric field is downward:
$$F_e - mg - F_d = ma_y$$
or
$$F_e - mg - F_{d,terminal} = 0 \quad (\because a_y = 0)$$
Substitute for F_e and $F_{d,terminal}$ to get:
$$qE - mg - 6\pi\eta r v_u = 0$$
or
$$NeE - mg - 6\pi\eta r v_u = 0 \quad (\because q = Ne)$$
Solve for v_u to get:
$$v_u = \frac{NeE - mg}{6\pi\eta r} \quad (1)$$
With the field pointing upward, the electric force is downward and the application of Newton's second law to the microsphere yields:
$$F_{d,\,terminal} - F_e - mg = 0$$
or
$$6\pi\eta r v_d - NeE - mg = 0$$
Solve for v_d to obtain:
$$v_d = \frac{NeE + mg}{6\pi\eta r} \quad \cdots (2)$$
Add equations (1) and (2) and simplify to obtain:
$$u = v_u + v_d = \frac{NeE - mg}{6\pi\eta r} + \frac{NeE + mg}{6\pi\eta r} = \frac{NeE}{3\pi\eta r} = \frac{qE}{3\pi\eta r}$$
Measuring both v_u and v_d has the advantage that you don't need to know the mass of the microsphere.

(b) Let Δu represents the change in the terminal speed of the microsphere due to a gain (or loss) of one electron we have:
$$\Delta u = v_{N+1} - v_N$$
Noting that Δu will be the same whether the microsphere is moving upward or downward, express its terminal speed when it is moving upward with N electronic charges on it:
$$v_N = \frac{NeE - mg}{6\pi\eta r}$$
Express its terminal speed upward when it has $N + 1$ electronic charges:
$$v_{N+1} = \frac{(N + 1)eE - mg}{6\pi\eta r}$$
Substitute and simplify to obtain:
$$\Delta u = \frac{(N + 1)eE - mg}{6\pi\eta r} - \frac{NeE - mg}{6\pi\eta r} = \frac{eE}{6\pi\eta r}$$
Substitute numerical values and evaluate Δu:
$$\Delta u = \frac{\left(1.602 \times 10^{-19}\text{C}\right)\left(6.00 \times 10^4 \text{ N/C}\right)}{6\pi\left(1.8 \times 10^{-5}\text{ Pa}\cdot\text{m}\right)\left(5.50 \times 10^{-7}\text{ m}\right)} = 52 \text{ } \mu\text{m/s}$$

31. APPROACH Choose a coordinate system in which the $+x$ direction is perpendicular to the surfaces and points to the right. Let the charge densities on the two plates be σ_1 and σ_2 and denote the three regions of interest as 1, 2, and 3 [Fig.1.372]. Choose a coordinate system in which the positive x direction is to the right. We can apply the equation for $\vec{E}$ near an infinite plane of charge and the superposition of fields to find the field in each of the three regions.

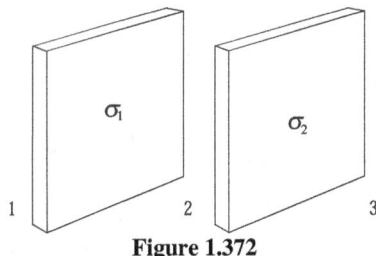

Figure 1.372

SOLUTION (a) Use the equation for $\vec{E}$ near an infinite plane of charge to express the field in region 1 when $\sigma_1 = \sigma_2 = +3.0 \, \mu C/m^2$
$$\vec{E}_1 = \vec{E}_{\sigma_1} + \vec{E}_{\sigma_2} = -2\pi k \sigma_1 \hat{i} - 2\pi k \sigma_2 \hat{i} = -4\pi k \sigma \hat{i}$$
Substitute numerical values and evaluate $\vec{E}_1$:
$$\vec{E}_1 = -4\pi \left(8.988 \times 10^9 \, N \cdot m^2/C^2\right) \left(3.0 \mu C/m^2\right) \hat{i}$$
$$= -\left(3.4 \times 10^5 \, N/C\right) \hat{i}$$
Proceed as above for region 2:
$$\vec{E}_2 = \vec{E}_{\sigma_1} + \vec{E}_{\sigma_2}$$
$$= 2\pi k \sigma_1 \hat{i} - 2\pi k \sigma_2 \hat{i} = 2\pi k \sigma \hat{i} - 2\pi k \sigma \hat{i} = 0$$
Proceed as above for region 3:
$$\vec{E}_3 = \vec{E}_{\sigma_1} + \vec{E}_{\sigma_2} = 2\pi k \sigma_1 \hat{i} + 2\pi k \sigma_2 \hat{i} = 4\pi k \sigma \hat{i}$$
$$= 4\pi \left(8.988 \times 10^9 \frac{N \cdot m^2}{C^2}\right) \left(3.0 \mu C/m^2\right) \hat{i}$$
$$= \left(3.4 \times 10^5 \, N/C\right) \hat{i}$$
(b) Use the equation for E near an infinite plane of charge to express and evaluate the field in region 1 when $\sigma_1 = +3.0 \, \mu C/m^2$ and $\sigma_2 = -3.0 \, \mu C/m^2$:
$$\vec{E}_1 = \vec{E}_{\sigma_1} + \vec{E}_{\sigma_2} = 2\pi k \sigma_1 \hat{i} - 2\pi k \sigma_2 \hat{i}$$
$$= 2\pi k \sigma \hat{i} - 2\pi k \sigma \hat{i} = 0$$
Proceed as above for region 2:
$$\vec{E}_2 = \vec{E}_{\sigma_1} + \vec{E}_{\sigma_2} = 2\pi k \sigma_1 \hat{i} + 2\pi k \sigma_2 \hat{i} = 4\pi k \sigma \hat{i}$$
$$= 4\pi \left(8.988 \times 10^9 \frac{N \cdot m^2}{C^2}\right) \left(3.0 \, \mu C/m^2\right) \hat{i}$$
$$= (3.4 \times 10^5 \, N/C) \hat{i}$$
Proceed as above for region 3:
$$\vec{E}_3 = \vec{E}_{\sigma_1} + \vec{E}_{\sigma_2} = 2\pi k \sigma_1 \hat{i} - 2\pi k \sigma_2 \hat{i}$$
$$= 2\pi k \sigma \hat{i} - 2\pi k \sigma \hat{i} = 0$$
(c) The electric field lines for (a) and (b) are shown in Fig.1.373.

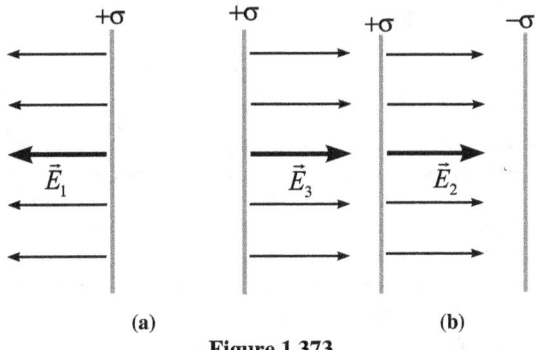

Figure 1.373

32. APPROACH The magnitude of the electric field at an axial distance z from a disk that has a uniform charge density σ and whose radius is R is given by
$$E_z = 2\pi k \sigma \left(1 - \frac{1}{\sqrt{1 + \frac{R^2}{z^2}}}\right)$$

We can equate this expression and $E_z = \frac{1}{2}\sigma/(2\epsilon_0)$ and solve for z.
Express the electric field on the axis of a disk charge:
$$E_z = 2\pi k \sigma \left(1 - \frac{1}{\sqrt{1 + \frac{R^2}{z^2}}}\right)$$
We're given that:
$$E_z = \frac{1}{2}\sigma/(2\epsilon_0) = \frac{\sigma}{4\epsilon_0}$$
Equating these expressions gives:
$$\frac{\sigma}{4\epsilon_0} = 2\pi k \sigma \left(1 - \frac{1}{\sqrt{1 + \frac{R^2}{z^2}}}\right)$$
Substituting for k yields:
$$\frac{\sigma}{4\epsilon_0} = 2\pi \left(\frac{1}{4\pi \epsilon_0}\right) \sigma \left(1 - \frac{1}{\sqrt{1 + \frac{R^2}{z^2}}}\right)$$
Solve for z to obtain: $z = R/\sqrt{3}$

33. The electric field at a distance z from the centre of a ring whose charge is Q and whose radius is a is given by
$$E_z = \frac{kQz}{(z^2 + a^2)^{3/2}}$$
(a) Evaluating $E_{z=0.2a}$ gives
$$E_{z=0.2a} = \frac{kQ(0.2a)}{[(0.2a)^2 + a^2]^{3/2}} = 0.189 \frac{kQ}{a^2}$$
(b) Evaluating $E_{z=0.5a}$ gives:
$$E_{z=0.5a} = \frac{kQ(0.5a)}{[(0.5a)^2 + a^2]^{3/2}} = 0.358 \frac{kQ}{a^2}$$
(c) Evaluating $E_{z=0.7a}$ gives:
$$E_{z=0.7a} = \frac{kQ(0.7a)}{[(0.7a)^2 + a^2]^{3/2}} = 0.385 \frac{kQ}{a^2}$$
(d) Evaluating $E_{z=a}$ gives:
$$E_{z=a} = \frac{kQa}{[a^2 + a^2]^{3/2}} = 0.354 \frac{kQ}{a^2}$$
(e) Evaluating $E_{z=2a}$ gives:
$$E_{z=2a} = \frac{2kQa}{[(2a)^2 + a^2]^{3/2}} = 0.179 \frac{kQ}{a^2}$$
(f) The field along the z axis is plotted in Fig.1.374. The z coordinates are in units of z/a and E is in units of kQ/a^2

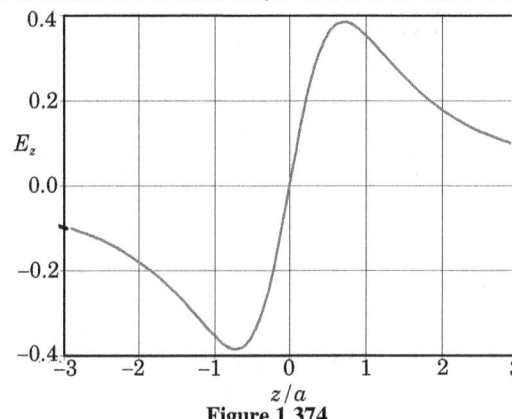

Figure 1.374

34. APPROACH The line charge and point $(0, y)$ are shown in the diagram 1.375. Also shown is a line element of length dx and the field $d\vec{E}$ its charge produces at $(0, y)$. We can find dE_x from $d\vec{E}$ and then integrate from $x = x_1$ to $x = x_2$.
SOLUTION Express the x component of $d\vec{E}$:

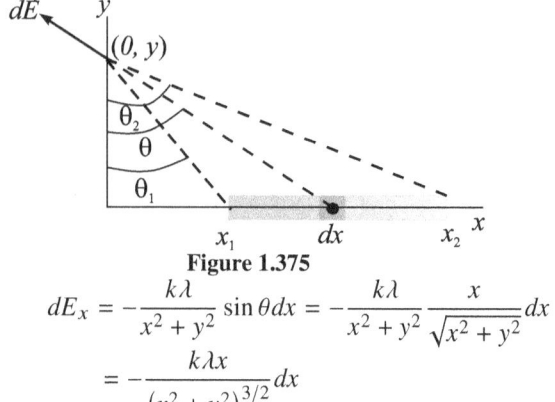

Figure 1.375

$$dE_x = -\frac{k\lambda}{x^2+y^2}\sin\theta dx = -\frac{k\lambda}{x^2+y^2}\frac{x}{\sqrt{x^2+y^2}}dx$$

$$= -\frac{k\lambda x}{(x^2+y^2)^{3/2}}dx$$

Integrate from $x = x_1$ to x_2 and simplify to obtain:

$$E_x = -k\lambda \int_{x_1}^{x_2} \frac{X}{(x^2+y^2)^{3/2}}dx = -k\lambda \left[-\frac{1}{\sqrt{x^2+y^2}}\right]_{x_1}^{x_2}$$

$$= -k\lambda\left[-\frac{1}{\sqrt{x_2^2+y^2}}+\frac{1}{\sqrt{x_1^2+y^2}}\right]$$

$$= -\frac{k\lambda}{y}\left[-\frac{y}{\sqrt{x_2^2+y^2}}+\frac{y}{\sqrt{x_1^2+y^2}}\right]$$

From the diagram we see that:

$$\cos\theta_2 = \frac{y}{\sqrt{x_2^2+y^2}} \quad \text{or} \quad \theta_2 = \tan^{-1}\left(\frac{x_2}{y}\right)$$

and

$$\cos\theta_1 = \frac{y}{\sqrt{x_1^2+y^2}} \quad \text{or} \quad \theta_1 = \tan^{-1}\left(\frac{x_1}{y}\right)$$

Substitute for $\frac{y}{\sqrt{x_2^2+y^2}}$ and $\frac{y}{\sqrt{x_1^2+y^2}}$ to Obtain:

$$E_x = -\frac{k\lambda}{y}[-\cos\theta_2+\cos\theta_1] = \frac{k\lambda}{y}[\cos\theta_2-\cos\theta_1]$$

35. APPROACH Consider the ring with its axis along the z direction shown in the Fig. 1.376. Its radius is $z = r\cos\theta$ and its width is $rd\theta$. We can use the equation for the field on the axis of a ring charge and then integrate to express the field at the center of the hemispherical shell.

SOLUTION Express the field on the axis of the ring of

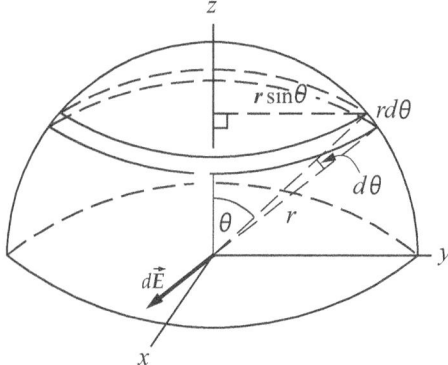

Figure 1.376

charge:

$$dE = \frac{kzdq}{(r^2\sin^2\theta + r^2\cos^2\theta)^{3/2}} = \frac{kzdq}{r^3}$$

(where $z = r\cos\theta$)

Express the charge dq on the ring:

$$dq = \sigma dA = \sigma(2\pi\cdot\sin\theta)rd\theta = 2\pi\sigma^2\sin\theta d\theta$$

Substitute this value of dq in above expression for dE to obtain:

$$dE = \frac{k(r\cos\theta)2\pi\sigma r^2\sin\theta d\theta}{r^3} = 2\pi k\sigma\sin\theta\cos\theta d\theta$$

Integrating dE from $\theta = 0$ to $\pi/2$ yields:

$$E = 2\pi k\sigma\int_0^{\pi/2}\sin\theta\cos\theta d\theta$$

$$= 2\pi k\sigma\left[\frac{1}{2}\sin^2\theta\right]_0^{\pi/2} = \pi k\sigma$$

36. APPROACH Because the cone encloses no charge, we know, from Gauss's law, that the net flux of the electric field through the cone's surface is zero [Fig.1.377]. Thus, the number of field lines penetrating the curved surface of the cone must equal the number of field lines penetrating the base and the entering flux must equal the exiting flux.

SOLUTION The flux penetrating the base of the cone is

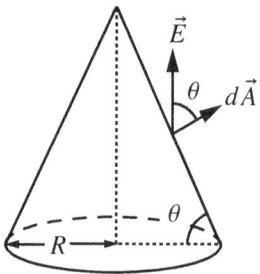

Figure 1.377

given by:

$$\phi_{\text{entering}} = EA_{\text{base}} \quad \cdots (1)$$

The flux penetrating the curved surface of the cone is given by:

$$\phi_{\text{exiting}} = \oint_S \vec{E}\cdot d\vec{A} = \oint_S E\cos\theta dA \quad \cdots (2)$$

Equating the fluxes given by Eq.(1) and (2) and simplifying yields:

$$A_{\text{base}} = \cos\theta\oint_S dA = (\cos\theta)A_{\text{curved surface}}$$

If n represents number of field lines crossing each surface, $N_{\text{curved surface}}$ represents field line density at curved surface, and N_{base} represents field line density at base, then
The ratio of the density of field lines is:

$$\frac{N_{\text{curved surface}}}{N_{\text{base}}} = \frac{n/A_{\text{curved surface}}}{n/A_{\text{base}}} = \frac{A_{base}}{A_{\text{curved surface}}} = \cos\theta$$

37. APPROACH We can find the total charge on the sphere by expressing the charge dq in a spherical shell and integrating this expression between $r = 0$ and $r = R$. By symmetry, the electric fields must be radial. To find E_r inside the charged sphere we choose a spherical Gaussian surface of radius $r < R$. To find E_r outside the charged sphere we choose a spherical Gaussian surface of radius $r > R$. On each of these surfaces, E_r is constant. Gauss's law then relates E_r to the total charge inside the surface.

SOLUTION (a) Express the charge dq in a shell of thickness dr and volume $4\pi r^2 dr$:

$$dq = 4\pi r^2\rho dr = 4\pi r^2(Ar)dr = 4\pi Ar^3 dr$$

Integrate this expression from $r = 0$ to R to find the total charge on the sphere:

$$Q = 4\pi A \int_0^R r^3 dr = \left[\pi A r^4\right]_0^R = \pi A R^4$$

(b) Apply Gauss's law to a spherical surface of radius $r > R$ that is concentric with the nonconducting sphere to obtain:

$$\oint_S E_r dA = \frac{1}{\epsilon_0} Q_{encl} \implies 4\pi r^2 E_r = \frac{Q_{encl}}{\epsilon_0}$$

Solving for E_r yields:

$$E_{r>R} = \frac{Q_{\text{inside}}}{4\pi\epsilon_0}\frac{1}{r^2} = \frac{kQ_{\text{inside}}}{r^2} = \frac{kA\pi R^4}{r^2} = \frac{AR^4}{4\epsilon_0 r^2}$$

Apply Gauss's law to a spherical surface of radius $r < R$ that is concentric with the non-conducting sphere to obtain:

$$\oint_S E_r dA = \frac{1}{\epsilon_0} Q_{\text{inside}} \implies 4\pi r^2 E_r = \frac{Q_{\text{inside}}}{\epsilon_0}$$

Solve for E_r and simplify to obtain:

$$E_{r<R} = \frac{Q_{\text{inside}}}{4\pi r^2 \epsilon_0} = \frac{\pi A r^4}{4\pi r^2 \epsilon_0} = \frac{Ar^2}{4\epsilon_0}$$

(c) E_r versus r/R graph, with E_r in units of $A/(4\varepsilon_0)$, is plotted in Fig.1.378.

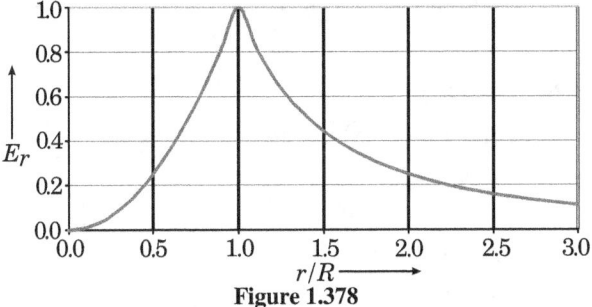

Figure 1.378

Remarks: Note that the results for (a) and (b) agree at $r = R$.

38. APPROACH By symmetry, the electric fields resulting from this charge distribution must be radial. To find E_r for $r < R_1$ we choose a spherical Gaussian surface of radius $r < R_1$. To find E_r for $R_1 < r < R_2$ we choose a spherical Gaussian surface of radius $R_1 < r < R_2$. To find E_r for $r > R_2$ we choose a spherical Gaussian surface of radius $r > R_2$. On each of these surfaces, E_r is constant. Gauss's law then relates E_r to the total charge inside the surface.

SOLUTION (a) The charge in an infinitesimal spherical shell of radius r and thickness dr is:

$$dQ = \rho dV = 4\pi \rho r^2 dr$$

Integrate dQ from $r = R_1$ to R_2 to find the total charge in the spherical shell in the interval $R_1 < r < R_2$:

$$Q_{\text{total}} = 4\pi\rho \int_{R_1}^{R_2} r^2 dr = 4\pi\rho \left[\frac{r^3}{3}\right]_{R_1}^{R_2} = \frac{4\pi\rho}{3}\left(R_2^3 - R_1^3\right)$$

(b) Apply Gauss's law to a spherical surface of radius r that is concentric with the non-conducting spherical shell to obtain:

$$\oint_S E_r dA = \frac{1}{\epsilon_0} Q_{\text{inside}} \implies 4\pi r^2 E_r = \frac{Q_{\text{inside}}}{\epsilon_0}$$

On solving for E_r, we get:

$$E_r = \frac{Q_{\text{inside}}}{4\pi\epsilon_0}\frac{1}{r^2} = \frac{kQ_{\text{inside}}}{r^2}$$

Evaluate $E_{r<R_1}$: $E_{r<R_1} = \frac{Q_{\text{inside}}}{4\pi\epsilon_0}\frac{1}{r^2} = \frac{kQ_{\text{inside}}}{r^2} = 0$

because $\rho_{r<R_1} = 0$ and, therefore, $Q_{\text{inside}} = 0$

Evaluate $E_{R_1<r<R_2}$:

$$E_{R_1<r<R_2} = \frac{kQ_{\text{inside}}}{r^2} = \frac{4\pi k\rho}{3r^2}\left(r^3 - R_1^3\right)$$
$$= \frac{\rho}{3\epsilon_0 r^2}\left(r^3 - R_1^3\right)$$

For $r > R_2$: $\quad Q_{\text{inside}} = \frac{4\pi\rho}{3}\left(R_2^3 - R_1^3\right)$

and $\quad E_{r>R_2} = \frac{4\pi k\rho}{3r^2}\left(R_2^3 - R_1^3\right) = \frac{\rho}{3\epsilon_0 r^2}\left(R_2^3 - R_1^3\right)$

Remark: Note that E is continuous at $r = R_2$.

39. APPROACH From symmetry, the field tangent to the surface of the cylinder must vanish. We can construct a Gaussian surface in the shape of a cylinder of radius r and length L and apply Gauss's law to find the electric field as a function of the distance from the centreline of the infinitely long non-conducting cylinder.

SOLUTION Apply Gauss's law to a cylindrical surface of radius r and length L that is concentric with the infinitely long non-conducting cylinder:

$$\oint_S E_n dA = \frac{1}{\epsilon_0} Q_{\text{inside}}$$

or

$$2\pi r L E_n = \frac{Q_{\text{inside}}}{\epsilon_0}$$

where we've neglected the end areas because there is no flux through them.

Due to symmetry, $E_n = E_r$. Solving for E_r yields:

$$E_r = \frac{Q_{\text{inside}}}{2\pi r L \epsilon_0} = \frac{2kQ_{\text{inside}}}{Lr}$$

Express Q_{inside} for $r < a$:

$$Q_{\text{inside}} = \rho(r)V = \rho_0 \left(\pi r^2 L\right)$$

Substitute to obtain:

$$E_{r<R} = \frac{2k\left(\pi\rho_0 L r^2\right)}{Lr} = \frac{\rho_0}{2\epsilon_0}r$$

since, $\lambda = \rho\pi a^2$, therefore

$$E_{r<R} = \frac{\lambda}{2\pi\epsilon_0 a^2}r$$

Express Q_{inside} for $r > a$:

$$Q_{\text{inside}} = \rho(r)V = \rho_0 \left(\pi a^2 L\right)$$

Substitute for Q_{inside} to obtain:

$$E_{r>a} = \frac{2k\left(\pi\rho_0 L a^2\right)}{Lr} = \frac{\rho_0 a^2}{2\epsilon_0 r}$$

Since, $\lambda = \rho\pi a^2$, therefore: $\quad E_{r>a} = \frac{\lambda}{2\pi\epsilon_0 r}$

40. APPROACH From symmetry; the field tangent to the surfaces of the cylindrical shells must vanish. We can construct a Gaussian surface in the shape of a cylinder of radius r and length L and apply Gauss's law to find the electric field as a function of the distance from the centreline of the infinitely long, uniformly charged cylindrical shells.

SOLUTION (a) Apply Gauss's law to the cylindrical surface of radius r and length L that is concentric with the infinitely long, uniformly charged cylindrical shell:

$$\oint_S E_n dA = \frac{1}{\varepsilon_0} Q_{\text{encl}} \implies 2\pi r L E_n = \frac{Q_{\text{encl}}}{\varepsilon_0}$$

where we've neglected the end areas because there is no flux through them.

Noting that, due to symmetry, $E_n = E_r$, Solve for E_r to obtain:

$$E_r = \frac{2kQ_{\text{encl}}}{Lr} \quad \cdots (1)$$

For $r < R_1, Q_{\text{encl}} = 0$, so: $\quad E_{r<R_1} = 0$

Express Q_{encl} for $a_1 < r < a_2$:

$$Q_{\text{encl}} = \sigma_1 A_1 = 2\pi\sigma_1 a_1 L$$

Substitute in equation (1) to obtain:

$$E_{a_1<r<a_2} = \frac{2k\left(2\pi\sigma_1 a_1 L\right)}{Lr} = \frac{\sigma_1 a_1}{\varepsilon_0 r}$$

(b) Set $E = 0$ for $r > a_2$ to obtain:

$$\frac{\sigma_1 a_1 + \sigma_2 a_2}{\varepsilon_0 r} = 0 \implies \frac{\sigma_2}{\sigma_1} = -\frac{a_1}{a_2}$$

(c) Because the electric field is determined by the enclosed charge within the Gaussian surface, the field under these conditions would be as given above:
$$E_{a_1 < r < a_2} = \sigma_1 a_1 / \varepsilon_0 r$$

41. APPROACH Because the penny is in an external electric field, it will have charges of opposite signs induced on its faces. The induced charge σ is related to the electric field by $E = \sigma \epsilon_0$. Once we know σ, we can use the definition of surface charge density to find the total charge on one face of the penny.

SOLUTION (a) Relate the electric field to the charge density on each face of the penny:
$$E = \frac{\sigma}{\epsilon_0} \implies \sigma = \epsilon_0 E$$

Substitute numerical values and evaluate σ:
$$\sigma = \left(8.854 \times 10^{-12} \text{C}^2/\text{N.m}^2\right)(1.60 \text{ kN/C})$$
$$= 14.17 \text{ nC/m}^2 = 14.2 \text{ nC/m}^2$$

(b) Use the definition of surface charge density to obtain:
$$\sigma = \frac{Q}{A} = \frac{Q}{\pi r^2} \implies Q = \sigma \pi r^2$$

Substitute numerical values and evaluate Q:
$$Q = \pi \left(14.17 \text{ nC/m}^2\right)(0.0100 \text{ m})^2 = 4.45 \text{ pC}$$

42. APPROACH Because the metal slab is in an external electric field, it will have charges of opposite signs induced on its faces. The induced charge σ is related to the electric field by $E = \sigma/\varepsilon_0$.

SOLUTION Relate the magnitude of the electric field to the charge density on the metal slab: $E = \dfrac{\sigma}{\epsilon_0}$

Use its definition to express σ: $\sigma = \dfrac{Q}{A} = \dfrac{Q}{L^2}$

Substitute for σ to obtain: $E = \dfrac{Q}{L^2 \varepsilon_0}$

Substitute numerical values and evaluate E:
$$E = \frac{1.2 \text{nC}}{(0.12 \text{ m})^2 \left(8.854 \times 10^{-12} \text{C}^2/\text{N} \cdot \text{m}^2\right)} = 9.4 \text{kN/C}$$

43. APPROACH From Gauss's law we know that the electric field at the surface of the charged sphere is given by $E = kQ/R^2$ where Q is the charge on the sphere and R is its radius. The minimum radius for dielectric breakdown corresponds to the maximum electric field at the surface of the sphere.

SOLUTION The relationship between E and R for dielectric breakdown:
$$E_{\max} = \frac{kQ}{R_{\min}^2} \implies R_{\min} = \sqrt{\frac{kQ}{E_{\max}}}$$

Substitute numerical values and evaluate $R_{\min}$:
$$R_{\min} = \sqrt{\frac{\left(8.988 \times 10^9 \text{ N} \cdot \text{m}^2/\text{C}^2\right)(18 \mu\text{C})}{3.0 \times 10^6 \text{ N/C}}} = 23 \text{ cm}$$

44. APPROACH Because the atom is uncharged, we know that the integral of the electron's charge distribution over all of space must equal its charge q_e. Evaluation of this integral will lead to an expression for ρ_0. In (b) we can express the resultant electric field at any point as the sum of the electric fields due to the proton and the electron cloud.

SOLUTION (a) Because the atom is uncharged, the integral of the electron's charge distribution over all of space must equal its charge e:
$$e = \int_0^\infty \rho(r) dV = \int_0^\infty \rho(r) 4\pi r^2 dr$$

Substitute for $\rho(r)$ and simplify to obtain:

$$e = -\int_0^\infty \rho_0 e^{-2r/a} 4\pi r^2 dr = -4\pi \rho_0 \int_0^\infty r^2 e^{-2r/a} dr$$

On integrating by parts, we get: $\displaystyle\int_0^\infty r^2 e^{-2r/a} dr = a^3/4$

On substituting for $\displaystyle\int_0^\infty r^2 e^{-2r/a} dr$, we get:
$$e = -4\pi \rho_0 a^3 / 3$$

Solving for ρ_0 yields: $e = -4\pi \rho_0 \left(\dfrac{a^3}{4}\right) = -\pi a^3 \rho_0$

Solving for ρ_0 yields: $\rho_0 = \dfrac{e}{\pi a^3}$

(b) The field will be the sum of the field due to the proton and that of the electron charge cloud:
$$E = E_p + E_{\text{cloud}}$$

Express the field due to the electron cloud:
$$E_{\text{cloud}}(r) = \frac{kQ(r)}{r^2}$$

where $Q(r)$ is the net negative charge enclosed a distance r from the proton. Substitute for E_p and E_{cloud} to obtain:
$$E(r) = \frac{ke}{r^2} + \frac{kQ(r)}{r^2} \qquad \cdots (1)$$

$Q(r)$ is given by:
$$Q(r) = \int_0^r 4\pi r'^2 \rho(r') dr' = 4\pi \int_0^r r'^2 \rho_0 e^{-2r'/a} dr'$$

From Part (a), $\rho_0 = \dfrac{-e}{\pi a^3}$:
$$Q(r) = 4\pi \left(\frac{-e}{\pi a^3}\right) \int_0^r r'^2 e^{-2r'/a} dr'$$
$$= \frac{-4e}{a^3} \int_0^r r'^2 e^{-2r'/a} dr'$$

Using the integration by parts, we get:
$$\int_0^r x^2 e^{-2x/a} dx = \frac{1}{4} e^{-2r/a} a \left[\left(e^{-2r/a} - 1\right) a^2 - 2ar - 2r^2\right]$$
$$= \frac{1}{4} e^{-2r/a} a^3 \left[\left(e^{-2r/a} - 1\right) - 2\frac{r}{a} - 2\frac{r^2}{a^2}\right]$$
$$= \frac{a^3}{4} \left[\left(1 - e^{-2r/a}\right) - 2e^{-2r/a} \left(\frac{r}{a} + \frac{r^2}{a^2}\right)\right]$$

Substituting for $\displaystyle\int_0^r r'^2 e^{-2r'/a} dr'$ in the expression for $Q(r)$ and simplifying yields:
$$Q(r) = \frac{-e}{4} \left[\left(1 - e^{-2r/a}\right) - 2e^{-2r/a} \left(\frac{r}{a} + \frac{r^2}{a^2}\right)\right]$$

Substitute for $Q(r)$ in equation (1) and simplify to obtain:
$$E(r) = \frac{ke}{r^2} - \frac{ke}{4r^2} \left[\left(1 - e^{-2r/a}\right) - 2e^{-2r/a} \left(\frac{r}{a} + \frac{r^2}{a^2}\right)\right]$$
$$= \frac{ke}{r^2} \left[1 - \frac{1}{4}\left(1 - e^{-2r/a}\right) - 2e^{-2r/a}\left(\frac{r}{a} + \frac{r^2}{a^2}\right)\right]$$

45. APPROACH See Fig.1.379. We can view the field in the hole as the sum of the field from a uniform spherical shell of charge Q plus the field due to a small patch with surface charge density equal but opposite to that of the patch cut out.

SOLUTION (a) Express the magnitude of the electric field at the centre of the hole:
$$E = E_{\substack{\text{spherical}\\\text{shell}}} + E_{\text{hole}}$$

Apply Gauss's law to a spherical gaussian surface just outside the given sphere:
$$E_{\substack{\text{spherical}\\\text{shell}}} \left(4\pi r^2\right) = \frac{Q_{\text{enclosed}}}{\varepsilon_0} = \frac{Q}{\varepsilon_0}$$

1.38. ANSWER KEYS AND SOLUTIONS

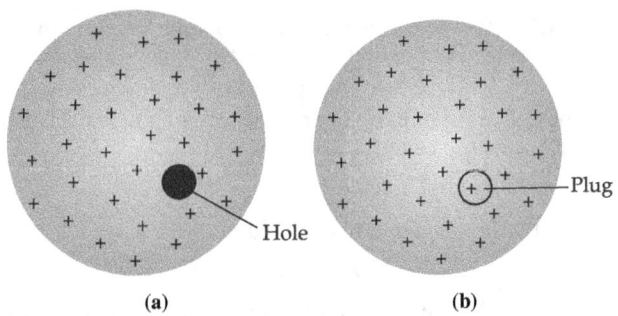

Figure 1.379

Solve for $E_{\text{spherical shell}}$ to obtain:
$$E_{\text{spherical shell}} = \frac{Q}{4\pi\epsilon_0 r^2}$$

The electric field due to the small hole (small enough so that we can treat it as a plane surface) is:
$$E_{\text{hole}} = \frac{-\sigma}{2\varepsilon_0}$$

Substitute for $E_{\text{spherical shell}}$ and E_{hole} and shell simplify to obtain:
$$E = \frac{Q}{4\pi\epsilon_0 r^2} + \frac{-\sigma}{2\varepsilon_0} = \frac{Q}{4\pi\epsilon_0 r^2} - \frac{Q}{2\varepsilon_0 (4\pi r^2)}$$
$$= \frac{Q}{8\pi\epsilon_0 r^2} \quad \text{(radially outward)}$$

(b) Express the force on the patch: $F = qE$
where q is the charge on the patch.
Assuming that the patch has radius a, express the proportion between its charge and that of the spherical shell:
$$\frac{q}{\pi a^2} = \frac{Q}{4\pi r^2} \quad \text{or} \quad q = \frac{a^2}{4r^2}Q$$

Substitute for q and E in the expression for F to obtain:
$$F = \left(\frac{a^2}{4r^2}Q\right)\left(\frac{Q}{8\pi\epsilon_0 r^2}\right) = \frac{Q^2 a^2}{32\pi\epsilon_0 r^4} \quad \text{(radially outward)}$$

(c) The pressure is the force exerted on the patch divided by the area of the patch:
$$P = \frac{\frac{Q^2 a^2}{32\pi \varepsilon_0 r^4}}{\pi a^2} = \frac{Q^2}{32\pi^2 \varepsilon_0 r^4}$$

46. **APPROACH** (a) We can apply Newton's second law to the particle to express its speed as a function of its mass m, charge q, and the radius of its path R, and the strength of the electric field due to the infinite line charge E [Fig.1.380].
(b) The period of the particle's motion is the ratio of the circumference of the circle in which it travels divided by its orbital speed.

SOLUTION (a) Apply Newton's second law to the particle

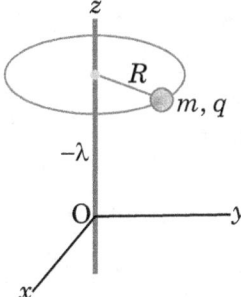

Figure 1.380

to obtain:
$$\sum F_{\text{radial}} = qE = m\frac{v^2}{R}$$
where the inward direction is positive.

Solving for v yields: $v = \sqrt{\frac{qRE}{m}}$

The strength of the electric field at a distance R from the infinite line charge is given by: $E = \frac{2k\lambda}{R}$

Substitute for E and simplify to obtain: $v = \sqrt{\frac{2kq\lambda}{m}}$

(b) The speed of the particle is equal to the circumference of its orbit divided by its period:
$$v = \frac{2\pi R}{T} \implies T = \frac{2\pi R}{v}$$

Substitute for v and simplify to obtain: $T = \pi R\sqrt{\frac{2m}{kq\lambda}}$

47. **APPROACH** In Part (a), you can apply Gauss's law to express $\vec{E}$ as a function of r for the uniformly charged non-conducting sphere with its center at the origin [Fig.1.381]. In Part (b), you can use the hint to express the field at a generic point $P(x, y)$ in the cavity as the sum of the fields due to equal positive and negative charge densities and then evaluate this expression at points 1 and 2.

SOLUTION (a) The electric field at a distance r from the

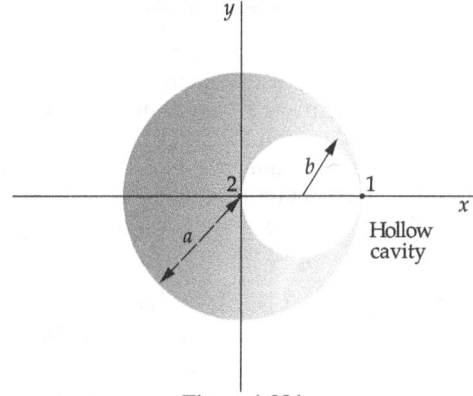

Figure 1.381

centre of the uniformly charged non-conducting sphere is given by:
$$\vec{E}_\rho = E\hat{r} \quad \cdots (1)$$
where $\hat{r}$ is a unit vector pointing radially outward. Apply Gauss's law to a spherical surface of radius r centered at the origin to obtain:
$$\oint_S E_n dA = E_\rho (4\pi r^2) = \frac{Q_{\text{inside}}}{\varepsilon_0}$$

Relate Q_{inside} to the charge density ρ:
$$\rho = \frac{Q_{\text{inside}}}{\frac{4}{3}\pi r^3} \implies Q_{\text{inside}} = \frac{4}{3}\rho\pi r^3$$

Substitute for Q_{inside}: $E_\rho (4\pi r^2) = \frac{4\rho\pi r^3}{3\epsilon_0}$

Solve for E_ρ to obtain: $E_\rho = \frac{\rho r}{3\epsilon_0}$

Substitute for E in equation (1) to obtain:
$$\vec{E}_\rho = \frac{\rho}{3\epsilon_0} r\hat{r}$$

(b) The electric field at point $P(x, y)$ is the sum of the electric fields due to the two charge distributions:
$$\vec{E} = \vec{E}_\rho + \vec{E}_{-\rho} = E_\rho \hat{r} + E_{-\rho} \hat{r}' \quad \cdots (2)$$
where $\hat{r}'$ is a unit vector normal to a spherical Gaussian surface whose center is at $x = b$.
Apply Gauss's law to a spherical surface of radius r' centred

at $x = b = R/2$ to obtain:
$$\oint_S E_n dA = E_{-\rho}\left(4\pi r'^2\right) = \frac{Q_{\text{inside}}}{\varepsilon_0}$$
Relate Q_{inside} to the charge density $-\rho$:
$$-\rho = \frac{Q_{\text{inside}}}{\frac{4}{3}\pi r'^3} \Rightarrow Q_{\text{inside}} = -\frac{4}{3}\rho\pi r'^3$$
Substitute for Q_{inside} to obtain:
$$E_{-\rho}\left(4\pi r'^2\right) = -\frac{4\pi r'^3 \rho}{3\epsilon_0}$$
Solving for $E_{-\rho}$ yields: $E_{-\rho} = -\frac{\rho}{3\epsilon_0}r'$

Substitute for E_ρ and $E_{-\rho}$ in equation (2) to obtain:
$$\vec{E} = \frac{\rho}{3\epsilon_0}r\hat{r} - \frac{\rho}{3\epsilon_0}r'\hat{r}' \qquad \cdots (3)$$
The vectors $\vec{r} = r\hat{r}$ and $\vec{r}' = r'\hat{r}'$ given by:
$$\vec{r} = x\hat{i} + y\hat{j} \quad \text{and} \quad \vec{r}' = (x-b)\hat{i} + y\hat{j}$$
where x and y are the coordinates of any point in the cavity.
Substitute for $r\hat{r}$ and $r'\hat{r}'$ in Eq.(3) and simplify to obtain:
$$\vec{E} = \frac{\rho}{3\epsilon_0}(x\hat{i}+y\hat{j}) - \frac{\rho}{3\epsilon_0}[(x-b)\hat{i}+y\hat{j}] = \frac{\rho b}{3\epsilon_0}\hat{i}$$
Because $\vec{E}$ is independent of x and y:
$$\vec{E}_1 = \vec{E}_2 = \frac{\rho b}{3\epsilon_0}\hat{i}$$

48. APPROACH The electric field in the cavity is the sum of the electric field due to the uniform and positive charge distribution of the sphere whose radius is a and the electric field due to any charge in the spherical cavity whose radius is b. You can use the hint given in Problem 47 to express the field at a generic point $P(x, y)$ in the cavity as the sum of the fields due to equal positive and negative charge densities to show that $\vec{E} = \frac{\rho}{3\varepsilon_0}b\hat{i}$.

SOLUTION The electric field at point $P(x,y)$ is the sum of the electric fields due to the two charge distributions:
$$\vec{E} = \vec{E}_\rho + \vec{E}_{-\rho} = E_\rho\hat{r} + E_{-\rho}\hat{r}' \qquad \cdots (1)$$
where $\hat{r}'$ is a unit vector normal to a spherical Gaussian surface whose centre is at $x = b$.
Apply Gauss's law to a spherical surface of radius r' centred at $x = b = R/2$ to obtain:
$$\oint_S E_n dA = E_{-\rho}\left(4\pi r'^2\right) = \frac{Q_{\text{inside}}}{\epsilon_0}$$
Relate Q_{inside} to the charge density $-\rho$: Substitute for Q_{inside} to obtain:
$$-\rho = \frac{Q_{\text{inside}}}{\frac{4}{3}\pi r'^3} \Rightarrow Q_{\text{inside}} = -\frac{4}{3}\rho\pi r'^3$$
Solving for $E_{-\rho}$ yields:
$$E_{-\rho}\left(4\pi r'^2\right) = -\frac{4\pi r'^3 \rho}{3\epsilon_0}$$
From Problem 47: $E_{-\rho} = -\frac{\rho}{3\epsilon_0}r'$

Substitute for E_ρ and $E_{-\rho}$ in equation (1) to obtain:
$$E_\rho = \frac{\rho}{3\epsilon_0}r \qquad \cdots (2)$$
The vectors $\vec{r} = r\hat{r}$ and $\vec{r}' = r'\hat{r}'$ are given by: $\vec{r} = x\hat{i} + y\hat{j}$ and $\vec{r}' = (x-b)\hat{i} + \hat{j}$, where x and y are the coordinates of any point in the cavity.
Substitute for $r\hat{r}$ and $r'\hat{r}'$ in equation (2) and simplify to obtain:
$$\vec{E} = \frac{\rho}{3\epsilon_0}(x\hat{i}+y\hat{j}) - \frac{\rho}{3\epsilon_0}[(x-b)\hat{i}+y\hat{j}] = \frac{\rho}{3\epsilon_0}b\hat{i}$$

49. APPROACH The electric field at a point $P(x, y)$ in the cavity is the sum of the fields due to the positive charge density and the total charge Q.

SOLUTION The electric field at point $P(x,y)$ is the sum of the electric fields due to the two charge distributions:
$$\vec{E} = \vec{E}_\rho + \vec{E}_{-\rho} + \vec{E}_Q \qquad \cdots (1)$$
where $\hat{r}'$ is a unit vector normal to a spherical Gaussian surface whose center is at $x = b$.
From Problem 47: $\vec{E}_\rho + \vec{E}_{-\rho} = \frac{\rho b}{3\varepsilon_0}\hat{i}$

Substituting in equation (1) yields:
$$\vec{E} = \frac{\rho b}{3\epsilon_0}\hat{i} + \vec{E}_Q$$
Assuming that the cavity is filled with positive charge Q:
$$\vec{E} = \frac{\rho b}{3\epsilon_0}\hat{i} + \frac{Q}{4\pi\varepsilon_0 b^3}r'\hat{r}'$$
The vectors $\vec{r} = r\hat{r}$ and $\vec{r}' = r'\hat{r}'$ are given by:
$$\vec{r} = x\hat{i} + y\hat{j} \quad \text{and} \quad \vec{r}' = (x-b)\hat{i} + y\hat{j}$$
where x and y are the coordinates of any point in the cavity.
Substitute for $r\hat{r}$ and $r'\hat{r}'$ and simplify to obtain:
$$\vec{E} = \frac{\rho b}{3\epsilon_0}\hat{i} + \frac{Q}{4\pi\varepsilon_0 b^3}[(x-b)\hat{i}+\hat{y}]$$
At point 1, $x = 2b$ and $y = 0$:
$$\vec{E}(2b,0) = \frac{\rho b}{3\epsilon_0}\hat{i} - \frac{Q}{4\pi\epsilon_0 b^3}[(2b-b)\hat{i}] = \left(\frac{\rho b}{3\epsilon_0} + \frac{Q}{4\pi\epsilon_0 b^2}\right)\hat{i}$$
At point 2, $x = 0$ and $y = 0$:
$$\vec{E}(0,0) = \frac{\rho b}{3\epsilon_0}\hat{i} + \frac{Q}{4\pi\epsilon_0 b^3}[(-b)\hat{i}] = \left(\frac{\rho b}{3\epsilon_0} - \frac{Q}{4\pi\epsilon_0 b^2}\right)\hat{i}$$

50. APPROACH We can find the field due to the infinitely long line charge from $E = 2k\lambda/r$ and the force that acts on the dipole using $F = p\,dE/dr$.

SOLUTION Express the force acting on the dipole:
$$F = p(dE/dr) \qquad \cdots (1)$$
The electric field at the location of the dipole is given by: $E = 2k\lambda/r$
Substitute for E in equation (1) to obtain:
$$F = p\frac{d}{dr}\left[\frac{2k\lambda}{r}\right] = -\frac{2k\lambda p}{r^2}$$
where the minus sign indicates that the dipole is attracted to the line charge.

1.38.13 Multiple Choice Assignments
1.38.13.1 Level 1

Q.No.	1	2	3	4	5	6	7	8	9
Ans.	D	D	B	A	B	A	D	A	D
Q.No.	10	11	12	13	14	15	16	17	18
Ans.	D	C	A	A	A	A	C	D	A
Q.No.	19	20	21	22	23	24	25	26	27
Ans.	C	A	A	B	B	A	C	A	B
Q.No.	28	29	30	31	32	33	34	35	36
Ans.	D	B	A	A	C	C	B	B	A
Q.No.	37	38	39	40	41	42	43	44	45
Ans.	D	A	C	A	C	B	A	B	C
Q.No.	46	47	48	49	50	51	52	53	54
Ans.	C	B	A	B	B	C	C	D	D
Q.No.	55	56	57	58	59	60	61	62	63
Ans.	C	C	A	A	C	A	C	D	B
Q.No.	64	65	66	67	68	69	70	71	72
Ans.	B	A	B	D	A	A	A	A	D
Q.No.	73	74	75	76	77	78	79	80	81
Ans.	D	C	A	C	D	C	C	B	A

1.38. ANSWER KEYS AND SOLUTIONS

1.38.13.2 Level 2

Q.No.	1	2	3	4	5	6	7	8	9
Ans.	C	D	A	D	B	B	C	B	B
Q.No.	10	11	12	13	14	15	16	17	18
Ans.	A	C	B	D	A	D	A	A	A
Q.No.	19	20	21	22	23	24	25	26	27
Ans.	A	C	A	A	C	D			

1.38.13.3 Level 3

Q.No.	1	2	3	4	5	6	7	8	9
Ans.	B	D	C	A	D	C	C	C	C
Q.No.	10	11	12	13	14	15	16	17	18
Ans.	C	C	D	B	A	B	C	C	D

1.38.13.4 Level 4

Section A

Q.No.	1	2	3	4	5	6	7	8	9
Ans.	A	B	A	D	D	B	A	D	A
Q.No.	10	11	12	13	14	15	16	17	18
Ans.	A	A	D	A	A	C	A	C	B
Q.No.	19	20	21	22	23	24	25	26	27
Ans.	D	C	B	A	B	D	D	A	B
Q.No.	28	29	30	31	32	33	34	35	36
Ans.	A	B	D	D	B	B	B	C	C
Q.No.	37	38	39	40	41	42	43	44	45
Ans.	C	C	D	B	C	B	B	C	B
Q.No.	46	47	48	49	50	51	52	53	54
Ans.	A	−48	D	A	B	C	C	B	A
Q.No.	55	56	57	58	59	60	61	62	63
Ans.	A	D	C	A	B	D	A	A	24
Q.No.	64	65	66	67	68	69	70	71	72
Ans.	C	B	B	12	640	D	2	288	A
Q.No.	73	74	75	76	77	78			
Ans.	A	A	6	B	B	C			

Section B

Q.No.	1	2	3	4	5	6	7	8
Ans.	A	B	D	B	A	C	B	D
Q.No.	9	10	11	12	13	14	15	16
Ans.	C	D	D	A	A	AD	A	D
Q.No.	17	18	19	20	21	22	23	24
Ans.	C	ACD	BD	C	C	A	2	3
Q.No.	25	26	27	28	29	30	31	32
Ans.	6	AB	B	BC	ABC	AC	6.40	3

This page intentionally left blank

Chapter 2
Electric Potential

In the previous chapter, we saw that the electric field vector $\vec{E}$ allows us to understand the exertion of electric force. In this chapter, we will build on the mechanical concepts of work and energy, and define two new scalar quantities: electric potential energy and electric potential, in terms of the work done by the electric field.

One of the advantages of using potential instead of field is that potential is a scalar quantity, while the field is a vector. This makes potential easier to handle in many cases than the electric field. If we know the electric potential of a charge distribution, we can derive the electric field due to that distribution from it. Similarly, potential energy is also a scalar quantity. By knowing the potential energy, we can find the force by taking appropriate derivatives.

2.1 Electric Potential Energy and Potential Difference in Fields

From mechanics, we know that every conservative force field is a potential field, i.e., the term potential energy can be associated with every conservative force field. Like gravitational field, the electrostatic field is also conservative force field, so we can associate the potential energy (U) with it. The electric potential energy U like the gravitational potential energy, is a scalar that depends on both the source and the test object under consideration (for example, test charge). In the case of gravity, the important property of each object is its mass. In the case of electricity, the important property is the charge. Electric potential energy depends on the charge of both the source and the object. Also, like gravity, electric potential energy depends on the configuration of the system.

The motion of a particle with positive charge q_0 in a uniform electric field is analogous to the motion of a particle of mass m in the uniform gravitational field near the earth.

When you move a particle of mass m, against the gravitational field and increase it's height from position A to B (Fig.2.1a), you have to do a positive work W_{ext}. If your external force on the particle, is just equal and opposite to the force due to the gravitational field, then it will move with a constant velocity v (say) and the kinetic energy of the particle will not change. In this case, all the external work (W_{ext}) increases only the gravitational potential energy of the earth-particle system, i.e.,

$$\boxed{W_{ext} = +\Delta U = U_f - U_i \quad (v \to \text{constant})} \quad (2.1)$$

where U_f and U_i are the final and initial potential energies respectively.

Therefore, when an external force moves a particle slowly (i.e., without acceleration), the work done by the force is equal to the change in potential energy. However, it's important to note that during accelerated motion of the particle, kinetic energy also changes. Consequently, in accelerated motion, the work done by the external force is equal to the change in mechanical energy (the sum of potential and kinetic energy), not just the change in potential energy. Therefore, Equation 2.1 is not valid for accelerated motion.

We can also write the change in potential energy in terms of work done by system's net internal conservative force acting on the par-

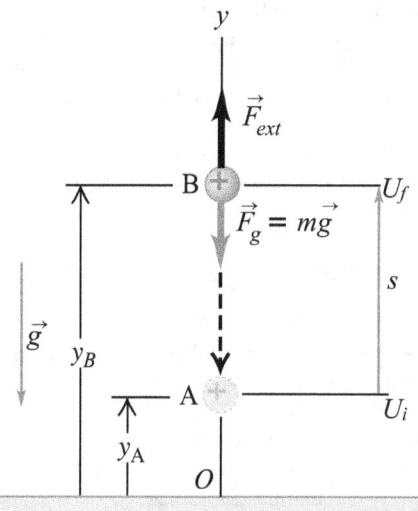

(a) Motion of a point mass in a gravitational field ($\vec{g}$).

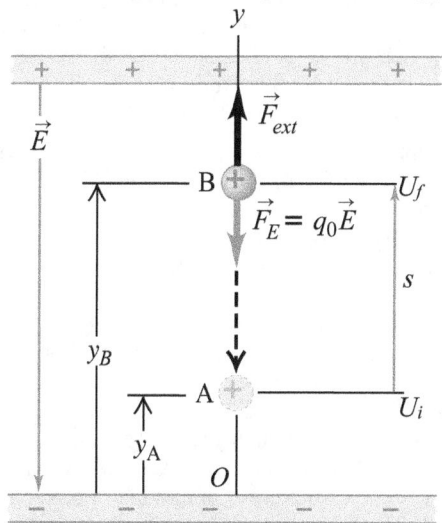

(b) Motion of a test charge in an electric field ($\vec{E}$)

Figure 2.1

ticle. If W_c is the work done by gravitational field (conservative force field), then by work kinetic energy theorem-

$$W_{net} = \underbrace{K_f - K_i}_{\text{Change in KE}}$$

Here, K_i and K_f are respectively the initial and final kinetic energies of the particle.

If W_{ext} and W_c represent work done by non conservative and conservative forces respectively, then net work (W_{net}) can be written as-

$$W_{net} = W_{ext} + W_c$$

and for constant velocity motion, $K_f = K_i$, so-

$$W_{ext} + W_c = 0 \quad \implies \quad W_{ext} = -W_c$$

Now, if we put $W_{ext} = -W_c$ in Eq.2.1, we get-

$$\boxed{\Delta U = U_f - U_i = -W_c} \quad (2.2)$$

Equation 2.2 defines the 'potential energy difference' in terms of the physically meaningful quantity 'work' done by a conservative force (here it is the gravitational force): The change in the system's potential energy is equal to the negative of the work done by the net conservative force acting on the particle. Importantly, Eq.2.2 represents the general definition of the change in potential energy and is valid for both accelerated and non-accelerated motion. This is why we typically define the change in potential energy in terms of the work done by the system's net conservative force, rather than in terms of the work done by an external agent (Eq.2.1), which is valid only for non-accelerated, i.e., slow motion of the particle.

Clearly, potential energy of a system decreases if the net work done by all conservative forces of the system is positive, i.e., $W_c = +ve$. In other words, system does positive work at the cost of it's potential energy.

But if the net work done by all conservative forces of the system is negative, i.e., $W_c = -ve$, then potential energy of the system increases because in this case, $\Delta U = -W_c = +ve$.

Potential energy so defined is undetermined to within an additive constant. What this means is that the actual value of potential energy is not physically significant; it is only the difference of potential energy that is significant. We can always add an arbitrary constant α to potential energy at every point, since this will not change the potential energy difference:

$$\Delta U = (U_f + \alpha) - (U_i + \alpha) = U_f - U_i \quad (2.3)$$

Put it differently, there is a freedom in choosing the point where potential energy is zero. A convenient choice is to have potential energy zero at infinity. With this choice, if we take the initial position i at infinity, we get from Eq.2.3-

$$U_f - U_\infty = U_f - 0 = U_f = -W_{\infty f} \quad (2.4)$$

here, $W_{\infty f}$ represents the work done by the conservative force in moving the particle from its initial position at infinity to the given final position.

Now, if we move a point mass m from the surface of earth to height y, the work done by Earth's gravitational force (a conservative force),

$$W_c = (mg)(y)\cos 180° = -mgy$$

Therefore, from Eq.2.3, change in gravitational potential energy, $\Delta U = -W_c = mgy$.

We can obtain a function that does not depend on m by defining the gravitational potential change as the potential energy change per unit mass: $\Delta V = \Delta U/m = gy$. Clearly, ΔV increases with increase in y, i.e., ΔV increases in moving opposite to the gravitational field. The SI unit of ΔV is J/kg.

Thus, gravitational potential difference can be given as-

$$\Delta V = V_f - V_i = \frac{\Delta U}{m} = -\frac{W_c}{m} \quad (2.5)$$

Clearly, ΔV is +ve corresponding to $-ve$ value of W_c. Thus, potential increases if work done by system's internal conservative force is $-ve$. If the motion of the particle is slow (i.e., non-accelerated), then $W_c = -W_{ext}$, therefore, for slow motion, Eq.2.5 can also be written as

$$\Delta V = V_f - V_i = \frac{\Delta U}{m} = -\frac{W_c}{m} = \frac{W_{ext}}{m} \quad (2.6)$$

If, we take initial position at zero potential level, i.e., $V_i = 0$, potential at final position $V_f = V$ and $m = 1$ kg, then Eq.2.5 gives-

$$\Delta V = V - 0 = -W_c \implies V = -W_c \quad (2.7)$$

If, in the above case, the motion of the particle is slow, then ($W_c = -W_{ext}$). Therefore, for particularly slow motion, the above expression can also be written in terms of external work (W_{ext}) as:

$$\Delta V = V - 0 = W_{ext} \implies V = -W_c = W_{ext} \quad (2.8)$$

From Eqs.2.7 and 2.8, we can define the gravitational potential at a given point in a gravitational field as follows:

The gravitational potential at a point is the negative of the work done by the gravitational field when we lift a unit mass from the zero level of potential to the given position, regardless of whether the motion of the mass is accelerated or non-accelerated. However, in terms of external work, it is the external work needed to lift a unit mass from the zero level of potential to the given position without a change in speed. A useful feature of the potential function is that it depends only on the source of the field (the earth) through the value of the gravitational field strength g, and not on the value of the "test mass", m.

Now, let us discuss the case of electric field instead of gravitational field. Fig.2.1b shows two points A and B in an electric field, and suppose the electric force applied by electric field on a positive test charge q_0 is $\vec{F}_E$ from point B to A. So, if we move this positive test charge infinitely slowly (without any acceleration) from A to B, we do work against the electric field $\vec{E}$.

Clearly, in the case of an electric field, the electric field $\vec{E}$ plays the role of the gravitational field $\vec{g}$, and the test charge q_0 plays the role of the point mass m.

Therefore, in electric field, the Eq.2.2 takes the form: $\Delta U = -W_c$, here W_c is the work done by electric force and similarly, Eq.2.5 takes the form-

$$\boxed{\Delta V = V_f - V_i = \frac{\Delta U}{q_0} = -\frac{W_c}{q_0}} \quad (2.9)$$

Here, ΔV is the electric potential difference between two positions in electric field, V_i is the electric potential at initial position, V_f is the electric potential at final position, W_c is the work done by electric field during this process.

Thus, when the test charge q_0 moves from one position to another in an electric field, the change in electric potential, ΔV, can be defined as the negative of work done by the external agent per unit charge or as the change in electric potential energy per unit charge.

From Eq.2.9, it is evident that potential increases corresponding to $-ve$ value of W_c.

If an external agent moves the test charge q_0 slowly in an electric field from one position to another and performs work W_{ext} on it during this process, then $W_c = -W_{ext}$. In this case Eq.2.9 can be written as-

$$\boxed{\Delta V = V_f - V_i = \frac{\Delta U}{q_0} = \frac{W_{ext}}{q_0}} \quad (v \to \text{constant}) \quad (2.10)$$

Here, it should be remembered that Eq.2.10 is valid for slow motion (i.e., for motion without acceleration) only.

From Fig. 2.1b, we observe that W_{ext} and hence ΔU are positive when q_0 moves against the electric field. Therefore, ΔV is also positive (i.e., the potential increases) when moving against the electric field. Conversely, if the test charge q_0 moves in the direction of the electric field, W_{ext} and hence ΔU are negative. Therefore, ΔV is also negative, signifying a decrease in electric potential when moving in the direction of the electric field.

The SI unit of electric potential is the volt (V), in honour of Alessandro Volta, inventor of the voltaic pile (the first primitive electric battery). Note that

$$\boxed{1\,\text{V} = 1\text{J/C}} \quad (2.11)$$

Definition of 1 volt: The potential difference between two points A and B is one volt if the work done by an external agent in moving one coulomb of positive charge slowly from A to B, is

2.1. ELECTRIC POTENTIAL ENERGY AND POTENTIAL DIFFERENCE IN FIELDS

one joule.

The quantity ΔV depends only on the field set up by the source charges, not on the test charge. Once the potential difference between two points is known, the external work needed to move a charge q_0, with no change in its speed, may be found from Eq.2.10 as:

$$\boxed{W_{ext} = \Delta U = q_0 \Delta V = q_0 (V_f - V_i) \quad (v \to \text{constant})}$$
(2.12)

The sign of this work depends on the sign of q_0 and the relative magnitudes of V_i and V_f. If $W_{ext} > 0$ [or $W_c < 0$], the positive test charge (q_0) is moving in the direction of force applied by external agent, i.e., opposite to the electric field.

Also, $W_{ext} > 0$ [or $W_c < 0$] $\implies q_0 (V_f - V_i) > 0$

Clearly, for +ve test charge (q_0), $V_f - V_i > 0 \implies V_f > V_i$.

i.e., corresponding to $W_{ext} > 0$ [or $W_c < 0$], the external force F_{ext} moves the positive test charge from lower to higher potential point.

And corresponding to $-$ ve test charge, for example, $q_0 = -q_0'$ (say), where, q_0' is a positive value,

$W_{ext} > 0 \implies q_0 (V_f - V_i) > 0 \implies -q_0' (V_f - V_i) > 0$
$\implies q_0' (V_f - V_i) < 0 \implies V_f - V_i < 0 \implies V_f < V_i$

i.e., corresponding to $W_{ext} > 0$, the external force F_{ext} moves the negative charge from higher to lower potential point.

If $W_{ext} < 0$ [or $W_c > 0$], the positive test charge q_0 is moving opposite to the force applied by external agent, i.e. the test charge q_0 is moving in the direction of electric field.

Also, $W_{ext} < 0 [\text{or} W_c > 0] \implies q_0 (V_f - V_i) < 0$

Therefore, corresponding to $q_0 > 0$, we have
$V_f - V_i < 0 \implies V_f < V_i$.

i.e., corresponding to negative external work or positive electric work, the external force moves the positive test charge from higher to lower potential point.

Corresponding to the negative test charge, as in the previous part, we can show that $V_f > V_i$, meaning the external force moves the negative test charge from a lower to a higher potential point.

From Eq.2.12, we see that only changes in potential, rather than the specific value of V_i and V_f, are significant. One can choose the reference point at which the potential is zero at some convenient point such as infinity. In electronic circuits it is convenient to choose the ground connection to earth as the zero of potential. If $V_i = 0$ and $V_f = V$, then Eq.2.9 gives-

$$V - 0 = -\frac{W_c}{q_0} \implies V = -\frac{W_c}{q_0}$$

> Thus, the electric potential at a point in an electric field is the negative of the work done by the electric field to bring a unit positive test charge from the reference point to that point.
> (Definition of Electric Potential)

If allowed to, positive charges tend to move "downhill" in potential in an electric field, just as do ordinary masses in a gravitational field. However, negative free charges tend to move "uphill" in potential. Alternatively, we can say that in an external electric field, both positive and negative charges move to decrease the electrostatic potential energy of the system.

From Eq.2.2, the potential energy difference, in terms of the work done (W_c) by the conservative force, is-

$$\Delta U = U_f - U_i = -W_c$$

The negative sign tells us that positive work done by the conservative force leads to a decrease in potential energy.

Now, consider a test charge q_0 moving in an external electric field created by a point source charge q. A and B are initial and final positions of q_0. Suppose, at any instant it's position is at P at distance r from q and the electric field at P is $\vec{E}$. At P, the conservative force (here it is electric force) on the test charge q_0, is $F_c = F_E = q_0 E$.

If dW_E is the work done by electric field in infinitesimal displacement $d\vec{s}$ of the charged particle q_0, then

$$dW_E = \vec{F}_e \cdot d\vec{s} = q_0 \vec{E} \cdot d\vec{s}$$

Therefore, the infinitesimal change in electric potential energy dU, associated with an infinitesimal displacement $d\vec{s}$, is-

$$dU = -dW_E = -q_0 \vec{E} \cdot d\vec{s}$$

So, the infinitesimal change in electric potential in displacement $d\vec{s}$, is -

$$dV = \frac{dU}{q_0} = -\vec{E} \cdot d\vec{s} \quad (2.13)$$

If V_A and V_B are electric potentials at A and B, the potential change from A to B is

$$\boxed{\Delta V = V_B - V_A = -\int_A^B \vec{E} \cdot d\vec{s}} \quad (2.14)$$

If θ is the angle between $\vec{E}$ and $d\vec{s}$, then

$$\vec{E} \cdot d\vec{s} = E ds \cos\theta = (E \cos\theta) ds = E_s ds$$

here, $E_s = E \cos\theta$ is the component of $\vec{E}$ in the direction of $d\vec{s}$.

Substituting this value in Eq.2.13, we get

$$\boxed{\Delta V = V_B - V_A = -\int_A^B \vec{E} \cdot d\vec{s} = -\int_A^B E_s ds} \quad (2.15)$$

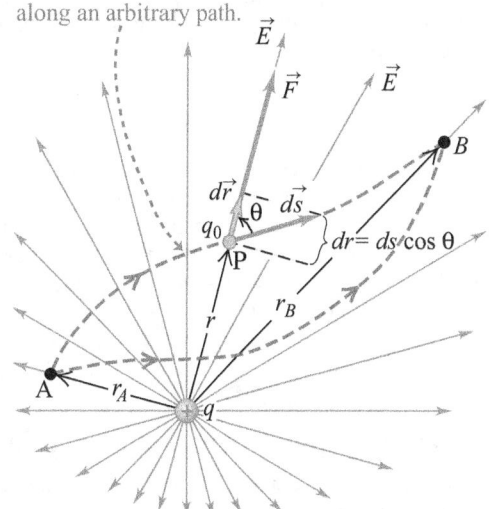

Figure 2.2: The work done on charge q_0 by the electric field of charge q does not depend on the path taken, but only on the distances r_A and r_B.

Thus, only the component of $\vec{E}$ along or opposite to the displacement is significant.

Since, the electrostatic field is a conservative field, the value of line integral $\int_A^B \vec{E} \cdot d\vec{s}$ depends only on the end points A and B, not on the path taken. The sign of the integral is determined (1) by the signs of the components of E, i.e., $E \cos\theta$ and (2) by the direction of the path taken-which is indicated by the limits.

If dr is the magnitude of projection of $d\vec{s}$ along $\vec{r}$, then from Fig. 2.2, we have-

$$ds \cos\theta = dr$$

Therefore, if $\vec{r}_A$ and $\vec{r}_B$ are the position vectors of points A and B respectively, then Eq.(2.14) can also be written as-

$$\Delta V = V_B - V_A = -\int_A^B E(ds\cos\theta) = -\int_{r_A}^{r_B} E\,dr \quad (2.16)$$

and change in electric potential energy-

$$\Delta U = U_B - U_A = q_0(V_B - V_A) = -q_0\int_{r_A}^{r_B} E\,dr \quad (2.17)$$

From Eq.2.17, we see that if q_0 is positive, then ΔU is negative. It shows that, **a system loses it's electric potential energy when the test charge moves in the direction of the field**. This means that an electric field does work on a positive charge when the charge moves in the direction of the electric field. (This is analogous to the work done by the gravitational field on a falling object, as shown in Figure 2.1a) If a positive test charge is released from rest in this electric field, it experiences an electric force $q_0\vec{E}$ in the direction of $\vec{E}$ (downwards in Fig.2.1b). Therefore, it accelerates downward, gaining kinetic energy. **As the charged particle gains kinetic energy, the charge in electric field loses an equal amount of potential energy**. This is accordance with the conservation of energy in an isolated system.

If q_0 is negative, then ΔU in Eq.(2.17) is positive and the situation is reversed: **A negative charge in an external electric field gains electric potential energy when the charge moves in the direction of the field.** If a negative charge is released from rest in an electric field, it accelerates in a direction opposite the direction of the field. In order for the negative charge to move in the direction of the field, an external agent must apply a force and do positive work on the charge.

☞ Since, ΔV and ΔU are path independent, therefore we can apply equations 2.16 and 2.17 without bothering about actual path of motion of test charge q_0.

Remarks

- From Eq.2.16, we have $\Delta V = V_B - V_A = -\int_{r_A}^{r_B} E\,dr$. Thus, it is evident that the negative area under the E vs r graph on the r-axis corresponds to the potential difference.
- CGS unit of electric potential is esu potential. At any point within the electric field, it is equal to the work done by external agent in bringing 1 esu of charge slowly (without any acceleration) from infinity to that point.

$$1 \text{ esu potential} = \frac{1 \text{ erg}}{1 \text{ esu charge}} = \frac{10^{-7} \text{ J}}{3.335641 \times 10^{-10} \text{C}}$$
$$= 299.79245 \text{ volt} \approx 300 \text{ volt}$$

2.1.1 Potential Difference in a Uniform Field

In a uniform field, $\vec{E}$ is constant, and therefore the integral in Eq.2.14 may be written as

$$\int \vec{E}\cdot d\vec{s} = \vec{E}\cdot\int d\vec{s} = \vec{E}\cdot\Delta\vec{s}$$

The finite change in potential ΔV associated with a finite displacement $\Delta\vec{s}$ takes the form

$$\boxed{\Delta V = -\vec{E}\cdot\Delta\vec{s} \qquad (\text{Uniform }\vec{E})} \quad (2.18)$$

Note that Δs and ΔV depend only on the initial and final positions, not on the path taken.

Figure 2.3a shows a uniform field $\vec{E} = E\hat{i}$. Let us find the change in potential in going from point A to point B, which are separated by a distance Δx along the lines. Since, from Fig.2.3a, $\vec{E} = E\hat{i}$ and $\Delta\vec{s} = \Delta x\hat{i} + \Delta y\hat{j}$, therefore,

$$\vec{E}\cdot\Delta\vec{s} = E\hat{i}\cdot(\Delta x\hat{i} + \Delta y\hat{j}) = E\Delta x$$

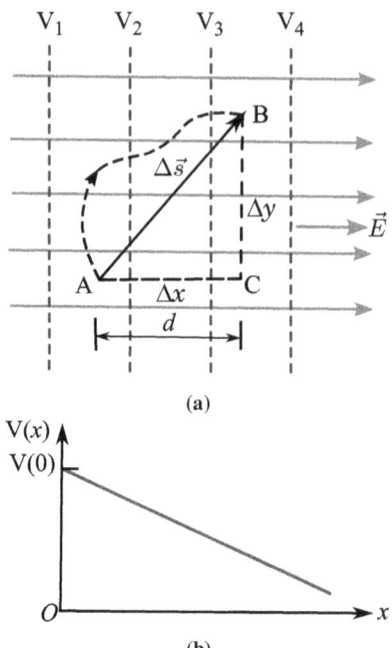

Figure 2.3

Putting this value in Eq.2.18, we get

$$\Delta V = -E\Delta x \qquad (\text{Uniform }\vec{E}) \quad (2.19)$$

If we write $\Delta x = +x$ in Eq.2.19, we get

$$\Delta V = V(x) - V(0) = -Ex \quad (2.20)$$

here, V_x is the potential at B and $V(0)$ is the potential at reference point A.

Thus, the potential decreases linearly along the x axis, as depicted in the graph of Fig. 2.3b. Notice that the field lines point from high potential to low potential.

Again from Fig.2.3a,

$$\Delta\vec{s} = \overrightarrow{AB} = \overrightarrow{AC} + \overrightarrow{CB}$$

Since, $\vec{E}$ is perpendicular to the displacement along $\overrightarrow{CB}$, no work will be done by field $\vec{E}$ on a test charge along this segment. Work is done only along the segment AC parallel to the field lines. Since only the component of the displacement along, or against, the field lines is significant, Eq.2.20 is often written in the form:

$$\Delta V = \pm Ed \qquad (\text{Uniform }\vec{E}) \quad (2.21)$$

where d is the magnitude of the component of the displacement along, or against, the field. The positive sign applies to a displacement opposite to the field. Thus, once again we see that V decreases in the direction of $\vec{E}$ and increases opposite to $\vec{E}$. Therefore, we can say that electric field is always directed from higher to lower potential.

From Eq.2.21 we see that an equivalent unit for electric field is V/m: $1 \text{V/m} = 1 \text{N/C}$

EXAMPLE 1. A point charge q moves from point P to point S along the path PQRS (Fig.2.5) in a uniform electric field $\vec{E}$ pointing parallel to the positive direction of the x-axis. The coordinates of points P, Q, R and S are $(a, b, 0), (2a, 0, 0), (a, -b, 0)$, $(0, 0, 0)$ respectively. Find the work done by the field in the above process.

APPROACH The work done by the conservative forces (electrostatic, gravitational, etc.) is independent of the path i.e., it depends only on the initial and final points. Thus, the work done by the field along path $P \to Q \to R \to S$ is same as the work done along the path $P \to T \to S$ as shown in Fig.2.6.

SOLUTION $W_{el}(\text{along PQRS}) = W_{el}(\text{along PTS})$

2.1. ELECTRIC POTENTIAL ENERGY AND POTENTIAL DIFFERENCE IN FIELDS

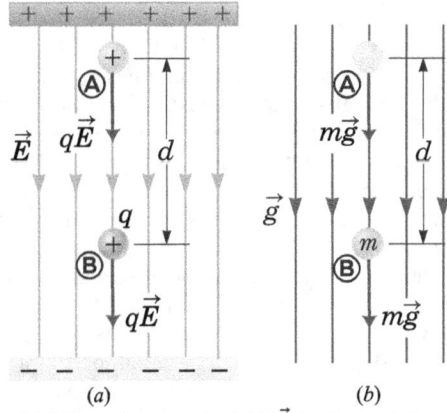

Figure 2.4: (a) When the electric field $\vec{E}$ is directed downward, point B is at a lower electric potential than point A. On a positive charge q electric force $q\vec{E}$ is acting in the direction of electric field $\vec{E}$, i.e., in the direction of lower electric potential. When a positive test charge moves from point A to point B, the charge configuration loses electric potential energy. (b) When an object of mass m moves downward in the direction of the gravitational field $\vec{g}$, the object-source system loses gravitational potential energy.

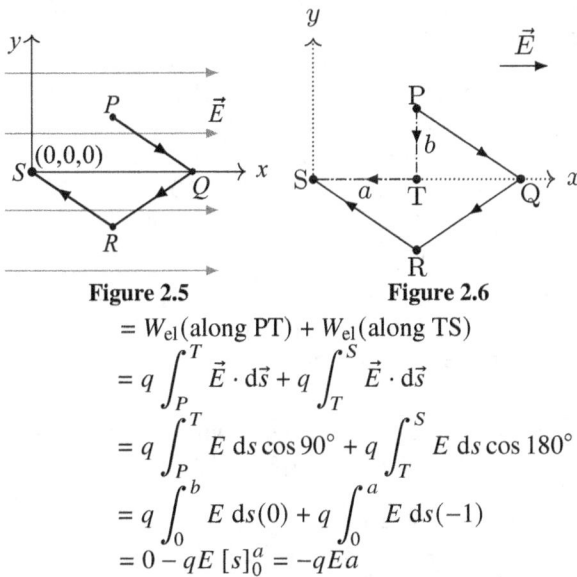

Figure 2.5   **Figure 2.6**

$$= W_{el}(\text{along PT}) + W_{el}(\text{along TS})$$
$$= q\int_P^T \vec{E}\cdot d\vec{s} + q\int_T^S \vec{E}\cdot d\vec{s}$$
$$= q\int_P^T E\,ds\cos 90° + q\int_T^S E\,ds\cos 180°$$
$$= q\int_0^b E\,ds(0) + q\int_0^a E\,ds(-1)$$
$$= 0 - qE\,[s]_0^a = -qEa$$

Aliter 1. The electric force on the point charge is $\vec{F} = qE\hat{\imath}$ and its displacement is
$$\vec{s} = \vec{r}_S - \vec{r}_P = 0 - (a\hat{\imath} + b\hat{\jmath}) = -a\hat{\imath} - b\hat{\jmath}.$$
Thus, $W_{el} = \vec{F}\cdot\vec{s} = (qE\hat{\imath})\cdot(-a\hat{\imath} - b\hat{\jmath}) = -qEa$.

Aliter 2. By using the definition of electric potential energy: Work done by electric field is equal to negative of change of electric potential energy. i.e.,
$$W_{el} = -\Delta U = -\left[-q\int_P^S \vec{E}\cdot d\vec{s}\right] = q\int_P^S \vec{E}\cdot d\vec{s}$$
Here, $\vec{E} = E\hat{\imath}$, and $d\vec{s} = dx\hat{\imath} + dy\hat{\jmath} + dz\hat{k}$
Therefore, $\vec{E}\cdot d\vec{s} = E\,dx$. So,
$$W_{el} = q\int_{-a}^0 E\,dx = -qEa$$

2.1.2 Electrostatic Potential Energy Difference and Potential Energy of a System of Two Point Charges

Let us consider the field $\vec{E}$ due to a point charge q placed at the origin. Now, imagine that we bring a test charge q_0 from a point A to a point B against the repulsive force on it due to the charge q. With reference to Fig. 2.7, this will happen if q and q_0 are both positive or both negative. For definiteness, let us take $q, q_0 > 0$. Two remarks may be made here. First, we

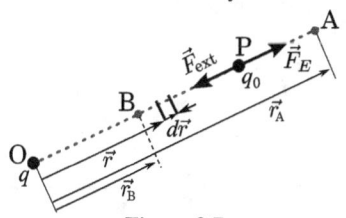

Figure 2.7

assume that the test charge q_0 is so small that it does not disturb the original configuration, namely the charge q at the origin (or else, we keep q fixed at the origin by some unspecified force). Second, in bringing the charge q_0 from A to B, we apply an external force F_{ext} just enough to counter the repulsive electric force F_E (i.e, $F_{ext} = -F_E$). This means there is no net force on or acceleration of charge q_0 when it is brought from A to B, meaning it is brought with infinitesimally slow constant speed. In this situation, the work done by the external force is the negative of the work done by the electric force, and it is fully stored as electrostatic potential energy of the charge q and q_0 system. If the external force is removed upon reaching B, the electric force will take the charge away from q. The stored energy (potential energy) for position B is used to provide kinetic energy to the test charge q_0 in such a way that the sum of the kinetic and potential energies is conserved.

☞ Electric potential energy of a system, containing the source (or sources) of electric field and the test charge q_0 at a given point within the electric field, is the work done by the external force (or −ve of work done by electric force) in bringing the test charge q_0 infinitely slowly[a] from infinity to that point.

[a]In the context of electric potential energy, "infinitely slowly" means moving charges very slowly. This is done to ensure that there is no acceleration, preventing any sudden changes in speed. The reason for this careful movement is to maintain forces in equilibrium, meaning a balance in the forces acting on the charges.

Now, from Eq.2.17, the change in potential energy when test charge q_0 moves from A to B is given as-

$$\Delta U = U_B - U_A = -q_0\int_{r_A}^{r_B} E\,dr \qquad (2.22)$$

Since, the magnitude of electric field at P, due to point charge q at O is given by-
$$E = k\frac{q}{r^2}, \quad \text{with} \quad k = \frac{1}{4\pi\epsilon_0}$$
Therefore, Eq.2.22, gives-
$$\Delta U = -q_0\int_{r_A}^{r_B} k\frac{q}{r^2}dr = -kqq_0\int_{r_A}^{r_B}\frac{dr}{r^2}$$
$$= -kqq_0\left[\frac{-1}{r}\right]_{r_A}^{r_B} = kqq_0\left(\frac{1}{r_B} - \frac{1}{r_A}\right)$$
$$\implies \Delta U = -kqq_0\left[\frac{-1}{r}\right]_{r_A}^{r_B} = kqq_0\left(\frac{1}{r_B} - \frac{1}{r_A}\right) \qquad (2.23)$$

here, $\vec{r}_A$ and $\vec{r}_B$ are the position vectors of points A and B, respectively, with respect to charge q at origin O.

Eq.2.23 shows that, the electric potential energy difference between two points A and B is path-independent and only depends on the positions of A and B. For example, consider two charges

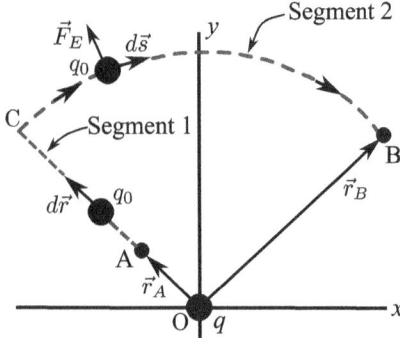

Figure 2.8: The two points A and B are not along the same radius. The path is taken to run radially outward from A to C then along circular arc from C to B. Position vectors of A and B with respect to O are $\vec{r}_A$ and $\vec{r}_B$ respectively.

q and q_0 as shown in Fig. 2.8. The source charge q is fixed at origin O and the test charge is initially at A having position vector $\overrightarrow{OA} = \vec{r}_A$. B is another point with position vector $\overrightarrow{OB} = \vec{r}_B$. The change in potential energy is still given by Eq.2.23, whether q_0 follows the dashed path or moves directly from point A to B.

Explanation: For segment 1 (Fig.2.8), which runs radially outward from A ($OA = r_A$) to C ($OC = r_B$), the result is identical to Eq.2.23. For segment 2, which follows a circumference at a distance r_B from the origin, the integral is zero because the electric force $\vec{F}_E$ is perpendicular to the path segment $d\vec{s}$ everywhere. So, the result for the change in potential energy-

$$\Delta U = \Delta U_{AC} + \Delta U_{CB} = \Delta U_{AC}$$
$$= kqq_0 \left(\frac{1}{r_B} - \frac{1}{r_A} \right)$$

It is the same result as given by Eq.2.23.

Note: Although we have shown the result for circular segment 2, the result holds for paths of any shape like one shown in Fig.2.2. Eq.2.23 shows that the change in electric potential energy is given by the difference of two functions, $U(r_B)$ and $U(r_A)$. We can therefore choose the zero of the potential energy function to be at whatever value of r we like. It is convenient and natural to choose zero potential energy to be at infinity. We can do this if we let $r_A \to \infty$ and let r_B take on a general value r in Eq.2.23:

$$\Delta U = U(r) - U(r_A)|_{r_A \to \infty} = k\frac{qq_0}{r}$$

We then say that the potential energy of a system having charge q_0 at a distance r from charge q is the difference in potential energy between that point and infinity. When we reverse the roles of q and q_0, the potential energy of q at a distance r from q_0 is again kqq_0/r. We can then say that the electric potential energy $U(r)$ for a system of two point charges q and q_0 separated by a distance r is-

$$\boxed{U(r) = k\frac{qq_0}{r}} \qquad (2.24)$$

It is indeed true that $U(r) = 0$ in the limit $r \to \infty$. Thus, the system has no potential energy when the two charges are infinitely far apart. Note that the potential energy of the two charges depends only on the distance r between them and on the magnitudes and signs of the charges.

Remarks

1. From Eq.2.24, we see that the electrical potential energy of system of two point charges is positive if the two charges are of the same type, either positive or negative, and negative if the two charges are of opposite types. This makes sense if you think of the change in the potential energy U as you bring the two charges closer or move them farther apart. Depending on the relative types of charges, you may have to work on the system or the system would do work on you, that is, your work is either positive or negative. If you have to do positive work on the system (actually push the charges closer), then the energy of the system should increase. If you bring two positive charges or two negative charges closer, you have to do positive work on the system, which raises their potential energy. Since potential energy is proportional to $1/r$, the potential energy goes up when r goes down between two positive or two negative charges.

   On the other hand, if you bring a positive and a negative charge nearer, you have to do negative work on the system (the charges are pulling you), which means that you take energy away from the system. This reduces the potential energy of the system. Since potential energy is negative in the case of a positive and a negative charge pair, the decrease in r (or increase in $1/r$) makes the potential energy more negative, which is the same as a reduction in potential energy.

2. Suppose, an electron with charge $q = e = 1.6 \times 10^{-19}$C is released in an electric field $\vec{E}$. This electric field applies an electric force $q\vec{E}$ i.e., $-e\vec{E}$ on it. Here, negative sign shows that electric force on electron is opposite to field $\vec{E}$. As a result, electron get accelerated opposite to $\vec{E}$ i.e., from lower potential point A (say) to higher potential point B (say). If V_A and V_B are potentials at A and B respectively and K_A and K_B are kinetic energies of electron at A and B respectively, then by work kinetic energy theorem, we have

   Work done by electric field = change in KE
   $$\Rightarrow \quad W_{el} = K_B - K_A = K_B - 0 = K_B$$

   Since, the work done by a conservative force field $\vec{E}$ is equal to negative of change of potential energy, therefore
   $$-\Delta U = K_B \quad \Rightarrow \quad -(-e)(V_B - V_A) = K_B$$
   If we write $V_B - V_A = \Delta V$, then
   $$e\Delta V = K_B \quad \Rightarrow \quad K_B = e\Delta V = 1.6 \times 10^{-19}\Delta V$$
   If we set $\Delta V = 1$ volt, then kinetic energy gained by electron
   $$K_B = e\Delta V = 1.6 \times 10^{-19} \text{J}$$
   This unit of energy is defined as 1 electron volt or 1 eV, i.e., 1 eV = 1.6×10^{-19}J.

   > The units based on eV are most commonly used in atomic, nuclear and particle physics. 1 keV = 10^3eV = 1.6×10^{-16} J,
   > 1 MeV = 10^6eV = 1.6×10^{-13} J,
   > 1 GeV = 10^9eV = 1.6×10^{-10} J and
   > 1 TeV = 10^{12}eV = 1.6×10^{-7} J

EXAMPLE 2. $\vec{E} = (100 \text{ V/m})\hat{i} - (50 \text{ V/m})\hat{j}$. Calculate potential difference between $(0,0)$ and $(3,4)$.

APPROACH From Eq.2.15, electric potential difference, between two points A and B, in terms of electric field, is given by-

$$\Delta V = V_B - V_A = -\int_A^B \vec{E} \cdot d\vec{s}$$

with $d\vec{s} = dx\hat{i} + dy\hat{j} + dz\hat{k}$.

Now, substitute the given values and integrate for given limits.

SOLUTION Given that $\vec{E} = (100\text{V/m})\hat{i} - (50\text{V/m})\hat{j}$, therefore-

$$\Delta V = -\int_{(0,0)}^{(3,4)} \vec{E} \cdot d\vec{s}$$
$$= -\int_{(0,0)}^{(3,4)} (100\hat{i} - 50\hat{j}) \cdot (dx\hat{i} + dy\hat{j} + dz\hat{k})$$
$$= -\int_0^3 100 dx + \int_0^4 50 dy = -100[x]_0^3 + 50[y]_0^4$$
$$= -100[3 - 0] + 50[4 - 0] = -100 \text{ V}$$

EXAMPLE 3. $\vec{E} = (10x\hat{i} - 30y^2\hat{j})$. Calculate potential difference between $(0,0)$ and $(3,4)$.

APPROACH From Eq.2.15, electric potential difference, between two points A and B, in terms of electric field, is given by-

$$\Delta V = V_B - V_A = -\int_A^B \vec{E}\cdot d\vec{s}$$

with $d\vec{s} = dx\hat{i} + dy\hat{j} + dz\hat{k}$.

Now, substitute the given values and integrate for given limits.

SOLUTION Given that $\vec{E} = 10x\hat{i} - 30y^2\hat{j}$, therefore-

$$\Delta V = -\int_{(0,0)}^{(3,4)} \vec{E}\cdot d\vec{s}$$

$$= -\int_{(0,0)}^{(3,4)} (10x\hat{i} - 30y^2\hat{j})\cdot(dx\hat{i} + dy\hat{j} + dz\hat{k})$$

$$= -\int_0^3 10x\,dx + \int_0^4 30y^2\,dy$$

$$= -10\left[\frac{x^2}{2}\right]_0^3 + 30\left[\frac{y^3}{3}\right]_0^4 = -10\left[\frac{3^2}{2}\right] + 30\left[\frac{4^3}{3}\right]$$

$$= -10\left[\frac{9}{2}\right] + 30\left[\frac{64}{3}\right] = -45 + 640 = 595\text{ V}$$

EXAMPLE 4. How much work is done by the electrical force when a point charge is brought from infinity to rest at a distance r from a fixed charge of the opposite sign? What is the meaning of the sign of your result?

SOLUTION Since, the charges are of opposite nature, therefore, the electric force between them is attractive and hence it is in the direction of displacement. So, if we bring a point charge from infinity to rest at a distance r from a fixed charge of the opposite sign with zero acceleration, the work done by electric force is positive.

Since, change in electric potential energy (ΔU) of the system is defined as the negative of work done by the electric force, so, it will be negative and hence system losses it's electric potential energy by doing positive work.

Calculation of Work done by Electric Force

Work done by electric force $W_{el} = -\Delta U$

$$\Rightarrow W_{el} = -(U_r - U_\infty) = -\left(k\frac{qq_0}{r} - 0\right) = -k\frac{qq_0}{r}$$

If, fixed charge $q = -Q$, and test charge is $q_0 = +Q_0$, then

$$\Rightarrow W_{el} = k\frac{QQ_0}{r}$$

EXAMPLE 5. A point charge 2×10^{-2} C is moved from P to S in a uniform electric field of 30 NC^{-1} directed along positive x-axis as shown in Fig.2.9. If coordinates of P and S are $(1\text{ m}, 2\text{ m}, 0\text{ m})$ and $(0\text{ m}, 0\text{ m}, 0\text{ m})$ respectively, find the work done by electric field during this displacement.

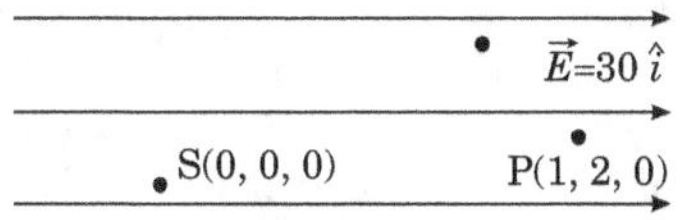

Figure 2.9

APPROACH Work done by electric force in displacement $\overrightarrow{PS}$ can be calculated by using work formula-

$$W_{el} = \vec{F}_{el}\cdot\overrightarrow{PS} = q\vec{E}\cdot\overrightarrow{PS}$$

SOLUTION The given situation is shown in Fig.2.9. The displacement,

$$\overrightarrow{PS} = \overrightarrow{OS} - \overrightarrow{OP} = (0\hat{i} + 0\hat{j} + 0\hat{k}) - (\hat{i} + 2\hat{j} + 0\hat{k}) = -\hat{i} - 2\hat{j}$$

Given electric field vector, $\vec{E} = 30$ NC$^{-1}\hat{i}$

Therefore, the work done by above electric field in displacing the charge 2×10^{-2} C by displacement $\overrightarrow{PS}$, is

$$W_E = q\vec{E}\cdot\overrightarrow{PS} = 2\times 10^{-2}[30\hat{i}\cdot(-\hat{i} - 2\hat{j})] = 2\times 10^{-2}(-30)$$
$$= -0.60\text{ J} = -600\text{ mJ}$$

☞ We encourage students to solve this problem by calculating the potential energy difference too.

2.2 Check Point 1

1. ••From the electric field $\vec{E}(r) = (kQ_s/r^2)\hat{r}$ produced by a positively charged particle, find the (usual) expression for its electric potential.

2. ••At a point (x, y, z), the electric field is given as-
$$\vec{E} = K\left[y^2\hat{i} + (2xy + z^2)\hat{j} + 2yz\hat{k}\right]$$
Here K is a constant with the appropriate units. Find the potential, using the origin as your reference point.

3. ••In some region of space, the electric field is given by $\vec{E} = Ax\hat{i} + By^2\hat{j}$. Find the electric potential difference between points whose positions are $(x_i, y_i) = (a, 0)$ and $(x_f, y_f) = (0, b)$. The constants A, B, a, and b have the appropriate SI units.

Multiple Choice Questions

4. ••The electric field in a region is given by $\vec{E} = (Ax + B)\hat{i}$, where E is in NC^{-1} and x is in metres. The values of constants are $A = 20$ SI unit and $B = 10$ SI unit. If the potential at $x = 1$ is V_1 and that at $x = -5$ is V_2, then $V_1 - V_2$ is:
   (A) 320 V (B) -48 V (C) 180 V (D) -520 V

5. ••On moving a charge of 20 C slowly by 2 cm, 2 J of external work is done, then the potential difference between the points is-
   (A) 0.1 V (B) 8 V (C) 2 V (D) 0.5 V

2.3 Electric Potential Energy of a System of Point Charges: Self Electric Energy

Suppose, the electric field $\vec{E}$ in which charge q_0 moves is caused by several point charges $q_1, q_2, q_3, \ldots$ at distances $r_1, r_2, r_3, \ldots$ from q_0, as shown in Fig.2.10. The total electric field at each point is the vector sum of the fields due to the individual charges, and the total work done on q_0 during any displacement is the sum of the contributions from the individual charges. Therefore, from Eq.2.24, we conclude that the potential energy associated with the test charge q_0 at point A in Fig. 2.10 is the algebraic sum:

$$U = \frac{q_0}{4\pi\epsilon_0}\left(\frac{q_1}{r_1} + \frac{q_2}{r_2} + \frac{q_3}{r_3} + \cdots\right) = \frac{q_0}{4\pi\epsilon_0}\sum_i \frac{q_i}{r_i} \quad (2.25)$$

The total electric potential energy of the system of point charges

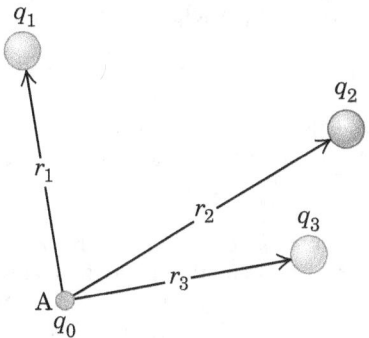

Figure 2.10: The potential energy associated with a charge q_0 at point A depends on the other charges q_1, q_2, and q_3 and on their distances r_1, r_2, and r_3 from point A.

is also referred to as the electric self-energy of the system.

In terms of the net electric field work, the net electric potential energy of the system of point charges, i.e., the self-energy of the system of charges, is defined as the negative of the net work done by the electric field in assembling the point charges from infinite separation to the given configuration.

We can also define self-energy in terms of the work done by an external agent: "The electric potential energy or self-energy of a system of point charges is the work done by an external force against the system's net electric force in assembling the charges from infinite separation to the present configuration without any change in the kinetic energy of any particle."

When q_0 is at a different point B (say), the potential energy is given by the same expression, but $r_1, r_2, \ldots$ are the respective distances from $q_1, q_2, \ldots$ to point B. The work done by electric field on test charge q_0, when it moves from A to B along any path is equal to the difference $(U_A - U_B)$ between the potential energies when q_0 is at A and at B respectively.

We can represent any charge distribution as a collection of point charges, so Eq.2.24 shows that we can always find a potential-energy function for any static electric field. It follows that for every electric field due to a static charge distribution, the force exerted by that field is conservative.

Equations 2.24 and 2.25 define U to be zero when distances $r_1, r_2, \ldots$ are infinite that is, when the test charge q_0 is very far away from all the charges that produce the field. As with any potential-energy function, the point where $U = 0$ is arbitrary. In electrostatics problems it's usually simplest to choose this point to be at infinity. When we analyse electric circuits, other choices will be more convenient.

Equation 2.25 gives the potential energy associated with the presence of the test charge q_0 in the $\vec{E}$ field produced by $q_1, q_2, q_3, \ldots$ But there is also potential energy involved in assembling these charges. If we start with charges $q_1, q_2, q_3, \ldots$ all separated from each other by infinite distances and then bring them together so that the distance between q_i and q_j is r_{ij}, the total potential energy U is the sum of the potential energies of interaction for each pair of charges. We can write this as

$$\boxed{U = \frac{1}{4\pi\epsilon_0} \sum_{i<j} \frac{q_i q_j}{r_{ij}}} \qquad (2.26)$$

This sum extends over all pairs of charges; we don't let $i = j$ (because that would be an interaction of a charge with itself), and we include only terms with $i < j$ to make sure that we count each pair only once. Thus, to account for the interaction between q_3 and q_4, we include a term with $i = 3$ and $j = 4$ but not a term with $i = 4$ and $j = 3$

Methods for Calculating U: We can find the electric potential energy of system of particles by any of the following three methods-

Method 1: You can directly apply Eq.2.26 to the given configuration.

Method 2: Initially, keep all the charges at ∞ separation from each other and then bring them one by one in present configuration and calculate the work done.
$$PE_{\text{sys}} = \Sigma W_i$$

Method 3: Find potential energies of each charge due to electric field of the other charges and then apply following Eq.2.27 to get the potential energy of the system of point charges-

$$PE_{\text{sys}} = \frac{PE_1 + PE_2 + PE_3 + \ldots}{2} \qquad (2.27)$$

where $PE_1 = PE_{12} + PE_{13} + \ldots$,
$PE_2 = P_{21} + PE_{23} + \ldots$, etc.

In the numerator of RHS of Eq.2.27, each pair of charge comes twice, that is why we divided it by 2. Eq.(2.27) is useful for symmetric charge arrangements.

EXAMPLE 6. Find the potential energy of the system shown in Fig.2.11. Each edge is of length a.

APPROACH I In this approach, we find the work done by an

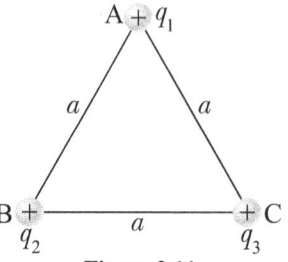

Figure 2.11

external agent in bringing each charge slowly, one by one, from infinity to shown positions. The work is done by the external agent against the electric field established by charges already reached to their position. This work is equal to change in potential energy of the system. Since, initially the charge separation is infinite, therefore, initial potential energy of the system of point charges is zero. If final potential energy is U, then
$$\Delta U = W_{\text{ext}} \implies U_f - U_i = W_{\text{ext}}$$
$$\implies U - 0 = W_{\text{ext}} \implies U = W_{\text{ext}}$$

SOLUTION Work done in bringing first charge q_1 from infinity to position A-
$$W_1 = 0$$
(in this case there was no electric field)

Now, this charge establishes an electric field in it's surroundings. So, work done in bringing second charge q_2 from infinity to position B against the electric field established by the charge q_1-
$$W_2 = \frac{kq_1q_2}{a}$$

Now, at position C, both q_1 and q_2 produce their electric fields. Therefore, in bringing, a third charge q_3 from infinity to position C, we have to do work against the electric field established by q_1 and q_2. This work done is given by-
$$W_3 = \frac{kq_1q_3}{a} + \frac{kq_2q_3}{a}$$

Net work done in forming the system ABC is stored in the form of the electric potential energy of the system. Some times we call it the self energy of the system.

So, self energy of the system ABC,
$$U = (W_{\text{ext}})_{\text{net}} = W_1 + W_2 + W_3$$
$$= 0 + \frac{kq_1q_2}{a} + \frac{kq_1q_3}{a} + \frac{kq_2q_3}{a}$$
$$\implies U = \frac{kq_1q_2}{a} + \frac{kq_1q_3}{a} + \frac{kq_2q_3}{a} \qquad (2.28)$$

So, we can say that the net potential energy of the system is equal to the sum of potential energy of each pair of charges.

APPROACH II There are $^3C_2 = 3$ pairs of charges. Find electrostatic potential energy for each pair by using Eq.2.24. Their sum will be the net electrostatic potential energy of the system.

SOLUTION Net electric potential energy of the given arrangement is-
$$U = U_{AB} + U_{AC} + U_{BC} = \frac{kq_1q_2}{a} + \frac{kq_1q_3}{a} + \frac{kq_2q_3}{a}$$
Here $k = 1/4\pi\epsilon_0$

☞ Student are advised to use approach II as a time saving trick in similar potential energy problems.

EXAMPLE 7. Determine the interaction energy of the system of point charges shown in Fig.2.12.

APPROACH There are $^4C_2 = 6$ pairs of charges. Find electrostatic potential energy for each pair by using Eq.2.24. Their sum

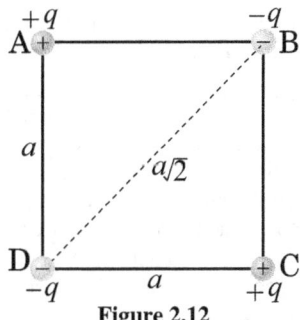

Figure 2.12

will be the net electric potential energy of the system.

SOLUTION Net electric potential energy of the arrangement is given by-

$$U = U_{AB} + U_{AC} + U_{AD} + U_{BC} + U_{BD} + U_{CD}$$
$$= -\frac{kq^2}{a} + \frac{kq^2}{(a\sqrt{2})} - \frac{kq^2}{a} - \frac{kq^2}{a} + \frac{kq^2}{(a\sqrt{2})} - \frac{kq^2}{a}$$
$$= -\frac{4kq^2}{a} + \frac{2kq^2}{a\sqrt{2}} = -\frac{kq^2}{a}\left(4 - \sqrt{2}\right)$$

EXAMPLE 8. Eight point charges are placed at the corners of a cube of edge a as shown in the Figure 2.13. Find the work done in disassembling this system of charges.

APPROACH If U_i is the initial electric potential energy of the

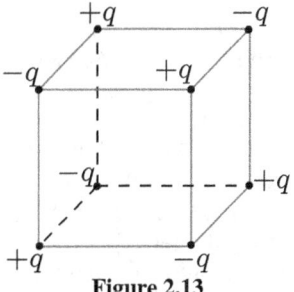

Figure 2.13

system and U_f is the final electric potential energy of the system, then work done by external agent in separating the charges slowly, is defined as-

$$W = \text{change in potential energy} = U_f - U_i \quad (1)$$

Disassembling the system means putting the charges at infinite separation. So, final electrostatic potential energy of the system $U_f = U_\infty = 0$

Now, calculate the initial potential energy (U_i) of the system by applying Eq.2.25 and substitute these values in above Eq.(1) and then solve for W.

SOLUTION In the given system, there are $^8C_2 = 28$ pairs of charges. The charge pairs on cube edges (twelve edges of length a each) have unlike charges. The charge pairs on face-diagonals (six faces, two diagonals per face, twelve face-diagonal of length $\sqrt{2}a$ each) have like charges.

The charge pairs on main-diagonals (four main-diagonal each with length $\sqrt{3}a$) have unlike charges. Thus, the initial potential energy of the system is

$$U_i = \frac{1}{4\pi\epsilon_0}\left[-12\frac{q^2}{a} + 12\frac{q^2}{\sqrt{2}a} - 4\frac{q^2}{\sqrt{3}a}\right] = -\frac{5.824}{4\pi\epsilon_0}\frac{q^2}{a}$$

Substituting these values of U_i and U_f in Eq.(1), we get-

$$W = 0 - U_i = \frac{5.824}{4\pi\epsilon_0}\frac{q^2}{a}$$

EXAMPLE 9. Three point charges q, $2q$ and $8q$ are to be placed on a 9 cm long straight line. Find the positions where the charges should be placed such that the potential energy of this system is minimum. In this situation, what is the electric field at the position of the charge q due to the other two charges?

APPROACH To obtain the minimum potential energy of the system, the charges, of larger magnitude, should be placed at the extreme positions. If the charge q is at distance x from $2q$, then corresponding to minimum potential energy, $dU/dx = 0$ and $d^2U/dx^2 = -ve$. By applying this concept, determine the value of x and then calculate the value of electric field at the position of q.

SOLUTION In Fig.2.14, the total potential energy of the given system is

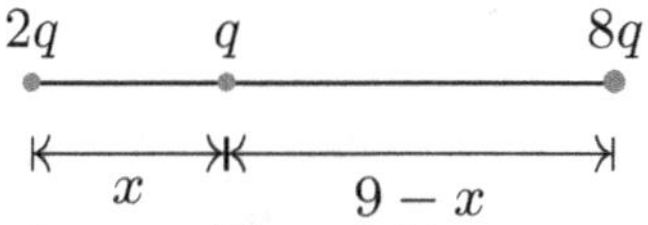

```
    2q         q              8q
    •──────────•───────────────•
    |←── x ──→|←──── 9 − x ───→|
```

Figure 2.14

$$U = \frac{1}{4\pi\epsilon_0}\left[\frac{(2q)(8q)}{9} + \frac{(2q)(q)}{x} + \frac{(q)(8q)}{9-x}\right]$$
$$= \frac{q^2}{2\pi\epsilon_0}\left[\frac{8}{9} + \frac{1}{x} + \frac{4}{9-x}\right]$$

The value of x for which potential energy is minimum is given by $dU/dx = 0$ i.e.,

$$\frac{dU}{dx} = \frac{q^2}{2\pi\epsilon_0}\left[-\frac{1}{x^2} + \frac{4}{(9-x)^2}\right] = 0$$

Simplifying, we get: $x = 3$, $x = -9$.

Now, $\frac{d^2U}{dx^2} = \frac{q^2}{2\pi\epsilon_0}\left[\frac{2}{x^3} - \frac{8}{(9-x)^3}\right]$

Corresponding to $x = 3$

$$\frac{d^2U}{dx^2} = \frac{q^2}{2\pi\epsilon_0}\left[\frac{2}{3^3} - \frac{8}{6^3}\right] = +ve$$

Since, corresponding to $x = 3$, $\frac{d^2U}{dx^2}$ is positive, therefore, potential energy is minimum at $x = 3$.

Now, corresponding to $x = -9$,

$$\frac{d^2U}{dx^2} = \frac{q^2}{2\pi\epsilon_0}\left[\frac{2}{(-9)^3} - \frac{8}{(18)^3}\right] = -ve$$

Since, corresponding to $x = -9$, $\frac{d^2U}{dx^2}$ is negative, therefore potential energy is maximum for $x = -9$.

Now, the electric field at the position of q due to the other two charges is

$$\vec{E} = \frac{1}{4\pi\epsilon_0}\left[\frac{2q}{x^2} - \frac{8q}{(9-x)^2}\right]\hat{\imath} = \frac{q}{4\pi\epsilon_0}\left[\frac{2}{(3)^2} - \frac{8}{(9-3)^2}\right]\hat{\imath} = \vec{0}$$

Note: Since, the force on the charge q is given by $\vec{F} = q\vec{E} = -dU/dx\,\hat{\imath}$. So, corresponding to minimum potential energy of the system, the charge q will be in the state of a stable equilibrium.

EXAMPLE 10. Two fixed, equal, positive charges, each of magnitude $q = 5 \times 10^{-5}$C are located at points A and B separated by a distance 6 m. An equal and opposite charge moves towards them along the line COD, the perpendicular bisector of the line AB. The moving charge, when reaches the point C at a distance of 4 m from O, has a kinetic energy of 4 J. Calculate the distance of the farthest point D which the negative charge will reach before returning towards C.

APPROACH At the farthest point, negative charge changes its direction of motion, so, the velocity of negative charge at this point will be zero. Since, there are only electrostatic forces on the charged particle, therefore, by applying the law of conservation of mechanical energy, we can find the position of farthest point.

SOLUTION If D is the farthest point at a distance r_{OD} from O, then the velocity of negative charge at D is $v_d = 0$. From Fig.2.16, we have-

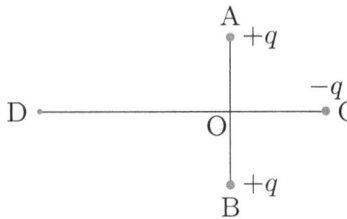

Figure 2.15

$r_{AC} = r_{BC} = \sqrt{3^2 + 4^2} = 5$ m

The electrostatic potential energies of the system for negative

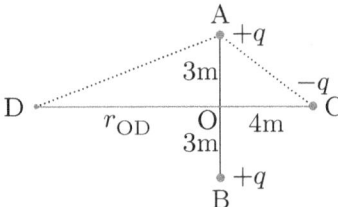

Figure 2.16

charge placed at C and D-

$$U_C = -\frac{q^2}{4\pi\epsilon_0}\left[\frac{1}{r_{AC}} + \frac{1}{r_{BC}}\right]$$

$$= -\frac{2(9 \times 10^9)(5 \times 10^{-5})^2}{5} = -9 \text{ J}$$

$$U_D = -\frac{q^2}{4\pi\epsilon_0}\left[\frac{1}{r_{AD}} + \frac{1}{r_{BD}}\right] = -\frac{45}{r_{AD}}\text{J}$$

The kinetic energy of the negative charge at C is $K_C = 4$ J and at D is $K_D = \frac{1}{2}mv_d^2 = 0$.

By conservation of mechanical energy, we have-

$$K_C + U_C = K_D + U_D$$

On substituting the values of K_C, U_C, K_D and U_D in above expression and simplifying for r_{AD}, we get

$$r_{AD} = 45/5 = 9 \text{ m}$$

Now, applying Pythagoras theorem in triangle AOD, we get-

$$r_{OD} = \sqrt{r_{AD}^2 - r_{AO}^2} = \sqrt{9^2 - 3^2} = \sqrt{72} = 8.48 \text{ m}$$

2.3.1 Potential Energy of a System of Two Charges in an External Field

In the previous section, we studied the potential energy of a system of point charges due to their own interactions. In this section, we will calculate the potential energy of a system of two charges q_1 and q_2 located at r_1 and r_2, respectively, in an external field. First, we calculate the work done by an external agent in bringing the charge q_1 slowly from infinity to r_1. The work done in this step is $q_1V(r_1)$. Next, we consider the work done by the external agent in bringing q_2 slowly to r_2. In this step, work is done not only against the external field $\vec{E}$ but also against the field due to q_1.

Work done on q_2 against the external field

$$= q_2V(r_2) \qquad (2.29)$$

Work done on q_2 against the field due to q_1

$$= \frac{q_1q_2}{4\pi\epsilon_0 r_{12}} \qquad (2.30)$$

where r_{12} is the distance between q_1 and q_2. By the superposition principle for fields, we add up the work done on q_2 against the two fields (E and that due to q_1):
Work done in bringing q_2 to r_2

$$= q_2V(r_2) + \frac{q_1q_2}{4\pi\epsilon_0 r_{12}} \qquad (2.31)$$

Thus, Potential energy of the system = the total work done in assembling the configuration

$$= q_1V(r_1) + q_2V(r_2) + \frac{q_1q_2}{4\pi\epsilon_0 r_{12}} \qquad (2.32)$$

EXAMPLE 11. (a) Determine the electrostatic potential energy of a system consisting of two charges $7\mu C$ and $-2\mu C$ (and with no external field) placed at $(-9$ cm, 0, 0$)$ and $(9$ cm, 0, 0$)$ respectively. (b) How much work is required to separate the two charges infinitely away from each other? (c) Suppose that the same system of charges is now placed in an external electric field $E = A(1/r^2); A = 9 \times 10^5 \text{Cm}^{-2}$. What would the electrostatic energy of the configuration be?

SOLUTION The given charges are $q_1 = 7 \times 10^{-6}$ C and $q_2 = -2 \times 10^{-6}$ C, located at $(-9$ cm, 0, 0$)$ and $(9$ cm, 0, 0 cm$)$ respectively. Thus, the distance between them is

$$r = 0.18 \text{ m}.$$

(a) Electrostatic potential energy of the two-charge system:
The potential energy of a pair of point charges is given by

$$U = k\frac{q_1q_2}{r}.$$

Substituting values:

$$U = \left(9 \times 10^9\right) \cdot \frac{(7 \times 10^{-6})(-2 \times 10^{-6})}{0.18}$$

$$= \left(9 \times 10^9\right) \cdot (-7.778 \times 10^{-11}) = -0.7 \text{ J}$$

(b) Work required to separate the charges to infinity:
At infinite separation, the potential energy becomes zero. Hence, the external work required to move the charges, without acceleration, from separation r to infinity is

$$W = U_\infty - U_{\text{initial}} = 0 - (-0.7) = 0.7 \text{ J}.$$

(c) Electrostatic energy in presence of external electric field:
In this case, the mutual interaction energy of the two charges remains unchanged. In addition, there is the energy of interaction of the two charges with the external electric field.

The external field is given by $E(r) = \dfrac{A}{r^2}$, which corresponds to a potential

$$V(r) = -\int_\infty^r E(r)\,dr = \frac{A}{r}.$$

Let r_1 is the distance of $(9$ cm, 0, 0 cm$)$ and r_2 is that of $(-9$ cm, 0, 0$)$ from origin, then

$$r_1 = \sqrt{(-0.09-0)^2 + (0-0)^2 + (0-0)^2} = 0.09 \text{ m}$$

and $\quad r_2 = \sqrt{(0.09-0)^2 + (0-0)^2 + (0-0)^2} = 0.09 \text{ m}$

Therefore, interaction energy with external electric field:

$$q_1V(\vec{r}_1) + q_2V(\vec{r}_2) = A\frac{7\mu C}{0.09 \text{ m}} + A\frac{-2\mu C}{0.09 \text{ m}}$$

$$= 70\text{ J} - 20\text{ J} = 50\text{ J}$$

Since, the interaction energy due to interaction between charges remains the same, therefore:

$$U = -0.7 \text{ J}.$$

Therefore, from (2.32), the net electrostatic energy of the system of charges in external electric field is:

$$q_1V(\vec{r}_1) + q_2V(\vec{r}_2) + \frac{q_1q_2}{4\pi\epsilon_0 r_{12}} = 50\text{ J} - 0.7\text{ J} = 49.3\text{ J}$$

2.3.2 Potential Energy of a Dipole in a Uniform Electric Field

Let us consider an electric dipole in a uniform external electric field as shown in Fig.2.17.

If τ is the magnitude of the torque applied by an external agent against the torque of the electric field for the shown position of the dipole, as depicted in Fig.2.17, then work done by external

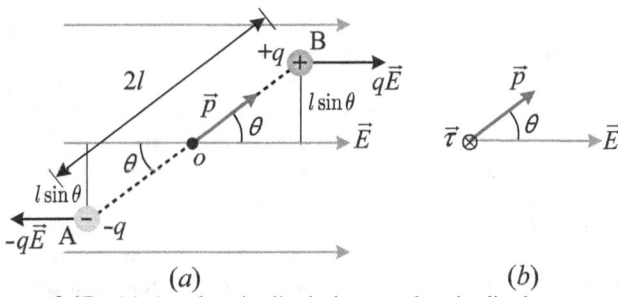

(a)   (b)

Figure 2.17: (a) An electric dipole has an electric dipole moment $\vec{p}$ in an external uniform electric field $\vec{E}$. The angle between $\vec{p}$ and $\vec{E}$ is θ. The line connecting the two charges represents their rigid connection and their centre of mass is assumed to be midway between them. (b) Representing the electric dipole moment by a vector $\vec{p}$ in the external electric field $\vec{E}$ and showing the direction of the torque $\vec{\tau}$ into the page by the symbol $\otimes$

force in rotating electric dipole anti-clock wise by small angle $d\theta$ against electric force

$$dW = \tau \, d\theta$$

Since, $\tau = pE \sin\theta$, therefore-

$$W_{\text{ext}} = \int_{\theta_1}^{\theta_2} \tau \, d\theta = \int_{\theta_1}^{\theta_2} pE \sin\theta \, d\theta$$

or $\boxed{W_{\text{ext}} = -pE \left(\cos\theta_2 - \cos\theta_1 \right)}$ (2.33)

In this case the work done by electric field

$$\boxed{W_{\text{el}} = -W_{\text{ext}} = pE \left[\cos\theta_2 - \cos\theta_1 \right]} \quad (2.34)$$

If $\theta_1 = 0$, $\theta_2 = \theta$, then

$$W_{\text{ext}} = -pE \left[\cos\theta - 1 \right]$$

and $\quad W_{\text{el}} = pE \left[\cos\theta - 1 \right]$

The work done by external force is stored in the form of electrostatic potential energy of the system. If we take zero potential energy at $\theta_1 = 90°$, then for $\theta_2 = \theta$, we have
Change in potential energy, $\Delta U = W_{\text{ext}} = -pE \cos\theta$
i.e., $\quad \Delta U = -\vec{p} \cdot \vec{E}$

When $\theta = 0°$, the dipole moment is in the direction of the field and the dipole is in *stable equilibrium*. If it is slightly displaced from its equilibrium position, it performs oscillations about the stable equilibrium position, which serves as the mean position. When $\theta = 180°$, the dipole moment is opposite to the direction of the field and the dipole is in *unstable equilibrium*.

2.3.3 The Electric Potential Due to a Point Charge

Let's calculate the electric potential of the simplest possible system: one point charge. The point charge q is the source of an electric potential. Consider a test charge q_0 separated by a distance r from a single point charge q. As Eq.2.24 shows, the potential energy of the system is $U(r) = q_0 q / 4\pi\epsilon_0 r$, and hence $U(r)/q_0 = q/4\pi\epsilon_0 r$. We have found the electric potential of a point charge q at a distance r from the charge:

$$\boxed{V(r) = \frac{U(r)}{q_0} = \frac{1}{4\pi\epsilon_0} \frac{q}{r}} \quad (2.35)$$

In Eq.2.35, we have assumed that zero potential energy is at infinity and, as a consequence, we have taken the electric potential due to a charge q to be zero at infinity. To emphasize this point, we might say that Eq.2.35 is the potential of a single charge with respect to infinity.
As for potential energy, the only physically relevant feature of the potential is how it differs between two points. The electric potential difference due to the charge q between the points A and B at locations $\vec{r}_A$ and $\vec{r}_B$ is given by (Fig.2.8):

$$\Delta V = V_B - V_A = \frac{U_B - U_A}{q_0} = \frac{q}{4\pi\epsilon_0}\left(\frac{1}{r_B} - \frac{1}{r_A}\right) \quad (2.36)$$

Here, we have abbreviated V as a function of r_A, or $V(r_A)$, as V_A, and so forth.

EXAMPLE 12. A conducting bubble of radius a and thickness t ($t \ll a$) has potential V. Now the bubble collapses into a droplet. Find the potential of the droplet.

SOLUTION The volume of liquid in a bubble of radius a and thickness t is given by

$$\mathcal{V}_b = \frac{4}{3}\pi(a+t)^3 - \frac{4}{3}\pi a^3 = \frac{4}{3}\pi a^3 \left(\left(1 + \frac{t}{a}\right)^3 - 1\right)$$

$$= \frac{4}{3}\pi a^3 \left(1 + 3\frac{t}{a} - 1\right) \quad \text{(since } t \ll a\text{)}$$

$$\approx 4\pi a^2 t$$

The volume of liquid in the droplet of radius r formed by collapsing the bubble, is-

$$\mathcal{V}_d = \frac{4}{3}\pi r^3$$

Since, mass of the material of bubble does not change during formation of drop from bubble, therefore

Mass of bubble = Mass of droplet

$\rho \mathcal{V}_b = \rho \mathcal{V}_d$,   [$\rho$ is the density of the material of bubble]

$\Longrightarrow \quad \frac{4}{3}\pi r^3 = 4\pi a^2 t \quad \Longrightarrow \quad r = \left(3a^2 t\right)^{1/3}$

The potential on a spherical shell of radius a and charge q is-

$$V = \frac{q}{(4\pi\epsilon_0 a)}$$

Therefore, the charge on bubble having potential V is-

$$q = 4\pi\epsilon_0 a V$$

So, by conservation of charge, the charge on the droplet will also be, $q = 4\pi\epsilon_0 a V$.
Therefore, the electric potential on the droplet is given by

$$V_d = \frac{q}{4\pi\epsilon_0 r} = \frac{4\pi\epsilon_0 a V}{(3a^2 t)^{1/3}} = \left(\frac{a}{3t}\right)^{1/3} V$$

EXAMPLE 13. Electric potential at a point 'P' due to a point charge of 5×10^{-9} C is 50 V. Calculate the distance of 'P' from the point charge. (Assume, $1/4\pi\epsilon_0 = 9 \times 10^{+9}$ Nm^{2}C^{-2})

APPROACH Apply relation, $V = kQ/r$ and then solve for r.

SOLUTION $V_P = \frac{kQ}{r} \quad \Longrightarrow \quad 50 = \frac{9 \times 10^9 \times 5 \times 10^{-9}}{r}$

$\Longrightarrow \quad r = \frac{45}{50} = \frac{9}{10} = 0.9$ m $= 90$ cm

EXAMPLE 14. Two equal point charges (each of magnitude q), are fixed at $x = -a$ and $x = +a$ on the x-axis. Another point charge Q is placed at the origin. Find the change in electrical potential energy of Q, when it is displaced by a small distance x along the x-axis.

SOLUTION Let O be the origin and O' be a point to the right of O at a distance x.

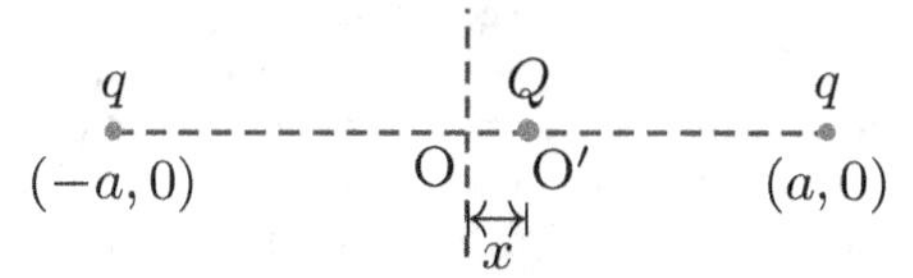

Figure 2.18

The potentials at O and O' respectively, due to charges at $(-a, 0)$ and $(a, 0)$ are

$$V_O = \frac{q}{4\pi\epsilon_0 a} + \frac{q}{4\pi\epsilon_0 a} = \frac{q}{2\pi\epsilon_0 a}$$

$$V_{O'} = \frac{q}{4\pi\epsilon_0(a+x)} + \frac{q}{4\pi\epsilon_0(a-x)} = \frac{q}{2\pi\epsilon_0}\left(\frac{a}{a^2 - x^2}\right)$$

The potential energy associated with the charge Q, when it is placed at a point where potential is V, is QV. Thus, the change in potential energy of charge Q when it is displaced by a small distance x is

$$\Delta U = QV_{O'} - QV_O = \frac{qQ}{2\pi\epsilon_0}\left[\frac{a}{a^2-x^2} - \frac{1}{a}\right]$$
$$= \frac{qQ}{2\pi\epsilon_0}\frac{x^2}{a(a^2-x^2)} \approx \frac{qQ}{2\pi\epsilon_0}\frac{x^2}{a^3} \quad (\text{for } x \ll a)$$

EXAMPLE 15. A uniform electric field pointing in positive x direction exists in a region. Let A be the origin, B be the point on the x-axis at $x = +1$ cm, and C be the point on the y-axis at $y = +1$ cm. Then the potentials at the points A, B and C satisfy
(A) $V_A < V_B$ (B) $V_A > V_B$ (C) $V_A < V_C$ (D) $V_A > V_C$

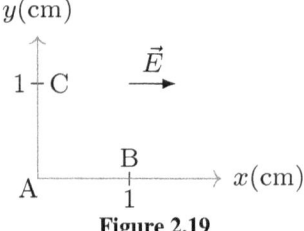

Figure 2.19

APPROACH Apply the relation, $V_B - V_A = -\int_i^f \vec{E} \cdot d\vec{r}$ along x and y directions.

SOLUTION (B) The uniform electric field in the region is $\vec{E} = E\hat{i}$. Let $d\vec{r}_x = dx\hat{i}$ and $d\vec{r}_y = dy\hat{j}$ be the small displacement vectors along x and y-axes.

The potentials at the point B and C relative to the point A are given by

$$V_B = V_A - \int \vec{E} \cdot d\vec{r}_x = V_A - \int_0^1 E\,dx = V_A - E$$
$$V_C = V_A - \int \vec{E} \cdot d\vec{r}_y = V_A \qquad (\because \vec{E} \perp d\vec{r}_y)$$

Note that the potential decreases along $\vec{E}$ but does not change in a direction perpendicular to $\vec{E}$.

2.4 Equipotential Surfaces

Any surface consisting of a continuous distribution of points having the same electric potential, is called an equipotential surface. It can be either an imaginary surface or a real, physical surface. Since, no point can be at two different potentials at the same time, so equipotential surfaces for different potentials can never touch or intersect. If a charged particle, having charge q_0, moves between two points A and B on the same equipotential surface (i.e., $V_A = V_B$), then from Eq.2.10:

$$V_B - V_A = \frac{W_{AB}}{q_0}$$

the net work done on the charged particle by an electric field:
$$W_{AB} = q_0(V_B - V_A) = q_0(0) = 0 \quad [\because V_A = V_B]$$

Thus, no net work W_{AB} is done on a charged particle by an electric field when it moves between two points A and B on the same equipotential surface. Furthermore, because of the path independence of potential, this result ($W_{AB} = 0$) holds for any two points A and B on the equipotential surface, even if the path between them does not lie entirely on the equipotential surface.

As change in potential energy is equal to negative of work done by electric field, i.e., $\Delta U = -W_{AB}$, so potential energy does not change (because, $W_{AB} = 0$) as a test charge moves over an equipotential surface, the electric field can do no work on such a charge. It follows that $\vec{E}$ must be perpendicular to the surface at every point so that the electric force $q_0\vec{E}$ is always perpendicular to the displacement of a charge moving on the surface. Field lines and equipotential surfaces are always mutually perpendicular. In general, field lines are curves, and equipotentials are curved surfaces. For the special case of a uniform field, in which the field lines are straight, parallel, and equally spaced, the equipotentials are parallel planes perpendicular to the field lines.

Figure 2.20 shows a family of equipotential surfaces associated with the electric field due to some distribution of charges. The work done by the electric field on a charged particle as the particle moves from one end to the other of paths I and II is zero because each of these paths begins and ends on the same equipotential surface and thus there is no net change in potential. The work done as the charged particle moves from one end to the other of paths III and IV is not zero but has the same value for both these paths because the initial and final potentials are identical for the two paths; that is, paths III and IV connect the same pair of equipotential surfaces.

☞ In diagrams we usually show only a few representative equipo-

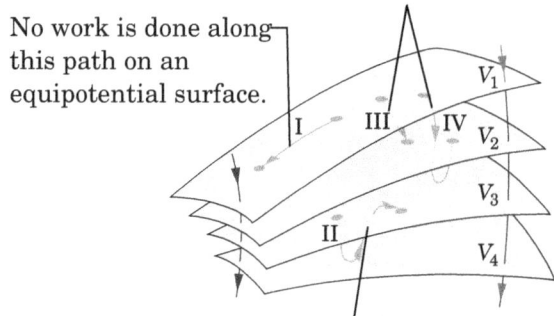

Figure 2.20: Portions of four equipotential surfaces at electric potentials V_1, V_2, V_3, and V_4 ($V_1 > V_2 > V_3 > V_4$). Four paths along which a test charge may move are shown. Two electric field lines are also indicated.

tentials, often with equal potential differences between adjacent surfaces.

2.4.1 Various Equipotential Surfaces

2.4.1.1 Equipotential Surfaces For a Point Charge

If a point charge q is fixed at origin O $(0,0,0)$ and P (x,y,z) be a point at distance $r\left(=\sqrt{x^2+y^2+z^2}\right)$ from O, then electric potential (V) at point P is-

$$V = k\frac{q}{r^2} = k\frac{q}{\sqrt{(x^2+y^2+z^2)}} = V_0 = \text{constant}$$
$$\implies x^2 + y^2 + z^2 = \frac{k^2q^2}{V_0^2} = \text{constant} \quad (2.37)$$

Equation 2.37 is of the form of a sphere with the center at $(0,0,0)$ and a radius of kq/V_0. Thus, the surfaces are concentric spheres with the origin (the location of the charge) as the centre and a radius (R) given by

$$R = \frac{kq}{V_0}$$

2.4.1.2 Equipotential Surfaces Due to an Electric Dipole

In Fig.2.22, electric field lines and equipotential lines are plotted for a set of two charges of equal magnitude, one positive and one negative. The field lines are closest together near the charges, indicating that the electric field is strongest there. Very

2.4. EQUIPOTENTIAL SURFACES

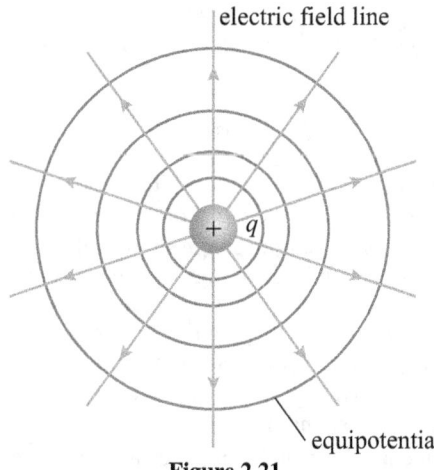

Figure 2.21

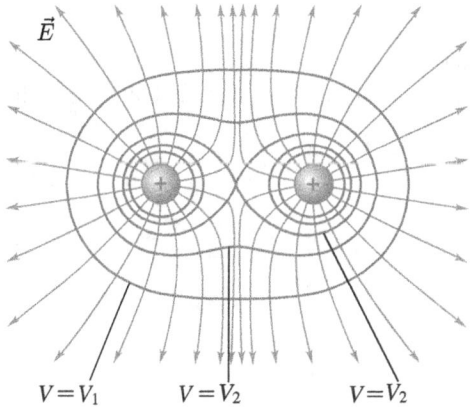

Figure 2.23: Two equal positive charges

close to each charge, the effect of that charge dominates on the equipotential plots, which are not strongly different from circles, but farther away the superposition of the two electric potentials produces more noticeable effects. Note that electric field lines always cross equipotential lines perpendicularly.

In Fig.2.22, electric field lines and equipotential lines are plotted for a pair of charges with equal magnitude, one positive and one negative. The field lines are closest together near the charges, indicating that the electric field is strongest there. In close proximity to each charge, the influence of that charge dominates on the equipotential plots, which appear not significantly different from circles. However, farther away, the superposition of the two electric potentials produces more noticeable effects. It is important to note that electric field lines always cross equipotential lines perpendicularly.

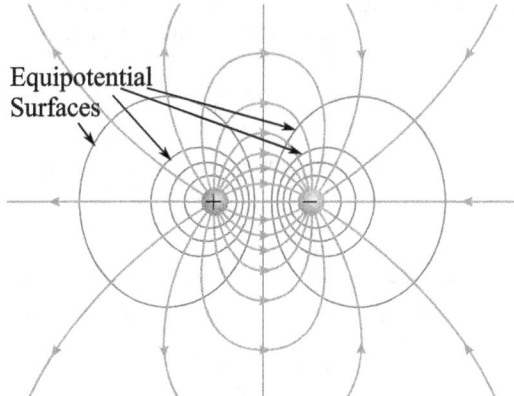

Figure 2.22: Equipotential lines and electric field lines for a pair of charges of equal magnitude but opposite sign.

2.4.1.3 Equipotential Surfaces Due to Two Identical Positive Charges:

Fig.2.23 displays equipotential surfaces and electric field lines resulting from two identical positive point charges. At first glance, it might seem that two equipotential surfaces intersect at the centre of Fig.2.23, contrary to the rule that prohibits such intersections. However, in reality, this is a single figure-eight-shaped equipotential surface.

2.4.1.4 Equipotential Surfaces of a Charged Wire of Infinite Length

Let us consider a line charge of infinite length having linear charge density λ. Suppose, it is placed along z-axis. The potential due to this line charge at a point P is given by Eq.2.60:

$$V(r) = -2k\lambda \ln r = -\frac{\lambda}{2\pi\epsilon_0} \ln r$$

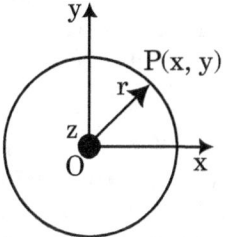

Figure 2.24: Cross-sectional view of cylindrical wire having axis along z-axis

where r is the distance of the point P from the line charge as shown in Fig.2.24.

Since, the line charge is along the z-axis, $r = \sqrt{x^2 + y^2}$, therefore,

$$V(r) = -\frac{\lambda}{4\pi\epsilon_0} \ln\left(x^2 + y^2\right)$$

Now, $V = \text{constant} = V_0$

$$\implies \ln\left(x^2 + y^2\right) = -\frac{4\pi\epsilon_0 V_0}{\lambda} \implies x^2 + y^2 = e^{-4\pi\epsilon_0 V_0/\lambda}$$

which represents cylinders with axis along the z-axis with radii $r = e^{-2\pi\epsilon_0 V_0/\lambda}$.

As V_0 increases, radius becomes smaller. Thus the cylinders are

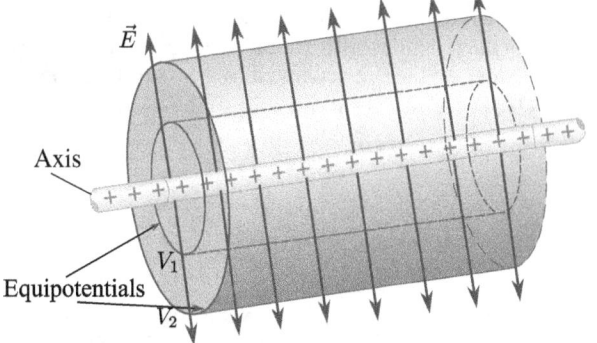

Figure 2.25

packed closer around the axis, showing that the field is stronger near the axis.

2.4.1.5 Uniformly Charged Plane Surface of Infinite Dimensions

Let us consider a plane sheet of charge, in x-y plane, having surface charge density σ. Consider a point P at distance z from this sheet. From Eq.2.62, the electric potential at P will be given by -

$$V = V_0 - \frac{\sigma}{2\epsilon_0}|z| \qquad (1)$$

Here, V_0 is the potential at $z = 0$.
Equipotential surface means

$$V = \text{constant} \implies V_0 - \frac{\sigma}{2\epsilon_0}|z| = C$$
$$\implies |z| = \text{constant}$$

Note that, the shapes of the equipotential surfaces in above di-

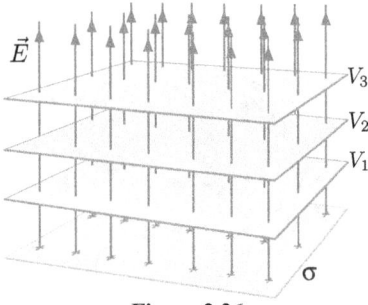

Figure 2.26

agrams will not change if the sign of each charge is reversed. If the positive charges in above diagrams, are replaced by negative charges, the equipotential surfaces will be the same but the sign of the potential will get reversed. For example, the surfaces with potentials $V = +30$ V and $V = -50$ V (say) will now have potentials $V = -30$ V and $V = +50$ V, respectively.

2.4.2 Properties of Equipotential surface

(i) The potential difference between any two points on an equipotential surface is always zero.
(ii) If a test charge q_0 moves from one point to the other on an equipotential surface, the electric potential energy $q_0 V$ remains constant.
(iii) No work is done by the electric force, when a charge moves along this surface.
(iv) Two equipotential surfaces can never intersect each other because in this scenario the point of intersection will have two potentials which is of course not acceptable.
(v) Field lines and equipotential surfaces are always mutually perpendicular.

2.5 Calculating the Field From the Potential: Potential Gradient

We can quantify the relation between potential and field by considering the potential difference dV between two nearby points A and B. Suppose they're separated by a small displacement ds in the 3-D space. Then, from Eq.2.13 we have-

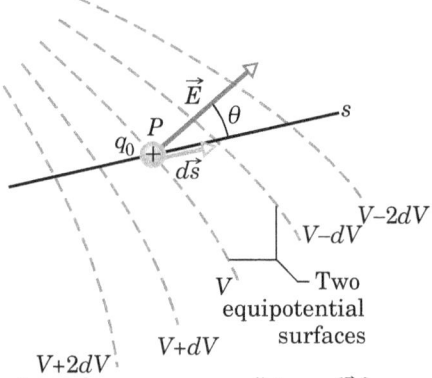

Figure 2.27: A test charge q_0 moves a distance $d\vec{s}$ from one equipotential surface to another. (The separation between the surfaces has been magnified for clarity.) The displacement $d\vec{s}$ makes an angle θ with the direction of the electric field $\vec{E}$.

$$dV = -\vec{E} \cdot d\vec{s} = -E \cos\theta \, ds \quad (2.38)$$

Here, θ is the angle between electric field vector $\vec{E}$ and displacement vector $d\vec{s}$.

Now, from Fig.2.27, we see that $E \cos\theta$ is the component of $\vec{E}$ in the direction of $d\vec{s}$. If we denote it by E_s, then above Eq.2.38 gives:

$$\boxed{E_s = E \cos\theta = -\frac{dV}{ds}} \quad (2.39)$$

In words, this equation states; *the negative of the rate of change of the potential with position in any direction is the component of $\vec{E}$ in that direction*. The minus sign implies that $\vec{E}$ points in the direction of decreasing V, as in Fig.2.27. It is clear from Eq.2.39 that an appropriate unit for E is the volt/meter (V/m).

According to Eq.2.39, the magnitude of $-dV/ds$ is maximum for $\theta = 0$. The quantity, E_s is also maximum for $\theta = 0$, i.e., in the direction of $\vec{E}$ which is always perpendicular to equipotential surface. Thus, corresponding to maximum magnitude of $-dV/ds$ the displacement $d\vec{s}$ should be in the direction of $\vec{E}$. Thus,

$$(E_s)_{\max} = E \cos 0 = -\left(\frac{dV}{ds}\right)_{\max}$$

or $$\boxed{E = -\left(\frac{dV}{ds}\right)_{\max}} \quad (2.40)$$

The maximum value of dV/ds at a given point is called the potential gradient at that point. The direction of $d\vec{s}$ for which dV/ds has its maximum value is always at right angles to the equipotential surface, corresponding to the direction of $\vec{E}$ as shown in Fig.2.27.

> **⚠ Caution**
>
> In practical physics usage, if you simply say *potential gradient* without specifying a direction, it is generally interpreted as the rate of change of the potential with position in the direction *perpendicular* to the equipotential surfaces, and is directly related to the electric field magnitude according to Eq. (2.40):
>
> $$E = -\left(\frac{dV}{ds}\right)_{\max}$$
>
> If you mean any other direction, it is clearer to say "rate of change of potential in that direction" rather than "potential gradient."

In a straightforward scenario where the electric field $\vec{E}$ is uniform, Eq.2.39 becomes

$$\boxed{E_s = -\frac{\Delta V}{\Delta s}} \quad (2.41)$$

where Δs is perpendicular to the equipotential surfaces. The component of the electric field is zero in any direction parallel to the equipotential surfaces because there is no change in potential along the surfaces.

If points A and B are separated by small displacement dx in the x-direction. Then similar to Eq.(2.38), we have-

$$dV = -E_x dx \quad (2.42)$$

where we again handled the dot product by considering only the component of $\vec{E}$ along the displacement.
Dividing both sides of Eq.2.42 by dx, we get-

$$E_x = -\frac{dV}{dx} \quad (2.43)$$

here, E_x is the electric-field component in the x-direction.
Similarly, we can write the expressions for y- and z-components

2.5. CALCULATING THE FIELD FROM THE POTENTIAL: POTENTIAL GRADIENT

as given below-
$$E_y = -\frac{dV}{dy}, \quad E_z = -\frac{dV}{dz} \tag{2.44}$$

When a function depends on more than one variable, as the potential generally does, we write derivatives with the partial derivative symbol ∂ instead of d to indicate the rate of change with respect to only one variable. Thus we have $E_x = -\partial V/\partial x$, $E_y = -\partial V/\partial y$, and $E_z = -\partial V/\partial z$.

Now, the entire electric-field vector can be written as-
$$\vec{E} = E_x \hat{i} + E_y \hat{j} + E_z \hat{k} \tag{2.45}$$

On substituting the values of E_x, E_y and E_z in above equation, we get-
$$\boxed{\vec{E} = -\left(\frac{\partial V}{\partial x}\hat{i} + \frac{\partial V}{\partial y}\hat{j} + \frac{\partial V}{\partial z}\hat{k}\right)} \tag{2.46}$$

Equation 2.46 confirms that the electric field is strong where the potential changes rapidly. The negative sign reflects that the electric field always points in the direction of decreasing potential, i.e., a positive test charge naturally moves from high to low potential.

In physics, we often use a shorthand called the gradient operator, written as:
$$\text{\bf grad} \text{ or } \vec{\nabla} = \frac{\partial}{\partial x}\hat{i} + \frac{\partial}{\partial y}\hat{j} + \frac{\partial}{\partial z}\hat{k}$$

It's just a compact way of writing the rate of change (i.e., derivative) of a quantity in all three directions.

Using this, we can write Eq.2.46 as-
$$\boxed{\vec{E} = -\left(\frac{\partial}{\partial x}\hat{i} + \frac{\partial}{\partial y}\hat{j} + \frac{\partial}{\partial z}\hat{k}\right)V = -\vec{\nabla}V \text{ or } -\text{\bf grad } V} \tag{2.47}$$

The quantity, $\vec{\nabla}V$ is called the potential gradient.

Clearly, the unit of potential gradient is volt per meter, while electric intensity or force per unit charge is expressed in newtons per coulomb. However,
$$\frac{\text{volt}}{\text{m}} = \frac{\text{J/C}}{\text{m}} = \frac{\text{N} \cdot \text{m}}{\text{C} \cdot \text{m}} = \frac{\text{N}}{\text{C}}$$

so, the volt/meter and the newton/coulomb are equivalent units.

Note: Potential is a scalar quantity but the gradient of potential is a vector quantity. In Cartesian co-ordinates, the potential gradient is defined by Eq.2.47.

Thus if V is known for all points of space, that is, if the function $V(x, y, z)$ is known, the components of $\vec{E}$, and thus E itself, can be found by taking partial derivatives*.

We therefore have two methods for calculating E for continuous charge distributions. One is based on integrating Coulomb's law, and the other is based on differentiating V (see Eq.2.47). In practice, the second method is often less difficult.

Special Case for Radial Fields:

In certain systems, the electric potential V depends only on the radial distance r from a central point, and not on the direction. Such systems are called *spherically symmetric*. This means that all points at the same distance from the centre have the same potential, and the electric field at any point is directed purely along the radial direction.

In Eq. (2.39), $E = -\frac{dV}{ds}$, the derivative is taken along the direction of the electric field. In a spherically symmetric system, the electric field at every point is purely radial, and the potential V depends only on the radial distance r. Therefore, when we move in the direction of the electric field, we are effectively moving along the radial direction itself. Hence, the displacement ds becomes identical to the radial displacement dr, and the relation simplifies to:
$$\boxed{E = -\frac{dV}{ds} = -\frac{dV}{dr}} \tag{2.48}$$

This form is valid only in spherically symmetric cases, where the direction of the field and the radial direction coincide.

Although $\vec{r}$ is a three-dimensional vector, the potential depends only on its magnitude r, and not on the angular coordinates. Hence, the variation of potential occurs only along the radial direction, and the electric field magnitude can be calculated using a single-variable derivative with respect to r.

This is the **scalar form** of the relation between electric field and potential in a radial field. Here, E represents the magnitude of the electric field, and $\frac{dV}{dr}$ is the rate at which the potential changes with distance from the centre.

The negative sign is important: it reflects the physical fact that the electric field always points in the direction of decreasing potential. Even though Eq. (2.48) is a scalar equation, the sign tells us about the relative direction between field and potential. For example, if V decreases with increasing r, then $\frac{dV}{dr} < 0$, making $E > 0$, which means the field points outward. On the other hand, if V increases with r, then $E < 0$, indicating an inward-pointing field. The negative sign, therefore, preserves the correct physical relationship between the electric field and the potential.

As an example, in Eq.2.35 we have shown that the potential at a point at the radial distance r from a source point charge q is
$$V = \frac{1}{4\pi\epsilon_0}\frac{q}{r}$$

To get E at the position of V, we consider the displacement $d\vec{s}$ in the direction of $\vec{E}$, i.e., radially away from point charge q.

Therefore, $E = -\frac{d}{dr}\left(\frac{1}{4\pi\epsilon_0}\frac{q}{r}\right) = \frac{1}{4\pi\epsilon_0}\frac{q}{r^2}$

which is in agreement with Coulomb's law.

Key Point

1. The component of the electric field $\vec{E}$ along any direction is related to the rate of change of potential with position by
$$E_s = -\frac{dV}{ds},$$
where ds is the infinitesimal displacement along that direction and dV is the corresponding change in potential.
If θ is the angle between $\vec{E}$ and $d\vec{s}$, then this component can also be written as
$$E_s = E\cos\theta.$$

2. For a spherically symmetric field, V depends only on the radial distance r from the source. In this case,
$$E = -\frac{dV}{dr},$$
where dr is the infinitesimal change in radial distance. The negative sign indicates that $\vec{E}$ points in the direction of decreasing potential.

EXAMPLE 16. Figure 2.28 shows lines of constant potential in a region in which an electric field is present. The values of the potential of each line is also shown. Of the points A, B and C, which one has the maximum magnitude of electric field $\vec{E}$

---

*The symbol $\partial V/\partial x$ denotes a partial derivative. In taking this derivative of the function $V(x, y, z)$, the quantity x is to be viewed as a variable and y and z are to be regarded as constants. Similar considerations hold for $\partial V/\partial y$ and $\partial V/\partial z$.

APPROACH From relation $E = -\partial V/\partial x$, we can say that the

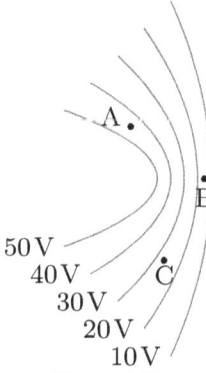

Figure 2.28

magnitude of $\vec{E}$ will be greater for greater value $\partial V/\partial x$.

SOLUTION The potential difference between all the successive lines of constant potential is $\Delta V = 10$ V. The perpendicular distances between successive lines at the point A and at the point C are almost equal but it is smaller at the point B i.e., $\Delta x_A = \Delta x_C > \Delta x_B$. Hence, $|E_B| = \Delta V/\Delta x_B > \Delta V/\Delta x_A = |E_A| = |E_C|$.

EXAMPLE 17. If $V = 3x^2y + y^2 + yz$, then find an expression for electric field $\vec{E}$

APPROACH From Eq.2.47, the electric field in terms of electric potential is defined as-
$$\vec{E} = -\left(\frac{\partial V}{\partial x}\hat{i} + \frac{\partial V}{\partial y}\hat{j} + \frac{\partial V}{\partial z}\hat{k}\right) \quad \ldots (1)$$

SOLUTION Given that, $V = 3x^2y + y^2 + yz$ therefore-
$$\frac{\partial V}{\partial x} = \frac{\partial}{\partial x}\left(3x^2y + y^2 + yz\right) = \frac{\partial}{\partial x}\left(3x^2y\right) = 3y\frac{d}{dx}\left(x^2\right) = 6xy$$

Similarly, $\frac{\partial V}{\partial y} = \frac{\partial}{\partial y}\left(3x^2y + y^2 + yz\right) = 3x^2 + 2y + z$,

and $\frac{\partial V}{\partial z} = \frac{\partial}{\partial z}\left(3x^2y + y^2 + yz\right) = \frac{\partial}{\partial z}yz = y$

Substituting these values in Eq.(1), we get-
$$\vec{E} = -6xy\hat{i} - (3x^2 + 2y + z)\hat{j} - y\hat{k}$$

EXAMPLE 18. If $V = -100/r$, then calculate intensity of electric field at point $(3, 4)$.

APPROACH Since, electric potential is given in terms of position r in 2-D, therefore, we use Eq.(2.48): $\left[E = -\dfrac{dV}{dr}\right]$ to get an expression for E. Here, $\vec{r} = x\hat{i} + y\hat{j}$ and $r = \sqrt{x^2 + y^2}$

SOLUTION $E = -\dfrac{dV}{dr} = -\dfrac{d}{dr}\left(-\dfrac{100}{r}\right) = -\dfrac{100}{r^2}$

The negative sign in the definition $E = -dV/dr$ indicates that the electric field vector points in the direction in which the potential decreases most rapidly.

In this problem, the potential is given by $V = -100/r$. Because of the negative sign in V, the potential *increases* with increasing r. Hence, $\vec{E}$ points opposite to the outward radial direction $\hat{r}$, i.e., towards the origin.

The magnitude of electric field is: $E = \dfrac{100}{r^2} > 0$

At $(3, 4)$, $r = \sqrt{3^2 + 4^2} = 5$, therefore, the magnitude of electric field:
$$E = \frac{100}{5^2} = 4 \text{ V/m}$$

In vector form the electric field can be written as
$$\vec{E} = -|E|\hat{r} = -4\hat{r} \quad (1)$$

where, $\vec{r} = 3\hat{i} + 4\hat{j}$, therefore, $\hat{r} = \dfrac{\vec{r}}{r} = \dfrac{3}{5}\hat{i} + \dfrac{4}{5}\hat{j}$

Substituting this value of $\hat{r}$ in Eq. (1), we get
$$\vec{E} = -\frac{12}{5}\hat{i} - \frac{16}{5}\hat{j} \text{ V/m}.$$

EXAMPLE 19. $V = 3x + 4y + 5z$. Find electric field intensity.

APPROACH Since, electric potential is given in terms of Cartesian coordinates, therefore, apply Eq. 2.46.

SOLUTION Given that- $V = 3x + 4y + 5z$, therefore-
$$\frac{\partial V}{\partial x} = 3, \frac{\partial V}{\partial y} = 4, \text{ and } \frac{\partial V}{\partial z} = 5$$

Substituting these values in Eq.2.46:
$$\vec{E} = -\left(\frac{\partial V}{\partial x}\hat{i} + \frac{\partial V}{\partial y}\hat{j} + \frac{\partial V}{\partial z}\hat{k}\right),$$

we get-
$$\vec{E} = -\left(\frac{\partial V}{\partial x}\hat{i} + \frac{\partial V}{\partial y}\hat{j} + \frac{\partial V}{\partial z}\hat{k}\right) = -\left(3\hat{i} + 4\hat{j} + 5\hat{k}\right) \text{ V/m}$$

Therefore, the magnitude of electric field is-
$$E = \sqrt{(-3)^2 + (-4)^2 + (5)^2} = \sqrt{50} = 5\sqrt{2} \text{ V/m}$$

EXAMPLE 20. $V = x^2y + y^2z + z^2x$. Find electric field intensity at $(1, 2, 3)$.

APPROACH Here, electric potential is given in terms of Cartesian coordinates, therefore, apply Eq.2.46.

SOLUTION Given that- $V = x^2y + y^2z + z^2x$, therefore-
$$\frac{\partial V}{\partial x} = 2xy + z^2, \frac{\partial V}{\partial y} = x^2 + 2zy, \text{ and } \frac{\partial V}{\partial z} = y^2 + 2zx$$

Substituting these values in Eq.2.46, we get-
$$\vec{E} = -\left(\frac{\partial V}{\partial x}\hat{i} + \frac{\partial V}{\partial y}\hat{j} + \frac{\partial V}{\partial z}\hat{k}\right)$$
$$= -\left((2xy + z^2)\hat{i} + (x^2 + 2zy)\hat{j} + (y^2 + 2zx)\hat{k}\right) \text{ V/m}$$

Therefore, at point $(1, 2, 3)$
$$\vec{E} = -((2(1)(2) + 3^2)\hat{i} + (1^2 + 2(3)(2))\hat{j} + (3^2 + 2(3)(1))\hat{k}) \text{ V/m}$$
$$= -\left(13\hat{i} + 13\hat{j} + 10\hat{k}\right) \text{ V/m}$$

and $|\vec{E}| = \sqrt{13^2 + 13^2 + 10^2} = \sqrt{438}$ V/m

EXAMPLE 21. If the electric potential at any point (x, y, z) m space is given by $V = 3x^2$ volt, find the electric field at the point $(1, 0, 3)$ m

SOLUTION Given: $V = 3x^2$, therefore, x, y and z components of electric field are-
$$E_x = -\frac{\partial V}{\partial x} = -6x, E_y = -\frac{\partial V}{\partial y} = 0, E_z = -\frac{\partial V}{\partial z} = 0,$$

respectively.

Now, net electric field,
$$\vec{E} = E_x\hat{i} + E_y\hat{j} + E_z\hat{k} = -6x\,\hat{i} + 0\hat{j} + 0\hat{k} = -6x\,\hat{i}$$

Therefore, at position $(1 \text{ m}, 0 \text{ m}, 3 \text{ m})$, $\vec{E} = -6(1) = -6$ V/m

2.6 Electric Potential For a System of Charges

For a total of N point charges, the potential V at any point P can be derived from the principle of superposition.

Recall that potential due to q_1 at point P:
$$V_1 = \frac{1}{4\pi\epsilon_0}\frac{q_1}{r_1}$$

Therefore, the total potential at point P due to all N point charges:
$$V = V_1 + V_2 + V_3 + \cdots + V_N \quad \text{(principle of superposition)}$$

2.6. ELECTRIC POTENTIAL FOR A SYSTEM OF CHARGES

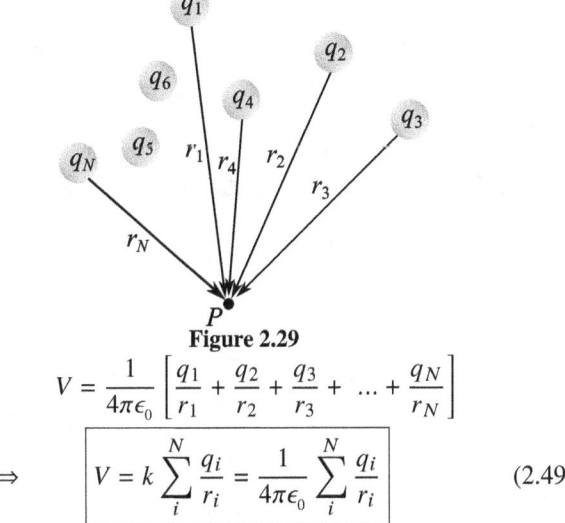

Figure 2.29

$$V = \frac{1}{4\pi\epsilon_0}\left[\frac{q_1}{r_1} + \frac{q_2}{r_2} + \frac{q_3}{r_3} + \ldots + \frac{q_N}{r_N}\right]$$

$$\Longrightarrow \boxed{V = k\sum_i^N \frac{q_i}{r_i} = \frac{1}{4\pi\epsilon_0}\sum_i^N \frac{q_i}{r_i}} \quad (2.49)$$

When we have a continuous distribution of charge along a line, over a surface, or through a volume, we divide the charge into elements dq, and the sum in Eq.2.49 becomes an integral:

$$\boxed{V = \frac{1}{4\pi\epsilon_0}\int \frac{dq}{r}} \quad (2.50)$$

Note: Remember that there does not have to be a charge at a given point for a potential V to exist at that point. (In the same way, an electric field can exist at a given point even if there's no charge there to respond to it.)

EXAMPLE 22. Positive and negative point charges of equal magnitude are kept at $(0,0,a/2)$ and $(0,0,-a/2)$, respectively. Find the work done by the electric field when another positive point charge is moved from $(-a,0,0)$ to $(0,a,0)$.

APPROACH Work done by electric field, when a point charge Q moves from potential V_A to potential V_B, is:
$$W = -Q(V_B - V_A)$$

SOLUTION The charge configuration is shown in the Fig.2.30. The point A$(-a,0,0)$ is at a distance $r_A = \sqrt{5}a/2$ from both the charges. Also, the point B$(0,a,0)$ is at a distance $r_B = \sqrt{5}a/2$ from both the charges.

The potentials at the point A and B are given by

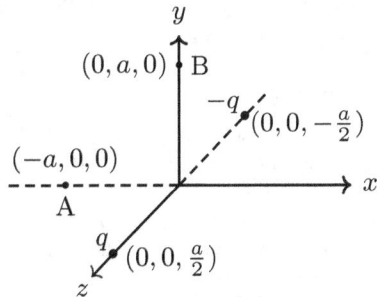

Figure 2.30

$$V_A = \frac{1}{4\pi\epsilon_0}\frac{q}{r_A} - \frac{1}{4\pi\epsilon_0}\frac{q}{r_A} = 0,$$
$$V_B = \frac{1}{4\pi\epsilon_0}\frac{q}{r_B} - \frac{1}{4\pi\epsilon_0}\frac{q}{r_B} = 0$$

Since $V_A = V_B$, therefore the work done in taking a point charge from A to B is
$$W = Q(V_B - V_A) = 0.$$

EXAMPLE 23. A charge $+q$ is fixed at each of the points $x = x_0, x = 3x_0, x = 5x_0, \cdots, \infty$ on the x-axis and a charge $-q$ is fixed at each of the points $x = 2x_0, x = 4x_0, x = 6x_0, \cdots, \infty$.

Here x_0 is a positive constant. Take the electric potential at a point due to a charge Q at a distance r from it to be $Q/(4\pi\epsilon_0 r)$. Find the potential at the origin due to the above system of charges.

APPROACH Due to several charge, the electric potential at any point is given by Eq.2.49, so apply it and solve for V at origin.

SOLUTION Using Eq.2.49, the electric potential at the origin due to the given system of charges is

$$V = \frac{q}{4\pi\epsilon_0 x_0}\left[\frac{1}{1} - \frac{1}{2} + \frac{1}{3} - \frac{1}{4} + \ldots\right] = \frac{q}{4\pi\epsilon_0 x_0}\ln 2$$

EXAMPLE 24. In Figure 2.31, six point charges are kept at the vertices of a regular hexagon of side L and centre O. Given that $k = \dfrac{1}{4\pi\epsilon_0}\dfrac{q}{L^2}$, which of the following statement(s) is (are) correct?

(A) The electric field at O is $6k$ along OD.
(B) The potential at O is zero.
(C) The potential at all points on the line PR is same.
(D) The potential at all points on the line ST is same.

APPROACH Apply the principle of superposition of electric

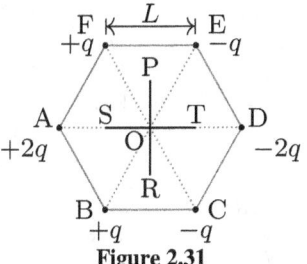

Figure 2.31

field as given in last chapter and principle of super position of electric potentials as given by Eq.2.49.

SOLUTION (A) (B) (C) The electric field at O due to the charges at A and D is $4K$ along OD, due to the charges at B and E is $2K$ along OE and due to the charges at C and F is $2K$ along OC. For the given geometry, resultant of these fields is $6K$ along OD.

The potential at O is
$$V_O = \sum \frac{1}{4\pi\epsilon_0}\frac{q_i}{L} = \frac{1}{4\pi\epsilon_0 L}\sum q_i = 0$$

For any point on PR, we have pairs of equal and opposite charges at the same distance making the potential at any point on PR zero. It may be seen that potential at points on OS is positive and that on OT is negative.

Comment: Students are advised to show that the potential on ST (at a distance x from O, taken positive towards the right) is

$$V(x) = \frac{q}{4\pi\epsilon_0}\left[\frac{2}{\sqrt{L^2 + x^2 + xL}} - \frac{2}{\sqrt{L^2 + x^2 - xL}} - \frac{4x}{L^2 - x^2}\right]$$

EXAMPLE 25. Six charges are placed around a regular hexagon of side length a as shown in the Fig.2.32. Five of them have charge q, and the remaining one has charge x. The perpendicular from each charge to the nearest hexagon side passes through the centre O of the hexagon and is bisected by the side. Which of the following statement(s) is (are) correct in SI units?

(A) When $x = q$, the magnitude of the electric field at O is zero.
(B) When $x = -q$, the magnitude of the electric field at O is $\dfrac{q}{6\pi\epsilon_0 a^2}$.
(C) When $x = 2q$, the potential at O is $\dfrac{7q}{4\sqrt{3}\pi\epsilon_0 a}$.
(D) When $x = -3q$, the potential at O is $\dfrac{3q}{4\sqrt{3}\pi\epsilon_0 a}$.

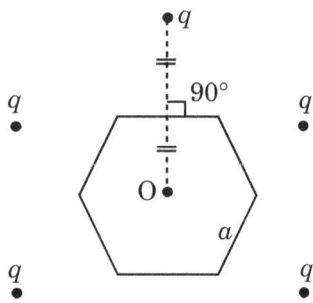

Figure 2.32

SOLUTION (A) When $x = q$, due to symmetry $\vec{E}_O = 0$

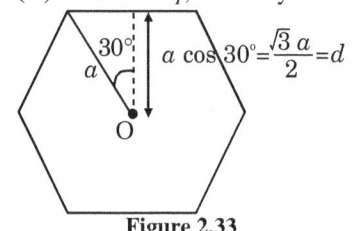

Figure 2.33

(B) When $x = -q$,
$$E_O = \frac{kq}{(2d)^2} \times 2 = \frac{2q \times 4}{4\pi\epsilon_0 \cdot 4 \cdot 3a^2} = \frac{q}{6\pi\epsilon_0 a^2}$$

(C) When $x = 2q$, the potential at O is
$$V_O = \frac{7kq}{2d} = \frac{7q}{4\pi\epsilon_0 \cdot \sqrt{3}a} = \frac{7q}{4\sqrt{3}\pi\epsilon_0 q}$$

(D) When $x = -3q$, the potential at O is
$$V_O = \frac{2kq}{2d} = \frac{2q}{4\pi\epsilon_0 \cdot \sqrt{3}a} = \frac{q}{2\sqrt{3}\pi\epsilon_0 q}$$

From above we see that the correct choices are- A, B and C.

EXAMPLE 26. In Fig.2.34a, four point charges $+8\ \mu C$, $-1\ \mu C$, $-1\ \mu C$, and $+8\ \mu C$ are fixed at the points $-\sqrt{27/2}$ m, $-\sqrt{3/2}$ m, $+\sqrt{3/2}$ m and $+\sqrt{27/2}$ m respectively on the y-axis. A particle of mass 6×10^{-4} kg and charge $+0.1\mu C$ moves along the x direction. Its speed at $x = +\infty$ is v_0. Find the least value of v_0 for which the particle will cross the origin. Also find the kinetic energy of the particle at the origin. Assume that space is gravity free. $\left[\dfrac{1}{4\pi\epsilon_0} = 9 \times 10^9\ \text{Nm}^2/\text{C}^2\right]$

APPROACH To cross the origin, the initial kinetic energy of the particle should be enough to cross the potential barrier applied by given four charges for any position $x = x_0$ (say) of the particle. Since, electric force is conservative, therefore, we can always apply the law of conservation of mechanical energy. If electric potential at any point on the x axis is V, then the position of potential barrier i.e., maximum potential, can be obtained by applying the relation $\partial V/\partial x = 0$.

Once you get the value of x_0, put $x = x_0$ in the expression for V and find net potential energy of the system for this position of moving charged particle. By conservation of energy, this energy will be equal to net initial kinetic energy of the particle for $x = \infty$. To get the kinetic energy of the particle at origin, again apply the conservation of mechanical energy principle and solve for KE.

SOLUTION Given $q = 1\mu C = 10^{-6}C$, $Q = 8\mu C = 8 \times 10^{-6}C = 8q$, $q_0 = 0.1\mu C = 10^{-7}C$, $m = 6 \times 10^{-4}$ kg and $a = \sqrt{3/2}$ m.
Consider a point P at a distance x from the origin.
The potential at P due to given charge distribution is

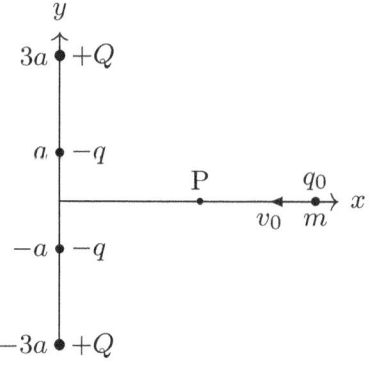

(a)

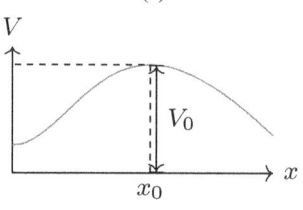

(b)

Figure 2.34

$$V(x) = \frac{1}{4\pi\epsilon_0}\left[\frac{2Q}{\sqrt{x^2 + 9a^2}} - \frac{2q}{\sqrt{x^2 + a^2}}\right]$$

The potential varies with x and attains its maximum at x_0 (Fig. 2.34b). For maximum value of $V(x)$, we have-

$$\frac{dV(x)}{dx} = -\frac{2x}{4\pi\epsilon_0}\left[\frac{Q}{(x^2 + 9a^2)^{3/2}} - \frac{q}{(x^2 + a^2)^{3/2}}\right] = 0 \quad \cdots (1)$$

On substituting, $Q = 8q$ in equation (1) and solving for x_0, we get $x_0 = a\sqrt{5/3} = \sqrt{5/2}$ m. The potential at x_0 is $V_0 = V(x_0) = 2.7 \times 10^4$ V.

Now, by conservation of mechanical energy, we can write-
$$\frac{1}{2}mv_0^2 = q_0 V_0, \quad \Longrightarrow \quad v_0 = \sqrt{2q_0 V_0/m} = 3 \text{ m/s}$$

The potential energy of the system for q_0 at the origin is
$$U = \frac{1}{4\pi\epsilon_0}\left[\frac{2Qq_0}{3a} - \frac{2qq_0}{a}\right] = 2.4 \times 10^{-3} \text{ J}$$

If K is the kinetic energy of q_0 at the origin, then the conservation of mechanical energy, $\frac{1}{2}mv_0^2 = K + U$, gives
$$K = \frac{1}{2}mv_0^2 - U = 3 \times 10^{-4} \text{ J}$$

EXAMPLE 27. Two fixed charges $-2Q$ and Q are located at the points with coordinates $(-3a, 0)$ and $(+3a, 0)$ respectively in the x-y plane.
(a) Show that all points in the $x - y$ plane where the electric potential due to the two charges is zero, lie on a circle. Find its radius and the location of its centre.
(b) Give the expression $V(x)$ at a general point on the x-axis and sketch the function $V(x)$ on the whole x-axis.
(c) If a particle of charge $+q$ starts from rest at the centre of the circle, show by a short quantitative argument that the particle eventually crosses the circle. Find its speed when it does so.

APPROACH For part (a), find the electric potential at any point $P(x, y)$ due to the charge $-2Q$ located at $(-3a, 0)$ and the charge Q located at $(3a, 0)$. For zero potential points, equate it to zero and find a relation between the coordinates x and y. The relation between x and y will give you the equation of path of zero potential points. For part (b), just find the electric potential at $(x, 0, 0)$ due to both the charges and apply the principle of superposition. To get speed of particle at the center of circle, apply the principle of conservation of energy.

SOLUTION (a) The net electric potential at the point $P(x, y)$ due to the charge $-2Q$ located at $(-3a, 0)$ and the charge Q located at $(3a, 0)$ is given by

$$V = \frac{Q}{4\pi\epsilon_0} \left[\frac{1}{\sqrt{(x-3a)^2 + y^2}} - \frac{2}{\sqrt{(x+3a)^2 + y^2}} \right]$$

For zero potential points: $V = 0$, therefore above expression gives-

$$\implies \frac{1}{\sqrt{(x-3a)^2 + y^2}} = \frac{2}{\sqrt{(x+3a)^2 + y^2}}$$

$$\implies (x-5a)^2 + y^2 = (4a)^2.$$

which is an equation of circle of radius $4a$ and centre $(5a, 0)$. **(b)**

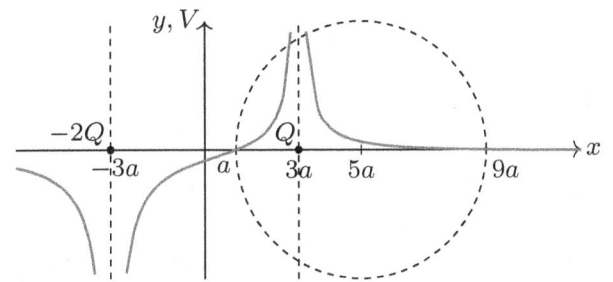

Figure 2.35

The potential on the x axis is given by

$$V(x) = \frac{Q}{4\pi\epsilon_0} \left(\frac{-2}{|x+3a|} + \frac{1}{|x-3a|} \right) \quad (2.51)$$

By definition of modulus function, we have-

$$|x| = \begin{cases} x, & \text{if } x \geq 0; \\ -x & \text{if } x < 0 \end{cases} \quad (2.52)$$

Using definition of modulus function (Eq.2.52), we can write Eq. 2.51 as -

$$V(x) = \begin{cases} \frac{Q}{4\pi\epsilon_0} \left(\frac{2}{3a+x} + \frac{1}{3a-x} \right), & \text{if } x \leq -3a; \\ \frac{Q}{4\pi\epsilon_0} \left(\frac{-2}{x+3a} + \frac{1}{3a-x} \right), & \text{if } -3a < x \leq 3a; \\ \frac{Q}{4\pi\epsilon_0} \left(\frac{-2}{x+3a} + \frac{1}{x-3a} \right), & \text{if } x > 3a. \end{cases} \quad (2.53)$$

From Eq.2.53, it is clear that- $V \to -\infty$ as $x \to -3a$ and $V \to \infty$ as $x \to 3a$. The potential is zero at $x = a$ and at $x = 9a$. V vs x graph is shown in Fig.2.35.

The potential at the centre of circle $(x = 5a)$ is

$$V = \frac{Q}{4\pi\epsilon_0} \left(\frac{-2}{8a} + \frac{1}{2a} \right) = \frac{Q}{16\pi\epsilon_0 a}$$

which has a positive value. The potential at the circumference of the circle is zero.

(c) A positive charge moves from a higher potential to a lower potential. By conservation of energy, decrease in the potential energy is equal to increase in kinetic energy i.e.,

$$\frac{1}{2}mv^2 = \frac{qQ}{16\pi\epsilon_0 a}, \quad \text{which gives} \quad v = \sqrt{\frac{Qq}{8\pi\epsilon_0 ma}}$$

2.6.1 Electric Potential at a Point $P(r, \theta)$ Due to an Electric Dipole

Let an electric dipole AB, of length $2l$, is placed along x-axis. The centre of the dipole is at origin O and charges $-q$ and $+q$ are situated at points A and B respectively (Fig.2.36a). $P(r, \theta)$ is a point where electric potential due to this electric dipole is required. If $r \gg l$, then we can assume $AP \| OP \| BP$. It is also shown in a separate Fig.2.36b.
In this case, $\angle PAB = \angle POB = \theta$.

Now, draw perpendiculars OC and BD from points O and B on the lines AP and BP respectively. From Fig. 2.36a, we have

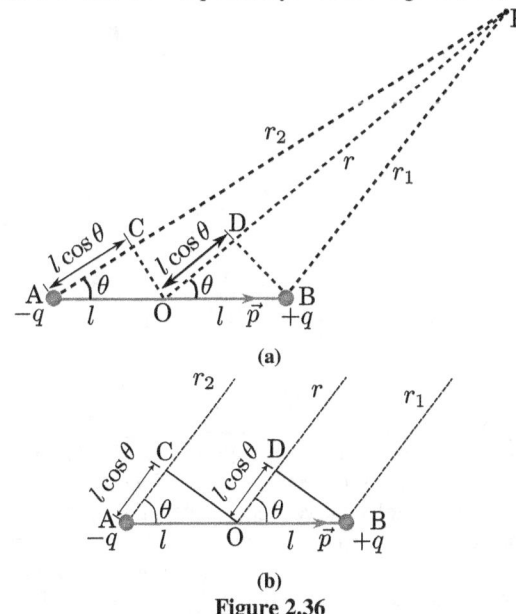

Figure 2.36

$AC = OD = l \cos\theta$ and $CP \approx OP$, $BP \approx DP$.
Therefore, $AP = AC + CP \approx AC + OP = l\cos\theta + r$
i.e., $r_2 \approx r + l\cos\theta$ and $BP \approx DP = OP - OD$
i.e., $r_1 \approx r - l\cos\theta$
Electric potential at point $P(r, \theta)$, due to charge $-q$ at A is-

$$V_{(-)} = \frac{kq}{r_2} \approx \frac{-kq}{r + l\cos\theta} \quad [\because r_2 \approx r + l\cos\theta]$$

And, the electric potential at point $P(r, \theta)$, due to charge $+q$ at B is -

$$V_{(+)} = \frac{kq}{r_1} \approx \frac{kq}{r - l\cos\theta} \quad [\because r_l \approx r - l\cos\theta]$$

Therefore, net electric potential at P due to both the charges of the electric dipole,

$$V = V_- + V_+ = kq \left[\frac{1}{r - l\cos\theta} - \frac{1}{r + l\cos\theta} \right]$$

$$= kq \left[\frac{r + l\cos\theta - r + l\cos\theta}{r^2 - l^2\cos^2\theta} \right] = \frac{kp\cos\theta}{[r^2 - l^2\cos^2\theta]}$$

For $r \gg l$

$$\boxed{V \approx \frac{kp\cos\theta}{r^2}} \quad (2.54)$$

here angle θ is measured from the dipole moment $\vec{p}$ to the position vector $\vec{r}$ that extends from the centre of the dipole to the point of interest. The sign of the electric potential depends on the angle θ (Fig. 2.37).

Special Cases

(i) End on Position: For End on positions, $\theta = 0$ or $\theta = \pi$ and then $\cos\theta = \pm 1$, therefore-

$$\boxed{V \approx \frac{kp\cos\theta}{r^2} = \pm\frac{kp}{r^2}} \quad (2.55)$$

(ii) Broadside on Position: For broadside on positions, $\theta = \pi/2$ and then $\cos\theta = 0$, therefore -

$$\boxed{V \approx \frac{kp\cos\theta}{r^2} = 0} \quad (2.56)$$

(iii) For $0 < \theta < \pi/2$, we have $\cos\theta > 0$. So-

$$\boxed{V \approx \frac{kp\cos\theta}{r^2} > 0} \quad (2.57)$$

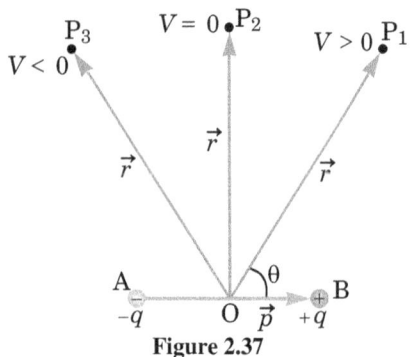

Figure 2.37

For $\pi/2 < \theta < \pi$, we have $\cos\theta < 0$. So -

$$\boxed{V \approx \frac{kp\cos\theta}{r^2} < 0} \qquad (2.58)$$

Figure 2.38 shows equipotential surfaces for the dipole in Figure 2.37. Most of the equipotential surfaces look like flattened spherical shells. Solid lines indicate positive electric potentials, and dashed lines indicate negative electric potentials. The solid black line in the centre of the Fig.2.38 represents the zero-potential surface, a flat plane. For a dipole, the electric potential is zero both at infinity and at any point on the plane shown in black. Notice that the sharpest change in the potential is between the two charges, near the plane. **EXAMPLE 28.** An electric dipole is formed

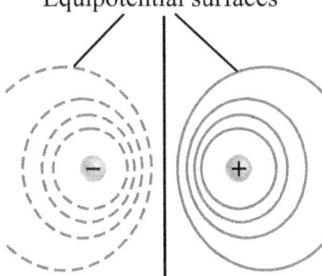

Figure 2.38: Equipotential surfaces for an electric dipole

by two charges $+q$ and $-q$ located in xy-plane at $(0, 2)$ mm and $(0, -2)$ mm, respectively, as shown in the Fig.2.39. The electric potential at point P$(100, 100)$ mm due to the dipole is V_0. The charges $+q$ and $-q$ are then moved to the points $(-1, 2)$ mm and $(1, -2)$ mm, respectively. What is the value of electric potential at P due to the new dipole?

(A) $V_0/4$ (B) $V_0/2$ (C) $V_0/\sqrt{2}$ (D) $3V_0/4$

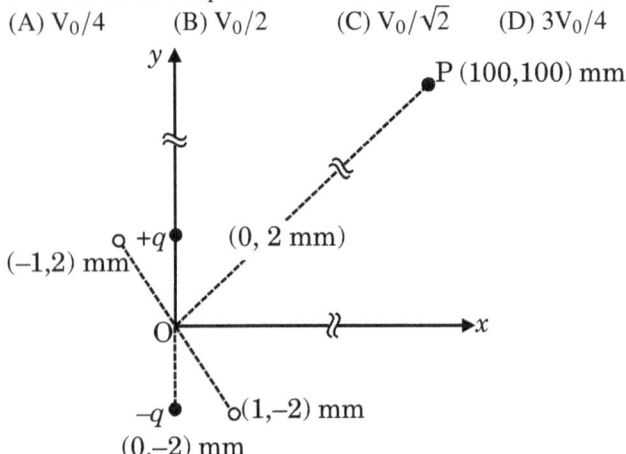

Figure 2.39

SOLUTION Initial separation between charges (Fig.2.40a):
$$2l = \sqrt{(0-0)^2 + (-2-2)^2} = 4\,\text{mm} = 4\times 10^{-3}\,\text{m}$$
Therefore, initial dipole moment,
$$\vec{p}_1 = p_1\hat{j} = q(2l)\,\hat{j} = q\left(4\times 10^{-3}\right)\hat{j},$$
Since, the centre of dipole divides it in the ratio of 1 : 1, therefore, the coordinates of centre of dipole will be $\left(\dfrac{0+0}{2}, \dfrac{2+(-2)}{2}\right)$, i.e., $(0, 0)$.
Position vector of point P $(100, 100)$ with respect to centre of the dipole-
$$\vec{r} = 100\hat{i} + 100\hat{j} = 100(\hat{i}+\hat{j}).$$
Therefore, magnitude of $\vec{r}$, i.e., $r = \sqrt{100^2 + 100^2} = 100\sqrt{2}\,\text{m}$
Initial voltage at point P,
$$V_0 = \frac{k\vec{p}_1\cdot\vec{r}}{r^3} = \frac{k(100\,p_1)}{(100\sqrt{2})^3} \qquad \cdots (1)$$
When the charges $+q$ and $-q$ are moved to the points $(-1, 2)$ mm

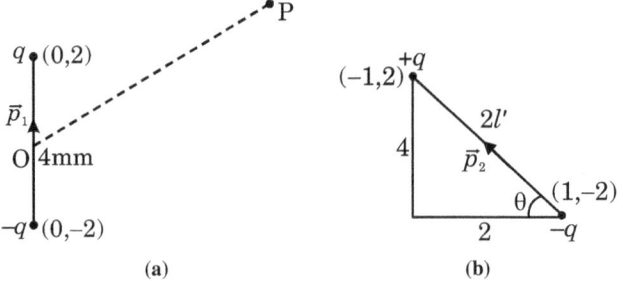

Figure 2.40

and $(1, -2)$ mm, respectively (Fig.2.40b), coordinates of new centre of dipole becomes $\left(\dfrac{-1+1}{2}, \dfrac{2+(-2)}{2}\right)$, i.e., $(0, 0)$.
New length of dipole is, $2l' = \sqrt{(1+1)^2 + (-2-2)^2} = 2\sqrt{5}\,\text{m}$
Therefore, magnitude of new dipole moment
$$p_2 = q(2l') = q\left(2\sqrt{5}\right)\times 10^{-3}\,\text{C.m}$$
New dipole moment vector,
$$\vec{p}_2 = -p_2\cos\theta\,\hat{i} + p_2\sin\theta\,\hat{j}$$
New position vector of point P with respect to centre of dipole
$$\vec{r}' = (100-0)\hat{i} + (100-0)\hat{j} = 100\hat{i} + 100\hat{j} = \vec{r} \text{ and } r' = r$$
New potential at point P
$$V = \frac{k\vec{p}_2\cdot\vec{r}'}{r'^3} = k\frac{-100 p_2\cos\theta + 100 p_2\sin\theta}{\left(100\sqrt{2}\right)^3}$$
$$\Rightarrow\quad V = k\frac{100}{\left(100\sqrt{2}\right)^3}(-p_2\cos\theta + p_2\sin\theta) \qquad \cdots (2)$$
Again from Fig.2.40b, $\cos\theta = \dfrac{2}{2\sqrt{5}} = \dfrac{1}{\sqrt{5}}$,
$$d\sin\theta = \frac{4}{2\sqrt{5}} = \frac{2}{\sqrt{5}}.$$
Therefore, eq.(2) gives
$$V = k\frac{100}{\left(100\sqrt{2}\right)^3} p_2\left(-\frac{1}{\sqrt{5}} + \frac{2}{\sqrt{5}}\right) = k\frac{100}{\left(100\sqrt{2}\right)^3}\frac{p_2}{\sqrt{5}}$$
Now, from Eq.(1) and (2), we get
$$\frac{V}{V_0} = \frac{p_2}{p_1\sqrt{5}} = \frac{q\left(2\sqrt{5}\right)\times 10^{-3}}{q\left(4\times 10^{-3}\right)\sqrt{5}} = \frac{1}{2} \Rightarrow V = \frac{V_0}{2}$$
Thus, the correct choice is B.

2.7 Check Point 2

1. ●●If $V = -5x + 3y + \sqrt{15}z$ then find magnitude of electric field at point (x, y, z).

2. ••Fig. 2.41 is a graph of the electric potential in a region of space where $\vec{E}$ is parallel to the x-axis. Draw a graph of E_x versus x.

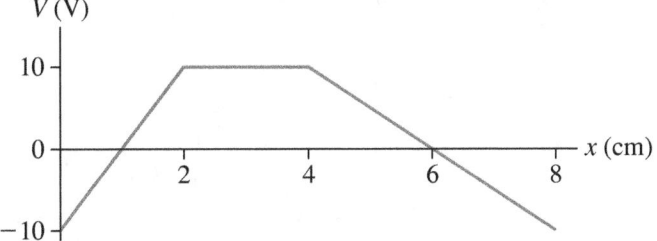

Figure 2.41

3. ••In Fig.2.42, a 1 cm × 1 cm grid is superimposed on a contour map of the potential. Estimate the strength and direction of the electric field at points 1, 2, and 3. Show your results graphically by drawing the electric field vectors on the contour map.

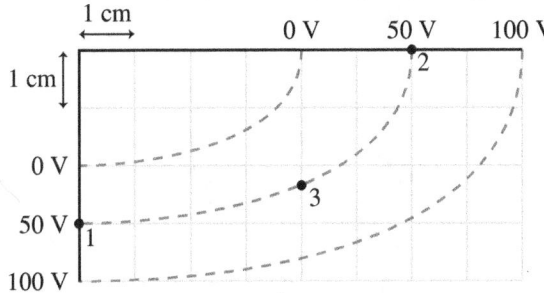

Figure 2.42: Equipotential lines

4. ••In a certain region of space, the electric potential is $V(x, y, z) = Axy - Bx^2 + Cy$, where A, B, and C are positive constants. (a) Calculate the x-, y-, and z-components of the electric field. (b) At which points is the electric field equal to zero?

5. ••In a certain region of space the electric potential is given by $V = +Ax^2y - Bxy^2$, where $A = 5.00$ V/m^3 and $B = 8.00$ V/m^3. Calculate the magnitude and direction of the electric field at the point in the region that has coordinates $x = 2.00$ m, $y = 0.400$ m, and $z = 0$

6. ••The electric potential V at any point (x, y, z) (all in metre) in space is given by $V = 4x^2$ volt. Find the electric field at the point $(1 \text{ m}, 0 \text{ m}, 2 \text{ m})$.

7. ••A metal sphere with radius r_a is supported on an insulating stand at the center of a hollow, metal spherical shell with radius r_b. There is charge $+q$ on the inner sphere and charge $-q$ on the outer spherical shell.
(a) Calculate the potential $V(r)$ for (i) $r < r_a$; (ii) $r_a < r < r_b$; (iii) $r > r_b$. (**Hint:** The net potential is the sum of the potentials due to the individual spheres.) Take V to be zero when r is infinite.
(b) Show that the potential of the inner sphere with respect to the outer is
$$V_{ab} = \frac{q}{4\pi\epsilon_0}\left(\frac{1}{r_a} - \frac{1}{r_b}\right)$$
(c) Use relation $E = -\partial V/\partial r$ and the result from part (a) to show that the electric field at any point between the spheres has magnitude
$$E(r) = \frac{V_{ab}}{(1/r_a - 1/r_b)}\frac{1}{r^2}$$
(d) Use Eq. $E = -\partial V/\partial r$ and the result from part (a) to find the electric field at a point outside the larger sphere at a distance r from the centre, where $r > r_b$.

(e) Suppose the charge on the outer sphere is not $-q$ but a negative charge of different magnitude, say $-Q$. Show that the answers for parts (b) and (c) are the same as before but the answer for part (d) is different.

8. ••If $\vec{E}$ is zero throughout a certain region of space, is the potential necessarily also zero in this region? Why or why not? If not, what can be said about the potential?

Multiple Choice Questions

9. ••In Figure 2.43, two points A and B are located within a region in which there is an electric field. The potential difference $\Delta V = V_B - V_A$ is-
(A) $+ve$ (B) $-ve$ (C) 0   (D) insufficient data

10. ••In Fig.2.44, a point charge $+q$ is placed at the origin O. Work done in taking another point charge $-Q$ from the point A [coordinates $(0, a)$] to another point B [coordinates $(a, 0)$] along the straight path AB, is-

(A) zero   (B) $\left(\frac{qQ}{4\pi\epsilon_0}\frac{1}{a^2}\right)\cdot\sqrt{2}a$

(C) $\left(\frac{-qQ}{4\pi\epsilon_0}\frac{1}{a^2}\right)\cdot\sqrt{2}a$ (D) $\left(\frac{qQ}{4\pi\epsilon_0}\frac{1}{a^2}\right)\cdot\frac{a}{\sqrt{2}}$

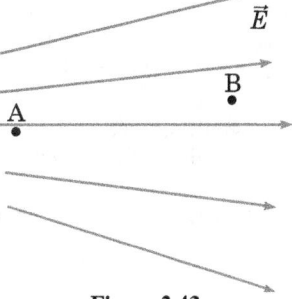

Figure 2.43 **Figure 2.44**

11. ••The electric potential at a point (x, y, z) is given by $V = -x^2y - xz^3 + 4$. The electric field at that point is
(A) $\vec{E} = \hat{i}2xy + \hat{j}\left(x^2 + y^2\right) + \hat{k}\left(3xz - y^2\right)$
(B) $\vec{E} = \hat{i}z^3 + \hat{j}xyz + \hat{k}z^2$
(C) $\vec{E} = \hat{i}\left(2xy - z^3\right) + \hat{j}xy^2 + \hat{k}3z^2x$
(D) $\vec{E} = \hat{i}\left(2xy + z^3\right) + \hat{j}x^2 + \hat{k}3xz^2$

12. ••If potential (in volts) in a region is expressed as $V(x,y,z) = 6xy - y + 2yz$, the electric field (in N/C) at point $(1, 1, 0)$ is
(A) $-(2\hat{i} + 3\hat{j} + \hat{k})$ (B) $-(6\hat{i} + 9\hat{j} + \hat{k})$
(C) $-(3\hat{i} + 5\hat{j} + 3\hat{k})$ (D) $-(6\hat{i} + 5\hat{j} + 2\hat{k})$

13. ••In a region, the potential is represented by $V(x, y, z) = 6x - 8xy - 8y + 6yz$, where V is in volts and x, y, z are in metres. The electric force experienced by a charge of 2 coulomb situated at point $(1, 1, 1)$ is
(A) $6\sqrt{5}$ N   (B) 30 N
(C) 24 N   (D) $4\sqrt{35}$ N

14. ••An alpha particle of energy 5 MeV is scattered through 180° by a fixed uranium nucleus. The distance of closest approach is of the order of
(A) 10^{-9} cm   (B) 10^{-10} cm
(C) 10^{-12} cm   (D) 10^{-15} cm

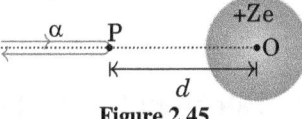

Figure 2.45

15. ••A dipole of dipole moment $\vec{p}$ is kept along an electric field $\vec{E}$ such that $\vec{E}$ and $\vec{p}$ are in the same direction. Find the

work done in rotating the dipole by an angle π.
(A) $W = 2Ep$ (B) $W = -2Ep$
(C) $W = Ep$ (D) $W = -Ep$

16. An electric dipole, made up of a positive and a negative charge, each of magnitude 1 μC and placed at a distance 2 cm apart, is placed in an electric field 10^5 N/C. Compute the maximum torque which the field can exert on the dipole, and the work that must be done to turn the dipole from a position $\theta = 0°$ to $\theta = 180°$
    (A) 6×10^{-3} N.m and 4×10^6 joule
    (B) 3×10^{-3} N.m and 4×10^9 joule
    (C) 4×10^{-3} N.m and $4 \times 10^{+6}$ joule
    (D) 2×10^{-3} N.m and 4×10^3 joule

2.7.1 Electric Potential due to a Line charge of Finite Length

EXAMPLE 29. Find the electric potential of a uniformly charged, non-conducting wire with linear density λ (in coulomb/meter) and length L at a point that lies on a line that divides the wire into two equal parts.

APPROACH We choose Cartesian coordinates in such a way as to exploit the symmetry in the problem as much as possible. We place the origin at the centre of the wire and orient the y-axis along the wire so that the ends of the wire are at $y = \pm L/2$. The field point P is in the xy-plane and since the choice of axes is up to us, we choose the x-axis to pass through the field point P, as shown in Figure 2.46.

SOLUTION Consider a small wire element dy of the charge

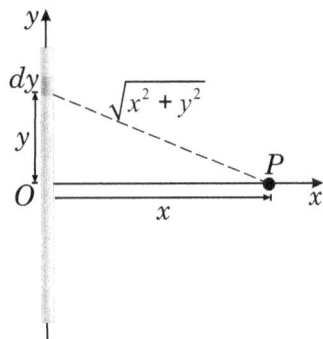

Figure 2.46: Electric potential due to a line of charge

distribution at distance y from O [Fig. 2.46]. The charge on the element is $dq = \lambda \, dy$ and the distance from the charge element to the field point P is $\sqrt{x^2 + y^2}$.
Therefore, the potential becomes

$$V = k \int \frac{dq}{r} = k \int_{-L/2}^{L/2} \frac{\lambda \, dy}{\sqrt{x^2 + y^2}}$$

$$= k\lambda \left[\ln\left(y + \sqrt{y^2 + x^2}\right) \right]_{-L/2}^{L/2}$$

$$= k\lambda \left[\ln\left(\left(\tfrac{L}{2}\right) + \sqrt{\left(\tfrac{L}{2}\right)^2 + x^2}\right) \right.$$
$$\left. - \ln\left(\left(-\tfrac{L}{2}\right) + \sqrt{\left(-\tfrac{L}{2}\right)^2 + x^2}\right) \right]$$

$$\Longrightarrow \boxed{V = k\lambda \ln\left[\frac{L + \sqrt{L^2 + 4x^2}}{-L + \sqrt{L^2 + 4x^2}}\right]} \quad (2.59)$$

2.7.2 Potential Due to an Infinite Charged Wire

In Example 29, we have already calculated the potential at any point due to a wire of finite length 'L.' Now, in Example 30, we will be considering a special case where the length of the wire is infinite.

EXAMPLE 30. Find the electric potential due to an infinitely long uniformly charged wire.

APPROACH We have already worked out the potential of a finite wire of length L in EXAMPLE 29. Now, taking $L \to \infty$ in Eq.2.59, we get

$$V = \lim_{L \to \infty} k\lambda \ln\left(\frac{L + \sqrt{L^2 + 4x^2}}{-L + \sqrt{L^2 + 4x^2}}\right)$$

$$= \lim_{L \to \infty} k\lambda \ln\left(\frac{1 + \sqrt{1 + \frac{4x^2}{L^2}}}{-1 + \sqrt{1 + \frac{4x^2}{L^2}}}\right) = k\lambda \ln\left(\frac{2}{0}\right)$$

Since, $\ln\left(\frac{2}{0}\right)$ is not defined, therefore this way of finding V of an infinite wire does not work. The reason for this problem may be traced to the fact that the charges are not localized in some space but continue to infinity in the direction of the wire. Hence, our (unspoken) assumption that zero potential must be an infinite distance from the wire is no longer valid.

To avoid this difficulty in calculating limits, we use the definition

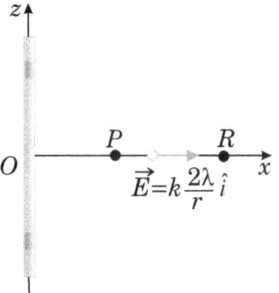

Figure 2.47: Electric potential of an infinite line of charge

of potential by integrating over the electric field, and the value of the electric field from this charge configuration from the previous chapter.

SOLUTION We use the integral

$$V_P - V_R = -\int_R^P \vec{E} \cdot d\vec{s}$$

where R is a finite distance from the line of charge, as shown in Figure 2.47.

With this setup, we use $\vec{E}_P = 2k\lambda \frac{1}{r}\hat{i}$ and $d\vec{s} = d\vec{r}$ to obtain

$$V_P - V_R = -\int_{r_R}^{r_P} 2k\lambda \frac{1}{r} dr = -2k\lambda \ln \frac{r_P}{r_R}$$

Now, if we define the reference potential $V_R = 0$ at $r_R = 1$m, this simplifies to

$$\boxed{V_P = -2k\lambda \ln r_P} \quad (2.60)$$

Note that this form of the potential is quite usable; it is 0 at 1m and is undefined at infinity, which is why we could not use the latter as a reference.

2.7.3 Electric Potential due to a Charged Thin Plane Sheet

Let's assume we have a charged thin plane sheet with surface charge density σ (charge per unit area) and it extends infinitely in both directions. The electric field due to an infinite plane sheet of charge is given by:

$$E = \frac{\sigma}{2\epsilon_0} \quad (2.61)$$

where ϵ_0 is the electric permittivity of free space.
Now, to find the potential at a point P at distance $|z|$ from the sheet, we integrate this electric field expression with respect to distance (from a reference point at $r = \infty$ to '$|z|$' from the sheet):

$$V = -\int_{\infty}^{|z|} E\,dr$$

Substitute the the value of E from Eq.2.61, we get

$$V = -\int_{\infty}^{|z|} \frac{\sigma}{2\epsilon_0}\,dr = -\frac{\sigma}{2\epsilon_0}\int_{\infty}^{|z|} dr$$

Integrating $\int_{\infty}^{|z|} dr$ is problematic as it gives undefined result. Instead, we should choose a reference point at a finite distance. Let's set the reference point at $r = 0$ (the location of the charged sheet). The potential at this reference point is V_0. In this scenario, the electric potential can be calculates as follows:

$$\int_{V_0}^{V} dV = -\int_{0}^{|z|} \frac{\sigma}{2\epsilon_0}\,dr \implies V - V_0 = -\frac{\sigma}{2\epsilon_0}|z|$$

$$\implies \boxed{V = V_0 - \frac{\sigma}{2\epsilon_0}|z|} \qquad (2.62)$$

This is the expression for the electric potential due to a charged thin plane sheet.

2.7.4 Electric Potential Due to a Charged Ring

In Fig.2.48, a ring of radius a is shown. A charge Q is uniformly distributed over the circumference of of this ring. P is a point on the axis of the ring at a distance x from the centre of the ring, where electric potential is to be calculated.

Notice that no vector considerations are necessary here because electric potential is a scalar.

☞ Because the ring consists of a continuous distribution of charge rather than a set of discrete charges, we must use the integration technique to calculate electric potential.

We divide the ring into infinitesimal segments and use Eq.2.50 to find V. All parts of the ring (and therefore all elements of the charge distribution) are at the same distance from P.

The electric potential at P due to the charge element dq of the ring is given by

$$dV = \frac{1}{4\pi\epsilon_0}\frac{dq}{r} = \frac{1}{4\pi\epsilon_0}\frac{dq}{(a^2 + x^2)^{1/2}}$$

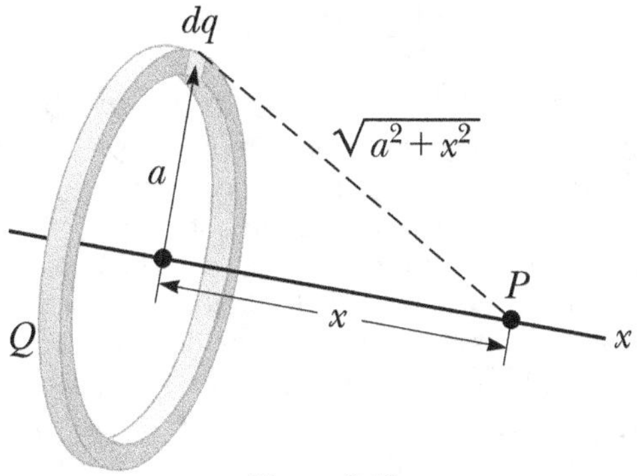

Figure 2.48

Hence, the electric potential at P due to the uniformly charged ring is given by-

$$V = \int_{0}^{Q} \frac{1}{4\pi\epsilon_0}\frac{dq}{(a^2 + x^2)^{1/2}}$$

Figure 2.48 shows that the distance from each charge element dq to P is $r = \sqrt{x^2 + a^2}$. Therefore, on taking the factor $1/\sqrt{x^2 + a^2}$ outside the integral in above equation, we get-

$$V = \frac{1}{4\pi\epsilon_0}\frac{1}{(a^2 + x^2)^{1/2}}\int_{0}^{Q} dq$$

or $\boxed{V = \frac{1}{4\pi\epsilon_0}\frac{Q}{\sqrt{(a^2 + x^2)}}} \qquad (2.63)$

☞ Eq.2.63 is also valid for a ring segment having radius a and charge Q.

Note: When x is much larger than a, our expression for V becomes approximately $V = Q/4\pi\epsilon_0 x$, which is the potential at a distance x from a point charge Q. Very far from a charged ring, its electric potential looks like that of a point charge.

EXAMPLE 31. What is the potential on the axis of a non-uniform ring of charge, where the charge density is $\lambda(\theta) = \lambda\cos\theta$?

SOLUTION Since, $\cos\theta$ is positive in first and fourth quadrants whereas it is negative in second and third quadrants, therefore for all points on the axis, there are equal and opposite charges equidistant from the point of interest. So, electric potential at all points on the axis will be zero. Note that this distribution will, in fact, have a dipole moment.

EXAMPLE 32. A circular ring of radius R with uniform positive charge density λ per unit length is located in the y-z plane with its centre at the origin O. A particle of mass m and positive charge q is projected from the point $P(R\sqrt{3}, 0, 0)$ on the positive x-axis directly towards O, with an initial speed v. Find the smallest (non-zero) value of the speed v such that the particle does not return to P.

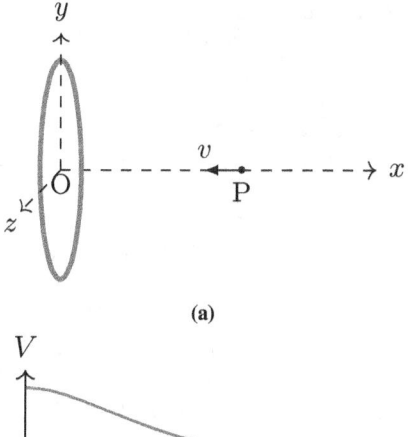

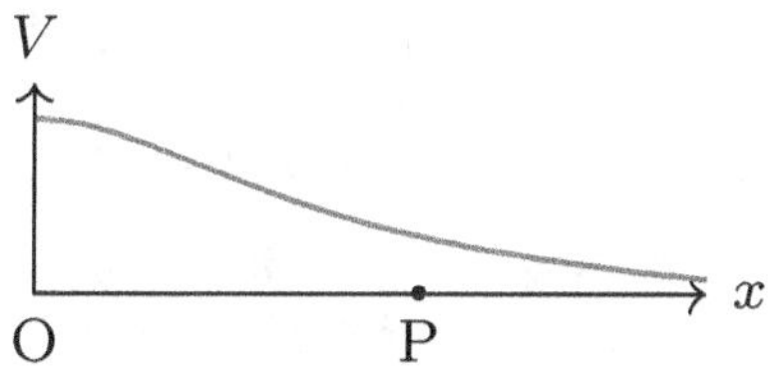

Figure 2.49

APPROACH From Eq.2.63, the electric potential at any point P $(x, 0, 0)$ on the axis of charged ring, is given by-

or $\boxed{V = \frac{1}{4\pi\epsilon_0}\frac{Q}{\sqrt{(a^2 + x^2)}}} \qquad (i)$

The potential decreases monotonically from centre O $(0, 0, 0)$ to P $(x, 0, 0)$ and it is maximum for $x = 0$, i.e., at the centre O of the ring. So, the particle cannot come back to P if it just crosses O. Since, the force acting on the particle is only electrostatic force, which is conservative in nature, therefore, the total mechanical energy of the ring-particle system will remain conserved. Now, apply conservation of mechanical energy to get the velocity of particle at point P.

SOLUTION The charge on the ring is $Q = 2\pi R\lambda$. Therefore, from Eq.(i), the electric potential at a point P $(R\sqrt{3}, 0, 0)$-

$$V_P = \frac{1}{4\pi\epsilon_0}\frac{Q}{2R} \qquad (ii)$$

The electric potentials at the point $O(0, 0, 0)$-

$$V_O = \frac{1}{4\pi\epsilon_0}\frac{Q}{R}, \qquad (iii)$$

Net mechanical energy of the particle-ring system at point P $(R\sqrt{3}, 0, 0)$-

$$E_P = K_P + U_{el} = \frac{1}{2}mv^2 + qV_P \qquad \text{(iv)}$$

here, v is the speed of the particle at point P.

If the charged particle just crosses the centre O of the ring, then it's velocity and hence it's kinetic energy at O will be zero. So, net mechanical energy of the particle ring system at point O $(0, 0, 0)$-

$$E_O = qV_O \qquad \text{(v)}$$

Applying, the principle of conservation of mechanical energy for points P and O, we get-

$$E_P = E_O$$

Substituting the values from Eq.v and iv in above, we get-

$$\frac{1}{2}mv^2 + qV_P = qV_O$$

or $\quad \frac{1}{2}mv^2 = q(V_O - V_P) \qquad \text{(vi)}$

Substituting the values of V_P and V_O, from Eq.ii and iii in Eq.vi, we get

$$\frac{1}{2}mv^2 = q\left(\frac{1}{4\pi\epsilon_0}\frac{Q}{R} - \frac{1}{4\pi\epsilon_0}\frac{Q}{2R}\right) = \frac{1}{4\pi\epsilon_0}\frac{qQ}{2R}$$

Since, $Q = \lambda(2\pi R)$, therefore-

$$\frac{1}{2}mv^2 = \frac{1}{4\pi\epsilon_0}\frac{q\lambda(2\pi R)}{2R} = \frac{q\lambda}{4\epsilon_0}$$

Solving for v, we get

$$v = \sqrt{q\lambda/(2\epsilon_0 m)}$$

EXAMPLE 33. Two identical thin rings, each of radius R, are co-axially placed a distance R apart. If Q_1 and Q_2 are respectively the charges uniformly spread on the two rings, the work done in moving a charge q slowly from the centre of one ring to that of the other is:

(A) zero $\qquad$ (B) $q(Q_1 - Q_2)\left(\dfrac{\sqrt{2}-1}{\sqrt{2}\pi\epsilon_0 R}\right)$

(C) $q\sqrt{2}\left(\dfrac{Q_1 + Q_2}{4\pi R}\right)$ $\qquad$ (D) $q(Q_1/Q_2)\left(\dfrac{\sqrt{2}+1}{\sqrt{2}\pi\epsilon_0 R}\right)$

APPROACH To find the work done in moving a charge q from the centre of one ring to that of the other, apply relation-

$$W = \Delta U = U_B - U_A$$

here U_A is the electric potential energy of the system when the charged particle was at any point on ring 1 and U_B is the potential energy of the system for the position of the charged particle on ring 2.

SOLUTION (B) The potential at A due to the charge Q_1 on the ring 1 is given as:

$$V_{A1} = k\frac{Q_1}{R}$$

The potential at A due to the charge Q_2 on the ring 2 is given as:

$$V_{A2} = k\frac{Q_2}{\sqrt{R^2 + R^2}} = k\frac{Q_2}{R\sqrt{2}}$$

Total potential at A is

$$V_A = V_{A1} + V_{A2} = k\frac{1}{R}\left(Q_1 + \frac{Q_2}{\sqrt{2}}\right)$$

The potential energy of charge q at A is

$$U_A = qV_A = k\frac{q}{R}\left(Q_1 + \frac{Q_2}{\sqrt{2}}\right)$$

Similarly, the potential energy of charge q at B is

$$U_B = qV_B = k\frac{q}{R}\left(Q_2 + \frac{Q_1}{\sqrt{2}}\right)$$

The work done in moving a charge q from point A to B is:

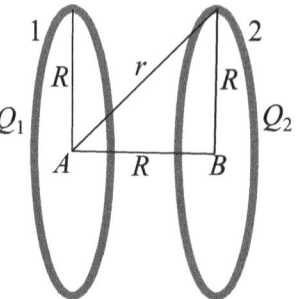

Figure 2.50

$$W = \Delta U = U_B - U_A = k\frac{q}{R}\left(Q_2 + \frac{Q_1}{\sqrt{2}} - Q_1 - \frac{Q_2}{\sqrt{2}}\right)$$

$$= \frac{1}{4\pi\epsilon_0 R}(Q_2 - Q_1)\frac{\sqrt{2}-1}{\sqrt{2}}$$

EXAMPLE 34. Calculate the electric potential at the centre of two concentric half rings of radii R_1 and R_2, having same linear charge density λ.

SOLUTION The situation is shown in Fig.2.51. The net potential (V_{centre}) at the centre will be equal to algebraic sum of potentials produced due to both semicircular concentric half rings.

$$V_{\text{centre}} = \frac{kQ_1}{R_1} + \frac{kQ_2}{R_2} \qquad \cdots (1)$$

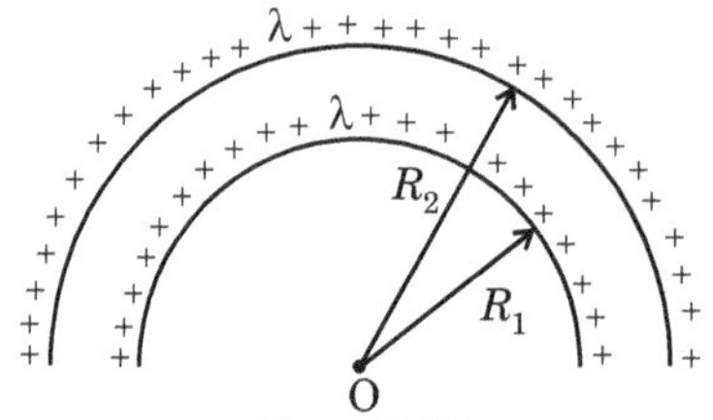

Figure 2.51

here, Q_1 and Q_2 are charges on semicircular rings of radii R_1 and R_2 respectively.

Since, charge density on each semicircular ring is λ, therefore

$$Q_1 = \lambda.\pi R_1 \text{ and } Q_2 = \lambda.\pi R_2$$

Substituting these values in Eq.(1), we get-

$$V_{\text{centre}} = \frac{(\lambda \cdot \pi R_1)}{4\pi\epsilon_0 R_1} + \frac{(\lambda \cdot \pi R_2)}{4\pi\epsilon_0 R_2} = \frac{\lambda}{2\epsilon_0}$$

2.7.5 Electric Potential Due to a Charged Disc at a Point on it's Geometric Axis

A non-conducting disc of radius R has a uniform surface charge density σ C/m^2. Let us calculate the potential at a point on the axis of the disc at a distance x from its centre. The symmetry of the disc tells us that the appropriate choice of element is a ring of radius x and thickness dx. All points on this ring are at the same distance $z = \sqrt{x^2 + r^2}$, from the point P. The charge on the ring is $dq = \sigma dA = \sigma(2\pi r dr)$ and so the potential due to the ring is

$$dV = \frac{1}{4\pi\epsilon_0}\frac{dq}{z} = \frac{1}{4\pi\epsilon_0}\frac{(\sigma 2\pi r dr)}{\sqrt{r^2 + x^2}}$$

Since potential is scalar, the potential due to the whole disc is given by

$$V = \frac{\sigma}{2\epsilon_0}\int_0^R \frac{r dr}{\sqrt{r^2 + x^2}}$$

$$= \frac{\sigma}{2\epsilon_0}[(r^2 + x^2)^{1/2}]_0^R = \frac{\sigma}{2\epsilon_0}[(R^2 + x^2)^{1/2} - x]$$

or $\quad \boxed{V = \frac{\sigma}{2\epsilon_0}[(R^2 + x^2)^{1/2} - x]} \qquad (2.64)$

Let us see this expression at large distance when $x \gg R$. In this case, $R^2 + x^2 \approx x^2$, therefore, corresponding to $x \gg R$, we have-

2.7. CHECK POINT 2

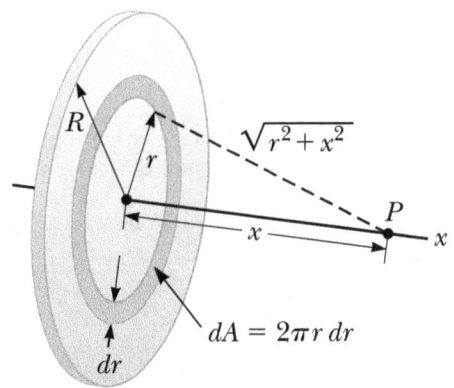

Figure 2.52

$$V = \frac{1}{4\pi\epsilon_0} \frac{Q}{x}$$

where $Q = \pi R^2 \sigma$ is the total charge on the disc.

Thus, we conclude that at large distance, the potential due to the disc is the same as that of a point charge Q.

At the centre of the disc $x = 0$, therefore

$$V = \frac{\sigma}{2\epsilon_0} R$$

EXAMPLE 35. A non-conducting disc of radius a and uniform positive surface charge density σ is placed on the ground with its axis vertical. A particle of mass m and positive charge q is dropped, along the axis of the disc from a height H with zero initial velocity. The particle has $q/m = 4\epsilon_0 g/\sigma$

(a) Find the value of H if the particle just reaches the disc.
(b) Sketch the potential energy of the particle as a function of its height and find its equilibrium position.

(a) APPROACH There are two forces on the charged particle:
1. Downward gravitational force
2. Upward electrostatic repulsive force due to electric field produced by the disc.

Both forces are conservative in nature, therefore, total mechanical energy of the earth-disc-particle system, will always remain conserved.

To find the height H, from where the particle is released, we apply the principle of conservation of mechanical energy.

Electric potential due to disc of radius R at distance x from it's centre is given by Eq.2.64-

$$V = \frac{\sigma}{2\epsilon_0}[(R^2 + x^2)^{1/2} - x] \quad \text{(i)}$$

In this problem, radius, $R = a$ and $x = H$, therefore V at height H-

$$V_P = \frac{\sigma}{2\epsilon_0}[(a^2 + H^2)^{1/2} - H]$$

Considering the position of disc as reference level, the gravitational potential energy of the earth-particle system for the particle at height H is-

$$[U_{\text{grav}}]_P = mgH$$

The electric potential energy of the disc particle system for the position of particle at height H-

$$[U_{\text{el}}]_P = qV_P = \frac{q\sigma}{2\epsilon_0}[(a^2 + H^2)^{1/2} - H]$$

Since, the particle is released from rest, therefore it's kinetic energy at height H-

$$K_P = 0$$

So, net mechanical energy of the disc-particle-earth system for the position of particle at height H, is given by-

$$E_P = [U_{\text{grav}}]_i + [U_{\text{el}}]_i + K_i$$
$$= mgH + \frac{q\sigma}{2\epsilon_0}[(a^2 + H^2)^{1/2} - H] + 0$$

$$E_P = mgH + \frac{q\sigma}{2\epsilon_0}[(a^2 + H^2)^{1/2} - H] \quad \text{(iii)}$$

Electric potential at the centre O of disc is-

$$V_O = \frac{\sigma a}{2\epsilon_0}$$

Therefore, electrostatic potential energy of the disc-particle system for particle at O is

$$[U_{\text{el}}]_O = qV_O = q\frac{\sigma a}{2\epsilon_0}$$

Since, disc surface is selected as reference level for gravitational potential energy, therefore at O-

$$[U_{\text{grav}}]_O = 0$$

Since, the particle just reaches the disc, so its kinetic energy at O will be zero.

$$K_O = 0$$

So, net mechanical energy of the earth-disc-particle system for position of particle at O, is given by-

$$E_O = [U_{\text{grav}}]_f + [U_{\text{el}}]_O + [K]_O$$
$$\Rightarrow \quad E_O = q\frac{\sigma a}{2\epsilon_0} + 0 + 0 = q\frac{\sigma a}{2\epsilon_0} \quad \text{(iv)}$$

Now, apply conservation of mechanical energy and solve for H

SOLUTION By conservation of mechanical energy, we have

$$E_P = E_O \quad \text{(v)}$$

Substituting the values of E_P and E_O in Eq.(v), we get

$$mgH + \frac{q\sigma}{2\epsilon_0}[(a^2 + H^2)^{1/2} - H] = q\frac{\sigma a}{2\epsilon_0}$$

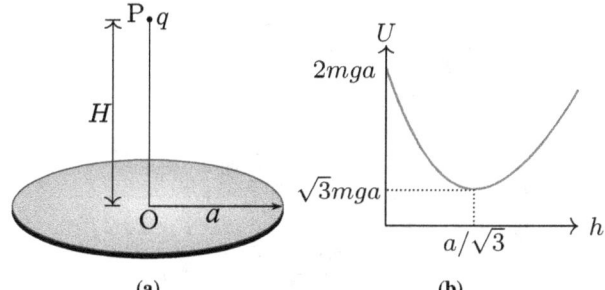

Figure 2.53

$$\Rightarrow \quad mgH = \frac{q\sigma}{2\epsilon_0}\left[a + H - \sqrt{a^2 + H^2}\right]$$

$$\Rightarrow \quad H = \frac{q\sigma}{2\epsilon_0 mg}\left[a + H - \sqrt{a^2 + H^2}\right]$$

On substitute $q/m = 4\epsilon_0 g/\sigma$, we get

$$H = 2\left[a + H - \sqrt{a^2 + H^2}\right]$$

$$\Rightarrow \quad \sqrt{a^2 + H^2} = a + H/2 \quad \Rightarrow \quad H = 4a/3$$

(b) APPROACH Net conservative force acting on the particle, is equal to $-ve$ of potential energy gradient, i.e.,

$$F = -\frac{\partial U}{\partial H} \quad \text{(vi)}$$

At equilibrium position, $F = 0$, so from Eq.(vi), we have-

$$\frac{\partial U}{\partial H} = 0$$

Therefore, the potential energy attains extremum at the equilibrium position.

SOLUTION Total potential energy (U) of the particle at point P is the sum of its gravitational and electrostatic potential energies i.e.,

$$U = mgH + \frac{q\sigma}{2\epsilon_0}\left[\sqrt{a^2 + H^2} - H\right]$$
$$= mg\left[2\sqrt{a^2 + H^2} - H\right]$$

At equilibrium position, we have

$$\frac{dU}{dH} = mg\left[\frac{2H}{\sqrt{a^2+H^2}} - 1\right] = 0$$

which gives $H_{\min} = a/\sqrt{3}$ and $U_{\min} = \sqrt{3}mga$. The Fig. 2.53 shows the variation of U with height H. Note that the equilibrium is stable (i.e., $d^2U/dH^2 > 0$).

EXAMPLE 36. A disc of radius R with uniform positive charge density σ is placed on the xy plane with its centre at the origin. The Coulomb potential along the z-axis is

$$V(z) = \frac{\sigma}{2\epsilon_0}\left(\sqrt{R^2 + z^2} - z\right)$$

A particle of positive charge q is placed initially at rest at a point on the z axis with $z = z_0$ and $z_0 > 0$. In addition to the Coulomb force, the particle experiences a vertical force $\vec{F} = -c\hat{k}$ with $c > 0$. Let $\beta = \dfrac{2c\epsilon_0}{q\sigma}$. Which of the following statement(s) is (are) correct?

(A) For $\beta = 1/4$ and $z_0 = 25R/7$, the particle reaches the origin.
(B) For $\beta = 1/4$ and $z_0 = 3R/7$, the particle reaches the origin.
(C) For $\beta = 1/4$ and $z_0 = R/\sqrt{3}$, the particle returns back to $z = z_0$.
(D) For $\beta > 1$ and $z_0 > 0$, the particle always reaches the origin.

SOLUTION By work–kinetic energy theorem, we have-
$$W_{el} + W_{ext} = K_f - K_i$$
$$\Rightarrow qV_i - qV_f + W_{ext} = K_f - K_i$$

Here, $V_i = \dfrac{\sigma}{2\epsilon_0}\left(\sqrt{R^2 + z_0^2} - z_0\right)$,

$V_f = \dfrac{\sigma}{2\epsilon_0}\left(\sqrt{R^2 + 0^2} - 0\right) = \dfrac{\sigma R}{2\epsilon_0}$,

$W_{ext} = cz = \dfrac{q\sigma\beta z}{2\epsilon_0}$ $\left(\because \beta = \dfrac{2c\epsilon_0}{q\sigma}\right)$ and $K_i = 0$

Substituting these values and solving for K_f, we get
$$\Rightarrow \frac{q\sigma}{2\epsilon_0}\left[\sqrt{R^2 + z^2} - z\right] - \frac{q\sigma R}{2\epsilon_0} + \frac{q\sigma\beta z}{2\epsilon_0} = K_f - 0$$
$$\Rightarrow K_f = \frac{q\sigma}{2\epsilon_0}\left[\sqrt{R^2 + z^2} - z\right] - \frac{q\sigma R}{2\epsilon_0} + \frac{q\sigma\beta z}{2\epsilon_0}$$

Now, substitute the values of β and z_0 from given options and solve for final kinetic energy K_f.
If kinetic energy is positive, then particle will reach at origin
If kinetic energy is negative, then particle will not reach at origin.
Thus correct choices are A, C and D.

2.7.6 Electric Potential Due to a Charged Conducting Shell or Sphere

Suppose, charge Q is uniformly distributed over the surface of a conducting shell of radius R. We will calculate the electric potential at a point-
(a) outside the shell; $(r > R)$
(b) on the surface of the shell $(r = R)$
(c) inside the shell $(r < R)$.

(a) At points outside a uniform spherical distribution, the electric field is
$$\vec{E} = \frac{1}{4\pi\epsilon_0}\frac{Q}{r^2}\hat{r}$$

Since, $\vec{E}$ is radially outward, therefore-
$$\vec{E}\cdot d\vec{s} = E\,dr\cos 0 = E\,dr$$

Since, $V(\infty) = 0$, therefore we have-
$$V(r) - V(\infty) = -\int_\infty^r \vec{E}\cdot d\vec{r}$$
$$\Rightarrow V - 0 = -\int_\infty^r \frac{Q}{4\pi\epsilon_0 r^2}dr$$

Integrating right hand side, we get-
$$V = \frac{1}{4\pi\epsilon_0}\frac{Q}{r} \quad (2.65)$$

From Eq.2.65, we see that the potential due to a uniformly charged shell is the same as that due to a point charge Q at the center of the shell.

(b) For a point on the surface of the shell, $r = R$, therefore Eq.2.65, we have-
$$V = \frac{1}{4\pi\epsilon_0}\frac{Q}{R} \quad (2.66)$$

(c) At points inside the shell, $E = 0$. So, the work done in bringing a unit positive charge from a point on the surface to any point inside the shell is zero. Thus, the potential has a fixed value at all points within the spherical shell and is equal to the potential at the surface.

$$V = \frac{1}{4\pi\epsilon_0}\frac{Q}{R} \quad (2.67)$$

Variation of electric potential with the distance from the centre (r) is given in Fig.2.54.

☞ *Note that, all above results also hold for a conducting sphere*

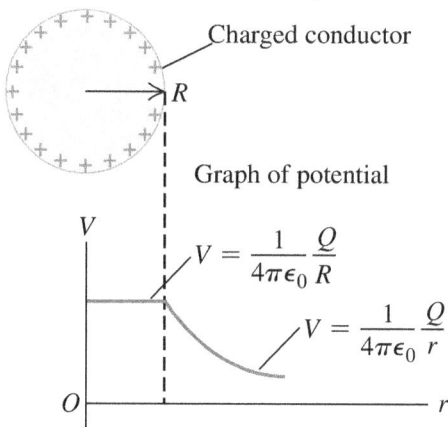

Figure 2.54: Electric potential V at points inside and outside a positively charged spherical shell/spherical conductor

whose charge lies entirely on the outer surface.

2.7.7 Electric Potential Due to a Non-Conducting Charged Sphere

Suppose a charge Q is uniformly distributed throughout a non-conducting spherical volume of radius R. Let the volume charge density of this non-conducting sphere is ρ. We have to find expressions for the potential at-
(a) an external point $(r > R)$; (b) on the surface $(r = R)$;
(c) an internal point $(r < R)$
where r is the distance of the point from the centre of the sphere.

(a) Electric Potential at an External Point of Uniformly Charged Sphere: Let O be the centre of a non-conducting sphere of radius R, have a charge Q distributed uniformly over its entire volume.

For $r \geq R$, the electric field E is radial and has a magnitude:
$$E_r = k\frac{Q}{r^2} \quad (r \geq R) \quad (2.68)$$

This is the same as the electric field due to a point charge, and hence the electric potential at any point of radius r in this region is given by:
$$V_r = -\int_\infty^r E_r\,dr = -\int_\infty^r \frac{kQ}{r^2}dr$$

Integrating right hand side, we get

$$V_r = k\frac{Q}{r} \quad (r \geq R) \tag{2.69}$$

(b) Electric Potential at a Point on the Surface of the Uniformly Charged Non-conducting Sphere: For a point on the surface of the non-conducting sphere, $r = R$, therefore Eq.(2.69) gives-

$$V_R = k\frac{Q}{R} \tag{2.70}$$

(c) Electric Potential at an internal Point of Uniformly Charged Sphere ($r < R$): To find the potential at some point r inside the sphere, we use Eq.2.16,

$$V_B - V_A = -\int_{r_A}^{r_B} E\, dr$$

along a radial path that begins at some point where the potential is known. We choose a point on the surface of the sphere, and integrate the electric field from that point (at $r_A = R$) to a point inside where we want to find the potential (at $r_B = r$). From Eq.1.145, the electric field at any point inside the non conducting sphere of radius R at $r = r$, is given by $E_r = k\frac{Qr}{R^3}$ and electric potential at $r = R$ is $V_R = k\frac{Q}{R}$, therefore-

$$V_r - V_R = -\int_R^r E_r\, dr = -k\frac{Q}{R^3}\int_R^r r\, dr$$

$$= -k\frac{Q}{R^3}\left|\frac{r^2}{2}\right|_{r=R}^{r=r} = k\frac{Q}{2R^3}(R^2 - r^2)$$

$$\implies V_r - V_R = k\frac{Q}{2R^3}(R^2 - r^2) \tag{2.71}$$

Using $V_R = kQ/R$ in the last result we reach to the following relation:

$$V_r = k\frac{Q}{2R}\left(3 - \frac{r^2}{R^2}\right) \quad (0 \leq r \leq R) \tag{2.72}$$

At $r = 0$, we have $V_0 = 3kQ/2R$, and at $r = R$, we get $V_R = kQ/R$ as expected. Figure 2.55 shows the electric potential in the two regions $0 \leq r \leq R$ and $r \geq R$.

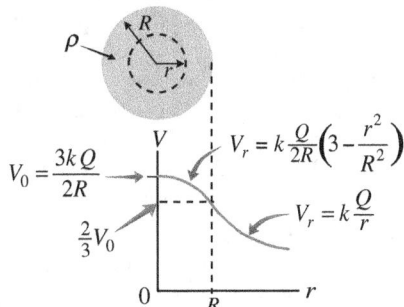

Figure 2.55: A sketch of the electric potential $V(r)$ as a function of r in the two regions $0 \leq r \leq R$ and $r \geq R$. The curve for the region $0 \leq r \leq R$ is parabolic and joins smoothly with the curve for the region $r \geq R$, which is hyperbola

☞ A point charge q can be considered to be the limiting case of a small spherical conductor whose radius tends to zero and potential to infinity.

EXAMPLE 37. In the Fig.2.56, the inner (shaded) region A represents a sphere of radius $r_A = 1$, within which the electrostatic charge density varies with the radial distance r from the centre as $\rho_A = kr$, where k is positive. In the spherical shell B of outer radius r_B, the electrostatic charge density varies as $\rho_B = 2k/r$. Assume that dimensions are taken care of. All physical quantities are in their SI units.

Which of the following statement(s) is(are) correct?

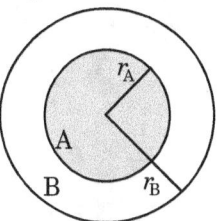

Figure 2.56

(A) If $r_B = \sqrt{3/2}$, then the electric field is zero everywhere outside B.
(B) If $r_B = 3/2$, then the electric potential just outside B is k/ϵ_0.
(C) If $r_B = 2$, then the total charge of the configuration is $15\pi k$.
(D) If $r_B = 5/2$, then the magnitude of the electric field just outside B is $13\pi k/\epsilon_0$.

SOLUTION $q_1 = \int_0^1 kr 4\pi r^2\, dr = \frac{4\pi k}{4} = \pi k$,

$$q_2 = \int_1^r \frac{2k}{r} 4\pi r^2\, dr = \frac{8\pi k (r^2 - 1^2)}{2},$$
$$= 4\pi k [r^2 - 1] = 4\pi k r^2 - 4\pi k$$

Therefore, $q_{\text{net}} = q_1 + q_2$
$$= 4\pi k r^2 - 3\pi k = \pi k [4r^2 - 3]$$

(A) $E_{\text{net}} = 0 \implies q_{\text{net}} = 0 \implies r = \frac{\sqrt{3}}{2}$

(B) $V = \frac{kQ_{\text{net}}}{r} = \frac{1}{4\pi\epsilon_0}\frac{\pi k (4r^2 - 3)}{r} = \frac{k}{4\epsilon_0}\left[4r - \frac{3}{r}\right]$

$$= \frac{k}{4\epsilon_0}\left[4 \times \frac{3}{2} - \frac{3 \times 2}{3}\right] = \frac{k}{\epsilon_0}$$

(C) $q_{\text{sat}} = \pi k [4(2)^2 - 3] = 13\pi k$

(D) $E_2 = \frac{kQ}{r^2} = \frac{1}{4\pi\epsilon_0}\frac{\pi k (4r^2 - 3)}{r^2} = \frac{k}{4\epsilon_0}\left[\frac{4\left(\frac{5}{2}\right)^2 - 3}{(5/2)^2}\right]$

$$= \frac{k}{25\epsilon_0}[25 - 3] = \frac{22}{25}\frac{k}{\epsilon_0}$$

From above it is evident that the correct choice is B.

2.7.8 Potential on the Edge of a Uniformly Charged Disc:

To calculate the potential at point O on the circumference of the disc, let us divide the disc in large number of rings with O as centre. ABD and EFG are the arcs of two such concentric rings having common centre O. The potential due to one segment between r and $r + dr$ is given by-

$$dV = \frac{1}{4\pi\epsilon_0} \cdot \frac{dq}{r}$$

Here, $dq = \sigma$ (Area of ring) = σ (arc $ABD \times dr$)
$= \sigma (r \times 2\theta \times dr) = \sigma (2r\theta)\, dr$

$$\therefore \quad dV = \frac{1}{4\pi\epsilon_0}\frac{\sigma (2r\theta)\, dr}{r} = \frac{\sigma}{2\pi\epsilon_0}\theta\, dr$$

Further, $r = OH \cos\theta = 2R \cos\theta$
$$\therefore \quad dr = -2R \sin\theta\, d\theta$$

Hence, $dV = -\frac{\sigma}{2\pi\epsilon_0} 2R\theta \sin\theta\, d\theta$

$$\therefore \quad V = \int_{\pi/2}^0 dV = \frac{\sigma R}{\pi\epsilon_0}\int_0^{\pi/2}\theta \sin\theta\, d\theta$$

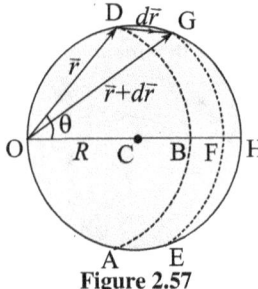

Figure 2.57

Simplifying it, we get -

$$V = \frac{\sigma R}{\pi \epsilon_0} \qquad (2.73)$$

EXAMPLE 38. A point charge q_1 is at the origin, and a second point charge q_2 is on the x axis at $x = a$. Find an expression for the electric potential everywhere on the x axis as a function of x.
APPROACH Find electric potentials due to each charge sepa-

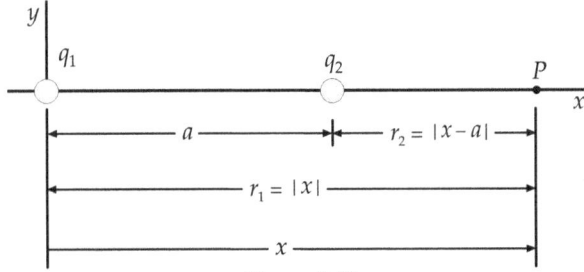

Figure 2.58

rately at a point on the x-axis and add them to find the net electric potential at that point.
SOLUTION Fig.2.58 shows the x axis and the two q_1, q_2 charges on it. Let r_1 be the distance from q_1 to an arbitrary field point P at position x on the x axis, that is, $r_1 = |x|$. Let r_2 be the distance from q_2 to P, that is, $r_2 = |x - a|$.
The potential as a function of the distances to the two charges, is given as-

$$V = \frac{kq_1}{r_1} + \frac{kq_2}{r_2} = \frac{kq_1}{|x|} + \frac{kq_2}{|x-a|} \qquad (x \neq 0, x \neq a)$$

Note that, $V \to \infty$ both as $x \to 0$ and as $x \to a$, and $V \to 0$ both

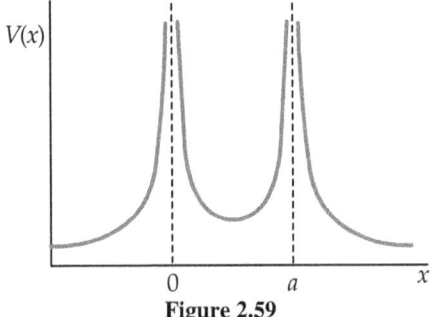

Figure 2.59

as $x \to -\infty$ and as $x \to \infty$ as one would expect.
Figure 2.59 shows V versus x on the x axis for $q_1 = q_2 > 0$.
EXAMPLE 39. 64 identical drops each charged up to potential of 10 mV are combined to form a bigger drop. Find the potential of the bigger drop.

SOLUTION If q is the charge on each drop and r is it's radius, then electric potential at the surface of the drop-

$$V = k\frac{q}{r} = 10\,\text{mV} \qquad \cdots (1)$$

If the radius of above 64 combined drops is R, then

$$64 \times (\text{volume})_{\substack{\text{smaller} \\ \text{drop}}} = (\text{volume})_{\substack{\text{bigger} \\ \text{drop}}}$$

$$\implies 64 \times \frac{4}{3}\pi r^3 = \frac{4}{3}\pi R^3 \implies R = 4r$$

And by charge conservation, the net charge on bigger drop, $Q = 64q$. Therefore, surface potential of the bigger drop-

$$V_{\text{bigger drop}} = \frac{kQ}{R} = \frac{k(64q)}{4r} = 16\frac{kq}{r} = 16 \times 10\,\text{mV}$$
$$= 160\,\text{mV} \qquad \text{(since, from Eq.(1), } kq/r = 10\,\text{mV)}$$

2.8 Equipotentials and Conductors

☞ When all charges are at rest, the surface of a conductor is always an equipotential surface.

The electric field $\vec{E}$ is always perpendicular to an equipotential

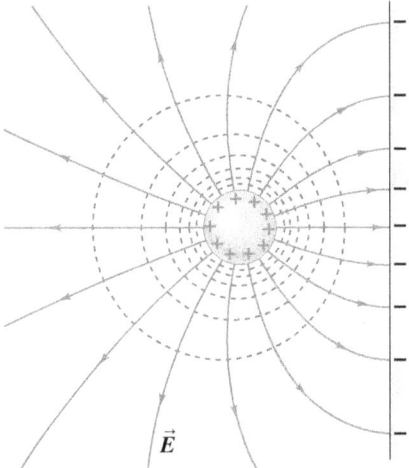

---- Cross sections of equipotential surfaces
——— Electric field lines

Figure 2.60: When charges are at rest, a conducting surface is always an equipotential surface. Field lines are perpendicular to a conducting surface.

surface.
To prove it by conservative nature of electric field, let us transport a test charge q_0 around the loop $abcda$ as shown in Fig. 2.61a. The segments bc and da (and the associated work) may be made arbitrarily small (because the length is very small. Think $\vec{F} \cdot \vec{s}$ i.e., work), and no work is done in the segment cd because, as already shown, the field is zero everywhere inside the conductor. If the field just outside the conductor has a component $E_\parallel$, parallel to the surface, this component does work equal to $q_0 E_\parallel$ that is the parallel component of $\vec{E}$ vector at the surface (if any) does work on the test charge q_0 equal to q_0 times l (length of ab) times parallel component of $\vec{E}$ at surface, i.e., $E_\parallel$.
So, the net work done in transporting the test charge q_0 along a closed path $abcda$,

$$W_{abcda} = W_{ab} + W_{bc} + W_{cd} + W_{da} \neq 0$$

i.e., the net work done is non zero. It is impossible, because, it shows that the electric force field is non-conservative. To avoid this contradiction, we must conclude that there cannot be a component of $\vec{E}$ parallel to the surface, and that $\vec{E}$ is therefore perpendicular to the surface.
Thus, $\vec{E}$ is perpendicular to the surface at each point, proving our statement.
It also follows that when all charges are at rest, the entire solid volume of a conductor is at the same potential. Eq.2.14: $\left[V_B - V_A = \frac{W_{AB}}{q_0} = -\int_A^B \vec{E} \cdot d\vec{s}\right]$, states that the potential

2.8. EQUIPOTENTIALS AND CONDUCTORS

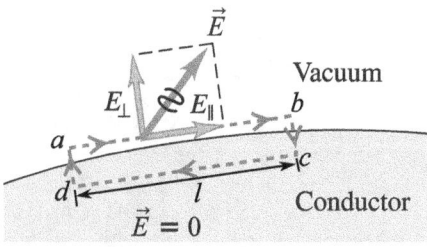

(a) Electric field at the surface of conductor.

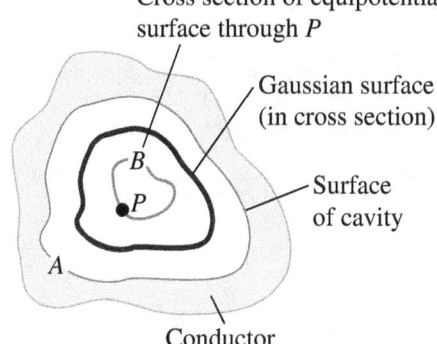

(b) Cavity in a conductor.

Figure 2.61: (a) At all points on a conductor's surface, the electric field must be perpendicular to the surface. If $\vec{E}$ had a tangential component, a net amount of work would be done on a test charge by moving it around a loop as shown-which is impossible because the electric force is conservative. (b) If the cavity contains no charge, every point in the cavity is at the same potential, the electric field is zero everywhere in the cavity, and there is no charge anywhere on the surface of the cavity.

difference between two points A and B within the conductor's solid volume, $V_B - V_A$, is equal to the line integral $\int_A^B \vec{E} \cdot d\vec{s}$ of the electric field from A to B. Since $\vec{E} = 0$ everywhere inside the conductor, the integral is guaranteed to be zero for any two such points A and B. Hence the potential is the same for any two points within the solid volume of the conductor. We describe this by saying that the solid volume of the conductor is an equipotential volume.

Theorem 1: *In an electrostatic situation, if a conductor contains a cavity and if no charge is present inside the cavity, then there can be no net charge anywhere on the surface of the cavity.*

Proof: To prove this theorem, we first prove that every point in the cavity is at the same potential. In Fig.2.61b, the conducting surface A of the cavity is an equipotential surface. Suppose point P in the cavity is at a different potential; then we can construct a different equipotential surface B including point P.

Now consider a Gaussian surface, shown in Fig.2.61b, between the two equipotential surfaces. Because of the relationship between $\vec{E}$ and the equipotentials, we know that the field at every point between the equipotentials is from A towards B, or conversely, at every point, it is from B towards A, depending on which equipotential surface is at higher potential.

In either case the flux through this Gaussian surface is certainly not zero. But then Gauss's law says that the charge enclosed by the Gaussian surface cannot be zero. This contradicts our initial assumption that there is no charge in the cavity. So the potential at P cannot be different from that at the cavity wall.

The entire region of the cavity must therefore be at the same potential. But for this to be true, the electric field inside the cavity must be zero everywhere. Finally, Gauss's law shows that the electric field at any point on the surface of a conductor is

proportional to the surface charge density σ at that point. We conclude that the surface charge density on the wall of the cavity is zero at every point.

☞ Don't confuse equipotential surfaces with the Gaussian surfaces. Gaussian surfaces have relevance only when we are using Gauss's law, and we can choose any Gaussian surface that's convenient. We cannot choose equipotential surfaces; the shape is determined by the charge distribution.

EXAMPLE 40. A solid conducting sphere, having a charge Q, is surrounded by an uncharged conducting hollow spherical shell. Let the potential difference between the surface of the solid sphere and that of the outer surface of the hollow shell be V. If the shell is now given a charge of $-4Q$, calculate the new potential difference between the same two surfaces.

APPROACH First indicate the charge distribution and then find the potential difference by applying the formula of potential at any point due to charged sphere/spherical shell.

SOLUTION Given situation is shown in Fig.2.62. Since, the

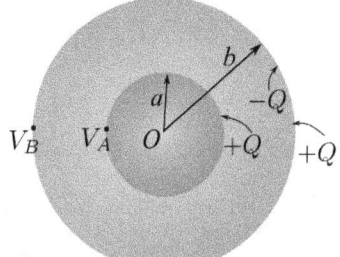

Figure 2.62

uncharged shell initially encloses charge Q, therefore, due to induction, the charge on inner surface of the shell will be $-Q$. By charge conservation, the charge on the outer surface of the shell will be $+Q$.

The electric potential at any point A on the inner surface of shell is given by-
$$V_A = \frac{kQ}{a} + \frac{k(-Q)}{b} + \frac{kQ}{b} \quad (2.74)$$
where, $k = \frac{1}{4\pi\epsilon_0} = 9.0 \times 10^9 \text{N.m}^2/\text{C}^2$

Electric potential on surface of outer shell is
$$V_B = \frac{kQ}{b} + \frac{k(-Q)}{b} + \frac{kQ}{b} \quad (2.75)$$
Therefore, the potential difference is
$$\Delta V_{AB} = V_A - V_B = kQ\left(\frac{1}{a} - \frac{1}{b}\right)$$
Given, that $\Delta V_{AB} = V$, so-
$$kQ\left(\frac{1}{a} - \frac{1}{b}\right) = V \quad (2.76)$$
Now, if the shell is given an extra charge $-4Q$, it will uniformly get spread on the outer surface of the shell. In this case, the new potential difference will be
$$\Delta V_{AB} = V_A - V_B$$
$$= \left(\frac{kQ}{a} + \frac{k(-4Q)}{b}\right) - \left(\frac{kQ}{b} + \frac{k(-4Q)}{b}\right)$$
$$= kQ\left(\frac{1}{a} - \frac{1}{b}\right) = V \quad [\text{from Eq.}(2.76)]$$

Remark: It is a very important result which shows that the potential difference between the outer surface of the spherical shell and inner sphere is independent on the charge of the shell.

EXAMPLE 41. Two conducting spheres 1 and 2, having radii a and b charged to q_1 and q_2 respectively. Find the potential difference between 1 and 2.

SOLUTION The potential on the surface of the sphere 1 is given

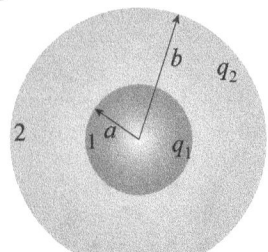

Figure 2.63

by -
$$V_1 = \frac{1}{4\pi\epsilon_0}\frac{q_1}{a} + \frac{1}{4\pi\epsilon_0}\frac{q_2}{b} \quad (2.77)$$

The potential on the surface of the sphere 2 is given by,
$$V_2 = \frac{1}{4\pi\epsilon_0}\frac{q_1}{b} + \frac{1}{4\pi\epsilon_0}\frac{q_2}{b} \quad (2.78)$$

$$\Rightarrow \quad V_1 - V_2 = \frac{1}{4\pi\epsilon_0}\frac{q_1}{a} - \frac{1}{4\pi\epsilon_0}\frac{q_1}{b}$$
$$= \frac{q_1}{4\pi\epsilon_0}\left(\frac{1}{a} - \frac{1}{b}\right)$$

EXAMPLE 42. A point charge q is kept at a distance l from a charged conducting sphere of radius R ($l > R$), having charge Q over its surface. Find electric potential due to induced charges at any point P inside the sphere.

SOLUTION The given situation is shown in Fig.2.64. Potential due to induced charges is zero as centre C is equidistant from all induced charges. Since, every point inside this conducting sphere will be at equal potential, therefore-
$$V_C = V_P = \frac{kq}{l} + \frac{kQ}{R}$$

Also at point P, we can write

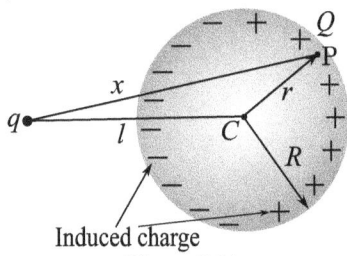

Figure 2.64

$V_P = V_{\text{due to }q} + V_{\text{due to induced charges}} + V_{\text{due to }Q}$

Comparing above two values of V_P, we get-
$$\frac{kq}{l} + \frac{kQ}{R} = \frac{kq}{x} + \frac{kQ}{R} + V_{\text{due to induced charges}}$$
$$\Longrightarrow \quad V_{\text{due to induced charges}} = \frac{kq}{l} - \frac{kq}{x}$$

Also, $\vec{E}_{\text{external}} + \vec{E}_{\text{conductor}} + \vec{E}_{\text{induced}} = 0$

Resultant electric field inside material is zero

EXAMPLE 43. A metal sphere of radius R, carrying charge q, is surrounded by a thick concentric metal shell (inner radius a, outer radius b, as in Fig.2.65). The shell carries no net charge.
(a) Find the surface charge density σ at R, at a, and at b.
(b) Find the potential at the center, using infinity as the reference point.
(c) Now the outer surface is touched to a grounding wire, which lowers its potential to zero (same as at infinity). How do your answers to (a) and (b) change?

SOLUTION (a) $\sigma_R = \frac{q}{4\pi R^2}$; $\sigma_a = \frac{-q}{4\pi a^2}$; $\sigma_b = \frac{-q}{4\pi b^2}$

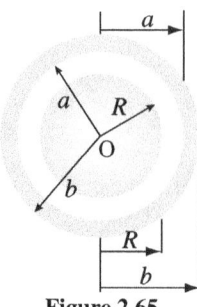

Figure 2.65

(b) $V(0) = -\int_{\infty}^{0} E\, dr$

To simplify right hand side of above expression, we break the limit ∞ to 0 into pieces: $(\infty \to b) + (b \to a) + (a \to R) + (R \to 0)$ and then apply corresponding E values:

$$V(0) = -\int_{\infty}^{b}\left(\frac{1}{4\pi\epsilon_0}\frac{q}{r^2}\right)dr - \int_{b}^{a}(0)dr$$
$$-\int_{a}^{R}\left(\frac{1}{4\pi\epsilon_0}\frac{q}{r^2}\right)dr - \int_{R}^{0}(0)dr$$
$$= \frac{1}{4\pi\epsilon_0}\left(\frac{q}{b} + \frac{q}{R} - \frac{q}{a}\right)$$

(c) $\sigma_b \to 0$ (the charge "drains off");

$$V(0) = -\int_{\infty}^{a}(0)dr - \int_{a}^{R}\left(\frac{1}{4\pi\epsilon_0}\frac{q}{r^2}\right) - \int_{R}^{0}(0)dr$$
$$= \frac{1}{4\pi\epsilon_0}\left(\frac{q}{R} - \frac{q}{a}\right)$$

EXAMPLE 44. A total charge q is spread uniformly over the inner surface of a non-conducting hemispherical cup of inner radius a. Calculate (a) the electric field and (b) the electric potential at the centre of the hemisphere (Consider the cup as a stack of rings).

SOLUTION Consider a circular strip symmetric about z-axis of radius r and width $a\,d\theta$ [Fig. 2.66]. The charge on the strip is
$$dq = q\,\frac{2\pi r\, ad\theta}{2\pi a^2} = \frac{qr\,d\theta}{a} = q\sin\theta\, d\theta$$

(a) At the centre of the hemisphere, the x-component of the field

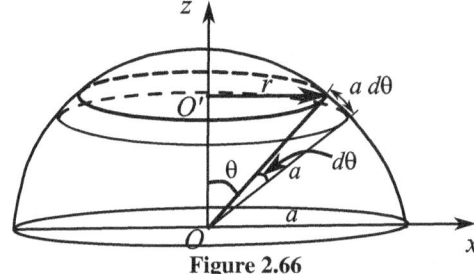

Figure 2.66

will be cancelled for reasons of symmetry. The entire field will be contributed by the z-component alone.
$$dE = dE_z = \frac{q\sin\theta\, d\theta \cos\theta}{4\pi\epsilon_0 a^2}$$
$$\therefore \quad E = \int dE_z = \frac{q}{4\pi\epsilon_0 a^2}\int_0^{\pi/2}\sin\theta\cos\theta\, d\theta = \frac{q}{8\pi\epsilon_0 a^2}$$

(b) $dV = \frac{q\sin\theta\, d\theta}{4\pi\epsilon_0 a}$;
$$\Rightarrow V = \int dV = \frac{q}{4\pi\epsilon_0 a}\int_0^{\pi/2}\sin\theta\, d\theta = \frac{q}{4\pi\epsilon_0 a}$$

EXAMPLE 45. An alpha particle (two protons, two neutrons) moves into a stationary gold atom (79 protons, 118 neutrons), passing through the electron region that surrounds the gold nucleus like a shell and headed directly toward the nucleus (Fig.

2.67). The alpha particle slows until it momentarily stops when its centre is at radial distance $r = 9.23$ fm from the nuclear centre. Then it moves back along its incoming path. (Because the gold nucleus is much more massive than the alpha particle, we can assume the gold nucleus does not move.) What was the kinetic energy K_i of the alpha particle when it was initially far away (hence external to the gold atom)? Assume that the only force acting between the alpha particle and the gold nucleus is the (electrostatic) Coulomb force and treat each as a single charged particle.

APPROACH During the entire process, the mechanical energy

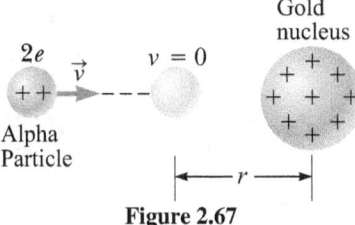

Figure 2.67

of the *alpha particle* + *gold atom* system is conserved. When the alpha particle is outside the atom, the system's initial electric potential energy U_i is zero because the atom has an equal number of electrons and protons, which produce a *net* electric field of zero. However, once the alpha particle passes through the electron region surrounding the nucleus on its way to the nucleus, the electric field due to the electrons goes to zero. The reason is that the electrons act like a closed spherical shell of uniform negative charge and such a shell produces zero electric field in the space it encloses. The alpha particle still experiences the electric field of the protons in the nucleus, which produces a repulsive force on the protons within the alpha particle.

As the incoming alpha particle is slowed by this repulsive force, its kinetic energy is transferred to electric potential energy of the system. The transfer is complete when the alpha particle momentarily stops and the kinetic energy is zero, i.e., $K_f = 0$.

Since, there is only electrostatic force between α particle and gold nucleus which is conservative, therefore, we can apply the principle of conservation of mechanical energy.

SOLUTION According to principle of conservation of mechanical energy, we have-
$$K_i + U_i = K_f + U_f$$
Here, $U_i = 0$ and $K_f = 0$, therefore-
$$K_i = k\frac{(2e)(79e)}{9.23 \times 10^{-15}} = 3.94 \times 10^{-12} \text{J} = 24.6 \text{ MeV}$$

EXAMPLE 46. Two particles of mass m and $2m$ carry a charge q each. Initially the heavier particle is at rest on a smooth horizontal plane and the other is projected along the plane directly towards the first from a distance d with speed u. Find the closest distance of approach.

APPROACH As the mass $2m$ is not fixed, it will also move away from m due to repulsion. The distance between the particles is minimum when their relative velocity is zero i.e., when they have equal velocities.

If v_1 and v_2 are the velocities of m and $2m$ respectively at closest distance of approach, then $v_1 = v_2$.

Additionally, the electrostatic force is the only force acting between the masses, therefore you can apply the principle of conservation of mechanical energy.

SOLUTION By conservation of linear momentum, we have-
$$mu = mv_1 + 2mv_2 \implies u = v_1 + 2v_2$$
Since, $v_1 = v_2$, therefore,
$$u = v_1 + 2v_1 \implies v_1 = u/3$$
$$\implies v_2 = v_1 = u/3$$

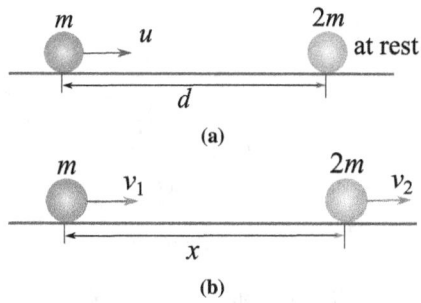

Figure 2.68

By conservation of mechanical energy, we have-
Loss in KE = gain in PE
$$\frac{1}{2}mu^2 - \left(\frac{1}{2}mv_1^2 + \frac{1}{2}2mv_2^2\right) = \frac{q^2}{4\pi\epsilon_0}\left(\frac{1}{x} - \frac{1}{d}\right)$$
$$\implies \frac{1}{2}mu^2 - \frac{1}{2}m\frac{u^2}{9}(1+2) = \frac{q^2}{4\pi\epsilon_0}\left(\frac{1}{x} - \frac{1}{d}\right)$$
$$\implies \frac{1}{3}mu^2 = \frac{q^2}{4\pi\epsilon_0}\left(\frac{1}{x} - \frac{1}{d}\right)$$
$$\implies \frac{1}{x} = \frac{1}{d} + \frac{4\pi\epsilon_0 mu^2}{3q^2} \implies x = \frac{3q^2 d}{3q^2 + 4\pi\epsilon_0 mu^2 d}$$

2.9 Connected Conducting Spheres

Now consider two charged conducting spheres with different radii connected with a long conducting wire, as shown in Figure 2.69. We assume that the spheres are far enough apart that the charge distribution on one does not directly affect the other. We already know that the electric field is stronger where the electric field lines are closer together and weaker where the electric field lines are widely separated. Electric charge is placed on one of the spheres. Some of that charge will then flow through the conducting wire so both spheres are charged. In electric equilibrium, the charge must be distributed so the electric potential is the same on the two spheres (since otherwise charges would flow in the conducting wire and the system would not be in electrical equilibrium as assumed). Two conductors connected can be seen as

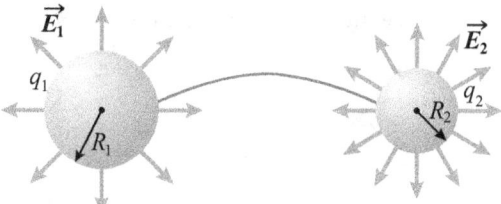

Figure 2.69

a single conductor. Therefore, electric potential will be identical everywhere on the connected spheres.

Potential at any point on the surface of charged sphere having charge q_1 and radius R_1 is-
$$V_1 = \frac{q_1}{4\pi\epsilon_0 R_1}$$
Potential at any point on the surface of charged sphere having charge q_2 and radius R_2 is-
$$V_2 = \frac{q_2}{4\pi\epsilon_0 R_2}$$
Therefore, in electrostatic equilibrium:
$$V_1 = V_2 \implies \frac{q_1}{R_1} = \frac{q_2}{R_2} \implies \frac{q_1}{q_2} = \frac{R_1}{R_2}$$
Surface charge density of the sphere of radius R_1, is-
$$\sigma_1 = \frac{q_1}{4\pi R_1^2}$$
Surface charge density of the sphere of radius R_2, is

$$\sigma_2 = \frac{q_2}{4\pi R_2^2}$$

$$\therefore \boxed{\frac{\sigma_1}{\sigma_2} = \frac{q_1}{q_2} \cdot \frac{R_2^2}{R_1^2} = \frac{R_1}{R_2} \cdot \frac{R_2^2}{R_1^2} = \frac{R_2}{R_1}} \quad (2.79)$$

So, if $R_1 < R_2$, then $\sigma_1 > \sigma_2$ and the surface electric field $E_1 > E_2$.

☞ Charge distribution on a conductor not necessarily have to be uniform.

EXAMPLE 47. Two isolated metallic solid spheres of radii R and $2R$ are charged such that both have same charge density σ. The spheres are then connected by a thin conducting wire. If the new charge density of the bigger sphere is σ', find the ratio σ'/σ

SOLUTION The initial situation is shown in Fig.2.70. Charge on sphere of radius R is

$$Q_1 = \sigma\left(4\pi R^2\right) = 4\pi R^2 \sigma$$

and charge on sphere of radius $2R$ is

$$Q_2 = \sigma\left(4\pi (2R)^2\right) = 16\pi R^2 \sigma$$

Therefore, net charge on both spheres

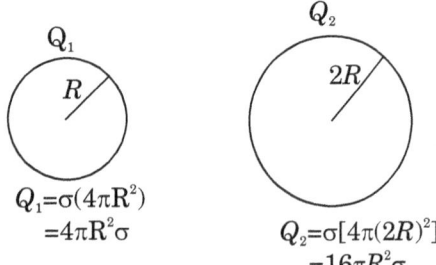

Figure 2.70

$$Q_1 + Q_2 = 4\pi R^2 \sigma + 16\pi R^2 \sigma = 20\pi R^2 \sigma \quad \cdots (1)$$

When both spheres are connected, the redistribution of charge starts until they both reach a common potential. Let their common potential be 'V,' and the charge on the sphere with a radius of R be Q_1' and the charge on the sphere with a radius of $2R$ be Q_2'. This situation is shown in Fig.2.71.

The common potential, of both the spheres, is given as

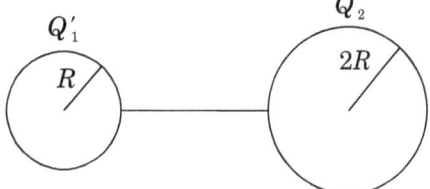

Figure 2.71

$$V = \frac{Q_1'}{4\pi\epsilon_0 R} = \frac{Q_2'}{4\pi\epsilon_0 (2R)} = \frac{Q_1' + Q_2'}{4\pi\epsilon_0 R + 4\pi\epsilon_0 (2R)}$$

By charge conservation,

$$Q_1' + Q_2' = Q_1 + Q_2 = 20\pi R^2 \sigma \text{ (By Eq.(1))}$$

Therefore, $V = \dfrac{Q_1'}{4\pi\epsilon_0 R} = \dfrac{Q_2'}{4\pi\epsilon_0 (2R)} = \dfrac{20\pi R^2 \sigma}{12\pi\epsilon_0 R} = \dfrac{5R\sigma}{3\epsilon_0}$

$$\implies Q_1' = \frac{5R\sigma}{3\epsilon_0} \times 4\pi\epsilon_0 R = \frac{20}{3}\pi\sigma R^2$$

and $Q_2' = \dfrac{5R\sigma}{3\epsilon_0} \times 8\pi\epsilon_0 R = \dfrac{40}{3}\pi\sigma R^2$

Clearly, $Q_2' = 2Q_1'$

If new charge density of bigger sphere is σ', then

$$\sigma' = \frac{Q_2'}{4\pi(2R)^2} = \frac{40\pi\sigma R^2}{3(16\pi R^2)} = \frac{5}{6}\sigma \implies \frac{\sigma'}{\sigma} = \frac{5}{6}$$

2.10 Earthing

Earth is very big conducting sphere. We assume potential of earth is zero because its size is very large.

When a conducting body is joined to earth through a conducting wire. Potential of the body also becomes zero.

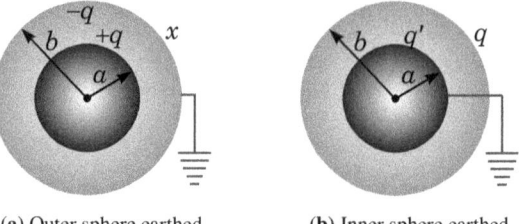

(a) Outer sphere earthed (b) Inner sphere earthed

Figure 2.72

In Fig.2.72a, charge $+q$ is given to inner sphere of radius a placed inside a larger concentric sphere of radius $b(>a)$. By using Gausses' law, we can show that the induced charge produced at inner surface of bigger sphere is $-q$. Let the charge on the outer surface of larger sphere is x.

Since, the outer sphere is earthed, therefore, the net electric potential at any point on the surface of outer sphere must be zero, i.e.,-

$$V_{\text{outer}} = 0 \implies k\frac{q}{b} + k\frac{(-q+x)}{b} = 0 \implies x = 0$$

So, there will not be any charge on the outer surface of the larger sphere, only $-q$ charge will be induced on the inner surface of the outer sphere.

In Fig. 2.72b the inner sphere is earthed and the net given charge on the outer sphere is q. If induced charge on outer surface of inner sphere is q', then -

$$V_{\text{inner}} = 0 \implies \frac{kq'}{a} + \frac{kq}{b} = 0 \implies \boxed{q' = -q\left(\frac{a}{b}\right)}$$

2.11 Corona Discharge

Let us consider an arbitrarily shaped charged conductor, as shown in Fig.2.73. Now, if we fit circles at different positions on the surface of the conductor, then the radii of these circles will be the radii of curvatures at these positions, respectively. Since the radius of curvature is smallest at the sharpest end of the conductor, therefore, from Eq.2.79, it is clear that the charge density is greatest for the sharpest regions. As a result of this, the electric field ($\propto$ charge density) is much greater in sharply pointed regions, and it may become large enough for the air or other gas to break down and conduct a current because we have ionized material near those points. *Corona discharge is an electric discharge due to the ionization of material in a fluid (such as air) surrounding a conductor. This discharge happens near pointed regions on conductors.* The minimum electric field needed for breakdown is called the breakdown field and is typically 3MV/m for fairly dry air. Corona discharge is used in ozone generators, in electrical precipitators that remove particulates in smokestacks, and in forced-air heating systems. It is also important in many other applications, including lightning protection rods.

EXAMPLE 48. Electric charge can accumulate on an airplane in flight. You may have observed needle-shaped metal extensions on the wing tips and tail of an airplane. Their purpose is to allow charge to leak off before much of it accumulates. The electric field around the needle is much larger than the field around the body of the airplane, and can become large enough to produce dielectric breakdown of the air, discharging the airplane. To model this process, assume that two charged spherical conductors are connected by a long conducting wire, and a charge of 1.20 μC is placed on the combination. One sphere, representing the body

2.11. CORONA DISCHARGE

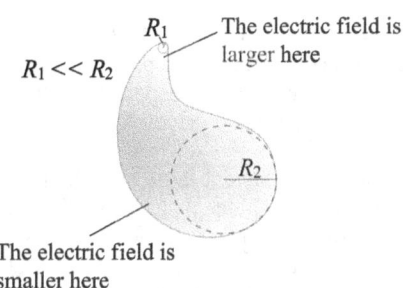

Figure 2.73

of the airplane, has a radius of 6.00 cm, and the other, representing the tip of the needle, has a radius of 2.00 cm. (a) What is the electric potential of each sphere? (b) What is the electric field at the surface of each sphere?

SOLUTION (a) Let q_1 and q_2 are charges on larger and smaller spheres. Radii of these spheres are $r_1 = 6.00$ cm and $r_2 = 2.00$ cm respectively.

Both spheres are connected with each other with metal body, therefore they must be at the same potential, i.e.,
$$\frac{kq_1}{r_1} = \frac{kq_2}{r_2} \implies q_1 = \frac{q_2 r_1}{r_2}$$
where also the total charge is-
$$q_1 + q_2 = 1.20 \times 10^{-6} C$$
Substituting the value of q_1 in above expression, we get
$$\frac{q_2 r_1}{r_2} + q_2 = 1.20 \times 10^{-6} C$$
$$\implies q_2 \left(\frac{r_1}{r_2} + 1\right) = 1.20 \times 10^{-6} C \implies q_2 = \frac{1.20 \times 10^{-6} C}{\frac{r_1}{r_2} + 1}$$
Now, substituting the values of r_1, r_2 in above expression and solving for q_2, we get
$$q_2 = \frac{1.20 \times 10^{-6} C}{1 + 6\text{ cm}/2\text{ cm}} = 0.300 \times 10^{-6} C$$
Thus, charge on smaller sphere is $0.300 \times 10^{-6} C$.
Since, $q_1 + q_2 = 1.20 \times 10^{-6} C$, therefore charge on larger sphere,
$$q_1 = 1.20 \times 10^{-6} C - q_2 = 1.20 \times 10^{-6} C - 0.300 \times 10^{-6} \text{ C}$$
$$= 0.900 \times 10^{-6} C$$
Now, electric potential
$$V = \frac{kq_1}{r_1} = \frac{(8.99 \times 10^9 \text{ N} \cdot \text{m}^2/\text{C}^2)(0.900 \times 10^{-6} C)}{6 \times 10^{-2} \text{ m}}$$
$$= 1.35 \times 10^5 \text{ V}$$
(b) Outside the larger sphere,
$$E_1 = \frac{kq_1}{r_1^2}\hat{r} = \frac{V_1}{r_1}\hat{r} = \frac{1.35 \times 10^5 \text{ V}}{0.06 \text{ m}}\hat{r}$$
$$= 2.25 \times 10^6 \text{ V/m away}$$
Outside the smaller sphere,
$$E_2 = \frac{1.35 \times 10^5 \text{ V}}{0.02 \text{ m}}\hat{r} = 6.74 \times 10^6 \text{ V/m away}$$
The smaller sphere carries less charge but creates a much stronger electric field than the larger sphere.

2.11.1 Electric Potential Energy for Continuous Charge System

Electric potential energy of a continuously charged object is also called self-energy* or internal energy of the object.

---

*For a system of stationary or moving discrete charges, the net electrostatic potential energy can be calculated as discussed in section 2.3 but for system of objects having a continuous charge distribution, we must also consider the self-energy of each object in the system.

2.11.1.1 Self (or Internal) Energy of a Uniformly Charged Conducting Spherical Shell or Conducting Sphere

For a conducting spherical shell or a sphere, charges are stored only on the outer surface, so the calculation of potential energy for the conducting sphere will be identical to that of a conducting spherical shell. Here, we will focus on discussing the conducting spherical shell.

Take an uncharged shell. Now bring charges one by one from infinity to the surface of the shell. The work required in this process will be stored as potential Energy.

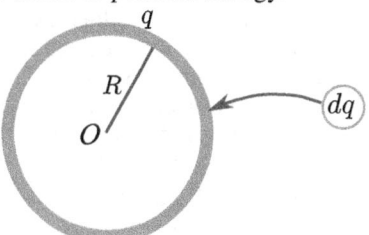

Figure 2.74

Suppose, at any instant the given charge on the conducting spherical shell is q and now we are giving extra charge dq to it. The work required to bring this extra charge dq from infinity to the shell is -
$$dW = (dq)(V_f - V_i) = dq\left(\frac{kq}{R} - 0\right) = \frac{kq}{R}dq$$
Therefore, net work done in charging the shell up to charge Q, is given by-
$$W = \int_0^Q \frac{kq}{R}dq = \frac{kQ^2}{2R}$$
This work will be stored in the form of electrostatic potential energy (self energy). So, the electric potential energy of the charged spherical shell-
$$\boxed{U = \frac{kQ^2}{2R}} \qquad (2.80)$$
Eq.2.80 gives the self energy (or electric potential energy) of a uniformly charged conducting shell (or sphere).

2.11.1.2 Self Energy of a Uniformly Charged Non Conducting Solid Sphere

To make a non conducting charged solid sphere, we have to assemble charged particles (each of infinitely small step-sized charge dq) one over the other until a sphere of required charge and radius is formed. For it, we bring the charged particles one by one from infinity to the sphere so that the size of the sphere increases. Suppose, at any instant when the collected charge is q,

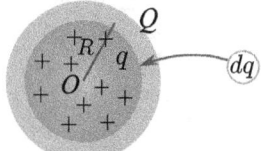

Figure 2.75

then the radius of sphere is r. Now, let us giving extra charge dq to it, which increases it's radius by dr.
∴ Work required to bring charge dq from infinity to the sphere
$$dW = dq(V_f - V_i) = dq\left(\frac{kq}{r} - 0\right) = \frac{kq\,dq}{r}$$
here, $q = \rho\left(\frac{4}{3}\pi r^3\right) \implies dq = \rho(4\pi r^2 dr)$
∴ Total work required to give charge Q to the sphere
$$W = \int_0^R k\frac{\rho\left(\frac{4}{3}\pi r^3\right)\rho(4\pi r^2 dr)}{r}$$

On simplifying it, we get:
$$W = \frac{3}{5}\frac{kQ^2}{R} = U_{\text{self}}$$
(for a non conducting solid sphere)

◊ The total electrical energy of a system of objects is always equal to the sum of self energies of each object and the electrical potential energy due to mutual interaction between the objects. If the objects are charged particles (i.e., point charges), then the self energy term will be zero.

EXAMPLE 49. A uniformly charged spherical shell of radius R and charge q, is expanded to a radius $2R$. Find the work performed by the external agent against electric forces and work done by electric forces, in this process.

APPROACH The work done by electric forces is always equal to the $-$ve of change in electric potential energy of the system, i.e.,
$$W_{\text{el}} = -\Delta U = -(U_f - U_i)$$
Here, U_i and U_f are respectively the initial and final electric potential energies of the spherical shell.

SOLUTION $W_{\text{el}} = -(U_f - U_i) = -\left[\frac{q^2}{16\pi\epsilon_0 R} - \frac{q^2}{8\pi\epsilon_0 R}\right]$
$$= \frac{q^2}{16\pi\epsilon_0 R}$$
Now, work done by an external agent against electric forces-
$$W_{\text{ext}} = -W_{\text{el}} = -\frac{q^2}{16\pi\epsilon_0 R}$$

☞ +ve work of electric field or negative work of an external agent shows the loss of potential energy of the system.

EXAMPLE 50. Two non-conducting hollow uniformly charged spheres of radii R_1 and R_2 with charge Q_1 and Q_2 respectively are placed at a distance r. Find out total energy of the system.

APPROACH To find the total energy of the system, add the self

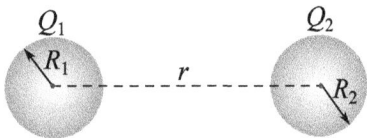

Figure 2.76

energies of both spheres and the potential energy due to mutual interaction.

SOLUTION $U_{\text{total}} = U_{\text{self}} + U_{\text{interaction}}$
$$= \frac{Q_1^2}{8\pi\epsilon_0 R_1} + \frac{Q_2^2}{8\pi\epsilon_0 R_2} + \frac{Q_1 Q_2}{4\pi\epsilon_0 r}$$

EXAMPLE 51. Two uniformly charged concentric spherical shells of radii R_1 and R_2 ($R_2 > R_1$) have charges Q_1 and Q_2 respectively. Find out total energy of the system.

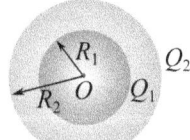

Figure 2.77

APPROACH To find the total energy of the system, add the self energies of both spherical shells and the potential energy due to mutual interaction.

SOLUTION $U_{\text{total}} = U_{\text{self1}} + U_{\text{self2}} + U_{\text{interaction}}$
$$= \frac{Q_1^2}{8\pi\epsilon_0 R_1} + \frac{Q_2^2}{8\pi\epsilon_0 R_2} + \frac{Q_1 Q_2}{4\pi\epsilon_0 R_2}$$

2.12 The Van de Graaff Generator

This is a machine that can build up high voltages of the order of a few million volts and electric field close to breakdown field of air which is about 3×10^6 V/m. The resulting large electric fields are used to accelerate charged particles (electrons, protons, ions) to high energies needed for experiments to probe the small scale structure of matter. The principle underlying the machine is as follows:

Principle: Suppose we have a large spherical conducting shell of radius R, on which we place a charge Q. This charge spreads itself uniformly over the entire surface of the shell. The field outside the sphere is like that of a point charge Q at the centre; while inside the sphere, the field vanishes. The potential outside is that of a point charge; whereas inside, it remains constant and equal to the value of potential at the surface of the conducting shell.

Thus, the potential inside conducting spherical shell of radius R carrying charge Q,
$$V_{\substack{\text{due to charge } Q \\ \text{on spherical shell}}} = \frac{1}{4\pi\epsilon_0}\frac{Q}{R} \quad (2.81)$$

Now, as shown in Fig. 2.78, let us introduce a small sphere of radius r, carrying some charge q, into the large one, and place it at the centre.

Now, the potential due to small sphere of radius r carrying charge q:

1. At surface of small sphere $= \dfrac{1}{4\pi\epsilon_0}\dfrac{q}{r}$

2. At large shell of radius $R = \dfrac{1}{4\pi\epsilon_0}\dfrac{q}{R}$

Taking both charges q and Q into account, the net potential at any point on the surface of conducting shell:
$$V(R) = \frac{1}{4\pi\epsilon_0}\left(\frac{Q}{R} + \frac{q}{R}\right) \quad (2.82)$$

And net potential at any point on the surface of inner sphere:
$$V(r) = \frac{1}{4\pi\epsilon_0}\left(\frac{Q}{R} + \frac{q}{r}\right) \quad (2.83)$$

Therefore, the potential difference between the surface of inner sphere and spherical shell:
$$V(r) - V(R) = \frac{q}{4\pi\epsilon_0}\left(\frac{1}{r} - \frac{1}{R}\right) \quad (2.84)$$

Because $r < R$, therefore, if q is positive then $V(r) - V(R)$ will also be positive.

From Eq.2.84, we see that, $V(r) - V(R)$ is independent of charge

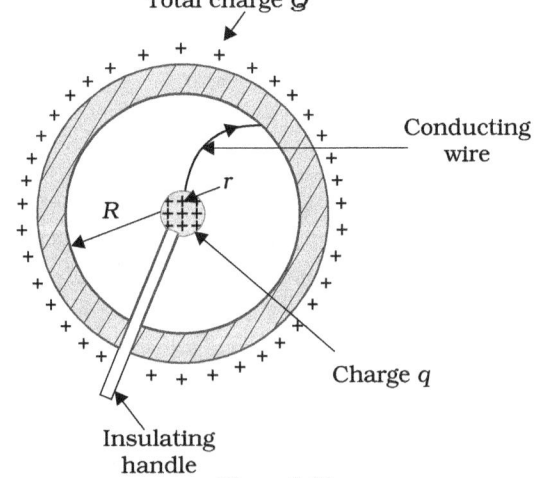

Figure 2.78

Q that may have accumulated on the larger sphere and even if it is positive, the inner sphere is always at a higher potential for +ve value of q (because, the difference $V(r) - V(R)$ is positive). The potential due to Q is constant up to radius R and so cancels out in the difference.

This means that if we now connect the smaller and larger sphere

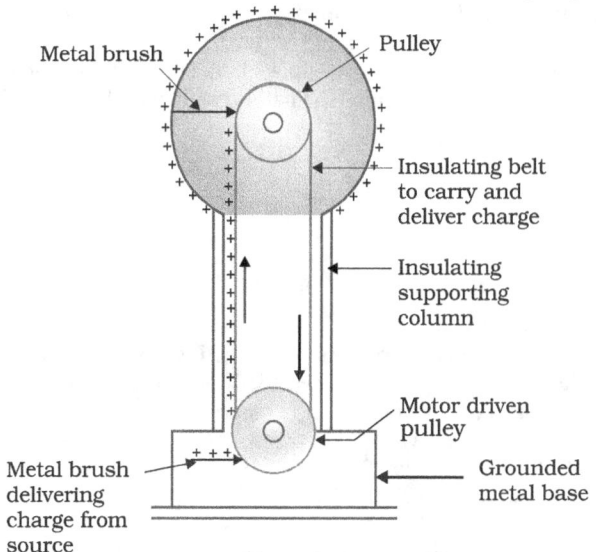

Figure 2.79

by a wire, the charge q on the smaller will immediately flow onto the larger, even though the charge Q may be quite large. The natural tendency is for positive charge to move from higher to lower potential. Thus, provided we are somehow able to introduce the small charged sphere into the larger one, we can in this way keep piling up larger and larger amount of charge on the latter. The potential (Eq.2.82) at the outer sphere would also keep rising, at least until we reach the breakdown field of air.

This is the principle of the van de Graaff generator.

Construction and Working: A schematic diagram of the van de Graaff generator is given in Fig.2.79. A large spherical conducting shell (of few meters radius) is supported at a height several meters above the ground on an insulating column. A long narrow endless belt insulating material, like rubber or silk, is wound around two pulleys – one at ground level, one at the center of the shell. This belt is kept continuously moving by a motor driving the lower pulley. It continuously carries positive charge, sprayed on to it by a brush at ground level, to the top. There it transfers its positive charge to another conducting brush connected to the large shell. Thus positive charge is transferred to the shell, where it spreads out uniformly on the outer surface. In this way, voltage differences of as much as 6 or 8 million volts (with respect to ground) can be built up.

Note: It is possible to increase the potential of the dome until electrical ionization occurs in the air. Since the ionization breakdown of air occurs at an electric field of about 3×10^6 V/m, a sphere of 1m can be raised to maximum of $V_{max} = ER = (3 \times 10^6 \text{ V/m})(1 \text{ m}) = 3 \times 10^6$ V. The dome's electric potential can be increased further by placing the dome in vacuum and by increasing the radius of the sphere.

2.13 The Millikan Oil-Drop Experiment

Robert Millikan conducted a series of experiments during 1909-1913 in which he measured e, the magnitude of the elementary charge on an electron, and demonstrated the quantized nature of this charge. His apparatus, diagrammed in Figure 2.80, contains two parallel metallic plates separated by distance d. Oil droplets from an atomizer are allowed to pass through a small hole in the upper plate. Millikan used x-rays to ionize the air in the chamber so that freed electrons would adhere to the oil drops, giving them a negative charge. A horizontally directed light beam is used to illuminate the oil droplets, which are viewed through a telescope whose long axis is perpendicular to the light beam. When viewed in this manner, the droplets appear as shining stars against a dark background and the rate at which individual drops fall can be determined.

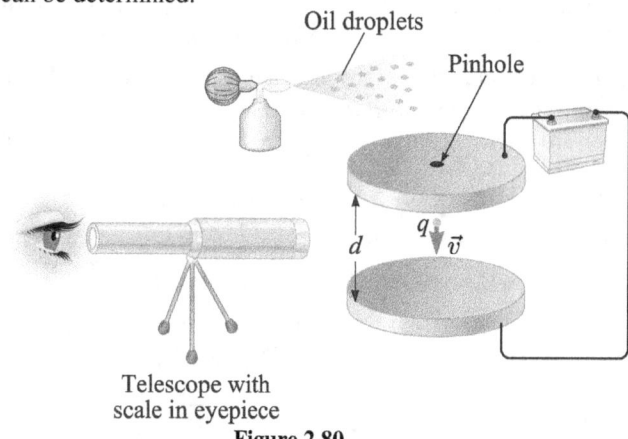

Figure 2.80

Let's assume a single drop having a mass m and carrying a charge q is being viewed and its charge is negative. If no electric field is present between the plates (Fig.2.81a), the two forces acting on the charge are-
1. Gravitational force $m\vec{g}$ (acting downwards)
2. Viscous drag force $\vec{F}_D$ (acting upwards)

Note that there is also a buoyant force on the oil drop due to the surrounding air. This force can be taken as a correction term in the gravitational force $m\vec{g}$ on the drop. For now, we will not consider it in our analysis.

The drag force is proportional to the drop's speed. When the drop reaches its terminal speed (v_T) the two forces balance each other, i.e.
$$mg = F_D$$
At terminal speed of the drop, the viscous drag force is given by-
$$F_D = 6\pi r \eta v_T$$
where r is the radius of the oil drop, η is the viscosity of the air, and v_T is the terminal velocity of the falling drop.

From above two equations, we can write-

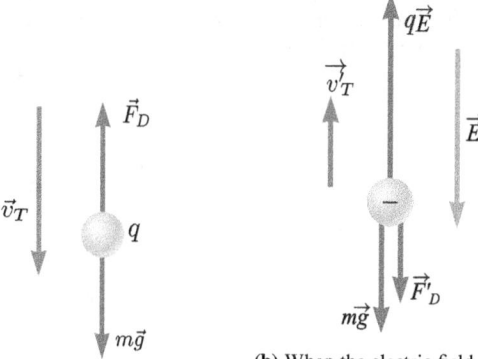

(a) With the electric field off, the droplet falls at terminal velocity $\vec{v}_T$ under the influence of the gravitational and drag forces.

(b) When the electric field is turned on, the droplet moves upward at terminal velocity $\vec{v}'_T$ under the influence of the electric, gravitational, and drag forces.

Figure 2.81: The forces acting on a negatively charged oil droplet in the Millikan experiment.

$$mg = 6\pi r \eta v_T$$
$$\implies r = \frac{mg}{6\pi \eta v_T} \qquad (2.85)$$

Now suppose a battery connected to the plates sets up an electric field between the plates such that the upper plate is at the higher electric potential. In this case, a third force $q\vec{E}$ acts on the charged drop. Since, q is negative and $\vec{E}$ is directed downward, this electric force is directed upward as shown in Figure 2.81b. If this upward force is strong enough, the drop moves upward and the drag force $\vec{F}'_D$ acts downward. When the upward electric force $q\vec{E}$ balances the sum of the gravitational force and the downward drag force $\vec{F}'_D$, the drop reaches a new terminal speed v'_T in the upward direction. In this situation-
$$qE = mg + F_D$$
$$\Rightarrow \quad qE = mg + 6\pi r \eta v'_T \quad (2.86)$$
Substituting the value of r from Eq.2.85 to Eq.2.86, we get-
$$qE = mg + 6\pi \left(\frac{mg}{6\pi \eta v_T}\right) \eta v'_T = mg + \frac{mg}{v_T} v'_T$$
$$\therefore \quad \boxed{q = \frac{mg}{E}\left(1 + \frac{v'_T}{v_T}\right)} \quad (2.87)$$
Since the electrodes are parallel plates, the magnitude of the electric field can be readily calculated from the potential difference (V) between the plates and the plate separation (d). Thus, by measuring the terminal velocities both with the field off and with the field on, it is possible to calculate the charge on a drop with the help of Eq.2.87.

Millikan made the necessary measurements on a number of drops and calculated the charge on each drop. He showed that all the drops had charges that are integer multiples of a fundamental unit of charge (e), i.e.,
$$q = ne \quad n = 0, \pm 1, \pm 2, \pm 3, \ldots$$
In his initial results, published in 1910, he reported a value of 1.63×10^{-19} C for the fundamental charge. After modifying his equipment and measuring a larger number of drops, he published a revised value of 1.59×10^{-19} C in 1913. The currently accepted value for the fundamental (or elementary) charge, e, is approximately 1.602×10^{-19} C.

2.14 Check Point 3

1. ••The labelled points in Figure 2.82 are on a series of equipotential surfaces associated with an electric field. Rank (from greatest to least) the work done by the electric field on a positively charged particle that moves from A to B; from B to C; from C to D; from D to E.

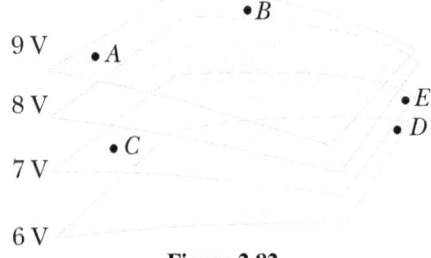

Figure 2.82

Multiple Choice Questions

2. ••Equipotential surfaces associated with an electric field which is increasing in magnitude along the X-direction are-
   (A) planes parallel to YZ-plane
   (B) planes parallel to XY-plane
   (C) planes parallel to XZ-plane
   (D) coaxial cylinders of increasing radii around the x-axis

3. ••For the equipotential surfaces in Figure 2.82, what is the approximate direction of the electric field?
   (A) Out of the page
   (B) Into the page
   (C) Toward the top of the page
   (D) Toward the bottom of the page.

4. ••A metallic solid sphere is placed in a uniform electric field. The lines of force follow the path(s) shown in the Fig.2.83

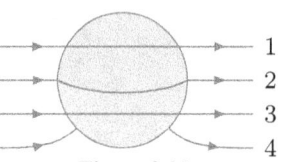

Figure 2.83

   (A) 1   (B) 2   (C) 3   (D) 4

5. ••A uniformly charged non-conducting solid sphere of radius R has potential V_0 (measured with respect to ∞) on its surface. For this sphere, the equipotential surfaces with potentials $3V_0/2$, $5V_0/4$, $3V_0/4$ and $V_0/4$ have radii R_1, R_2, R_3, and R_4 respectively. Then,
   (A) $R_1 = 0$ and $R_2 > (R_4 - R_3)$
   (B) $R_1 \neq 0$ and $(R_2 - R_1) > (R_4 - R_3)$
   (C) $R_1 = 0$ and $R_2 < (R_4 - R_3)$
   (D) $2R < R_4$

6. ••A hollow metal sphere of radius 5 cm is charged such that the potential on its surface is 10 V. The potential at the centre of the sphere is
   (A) zero
   (B) 10 V
   (C) same as at a point 5 cm away from the surface
   (D) same as at a point 25 cm away from the surface.

7. ••The arc AB with the centre C and the infinitely long wire having linear charge density λ are lying in the same plane as shown in Fig.2.84. The minimum amount of work to be expended to move a point charge q_0 from point A to B through a circular path AB of radius a is equal to:
   (A) $\frac{q_0}{2\pi \epsilon_0} \ln \frac{2}{3}$ (B) $\frac{q_0 \lambda}{2\pi \epsilon_0} \ln \frac{3}{2}$
   (C) $\frac{q_0 \lambda}{2\pi \epsilon_0} \ln \frac{2}{3}$ (D) $\frac{q_0 \lambda}{\sqrt{2}\pi \epsilon_0}$

Figure 2.84

8. ••• Consider a system of three charges $q/3$, $q/3$ and $-2q/3$ placed at points A, B and C, respectively, as shown in the Fig.2.85. Take O to be the centre of the circle of radius R and angle CAB = 60°.
   (A) The electric field at point O is $\frac{q}{8\pi \epsilon_0 R^2}$ directed along the negative x-axis
   (B) The potential energy of the system is zero.
   (C) The magnitude of the force between the charge C and B is $\frac{q^2}{54\pi \epsilon_0 R^2}$
   (D) The potential at point O is $\frac{q}{12\pi \epsilon_0 R}$.

2.15 Questions and Exercises

2.15.1 Conceptual Questions

1. A negative charge moves in the direction of a uniform electric field. Does the potential energy of the charge–field system

2.15. QUESTIONS AND EXERCISES

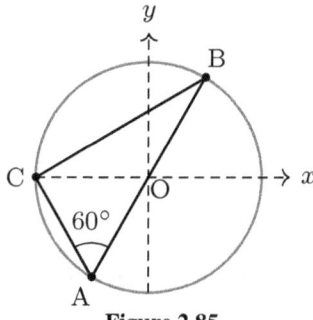

Figure 2.85

increase or decrease? Does the charge move to a position of higher or lower potential?

2. A proton is moved to the left in a uniform electric field that points to the right. Is the proton moving in the direction of increasing or decreasing electric potential? Is the electrostatic potential energy of the proton, in presence of field, increasing or decreasing?

3. An electron is moved to the left in a uniform electric field that points to the right. Is the electron moving in the direction of increasing or decreasing electric potential? Is the electrostatic potential energy of the electron, in presence of field, is increasing or decreasing?

4. If the electric potential is uniform throughout a region of space, what can be said about the electric field in that region?

5. Can there be a potential difference between two conductors that carry like charges of the same magnitude?

6. If V is known at only a single point in space, can $\vec{E}$ be found at that point? Explain your answer.

7. If two points are at the same potential, does this mean that no work is done in moving a test charge from one point to the other? Does this imply that no force must be exerted? Explain.

8. If a negative charge is initially at rest in an electric field, will it move toward a region of higher potential or lower potential? What about a positive charge? How does the potential energy of the charge change in each instance?

9. Figure 2.86 shows a point particle that has a positive charge $+Q$ and a metal sphere that has a charge $-Q$. Sketch the electric field lines and equipotential surfaces for this system of charges.

Figure 2.86

10. Can a free particle ever move from a region of low electric potential to one of high potential and yet have its electric potential energy decrease? Explain.

11. Figure 2.87 shows a point particle and metal sphere. Both have equal charge $+Q$. Sketch the electric field lines and equipotential surfaces for this system of charges.

12. Two equal positive point charges are separated by a finite distance. Sketch the electric field lines and the equipotential surfaces for this system.

13. Two point charges are fixed on the x-axis. (a) Each has a positive charge q. One is at $x = -a$ and the other is at $x = +a$. At the origin, which of the following is true?
(A) $\vec{E} = 0$ and $V = 0$,

Figure 2.87

(B) $\vec{E} = 0$ and $V = 2kq/a$,
(C) $\vec{E} = (2kq/a^2)\,\hat{\imath}$ and $V = 0$,
(D) $\vec{E} = (2kq/a^2)\,\hat{\imath}$ and $V = 2kq/a$,

(b) One point charge has a positive charge $+q$ and the other has a negative charge $-q$. The positive point charge is at $x = -a$ and the negative point charge is at $x = +a$. At the origin, which of the following is true?
(A) $\vec{E} = 0$ and $V = 0$,
(B) $\vec{E} = 0$ and $V = 2kq/a$,
(C) $\vec{E} = (2kq/a^2)\,\hat{\imath}$ and $V = 0$,
(D) $\vec{E} = (2kq/a^2)\,\hat{\imath}$ and $V = 2kq/a$.

14. A uniform electric field is parallel to the x axis. In what direction can a charge be displaced in this field without any external work being done on the charge?

15. The electric potential (in volts) is given by $V(x, y, z) = 4.00|x| + V_0$, where V_0 is a constant, and x is in meters. (a) Sketch the electric field for this potential. (b) Which of the following charge distributions is most likely responsible for this potential:
(A) A negatively charged flat sheet in the $x = 0$ plane,
(B) a point charge at the origin,
(C) a positively charged flat sheet in the $x = 0$ plane,
(D) a uniformly charged sphere centered at the origin?
Explain your answer.

16. The electric potential is the same everywhere on the surface of a conductor. Does this mean that the surface charge density is also the same everywhere on the surface? Explain your answer.

17. If $V = 0$ at a point in space, must $\vec{E} = 0$? If $\vec{E} = 0$ at some point, must $V = 0$ at that point? Explain. Give examples for each.

18. When dealing with practical devices, we often take the ground (the Earth) to be 0 V. (a) If instead we said the ground was -10 V, how would this affect V and E at other points? (b) Does the fact that the Earth carries a net charge affect the choice of V at its surface?

19. Explain why electric field lines are always perpendicular to equipotential surfaces.

20. Can two equipotential lines cross? Explain.

21. A satellite orbits the Earth along a gravitational equipotential line. What shape must the orbit be?

22. Suppose a charged ring is non-uniformly charged, so that the density of charge was twice as great near the top as near the bottom. Assuming the total charge on the ring is Q. Would this affect the potential at point P on the axis (Fig.2.88) as obtained for a uniformly charged ring? Would it affect the value of $\vec{E}$ at that point obtained for uniformly charged ring? Is there a discrepancy here? Explain.

23. Consider a metal conductor in the shape of a football. If it carries a total charge Q, where would you expect the charge density σ to be greatest, at the ends or along the flatter sides? Explain. [Hint: Near the surface of a conductor, $E = \sigma/\epsilon_0$.]

24. If you know V at a point in space, can you calculate $\vec{E}$ at that point? If you know $\vec{E}$ at a point can you calculate V at that

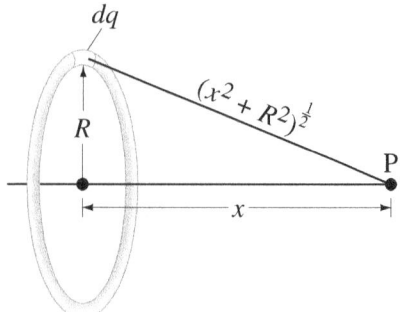

Figure 2.88

point? If not, what else must be known in each case?

25. A conducting sphere carries a charge Q and a second identical conducting sphere is neutral. The two are initially isolated, but then they are placed in contact. (a) What can you say about the potential of each when they are in contact? (b) Will charge flow from one to the other? If so, how much? (c) If the spheres do not have the same radius, how are your answers to parts (a) and (b) altered?

26. At a particular location, the electric field points due north. In what direction(s) will the rate of change of potential be (a) greatest, (b) least, and (c) zero?

27. Why is it important, when soldering connectors onto a piece of electronic circuitry, to leave no pointy protrusions from the solder joints?

28. Equipotential lines are spaced 1.00 V apart. Does the distance between the lines in different regions of space tell you anything about the relative strengths of $\vec{E}$ in those regions? If so, what?

29. If the electric field $\vec{E}$ is uniform in a region, what can you infer about the electric potential V? If V is uniform in a region of space, what can you infer about $\vec{E}$?

30. Is the electric potential energy of two unlike charges positive or negative? What about two like charges? What is the significance of the sign of the potential energy in each case?

31. Fig.2.89 shows the x-component of $\vec{E}$ as a function of x. Draw a graph of V versus x in this same region of space. Let $V = 0$ V at $x = 0$ m and include an appropriate vertical scale.

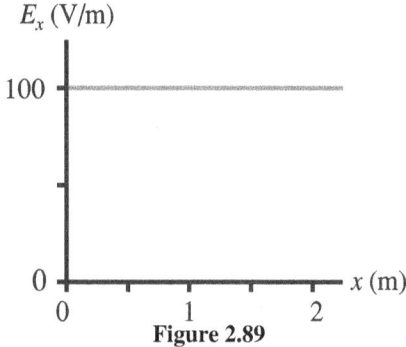

Figure 2.89

32. Fig.2.90 shows the electric potential as a function of x. Draw a graph of E_x versus x in this same region of space.

33. For each contour map in Fig.2.91, estimate the electric fields $\vec{E}_1$ and $\vec{E}_2$ at points 1 and 2.

34. An electron is released from rest at $x = 2$ m in the potential shown in Fig.2.92. Does it move? If so, to the left or to the right? Explain.

35. Fig.2.93 shows an electric field diagram. Dashed lines 1 and 2 are two surfaces in space, not physical objects.
(a) Is the electric potential at point a higher than, lower than, or equal to the electric potential at point b? Explain.

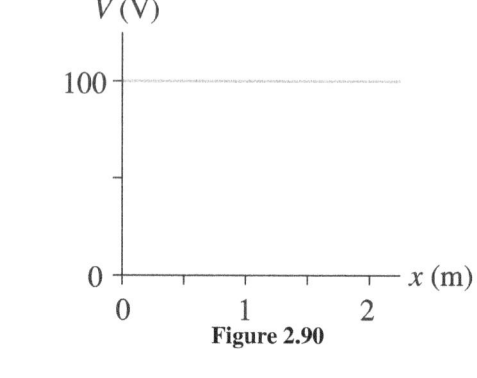

Figure 2.90

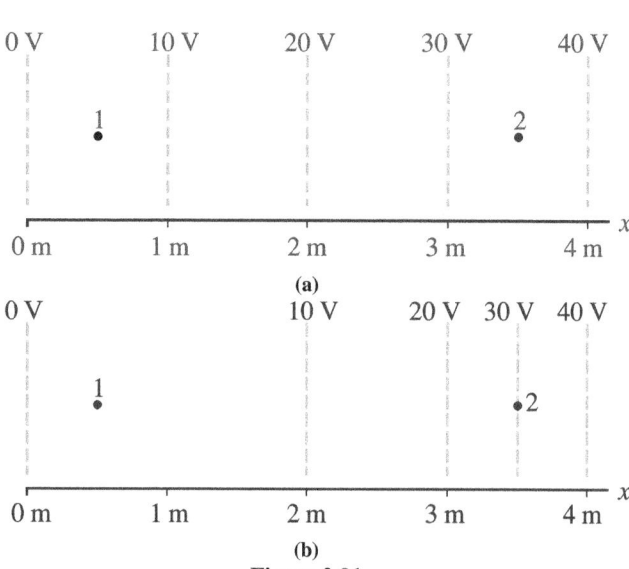

Figure 2.91

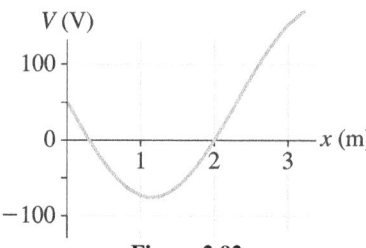

Figure 2.92

(b) Rank in order, from largest to smallest, the magnitudes of the potential differences ΔV_{ab}, ΔV_{cd}, and ΔV_{ef}.
(c) Is surface 1 an equipotential surface? What about surface 2? Explain why or why not.

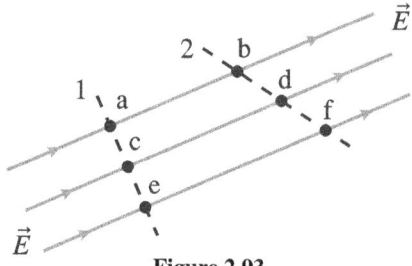

Figure 2.93

36. Fig.2.94 shows a negatively charged electroscope. The gold leaf stands away from the rigid metal post. Is the electric potential of the leaf higher than, lower than, or equal to the potential of the post? Explain.

37. The two metal spheres in Fig.2.95 are connected by a metal wire with a switch in the middle. Initially the switch is open. Sphere 1, with the larger radius, is given a positive charge.

2.15. QUESTIONS AND EXERCISES

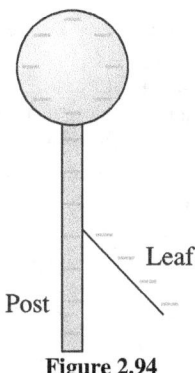

Figure 2.94

Sphere 2, with the smaller radius, is neutral. Then the switch is closed. Afterward, sphere 1 has charge Q_1, is at potential V_1, and the electric field strength at its surface is E_1. The values for sphere 2 are Q_2, V_2, and E_2

(a) Is V_1 larger than, smaller than, or equal to V_2? Explain.

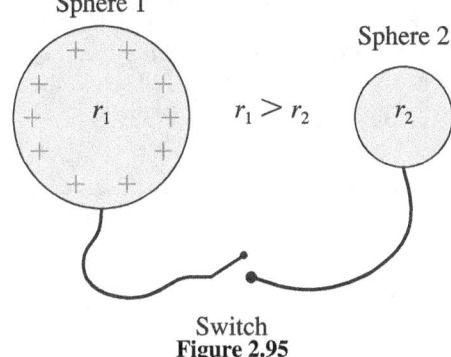

Figure 2.95

(b) Is Q_1 larger than, smaller than, or equal to Q_2? Explain.
(c) Is E_1 larger than, smaller than, or equal to E_2? Explain.

38. Fig.2.96 shows a graph of E_x vs x. What is the potential difference between $x_i = 1.0$ m and $x_f = 3.0$ m?

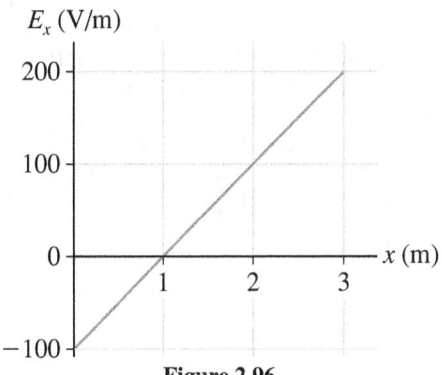

Figure 2.96

39. Fig.2.97 is a graph of E_x vs x. The potential at the origin is -50 V. What is the potential at $x = 3.0$ m?

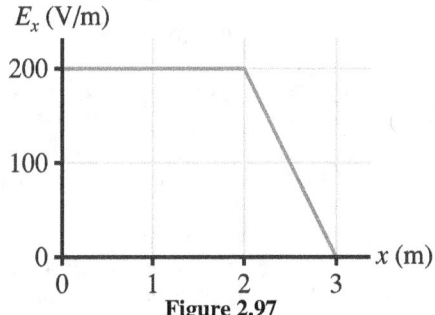

Figure 2.97

40. What are the magnitude and direction of the electric field at the dot in Fig.2.98?

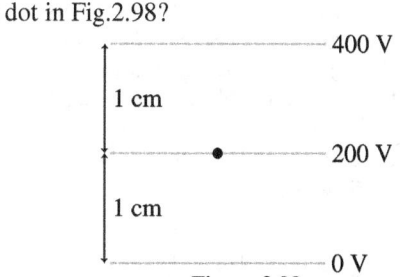

Figure 2.98

41. What are the magnitude and direction of the electric field at the dot in Fig.2.99?

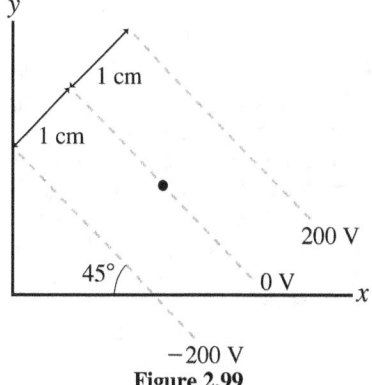

Figure 2.99

42. Fig.2.100 shows V versus x graphs. Draw the corresponding graphs of E_x versus x.

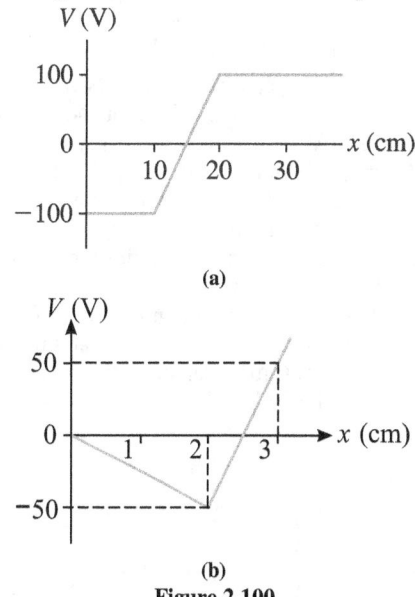

(a)

(b)
Figure 2.100

43. The electric potential in a region of uniform electric field is -1000 V at $x = -1.0$ m and $+1000$ V at $x = +1.0$ m. What is E_x?

44. The electric potential along the x-axis is $V = 100x^2$ V, where x is in meters. What is E_x at (a) $x = 0$ m and (b) $x = 1$ m?

45. The electric potential along the x-axis is $V = 100\,e^{-2x}$ V, where x is in meters. What is E_x at (a) $x = 1.0$ m and (b) $x = 2.0$ m?

2.15.2 Problems

In all the problems in this section, assume that the electric potential is zero at distances far from all charges unless otherwise stated.

Potential Differences in a Uniform Electric Field

1. ••When an electron moves from A to B along an electric field line in Fig.2.101, the electric field does 3.94×10^{-19} J of work on it. What are the electric potential differences (a) $V_B - V_A$ (b) $V_C - V_A$, and (c) $V_C - V_B$?

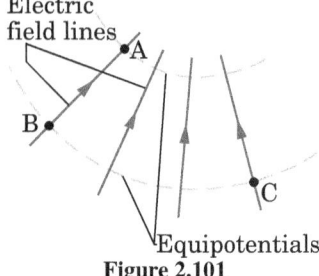
Figure 2.101

2. ••A graph of the x component of the electric field as a function of x in a region of space is shown in Fig. 2.102. The scale of the vertical axis is set by $E_{xs} = 20.0$ N/C. The y and z components of the electric field are zero in this region. If the electric potential at the origin is 10 V, (a) what is the electric potential at $x = 2.0$ m, (b) what is the greatest positive value of the electric potential for points on the x axis for which $0 \leq x \leq 6.0$ m, and (c) for what value of x is the electric potential zero?

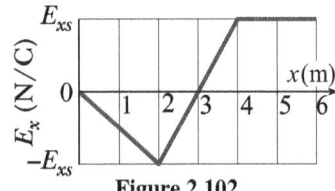
Figure 2.102

3. ••What are (a) the charge and (b) the charge density on the surface of a conducting sphere of radius 0.15 m whose potential is 200 V (with $V = 0$ at infinity)?

4. ••The difference in potential between the accelerating plates in the electron gun of a TV picture tube is about 25000 V. If the distance between these plates is 1.50 cm, what is the magnitude of the uniform electric field in this region?

5. ••A uniform electric field of magnitude 325 V/m is directed in the negative y direction in Figure 2.103. The coordinates of point A are $(-0.200$ m, -0.300 m$)$, and those of point B are $(0.400$ m, 0.500 m$)$. Calculate the potential difference $V_B - V_A$, using the path ACB.

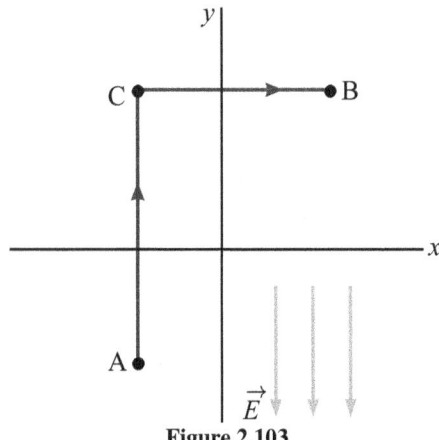
Figure 2.103

6. ••A block having mass m and charge $+Q$ is connected to a spring having spring constant k. The block lies on a frictionless horizontal track, and the system is immersed in a uniform electric field of magnitude E, directed as shown in Fig.2.104. If the block is released from rest when the spring is unstretched (at $x = 0$), (a) by what maximum amount does the spring expand? (b) What is the equilibrium position of the block? (c) Show that the block's motion is simple harmonic, and determine its period. (d) What If? Repeat part (a) if the coefficient of kinetic friction between block and surface is μ_k.

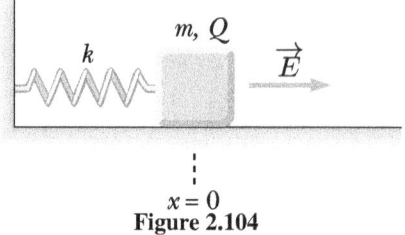
Figure 2.104

7. ••An insulating rod having linear charge density $\lambda = 40.0$ μC/m and linear mass density $\mu = 0.100$ kg/m is released from rest in a uniform electric field $E = 100$ V/m directed perpendicular to the rod (Fig. 2.105). (a) Determine the speed of the rod after it has travelled 2.00 m. (b) How does your answer to part (a) change if the electric field is not perpendicular to the rod? Explain.

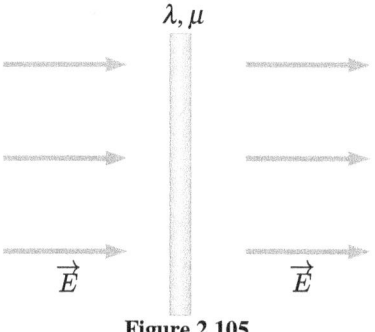
Figure 2.105

8. ••A particle having charge $q = +2.00$ μC and mass $m = 0.0100$ kg is connected to a string that is $L = 1.50$ m long and is tied to the pivot point P in Figure 2.106. The particle, string and pivot point all lie on a frictionless horizontal table. The particle is released from rest when the string makes an angle $\theta = 60.0°$ with a uniform electric field of magnitude $E = 300$ V/m. Determine the speed of the particle when the string is parallel to the electric field (point a in Fig.2.106).

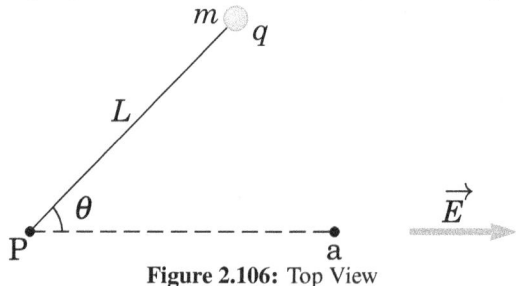
Figure 2.106: Top View

Electric Potential and Potential Energy Due to Point Charges

9. •(a) Find the potential at a distance of 1.00 cm from a proton. (b) What is the potential difference between two points that are 1.00 cm and 2.00 cm from a proton? (c) What If? Repeat parts (a) and (b) for an electron.

10. ••Consider a particle with charge $q = 1.0$ μC, point A at distance $d_1 = 2.0$ m from q, and point B at distance $d_2 = 1.0$ m. (a) If A and B are diametrically opposite to each other, as in Fig.2.107a, what is the electric potential difference $V_A - V_B$? (b) What is that electric potential difference if A

and B are located as in Fig. 2.107b?

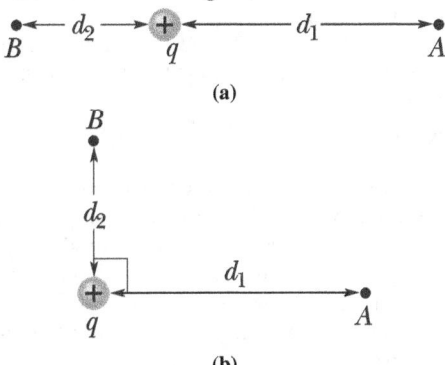

Figure 2.107

11. ••A spherical drop of water carrying a charge of 30 pC has a potential of 500 V at its surface (with $V = 0$ at infinity). (a) What is the radius of the drop? (b) If two such drops of the same charge and radius combine to form a single spherical drop, what is the potential at the surface of the new drop?

12. ••Figure 2.108 shows a rectangular array of charged particles fixed in place, with distance $a = 39.0$ cm and the charges shown as integer multiples of $q_1 = 3.40$ pC and $q_2 = 6.00$ pC. With $V = 0$ at infinity, what is the net electric potential at the rectangle's centre? (Hint: Thoughtful examination of the arrangement can reduce the calculation.)

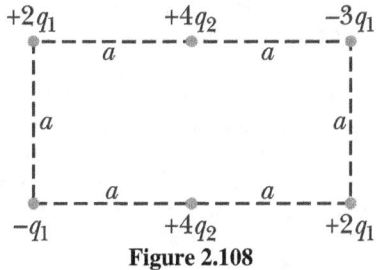

Figure 2.108

13. ••In Fig.2.109, what is the net electric potential at point P due to the four particles if $V = 0$ at infinity, $q = 5.00$ fC, and $d = 4.00$ cm?

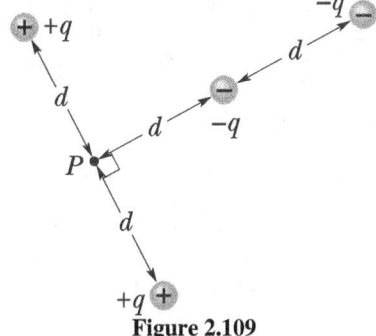

Figure 2.109

14. ••In Fig.2.110, particles with the charges $q_1 = +5e$ and $q_2 = -15e$ are fixed in place with a separation of $d = 24.0$ cm. With electric potential defined to be $V = 0$ at infinity, what are the finite (a) positive and (b) negative values of x at which the net electric potential on the x axis is zero?

15. ••Two particles, of charges q_1 and q_2, are separated by distance d in Fig. 2.110. The net electric field due to the particles is zero at $x = d/4$. With $V = 0$ at infinity, locate (in terms of d) any point on the x axis (other than at infinity) at which the electric potential due to the two particles is zero.

16. ••Given two 2.00 μC charges, as shown in Figure 2.111, and

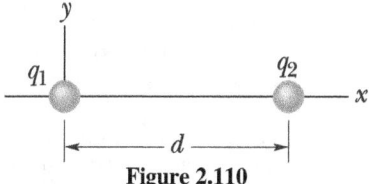

Figure 2.110

a positive test charge $q = 1.28 \times 10^{-18}$C at the origin, (a) what is the net force exerted by the two 2.00 μC charges on the test charge q? (b) What is the electric field at the origin due to the two 2.00 μC charges? (c) What is the electric potential at the origin due to the two 2.00 μC charges?

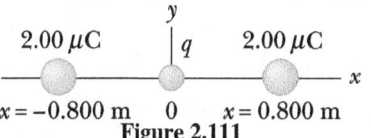

Figure 2.111

17. ••A charge $+q$ is at the origin. A charge $-2q$ is at $x = 2.00$ m on the x axis. For what finite value(s) of x is (a) the electric field zero? (b) the electric potential zero?

18. ••The three charges in Figure 2.112 are at the vertices of an isosceles triangle. Calculate the electric potential at the midpoint of the base, taking $q = 7.00$ μC.

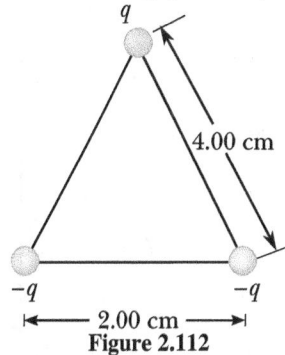

Figure 2.112

19. ••Two point charges, $Q_1 = +5.00$ nC and $Q_2 = -3.00$ nC, are separated by 35.0 cm. (a) What is the potential energy of the pair? What is the significance of the algebraic sign of your answer? (b) What is the electric potential at a point midway between the charges?

20. ••Show that the amount of work required to assemble four identical point charges of magnitude Q at the corners of a square of side s is $5.41kQ^2/s$

21. •••Two insulating spheres have radii 0.300 cm and 0.500 cm, masses 0.100 kg and 0.700 kg, and uniformly distributed charges of -2.00 μC and 3.00 μC. They are released from rest when their centres are separated by 1.00 m. (a) How fast will each be moving when they collide? (b) What If? If the spheres were conductors, would the speeds be greater or less than those calculated in part (a)? Explain.

22. •••Two insulating spheres have radii r_1 and r_2, masses m_1 and m_2, and uniformly distributed charges $-q_1$ and q_2. They are released from rest when their centres are separated by a distance d. (a) How fast is each moving when they collide? (Suggestion: consider conservation of energy and conservation of linear momentum.) (b) If the spheres were conductors, would their speeds be greater or less than those calculated in part (a)? Explain.

23. ••Two particles, with charges of 20.0 nC and -20.0 nC, are placed at the points with coordinates (0, 4.00 cm) and (0, -4.00 cm), as shown in Figure 2.113. A particle with charge 10.0 nC is located at the origin. (a) Find the electric potential energy of the configuration of the three fixed

charges. (b) A fourth particle, with a mass of 2.00×10^{-13} kg and a charge of 40.0 nC, is released from rest at the point $(3.00 \text{ cm}, 0)$. Find its speed after it has moved freely to a very large distance away.

24. ••A light unstressed spring has length d. Two identical particles, each with charge q, are connected to the opposite ends of the spring. The particles are held stationary a distance d apart and then released at the same time. The system then oscillates on a horizontal frictionless table. The spring has a bit of internal kinetic friction, so the oscillation is damped. The particles eventually stop vibrating when the distance between them is $3d$. Find the increase in internal energy that appears in the spring during the oscillations. Assume that the system of the spring and two charges is isolated.

25. ••Two point charges of equal magnitude are located along the y axis at equal distances above and below the x axis, as shown in Figure 2.114. (a) Plot a graph of the potential at points along the x axis over the interval $-3a < x < 3a$. (**Hint:** You should plot the potential in units of kQ/a.) (b) If the charge located at $-a$ be negative then plot the potential along the y axis over the interval $-4a < y < 4a$.

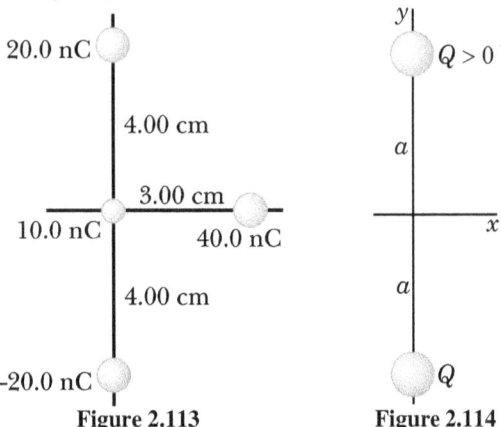

Figure 2.113   **Figure 2.114**

26. •••An alpha particle, has charge $+2e$ and mass 6.64×10^{-27} kg. Assume an alpha particle, initially very far from a gold nucleus, is fired with a velocity of 2.00×10^7 m/s directly toward the nucleus (charge $+79\ e$). How close does the alpha particle get to the nucleus before turning around? Assume the gold nucleus remains stationary.

27. ••An electron starts from rest 3.00 cm from the centre of a uniformly charged insulating sphere of radius 2.00 cm and total charge 1.00 nC. What is the speed of the electron when it reaches the surface of the sphere?

28. ••Calculate the energy required to assemble the array of charges shown in Fig.2.115, where $a = 0.200$ m, $b = 0.400$ m, and $q = 6.00\mu$C.

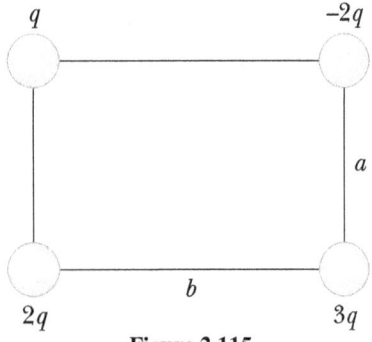

Figure 2.115

29. ••Four identical particles each have charge q and mass m. They are released from rest at the vertices of a square of side L. How fast is each charge moving when their distance from the centre of the square doubles?

30. ••How much work is required to assemble eight identical point charges, each of magnitude q, at the corners of a cube of side s?

31. ••A positive point charge $+Q$ is located on the x axis at $x = -a$. (a) How much work is required to bring an identical point charge from infinity to the point on the x axis at $x = +a$? (b) With the two identical point charges in place at $x = -a$ and $x = +a$, how much work is required to bring a third point charge $-Q$ from infinity to the origin? (c) How much work is required to move the charge $-Q$ from the origin to the point on the x axis at $x = 2a$ along the semicircular path shown in Fig.2.116?

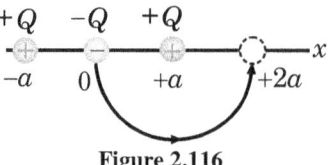

Figure 2.116

Obtaining the Value of the Electric Field from the Electric Potential

32. ••An ion accelerated through a potential difference of 115 V experiences an increase in kinetic energy of 7.37×10^{-17} J. Calculate the charge on the ion.

33. ••(a) Calculate the speed of a proton that is accelerated from rest through a potential difference of 120 V. (b) Calculate the speed of an electron that is accelerated through the same potential difference.

34. ••What potential difference is needed to stop an electron having an initial speed of 4.20×10^5 m/s?

35. ••The potential in a region between $x = 0$ and $x = 6.00$ m is $V = a + bx$, where $a = 10.0$ V and $b = -7.00$ V/m. Determine (a) the potential at $x = 0$, 3.00 m, and 6.00 m, and (b) the magnitude and direction of the electric field at $x = 0$, 3.00 m, and 6.00 m.

36. ••The electric potential inside a charged spherical conductor of radius R is given by $V = kQ/R$, and the potential outside is given by $V = kQ/r$. Using $E_r = -dV/dr$, derive the electric field (a) inside and (b) outside this charge distribution.

37. ••Over a certain region of space, the electric potential is $V = 5x - 3x^2y + 2yz^2$. Find the expressions for the x, y and z components of the electric field over this region. What is the magnitude of the field at the point P that has coordinates $(1, 0, -2)$m?

38. ••Fig.2.117 shows several equipotential lines each labelled by its potential in volts. The distance between the lines of the square grid represents 1.00 cm. (a) Is the magnitude of the field larger at A or at B? Why? (b) What is E at B? (c) Represent what the field looks like by drawing at least eight field lines.

Electric Potential Due to Continuous Charge Distributions

39. ••Consider a ring of radius R with the total charge Q spread uniformly over its perimeter. What is the potential difference between the point at the centre of the ring and a point on its axis a distance $2R$ from the centre?

40. ••In Fig.2.118, a plastic rod having a uniformly distributed charge $Q = -25.6$ pC has been bent into a circular arc of radius $R = 3.71$ cm and central angle $\phi = 120°$. With $V = 0$ at infinity, what is the electric potential at P, the centre of curvature of the rod?

41. ••A plastic rod has been bent into a circle of radius $R = 8.20$ cm. It has a charge $Q_1 = +4.20$ pC uniformly dis-

2.15. QUESTIONS AND EXERCISES

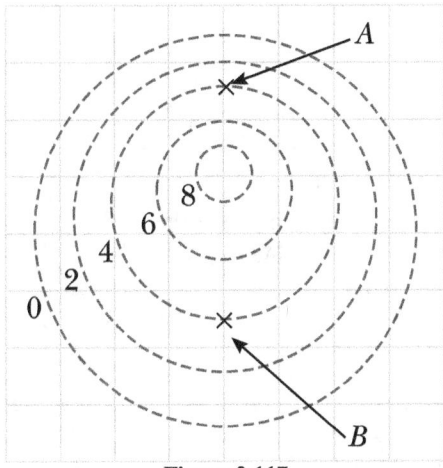

Figure 2.117

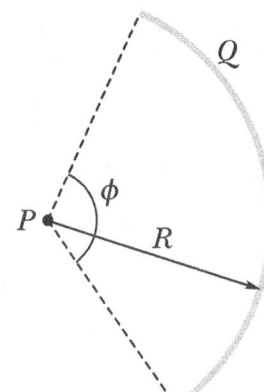

Figure 2.118

tributed along one-quarter of its circumference and a charge $Q_2 = -6Q_1$ uniformly distributed along the rest of the circumference (Fig. 2.119). With $V = 0$ at infinity, what is the electric potential at (a) the center C of the circle and (b) point P, on the central axis of the circle at distance $D = 6.71$ cm from the center?

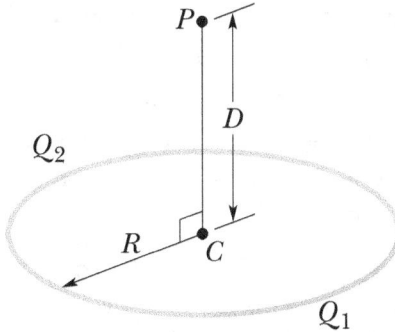

Figure 2.119

42. ••In Fig.2.120, three thin plastic rods form quarter-circles with a common center of curvature at the origin. The uniform charges on the three rods are $Q_1 = +30$ nC, $Q_2 = +3.0Q_1$, and $Q_3 = -8.0Q_1$. What is the net electric potential at the origin due to the rods?
43. ••A rod of length L (Fig.2.121) lies along the x axis with its left end at the origin. It has a non-uniform charge density $\lambda = \alpha x$, where α is a positive constant. (a) What are the units of α? (b) Calculate the electric potential at A.
44. ••For the arrangement described in the previous problem, calculate the electric potential at point B, which lies on the perpendicular bisector of the rod a distance b above the x axis.

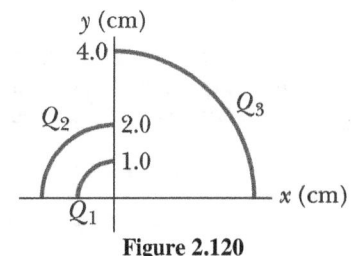

Figure 2.120

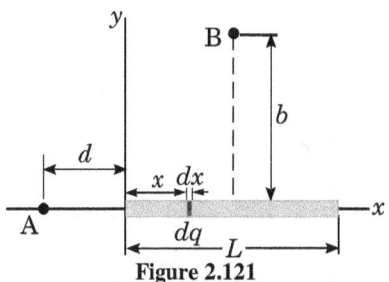

Figure 2.121

45. ••Calculate the electric potential at point P on the axis of the annulus shown in Fig.2.122, which has a uniform charge density σ.

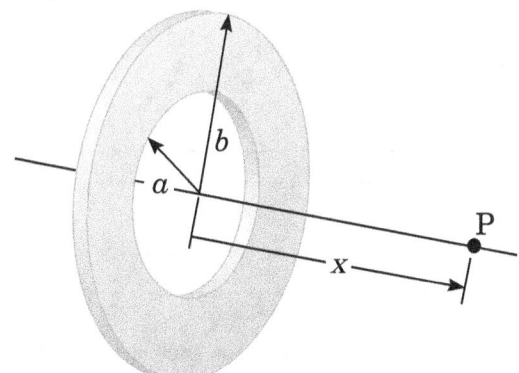

Figure 2.122

46. ••A wire having a uniform linear charge density λ is bent into the shape shown in Figure 2.123. Find the electric potential at point O.

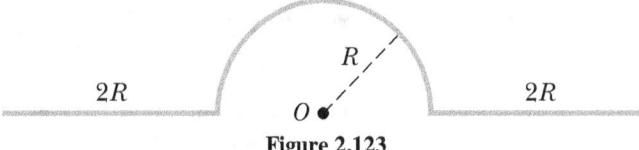

Figure 2.123

Electric Potential Due to a Charged Conductor

47. ••How many electrons should be removed from an initially uncharged spherical conductor of radius 0.300 m to produce a potential of 7.50 kV at the surface?
48. ••A spherical conductor has a radius of 14.0 cm and charge of 26.0 μC. Calculate the electric field and the electric potential at (a) $r = 10.0$ cm, (b) $r = 20.0$ cm, and (c) $r = 14.0$ cm from the center.

Applications of Electrostatics

49. ••An electron is released from rest on the axis of a uniform positively charged ring, 0.100 m from the ring's centre. If the linear charge density of the ring is +0.100 μC/m and the radius of the ring is 0.200 m, how fast will the electron be moving when it reaches the centre of the ring?
50. ••Four balls, each with mass m, are connected by four non-conducting strings to form a square with side a, as shown

in Figure 2.124. The assembly is placed on a horizontal non-conducting frictionless surface. Balls 1 and 2 each have charge q, and balls 3 and 4 are uncharged. Find the maximum speed of balls 1 and 2 after the string connecting them is cut.

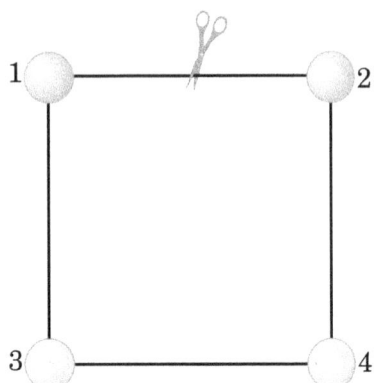

Figure 2.124

51. ●●A disc of radius R (Fig.2.125) has a non-uniform surface charge density $\sigma = Cr$, where C is a constant and r is measured from the centre of the disk. Find (by direct integration) the potential at P.

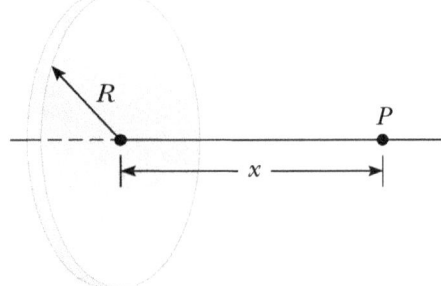

Figure 2.125

52. ●●Suppose, an alpha particle is shot at a gold foil from a large distance with kinetic energy of 5.0 MeV. If the alpha particle is aimed directly at the gold nucleus, and the only force acting on it is the electric force of repulsion exerted on it by the gold nucleus, how close will it approach the gold nucleus before turning back?

2.15.3 Multiple Choice Questions
Level 1
Potential and Potential Difference

1. When charge of 3 coulomb is placed in a Uniform electric field, it experiences a force of 3000 newton, within this field, potential difference between two points separated by a distance of 1 cm is-
   (A) 10 V (B) 90 V (C) 1000 V (D) 3000 V

2. A uniform electric field having a magnitude E_0 and direction along positive x-axis exists. If the electric potential (V) is zero at $x = 0$ then its value at $x = +x$ will be-
   (A) $V_x = xE_0$ (B) $V_x = -x \cdot E_0$
   (C) $V_x = x^2 E_0$ (D) $V_x = x^2 E_0$

3. The dimensions of potential difference are -
   (A) $[ML^2 T^{-2} Q^{-1}]$ (B) $[MLT^{-2} Q^{-1}]$
   (C) $[MT^{-2} Q^{-2}]$ (D) $[ML^2 T^{-1} Q^{-1}]$

4. Three equal charges are placed at the three corners of an isosceles triangle as shown in the Figure 2.126. The statement which is true for electric potential V and the field intensity E at the centre of the triangle is-
   (A) $V = 0, E = 0$ (B) $V = 0, E \neq 0$
   (C) $V \neq 0, E = 0$ (D) $V \neq 0, E \neq 0$

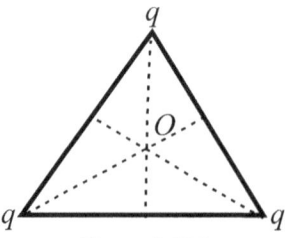

Figure 2.126

5. 1 esu of potential is equal to-
   (A) 1/300 volt (B) 8×10^{10} volt
   (C) 300 volt (D) 3 volt

6. The earth's surface is considered to be at -
   (A) Zero potential (B) Negative Potential
   (C) Infinite Potential (D) Positive Potential

7. Electric potential is a -
   (A) vector quantity (B) scalar quantity
   (C) fictious quantity (D) none of above

8. The electric potential V at any point (x, y, z) in space is given by $V = 4x^2$ volt. The electric field E in V/m at the point $(1, 0, 2)$ is
   (A) +8 in x direction (B) 8 in $-x$ direction
   (C) 16 in +x direction (D) 16 in $-x$ direction

9. ABC is equilateral triangle of side 1 m. Charges are placed at its corners as shown in Fig.2.127. O is the mid point of side BC. The potential at point O, is-
   (A) 2.7×10^3 V (B) 1.52×10^5 V
   (C) 1.3×10^3 V (D) -1.52×10^5 V

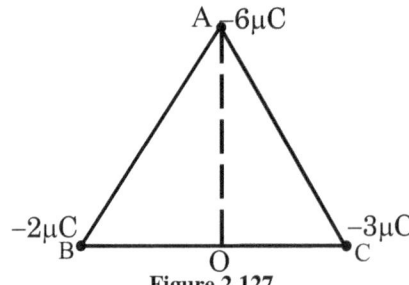

Figure 2.127

10. In a region where $E = 0$, the potential (V) varies with distance r as -
    (A) $V \propto 1/r$ (B) $V \propto r$
    (C) $V \propto 1/r^2$ (D) V is independent on r

11. Charges of $+\left(\frac{10}{3}\right) \times 10^{-9}$ are placed at each of the four corners of a square of side 8 cm. The potential at the intersection of the diagonals is
    (A) $150\sqrt{2}$ V (B) $1500\sqrt{2}$ V
    (C) $900\sqrt{2}$ V (D) 900 V

12. An equipotential surface is that surface-
    (A) On which each and every point has the same potential
    (B) Which has negative potential
    (C) Which has positive potential
    (D) Which has zero potential

13. The surface of a conductor -
    (A) is a non-equipotential surface
    (B) has all the points at the same potential
    (C) has different points at different potential
    (D) has at least two points at the same potential

14. The electron potential (V) as a function of distance (x) [in meters] is given by $V = (5x^2 + 10x - 9)$ volt. The value of electric field at x = 1 m would be-
    (A) 20 volt/m (B) 6 volt/m
    (C) 11 volt/m (D) -23 volt/m

15. Some equipotential lines are as shown is Fig.2.128 E_1, E_2 and E_3 are the electric fields at points 1, 2 and 3 then-
    (A) $E_1 = E_2 = E_3$ (B) $E_1 > E_2 > E_3$
    (C) $E_1 > E_2, E_2 < E_3$ (D) $E_1 < E_2 < E_3$

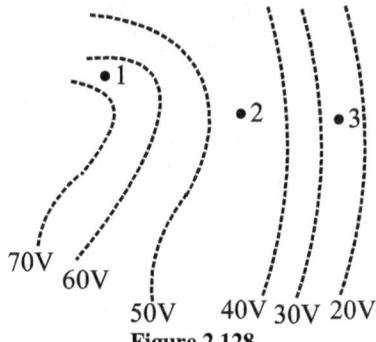

Figure 2.128

16. Three charges $2q, -q, -q$ are located at the vertices of an equilateral triangle. At the circumcenter of the triangle-
    (A) the field is zero but potential is not zero.
    (B) the field is non-zero but the potential is zero.
    (C) both, field and potential are zero.
    (D) both, field and potential are non- zero

Electric Potential Energy and Work Done

17. A positive point charge of Q' units is moved round another point positive charge of Q units in circular path. If the radius of the circle is r, then the work done on the charge Q' in making one complete revolution is-
    (A) $\dfrac{Q}{4\pi \epsilon_0 r}$ (B) $\dfrac{QQ'}{4\pi \epsilon_0 r}$ (C) $\dfrac{Q'}{4\pi \epsilon_0 r}$ (D) 0

18. A proton is projected with velocity 7.45×10^5 m/s towards an other proton which is at rest. The minimum distance of approach is-
    (A) 10^{-12} m   (B) 10^{-14} m
    (C) 10^{-10} m   (D) 10^{-8} m

19. Three charges are placed as shown in Fig.2.129. If the electric potential energy of system is zero, then $Q : q$ -
    (A) $Q/q = -2/1$ (B) $Q/q = 2/1$
    (C) $Q/q = -1/2$ (D) $Q/q = 1/4$

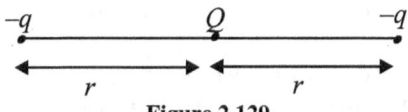

Figure 2.129

20. If a unit charge is taken from one point to another over an equipotential surface then-
    (A) work is done on the charge
    (B) work is done by the charge
    (C) work on the charge is constant
    (D) no work is done

21. In an electric field the work done in moving a unit positive charge between two points is the measures of -
    (A) Resistance
    (B) Potential difference
    (C) Intensity of electric field
    (D) Capacitance

22. State which one of the following is correct?
    (A) joule = coulomb × volt
    (B) joule = coulomb / volt
    (C) joule = volt / ampere
    (D) joule = volt × ampere

23. One electron volt (eV) of energy is equal to -
    (A) 1.6×10^{-12} ergs   (B) 4.8×10^{-10} ergs.
    (C) 9×10^{11} ergs.   (D) 3×10^9 ergs.

24. The kinetic energy in electron volt gained by an $\alpha-$ particle when it moves from rest at point where its potential is 70 volt to a point where potential is 50 volts, is -
    (A) 20 eV   (B) 20 MeV
    (C) 40 eV   (D) 40 MeV

25. An α - particle moves towards a nucleus at rest, if kinetic energy of α-particle is 10 MeV and atomic number of nucleus is 50. The distance of closest approach will be -
    (A) 1.44×10^{-14} m   (B) 2.88×10^{-14} m
    (C) 1.44×10^{-10} m   (D) 2.88×10^{-10} m

26. A point charge q moves form point P to point S along the path PQRS as shown in Fig.2.130 in a uniform electric field $\vec{E}$, pointing co-parallel to the positive direction of the x-axis. The coordinates of the points P, Q, R and S are $(a, b, 0)$, $(2a, 0, 0)$, $(a, -b, 0)$ and $(0, 0, 0)$ respectively. The work done by the field in the above process is given by the expression
    (A) qEq (B) $-qEa$
    (C) $q E a\sqrt{2}$ (D) $qE\sqrt{[(2a)^2 + b^2]}$

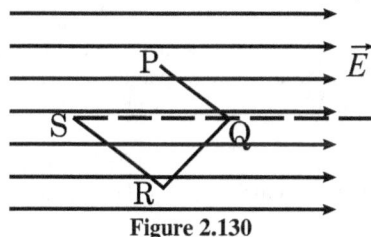

Figure 2.130

27. Two identical thin rings, each of radius R metres, are coaxially placed at a distance (R) metres apart. Q_1 and Q_2 respectively are the charges on two rings. The charge distribution on both the rings is uniform. The work done in moving a charge q from the centre of one ring to that of other is-
    (A) zero
    (B) $\dfrac{q(Q_1-Q_2)(\sqrt{2}-1)}{(4\sqrt{2}\epsilon_0 \pi R)}$
    (C) $\dfrac{q\sqrt{2}(Q_1+Q_2)}{(4\epsilon_0 \pi R)}$
    (D) $\dfrac{q(Q_1+Q_2)(\sqrt{2}+1)}{(4\sqrt{2}\epsilon_0 \pi R)}$

Applications of Gauss's Law

28. A solid conducting sphere having a charge Q is surrounded by an uncharged concentric conducting hollow spherical shell. Let the potential difference between the surface of the solid sphere and that of the outer surface of the hollow shell be V. If the shell is now given a charge of $3Q$ the new potential difference between the same two surfaces is
    (A) V   (B) 2 V   (C) 4 V   (D) -2 V

29. The dependence of electric potential V on the distance 'r' from the centre of a charged spherical shell is shown by which of the graph in Fig.??-

Electric dipole

30. The electric potential at a point due to an electric dipole will be-
    (A) $k\dfrac{\vec{p} \cdot \vec{r}}{r^3}$ (B) $k\dfrac{\vec{p} \cdot \vec{r}}{r^2}$
    (C) $\dfrac{k(\vec{p} \times \vec{r})}{r}$ (D) $\dfrac{k(\vec{p} \times \vec{r})}{r^2}$

31. An electric dipole of dipole moment p is aligned parallel to a uniform electric field E. The energy required to rotate the dipole by 90° is-
    (A) p^2E (B) pE (C) ∞ (D) pE^2

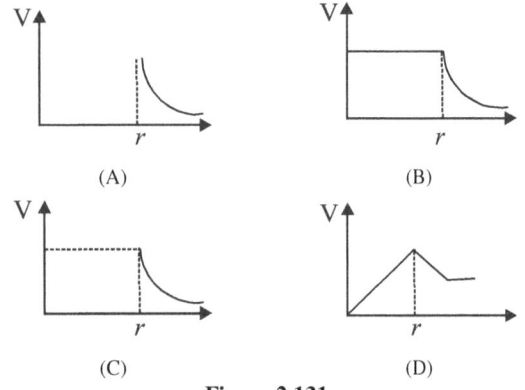

Figure 2.131

32. An electric dipole of moment p is placed in an electric field of intensity E. The dipole acquires a position such that the axis of the dipole makes an angle θ with direction of the field. Assuming that the potential energy of the dipole to be zero when $\theta = 90°$, the torque and the potential energy of the dipole will respectively be-
    (A) $pE\sin\theta, -pE\cos\theta$
    (B) $pE\sin\theta, -2pE\cos\theta$
    (C) $pE\sin\theta, 2pE\cos\theta$
    (D) $pE\cos\theta, -pE\sin\theta$

Miscellaneous

33. A proton is first placed at A and then at B between the two plates of a parallel plate capacitor charged to a potential difference of V volt as shown in Fig.2.132. The force on proton at A is-
    (A) more than at B
    (B) less than at B
    (C) equal to that at B
    (D) nothing can be said

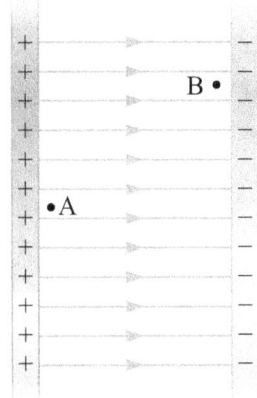

Figure 2.132

34. In an electric field, a $6.75\mu C$ charge experiences 2.5 N force, when placed at distance of 5 m from the origin. Then potential gradient at this point will be- (in MKS)
    (A) 5.71×10^5
    (B) 3.71×10^5
    (C) 18.81×10^5
    (D) 1.881×10^5

35. The electric flux coming out of the equipotential surface is-
    (A) perpendicular to the surface
    (B) parallel to the surface
    (C) in all directions
    (D) zero

36. A charge Q is distributed over two concentric hollow spheres of radii r and $R(>r)$ such that the surface densities are equal. Find the potential at the common centre-
    (A) $\frac{Q}{4\pi\epsilon_0} \times \frac{(r+R)}{(R+r)^2}$
    (B) $\frac{Q(R^2+r)^2}{4\pi\epsilon_0(r+R)}$
    (C) $\frac{Q(r+R)}{4\pi\epsilon_0(R^2+r^2)}$
    (D) none of these

37. The earth had a net charge equivalent to 1 electron per square metre of surface area of radius 6.4×10^6 m. Its potential would be-
    (A) $+0.12$ V (B) -0.12 V (C) $+1.2$ V (D) -1.2 V

38. The electric potential at the surface of an atomic nucleus ($Z = 50$) of radius 9×10^{-15} m is -
    (A) 80 V
    (B) 8×10^6 V
    (C) 8×10^4 V
    (D) 8×10^2 V

39. A hollow metal sphere of radius 5 cm is charged such that the potential on its surface is 10 V. The potential at the centre of the sphere is -
    (A) 0 V
    (B) 10 V
    (C) same as at point 5 cm away from the surface
    (D) same as at point 25 cm away from the surface

2.15.3.1 Level 2

1. Electric potential in an electric field is given as $V = K/r$, (K being constant), if position vector $\vec{r} = 2\hat{i} + 3\hat{j} + 6\hat{k}$ then electric field will be
    (A) $\frac{(2\hat{i}+3\hat{j}+6\hat{k})K}{243}$
    (B) $\frac{(2\hat{i}+3\hat{j}+6\hat{k})K}{343}$
    (C) $\frac{(3\hat{i}+2\hat{j}+6\hat{k})K}{243}$
    (D) $\frac{(6\hat{i}+2\hat{j}+3\hat{k})K}{343}$

2. At any point $(x, 0, 0)$ the electric potential V is $\left(\frac{1000}{x} + \frac{1500}{x^2} + \frac{500}{x^3}\right)$ volt, then electric field at $x = 1$ m
    (A) $5500(\hat{j}+\hat{k})V/m$
    (B) $5500\hat{i} V/m$
    (C) $\frac{5500}{\sqrt{2}}(\hat{j}+\hat{k})V/m$
    (D) $\frac{5500}{\sqrt{2}}(\hat{i}+\hat{k})V/m$

3. Potential difference between centre and the surface of a sphere of radius R with uniform charge density σ with in it will be
    (A) $\sigma R^2/6\epsilon_0$
    (B) $\sigma R^2/4\epsilon_0$
    (C) zero
    (D) $\sigma R^2/2\epsilon_0$

4. Two conducting spheres of radii r_1 and r_2 are equally charged. The ratio of their potential is-
    (A) r_1^2/r_2^2 (B) r_2^2/r_1^2 (C) r_1/r_2 (D) r_2/r_1

5. Two similar rings P and Q (radius = 0.1 m) are placed co-axially at a distance 0.5 m apart. The charge on P and Q is 2 μC and 4 μC respectively. Work done in moving a 5 μC charge from center of P to the center of Q is -
    (A) 1.28 J (B) 0.72 J (C) 0.144 J (D) 1.44 J

6. Three point charge $-q$, $+q$ and $-q$ are placed along a straight line at equl distances (say r meter). Electric potential energy of this system of charges will be if $+q$ charge is in the middle-
    (A) $\frac{-3q^2}{4\pi\epsilon_0 r}$
    (B) $\frac{-8q^2}{3\pi\epsilon_0 r}$
    (C) $\frac{-3q^2}{8\pi\epsilon_0 r}$
    (D) $\frac{-q^2}{8\pi\epsilon_0 r}$

7. Four equal charges of charge q are placed at corner of a square of side a. Potential energy of the whole system is-
    (A) $\frac{4kq^2}{a}$
    (B) $\frac{4kq^2}{a}\left(1 + \frac{1}{2\sqrt{2}}\right)$
    (C) $\frac{1}{2\sqrt{2}}\frac{kq^2}{a}$
    (D) $\frac{kq^2}{a}\left(4 + \frac{1}{2\sqrt{2}}\right)$

8. The potential of a charged drop is v. This is divided into n smaller drops, then each drop will have the potential as;
    (A) $n^{-1}v$ (B) $n^{2/3}v$ (C) $n^{3/2}v$ (D) $n^{-2/3}v$

9. 8 small droplets of water of same size and same charge form a large spherical drop. The potential of the large drop, in comparison to potential of a small drop will be-
    (A) 2 times
    (B) 4 times
    (C) 8 times
    (D) same

10. In Millikan's oil drop experiment an oil drop carrying a charge Q is held stationary by a potential difference 2400 V between the plates. To keep a drop of half the radius

stationary the potential difference had to be made 600 V. What is the charge on the second drop-
(A) $Q/4$ (B) $Q/2$ (C) Q (D) $3Q/2$

11. There is an electric field E in X-direction. If the work done on moving a charge 0.2 C through a distance of 2 m along a line making an angle 60° with the X-axis, is 4.0 J, what is the value of E
    (A) $\sqrt{3}$ N/C (B) 4 N/C
    (C) 5 N/C (D) None of these

12. A ball of mass 1 g and charge 10^{-3}C moves from a point A. where potential is 800 volt to the point B where potential is zero. Velocity of the ball at the point B is 20 cm/s. The velocity of the ball at the point A will be-
    (A) 22.8 cm/s (B) 228 cm/s
    (C) 16.8 cm/s (D) 168 m/s

13. An electric dipole is placed along the x-axis at the origin O. A point P is at a distance of 20 cm from this origin such that OP makes an angle $\pi/3$ with the x-axis. If the electric field at P makes an angle θ with the x-axis. the value of θ would be
    (A) $\pi/3$ (B) $\frac{\pi}{3} + \tan^{-1}\left(\frac{\sqrt{3}}{2}\right)$
    (C) $2\pi/3$ (D) $\tan^{-1}\left(\frac{\sqrt{3}}{2}\right)$

14. An electric dipole of moment $\vec{p}$ is placed normal to the lines of force of electric intensity $\vec{E}$, then the work done in deflecting it through an angle of 180° is
    (A) pE (B) $+2pE$ (C) $-2pE$ (D) Zero

15. An electric dipole of moment $\vec{p}$ placed in a uniform electric field $\vec{E}$ has minimum potential energy when the angle between $\vec{p}$ and $\vec{E}$ is-
    (A) Zero (B) $\pi/2$ (C) π (D) $3\pi/2$

16. An electric dipole has the magnitude of its charge as q and its dipole moment is p. It is placed in a uniform electric field $\vec{E}$. If its dipole moment is along the direction of the field, the force on it and its potential energy are respectively-
    (A) $2qE$, minimum (B) qE, pE
    (C) 0, minimum (D) qE, maximum

2.15.3.2 Level 3

1. A conducting sphere of radius R is charged to a potential of V volt. Then the electric field at a distance $r(> R)$ from the centre of the sphere would be -
    (A) RV/r^2 (B) rV/R^2 (C) V/r (D) R^2V/r^3

2. The variation of electric potential with distance from a fixed point is shown in Fig.2.133. The value of electric field at $x = 2$ m -
    (A) 0 (B) 6/2 (C) 6/1 (D) 6/3

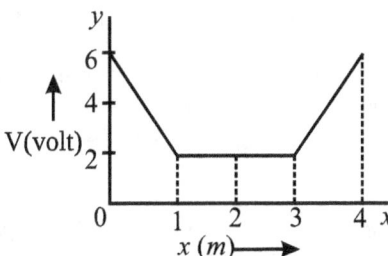

Figure 2.133

3. A positive point charge q is carried from a point B to a point A in the electric field of a point charge $+Q$ at origin O. If the permittivity of free space is ϵ_0, the work done in the process is given by (where $a = OA$ and $b = OB$)

(A) $\frac{qQ}{4\pi \epsilon_0}\left(\frac{1}{a} + \frac{1}{b}\right)$ (B) $\frac{qQ}{4\pi \epsilon_0}\left(\frac{1}{a} - \frac{1}{b}\right)$
(C) $\frac{qQ}{4\pi \epsilon_0}\left(\frac{1}{a^2} - \frac{1}{b^2}\right)$ (D) $\frac{qQ}{4\pi \epsilon_0}\left(\frac{1}{a^2} + \frac{1}{b^2}\right)$

4. Two conducting spheres each of radius R carry charge q. They are placed at a distance r from each other, where $r > 2R$. The neutral point lies at a distance $r/2$ from either sphere. If the electric field at the neutral point due to either sphere be E, then the total electric potential at that point will be -
    (A) $rE/2$ (B) rE (C) $RE/2$ (D) RE

5. A ring of radius R carries a charge $+q$. A test charge $-q_0$ is released on its axis at a distance $\sqrt{3}R$ from its center. How much kinetic energy will be acquired by the test charge when it reaches the centre of the ring-
    (A) $\frac{1}{4\pi\epsilon_0}\frac{qq_0}{R}$ (B) $\frac{1}{4\pi\epsilon_0}\frac{qq_0}{2R}$
    (C) $\frac{1}{4\pi\epsilon_0}\frac{qq_0}{\sqrt{3}R}$ (D) $\frac{1}{4\pi\epsilon_0}\frac{qq_0}{3R}$

6. Two spheres of radii r_1 and r_2 are at the same potentials. If their surface densities of charges be σ_1 and σ_2 respectively, then σ_1/σ_2-
    (A) r_1/r_2 (B) r_2/r_1 (C) $(r_1/r_2)^2$ (D) $(r_2/r_1)^2$

7. A proton and an electron are released from an infinite distance apart and are attracted to each other. Which of the following statements about their kinetic energy is true?
    (A) Kinetic energy of electron is more than that of proton
    (B) Kinetic energy of electron is less than that of proton
    (C) Kinetic energy of electron = kinetic energy of proton
    (D) None of the above is true as it depending on the distance between the particles

8. Two conducting spheres of radii r_1 and r_2 are charged such that they have the same electric field on their surfaces. The ratio of the electric potential at their centres is -
    (A) $\sqrt{r_1/r_2}$ (B) r_1/r_2
    (C) r_1^2/r_2^2 (D) None of the above

9. A charge $+Q$ at A (Figure 2.134) produces electric field E and electric potential V at D. If we now put charges $-2Q$ and $+Q$ at B and C respectively, then the electric field and potential at D will be -
    (A) E and 0 (B) 0 and V
    (C) $\sqrt{2}E$ and $\frac{V}{\sqrt{2}}$ (D) $\frac{E}{\sqrt{2}}$ and $\frac{V}{\sqrt{2}}$

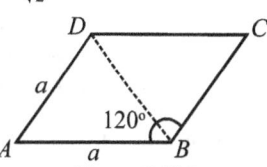

Figure 2.134

10. If a positive charge is shifted from a low-potential region to a high-potential region, the electric potential energy-
    (A) increases
    (B) decreases
    (C) remains the same
    (D) may increase or decrease

11. A particle of mass 0.002 kg and a charge 1 μC is held at rest on a frictionless horizontal surface at a distance of 1 m from a fixed charge of 1 mC. If the particle is released, it will be repelled. The speed of the particle when it is at a distance of 10 m from the fixed charge is-
    (A) 60 ms^{-1} (B) 75 ms^{-1} (C) 90 ms^{-1} (D) 100 ms^{-1}

12. Electric potential is given by: $V = 6x - 8xy^2 - 8y + 6yz - 4z^2$ Electric field at the origin is-
    (A) $-6\hat{i} + 8\hat{j}$ (B) $6\hat{i} - 8\hat{j}$ (C) $\hat{i} + \hat{j}$ (D) Zero

13. A hollow conducting sphere of radius R has charge $(+Q)$ on its surface. The electric potential within the sphere at a

distance $r = R/3$ from the centre is -
(A) 0 (B) $\frac{1}{4\pi\epsilon_0}\frac{Q}{r}$ (C) $\frac{1}{4\pi\epsilon_0}\frac{Q}{R}$ (D) $\frac{1}{4\pi\epsilon_0}\frac{Q}{r^2}$

14. A particle has a mass 400 times than that of the electron and charge is double than that of a electron. It is accelerated by 5 V of potential difference. Initially the particle was at rest. Then its final kinetic energy will be-
(A) 5 eV (B) 10 eV (C) 100 eV (D) 2000 eV

15. Two equal positive charges are kept at points A and B. The electric potential at the points between A and B (excluding these points) is studied while moving from A to B. The potential-
(A) continuously increases
(B) continuously decreases
(C) increases then decreases
(D) decreases then increases

Passage Type Questions (Q.16 and 18)
In the diagram shown in Fig.2.135, the broken lines represent the paths followed by particles W, X, Y and Z respectively through the constant field E. The numbers below the field represents meters.

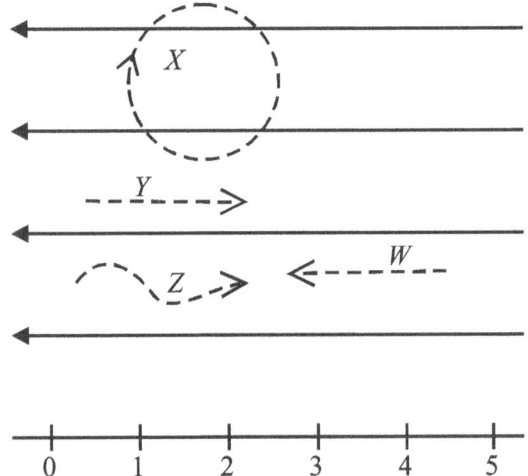

Figure 2.135

16. If the particles begin and end at rest, and all are positively charged, the same amount of work was done on which particles?
(A) W and Z (B) W, Y and Z
(C) Y and Z (D) W, X, Y and Z

17. If the particles started from rest and all are positively charged which particles must have been acted upon by a force other than that produced by the electric field.
(A) W and Y (B) X and Z
(C) X, Y and Z (D) W, X, Y and Z

18. If the particles are positively charged, which particles increased their electrical potential energy -
(A) X and Z
(B) Y and Z
(C) W, X, Y and Z
(D) Since the electric field is constant none of the particles increased their electrical potential energy.

19. Five equal and similar charges are placed at the corners of a regular hexagon as shown in the Fig.2.136. What is the electric field and potential at the centre of the hexagon -
(A) $\frac{5}{4\pi\epsilon_0}\frac{q}{l}, \frac{5}{4\pi\epsilon_0}\frac{q}{l^2}$ (B) $\frac{1}{4\pi\epsilon_0}\frac{q}{l}, \frac{5}{4\pi\epsilon_0}\frac{q}{l^2}$
(C) $\frac{1}{4\pi\epsilon_0}\frac{q}{l^2}, \frac{5}{4\pi\epsilon_0}\frac{q}{l}$ (D) $\frac{1}{4\pi\epsilon_0}\frac{q}{l}, \frac{1}{4\pi\epsilon_0}\frac{q}{l^2}$

Statements Type Question:-
Each of the questions given below consist of Statement-I and Statement- II. Use the following Key to choose the

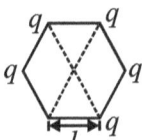

Figure 2.136

appropriate answer.
(A) If both Statement-I and Statement-II are true, and Statement-II is the correct explanation of Statement-I.
(B) If both Statement-I and Statement-II are true but Statement-II is not the correct explanation of Statement $-I$.
(C) If Statement-I is true but Statement-II is false.
(D) If Statement-I is false but Statement-II is true.

20. **Statement I:** If a point charge q is placed in front of an infinite grounded conducting plane surface, the point charge will experience a force.
Statement II: The force is due to the induced charge on the conducting surface which is at zero potential.

21. **Statement I:** Electrons move away from a region of lower potential to a region of higher potential.
Statement II: An electron has negative charge.

22. **Statement I:** Work done in moving a charge between any two points in an electric field is independent of the path followed by the charge, between these points.
Statement II: Electrostatic forces are non conservative.

2.15.3.3 Level 4 (Previous Years JEE Main and Advanced Questions)

Section A: JEE Main

1. On moving a charge of 20 coulombs by 2 cm, 2J of work is done, then the potential difference between the points is –
[AIEEE-2002]
(A) 0.1 V (B) 8 V (C) 2 V (D) 0.5 V

2. A thin spherical conducting shell of radius R has a charge q. Another charge Q is placed at the centre of the shell. The electrostatic potential at a point P at a distance $R/2$ from the centre of the shell is- [AIEEE 2003]
(A) $\frac{2Q}{4\pi\epsilon_0 R} - \frac{2q}{4\pi\epsilon_0 R}$ (B) $\frac{2Q}{4\pi\epsilon_0 R} + \frac{q}{4\pi\epsilon_0 R}$
(C) $\frac{(q+Q)}{4\pi\epsilon_0}\frac{2}{R}$ (D) $\frac{2Q}{4\pi\epsilon_0 R}$

3. A charged particle 'q' is shot towards another charged particle 'Q', which is fixed, with a speed v. It approaches Q upto a closest distance r and then returns. If q were given a speed of $2v$, the closest distances of approach would be-
[AIEEE-2004]
(A) r (B) $2r$ (C) $r/2$ (D) $r/4$

4. Two thin wire rings each having a radius R are placed at a distance d apart with their axes coinciding. The charges on the two rings are $+q$ and $-q$. The potential difference between the centres of the two rings is- [AIEEE-2005]
(A) $QR/4\pi\epsilon_0 d^2$ (B) $\frac{Q}{2\pi\epsilon_0}\left[\frac{1}{R} - \frac{1}{\sqrt{R^2+d^2}}\right]$
(C) zero (D) $\frac{Q}{4\pi\epsilon_0}\left[\frac{1}{R} - \frac{1}{\sqrt{R^2+d^2}}\right]$

5. Two insulating plates are both uniformly charged, as shown in Fig.2.137, in such a way that the potential difference between them is $V_2 - V_1 = 20$ V (i.e. plate 2 is at a higher potential). The plates are separated by $d = 0.1$ m and can be treated as infinitely large. An electron is released from rest on the inner surface of plate 1. What is its speed when it hits plate 2? $(e = 1.6 \times 10^{-19}$C$, m_e = 9.11 \times 10^{-31}$ kg$)$ –
[AIEEE 2006]

2.15. QUESTIONS AND EXERCISES

(A) 1.87×10^6 m/s (B) 32×10^{-19} m/s
(C) 2.65×10^6 m/s (D) 7.02×10^{12} m/s

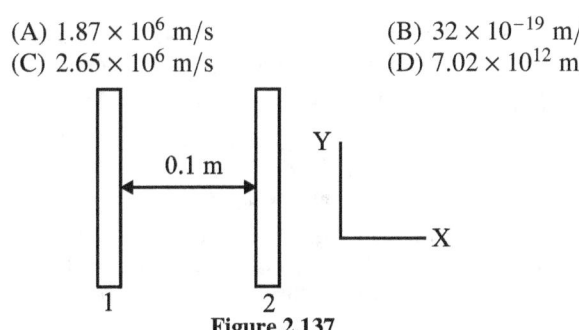

Figure 2.137

6. Two spherical conductors A and B of radii 1 mm and 2 mm are separated by a distance of 5 cm and are uniformly charged. If the sphere are connected by a conducting wire then in equilibrium condition, the ratio of the magnitude of the electric fields at the surfaces of spheres A and B is - [AIEEE 2007]
(A) 2 : 1 (B) 1 : 4 (C) 4 : 1 (D) 1 : 2

7. An electric charge $10^{-3} \mu C$ is placed at the origin $(0, 0)$ of X-Y co-ordinate system. Two points A and B are situated at $(\sqrt{2}, \sqrt{2})$ and $(2, 0)$ respectively. The potential difference between the points A and B will be- [2006]
(A) 9 V (B) 0 V (C) 2 V (D) 4.5 V

8. Charges are placed on the vertices of a square as shown. Let $\vec{E}$ be the electric field and V the potential at the centre. If the charges on A and B are interchanged with those on D and C respectively, then [AIEEE 2007]

(A) $\vec{E}$ remains unchanged, V changes
(B) Both $\vec{E}$ and V change
(C) $\vec{E}$ and V remain unchanged
(D) $\vec{E}$ changes, V remains unchanged

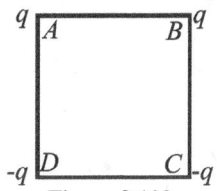

Figure 2.138

9. The potential at a point x (measured in μm) due to some changes situated on the x-axis is given by $V(x) = 20/(x^2 - 4)$ volts. The electric field E at $x = 4$ μm is given by [2007]
(A) 5/3 volt /μm and in the $-$ve x direction
(B) 5/3 volt /μm and in the $+$ve x direction
(C) 10/9 volt μm and in the $-$ve x direction
(D) 10/9 volt μm and in the $+$ve x direction

10. Two points P and Q are maintained at the potentials of 10 V and -4 V, respectively. The work done in moving 100 electrons from P to Q is - [AIEEE-2009]
(A) -9.60×10^{-17} J (B) 9.60×10^{-17} J
(C) -2.24×10^{-16} J (D) 2.24×10^{-16} J

11. The electrostatic potential inside a charged spherical ball is given by $\phi = ar^2 + b$ where r is the distance from the centre; a, b are constants. Then the charge density inside ball is [AIEEE 2011]
(A) $-6a\epsilon_0 r$ (B) $-24\pi a\epsilon_0 r$
(C) $-6a\epsilon_0$ (D) $-24\pi a\epsilon_0$

12. This question has statement 1 and statement 2. Of the four choices given after the statements, choose the one that best describes the two statements.
An insulating solid sphere of radius R has a uniformly positive charge density ρ. As a result of this uniform charge distribution there is a finite value of electric potential at the centre of the sphere, at the surface of the sphere and also at a point outside the sphere. The electric potential at infinity is zero. [2012]
Statement 1: When a charge q is taken from the centre to the surface of the sphere, its potential energy changes by $q\rho/(3\epsilon_0)$
Statement 2: The electric field at a distance $r(r < R)$ from the centre of the sphere is $\rho r / 3\epsilon_0$
(A) Statement 1 is true, Statement 2 is true, Statement 2 is not the correct explanation for statement 1.
(B) Statement 1 is true, Statement 2 is false
(C) Statement 1 is false, Statement 2 is true
(D) Statement 1 is true, Statement 2 is the correct explanation for statement 1

13. A charge Q is uniformly distributed over a long rod AB of length L as shown in the Fig.2.139 The electric potential at the point O lying at distance L from the end A is: [2013]
(A) $\frac{Q}{8\pi\epsilon_0 L}$ (B) $\frac{3Q}{4\pi\epsilon_0 L}$
(C) $\frac{Q}{4\pi\epsilon_0 L \ln 2}$ (D) $\frac{Q \ln 2}{4\pi\epsilon_0 L}$

Figure 2.139

14. Assume that an electric field $\vec{E} = 30x^2 \hat{i}$ exists in space. Then the potential difference $V_A - V_0$, where V_0 is the potential at the origin and V_A the potential at $x = 2$ m is: [2014]
(A) -80 J (B) 80 J (C) 120 J (D) -120 J

15. Three concentric metal shells A, B and C of respective radii a, b and c $(a < b < c)$ have surface charge densities $+\sigma, -\sigma$ and $+\sigma$ respectively. The potential of shell B is [2018]
(A) $\frac{\sigma}{\epsilon_0}\left[\frac{a^2-b^2}{a}+c\right]$ (B) $\frac{\sigma}{\epsilon_0}\left[\frac{a^2-b^2}{b}+c\right]$
(C) $\frac{\sigma}{\epsilon_0}\left[\frac{b^2-c^2}{b}+a\right]$ (D) $\frac{\sigma}{\epsilon_0}\left[\frac{b^2-c^2}{c}+a\right]$

16. A charge Q is distributed over two concentric conducting thin spherical shells of radii r and $R(R > r)$ (Fig.2.140). If the surface charge densities on the two shells are equal, the electric potential at the common centre is: [2019]

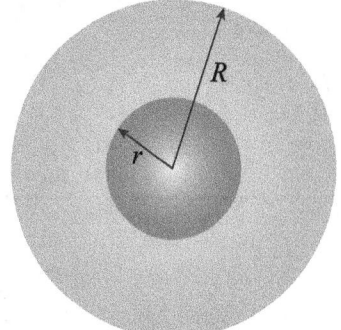

Figure 2.140

(A) $\frac{1}{4\pi\epsilon_0}\frac{2(R+r)}{(R^2+r^2)}Q$ (B) $\frac{1}{4\pi\epsilon_0}\frac{(R+r)}{2(R^2+r^2)}Q$
(C) $\frac{1}{4\pi\epsilon_0}\frac{(R+2r)Q}{2(R^2+r^2)}$ (D) $\frac{1}{4\pi\epsilon_0}\frac{(R+r)}{2(R^2+r^2)}Q$

17. A solid conducting sphere, having a charge Q, is surrounded by an uncharged conducting hollow spherical shell. Let the potential difference between the surface of the solid sphere and that of the outer surface of the hollow shell be V. If

the shell is now given a charge of $-4Q$, the new potential difference between the same two surfaces is: [2019]
(A) $-2V$ (B) $2V$ (C) $4V$ (D) V

18. A positive point charge is released from rest at a distance r_o from a positive line charge with uniform density. The speed (v) of the point charge, as a function of instantaneous distance r from line charge, is proportional to: [2019]

(A) $v \propto e^{+r/r_o}$ (B) $v \propto \sqrt{\ln\left(\frac{r}{r_o}\right)}$
(C) $v \propto \ln\left(\frac{r}{r_o}\right)$ (D) $v \propto \left(\frac{r}{r_o}\right)$

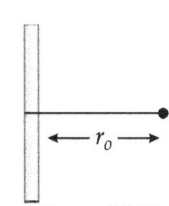

Figure 2.141

19. The electric field in a region is given by $\vec{E} = (Ax + B)\hat{i}$, where E is in NC^{-1} and x is in metres. The values of constants are $A = 20$ SI unit and $B = 10$ SI unit. If the potential at $x = 1$ is V_1 and that at $x = -5$ is V_2, then $V_1 - V_2$ is: [2019]
(A) $320\,V$ (B) $-48\,V$ (C) $180\,V$ (D) $-520\,V$

20. A system of three charges are placed as shown in the Fig.2.142. If $D \gg d$, the potential energy of the system is best given by: [2019]

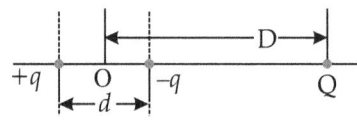

Figure 2.142

(A) $\frac{1}{4\pi\epsilon_o}\left[-\frac{q^2}{d} - \frac{qQd}{2D^2}\right]$ (B) $\frac{1}{4\pi\epsilon_o}\left[-\frac{q^2}{d} + \frac{2qQd}{D^2}\right]$
(C) $\frac{1}{4\pi\epsilon_o}\left[+\frac{q^2}{d} + \frac{qQd}{D^2}\right]$ (D) $\frac{1}{4\pi\epsilon_o}\left[-\frac{q^2}{d} - \frac{qQd}{D^2}\right]$

21. In free space, a particle A of charge $1\,\mu C$ is held fixed at a point P. Another particle B of the same charge and mass $4\,\mu g$ is kept at a distance of $1\,mm$ from P. If B is released, then its velocity at a distance of $9\,mm$ from P is (Take $\frac{1}{4\pi\epsilon_o} = 9 \times 10^9 Nm^2C^{-2}$): [2019]

(A) $1.0\,m/s$ (B) $3.0 \times 10^4\,m/s$
(C) $2.0 \times 10^3\,m/s$ (D) $1.5 \times 10^2\,m/s$

22. A point dipole $\vec{p} = -p_0\hat{x}$ is kept at the origin. The potential and electric field due to this dipole on the y-axis at a distance d are, respectively: (Take $V = 0$ at infinity) [2019]
(A) $\frac{|\vec{p}|}{4\pi\epsilon_0 d^2}, \frac{\vec{p}}{4\pi\epsilon_0 d^3}$ (B) $0, \frac{-\vec{p}}{4\pi\epsilon_0 d^3}$
(C) $0, \frac{\vec{p}}{4\pi\epsilon_0 d^3}$ (D) $\frac{|\vec{p}|}{4\pi\epsilon_0 d^2}, \frac{-\vec{p}}{4\pi\epsilon_0 d^3}$

23. A charge Q is distributed over three concentric spherical shells of radii a, b, c ($a < b < c$) such that their surface charge densities are equal to one another. The total potential at a point at distance r from their common centre, where $r < a$, would be: [2019]
(A) $\frac{Q}{12\pi\epsilon_0}\frac{ab+bc+ca}{abc}$ (B) $\frac{Q(a^2+b^2+c^2)}{4\pi\epsilon_0(a^3+b^3+c^3)}$
(C) $\frac{Q}{4\pi\epsilon_0(a+b+c)}$ (D) $\frac{Q(a+b+c)}{4\pi\epsilon_0(a^2+b^2+c^2)}$

24. Two electric dipoles, A, B with respective dipole moments $\vec{d}_A = -4qa\hat{i}$ and $\vec{d}_B = -2qa\hat{i}$ are placed on the x-axis with a separation R, as shown in the Fig.2.143 The distance from A at which both of them produce the same potential is: [2019]
(A) $\frac{R}{\sqrt{2}+1}$ (B) $\frac{\sqrt{2}R}{\sqrt{2}+1}$ (C) $\frac{R}{\sqrt{2}-1}$ (D) $\frac{\sqrt{2}R}{\sqrt{2}-1}$

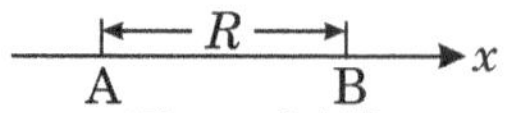

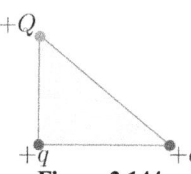

Figure 2.143

25. Four equal point charges Q each are placed in the xy plane at $(0,2), (4,2), (4,-2)$ and $(0,-2)$. The work required to put a fifth charge Q at the origin of the coordinate system will be: [2019]
(A) $\frac{Q^2}{4\pi\epsilon_0}\left(1+\frac{1}{\sqrt{3}}\right)$ (B) $\frac{Q^2}{4\pi\epsilon_0}\left(1+\frac{1}{\sqrt{5}}\right)$
(C) $\frac{Q^2}{2\sqrt{2}\pi\epsilon_0}$ (D) $\frac{Q^2}{4\pi\epsilon_0}$

26. Three charges Q, $+q$ and $+q$ are placed at the vertices of a right-angle isosceles triangle as shown in Fig.2.144. The net electrostatic energy of the configuration is zero, if the value of Q is: [2019]
(A) $\frac{-q}{1+\sqrt{2}}$ (B) $+q$ (C) $-2q$ (D) $\frac{-\sqrt{2}q}{\sqrt{2}+1}$

Figure 2.144

27. The given graph in Fig.2.145, shows variation (with distance r from centre) of: [2019]
(A) Electric field of a uniformly charged sphere
(B) Potential of a uniformly charged spherical shell
(C) Potential of a uniformly charged sphere
(D) Electric field of a uniformly charged spherical shell

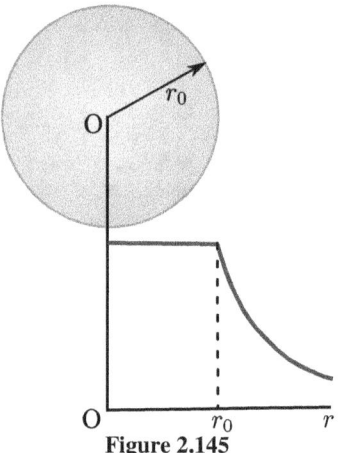

Figure 2.145

28. An electric field of $1000\,V/m$ is applied to an electric dipole at angle of $45°$. The value of electric dipole moment is 10^{-29} C.m. What is the potential energy of the electric dipole? [2019]
(A) -20×10^{-18} J (B) -7×10^{-27} J
(C) -10×10^{-29} J (D) -9×10^{-20} J

29. Determine the electric dipole moment of the system of three charges, placed on the vertices of an equilateral triangle, as shown in the Fig.2.146 [2019]

(A) $\sqrt{3}ql\left(\frac{\hat{j}-\hat{i}}{\sqrt{2}}\right)$ (B) $(ql)\left(\frac{\hat{i}+\hat{j}}{\sqrt{2}}\right)$
(C) $2ql\hat{j}$ (D) $-\sqrt{3}ql\hat{j}$

30. There is a uniform spherically symmetric surface charge density at a distance R_0 from the origin. The charge dis-

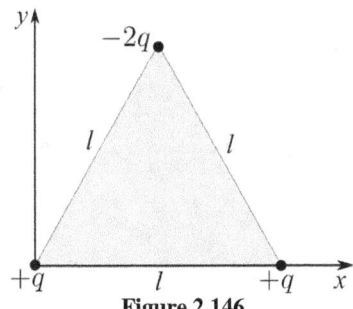

Figure 2.146

tribution is initially at rest and starts expanding because of mutual repulsion. The choice of Fig.2.147 that represents best the speed $v(R(t))$ of the distribution as a function of its instantaneous radius $R(t)$ is: [2019]

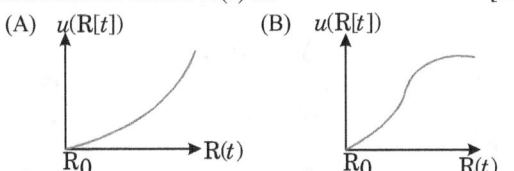

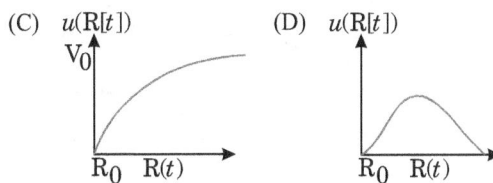

Figure 2.147

31. An electric field of 1000 V/m is applied to an electric dipole at angle of 45°. The value of electric dipole moment is 10^{-29} C·m. What is the potential energy of the electric dipole? [11 Jan 2019, (II)]
    (A) -20×10^{-18} J (B) -10×10^{-29} J
    (C) -7×10^{-27} J (D) -9×10^{-20} J

32. Two isolated conducting spheres S_1 and S_2 of radius $2R/3R$ and $R/3$ have 12 μC and −3 μC charges, respectively, and are at a large distance from each other. They are now connected by a conducting wire. A long time after this is done the charges on S_1 and S_2 are respectively: [2020]
    (A) 6 μC and 3 μC   (B) 4.5 μC on both
    (C) +4.5 μC and −4.5 μC   (D) 3 μC and 6 μC

33. Concentric metallic hollow spheres of radii R and $4R$ hold charges Q_1 and Q_2 respectively. Given that surface charge densities of the concentric spheres are equal, the potential difference $V(9R) - V(4R)$ is: [2020]
    (A) $\frac{Q_2}{4\pi\epsilon_0 R}$ (B) $\frac{3Q_2}{4\pi\epsilon_0 R}$
    (C) $\frac{3Q_1}{16\pi\epsilon_0 R}$ (D) $\frac{3Q_1}{4\pi\epsilon_0 R}$

34. Two point charges $4q$ and $-q$ are fixed on the x-axis at $x = -d/2$ and $x = d/2$, respectively. If the third point charge 'q' is taken from the origin to $x = d$ along the semicircle as shown in the Fig.2.148, the energy of the charge will: [2020]
    (A) increase by $\frac{2q^2}{3\pi\epsilon_0 d}$ (B) increase by $\frac{2q^2}{4\pi\epsilon_0 d}$
    (C) decrease by $\frac{q^2}{4\pi\epsilon_0 d}$ (D) decrease by $\frac{4q^2}{3\pi\epsilon_0 d}$

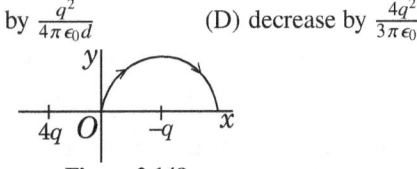

Figure 2.148

35. A particle of charge q and mass m is subjected to an electric field $E = E_0(1 - ax^2)$ in the x-direction, where a and E_0 are constants. Initially the particle was at rest at $x = 0$. Other than the initial position the kinetic energy of the particle becomes zero when the distance of the particle from the origin is: [2020]
    (A) a (B) $\sqrt{1/a}$ (C) $\sqrt{3/a}$ (D) $\sqrt{2/a}$

36. A solid sphere of radius R carries a charge $Q + q$ distributed uniformly over its volume. A very small point like piece of it of mass m gets detached from the bottom of the sphere and falls down vertically under gravity. This piece carries charge q. If it acquires a speed v when it has fallen through a vertical height y (Fig.2.149), then (assume the remaining portion to be spherical) [2020]
    (A) $v^2 = 2y\left[\frac{QqR}{4\pi\epsilon_0(R+y)^3 m} + g\right]$
    (B) $v^2 = y\left[\frac{qQ}{4\pi\epsilon_0 R^2 ym} + g\right]$
    (C) $v^2 = y\left[\frac{qQ}{4\pi\epsilon_0 R(R+y)m} + g\right]$
    (D) $v^2 = 2y\left[\frac{qQ}{4\pi\epsilon_0 R(R+y)m} + g\right]$

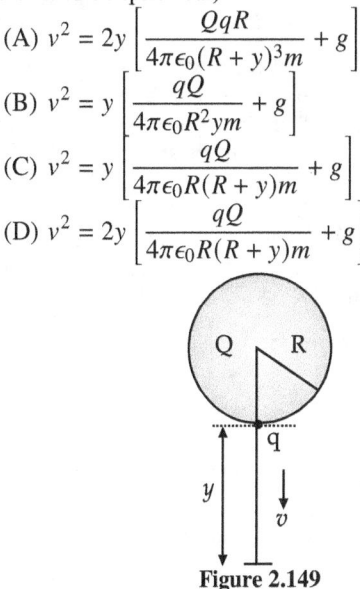

Figure 2.149

37. Ten charges are placed on the circumference of a circle of radius R with constant angular separation between successive charges. Alternate charges 1, 3, 5, 7, 9 have charge $(+q)$ each, while 2, 4, 6, 8, 10 have charge $(-q)$ each. The potential V and the electric field E at the centre of the circle are respectively: [Take $V = 0$ at infinity] [2020]
    (A) $V = 0$; $E = 0$ (B) $V = \frac{10q}{4\pi\epsilon_0 R}$; $E = 0$
    (C) $V = \frac{10q}{4\pi\epsilon_0 R}$; $E = \frac{10q}{4\pi\epsilon_0 R^2}$ (D) $V = 0, E = \frac{10q}{4\pi\epsilon_0 R^2}$

38. Two identical electrical point dipoles have dipole moments $\vec{p}_1 = p\hat{i}$ and $\vec{p}_2 = -p\hat{i}$ and held on the x-axis at distance 'a' from each other. When released, they move along the x-axis with the direction of their dipole moments remaining unchanged. If the mass of each dipole is 'm', their speed when they are infinitely far apart is: [2020]
    (A) $\frac{p}{a}\sqrt{\frac{3}{2\pi\epsilon_0 ma}}$ (B) $\frac{pP}{a}\sqrt{\frac{1}{2\pi\epsilon_0 ma}}$
    (C) $\frac{p}{a}\sqrt{\frac{1}{\pi\epsilon_0 ma}}$ (D) $\frac{p}{a}\sqrt{\frac{2}{\pi\epsilon_0 ma}}$

39. Consider two charged metallic spheres S_1 and S_2 of radii R_1 and R_2, respectively. The electric fields E_1 (on S_1) and E_2 (on S_2) on their surfaces are such that $\frac{E_1}{E_2} = \frac{R_1}{R_2}$. Then the ratio $\frac{V_1(\text{on } S_1)}{V_2(\text{on } S_2)}$ of the electrostatic potentials on each sphere is: [2020]
    (A) $\left(\frac{R_1}{R_2}\right)^3$ (B) $\left(\frac{R_2}{R_1}\right)$ (C) $\frac{R_1}{R_2}$ (D) $\left(\frac{R_1}{R_2}\right)^2$

40. The two thin coaxial rings, each of radius 'a' and having charges $+Q$ and $-Q$ respectively are separated by a distance of 's'. The potential difference between the centres of the

two ring is: [2021]

(A) $\frac{Q}{4\pi\epsilon_0}\left[\frac{1}{a}-\frac{1}{\sqrt{s^2+a^2}}\right]$ (B) $\frac{Q}{2\pi\epsilon_0}\left[\frac{1}{a}-\frac{1}{\sqrt{s^2+a^2}}\right]$

(C) $\frac{Q}{4\pi\epsilon_0}\left[\frac{1}{a}+\frac{1}{\sqrt{s^2+a^2}}\right]$ (D) $\frac{Q}{2\pi\epsilon_0}\left[\frac{1}{a}+\frac{1}{\sqrt{s^2+a^2}}\right]$

41. If the electric potential at any point (x, y, z) m space is given by $V = 3x^2$ volt. The electric field at the point $(1, 0, 3)$ m will be: [29 June 22 Shift II]
    (A) 3 Vm^{-1}, directed along positive x-axis.
    (B) 3 Vm^{-1}, directed along negative x-axis.
    (C) 6 Vm^{-1}, directed along positive x-axis.
    (D) 6 Vm^{-1}, directed along negative x-axis.

42. The electric potential at the centre of two concentric half rings of radii R_1 and R_2, having same linear charge density λ is- [24 Jan 2023 Shift II]
    (A) $\frac{2\lambda}{\epsilon_0}$ (B) $\frac{\lambda}{2\epsilon_0}$ (C) $\frac{\lambda}{4\epsilon_0}$ (D) $\frac{\lambda}{\epsilon_0}$

43. Two isolated metallic solid spheres of radii R and $2R$ are charged such that both have same charge density σ. The spheres are then connected by a thin conducting wire. If the new charge density of the bigger sphere is σ'. The ratio σ'/σ is: [30 Jan 2023 Shift I]
    (A) 9/4 (B) 4/3 (C) 5/3 (D) 5/6

44. Which of the graphs shown in Fig.2.150 correctly represents the variation of electric potential (V) of a charged spherical conductor of radius (R) with radial distance (r) from the centre? [31 Jan 2023 Shift I]

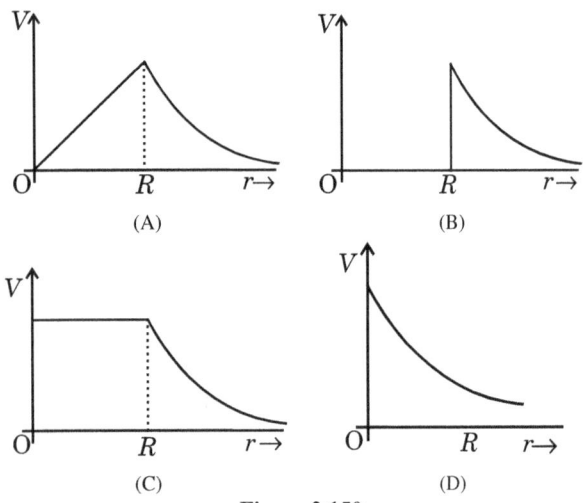

Figure 2.150

45. Which of the graphs shown in Fig.2.152 represents the electric potential (V) radially away from the centre 'O' of a uniformly charged thin spherical shell as shown in Fig.2.151? [06 April 23 Shift I]

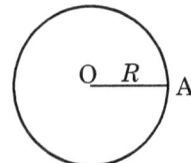

Figure 2.151

46. Electric potential at a point 'P' due to a point charge of 5×10^{-9} C is 50 V. The distance of 'P' from the point charge is: [08 April 23 Shift II]
    (Assume, $1/4\pi\epsilon_0 = 9 \times 10^{+9}$ Nm^{2}C^{-2})
    (A) 3 cm (B) 9 cm (C) 90 cm (D) 0.9 cm

47. Three concentric spherical metallic shells X, Y and Z of

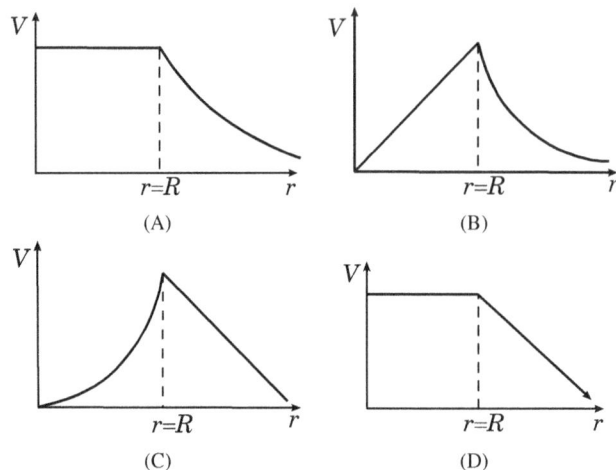

Figure 2.152

radius a, b and c respectively [$a < b < c$] (Fig.2.153) have surface charge densities σ, $-\sigma$ and σ, respectively. The shells X and Z are at same potential. If the radii of X & Y are 2 cm and 3 cm, respectively. The radius of shell Z is _____ cm. [10 April 23 Shift I]

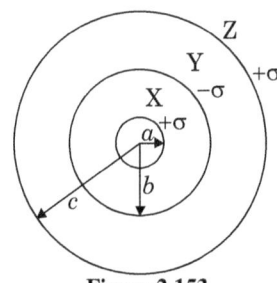

Figure 2.153

48. 64 identical drops each charged up to potential of 10 mV are combined to form a bigger drop. The potential of the bigger drop will be _____ mV. [12 April 23 Shift I]

Section - B (JEE Advanced)

1. Two identical thin rings, each of radius R, are coaxially placed a distance R apart. If Q_1 and Q_2 are respectively the charges uniformly spread on the two rings, the work done in moving a charge q from the centre of one ring to that of the other is [1992]
    (A) zero (B) $\frac{q(Q_1-Q_2)(\sqrt{2}-1)}{\sqrt{2}4\pi\epsilon_0 R}$
    (C) $\frac{q\sqrt{2}(Q_1+Q_2)}{4\pi\epsilon_0 R}$ (D) $\frac{q(Q_1/Q_2)(\sqrt{2}+1)}{\sqrt{2}4\pi\epsilon_0 R}$

2. The electric potential V at any point x, y, z in space is given by $V = 4x^2$ V/m^2. The electric field at the point $(1\,\text{m}, 0, 2\,\text{m})$ is [1992]
    (A) 8 V/m (B) 4 V/m (C) 16 V/m (D) $\frac{4}{3}$ V/m

3. A non-conducting ring of radius 0.5 m carries a total charge of 1.11×10^{-10} C distributed non-uniformly on its circumference producing an electric field E everywhere in space. The value of the line integral $\int_{l=\infty}^{l=0} -\vec{E} \cdot d\vec{l}$ ($l = 0$ being centre of the ring) in volts is [1994]
    (A) +2 (B) −1 (C) −2 (D) 0

4. A charge $+q$ is fixed at each of the point $x = x_0$, $x = 3x_0$, $x = 5x_0, \ldots$ ad inf. on the x-axis, and charges $-q$ is fixed at each of the point $x = 2x_0$, $x = 4x_0$, $x = 6x_0, \ldots$ ad inf. Here x_0 is a positive constant. Take the electric potential at a point due to a charge Q at a distance r from it to be $Q/(4\pi\epsilon_0 r)$. Then the potential at the origin due to the above system of charges is [1998]

(A) −1 (B) $\frac{q}{8\pi\epsilon_0 x_0/n2}$
(C) ∞ (D) $\frac{q\ln 2}{4\pi\epsilon_0 x_0}$

5. An ellipsoidal cavity is carved within a perfect conductor. A positive charge q is placed at the centre of the cavity. The points A and B are on the cavity surface as shown in the Fig.2.154. Then [1999]
   (A) Electric field near A in the cavity = electric field near B in the cavity
   (B) Charge density at A = charge density at B
   (C) Potential at $A \neq$ potential at B
   (D) Total electric field flux through the surface of the cavity is q/ϵ_0.

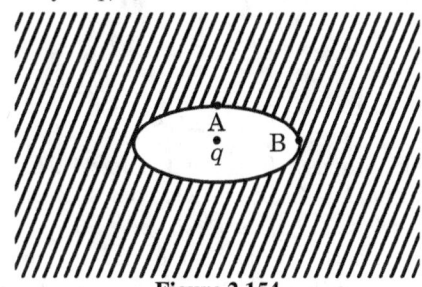

Figure 2.154

6. Three charges $Q, +q$ and $+q$ are placed at the vertices of a right-angled isosceles triangle as shown in fig. The net electrostatics energy of the configuration is zero if Q is equal to [2000]
   (A) $\frac{-q}{1+\sqrt{2}}$ (B) $\frac{-2q}{2+\sqrt{2}}$ (C) $-2q$ (D) $+q$

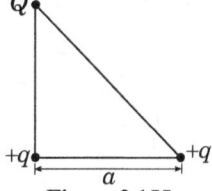

Figure 2.155

7. A uniform electric field pointing in positive x-direction exists in a region. Let A be the origin, B be the point on the x-axis at $x = +1$ cm and C be the point on the y axis at $y = +1$ cm. Then the potentials at the points A, B, and C satisfy. [2001]
   (A) $V_A < V_B$ (B) $V_A > V_B$
   (C) $V_A < V_C$ (D) $V_A > V_C$

8. Two equal point charges are fixed at $x = -a$ and $x = +a$ on the x-axis. Another point charge Q is placed at the origin. The change in the electrical potential energy of Q, when it is displaced by a small distance x along the x-axis, is approximately proportional to - [2002]
   (A) x (B) x^2 (C) x^3 (D) $1/x$

9. Positive and negative point charges of equal magnitude are kept at $(0, 0, a/2)$ and $(0, 0, -a/2)$, respectively. The work done by the electric field when another positive point charge is moved from $(-a, 0, 0)$ to $(0, a, 0)$ is [2007]
   (A) positive
   (B) negative
   (C) zero
   (D) depends on the path connecting the initial and final positions

10. Consider a system of three charges $q/3, q/3$ and $-2q/3$ placed at point A, B and C, respectively, as shown in the Fig.2.156. Take O to be the centre of the circle of radius R and angle CAB = 60°. [2008]
    (A) The electric field at point O is $\frac{q}{8\pi\epsilon_0 R^2}$ direction along the negative x-axis.
    (B) The potential energy of the system is zero.
    (C) The magnitude of the force between the charges at C and B is $\frac{q^2}{54\pi\epsilon_0 R^2}$.
    (D) The potential at point O is $\frac{q}{12\pi\epsilon_0 R}$.

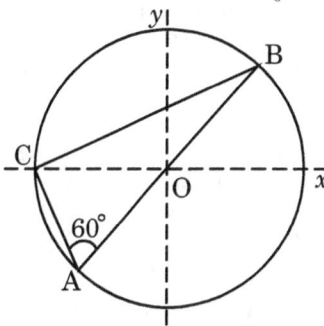

Figure 2.156

11. **STATEMENT 1:** For practical purposes the earth is used as a reference at zero potential in electrical circuits.
    STATEMENT 2: The electrical potential of a sphere of radius R with charge Q uniformly distributed on the surface is given by $\frac{Q}{4\pi\epsilon_0 R}$.
    (A) STATEMENT-1 is True, STATEMENT-2 is True; STATEMENT-2 is a correct explanation for STATEMENT-1
    (B) STATEMENT-1 is True, STATEMENT-2 is True; STATEMENT-2 is NOT a correct explanation for STATEMENT-1
    (C) STATEMENT-1 is True, STATEMENT-2 is False
    (D) STATEMENT-1 is False, STATEMENT-2 is True

12. Three concentric metallic spherical shells of radii $R, 2R, 3R$ are given charges Q_1, Q_2, Q_3 respectively. It is found that the surface charge densities on the outer surfaces of the shells are equal. Then, the ratio of the charges given to the shells, $Q_1 : Q_2 : Q_3$ is- [2009]
    (A) 1 : 2 : 3 (B) 1 : 3 : 5
    (C) 1 : 4 : 9 (D) 1 : 8 : 18

13. A spherical metal shell A of radius R_A and a solid metal sphere B of radius $R_B (< R_A)$ are kept far apart and each is given charge $+Q$. Now they are connected by a thin metal wire. Then [2011]
    (A) $E_A^{\text{inside}} = 0$ (B) $Q_A > Q_B$
    (C) $\sigma_A/\sigma_B = R_B/R_A$ (D) $E_A^{\text{surface}} < E_B^{\text{surface}}$

Paragraph 14-15

A dense collection of equal number of electrons and positive ions is called neutral plasma. Certain solids containing fixed positive ions surrounded by free electrons can be treated as neutral plasma. Let 'N' be the number density of free electrons, each of mass 'm'. When the electrons are subjected to an electric field, they are displaced relatively away from the heavy positive ions. If the electric field becomes zero, the electrons begin to oscillate about the positive ions with a natural angular frequency 'ω_p' which is called the plasma frequency. To sustain the oscillations, a time varying electric field needs to be applied that has an angular frequency ω, where a part of the energy is absorbed and a part of it is reflected. As ω approaches ω_p all the free electrons are set to resonance together and all the energy is reflected. This is the explanation of high reflectivity of metals.

14. Taking the electronic charge as 'e' and the permittivity as 'ϵ_0'. Use dimensional analysis to determine the correct

expression for ω_p. [2011]
(A) $\sqrt{Ne/me_e}$
(B) $\sqrt{me_0/Ne}$
(C) $\sqrt{Ne^2/m\epsilon_0}$
(D) $\sqrt{m\epsilon_0/Ne^2}$

15. Estimate the wavelength at which plasma reflection will occur for a metal having the density of electrons $N \approx 4 \times 10^{27}$ m^{-3}. Taking $e_0 = 10^{-11}$ and $m \approx 10^{-30}$, where these quantities are in proper SI units. [2011]
(A) 800 nm (B) 600 nm (C) 300 nm (D) 200 nm

16. Two large vertical and parallel metal plates having a separation of 1 cm are connected to a DC voltage source of potential difference X. A proton is released at rest midway between the two plates. It is found to move at 45° to the vertical just after release. Then X is nearly. [2012]
(A) 1×10^{-5} V
(B) 1×10^{-7} V
(C) 1×10^{-9} V
(D) 1×10^{-10} V

17. Consider a thin spherical shell of radius R with centre at the origin, carrying uniform positive surface charge density. The variation of the magnitude of the electric field $|\vec{E}(r)|$ and the electric potential $V(r)$ with the distance r from the centre, is best represented by which of the graph shown in Fig.2.157? [2012]

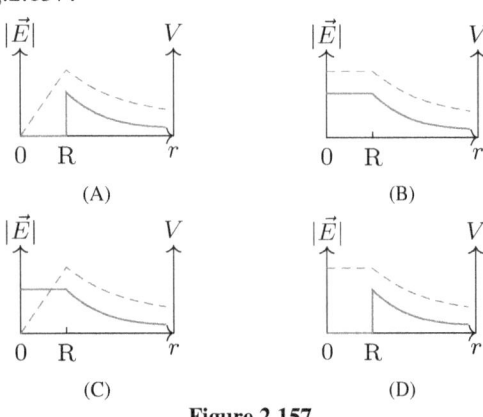

Figure 2.157

18. Two non-conducting spheres of radii R_1 and R_2 and carrying uniform volume charge densities $+\rho$ and $-\rho$, respectively, are placed such that they partially overlap, as shown in the Fig.2.158. At all points in the overlapping region, [2013]
(A) The electrostatic field is zero
(B) The electrostatic potential is constant
(C) The electrostatic field is constant in magnitude
(D) The electrostatic field has same direction

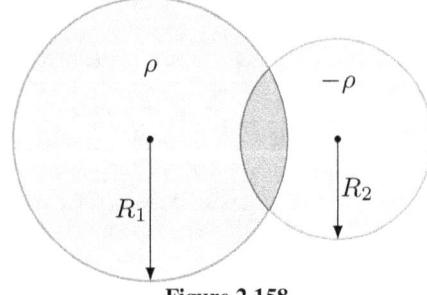

Figure 2.158

19. A point charge $+Q$ is placed just outside an imaginary hemispherical surface of radius R as shown in the Fig.2.159. Which of the following statements is/are correct? [2017]

(A) The circumference of the flat surface is an equipotential
(B) The electric flux passing through the curved surface of the hemisphere is $-\frac{Q}{2\epsilon_0}\left(1 - \frac{1}{\sqrt{2}}\right)$
(C) Total flux through the curved and the flat surfaces is Q/ϵ_0
(D) The component of the electric field normal to the flat surface is constant over the surface.

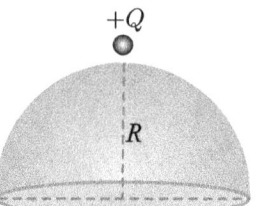

Figure 2.159

20. A particle, of mass 10^{-3} kg and charge 1.0 C, is initially at rest. At time $t = 0$, the particle comes under the influence of an electric field $\vec{E}(t) = E_0 \sin \omega t \,\hat{i}$, where $E_0 = 1.0$ NC^{-1} and $\omega = 10^3$ rads^{-1}. Consider the effect of only the electrical force on the particle. Then the maximum speed, in ms^{-1}, attained by the particle at subsequent times is ____. [2018]

21. Two large circular discs separated by a distance of 0.01 m are connected to a battery via a switch as shown in the Fig.2.160. Charged oil drops of density 900 kg m^{-3} are released through a tiny hole at the centre of the top disc. Once some oil drops achieve terminal velocity, the switch is closed to apply a voltage of 200 V across the discs. As a result, an oil drop of radius 8×10^{-7} m stops moving vertically and floats between the discs. The number of electrons present in this oil drop is ____ (neglect the buoyancy force, take acceleration due to gravity = 10 ms^{-2} and charge on an electron $(e) = 1.6 \times 10^{-19}$ C) [2020]

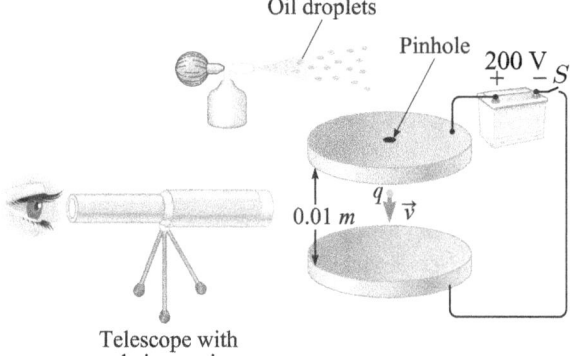

Figure 2.160

22. A point charge q of mass m is suspended vertically by a string of length l. A point dipole of dipole moment $\vec{p}$ is now brought towards q from infinity so that the charge moves away. The final equilibrium position of the system including the direction of the dipole, the angles and distances is shown in Fig.2.161. If the work done in bringing the dipole to this position is $N \times (mgh)$, where g is the acceleration due to gravity, then find the value of N.
(Note that for three coplanar forces keeping a point mass in equilibrium, $F/\sin\theta$ is the same for all forces, where F is any one of the forces and θ is the angle between the other two forces) [2020]

Paragraph for Q.23-24
Two point charges $-Q$ and $+Q/\sqrt{3}$ are placed in the xy-plane

2.16. ANSWER KEYS AND SOLUTIONS

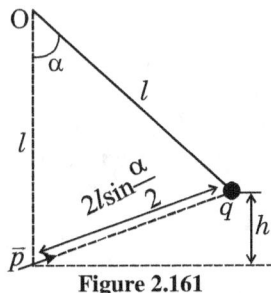

Figure 2.161

at the origin $(0, 0)$ and a point $(2, 0)$, respectively, as shown in the Fig.2.162. This results in an equipotential circle of radius R and potential $V = 0$ in the xy-plane with its centre at $(b, 0)$. All lengths are measured in meters. [2021]

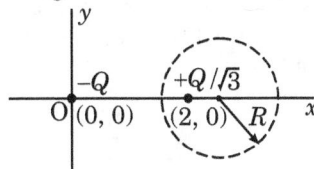

Figure 2.162

23. The value of R is ____ meter.
24. The value of b is ____ meter.
25. Six charges are placed around a regular hexagon of side length a as shown in the Fig.2.163. Five of them have charge q, and the remaining one has charge x. The perpendicular from each charge to the nearest hexagon side passes through the centre O of the hexagon and is bisected by the side. Which of the following statement(s) is (are) correct in SI units? [2022]
    (A) When $x = q$, the magnitude of the electric field at O is zero.
    (B) When $x = -q$, the magnitude of the electric field at O is $\dfrac{q}{6\pi\epsilon_0 a^2}$.
    (C) When $x = 2q$, the potential at O is $\dfrac{7q}{4\sqrt{3}\pi\epsilon_0 a}$.
    (D) When $x = -3q$, the potential at O is $\dfrac{3q}{4\sqrt{3}\pi\epsilon_0 a}$.

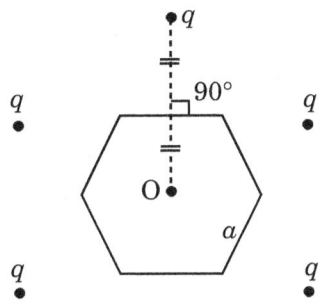

Figure 2.163

26. A disc of radius R with uniform positive charge density σ is placed on the xy plane with its centre at the origin. The Coulomb potential along the z-axis is
$$V(z) = \dfrac{\sigma}{2\epsilon_0}\left(\sqrt{R^2 + z^2} - z\right)$$
A particle of positive charge q is placed initially at rest at a point on the z axis with $z = z_0$ and $z_0 > 0$. In addition to the Coulomb force, the particle experiences a vertical force $\vec{F} = -c\hat{k}$ with $c > 0$. Let $\beta = \dfrac{2c\epsilon_0}{q\sigma}$. Which of the following statement(s) is(are) correct? [2022]
    (A) For $\beta = 1/4$ and $z_0 = 25R/7$, the particle reaches the origin.
    (B) For $\beta = 1/4$ and $z_0 = 3R/7$, the particle reaches the origin.
    (C) For $\beta = 1/4$ and $z_0 = R/\sqrt{3}$, the particle returns back to $z = z_0$.
    (D) For $\beta > 1$ and $z_0 > 0$, the particle always reaches the origin.

27. An electric dipole is formed by two charges $+q$ and $-q$ located in xy-plane at $(0, 2)$ mm and $(0, -2)$ mm, respectively, as shown in the Fig.2.164. The electric potential at point P $(100, 100)$ mm due to the dipole is V_0. The charges $+q$ and $-q$ are then moved to the points $(-1, 2)$ mm and $(1, -2)$ mm, respectively. What is the value of electric potential at P due to the new dipole? [2023]
    (A) $V_0/4$ (B) $V_0/2$ (C) $V_0/\sqrt{2}$ (D) $3V_0/4$

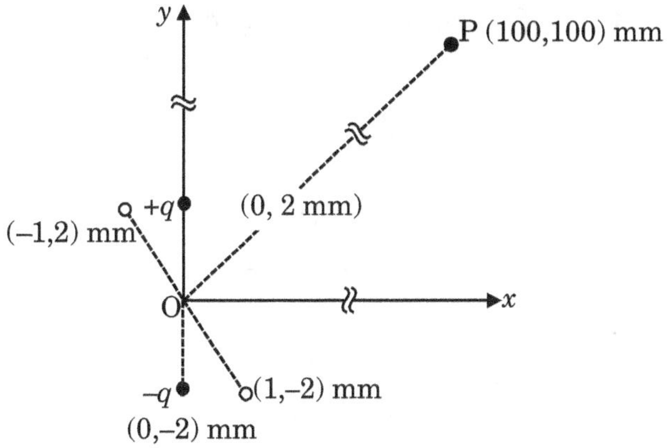

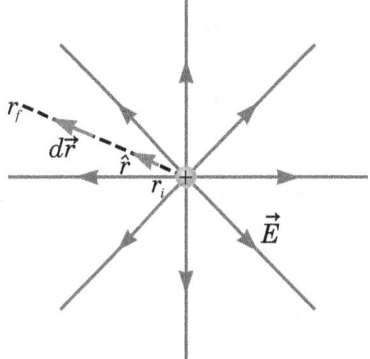

Figure 2.164

2.16 Answer Keys and Solutions
2.16.1 Check Point 1
1. **APPROACH** Sketch the electric field and include a path from r_i to r_f (Fig. 2.165).
To get potential difference, apply Eq. 2.16:

Figure 2.165

$$\Delta V = -\int_{r_A}^{r_B} E\, dr \qquad \cdots (1)$$

from limit r_i to r_f.
SOLUTION From Eq.(1), we can write-
$$\Delta V = -\int_{r_i}^{r_f} E\, dr = -\int_{r_i}^{r_f}\left(k\dfrac{Q}{r^2}\right)dr$$
$$= -kQ\int_{r_i}^{r_f}\left(\dfrac{1}{r^2}\right)dr = -kQ\left[-\dfrac{1}{r}\right]_{r_i}^{r_f} = kQ\left[\dfrac{1}{r}\right]_{r_i}^{r_f}$$

Evaluate between limits and reduce.
$$\Delta V = \frac{kQ}{r_f} - \frac{kQ}{r_i}$$
By convention, we can take the electrostatic potential zero at infinity, therefore, the expression for the electric potential at a distance r from the source will become-
$$\Delta V = V(r) - V(\infty) = V(r) - 0 = V$$
or
$$\boxed{V = \frac{kQ}{r}}$$

2. **APPROACH** The potential difference between any two points in an electric field is given by Eq.2.14:
$$\Delta V = V_B - V_A = -\int_A^B \vec{E} \cdot d\vec{s} \quad (1)$$
here, $d\vec{s} = dx\hat{i} + dy\hat{j} + dz\hat{k}$
You must select a specific path to integrate along. It doesn't matter what path you choose, since the answer is path-independent, but you simply cannot integrate unless you have a definite path in mind.
SOLUTION Let's go by the indicated path:

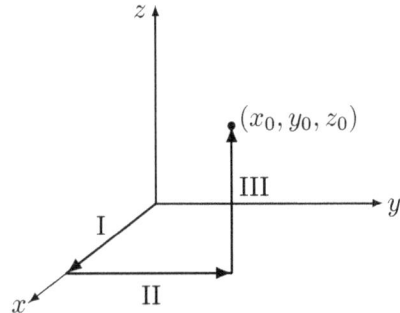

Figure 2.166

$\vec{E} \cdot d\vec{s} = \left(y^2 dx + \left(2xy + z^2\right) dy + 2yz\,dz\right) K$

Along Path I: $y = z = 0;\, dy = dz = 0$. $\vec{E} \cdot d\vec{s} = Ky^2 dx = 0$.
$\therefore \int_I \vec{E} \cdot d\vec{s} = 0$.

Along Path II: $x = x_0,\, y: 0 \to y_0,\, z = 0$. $dx = dz = 0$.
$\vec{E} \cdot d\vec{s} = K\left(2xy + z^2\right) dy = 2Kx_0 y\,dy$.
$\therefore \int_{II} \vec{E} \cdot d\vec{s} = 2Kx_0 \int_0^{y_0} y\,dy = Kx_0 y_0^2$.

Along Path III: $x = x_0,\, y = y_0,\, z: 0 \to z_0;\, dx = dy = 0$.
$\vec{E} \cdot d\vec{s} = 2Kyz\,dz = 2Ky_0 z\,dz$.
$\therefore \int_{III} \vec{E} \cdot d\vec{s} = 2y_0 K \int_0^{z_0} z\,dz = Ky_0 z_0^2$.

Therefore, potential at point (x_0, y_0, z_0)-
$$V(x_0, y_0, z_0) = -\int_I \vec{E} \cdot d\vec{s} - \int_{II} \vec{E} \cdot d\vec{s} - \int_{III} \vec{E} \cdot d\vec{s}$$
$$= -K\left(x_0 y_0^2 + y_0 z_0^2\right)$$

Generalizing the result, we get
$$V(x, y, z) = -K\left(xy^2 + yz^2\right)$$

3. $\left(Aa^2/2\right) - \left(Bb^3/3\right)$

4. From Eq.2.14, electric potential difference, between two points A and B, in terms of electric field, is given by-
$$\Delta V = V_B - V_A = -\int_A^B \vec{E}.d\vec{s} \quad (1)$$
with $d\vec{s} = dx\hat{i} + dy\hat{j} + dz\hat{k}$.
Now, substitute the given values and integrate within the given limits.

SOLUTION (C) Given that- $\vec{E} = (Ax + B)\hat{i}$, with A = 20, B = 10, therefore:
$\vec{E}.d\vec{s} = (Ax + B)\hat{i}.(dx\hat{i} + dy\hat{j} + dz\hat{k}) = (Ax + B)dx$
potential at $x = 1$ is V_1, potential at $x = -5$ is V_2, all the values are in SI units. Therefore, equation 1, gives-
$$V_2 - V_1 = -\int_{x=1}^{x=-5} (Ax + B)dx$$
Put values of A and B.
$$V_2 - V_1 = -\frac{20}{2}\left[x^2\right]_{x=1}^{x=-5} - 10[x]_{x=1}^{x=-5}$$
$$= -10(24) - 10(-6) = -240 + 60 = -180 \text{ V}$$
So, $V_1 - V_2 = 180$ V

5. **APPROACH** From Eq.2.9, the potential difference between two points and the work done by external force W_{ext}, in moving a point charge q_0 slowly, from one point to other are related as-
$$\Delta V = \frac{W_{\text{ext}}}{q_0}$$
Substitute the given values in above equation and solve for ΔV.
SOLUTION (A) Put the given values of W_{ext} and q_0 and solve for $V_B - V_A$
SOLUTION Given: $W_{\text{ext}} = 2$ J, $q_0 = 20$ C
So, $\Delta V = \dfrac{W_{\text{ext}}}{q_0} = \dfrac{2}{20} = 0.1$ V

2.16.2 Check Point 2

1. The electric field is the negative of the slope of the potential graph.
   There are three regions of different slope:
   $0 < x < 2$ cm
   $\quad \begin{cases} \Delta V/\Delta x = (20 \text{ V})/(0.020 \text{ m}) = 1000 \text{ V/m} \\ E_x = -1000 \text{ V/m} \end{cases}$
   $2 < x < 4$ cm
   $\quad \begin{cases} \Delta V/\Delta x = 0 \text{ V/m} \\ E_x = 0 \text{ V/m} \end{cases}$
   $4 < x < 8$ cm
   $\quad \begin{cases} \Delta V/\Delta x = (-20 \text{ V})/(0.040 \text{ m}) = -500 \text{ V/m} \\ E_x = 500 \text{ V/m} \end{cases}$
   The results are shown in Fig.2.167.

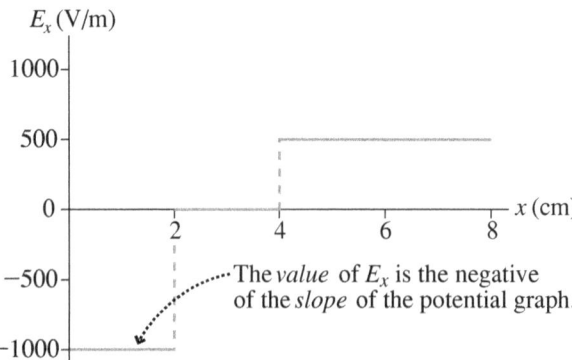

Figure 2.167

2. **APPROACH** The electric field is perpendicular to the equipotential lines, points "downhill", and depends on the slope of the potential hill.
   The potential is highest on the bottom and the right. An elevation graph of the potential would look like the lower-right quarter of a bowl or a football stadium.
   SOLUTION Some distant but unseen source charges have created an electric field and potential. We do not need to

see the source charges to relate the field to the potential. Because $E \approx -\Delta V/\Delta s$, the electric field is stronger where the equipotential lines are closer together and weaker where they are farther apart. If Figure 2.42 were a topographic map, you would interpret the closely spaced contour lines at the bottom of the figure as a steep slope.

Fig.2.168 shows how measurements of Δs from the grid are combined with values of ΔV to determine $\vec{E}$. Point 3 requires an estimate of the spacing between the 0 V and the 100 V surfaces. Notice that we're using the 0 V and 100 V equipotential surfaces to determine $\vec{E}$ at a point on the 50 V equipotential.

Note: The directions of $\vec{E}$ are found by drawing downhill

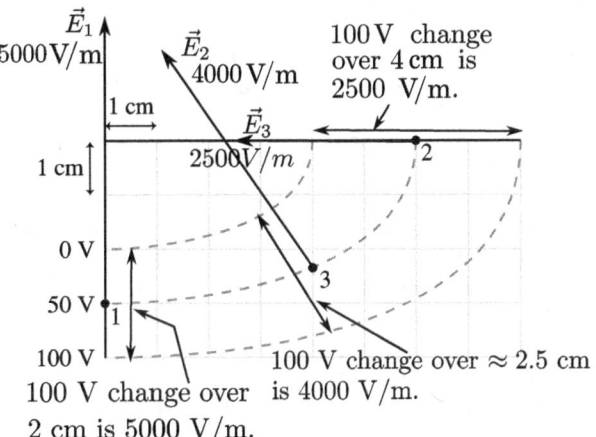

Figure 2.168

vectors perpendicular to the equipotentials. The distances between the equipotential surfaces are needed to determine the field strengths.

3. **APPROACH** Components of $\vec{E}$, are given by-
$$E_x = -\frac{\partial V}{\partial x}, \quad E_y = -\frac{\partial V}{\partial y} \text{ and } E_z = -\frac{\partial V}{\partial z}$$
SOLUTION $V = Axy - Bx^2 + Cy$
(a) $E_x = -\frac{\partial V}{\partial x} = -Ay + 2Bx$, $E_y = -\frac{\partial V}{\partial y} = -Ax - C$,
$E_z = \frac{\partial V}{\partial z} = 0$
(b) $E = 0$ requires that $E_x = E_y = E_z = 0$. $E_z = 0$ everywhere. $E_y = 0$ at $x = -C/A$. And E_x is also equal to zero for this x, any value of z and $y = 2Bx/A = (2B/A)(-C/A) = -2BC/A^2$.
Since, V doesn't depend on z, so $E_z = 0$ everywhere.

4. **APPROACH** Components of $\vec{E}$, are given by-
$$E_x = -\frac{\partial V}{\partial x}, \quad E_y = -\frac{\partial V}{\partial y} \text{ and } E_z = -\frac{\partial V}{\partial z}$$
SOLUTION The electric field at distance r due to a point charge q is given by:
$$\vec{E} = \frac{1}{4\pi\epsilon_0}\frac{q}{r^2}\hat{r}$$
(a) $E_x = -\frac{\partial V}{\partial x} = -\frac{\partial}{\partial x}\left(\frac{kQ}{\sqrt{x^2+y^2+z^2}}\right)$
$= \frac{kQx}{(x^2+y^2+z^2)^{3/2}} = \frac{kQx}{r^3}$
Similarly, $E_y = \frac{kQy}{r^3}$ and $E_z = \frac{kQz}{r^3}$.
(b) From part (a), $E = \frac{kQ}{r^2}\left(\frac{x\hat{i}}{r}+\frac{y\hat{j}}{r}+\frac{z\hat{k}}{r}\right) = \frac{kQ}{r^2}\hat{r}$

Note: V is a scalar. $\vec{E}$ is a vector and has components.

5. The electric field is given by
$$\vec{E} = -\frac{\partial V}{\partial x}\hat{i} - \frac{\partial V}{\partial y}\hat{j} - \frac{\partial V}{\partial z}\hat{k} = -8x\hat{i}$$
Therefore, $\vec{E}_{(1m, 0m, 2m)} = -8\hat{i}$.

6. **APPROACH** For a solid metal sphere or for a spherical shell, $V = kq/r$ outside the sphere and $V = kq/R$ at all points inside the sphere, where R is the radius of the sphere. When the electric field is radial, $E = -\frac{\partial V}{\partial r}$

SOLUTION (a) (i) $r < r_a$: This region is inside both spheres.
$$V = \frac{kq}{r_a} - \frac{kq}{r_b} = kq\left(\frac{1}{r_a} - \frac{1}{r_b}\right)$$
(ii) $r_a < r < r_b$: This region is outside the inner shell and inside the outer shell.
$$V = \frac{kq}{r} - \frac{kq}{r_b} = kq\left(\frac{1}{r} - \frac{1}{r_b}\right)$$
(iii) $r > r_b$: This region is outside both spheres and $V = 0$ since outside a sphere the potential is the same as for a point charge. Therefore the potential is the same as for two oppositely charged point charges at the same location. These potentials cancel.

(b) $V_a = k\left(\frac{q}{r_a} - \frac{q}{r_b}\right)$ and $V_b = 0$,
so, $V_{ab} = kq\left(\frac{1}{r_a} - \frac{1}{r_b}\right)$

(c) Between the spheres $r_a < r < r_b$ and $V = kq\left(\frac{1}{r} - \frac{1}{r_b}\right)$
So, $E = -\frac{\partial V}{\partial r} = -kq\frac{\partial}{\partial r}\left(\frac{1}{r} - \frac{1}{r_b}\right)$
$= +k\frac{q}{r^2} = \frac{V_{ab}}{\left(\frac{1}{r_a} - \frac{1}{r_b}\right)}\frac{1}{r^2}$
$\left[\text{Since, from part (b) } kq = \frac{V_{ab}}{\left(\frac{1}{a} - \frac{1}{b}\right)}\right]$

(d) Since, V is constant, therefore $E = -\partial V/\partial r = 0$, outside the spheres.

(e) If the outer charge is different, then outside the outer sphere the potential is no longer zero but is $V = k\frac{q}{r} - k\frac{Q}{r} = k\frac{(q-Q)}{r}$. All potentials inside the outer shell are just shifted by an amount $V = kQ/r_b$. Therefore relative potentials within the shells are not affected. Thus (b) and (c) do not change. However, now that the potential does vary outside the spheres, there is an electric field there:
$$E = -\frac{\partial V}{\partial r} = -\frac{\partial}{\partial r}\left(\frac{kq}{r} + \frac{-kQ}{r}\right) = \frac{kq}{r^2}\left(1 - \frac{Q}{q}\right) = \frac{k}{r^2}(q-Q)$$
Note: In part (a) the potential is greater than zero for all $r < r_b$.

7. Since, $E = -\partial V/\partial r$, therefore, $\vec{E} = 0$ everywhere implies a constant potential and not necessarily zero potential.

8. (B). When moving straight from A to B, $\vec{E}$ and $d\vec{s}$ both point toward the right. Thus, the dot product $\vec{E} \cdot d\vec{s}$ in equation $\Delta V = V_B - V_A = -\vec{E} \cdot d\vec{s}$ is positive and so ΔV is negative.

9. (A) Work done is equal to zero because the potential of A and B are the same $= \frac{1}{4\pi\epsilon_0}\frac{q}{a}$
No work is done if a particle does not change its potential

energy. i.e. initial potential energy = final potential energy.

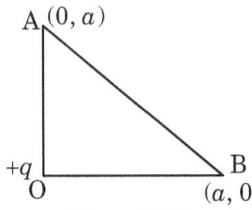

Figure 2.169

10. $\vec{E} = -\left[\dfrac{\partial V}{\partial x}\hat{i} + \dfrac{\partial V}{\partial y}\hat{j} + \dfrac{\partial V}{\partial z}\hat{k}\right] = -(-5\hat{i} + 3\hat{j} + \sqrt{15}\hat{k})$

[since, $V = -5x + 3y + \sqrt{15}z$]

$\Rightarrow \quad |\vec{E}| = \sqrt{25 + 9 + 15} = \sqrt{49} = 7$ unit

11. (D): The electric potential at a point,
$$V = -x^2y - xz^3 + 4$$
The field $\vec{E} = -\vec{\nabla}V = -\left(\dfrac{\partial V}{\partial x}\hat{i} + \dfrac{\partial V}{\partial y}\hat{j} + \dfrac{\partial V}{\partial z}\hat{k}\right)$

$\therefore \quad \vec{E} = \hat{i}(2xy + z^3) + \hat{j}x^2 + \hat{k}(3xz^2)$

12. (D): The electric field $\vec{E}$ and potential V in a region are related as $\vec{E} = -\left[\dfrac{\partial V}{\partial x}\hat{i} + \dfrac{\partial V}{\partial y}\hat{j} + \dfrac{\partial V}{\partial z}\hat{k}\right]$

Here, $V(x, y, z) = 6xy - y + 2yz$

$\therefore \quad \vec{E} = -\left[\dfrac{\partial}{\partial x}(6xy - y + 2yz)\hat{i} + \dfrac{\partial}{\partial y}(6xy - y + 2yz)\hat{j}\right.$
$\left. + \dfrac{\partial}{\partial z}(6xy - y + 2yz)\hat{k}\right]$

$= -[(6y)\hat{i} + (6x - 1 + 2z)\hat{j} + (2y)\hat{k}]$

At point $(1, 1, 0)$,
$\vec{E} = -[(6(1))\hat{i} + (6(1) - 1 + 2(0))\hat{j} + (2(1))\hat{k}]$
$= -(6\hat{i} + 5\hat{j} + 2\hat{k})$

13. (D): Here, $V(x, y, z) = 6x - 8xy - 8y + 6yz$

The x, y and z components of electric field are

$E_x = -\dfrac{\partial V}{\partial x} = -\dfrac{\partial}{\partial x}(6x - 8xy - 8y + 6yz)$
$= -(6 - 8y) = -6 + 8y,$

$E_y = -\dfrac{\partial V}{\partial y} = -\dfrac{\partial}{\partial y}(6x - 8xy - 8y + 6yz)$
$= -(-8x - 8 + 6z) = 8x + 8 - 6z,$

$E_z = -\dfrac{\partial V}{\partial z} = -\dfrac{\partial}{\partial z}(6x - 8xy - 8y + 6yz) = -6y$

So, $\vec{E} = E_x\hat{i} + E_y\hat{j} + E_z\hat{k}$
$= (-6 + 8y)\hat{i} + (8x + 8 - 6z)\hat{j} - 6y\hat{k}$

At point $(1, 1, 1)$
$\vec{E} = (-6 + 8)\hat{i} + (8 + 8 - 6)\hat{j} - 6\hat{k} = 2\hat{i} + 10\hat{j} - 6\hat{k}$

The magnitude of electric field $\vec{E}$ is

$\vec{E} = \sqrt{E_x^2 + E_y^2 + E_z^2} = \sqrt{(2)^2 + (10)^2 + (-6)^2}$
$= \sqrt{140} = 2\sqrt{35} \text{ NC}^{-1}$

Electric force experienced by the charge
$$F = qE = 2\text{C} \times 2\sqrt{35} \text{ NC}^{-1} = 4\sqrt{35} \text{ N}$$

14. **APPROACH** If we assume the space as gravity free, then the force acting on the alpha particle is electrostatic in nature, which is conservative. So, we can apply the principle of conservation of mechanical energy. According to this principle,
$$K_i + U_i = K_f + U_f$$
Here letters have their usual meanings.

SOLUTION (C) Initially, kinetic energy of the alpha particle is $K_i = 5$ MeV and its potential energy is $U_i = 0$ (because it is far away from the nucleus). The charge of the uranium nucleus is $Q = Ze = 92e$ and charge on the alpha particle is $q = 2e$. Let O be the centre of the uranium nucleus. The alpha particle starts moving towards O and is scattered by an angle 180° at P.

The distance of the closest approach is $OP = d$. The kinetic energy of the alpha particle at P is $K_f = 0$ (since its velocity is zero) and its potential energy is
$$U_f = 2Ze^2/(4\pi\epsilon_0 d)$$
Apply conservation of energy, $K_i + U_i = K_f + U_f$, to get

$d = \dfrac{2Ze^2}{4\pi\epsilon_0 K_i} = \dfrac{(9 \times 10^9)(2)(92)(1.6 \times 10^{-19})^2}{(5 \times 10^6)(1.6 \times 10^{-19})}$

$= 5.3 \times 10^{-14}$ m

15. (A) From Eq.2.33,
$W_{ext} = -pE(\cos\theta_2 - \cos\theta_1) = -pE(\cos\pi - \cos 0)$
$= -pE(-1 - 1) = 2pE$

16. (D) The torque exerted by an electric field E on a dipole of moment p is given by-
$$\tau = pE\sin\theta \qquad \cdots (1)$$
where θ is the angle which the dipole is making with the electric-field.

Corresponding to maximum torque, $\theta = 90°$, therefore from Eq.(1), we get
$$\tau_{max} = pE$$
Here, $p = q(2l) = 1 \times 10^{-6} \times 0.02$ C/m and $E = 10^5$ N/C.
Therefore, $\tau_{max} = 1 \times 10^{-6} \times 0.02 \times 10^5 = 2 \times 10^{-3}$ N.m
The work done in rotating the dipole from an angle $\theta = 0°$ to 180° is given by-
$$W = \int_{\theta_o}^{\theta} pE\sin\theta\, d\theta = pE(\cos\theta_0 - \cos\theta)$$
Here, $\theta_0 = 0°$ and $\theta = 180°$
Therefore, $W = pE(\cos 0° - \cos 180°)$
$= 2pE = 4 \times 10^{-3}$ joule

2.16.3 Check Point 3

1. B → C, C → D, A → B, D → E Moving from B to C decreases the electric potential by 2 V, so the electric field performs 2 J of work on each coulomb of positive charge that moves. Moving from C to D decreases the electric potential by 1 V, so 1 J of work is done by the field. It takes no work to move the charge from A to B because the electric potential does not change. Moving from D to E increases the electric potential by 1 V, and thus the field does −1 J of work per unit of positive charge that moves.

2. (A) Any surface which has same electrostatic potential at every point is called an equipotential surface. Electric field is always perpendicular to an equipotential surface. Therefore, X-direction is perpendicular only to YZ-plane.

3. (D) The electric field points in the direction of decreasing electric potential.

4. (D) In electrostatics (i.e., when charges are stationary and charge density does not vary with the time), electric field inside a conductor remains zero and the electric field lines remain normal to the surface and never enter inside the conductor.

5. **APPROACH** Electric potential at a distance r from the centre of a uniformly charged non-conducting sphere of radius R, as shown in Fig.2.170, is given by-
$$V_r = k\dfrac{Q}{r} \quad (r \geq R) \quad [\text{see Eq.(2.69)}] \qquad (1)$$
At the surface of non-conducting uniformly charged sphere,

2.16. ANSWER KEYS AND SOLUTIONS

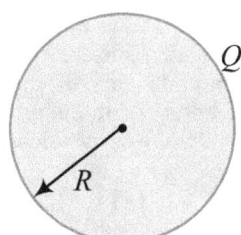

Figure 2.170

electric potential is given as:
$$V_R = k\frac{Q}{R} \quad \text{[see Eq.(2.70)]} \quad (2)$$

And inside the non-conducting uniformly charged sphere, the electric potential is given by:
$$V_r = k\frac{Q}{2R}\left(3 - \frac{r^2}{R^2}\right) \quad (0 \leq r \leq R) \quad \text{[see Eq.(2.52)]} \quad (3)$$

Substitute the given values of V in above appropriate equations and simplify for r.

SOLUTION (C, D) Given that the electric potential of the sphere with respect to ∞, is V_0, therefore, from Eq.(2)-
$$V_0 = \frac{kQ}{R} \implies \frac{kQ}{R} = V_0 \quad (4)$$

At the centre of sphere, $r = 0$, therefore, from Eq.(3), we have-
$$V_C = k\frac{Q}{2R}\left(3 - \frac{0^2}{R^2}\right) = \frac{3kQ}{2R} = \frac{3}{2}V_0$$
$$\text{or} \quad V_C = \frac{3}{2}V_0 \quad (5)$$

Calculation of distances from the centre of the sphere for given potentials:

(i) For potential $V_{R_1} = 3V_0/2$
From Eq.(5), it is clear that-
$$V_{R_1} = V_C \implies R_1 = 0$$

(ii) For potential $V_{R_2} = 5V_0/4$
For a non-conducting sphere of radius R, potential $V < V_0$ for $r > R$ and the point for which $V_{R_2} = 5V_0/4 > V_0$ lies inside the sphere. Therefore, we use Eq.(3) for it.
$$V_{R_2} = \frac{5V_0}{4} = k\frac{Q}{2R}\left(3 - \frac{R_2^2}{R^2}\right)$$
$$V_{R_2} = \frac{5V_0}{4} = \frac{kQ}{2R^3}\left(3R^2 - R_2^2\right)$$
or $\quad \frac{5V_0}{4} = \frac{V_0}{2R^2}\left(3R^2 - R_2^2\right) \implies \frac{5}{2} = 3 - \left(\frac{R_2}{R}\right)^2$
$$\implies \left(\frac{R_2}{R}\right)^2 = 3 - \frac{5}{2} = \frac{1}{2} \implies R_2 = \frac{R}{\sqrt{2}}$$

Similarly, (iii) For potential $V_{R_3} = 3V_0/4$
$$V_{R_3} = \frac{3V_0}{4} \implies \frac{kQ}{R_3} = \frac{3}{4} \times \frac{kQ}{R} \implies R_3 = 4R/3$$

(iv) Now, for potential $V = V_0/4$
$$V_{R_4} = \frac{kQ}{R_4} = \frac{V_0}{4} \implies \frac{kQ}{R_4} = \frac{1}{4} \times \frac{kQ}{R} \implies R_4 = 4R$$

6. **(B)** Note that the potential inside a hollow conducting sphere is always constant and its value is equal to the potential at it's surface. So, the potential at the center will be 10 V. Note that the electric field inside the hollow conducting sphere is zero.

7. **(B)** $E = \frac{\lambda}{2\pi\epsilon_0 x}$
$$\int_{V_A}^{V_B} dV = -E\,dx = -\frac{\lambda}{2\pi\epsilon_0}\int_{3a}^{2a}\frac{dx}{x}$$

$$\implies V_B - V_A = \frac{\lambda}{2\pi\epsilon_0}\ln\frac{3}{2}$$

Therefore, the work done by external agent = $\frac{q_0\lambda}{2\pi\epsilon_0}\ln\frac{3}{2}$

8. **(C)** The charges at $A, B,$ and C are $q_A = q/3, q_B = q/3$, and $q_C = -2q/3$. The electric fields at O due to q_A and q_B are equal in magnitude but opposite in direction. Thus, the resultant electric field at O is only due to charge q_C and is given by
$$\vec{E}_O = -\frac{q}{6\pi\epsilon_0 R^2}\hat{i}$$

The triangle ABC is right-angled with $\angle A = 60°, \angle C = 90°$, and $r_{AB} = 2R$. Thus, $r_{AC} = R$ and $r_{BC} = \sqrt{3}R$. The potential energy for the given charge distribution is
$$U = \frac{1}{4\pi\epsilon_0}\left[\frac{q_A q_B}{r_{AB}} + \frac{q_A q_C}{r_{AC}} + \frac{q_B q_C}{r_{BC}}\right]$$
$$= \frac{1}{4\pi\epsilon_0}\left[\frac{q^2}{18R} - \frac{2q^2}{9R} - \frac{2q^2}{9\sqrt{3}R}\right] \neq 0$$

The magnitude of force between q_C and q_B is
$$F_{BC} = \frac{1}{4\pi\epsilon_0}\frac{q_B q_C}{r_{BC}^2} = \frac{q^2}{54\pi\epsilon_0 R^2}$$

The potential at O is
$$V = \frac{1}{4\pi\epsilon_0}(q_A/R + q_B/R + q_C/R) = 0$$

2.16.4 Conceptual Questions

1. The potential energy increases. When an outside agent makes it a negative charge to move in the direction of the field, the charge moves to a region of lower electric potential. The product of its negative charge with a lower number of volts gives a higher number of joules. Keep in mind that a negative charge feels an electric force in the opposite direction to the field, while the potential is the work done on the charge to move it in a field per unit charge.

2. Given that the electric field points to the right and the proton is moving to the left, it means the proton is moving in the opposite direction of the electric field. Since an electric field always directs from higher potential to lower potential, we can conclude that the proton is moving toward a region of higher potential. We have already studied that if a charged particle is free to move in an electric field, it moves in a direction that minimizes its potential energy. Now, if the proton is free to move, then to minimize its electric potential energy in the electric field, it would move in the direction of the electric field, i.e., towards a lower potential point. So, it is clear that there must be some external agent to transport the proton from a lower potential point to a higher one. As it moves toward the higher potential point, the electric potential energy of the proton in the electric field increases. This increase occurs at the cost of work done by the external agent against the electric force.

3. Since the electron is moving toward a region of higher electric potential, therefore, its electric potential energy in the external electric field is decreasing.

4. If V is constant, its gradient is zero; consequently the electric field is zero throughout the region.

5. Yes.

6. No. $\vec{E}$ cannot be determined without knowing V at a continuum of points.

7. Not necessarily. If two points are at the same potential, then no net work is done in moving a charge from one point to the other, but work (both positive and negative) could be done at different parts of the path. No. It is possible

that positive work was done over one part of the path (for example accelerated motion), and negative work done over another part of the path (decelerated or retarded motion), so that these two contributions to the net work sum to zero. In this case, a non-zero force would have to be exerted over both parts of the path.

8. The negative charge will move toward a region of higher potential and the positive charge will move toward a region of lower potential. In both cases, the potential energy of the charge will decrease.

9. The electric field lines, shown as solid lines, and the equipotential surfaces (intersecting the plane of the paper), shown as dashed lines, are sketched in the Figure 2.171. The point charge $+Q$ is the point at the right, and the metal sphere with charge $-Q$ is at the left. Near the two charges the equipotential surfaces are spheres, and the field lines are normal to the metal sphere at the sphere's surface,

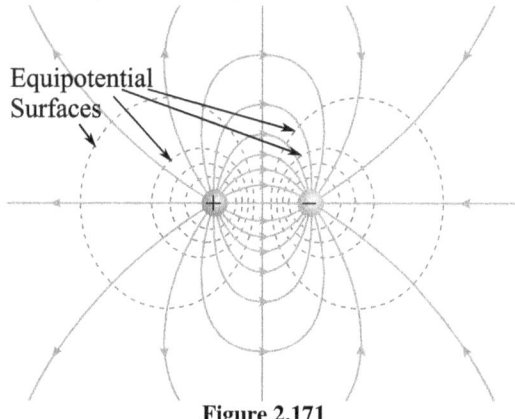

Figure 2.171

10. Yes. If the charge on the particle is negative then under the electric force only, it moves from a region of low electric potential to a region of high electric potential, its electric potential energy will decrease.

11. The electric field lines, shown as solid lines, and the equipotential surfaces (intersecting the plane of the paper), shown as dashed lines, are sketched in the adjacent Figure 2.172. The point charge $+Q$ is the point at the right, and the metal sphere with charge $+Q$ is at the left. Near the two charges the equipotential surfaces are spheres, and the field lines are normal to the metal sphere at the sphere's surface. Very far from both charges, the equipotential surfaces and field lines approach those of a point charge $2Q$ located at the midpoint.

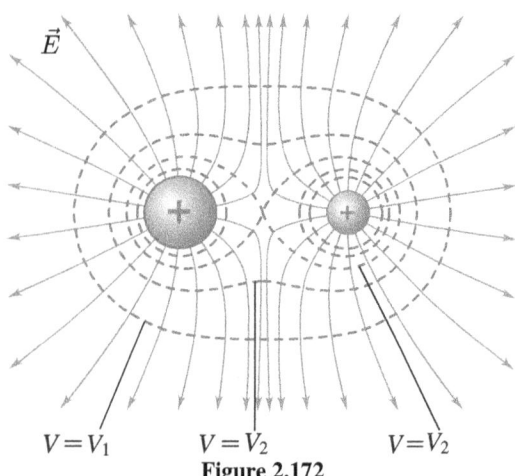

Figure 2.172

12. The equipotential surfaces are shown with dashed lines, the electric field lines are shown with solid lines. Near each charge, the equipotential surfaces are spheres centered on each charge; far from the charges, the equipotential surface is a sphere centered at the midpoint between the charges. The electric field lines are perpendicular to the equipotential surfaces.

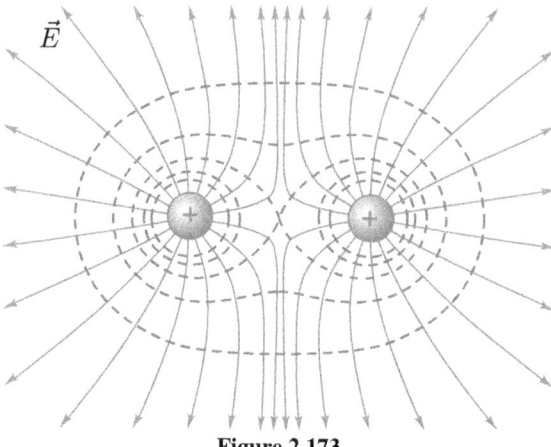

Figure 2.173

13. **APPROACH** We can use Coulomb's law and the superposition of fields to find $\vec{E}$ at the origin and the definition of the electric potential due to a point charge to find V at the origin.

SOLUTION (a) Apply Coulomb's law and the superposition of fields to find the electric field $\vec{E}$ at the origin:

$$\vec{E} = \vec{E}_{+q \text{ at } -a} + \vec{E}_{+q \text{ at } +a} = \frac{kq}{a^2}\hat{i} - \frac{kq}{a^2}\hat{i} = 0$$

The potential V at the origin is given by:

$$V = V_{+q \text{ at } -a} + V_{+q \text{ at } +a} = \frac{kq}{a} + \frac{kq}{a} = \frac{2kq}{a}$$

and (B) is correct.

(b) Apply Coulomb's law and the superposition of fields to find the electric field $\vec{E}$ at the origin:

$$\vec{E} = \vec{E}_{+q \text{ at } -a} + \vec{E}_{-q \text{ at } +a} = \left(\frac{kq}{a^2}\right)\hat{i} + \left(\frac{kq}{a^2}\right)\hat{i} = \left(\frac{2kq}{a^2}\right)\hat{i}$$

The potential V at the origin is given by:

$$V = V_{+q \text{ at } -a} + V_{-q \text{ at } +a} = \frac{kq}{a} + \frac{k(-q)}{a} = 0$$

and (C) is correct.

14. The charge can be moved along any path parallel to the y-z plane, namely perpendicular to the field.

15. **APPROACH** Use $\vec{E} = -\frac{\partial V}{\partial x}\hat{i}$ to find the electric field corresponding to the given potential and then compare its form to those produced by the four alternatives listed.

SOLUTION (a) Find the electric field corresponding to the given potential function: $V = 4.00|x| + V_0$

$$\vec{E} = -\frac{\partial V}{\partial x}\hat{i} = -\frac{\partial}{\partial x}[4.00|x| + V_0]\hat{i} = -4.00\frac{\partial}{\partial x}[|x|]\hat{i}$$

Note that $\dfrac{d}{dx}|x| = \begin{cases} 1 & \text{if } x > 0 \\ -1 & \text{if } x < 0 \end{cases}$

If $x > 0$, then $\dfrac{\partial}{\partial x}|x| = 1$ therefore-

$$\vec{E}_{x>0} = \left(-4.00\frac{\text{V}}{\text{m}}\right)\hat{i}$$

If $x < 0$, then $\dfrac{\partial}{\partial x}|x| = -1$ and hence-

$$\vec{E}_{x<0} = \left(4.00\frac{\text{V}}{\text{m}}\right)\hat{i}$$

A sketch of the electric field in this region is shown in Fig. 2.174.

(b) (1) is correct because field lines end on negative

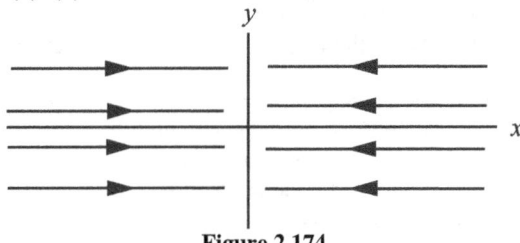

Figure 2.174

charges.

16. No. The local surface charge density is proportional to the normal component of the electric field ($E = \sigma/\epsilon_0 \implies \sigma = \epsilon_0 E$), not the potential on the surface.

17. No. Electric potential is the potential energy per unit charge at a point in space and electric field is the electric force per unit charge at a point in space. If one of these quantities is zero, the other is not necessarily zero. For example, the point exactly between two charges with equal magnitudes and opposite signs will have a zero electric potential because the contributions from the two charges will be equal in magnitude and opposite in sign. (Net electric potential is a scalar sum.) This point will not have a zero electric field, however, because the electric field contributions will be in the same direction (towards the negative and away from the positive) and so will add. (Net electric field is a vector sum.) As another example, consider the point exactly between two equal positive point charges. The electric potential will be positive since it is the sum of two positive numbers, but the electric field will be zero since the field contributions from the two charges will be equal in magnitude but opposite in direction.

18. (a) V at other points would be lower by 10 V. E would be unaffected, since E is the negative gradient of V, and a change in V by a constant value will not change the value of the gradient.
(b) If V represents an absolute potential, then yes, the fact that the Earth carries a net charge would affect the value of V at the surface. If V represents a potential difference, then no, the net charge on the Earth would not affect the choice of V.

19. If the field lines are not perpendicular to the surface, then there is a component of electric field parallel to the surface. If there is an electric field component parallel to the surface, then there must be a potential difference along the surface too.
But the surface is an equipotential surface, so, there is no potential difference along it. This gives a contradiction which means the initial assumption of field lines being not perpendicular to the surface is wrong.

20. No. An equipotential line is a line connecting points of equal electric potential. If two equipotential lines crossed, it would indicate that their intersection point has two different values of electric potential simultaneously, which is impossible. As an analogy, imagine contour lines on a topographic map. They also never cross because one point on the surface of the Earth cannot have two different values for elevation above sea level.

21. The Earth's gravitational equipotential lines are roughly circular, so the orbit of the satellite would have to be roughly circular.

22. The potential at point P would be unchanged. Each bit of positive charge will contribute an amount to the potential based on its charge and its distance from point P. Moving charges to different locations on the ring does not change their distance from P, and hence does not change their contributions to the potential at P.
The value of the electric field will change. The electric field is the vector sum of all the contributions to the field from the individual charges. When the charge Q is distributed uniformly about the ring, the y-components of the field contributions cancel, leaving a net field in the x-direction. When the charge is not distributed uniformly, the y-components will not cancel, and the net field will have both x- and y-components, and will be larger than for the case of the uniform charge distribution. There is no discrepancy here, because electric potential is a scalar and electric field is a vector.

23. The charge density and the electric field strength will be greatest at the pointed ends of the football because at the pointed surface there has a smaller radius of curvature than the middle.

24. No. You cannot calculate electric potential knowing only electric field at a point and you cannot calculate electric field knowing only electric potential at a point. As an example, consider the uniform field between two charged, conducting plates. If the potential difference between the plates is known, then the distance between the plates must also be known in order to calculate the field. If the field between the plates is known, then the distance to a point of interest between the plates must also be known in order to calculate the potential there. In general, to find V, you must know E and be able to integrate it. To find E, you must know V and be able to take its derivative. Thus you need E or V in the region around the point, not just at the point, in order to be able to find the other variable.

25. (a) Once the two spheres are placed in contact with each other, they effectively become one larger conductor. They will have the same potential because the potential everywhere on a conducting surface is constant.
(b) Because the spheres are identical in size, an amount of charge $Q/2$ will flow from the initially charged sphere to the initially neutral sphere so that they will have equal charges.
(c) Even if the spheres do not have the same radius, they will still be at the same potential once they are brought into contact because they still create one larger conductor. However, the amount of charge that flows will not be exactly equal to half the total charge. The larger sphere will end up with the larger charge.

26. If the electric field points due north, the change in the potential will be (a) greatest in the direction opposite the field, south; (b) least in the direction of the field, north; and (c) zero in a direction perpendicular to the field, east and west.

27. In the vicinity of a pointy protrusion, the electric field can be very high. This can lead to a spark inside an electronic device which can make the device to stop functioning.

28. Yes. In regions of space where the equipotential lines are closely spaced, the electric field is stronger than in regions of space where the equipotential lines are farther apart.

29. If the electric field in a region of space is uniform, then you can infer that the electric potential is increasing or decreasing uniformly in that region. For example, if the electric field is 10 V/m in a region of space then you can infer that the potential difference between two points 1 meter apart (measured parallel to the direction of the field) is 10 V. If the electric potential in a region of space is uniform, then

you can infer that the electric field there is zero.
30. The electric potential energy of two unlike charges is negative. The electric potential energy of two like charges is positive. In the case of unlike charges, work must be done to separate the charges. In the case of like charges, work must be done to move the charges together.
31. The electric field is the negative of the slope of the V vs. x graph [Fig. 2.175].

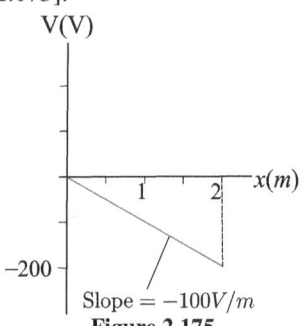

Figure 2.175

32. The electric field is the negative of the slope of the V vs. x graph. If $V = $ constant in some region then $E = 0$ in that region.

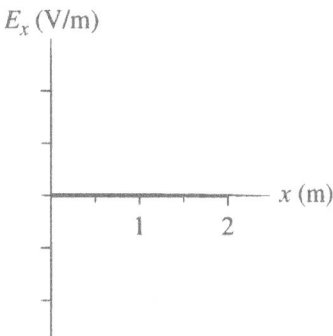

Figure 2.176

33. For contour map shown in Fig.2.91a, electric field at point ①;
$$E_1 = -\left(\frac{10-0}{1}\right)\hat{i} = -10 \text{ V/m } \hat{i}$$
and electric field at point ②;
$$E_2 = -\left(\frac{40-30}{1}\right)\hat{i} = -10 \text{ V/m } \hat{i}$$
here, negative sign shows that electric field is directed along $-$ve direction of x axis.
Now, for contour map shown in Fig.2.91b, electric field at point ①;
$$E_1 = -\left(\frac{10-0}{2}\right)\hat{i} = -5 \text{ V/m } \hat{i}$$
and electric field at point ②;
$$E_2 = -\left(\frac{40-20}{1}\right)\hat{i} = -20 \text{ V/m } \hat{i}$$
here, again negative sign shows that electric field is directed along $-$ve direction of x axis.

34. The electron moves to the right. The electric field is the negative of the slope of the electric potential, so the electric field will be non-zero at $x = 2$ m and points to the left. The electron will therefore experience a net force to the right and accelerate to the right.

35. **(a)** The electric field vector points in the direction of decreasing potential. Therefore $V_a > V_b$.
(b) $|\Delta V_{ab}| < |\Delta V_{cd}| < |\Delta V_{ef}|$. For a uniform electric field $|\Delta V| = |-E\Delta s| = |E\Delta s|$. The order is determined by the fact that $\Delta s_{ab} < \Delta s_{cd} < \Delta s_{ef}$
(c) Surface 1 is an equipotential surface because it is perpendicular to the electric field vectors. Surface 2 is not perpendicular to the electric field so it is not an equipotential surface.

36. Equal. When a conductor is in electrostatic equilibrium, the entire conductor is at the same potential.

37. **(a)** $V_1 = V_2$. Both spheres and the wire become one conductor and so are all at the same potential.
(b) Since $V_1 = V_2$, $\frac{1}{4\pi\epsilon_0}\frac{Q_1}{r_1} = \frac{1}{4\pi\epsilon_0}\frac{Q_2}{r_2}$, thus $\frac{Q_1}{r_1} = \frac{Q_2}{r_2}$. Since $r_1 > r_2$, $Q_1 > Q_2$.
(c) Recall $E_1 = \frac{1}{4\pi\epsilon_0}\frac{Q_1}{r_1^2} = \frac{V_1}{r_1}$ and $E_2 = \frac{1}{4\pi\epsilon_0}\frac{Q_2}{r_2^2} = \frac{V_2}{r_2}$. Since $r_1 > r_2$, $E_1 < E_2$.

38. **APPROACH**: The potential difference is the negative of the area under the E_x vs. x curve.
SOLUTION The potential difference between $x = 1.0$ m and $x = 3.0$ m (Fig.2.96) is-
$$\Delta V = -\frac{1}{2}(200 \text{ V})(3.0 \text{ m} - 1.0 \text{ m}) = -200 \text{ V}$$
The potential difference is negative since the electric field points in the direction of decreasing potential.

39. **APPROACH** The potential difference is the negative of the area under the E_x vs. x curve.
SOLUTION The potential difference between the origin and $x = 3.0$ m is
$$\Delta V = V(x = 3.0 \text{ m}) - V(x = 0.0 \text{ m})$$
$$= -\frac{1}{2} \times (3 \text{ m} + 2 \text{ m}) \times 200 \text{ V/m} = -500 \text{ V}$$
Thus, $V(3.0 \text{ m}) = V(0 \text{ m}) - 500 \text{ V}$
$$= -50 \text{ V} - 500 \text{ V} = -550 \text{ V} = -5.50 \text{ kV}.$$
The potential decreases from the origin to $x = 3.0$ m.

40. **APPROACH** The electric field points in the direction of decreasing potential and is perpendicular to the equipotential lines.
SOLUTION Please refer to Figure 2.98. The three equipotential surfaces correspond to potentials of 0 V, 200 V, and 400 V.
The electric field component along a direction of constant potential is $E_s = -dV/ds = 0$ V/m. But, the electric field component perpendicular to the equipotential surface is
$$|\vec{E}| = \frac{\Delta V}{\Delta s} = \frac{400 \text{ V}}{0.02 \text{ m}} = 20 \text{ kV/m}$$
The direction of the electric field vector is "downhill," perpendicular to the equipotential surfaces. That is, the electric field is 20 kV/m downward or $\vec{E} = -20\,\hat{j}$ kV/m

41. **APPROACH** The electric field is perpendicular to the equipotential lines and points "downhill".
SOLUTION Refer to Figure 2.99. Three equipotential surfaces at potentials of -200V, 0 V, and 200 V are shown.
The electric field is perpendicular to the equipotential surfaces:
$$E = -\frac{\Delta V}{\Delta s} = -\frac{400 \text{ V}}{0.02 \text{ m}} = -20 \text{ kV/m}$$
The electric field vector is in the third quadrant, 45° below the negative x-axis. That is,
$$\vec{E} = (20 \text{ kV/m}, 45° \text{ below } x \text{ axis })$$
In component form, this would be
$$\vec{E} = 20\left[-\left(\frac{\hat{i}+\hat{j}}{\sqrt{2}}\right)\right] \text{kV/m} = \left(-\frac{20}{\sqrt{2}}\hat{i} - \frac{20}{\sqrt{2}}\hat{j}\right) \text{kV/m}$$

42. **APPROACH** The electric field is the negative of the slope of the graph of the potential function.
SOLUTION (a) Please refer to Figure 2.100(a). There are

three regions of different slope. For 0 cm ≤ x < 10 cm and 20 cm ≤ x < 30 cm,
$$\frac{\Delta V}{\Delta x} = 0 \text{ V/m} \implies E_x = 0 \text{ V/m}$$
For 10 cm ≤ x < 20 cm,
$$\frac{\Delta V}{\Delta x} = \frac{100 \text{ V} - (-100 \text{ V})}{0.20 \text{ m} - 0.10 \text{ m}} = 2000 \text{V/m}$$
$$\implies E_x = -2000 \text{V/m}$$

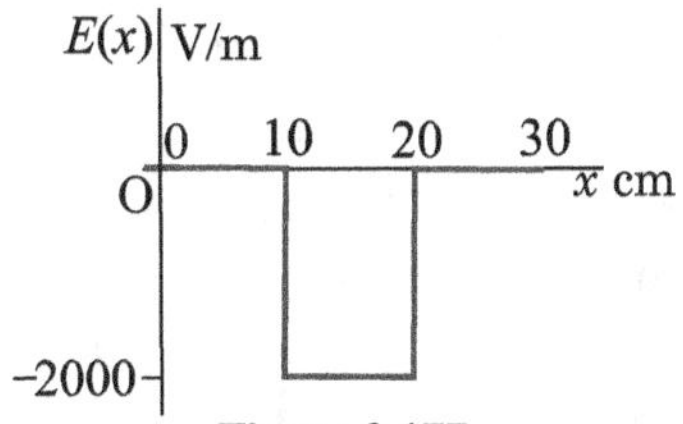

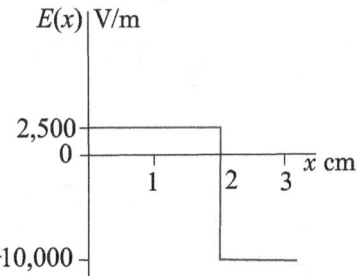
Figure 2.177

Because $E_s = -dV/ds$, the electric field is zero where the potential is not changing.

(b) Now, refer to Figure 2.100(b). There are three regions of different slope. For 0 cm ≤ x < 2 cm,
$$\frac{\Delta V}{\Delta x} = \frac{-50 \text{ V} - 0 \text{ V}}{0.02 \text{ m} - 0 \text{ m}} = -2500 \text{ V/m} \implies E_x = 2500 \text{ V/m}$$
For $x \geq 2$ cm
$$\frac{\Delta V}{\Delta x} = \frac{50 \text{ V} - (-50 \text{ V})}{0.03 \text{ m} - 0.02 \text{ m}} = 10{,}000 \text{ V/m}$$
$$\implies E_x = -10{,}000 \text{ V/m}$$
$\Delta V/\Delta x$ and E_x are the negative of each other.

Figure 2.178

43. The electric potential difference ΔV between two points in a uniform electric field is

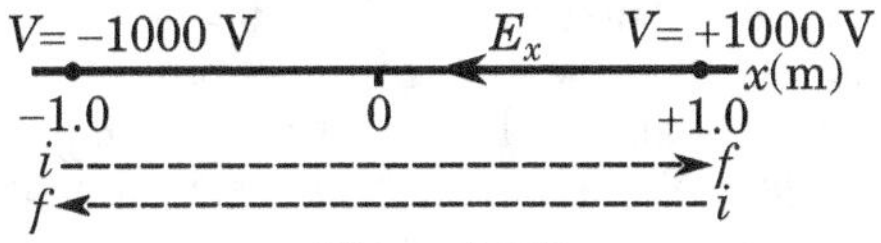

Figure 2.179

$$V(x_f) - V(x_i) = -\int E_x dx = -E_x(x_f - x_i)$$
Choosing $x_i = -1.0$ m and $x_f = +1.0$ m,
$$+1000 \text{ V} - (-1000 \text{ V}) = -E_x[1.0 \text{ m} - (-1.0 \text{ m})]$$
$$\implies E_x = -1.0 \text{ kV/m}$$
Alternatively, $x_i = -1.0$ m and $x_f = -1.0$ m. For this choice,
$$-1000 \text{ V} - (+1000 \text{ V}) = -E_x[-1.0 \text{ m} - (1.0 \text{ m})]$$
$$\implies E_x = -1.0 \text{ kV/m}$$
The choice of initial and final positions does not change the physical nature of the electric field or the potential difference.

44. APPROACH The electric field is the negative of the derivative of the potential function.
SOLUTION (a) The component of the electric field in the s-direction is $E_s = -dV/ds$. For the given potential,
$$\frac{dV}{dx} = \frac{d}{dx}\left(100x^2 \text{ V}\right) = 200x \text{ V/m} \implies E_x = -200x \text{ V/m}$$

(b) At $x = 1$ m, $E_x = -200 \times 1 \text{ V/m} = -0.2 \text{ kV/m}$.
The potential increases with x, so the electric field must point in the $-x$-direction.

45. (a) Since $E_x = -dV/dx$, we have
$$E_x = -\frac{d}{dx}\left(100\,e^{-2x}\right) \text{ V/m} = 200\,e^{-2x} \text{ V/m}$$
At $x = 1.0$ m, $E_x = \left(200\,e^{-2(1.0)}\right) \text{ V/m} = 27 \text{ V/m}$

(b) At $x = 2.0$ m, $E_x = \left(200\,e^{-2(2.0)}\right) \text{ V/m} = 3.7 \text{ V/m}$

2.16.5 Problems

1. (a) $V_B - V_A = \Delta U/q = -W/(-e)$
$= -\left(3.94 \times 10^{-19} \text{ J}\right)/\left(-1.60 \times 10^{-19} \text{C}\right)$
$= 2.46$ V.
(b) $V_C - V_A = V_B - V_A = 2.46$ V.
(c) $V_C - V_B = 0$ (since C and B are on the same equipotential line).

2. (a) The change in potential is the negative of the "area" under the curve. Thus, using the area of a triangle formula, we have
$$V - V_0 = -\int_0^{x=2} \vec{E} \cdot d\vec{s} = -\int_0^{x=2} E\,dx = -\frac{1}{2}(2)(-20)$$
here, $V_0 = 10$ V is the potential at $x = 0$. Substituting this value and solving for V yields $V = 30$ V.

(b) For any region within $0 < x < 3$ m, the area under E_x vs x graph is negative, therefore $-\int \vec{E} \cdot d\vec{s}$ is positive, but for any region for which $x > 3$ m, the area is positive therefore $-\int \vec{E} \cdot d\vec{s}$ is negative. Therefore, $V = V_{\max}$ occurs at $x = 3$ m.
If V_0 is the electric potential at $x = 0$ and $V_{\max}$ at $x = 3$ m, then
$$V_{\max} - V_0 = -\int_0^{x=3} \vec{E} \cdot d\vec{s} = -\int_0^{x=3} E\,dx = -\frac{1}{2}(3)(-20)$$
Since, $V_0 = 10$ V, therefore, above equation gives-
$$V_{\max} - 10 \text{ V} = 30 \text{ V} \implies V_{\max} = 40 \text{ V}$$

(c) From $x = 0$ to $x = 3$ m, the area under E vs x graph i.e., $\int_{x=0}^{x=3} E\,dx$ is negative, therefore $V - V_0 = -\int_{x=0}^{x=3} E\,dx$ will be positive. Since, $V_0 = 10$ V, therefore V can not be zero anywhere in this region. Furthe from part (b), we have seen that the voltage at $x = 3$ m is 40 V. Now, further the area under the E vs x graph from $x = 3$ to $x = 4$ is only $\frac{1}{2}(4-3)(20) = +10$ V, therefore the change in potential corresponding to this area is -10 V. So from $x = 0$ to $x = 3$ m potential increases from 10 V to 40 V, then from $x = 3$ m to $x = 4$ m there is a decrease of 10 V and voltage reaches to 40 V − 10 V, i.e., 30 V. Thus, potential can not be zero for any value of $x \leq 4$.
Now, suppose at some $x = x'(> 4$ m), the electric potential is zero, then
$$V - V_0 = -\int_{x=0}^{x=x'} \vec{E} \cdot d\vec{s} = -\int_{x=0}^{x=x'} E\,dx$$
$$\implies V - V_0 = -\left[\int_{x=0}^{x=3} E\,dx + \int_{x=3}^{x=4} E\,dx + \int_{x=4}^{x=x'} E\,dx\right]$$
$$= -\left[\frac{1}{2}(3)(-20) + \frac{1}{2}(4-3)(20) + (x'-4)(20)\right]$$
$$= -[-30 + 10 + 20x' - 80] = 100 - 20x'$$
Given that $V_0 = 10$ V, therefore corresponding to $V = 0$, we have
$$0 - 10 = 100 - 20x' \implies x' = 5.5 \text{ m}$$

3. (a) The charge on the sphere is
$$q = 4\pi\epsilon_0 VR = \frac{(200 \text{ V})(0.15 \text{ m})}{8.99 \times 10^9 \text{ N}\cdot\text{m}^2/\text{C}^2} = 3.3 \times 10^{-9} \text{C}$$
(b) The (uniform) surface charge density (charge divided by the area of the sphere) is
$$\sigma = \frac{q}{4\pi R^2} = \frac{3.3 \times 10^{-9} \text{C}}{4\pi(0.15 \text{ m})^2} = 1.2 \times 10^{-8} \text{C/m}^2$$

4. $E = \frac{|\Delta V|}{d} = \frac{25.0 \times 10^3 \text{ J/C}}{1.50 \times 10^{-2} \text{ m}}$
$= 1.67 \times 10^6 \text{ N/C} = 1.67 \text{ MN/C}$

5. From Fig.2.180, we can write-
$$V_B - V_A = -\int_A^B \vec{E}\cdot d\vec{s}$$
$$= -\int_A^C \vec{E}\cdot d\vec{s} - \int_C^B \vec{E}\cdot d\vec{s}$$
$$V_B - V_A = (-E\cos 180°)\int_{-0.300}^{0.500} dy$$
$$- (E\cos 90.0°)\int_{-0.200}^{0.400} dx$$
$$V_B - V_A = (325)(0.800) = +260 \text{ V}$$

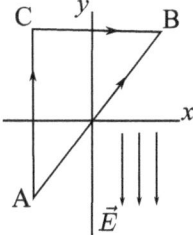

Figure 2.180

6. (a) In Fig. 2.181, a charged block with mass m and charge Q is attached to a spring with a spring constant k in an electric field $\vec{E}$ on a smooth surface. Since the block is positively charged, it moves toward a lower electric potential when released from rest, due to the electric force exerted by the electric field.
If at any instant, x represents the extension in spring, then the corresponding decrement in electric potential is Ex. Let's arbitrarily choose $V = 0$ at $x = 0$. For an extension x in the spring, the electric potential decreases to $V = -Ex$ and electric potential energy becomes $U_e = QV = -QEx$.
Between the endpoints of the motion,

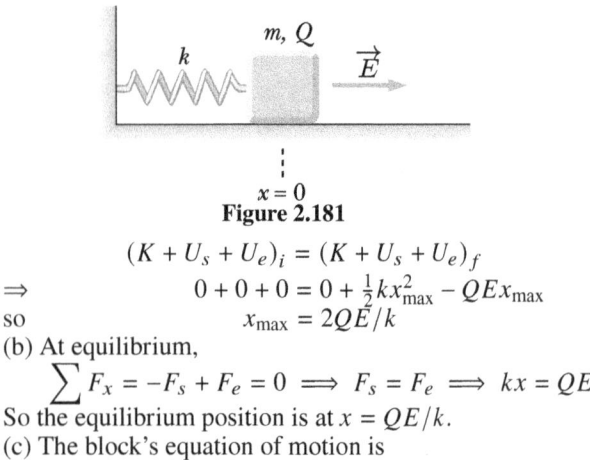

Figure 2.181

$(K + U_s + U_e)_i = (K + U_s + U_e)_f$
$\Rightarrow \quad 0 + 0 + 0 = 0 + \frac{1}{2}kx_{\max}^2 - QEx_{\max}$
so $\quad x_{\max} = 2Q\vec{E}/k$

(b) At equilibrium,
$$\sum F_x = -F_s + F_e = 0 \Rightarrow F_s = F_e \Rightarrow kx = QE$$
So the equilibrium position is at $x = QE/k$.

(c) The block's equation of motion is
$$\sum F_x = -kx + QE = m\frac{d^2x}{dt^2}.$$

Let $x' = x - \frac{QE}{k}$ or $x = x' + \frac{QE}{k}$, so, the equation of motion becomes:
$$-k\left(x' + \frac{QE}{k}\right) + QE = m\frac{d^2(x + QE/k)}{dt^2}$$
$$\Rightarrow \frac{d^2x'}{dt^2} = -\left(\frac{k}{m}\right)x'$$
This equation has the form of SHM:
$$\frac{d^2x'}{dt^2} = -\omega^2 x'$$
with angular frequency, $\omega = \sqrt{\frac{k}{m}}$
The period of the motion is then
$$T = \frac{2\pi}{\omega} = 2\pi\sqrt{\frac{m}{k}}$$

(d) $(K + U_s + U_e)_i + \Delta E_{\text{mech}} = (K + U_s + U_e)_f$
$\Rightarrow \quad 0 + 0 + 0 - \mu_k mg x_{\max} = 0 + \frac{1}{2}kx_{\max}^2 - QEx_{\max}$
$\Rightarrow \quad x_{\max} = \frac{2(QE - \mu_k mg)}{k}$

7. Total charge of the rod of length L: $q = \lambda L$, and total mass of the rod: $m = \mu L$. Arbitrarily take $V = 0$ at the initial point. Then at distance d downfield the electric potential V is given by-
$V = -Ed$ and potential energy $U_e = qV = -\lambda LEd$.

(a) $(K + U)_i = (K + U)_f \Rightarrow 0 + 0 = \frac{1}{2}\mu Lv^2 - \lambda LEd$
Solving for v, we get
$$v = \sqrt{\frac{2\lambda Ed}{\mu}} = \sqrt{\frac{2(40.0 \times 10^{-6} \text{C/m})(100 \text{ N/C})(2.00 \text{ m})}{(0.100 \text{ kg/m})}}$$
$$= 0.400 \text{ m/s}$$

(b) The same.

8. Arbitrarily take $V = 0$ at point P. Then from Fig.2.182a, the potential at the original position of the charged particle is $-\vec{E}\cdot\vec{s} = Es\cos\theta = -EL\cos\theta$ ($\because |\vec{s}| = L$). From Fig.2.182b, the potential at final position a, $V = -EL$. Given that, the table is frictionless, therefore the forces acting on the particles are–

(a) Normal reaction $\vec{N}$ of table acting perpendicularly outward to the plane of page. It is not shown in diagram.
(b) gravitational force $m\vec{g}$ acting perpendicularly inward to the plane of page. It is also not shown in diagram.
(c) Electric force $q\vec{E}$ in the direction of electric field $\vec{E}$.

Since $\vec{N}$ and $m\vec{g}$ both are perpendicular to the plane of displacement of the particle, therefore, the work done by conservative force $m\vec{g}$ and non-conservative force $\vec{N}$ both are zero. The force which is doing work in this case is the electric force $q\vec{E}$ which is a conservative force in nature, therefore, net mechanical energy of the charged particle in electric field will remain conserved. So, by principle of conservation of mechanical energy between original and final positions of charged particle, we can write

$(K + U)_i = (K + U)_f \Rightarrow 0 - qEL\cos\theta = \frac{1}{2}mv^2 - qEL$
Solving for v, we get
$$v = \sqrt{\frac{2qEL(1 - \cos\theta)}{m}}$$
$$= \sqrt{\frac{2(2.00 \times 10^{-6}\text{C})(300 \text{ N/C})(1.50 \text{ m})(1 - \cos 60.0°)}{0.0100 \text{ kg}}}$$
$$= 0.300 \text{ m/s}$$

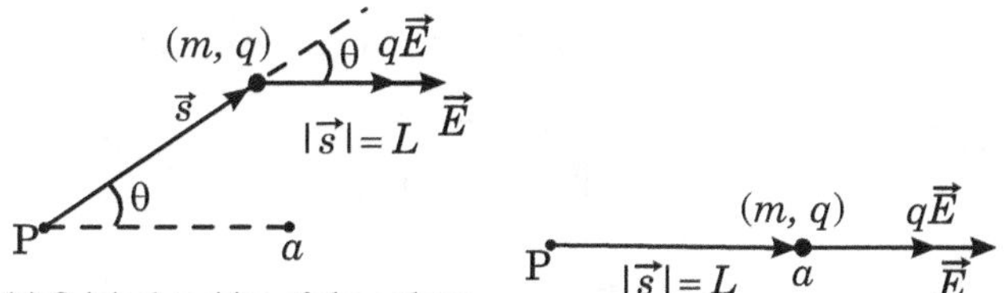

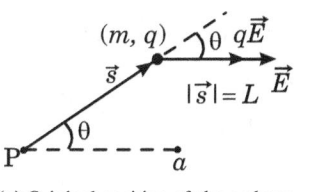

(a) Original position of charged particle

(b) Final position of charged particle

Figure 2.182

9. (a) The potential at 1.00 cm is
$$V_1 = k\frac{q}{r} = \frac{(8.99 \times 10^9 \text{ N}\cdot\text{m}^2/\text{C}^2)(1.60 \times 10^{-19}\text{C})}{1.00 \times 10^{-2} \text{ m}}$$
$$= 1.44 \times 10^{-7} \text{ V}$$

(b) The potential at 2.00 cm is
$$V_2 = k\frac{q}{r} = \frac{(8.99 \times 10^9 \text{ N}\cdot\text{m}^2/\text{C}^2)(1.60 \times 10^{-19}\text{C})}{2.00 \times 10^{-2} \text{ m}}$$
$$= 0.719 \times 10^{-7} \text{ V}.$$

Thus, the difference in potential between the two points is
$$\Delta V = V_2 - V_1 = -7.19 \times 10^{-8} \text{ V}.$$

(c) The approach is the same as above except the charge is -1.60×10^{-19}C. This changes the sign of each answer, with its magnitude remaining the same.
That is, the potential at 1.00 cm is -1.44×10^{-7} V.
The potential at 2.00 cm is -0.719×10^{-7} V,
so, $\Delta V = V_2 - V_1 = 7.19 \times 10^{-8}$ V.

10. (a) The potential difference is
$$V_A - V_B = \frac{q}{4\pi\epsilon_0 r_A} - \frac{q}{4\pi\epsilon_0 r_B}$$
$$= (1.0 \times 10^{-6}\text{C})(8.99 \times 10^9 \text{ N}\cdot\text{m}^2/\text{C}^2)$$
$$\left(\frac{1}{2.0 \text{ m}} - \frac{1}{1.0 \text{ m}}\right)$$
$$= -4.5 \times 10^3 \text{ V}$$

(b) Since $V(r)$ depends only on the magnitude of $\vec{r}$, the result is unchanged.

11. **APPROACH** The electric potential for a spherically symmetric charge distribution falls off as $1/r$, where r is the radial distance from the centre of the charge distribution. The electric potential V at the surface of a drop of charge q and radius R is given by $V = q/4\pi\epsilon_0 R$.
SOLUTION (a) With $V = 500$ V and $q = 30 \times 10^{-12}$C, we find the radius to be
$$R = \frac{q}{4\pi\epsilon_0 V} = \frac{(8.99 \times 10^9 \text{ N}\cdot\text{m}^2/\text{C}^2)(30 \times 10^{-12}\text{C})}{500 \text{ V}}$$
$$= 5.4 \times 10^{-4} \text{ m}$$

(b) After the two drops combine to form one big drop, the total volume is twice the volume of an original drop, so the radius R' of the combined drop is given by $(R')^3 = 2R^3$ and $R' = 2^{1/3}R$. The charge is twice the charge of the original drop: $q' = 2q$. Thus,
$$V' = \frac{1}{4\pi\epsilon_0}\frac{q'}{R'} = \frac{1}{4\pi\epsilon_0}\frac{2q}{2^{1/3}R} = 2^{2/3}V = 2^{2/3}(500 \text{ V})$$
$$\approx 790 \text{ V}$$

☞ A positively charged configuration produces a positive electric potential, and a negatively charged configuration produces a negative electric potential. Adding more charge increases the electric potential.

12. The electric potential is given by-
$$V = \frac{1}{4\pi\epsilon_0}\frac{q}{r} \quad (1)$$

In applying Eq.1, we are assuming $V \to 0$ as $r \to \infty$. All corner particles are equidistant from the centre, and since their total charge is
$$2q_1 - 3q_1 + 2q_1 - q_1 = 0$$
therefore, their contribution to Eq.1 vanishes. The net potential is due, then, to the two $+4q_2$ particles, each of which is a distance of $a/2$ from the centre:
$$V = \frac{1}{4\pi\epsilon_0}\frac{4q_2}{a/2} + \frac{1}{4\pi\epsilon_0}\frac{4q_2}{a/2} = \frac{16q_2}{4\pi\epsilon_0 a}$$
$$= \frac{16(8.99 \times 10^9 \text{ N}\cdot\text{m}^2/\text{C}^2)(6.00 \times 10^{-12}\text{C})}{0.39 \text{ m}}$$
$$= 2.21 \text{ V}$$

13. A charge $-5q$ is at a distance $2d$ from P, a charge $-5q$ is at a distance d from P, and two charges $+5q$ are each at a distance d from P, so the electric potential at P is
$$V = \frac{q}{4\pi\epsilon_0}\left[-\frac{1}{2d} - \frac{1}{d} + \frac{1}{d} + \frac{1}{d}\right]$$
$$= \frac{q}{8\pi\epsilon_0 d} = \frac{(8.99 \times 10^9 \text{ N}\cdot\text{m}^2/\text{C}^2)(5.00 \times 10^{-15}\text{C})}{2(4.00 \times 10^{-2} \text{ m})}$$
$$= 5.62 \times 10^{-4} \text{ V}$$

The zero of the electric potential was taken to be at infinity.

14. First, we observe that $V(x)$ cannot be equal to zero for $x > d$. In fact $V(x)$ is always negative for $x > d$. Now we consider the two remaining regions on the x axis: $x < 0$ and $0 < x < d$.
(a) For $0 < x < d$ we have $d_1 = x$ and $d_2 = d - x$. Now,
$$V(x) = k\left(\frac{q_1}{d_1} + \frac{q_2}{d_2}\right) = \frac{q}{4\pi\epsilon_0}\left(\frac{1}{x} + \frac{-3}{d-x}\right) = 0$$
$$\Longrightarrow x = d/4.$$
Since, $d = 24.0$ cm, therefore, $x = d/4 = 6.00$ cm.
(b) Similarly, for $x < 0$ the separation between q_1 and a point on the x axis whose coordinate is x is given by $d_1 = -x$; while the corresponding separation for q_2 is $d_2 = d - x$. Now,
$$V(x) = k\left(\frac{q_1}{d_1} + \frac{q_2}{d_2}\right) = \frac{q}{4\pi\epsilon_0}\left(\frac{1}{-x} + \frac{-3}{d-x}\right) = 0$$
$$\Longrightarrow x = -\frac{d}{2}$$
Since, $d = 24.0$ cm, therefore, $x = -d/2 = -12.0$ cm.

15. Since according to the problem statement there is a point in between the two charges on the x axis where the net electric field is zero, the fields at that point due to q_1 and q_2 must be directed opposite to each other. This means that q_1 and q_2 must have the same sign (i.e., either both are positive or both negative). Thus, the potentials due to either of them must be of the same sign. Therefore, the net electric potential cannot possibly be zero anywhere except at infinity.

16. (a) Since the charges are equal and placed symmetrically, $F = 0$
(b) $F = qE = 0, E = 0.$

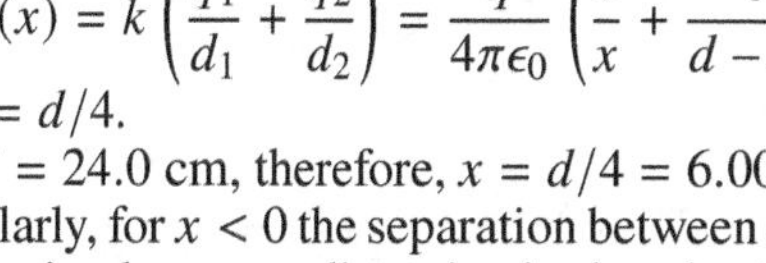

Figure 2.183

(c) $V = 2k\frac{q}{r}$
$$= 2(8.99 \times 10^9 \text{ N}\cdot\text{m}^2/\text{C}^2)\left(\frac{2.00 \times 10^{-6}\text{C}}{0.800 \text{ m}}\right)$$
$$= 4.50 \times 10^4 \text{ V} = 45.0 \text{ kV}$$

17. (a) $E_x = \frac{kq_1}{x^2} + \frac{kq_2}{(x-2.00)^2} = 0$
here, $q_1 = q$ and $q_2 = -2q$. Therefore, corresponding to $E_x = 0$ above equation becomes-

$$E_x = k\left(\frac{+q}{x^2} + \frac{-2q}{(x-2.00)^2}\right) = 0.$$

Dividing by k, we get-
$$2qx^2 = q(x-2.00)^2 \implies x^2 + 4.00x - 4.00 = 0.$$
Therefore $E_x = 0$ when
$$x = \frac{-4.00 \pm \sqrt{16.0 + 16.0}}{2} = -4.83 \text{ m}$$

(Note that the positive root does not correspond to a physically valid situation.)

(b) Since, zero potential points are always closer to that charge which is smaller in magnitude, therefore either zero potential point lies between the charges q and $-2q$ or at the left of q.

Now, corresponding to zero potential, we have
$$V = \frac{kq_1}{|x|} + \frac{kq_2}{|2.00-x|} = 0 \implies k\left(\frac{+q}{|x|} - \frac{2q}{|2.00-x|}\right) = 0.$$
$$\implies \frac{1}{|x|} = \frac{2}{|2.00-x|} \implies 2|x| = |2.00-x|$$
$$\implies 2x = \pm(2.00-x) \quad (\because \text{ if } |x| = |y|, \text{ then } x = \pm y)$$
First considering positive sign, we get
$$2x = (2.00 - x) \implies x = 0.667 \text{ m}$$
Now, considering negative sign, we get
$$2x = -(2.00 - x) \implies x = -2.00 \text{ m}$$
Thus, $V = 0$ at $x = \begin{cases} 0.667 \text{ m} & \text{when } 0 \le x \le 2.00 \\ -2.00 \text{ m} & \text{when } x < 0 \end{cases}$

18. $V = \sum_i k\frac{q_i}{r_i}$
$= (8.99 \times 10^9)(7.00 \times 10^{-6})\left[\frac{-1}{0.0100} - \frac{1}{0.0100} + \frac{1}{0.0387}\right]$
$= -1.10 \times 10^7 \text{ V} = -11.0 \text{ MV}$

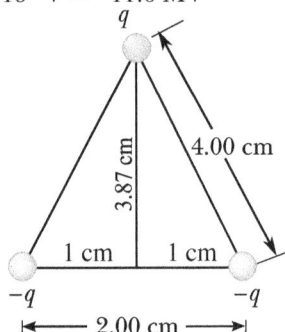

Figure 2.184

19. **(a)** $U = \frac{qQ}{4\pi\epsilon_0 r}$
$= \frac{(5.00\times 10^{-9}\text{C})(-3.00\times 10^{-9}\text{C})(8.99\times 10^9 \text{ V·m/C})}{(0.350 \text{ m})}$
$= -3.86 \times 10^{-7} \text{ J}$

The minus sign means it takes 3.86×10^{-7} J to pull the two charges apart from 35 cm to a much larger separation.

(b) $V = \frac{Q_1}{4\pi\epsilon_0 r_1} + \frac{Q_2}{4\pi\epsilon_0 r_2}$
$= \frac{(5.00 \times 10^{-9}\text{C})(8.99 \times 10^9 \text{ V·m/C})}{0.175 \text{ m}}$
$+ \frac{(-3.00 \times 10^{-9}\text{C})(8.99 \times 10^9 \text{ V.m/C})}{0.175 \text{ m}} = 103 \text{ V}$

20. From Fig.2.185, we have-
$U = U_1 + U_2 + U_3 + U_4$
$U = 0 + U_{12} + (U_{13} + U_{23}) + (U_{14} + U_{24} + U_{34})$
$U = 0 + \frac{kQ^2}{s} + \frac{kQ^2}{s}\left(\frac{1}{\sqrt{2}} + 1\right) + \frac{kQ^2}{s}\left(1 + \frac{1}{\sqrt{2}} + 1\right)$
$U = \frac{kQ^2}{s}\left(4 + \frac{2}{\sqrt{2}}\right) = 5.41\frac{kQ^2}{s}$

An alternate way to get the term $\left(4 + \frac{2}{\sqrt{2}}\right)$ is to recognize

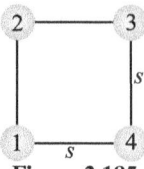

Figure 2.185

that there are 4 side pairs and 2 face diagonal pairs.

21. Consider the two spheres as a system.
(a) Conservation of momentum:
$$0 = m_1 v_1 \hat{i} + m_2 v_2(-\hat{i}) \implies v_2 = \frac{m_1 v_1}{m_2}$$
By conservation of energy,
$$0 = \frac{k(-q_1)q_2}{d} = \frac{1}{2}m_1 v_1^2 + \frac{1}{2}m_2 v_2^2 + \frac{k(-q_1)q_2}{r_1 + r_2}$$
and
$$\frac{kq_1q_2}{r_1+r_2} - \frac{kq_1q_2}{d} = \frac{1}{2}m_1 v_1^2 + \frac{1}{2}\frac{m_1^2 v_1^2}{m_2}$$
Solving for v_1, we get
$$v_1 = \sqrt{\frac{2m_2 kq_1 q_2}{m_1(m_1 + m_2)}\left(\frac{1}{r_1+r_2} - \frac{1}{d}\right)}$$
Substituting the given value in above expression, we get-
$$v_1 = 10.8 \text{ m/s},$$
and $v_2 = \frac{m_1 v_1}{m_2} = \frac{0.100 \text{ kg}(10.8 \text{ m/s})}{0.700 \text{ kg}} = 1.55 \text{ m/s}$

(b) If the spheres are conductor, electrons will move around on them with negligible energy loss to place the centers of excess charge on the insides of the spheres. Then just before they touch, the effective distance between charges will be less than $r_1 + r_2$ and the spheres will really be moving faster than calculated in (a).

22. Consider the two spheres as a system.
(a) By conservation of momentum:
$$0 = m_1 v_1 \hat{i} + m_2 v_2(-\hat{i}) \implies v_2 = \frac{m_1 v_1}{m_2}$$
and by conservation of mechanical energy,
$$0 + \frac{k(-q_1)q_2}{d} = \frac{1}{2}m_1 v_1^2 + \frac{1}{2}m_2 v_2^2 + \frac{k(-q_1)q_2}{r_1 + r_2}$$
Substituting the value of v_2 and then rearranging terms, we get
$$\frac{kq_1q_2}{r_1+r_2} - \frac{kq_1q_2}{d} = \frac{1}{2}m_1 v_1^2 + \frac{1}{2}\frac{m_1^2 v_1^2}{m_2}$$
Solving for v_1, we get
$$v_1 = \sqrt{\frac{2m_2 kq_1 q_2}{m_1(m_1 + m_2)}\left(\frac{1}{r_1+r_2} - \frac{1}{d}\right)}$$
Now, substituting this value of v_1 in expression for v_2, we get
$$v_2 = \left(\frac{m_1}{m_2}\right)v_1 = \sqrt{\frac{2m_1 kq_1 q_2}{m_2(m_1 + m_2)}\left(\frac{1}{r_1+r_2} - \frac{1}{d}\right)}$$

(b) If the spheres are conductor, electrons will move around on them with negligible energy loss to place the centers of excess charge on the insides of the spheres. Then just before they touch, the effective distance between charges will be less than $r_1 + r_2$ and the spheres will really be moving faster than calculated in (a).

23. **(a)** In an empty universe, the 20 nC charge can be placed at its location with no energy investment. At a distance of 4 cm, it creates a potential
$$V_1 = \frac{kq_1}{r} = \frac{(8.99 \times 10^9 \text{ N·m}^2/\text{C}^2)(20 \times 10^{-9}\text{C})}{0.04 \text{ m}}$$
$= 4.50 \text{kV}$

2.16. ANSWER KEYS AND SOLUTIONS

To place the 10 nC charge there we must put in energy
$$U_{12} = q_2 V_1 = \left(10 \times 10^{-9} \text{C}\right)\left(4.5 \times 10^3 \text{ V}\right)$$
$$= 4.50 \times 10^{-5} \text{ J}$$
Next, to bring up the -20 nC charge requires energy
$$U_{23} + U_{13} = q_3 V_2 + q_3 V_1 = q_3 (V_2 + V_1)$$
$$= -20 \times 10^{-9} \text{C} \left(8.99 \times 10^9 \text{ N} \cdot \text{m}^2/\text{C}^2\right)$$
$$\left(\frac{10 \times 10^{-9} \text{C}}{0.04 \text{ m}} + \frac{20 \times 10^{-9} \text{C}}{0.08 \text{ m}}\right)$$
$$= -4.50 \times 10^{-5} \text{ J} - 4.50 \times 10^{-5} \text{ J} = -9.0 \times 10^{-5} \text{ J}$$
The total energy of the three charges is
$$U_{12} + U_{23} + U_{13} = -4.50 \times 10^{-5} \text{ J}$$
(b) The three fixed charges create this potential at the location where the fourth is released:
$$V = V_1 + V_2 + V_3$$
$$= \left(8.99 \times 10^9 \text{ N} \cdot \text{m}^2/\text{C}^2\right)$$
$$\left(\frac{20 \times 10^{-9}}{\sqrt{0.04^2 + 0.03^2}} + \frac{10 \times 10^{-9}}{0.03} - \frac{20 \times 10^{-9}}{0.05}\right) \text{C/m}$$
$$= 3.00 \times 10^3 \text{ V}$$ Energy of the system of four charged objects is conserved as the fourth charge flies away:
$$\left(\frac{1}{2}mv^2 + qV\right)_i = \left(\frac{1}{2}mv^2 + qV\right)_f$$
$$\implies 0 + \left(40 \times 10^{-9} \text{C}\right)\left(3.00 \times 10^3 \text{ V}\right)$$
$$= \frac{1}{2}\left(2.00 \times 10^{-13} \text{ kg}\right) v^2 + 0$$
Solving for v, we get
$$v = \sqrt{\frac{2 \left(1.20 \times 10^{-4} \text{ J}\right)}{2 \times 10^{-13} \text{ kg}}} = 3.46 \times 10^4 \text{ m/s}$$

24. The original electrical potential energy is
$$U = qV = q\frac{kq}{d} = \frac{kq^2}{d}$$
In the final configuration we have mechanical equilibrium. If k_s is the spring constant, then the spring and electrostatic forces on each charge are $-k_s(2d)$ and $q\frac{kq}{(3d)^2}$ respectively.
In mechanical equilibrium, we have-
$$-k_s(2d) + q\frac{kq}{(3d)^2} = 0 \implies k_s = \frac{kq^2}{18d^3}$$
In the final configuration the total potential energy is-
$$\frac{1}{2}k_s x^2 + qV = \frac{1}{2}\frac{kq^2}{18d^3}(2d)^2 + q\frac{kq}{3d} = \frac{4}{9}\frac{kq^2}{d}$$
The missing energy must have become internal energy, as the system is isolated:
$$\frac{kq^2}{d} = \frac{4kq^2}{9d} + \Delta E_{int} \implies \Delta E_{int} = \frac{5}{9}\frac{kq^2}{d}$$

25. (a) **Calculation of potential on x-axis**
Due to both charges, the net potential $V(x)$ on the x-axis at distance x from the origin is given by
$$V(x) = \frac{kQ_1}{r_1} + \frac{kQ_2}{r_2} = \frac{k(+Q)}{\sqrt{x^2 + a^2}} + \frac{k(+Q)}{\sqrt{x^2 + (-a)^2}}$$
$$= \frac{2kQ}{\sqrt{x^2 + a^2}} = \frac{kQ}{a}\left(\frac{2}{\sqrt{(x/a)^2 + 1}}\right)$$
$$\implies \frac{V(x)}{(kQ/a)} = \frac{2}{\sqrt{(x/a)^2 + 1}}$$
Potential $V(x)$ vs x graph is plotted in Fig.2.186.

(b) Similarly, the net potential $V(y)$ on y-axis at distance y from origin is given by

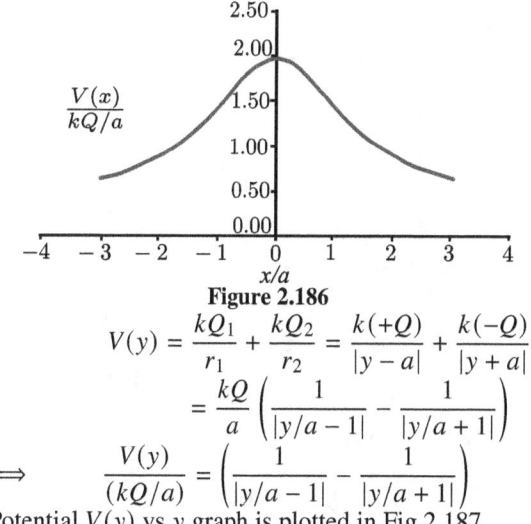

Figure 2.186

$$V(y) = \frac{kQ_1}{r_1} + \frac{kQ_2}{r_2} = \frac{k(+Q)}{|y-a|} + \frac{k(-Q)}{|y+a|}$$
$$= \frac{kQ}{a}\left(\frac{1}{|y/a - 1|} - \frac{1}{|y/a + 1|}\right)$$
$$\implies \frac{V(y)}{(kQ/a)} = \left(\frac{1}{|y/a - 1|} - \frac{1}{|y/a + 1|}\right)$$
Potential $V(y)$ vs y graph is plotted in Fig.2.187.

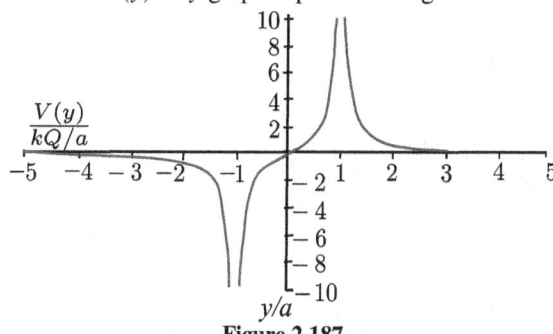

Figure 2.187

26. Using conservation of energy for the alpha particle-nucleus system, we have
$$K_f + U_f = K_i + U_i$$
But $U_i = \frac{kq_\alpha q_{\text{gold}}}{r_i}$ and $r_i \approx \infty$.
Thus, $U_i = 0$, $K_f = 0$ (since, at turning point, $v_f = 0$)
Therefore, $U_f = K_i \implies \frac{kq_\alpha q_{\text{gold}}}{r_{\min}} = \frac{1}{2}m_\alpha v_\alpha^2$
Solving for $r_{\min}$, we get
$$r_{\min} = \frac{2kq_\alpha q_{\text{gold}}}{m_\alpha v_\alpha^2}$$
$$= \frac{2\left(8.99 \times 10^9 \text{ N} \cdot \text{m}^2/\text{C}^2\right)(2)(79)\left(1.60 \times 10^{-19}\text{C}\right)^2}{\left(6.64 \times 10^{-27} \text{ kg}\right)\left(2.00 \times 10^7 \text{ m/s}\right)^2}$$
$$= 2.74 \times 10^{-14} \text{ m} = 27.4 \text{fm}$$

27. Using conservation of mechanical energy, we have:
$$\frac{keQ}{r_1} = \frac{keQ}{r_2} + \frac{1}{2}mv^2$$
which gives: $v = \sqrt{\frac{2keQ}{m}\left(\frac{1}{r_1} - \frac{1}{r_2}\right)}$
$$= \left[\frac{(2)\left(8.99 \times 10^9 \text{ N} \cdot \text{m}^2/\text{C}^2\right)\left(-1.60 \times 10^{-19}\text{C}\right)\left(10^{-9}\text{C}\right)}{9.11 \times 10^{-31} \text{ kg}}\right.$$
$$\left.\left(\frac{1}{0.0300 \text{ m}} - \frac{1}{0.0200 \text{ m}}\right)\right]^{1/2}$$
$$= 7.26 \times 10^6 \text{ m/s}$$

28. $U = \sum \frac{kq_i q_j}{r_{ij}}$, summed over all pairs of (i, j) where $i \neq j$
Fig.2.188.
$$U = k\left[\frac{q(-2q)}{b} + \frac{(-2q)(3q)}{a} + \frac{(2q)(3q)}{b} + \frac{q(2q)}{a}\right]$$

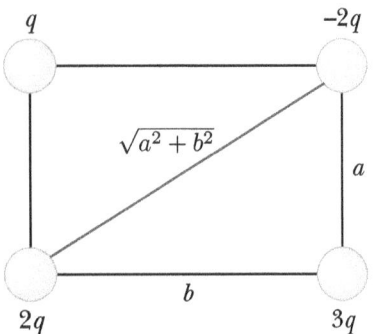

Figure 2.188

$$+ \frac{q(3q)}{\sqrt{a^2+b^2}} + \frac{2q(-2q)}{\sqrt{a^2+b^2}}\right]$$

Substituting the given values, we get

$$U = kq^2 \left[\frac{-2}{0.400} - \frac{6}{0.200} + \frac{6}{0.400} + \frac{2}{0.200}\right.$$
$$\left. + \frac{3}{0.447} - \frac{4}{0.447}\right]$$

$$U = (8.99 \times 10^9)(6.00 \times 10^{-6})^2 \left[\frac{4}{0.400} - \frac{4}{0.200} - \frac{1}{0.447}\right]$$

$$= -3.96 \text{ J}$$

29. Since, the effective force on each charge is diagonally away, therefore each charge moves off on its diagonal line. All charges have equal speeds.
By conservation of mechanical energy, we have

$$\sum (K + U)_i = \sum (K + U)_f$$

$$\Rightarrow \quad 0 + \frac{4kq^2}{L} + \frac{2kq^2}{\sqrt{2}L} = 4\left(\frac{1}{2}mv^2\right) + \frac{4kq^2}{2L} + \frac{2kq^2}{2\sqrt{2}L}$$

$$\Rightarrow \quad \left(2 + \frac{1}{\sqrt{2}}\right)\frac{kq^2}{L} = 2mv^2 \Rightarrow v = \sqrt{\left(1 + \frac{1}{\sqrt{8}}\right)\frac{kq^2}{mL}}$$

30. A cube has 12 edges and 6 faces. Consequently, there are 12 edge pairs separated by s, $2 \times 6 = 12$ face diagonal pairs separated by $\sqrt{2}s$ and 4 interior diagonal pairs separated $\sqrt{3}s$.

$$U = \frac{kq^2}{s}\left[12 + \frac{12}{\sqrt{2}} + \frac{4}{\sqrt{3}}\right] = 22.8\frac{kq^2}{s}$$

31. We can use $W_{q \to \text{final position}} = Q\Delta V_{i \to f}$ to find the work required to move these charges between the given points.
(a) Express the required work in terms of the charge being moved and the potential due to the charge at $x = +a$ and simplify to obtain:

$$W_{+Q \to +a} = Q\Delta V_{\infty \to +a} = Q[V(a) - V(\infty)]$$
$$= QV(a) = Q\left(\frac{kQ}{2a}\right) = \frac{kQ^2}{2a}$$

(b) Express the required work in terms of the charge being moved and the potentials due to the charges at $x = +a$ and $x = -a$ and simplify to obtain:

$$W_{-Q \to 0} = -Q\Delta V_{\infty \to 0} = -Q[V(0) - V(\infty)]$$
$$= -QV(0) = -Q\left[V_{\text{due to charge} \atop \text{at } x=-a} + V_{\text{due to charge} \atop \text{at } x=+a}\right]$$
$$= -Q\left(\frac{kQ}{at} + \frac{kQ}{a}\right) = \frac{-2kQ^2}{a}$$

(c) Express the required work in terms of the charge being moved and the potentials due to the charges at $x = +a$ and $x = -a$ and simplify to obtain:

$$W_{-Q \to 2a} = -Q\Delta V_{0 \to 2a} = -Q[V(2a) - V(0)]$$
$$= -Q\left[V_{\text{due to charge} \atop \text{at } x=-a} + V_{\text{due to charge} \atop \text{at } x=+a} - V(0)\right]$$

$$= -Q\left(\frac{kQ}{3a} + \frac{kQ}{a} - \frac{2kQ}{a}\right) = \frac{2kQ^2}{3a}$$

32. $\Delta K = q|\Delta V| \Rightarrow 7.37 \times 10^{-17} = q(115)$
$\Rightarrow q = 6.41 \times 10^{-19}$ C

33. (a) Energy of the proton-field system is conserved as the proton moves from high to low potential, which can be defined for this problem as moving from 120 V down to 0 V.

$$K_i + U_i + \Delta E_{\text{mech}} = K_f + U_f$$

$$\Rightarrow \quad 0 + qV + 0 = \frac{1}{2}mv_p^2 + 0$$

$$\Rightarrow (1.60 \times 10^{-19}\text{C})(120 \text{ V}) = \frac{1}{2}(1.67 \times 10^{-27} \text{ kg})v_p^2$$

$$\Rightarrow \quad v_p = 1.52 \times 10^5 \text{ m/s}$$

(b) The electron will gain speed in moving the other way, from $V_i = 0$ to $V_f = 120$ V :

$$K_i + U_i + \Delta E_{\text{mech}} = K_f + U_f$$

$$\Rightarrow \quad 0 + 0 + 0 = \frac{1}{2}mv_e^2 + qV$$

$$\Rightarrow \quad 0 = \frac{1}{2}(9.11 \times 10^{-31} \text{ kg})v_e^2$$
$$+ (-1.60 \times 10^{-19}\text{C})(120 \text{ J/C})$$

$$\Rightarrow \quad v_e = 6.49 \times 10^6 \text{ m/s}$$

34. By work-kinetic energy theorem, we have

$$W = \Delta K \Rightarrow -q\Delta V = \frac{1}{2}mv_2^2 - \frac{1}{2}mv_1^2$$

$$\Rightarrow \quad -(-1.60 \times 10^{-19}\text{C})\Delta V$$
$$= 0 - \frac{1}{2}(9.11 \times 10^{-31} \text{ kg})(4.20 \times 10^5 \text{ m/s})^2$$

From which we get, $\Delta V = -0.502$ V.

35. $V = a + bx = 10.0 \text{ V} + (-7.00 \text{ V/m})x$
(a) At $x = 0$, $V = 10.0$ V
At $x = 3.00$ m, $V = -11.0$ V
At $x = 6.00$ m, $V = -32.0$ V
(b) $E = -\frac{dV}{dx} = -b = -(-7.00 \text{ V/m}) = 7.00$ N/C in the $+x$ direction

36. (a) For $r < R$, $V = \frac{kQ}{R}$ and $E_r = -\frac{dV}{dr} = 0$
(b) For $r \geq R$, we have $V = \frac{kQ}{r}$
and $E_r = -\frac{dV}{dr} = -\left(-\frac{kQ}{r^2}\right) = \frac{kQ}{r^2}$

37. $V = 5x - 3x^2y + 2yz^2$
Evaluate E at $(1, 0, -2)$

$$E_x = -\frac{\partial V}{\partial x} = -5 + 6xy = -5 + 6(1)(0) = -5$$

$$E_y = -\frac{\partial V}{\partial y} = +3x^2 - 2z^2 = 3(1)^2 - 2(-2)^2 = -5$$

$$E_z = -\frac{\partial V}{\partial z} = -4yz = -4(0)(-2) = 0$$

$$E = \sqrt{E_x^2 + E_y^2 + E_z^2} = \sqrt{(-5)^2 + (-5)^2 + 0^2} = 7.07 \text{ N/C}$$

38. (a) $E_A > E_B$ since $E = \frac{\Delta V}{\Delta s}$

(b) $E_B = -\frac{\Delta V}{\Delta s} = -\frac{(6-2)\text{V}}{2 \text{ cm}} = 200$ N/C down

(c) Electric field lines are shown in Fig.2.189.

39. $\Delta V = V_{2R} - V_0$

$$= \frac{kQ}{\sqrt{R^2 + (2R)^2}} - \frac{kQ}{R} = \frac{kQ}{R}\left(\frac{1}{\sqrt{5}} - 1\right) = -0.553\frac{kQ}{R}$$

40. The potential is

$$V_P = \frac{1}{4\pi\epsilon_0}\int_{\text{rod}}\frac{dq}{R} = \frac{1}{4\pi\epsilon_0 R}\int_{\text{rod}}dq = \frac{-Q}{4\pi\epsilon_0 R}$$

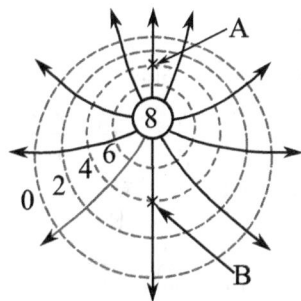

Figure 2.189

$$= -\frac{(8.99 \times 10^9 \text{ N} \cdot \text{m}^2/\text{C}^2)(25.6 \times 10^{-12}\text{C})}{3.71 \times 10^{-2} \text{ m}} = -6.20 \text{ V}$$

We note that the result is exactly what one would expect for a point-charge $-Q$ at a distance R. This "coincidence" is due, in part, to the fact that V is a scalar quantity.

41. (a) All the charge is at the same distance R from C, so the electric potential at C is
$$V = \frac{1}{4\pi\epsilon_0}\left(\frac{Q_1}{R} - \frac{6Q_1}{R}\right) = -\frac{5Q_1}{4\pi\epsilon_0 R}$$
$$= -\frac{5(8.99 \times 10^9 \text{ N} \cdot \text{m}^2/\text{C}^2)(4.20 \times 10^{-12}\text{C})}{8.20 \times 10^{-2} \text{ m}} = -2.30 \text{ V}$$
where the zero was taken to be at infinity.

(b) All the charge is at the same distance from P. That distance is $\sqrt{R^2 + D^2}$, so the electric potential at P is
$$V = \frac{1}{4\pi\epsilon_0}\left[\frac{Q_1}{\sqrt{R^2+D^2}} - \frac{6Q_1}{\sqrt{R^2+D^2}}\right] = -\frac{5Q_1}{4\pi\epsilon_0\sqrt{R^2+D^2}}$$
$$= -\frac{5(8.99 \times 10^9 \text{ N} \cdot \text{m}^2/\text{C}^2)(4.20 \times 10^{-12}\text{C})}{\sqrt{(8.20 \times 10^{-2} \text{ m})^2 + (6.71 \times 10^{-2} \text{ m})^2}} = -1.78 \text{ V}$$

42. Letting d denote 0.010 m, we have
$$V = \frac{Q_1}{4\pi\epsilon_0 d} + \frac{3Q_1}{8\pi\epsilon_0 d} - \frac{3Q_1}{16\pi\epsilon_0 d} = \frac{Q_1}{8\pi\epsilon_0 d}$$
$$= \frac{(8.99 \times 10^9 \text{ N} \cdot \text{m}^2/\text{C}^2)(30 \times 10^{-9}\text{C})}{2(0.01 \text{ m})} = 1.3 \times 10^4 \text{ V}$$

43. (a) $[\alpha] = \left[\frac{\lambda}{x}\right] = \frac{C}{m} \cdot \left(\frac{1}{m}\right) = \frac{C}{m^2}$

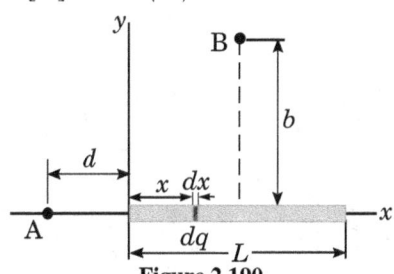

Figure 2.190

(b) $V = k\int \frac{dq}{r} = k\int \frac{\lambda dx}{d+x} = k\alpha \int_0^L \frac{xdx}{d+x}$
$$= k\alpha\left[L - d\ln\left(1 + \frac{L}{d}\right)\right]$$

44. $V = \int \frac{kdq}{r} = k\int \frac{\alpha x dx}{\sqrt{b^2 + (L/2 - x)^2}}$

Let $z = \frac{L}{2} - x$, then $x = \frac{L}{2} - z$, and $dx = -dz$, therefore
$$V = k\alpha \int \frac{(L/2 - z)(-dz)}{\sqrt{b^2 + z^2}} = -\frac{k\alpha L}{2}\int \frac{dz}{\sqrt{b^2 + z^2}}$$
$$+ k\alpha \int \frac{zdz}{\sqrt{b^2 + z^2}}$$

$$= -\frac{k\alpha L}{2}\ln\left(z + \sqrt{z^2 + b^2}\right) + k\alpha\sqrt{z^2 + b^2}$$

$$V = -\frac{k\alpha L}{2}\ln\left[\left(\frac{L}{2} - x\right) + \sqrt{\left(\frac{L}{2} - x\right)^2 + b^2}\right]_0^L$$

$$+ k\alpha\sqrt{\left(\frac{L}{2} - x\right)^2 + b^2}\Bigg|_0^L$$

$$V = -\frac{k\alpha L}{2}\ln\left[\frac{L/2 - L + \sqrt{(L/2)^2 + b^2}}{L/2 + \sqrt{(L/2)^2 + b^2}}\right]$$

$$+ k\alpha\left[\sqrt{\left(\frac{L}{2} - L\right)^2 + b^2} - \sqrt{\left(\frac{L}{2}\right)^2 + b^2}\right]$$

$$V = \left[-\frac{k\alpha L}{2}\ln\left[\frac{\sqrt{b^2 + (L^2/4)} - L/2}{\sqrt{b^2 + (L^2/4)} + L/2}\right]\right]$$

45. $dV = \frac{kdq}{\sqrt{r^2 + x^2}}$, where $dq = \sigma dA = \sigma 2\pi rdr$
$$V = 2\pi\sigma k \int_a^b \frac{rdr}{\sqrt{r^2 + x^2}} = 2\pi k\sigma\left[\sqrt{x^2 + b^2} - \sqrt{x^2 + a^2}\right]$$
[see Fig.2.191]

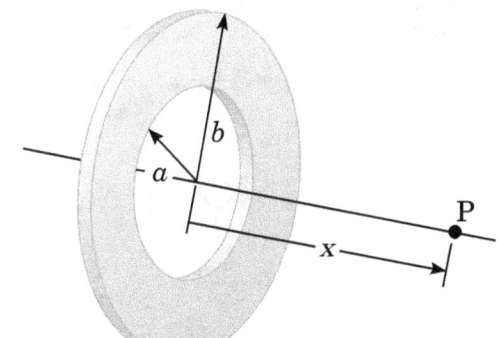

Figure 2.191: Problem 45

46. $V = k\int_{\text{all charge}} \frac{dq}{r} = k\int_{-3R}^{-R} \frac{\lambda dx}{-x} + k\int_{\text{semicircle}} \frac{\lambda ds}{R}$
$$+ k\int_R^{3R} \frac{\lambda dx}{x}$$
$$= -k\lambda \ln(-x)\big|_{-3R}^{-R} + \frac{k\lambda}{R}\pi R + k\lambda \ln x\big|_R^{3R}$$
$$\implies V = k\ln\frac{3R}{R} + k\lambda\pi + k\ln 3 = k\lambda(\pi + 2\ln 3)$$

47. Substituting given values into $V = kq/r$, we get
$$7.50 \times 10^3 \text{ V} = \frac{(8.99 \times 10^9 \text{ N} \cdot \text{m}^2/\text{C}^2) q}{0.300 \text{ m}}$$
$$\implies q = 2.50 \times 10^{-7}\text{C}$$

Now, from $q = Ne$, number of electrons removed from spherical conductor:
$$N = \frac{q}{e} = \frac{2.50 \times 10^{-7}\text{C}}{1.60 \times 10^{-19}\text{C}} = 1.56 \times 10^{12} \text{ electrons}$$

48. (a) $E_{max} = 3.00 \times 10^6 \text{ V/m} = \frac{kQ}{r^2} = \frac{kQ}{r}\left(\frac{1}{r}\right) = V_{max}\left(\frac{1}{r}\right)$

$V_{max} = E_{max}r = 3.00 \times 10^6(0.150) = 450$ kV

(b) $\frac{kQ_{max}}{r^2} = E_{max}$ { or $\frac{kQ_{max}}{r} = V_{max}$}

or $Q_{max} = \frac{E_{max}r^2}{k} = \frac{3.00 \times 10^6(0.150)^2}{8.99 \times 10^9} = 7.51$ μC

49. The potential created by the ring at the electron's starting

point is
$$V_i = \frac{kQ}{\sqrt{x_i^2 + a^2}} = \frac{k(2\pi\lambda a)}{\sqrt{x_i^2 + a^2}}$$
while at the centre, it is $V_f = 2\pi k\lambda$. From conservation of mechanical energy, we have
$$0 + (-eV_i) = \frac{1}{2}m_e v_f^2 + (-eV_f)$$
$$\implies v_f^2 = \frac{2e}{m_e}(V_f - V_i) = \frac{4\pi e k\lambda}{m_e}\left(1 - \frac{a}{\sqrt{x_i^2 + a^2}}\right)$$
$$\implies v_f^2 = \frac{4\pi (1.60 \times 10^{-19})(8.99 \times 10^9)(1.00 \times 10^{-7})}{9.11 \times 10^{-31}}$$
$$\left(1 - \frac{0.200}{\sqrt{(0.100)^2 + (0.200)^2}}\right)$$
$$\implies v_f = \sqrt{1.45 \times 10^7} \text{ m/s}$$

50. Take the illustration presented with the problem as an initial picture (Fig.2.124). No external horizontal forces act on the set of four balls, so its centre of mass stays fixed at the location of the centre of the square. As the charged balls 1 and 2 swing out and away from each other, balls 3 and 4 move up with equal y-components of velocity. The maximum-kinetic energy state is shown in Fig.2.192. By conservation of mechanical energy, we have-
$$\frac{kq^2}{a} = \frac{kq^2}{3a} + \frac{1}{2}mv^2 + \frac{1}{2}mv^2 + \frac{1}{2}mv^2 + \frac{1}{2}mv^2$$

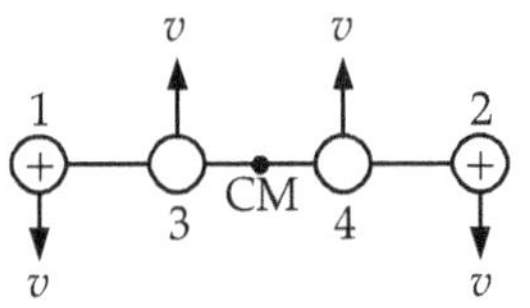

Figure 2.192

$$\implies \frac{2kq^2}{3a} = 2mv^2 \implies v = \sqrt{\frac{kq^2}{3am}}$$

51. For an element of area which is a ring of radius r and width dr,
$$dV = \frac{k\,dq}{\sqrt{r^2 + x^2}}$$
Here, $dq = \sigma dA = Cr(2\pi r\,dr)$, therefore-
$$V = C(2\pi k)\int_0^R \frac{r^2\,dr}{\sqrt{r^2 + x^2}}$$
$$= C(\pi k)\left[R\sqrt{R^2 + x^2} + x^2 \ln\left(\frac{x}{R + \sqrt{R^2 + x^2}}\right)\right]$$

52. **APPROACH** The work done by the electric field of the gold nucleus changes the kinetic energy of the alpha particle-eventually bringing it to rest. We can apply the work-kinetic energy theorem to derive an expression for the distance of closest approach. Because the repulsive Coulomb force $\vec{F}_e$ varies with distance, we'll have to evaluate $\int \vec{F}_e \cdot d\vec{r}$ in order to find the work done on the alpha particles by this force.
SOLUTION Apply the work-kinetic energy theorem to the alpha particle to obtain:
$$W_{net} = \int_\infty^{r_{min}} \vec{F}_e \cdot d\vec{r} = \Delta K$$
since, $\int_\infty^{r_{min}} \vec{F}_e \cdot d\vec{r} = -\int_\infty^{r_{min}} \frac{k(2e)(79e)}{r^2}dr$ and $K_f = 0$

Therefore, $-158ke^2 \int_\infty^{r_{min}} \frac{dr}{r^2} = -K_i$
Evaluating the integral yields:
$$-158ke^2\left[\frac{1}{r}\right]_\infty^{r_{min}} = -\frac{158ke^2}{r_{min}} = -K_i$$
Solving for r_{min} we get:
$$r_{min} = \frac{158ke^2}{K_i}$$
Substituting numerical values in above expression, we get
$$r_{min} = \frac{158\left(8.988 \times 10^9 \frac{N\cdot m^2}{C^2}\right)\left(1.602 \times 10^{-19}C\right)^2}{5.0 \text{ Mev} \times \frac{1.602 \times 10^{-19} J}{eV}}$$
$$= 4.6 \times 10^{-14} \text{ m}$$

2.16.6 Multiple Choice Assignments

2.16.6.1 Level 1

Q.No.	1	2	3	4	5	6	7	8	9
Ans.	A	B	A	C	C	A	B	B	D
Q.No.	10	11	12	13	14	15	16	17	18
Ans.	D	B	A	B	A	C	B	D	A
Q.No.	19	20	21	22	23	24	25	26	27
Ans.	D	D	B	A	A	C	A	B	B
Q.No.	28	29	30	31	32	33	34	35	36
Ans.	A	B	A	B	A	C	B	A	C
Q.No.	37	38	39						
Ans.	B	B	B						

2.16.6.2 Level 2

Q.No.	1	2	3	4	5	6	7	8	9
Ans.	B	B	C	D	B	C	B	D	B
Q.No.	10	11	12	13	14	15	16	17	18
Ans.	B	D	C	B	D	A	C		

2.16.6.3 Level 3

Q.No.	1	2	3	4	5	6	7	8	9
Ans.	A	A	B	B	B	B	A	B	A
Q.No.	10	11	12	13	14	15	16	17	18
Ans.	A	C	A	C	B	D	B	C	B
Q.No.	19	20	21	22	23	24	25	26	27
Ans.	C	A	A	C					

2.16.6.4 Level 4
Section A

Q.No.	1	2	3	4	5	6	7	8	9
Ans.	A	B	D	B	C	A	B	D	D
Q.No.	10	11	12	13	14	15	16	17	18
Ans.	D	C	C	D	A	B	D	D	B
Q.No.	19	20	21	22	23	24	25	26	27
Ans.	C	D	C	B	D	B	B	D	B
Q.No.	28	29	30	31	32	33	34	35	36
Ans.	B	D	C	B	A	C	D	C	D
Q.No.	37	38	39	40	41	42	43	44	45
Ans.	A	B	D	B	D	B	D	C	A
Q.No.	46	47	48						
Ans.	C	5	160						

Section B

Q.No.	1	2	3	4	5	6	7	8
Ans.	B	A	A	D	D	B	B	B
Q.No.	9	10	11	12	13	14	15	16
Ans.	C	C	A	B	ABCD	C	B	C

2.16. ANSWER KEYS AND SOLUTIONS

Q.No.	17	18	19	20	21	22	23	24
Ans.	D	CD	AB	2	6	2	1.73	3.00
Q.No.	25	26	27					
Ans.	ABC	ACD	B					

This page intentionally left blank

Chapter 3
Capacitance and Dielectrics

3.1 Capacitor

A **capacitor** is a device that stores energy in the electrostatic field between its plates. When connected to a battery, it charges up by drawing energy relatively slowly because the charging current is limited by the internal resistance of the battery and the circuit. However, once charged, it can release this energy extremely quickly — sometimes in just a few milliseconds — if connected across a low-resistance path such as a bulb. The smaller the resistance in the discharge path, the smaller the time constant $\tau = RC$*, and the faster the capacitor releases its stored energy. This is why even a small capacitor can produce a brief but intense burst of power.

This property of rapid energy release has many important applications. In laboratories, very large capacitors are used to produce short, intense laser pulses. In some fusion experiments, they deliver power levels of the order of 10^{14} W, although the pulses last for only about 10^{-9} s. At the other extreme, very tiny capacitors are used in the memory cells of computers to store the binary "ones" and "zeros" in random access memory (RAM). Capacitors are also used to create nearly uniform electric fields, such as those that deflect electron beams in television or oscilloscope tubes. In electrical circuits, they help smooth out sudden voltage fluctuations, reduce electrical noise, and protect sensitive components from damaging surges.

3.2 Capacitance of an Isolated Spherical Conductor

Let us consider an isolated conductor, i.e. the conductor removed from other conductors, bodies, and charges. Experiments show that the charge q of this conductor is directly proportional to its potential V (we assumed that at infinity potential is equal to zero). Consequently, the ratio q/V does not depend on the charge q and has a certain value for each isolated conductor. So, we can write:

$$V \propto q \implies q = CV$$

$$\text{or} \quad \boxed{C = \frac{q}{V}} \quad (3.1)$$

The quantity C is called the electrostatic capacitance of an isolated conductor (or simply capacitance). It is numerically equal to the charge that must be supplied to the conductor in order to increase its potential by unity.

For a spherical conductor having radius R and charge q, the potential at any point on its surface is

$$V = \frac{1}{4\pi\epsilon_0}\frac{q}{R}$$

Therefore, $C = \dfrac{q}{V} = \dfrac{q}{\frac{1}{4\pi\epsilon_0}\frac{q}{R}} = 4\pi\epsilon_0 R$

$$\implies \boxed{C = 4\pi\epsilon_0 R} \quad (3.2)$$

Here, ϵ_0 is the electric permittivity of free space.
But if the spherical conductor is placed in a dielectric medium of dielectric constant κ, then we have

$$\implies \boxed{C = 4\pi\epsilon R = 4\pi\epsilon_0 R} \quad (3.3)$$

here, $\epsilon = \epsilon_0 \kappa$ is the electric permittivity of the dielectric medium.

The capacitance is dependent on the size, shape, and medium in which the conductor is placed.
C is always taken to be positive. Furthermore, the charge q and the potential V are always expressed in Eq.3.1 as positive quantities. The unit of capacitance is coulomb per volt (C/V), but capacitance occurs so frequently that it has been given its own SI unit, the farad (F), in honour of Michael Faraday:

$$\boxed{1\,\text{F} = 1\,\text{C/V}} \quad (3.4)$$

In practice, the farad is inconveniently large unit (spherical conductor of 1 farad must have a radius of 9×10^8 m), and for practical use, units — μF ("micro farad"), nF ("nano farad"), and pF ("pico farad")— are more common.
$1\,\mu\text{F} = 1 \times 10^{-6}$ F, $1\,\text{nF} = 1 \times 10^{-9}$ F, $1\,\text{pF} = 1 \times 10^{-12}$ F.
Sometimes we also use unit, "fF (femtofarad)" for a relatively smallar capacitance: $1\,\text{fF} = 1 \times 10^{-15}$ F.
CGS unit: CGS unit of capacitance is stat-farad.
$\quad 1\,\text{Farad} = 9 \times 10^{11}$ stat-farad
Dimensional Formula: Dimensional formula of capacitance is $[M^{-1}L^{-2}T^4A^2]$

☞ For practical purposes and as a simplification, we consider the surface of Earth as an equipotential surface at zero potential[a]. Therefore, the capacity of a conductor connected to earth will be infinite

$$C = \frac{q}{V} = \frac{q}{0} = \infty$$

In this case, the capacitance of Earth (if conductor-Earth system is assumed to be isolated) will also be infinite because it is connected with the conductor.

[a]We consider Earth's surface to have zero electric potential because we live on Earth and we measure the electric potential with respect to Earth. It is like using "zero" on a ruler to measure distances. This doesn't mean there are no electric fields or charges; it's just a helpful starting point for our calculations. Also note that due to various factors like atmospheric charges, weather conditions, and Earth's magnetic field, and human-made electric systems, Earth's surface is also not an exactly equipotential surface.

EXAMPLE 1. A conductor gets a charge of 50.0 μC when it is connected to a battery of emf 5.0 V. Calculate capacity of the conductor.
APPROACH From Eq.3.1, the capacitance of a conductor, is given by-

$$C = \frac{q}{V} \quad \ldots (1)$$

So, substitute the given values of charge q and voltage (V) [here emf] in above expression and simplify for C.
SOLUTION Given that: $q = 50.0\,\mu$C, $V = 5.0$ V, therefore from Eq.(1), we get -

$$C = \frac{q}{V} = \frac{50.0 \times 10^{-6}}{5.0} = 10.0\,\mu\text{F}$$

EXAMPLE 2. Find the capacitance of an isolated conductor which has the shape of a sphere of radius R (Fig.3.1).
APPROACH It can be seen from Eq.3.1 that for this purpose we

---
*The time constant ($\tau = RC$) would be discussed in subsection 4.30.2.1 of page 371 of chapter "Current Electricity and DC Circuits"

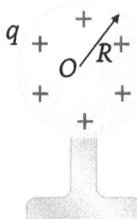

Figure 3.1: An isolated metallic sphere

must mentally charge the conductor by a charge q and calculate its potential V. The electric potential of a metal sphere, having charge q, is given by-

$$V = -\int_\infty^R E_r dr = -\frac{1}{4\pi\epsilon_0}\int_\infty^R \frac{q}{r^2}dr = \frac{1}{4\pi\epsilon_0}\frac{q}{R} \quad (3.5)$$

SOLUTION Substituting this result into Eq.3.1, we get

$$\boxed{C = \frac{q}{V} = 4\pi\epsilon_0 R} \quad (3.6)$$

SI unit of capacitance is 'farad (F)'.
1 F is the capacitance of a conductor whose potential changes by 1V when a charge of 1C is supplied to it.

From Eq.3.6, $R = \dfrac{C}{4\pi\epsilon_0}$ ... (1)

Here, $\dfrac{1}{4\pi\epsilon_0} = 9.0 \times 10^9 \text{ Nm}^2/\text{C}^2$.

For $C = 1$ F, Eq.(1), gives
$$R = 9.0 \times 10^9 \text{Nm}^2/\text{C}^2 \times 1\text{ F} = 9.0 \times 10^9 \text{ m}$$

Thus, a farad is a very large quantity. It corresponds to the capacitance of an isolated sphere of radius 9×10^6 km, which is 1500 times the radius of the Earth (the capacitance of the Earth is 0.7 mF). In actual practice, we encounter capacitances between 1 μF and 1 pF.

☞ **The capacity of a spherical conductor having same radius as that of Earth $(6.4 \times 10^6$ m)**

$$C = 4\pi\epsilon_0 R = \frac{1}{(9.0\times 10^9)}\frac{\text{C}^2}{\text{N.m}^2} \times 6.4 \times 10^6 \text{ m} \approx 711\ \mu\text{F}$$

Note that this is the result for an isolated spherical conductor not for Earth. Earth's capacitance is influenced by various factors like the height of the ionosphere, the distribution of charges and more.

EXAMPLE 3. Capacitance of the Echo Satellite: The Echo I satellite was a metal-coated plastic sphere, 30 m in diameter, that orbited the earth at a mean altitude of 1600 km. Because its altitude was large compared to its radius, it was essentially isolated. Find it's capacitance.

APPROACH From Eq.3.6, the capacitance of an isolated spherical conductor of radius R, is given by-
$$C = 4\pi\epsilon_0 R$$
Now, substitute the values of $4\pi\epsilon_0$ and R in above equation and simplify for C.

SOLUTION Required capacitance of the Echo I satellite -
$$C = 4\pi\epsilon_0 R = 4\pi\epsilon_0 \cdot 15 = \frac{1}{9.0 \times 10^9} \times 15 \text{ F}$$
$$= 1.7 \times 10^{-9} \text{ F}$$

3.2.1 q Versus V Graph

From Eq.3.1, we have:
$$q = CV$$
It is of the form of straight line $y = mx$. So, a graph between V and q will be a straight line with slope $\tan\theta = \dfrac{q}{V} = C$ (see Fig.3.2)

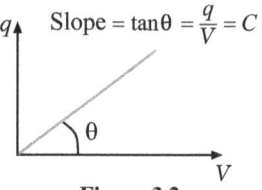

Figure 3.2

EXAMPLE 4. Charge vs voltage graphs for three conductors A, B and C, are shown in Fig.3.3. Which has the greatest capacitance?

SOLUTION From q-V graphs shown in Fig.3.3, it is clear that,

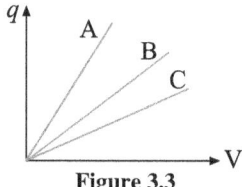

Figure 3.3

the slope of conductor A is greater than the slope of conductor B, which, in turn, is greater than the slope of conductor C. Therefore, the decreasing order of capacitances is $C_A > C_B > C_C$. Thus the conductor A has greatest capacitance.

3.3 Capacitance of System of Two Conductors

If a conductor is not isolated, its capacitance will considerably increase as other bodies approach it. This is due to the fact that the field of the given conductor causes a redistribution of charges on the surrounding bodies, i.e., induces charges on them. Let the charge of the conductor be $q > 0$. Then negative induced charges will be nearer to the conductor than the positive charges. For this reason, the potential of the conductor, which is the algebraic sum of the potentials of its own charge and of the charges induced on other bodies will decrease when other uncharged bodies approach it. This means that its capacitance increases.

This circumstance made it possible to create the system of conductors, which has a considerably higher capacitance than that of an isolated conductor. Moreover, the capacitance of this system does not depend on surrounding bodies. Such a system is called a capacitor (sometimes called condensers). The simplest capacitor consists of two conductors (plates) separated by a small distance. In order to exclude the effect of external bodies on the capacitance of a capacitor, its plates are arranged with respect to one another in such a way that the field created by the charges accumulated on them is concentrated almost completely inside the capacitor. This means that the lines of E emerging on one plate must terminate on the other, i.e. the charges on the plates must be equal in magnitude and opposite in sign (q and $-q$).

The basic characteristic of a capacitor is its capacitance. Unlike the capacitance of an isolated conductor, the capacitance of a capacitor is defined as the ratio of its charge to the potential difference between the plates (this difference is called the voltage):

$$\boxed{C = q/V} \quad (3.7)$$

Here, V is the potential difference between the plates of the capacitor, whereas for an isolated conductor it is just a potential of the conductor. Note that potential is also a potential difference between the potential of the isolated conductor and zero potential at infinity.

The charge q of a parallel plate capacitor is the magnitude of charge on the facing surface of either plate.

The capacitance of a capacitor is also measured in farads.

The capacitance of a capacitor depends on its geometry (size and shape of its plates), the gap between the plates, and the material that fills the capacitor.

EXAMPLE 5. Two capacitors, A and B, are each connected to a 9 V battery. If $C_A > C_B$, which capacitor stores the greater amount of charge?

SOLUTION C represents the capacitor's capacity to store charge: The greater C, the greater the amount of charge stored for a given value of V. So, A stores the greater amount of charge.
Mathematically, from Eq.3.7, we have
$$q = CV.$$
Given that $V = 9$ volt, so, $q = 9C$ and hence,
$$q_A = 9C_A, q_B = 9C_B.$$
Since, $C_A > C_B$, therefore $q_A > q_B$.
Also, $q_A : q_B = C_A : C_B$

3.4 Parallel-plate Capacitor

A parallel-plate capacitor consists of two plane-parallel thin conductor plates of equal area separated by an insulator (Fig.3.4a). Often the two plates are rolled into the form of a cylinder with plastic, paper, or other insulator separating the plates (Fig.3.4b). For simplicity, let's assume that the insulator is a vacuum. If the capacitor plates were of infinite size, the electric field between the plates would be $E = \sigma/\epsilon_0$ (see Eq.1.66 of chapter 1) where σ is the surface charge density of plates. Practically, parallel-plate capacitors are not infinitely large, but the dimensions of the area A of the plates are usually very large as compared to the plate separation d. When this is the case, the electric field can be approximated by the infinite-plate expression.

Using this expression, we obtain for the potential difference

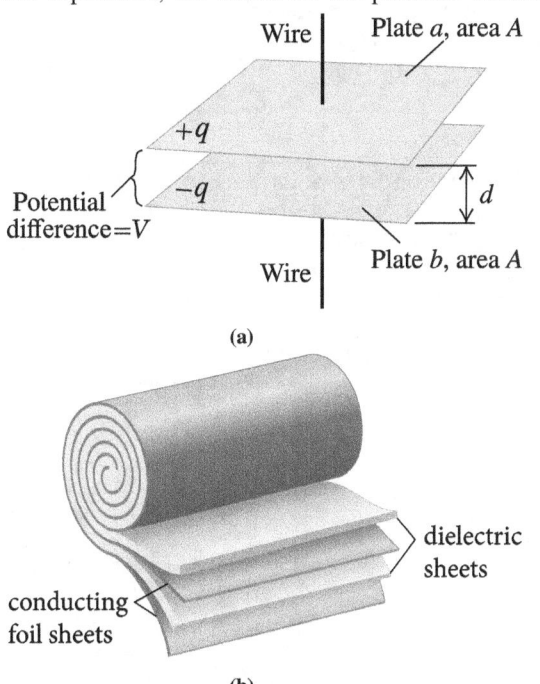

(b)

Figure 3.4: (a) Arrangement of the capacitor plates (b) One way to design a compact capacitor with a large surface area

between the plates of the capacitor in Figure 3.4a.

Figure 3.4a shows such a two-conductor capacitor consisting of two large, parallel thin metallic plates, each of area A, separated by a distance d. The plates carry charges $+q$ and $-q$, respectively, on their facing surfaces. The electric field in the region between the plates is-
$$E = \frac{\sigma}{\epsilon_0}$$

here, σ is the surface charge density of the plates.
Since, $\sigma = \frac{q}{A}$, therefore-
$$E = \frac{q}{\epsilon_0 A}$$
Consequently, the voltage between the plates is-
$$V = Ed = \frac{q}{\epsilon_0 A}d$$
Substituting this expression into Eq.3.7, we obtain-
$$\boxed{C = \frac{\epsilon_0 A}{d}} \qquad (3.8)$$

This is the capacitance of a parallel-plate capacitor. This calculation was made without taking into account field distortions near the edges of the plates (edge effects)[Fig.3.5]. The capacitance of a real plane capacitor determined by this formula is more accurate for smaller values of d in comparison with the linear dimensions of the plates.

From Eq.3.8, C depends on only the geometry of the capacitor;

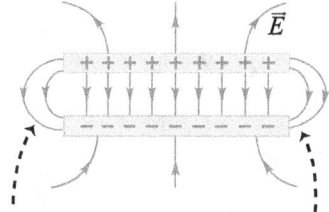

When the separation of the plates is small compared to their size, the fringing of the field is slight.

Figure 3.5: Side view of the electric field $\vec{E}$

it is directly proportional to the area A of each plate and inversely proportional to their separation d. The quantities A and d are constants for a given capacitor, and ϵ_0 is a universal constant. Thus in vacuum the capacitance C is a constant independent of the charge on the capacitor or the potential difference between the plates. If one of the capacitor plates is flexible, C changes as the plate separation d changes. This is the operating principle of a condenser microphone (Fig.3.7).

When matter is present between the plates, its properties affect the capacitance. If the space contains air at atmospheric pressure instead of vacuum, the capacitance differs from the prediction of Eq.3.8 by less than 0.06%. From Eq.(3.8), we can also show that alternate units for ϵ_0 are farads per meter. So,
$$\epsilon_0 = 8.85 \times 10^{-12} \text{ F/m} \qquad (3.9)$$
This relationship is useful in capacitance calculations, and it also helps us to verify that Eq.3.8 is dimensionally consistent.

Although Eq.3.8 pertains only to parallel-plate capacitors, its implications are much broader. The capacitance of any capacitor increases as the distance between the conductors decreases. Decreasing d is an effective way of increasing capacitance.

The proportionality, $C \propto A/d$ in Eq.3.8, is also valid for a parallel-plate capacitor even if it is rolled up into a spiral cylinder, as in Fig.3.4b. However, the constant factor, ϵ_0, must be replaced by electric permittivity of the insulator used to separate the plates (It will be discussed in next section).

A particular computer keyboard operates by capacitance. As shown in Fig.3.6, each key is connected to the upper plate of a capacitor. The upper plate moves down when the key is pressed, reducing the spacing between the capacitor plates, and increasing the capacitance (Eq.3.8: $C = \epsilon_0 A/d \implies$ with a smaller d, the capacitance C becomes larger). The change in capacitance results in an electric signal that is detected by an electronic circuit.

One type of microphone is a condenser, or capacitor, micro-

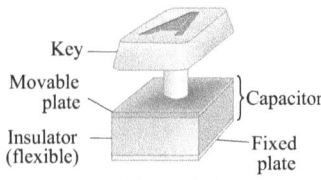

Figure 3.6

phone, diagrammed in Fig.3.7. The changing air pressure in a sound wave causes one plate of the capacitor C to move back and forth, and that plate is isolated so the charge on it doesn't change. Hence the voltage across the capacitor changes at the same frequency as the sound wave.

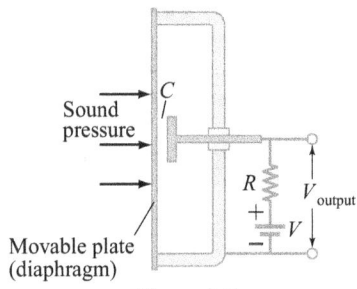

Figure 3.7

Important Points

☞ When we say that a capacitor has charge q or that a charge q is stored on the capacitor, we mean that the conductor (or capacitor plate) at higher potential has charge $+q$ and the conductor (or capacitor plate) at lower potential has charge $-q$ (*assuming that q is positive*).

☞ If one of the plates of parallel plate capacitor slides relatively to other, then C decreases (As overlapping area decreases (Fig.3.8)).

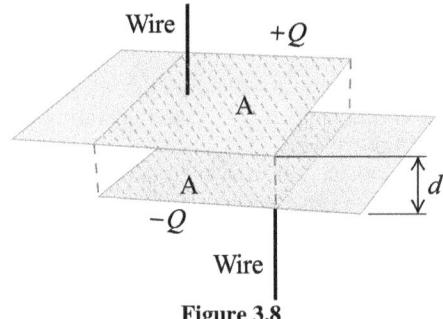

Figure 3.8

☞ If both the plates of parallel plate capacitor are connected to each other (Fig.3.9), then the potential of both the plates will be same and therefore potential difference (V) between them will be zero. So, from $C = q/V$ capacitance would be indeterminate and charge stored on them would be zero.

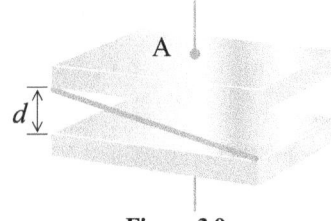

Figure 3.9

EXAMPLE 6. Size of a 1 F Capacitor: A parallel-plate capacitor has a capacitance of 1.0 F. If the plates are 1.0 mm apart, what is the effective area of the plates?
APPROACH The effective area of plates of a parallel plate capacitor, can be calculated by using Eq.3.8: $C = \epsilon_0 A/d$ -

SOLUTION $C = \dfrac{\epsilon_0 A}{d} \implies A = \dfrac{Cd}{\epsilon_0}$

Here, $C = 1\,F$, $d = 1.0\,mm$, $\epsilon_0 = 8.85 \times 10^{-12}\,F/m$

$\therefore \quad A = \dfrac{(1.0\,F)\left(10^{-3}\,m\right)}{8.85 \times 10^{-12}\,F/m} = 1.1 \times 10^8\,m^2$

This corresponds to a square about 10 km on a side. Clearly this is not a very practical design for a capacitor. However, the plate area can be increased by rolling up into a spiral cylinder, as in Fig.3.4b.

EXAMPLE 7. A parallel-plate capacitor has square metallic plates of edge length 10 cm separated by 1.0 mm. (a) Calculate the capacitance of this device. (b) As this capacitor is charged to 12 V, how much charge is transferred from one plate to another?
APPROACH The capacitance C is determined by the area and the separation of the plates (Eq.3.8). Once C is found, the charge for a given voltage V is found from the definition of capacitance $C = q/V$.
(a) We find the capacitance using Eq.3.8: $C = \epsilon_0 A/d$.

$$C = \dfrac{\epsilon_0 A}{d} = \dfrac{(8.85\,pF/m)(0.10\,m)^2}{0.0010\,m} = 88.5\,pF$$

(b) The charge transferred is found from $Q = CV$ (the definition of capacitance):

$$Q = CV = (88.5\,pF)(12\,V) = 1.06 \times 10^{-9}\,C = 1.1\,nC$$

3.4.1 Parallel Plate Capacitor with With Unequal Plate Charges

In Fig.3.10 two parallel metal plates having charges Q_1 and Q_2 are shown. From **EXAMPLE 101** of chapter 1, the surface charges on the facing surfaces of the plates having charges Q_1 and Q_2 are $(Q_1 - Q_2)/2$ and $-(Q_1 - Q_2)/2$ respectively. Thus the magnitude of charge on the facing surface of either plate-

$$\boxed{q = \dfrac{|Q_1 - Q_2|}{2}} \quad (3.10)$$

and the charge on the back surface of plate having net charge Q_1 is $Q_1 - (Q_1 - Q_2)/2$, i.e., $(Q_1 + Q_2)/2$ and the charge on the back surface of plate having net charge Q_2 is $Q_2 - (-(Q_1 - Q_2)/2)$, i.e., $(Q_1 + Q_2)/2$.

Special Case: If two plates of a parallel plate capacitor have

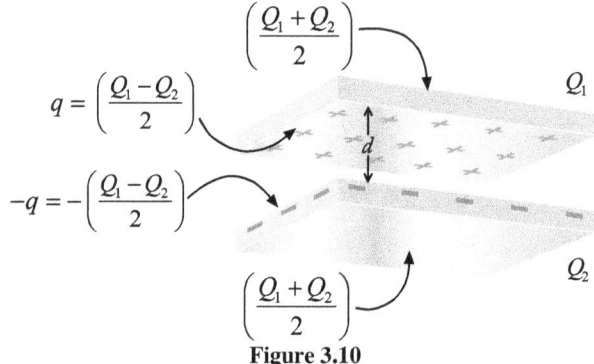

Figure 3.10

charges $+Q$ and $-Q$ respectively, then the magnitude of charge on the facing surface of either plate will be

$$\boxed{q = \dfrac{|Q_1 - Q_2|}{2} = \dfrac{|Q - (-Q)|}{2} = Q}$$

which is the magnitude of net charge on any one of the two charged plates.
The charge on the back surfaces of each plate,

$$q' = \dfrac{Q_1 + Q_2}{2} = \dfrac{Q - Q}{2} = 0$$

3.4. PARALLEL-PLATE CAPACITOR

In this case, due to electrostatic attraction, the net charge on each plate remains confined to its facing surface only.

EXAMPLE 8. Charges of 1 μC and 2 μC are given to two plates of a parallel-plate capacitor of capacitance 0.1 μF. Find the potential difference developed between the plates.

APPROACH In this case, the magnitudes of charges on the plates of a capacitor are not equal, therefore, to find the magnitude of charges on facing surfaces of either plate, we use Eq.3.10 and then for voltage, apply the relation $C = q/V$.

SOLUTION From Eq.3.10, the magnitude of charge on facing surface of either plate is-
$$q = \frac{|Q_1 - Q_2|}{2} = \frac{|2\ \mu\text{C} - 1\ \mu\text{C}|}{2} = 0.5\ \mu\text{C}$$
Given that, $C = 0.1\ \mu$F, therefore $C = q/V$, gives-
$$V = q/C = \frac{0.5\ \mu\text{C}}{0.1\ \mu\text{F}} = 5\ \text{volt}$$

EXAMPLE 9. A charge of 20 μC is placed on a plate of a parallel-plate capacitor of capacitance 10 μF. It's other plate does not have any net charge. Calculate the potential difference developed between the plates.

APPROACH In this case too, the magnitudes of charges on the plates of a capacitor are not equal, therefore, to find the magnitude of charges on facing surfaces of either plate, we use Eq.3.10 and then for voltage, apply the relation $C = q/V$.

SOLUTION From Eq.3.10, the magnitude of charge on facing surface of either plate is-
$$q = \frac{|Q_1 - Q_2|}{2} = \frac{|20\ \mu\text{C} - 0\ \mu\text{C}|}{2} = 10\ \mu\text{C}$$
Given that, $C = 10\ \mu$F, therefore $C = q/V$, gives-
$$V = q/C = \frac{10\ \mu\text{C}}{10\ \mu\text{F}} = 1\ \text{volt}$$

EXAMPLE 10. Two identical metal plates are given positive charges q_1 and q_2 ($< q_1$) respectively. If they are now brought close together to form a parallel plate capacitor with capacitance C, calculate the potential difference between them.

SOLUTION Although, this problem too can be easily solved

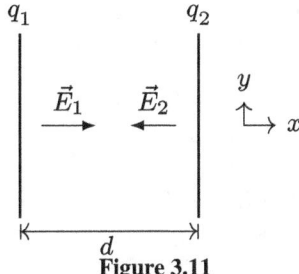

Figure 3.11

like previous, we solve it by calculating electric field between the plates. Let A be the area of the plates and d be the separation between them.

The electric field, in the region between the plates, due to the left plate is
$$\vec{E}_1 = \frac{\sigma_1}{2\epsilon_0}\hat{i} = \frac{q_1}{2\epsilon_0 A}\hat{i},$$
and due to the right plate is
$$\vec{E}_2 = -\frac{\sigma_2}{2\epsilon_0}\hat{i} = -\frac{q_2}{2\epsilon_0 A}\hat{i}$$
The resultant field $(\vec{E})$ and potential difference between the two plates (V) are respectively-
$$\vec{E} = \vec{E}_1 + \vec{E}_2 = \frac{q_1 - q_2}{2\epsilon_0 A}\hat{i}$$
and
$$V = Ed = \frac{q_1 - q_2}{2(\epsilon_0 A/d)} = \frac{q_1 - q_2}{2C}.$$

Paragraph Type Questions Paragraph for Examples 11 and 12 Consider an evacuated cylindrical chamber of height h with rigid conducting plates at the ends and an insulating curved surface, as shown in Fig.3.12. Several spherical balls made of lightweight and soft material, coated with a conducting material, are placed on the bottom plate. The balls have a radius $r \ll h$. Now, connect a high voltage source (HV) across the conducting plates so that the bottom plate is at $+V_0$ and the top plate is at $-V_0$. Due to their conducting surface, the balls will become charged, reach equipotential with the plate, and be repelled by it. Eventually, the balls will collide with the top plate, where the coefficient of restitution can be taken as zero due to the soft nature of the material. The electric field in the chamber can be considered as that of a parallel plate capacitor. Assume there are no collisions between the balls, and the interaction between them is negligible. [Ignore gravity.]

EXAMPLE 11. Which one of the following statements is

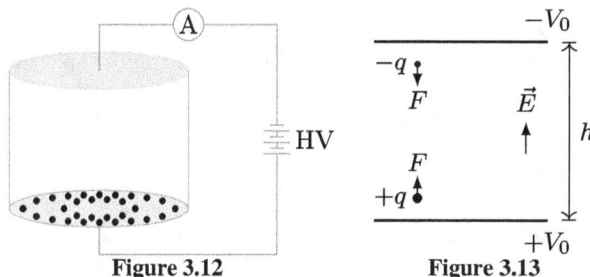

Figure 3.12   Figure 3.13

correct?
(A) The balls will stick to the top plate and remain there.
(B) The balls will bounce back to the bottom plate carrying the same charge they went up with.
(C) The balls will bounce back to the bottom plate carrying the opposite charge they went up with.
(D) The balls will execute simple harmonic motion between the two plates.

SOLUTION (C) The distance between the two plates is h. The potential of the bottom plate is V_0 and that of the top plates is $-V_0$. It is just like a parallel plate capacitor as shown in Fig.3.13. The electric field between the plates is $E = 2V_0/h$ (directed upwards). The radius of each ball is $r (\ll h)$. Let m be the mass and C be the capacitance of each ball.

When the ball touches the bottom plate, it gets a positive charge $q = CV_0$ (we assume that the charge transfer is instantaneous). This positively charged ball experiences an upward force, $F = qE = 2CV_0^2/h$, which accelerates the ball upwards. Since the force is constant, the ball cannot do SHM (for SHM, the force should be proportional to the displacement and directed towards the centre).

When the ball hits the top plate, it transfers the positive charge to the plate and gets negative charge $q = -CV_0$. This negatively charged ball again experience a force $F = qE$ (downward) and starts accelerating downwards. Thus, the ball keeps moving between the bottom and the top plates carrying a charge $+q$ when moving upwards and $-q$ when moving downwards.

EXAMPLE 12. The average current in the steady state registered by the ammeter in the circuit will be
(A) 0   (B) $\propto V_0$ (C) $\propto V_0^{1/2}$ (D) $\propto V_0^2$

SOLUTION (D) The average current when a charge q moves from the bottom plate to the top plate in time T is $i = q/T$. The ball moves with a uniform acceleration $a = F/m = 2CV_0^2/(mh)$. The distance travelled in time T is
$$h = \frac{1}{2}aT^2 = \frac{1}{2}\frac{2CV_0^2}{mh}T^2, \quad \text{i.e.,} \quad T = \frac{h}{V_0}\sqrt{\frac{m}{C}}$$
The average current carried by each ball is

$$i = \frac{q}{T} = \frac{C}{h}\sqrt{\frac{C}{m}}V_0^2$$

3.5 Electric Force Between the Plates of a Parallel Plate Capacitor

If σ is the charge density and A is the facing area of each plate, then, the electric field on the facing surface of each plate due to other is,
$$E = \frac{\sigma}{2\epsilon_0} = \frac{q}{2A\epsilon_0}$$
Therefore, electrostatic attraction due to charge q on the facing surface of the plate is given by-
$$\boxed{F = qE = \frac{q^2}{2A\epsilon_0}} \quad (3.11)$$

3.5.1 Spherical Capacitor

A spherical capacitor consists of a solid or hollow spherical conductor of a certain radius surrounded by another hollow concentric spherical shell of different radius as illustrated in Figure 3.14.

EXAMPLE 13. A spherical capacitor consists of two concentric spherical conducting shells, of different radii a and b respectively (Figure 3.14). The inner shell carries a uniformly distributed charge Q on its surface, and the outer surface of outer shell is grounded. Determine the capacitance of the two shells.

APPROACH Suppose, charge $+Q$ is given to inner shell and the

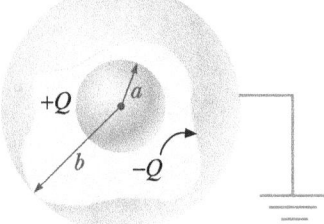

Figure 3.14: A spherical capacitor consists of an inner sphere of radius a surrounded by a concentric spherical shell of radius b. The diagram shows the capacitor carrying a charge Q. The electric field between the spheres is directed radially outward when the inner sphere is positively charged.

outer surface of outer shell is earthed. In this case, the induced charge on the inner surface of outer shell will be $-Q$.

By definition, the capacitance C is the magnitude q of the charge

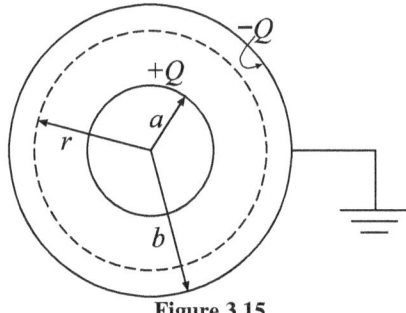

Figure 3.15

on either sphere divided by the potential difference V between the spheres. We first find V, and then find the capacitance by using expression $C = q/V$, with $q = Q$.

SOLUTION By Gauss's theorem, we can show that the charge on a conducting shell produces zero field inside the it, so the outer shell makes no contribution to the field between the spherical shells. Therefore, the electric field between the shells are the same as those outside a charged conducting sphere with charge $+Q$.

In Eq.1.141 of chapter 1, we used Gauss's law to show that the electric field outside a uniformly charged conducting sphere is $E = Q/4\pi\epsilon_0 r^2$ as if all the charge were concentrated at the centre. Therefore, the potential difference between shells-
$$V = \int_-^+ \vec{E} \cdot d\vec{s} = -\frac{Q}{4\pi\epsilon_0}\int_b^a \frac{dr}{r^2}$$
$$= \frac{Q}{4\pi\epsilon_0}\left(\frac{1}{a} - \frac{1}{b}\right) = \frac{Q}{4\pi\epsilon_0}\left(\frac{b-a}{ab}\right)$$
Therefore,
$$\boxed{C = Q/V = 4\pi\epsilon_o\left(\frac{ab}{b-a}\right)} \quad (3.12)$$

☞ It is interesting that when the separation $d = b - a$ is very small, i.e., $b - a \ll a$ (or b), then $C = 4\pi\epsilon_0 a^2/d \approx \epsilon_0 A/d$ (since $A = 4\pi a^2$), which is the parallel-plate formula, Eq.3.8.

☞ Equation 3.12 can also be used to find the capacitance of an isolated spherical conductor. An isolated spherical conductor with a radius 'R' can be considered as a capacitor, where the outer spherical shell approaches infinity (i.e., $b \to \infty$) and the radius of the inner shell is $a = R$. Therefore, for this isolated spherical conductor, Equation 3.12 yields-
$$C = \lim_{b\to\infty} 4\pi\epsilon_0\frac{Rb}{(b-R)} = \lim_{b\to\infty} 4\pi\epsilon_0 \frac{R}{\left(1-\frac{R}{b}\right)} = 4\pi\epsilon_0 R$$
which is same as Eq.3.6.

EXAMPLE 14. The stratosphere acts as a conducting layer for the earth. If the stratosphere extends beyond 50 km from the surface of earth, then calculate the capacitance of the spherical capacitor formed between stratosphere and earth's surface. Take radius of earth as 6400 km.

SOLUTION From Eq.3.12, the capacitance of a spherical capacitor is
$$C = 4\pi\epsilon_0\left(\frac{ab}{b-a}\right)$$
here, b = radius of the top of stratosphere layer
= 6400 km + 50 km = 6450 km = 6.45×10^6 m,
and a = radius of earth
= 6400 km = 6.4×10^6 m, therefore,-
$$C = \frac{1}{9.0 \times 10^9} \times \frac{6.45 \times 10^6 \times 6.4 \times 10^6}{6.45 \times 10^6 - 6.4 \times 10^6} = 0.092 \text{ F}$$

3.5.2 The Cylindrical Capacitor

EXAMPLE 15. A solid cylindrical conductor of radius a is coaxial with a cylindrical shell of negligible thickness and radius $b > a$ (Fig. 3.16a). Find the capacitance of this cylindrical capacitor if its length is $l (\gg b)$.

APPROACH Because of the cylindrical symmetry of the system, we can use results from previous studies of cylindrical systems to find the capacitance.

Suppose, the capacitor carries a charge Q and it's length l is much greater than a and b, we can neglect end effects. In this case, the electric field is perpendicular to the long axis of the cylinders and is confined to the region between them (Fig.3.16b). The potential difference between the two charged cylinders-
$$V = V_a - V_b = -\int_b^a \vec{E} \cdot d\vec{s}$$
Notice from Figure 3.16b that $\vec{E}$ is parallel to $d\vec{s}$ along a radial line and apply Eq.1.128: $E = 2k\lambda/r$ for the electric field outside a cylindrically symmetric charge distribution:
$$V = -\int_b^a E_r dr = -2k\lambda \int_b^a \frac{dr}{r} = 2k\lambda \ln\left(\frac{b}{a}\right)$$
Substitute the absolute value of V into equation $C = Q/V$ and

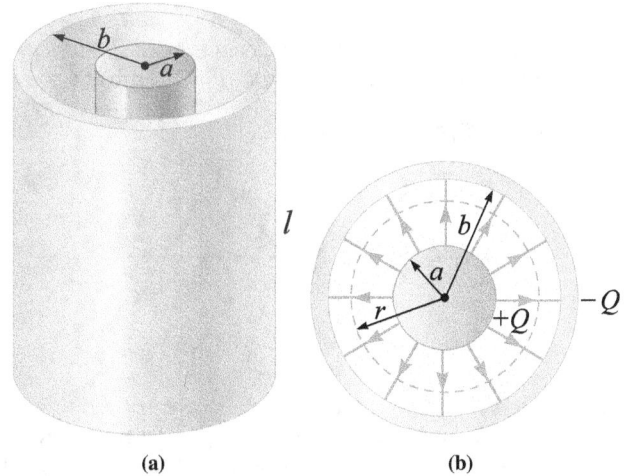

Figure 3.16: (a) A cylindrical capacitor (b) End view of the capacitor if it is charged. The electric field lines are radial.

use $\lambda = Q/l$:

$$C = \frac{Q}{V} = \frac{Q}{(2kQ/l)\ln(b/a)} = \frac{l}{2k\ln(b/a)} \quad (3.13)$$

Since, $k = \dfrac{1}{4\pi\epsilon_0}$, therefore above equation can also be written as-

$$C = \frac{2\pi\epsilon_0 l}{\ln(b/a)} \quad (3.14)$$

This result shows that the capacitance is proportional to l, which makes sense: The longer the coaxial cylinders, the greater the quantity of charge that can be stored on them. Decreasing a or increasing b is equivalent to increasing the plate separation distance d in a parallel-plate capacitor, which decreases the capacitance. Indeed, this result shows a decreasing capacitance for decreasing a or increasing b. (The dependence on a and b is a bit more complicated than the dependence on d in a parallel-plate capacitor because the electric field in the coaxial capacitor is nonuniform and because changing the radii of the cylinders affects their surface areas.)

Equation 3.13 shows that the capacitance per unit length of a combination of concentric cylindrical conductors is

$$\frac{C}{l} = \frac{2\pi\epsilon_0}{\ln(b/a)} \quad (3.15)$$

An example of this type of geometric arrangement is a coaxial cable, which consists of two concentric cylindrical conductors separated by an insulator. The coaxial cable is especially useful for shielding electrical signals from any possible external influences.

EXAMPLE 16. A cylindrical capacitor has two coaxial cylinders of length 15 cm and radii 1.5 cm and 1.4 cm. The outer cylinder is earthed and the inner cylinder is given a charge of 3.5 μC. Determine the capacitance of the system and the potential of the inner cylinder.

SOLUTION Given that: $l = 15$ cm $= 15 \times 10^{-2}$ m; $a = 1.4$ cm $= 1.4 \times 10^{-2}$ m; $b = 1.5$ cm $= 1.5 \times 10^{-2}$ m; $q = 3.5$ μC $= 3.5 \times 10^{-6}$ C.

Capacitance $C = \dfrac{2\pi\epsilon_0 l}{\log_{10}\left(\frac{b}{a}\right)} = \dfrac{2\pi\epsilon_0 l}{2.303\log_{10}\left(\frac{b}{a}\right)}$

$= \dfrac{2\pi \times 8.854 \times 10^{-12} \times 15 \times 10^{-2}}{2.303 \log_{10}\frac{1.5\times 10^{-2}}{1.4\times 10^{-2}}}$

$= 1.21 \times 10^{-8}$ F

Since the outer cylinder is earthed, so it is at zero potential, therefore the potential of the inner cylinder will be equal to the potential difference between them. Potential of inner cylinder, is

$$V = \frac{q}{C} = \frac{3.5 \times 10^{-6}}{1.2 \times 10^{-10}} = 2.89 \times 10^{4} \text{V}$$

EXAMPLE 17. How the magnitude of electric field $\vec{E}$ in the annular region of a charged cylindrical capacitor depend on the position from the central axis?

SOLUTION Consider a cylindrical Gaussian surface of length

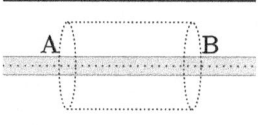

Figure 3.17

l and radius r as shown in Fig.3.17.

By symmetry, electric field $\vec{E}$ is radial and is of same magnitude at all points at a radial distance r. Since, area vectors at end surfaces 'A' and 'B' are perpendicular to $\vec{E}$, therefore the fluxes through the end surfaces A and B of the cylinder are zero. Hence, the flux through Gaussian surface is

$$\oint \vec{E} \cdot d\vec{S} = 2\pi r l E$$

If λ is the charge per unit length on the inner cylinder then the charge enclosed by Gaussian surface is

$$q_{\text{encl}} = \lambda l$$

Gauss's law, $\oint \vec{E} \cdot d\vec{S} = q_{\text{encl}}/\epsilon_0$, gives

$$E = \frac{\lambda}{2\pi\epsilon_0 r}$$

3.5.2.1 Capacitance of two long parallel wires

EXAMPLE 18. Estimate the capacitance per unit length of two very long straight parallel wires, each of radius R, carrying uniform charges $+Q$ and $-Q$, and separated by a distance d which is large compared to $R(d \gg R)$, as shown in Fig.3.18.

APPROACH First of all calculate the potential difference be-

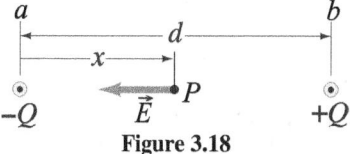

Figure 3.18

tween the wires by treating the electric field at any point between them as the superposition of the two fields created by each wire. (The electric field inside each wire conductor is zero because the charge is in static equilibrium.) Now, the capacitance is given as: $C = Q/V$, therefore,

$$\frac{C}{l} = \frac{1}{l}\left(\frac{Q}{V}\right)$$

here, l is the length of each wire under consideration.

SOLUTION The electric field outside of a long straight conductor was found in Eq.1.128 (chapter 1) to be radial and given by $E = \lambda/(2\pi\epsilon_0 r)$ where λ ($= Q/l$) is the charge per unit length and r is the distance of field point from the central axis of wire ($r > R$). Since, wires are oppositely charged, therefore, fields due to each wire will be in the same direction. So, the net electric field due to both wires at distance x from the left-hand wire in Fig.3.18 has magnitude-

$$E = \frac{\lambda}{2\pi\epsilon_0 x} + \frac{\lambda}{2\pi\epsilon_0(d-x)}$$

and points to the left (from + to −). Now, we find the potential difference between the two wires using Eq.2.14: $V = V_b - V_a = -\int_a^b \vec{E} \cdot d\vec{s}$ and integrating along the straight line from the surface of the negative wire to the surface of the positive wire, noting that $\vec{E}$ and $d\vec{s}$ point in opposite directions ($\vec{E} \cdot d\vec{s} < 0$):

$$V = V_b - V_a = -\int_a^b \vec{E} \cdot d\vec{s}$$
$$= \left(\frac{\lambda}{2\pi\epsilon_0}\right) \int_R^{d-R} \left[\frac{1}{x} + \frac{1}{(d-x)}\right] dx$$
$$= \left(\frac{\lambda}{2\pi\epsilon_0}\right) [\ln(x) - \ln(d-x)]\Big|_R^{d-R}$$
$$= \left(\frac{\lambda}{2\pi\epsilon_0}\right) [\ln(d-R) - \ln R - \ln R + \ln(d-R)]$$
$$= \left(\frac{\lambda}{\pi\epsilon_0}\right) [\ln(d-R) - \ln(R)] \approx \left(\frac{\lambda}{\pi\epsilon_0}\right) [\ln(d) - \ln(R)]$$

Since, $d \gg R$, therefore $V \approx \left(\frac{Q}{\pi\epsilon_0 l}\right) \left[\ln\left(\frac{d}{R}\right)\right]$

The capacitance from Eq.(3.7) is,

$$\boxed{C = Q/V \approx (\pi\epsilon_0 l)/\ln(d/R)} \qquad (3.16)$$

So, the capacitance per unit length is approximately-

$$\boxed{\frac{C}{l} \approx \frac{\pi\epsilon_0}{\ln(d/R)}}$$

3.5.2.2 Circuit Symbols of Capacitors

Fig.3.19a and 3.19b represent the circuit symbols of a capacitor. It is similar to a battery symbol except that a capacitor has equal arms unlike to a battery which has unequal arms as shown in Fig.3.19c.

A battery, which is a source of voltage, is indicated by the symbol

Figure 3.19: (a) and (b) represent a capacitor symbols whereas (c) shows a battery symbol

shown in Fig. 3.19c with unequal arms.

3.6 Check Point 1

1. •An isolated conducting sphere that has a 10.0 cm radius has an electric potential of 2.00 kV (the potential far from the sphere is zero). (a) How much charge is on the sphere? (b) What is the self-capacitance of the sphere? (c) By how much does the selfcapacitance change if the sphere's electric potential is increased to 6.00 kV?
2. •How much charge flows from a 12.0 V battery when it is connected to a 12.6 μF capacitor?
3. ••When 1.0×10^{12} electrons are transferred from one conductor to another, a potential difference of 10 V appears between the conductors. Calculate the capacitance of the two-conductor system.
4. ••At constant temperature, two uniformly charged spherical drops at potential V coalesce to form a larger drop. If capacity of each smaller drop is C then find capacity and potential of larger drop.
5. •A charge of $+2.0 \times 10^{-8}$ C is placed on the positive plate and a charge of -1.0×10^{-8} C on the negative plate of a parallel-plate capacitor of capacitance $1.5 \times 10^{-3} \mu$F. Calculate the potential difference developed between the plates.
6. ••The charge on a capacitor increases by 26 μC when the voltage across it increases from 28 V to 78 V. What is the capacitance of the capacitor?

7. ••A 7.7 μF capacitor is charged by a 125 V battery (Fig.3.20a) and then is disconnected from the battery. When this capacitor (C_1) is connected (Fig. 3.20b) to a second (initially uncharged) capacitor, C_2, the final voltage on each capacitor is 15 V. What is the value of C_2?

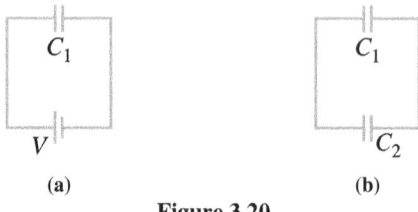

Figure 3.20

8. ••Two isolated conducting spheres of equal radius R have charges $+Q$ and $-Q$, respectively. Their centres are separated by a distance d that is large compared to their radii. Estimate the capacitance of this unusual capacitor.
9. ••Consider two different capacitors, A and B. Figure 3.21 shows a graph of the potential difference V between the two plates of each capacitor versus the charge q on the plates. Use these graphs to find the capacitance of each capacitor.

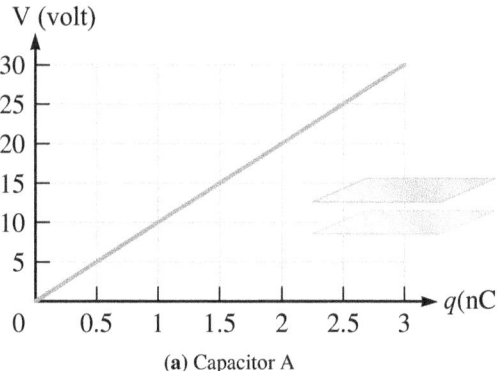

(a) Capacitor A

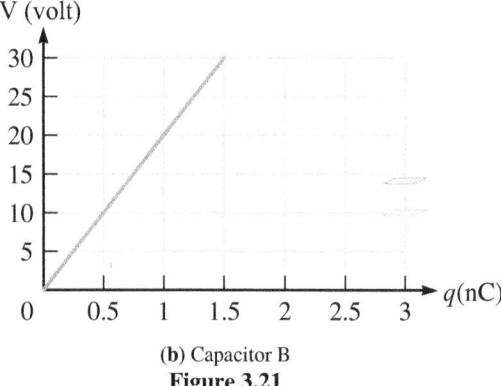

(b) Capacitor B
Figure 3.21

10. ••The ground and the oceans are conductors, and the Earth can therefore be regarded as a conducting sphere. What is its capacitance? How much charge is needed to alter the potential of earth by 1 V
11. •••A radio active source in the form of a metal sphere of diameter 10^{-3} m emits β particles at a constant rate of 6.25×10^{10} particles per second. If the source is electrically insulated, how long will it take for its potential to rise by 1.0 volt, assuming that 80% of emitted β particles escape from the surface.

3.7 Capacitors in Series and in Parallel

When there is a combination of capacitors in a circuit*, we can sometimes describe its behaviour as an equivalent capacitor – a single capacitor that has the same capacitance as the actual combination of capacitors. In this way, we can simplify the circuit and make it easier to analyse. There are two basic configurations of capacitors that allow such a replacement.

3.7.1 Capacitors in Series

In Figure 3.22, three capacitors of capacitances C_1, C_2 and C_3 are connected between terminals a and b by conducting wires in series (one after the other). The left plate of capacitor C_1 and the right plate of capacitor C_3 are connected to the terminals of a battery. The other plates are connected to each other and to nothing else; hence, they form an isolated system that is initially uncharged and must continue to have zero net charge.

When a constant positive potential difference $V_{ab} = V$ is applied

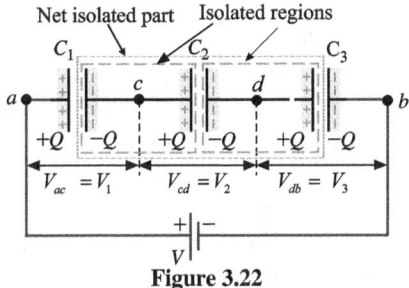

Figure 3.22

between terminals a and b, the negative charge flows from left plate of the capacitor C_1 to positive terminal of battery and the plate acquires a positive charge $+Q$. Equivalently, we can say that a charge $+Q$ flows from the battery to left plate of C_1. The electric field of this positive charge attracts negative charge up to the right plate of C_1 until all of the field lines that begin on the left plate of C_1 end on it's right plate. This requires that the right plate have charge $-Q$. Since, the region 'c' is isolated therefore, these negative charges had to come from the left plate of next capacitor C_2, which becomes positively charged with charge $+Q$. Similarly, this charge $+Q$ of the left plate of capacitor C_2 pulls a charge of $-Q$ on the opposite plate. Now, next region 'd' is also isolated, so, it must also have a zero net charge. Therefore, there is a charge $+Q$ on the left plate of C_3. This positive charge then pulls negative charge $-Q$ from the connection at point b onto the right plate of C_3. The total charge on the net isolated part shown in diagram 3.22 must always be zero because the plates in this region aren't connected to anything except each other. Thus, **in a series connection the magnitude of charge on all plates is the same.**

Now, from Fig.3.22, we have-
$$V_{ab} = V = V_{ac} + V_{cd} + V_{db} = V_1 + V_2 + V_3$$
Here, $V_{ac} = V_1 = Q/C_1$, $V_{cd} = V_2 = Q/C_2$,
and $V_{db} = V_3 = Q/C_3$

Therefore, $V_{ab} = V = V_1 + V_2 + V_3 = Q\left(\dfrac{1}{C_1} + \dfrac{1}{C_2} + \dfrac{1}{C_3}\right)$

or, $\dfrac{V}{Q} = \dfrac{1}{C_1} + \dfrac{1}{C_2} + \dfrac{1}{C_3}$ ($\because V_{ab} = V$) ... (i)

The equivalent capacitance C_{eq} of the series combination is defined as the capacitance of a single capacitor for which the charge Q is the same as for the combination, when the potential difference V is the same. In other words, the combination can be replaced by an equivalent capacitor of capacitance C_{eq}. For such a capacitor, shown in Fig.3.22,

$$C_{eq} = \dfrac{Q}{V} \quad \text{or} \quad \dfrac{1}{C_{eq}} = \dfrac{V}{Q} \quad \text{... (ii)}$$

Comparing equations (i) and (ii), we get-

$$\boxed{\dfrac{1}{C_{eq}} = \dfrac{1}{C_1} + \dfrac{1}{C_2} + \dfrac{1}{C_3}} \quad \text{[series]} \quad (3.17)$$

Similarly for more than three capacitors in series, the equivalent capacitance is given by-

$$\boxed{\dfrac{1}{C_{eq}} = \dfrac{1}{C_1} + \dfrac{1}{C_2} + \dfrac{1}{C_3} + \cdots} \quad \text{[series]} \quad (3.18)$$

The reciprocal of the equivalent capacitance of a series combination equals the sum of the reciprocals of the individual capacitances.

☞ Notice that the equivalent capacitance C_{eq} is smaller than the smallest contributing capacitance.

★ **Important Point:** The magnitude of charge is the same on all plates of all the capacitors in a series combination. However, from the relation $V = Q/C$ for each capacitor in series (i.e., with the same Q), we can infer that the potential differences of the individual capacitors are not the same unless their individual capacitances are equal. The potential differences of the individual capacitors add up to give the total potential difference across the series combination:

$$\boxed{V_{\text{total}} = V_1 + V_2 + V_3 + \cdots}$$

EXAMPLE 19. An infinite number of capacitors of capacitance C, $4C$, $16C$, ... ∞ are connected in series. Find their equivalent capacitance?

APPROACH For equivalent capacitance in series, apply Eq.3.18-

SOLUTION The equivalent capacitance (C_{eq}) of the given combination in series, is-
$$\dfrac{1}{C_{eq}} = \dfrac{1}{C} + \dfrac{1}{4C} + \dfrac{1}{16C} + \ldots \infty$$
$$= \left[1 + \dfrac{1}{4} + \dfrac{1}{16} + \ldots \infty\right]\dfrac{1}{C} \quad (1)$$

The right hand side part, in square bracket, represents geometric progression†.

For the square bracket part of right hand side, the first term $a = 1$, common ratio, $r = \dfrac{1}{4}$, therefore-

$$\dfrac{1}{C_{eq}} = \left[1 + \dfrac{1}{4} + \dfrac{1}{16} + \ldots \infty\right]\dfrac{1}{C} = \dfrac{1}{1 - \frac{1}{4}} \times \dfrac{1}{C} \implies C_{eq} = \dfrac{3}{4}C$$

EXAMPLE 20. What is the charge on capacitor of capacitance $15\ \mu F$ in the Fig.3.23?

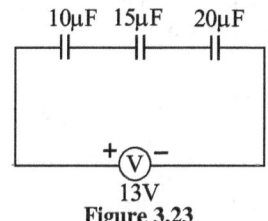

Figure 3.23

SOLUTION Equivalent capacitance is given by-
$$\dfrac{1}{C_{eq}} = \dfrac{1}{10} + \dfrac{1}{15} + \dfrac{1}{20} = \dfrac{26}{120} \implies C_{eq} = \dfrac{60}{13}\ \mu F$$

Now, we can replace all these capacitors by above equivalent ca-

---

*By electric circuit we mean a closed path of conductors, usually wires connecting capacitors and/or other devices, in which charge can flow and which includes a source of voltage such as a battery. The battery voltage is usually given the symbol V, which means that V represents a potential difference.

†The sum of infinite terms in a geometric progression, is given by
$$a + ar + ar^2 + ar^3 + \cdots = \dfrac{a}{1-r} \quad \text{(for } r < 1\text{)}$$

pacitor. This equivalent capacitor has terminal voltage of 13 V, therefore the charge supplied by the battery
$$Q = C_{eq}V = \frac{60}{13} \mu F \times 13 \text{ V} = 60 \mu C$$
As all capacitors are in series, therefore charge on each capacitor will be the same and each of them has charge of 60 μC

EXAMPLE 21. As shown in the Fig.3.24, two parallel plate capacitors having equal plate area of 200 cm^2 are joined in such a way that $a \neq b$. Find the equivalent capacitance of the combination.

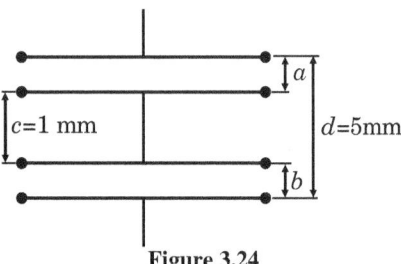

Figure 3.24

SOLUTION Combination can be considered as two capacitors in series. If C_1 is the capacitance of capacitor having separation a and C_2 is that of having separation b, then equivalent capacitance
$$C = \frac{C_1 C_2}{C_1 + C_2} = \frac{\left(\frac{\epsilon_0 A}{a}\right)\left(\frac{\epsilon_0 A}{b}\right)}{\frac{\epsilon_0 A}{a} + \frac{\epsilon_0 A}{b}} = \frac{\epsilon_0 A}{ab\left(\frac{1}{a} + \frac{1}{b}\right)} = \frac{\epsilon_0 A}{a+b}$$
Also from Fig.3.24, $a + b = d - c$, therefore
$$C = \frac{\epsilon_0 A}{(d-c)} = \frac{\epsilon_0 \times 200 \times 10^{-4}}{4 \times 10^{-3}} = 5\epsilon_0 \text{ F}$$
Comparing this value of capacitance with given capacitance $C = x\epsilon_0$ F, we get, $x = 5$.

3.7.2 Capacitors in Parallel

In Fig.3.25, we have three capacitors C_1, C_2 and C_3. The left plate of each capacitor is connected by conducting wires to a common point a at potential V_a and the right plate of each capacitor is connected by conducting wires to another common point b at potential V_b. Such type af arrangement of capacitors in which one plate of each capacitor is at one common potential and other plate of each capacitor is at some other common potential, is called a parallel arrangement of capacitors.

In a parallel arrangement, the same potential difference $V = V_{ab} = V_a - V_b$ exists across each capacitor. Positive plates of capacitors C_1, C_2 and C_3 acquire charges $Q_1 = C_1V$, $Q_2 = C_2V$, and $Q_3 = C_3V$ respectively. The total charge Q that must leave the battery is then
$$Q = Q_1 + Q_2 + Q_3 = C_1V + C_2V + C_3V \quad \ldots \text{(i)}$$
Let us try to find a single equivalent capacitor that will hold the same charge Q at the same voltage $V = V_{ab} = V_a - V_b$. It will have a capacitance C_{eq} given by
$$Q = C_{eq}V \quad \ldots \text{(ii)}$$
On comparing equations (i) and (ii), we get-
$$C_{eq}V = C_1V + C_2V + C_3V = (C_1 + C_2 + C_3)V$$
$$\boxed{C_{eq} = C_1 + C_2 + C_3 \quad \text{[parallel]}} \quad (3.19)$$
Similarly, for more than three capacitors in parallel, we can write-
$$\boxed{C_{eq} = C_1 + C_2 + C_3 + \cdots \quad \text{[parallel]}} \quad (3.20)$$
The equivalent capacitance of a parallel combination equals the sum of the individual capacitances.

☞ Notice that in a parallel connection the equivalent capacitance is always greater than the largest contributing capacitance.
The net effect of connecting capacitors in parallel is thus to increase the capacitance. This makes sense because we are essentially increasing the area of the plates where charge can accumulate.

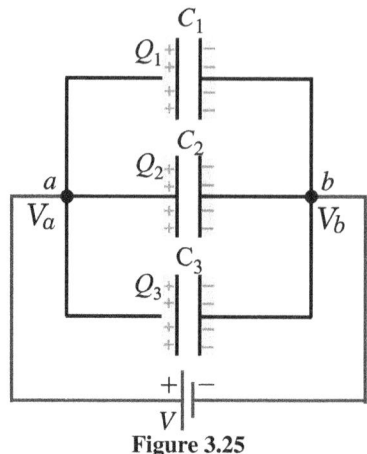

Figure 3.25

☞ In parallel connection,
$$Q_1 : Q_2 : Q_3 = C_1 : C_2 : C_3$$
The charge on the capacitor is proportional to its capacitance
$$Q_1 = C_1 V = \frac{C_1}{C_1 + C_2 + C_3}Q, \, Q_2 = C_2V = \frac{C_2}{C_1 + C_2 + C_3}Q \text{ and}$$
$$Q_3 = C_3V = \frac{C_3}{C_1 + C_2 + C_3}Q$$
$$(\because \text{from Eq. (i)}, V = Q/(C_1 + C_2 + C_3))$$
We can also analyse the other connections of capacitors by using charge conservation, and often simply in terms of series and parallel connections.

★ **Important Point:** The potential differences are the same for all capacitors in a parallel combination; however from relation $Q = CV$, the charges on individual capacitors are not the same unless their individual capacitances are the same. The charges on the individual capacitors add to give the total charge on the parallel combination:
$$Q_{\text{total}} = Q_1 + Q_2 + Q_3 + \cdots$$
Results 3.18 and 3.20 will be used later in circuit analysis.

EXAMPLE 22. Capacitors of capacitances $C, 2C, 4C, \ldots \infty$ are connected in parallel. Find their effective capacitance?
APPROACH For effective or equivalent capacitance (C_{eq}) of capacitors connected in parallel, apply Eq.3.20.
SOLUTION From Eq.3.20, the equivalent or effective capacitance-
$$C_{eq} = C + 2C + 4C + \ldots \infty$$
$$= C[1 + 2 + 4 + \ldots \infty] = C(\infty) = \infty$$

☞ **Important Points**
- If N identical capacitors each having breakdown voltage* V are joined in-
  1. series, then the break down voltage of the combination will be equal to NV, because in series, voltages across capacitors are additive, i.e.,
  $$V_{\text{total}} = V_1 + V_2 + V_3 + \cdots + V_N$$
  2. parallel, then the breakdown voltage of the combination is equal to V because in parallel, the voltage across all capacitos is the same, i.e.,
  $$V_{\text{total}} = V_1 = V_2 = \cdots = V_N$$
- Two capacitors are connected in series with a battery. Now, if battery is removed and loose wires connected together then by charge conservation, the final charge on each capacitor would be zero.
- If N identical capacitors are connected then

---
*Breakdown voltage of a medium is the minimum required voltage to ionize it so that it conducts charge

3.7. CAPACITORS IN SERIES AND IN PARALLEL

$$(C_{eq})_{series} = \frac{C}{N}, \quad (C_{eq})_{parallel} = NC$$

- In DC, all capacitor's offer infinite resistance in steady state, so there would not be any current through a branch containing capacitor.

EXAMPLE 23. In a spherical capacitor shown in Fig.3.26, the inner sphere of radius a, is earthed. The inner and outer radii of outer sphere are b and c respectively. Determine the capacitance of the system. If charge Q is given to outer sphere, then find the charge stored on each surface.

APPROACH In this case, there are two capacitors. One capacitor is formed by inner sphere (radius = a) and inner surface (radius = b) of outer sphere. Suppose it's capacitance is C_1. The other capacitor is formed by outer surface (radius = c) of outer sphere having it's second plate at infinite (i.e., at zero potential). It can be considered as an isolated spherical conductor. Suppose, it's capacitance is C_2.

The capacitances of inner and outer capacitors are —

$$C_1 = 4\pi\epsilon_0 \left(\frac{ab}{b-a}\right) \quad \text{and} \quad C_2 = 4\pi\epsilon_0 c \text{ respectively.}$$

The inner sphere is grounded, so it's potential is zero. Suppose, the potential of outer sphere is V. Now, for inner capacitor C_1, one plate is at zero potential and other is at V, where as for outer capacitor C_2 one plate is at V and other is zero potential (because second plate is at infinite). So, these capacitors are connected between same voltage and hence regarded in parallel (Fig.3.27). Therefore, equivalent capacitance is given by-

$$C = C_1 + C_2 \qquad (3.21)$$

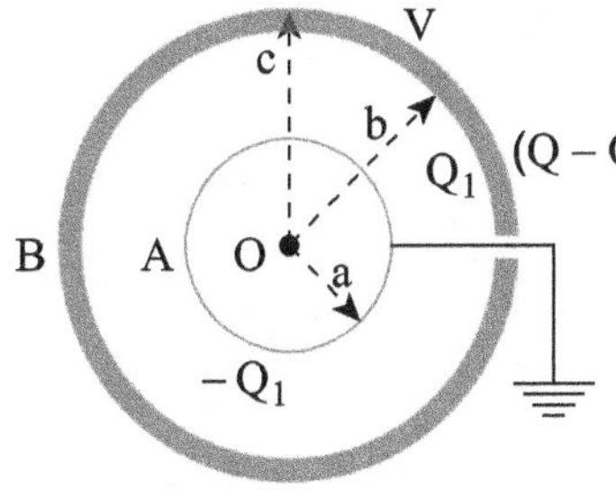

Figure 3.26 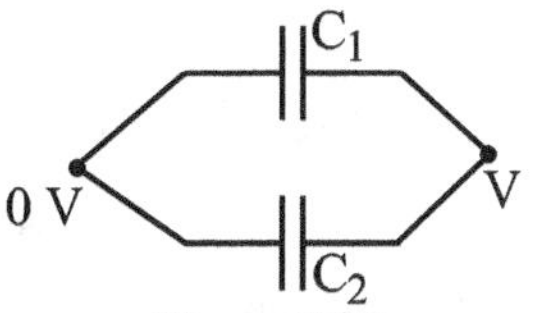 **Figure 3.27**

SOLUTION Substituting the values of C_1 and C_2 in Eq.3.21, we get-

$$\boxed{C = 4\pi\epsilon_0 \frac{ab}{b-a} + 4\pi\epsilon_0 c} \qquad (3.22)$$

If outer shell is thin, then $c = b$, in this case above equation becomes-

$$\boxed{C = 4\pi\epsilon_0 \left[\frac{ab}{b-a} + b\right] = 4\pi\epsilon_0 \left[\frac{b^2}{b-a}\right]} \qquad (3.23)$$

Calculation of Charge on Each Surface: Net charge on outer sphere is Q. Suppose, the charge on inner surface of outer sphere is Q_1, then by conservation of charge, the charge on it's outer surface will be $Q - Q_1$. Now, by Gauss's theorem, we can show that the charge on inner sphere will be $-Q_1$.

Net electric potential at any point on the surface of the inner sphere will be equal to sum of potentials produced by all plate charges, i.e.,

$$V = k\frac{-Q_1}{a} + k\frac{Q_1}{b} + k\frac{Q - Q_1}{c}$$

Since, it is grounded, i.e., $V = 0$, therefore –

$$\frac{-Q_1}{a} + \frac{Q_1}{b} + \frac{Q - Q_1}{c} = 0$$

or

$$-Q_1\left(\frac{1}{a} + \frac{1}{c} - \frac{1}{b}\right) + Q\left(\frac{1}{c}\right) = 0$$

or

$$Q_1\left(\frac{1}{a} + \frac{1}{c} - \frac{1}{b}\right) = Q\left(\frac{1}{c}\right)$$

or

$$Q_1 = \left(\frac{ab}{bc + ab - ac}\right)Q$$

So, the charge on inner surface of outer sphere,

$$Q_1 = \left(\frac{ab}{bc + ab - ac}\right)Q$$

Charge on inner sphere,

$$= -Q_1 = -\left(\frac{ab}{bc + ab - ac}\right)Q$$

And charge on outer surface of outer sphere

$$= Q - Q_1 = Q - \left(\frac{ab}{bc + ab - ac}\right)Q$$

$$= Q\left[1 - \left(\frac{ab}{bc + ab - ac}\right)\right] = Q\left[\frac{c(b-a)}{bc + ab - ac}\right]$$

If outer sphere is thin, i.e., $b = c$, then $Q_1 = \left(\frac{a}{b}\right)Q$

Therefore, charge on inner sphere $= -Q_1 = -\left(\frac{a}{b}\right)Q$,

Charge on inner surface of outer sphere $= Q_1 = \left(\frac{a}{b}\right)Q$,

And charge on the outer surface of outer sphere $= \left(\frac{b-a}{b}\right)Q$

Remark

☞ If there is no outer sphere and inner sphere is connected to Earth as shown in Fig.3.28, then we can take $\lim b \to \infty$. In this case, Eq.3.23 becomes:

$$\boxed{C = \lim_{b \to \infty} 4\pi\epsilon_0 \left[\frac{b^2}{b-a}\right] \to \infty} \qquad (3.24)$$

So, once again we can say that the capacitance of a conductor

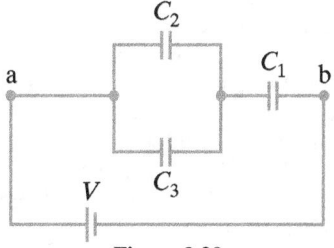

Figure 3.28

connected to Earth is approximately infinite.

EXAMPLE 24. (a) Determine the capacitance of a single capacitor that will have the same effect as the combination shown in Fig.3.29. Take $C_1 = C_2 = C_3 = C$.

(b) Determine the charge on each capacitor in Fig.3.29 and the voltage across each capacitor, assuming $C = 3.0 \ \mu F$ and the battery voltage is $V = 4.0$ V.

(a) **APPROACH** The terminals of capacitors C_2 and C_3 are at

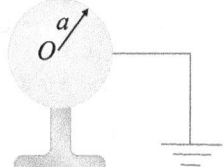

Figure 3.29

common potentials. So, these are in parallel. The equivalent of C_2 and C_3 is in series with C_1.

Therefore, first we find the equivalent capacitance of C_2 and C_3 in parallel, and then consider that capacitance in series with C_1 (Fig.3.30).

SOLUTION Capacitors C_2 and C_3 are connected in parallel, so

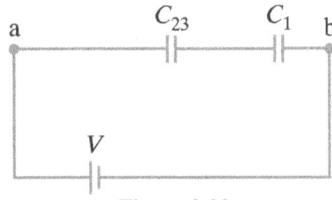

Figure 3.30

they are equivalent to a single capacitor having capacitance C_{23} given as-
$$C_{23} = C_2 + C_3 = C + C = 2C$$
This C_{23} is in series with C_1, Fig.3.30, so the equivalent capacitance of the entire circuit, C_{eq}, is given by
$$\frac{1}{C_{eq}} = \frac{1}{C_1} + \frac{1}{C_{23}} = \frac{1}{C} + \frac{1}{2C} = \frac{3}{2C}$$
Hence the equivalent capacitance of the entire combination is $C_{eq} = \frac{2}{3}C$, and it is smaller than any of the contributing capacitors, $C_1 = C_2 = C_3 = C$.

(b) APPROACH We first find the charge Q that leaves the battery, using the equivalent capacitance. Then we find the charge on each separate capacitor and the voltage across each. Each step uses: $Q = CV$.

SOLUTION The 4.0 V battery behaves as if it is connected to a capacitance $C_{eq} = \frac{2}{3}C = \frac{2}{3}(3.0\ \mu F) = 2.0\ \mu F$. Therefore the charge Q that leaves the battery is-
$$Q = CV = (2.0\ \mu F)(4.0\ V) = 8.0\ \mu C$$
From Fig.3.29, this charge arrives at the negative plate of C_1, so $Q_1 = 8.0\ \mu C$. The charge Q that leaves the positive plate of the battery is split evenly between C_2 and C_3 (symmetry: $C_2 = C_3$) and is $Q_2 = Q_3 = \frac{1}{2}Q = 4.0\ \mu C$. Next, the voltages across C_2 and C_3 have to be the same. The voltage across each capacitor is obtained using $V = Q/C$. So–
$$V_1 = Q_1/C_1 = (8.0\ \mu C)/(3.0\ \mu F) = 2.7\ V,$$
$$V_2 = Q_2/C_2 = (4.0\ \mu C)/(3.0\ \mu F) = 1.3\ V,$$
and
$$V_3 = Q_3/C_3 = (4.0\ \mu C)/(3.0\ \mu F) = 1.3\ V$$

EXAMPLE 25. The connections shown in Fig.3.31 are established with the switch S open. How much charge will flow through the switch if it is closed?

APPROACH Given diagram is again shown in Fig.3.32 with

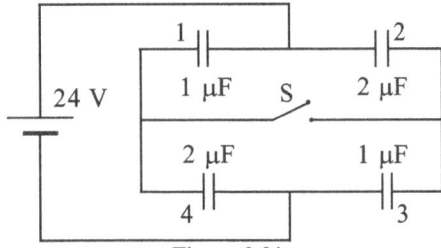

Figure 3.31

capacitors plates marked by letters a, b, c, d, e, f, g and h. Since, switch S is open, therefore, the system containing plates a, g, is isolated and the system of plates d, f is also isolated. These isolated parts are shown by dashed closed curves in Fig.3.32. Since, switch S is open, therefore, we can remove the branch containing switch S and it won't effect the net capacitance of circuit between the terminals connected with battery. The circuit without switch S branch is shown in Fig.3.33. In this case the net charge on isolated plates $a + g$ will be zero and on plates $d + f$ will also be zero.

From Fig.3.33, it is clear that capacitors 1, 4 are in series and 2, 3 are also in another series. Their series equivalents are in parallel.

When switch S is turned ON, the plates a and d of capacitors 1 and 2 respectively comes at a common potential V_0 (say) through switch S whereas plate b of capacitor 1 and plate c of capacitor

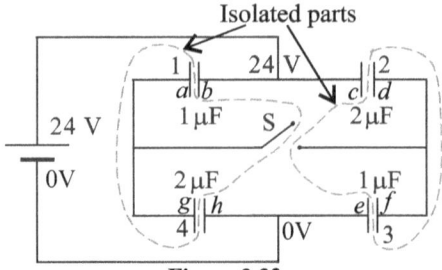

Figure 3.32

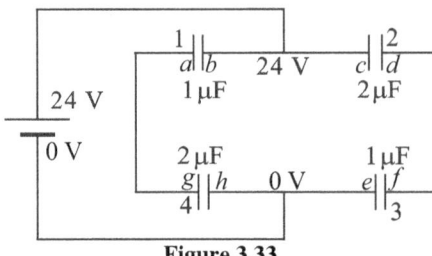

Figure 3.33

2 is at other common potential of 24 V. So, these are in parallel. Similarly, the plates e and h of capacitors 3 and 4 respectively are at a common potential 0 V whereas plates f and g of capacitors 3 and 4 respectively, are at same potential V_0, so capacitors 3 and 4 are also in parallel. The equivalent capacitance of 1 and 2 is in series with equivalent capacitance of capacitors 3 and 4.

It's simplified circuit diagram is again shown in Fig.3.35. Now,

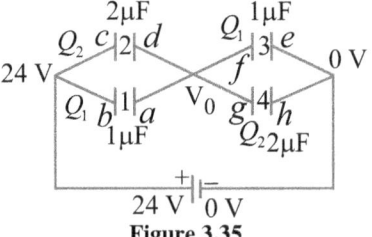

Figure 3.34

find net capacitance of the circuit, then find the charge supplied by battery. After that find the sum of charges either on plates a and g or on plates d and f by using $Q = CV$ for each capacitor. The difference in charge, before and after switch S is turned ON, would have passed through switch S.

SOLUTION When switch S was open:

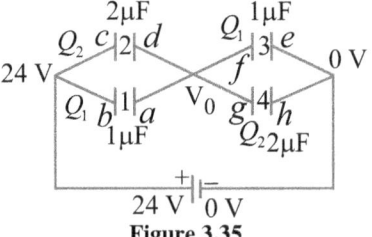

Figure 3.35

In Fig.3.33, the series equivalent capacitance of capacitors 2 and 3 is, $\frac{2 \times 1}{2 + 1} = \frac{2}{3}\ \mu F$, similarly, the series equivalent capacitance of capacitors 1 and 2 is $\frac{2}{3}\ \mu F$. The net capacitance of the circuit connected with battery, is -
$$C_0 = \frac{2}{3} + \frac{2}{3} = \frac{4}{3}\ \mu F$$
In this case, the net charge supplied by the battery-
$$Q_0 = C_0 V = \frac{4}{3}\ \mu F \times 24\ V = 32\ \mu C$$
Since, each branch of Fig.3.33, has equal capacitance of $\frac{2}{3}\ \mu F$,

3.7. CAPACITORS IN SERIES AND IN PARALLEL

therefore charge supplied to each branch:
$$Q_{01} = \frac{2}{3} \mu F \times 24 \text{ V} = 16 \ \mu C.$$
In series, the charge on each capacitor remains same, therefore the charge on each capacitor is $16 \ \mu C$
Therefore, the sum of charges on plates a and g = the sum of charges on plates d and $f = -16 \ \mu C + 16 \ \mu C = 0$

When switch S is closed: The net capacitance of the circuit connected with the terminals of battery [Fig.3.35] is given by-
$$\frac{1}{C} = \frac{1}{1 \ \mu F + 2 \ \mu F} + \frac{1}{1 \ \mu F + 2 \ \mu F} = \frac{2}{3 \ \mu F} \implies C = \frac{3}{2} \ \mu F$$
In this case, net charge supplied by battery-
$$Q = CV = \frac{3}{2} \mu F \times 24 \text{ V} = 36 \ \mu C$$
Now, from Fig.3.35,
$$V - V_0 = \frac{Q_1}{1 \ \mu F} = \frac{Q_2}{2 \ \mu F} = \frac{Q_1 + Q_2}{1 \ \mu F + 2 \ \mu F}$$
$$= \frac{36 \ \mu C}{3 \ \mu F} = 12 \text{ volt}$$
$$\implies V_0 = V - 12 \text{ volt} = 24 \text{ volt} - 12 \text{ volt} = 12 \text{ volt}$$
Now, charge on +ve plate of capacitor 1 -
$$Q_1 = 1 \ \mu F \times 12 \text{ V} = 12 \ \mu C$$
Charge on +ve plate of capacitor 4 is
$$Q_2 = (2 \ \mu F)(V_0) = 2 \ \mu F \times 12 \text{ V} = 24 \ \mu C$$
Therefore, net charge on plates a and g is
$$(-Q_1) + Q_2 = -12 \ \mu C + 24 \ \mu C = +12 \ \mu C$$
When the switch was open, this charge was zero. Thus, $12 \mu C$ of charge has passed from plates $d + f$ to plates $a + g$ through the switch after it was closed.

EXAMPLE 26. Seven capacitors each of capacitance $2\mu F$ are connected in a configuration to obtain an effective capacitance $\frac{10}{11} \ \mu F$. Which of the combination shown in Fig.3.36, will achieve the desired result?

SOLUTION In option A, equivalent capacitance of five capaci-

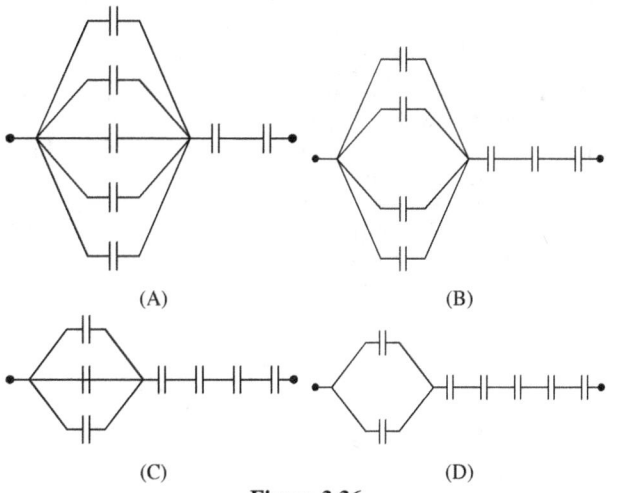

Figure 3.36

tors connected in parallel is $C_1 = 10 \mu F$ and that of two capacitors connected in series is $C_2 = 1 \mu F$. The C_1 and C_2 are connected in series so equivalent capacitance is
$$C_{eq} = \frac{C_1 C_2}{C_1 + C_2} = \frac{10}{11} \ \mu F.$$

EXAMPLE 27. For the circuit shown in Fig.3.37, which of the following statements is true?
(A) with S_1 closed, $V_1 = 15$ V, $V_2 = 20$ V.
(B) with S_3 closed, $V_1 = V_2 = 25$ V.
(C) with S_1 and S_2 closed, $V_1 = V_2 = 0$.
(D) with S_1 and S_3 closed, $V_1 = 30$ V, $V_2 = 20$ V.

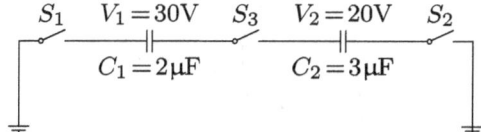

Figure 3.37

SOLUTION (D) The charge on the left capacitor is $Q_1 = C_1 V_1 = 60 \ \mu C$ and that on the right capacitor is $Q_2 = C_2 V_2 = 60 \ \mu C$. The potential or charge on the capacitor does not change in any case.

EXAMPLE 28. In the circuit shown in Fig.3.38, there are two parallel plate capacitors each of capacitance C. The switch S_1 is pressed first to fully charge the capacitor C_1 and then released. The switch S_2 is then pressed to charge the capacitor C_2. After some time, S_2 is released and then S_3 is pressed. After some time, find the charges on upper plates of C_1 and C_2.

SOLUTION Pressing switch S_1 charges C_1 to a charge $Q =$

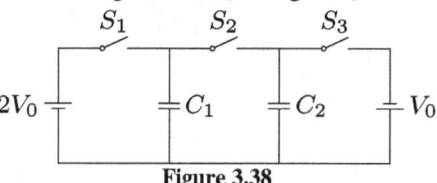

Figure 3.38

$C_1 V = 2CV_0$ with its upper plate (attached to positive terminal) having charge $+2CV_0$. Releasing switch S_1 and pressing switch S_2 equally distributes the charge on C_1 and C_2 (identical capacitors). Thus, the charge on the upper plate of C_1 and C_2 is $+CV_0$. When switch S_2 is released and switch S_3 is pressed, charge on the upper plate of C_1 remains $+CV_0$ but charge on C_2 changes. The battery attached to C_2 charges it to charge CV_0 with upper plate (attached to negative terminal) having charge $-CV_0$.

EXAMPLE 29. In the circuit shown in Fig.3.39, a charge of $+80 \ \mu C$ is given to the upper plate of the $4\mu F$ capacitor. In steady state, find the charge on the upper plate of the $3 \ \mu F$ capacitor.

SOLUTION Let the charges on $3 \ \mu F$ and $2 \ \mu F$ capacitors be q

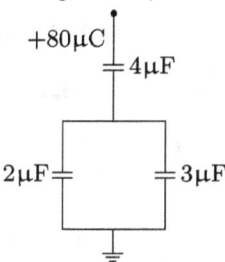

Figure 3.39

and q' respectively. Charge on lower plate of $4 \ \mu F$ capacitor is $-80 \ \mu C$. Lower plate of $4 \ \mu F$ capacitor and upper plates of $2 \ \mu F$ and $3 \ \mu F$ capacitors form an isolated system. Hence, net charge on this system must be zero i.e.,
$$q + q' - 80 \ \mu C = 0 \qquad \ldots (1)$$
The potentials $(V = q/C)$ across these two capacitors are equal which gives $q/3 = q'/2$ $\qquad \ldots (2)$
On eliminating q' from equations (1) and (2), we get $q = 48 \ \mu C$.

EXAMPLE 30. Five identical capacitor plates, each of area A, are arranged such that adjacent plates are at a distance d apart, the plates are connected to a source of emf V as shown in the Fig.3.40. Find the charge on plates 1 and 4.

SOLUTION Given circuit is redrawn in Fig.3.41a. This time we have marked each plate surface with a lowercase English letter. Surfaces a, b, e, f, i and j are connected to the positive terminal of the battery of emf V and surfaces c, d, g and h are connected to the negative terminal of the battery. The equivalent circuit is shown in the Fig.3.41b

There are four capacitors, each of capacity $C = \epsilon_0 A/d$, connected

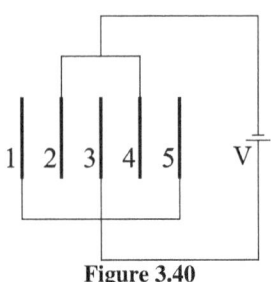

Figure 3.40

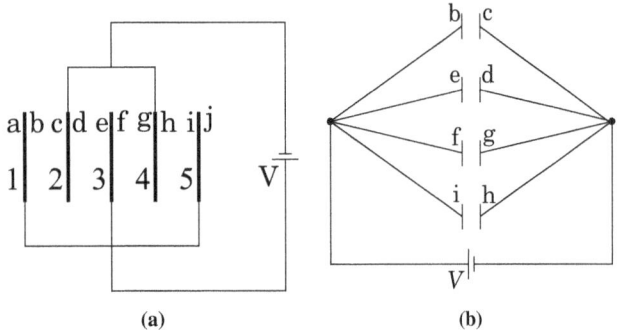

Figure 3.41

in parallel. The charge on each capacitor is $q = CV = \epsilon_0 AV/d$. Therefore, each surface b, e, f and i has charge $\epsilon_0 AV/d$ and each surface c, d, g and h has charge $-\epsilon_0 AV/d$.

Calculation of Charge on Each Plate

Charge on Plate 1: It has two surfaces a and b. Surface a is not a part of any capacitor so it does not have any charge. The charge on surface b is $\epsilon_0 AV/d$. Therefore total charge on plate 1:
$$Q_1 = 0 + \epsilon_0 AV/d = \epsilon_0 AV/d$$

Charge on Plate 2: It has two surfaces c and d. The charge on each surface c and d is $-\epsilon_0 AV/d$. Therefore total charge on plate 2:
$$Q_2 = -\epsilon_0 AV/d - \epsilon_0 AV/d = -2\epsilon_0 AV/d$$

Charge on Plate 3: It has two surfaces e and f. The charge on each surface e and f is $\epsilon_0 AV/d$. Therefore total charge on plate 3:
$$Q_3 = \epsilon_0 AV/d + \epsilon_0 AV/d = 2\epsilon_0 AV/d$$

Charge on Plate 4: It has two surfaces g and h. The charge on each surface g and h is $-\epsilon_0 AV/d$. Therefore total charge on plate 4:
$$Q_4 = -\epsilon_0 AV/d - \epsilon_0 AV/d = -2\epsilon_0 AV/d$$

Charge on Plate 5: It has two surfaces i and j. Surface j is not a part of any capacitor so it does not have any charge. The charge on surface i is $\epsilon_0 AV/d$. Therefore total charge on plate 5:
$$Q_5 = \epsilon_0 AV/d + 0 = \epsilon_0 AV/d$$

EXAMPLE 31. What is the equivalent capacitance between points A and B in the circuit of shown Fig.3.42?

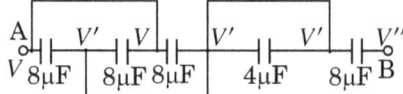

Figure 3.42

SOLUTION In Fig.3.42, points A and D are connected by a conducting wire, therefore points A and D are at same potential. Similarly points C, E and P are also at same potential. Between points E and P, there are two capacitors, each with a capacitance of 8 μF, connected in series. The equivalent capacitance of these two capacitors is $\dfrac{8 \times 8}{8+8}$ μF, i.e., 4 μF. In next Fig.3.43, we have replaced these capacitors by their equivalent capacitor. Now, in Fig.3.43 we see that the 4 μF capacitor is short circuited, therefore it's both plates are at same potential and hence we can replace it by a conducting wire.

Suppose V, V' and V'' are the potentials of points A, P and

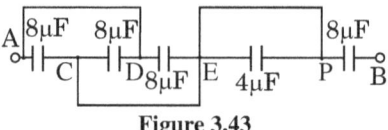

Figure 3.43

B respectively. The circuit diagram showing potentials of all points, is given in Fig.3.44. Next simplified diagram is shown in

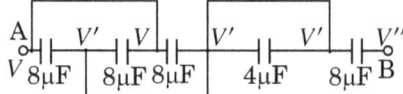

Figure 3.44

Fig.3.45.

In Figure 3.45, all three capacitors, each with a magnitude

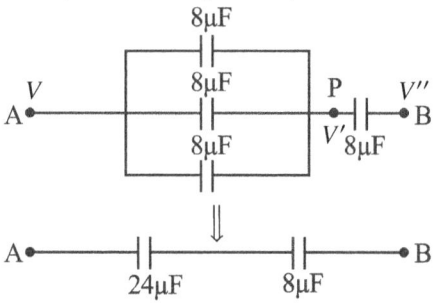

Figure 3.45

of 8 μF between terminals A and P, are connected in parallel. Their equivalent (8 μF + 8 μF + 8 μF = 24 μF) is in series with the 8 μF capacitor connected between points P and B.

Finally, the equivalent capacitance between the terminals A and B is
$$C_{eq} = \frac{24 \times 8}{24 + 8} = \frac{24 \times 8}{32} = 6 \ \mu F$$

EXAMPLE 32. A parallel plate capacitor is made up of stair like structure with a plate area A of each stair and that is connected with a wire of length b, as shown in the Fig.3.46. Find the capacitance of the arrangement.

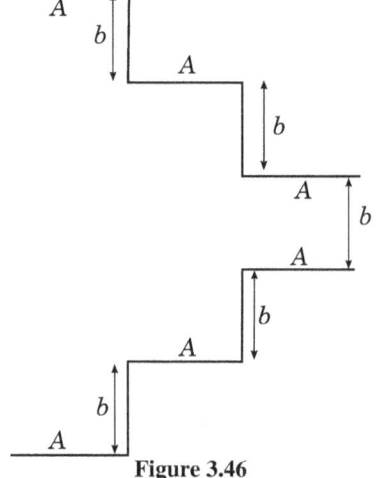

Figure 3.46

SOLUTION Equivalent capacitance of given parallel combination is
$$C_{eq} = \epsilon_0 A \left[\frac{1}{5b} + \frac{1}{3b} + \frac{1}{b} \right] = \frac{23}{15} \frac{\epsilon_0 A}{b}$$

EXAMPLE 33. In the given circuit in Fig.3.47, $C_1 = 2 \ \mu F$,

$C_2 = 0.2\ \mu F$, $C_3 = 2\ \mu F$, $C_4 = 4\ \mu F$, $C_5 = 2\ \mu F$, $C_6 = 2\ \mu F$, calculate the charge stored on capacitor C_4

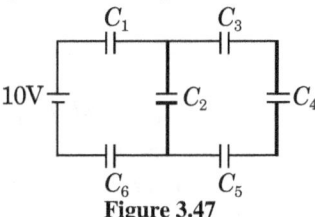

Figure 3.47

SOLUTION C_2, C_4 and C_5 are in series. If their equivalent capacitance is C', then
$$\frac{1}{C'} = \frac{1}{C_3} + \frac{1}{C_4} + \frac{1}{C_5} = \frac{1}{2} + \frac{1}{4} + \frac{1}{2} = \frac{5}{4}$$
Therefore, $C' = \frac{4}{5}\ \mu F = 0.8\ \mu F$.

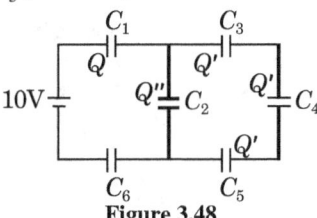

Figure 3.48

This is in parallel with C_2, therefore, if C_1' is the equivalent of C' and C_2, then
$$C_1' = C' + C_2 = 0.8 + 0.2 = 1\ \mu F$$
Again, C_1' is in series with C_1 and C_6. Therefore, the equivalent capacitance (C_{eq}) connected between the terminals of the battery is given as:
$$\frac{1}{C_{eq}} = \frac{1}{C_1} + \frac{1}{C_1'} + \frac{1}{C_6} = \frac{1}{2} + \frac{1}{1} + \frac{1}{2} = 2$$
Therefore, $C_{eq} = \frac{1}{2}\ \mu F = 0.5\ \mu F$
Therefore, charge supplied by battery,
$$Q = C_{eq} V = 0.5 \times 10 = 5\ \mu C$$
Thus, charge on each of capacitors C_1 and C_6 is $5\ \mu C$. Now, at the junction of C_1 and C_3, the charge Q divided to two parts: Q' and Q'' as shown in Fig.3.48.
As the potential difference between C_2 and C' (which is a series equivalent of C_3, C_4 and C_5) is same, therefore
$$\frac{Q'}{C'} = \frac{Q''}{C_2} = \frac{Q' + Q''}{C' + C_2} = \frac{Q}{C' + C_2} = \frac{5}{0.8 + 0.2} = 5$$
Therefore, $Q' = 5 \times C' = 5 \times 0.8 = 4\ \mu C$.
Since, C_3, C_4 and C_5 are in series, therefore same charge of $4\ \mu C$ goes to these capacitors.

3.8 An Infinite Network

In this section, we describe a method to find the equivalent capacitance of a circuit that has infinite identical stages. This type of network is also known as an infinite ladder. In such problems, we start by guessing a finite and non zero value (C_{eq}) for the equivalent capacitance. For this finite and non-zero equivalent capacitance, adding or removing one or more stages to the ladder will not change the capacitance of the network. Therefore, to simplify the circuit we can break it in a manner that only one stage is left with us and in place of the remaining portion we connect a capacitor of magnitude C_{eq}. Now, we calculate the equivalent capacitance of the simplified circuit and put it equal to the guessed value i.e., C_{eq}. This process involves solving a quadratic equation. By solving this equation, we can find the desired value of the equivalent capacitance. See Example 35 for clear understanding of such types of problems.

☞ Note that an infinite ladder should be broken from those points from where the broken part resembles with the original ladder.

EXAMPLE 34. (a) Calculate the equivalent capacitance (in terms of C, the capacitance of each individual capacitor) between points A and B for the infinite ladder of capacitors shown in Figure 3.49 assuming the capacitors are identical. That is assuming $C = C_1 = C_2$. (b) Repeat Part (a) but do not assume that $C_1 = C_2$ and express your answer in terms of C_1 and C_2. (c) Check your results by showing that your result from Part (b) agrees with your result from part (a) if you substitute C for both C_1 and C_2.

APPROACH Let C be the capacitance of each capacitor in the

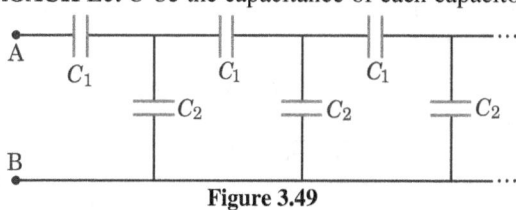

Figure 3.49

ladder and let C_{eq} be the equivalent capacitance of the infinite ladder between terminals A and B. If the capacitance is finite and non-zero, then removing one stage to the ladder will not change the capacitance of the network. In Fig.3.50a, we have drawn a dashed line XY which separates the first stage from rest of the part. So, the equivalent capacitance of rest of the part will still be C_{eq}. Therefore, the equivalent circuit diagram will be similar to Fig.3.50b. We can apply the rules for capacitance combination to the diagram shown in Fig.3.50b to obtain a quadratic equation in C_{eq} that we can solve for the equivalent capacitance between points A and B.

SOLUTION (a) The equivalent capacitance of the series com-

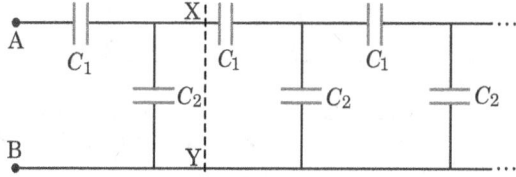

(a) Equivalent capacitance between terminals X any Y will be same as that of between terminals A and B

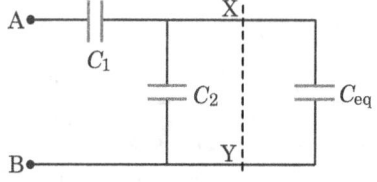

(b) Equivalent Circuit

Figure 3.50

bination of C_1 and $(C_2 \| C_{eq})$ is C_{eq}, so:
$$C_{eq} = \frac{C_1(C_2 + C_{eq})}{C_1 + (C_2 + C_{eq})} = \frac{C(C + C_{eq})}{C + (C + C_{eq})} \quad (\because C_1 = C_2 = C)$$
$$\Rightarrow \quad 2CC_{eq} + C_{eq}^2 = C^2 + CC_{eq}$$
$$\Rightarrow \quad C_{eq}^2 + CC_{eq} - C^2 = 0$$
Solving for C_{eq}, we get
$$C_{eq} = \frac{-C \pm \sqrt{C^2 + 4C^2}}{2} = \frac{-C \pm \sqrt{5C^2}}{2} = \left(\frac{-1 \pm \sqrt{5}}{2}\right)C$$
Since, capacitance of any circuit can never be negative, therefore, the only acceptable value of C_{eq} is-
$$C_{eq} = \left(\frac{-1 + \sqrt{5}}{2}\right)C \approx 0.618\ C$$

(b) The equivalent capacitance of the series combination of C_1 and $(C_2 \| C_{eq})$ is C_{eq}, so:

$$C_{eq} = \frac{C_1(C_2 + C_{eq})}{C_1 + (C_2 + C_{eq})}$$

$$\Rightarrow \quad C_{eq}^2 + (C_1 + C_2)C_{eq} = C_1C_2 + C_1C_{eq}$$

$$\Rightarrow \quad C_{eq}^2 + C_2C_{eq} - C_1C_2 = 0$$

Solving for C_{eq}, we get

$$C_{eq} = \frac{-C_2 \pm \sqrt{C_2^2 + 4C_1C_2}}{2}$$

Since, C_{eq} is a finite and positive value, therefore, ignoring negative value, we get

$$C_{eq} = \frac{-C_2 + \sqrt{C_2^2 + 4C_1C_2}}{2}$$

(c) If $C_1 = C_2 = C$, then:

$$C_{eq} = \frac{-C + \sqrt{C^2 + 4CC}}{2} = \left(\frac{-1+\sqrt{5}}{2}\right)C = 0.618\,C$$

It is in agreement with Part (a).

EXAMPLE 35. Determine the equivalent capacitance C between terminals A and B of the infinite set of capacitors represented in Fig.3.51. Each capacitor has capacitance C_0.

APPROACH Let C_{eq} be the equivalent capacitance of the

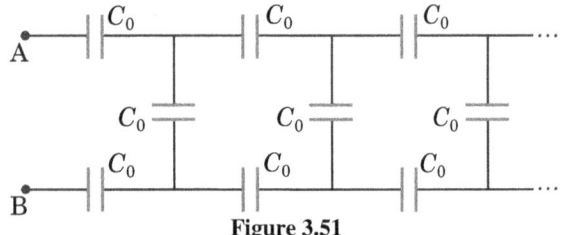
Figure 3.51

infinite ladder between terminals A and B. If the capacitance is finite and non-zero, then removing one or more stages to the ladder will not change the capacitance of the network.

Imagine that the ladder is cut at the dashed line XY [Fig.3.52]

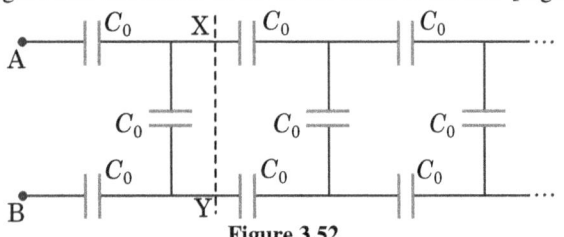
Figure 3.52

and note that the equivalent capacitance of the infinite section to the right of XY is still C_{eq}. The combination of capacitors shown is equivalent to Fig.3.53a

In Fig.3.53b, we have connected the the equivalent capacitance

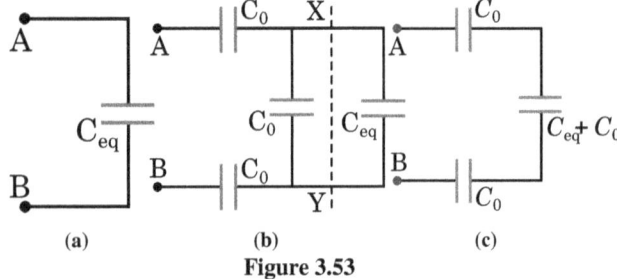
Figure 3.53

(C_{eq}) of right side of line AB in parallel with C_0. Equivalent capacitance of these two parallel capacitors is $C_{eq} + C_0$. It is shown in Fig.3.53c.

SOLUTION From Fig.3.53c, the equivalent capacitance between terminals A and B is given as-

$$\frac{1}{C_{eq}} = \frac{1}{C_0} + \frac{1}{C_{eq} + C_0} + \frac{1}{C_0} = \frac{C_{eq} + C_0 + C_0 + C_{eq} + C_0}{C_0(C_{eq} + C_0)}$$

$$C_0 C_{eq} + C_0^2 = 2C_{eq}^2 + 3C_0 C_{eq}$$

$$\Rightarrow \quad 2C_{eq}^2 + 2C_0 C_{eq} - C_0^2 = 0$$

$$\Rightarrow \quad C_{eq} = \frac{-2C_0 \pm \sqrt{4C_0^2 + 4\left(2C_0^2\right)}}{4}$$

Since, the capacitance can never be $-$ve, therefore, only the positive root is physical. Thus,

$$C_{eq} = \frac{C_0}{2}(\sqrt{3} - 1)$$

3.9 Wheatstone's Balanced Capacitance Bridge

Fig.3.54 shows a Wheatstone's capacitance bridge. V_0 is the applied voltage across it's terminals A and B. If capacitors are so adjusted that there is a zero voltage between points C and D (No charge flows through the voltmeter when it reads zero.), then-

$$\boxed{\frac{C_1}{C_2} = \frac{C_3}{C_4}} \quad (3.25)$$

In this situation, the bridge is said to be balanced.

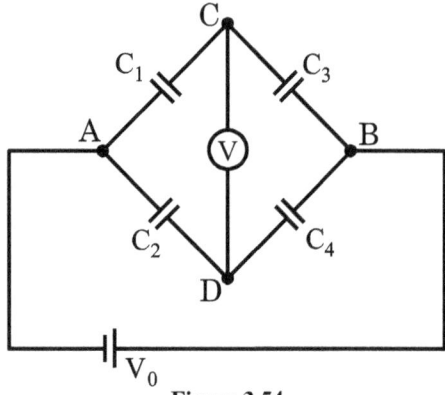
Figure 3.54

Proof: Since, there is no voltage across points C and D, these points will be at the same potential. In this situation, we can imagine there being a connecting wire between points C and D. Then capacitors C_1 and C_2 are in parallel, and so have the same voltage. Also capacitors C_3 and C_4 are in parallel, and so have the same voltage.

If in balanced Wheatstone's capacitance bridge, Q_1, Q_2, Q_3 and Q_4 are the charges on capacitors C_1, C_2, C_3 and C_4 respectively, then

$$V_{AC} = V_{AD} \implies \frac{Q_1}{C_1} = \frac{Q_2}{C_2}$$

$$\implies \frac{Q_1}{Q_2} = \frac{C_1}{C_2} \quad (3.26)$$

and $\quad V_{CB} = V_{DB} \implies \frac{Q_3}{C_3} = \frac{Q_4}{C_4}$

$$\implies \frac{Q_3}{Q_4} = \frac{C_3}{C_4} \quad (3.27)$$

Since, no charge flows through the voltmeter, we could also remove it from the circuit and have no change in the capacitance of the circuit. In that case, capacitors C_1 and C_3 are in series and so have the same charge, i.e., $Q_1 = Q_3$. Likewise capacitors C_2 and C_4 are in other series, and so have the same charge, i.e., $Q_2 = Q_4$.

3.9. WHEATSTONE'S BALANCED CAPACITANCE BRIDGE

Thus,
$$\Rightarrow \frac{Q_1}{Q_2} = \frac{Q_3}{Q_4} \quad (3.28)$$

Substituting the values of Q_1/Q_2 and Q_3/Q_4 from Eq:(3.26) and (3.27) respectively in Eq.(3.28), we get-
$$\boxed{\frac{C_1}{C_2} = \frac{C_3}{C_4}} \quad (3.29)$$

which is the required condition for the bridge to be balanced. In this case, if a fifth capacitor is connected between terminals C and D, then net capacitance of the system does not change. In other words: adding or removing any capacitor between terminals C and D does not change the capacitance of the circuit.

EXAMPLE 36. In Fig.4.172, capacitance of each capacitor is C. Find the equivalent capacitances of the combinations between terminals A and B.

SOLUTION (a) Given circuit diagram is redrawn in Fig.3.56.

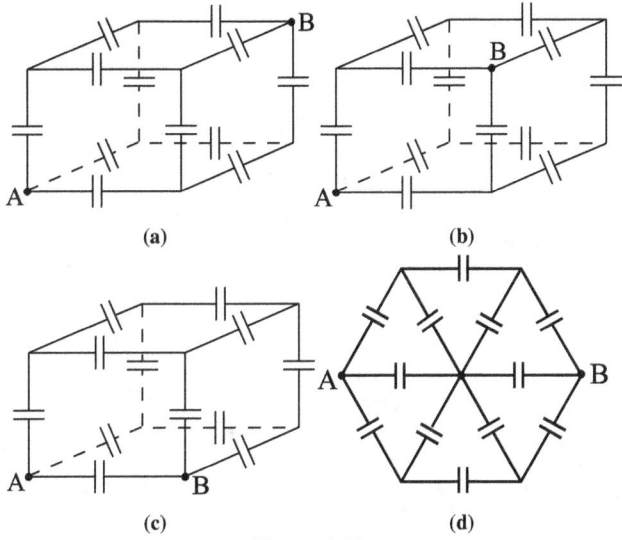

Figure 3.55

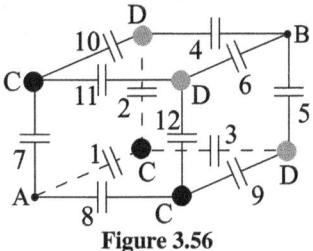

Figure 3.56

Symmetry Observation: All symmetrical points are marked by same letters. These are C and D.

Point C: Between each point C and terminal point A, there is exactly one capacitor of identical capacitance from all directions. These are- 1, 7 and 8 from each direction respectively. Between terminal points B and C there are two similar capacitors from each sides. Theses are- (4, 10), (6, 11), (5, 9), (6, 12), (5, 3), (4, 2).

Point D: Between points D and A, there are exactly two identical capacitors from each direction. These are- (1, 3), (8, 9), (1, 2), (7, 10), (8, 12), (7, 11). Between terminal point B and D there is one capacitor of same magnitude from each direction. These are-4, 5 and 6.

So, all points marked by same letters are symmetrical. At all symmetrical points, electric potential will be same. Therefore, all capacitors attached between terminal A and point B will be in parallel and between terminal points B and D the attached capacitors are also in parallel combination. It's new simplified diagram is shown in Fig.3.57.

In Fig.3.57, equivalent capacitance between terminals A and C

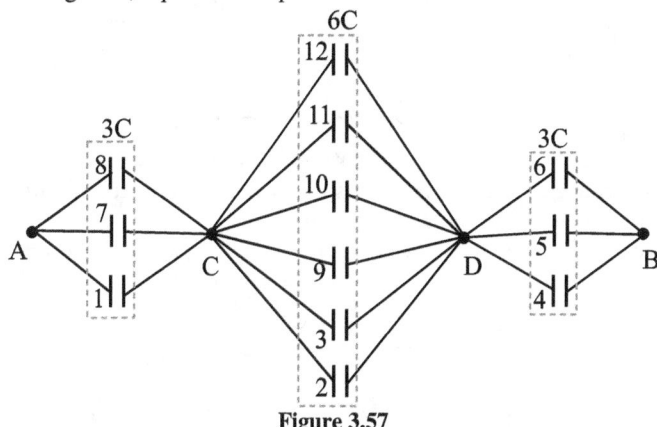

Figure 3.57

is $3C$; between terminals C and D is $6C$ and between terminals D and B it is $3C$. It's next simplified circuit diagram is shown in Fig.3.58.

In Fig.3.58, all capacitors are in series, their equivalent

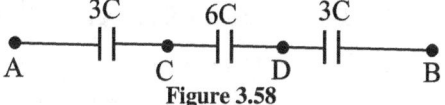

Figure 3.58

capacitance is given by-
$$\frac{1}{C_{eq}} = \frac{1}{3C} + \frac{1}{6C} + \frac{1}{3C} \Rightarrow C_{eq} = 6C/5$$

(b) Fig.3.59 shows circuit diagram with two symmetrical points each marked by C and two other symmetrical points each marked by D.

Symmetry observations: Between terminals A and B, there are

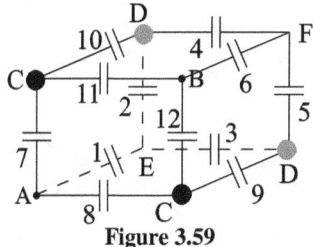

Figure 3.59

some mirror symmetric points. Two mirror symmetric points are marked by C and two other mirror symmetric points are marked by letter D. All mirror symmetric points are equipotential points. It's simplified circuit diagram is shown in Fig.3.60.

Capacitors connected in parallel are shown by dashed boxes.

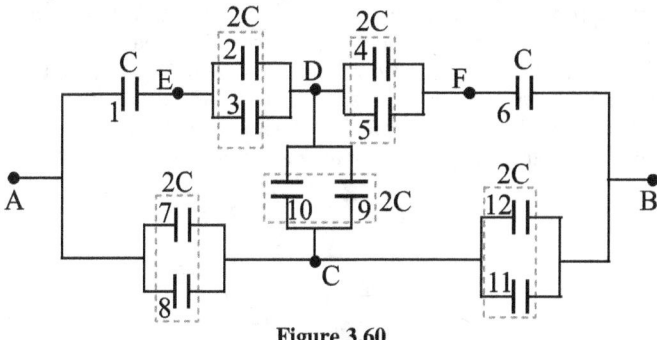

Figure 3.60

Their equivalent capacitances are also indicated with respective boxes. Next simplified diagram is shown in Fig.3.61.

In Fig.3.61, the capacitors in series, with their resultant

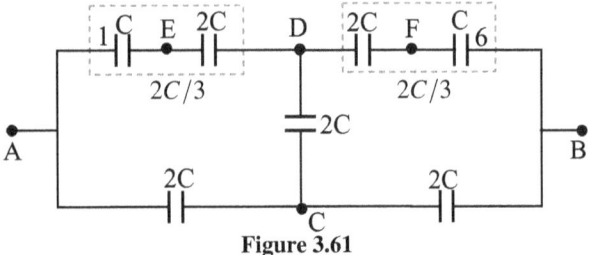

Figure 3.61

capacitances, are shown in dashed boxes. The next simplified circuit diagram is shown in Fig.3.62

In Fig.3.65, we observe that the ratio of capacitance between

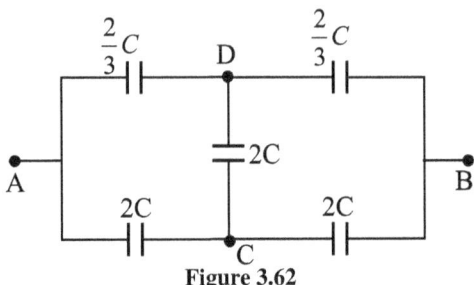

Figure 3.62

terminals A and D with capacitance between terminals A and C i.e., $\dfrac{2C/3}{2C}$ is equal to the ratio of capacitance between terminals B and D with capacitance between terminals B and C, i.e., $\dfrac{2C/3}{2C}$. So, it satisfies the Wheatstone's balanced bridge condition (Eq.3.25). In this situation, point C and D will be equipotential points and so we can remove the capacitor between terminals C and D. It's next simplified circuit diagram is shown in Fig.3.65. Capacitors in series with their equivalent capacitances, are shown in dashed boxes. Their equivalent capacitances are in parallel. The net equivalent capacitance between terminals A and B is given by-

$$C_{eq} = C + \dfrac{C}{3} = \dfrac{4C}{3}$$

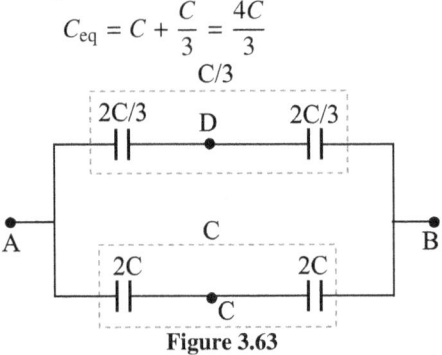

Figure 3.63

(c) Symmetry observation: The circuit diagram is shown in Fig.3.64. There are two mirror symmetric points marked by letter C and two other symmetric points marked by D. So, potentials at both C's will be same and at both D's it will also be the same. It's first simplified circuit diagram is shown in Fig.3.65

All capacitors in parallel with their resultants are shown by

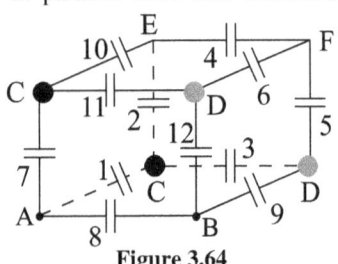

Figure 3.64

dashed boxes. Next simplified diagram is shown in Fig.3.66. Capacitors in series with their equivalent capacitance $(C/2)$ is shown by dashed box. This series equivalent capacitance is in parallel order with the capacitor $2C$. Their equivalent capacitance is $2C + (C/2) = 5C/2$

Next simplified circuit diagram is shown in Fig.3.67. Capacitors

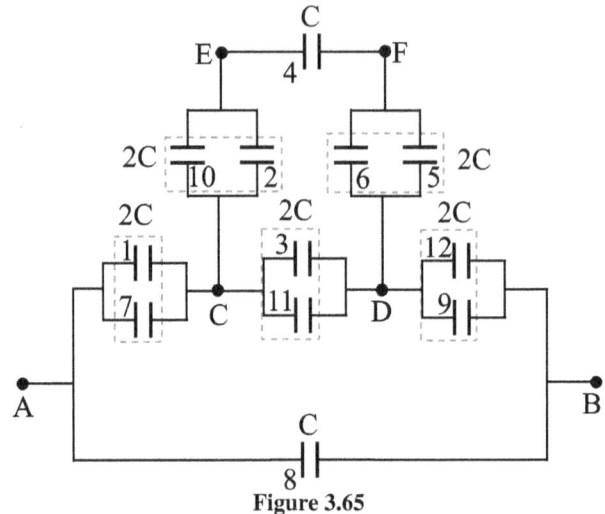

Figure 3.65

connected in branches AC, CD and DB are in series. Their equivalent capacitance (C_1), is given by-

$$\dfrac{1}{C_1} = \dfrac{1}{2C} + \dfrac{1}{5C/2} + \dfrac{1}{2C} = \dfrac{1}{C}\left(1 + \dfrac{2}{5}\right) = \dfrac{7}{5C}$$

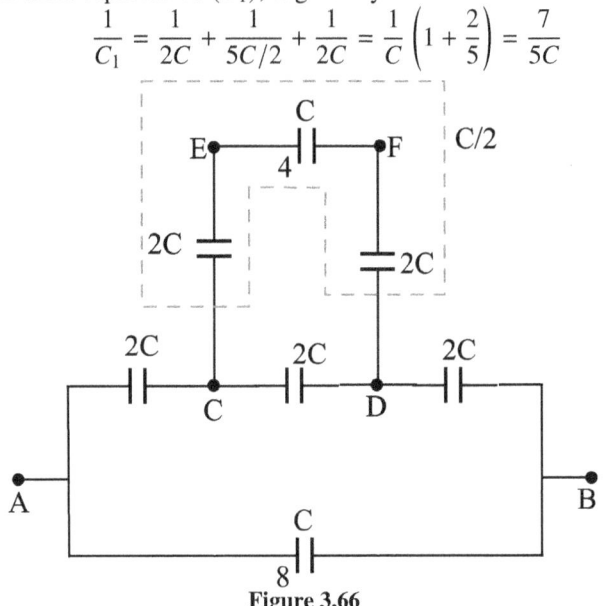

Figure 3.66

Therefore, $\qquad C_1 = 5C/7$

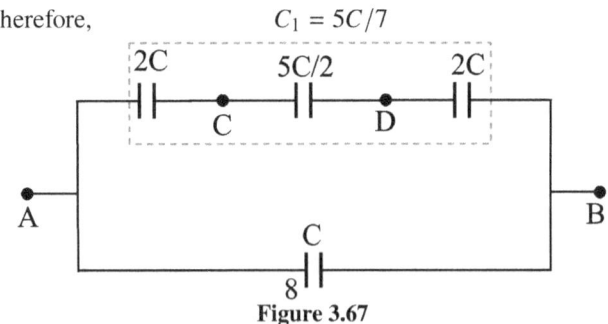

Figure 3.67

This equivalent capacitance C_1 is in parallel with capacitor number 8.

Equivalent capacitance between terminals A and B:

$$C_{eq} = \dfrac{5C}{7} + C = \dfrac{12}{7}C$$

(d) Given circuit diagram is again shown in Fig.3.68a. All symmetric points are shown by same letter. There are two symmetrical points marked by C and two other symmetrical points are marked by D. Both C have same potential and both D also have same potential. So, charges on capacitors connected between terminal A and all terminal points C will be same, i.e., each capacitor 1 and 2 have same charge. Let this charge is q, and the charge on capacitor 7 is q_1. Points marked by letter D are also mirror symmetric to points C, therefore, charges on capacitors 3 and 6 and 8 are similar to 1, 6 and 7 respectively (Fig.3.68b). The net charge on plates inside the dashed rectangle (Fig.3.68b) is zero, therefore, we can isolate this branch from upper and lower branches. Next simplified circuit diagram is shown in Fig.4.186a.

Now, in circuit diagram 4.186a, capacitors 9 and 11 are in series;

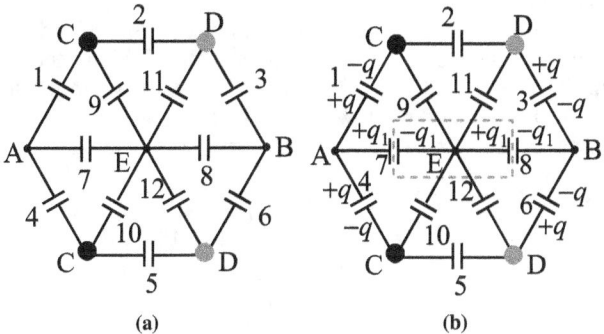

Figure 3.68

similarly, 10 and 12 are also in series. If C_1 is the equivalent capacitance of capacitors 9 and 11, then-
$$\frac{1}{C_1} = \frac{1}{C} + \frac{1}{C} \implies C_1 = \frac{C}{2}$$
Similarly, the equivalent capacitance of 10 and 12 is also $C/2$. These equivalent capacitors are in parallel with capacitors 2 and 5 respectively (Fig.3.69b).

In Fig.3.69b, net capacitance between upper terminals C and D is given by-
$$C_2 = \frac{C}{2} + C = \frac{3}{2}C$$
Similarly, net capacitance between lower terminals C and D is also $3C/2$.

Next simplified circuit diagram is shown in Fig.3.70a.

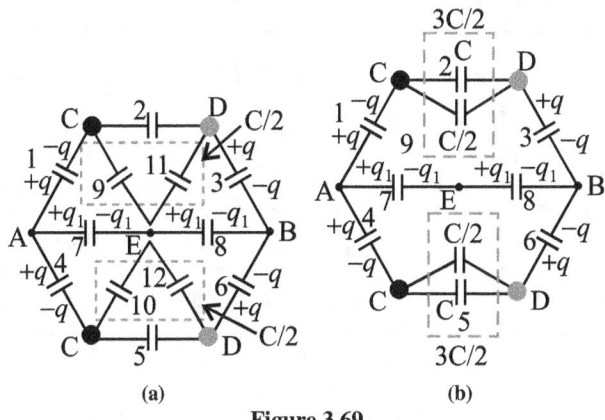

Figure 3.69

The capacitors in series are rounded by dashed rectangles. Series equivalent capacitance of upper rectangle is given by-
$$\frac{1}{C_3} = \frac{1}{C} + \frac{2}{3C} + \frac{1}{C} \implies C_3 = \frac{3C}{8}$$
Similarly, the series equivalent of lower rectangle is also $3C/8$. Series equivalent of middle rectangle is-
$$C_4 = \frac{C \cdot C}{C + C} = \frac{C}{2}$$
Next simplified circuit diagram is shown in Fig.3.70b. All capacitors connected between terminals A and B are in parallel. Their equivalent capacitance is given by-
$$C_{eq} = \frac{3C}{8} + \frac{C}{2} + \frac{3C}{8} = \frac{5C}{4}$$

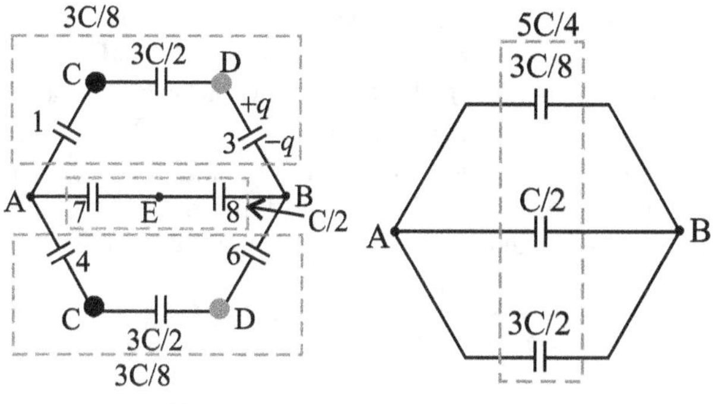

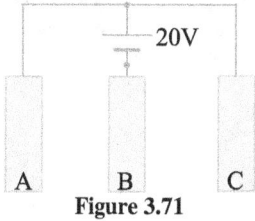

Figure 3.70

EXAMPLE 37. Each of the three plates shown in Fig.3.71 has an area of 200 cm^2 on one side and the gap between the adjacent plates is 0.2 mm. The emf of the battery is 20 V. What is the equivalent capacitance of the system between the terminal points? Find the distribution of charge on various surfaces of the plates.

SOLUTION All surfaces of given plates are numbered as shown

Figure 3.71

in Fig 3.72a. Let the +ve terminal of battery is at +20 V and it's −ve terminal at 0 V. Surfaces ①, ②, ⑤ and ⑥ are directly connected to the +ve terminals of battery, therefore they all are at +20 V potential. Surfaces ③ and ④ are connected with −ve terminals of battery at 0 V so they are at 0 V potential.

Surfaces ②, ③ and surfaces ④, ⑤ form two parallel plate

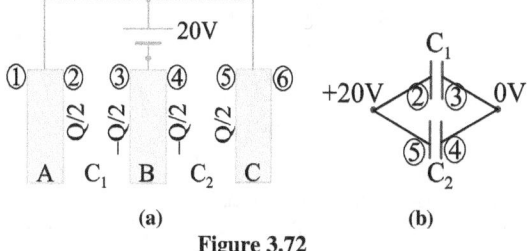

Figure 3.72

capacitors of capacitances C_1 and C_2 respectively connected in parallel as shown in Fig.3.72b. For surfaces ① and ⑥ second capacitor plates can be assumed at infinite so for each of them the plate separation will be infinite and hence the capacitances of these capacitors will be zero.

Therefore, equivalent capacitance of the system -
$$C = C_1 + C_2 \quad \text{(capacitors are in parallel)}$$
Here, $C_1 = C_2 = \frac{\epsilon_0 A}{d}$
$$= \frac{8.85 \times 10^{-12} \text{Fm}^{-1} \times 200 \times 10^{-4} \text{ m}^2}{2 \times 10^{-4} \text{ m}}$$

$= 8 \cdot 85 \times 10^{-10} \text{F} = 0.885 \text{ nF}$
Therefore, $C = C_1 + C_2 = 2 \times 0.885 \text{ nF} = 1.77 \text{ nF}$
Calculation of charge on each plate surface: Net charge supplied by the battery:
$Q = CV = (1.77 \text{ nF}) \times 20 \text{ V} = 35.4 \text{ nC}$
Let charge on capacitor C_1 is Q_1, then-
$$Q_1 = C_1 V = (0.885 \text{ nF}) \times 20 \text{ V} = 17.7 \text{ nC}$$
Similarly, the charge Q_2 on capacitor C_2 is
$$Q_2 = C_2 V = (0.885 \text{ nF}) \times 20 \text{ V} = 17.7 \text{ nC}$$
Therefore, electric charge on each of the surfaces ② and ⑤ is 17.7 nC whereas on each of the surfaces ③ and ④ it is -17.7 nC
From symmetry it is clear that surfaces ① and ⑥ are symmetrical. So charges on them will also be similar. Let charge on each of the surface ① and ⑥ is q, then by conservation of charge, we have-
$$Q_1 + Q_2 + 2q = Q \implies 2q = Q - (Q_1 + Q_2)$$
$$\implies 2q = 35.4 \text{ nC} - (17.7 \text{ nC} + 17.7 \text{ nC}) = 0$$
$$\implies q = 0$$
So, surfaces ① and ⑥ i.e., outer surfaces of plates A and C will not have any charge.

EXAMPLE 38. A parallel-plate capacitor has a plate separation d and plate area A. An uncharged metallic slab of thickness t is inserted midway between the plates.
(a) Find the capacitance of the device.
(b) Show that the capacitance of the original capacitor is unaffected by the insertion of the metallic slab if the slab is infinitesimally thin.
(c) What if the metallic slab in part (a) is not midway between the plates? How would that affect the capacitance?

(a) APPROACH Figure 3.73a shows the metallic slab between the plates of the capacitor. Any charge that appears on one plate of the capacitor must induce a charge of equal magnitude and opposite sign on the near side of the slab as shown in Figure 3.73a. Consequently, the net charge on the slab remains zero and the electric field inside the slab is zero.
The planes of charge on the metallic slab's upper and lower edges are identical to the distribution of charges on the plates of a capacitor. The metal between the slab's edges serves only to make an electrical connection between the edges. Therefore, we can model the edges of the slab as conducting planes and the bulk of the slab as a wire. As a result, the capacitor in Figure 3.73a is equivalent to two capacitors in series, each having a plate separation $(d - t)/2$ as shown in Figure 3.73b.
To find net capacitance, we use Eq.3.8 and the rule for adding two capacitors in series (Eq.3.17) to find the equivalent capacitance in Figure 3.73b:
$$\frac{1}{C} = \frac{1}{C_1} + \frac{1}{C_2} = \frac{1}{\frac{\epsilon_0 A}{(d-t)/2}} + \frac{1}{\frac{\epsilon_0 A}{(d-t)/2}} \implies C = \frac{\epsilon_0 A}{d - t}$$

(b) In the result for part (a), let $t \to 0$:
$$C = \lim_{t \to 0} \left(\frac{\epsilon_0 A}{d - t} \right) = \frac{\epsilon_0 A}{d}$$
The result of part (b) is the original capacitance before the slab is inserted, which tells us that we can insert an infinitesimally thin metallic sheet between the plates of a capacitor without affecting the capacitance.

(c) Let's imagine moving the slab in Figure 3.73a upward so that the distance between the upper edge of the slab and the upper plate is x (see Fig.3.74). Then, the distance between the lower edge of the slab and the lower plate is $d - x - t$. As in part (a), we find the total capacitance of the series combination:
$$\frac{1}{C} = \frac{1}{C_1} + \frac{1}{C_2} = \frac{1}{\epsilon_0 A / x} + \frac{1}{\epsilon_0 A / (d - x - t)}$$

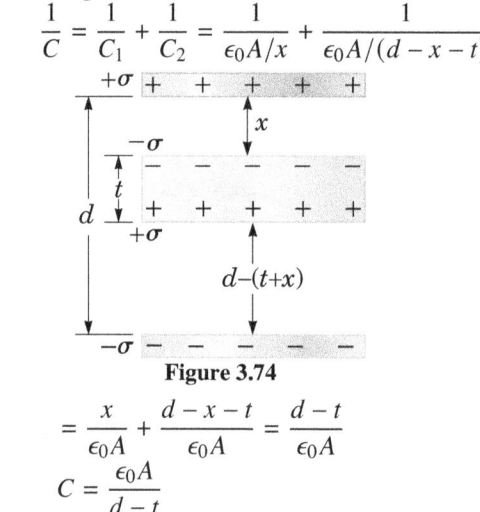

Figure 3.74

$$= \frac{x}{\epsilon_0 A} + \frac{d - x - t}{\epsilon_0 A} = \frac{d - t}{\epsilon_0 A}$$
$$\implies C = \frac{\epsilon_0 A}{d - t}$$
which is the same result as found in part (a). Thus, capacitance is independent of the value of x, so it does not matter where the slab is located. In Figure 3.73b, when the central structure is moved up or down, the decrease in plate separation of one capacitor is compensated by the increase in plate separation for the other.

3.10 Energy Stored in the Capacitor's Electric Field

In order to charge a capacitor, some charge-separating device (a cell or battery) must transfer positive charge from one plate to the other*(Fig.3.75b). During this transfer, the charge-separating device does work on the capacitor. The total work done by this device in charging the capacitor up to a final amount Q is stored in the form of electric potential energy in it.

A capacitor does not become charged instantly. It takes some time. Initially, when the capacitor is uncharged, no work is required to move the first bit of charge over. When some charge is transferred from one plate to other, it requires work to add more charge of the same sign because of the electric repulsion. The more charge already on a plate, the more will be potential difference between the plates and the more work required to add additional charge.

Suppose that at some intermediate stage of charging, the charge

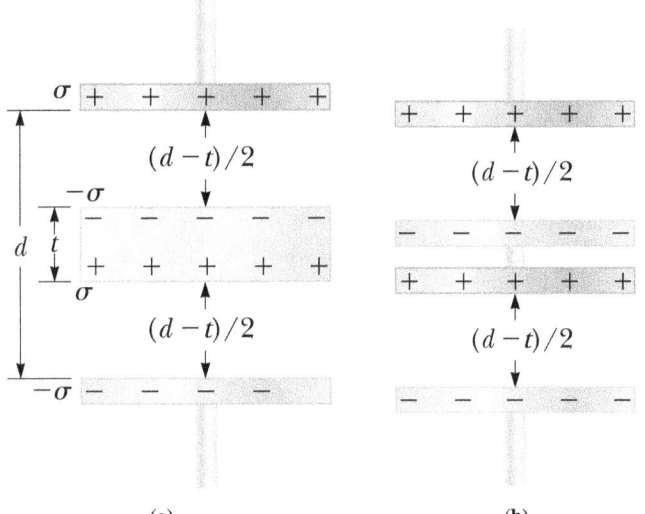

(a)   **(b)**
Figure 3.73: (a) A parallel-plate capacitor of plate separation d partially filled with a metallic slab of thickness t. (b) The equivalent circuit of the device in (a) consists of two capacitors in series, each having a plate separation $(d - t)/2$.

---
*Actually, it is a transfer of electrons in opposite direction which is equivalent to transfer of same amount of positive charge in the given direction

3.10. ENERGY STORED IN THE CAPACITOR'S ELECTRIC FIELD

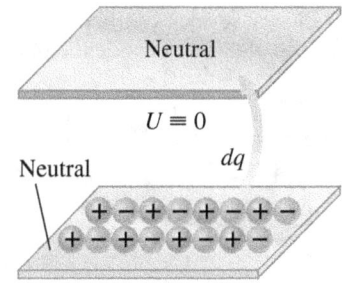

(a) Initial uncharged stage

The instantaneous charge on the plates is $\pm q$

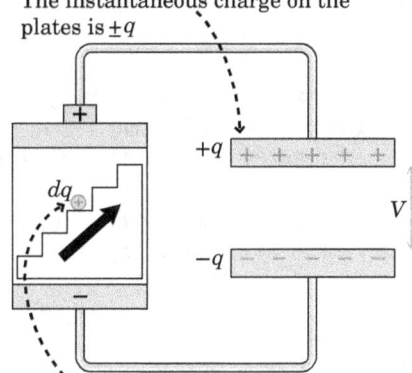

The battery does work $V\,dq$ to move charge dq from the negative plate to the positive plate.

(b) Side view with battery when potential difference between the plates is V

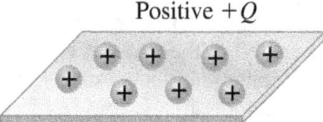

Positive $+Q$

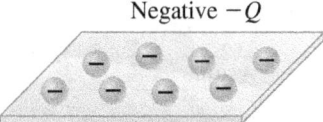

Negative $-Q$

(c) Final charged stage

Figure 3.75

on the positive plate is q, so that the potential difference across the plates, is q/C. The work needed to transport the next piece of charge dq from $-$ve plate to $+$ve plate, is-

$$dW = V\,dq = \frac{q}{C}dq$$

The total work necessary, then, to go from $q = 0$ to $q = Q$, is

$$W = \int dW = \frac{1}{C}\int_0^Q q\,dq = \frac{Q^2}{2C}$$

In practice, this work done is supplied by a battery.
Thus, we can say that the energy "stored" in a capacitor is

$$U = \frac{1}{2}\frac{Q^2}{C} \qquad (3.30)$$

Since, $Q = CV$, therefore, in terms of C and V Eq.3.30, becomes-

$$U = \frac{1}{2}CV^2$$

Again using, $C = Q/V$, in above expression, we get

$$U = \frac{1}{2}QV$$

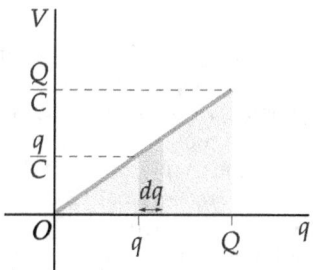

Figure 3.76: The work needed to charge a capacitor is the integral of $V\,dq$ from the original charge of $q = 0$ to the final charge of $q = Q$. This work is equal to area under the curve. That is, the work equals the area of the triangle of height Q/C and width Q.

Therefore, we can write-

$$\boxed{U = \frac{Q^2}{2C} = \frac{1}{2}CV^2 = \frac{1}{2}QV} \qquad (3.31)$$

A potential energy value is relative to an assumed zero reference. For capacitors, the potential energy is zero when the capacitor is completely uncharged.

The total work required to charge the capacitor from original charge $q = 0$ to a final charge of $q = Q$, is the shaded area under the straight line in q - V graph[Fig.3.76].

Total Work done by a Battery in charging a Capacitor

Suppose we charge a capacitor by connecting it to a battery. The potential difference V when the capacitor is fully energized with charge $+Q$ on one conductor and charge $-Q$ on the other is just the potential difference between the terminals of the battery before they were connected to the capacitor. The total work done by the battery in charging the capacitor is QV, which is twice the amount of energy stored in the capacitor. The additional work done by the battery is either dissipated as thermal energy in the battery and in the connecting wires or radiated as electromagnetic waves. Thus,

Energy stored in capacitor = $\frac{1}{2} \times$ Energy given by battery

☞ Thus, the battery loses 50% of its energy in the form of thermal energy, sound energy and electromagnetic radiation.

EXAMPLE 39. Charge on capacitor is increased by 2 C and energy stored becomes 44%. Find initial charge.

SOLUTION If Q is the initial charge on the positive plate of capacitor and C is it's capacitance, then initial value of electric energy stored in capacitor is given by

$$U_i = \frac{Q_i^2}{2C}$$

Now, if charge is increased to Q_f, then final value of stored energy-

$$U_f = \frac{Q_f^2}{2C}$$

Therefore, % increase in energy

$$= \frac{U_f - U_i}{U_i} \times 100 = \frac{\frac{Q_f^2}{2C} - \frac{Q_i^2}{2C}}{\frac{Q_i^2}{2C}} \times 100 = 44$$

$$\implies \left(Q_f^2 - Q_i^2\right) = 0.44\,Q_i^2$$

$$\implies Q_f^2 = 1.44\,Q_i^2 \qquad \cdots (1)$$

Given: $\qquad Q_f = Q_i + 2 \qquad \cdots (2)$

Putting this value of Q_f from Eq.(2) in Eq.(1), we get

$(Q_i + 2)^2 = 1.44 Q_i^2 \implies 0.44 Q_i^2 - 4Q_i - 4 = 0$

$\implies 11Q_i^2 - 100Q_i - 100 = 0$

$\implies 11Q_i^2 - 110Q_i + 10Q_i - 100 = 0$

$\implies 11Q_i(Q_i - 10) + 10(Q_i - 10) = 0$

$\implies (Q_i - 10)(10Q_i + 10) = 0$

$\implies Q_i = 10$ C and $Q_i = -1$ C
Corresponding to the root $Q_i = -1$ C, Eq.(2) gives
$$Q_f = Q_i + 2 = +1 \text{ C}.$$
These values of Q_i and Q_f do not satisfy Eq.(1), therefore $Q_i = -1$ C is not an acceptable root.
Now, corresponding to the root $Q_i = 10$ C, Eq.(2) gives
$$Q_f = Q_i + 2 = +12 \text{ C}.$$
These values satisfy Eq.(1), therefore $Q_i = 10$ C is the only acceptable value of initial charge on capacitor.

3.10.1 Electric Field Energy Density

During the process of charging a capacitor, an electric field is produced between the plates. The work required to charge the capacitor can be thought of as the work required to establish the electric field. That is, we can think of the energy stored in a capacitor as energy stored in the electric field, called electrostatic field energy.

Consider a parallel-plate capacitor. We can relate the energy stored in the capacitor to the electric field strength E between the plates. The potential difference between the plates is related to the electric field by $V = Ed$, where d is the separation between the plates. The capacitance is given by-
$$C = \epsilon_0 A/d$$
therefore, the electric potential energy can be written as-
$$U = \frac{1}{2}CV^2 = \frac{\epsilon_0 A}{2d}V^2 = \frac{\epsilon_0 A d}{2}\left(\frac{V}{d}\right)^2 = \frac{1}{2}\epsilon_0(Ad)E^2,$$
where $E(= V/d)$ is the electric field strength between the plates. The quantity Ad is the volume of the space between the plates of the capacitor. This volume is the volume of the region containing the electric field. The energy per unit volume is called the energy density u_e. The energy density in an electric field of strength E is thus,
$$\boxed{u = \frac{U}{Ad} = \frac{1}{2}\epsilon_0 E^2} \quad (3.32)$$
Note that this result, which is derived for the special case of a parallel-plate capacitor, is valid for any capacitors.

This is a viewpoint which is very different from the conventional viewpoint of Newtonian mechanics, where potential energy is associated with the position of a particle (matter), but not associated with a force field. The viewpoint that energy is associated with a force field, if true, suggests that electric field has a physical significance that is more than representing force between charges. It is also a kind of matter because energy and matter are equivalent.

3.10.1.1 Alternative Method

In a parallel plate capacitor, two oppositely charged plates are separated by some distance and there is an electrostatic attraction between them. If we imagine that, initially both oppositely charged plates were infinitely close (without touching) to each other and we are going to make a capacitor by separating them slowly at a distance d from each other. For it we have to apply an external force against the electrostatic attraction of the plates. This work creates an electric field in the space between the plates and stored in the form of electrostatic energy.

Let us consider two infinitely close (without touching) oppositely charged parallel plates A and B having charges $+Q$ and $-Q$ respectively. Their surface charge densities are $+\sigma$ and $-\sigma$ respectively. Suppose, plate B is fixed whereas the plate A is free to move in a perpendicular direction of B. Now, we move plate A without any acceleration away from B. Suppose at any instant it's distance from B is x. The intensity of electric field between the plates is σ/ϵ_0 in downward direction and the electric field inside the metal plates A is always zero. All the charge resides on the inner surface of plates. The intensity of electric field at the surface where charge resides can be taken as the average of intensities between the plates and inside the plates. i.e.,
$$E_{av} = \frac{(\sigma/\epsilon_0) + 0}{2} = \frac{\sigma}{2\epsilon_0}$$
Now, if we displace the plate A with zero acceleration, in upward

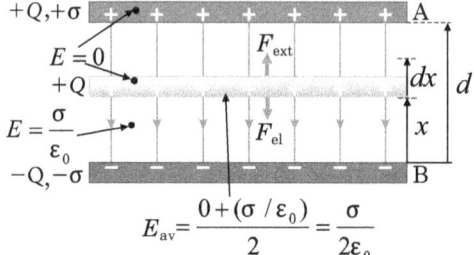

$$E_{av} = \frac{0 + (\sigma/\epsilon_0)}{2} = \frac{\sigma}{2\epsilon_0}$$

Figure 3.77: Parallel plate capacitor

direction, by an extra infinitesimally small displacement dx, then work done by external force against electric force is given by
$$dW = \vec{F}_{ext}\cdot d\vec{x}$$
Since, corresponding to zero acceleration, the external force should be exactly equal to electrostatic force (F_{el}) applied by lower plate, i.e., $F_{ext} = F_{el} = QE_{av}$, therefore-
$$dW = \vec{F}_{ext}\cdot d\vec{x} = F_{ext}\,dx = F_{el}\,dx = QE_{av}dx$$
$$= Q\frac{\sigma}{2\epsilon_0}dx = (\sigma A)\frac{\sigma}{2\epsilon_0}dx = \frac{1}{2}\left(\frac{\sigma}{\epsilon_0}\right)^2 \epsilon_0 A\,dx$$
Total external work done in displacing the plate A by distance d is-
$$\boxed{W = \frac{1}{2}\epsilon_0 E^2 A \int_0^d dx = \frac{1}{2}\epsilon_0 E^2(Ad)} \quad (3.33)$$
Here, Ad is the space volume between the plates of the capacitor. This work done is stored in the form of electric potential energy in the two oppositely charged plate system (i.e., capacitor). Therefore, electrostatic potential energy,
$$U = \frac{1}{2}\epsilon_0 E^2(Ad)$$
Therefore, energy density-
$$u = \frac{U}{\text{volume}} = \frac{U}{Ad} = \frac{1}{2}\epsilon_0 E^2$$
i.e., $\boxed{u = \frac{1}{2}\epsilon_0 E^2} \quad (3.34)$

EXAMPLE 40. A parallel plate capacitor is charged and the charging battery is then disconnected. If the plates of the capacitor are moved farther apart by means of insulating handles, then which of the following option(s), correct?
(A) the charge on the capacitor increases.
(B) the voltage across the plates increases.
(C) the capacitance increases.
(D) the electrostatic energy stored in the capacitor increases.

SOLUTION (B, D) The capacitance of a parallel plate capacitor of plate area A and plate separation d is $C = \epsilon_0 A/d$. On increasing the plate separation from d to d', the capacitance decreases to $C' = \epsilon_0 A/d' < C$. The charge Q on the capacitor does not change because it is an isolated system (battery is removed and insulated handles are used to increase the plate separation). The voltage across the plates, $V' = Q/C'$, and the electrostatic energy stored in the capacitor, $U = Q^2/(2C')$, both increase due to decrease in the capacitance.

☞ We can also find the work done by the external force in increasing the plate separation from d to d', as given in Eq. 3.33 — except the lower limit of integration is d, and the upper limit is d' this time

$$W = \frac{1}{2}\epsilon_0 E^2 A \int_d^{d'} dx = \frac{1}{2}\epsilon_0 E^2 A (d' - d)$$

3.10.2 Redistribution of Charges

Suppose two isolated spherical conductors of capacitances C_1 and C_2 are given charges q_1 and q_2 and raised to potential V_1 and V_2 respectively. Then,

Initial Charges: q_1 q_2

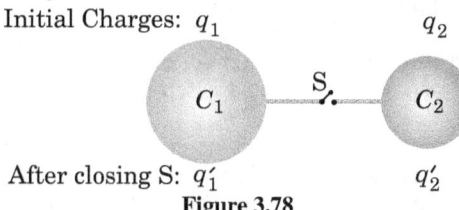

After closing S: q_1' q_2'

Figure 3.78

$$q_1 = C_1 V_1 \quad \text{and} \quad q_2 = C_2 V_2 \quad (3.35)$$

Now, if we join these two conductors by a thin conducting wire as shown in Fig.3.78, then on closing the switch 'S', the positive charge will flow from the conductor at higher potential to conductor at lower potential till their potentials become equal. During this process, the net charge always remains conserved. So, if q_1' and q_2' are final charges after redistribution and V the common potential difference of each conductor, then

$$q_1' = C_1 V \quad \text{and} \quad q_2' = C_2 V \quad (3.36)$$

Therefore, from conservation of charge, we have-

$$q_1 + q_2 = q_1' + q_2' \quad (3.37)$$

Substituting the values of q_1, q_2, q_1' and q_2' from Eq.3.35 and 3.36 in Eq.3.37, we get

$$C_1 V_1 + C_2 V_2 = C_1 V + C_2 V$$

Solving above equation for V, we get-

$$\boxed{V = \frac{C_1 V_1 + C_2 V_2}{C_1 + C_2}} \quad (3.38)$$

Again from Eq.(3.36) and (3.37), we have

$$V = \frac{q_1'}{C_1} = \frac{q_2'}{C_2} = \frac{q_1' + q_2'}{C_1 + C_2} = \frac{q_1 + q_2}{C_1 + C_2}$$

Solving for q_1' and q_2', we get

$$q_1' = \left(\frac{C_1}{C_1 + C_2}\right)(q_1 + q_2), \quad q_2' = \left(\frac{C_2}{C_1 + C_2}\right)(q_1 + q_2)$$

$$\Rightarrow \quad \boxed{\frac{q_1'}{q_2'} = \frac{C_1}{C_2}} \quad (3.39)$$

From Eq.3.39, we can say that in electrostatic equilibrium, the charges on capacitors (or conductors) are in the ratio of their capacitances.

☞ Students are advised to derive Eq.3.39 by also considering two parallel plate capacitors.

EXAMPLE 41. Two capacitors having capacitances C_1 and C_2 respectively are connected as shown in Fig.3.79. Initially, capacitor C_1 is charged to a potential difference V volt by a battery. The battery is then removed and the charged capacitor C_1 is now connected to uncharged capacitor C_2 by closing the switch S. Calculate the amount of charge on the capacitor C_2, after equilibrium is reached.

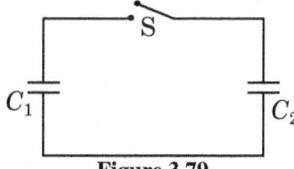

Figure 3.79

SOLUTION Initially, capacitor C_1 is charged to a potential difference V, therefore magnitude of charge stored on each plate of this capacitor is

$$Q_0 = C_1 V \quad \cdots (1)$$

When both capacitors are connected by switch S as shown in Fig.3.79, the distribution of charge started until both capacitors reach to a common potential V' (say). If Q_1 and Q_2 are charges on capacitors C_1 and C_2 respectively when the reach to a common potential V', then

$$V' = \frac{Q_1}{C_1} = \frac{Q_2}{C_2} = \frac{Q_1 + Q_2}{C_1 + C_2} \quad \cdots (2)$$

By charge conservation, $Q_1 + Q_2 = Q_0 = C_1 V$ (from Eq.(1)), therefore above Eq.(2) gives

$$V' = \frac{Q_1}{C_1} = \frac{Q_2}{C_2} = \frac{Q_1 + Q_2}{C_1 + C_2} = \frac{C_1 V}{C_1 + C_2}$$

Solving for Q_2, we get

$$Q_2 = \frac{C_1 C_2}{C_1 + C_2} V$$

EXAMPLE 42. A capacitor C_1 of capacitance 5 μF is charged to a potential of 30 V using a battery. The battery is then removed and the charged capacitor is connected to an uncharged capacitor C_2 of capacitance 10 μF as shown in Fig.3.80. When the switch is closed charge flows between the capacitors. Calculate the charge stored on the capacitor C_2 after equilibrium is reached.

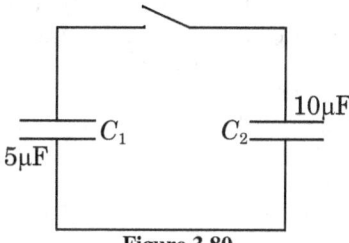

Figure 3.80

APPROACH This problem is similar to Example 41 with $C_1 = 5\,\mu$F, $C_2 = 10\,\mu$F and $V = 30$ V.

SOLUTION Substituting above values in the result of Example 41: $Q_2 = \dfrac{C_1 C_2}{C_1 + C_2} V$, we get

$$Q_2 = \frac{C_1 C_2}{C_1 + C_2} V = \frac{(5\,\mu F)((10\,\mu F))}{(5\,\mu F) + (10\,\mu F)} \times 30\,V = 100\,\mu C.$$

EXAMPLE 43. Two capacitors, $C_1 = 2.2\,\mu$F and $C_2 = 1.2\,\mu$F, are connected in parallel to a 24 V source as shown in Fig.3.81a. After they are charged they are disconnected from the source and from each other, and then reconnected directly to each other with plates of opposite sign connected together as shown in Fig.3.81b. Find the charge on each capacitor and the potential across each after equilibrium is established.

APPROACH Initially, both capacitors C_1 (= 2.2 μF) and C_2 (=

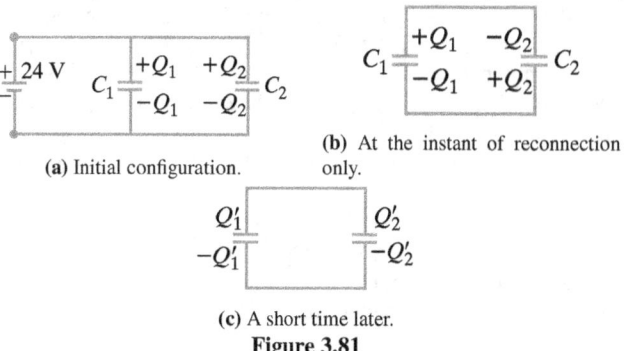

Figure 3.81

1.2 μF) are are connected with a 24 V battery in parallel as shown in Fig.3.81a. Therefore, the initial charge on each capacitor can be obtained by using expression: $Q = CV$. After disconnecting from the source, when they are directly connected with each other in opposite polarities, the redistribution of charge takes place until both capacitors reach the same potential. During this charge will remain conserved.

SOLUTION First we calculate how much charge has been placed on each capacitor after the power source has charged them fully, using Eq. $Q = CV$:
$$Q_1 = C_1V = (2.2\ \mu F)(24\ V) = 52.8\ \mu C,$$
$$Q_2 = C_2V = (1.2\ \mu F)(24\ V) = 28.8\ \mu C$$

Next the capacitors are connected in parallel, Fig.3.81b, and the potential difference across each must quickly equalize. Thus, the charge cannot remain as shown in Fig.3.81b, but the charge must rearrange itself so that the upper plates have the same sign of charge, with the lower plates having the opposite charge as shown in Fig.3.81c. Relation : $Q = CV$ can be applied for each capacitor:
$$Q'_1 = C_1V' \quad \text{and} \quad Q'_2 = C_2V'$$
where V' is the voltage across each capacitor after the charges have rearranged themselves. We don't know Q'_1, Q'_2, or V', so we need a third equation. This is provided by charge conservation. The charges have rearranged themselves between Figs.3.81b and 3.81c. The total charge on the upper plates in those two Figures must be the same, so we have
$$Q'_1 + Q'_2 = Q_1 - Q_2 = (52.8 - 28.8)\ \mu C = 24.0\ \mu C$$
On combining the last three equations, we get:
$$Q'_1 + Q'_2 = C_1V' + C_2V' = (C_1 + C_2)\ V'$$
or $\quad V' = (Q'_1 + Q'_2)/(C_1 + C_2)$
$$= 24.0\ \mu C/3.4\ \mu F = 7.06\ V \approx 7.1\ V,$$
$$Q'_1 = C_1V' = (2.2\ \mu F)(7.06\ V) = 15.5\ \mu C \approx 16\ \mu C,$$
$$Q'_2 = C_2V' = (1.2\ \mu F)(7.06\ V) = 8.5\ \mu C$$
where we have kept only two significant figures in our final answers.

EXAMPLE 44. Three capacitors of capacitances $2\ \mu F$, $3\ \mu F$ and $6\ \mu F$ are connected in series with a 12 V battery. All the connecting wires are disconnected, the three positive plates are connected together and the three negative plates are connected together. Find the charges on the three capacitors after the reconnection.

APPROACH First find the equivalent capacitance in series combination by applying Eq.3.18:
$$\frac{1}{C} = \frac{1}{C_1} + \frac{1}{C_2} + \frac{1}{C_3}$$
Here, C is the equivalent capacitance of the combination.
After finding C, find charge supplied by the battery by applying Eq.3.1: $Q = CV$
Since, all capacitors are in series, so equal charge ($= Q$) will appear on each capacitor.
When all the three positively charged plates connected together and all the three negative plates are connected together, they comes to same potential difference. If Q_1, Q_2 and Q_3 were the charges on capacitors of capacitances $2\ \mu F$, $3\ \mu F$, and $6\ \mu F$ respectively when they were going to be connected in parallel and Q'_1, Q'_2 and Q'_3 are the charges after reconnecting the capacitors, then common potential difference between the terminals of all the capacitors-
$$V' = \frac{Q'_1}{C_1} = \frac{Q'_2}{C_2} = \frac{Q'_3}{C_3} = \frac{Q'_1 + Q'_2 + Q'_3}{C_1 + C_2 + C_3}$$
By charge conservation, $Q'_1 + Q'_2 + Q'_3 = Q$.
Therefore, from above equation,
$$Q'_1 = \frac{Q}{C_1 + C_2 + C_3}C_1 \quad \ldots \text{(i)}$$
$$Q'_2 = \frac{Q}{C_1 + C_2 + C_3}C_2 \quad \ldots \text{(ii)}$$
and $\quad Q'_1 = \frac{Q}{C_1 + C_2 + C_3}C_3 \quad \ldots \text{(iii)}$

Now simplify for Q'_1, Q'_2 and Q'_3.

SOLUTION The equivalent capacitance of the three capacitors connected in series is given by
$$\frac{1}{C} = \frac{1}{2\mu F} + \frac{1}{3\mu F} + \frac{1}{6\mu F}$$
$$C = 1\mu F$$
The charge supplied by the battery,
$$Q = CV = 1\mu F \times 12\ V = 12\ \mu C$$
As the capacitors are connected in series, $+12\ \mu C$ charge appears on each of the positive plates and $-12\ \mu C$ on each of the negative plates, i.e., $Q_1 = Q_2 = Q_3 = 12\ \mu C$.
When the charged capacitors are connected in parallel as shown in Fig.3.82, the total positive charge $Q_1 + Q_2 + Q_3 = +36\ \mu C$ and total negative charge $-Q_1 - Q_2 - Q_3 = -36\ \mu C$ redistribute until all capacitors reach to a common potential.
Now, from Eq.(i), (ii) and (iii), we have-

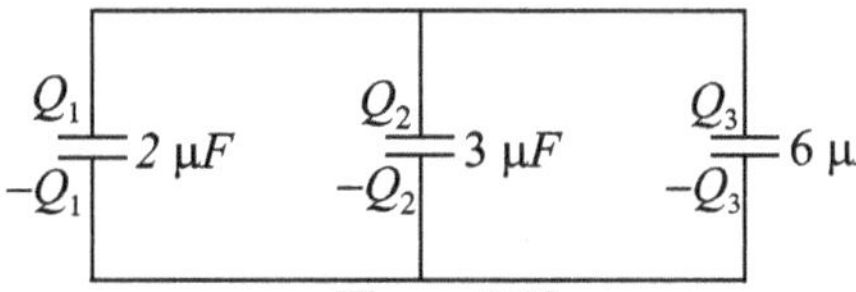

Figure 3.82

$$Q'_1 = \frac{Q}{C_1 + C_2 + C_3}C_1$$
$$= \frac{36\ \mu C}{2\ \mu F + 3\ \mu F + 6\ \mu F}(2\ \mu F) = \frac{72}{11}\ \mu C$$
$$Q'_2 = \frac{Q}{C_1 + C_2 + C_3}C_2$$
$$= \frac{36\ \mu C}{2\ \mu F + 3\ \mu F + 6\ \mu F}(3\ \mu F) = \frac{108}{11}\ \mu C$$
$$Q'_3 = \frac{Q}{C_1 + C_2 + C_3}C_1$$
$$= \frac{36\ \mu C}{2\ \mu F + 3\ \mu F + 6\ \mu F}(6\ \mu F) = \frac{216}{11}\ \mu C$$

3.10.2.1 Energy Loss in Redistribution of Charges

Initial electric potential energy of two conductor system is -
$$U_{\text{initial}} = \frac{1}{2}C_1V_1^2 + \frac{1}{2}C_2V_2^2$$
Final electric potential energy after redistribution of charge is -
$$U_{\text{final}} = \frac{1}{2}C_1V^2 + \frac{1}{2}C_2V^2$$
Therefore, the loss in electrostatic potential energy-
$$\Delta U = U_{\text{initial}} - U_{\text{final}}$$
$$= \left[\tfrac{1}{2}C_1V_1^2 + \tfrac{1}{2}C_2V_2^2\right] - \left[\tfrac{1}{2}C_1V^2 + \tfrac{1}{2}C_2V^2\right]$$
Now, substituting the value of V from Eq.3.38 in above expression and simplifying for ΔU, we get
$$\boxed{\Delta U = \frac{1}{2}\frac{C_1C_2}{C_1 + C_2}(V_1 - V_2)^2} \quad (3.40)$$

Special case: Suppose, second capacitor was initially uncharged, i.e., $q_2 = 0$, $V_2 = 0$. Then, After connection, the common equilibrium potential,
$$V = \frac{q_1}{C_1 + C_2} = \frac{C_1V_1}{C_1 + C_2}$$
The loss of energy now becomes
$$\Delta U = \frac{1}{2}\frac{C_1C_2}{C_1 + C_2}V_1^2$$
Therefore, the % loss in energy, $= \left(\frac{\Delta U}{U_i}\right) \times 100\%$

This energy appears partly as heat in the connecting wire and partly as light and sound if sparking occurs.

Note: If $V_1 = V_2$, then from Eq.3.40, $\Delta U = 0$, i.e. there is no loss of energy on connecting two capacitors (conductors) at the same potential.

EXAMPLE 45. The plates of a capacitor are charged to a potential difference of 100 V and then connected across a resister. The potential difference across the capacitor decays exponentially with respect to time. After one second the potential difference between the plates of the capacitor is 80 V. What is the fraction of the stored energy which has been dissipated?

APPROACH From Eq.3.31, the energy stored in a capacitor (capacitance = C), charged to voltage V, is given by-

$$U = \frac{1}{2}CV^2 \qquad (1)$$

If U_i and U_f are initial and final potential energies respectively, then from above Eq.(1), loss in potential energy-

$$\Delta U = U_i - U_f = \frac{1}{2}CV_i^2 - \frac{1}{2}CV_f^2 \qquad (2)$$

Here, V_i and V_f are initial and final voltages across the capacitor. Therefore, fractional energy loss,

$$\frac{\Delta U}{U_i} = \frac{\frac{1}{2}CV_i^2 - \frac{1}{2}CV_f^2}{\frac{1}{2}CV_i^2} = \frac{V_i^2 - V_f^2}{V_i^2} \qquad (3)$$

Now, substitute the given values in Equation (3) and solve for $\frac{\Delta U}{U_i}$.

SOLUTION Given that: $V_i = 100$ V, $V_f = 80$ V, therefore from Eq.(3), we get

Fractional energy loss $\frac{\Delta U}{U_i} = \frac{V_i^2 - V_f^2}{V_i^2}$

$$= \frac{(100)^2 - (80)^2}{(100)^2} = \frac{20 \times 180}{(100)^2} = \frac{9}{25}$$

EXAMPLE 46. Two capacitors A and B with capacities 3 μF and 2 μF are charged to a potential difference of 100 V and 180 V respectively. The plates of the capacitors are connected as shown in the Fig.3.83a with one wire of each capacitor free. The upper plate of A is positive and that of B is negative. An uncharged 2 μF capacitor C with lead wires falls on the free ends to complete the circuit. Calculate,
(a) the final charge on the three capacitors.
(b) the amount of electrostatic energy stored in the system before and after completion of the circuit.

SOLUTION If q_{A0} and q_{B0} are initial charges on the capacitors

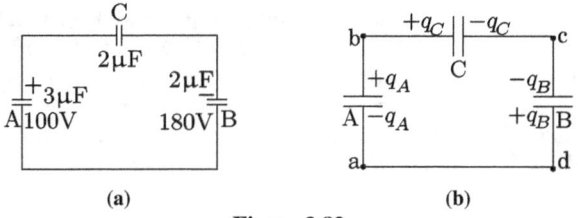

Figure 3.83

A and B respectively, then

$$q_{A0} = C_A V_A = 3 \times 100 = 300 \mu C,$$
$$q_{B0} = C_B V_B = 2 \times 180 = 360 \mu C$$

Let q_A, q_B, and q_C be the final charges on the three capacitors. The charge on each plate is shown in Fig.3.83b
By charge conservation, total charge on the upper plate of A and the left plate of C is equal to the initial charge on the upper plate of A i.e.,

$$q_A + q_C = q_{A0} = 300 \mu C \qquad (1)$$

Similarly, total charge on the upper plate of B and the right plate of C is equal to the initial charge on the upper plate of B i.e.,

$$q_B + q_C = q_{B0} = 360 \mu C \qquad \ldots (2)$$

If V_a, V_b, V_c and V_d are potentials of points a, b, c and d respectively in Fig.3.83b, then starting from point a in clock wise direction we can write-

$$(V_b - V_a) + (V_c - V_b) + (V_d - V_c) + (V_a - V_d) = 0 \quad \cdots (3)$$

It is also known as Kirchhoff's voltage or loop law. It will be discussed in next chapter "Current Electricity".
Here,

$V_b - V_a = +\frac{q_A}{C_A}$: +ve sign in right side shows voltage gain as we move from −ve plate to +ve plate of capacitor A.

$V_c - V_b = -\frac{q_C}{C_C}$: −ve sign in right side shows voltage drop as we move from +ve plate to −ve plate of capacitor C.

$V_d - V_c = +\frac{q_B}{C_B}$: +ve sign in right side shows voltage gain as we move from −ve plate to +ve plate of capacitor B.

$V_a - V_d = 0$: zero in right side shows that there is no voltage change as we move from a to b across the conducting wire da.

Substituting these values of potential differences in equation (3), we get

$$\frac{q_A}{C_A} - \frac{q_C}{C_C} + \frac{q_B}{C_B} + 0 = 0,$$

i.e., $\frac{q_A}{3} - \frac{q_C}{2} + \frac{q_B}{2} = 0 \qquad \cdots (4)$

Solving equations (1), (2) and (4), for q_A, q_B and q_C, we get
$q_A = 90 \mu C$, $q_B = 150 \mu C$, and $q_C = 210 \mu C$.
Stored electrostatic energy in the initial configuration:

$$U_i = \frac{q_{A0}^2}{2C_A} + \frac{q_{B0}^2}{2C_B}$$

$$= \frac{(300 \times 10^{-6})^2}{2(3 \times 10^{-6})} + \frac{(360 \times 10^{-6})^2}{2(2 \times 10^{-6})} = 47.4 \times 10^{-3} J,$$

Stored electrostatic energy in final configuration:

$$U_f = \frac{q_A^2}{2C_A} + \frac{q_B^2}{2C_B} + \frac{q_C^2}{2C_C}$$

$$= \frac{(90 \times 10^{-6})^2}{2(3 \times 10^{-6})} + \frac{(150 \times 10^{-6})^2}{2(2 \times 10^{-6})} + \frac{(210 \times 10^{-6})^2}{2(2 \times 10^{-6})}$$

$$= 18.0 \times 10^{-3} J$$

EXAMPLE 47. A 2 μF capacitor is charged as shown in the Fig.3.84. Calculate the percentage of it's stored energy dissipated after the switch S is turned to position 2.

SOLUTION Let $C_1 = 2 \mu F$ and $C_2 = 8 \mu F$ be the capacitances of

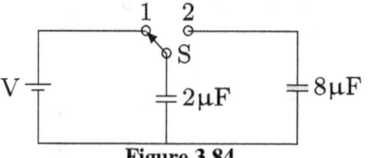

Figure 3.84

given capacitors. Initially, charge on the capacitor C_1 is $Q_i = C_1 V$ and its electrostatic energy is $U_i = Q_i^2 / 2C_1$.
Let the final (after switch is turned to position 2) charges on the two capacitors be Q_{f_1} and Q_{f_2}. The charge conservation gives

$$Q_{f_1} + Q_{f_2} = Q_i \qquad (1)$$

The capacitors C_1 and C_2 are connected in parallel. Thus, the voltages across both the capacitors are equal i.e.,

$$\frac{Q_{f_1}}{C_1} = \frac{Q_{f_2}}{C_2} \qquad (2)$$

Solve equations (1) and (2) to get the charges

$$Q_{f_1} = \frac{C_1}{C_1 + C_2} Q_i = \frac{1}{5} Q_i,$$

$$Q_{f_2} = \frac{C_2}{C_1 + C_2} Q_i = \frac{4}{5} Q_i,$$

and the electrostatic potential energies
$$U_{f_1} = \frac{Q_{f_1}^2}{2C_1} = \frac{1}{25}\left(\frac{Q_i^2}{2C_1}\right) = \frac{1}{25}U_i$$
$$U_{f2} = \frac{Q_{f2}^2}{2C_2}$$
$$= \frac{Q_{f2}^2}{2(Q_{f2}/Q_{f1})C_1} \quad \text{(using Eq.(2))}$$
$$= \frac{Q_{f2}Q_{f1}}{2C_1} = \frac{\frac{4}{5}Q_1\frac{1}{5}Q_1}{2C_1} = \frac{4}{25}\left(\frac{Q_i^2}{2C_1}\right) = \frac{4}{25}U_i$$

Therefore, the energy loss is $U_i - (U_{f_1} + U_{f_2}) = 0.80U_i$ i.e., 80%. This energy is lost in the form of heat, light and sound.

EXAMPLE 48. Two identical capacitors have the same capacitance C. One of them is charged to potential V_1 and the other to V_2. The negative ends of the capacitors are connected together. When the positive ends are also connected, calculate the decrease in energy of the combined system.

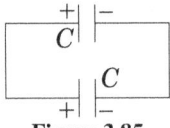

Figure 3.85

SOLUTION Initially, the total electrostatic energy stored in the two capacitors is $U_i = \frac{1}{2}CV_1^2 + \frac{1}{2}CV_2^2$. Initial charges on the two capacitors are $q_1 = CV_1$ and $q_2 = CV_2$. Let q_1' and q_2' be the charges on the capacitors when same terminals of the two capacitors are connected to each other.
The charge conservation gives -
$$q_1' + q_2' = q_1 + q_2 \quad (1)$$
The final potentials V' across the two capacitors are equal, i.e.,
$$q_1'/C = q_2'/C \quad (2)$$
On solving equations (1) and (2), we get
$$q_1' = q_2' = (q_1 + q_2)/2, \quad \text{and} \quad V' = (V_1 + V_2)/2.$$
Thus, the final electrostatic energy stored in the two capacitors is
$$U_f = \frac{1}{2}CV'^2 + \frac{1}{2}CV'^2 = \frac{1}{4}C\left(V_1^2 + V_2^2 + 2V_1V_2\right)$$
and loss in energy is -
$$U_i - U_f = \frac{1}{4}C(V_1 - V_2)^2.$$

EXAMPLE 49. Consider the situation shown in the Fig.3.86. The capacitor A has a charge q on it whereas B is uncharged. Calculate the charge appearing on the capacitor B a long time after the switch S is closed.

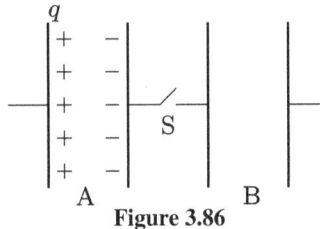

Figure 3.86

SOLUTION Initially, charge on the right plate of capacitor A is $-q$ and that on the left plate of capacitor B is zero. The electric field inside the two capacitors and energy stored in them are
$$E_A = \frac{q/A}{2\epsilon_0} - \frac{-q/A}{2\epsilon_0} = \frac{q}{A\epsilon_0}, \quad E_B = 0;$$
$$U_A = \frac{1}{2}\epsilon_0 E_A^2(Ad) = \frac{q^2}{2\epsilon_0}\frac{d}{A}, \quad \left[\because E_A = \frac{q}{A\epsilon_0}\right]$$
$$U_B = 0.$$

Let the charge q' moves from the right plate of A to the left plate of B when the switch S is closed. By charge conservation, charge remaining on the right plate of A is $-q - q'$ and charge on the left plate of B is q'. Now, the electric field inside the two capacitors and energy stored in them are
$$E_A' = \frac{q/A}{2\epsilon_0} - \frac{(-q - q')/A}{2\epsilon_0} = \frac{q}{A\epsilon_0}\left(1 + \frac{q'}{2q}\right),$$
$$E_B' = \frac{q'/A}{2\epsilon_0},$$
$$U_A' = \frac{1}{2}\epsilon_0 E_A'^2(Ad) = \frac{q^2 d}{2\epsilon_0 A}\left(1 + \frac{q'}{2q}\right)^2$$
$$\text{[Substituting the value of } E_A']$$
$$U_B' = \frac{q'^2}{8\epsilon_0}\frac{d}{A}$$

By conservation of energy, we have
$$U_A + U_B = U_A' + U_B'$$
$$\Rightarrow \frac{q^2}{2\epsilon_0}\frac{d}{A} + 0 = \frac{q^2 d}{2\epsilon_0 A}\left(1 + \frac{q'}{2q}\right)^2 + \frac{q'^2}{8\epsilon_0}\frac{d}{A}$$

Now, solving for q', we get-
$$q^2 = q^2\left(1 + \frac{q'^2}{4q^2} + \frac{q'}{q}\right) + \frac{q'^2}{4}$$
$$\Rightarrow 0 = \left(\frac{q'^2}{4} + qq'\right) + \frac{q'^2}{4} \Rightarrow 0 = \frac{q'^2}{2} + qq'$$
$$\Rightarrow 0 = q'\left(\frac{q'}{2} + q\right) \Rightarrow q' = 0, q' = -2q$$

Since the negative charge on the right plate of capacitor A is confined within it due to the positive charge q on the left plate of A, the acceptable conclusion is $q' = 0$. Furthermore, a charge of $-2q$ on the left plate of capacitor B implies a $+2q$ charge on its right plate. However, as the right plate is isolated, no charge can be transferred to it from elsewhere. Consequently, the presence of a $-2q$ charge on the left plate of capacitor B is impossible. Considering charge conservation, a $-2q$ charge on the left plate of the capacitor B would signify a $+q$ charge on the right plate of capacitor A. This scenario would further entail a $-q$ charge on the left plate of capacitor A, which contradicts its isolation and existing charge of $+q$.

EXAMPLE 50. A parallel plate capacitor of capacitance C is connected to a battery and is charged to a potential difference V. Another capacitor of capacitance $2C$ is similarly charged to a potential difference $2V$. The charging battery is now disconnected and the capacitors are connected in parallel to each other in such a way that the positive terminal of one is connected to the negative terminal of the other. Find the final energy of the configuration.

SOLUTION Initially, charge on the capacitor of capacitance

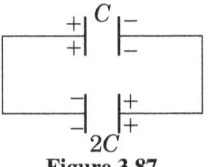

Figure 3.87

C is $q_1 = CV$ and that on the capacitor of capacitance $2C$ is $q_2 = (2C)(2V) = 4CV$. Let q_1' and q_2' be the charges on the two capacitors when the positive terminal of one is connected to the negative terminal of the other (Fig.3.87).
The charge conservation gives
$$q_1' + q_2' = -q_1 + q_2 = 3CV \quad (1)$$
Also, the potentials across the two capacitors are equal
$$q_1'/C = q_2'/(2C). \quad (2)$$
Solving equations (1) and (2) for q_1' and q_2', we get:

3.11. DIELECTRICS

$$q'_1 = CV \text{ and } q'_2 = 2CV$$

Therefore, final energy of the configuration is

$$U = \frac{q_1'^2}{2C} + \frac{q_2'^2}{2(2C)} = \frac{3}{2}CV^2$$

EXAMPLE 51. A 600 pF capacitor is charged by 200 V supply. It is then disconnected from the supply and is connected to another uncharged 600 pF capacitor. Calculate the electrostatic energy loss in this process.

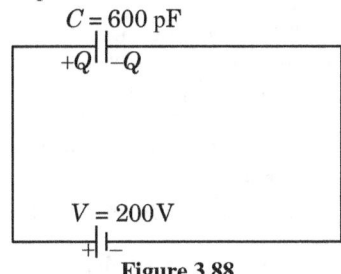

Figure 3.88

SOLUTION $Q = CV = 600 \times 10^{-12} \times 200 = 12 \times 10^{-8}$ C

Initial energy $= \frac{1}{2}CV^2 = \frac{1}{2} \times 600 \times 10^{-12} \times (200)^2 = 12\,\mu$J

When connected to another uncharged identical capacitor, the re-arrangement of charge started until both capacitors reach to same potential. If Q_1 and Q_2 are final charges on initially charged and uncharged capacitors respectively and V is the common potential, then

$$V = \frac{Q_1}{C_1} = \frac{Q_2}{C_2} = \frac{Q_1+Q_2}{C_1+C_2} = \frac{Q}{C+C}$$

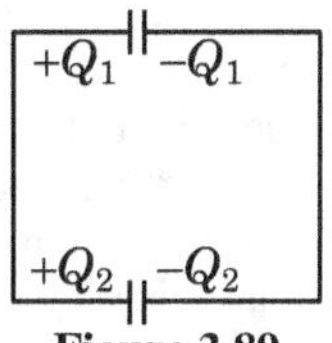

Figure 3.89

$$\Rightarrow \quad \frac{Q_1}{C} = \frac{Q_2}{C} = \frac{Q}{2C} \quad \Rightarrow \quad Q_1 = Q_2 = \frac{Q}{2}$$

Thus, the charge will be equally distributed on identical capacitor

$$Q_1 = Q_2 = \frac{Q}{2} = 6 \times 10^{-8}$$

Final energy $= \frac{Q_1^2}{2C} + \frac{Q_2^2}{2C} = \frac{1}{2C}\left[\left(\frac{Q}{2}\right)^2 + \left(\frac{Q}{2}\right)^2\right]$

$$= \frac{1}{2C}\left[2\left(\frac{Q}{2}\right)^2\right] = \frac{1}{C}\left(\frac{Q}{2}\right)^2$$

$$= \frac{(6 \times 10^8)^2}{600 \times 10^{-12}} = 6\,\mu\text{J} \quad \left[\text{since}, \frac{Q}{2} = 6 \times 10^{-8}\,\text{C}\right]$$

Therefore, Energy lost = Initial energy − Final energy
$= (12 - 6)\,\mu\text{J} = 6\,\mu\text{J}$

3.11 Dielectrics

An insulator or a non-conducting material (for example, air, glass, paper, or wood) is called a **dielectric**. These substances, practically do not conduct electric current. This means that in contrast, for example, to conductors dielectrics do not contain charges that can move over considerable distances and create electric current.

When even a neutral dielectric is introduced into an external electric field, appreciable changes are observed in the field and in the dielectric itself. Faraday discovered that the capacitance of a capacitor is increased when an insulator is put between the plates. If the insulator completely fills the space between the plates, the capacitance is increased by a factor κ (kappa) which depends only on the nature of the dielectric; the factor κ is then a property of the dielectric, and is called the dielectric constant. The reason for this increase is that the electric field between the plates of a capacitor is weakened by the dielectric. Thus, for a given charge on the plates, the potential difference V is reduced and the capacitance ($C = Q/V$) is increased. The dielectric constant of vacuum is unity. κ is a dimensionless factor.

Suppose, a capacitor has a capacitance C_0 when there is only vacuum between the plates. When a dielectric material is inserted to completely fill the space between the plates, the capacitance becomes C, then-

$$\boxed{C = \kappa C_0} \quad (3.41)$$

Experiments indicate that all dielectric materials have $\kappa > 1$.
On substituting the value of C_0 in Eq.3.41, we get-

$$\boxed{C = \kappa\frac{\epsilon_0 A}{d}} \quad (3.42)$$

The quantity $\kappa\epsilon_0$ appears so often in formulas that we define a new quantity-

$$\boxed{\epsilon = \epsilon_0 \kappa} \quad (3.43)$$

called the electric permittivity of a material. Then the capacitance of a parallel-plate capacitor becomes in general:

$$\boxed{C = \frac{\epsilon A}{d}} \quad (3.44)$$

Note that ϵ_0 represents the electric permittivity of free space (a vacuum).

EXAMPLE 52. The capacity of a spherical capacitor in air is 50 μF and on immersing it into oil it becomes 110 μF. Calculate the dielectric constant of oil.

APPROACH If C_0 and C are the capacitances of the spherical capacitor in air and in oil respectively, then from Eq.3.41, the dielectric constant of the oil is given by-

$$\kappa = \frac{C}{C_0} \quad (1)$$

Now, substitute the given values of C and C_0 and solve for κ.

SOLUTION Given that, $C_0 = 50\,\mu$F and $C = 110\,\mu$F, therefore from Eq.(1), we have-

$$\kappa = \frac{C}{C_0} = \frac{110}{50} = 2.2$$

EXAMPLE 53. A parallel plate capacitor is formed by two plates each of area 30π cm^2 separated by 1 mm. A material of dielectric strength 3.6×10^7 Vm^{-1} is filled between the plates. The maximum charge that can be stored on the capacitor without causing any dielectric breakdown is 7×10^{-6} C. What is the value of dielectric constant of the material? (Use: $1/4\pi\epsilon_0 = 9 \times 10^9$ Nm^{2}C^{-2})

SOLUTION The magnitude of charge that can be stored on capacitor's plate is:

$$q = CV = \left(\frac{\epsilon_0 \kappa A}{d}\right)(Ed) \quad \Rightarrow \quad \kappa = \frac{q}{\epsilon_0 A V}$$

Substituting the given values, we get-

$$\kappa = \frac{7 \times 10^{-6}}{\frac{1}{4\pi \times 9 \times 10^9} \times 30\pi \times 10^{-4} \times 3.6 \times 10^7} = \frac{36 \times 7}{30 \times 3.6} = 2.33$$

EXAMPLE 54. A parallel plate capacitor with air between the plate has a capacitance of 15 pF. The separation between the plates becomes twice and the space between them is filled with a medium of dielectric constant 3.5. Then the capacitance becomes $\frac{x}{4}$ pF. Find the value of x.

SOLUTION Capacitance of the parallel plate capacitor, in free space, is given as

$$C_0 = \frac{\epsilon_0 A}{d} = 15 \text{ pF}$$

When the plates separation is doubled and the medium is filled with a dielectric of dielectric constant κ, the new capacitance becomes

$$C = \frac{\kappa \epsilon_0 A}{2d} = \frac{3.5}{2} \times 15 \text{ pF} = \frac{105}{4} \text{ pF}$$

But according to given problem, $C = \frac{x}{4}$ pF, therefore comparing it with obtained capacitance, we get- $x = 105$

3.11.1 Molecular View of a Dielectric

To develop a conceptual understanding of Eq.3.41, we must take into consideration that dielectrics consist either of neutral molecules or of charged ions located at the sites of a crystal lattice (ionic crystals, for example, of the NaCl type). The molecules can be either polar or nonpolar. In a polar molecule (for example, HCl, H_2O), the "centre of the negative charge" is displaced relative to the "centre of the positive charge" (Fig.3.90a). As a result, the molecule acquires an intrinsic dipole moment $\vec{p}$. Non-polar molecules (Fig.3.90b) do not have intrinsic dipole moments, since the centres of the positive and negative charges in them coincide.

Polarization Under the action of an external electric field, the di-

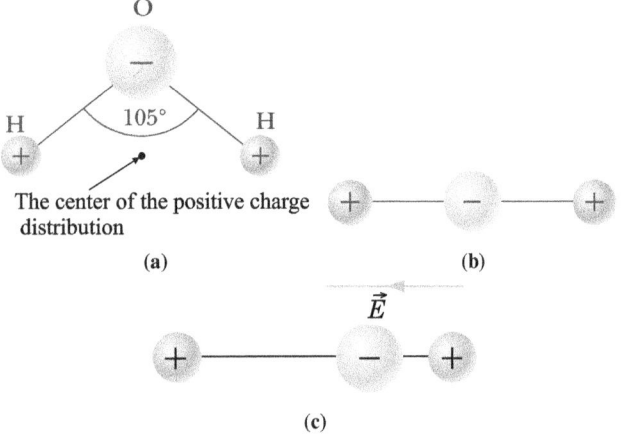

Figure 3.90: (a) The water molecule, H_2O, has a permanent polarization resulting from its nonlinear geometry. (b) A linear symmetric molecule has no permanent polarization. (c) An external electric field induces a polarization in the molecule.

electric is polarized. Let us first consider monoatomic dielectrics. Monoatomic dielectrics are made of atoms. We can think of an atom as a very small, positively charged nucleus surrounded by a negatively charged electron cloud (Fig.3.91a). Generally, in an atom, the charge configuration is sufficiently symmetric so that the centre of negative charge coincides with the centre of positive charge. An atom or molecule that has this symmetry has zero dipole moment and is said to be nonpolar. In the presence of an external electric field, however, the positive and negative charges experience forces in opposite directions, so the positive and negative charges then separate until the attractive force they exert on each other balances the forces due to the external electric field (Figure 3.91b). The molecule is then said to be polarized and it behaves like the electric dipole it is.

Now, consider the case in which, polyatomic dielectrics are made up by permanent polar molecules. Polar molecules have different directions of the dipole moment because of the random thermal agitation in the material. In any volume containing a large number of molecules (say more than a thousand), the net dipole moment is zero. If such a material is placed in an electric field, the individual dipoles experience torque due to the field and they try to align along the field. On the other hand, thermal agitation tries to randomise the orientation and hence, there is a

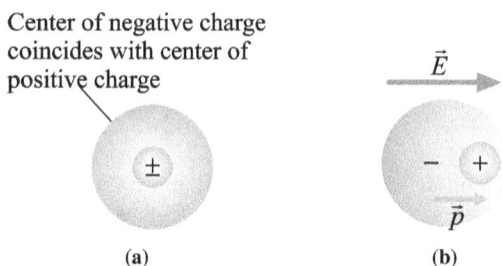

Figure 3.91: Schematic diagrams of the charge distributions of an atom or nonpolar molecule. (a) In the absence of an external electric field, the centre of positive charge coincides with the centre of negative charge. (b) In the presence of an external electric field, the centres of positive and negative charge are displaced, producing an induced dipole moment in the direction of the external field.

partial alignment. As a result, we get a net dipole moment in any volume of the material.

If a dielectric is made up by nonpolar polyatomic molecules (Fig.3.90b), the positive charge in each molecule is shifted along the field and the negative, in the opposite direction (Fig.3.90c). This induces dipole moment in each atom or molecule and thus, we get a dipole moment in any volume of the material.

Finally, in ionic dielectric molecules (Fig.3.90b), an external field displaces all the positive ions along the field and the negative ions, against the field.[*]

Thus, when a dielectric material is placed in an electric field, dipole moment appears in any volume in it. This fact is known as "polarization" of the material. The dipole moment per unit volume of the dielectric is called the polarization vector $\vec{P}$. It's magnitude P is often referred to as the polarization.

So, we can say that, the mechanism of polarization depends on the structure of a dielectric. For further discussion it is only important that regardless of the polarization mechanism, all the positive charges during this process are displaced along the field, while the negative charges, against the field. It should be noted that under normal conditions the displacements of charges are very small even in comparison with the dimensions of the molecules. This is due to the fact that the intensity of the external field acting on the dielectric is considerably lower than the intensities of internal electric fields in the molecules.

When a dielectric is placed in the strong field of a charged capacitor, its molecules are polarized in such a way that there is a net dipole moment parallel to the field. If the molecules are polar, their dipole moments (originally oriented at random) tend to become aligned due to the torque exerted by the field[†]. If the molecules are nonpolar, the field induces dipole moments that are parallel to the field. In either case, the molecules in the dielectric are polarized in the direction of the external field (Figure 3.92). The net effect of the polarization of a homogeneous dielectric in a parallel-plate capacitor is the creation of surface charges on the dielectric faces near the plates, as shown in Figure 3.93. This charge is called the **induced charge**. The induced surface charge on the dielectric is also called the **bound charge**, because the surface charge is bound to the surface molecules of the dielectric and can not move about like the free charge on the conducting capacitor plates. This induced charge produces an induced electric field ($\vec{E}_{\text{ind}}$) opposite in direction to the electric field $\vec{E}_0$ produced by the free charge on the conductors. Thus, the resultant electric

---

[*]There exist ionic crystals polarized even in the absence of an external field. This property is inherent in dielectrics which are called *electrets* (they resemble permanent magnets).

[†]The degree of alignment depends on the external field and on the temperature. It is approximately proportional to $pE/(kT)$, where pE is the maximum energy of a dipole in a field E, and kT is the characteristic thermal energy.

3.11. DIELECTRICS

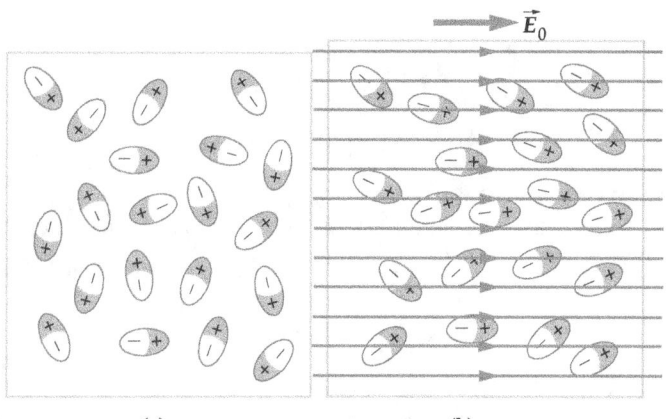

Figure 3.92: (a) The randomly oriented electric dipoles of a polar dielectric in the absence of an external electric field. (b) In the presence of an external electric field, the dipoles are partially aligned parallel to the field.

field between the plates in dielectric is-

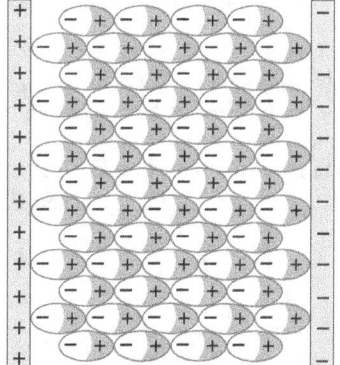

Figure 3.93: When a dielectric is placed between the plates of a capacitor, the electric field of the capacitor polarizes the molecules of the dielectric. The result is a bound charge on the surface of the dielectric that produces its own electric field; this field opposes the external field. The field of the bound surface charges thus weakens the electric field within the dielectric.

$$\vec{E} = \vec{E}_0 + \vec{E}_{\text{ind}} \quad (3.45)$$

For homogeneous and isotropic dielectrics, the direction of $\vec{E}_{\text{ind}}$ is opposite to the direction of $\vec{E}_0$. The resultant field $\vec{E}$ is in the same direction as the applied field $\vec{E}_0$ but its magnitude is reduced. Therefore, from Eq.3.45, we can write-

$$E = E_0 - E_{\text{ind}} \quad (3.46)$$

Here, E_{ind} is the electric field produced by induced charge
Clearly, $E < E_0$, as illustrated in Figure 3.94.

The surface charge density of the induced charge has a simple relationship with the polarization P. Suppose, the rectangular slab of Figure 3.94b has a length l and area of cross-section A. Let σ_i be the magnitude of the induced charge per unit area on the faces.

The dipole moment of the slab is then-

$$Q_i \times l = (\sigma_i A) l = \sigma_i (Al)$$

Therefore, the net polarization dipole moment induced per unit volume-

$$\boxed{P = \frac{\sigma_i(Al).}{Al} = \sigma_i} \quad (3.47)$$

Although this result is deduced for a rectangular slab, it is true in general. The induced surface charge density is equal in magnitude to the polarization P.

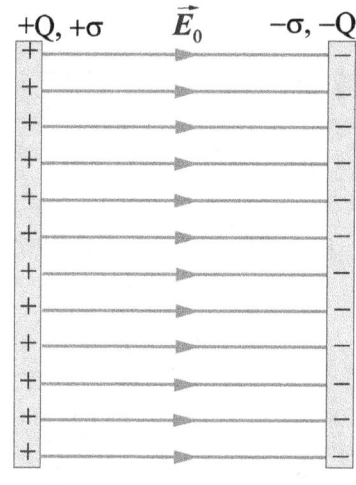

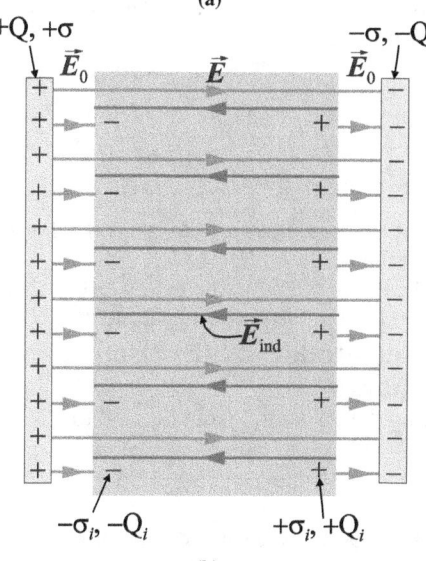

Figure 3.94: The electric field between the plates of a capacitor that has (a) no dielectric and (b) a dielectric. The surface charge on the dielectric weakens the original field between the plates.

3.11.1.1 Mathematical Description of Dielectric Constant

If σ and σ_i are free and bound charge densities, then
$$E_0 = \sigma/\epsilon_0, \text{ and } E_{\text{ind}} = \sigma_i/\epsilon_0.$$

Substituting these values in equation 3.46, we get-

$$E = \frac{\sigma}{\epsilon_0} - \frac{\sigma_i}{\epsilon_0} \quad (3.48)$$

Surface density of bound charge (σ_i) is directly proportional to free surface charge density σ, i.e.,

$$\sigma_i \propto \sigma$$
$$\implies \quad \sigma_i = b\sigma \quad (3.49)$$

here b is the bound charge density factor. It's value lies in the range 0 to 1, i.e., $0 \leq b \leq 1$.
For free space or vacuum, $b = 0$ and for conducting medium, $b = 1$.
Substituting this value of σ_i in Eq.3.48, we get-

$$E = \frac{\sigma}{\epsilon_0} - \frac{b\sigma}{\epsilon_0}$$
$$\implies \quad E = \frac{\sigma}{\epsilon_0}(1 - b) \quad (3.50)$$

The reciprocal of $(1 - b)$ is denoted by κ and it is called the dielectric constant for given dielectric material, i.e.,

Therefore,
$$1 - b = \frac{1}{\kappa}$$
$$b = 1 - \frac{1}{\kappa} \tag{3.51}$$

Since, $0 \le b \le 1$, therefore-
$$0 < 1 - \frac{1}{\kappa} < 1$$

On subtracting 1 from all sides of this inequality, we get
$$-1 \le -\frac{1}{\kappa} \le 0$$
or
$$1 \ge \frac{1}{\kappa} \ge 0$$
or
$$1 \le \kappa \le \infty$$

For free space or air, $\kappa = 1$, for conductors $\kappa = \infty$ and for dielectric medium, $\boxed{1 < \kappa < \infty}$

Note that κ is a material property of the dielectrics, therefore its value depends on the type of dielectric used in capacitor.

Again from Eq.3.50,
$$E = \frac{\sigma}{\epsilon_0}(1-b)$$

Substituting, $1 - b = 1/\kappa$, in above equation, we get-
$$\boxed{E = \frac{\sigma}{\epsilon_0 \kappa} = \frac{E_0}{\kappa}} \tag{3.52}$$

Since, $\kappa > 1$, therefore, electric field inside the dielectric material is always less than E_0, i.e., $E < E_0$.

Thus, in a dielectric medium, the field is everywhere smaller by the factor $1/\kappa$, than in the case without the dielectric.

3.11.2 Magnitude of the Bound or Induced Charge

Substituting the value of b from Eq.3.50 to 3.49, we get-
$$\boxed{\sigma_i = \left(1 - \frac{1}{\kappa}\right)\sigma} \tag{3.53}$$

This is the relation between induced (or bound) surface charge density and free surface charge density.

If Q, Q_i are free and induced (i.e., bound) charges respectively, then $\sigma = Q/A$ and $\sigma_i = Q_i/A$.

Here, A is the effective facing area of each capacitor plate.

On substituting these values of σ and σ_i in Eq.3.53, we get-
$$\boxed{Q_i = \left(1 - \frac{1}{\kappa}\right)Q} \tag{3.54}$$

Eq.3.54 gives the required relation between induced and free surface charges. For a dielectric $1 - \frac{1}{\kappa} < 1$, therefore, $Q_i < Q$. Thus, the bound charge is always less than the free charge that produces it.

3.11.3 Electric Displacement Vector

☞This topic is out of the JEE and NEET syllabi.

The field due to the polarization is
$$E_{\text{ind}} = \frac{\sigma_i}{\epsilon_0} = \frac{P}{\epsilon_0} \quad \text{(Using Eq.3.47)}$$

where, P is the polarization (the dipole moment per unit volume). As the direction of E_{ind} is opposite to the polarization vector $\vec{P}$, we can write-
$$\vec{E}_{\text{ind}} = -\frac{\vec{P}}{\epsilon_0}$$

Substituting the value of $\vec{E}_{\text{ind}}$ in Eq.3.45:
$$\vec{E} = \vec{E}_0 + \vec{E}_{\text{ind}}$$

we get-

$$\vec{E} = \vec{E}_0 - \frac{\vec{P}}{\epsilon_0}$$
$$\implies \quad \epsilon_0 \vec{E} + \vec{P} = \epsilon_0 \vec{E}_0 \tag{3.55}$$
or,
$$\oint \left(\epsilon_0 \vec{E} + \vec{P}\right).d\vec{A} = \oint \epsilon_0 \vec{E}_0.d\vec{A}$$

over any closed surface.

As $\vec{E}_0$ is the field produced by the free charge Q, therefore,
$$\oint \vec{E}_0.d\vec{A} = \frac{Q}{\epsilon_0} \quad \text{(from Gauss's law)}$$
or
$$\oint \epsilon_0 \vec{E}_0.d\vec{A} = Q$$

Thus,
$$\oint \left(\epsilon_0 \vec{E} + \vec{P}\right).d\vec{A} = Q \tag{3.56}$$

The quantity $\epsilon_0 \vec{E} + \vec{P}$ is known as the electric displacement vector $\vec{D}$. Therefore, Eq.3.56 can also be written as-
$$\boxed{\oint \vec{D}.d\vec{A} = Q} \tag{3.57}$$

which is another form of Gauss's law.

If there is no polarization (i.e. $P = 0$), then $\vec{D} = \epsilon_0 \vec{E}$ and Q is equal to the total charge inside the Gaussian surface. Eq.3.57 then reduced to the usual form of Gauss's law.

In case of homogeneous and isotropic dielectrics, $\vec{E}_0 = \kappa \vec{E}$ so that Eq.3.55 gives $\vec{D} = \epsilon_0 \kappa \vec{E}$ and Eq.3.57 reduces to
$$\epsilon_0 \kappa \oint \vec{E}.d\vec{A} = Q$$
$$\implies \boxed{\oint \vec{E}.d\vec{A} = \frac{Q}{\epsilon_0 \kappa}} \tag{3.58}$$

Eq.3.58 is the integral form of Gauss's law in dielectric medium. This law, in a dielectric medium, is separately derived in section 3.12. It's generalized form is-

$$\implies \boxed{\oint \vec{E}.d\vec{A} = \frac{Q_{\text{encl}}}{\epsilon_0 \kappa}} \tag{3.59}$$

Here, Q_{encl} is the net charge enclosed within the Gaussian surface in dielectric.

3.11.4 Dielectric Breakdown and Dielectric Strength

As we know that, any insulating material, when subjected to a sufficiently large electric field, experiences a partial ionization. In ionization, in a large electric field, the electrons detached from outer orbits of their parent atoms, may crash into other molecules, liberating even more electrons. This avalanche of moving charge forms a spark or arc discharge. The dielectric then behaves like a conductor and permits conduction through it. This phenomenon is known as dielectric breakdown.

Many dielectric materials can tolerate stronger electric fields without breakdown than can air. Thus, using a dielectric allows a capacitor to sustain a higher potential difference and so store greater amounts of charge and energy.

The maximum value of electric field before breakdown occurs and charges begin to flow, is called the 'dielectric strength' or 'breakdown limit' of the material.

Since, voltage across dielectric $V = Ed$ for a uniform field, the dielectric strength determines the maximum potential difference that can be applied across a capacitor per meter of plate spacing. Lightning is an example of dielectric breakdown in air. The di-

3.11. DIELECTRICS

electric strength of dry air is about 3×10^6 V/m.
The values of the dielectric constant and dielectric strength for various materials are given in Table 3.1. Three basic functions of

Table 3.1: Dielectric constants for some Dielectrics at $20°C$

Material	Dielectric Constant κ	Dielectric Strength (V/m)
Vacuum	1.0000	
Air (1 atm)	1.0006	3×10^6
Paraffin	2.2	10×10^6
Polystyrene	2.6	24×10^6
Vinyl (plastic)	2 – 4	50×10^6
Paper	3.7	15×10^6
Quartz	4.3	8×10^6
Oil	4	12×10^6
Glass, Pyrex	5	14×10^6
Rubber, neoprene	6.7	20×10^6
Porcelain	6 – 8	5×10^6
Mica	7	120×10^6
Water (liquid)	80	70×10^6
Strontium titanate	300	8×10^6

a solid Dielectric between the plates of a Capacitor are-
1. It provides a mechanical support between the plates, which allows the plates to be close together without touching, thereby decreasing d and increasing C.
2. It increases the maximum operating voltage without breakdown.
3. It increases the capacitance.

Note: Because of dielectric breakdown, capacitors always have maximum voltage ratings. When a capacitor is subjected to excessive voltage, an arc may form through a layer of dielectric, burning or melting a hole in it. This arc creates a conducting path (a short circuit) between the conductors. If a conducting path remains after the arc is disappears, the device is rendered permanently useless as a capacitor.

☞ Do not confuse dielectric constant and dielectric strength; they are not related. The dielectric constant determines how much charge can be stored for a given potential difference, while dielectric strength determines how large a potential difference can be applied to a capacitor before dielectric breakdown occurs.

EXAMPLE 55. A parallel-plate capacitor is formed by two plates, each of area 100 cm^2, separated by a distance of 1 mm. A dielectric of dielectric constant 5.0 and dielectric strength 1.9×10^7 V/m is filled between the plates. Find the maximum charge that can be stored on the capacitor without causing any dielectric breakdown.

SOLUTION If the charge on the capacitor = Q
the surface charge density $\sigma = \dfrac{Q}{A}$

and the electric field $= \dfrac{\sigma}{\epsilon} = \dfrac{Q/A}{\kappa \epsilon} = \dfrac{Q}{\kappa A \epsilon_0}$.

This electric field should not exceed the dielectric strength 1.9×10^7 V/m.

Therefore, $\dfrac{Q}{\kappa A \epsilon_0} \leq 1.9 \times 10^7$ V/m

$\implies Q \leq \kappa A \epsilon_0 \left(1.9 \times 10^7 \text{ V/m}\right)$

Given that $A = 100$ cm$^2 = 10^{-2}$ m^2, therefore-

$Q \leq (5.0) \times \left(10^{-2}\right) \times \left(8.85 \times 10^{-12}\right) \times \left(1.9 \times 10^7\right)$

$= 8.4 \times 10^{-6}$ C

Therefore, the maximum possible charge (Q) on plate (without dielectric breakdown) is-

$Q = 8.4 \times 10^{-6}$ C

3.11.5 Change in Various Factors when a Dielectric Slab is Introduced Between the Plates of a Capacitor

Let us consider a charged capacitor in free space. Suppose, Q_0, C_0, V_0, E_0 and U_0 represent it's charge, capacity, potential difference, electric field and energy respectively. If the space between the plates, is completely filled with a dielectric slab of dielectric constant κ, then it's capacitance increases to κC_0. Now, we discuss following two different possible cases-

3.11.5.1 When the Battery is Disconnected

This situation is shown in Fig.3.95a. When the battery disconnected, the capacitor becomes isolated. Now, a dielectric slab of a dielectric constant κ is introduced to completely fill the space between the plates of the capacitor as shown in Fig.3.95b. In this case [Fig.3.95]-

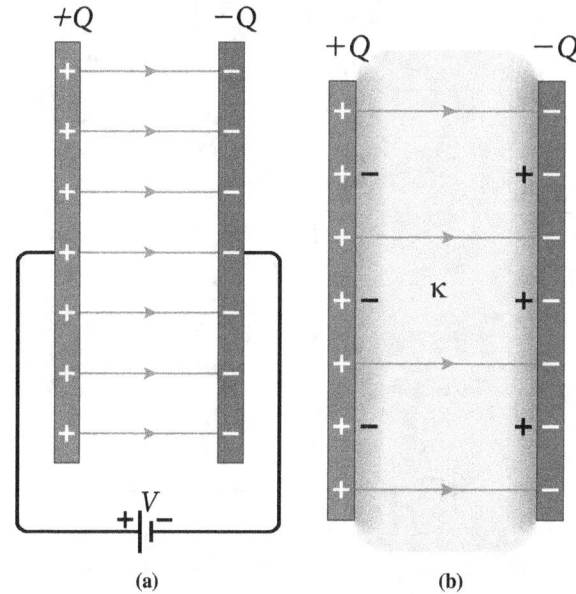

Figure 3.95: The presence of a polarized dielectric reduces the strength of the electric field between the plates of a capacitor.

1. **Effect on Capacitance:** Since, in the presence of a dielectric slab capacity becomes κ times i.e., $\boxed{C = \kappa C_0}$ and for a dielectric, $\kappa > 1$, so capacitance increases to κ times.
2. **Effect on Charge:** Since, battery is disconnected, so both plates of capacitor becomes isolated and therefore, charge remains constant i.e., $\boxed{Q = Q_0}$.
3. **Effect on Potential Difference** V: Potential difference between the capacitor plates is-

$$V = \dfrac{Q}{C} = \dfrac{Q_0}{\kappa C_0} = \dfrac{V_0}{\kappa}$$

$[\because \; Q = Q_0 \text{ and } C = \kappa C_0, V_0 = Q_0/C_0]$

Since, corresponding to a dielectric medium, $\kappa > 1$, the potential difference between the plates decreases and becomes $1/\kappa$ times its original value, i.e., $\boxed{V = \dfrac{V_0}{\kappa}}$.

4. **Effect on Electric Field between the Plates:** In presence of dielectric slab, the electric field between the plates of capacitor is given by-

$$E = \dfrac{V}{d} = \dfrac{V_0}{\kappa d} = \dfrac{E_0}{\kappa} \quad \left[\text{as } V = \dfrac{V_0}{\kappa}, \text{ and } E_0 = \dfrac{V_0}{d}\right]$$

So, electric field between the plates decreases and becomes $1/\kappa$ times of the original value, i.e., $\boxed{E = \dfrac{E_0}{\kappa}}$.

5. **Effect on Energy Stored in Capacitor:** In dielectric medium, the electrostatic potential energy stored in a capacitor is given by
$$U = \dfrac{Q^2}{2C} = \dfrac{Q_0^2}{2\kappa C_0} = \dfrac{U_0}{\kappa} \quad (\text{as } Q = Q_0 \text{ and } C = \kappa C_0)$$
So, $\boxed{U = \dfrac{U_0}{\kappa}}$

Thus, the energy stored in the capacitor decreases and becomes $1/\kappa$ times of it's value in free space or vacuum.

3.11.5.2 When the Battery Remains Connected

This situation is shown in Fig.3.96. In this case, the capacitor is not isolated. The battery keeps the potential difference between the capacitor plates the same regardless of the presence of the dielectric [Fig.3.96b].

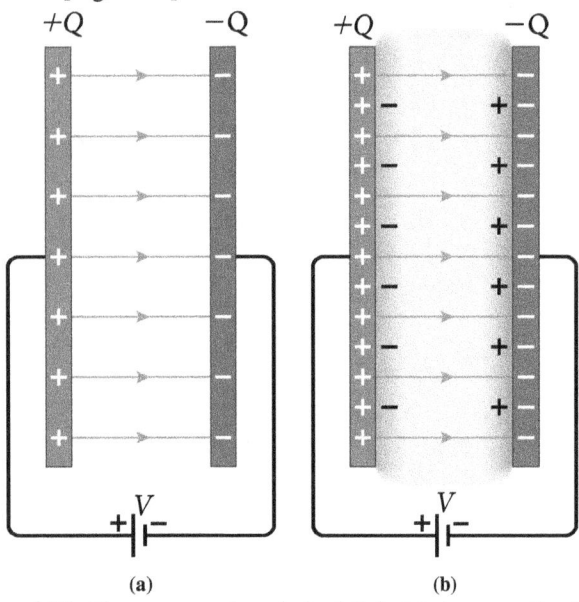

Figure 3.96: The presence of a polarized dielectric increases the charge on the plates of a capacitor connected to a battery.

1. **Effect on Capacitance:** Since, in the presence of a dielectric slab capacity becomes κ times i.e., $\boxed{C = \kappa C_0}$ and for a dielectric, $\kappa > 1$, so capacitance increases to κ times.
2. **Effect on Potential Difference V:** Since, battery is a source of constant potential difference, therefore potential difference between the plates remains constant i.e.,
$$\boxed{V = V_0}$$
3. **Effect on Charge:** Since, $Q = CV = (\kappa C_0) V = \kappa Q_0$
   [because $Q_0 = C_0 V$]
   Therefore, charge increases to κ times and becomes
   $$\boxed{Q = \kappa Q_0}$$
   So, dielectric causes plates to carry free charge greater than Q. The extra charge is provided by battery connected to the plates.
4. **Effect on Electric Field between the Plates:** Since,
$$E = \dfrac{V}{d} = \dfrac{V_0}{d} = E_0 \quad \left[\text{as } V = V_0, \text{ and } \dfrac{V_0}{d} = E_0\right]$$
It follows that, as long as the capacitor is connected to the battery, the electric field must be the same regardless of the dielectric, i.e., $E = E_0$
5. **Effect on Energy Stored in Capacitor:** In presence of dielectric material the electrostatic energy stored in the capacitor becomes-
$$U = \tfrac{1}{2}CV^2 = \tfrac{1}{2}(\kappa C_0) V_0^2 = \kappa U_0$$
$$[\because C = \kappa C_0 \text{ and } U_0 = \tfrac{1}{2}C_0 V_0^2]$$
or, $\boxed{U = \kappa U_0}$,

So, the electric energy stored in the capacitor increases to κ times.

Key Points:
☞ A dielectric increases the capacitance of a capacitor by a factor κ (the dielectric constant) over its capacitance when free space or air is between the plates.
☞ For problems in which a battery is being connected or disconnected, note whether modifications to the capacitor are made while it is connected to the battery or after it has been disconnected. If the capacitor remains connected to the battery, the voltage across the capacitor remains unchanged (equal to the battery voltage), and the charge is proportional to the capacitance, although it may be modified (for instance, by the insertion of a dielectric). If you disconnect the capacitor from the battery before making any modification to the capacitor, then its charge remains fixed. In this case, as you vary the capacitance, the voltage across the plates changes according to the expression $V = Q/C$.

EXAMPLE 56. A parallel-plate capacitor is charged with a battery to a charge Q_0, as shown in Figure 3.97a. The battery is then removed, and a slab of material that has a dielectric constant κ is inserted between the plates, as shown in Figure 3.97b. Find the energy stored in the capacitor before and after the dielectric is inserted.

SOLUTION From Eq.3.31, the energy stored in the capacitor in

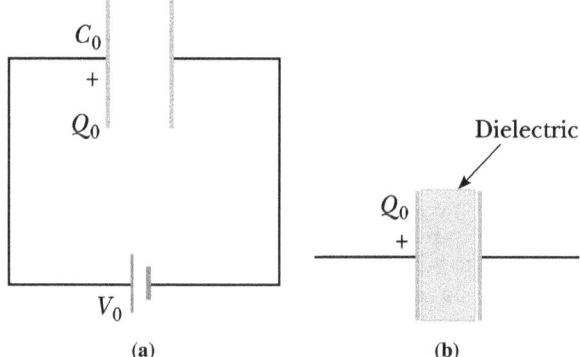

Figure 3.97: (a) A battery charges up a parallel-plate capacitor. (b) The battery is removed and a slab of dielectric material is inserted between the plates.

the absence of the dielectric is-
$$U_0 = \dfrac{Q_0^2}{2C_0}$$
After the battery is removed and the dielectric is inserted between the plates, the charge on the capacitor remains the same because the unconnected capacitor is an isolated system. Hence, the energy stored in the presence of the dielectric, is
$$U = \dfrac{Q_0^2}{2C}$$
The capacitance in the presence of the dielectric, however, is given by $C = \kappa C_0$, so U becomes
$$U = \dfrac{Q_0^2}{2\kappa C_0} = \dfrac{U_0}{\kappa}$$
Because $\kappa > 1$, we see that the final energy is less than the initial energy by the factor $1/\kappa$.

Where does the energy go? The answer lies in the fringing field

3.12 Guass's Law in Dielectrics

Fig.3.99a shows a capacitor filled with a dielectric material of dielectric constant κ. Figure 3.99b is a close-up view of the left capacitor plate and left surface of the dielectric. Let's apply Gauss's law to the rectangular box shown in cross section; the surface area of the left and right sides is A. The left side is embedded in the conductor that forms the left capacitor plate, and so the electric field everywhere on that surface is zero. The right side is embedded in the dielectric, where the electric field has magnitude E, and $E_\perp = 0$ everywhere on the other four sides. The total charge enclosed, including both the charge on the capacitor plate and the induced charge on the dielectric surface, is

$$Q_{\text{encl}} = (\sigma - \sigma_i) A = Q - Q_i,$$

so by Gauss's law, we have-

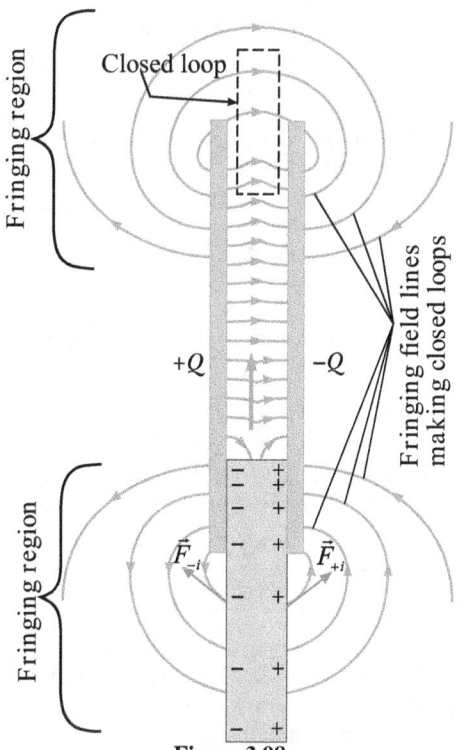

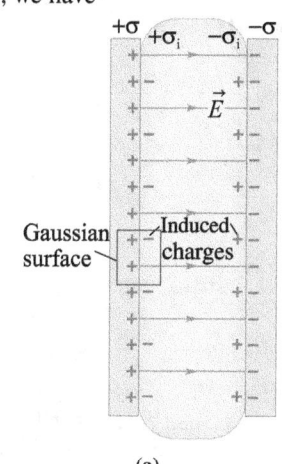

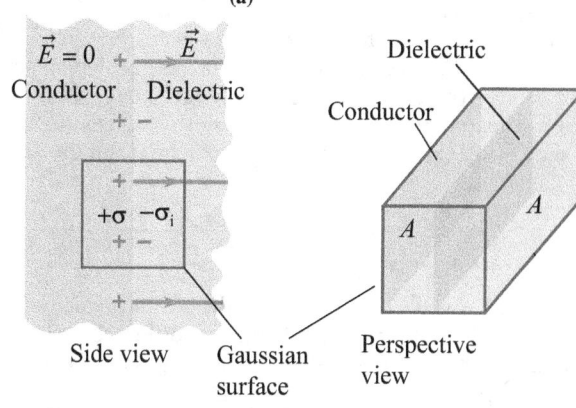

Figure 3.98

at the edges of a real parallel-plate capacitor as shown in Fig.3.98. Actually, there exists a non-uniform electric field around the edges of the capacitor, which, for most purposes, could be ignored. However, in this case, it is responsible for the attraction on the dielectric. This non uniform electric field near the edges of the capacitor, is called fringing field*. Any fringing field line cannot terminate abruptly at the edge of the capacitor, because if it terminates, the line integral $\oint \vec{E} \cdot d\vec{l}$ around the closed loop shown by dashed rectangle in Fig. 3.98 would not be zero†. These field lines extend beyond the edges to form the closed loops. It is this non-uniform fringing field that pulls the dielectric into the capacitor.

Explanation for Attraction: When a dielectric slab is brought near the plates of a capacitor, the fringing field of charged plates induce a polarization in the dielectric. As a result of which the bound or induced charges developed over the surface of the dielectric in opposite sense of free charges of the plates. This bound charge tends to accumulate near the free charge of the opposite sign and consequently it is pulled into the space between the plates (Fig.3.98).

Since, the field inside the capacitor does positive work in pulling the dielectric, the field energy density decreases as the dielectric enters the space between the plates. This potential energy loss ($\Delta U = U_0 - U$) is converted to kinetic energy of dielectric slab. Therefore, the unconnected capacitor and dielectric form an isolated system for electric charge, but when considering energy, it is a nonisolated system.

The electric field deep inside a charged capacitor is almost constant, so a neutral dielectric will not be attracted to either plate inside the parallel plates of a capacitor. The variation of the field, near the edges, is an essential part of the attraction mechanism. If the dielectric were released, it would be pulled into the plates and would pass through the plates with a kinetic energy, emerging on the other side and exhibiting oscillatory motion.

Figure 3.99: (a) Dielectric between the plates. (b) Close-up of the left-hand capacitor plate. The Gaussian surface is a rectangular box that lies half in the conductor and half in the dielectric.

$$\oint \vec{E}.d\vec{A} = \frac{Q_{\text{encl}}}{\epsilon_0} = \frac{Q - Q_i}{\epsilon_0} = \frac{1}{\epsilon_0}\left[Q - Q\left(1 - \frac{1}{\kappa}\right)\right] = \frac{Q}{\epsilon_0 \kappa}$$

or
$$\oint \vec{E}.d\vec{A} = \frac{Q_{\text{encl-free}}}{\epsilon_0 \kappa}$$

Here, $Q_{\text{encl-free}}$ does not include the bound charge due to polarization. So, while applying Gauss's law in a dielectric we can use either of the two equations given below,

$$\boxed{\oint \vec{E}.d\vec{A} = \frac{Q_{\text{encl}}}{\epsilon_0}} \quad (3.60a)$$

---

*If we consider a uniform electric field inside the parallel-plate capacitor, and zero outside, then there would be no net force on the dielectric at all, since the field everywhere would be perpendicular to the plates.

†The electric field is a conservative force field, so the line integral of the field around a closed loop should always be zero, i.e., $\oint \vec{E} \cdot d\vec{l} = 0$

or $\oint \vec{E}.d\vec{A} = \dfrac{Q_{\text{encl-free}}}{\epsilon_0 \kappa}$ (3.60b)

where $Q_{\text{encl-free}}$ is the total free charge (not bound charge) enclosed by the Gaussian surface. The significance of this [Eq.(3.60b)] results is that the right side contains only the free charge on the conductor, not the bound (induced) charge.

This form of Gauss's law is very general: Even though we derived it for the special case of a parallel-plate capacitor, it holds in any situation, even one without a dielectric. In the absence of matter (that is, in vacuum), $\kappa = 1$, and because there is no bound charge we have $Q_{\text{encl, free}} = Q_{\text{encl}}$. Then Eq.3.60b becomes identical to the familiar form of Gauss's law given by Eq.3.60a

3.12.1 Capacitor with Partially Filled Dielectric Medium

Let a non-conducting slab of thickness t, area A and dielectric constant κ is inserted into the space between the plates of a parallel-plate capacitor with spacing d, charge Q and facing area A, as shown in Fig.3.100. The slab is at distance x from the positively charged plate.

To find the capacitance C, we first calculate the potential differ-

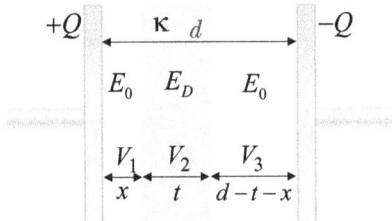

Figure 3.100: Partially filled Capacitor

ence between the plates. If E_0 is the magnitude of intensity of electric field in free space, then the electric field intensity between the plates, in dielectric medium is $E = E_0/\kappa$ (Eq.3.52), where, κ is the dielectric constant of the medium.

The potential difference across the plates, can be found by integrating the electric field along a perpendicular straight line starting from the left plate to the right plate. If V_1, V_2 and V_3 are potential differences across regions 1, 2 and 3 respectively as shown in Fig.3.100, then the net potential difference across the plates-

$V = -\int_{+}^{-} E dl = V_1 + V_2 + V_3 = -E_0 x - E_D t - E_0(d - t - x)$

or $V = -E_0(d-t) - E_D t$ (3.61)

where, E_D is the electric field in the dielectric medium. Since, V is free from x, therefore, net potential difference across the plates of a dielectric is independent on the position of the dielectric slab.

Since, $E_0 = \dfrac{Q}{\epsilon_0 A}$, $E_D = \dfrac{E_0}{\kappa} = \dfrac{Q}{\kappa \epsilon_0 A}$, therefore, Eq.3.61 gives-

$V = -\dfrac{Q}{\epsilon_0 A}(d-t) - \dfrac{Q}{\kappa \epsilon_0 A} t$

or $V = -\dfrac{Q}{\epsilon_0 A}\left[d - t\left(1 - \dfrac{1}{\kappa}\right)\right]$ (3.62)

Therefore, required capacitance,

$C = \dfrac{Q}{|V|} = \dfrac{\epsilon_0 A}{d - t\left(1 - \dfrac{1}{\kappa}\right)}$ (3.63)

It is useful to check the following limits:
1. As $t \to 0$, i.e., the thickness of the dielectric approaches zero, in this case-

$C \to \dfrac{\epsilon_0 A}{d} = C_0$

which is an expected result for no dielectric.
2. As $\kappa \to 1$, we again have

$C \to \dfrac{\epsilon_0 A}{d} = C_0$

and the situation also correspond to the case where the dielectric is absent.
3. In the limit where the complete space is filled with dielectric, i.e., $t = d$, Eq.3.63 gives-

$C = \dfrac{\epsilon_0 A}{d - d\left(1 - \dfrac{1}{\kappa}\right)} = \dfrac{\kappa \epsilon_0 A}{d} = \kappa C_0$

For a metal slab $\kappa = \infty$, therefore from Eq.3.63, we get

$C = \dfrac{\epsilon_0 A}{d - t}$ (3.64)

Eq.3.64 gives the capacitance of a capacitor partially filled with a metal plate of thickness t. Eq.3.64 shows that capacitance reduces on inserting a metal slab of thickness $t(< d)$ between the plates of the capacitor.

If $t \to 0$, then from Eq.3.64,

$C \to \dfrac{\epsilon_0 A}{d} = C_0$

✓ So, if required, we can insert a thin ($t \to 0$) hypothetical parallel metal plate between the parallel plates of a capacitor. This result is useful when we deal a capacitor having two or more dielectric mediums in between the plates. In this case we can assume an imaginary metal plate of thickness zero at each boundary of dielectrics.

An Important Outcome

As we have found that an infinitesimally thin metallic sheet inserted between the plates of a capacitor does not affect the capacitance. So, we slide an infinitesimally thin metallic slab along the boundary face of the dielectric as shown in Fig.3.101a. Now, we can model this system as a combination of three capacitors C_1, C_2 and C_3 connected in series as shown in Fig.3.101b. Air capacitor C_1 has a plate separation x, Capacitor C_2 with dielectric of dielectric constant κ has plate separation t and air capacitor C_3 has plate separation $(d - t - x)$.

The equivalent capacitance (C) of these three capacitors, is given

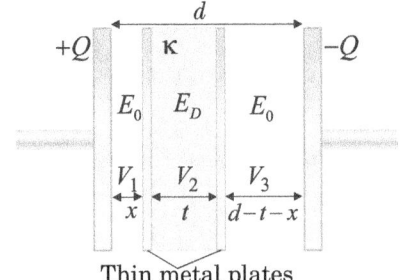

Figure 3.101

by -

3.12. GUASS'S LAW IN DIELECTRICS

$$\frac{1}{C} = \frac{1}{C_1} + \frac{1}{C_2} + \frac{1}{C_3} = \frac{x}{\epsilon_0 A} + \frac{t}{\epsilon_0 \kappa A} + \frac{d-t-x}{\epsilon_0 A}$$

$$= \frac{1}{\epsilon_0 A}\left(d - t + \frac{t}{\kappa}\right)$$

$$\implies \boxed{C = \frac{\epsilon_0 A}{d - t\left(1 - \frac{1}{\kappa}\right)}}$$

This value is same as obtained in Eq.(3.63).
If there are number of slabs having dielectric constants $\kappa_1, \kappa_2, \kappa_3, \cdots$ and thickness $t_1, t_2, t_3, \cdots$ respectively between the plates of a capacitor as described above, then

$$\boxed{C = \frac{\epsilon_0 A}{d - (t_1 + t_2 + t_3 + \ldots) + \left(\frac{t_1}{\kappa_1} + \frac{t_2}{\kappa_2} + \frac{t_3}{\kappa_3} + \ldots\right)}} \quad (3.65)$$

Note: From Eq.3.63, the magnitude of potential difference between the plates in presence of the dielectric slab is-

$$V = \frac{Q}{C} = \frac{Q}{\epsilon_0 A}\left[d - t\left(1 - \frac{1}{\kappa}\right)\right]$$

But in the absence of slab it is

$$V_0 = \frac{Q}{C_0} = \frac{Q}{\epsilon_0 A} d$$

Therefore, the decrease in potential difference is-

$$\Delta V = V_0 - V = \frac{Q}{\epsilon_0 A}\left[t\left(1 - \frac{1}{\kappa}\right)\right]$$

Now, suppose we want to increase the potential difference again up to its original value by increasing the separation of plates by a distance t', then

$$\frac{Q}{\epsilon_0 A}\left[t\left(1 - \frac{1}{\kappa}\right)\right] = \frac{Q}{\epsilon_0 A} t'$$

or $\boxed{t\left(1 - \frac{1}{\kappa}\right) = t'}$ \quad (3.66)

Thus, to get initial voltage after inserting a dielectric slab between the plates of an isolated charged capacitor, we have to increase the separation between the plates by t' as given by Eq.3.66.

EXAMPLE 57. In Fig.3.102, a parallel plate capacitor has plate area 40 cm² and plates separation 2 mm. The space between the plates is filled with a dielectric medium of a thickness 1 mm and dielectric constant 5. What is the capacitance of the system?

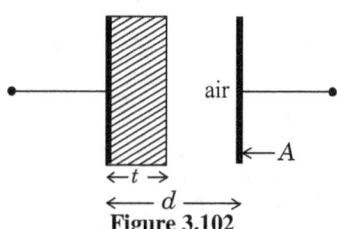

Figure 3.102

SOLUTION If we insert an infinitely thin metal plate (parallel to capacitor plates) at the dielectric-free space interface, the given situation will be equivalent to two capacitors in series: one filled with a dielectric and the other filled with air. If C_1 and C_2 represent the capacitances of the capacitors with dielectric and air, respectively, then the equivalent capacitance C_{eq} is determined by

$$\frac{1}{C_{eq}} = \frac{1}{C_1} + \frac{1}{C_2} = \frac{1}{\kappa \epsilon_0 A/t} + \frac{1}{\epsilon_0 A/(d-t)} = \frac{t}{\kappa \epsilon_0 A} + \frac{d-t}{\epsilon_0 A}$$

$$= \frac{1 \times 10^{-3}}{5\epsilon_0 \times 40 \times 10^{-4}} + \frac{1 \times 10^{-3}}{\epsilon_0 40 \times 10^{-4}}$$

$$\implies \frac{1}{C_{eq}} = \frac{1}{20\epsilon_0} + \frac{1}{4\epsilon_0} \implies C_{eq} = \frac{20 \times 4\epsilon_0}{24} = \frac{10\epsilon_0}{3} \text{ F}$$

EXAMPLE 58. A parallel plate capacitor with plate area A and plate separation d is filled with a dielectric material of dielectric constant $\kappa = 4$. The thickness of the dielectric material is t, where $t < d$. Let C_1 and C_2 be the capacitance of the system for $t = d/3$ and $t = 2d/3$, respectively. If $C_1 = 2\ \mu F$, find the value of C_2.

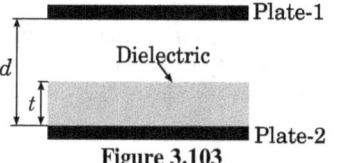

Figure 3.103

SOLUTION For $t = d/3$, Eq.3.63 gives

$$C_1 = \frac{\epsilon_0 A}{d - t\left(1 - \frac{1}{\kappa}\right)} = \frac{\epsilon_0 A}{d - \frac{d}{3}\left(1 - \frac{1}{\kappa}\right)} = \frac{\epsilon_0 A}{d - \frac{d}{3}\left(1 - \frac{1}{4}\right)} = \frac{4}{3}\frac{\epsilon_0 A}{d}$$

It is given that, $C_1 = 2\mu F$, therefore-

$$C_1 = \frac{4}{3}\frac{\epsilon_0 A}{d} = 2\mu F \implies \frac{\epsilon_0 A}{d} = \frac{3}{2}\mu F \quad \cdots (1)$$

For $t = 2d/3$, Eq.3.63 gives-

$$C_2 = \frac{\epsilon_0 A}{d - \frac{2d}{3}\left(1 - \frac{1}{4}\right)} = \frac{\epsilon_0 A}{d/2} = 2\left(\frac{\epsilon_0 A}{d}\right) \quad \cdots (2)$$

Substituting the value of $\frac{\epsilon_0 A}{d}$ from Eq.(1) in Eq.(2), we get

$$C_2 = 2\left(\frac{\epsilon_0 A}{d}\right) = 2\left(\frac{3}{2}\mu F\right) = 3\ \mu F$$

EXAMPLE 59. The distance between two plates of a capacitor is d and its capacitance is C_1, when air is the medium between the plates. If a metal sheet of thickness $2d/3$ and of same area as plate is introduced between the plates, the capacitance of the capacitor becomes C_2. Calculate the ratio C_2/C_1.

SOLUTION When air is the medium between the plates of given capacitor, the its capacitance-

$$C_1 = \frac{\epsilon_0 A}{d}$$

In presence of a metal sheet between the plates, the capacitance can be calculated by applying Eq.3.63:

$$C = \frac{\epsilon_0 A}{d - t\left(1 - \frac{1}{\kappa}\right)}$$

Here, $\kappa_{\text{metal sheet}} = \infty$, $t = \frac{2d}{3}$, therefore

$$C_2 = \frac{\epsilon_0 A}{d - \frac{2d}{3}\left(1 - \frac{1}{\infty}\right)} = 3\left(\frac{\epsilon_0 A}{d}\right) = 3C_1 \implies \frac{C_2}{C_1} = 3$$

EXAMPLE 60. A parallel plate capacitor with plate area A and plate separation $d = 2$ m has a capacitance of $4\ \mu F$. Find the new capacitance of the system if half of the space between them is filled with a dielectric material of dielectric constant $\kappa = 3$ as shown in Fig.3.104.

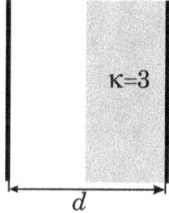

Figure 3.104

SOLUTION The capacitance of empty capacitor originally,

$$C_0 = \frac{\epsilon_0 A}{d} = 4\ \mu F \quad \text{(Given)} \quad \cdots (1)$$

Now, half of the space is filled with a dielectric medium of dielectric constant $\kappa = 3$, as shown in Fig.3.104. At the vacuum-dielectric interface, if we introduce an infinitely thin metal plate parallel to the given capacitor plates, the structure get converted into two capacitors in series–one with vacuum and the other filled with a dielectric medium. Both capacitors have the same surface area and plate separation.

Suppose, the capacitance of the capacitor having vacuum is C_1 and that of the other capacitor having dielectric medium, is C_2. The equivalent capacitance of this newly formed system of capacitors is given as-

$$\frac{1}{C_{eq}} = \frac{1}{C_1} + \frac{1}{C_2}$$

$$\implies \frac{1}{C_{eq}} = \frac{d/2}{\epsilon_0 A} + \frac{d/2}{\epsilon_0 \kappa A} = \frac{d}{2\epsilon_0 A}\left(1 + \frac{1}{\kappa}\right)$$

$$= \frac{d}{2\epsilon_0 A}\left(1 + \frac{1}{3}\right) = \frac{2d}{3\epsilon_0 A} = \frac{2}{3C_0} \quad (\because \kappa = 3)$$

Therefore, $C_{eq} = \frac{3}{2}C_0 = \frac{3}{2}(4\,\mu F) = 6\,\mu F$

(Since, from Eq.(1), $C_0 = 4\,\mu F$)

EXAMPLE 61. The plates of a parallel plate capacitor are charged upto 100 volt. A 2 mm thick dielectric plate is inserted between the plates, then to maintain the same potential difference, the distance between the capacitor plates is increased by 1.6 mm. What is the dielectric constant of the plate?

APPROACH In this problem, to maintained the original potential difference, after inserting the dielectric plate, the separation between the plates is increased. Therefore, we use Eq.3.66

SOLUTION From Eq.3.66, we have-

$$t\left(1 - \frac{1}{\kappa}\right) = t' \implies 2\left(1 - \frac{1}{\kappa}\right) = 1.6 \implies \kappa = 5$$

EXAMPLE 62. The distance between the plates of a parallel-plate capacitor is 0.05 m. It is connected to a battery which establishes a field of 3×10^4 V/m between the plates. Now, it is disconnected from the battery and an uncharged metal plate of thickness 0.01 m is inserted into it. Find the voltage between the plates- (i) before the introduction of the metal plate and (ii) after its introduction. What would be the voltage if we replace the metal plate by a dielectric plate with dielectric constant $\kappa = 2$?

SOLUTION (i) Before introducing the metal plate: For a parallel plate capacitor, the electric field E and voltage V are related as-

$$V = Ed$$

Given values: $d = 0.05$ m, $E = 3 \times 10^4$ V/m
On substituting these values in $V = ED$, we get

$$V = 3 \times 10^4 \times 0.05 = 1.5\,\text{kV}$$

(ii) After introducing the metal plate: When the battery is disconnected, capacitor becomes isolated. For an isolated capacitor, the charge on its plates remains constant, i.e., q = constant.
If after introducing the metal plate of thickness t, the capacitance becomes C' and voltage becomes V', then-

$$q = CV = C'V' \implies V' = \frac{C}{C'}V$$

Here, $C = \frac{\epsilon_0 A}{d}$, $C' = \frac{\epsilon_0 A}{(d-t)+(t/\kappa)}$ (Eq.3.63), therefore-

$$V' = \frac{C}{C'}V = \frac{(d-t)+(t/\kappa)}{d} \times V$$

Given that, $d = 0.05$ m, $t = 0.01$ m and for a metal $\kappa = \infty$, therefore-

$$V' = \left[\frac{(0.05-0.01)+(0.01/\infty)}{0.05}\right] \times 1.5 = 1.2 \times 10^3\,\text{V} = 1.2\,\text{kV}$$

If instead of metal plate, we introduce a dielectric plate ($\kappa = 2$) of same thickness, then-

$$V' = \left[\frac{(0.05-0.01)+(0.01/2)}{0.05}\right] \times 1.5$$

$$= 1.35 \times 10^3\,\text{V} = 1.35\,\text{kV}$$

3.12.2 Capacitor Filled with a Variable Dielectric Medium

Now, we consider a special case in which the magnitude of the dielectric constant varies according to certain pattern, from positive plate to negative plate. In such cases, always divide the complete space between the plates into elementary parallel strips. Now, consider such a strip of thickness dx at a distance x from the positive plate and insert two infinitely thin parallel metal plates along the boundary of this elementary strip. It forms an elementary capacitor of plate separation dx. If κ is the dielectric constant at this position, then capacitance of this elementary strip capacitor will be-

$$dC = \frac{\epsilon_0 \kappa A}{dx}$$

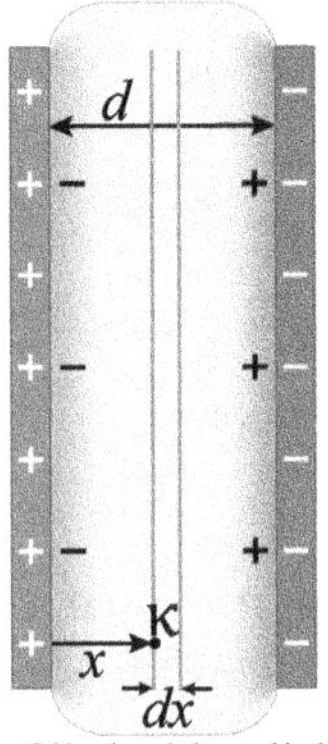

Figure 3.105: Capacitor filled with a dielectric having variable κ.

The net capacitance of the capacitor can be obtained by considering all such capacitors into series.
If net capacitance of the capacitor is C, then-

$$\frac{1}{C} = \int_{x=0}^{d} \frac{1}{dC} = \int_{x=0}^{d} \frac{1}{\frac{\epsilon_0 \kappa A}{dx}} = \int_{x=0}^{d} \frac{dx}{\epsilon_0 \kappa A}$$

$$\implies \boxed{\frac{1}{C} = \int_{x=0}^{d} \frac{dx}{\epsilon_0 \kappa A}} \qquad (3.67)$$

Substituting the value of κ in terms of x in Eq.(3.67) and solving for C gives the required capacitance.
For example, let $\kappa = ax + b$, then from Eq.3.67, we get-

$$\frac{1}{C} = \frac{1}{\epsilon_0 A}\int_{x=0}^{d} \frac{dx}{ax+b} = \frac{1}{\epsilon_0 Aa}\ln\left[\frac{a}{b}d+1\right]$$

or

$$C = \frac{\epsilon_0 Aa}{\ln\left[\frac{a}{b}d+1\right]}$$

This is the required capacitance of the given capacitor.

EXAMPLE 63. Two parallel plate capacitors A and B have the same separation $d = 8.85 \times 10^{-4}$ m between the plates. The plate areas of A and B are 0.04 m^2 and 0.02 m^2, respectively. A slab of dielectric constant (relative permittivity) $\kappa = 9$ has dimensions such that it can exactly fill the space between the plates of capacitor B.

(a) The dielectric slab is placed inside A as shown in the Fig.3.106a. A is then charged to a potential difference of 110 V. Calculate the capacitance of A and the energy stored in it.

(b) The battery is disconnected and then the dielectric slab is removed from A. Find the work done by the external agency in removing the slab from A.

(c) The same dielectric slab is now placed inside B, filling it completely. The two capacitors A and B are then connected as shown in the Fig.3.106c. Calculate the energy stored in the system.

SOLUTION The plate area of the capacitor B is $A = 0.02\,\text{m}^2$. The area of dielectric slab is $A = 0.02\,\text{m}^2$ and the plate area of

3.12. GAUSS'S LAW IN DIELECTRICS

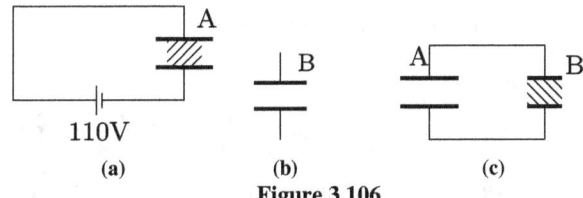

Figure 3.106

the capacitor A is $2A = 0.04 \text{ m}^2$. The dielectric slab fills half portion of the capacitor A. The capacitor A with dielectric slab inside it is equivalent to two capacitors of capacitance $\epsilon_0 A/d$ and $\kappa \epsilon_0 A/d$ connected in parallel. Thus, the equivalent capacitance of A with dielectric slab is

$$C_A = \frac{\epsilon_0 A}{d} + \frac{\kappa \epsilon_0 A}{d} = \frac{(1+\kappa)\epsilon_0 A}{d}$$
$$= \frac{(1+9)(8.85 \times 10^{-12})(0.02)}{8.85 \times 10^{-4}} = 2 \times 10^{-9} \text{ F}$$

The charge on capacitor A and electrostatic energy stored in it are

$$q_A = C_A V = (2 \times 10^{-9})(110) = 2.2 \times 10^{-7} C,$$
$$U_A = \frac{1}{2}C_A V^2 = \frac{1}{2} \times (2 \times 10^{-9})(110)^2 = 1.21 \times 10^{-5} \text{ J}$$

The charge on A remains same after the dielectric is removed i.e., $q'_A = q_A$. The capacitance of A after the removal of dielectric is

$$C'_A = \epsilon_0 (2A)/d = 0.4 \times 10^{-9} \text{ F},$$

and the stored energy after the removal of dielectric is

$$U'_A = \frac{q'^2_A}{2C'_A} = \frac{(2.2 \times 10^{-7})^2}{2(0.4 \times 10^{-9})} = 6.05 \times 10^{-5} \text{ J}$$

The stored energy increases because the work is done by the external source. By energy conservation, the work done by the external source is

$$W = U'_A - U_A = 4.84 \times 10^{-5} \text{ J}.$$

The capacitance of the capacitor B with dielectric slab is $C_B = \kappa \epsilon_0 A/d = 1.8 \times 10^{-9}$ F. The capacitors A and B are connected in parallel giving equivalent capacitance

$$C = C'_A + C_B = 0.4 \times 10^{-9} + 1.8 \times 10^{-9} = 2.2 \times 10^{-9} \text{ F}$$

Total charge on the equivalent capacitor is $q = q_A$ and the stored energy is

$$U = \frac{q^2}{2C} = \frac{(2.2 \times 10^{-7})^2}{2(2.2 \times 10^{-9})} = 1.1 \times 10^{-5} \text{ J}$$

EXAMPLE 64. Figure 3.107 shows two identical parallel plate capacitors connected to a battery with the switch S closed. The switch is now opened and the free space between the plates of the capacitors is filled with a dielectric of dielectric constant (or relative permittivity) 3. Find the ratio of the total electrostatic energy stored in both capacitors before and after the introduction of the dielectric.

SOLUTION Let $C_A = C$ and $C_B = C$ be the initial capaci-

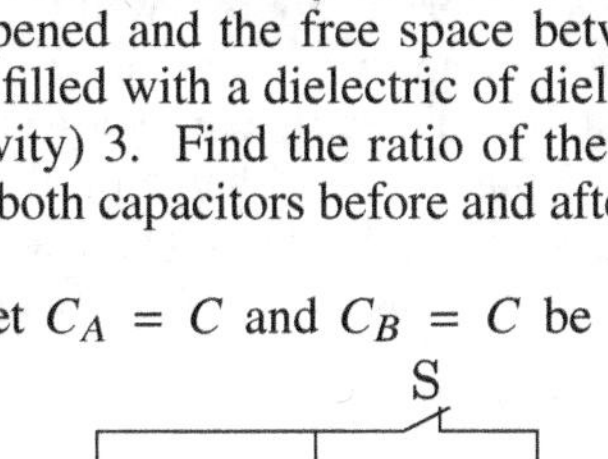

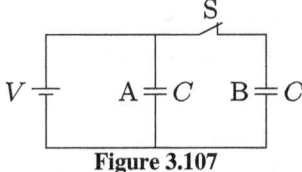

Figure 3.107

tances of the capacitors A and B, respectively. Initially, potential difference across both the capacitors is V and electrostatic energies stored in two capacitors are $U_A = \frac{1}{2}C_A V^2 = \frac{1}{2}CV^2$ and $U_B = \frac{1}{2}C_B V^2 = \frac{1}{2}CV^2$ (Fig.3.108). Thus, the total energy stored in two capacitors is

$$U = U_A + U_B = CV^2$$

When a dielectric of dielectric constant $\kappa = 3$ is introduced in the two capacitors, their capacitances become $C'_A = \kappa C_A = 3C$ and $C'_B = \kappa C_B = 3C$. The potential difference across capacitor A does not change when the switch is opened. Thus, the electrostatic energy stored in capacitor A is -

$$U'_A = \frac{1}{2}C'_A V^2 = \frac{1}{2}(3C)V^2 = \frac{3}{2}CV^2$$

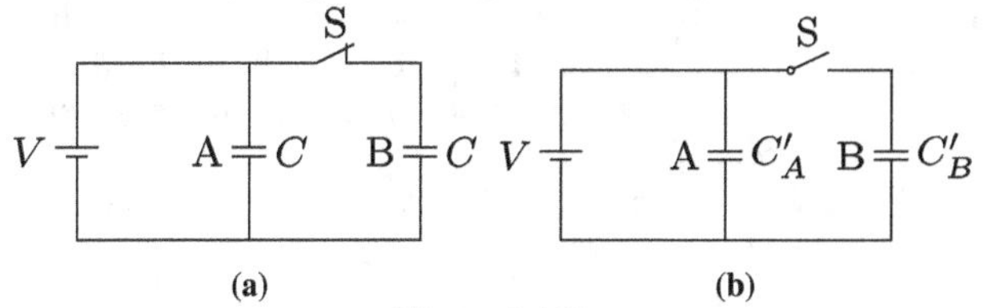

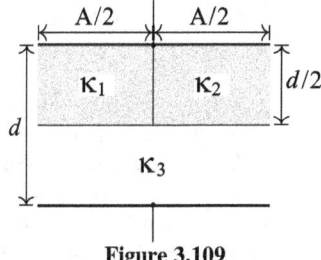

Figure 3.108

The charge on capacitor B does not change when the switch is opened, i.e., charge remains equal to its initial value of $Q = V/C_A = V/C$. Thus, the electrostatic energy stored in capacitor B is

$$U'_B = \frac{Q'^2}{2C'_B} = \frac{Q^2}{2(3C)} = \frac{(V/C)^2}{6C} = \frac{1}{6}CV^2$$

Total energy stored in the two capacitors after the introduction of dielectric is

$$U' = U'_A + U'_B = \frac{3}{2}CV^2 + \frac{1}{6}CV^2 = \frac{5}{3}CV^2$$

Hence, the ratio of the total electrostatic energy stored in both capacitors before and after the introduction of the dielectric is $U_i/U_f = 3/5$.

EXAMPLE 65. A parallel plate capacitor of area A, plate separation d and capacitance C is filled with three different dielectric materials having dielectric constants κ_1, κ_2 and κ_3 as shown in Fig.3.109. If a single dielectric material is to be used to have the same capacitance C in this capacitor then prove that-

$$\frac{1}{\kappa} = \frac{1}{2\kappa_3} + \frac{1}{\kappa_1 + \kappa_2}$$

SOLUTION The capacitance of a parallel plate capacitor of

Figure 3.109

plate area A, plate separation d, and dielectric constant κ is given by $C = \frac{\kappa \epsilon_0 A}{d}$. Given configuration can be considered as a combination of three capacitors having capacitances

$$C_1 = \frac{\kappa_1 \epsilon_0 (A/2)}{d/2} = \kappa_1 \frac{\epsilon_0 A}{d} = \kappa_1 C_0,$$
$$C_2 = \frac{\kappa_2 \epsilon_0 (A/2)}{d/2} = \kappa_2 C_0,$$
$$C_3 = \frac{\kappa_3 \epsilon_0 A}{d/2} = 2\kappa_3 C_0$$

where $C_0 = \epsilon_0 A/d$. The capacitor C_1 and C_2 are connected in parallel with their effective capacitance:

$$C_{12} = C_1 + C_2 = (\kappa_1 + \kappa_2)C_0$$

The C_{12} is combined in series with C_3 giving effective capacitance of the three capacitors as:

$$C_{123} = \frac{C_{12}C_3}{C_{12} + C_3} = \frac{(\kappa_1 + \kappa_2)2\kappa_3}{\kappa_1 + \kappa_2 + 2\kappa_3}C_0 = \frac{\kappa \epsilon_0 A}{d}$$

On simplifying it, we get

$$\frac{1}{\kappa} = \frac{1}{2\kappa_3} + \frac{1}{\kappa_1 + \kappa_2}.$$

EXAMPLE 66. A parallel plate capacitor has a dielectric slab of dielectric constant κ between its plates that covers 1/3 of the area of its plates, as shown in the Figure 3.110. The total capacitance of the capacitor is C while that of the portion with dielectric in between is C_1. When the capacitor is charged, the plate area covered by the dielectric gets charge Q_1 and the rest of the area gets charge Q_2. The electric field in the dielectric is E_1 and that in the other portion is E_2. Ignoring edge effects, calculate (a) total capacitance, (b) E_1/E_2 and (c) Q_1/Q_2.

SOLUTION Let A be the total plate area of the capacitor and

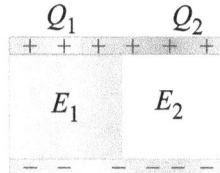

Figure 3.110

d be the separation between the plates. The plate area for the portion with dielectric is $A_1 = A/3$ and that without dielectric is $A_2 = 2A/3$. The capacitances of the portions with dielectric and without dielectric are

$$C_1 = \frac{\kappa\epsilon_0(A/3)}{d} = \frac{\kappa\epsilon_0 A}{3d} \quad \text{and}$$
$$C_2 = \frac{\epsilon_0(2A/3)}{d} = \frac{2\epsilon_0 A}{3d}$$

respectively.

These two portions are connected in parallel. Hence, the total capacitance is

$$C = C_1 + C_2 = \frac{(2+\kappa)\epsilon_0 A}{3d}$$

In parallel connection, potential between the plates is equal which gives

$$\frac{E_1}{E_2} = \frac{V/d}{V/d} = 1.$$

The ratio of the charges on the two portions is

$$\frac{Q_1}{Q_2} = \frac{C_1 V}{C_2 V} = \frac{\kappa}{2}.$$

EXAMPLE 67. A dielectric slab of thickness d is inserted in a parallel plate capacitor whose negative plate is at $x = 0$ and positive plate is at $x = 3d$. The slab is equidistant from the plates. The capacitor is given some charge. As x goes from 0 to $3d$,
(A) the magnitude of the electric field remains the same.
(B) the direction of the electric field remains the same.
(C) the electric potential increases continuously.
(D) the electric potential increases at first, then decreases and again increases.

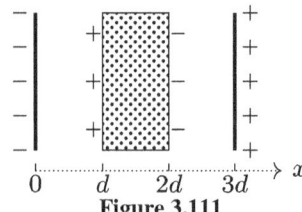

Figure 3.111

SOLUTION (B, C) Let $\sigma > 0$ be the charge density of the plate at $x = 3d$.
The electric field in the various regions inside the capacitor is given by

$$\vec{E}(x) = \begin{cases} -\frac{\sigma}{\epsilon_0}\hat{\imath}, & \text{if } 0 < x \le d \\ -\frac{\sigma}{\epsilon_0\kappa}\hat{\imath}, & \text{if } d < x \le 2d \\ -\frac{\sigma}{\epsilon_0}\hat{\imath}, & \text{if } 2d < x \le 3d \end{cases}$$

The electric potential is given by $V(x) = -\int E dx$. If the plate at $x = 0$ is assumed to be at zero potential then the potential in the various regions is

$$V(x) = \begin{cases} \frac{\sigma x}{\epsilon_0}, & \text{if } 0 < x \le d \\ \frac{\sigma d}{\epsilon_0} + \frac{\sigma(x-d)}{\epsilon_0\kappa}, & \text{if } d < x \le 2d \\ \frac{\sigma d}{\epsilon_0} + \frac{\sigma d}{\epsilon_0\kappa} + \frac{\sigma(x-2d)}{\epsilon_0}, & \text{if } 2d < x \le 3d \end{cases}$$

EXAMPLE 68. A parallel plate capacitor of plate area A and plate separation d is charged to potential difference V and then the battery is disconnected. A slab of dielectric constant κ is then inserted between the plates of the capacitor so as to fill the space between the plates. If Q, E and W denote respectively, the magnitude of charge on each plate, the electric field between the plates (after the slab is inserted), and work done on the system in question, in the process of inserting the slab, then, find the values of Q, W and E.

SOLUTION Initially, the capacitance of the capacitor, charge on its plates, and electrostatic energy stored in it are

$$C = \epsilon_0 A/d,$$
$$Q = CV = \epsilon_0 AV/d,$$
and $$U = \frac{1}{2}CV^2 = \epsilon_0 AV^2/(2d) \text{ respectively.}$$

The charge on the capacitor plates does not change if the dielectric slab is inserted after the battery disconnection. Finally, the capacitance, charge, and stored energy are

$$C' = \kappa\epsilon_0 A/d$$
$$Q' = Q = \epsilon_0 AV/d$$
and $$U' = Q'^2/(2C') = \epsilon_0 AV^2/(2d\kappa) \text{ respectively.}$$

Since, work done (W) on the system is equal to change in it's potential energy, therefore:

$$W = U' - U = -\frac{\epsilon_0 AV^2}{2d}\left[1 - \frac{1}{\kappa}\right]$$

The potential difference across the capacitor plates becomes $V' = Q'/C' = V/\kappa$ and electric field between the plates is $E' = V'/d = V/(\kappa d)$.

EXAMPLE 69. A parallel plate air capacitor is connected to a battery. The quantities charge, voltage, electric field and energy associated with this capacitor are given by Q_0, V_0, E_0 and U_0, respectively. A dielectric slab is now introduced to fill the space between the plates with the battery still in connection. The corresponding quantities now given by Q, V, E and U respectively. Find relations between new and original quantities.

SOLUTION The potential across the capacitor is equal to the battery voltage if the battery is not removed i.e., $V = V_0$.
The capacitance of the capacitor increases from C_0 to $C = \kappa C_0$ when a dielectric slab of dielectric constant κ is introduced to fill the gap between the plates. The charge on the capacitor plates become $Q = CV = (\kappa C_0) V_0 = \kappa Q_0 > Q_0$ (because $\kappa > 1$).
The electric field between the plates separated by a distance d is given by

$$E = V/d = V_0/d = E_0.$$

The electrostatic energy stored in the capacitor is given by

$$U = \frac{1}{2}CV^2 = \frac{1}{2}(\kappa C_0) V_0^2 = \kappa\left(\frac{1}{2}C_0 V_0^2\right)$$
$$= \kappa U_0 > U_0$$

EXAMPLE 70. Two parallel plate capacitors of capacitance C and $2C$ are connected in parallel and charged to a potential difference V. The battery is then disconnected and the region between the plates of capacitor C is completely filled with a material of dielectric constant κ. Find the new potential difference across the capacitors.

SOLUTION The capacitors arrangement is shown Fig.3.112. In Fig.3.112a two capacitors of capacitances C and $2C$ respectively, are connected in parallel and charged to a potential difference V. In Fig.3.112b, the battery is disconnected and the region between

3.12. GUASS'S LAW IN DIELECTRICS

the plates of capacitor C is completely filled with a material of dielectric constant κ.

The charges on the plates connected to the positive terminal

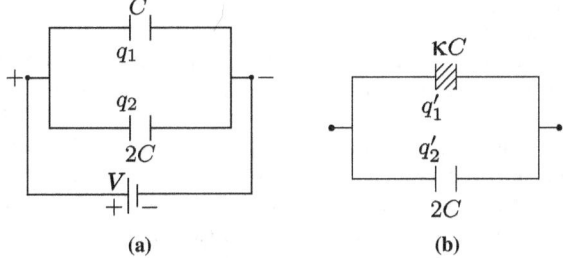

Figure 3.112

of the battery of potential V are $q_1 = CV$ and $q_2 = 2CV$. Let q'_1 and q'_2 be the charges on these plates after the battery is disconnected and the dielectric is inserted. The capacitance of the first capacitor becomes κC when a material of dielectric constant κ is inserted into it. By charge conservation, total charge on the plates connected to positive terminal remains constant, once the battery is disconnected (isolated system), i.e.,

$$q'_1 + q'_2 = q_1 + q_2 = 3CV \quad (1)$$

The potential across the two capacitors after insertion of dielectric is given by

$$V' = \frac{q'_1}{\kappa C} = \frac{q'_2}{2C} = \frac{q'_1 + q'_2}{\kappa C + 2C} \quad (2)$$

Since from Eq.(1), $q'_1 + q'_2 = 3CV$, therefore Eq.(2) gives-

$$V' = \frac{3CV}{C(\kappa + 2)} = \frac{3V}{\kappa + 2}$$

EXAMPLE 71. Three identical capacitors C_1, C_2 and C_3 have a capacitance of $1.0\ \mu F$ each and they are uncharged initially. They are connected in a circuit as shown in the Fig.3.113 and C_1 is then filled completely with a dielectric material of relative permittivity ϵ_r. The emf of battery is $V_0 = 8$ V. First the switch S_1 is closed while the switch S_2 is kept open. When the capacitor C_3 is fully charged, S_1 is opened and S_2 is closed simultaneously. When all the capacitors reach equilibrium, the charge on C_3 is found to be $5\ \mu C$. Find the value of ϵ_r.

SOLUTION When the switch S_1 is closed and the switch S_2 is

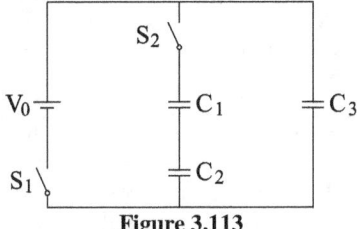

Figure 3.113

kept open, the capacitor $C_3 (= 1\ \mu F)$ gets fully charged by a battery of emf $V_0 (= 8\ V)$ (Fig.3.114a). On complete charging, the charge on the capacitor C_3 becomes $q_3 = C_3 V_0 = (1)(8) = 8\ \mu C$ (i.e., charge $+8\ \mu C$ on the plate connected to the positive terminal of the battery and $-8\ \mu C$ on the plate connected to the negative terminal of the battery).

When the switch S_1 is opened and the switch S_2 is closed the situation becomes similar to Fig.3.114b. In this situation, the positive charge starts flowing from the upper plate of C_3 to the upper plate of C_1. Similarly, the negative charge starts flowing from the lower plate of C_3 to the lower plate of C_2. An equal but opposite charge is induced on the other plates of C_1 and C_2. In equilibrium, charge on the upper plate of C_3 is $+5\mu C$ and charge on the lower plate of C_3 is $-5\mu C$. By conservation of charge, charge on the upper plate of C_1 is $+3\ \mu C$ and charge on the lower plate of C_2 is $-3\ \mu C$.

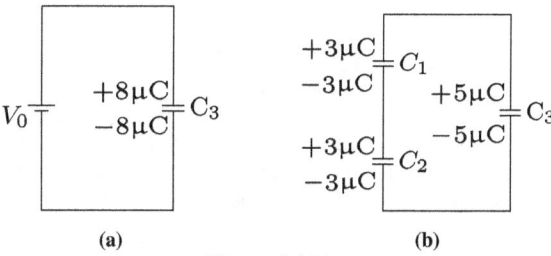

Figure 3.114

The potential difference across a capacitor of capacitance C having a charge q is given by $V = q/C$. When capacitor C_1 is filled with a dielectric of relative permittivity ϵ_r, its capacitance increases from C_1 to $C'_1 = \epsilon_r C_1$.

The potential difference across C_3 is equal to the potential difference across C_1 and C_2 together i.e.,

$$\frac{q_1}{C'_1} + \frac{q_2}{C_2} = \frac{q_3}{C_3}$$

$$\Rightarrow \quad \frac{3\mu C}{\epsilon_r (1\mu F)} + \frac{3\mu C}{1\mu F} = \frac{5\mu C}{1\mu F}$$

On simplifying it, we get $\epsilon_r = 1.5$.

EXAMPLE 72. The capacitance of a parallel plate capacitor with plate area A and separation d is C. The space between the plates is filled with two wedges of dielectric constants κ_1 and κ_2 respectively as shown in Fig.3.115. Find the capacitance of the resulting capacitor.

SOLUTION Let length and breadth of the capacitor be l and b

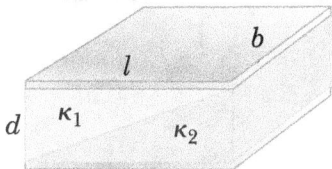

Figure 3.115

respectively and d be the distance between the plates as shown in Fig.3.116. Now, consider a strip of width dx at a distance x from left end.

From Fig.3.116, $QR = x \tan \theta$ and $PQ = d - x \tan \theta$

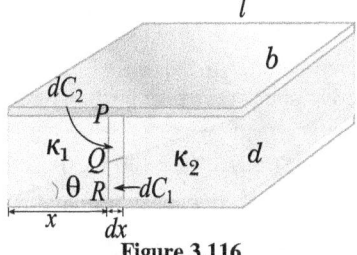

Figure 3.116

where $\tan \theta = d/l$.
Capacitance of portion PQ

$$dC_1 = \frac{\kappa_1 \epsilon_0 (b\,dx)}{\text{distance PQ}} = \frac{\kappa_1 \epsilon_0 (b\,dx)}{d - x \tan \theta} = \frac{\kappa_1 \epsilon_0 (b\,dx)}{d - \frac{xd}{l}}$$

or $$dC_1 = \frac{\kappa_1 \epsilon_0 bl\,dx}{d(l-x)} = \frac{\kappa_1 \epsilon_0 A(dx)}{d(l-x)}$$

and capacitance of QR

$$dC_2 = \frac{\kappa_2 \epsilon_0 (b\,dx)}{\text{distance QR}} = \frac{\kappa_2 \epsilon_0 b(dx)}{x \tan \theta} = \frac{\kappa_2 \epsilon_0 A(dx)}{x\,d}$$

$$[\because \tan \theta = d/l]$$

Now dC_1 and dC_2 are in series. Therefore, their resultant capacity dC will be given by

$$\frac{1}{dC} = \frac{1}{dC_1} + \frac{1}{dC_2} = \frac{d(l-x)}{\kappa_1 \epsilon_0 A(dx)} + \frac{x \cdot d}{\kappa_2 \epsilon_0 A(dx)}$$

$$\Longrightarrow \quad \frac{1}{dC} = \frac{d}{\epsilon_0 A(dx)}\left(\frac{1-x}{\kappa_1}+\frac{x}{\kappa_2}\right) = \frac{d[\kappa_2(l-x)+\kappa_1 x]}{\epsilon_0 A \kappa_1 \kappa_2 (dx)}$$

$$\Longrightarrow \quad dC = \frac{\epsilon_0 A \kappa_1 \kappa_2}{d[\kappa_2(l-x)+\kappa_1 x]}dx = \frac{\epsilon_0 A \kappa_1 \kappa_2}{d[\kappa_2 l + (\kappa_1-\kappa_2)x]}dx$$

All such elemental capacitor dC are connected in parallel. Now the capacitance of the given parallel plate capacitor is obtained by adding such infinitesimal capacitors parallel from $x = 0$ to $x = l$. i.e.,

$$C = \int_{x=0}^{x=l} dC = \int_0^l \frac{\epsilon_0 A \kappa_1 \kappa_2}{d[\kappa_2 l + (\kappa_1-\kappa_2)x]}dx$$

$$\Longrightarrow \quad C = \frac{\kappa_1 \kappa_2 \epsilon_0 A}{(\kappa_1-\kappa_2)d}\ln\frac{\kappa_2}{\kappa_1}$$

EXAMPLE 73. A container has a base of $50\,\text{cm} \times 5\,\text{cm}$ and height 50 cm, as shown in the Fig.3.117. It has two parallel electrically conducting walls each of area $50\,\text{cm} \times 50\,\text{cm}$. The remaining walls of the container are thin and non-conducting. The container is being filled with a liquid of dielectric constant 3 at a uniform rate of $250\,\text{cm}^3\text{s}^{-1}$. What is the value of the capacitance of the container after 10 seconds? [Given: Permittivity of free space $\epsilon_0 = 9\times 10^{-12}\,\text{C}^2\text{N}^{-1}\text{m}^{-2}$, the effects of the non-conducting walls on the capacitance are negligible]

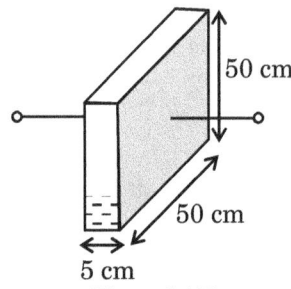

Figure 3.117

SOLUTION In $t = 10$ sec volume of liquid, $V = 2500\,\text{cm}^3$,

Therefore, height of the liquid column, $h = \dfrac{2500}{50\times 5} = 10\,\text{cm}$,

Therefore, effective plate area of the portion of capacitor containing liquid dielectric

$$A_d = \left(50\times 10^{-2}\right)\times\left(10\times 10^{-2}\right)$$

The height of portion of air in capacitor $= 50 - 10 = 40\,\text{cm}$

Therefore, effective plate area of the portion of capacitor containing air

$$A_a = \left(50\times 10^{-2}\right)\times\left(40\times 10^{-2}\right)$$

Capacitance of the portion of capacitor containing liquid dielectric,

$$C_d = \frac{A_d \epsilon_0 \kappa}{d} = \frac{50\times 10^{-2}\times 10\times 10^{-2} \epsilon_0 \times 3}{5\times 10^{-2}} = 3\epsilon_0,$$

Capacitance of the portion of capacitor containing air,

$$C_a = \frac{A_a \epsilon_0}{d} = \frac{50\times 10^{-2}\times 40\times 10^{-2}\epsilon_0}{5\times 10^{-2}} = 4\epsilon_0,$$

These two capacitors can be considered in parallel, therefore net capacitance of the system

$$C = C_a + C_d = 7\epsilon_0 = 7\times 9\times 10^{-12} = 63\,\text{pF}$$

3.13 Effects of Nonuniform Electric Fields on Dielectrics

Experiments show that a dielectric in an electric field experiences the action of forces. These forces are called **ponderomotive forces**. These forces appear when the dielectric is neutral as a whole. Ponderomotive forces appear due to the action of a nonuniform electric field on dipole molecules of the polarized dielectric. In this case, the forces are caused by the non-uniformity of not only the macroscopic field but the microscopic field as well, which is created mainly by the nearest molecules of the polarized dielectric.

Under the action of these electric forces, the polarized solid dielectric is deformed. This phenomenon is called **electrostriction**. As a result of electrostriction, mechanical stresses appear in the dielectric.

Due to electrostriction, not only the electric force (which depends on the charges) acts on a conductor in a polarized dielectric, but also an additional mechanical force caused by the dielectric. In the general case, the effect of a dielectric on the resultant force acting on a conductor cannot be taken into account by any simple relations, and the problem of calculating the forces with simultaneous analysis of the mechanism of their appearance is, very complicated. However, in many cases, [like force on dielectric slab placed near the edge of capacitors (subsection 3.13.1.1)], these forces can be calculated in a sufficiently simple way without a detailed analysis of their origin by using the law of conservation of energy.

3.13.1 Energy Method for Calculating the Force Between Dielectrics and Conductors

This method is the most general and allows us to take into account automatically all force interactions (both electric and mechanical) ignoring their origin, and hence leads to a correct result.

In Example 56, we explained why a dielectric slab placed near

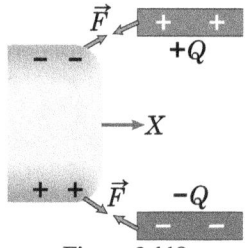

Figure 3.118

the edges of a charged parallel plate capacitor is attracted in the space between the plates, similar to how conductors are attracted towards larger electric fields. Now, we will discuss the energy method used to calculate the force acting between the plates and the dielectric slab.

Figure 3.118 illustrates a situation where charged conducting plates are disconnected from the power supply and then placed near a dielectric slab. Both the dielectric slab and the plates are free to move. In this case, the charges on the plates remain unchanged, and we can state that the work W of all internal forces within the system during a slow relative displacement between the charged plates and the dielectric slab is done entirely at the expense of a decrease in the electric energy U of the system (or its field). We assume that these displacements do not cause the transformation of electric energy into other forms of energy, or if they do, the transformations are infinitesimally small. Thus, for an infinitesimal displacement, we can express it as:

$$dW = -\,dU|_Q \qquad (3.68)$$

Here, the symbol Q emphasizes that the decrease in the energy of the system must be calculated when charges on the conductors are constant. If F_x is the x-component of the electric force during an infinitesimal displacement dx, then:

$$dW = F_x\,dx$$

By substituting this value for dW into Eq.3.68 and dividing both sides of Eq. 3.68 by dx, we obtain:

$$F_x = -\left.\frac{\partial U}{\partial x}\right|_Q \quad (3.69)$$

The force depends solely on the positions of the plates and the dielectric slab, as well as on the distribution of charges at a given instant. It does not rely on how the energy process will evolve if the system starts to move under the influence of forces. This implies that, to calculate F_x using Eq.3.69, we do not need to specify conditions under which all the charges on the conductor are necessarily constant (Q = constant). Instead, we only need to determine the change dU under the condition that Q = constant, which is a purely mathematical operation.

It should be noted that if a displacement is performed at constant potential difference between the plates, the corresponding calculation leads to another expression for the force:

$$F_x = +\left.\frac{\partial U}{\partial x}\right|_V \quad (3.70)$$

However, the result of the calculation of F_x with the help of this formula or Eq.3.69 is the same, as expected. Henceforth, we shall confine ourselves to the application of only Eq.3.69 and will use it for any conditions, including those where $Q \neq$ constant upon small displacements. It's important to note that, in such cases, the derivative $\partial U / \partial x$ will be consistently calculated at Q = constant.

3.13.1.1 Calculation of Force on a Dielectric Slab in a Fringing Field of a Capacitor

As given above, we find the force on a dielectric slab under following two conditions:
(i) when battery is disconnected after charging the capacitor (i.e., charge Q = constant)
(ii) When battery remains connected (i.e., V = constant)

(i) When Battery is Disconnected after Charging the Capacitor (i.e., Charge Q = constant)

Let us consider a parallel plate capacitor with plates of width b and length l. The distance between the plates is d ($\ll l$). Suppose, the charge on the positive plate of the capacitor is Q. If at any instant, the dielectric slab of dielectric constant κ, is pulled in by the fringing field up to a distance x inside the capacitor, then the system can be considered to be two capacitors in parallel, one with plate area $b(l - x)$ and air between the plates and one with area bx and dielectric filling the space between the plates (see Fig.3.119).

The formula for capacitance of a parallel-plate capacitor with

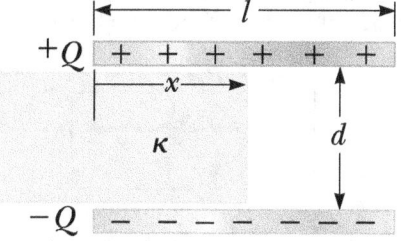

Figure 3.119: A charged capacitor disconnected from battery: Q = constant.

plate area A is given by-

$$C = \frac{\kappa \epsilon_0 A}{d}$$

The capacitance of air capacitor-

$$C_1 = \frac{\epsilon_0 b (l - x)}{d}$$

and the capacitance of capacitor with dielectric slab-

$$C_2 = \frac{\kappa \epsilon_0 b x}{d}$$

Since, the potential difference across both capacitors, is same, therefore these capacitors can be regarded in parallel and hence, the net capacitance of the system-

$$C = C_1 + C_2 = \frac{\epsilon_0 b}{d}[l + x(\kappa - 1)] \quad (3.71)$$

In this case, the plates - dielectric slab system is isolated, so the plate charge Q will remain constant and confined to the plates of the capacitor.

Let U be the electric energy of the system and F_x be the pulling force of fringing field on the dielectric slab when x length of the slab is inside the plates. If the fringing field pulls the dielectric by an additional infinitesimal displacement δx, then capacitance C increases and therefore according to relation $U = Q^2/2C$, the electric potential energy decreases by an amount equal to the work done by the electric field, i.e.

$$-\delta U = \delta W = F_x \delta x \quad (3.72)$$

Now, the electric force (F_x) on the slab is

$$F_x = -\frac{\partial U}{\partial x}$$

The electric potential energy stored in the capacitor is given by-

$$U = \frac{1}{2}\frac{Q^2}{C} = \frac{1}{2}\frac{Q^2 d}{\epsilon_0 [l + x(\kappa - 1)] b}$$

Therefore, electrostatic force on the dielectric slab-

$$F_x = -\frac{\partial U}{\partial x} = -\frac{\partial}{\partial x}\left(\frac{1}{2}\frac{Q^2}{C}\right) = \frac{Q^2}{2C^2}\frac{\partial C}{\partial x} \quad (3.73)$$

Now, substituting the value of C from Eq.3.71 in above expression, we get-

$$F_x = -\frac{Q^2 d}{2\epsilon_0 b}\frac{(\kappa - 1)(-1)}{[l + x(\kappa - 1)]^2}$$

$$\Rightarrow \quad F_x = \frac{Q^2 d (\kappa - 1)}{2\epsilon_0 b [l + x(\kappa - 1)]^2} \quad (3.74)$$

(ii) When Battery Remains Connected (i.e., V = constant)

When the plates are connected to the battery of voltage V, the plates-dielectric system is not an isolated system.

In Fig.3.120, the dielectric slab with dielectric constant κ is

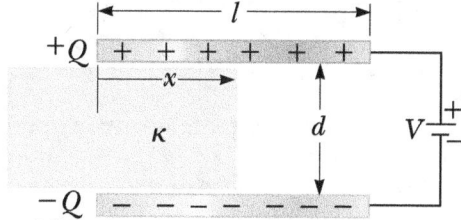

Figure 3.120: Capacitor connected to battery: V = constant

pulled in by the fringing field up to a distance x. At this instant, from Eq.3.71, the net capacitance of the capacitor is given by

$$C = \frac{\epsilon_0 b}{d}[l + x(\kappa - 1)]$$

Therefore, the total electric energy stored in the capacitor is given by the expression

$$U = \frac{1}{2}CV^2 = \frac{1}{2}\frac{\epsilon_0 b}{d}[l + x(\kappa - 1)]V^2 \quad (3.75)$$

If F_x is the pulling force applied by the electric field at this instant, and dW_{el} is the work done by the electric field during an additional infinitesimally small displacement δx, then

$$W_{el} = F_x \, dx \quad (3.76)$$

If change in potential energy due to this work is δU_{el}, then

$$\delta U_{el} = -W_{el} = -F_x \, dx \quad (3.77)$$

We must be careful here. During the inward displacement of the dielectric slab, the capacitance of the capacitor increases while the potential difference between the plates remains constant. From $Q = CV$, to increase the capacitance at a constant potential difference (V), the charge must be supplied by the battery. During the additional inward displacement δx of the slab, the battery transfers an additional charge $+\delta Q$ from the negative plate to the positive plate of the capacitor. In this process, the battery loses energy, and the capacitor gains energy. If δU_b is the change in potential energy of the capacitor due to the work done by the battery (W_b), then

$$\delta U_b = \delta W_b = V \delta Q \quad (3.78)$$

☞ Note that the potential energy of the voltage source and the potential energy of the capacitor both change as x changes.

From Eq.3.77 and Eq. 3.78, the net change in potential energy of the capacitor during this process is

$$\delta U = \delta U_{el} + \delta U_b = -F_x \, dx + V \, \delta Q$$

Rearranging terms, we get

$$F_x \, \delta x = -\delta U + V \, \delta Q \quad (3.79)$$

Solving Eq.3.79 for F_x, we get

$$\boxed{F_x = -\frac{\partial U}{\partial x} + V \frac{\partial Q}{\partial x}} \quad (3.80)$$

Since, $U = \frac{1}{2} CV^2$ and $Q = CV$, therefore, from above Eq.3.80, we can write-

$$\boxed{F_x = -\frac{1}{2} V^2 \frac{\partial C}{\partial x} + V^2 \frac{\partial C}{\partial x} = \frac{1}{2} V^2 \frac{\partial C}{\partial x}} \quad (3.81)$$

Substituting the value of C from Eq.3.71 to Eq.3.81, we get:

$$\boxed{F_x = \frac{1}{2} \frac{\epsilon_0 b V^2}{d} (\kappa - 1)} \quad (3.82)$$

We can also find electric force by just using the previous approach as follows-

If the instantaneous charge on capacitor is Q, then from Eq.3.69, the pulling force applied by capacitor on the dielectric slab is given by:

$$F_x = -\left.\frac{\partial U}{\partial x}\right|_Q = -\frac{\partial}{\partial x}\left(\frac{Q^2}{2C}\right) = -\frac{Q^2}{2} \frac{\partial}{\partial x}\left(\frac{1}{C}\right) = \frac{Q^2}{2C^2} \frac{\partial C}{\partial x}$$

$$\Rightarrow \quad F_x = -\left.\frac{\partial U}{\partial x}\right|_Q = \frac{Q^2}{2C^2} \frac{\partial C}{\partial x} \quad (3.83)$$

Since, the capacitor is connected with voltage source (battery) of voltage V, therefore, $Q = CV$.

Substituting this value in Eq.3.83, we get-

$$\boxed{F_x = -\frac{\partial U}{\partial x} = \frac{1}{2} V^2 \frac{\partial C}{\partial x}} \quad (3.84)$$

It is same as Eq.3.81.

CAUTION! It is a common error to use equation,

$$F_x = -\left.\frac{\partial U}{\partial x}\right|_V \quad \text{(with } V \text{ constant)},$$

rather than Eq.3.70 or Eq.3.80 or Eq.3.83 (with Q constant), in computing the force. One then obtains-

$$F_x = -\frac{\partial}{\partial x}\left(\frac{1}{2} CV^2\right) = -\frac{1}{2} V^2 \frac{dC}{dx},$$

which is off by a sign. So, if you want to use constant voltage equation directly as a time saving trick for simplifying the problems, then use Eq.3.70, i.e.,-

$$F_x = +\left.\frac{\partial U}{\partial x}\right|_V$$

Now, put the value $U = \frac{1}{2} CV^2$ in above expression, you get the correct expression for F_x as-

$$F_x = +\left.\frac{\partial U}{\partial x}\right|_V = \frac{1}{2} V^2 \frac{\partial C}{\partial x}$$

the same as before (Eq. 3.81), with the correct sign.

☞ The force on the dielectric cannot possibly depend on whether you plan to hold Q constant or V constant-it is determined entirely by the distribution of charge, free and bound. It's simpler to calculate the force assuming constant Q, because then you don't have to worry about work done by the battery; but if you require, it can be done correctly either way.

EXAMPLE 74. Two long coaxial cylindrical metal tubes (inner radius a, outer radius b and length l) stand vertically in a tank of dielectric oil (dielectric constant κ and and mass density ρ). The inner one is maintained at potential V, and the outer one is grounded (Fig.3.121). To what height (x) does the oil rise, in the space between the tubes?

APPROACH First of all find net capacitance of the capacitor

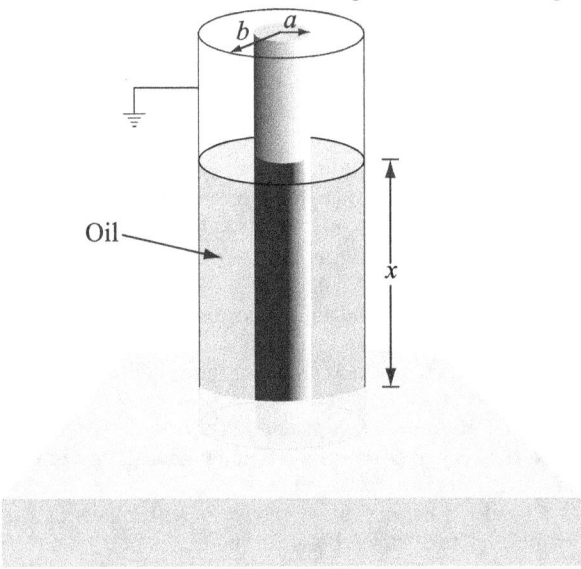

Figure 3.121

with dielectric oil up to height x and air in remaining length $(l - x)$. The system is connected with constant voltage supply, therefore, to find upward electric force apply Eq.3.70

In equilibrium, this electric force just balances the downward gravitational force.

SOLUTION We consider the cylinder as two cylindrical capacitors in parallel. Schematically, it is like Figure 3.122.

The capacitance of air part of the cylindrical capacitor is given

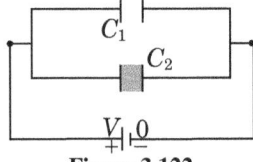

Figure 3.122

by Eq:3.14:

$$C_1 = \frac{2\pi \epsilon_0 (l - x)}{\ln(b/a)} \quad \ldots (1)$$

The capacitance of oil part of the cylindrical capacitor is Eq.3.14:

$$C_2 = \frac{2\pi \epsilon_0 \kappa (x)}{\ln(b/a)} \quad \ldots (2)$$

Therefore, net capacitance

$$C = C_1 + C_2 = \frac{2\pi \epsilon_0 (l - x)}{\ln(b/a)} + \frac{2\pi \epsilon_0 \kappa x}{\ln(b/a)}$$

3.13. EFFECTS OF NONUNIFORM ELECTRIC FIELDS ON DIELECTRICS

$$\Rightarrow \quad C = \frac{2\pi\epsilon_0}{\ln(b/a)}[l + (\kappa - 1)x] \quad \ldots (3)$$

The potential difference between the plates is V, therefore the electric energy stored in capacitor-

$$U = \frac{1}{2}CV^2$$

Therefore, from Eq.3.70, net electric force acting on dielectric is given by-

$$F = +\frac{\partial U}{\partial x} \quad \text{(constant } V \text{ condition)}$$

$$\Rightarrow \quad F = \frac{\partial}{\partial x}\left(\frac{1}{2}CV^2\right) = \frac{1}{2}V^2\frac{\partial C}{\partial x} \quad \ldots (4)$$

Substituting the value of C from Eq.(3) in (4) we get

$$F = \frac{1}{2}V^2\frac{2\pi\epsilon_0}{\ln(b/a)}(\kappa - 1) \quad \ldots (5)$$

Downward gravitational force, on the oil:

$$F_g = mg = \rho\pi\left(b^2 - a^2\right)gx \quad \ldots (6)$$

In equilibrium, Eq. (5) and (6) give-

$$\rho\pi\left(b^2 - a^2\right)gx = \frac{1}{2}V^2\frac{2\pi\epsilon_0}{\ln(b/a)}(\kappa - 1)$$

or $$x = \frac{2\epsilon_0 V^2}{\rho g(b^2 - a^2)\ln(b/a)}(\kappa - 1)$$

EXAMPLE 75. A parallel-plate capacitor is placed in such a way that its plates are horizontal and the lower plate is dipped into an oil of dielectric constant κ and density ρ. Each plate has an area A. The plates are now connected to a battery which supplies a positive charge of magnitude Q to the upper plate. Find the rise in the level of the liquid in the space between the plates.

SOLUTION Fig.3.123 shows an oil-capacitor system given in this problem. The induced charge produced on the upper surface of the liquid is $-Q\left(1 - \frac{1}{\kappa}\right)$ whereas induced charge in the oil at contact surface of lower plate is $Q\left(1 - \frac{1}{\kappa}\right)$.

Therefore, the net charge on the lower plate,

$$Q_L = -Q + Q\left(1 - \frac{1}{\kappa}\right) = -\frac{Q}{\kappa}.$$

Now, consider the equilibrium of the liquid in the volume $ABCD$. The forces on this liquid are
(a) the force due to the electric field at CD,

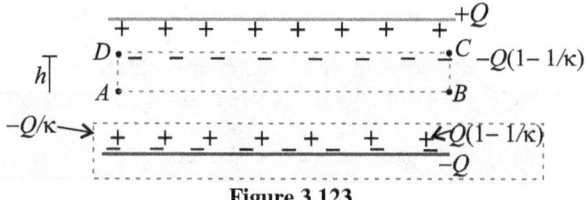

Figure 3.123

(b) the weight of the liquid,
(c) the force due to atmospheric pressure and
(d) the force due to the pressure of the liquid below AB.

As AB is in the same horizontal level as the outside surface, the pressure here is the same as the atmospheric pressure. The forces in (c) and (d), therefore, balance each other. Hence, for equilibrium, the forces in (a) and (b) should also balance each other.

The electric field at CD due to the charge Q is

$$E_1 = \frac{Q}{2A\epsilon_0}$$

in the downward direction.
The field at CD due to the charge $Q_L = -Q/\kappa$, is

$$E_2 = \frac{Q}{2A\epsilon_0\kappa}$$

also in the downward direction.
The net field at CD is

$$E_1 + E_2 = \frac{(\kappa + 1)Q}{2A\epsilon_0\kappa}$$

Therefore, the force on the charge $-Q\left(1 - \frac{1}{\kappa}\right)$ at CD is

$$F = Q\left(1 - \frac{1}{\kappa}\right)\frac{(\kappa + 1)Q}{2A\epsilon_0\kappa} = \frac{(\kappa^2 - 1)Q^2}{2A\epsilon_0\kappa^2}$$

in the upward direction. The weight of the liquid considered is $hA\rho g$. Thus,

$$hA\rho g = \frac{(\kappa^2 - 1)Q^2}{2A\epsilon_0\kappa^2} \quad \Rightarrow \quad h = \frac{(\kappa^2 - 1)Q^2}{2A^2\kappa^2\epsilon_0\rho g}$$

EXAMPLE 76. Two square metal plates of side 1 m are kept 0.01 m apart like a parallel plate capacitor in air in such a way that one of their edges is perpendicular to an oil surface in a tank filled with an insulating oil. The plates are connected to a battery of emf 500 V. The plates are then lowered vertically into the oil at a speed of 0.001 m/s. Calculate the current drawn from the battery during the process. [Dielectric constant of oil = 11, $\epsilon_0 = 8.85 \times 10^{-12} C^2 N^{-1} m^{-2}$.]

APPROACH If q is the instantaneous charge on the capacitor

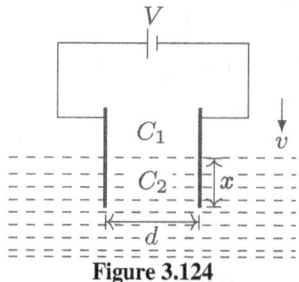

Figure 3.124

plate, then instantaneous current provided by battery is given by, $i = dq/dt$. So, we first assume that at any instant t, plate length x is inside the oil and $(l - x)$ in air, then at this instant the configuration is equivalent to air capacitor C_1 and oil filled capacitor C_2 connected in parallel (Fig.3.124). Here,

$$C_1 = \epsilon_0 l(l - x)/d, \quad \text{and} \quad C_2 = \kappa\epsilon_0 lx/d$$

Now, find parallel equivalent capacitance $C = C_1 + C_2$, then apply the relation $q = CV$ to get charge q in terms of x. Apply $i = dq/dt$ and substitute v for dx/dt to get i in terms of x. Finally substitute the given values and simplify for numeric value of i.

SOLUTION Equivalent capacitance of the system is given by-

$$C = C_1 + C_2 = \frac{\epsilon_0 l}{d}[l + (\kappa - 1)x]$$

The charge on the capacitor is $q = CV$ and current through the circuit is

$$i = \frac{dq}{dt} = V\frac{dC}{dt} = \frac{\epsilon_0 l}{d}(\kappa - 1)\frac{dx}{dt} = V\left[\frac{\epsilon_0 l(\kappa - 1)v}{d}\right]$$

where $v = dx/dt$ is the speed at which the plates are being lowered.

Given that, side of the square plate $l = 1$ m, separation between the plates $d = 0.01$ m, dielectric constant of oil $\kappa = 11$ and emf of the battery $V = 500$ V. Substituting these values, we get:

$$i = 500\left[\frac{(8.85 \times 10^{-12})(1)(11 - 1)(0.001)}{0.01}\right] = 4.425 \times 10^{-9} \text{ A}$$

EXAMPLE 77. A parallel plate capacitor having plates of area A and plate separation d, has capacitance C_1 in air. When two dielectrics of different relative permitivities ($\epsilon_1 = 2$ and $\epsilon_2 = 4$) are introduced between the two plates as shown in the Figure 3.125, the capacitance becomes C_2. Find the ratio C_2/C_1.

APPROACH The capacitance of an air filled parallel plate capacitor having plates of area A and plate separation d is given by

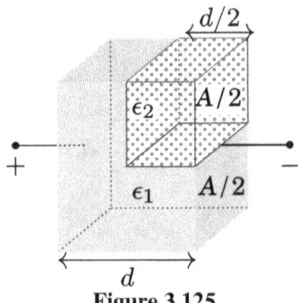

Figure 3.125

$C_1 = \epsilon_0 A/d$. Consider the case when two dielectrics of relative permittivities ($\epsilon_1 = 2$ and $\epsilon_2 = 4$) are introduced between the two plates as shown in the Figure 3.125.

Now, divide the given configuration into three parts- as follows-
(a) Capacitor of plate area $A/2$, plate separation d, and filled with a dielectric of relative permittivity $\epsilon_1 = 2$ (lower half). The capacitance of this part is

$$C_a = \epsilon_1 \epsilon_0 \frac{A/2}{d} = \epsilon_0 \frac{A}{d} = C_1$$

(b) Capacitor of plate area $A/2$, plate separation $d/2$, and filled with a dielectric of relative permittivity $\epsilon_1 = 2$ (left upper half). The capacitance of this part is

$$C_b = \epsilon_1 \epsilon_0 \frac{A/2}{d/2} = 2\epsilon_0 \frac{A}{d} = 2C_1$$

(c) Capacitor of plate area $A/2$, plate separation $d/2$, and filled with a dielectric of relative permittivity $\epsilon_2 = 4$ (right upper half). The capacitance of this part is

$$C_c = \epsilon_2 \epsilon_0 \frac{A/2}{d/2} = 4\epsilon_0 \frac{A}{d} = 4C_1$$

The capacitors C_b and C_c are connected in series and their equivalent capacitor C_{bc} (say) is connected in parallel with with C_a. Now find their equivalent capacitance.

SOLUTION Since, C_b and C_c are in series, therefore their equivalent capacitance:

$$C_{bc} = \frac{C_b C_c}{C_b + C_c} = \frac{(2C_1)(4C_1)}{(2C_1)+(4C_1)} = \frac{4}{3}C_1$$

The capacitor C_{bc} is connected in parallel with C_a. Therefore, the equivalent capacitance is

$$C_2 = C_{abc} = C_a \| C_{bc} = C_1 + \frac{4}{3}C_1 = \frac{7}{3}C_1$$

3.14 Check Point 2

1. ●●A capacitor is made of two circular plates as shown in Fig.3.126. Radius of each plate is 8.0 cm and distance between them is 1.0 mm. When mica (dielectric constant, $\kappa = 6.0$) is placed between the plates, calculate the capacitance of this capacitor and the energy stored when it is given a potential of 150.0 volt.

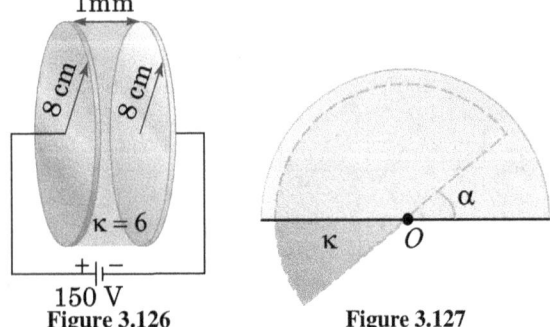

Figure 3.126   **Figure 3.127**

2. ●●A cylindrical layer of a homogeneous dielectric with the dielectric constant κ is introduced into a cylindrical capacitor so that the layer fills the gap of width d between the plates. The mean radius of the plates is R such that $R \gg d$. The capacitor is connected to a source of a permanent voltage V. Find the force pulling the dielectric inside the capacitor.

3. ●●●A capacitor consists of two fixed plates in the form of a semicircle of radius R and a movable plate of thickness d made of a dielectric with the dielectric constant κ, placed between them. The latter plate can freely rotate about an axis perpendicular to plane of plate and passing through O (Fig.3.127) and practically fills the entire gap between the fixed plates. A constant voltage V is maintained between the plates. Find the moment or torque τ of forces acting on the movable plate about above axis, when it is placed as shown in the Fig.3.127.

4. ●●Fig.3.128 shows a parallel plate capacitor inside a liquid dielectric. Find the force acting on one of the plates of the capacitor, if the distance between the plates is h, the capacitance of the capacitor under given conditions is C and the voltage V is maintained across its plates.

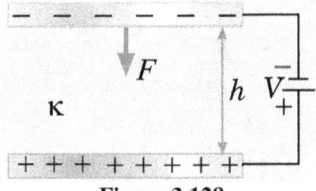

Figure 3.128

5. ●●●**The work done upon moving capacitor plates apart:** A parallel-plate air capacitor has the plates of area A each. Find the work W', against the electric forces, done to increase the distance between the plates from x_1 to x_2, if (1) the charge q of the capacitor and (2) its voltage V are maintained constant. Find the increments of the electric energy of the capacitor in the two cases.

6. ●●**Forces acting between conductors in a dielectric:** A parallel-plate capacitor is immersed, in the horizontal position, into a liquid dielectric with the dielectric constant κ, filling the gap of width d between the plates. Then the capacitor is connected to a source of permanent voltage V. Find the force F' acting on a unit surface of the plate from the dielectric.

7. ●●●A parallel plate capacitor has length $2l$, breadth b and plate separation d (Fig.3.129). It is half filled with a dielectric (dielectric constant κ) of mass M. Capacitor is attached with a battery of voltage V. Plates are held fixed on smooth insulating horizontal surface. A bullet of mass M hits the dielectric elastically and it is found that dielectric just leaves out the capacitor. Find speed of bullet.

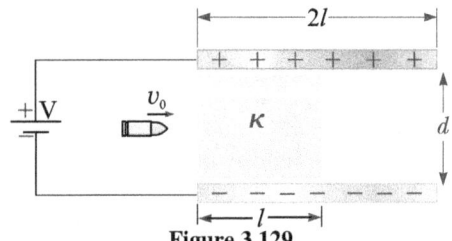

Figure 3.129

8. ●●●In Fig.3.130, a block A of mass M, is kept on a rough horizontal surface. It is connected to a dielectric slab B of dielectric constant κ) and mass $M/6$ by means of a light and inextensible string passing over a smooth pulley. The dielectric can completely fill the space between the parallel plate capacitor of plate length l and breadth b. Plates are separated by distance d in vertical position. Initially switch S is open and length of the dielectric inside the capacitor is

l_0. The coefficient of kinetic friction between the block A and the surface of table is $\mu = 1/4$. Ignore any other friction.
(a) When switch S is closed, find the minimum value of emf V so that the block A starts motion with zero acceleration.
(b) If we apply voltage $V' = 2V$, find the speed of the block A when the dielectric completely fills the space between the plates of the capacitor.

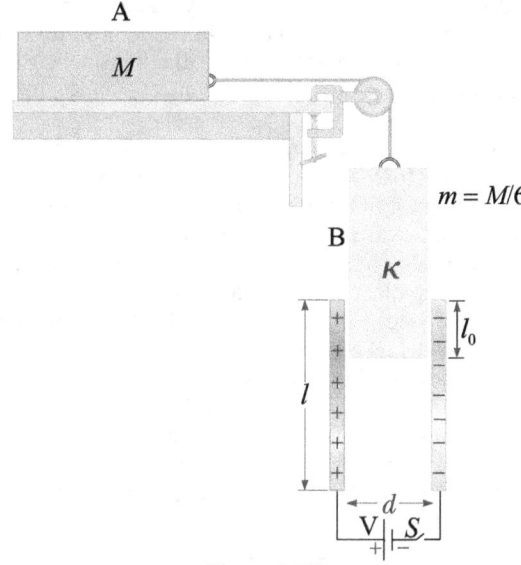

Figure 3.130

3.15 Types of Commercial Capacitors

Commercial capacitors are often made from metallic foil interlaced with thin sheets of either paraffin-impregnated paper or Mylar as the dielectric material. These alternate layers of metallic foil and dielectric are rolled into a cylinder to form a small package (Fig. 3.131a). High-voltage capacitors commonly consist of a number of interwoven metallic plates immersed in silicone oil (Fig. 3.131b). Small capacitors are often constructed from ceramic materials.

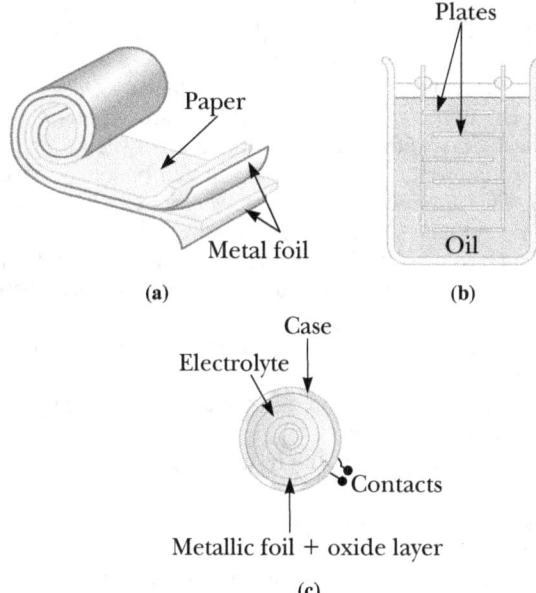

Figure 3.131: Three commercial capacitor designs

Often, an electrolytic capacitor (Fig.3.131c) is used to store large amounts of charge at relatively low voltages. It consists of a metallic foil in contact with an electrolyte. When a voltage is applied between the foil and the electrolyte, a thin layer of metal oxide (an insulator) is formed on the foil, and this layer serves as the dielectric. Very large values of capacitance can be obtained in an electrolytic capacitor because the dielectric layer is very thin, and thus the plate separation is very small.

Electrolytic capacitors are not reversible as are many other capacitors-they have a polarity, which is indicated by positive and negative signs marked on the device. When electrolytic capacitors are used in circuits, the polarity must be aligned properly. If the polarity of the applied voltage is opposite that which is intended, the oxide layer is removed and the capacitor conducts electricity instead of storing charge.

Variable capacitors (typically 10 to 500 pF) usually consist of two interwoven sets of metallic plates, one fixed and the other movable, and contain air as the dielectric (Fig. 3.131). These types of capacitors are often used in radio tuning circuits.

3.16 Solved Examples

EXAMPLE 1. Seven capacitors, each of capacitance 2 μF are to be connected to obtain a capacitance of $10/11$ μF. Which of the following combinations is possible?
(A) 5 in parallel 2 in series   (B) 4 in parallel 3 in series
(C) 3 in parallel 4 in series   (D) 2 in parallel 5 in series
SOLUTION (A) 5 capacitors each of magnitude 2 μF in parallel give equivalent capacitance of 5×2 μF $= 10$ μF.
2 capacitors in series give equivalent capacitance of 1 μF.
Now, the series combination of these two equivalent capacitors give net equivalent capacitance
$$\frac{10 \times 1}{10 + 1} = \frac{10}{11} \ \mu F$$
EXAMPLE 2. In the circuit 3.132a, if potential of A is 10 V, then find the potential of B.
APPROACH To find potential of point B, we use charge,

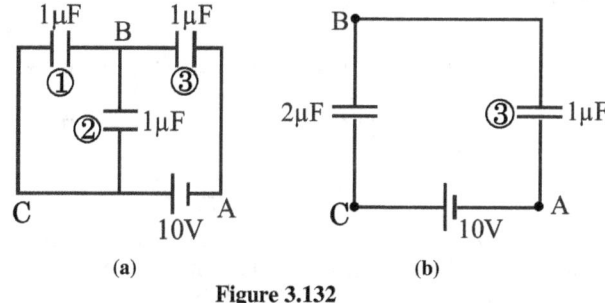

Figure 3.132

capacitance and voltage relation between terminals A and B:
$V_B - V_A = Q/C$
Here, $V_A = 10$ V and Q is the charge given by battery to capacitor ③. Q is unknown to us. So, to find Q, we have to know the total capacitance of the circuit connected with the terminal of the battery.

In Fig.3.132a, Capacitors ① and ② both have their one terminal at one common potential of B, whereas second terminal at other common potential of C. Therefore, capacitors ① and ② are in parallel order. Their equivalent is in series with the capacitor ③.
SOLUTION The equivalent capacitance of capacitors ① and ② is-
$$C_{12} = 1 \ \mu F + 1 \ \mu F = 2 \ \mu F$$
Now, this is in series with capacitor ③. It's new circuit diagram is shown in Fig.3.132b.
The series equivalent capacitance of circuit 3.132b is-
$$C = \frac{2 \ \mu F \times 1 \ \mu F}{2 \ \mu F + 1 \ \mu F} = \frac{2}{3} \ \mu F$$
Charge given by battery of 10 V is

$$Q = CV = \frac{2}{3} \mu F \times 10 \text{ V} = \frac{20}{3} \mu C$$

In series, the charges on capacitor remain same, therefore, in Fig.3.132b, the same charge of magnitude $\frac{20}{3}$ μC will appear on capacitors of 1 μF and 2 μF.

Now, the potential difference between terminal A and B is-

$$V_B - V_A = \frac{Q}{C_{(3)}} = \frac{(20/3) \ \mu C}{1 \ \mu F} = \frac{20}{3} \text{ V}$$

$$\Rightarrow \qquad V_B = \frac{20}{3} \text{ V} + V_A \qquad \text{(Given that, } V_A = 10 \text{ V)}$$

$$\Rightarrow \qquad V_B = \frac{20}{3} \text{ V} + 10 \text{ V} = \frac{50}{3} \text{ V}$$

EXAMPLE 3. Dimensions and separation (d) between the plates of a parallel plate capacitor, are very small. The charges on positive and negative plates are $+Q$ and $-Q$ respectively(Fig.3.133). What is the force on the charged particle of charge q at a distance l from the capacitor? (Assume that the distance between the plates is $d \ll l$)

APPROACH Since, dimensions of plates are very small, there-

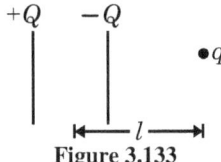

Figure 3.133

fore, we can treat them like point charges. Now, it is also given that the separation between plates is very small as compared to the the position of charged particle of charge q from plates, i.e., $d \ll l$, therefore we can consider the two plate system as an electric dipole for the given position of charge q (Fig.3.133).

If $p (= Qd)$ is the magnitude of the dipole moment of considered dipole, then electric field of it at axial position of q-

$$E = k \frac{2p}{l^3} \qquad (1)$$

Now electric force on charged particle can be written as-

$$F = qE \qquad (2)$$

SOLUTION From Equations (1) and (2), the electric force on the charged particle,

$$F = qE = qk\frac{2p}{l^3} = \frac{1}{2\pi\epsilon_0} \frac{qQd}{l^3}$$

EXAMPLE 4. A varying voltage is applied between the terminals A, B (Fig.4.224a) so that the voltage across the capacitor varies as shown in the Fig.4.224b. Then-
 (A) The voltage between the terminals C and D is constant between $2t_0$ and $3t_0$
 (B) The current in the resistor is 0 between $2t_0$ and $3t_0$
 (C) The current in the resistor between t_0 and $2t_0$ is twice the current between $3t_0$ and $5t_0$
 (D) None of these

APPROACH Charge stored, on the positive plate of a capacitor,

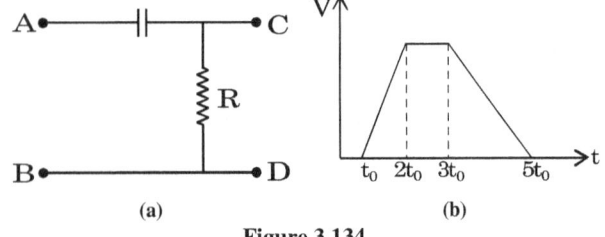

Figure 3.134

is given by-

$$q = CV \qquad (1)$$

So, when the voltage across the capacitor is constant, charge stored on it will also be constant. In this case, there will not be any current in capacitor branch. So, in this case, (Fig.4.224b), there will not be any current in resistor R.

On differentiating Eq.(1), with respect to t, we get-

$$\frac{dq}{dt} = C\frac{dV}{dt} \implies i = C\frac{dV}{dt} \qquad (2)$$

So, current in circuit is equal to C times of slope of V-t graph. and for constant value of V, current i will be zero.

SOLUTION (ABC) From Fig.4.224b, voltage V across C is constant in time interval 0 to t_0 and again constant in time interval $2t_0$ to $3t_0$, so in these intervals, there will not be any current in R and hence voltage across C and D will also be constant.
Therefore, options (A) and (B) are correct.

From Fig.4.224b, the magnitude of slope of V-t graph is double in time interval t_0 to $2t_0$ as compared to that in interval $3t_0$ to $5t_0$, therefore the current in time interval t_0 to $2t_0$ will also be double that of in time interval $3t_0$ to $5t_0$. So option (C) is also correct.

EXAMPLE 5. In Fig.3.135, A, B and C are three large, parallel conducting plates, placed horizontally. A and C are rigidly fixed and earthed. B is given some charge. Under electrostatic and gravitational forces, B may be-
 (A) in equilibrium midway between A and C.
 (B) in equilibrium if it is closer to A than to C.
 (C) in equilibrium if it is closer to C than to A.
 (D) B can never be in stable equilibrium.

APPROACH Since, plates A and C are earthed, therefore they

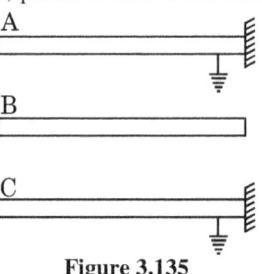

Figure 3.135

are at same potential. So, the potential difference between plates A an B will be same as that between B and C . Now, the system can be considered as two capacitors in parallel-

One capacitor is formed by lower surface of plate A and upper surface of plate B , whereas second capacitor is formed between the lower surface of plate B and upper surface of plate C.

If B is closer to A than to C then the capacitance C_{AB} is greater than C_{BC}. The upper surface of B will have greater charge than the lower surface. As the force of attraction between the plates of a capacitor is proportional to Q^2, there will be a net upwards force on B. This force can balance its weight. If we shift plate B towards lower plate C, the charge on lower surface of plate B and upper surface of plate C increase, so electrostatic attraction between them also increase. As a result, plate B moves and stick on plate C.

ANSWER (B, D) From above discussion, options (B) and (D) are correct.

EXAMPLE 6. Study the circuit diagram shown in Fig.4.225 and mark the correct option(s)
 (A) The potential of point a with respect to point b when switch S is open, is -6 V.
 (B) The points a and b, are at the same potential, when S is opened.
 (C) The charge flows through switch S when it is closed, is 54 μC
 (D) The final potential of b with respect to ground when switch S is closed is 8 V

SOLUTION (A, C) In given circuit diagram Fig. 4.225, terminal d is grounded. So, d can be considered at zero potential. The

3.16. SOLVED EXAMPLES

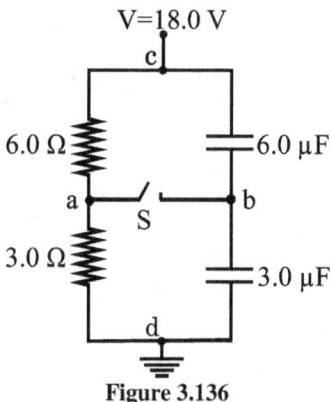

Figure 3.136

potential of terminal c is 18.0 V. Therefore, potential difference between terminals d and c is
$$V_{cd} = V_c - V_d = 18\,V - 0\,V = 18\,V$$

When switch S is opened: When switch S is opened, capacitors $3.0\,\mu F$ and $6\,\mu F$ will be in series. Equivalent capacitance between terminals c and d is-
$$C_{cd} = \frac{3.0\,\mu F \times 6.0\,\mu F}{3.0\,\mu F + 6.0\,\mu F} = 2\,\mu F$$
So, charge on each capacitor (Fig.4.226a),
$$q = C_{cd} V_{cd} = 2\,\mu F \times 18\,V = 36\,\mu C$$
Therefore, net charge on capacitor plates connected with point b
$$q_b = q + (-q) = 36\,\mu C + (-36)\,\mu C = 0\,\mu C$$
If potential at terminal b is V_b, then from Fig.4.225, potential difference between terminals b and c is
$$V_c - V_b = \frac{q}{C_{bc}} = \frac{36\,\mu C}{6\,\mu F} = 6\,V$$
or $\quad V_b = V_c - 6\,V = 18\,V - 6\,V = 12\,V$
Net resistance between terminals c and d is

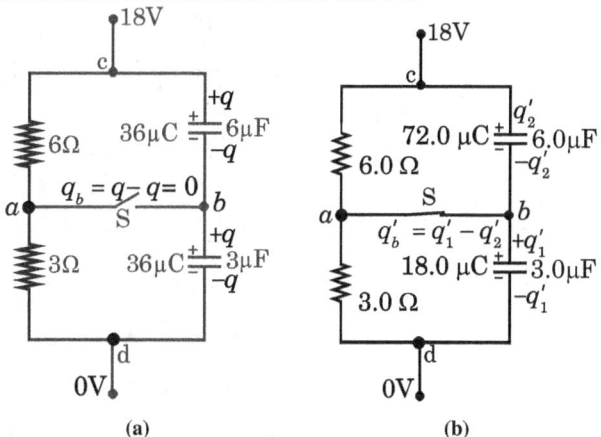

Figure 3.137

$$R_{cd} = R_{ca} + R_{ac} = 3.0\,\Omega + 6.0\,\Omega = 9.0\,\Omega$$
Therefore, by Ohm's law, the electric current in branch cad is -
$$I = \frac{V_{cd}}{R_{cd}} = \frac{18.0\,V}{9.0\,\Omega} = 2.0\,A$$
If electric potential at point a is V_a, then by Ohm's law between terminals a and c, we can write-
$$V_c - V_a = IR_{ac}$$
or $\quad V_a = V_c - IR_{ac} = 18\,V - (2.0\,A \times 6.0\,\Omega) = 6\,V$
Therefore, potential difference between terminals a and b is
$$V_{ba} = V_b - V_a = 12\,V - 6\,V = 6\,V$$
So, option (B) is incorrect whereas option (A) is correct, i.e., potential of terminal a with respect to b is $-6\,V$.
When switch S is closed: In this case, the potential of point a will also be the potential of b.
So, the potential of point b is $V_b = 6\,V$
Therefore, option (D) is incorrect.

Now, charge stored on $3\,\mu F$ capacitor-
$$q'_1 = (3\,\mu F)(V_b - V_a) = (3\,\mu F)(6\,V - 0) = 18\,\mu C$$
and charge stored on $6\,\mu F$ capacitor-
$$q'_2 = (6\,\mu F)(V_c - V_b) = (6\,\mu F)(18\,V - 6\,V) = 72\,\mu C$$
Therefore, after switch S is closed (Fig.4.226b), net charge on capacitors plates connected with point b
$$q'_b = 18\,\mu C + (-72)\,\mu C = -54\,\mu C$$
Therefore, charges flown through S after it is closed: $54 - 0 = 54\,\mu C$.
So, option (C) is also correct.

EXAMPLE 7. Fig.3.138 shows the initial charges on the capacitor. Now, switch S is closed at $t = 0$. Find-
(a) Charge on capacitor C_1, at $t = 0$.

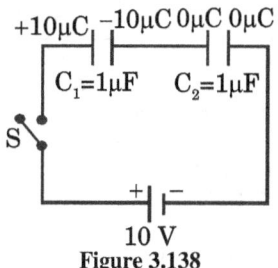

Figure 3.138

(b) Charge on capacitor C_2, at $t = 0$.
(c) Work done by battery in redistribution of charges.

SOLUTION When switch S is open, the charge of the left plate of capacitor C_1 cannot go anywhere, so the charge of the right plate will also remain confined to it.
Therefore, charge on C_1 at $t = 0$ is $10\,\mu C$ and charge on capacitor C_2 will be zero **(answers of part (a) and (b))**.
After switch S is closed, the equivalent capacitance of capacitors across the battery [Fig.3.138],-
$$C = \frac{C_1 C_2}{C_1 + C_2} = \frac{(1\,\mu F)(1\,\mu F)}{1\,\mu F + 1\,\mu F} = \frac{1}{2}\,\mu F$$
After long time, the charge given by battery-
$$q = CV = \frac{10}{2}\,\mu C = 5\,\mu C$$
Therefore, $+5\,\mu C$ charge flows from left plate of capacitor C_1 to positive terminal of battery and also $+5\,\mu C$ charge flows from right plate of capacitor C_2 to negative terminal of battery. So, net charge crossing the battery-
$$q_b = 5\,\mu C - 5\,\mu C = 0$$
Therefore, work done by battery-
$$W_{\text{battery}} = q_b V = 0(10\,V) = 0\,J \text{ (answer of part (c))}$$

EXAMPLE 8. In the circuit of Fig.3.139, capacitor A has capacitance $C_1 = 2\,\mu F$ when filled with dielectric slab ($\kappa = 2$). Capacitor B and C are air capacitors and have capacitances $C_2 = 3\,\mu F$ and $C_3 = 6\,\mu F$ respectively.
(a) Calculate the energy supplied by battery during process of

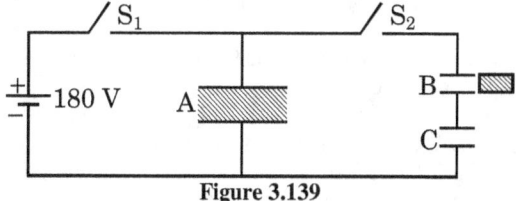

Figure 3.139

charging when switch S_1 is closed alone.
(b) Find the charge on capacitor B, when switch S_1 is opened and S_2 is closed.
(c) Now switch S_2 is opened, slab of A is removed. Another dielectric slab $\kappa = 2$ which can just fill the space in B, is inserted into it and then switch S_2 is closed. Find the charge on capacitor

B.

SOLUTION (a) When switch S_1 is closed and S_2 is open, the charge given by battery-
$$q = CV = 2 \ \mu\text{C} \times 180 \ \text{V} = 360 \ \mu\text{C}$$
Therefore, the energy supplied by battery-
$$E_b = qV = 360 \times 10^{-6} \ \text{C} \times 180 \ \text{V} = 0.0648 \ \text{J}.$$

(b) Capacitors B and C are in series, therefore, their equivalent capacitance-
$$C_{BC} = \frac{(3 \ \mu\text{F})(6 \ \mu\text{F})}{3 \ \mu\text{F} + 6 \ \mu\text{F}} = 2 \ \mu\text{F}$$

In Fig.3.140, the equivalent of B and C is denoted by D.
If V is the common potential and q'_1, q'_2 are final charges on capacitors A and D respectively, then-
$$V = \frac{q'_1}{C_1} = \frac{q'_2}{C_{BC}} = \frac{q'_1 + q'_2}{C_1 + C_{BC}} = \frac{360 \ \mu\text{C}}{2\text{F} + 2\text{F}} = 90 \ \text{V}$$
$$\Rightarrow \quad q'_1 = 90 \ \text{V} \times 2 \ \mu\text{F} = 180 \ \mu\text{C}$$
and $\quad q'_2 = 90 \ \text{V} \times 2 \ \mu\text{F} = 180 \ \mu\text{C}$

Therefore, equal charges of magnitude 180 μC stored on each capacitor A and D respectively.

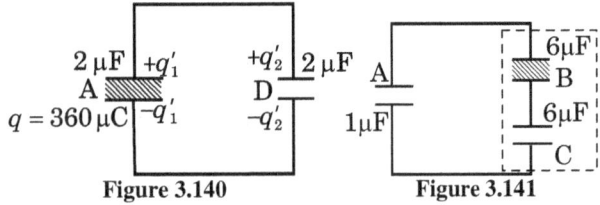

Figure 3.140   **Figure 3.141**

Since, D is the series equivalent of capacitors B and C, therefore, equal charges of magnitude 180 μC stored on the plates of capacitors B and C.
∴ charge q on B = $90 \times 2 \times 10^{-6} = 180 \ \mu$C.

(c) If switch S_2 is opened and then slab of A is removed, the charge of capacitor A does not change whereas it's capacitance reduced to $1/\kappa$ times, i.e., new capacitance of capacitor A is $1 \ \mu$F [Fig.3.141].

Now, another dielectric slab of dielectric constant $\kappa = 2$, is fully inserted in the space of B, so the capacitance of B increases to κ times of C_2 i.e, new capacitance of B is 6 μF, whereas it's charge remains unchanged.

In this case, new equivalent capacitance of B and C is-
$$C''_{BC} = \frac{(6 \ \mu\text{F})(6 \ \mu\text{F})}{6 \ \mu\text{F} + 6 \ \mu\text{F}} = 3 \ \mu\text{F}$$

If V' is the common potential and q''_1, q''_2 are final charges on capacitors A and the equivalent of B and C respectively, then-
$$V' = \frac{q''_1}{C'_1} = \frac{q''_2}{C''_{BC}} = \frac{q''_1 + q''_2}{C'_1 + C''_{BC}} = \frac{360 \ \mu\text{C}}{1 \ \text{F} + 3 \ \text{F}} = 90 \ \text{V}$$
$$\Rightarrow \quad q''_1 = 90 \ \text{V} \times 1 \ \mu\text{F} = 90 \ \mu\text{C}$$
and $\quad q''_2 = 90 \ \text{V} \times 3 \ \mu\text{F} = 270 \ \mu\text{C}$

EXAMPLE 9. **Column match:** All capacitors given in column-I have capacitance of 1 μF in Table 3.2. Match correct options.

APPROACH Students are advised to to identify series and parallel connection for each given circuit in Table 3.2.

ANSWER (A) → (S), (B) → (R), (C) → (R), (D) → (Q)

EXAMPLE 10. Six metal plates are with their separation are shown in Fig.3.142a. Effective area of each plate is A. Find the equivalent capacitance between points b and e.

SOLUTION The circuit is redrawn in Fig.3.142b. Each surface is labelled by numbers ①, ②, ... etc. Let terminals e and b are at voltages V_1 and V_2 respectively. Suppose, interconnected plates c and f are at common potential V_3 whereas plates a and d are at common potential V_3.

Parallel plate capacitors formed by labeled surfaces are shown in

Table 3.2: Column-I shows the capacitive circuits and column-II shows the capacitances between terminals A and B.

Column-I (Circuit)	Column-II (Capacitance)
(A) 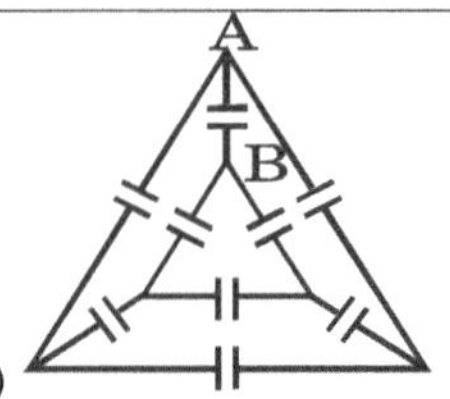	(P) $\frac{4}{3} \ \mu$F
(B) 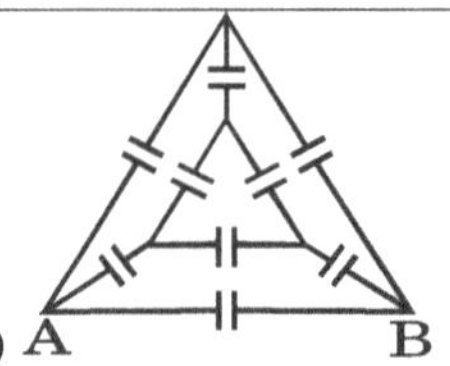	(Q) $\frac{3}{2} \ \mu$F
(C) 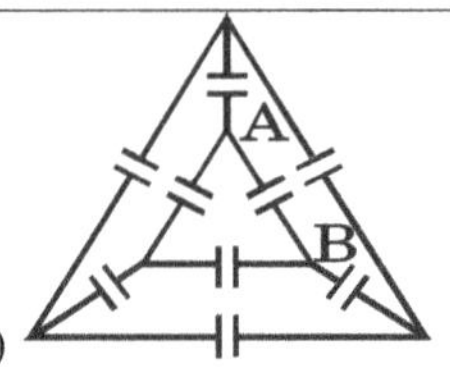	(R) $\frac{15}{8} \ \mu$F
(D) 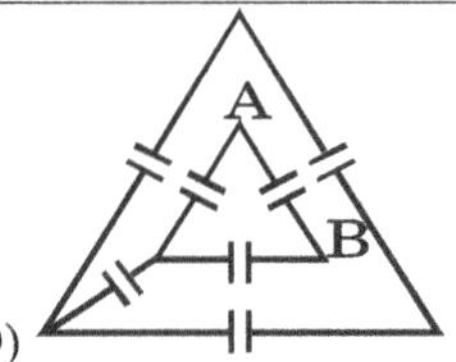	(S) $\frac{5}{3} \ \mu$F
	(T) None of these

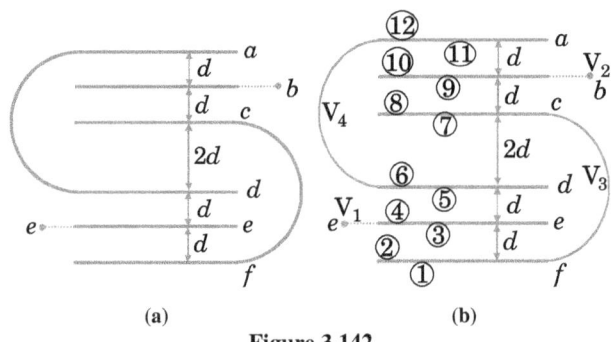

Figure 3.142

Fig.3.143a. If $\frac{\epsilon_0 A}{d} = C$, then capacitance of capacitor formed by surfaces ⑥ and ⑦ will be of magnitude $C/2$, because these surfaces have separation $2d$. All other surfaces have separation d.

Since, circuit 3.143a satisfies wheatstone balanced bridge

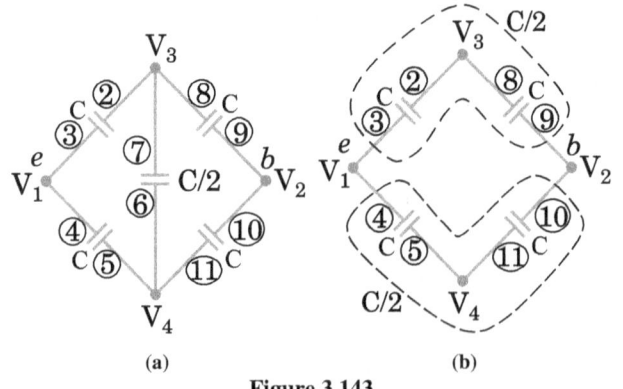

Figure 3.143

condition-
$$\frac{C_{2,3}}{C_{4,5}} = \frac{C_{8,9}}{C_{10,11}} \implies \frac{C}{C} = \frac{C}{C}$$
Therefore, $V_3 = V_4$, so we can remove the capacitor of capacitance $C/2$ connected between these terminals.

Now, next simplified circuit diagram is shown in Fig.3.143b. Capacitor's in series are enclosed within the dashed closed curves. The equivalent capacitance within each closed curve is $\frac{C \cdot C}{C+C} = C/2$. Now, these equivalents are in parallel order, so final capacitance of the circuit:
$$C_{eq} = C/2 + C/2 = C = \frac{\epsilon_0 A}{d}$$

EXAMPLE 11. In Fig.3.144, the charge on 3 μF capacitor is 3 μC. Find the charge on capacitor of capacitance C in μC.

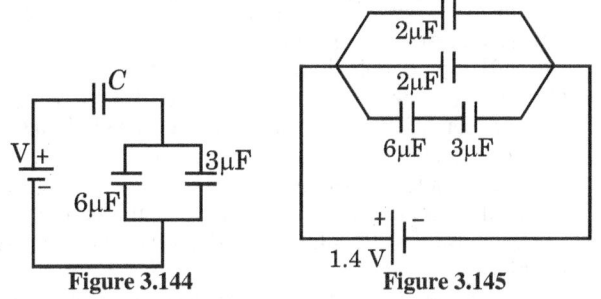

Figure 3.144   **Figure 3.145**

SOLUTION Charge on 3 μF capacitor is 3 μC, therefore voltage across 3 μF capacitor is $q/C = 3\ \mu C/3\ \mu F = 1$ V.

Now, from Fig.3.144, potential difference across 3 μF = P.D. across 6 μF = 1 V, therefore charge on 6 μF capacitor is-
$$q_1 = CV = (6\ \mu F)(1\ V) = 6\ \mu C$$
$\Rightarrow$ Total charge on combination of 6 μF and 3 μF = 9 μC
As capacitor C is in series with the equivalent of 6 μF and 3 μF, therefore charge on $C = 9\ \mu C$

EXAMPLE 12. Find total energy stored in circuit shown in Fig.3.145.

SOLUTION In circuit 3.145, 6 μF and 3 μF capacitors are in series, their equivalent capacitance is
$$C_1 = \frac{(6\ \mu F)(3\ \mu F)}{6\ \mu F + 3\ \mu F} = 2\ \mu F$$
Now, C_1 is in parallel with other two capacitors each of magnitude 2 μF.
Therefore, equivalent capacitance of the circuit-
$$C_{eq} = C_1 + 2\ \mu F + 2\ \mu F = 2\ \mu F + 2\ \mu F + 2\ \mu F = 6\ \mu F$$
Therefore, total energy stored in the circuit-
$$E = \frac{1}{2}C_{eq}V^2 = \frac{1}{2}\left(6 \times 10^{-3}\right)(1.4)^2 = 6\ mJ$$

EXAMPLE 13. Two parallel plate capacitors with area A are connected through a conducting spring of natural length l in series as shown in Fig.3.146a. Plates P and S have fixed positions at separation d. Now the plates are connected by a battery of emf $\mathcal{E}$ by closing switch S. If the extension in the spring in equilibrium is equal to the separation between the plates, find the spring constant κ.

SOLUTION Let charge on capacitors be q and separation be-

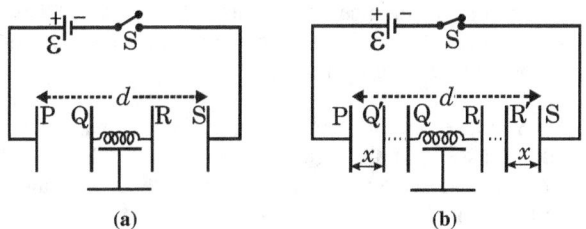

Figure 3.146

tween plates P and Q and R and S be x (at any time distance between plates P and Q and R and S is same because force acting on them is same.)

Capacitance of capacitor PQ, $C_1 = \dfrac{\epsilon_0 A}{x}$

Capacitance of capacitor RS, $C_2 = \dfrac{\epsilon_0 A}{x}$

If V_1 and V_2 are voltages across PQ and RS respectively, then
$$V_1 + V_2 = \mathcal{E} \implies \frac{q}{C_1} + \frac{q}{C_2} = \mathcal{E} \implies q = \frac{\epsilon_0 A \mathcal{E}}{2x}$$
Final length of spring $= d - 2x$
At this moment extension in spring, $y = (d - 2x) - l$.
Force on plate Q towards P,
$$F_1 = \frac{q^2}{2A\epsilon_0} = \frac{\epsilon_0^2 A^2 \mathcal{E}^2}{8Ax^2 \epsilon_0} = \frac{A\epsilon_0 \mathcal{E}^2}{8x^2} \quad \cdots (1)$$
Spring force on plate Q due to extension in spring, $F_2 = ky$.
As it is given that the equilibrium separation between plates = extension in the spring
Therefore, $x = y = d - 2x - l \implies x = \frac{d-l}{3} \quad \cdots (2)$
and $\quad F_1 = F_2 = \left(\dfrac{A\epsilon_0 \mathcal{E}^2}{8\kappa}\right) \quad \cdots (3)$
From Eq.(2) and (3), we have-
$$\frac{A\epsilon_0 \mathcal{E}^2}{8x^2} = \kappa y = \kappa x \implies x = \left(\frac{A\epsilon_0 \mathcal{E}^2}{8\kappa}\right)^{1/3} \quad \cdots (4)$$
From Eq.(2) and Eq.(4),
$$\left(\frac{d-l}{3}\right) = \left(\frac{A\epsilon_0 \mathcal{E}^2}{8\kappa}\right)^{1/3} \implies \kappa = \frac{A\epsilon_0 \mathcal{E}^2 27}{8(d-l)^3}$$

3.17 Questions and Exercises

3.17.1 Conceptual Questions

1. Must a capacitor's plates be made of conducting material? What would happen if two insulating plates were used instead of conducting plates?
2. (a) Why is it dangerous to touch the terminals of a high-voltage capacitor even after the voltage source that charged the capacitor is disconnected from the capacitor? (b) What can be done to make the capacitor safe to handle after the voltage source has been removed?
3. Suppose the separation of plates d in a parallel-plate capacitor is not very small compared to the dimensions of the plates. Would you expect Eq. $C = Q/V = \epsilon_0 A/d$ to give an overestimate or underestimate of the true capacitance? Explain.
4. Assume you want to increase the maximum operating voltage of a parallel-plate capacitor. Describe how you can do that with a fixed plate separation.
5. Liquid dielectrics having polar molecules (such as water) have dielectric constants that decrease with increasing temperature. Why?
6. When a battery is connected to a capacitor, why do the two plates acquire charges of the same magnitude? Will this be true if the two plates have different sizes or shapes?
7. You have an electric device containing a 10.0 μF capacitor, but an application requires an 18.0 μF capacitor. What modification can you make to your device to increase its capacitance to 18.0 μF?
8. True or false: The electrostatic energy density is uniformly distributed in the region between the conductors of a cylindrical capacitor.
9. Describe a simple method of measuring ϵ_0 using a capacitor.
10. Suppose, three identical capacitors are connected to a battery. Will they store more energy if connected in series or in

parallel?

11. A large copper sheet of thickness l is placed between the parallel plates of a capacitor, but does not touch the plates. How will this affect the capacitance?
12. Explain why a dielectric increases the maximum operating voltage of a capacitor even though the physical size of the capacitor doesn't change.
13. The parallel plates of an isolated capacitor carry opposite charges, Q. If the separation of the plates is increased, is a force required to do so? Is the potential difference changed? What happens to the work done in the pulling process?
14. An isolated charged capacitor has horizontal plates. If a thin dielectric is inserted a short way between the plates, Fig.3.147, will it move left or right when it is released?

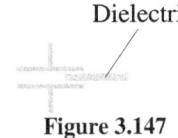

Figure 3.147

15. Explain why the work needed to move a particle with charge Q through a potential difference ΔV is $W = Q\Delta V$, whereas the energy stored in a charged capacitor is $U_E = \frac{1}{2}Q\Delta V$. Where does the factor $1/2$ come from?
16. Suppose a battery remains connected to the capacitor in Question 14. What then will happen when the dielectric is released?
17. How does the energy stored in a capacitor change when a dielectric is inserted if (a) the capacitor is isolated so Q does not change; (b) the capacitor remains connected to a battery so V does not change?
18. For dielectrics consisting of polar molecules, how would you expect the dielectric constant to change with temperature?
19. A dielectric is pulled out from between the plates of a capacitor which remains connected to a battery. What changes occur to the capacitance, charge on the plates, potential difference, energy stored in the capacitor, and electric field?
20. We have seen that the capacitance C depends on the size, shape, and position of the two conductors, as well as on the dielectric constant κ. What then did we mean when we said that C is a constant in Eq. $Q = CV$?
21. What value might we assign to the dielectric constant for a good conductor? Explain.
22. Because the charges on the plates of a parallel-plate capacitor are opposite in sign, they attract each other. Hence, it would take positive work to increase the plate separation. What type of energy in the system changes due to the external work done in this process?
23. Does it take more work to separate the plates of a charged parallel plate capacitor while it remains connected to the charging battery or after it has been disconnected from the charging battery?
24. A parallel plate capacitor is charged with a battery and then disconnected from the battery, leaving a certain amount of energy stored in the capacitor. The separation between the plates is then increased. What happens to the energy stored in the capacitor? Discuss your answer in terms of energy conservation.

3.18 Problems

Definition of Capacitance

1. •(a) How much charge is on each plate of a $4.00\,\mu F$ capacitor when it is connected to a 12.0 V battery? (b) If this same capacitor is connected to a 1.50 V battery, what charge is stored?
2. •Two conductors having net charges of $+10.0\,\mu C$ and $-10.0\,\mu C$ have a potential difference of 10.0 V between them. (a) Determine the capacitance of the system. (b) What is the potential difference between the two conductors if the charges on each are increased to $+100\,\mu C$ and $-100\,\mu C$?

Calculating Capacitance

3. •An isolated charged conducting sphere of radius 12.0 cm creates an electric field of 4.90×10^4 N/C at a distance 21.0 cm from its centre. (a) What is its surface charge density? (b) What is its capacitance?
4. •(a) If a drop of liquid has capacitance 1.00 pF, what is its radius? (b) If another drop has radius 2.00 mm, what is its capacitance? (c) What is the charge on the smaller drop if its potential is 100 V?
5. •An air-filled capacitor consists of two parallel plates, each with an area of $7.60\,cm^2$, separated by a distance of 1.80 mm. A 20.0 V potential difference is applied to these plates. Calculate (a) the electric field between the plates, (b) the surface charge density, (c) the capacitance, and (d) the charge on each plate.
6. ••When a potential difference of 150 V is applied to the plates of a parallel-plate capacitor, the plates carry a surface charge density of $30.0\,nC/cm^2$. What is the spacing between the plates?
7. •••A variable air capacitor used in a radio tuning circuit is made of N semicircular plates each of radius R and positioned a distance d from its neighbours, to which it is electrically connected. As shown in Figure 3.148, a second identical set of plates is enmeshed with its plates halfway between those of the first set. The second set can rotate as a unit. Determine the capacitance as a function of the angle of rotation θ, where $\theta = 0$ corresponds to the maximum capacitance.
8. ••A 50.0 m length of coaxial cable has an inner conductor that has a diameter of 2.58 mm and carries a charge of $8.10\,\mu C$. The surrounding conductor has an inner diameter of 7.27 mm and a charge of $-8.10\,\mu C$. (a) What is the capacitance of this cable? (b) What is the potential difference between the two conductors? Assume the region between the conductors is air.
9. ••A $20.0\,\mu F$ spherical capacitor is composed of two concentric metal spheres, one having a radius twice as large as the other. The region between the spheres is a vacuum. Determine the volume of this region.
10. •• An air-filled spherical capacitor is constructed with inner and outer shell radii of 7.00 and 14.0 cm, respectively. (a) Calculate the capacitance of the device. (b) What potential difference between the spheres results in a charge of $4.00\,\mu C$ on the capacitor?
11. •••A small object of mass m carries a charge q and is suspended by a thread between the vertical plates of a parallel-plate capacitor. The plate separation is d. If the thread makes an angle θ with the vertical, what is the potential difference between the plates?
12. •Find the capacitance of the Earth. (Suggestion: The outer conductor of the "spherical capacitor" may be considered as a conducting sphere of infinite radius where V approaches zero.)
13. •• Suppose one plate of a parallel-plate capacitor is tilted so it makes a small angle θ with the other plate, as shown in Fig.3.149. Determine a formula for the capacitance C in terms of A, d, and θ, where A is the area of each plate and θ is small. Assume the plates are square. [Hint: Imagine the

capacitor as many infinitesimal capacitors in parallel.]

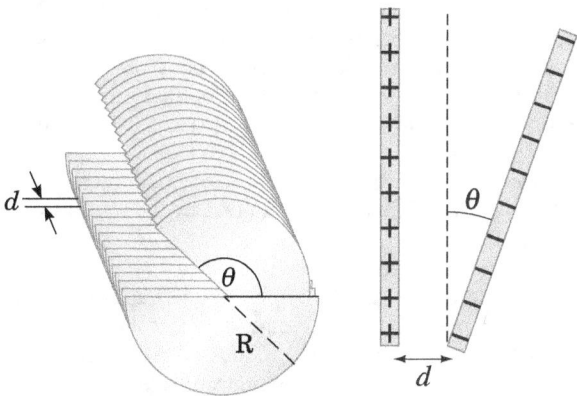

Figure 3.148: Problem 7   **Figure 3.149:** Problem 13

Combinations of Capacitors

14. ••Two capacitors, $C_1 = 5.00~\mu F$ and $C_2 = 12.0~\mu F$, are connected in parallel, and the resulting combination is connected to a 9.00 V battery. (a) What is the equivalent capacitance of the combination? What are (b) the potential difference across each capacitor and (c) the charge stored on each capacitor?

15. ••The two capacitors of Problem 14 are now connected in series and to a 9.00 V battery. Find (a) the equivalent capacitance of the combination, (b) the potential difference across each capacitor, and (c) the charge on each capacitor.

16. ••A capacitor is made of a flat plate of area A and a second plate having a stair-like structure as shown in Figure 3.150. The width of each stair is a and the height is b. Find the capacitance of the assembly.

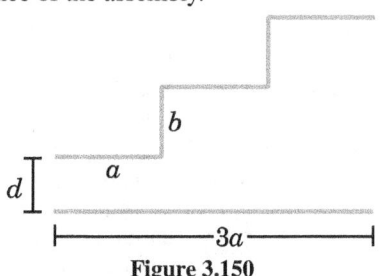

Figure 3.150

17. (a)••A voltage V is applied to the capacitor network shown in Fig.3.151. (a) What is the equivalent capacitance? Determine the equivalent capacitance if $C_1 = C_2 = 8.0~\mu F$ and $C_3 = C_4 = C_5 = 4.5~\mu F$.
    (b) What if, $C_2 = C_4 = 8.0~\mu F$ and $C_1 = C_3 = C_5 = 4.5~\mu F$ [Hint: Assume a potential difference V_{ab} exists across the network as shown; write potential differences for various pathways through the network from a to b in terms of the charges on the capacitors and the capacitances.]

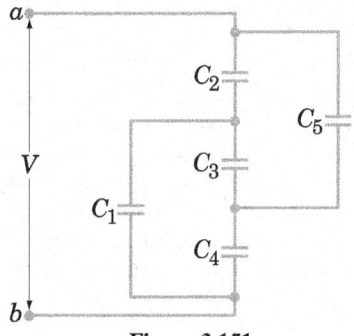

Figure 3.151

18. ••What is the equivalent capacitance between terminals a and b, (in terms of C, which is the capacitance of one of the capacitors) of the infinite ladder of capacitors shown in Figure 3.152?

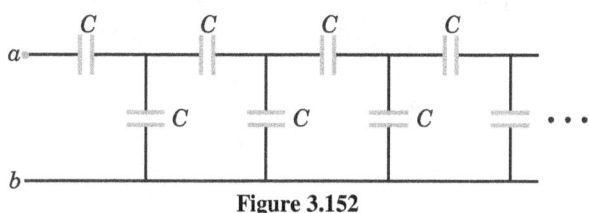

Figure 3.152

19. ••Take $C_1 = 4.0~\mu F$ and $C_2 = 6.0~\mu F$ in Figure 3.153. Calculate the equivalent capacitance of the combinations between the points A and B.

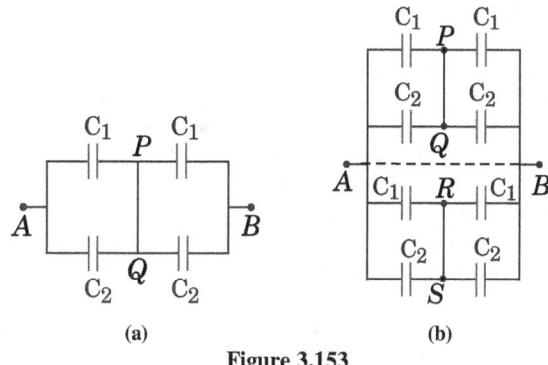

Figure 3.153

20. ••Two capacitors when connected in parallel give an equivalent capacitance of 9.00 pF and give an equivalent capacitance of 2.00 pF when connected in series. What is the capacitance of each capacitor?

21. ••Four capacitors are connected as shown in Figure 3.154. (a) Find the equivalent capacitance between points a and b. (b) Calculate the charge on each capacitor if $V_{ab} = 15.0$ V.

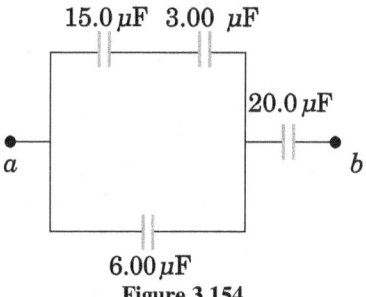

Figure 3.154

22. ••Consider the circuit shown in Figure 3.155, where $C_1 = 6.00~\mu F$, $C_2 = 3.00~\mu F$, and $V = 20.0$ volt. Capacitor C_1 is first charged by the closing of switch S_1. Switch S_1 is then opened, and the charged capacitor is connected to the uncharged capacitor by the closing of S_2. Calculate the initial charge acquired by C_1 and the final charge on each capacitor.

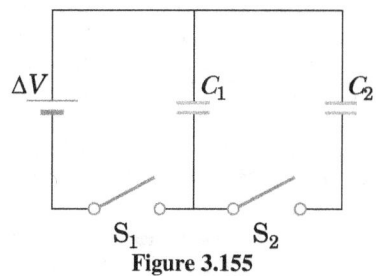

Figure 3.155

23. ••A group of identical capacitors is connected first in series

and then in parallel. The combined capacitance in parallel is 100 times larger than for the series connection. How many capacitors are in the group?

24. ••Find the equivalent capacitance between points a and b in the combination of capacitors shown in Figure 3.156.

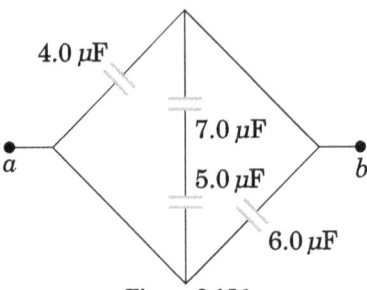

Figure 3.156

Energy Stored in a Charged Capacitor

25. •(a) A 3.00 μF capacitor is connected to a 12.0 V battery. How much energy is stored in the capacitor? (b) If the capacitor had been connected to a 6.00 V battery, how much energy would have been stored?

26. ••A parallel-plate capacitor is charged and then disconnected from a battery. By what fraction does the stored energy change (increase or decrease) when the plate separation is doubled?

27. ••A uniform electric field $E = 3000$ V/m exists within a certain region. What volume of space contains an energy equal to 1.00×10^{-7}J? Express your answer in cubic meters and in litres.

28. ••A parallel-plate capacitor has a charge Q and plates of area A. What force acts on one plate to attract it toward the other plate? Show that the force exerted on each plate is actually $F = Q^2/2\epsilon_0 A$. (Suggestion: Let $C = \epsilon_0 A/x$ for an arbitrary plate separation x; then the work done in separating the two charged plates be $W = \int F dx$.)

29. •••The circuit in Figure 3.157 consists of two identical parallel metal plates connected by identical metal springs to a 100 V battery. With the switch open, the plates are uncharged, are separated by a distance $d = 8.00$ mm, and have a capacitance $C = 2.00$ μF. When the switch is closed, the distance between the plates decreases by a factor of 0.500. (a) How much charge collects on each plate and (b) what is the spring constant (k_{sp}) for each spring? (Suggestion: Use the result of Problem 28)

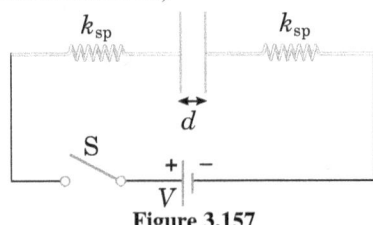

Figure 3.157

30. •••A certain storm cloud has a potential of 1.00×10^8 V relative to a tree. If, during a lightning storm, 50.0 C of charge is transferred through this potential difference and 1.00% of the energy is absorbed by the tree, how much sap in the tree can be boiled away? Model the sap as water initially at 30.0°C. Water has a specific heat of 4186 J/kg°C, a boiling point of 100°C, and a latent heat of vaporization of 2.26×10^6 J/kg.

31. ••Two identical parallel-plate capacitors, each with capacitance C, are charged to potential difference V and connected in parallel. Then the plate separation in one of the capacitors is doubled. (a) Find the total energy of the system of two capacitors before the plate separation is doubled. (b) Find the potential difference across each capacitor after the plate separation is doubled. (c) Find the total energy of the system after the plate separation is doubled. (d) Reconcile the difference in the answers to parts (a) and (c) with the law of conservation of energy.

32. ••Show that the energy associated with a conducting sphere of radius R and charge Q surrounded by a vacuum is $U = kQ^2/2R$

Capacitors with Dielectrics

33. ••(a) How much charge can be placed on a capacitor with air between the plates before it breaks down, if the area of each of the plates is 5.00 cm^2 ? (b) Find the maximum charge if polystyrene is used between the plates instead of air.

34. ••Two different dielectrics each fill half the space between the plates of a parallel-plate capacitor as shown in Fig.3.158a. Determine a formula for the capacitance in terms of κ_1, κ_2, the area A of the plates, and the separation d.

35. ••Two different dielectrics fill the space between the plates of a parallel plate capacitor as shown in Fig.3.158b. Determine a formula for the capacitance in terms of κ_1, κ_2 the area A, of the plates, and the separation $d_1 = d_2 = d/2$.

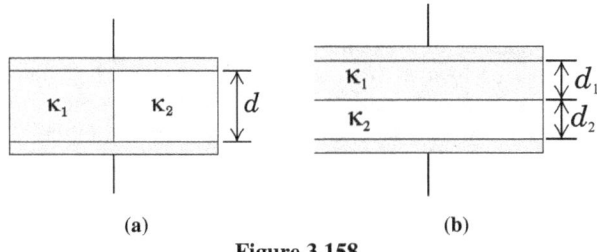

(a)   (b)

Figure 3.158

36. ••A wafer of titanium dioxide ($\kappa = 173$) of area 1.00 cm^2 has a thickness of 0.100 mm. Aluminum is evaporated on the parallel faces to form a parallel-plate capacitor. (a) Calculate the capacitance. (b) When the capacitor is charged with a 12.0 V battery, what is the magnitude of charge delivered to each plate? (c) For the situation in part (b), what are the free and induced surface charge densities? (d) What is the magnitude of the electric field?

37. ••Each capacitor in the combination shown in Figure 3.159 has a breakdown voltage of 15.0 V. What is the breakdown voltage of the combination?

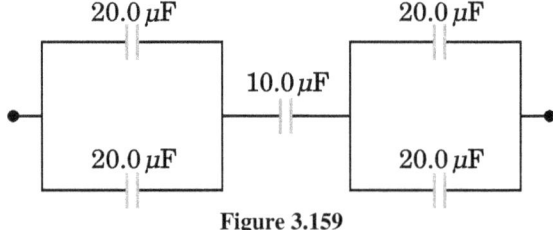

Figure 3.159

38. ••Consider the situation shown in Fig.3.160. The width of each plate is b. The capacitor plates are rigidly clamped in the laboratory and connected to a battery of emf $\mathcal{E}$. All surfaces are frictionless. Calculate the value of M for which the dielectric slab will stay in equilibrium.

39. ••Figure 3.161 shows two parallel plate capacitors with fixed plates and connected to two batteries. The separation between the plates is the same for the two capacitors. The plates are rectangular in shape with width b and lengths l_1 and l_2. The left half of the dielectric slab has a dielectric constant κ_1

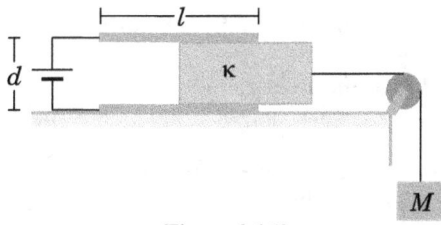

Figure 3.160

and the right half κ_2. Neglecting any friction, find the ratio of the emf of the left battery to that of the right battery for which the dielectric slab may remain in equilibrium.

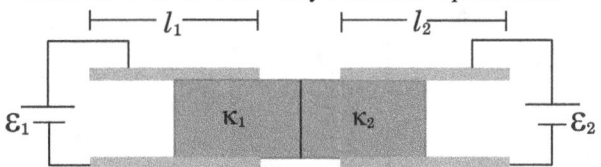

Figure 3.161

An Atomic Description of Dielectrics

40. ••The capacitor shown in Fig.3.162 is connected to a 90.0 V battery. Calculate (and sketch) the electric field everywhere between the capacitor plates. Find both the free charge on the capacitor plate and the induced charge on the faces of the glass dielectric plate.

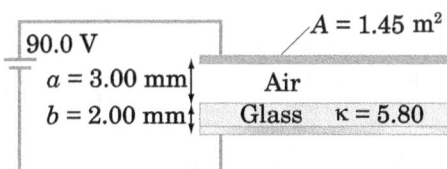

Figure 3.162

Additional Problems

41. ••A parallel-plate capacitor is constructed by filling the space between two square plates with blocks of three dielectric materials, as in Figure 3.163. You may assume that $l \gg d$. (a) Find an expression for the capacitance of the device in terms of the plate area A and d, κ_1, κ_2, and κ_3. (b) Calculate the capacitance using the values $A = 1.00$ cm^2, $d = 2.00$ mm, $\kappa_1 = 4.90$, $\kappa_2 = 5.60$, and $\kappa_3 = 2.10$.

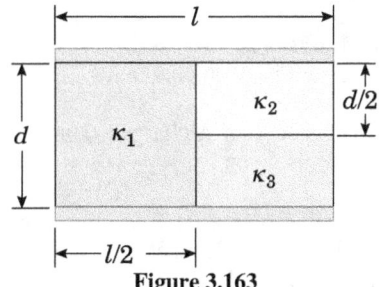

Figure 3.163

42. ••(a) Two spheres have radii a and b and their centres are a distance d apart. Show that the capacitance of this system is

$$C = \frac{4\pi\epsilon_0}{\frac{1}{a} + \frac{1}{b} - \frac{2}{d}}$$

provided that d is large compared with a and b. (Suggestion: Because the spheres are far apart, assume that the potential of each equals the sum of the potentials due to each sphere, and when calculating those potentials assume that $V = kQ/r$ applies.) (b) Show that as d approaches infinity the above result reduces to that of two spherical capacitors in series.

43. ••A vertical parallel-plate capacitor is half filled with a dielectric for which the dielectric constant is 2.00 (Fig.3.164a). When this capacitor is positioned horizontally, what fraction of it should be filled with the same dielectric (Fig.3.164b) in order for the two capacitors to have equal capacitance?

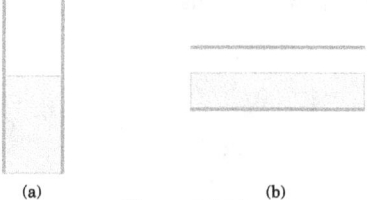

Figure 3.164

44. •••A capacitance balance is shown in Figure 3.165. The balance has a weight attached on one side and a capacitor that has a variable gap width on the other side. Assume the upper plate of the capacitor has negligible mass. When the capacitor potential difference between the plates is V_0, the attractive force between the plates balances the weight of the hanging mass. (a) Is the balance stable? That is, if we balance it out, and then move the plates a little closer together, will they snap shut or move back to the equilibrium point? (b) Calculate the value of V_0 required to balance an object of mass M, assuming the plates are separated by distance d_0 and have area A.

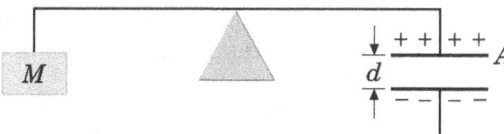

Figure 3.165

45. •••A fuel gauge uses a capacitor to determine the height of the fuel in a tank. The effective dielectric constant κ_{eff} changes from a value of 1 when the tank is empty to a value of κ, the dielectric constant of the fuel, when the tank is full. Each of the two rectangular plates has a width w and a length L (Fig.3.166). The height of the fuel between the plates is h. You can ignore any fringing effects. (a) Derive an expression for κ_{eff} as a function of h. (b) What is the effective dielectric constant for a tank 1/4 full, 1/2 full, and 3/4 full if the fuel is gasoline ($\kappa = 1.95$) ? (c) Repeat part (b) for methanol ($\kappa = 33.0$). (d) For which fuel is this fuel gauge more practical?

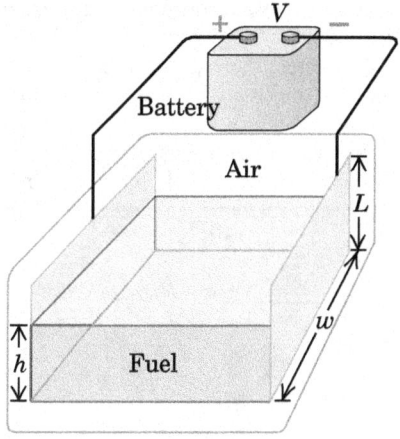

Figure 3.166

46. •••Three square metal plates A, B, and C, each 12.0 cm on a side and 1.50 mm thick, are arranged as in Fig.3.167. The plates are separated by sheets of paper 0.45 mm thick and with dielectric constant 4.2. The outer plates are connected together and connected to point b. The inner plate is

connected to point a. (a) Show by plus and minus signs the charge distribution on the plates when point a is maintained at a positive potential relative to point b. (b) What is the capacitance between points a and b?

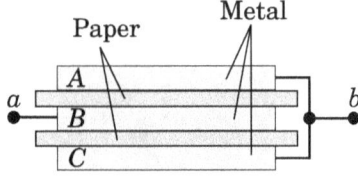

Figure 3.167

47. • • • An isolated spherical capacitor has charge $+Q$ on its inner conductor (radius r_a) and charge $-Q$ on its outer conductor (radius r_b). Half of the volume between the two conductors is then filled with a liquid dielectric of constant κ, as shown in cross section in Fig. 3.168. (a) Find the capacitance of the half-filled capacitor. (b) Find the magnitude of $\vec{E}$ in the volume between the two conductors as a function of the distance r from the centre of the capacitor. Give answers for both the upper and lower halves of this volume. (c) Find the surface density of free charge on the upper and lower halves of the inner and outer conductors. (d) Find the surface density of bound charge on the inner ($r = r_a$) and outer ($r = r_b$) surfaces of the dielectric. (e) What is the surface density of bound charge on the flat surface of the dielectric? Explain.

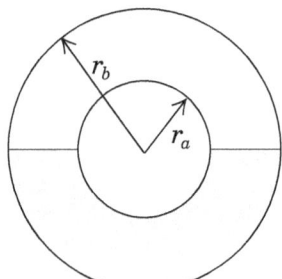

Figure 3.168

48. • • • Two identical metal plates with area A and mass m are kept separated with the help of three insulating springs as shown in the Fig.3.169. The equilibrium separation between the plates is $d_0 (\ll \sqrt{A})$ and force constant of each spring is κ. When a constant voltage source having emf V is connected to the plates, the equilibrium separation changes to d. Assume that the lower plate is fixed and the upper plate is free to move.
(a) Find V in terms of given parameters.
(b) If the upper disc is slightly displaced from its equilibrium position and released, calculate the time period of its oscillation.

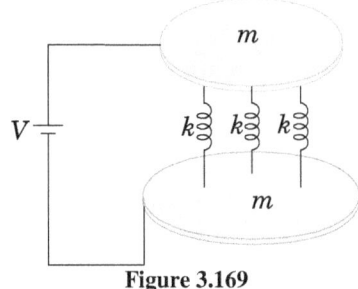

Figure 3.169

3.18.1 Multiple Choice Questions
3.18.1.1 Level 1
Capacitance and Energy

1. In Fig.3.170, what physical quantities may X and Y represent?
   (A) pressure and temperature respectively of a given gas (constant volume)
   (B) kinetic energy and velocity respectively of a particle
   (C) capacitance and charge respectively at a constant potential
   (D) potential and capacitance respectively at a constant charge

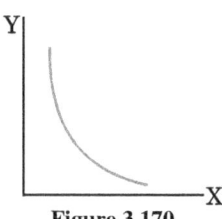

Figure 3.170

2. The capacitance of a spherical conductor of radius r is proportional to -
   (A) r (B) $1/r$ (C) r^2 (D) $1/r^3$

3. The capacitance of a metallic sphere will be $1\mu F$, if its radius is nearly
   (A) 9 km   (B) 10 m   (C) 1.11 m   (D) 1.11 cm

4. The energy of a charged conductor is given by the expression
   (A) $q^2/2C$ (B) q^2/C (C) $2qC$ (D) $q^2/2C^2$

5. No current flows between two charged bodies connected together when they have the same
   (A) capacitance or QN ratio
   (B) charge
   (C) resistance
   (D) potential or Q/C ratio

6. Which of the following expressions represents a farad?
   (A) joule/volt   (B) volt/coulomb
   (C) coulomb/volt   (D) coulomb/joule

7. Two spherical conductors A and B of radii R and $2R$ respectively, are each given a charge Q. When they are connected by a metallic wire. The charge will
   (A) flow from A to B
   (B) flow from B to A
   (C) remain stationary on conductor
   (D) none of these

8. The potential energy of a charged conductor of charge q and potential V, is given by
   (A) $\frac{1}{2}qV$ (B) $\frac{1}{2}q^2V$ (C) $\frac{1}{2}\frac{q}{V}$ (D) $\frac{1}{2}qV^2$

9. A conductor of capacitance 0.5 μF has been charged to 100 volts. It is now connected to uncharged conductor of capacitance 0.2 μF. The loss in potential energy is nearly:
   (A) 7×10^{-4}J   (B) 3.5×10^{-4}J
   (C) 14×10^{-4}J   (D) 7×10^{-3}J

10. Two spherical conductors of capacitance 3.0 μF and 5.0 μF are charged to potentials of 300 volt and 500 volt. The two are connected resulting in redistribution of charges. Then the final potential is:
    (A) 300 volt   (B) 500 volt
    (C) 425 volt   (D) 400 volt

11. N drops of mercury of equal radii and possessing equal charges combine to form a big spherical drop. Then the capacitance of the bigger drop compared to each individual drop is

3.18. PROBLEMS

(A) N times (B) $N^{2/3}$ times
(C) $N^{1/3}$ times (D) $N^{5/3}$ times

Parallel Plate Capacitor

12. The capacitance of a parallel plate condenser is C. Its capacitance when the separation between the plates is halved will be
    (A) $4C$ (B) $2C$ (C) $C/2$ (D) $C/4$

13. A parallel plate condenser has a capacitance 50 μF in air and 110 μF when immersed in an oil. The dielectric constant κ of the oil is:
    (A) 0.45 (B) 0.55 (C) 1.10 (D) 2.20

14. The capacity of a parallel plate condenser is 5 μF. When glass plate is placed between the plates of the conductor, its potential becomes 1/8 th of the original value. The value of dielectric constant will be-
    (A) 1.6 (B) 5 (C) 8 (D) 40

15. Capacitance of parallel plate condenser having a medium of dielectric constant κ is given by -
    (A) $C = QV$ F (B) $C = \frac{\kappa A}{9 \times 10^9 4\pi d}$ F
    (C) $C = \frac{\kappa d}{9 \times 10^9 4\pi A}$ F (D) $C = \frac{A}{9 \times 10^9 4\pi \kappa d}$ F

16. The capacity of a parallel plate air condenser is 2 $\mu\mu F$. If the distance between the plates is 4 cm and the area of each plate is 0.01 m^2, the value of permittivity of air is -
    (A) 8×10^{-12} Fm^{-1} (B) 5×10^{-13} Fm^{-1}
    (C) 8×10^{-14} Fm^{-1} (D) 5×10^{-14} Fm^{-1}

17. If the potential difference across the ends of a capacitor of 4 μF is 1.0 kilovolt. Then its electrical potential energy will be-
    (A) 4×10^{-3} ergs (B) 2 ergs
    (C) 2 joules (D) 4 joules

18. A 6μF capacitor charged from 10 volts to 20 volts. Increase in energy will be-
    (A) 18×10^{-4} joule (B) 9×10^{-4} joule
    (C) 4.5×10^{-4} joule (D) 9×10^{-9} joule

19. The energy of a charged capacitor resides in
    (A) the electric field only
    (B) the magnetic field only
    (C) both the electric and magnetic field
    (D) neither in electric nor magnetic field

20. The capacity and the energy stored in a parallel plate condenser with air between its plates are respectively C_0 and W_0. If the air is replaced by glass (dielectric constant = 5) between the plates, the capacity of the plates and the energy stored in it will respectively be -
    (A) $5C_0, 5W_0$ (B) $5C_0, \frac{W_0}{5}$
    (C) $\frac{C_0}{5}, 5W_0$ (D) $\frac{C_0}{5}, \frac{W_0}{5}$

21. By inserting a plate of dielectric material between the plates of a parallel plate capacitor, the energy is increased five times. The dielectric constant of the material is
    (A) 1/25 (B) 1/5 (C) 5 (D) 25

22. A capacitor of capacity C has charge Q and stored energy is W. If the charge is increased to $2Q$ the stored energy will be -
    (A) $2W$ (B) $W/2$ (C) $4W$ (D) $W/4$

23. A glass slab is put with in the plates of a charged parallel plate condenser. Which of the following quantities does not change?
    (A) energy of the condenser
    (B) capacity
    (C) intensity of electric field
    (D) charge

24. A parallel plate capacitor is connected to a battery and inserted a dielectric plate between the place of plates then which quantity increase.

    (A) potential difference (B) electric field
    (C) stored energy (D) emf of battery

25. A parallel plate capacitor is connected to a battery and decreased the distance between the plates then which quantity is same on the parallel plate capacitor
    (A) potential difference (B) capacitance
    (C) intensity of electric field (D) stored energy

26. A parallel plate capacitor is charged by a battery after charging the capacitor, battery is disconnected. Now, if a dielectric slab is inserted between the space of plates. Then which one of the following statements is not correct.
    (A) increase in the stored energy
    (B) decrease in the potential difference
    (C) decrease in the electric field
    (D) increase in the capacitance

27. A parallel plate capacitor has a capacity C. The separation between plates is doubled and a dielectric medium is inserted between plates. The new capacity is $3C$. The dielectric constant of medium is
    (A) 1.5 (B) 3.0 (C) 6.0 (D) 12.0

28. A parallel plate capacitor is charged by a battery. After charging the capacitor, battery is disconnected and decrease the distance between the plates then which following statement is correct?
    (A) electric field is not constant
    (B) potential difference is increased
    (C) decrease the capacitance
    (D) decrease the stored energy

29. The capacitance of a parallel plate condenser does not depend upon
    (A) the distance between the plates
    (B) area of the plates
    (C) medium between the plates
    (D) metal of the plates

30. A metallic plate of thickness 't' and face area of one side 'A' is inserted between the plates of a parallel plate air capacitor with a separation 'd' and face area 'A'. Then the equivalent capacitance is:
    (A) $\frac{\epsilon_0 A}{d}$ (B) $\frac{\epsilon_0 A}{dxt}$
    (C) $\frac{\epsilon_0 A}{d-t}$ (D) $\frac{\epsilon_0 A}{d+t}$

31. An air capacitor of 1 $\mu\mu F$ is immersed in a transformer oil of dielectric constant 3. The capacitance of the oil capacitor is
    (A) 1 $\mu\mu F$ (B) $\frac{1}{3} \mu\mu F$ (C) 3 $\mu\mu F$ (D) 2 $\mu\mu F$

32. Two metal plates form a parallel plate condenser. The distance between the plates in d. Now a metal plate of thickness $d/2$ and of same area is inserted completely between the plates, the capacitance -
    (A) remains unchanged (B) is doubled
    (C) is halved (D) reduced to one fourth

33. The capacity of a parallel plate capacitor with air as medium is 2 μF. After inserting a sheet of mica a equal air thickness, it becomes 5 μF. The dielectric constant of mica is -
    (A) 0.1 (B) 0.4 (C) 2.5 (D) 10

34. A parallel plate capacitor has rectangular plates of area 400 cm^2 and are separated by a distance of 2 mm with air as medium. If a 200 volt potential difference is applied across the condenser, the charge appeared the plates:
    (A) 3.54×10^{-6} C (B) 3.54×10^{-8} C
    (C) 3.54×10^{-10} C (D) 1770.8×10^{-13} C

35. A parallel plate condenser is immersed in an oil of dielectric constant 2. The field between the plates is

(A) increased proportional to 2
(B) decreased proportional to 1/2
(C) increased proportional to $\sqrt{2}$
(D) decreased proportional to $1/\sqrt{2}$

36. A parallel plate capacitor consists of two plates of 2 m × 1 m. The space between the plates is of 1 mm and filled with a dielectric of relative permittivity of 7. A potential difference of 300 volts is applied across the plates. Find the potential gradient
    (A) 6×10^5 N/C
    (B) 3×10^5 N/C
    (C) 18×10^5 N/C
    (D) 12×10^5 N/C

37. The energy of a charged capacitor resides in
    (A) the electric field only
    (B) the magnetic field only
    (C) both the electric and magnetic fields
    (D) neither in electric nor magnetic field

38. Two conductors insulated from each other, charged by transferring electrons from one conductor to the other. After 25×10^{12} electrons have been transferred. The potential difference between the conductors is found to be 16 V. The capacitance of the system is
    (A) 25 μF
    (B) 0.25 μF
    (C) 25 nF
    (D) 25 pF

39. The energy density in a parallel plate capacitor is given as 2.2×10^{-10} J/m^3. The value of the electric field in the region between the plates is-
    (A) 7 NC^{-1}
    (B) 3.6 NC^{-1}
    (C) 72 NC^{-1}
    (D) 8.4 NC^{-1}

40. If a 10 μF capacitor is to have an energy content of 1 Joule. It must be placed across a potential difference of (in volts)
    (A) 900
    (B) 450×10^8
    (C) 200
    (D) 450

41. A capacitor of capacitance $\frac{1}{3}\mu$F is connected to a battery of 300 volt and charged. Then the energy stored in capacitor is
    (A) 3×10^{-2} joule
    (B) 1.5×10^{-2} joule
    (C) 6×10^2 joule
    (D) 12×10^2 joule

42. The two parallel plates of a condenser have been connected to a battery of 300 V and the charge collected at each plate is 1 μC. The energy supplied by battery is -
    (A) 6×10^{-4} J
    (B) 3×10^{-4} J
    (C) 1.5×10^{-4} J
    (D) 4.5×10^{-4} J

43. The plates of a parallel plate capacitor are charged with a battery so that the plates of the capacitor have acquired the potential difference(pd) equal to emf of the battery. The ratio of the work done by the battery and the energy stored in capacitor is
    (A) 2 : 1
    (B) 1 : 1
    (C) 1 : 2
    (D) 1 : 4

44. A parallel plate condenser has plates of area 200 cm^2 and separation 0.05 cm has been filled with a dielectric having $\kappa = 8$ and then charged to 300 volts. The final energy of condenser is-
    (A) 1.6×10^{-5} J
    (B) 2.0×10^{-6} J
    (C) 12.8×10^{-5} J
    (D) 64×10^{-5} J

Combination of Capacitors

45. Three capacitors of capacity C_1, C_2, C_3 are connected in series. Their total capacity will be -
    (A) $C_1 + C_2 + C_3$
    (B) $\frac{1}{(C_1+C_2+C_3)}$
    (C) $(C_1^{-1} + C_2^{-1} + C_3^{-1})$
    (D) none of these

46. Three capacitors each of capacitance 1 μF are connected in parallel. To this combination a fourth capacitor of capacitance 1 μF connected in series. The resultant capacitance of the system is
    (A) 4 μF
    (B) 2 μF
    (C) 4/3 μF
    (D) 3/4 μF

47. Two capacitors of capacity C_1 and C_2 are connected in series and potential difference V is applied across it. Then the potential difference across C_1 will be
    (A) $V\frac{C_2}{C_1}$
    (B) $V\frac{C_1+C_2}{C_1}$
    (C) $V\frac{C_2}{C_1+C_2}$
    (D) $V\frac{C_1}{C_1+C_2}$

48. Two condensers of capacities 1 μF and 2 μF are connected in series and system charged to 120 volts. Then the potential difference across 1 μF capacitor (in volts) will be
    (A) 40
    (B) 60
    (C) 80
    (D) 120

49. Two condensers of capacity 0.3 μF and 0.6 μF respectively are connected in series. The combination is connected across a potential of 6 volts. The ratio of energies stored by the condensers will be-
    (A) 1/2
    (B) 2
    (C) 1/4
    (D) 4

50. Three capacitors $C_a < C_b < C_c$ are connected in series. Their equivalent capacitance will be
    (A) greater than C_c
    (B) less than C_c but greater then C_a
    (C) less than C_a
    (D) infinite

51. In the circuit (Fig.3.171) the resultant capacitance between A and B is 1 μF. Then value of C is

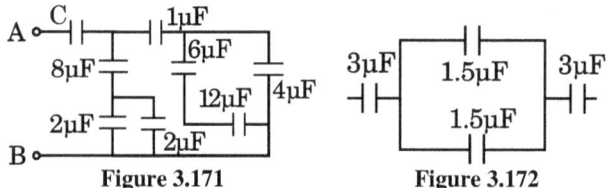

Figure 3.171   Figure 3.172

(A) 11 μF
(B) 11 μF
(C) $\frac{23}{32}$ μF
(D) $\frac{32}{23}$ μF

52. The condensers of capacity C_1 and C_2 are connected in parallel, then the equivalent capacitance is
    (A) $C_1 + C_2$
    (B) $\frac{C_1 C_2}{C_1+C_2}$
    (C) C_1/C_2
    (D) C_2/C_1

53. The equivalent capacity of the system of capacitors between free terminals, as shown in Fig.3.172, will be:
    (A) 1 μF
    (B) 2 μF
    (C) 1.5 μF
    (D) 3 μF

54. Three capacitors of capacitance 3 μF, 9 μF and 18 μF are connected once in series and another time in parallel. The ratio of equivalent capacitance in the two cases C_s/C_p will be-
    (A) 1 : 15
    (B) 15 : 1
    (C) 1 : 1
    (D) 1 : 3

55. Three equal capacitors, each with capacitance C are connected as shown in Fig.3.173. Then the equivalent capacitance between A and B is

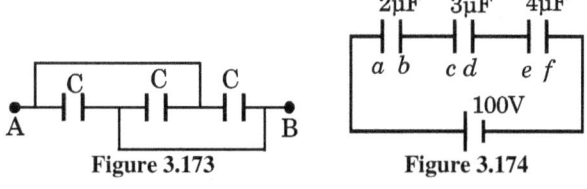

Figure 3.173   Figure 3.174

(A) C
(B) $3C$
(C) $C/3$
(D) $3C/2$

56. Three capacitors are connected to DC source of 100 volts as shown in the Figure 3.174. If the charge accumulated on plates of C_1, C_2 and C_3 are $q_a, q_b, q_c, q_d, q_e, q_f$ respectively then
    (A) $q_b + q_d + q_f = 100/9$ C
    (B) $q_b + q_d + q_f = 0$
    (C) $q_a + q_c + q_e = 50$ C
    (D) $q_b = q_d = q_f$

57. A capacitor $C_1 = 4\mu$F is connected in series with another capacitor $C_2 = 1\mu$F. The combination is connected across a DC source of voltage 200 V. The ratio of potential across C_1 and C_2 is -
    (A) 1 : 4
    (B) 4 : 1
    (C) 1 : 2
    (D) 2 : 1

58. Two condensers of 20 and 30 microfarad, are connected in

series across a 200 volt directional current (dc) supply. Find the charge on each condenser?
(A) 2400 μC (B) 4200 μC
(C) 2600 μC (D) 3000 μC

59. Two condensers of capacities C_1 and C_2 respectively are connected in parallel. The equivalent capacitance of the system is-
(A) $C_1 + C_2$ (B) $\frac{C_1 C_2}{C_1 + C_2}$
(C) $C_1 - C_2$ (D) $\frac{1}{C_1} + \frac{1}{C_2}$

60. The three condensers of capacitances 10 μF, 20 μF and 30 μF are first connected in series and then connected in parallel. The ratio of the resultant capacitance in the two cases is-
(A) 1 : 11 (B) 11 : 1 (C) 1 : 6 (D) 6 : 1

61. Five equal capacitors, each with capacitance (C) are connected as shown in the adjoining Fig. 3.175. Then the equivalent capacitance between A and B is:
(A) C (B) $5C$ (C) $C/5$ (D) $3C$

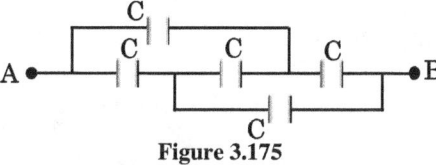

Figure 3.175

62. The total capacity of the system of capacitors shown in adjoining Figure 3.176 between the points A and B is:
(A) 1 μF (B) 2 μF (C) 3 μF (D) 4 μF

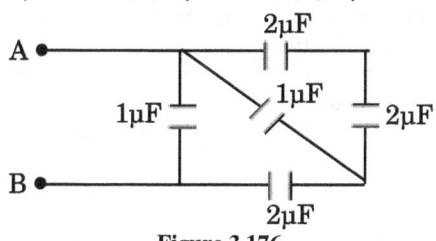

Figure 3.176

63. The equivalent capacitance between the points A and B in the Fig.3.177 is-
(A) 8 μF (B) 6 μF (C) $\frac{8}{3}$ μF (D) $\frac{3}{8}$ μF

64. Five capacitors of 10 μF capacitor each are connected to a DC potential of 100 volts as shown in Figure 3.178. The equivalent capacitance between the points A and B will be equal to

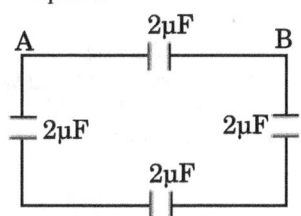 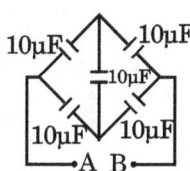

Figure 3.177: Question 63. **Figure 3.178:** Question 64.

(A) 40 μF (B) 20 μF (C) 30 μF (D) 10 μF

65. Three capacitors of capacity 10 μF, 5 μF and 5 μF are connected in parallel. The total capacity will be-
(A) 10 μF (B) 5 μF
(C) 20 μF (D) none

66. Two capacitors connected in parallel having the capacities C_1 and C_2 are given charge 'q', which is distributed among them. The ratio of the charge on C_1 and C_2 will be
(A) $\frac{C_1}{C_2}$ (B) $\frac{C_2}{C_1}$
(C) $C_1 C_2$ (D) $\frac{1}{C_1 C_2}$

67. Three capacitors C_1, C_2 and C_3 are joined to a battery as shown in Fig.3.179. The correct condition will be
(A) $Q_1 = Q_2 = Q_3$ and $V_1 = V_2 = V_3 = V$
(B) $Q_2 = Q_2 + Q_3$ and $V = V_1 + V_2 + V_3$
(C) $Q_1 = Q_2 + Q_3$ and $V = V_1 + V_2$
(D) $Q_2 = Q_3$ and $V_2 = V_3$

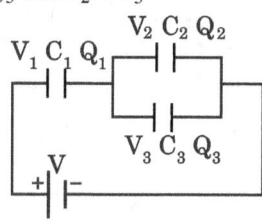

Figure 3.179

(Symbols have their usual meanings)

68. Two capacitors of equal capacity are first connected in parallel and then in series. The ratio of the total capacities in the two cases will be
(A) 2 : 1 (B) 1 : 2 (C) 4 : 1 (D) 1 : 4

69. A 4 μF condenser is connected in parallel to another condenser of 8 μF. both the condensers are then connected in series with a 12 μF condenser and charged to 20 volts. The charge on the plate of 4 μF condenser is
(A) 3.3 μC (B) 40 μC (C) 80 μC (D) 240 μC

70. If three capacitors each of capacity 1 μF are connected in such a way that the resultant capacity is 1.5 μF then-
(A) all the three are connected in series
(B) all the three are connected in parallel
(C) two of them are in parallel and then connected in series to the third
(D) two of them are in series and then connected in parallel to the third

71. Two capacitors each of capacity 2 μF are connected in parallel. This system is connected in series with a third capacitance of 12 μF capacity. The equivalent capacity of the system will be
(A) 16 μF (B) 13 μF (C) 4 μF (D) 3 μF

72. Seven capacitors each of capacitance 2 μF are to be so connected to have total capacity (10/11) μF. Which will be the necessary figure as shown in Fig.3.180

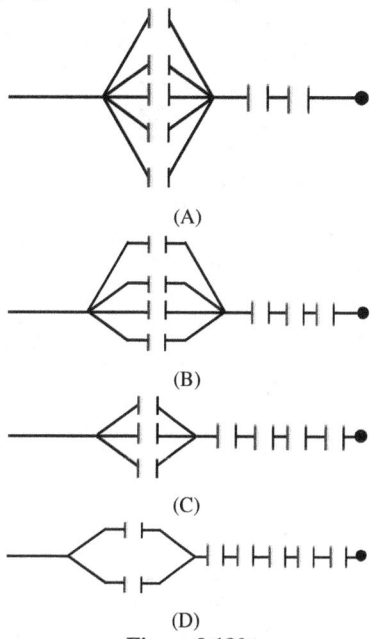

Figure 3.180

73. Two capacitor each of 1 μF capacitance are connected in parallel and are then charged by 200 volts dc supply. The total energy of their charges (in joules is)
    (A) 0.01   (B) 0.02   (C) 0.04   (D) 0.06
74. An infinite number of identical capacitors each of capacitance 1 μF are connected as in Fig.3.181. Then the equivalent capacitance between A and B is
    (A) 1 μF   (B) 2 μF   (C) 1/2 μF   (D) ∞

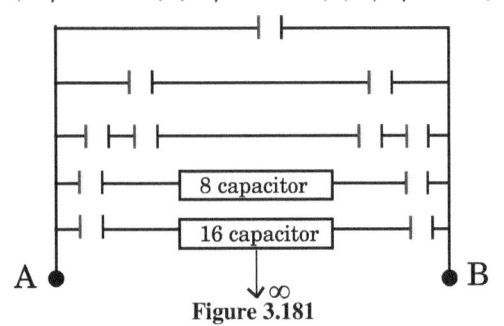

Figure 3.181

75. Four capacitors are connected as shown in the Fig.3.182. The equivalent capacitance between the points P and Q is
    (A) 4 μF   (B) 1/4 μF   (C) 3/4 μF   (D) 4/3 μF
76. The total capacity of the system of capacitors shown in the Fig.3.183 between the points A and B will be:
    (A) 1 μF   (B) 2 μF   (C) 3 μF   (D) 4 μF

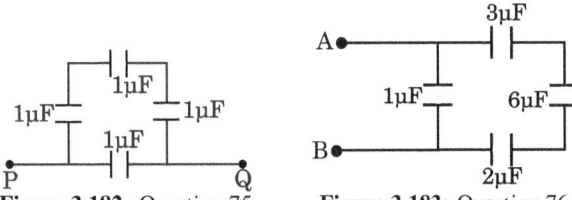

Figure 3.182: Question 75.   **Figure 3.183:** Question 76.

77. Two dielectric slabs of dielectric constants κ_1 and κ_2 have been filled in between the plates of a capacitor as shown in Fig.3.184. What will be the capacitance of the capacitor
    (A) $\frac{2\epsilon_0 A}{d}(\kappa_1 + \kappa_2)$
    (B) $\frac{2\epsilon_0 A}{d}\left(\frac{\kappa_1 + \kappa_2}{\kappa_1 \times \kappa_2}\right)$
    (C) $\frac{d}{2\epsilon_0 A}$
    (D) $\frac{2\epsilon_0 A}{d}\left(\frac{\kappa_1 \times \kappa_2}{\kappa_1 + \kappa_2}\right)$

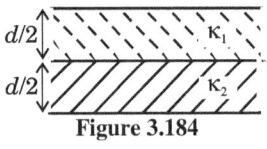

Figure 3.184

78. Five identical plates each of area A are joined as shown in the Fig.3.185. The distance between the plates is d. The plates are connected to battery of voltage V volts. The charges on plates 1 and 4 will be
    (A) $\frac{\epsilon_0 AV}{d}, \frac{2\epsilon_0 AV}{d}$
    (B) $\frac{-\epsilon_0 AV}{d}, \frac{2\epsilon_0 AV}{d}$
    (C) $\frac{\epsilon_0 AV}{d}, \frac{-2\epsilon_0 AV}{d}$
    (D) $\frac{-\epsilon_0 AV}{d}, \frac{-2\epsilon_0 AV}{d}$

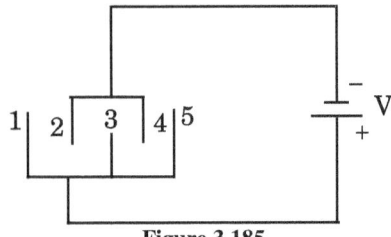

Figure 3.185

79. A parallel plate capacitor with air as medium between the plates has a capacitance of 10 μF. The area of capacitor is divided into two equal halves and filled with two media having dielectric constant $\kappa_1 = 2$ and $\kappa_2 = 4$. The capacitance of the system will now be-
    (A) 10 μF   (B) 20 μF   (C) 30 μF   (D) 40 μF
80. Separation between the plates of a parallel plate capacitor is d and the area of each plate is A. When a slab of material of dielectric constant κ and thickness $t(t < d)$ is introduced between the plates, its capacitance becomes
    (A) $\frac{\epsilon_0 A}{d+t\left(1-\frac{1}{\kappa}\right)}$
    (B) $\frac{\epsilon_0 A}{d+t\left(1+\frac{1}{\kappa}\right)}$
    (C) $\frac{\epsilon_0 A}{d-t\left(1-\frac{1}{\kappa}\right)}$
    (D) $\frac{\epsilon_0 A}{d-t\left(1+\frac{1}{\kappa}\right)}$
81. In Fig.3.186, the area of each plate is A and plates are separated from each other by a distance d then C_{eq} between A and B is -
    (A) $\frac{3\epsilon_0 A}{d}$ (B) $\frac{\epsilon_0 A}{3d}$ (C) $\frac{3\epsilon_0 A}{2d}$ (D) $\frac{2\epsilon_0 A}{3d}$
82. The capacitance of a capacitor, filled with two dielectrics of same dimensions but of dielectric constants κ_1 and κ_2 respectively as shown in Fig.3.187, will be:
    (A) $\frac{\epsilon_0 A}{2d}(\kappa_1 + \kappa_2)$
    (B) $\frac{\epsilon_0 A}{d}(\kappa_1 + \kappa_2)$
    (C) $\frac{\epsilon_0 A}{2d}\left(\frac{\kappa_1 \kappa_2}{\kappa_1 + \kappa_2}\right)$
    (D) $\frac{\epsilon_0 A}{d}\left(\frac{\kappa_1 \kappa_2}{\kappa_1 + \kappa_2}\right)$

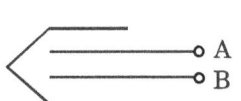 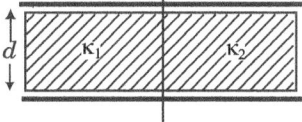

Figure 3.186: Question 81.   **Figure 3.187:** Question 82.

83. The capacitance of a parallel plate capacitor is 2.5 μF. When it is half filled with a dielectric as shown in the Fig.3.188, its capacitance becomes 5 μF, the dielectric constant of the dielectric is:
    (A) 7.5   (B) 3.0   (C) 0.33   (D) 4.0

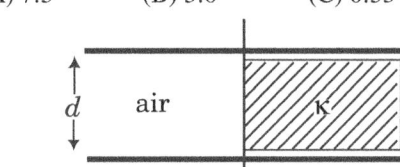

Figure 3.188

Spherical Capacitors

84. The capacitance of a spherical condenser is 1 μF. If the spacing between the two spheres is 1 mm, then the radius of the outer sphere is
    (A) 30 cm   (B) 6 m   (C) 5 cm   (D) 3 m
85. The capacitance (C) for an isolated conducting sphere of radius (a) is given by $4\pi\epsilon_0 a$. If the sphere is enclosed with an earthed concentric sphere. The ratio of the radii of the sphere being $(n/(n-1))$ then the capacitance of such a sphere will be increased by a factor-
    (A) n (B) $n/(n-1)$
    (C) $(n-1)/n$ (D) $a.n$
86. A_1 is a spherical conductor of radius (r) placed concentrically inside a thin spherical hollow conductor A_2 of radius R. A_1 is earthed and A_2 is given a charge $+Q$ then the charge on A_1 is-
    (A) $-Q$ (B) Qr/R
    (C) $-rQ/R$ (D) $-Q(R-r)/R$
87. Two spherical conductors A_1 and A_2 of radii r_1 and r_2 $(r_2 > r_1)$ are placed concentrically in air. A_1 is given a charge $+Q$ while A_2 in earthed. Then the capacitance of the system is-

(A) $4\pi\epsilon_0 \frac{r_1 \cdot r_2}{r_2-r_1}$ (B) $4\pi\epsilon_0 (r_1 + r_2)$

(C) $4\pi\epsilon_0 \cdot r_2$ (D) $4\pi\epsilon_0 \frac{r_2^2}{r_2-r_1}$

Connection of Charged Capacitors

88. Two condensers of capacitances $2C$ and C respectively, are joined in parallel and charged up to potential V. The battery is removed and the condenser of capacity C is filled completely with a medium of dielectric constant κ. The potential difference across the capacitors will now be-
    (A) $\frac{3V}{\kappa+2}$ (B) $\frac{3V}{\kappa}$ (C) $\frac{V}{\kappa+2}$ (D) $\frac{V}{\kappa}$

89. 0.2 F capacitor is charged to 600 V by a battery. After removing the battery, it is connected with another parallel plate condenser of magnitude 1.0 F. The common potential difference is-
    (A) 100 V   (B) 120 V   (C) 300 V   (D) 600 V

90. A 0.01 μF capacitor is charged to a potential of 500 V. It is then connected to an instrument of input capacitance 1.0 μF. The potential difference across the instrument in "volts" is now-
    (A) 1.00   (B) 4.95   (C) 5.00   (D) 50.0

91. A condenser of capacitance 10 μF has been charged to 100 V. It is now connected to another uncharged condenser. The common potential becomes 40 V. The capacitance of another condenser is-
    (A) 5 μF   (B) 10 μF   (C) 15 μF   (D) 20 μF

92. A capacitor having capacitance C is charged to a voltage V. It is then removed and connected in parallel with another identical capacitor which is uncharged. The new charge on each capacitor is now
    (A) CV (B) $CV/2$ (C) $2CV$ (D) $CV/4$

93. Two capacitors of capacities C_1 and C_2 are charged to voltages V_1 and V_2 respectively. There will be no exchange of energy in connecting them in parallel. If
    (A) $C_1 = C_2$ (B) $C_1V_1 = C_2V_2$
    (C) $V_1 = V_2$ (D) $C_1 = C_2$

3.18.1.2 Level 2

1. A parallel plate capacitor is charged and kept connected with the battery. If now a dielectric slab is inserted between the plates to fill the entire space between the plates then what will be the change in the charge, potential difference and electric field intensity between the plates respectively-
    (A) increases, constant, increases
    (B) increases, constant, constant
    (C) increases, constant, decreases
    (D) constant, decreases, decreases.

2. A parallel plate air capacitor is connected to a battery. The quantities charge, voltage electric field, and energy associated with this capacitor are given by Q_0, V_0, E_0 and U_0 respectively. A dielectric slab is now introduced to fill the space between the plates with battery still in connection. The corresponding quantities now given by Q, V, E, and U are related of the previous one as-
    (A) $Q > Q_0$ (B) $V > V_0$
    (C) $E > E_0$ (D) $U \geq U_0$

3. A battery charges a parallel plate capacitor of thickness d so that an energy U_0 is stored in the system. Now, if battery remains connected and a slab of dielectric constant κ and thickness d, is then introduced between the plates of the capacitor. The new energy of the system is given by-
    (A) κU_0 (B) $\kappa^2 U_0$ (C) U_0/κ (D) U_0/κ^2

4. Six identical capacitors, joined in parallel, are charged to a potential difference of 10 volt. They are now separated and then connected in series i.e., the positive plate of one is connected to the negative plate of the other. Then the potential difference between the free plates is-
    (A) 10 V   (B) 30 V   (C) 60 V   (D) 10/6 V

5. Two spheres of radii R_1 and R_2 have equal charge are joined together with a copper wire. If the potential on each sphere after they are separated to each other is V, then initial charge on any sphere was ($k = 1/4\pi\epsilon_0$):
    (A) $\frac{V}{k}(R_1 + R_2)$ (B) $\frac{V}{2k}(R_1 + R_2)$
    (C) $\frac{V}{3k}(R_1 + R_2)$ (D) $\frac{V}{k}\frac{(R_1R_2)}{(R_1+R_2)}$

6. In Fig.3.189, calculate the reading of voltmeter between X and Y i.e., $(V_X - V_Y)$ is equal to:

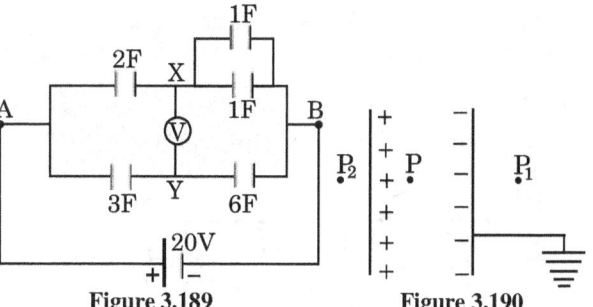

Figure 3.189   Figure 3.190

(A) 10 V   (B) 13.33 V   (C) 3.33 V   (D) 10.33 V

7. The capacitance of two capacitors was compared with the aid of an electrometer. The capacitors were charged to potential of $V_1 = 300$ V and $V_2 = 100$ V and were connected in parallel. The potential difference between the plates measured by the electrometer was 250 V. The capacitance ratio C_1/C_2 is-
    (A) 3 : 1   (B) 1 : 3   (C) 1 : 2.5   (D) 2.5 : 1

8. Three capacitors 2, 3 and 4 μF are connected in series with 6 V battery. When the current stops, the charge on the 3 μF capacitor is
    (A) 5.5 μC   (B) 4.4 μC   (C) 3.3 μC   (D) 2.2 μC

9. Fig.3.190 shows a parallel plate capacitor in which one plate is given a charge $+q$ while the other is earthed. There are three points P, P_1 and P_2. The electric intensity is not zero at:
    (A) P only   (B) P_1 only
    (C) P_2 only   (D) P, P_1 and P_2

10. The resultant capacitance between A and B in the Fig.3.191 is-
    (A) 1.0 μF   (B) 3 μF   (C) 2 μF   (D) 1.5 μF

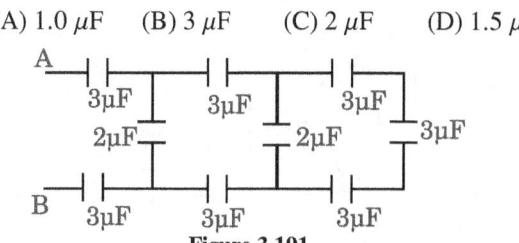

Figure 3.191

11. The diameter of the plate of a parallel plate condenser is 6 cm. If it's capacity is equal to that of a sphere of diameter 200 cm, the separation between the plates of the condenser is-
    (A) 4.5×10^{-4} m   (B) 2.25×10^{-4} m
    (C) 6.75×10^{-4} m   (D) 9×10^{-4} m

12. Four metallic plates of each with a surface area of one side A, are placed at a distance d from each other. The alternate plate are connected to point A and B as shown in the Fig.3.192. The capacitance of the system is
    (A) $\epsilon_0 A/d$ (B) $2\epsilon_0 A/d$
    (C) $3\epsilon_0 A/d$ (D) $4\epsilon_0 A/d$

13. A sheet of aluminium foil of negligible thickness is placed

between the plates of a capacitor of capacitance C as shown in the Figure 3.193, then capacitance of capacitor becomes
(A) $2C$ (B) C (C) $C/2$ (D) zero

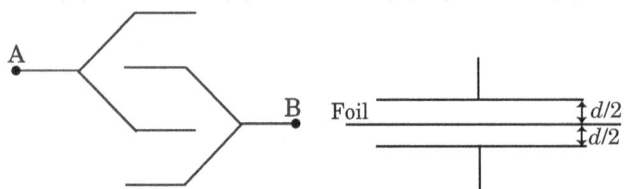

Figure 3.192: Question 12.   **Figure 3.193:** Question 13.

14. In above problem if foil is connected to any one plate of capacitor by means of conducting wire then capacitance of capacitor becomes -
(A) $2C$ (B) C (C) $C/2$ (D) zero

15. For circuit [Fig.3.194], the equivalent capacitance between P and Q is-
(A) $6C$ (B) $4C$ (C) $3C/2$ (D) $3C/4$

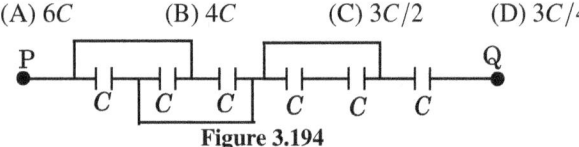

Figure 3.194

16. The Figure 3.195 shows a circuit consisting of four capacitors. The effective capacitance between A and B is-

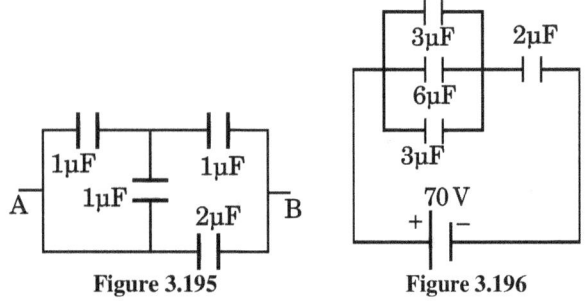

Figure 3.195   **Figure 3.196**

(A) $\frac{5}{6}\,\mu F$ (B) $\frac{7}{6}\,\mu F$ (C) $\frac{8}{3}\,\mu F$ (D) $1\,\mu F$

17. The potential across the capacitance of $2\,\mu F$ in the Figure 3.196 is -
(A) 10 V   (B) 60 V   (C) 28 V   (D) 56 V

18. A circuit is shown in the Figure 3.197. Find out the charge of the condenser having capacity $5\,\mu F$

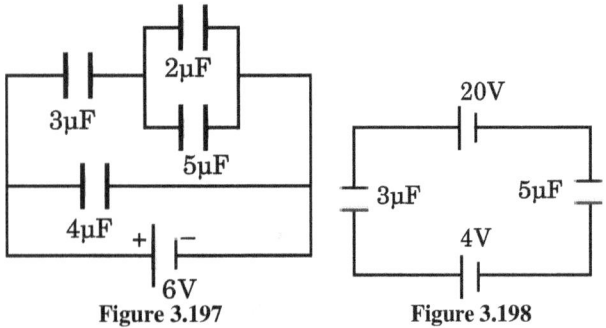

Figure 3.197   **Figure 3.198**

(A) $4.5\,\mu C$ (B) $6.0\,\mu C$ (C) $9.0\,\mu C$ (D) $30\,\mu C$

19. In the circuit shown in Fig.3.198, the potential difference across $3\,\mu F$ capacitor is-
(A) 4 V   (B) 6 V   (C) 10 V   (D) 16 V

20. Three capacitors of capacitors C_1, C_2, C_3 are connected as shown in the Figure 3.199. The points A, B and C are at potential V_1, V_2 and V_3 respectively. Then the potential at O will be-
(A) $\frac{V_1+V_2+V_3}{2}$ (B) $\frac{V_1V_2+V_2V_3+V_3V_1}{V_1+V_2+V_3}$
(C) $\frac{V_1C_1+V_2C_2+V_3C_3}{C_1+C_2+C_3}$ (D) zero

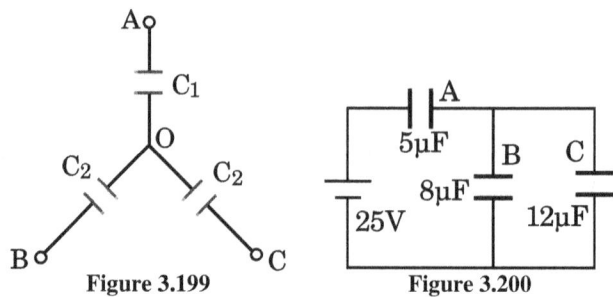

Figure 3.199   **Figure 3.200**

21. Three capacitors A, B and C are connected to a battery of 25 volt as shown in the Figure 3.200. The ratio of charges on capacitors A, B and C will be-
(A) 5 : 2 : 3   (B) 5 : 3 : 2
(C) 2 : 5 : 3   (D) 2 : 3 : 5

22. Four equal capacitors, each with a capacitance (C) are connected to a battery of emf 10 volts as shown in the Figure 3.201. The mid point of the capacitor system is connected to earth. Then the potentials of B and D are respectively-
(A) +10 volts, zero volts   (B) +5volts, −5 volts
(C) −5 volts, +5volts   (D) zero volts, 10 volts

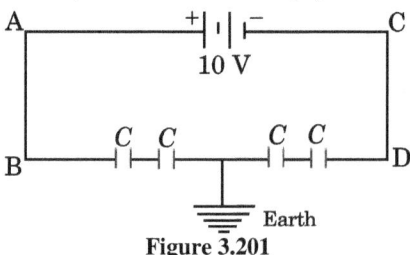

Figure 3.201

23. A circuit has a section AB as shown in the Fig.3.202, with $\mathcal{E} = 10$ V, $C_1 = 1.0\,\mu F$, $C_2 = 2.0\,\mu F$ and the potential difference $V_A - V_B = 5$ V. The voltage across C_1 is-
(A) zero   (B) 5 V   (C) 10 V   (D) 15 V

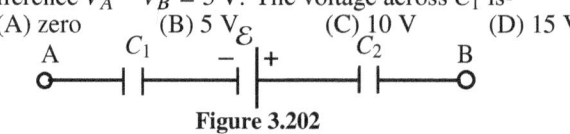

Figure 3.202

24. The potential difference between points P and Q of the circuit (Fig.3.203) is-

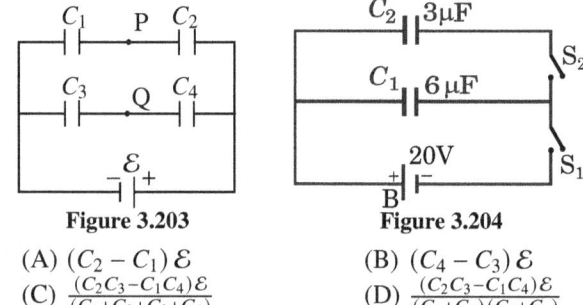

Figure 3.203   **Figure 3.204**

(A) $(C_2 - C_1)\mathcal{E}$ (B) $(C_4 - C_3)\mathcal{E}$
(C) $\frac{(C_2C_3 - C_1C_4)\mathcal{E}}{(C_1+C_2+C_3+C_4)}$ (D) $\frac{(C_2C_3 - C_1C_4)\mathcal{E}}{(C_1+C_2)(C_3+C_4)}$

25. In the circuit of Fig.3.204, $C_1 = 6\,\mu F$, $C_2 = 3\,\mu F$ and battery $B = 20$ V. The Switch S_1 is first closed. It is then opened and afterwards S_2 is closed. What is the charge finally on C_2?
(A) $120\,\mu C$ (B) $80\,\mu C$ (C) $40\,\mu C$ (D) $20\,\mu C$

26. A parallel plate capacitor has plate area A and separation d. It is charged to a potential difference V_0. The charging battery is disconnected and the plates are pulled apart to three times the initial separation. The work required to separate the plates is
(A) $\frac{3\epsilon_0 AV_0^2}{d}$ (B) $\frac{\epsilon_0 AV_0^2}{2d}$
(C) $\frac{\epsilon_0 AV_0^2}{3d}$ (D) $\frac{\epsilon_0 AV_0^2}{d}$

3.18. PROBLEMS

27. A capacitor of capacity C_1 is charged to the potential of V_0. On disconnecting with the battery, it is connected with a capacitor of capacity C_2 as shown in the Figure 3.205. The ratio of energies before and after the connection of switch S will be
    (A) $(C_1 + C_2)/C_1$
    (B) $C_1/(C_1 + C_2)$
    (C) $C_1 C_2$
    (D) C_1/C_2

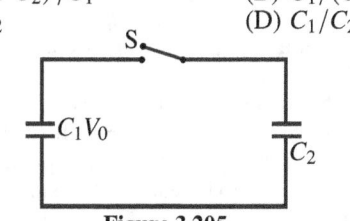

Figure 3.205

28. Condenser A has a capacity of 15 μF when it is filled with a medium of dielectric constant 15. Another condenser B has a capacity of 1 μF with air between the plates. Both are charged separately by a battery of 100 V. After charging, both are connected in parallel without the battery and the dielectric medium being removed. The common potential now is
    (A) 400 V (B) 800 V (C) 1200 V (D) 1600 V

3.18.1.3 Level 3

1. Two conducting spheres of radii 6 cm and 12 cm each having same charge of 3×10^{-8} C are kept very far apart. If the spheres are connected to each other by a conducting wire. the amount of charge transferred -
   (A) 1×10^{-8} C from smaller to bigger sphere
   (B) 1×10^{-8} C from bigger to smaller sphere
   (C) 2×10^{-8} C from bigger to smaller sphere
   (D) 2×10^{-8} C from smaller to bigger sphere

2. Two charged metal spheres of radii R and $2R$ are temporarily placed in contact and then separated. At the surface of each, the ratio of electric field will be-
   (A) 1 : 1 (B) 1 : 2 (C) 2 : 1 (D) 1 : 4

3. A 10 μF capacitor is charged by a battery of emf $100V$. The energy drawn from the battery and the energy stored in the capacitor are respectively -
   (A) 0.10 J and 0.05 J
   (B) 0.05 J and 0.10 J
   (C) 1.0 mJ and 0.5 mJ
   (D) 0.05 J and 0.05 mJ

4. Two identical capacitors A and B shown in the circuit in Fig.3.206 are joined in series with a battery. If a dielectric slab of dielectric constant κ is slipped between the plates of capacitor B and battery remain connected, then the energy of capacitor A will-
   (A) Decrease
   (B) Increase
   (C) Remain the same
   (D) Be zero since circuit will not work

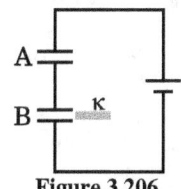

Figure 3.206

5. Two parallel plate air filled capacitors each of capacitance C, are joined in series to a battery of emf V. The space between the plates of one of the capacitors is then completely filled up with a uniform dielectric having dielectric constant κ. The quantity of charge which flows through the battery is -

(A) $\frac{CV}{2}\left(\frac{\kappa-1}{\kappa+1}\right)$
(B) $\frac{CV}{2}\left(\frac{\kappa+1}{\kappa-1}\right)$
(C) $CV\left(\frac{\kappa-1}{\kappa+1}\right)$
(D) $CV\left(\frac{\kappa+1}{\kappa-1}\right)$

6. A capacitor when filled with a dielectric $\kappa = 3$ has charge Q_0, voltage V_0 and Electric field E_0. If the dielectric is replaced with another one having $\kappa = 9$, the new value of charge, voltage and field will be respectively-
   (A) $3Q_0, 3V_0, 3E_0$
   (B) $Q_0, 3V_0, 3E_0$
   (C) $Q_0, V_0/3, 3E_0$
   (D) $Q_0, V_0/3, E_0/3$

7. Fig.3.207a shows two capacitors connected in series and joined to a battery. The graph in Fig.3.207b shows the variation in potential as one moves from left to right on the branch containing the capacitors if-
   (A) $C_1 > C_2$
   (B) $C_1 = C_2$
   (C) $C_1 < C_2$
   (D) The information is not sufficient to decide the relation between C_1 and C_2

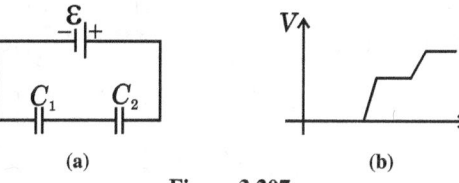

Figure 3.207

8. Four condensers are joined as shown in Fig.3.208, the capacity of each capacitor is 8 μF. The equivalent capacity between points A and B will be -

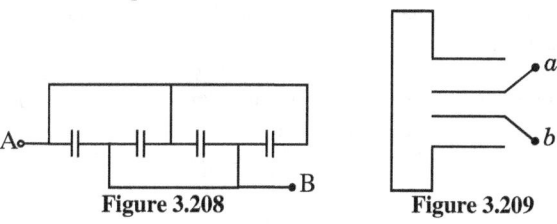

Figure 3.208   **Figure 3.209**

   (A) 32 μF (B) 2 μF (C) 8 μF (D) 16 μF

9. In a parallel plate capacitor, the separation between the plates is 3 mm with air between them. Now, a 1 mm thick layer of a material of dielectric constant 2 is introduced between the plates due to which the capacity increases. In order to bring its capacity of the original value, the separation between the plates must be made-
   (A) 1.5 mm (B) 2.5 mm (C) 3.5 mm (D) 4.5 mm

10. Four metallic plates each with a surface area A of one side and placed at a distance d from each other. the plates are connected as shown in the Fig.3.209. Then the capacitance of the system between a and b is -
    (A) $\frac{3\epsilon_0 A}{d}$
    (B) $\frac{2\epsilon_0 A}{d}$
    (C) $\frac{2\epsilon_0 A}{3d}$
    (D) $\frac{3\epsilon_0 A}{2d}$

11. Two identical parallel plate capacitors are placed in series and connected to a constant voltage source of V_0 volt. If one of the capacitors is completely immersed in a liquid with dielectric constant κ, the potential difference between the plates of the other capacitor will change to-
    (A) $\frac{\kappa+1}{\kappa}V_0$
    (B) $\frac{\kappa}{\kappa+1}V_0$
    (C) $\frac{\kappa+1}{2\kappa}V_0$
    (D) $\frac{2\kappa}{\kappa+1}V_0$

12. A number of capacitors each of capacitance 1 μF and each one of which get punctured if a potential difference just exceeding 500 volt is applied, are provided. Then an arrangement suitable for giving a capacitor of 2 μF across which 3000 volt may be applied requires at least-

(A) 18 component capacitors
(B) 36 component capacitors
(C) 72 component capacitors
(D) 144 component capacitors

13. A capacitor of capacitance 1 μF withstands a maximum voltage of 6 kV, while another capacitor of capacitance 2 μF, the maximum voltage 4 kV. If they are connected in series, the combination can withstand a maximum of-
    (A) 6 kV   (B) 4 kV   (C) 10 kV   (D) 9 kV

14. A capacitor of capacitance C_1 is charged to a potential V_0. The electrostatic energy stored in it is U_0. It is connected to another uncharged capacitor of capacitance C_2 in parallel. The energy dissipated in the process is-
    (A) $\frac{C_2}{C_1+C_2}U_0$
    (B) $\frac{C_1}{C_1+C_2}U_0$
    (C) $\left(\frac{C_1-C_2}{C_1+C_2}\right)^2$
    (D) $\frac{C_1 C_2}{2(C_1+C_2)}U_0$

15. The effective capacitance between A and B of an infinite chain of capacitors joined as shown in Fig.3.210-
    (A) $(\sqrt{5}-1)\frac{C}{2}$
    (B) $(\sqrt{5}+1)\frac{C}{2}$
    (C) $(\sqrt{3}-1)\frac{C}{2}$
    (D) $(\sqrt{5}-1)C$

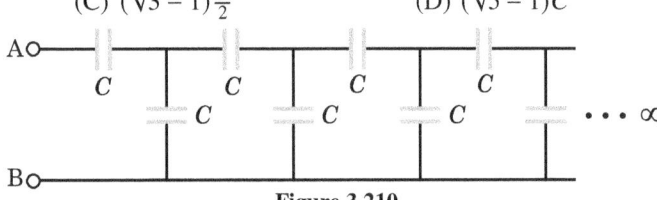

Figure 3.210

16. In the given circuit in Fig.3.211, $C_1 = C$, $C_2 = 2C$, $C_3 = 3C$. If charge at the capacitor C_2 is Q, then the charge at the capacitor C_3 will be -
    (A) $3Q/2$ (B) $9Q/2$ (C) $Q/3$ (D) $Q/6$

17. Consider the circuit shown in the Fig.3.212, the ratio of charge on the capacitors 2 μC to 6 μC will be-
    (A) 1/3   (B) 2/3   (C) 1/2   (D) 4/3

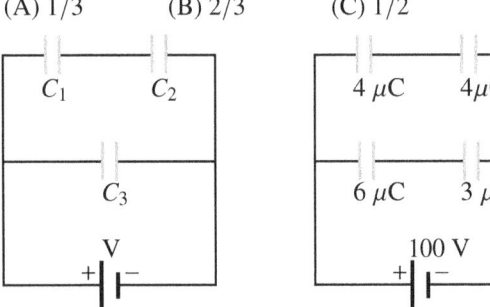

Figure 3.211: Question 17.   Figure 3.212: Question 17.

Passage Type Question:
In its simplest form a capacitor consists of two parallel plates of conducting material separated by an insulator called a dielectric. When the conductors are connected to the terminals of a voltage source, electrons move to one plate, giving a net negative charge. This forces us electrons to leave the opposite plate, giving it a net positive charge. When the capacitor is removed from the power source, charge can not leave the plate. The capacitor stores the electrical energy in the dielectric. Capacitance indicates the amount of charge. A particular capacitor can store per volt of potential difference across its plates.
(A) In terms of the charge, stored capacitance is the ratio of the charge q on either plate to the potential difference between the plates $C = q/V$ capacitance is measured in the unit, "farad = coulomb/ volt".
(B) In terms of geometry of the capacitor, the capacitance is-

$$C = \frac{\epsilon_0 \kappa A}{d}$$

here A is the area of one of the plates and d is the separation between the plates, κ is the unit less dielectric constant of the insulating material between the plates. The total charge on the plates increases by the factor κ.

$$q = \kappa C_0 V$$

where C_0 is the capacitance when the plates are separated by air. For air $\kappa = 1$
For capacitors connected in series, the reciprocal of the total capacitance is equal to the sum of the reciprocals of the all separate capacitances. For capacitors connected in parallel, the total capacitance is the sum of the individual capacitances.

18. A parallel plate capacitor has square plates separated by an air gap 0.012 mm wide. If the length of the sides of the plates are tripled, what is the separation required to keep the capacitance the same?
    (A) 0.0013 mm   (B) 0.004 mm
    (C) 0.036 mm   (D) 0.108 mm

19. A 10 V power source charges a 5 μF capacitor with air as its dielectric. The power source is removed and the air gap is carefully replaced with a material of dielectric constant $\kappa = 5$. What is the final charge on the capacitor?
    (A) 250 μF   (B) 50 μF   (C) 25 μF   (D) 5 μF

20. A parallel plate capacitor has a capacitance of 19.2 μF, when the insulator between the plates is glass, $\kappa = 8.0$. When the glass is replaced with rubber, the capacitance between the plates is 6.0 μF. What is the dielectric constant of the rubber?
    (A) 2.5   (B) 4.7   (C) 7.5   (D) 12.0

Statements Type Question:
Each of the questions given below consist of Statement - I and Statement - II. Use the following Key to choose the appropriate answers.
(A) If both Statement - I and Statement - II are true, and Statement - II is the correct explanation of Statement- I.
(B) If both Statement - I and Statement - II are true but Statement - II is not the correct explanation of Statement $-I$.
(C) If statement - I is true but statement - II is false.
(D) If statement $-I$ is false but statement - II is true.

21. **Statement I :** farad is a very large unit of capacity.
    Statement II: Capacity of Earth– which is the largest sphere, is in microfarad.

22. **Statement I:** Capacity of a parallel plate condenser increases on introducing a conducting or insulating slab between the plates.
    Statement II: In both the cases, electric field intensity between the plates reduces.

23. **Statement I:** When charges are shared between any two bodies, some charge is lost, and some loss of energy does occur.
    Statement II: Some energy disappears in the form of heat, sparking etc.

24. **Statement I:** The whole charge of a body can be transferred to another body.
    Statement II: Charge can not be transferred partially.

25. **Statement I:** In a series combination of capacitors, charge on each capacitor is same.
    Statement II: In such a combination, charge can move only along one route.

3.18.1.4 Level 4 (Previous Years JEE Main & Advanced Questions)
Section - A (JEE Main)

3.18. PROBLEMS

1. If n capacitor connected in series with a cell of emf V volt. The energy of system is - [2002]
   (A) $\frac{1}{2}nCV^2$
   (B) $\frac{1}{2}\frac{CV^2}{n}$
   (C) $\frac{1}{2}CV^2$
   (D) none of above

2. Capacitance (in F) of a spherical conductor with radius 1 m is - [2002]
   (A) 1.1×10^{-10}
   (B) 10^{-6}
   (C) 9×10^{-9}
   (D) 10^{-3}

3. A sheet of aluminium foil of negligible thickness is introduced between the plates of a capacitor. The capacitance of the capacitor - [2003]
   (A) Remains unchanged
   (B) Becomes infinite
   (C) Increases
   (D) Decreases

4. The work done in placing a charge of 8×10^{-18} C on a condenser of capacity 100 microfarad is - [2003]
   (A) 3.1×10^{-26} J
   (B) 4×10^{-10} J
   (C) 32×10^{-32} J
   (D) 16×10^{-32} J

5. A fully charged capacitor has a capacitance 'C'. It is discharged through a small coil of resistance wire embedded in a thermally insulated block of specific heat capacity 's' and mass 'm'. If the temperature of the block is raised by 'ΔT', the potential difference 'V' across the capacitance is - [2005]
   (A) $\sqrt{\frac{2mC\Delta T}{s}}$
   (B) $\frac{mC\Delta T}{s}$
   (C) $\frac{ms\Delta T}{C}$
   (D) $\sqrt{\frac{2ms\Delta T}{C}}$

6. A parallel plate capacitor is made by stacking n equally spaced plates connected alternatively. If the capacitance between any two adjacent plates is 'C' then the resultant capacitance is - [2005]
   (A) $(n-1)C$ (B) $(n+1)C$ (C) C (D) nC

7. A battery is used to charge a parallel plate capacitor till the potential difference between the plates becomes equal to the electromotive force of the battery. The ratio of the energy stored in the capacitor and the work done by the battery will be - [2007]
   (A) 1 (B) 2 (C) 1/4 (D) 1/2

8. A parallel plate condenser with a dielectric of dielectric constant κ between the plates has a capacity C and is charged to a potential V volts. The dielectric slab is slowly removed from between the plates and then reinserted. The net work done by the system in this process is - [2007]
   (A) $1/2(\kappa-1)CV^2$
   (B) $CV^2(\kappa-1)/\kappa$
   (C) $(\kappa-1)CV^2$
   (D) zero

9. A parallel plate capacitor with air between the plates has a capacitance of 9 pF. The separation between its plates is 'd'. The space between the plates is now filled with two dielectrics. One of the dielectrics has dielectric constant $\kappa_1 = 3$ and thickness $d/3$ while the other one has dielectric constant $\kappa_2 = 6$ and thickness $2d/3$. Capacitance of the capacitor is now [2008]
   (A) 45 pF
   (B) 40.5 pF
   (C) 20.25 pF
   (D) 1.8 pF

10. Let C be the capacitance of a capacitor discharging through a resistor R. Suppose t_1 is the time taken for the energy stored in the capacitor to reduce to half its initial value and t_2 is the time taken for the charge to reduce to one-fourth its initial value. Then the ratio t_1/t_2 will be [2010]
    (A) 1 (B) 1/2 (C) 1/4 (D) 2

11. Two circuits (a) and (b) have charged capacitors of capacitance C, $2C$ and $3C$ with open switches. Charges on each of the capacitor are as shown in the Figures 3.213. On closing the switches [2012]
    (A) No charge flows in (a) but charge flows from R to L in (B)
    (B) Charges flow from L to R in both (A) and (B)
    (C) Charges flow from R to L in (A) and from L to R in (B)
    (D) No charge flows in (A) but charge flows from L to R in (B)

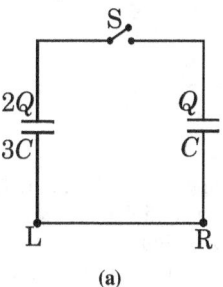
(a)
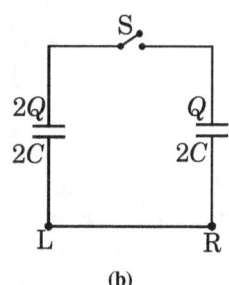
(b)

Figure 3.213

12. A series combination of n_1 capacitors, each of capacity C_1 is charged by source of potential difference $4V$. When another parallel combination of n_2 capacitors each of capacity C_2 is charged by a source of potential difference V, it has the same total energy stored in it as the first combination has. The value of C_2 in terms of C_1 is then [2012]
    (A) $16\frac{n_2}{n_1}C_1$
    (B) $\frac{2C_1}{n_1 n_2}$
    (C) $2\frac{n_2}{n_1}C_1$
    (D) $\frac{16C_1}{n_1 n_2}$

13. **Statement 1:** It is not possible to make a sphere of capacity 1 farad using a conducting material.
    Statement 2: It is possible for earth as its radius is 6.4×10^6 m. [2012]
    (A) Statement 1 is true, Statement 2 is true, Statement 2 is the correct explanation of Statement 1.
    (B) Statement 1 is false, Statement 2 is true.
    (C) Statement 1 is true, Statement 2 is true, Statement 2 is not the correct explanation of Statement 1.
    (D) Statement 1 is true, Statement 2 is false.

14. The capacitor of an oscillatory circuit is enclosed in a container. When the container is evacuated, the resonance frequency of the circuit is 10 kHz. When the container is filled with a gas, the resonance frequency changes by 50 Hz. The dielectric constant of the gas is [2012]
    (A) 1.001 (B) 2.001 (C) 1.01 (D) 3.01

15. The Figure 3.214 shows an experimental plot discharging of a capacitor in an RC circuit. The time constant τ of this circuit lies between : [2012]
    (A) 150 s and 200 s
    (B) 0 s and 50 s
    (C) 50 s and 100 s
    (D) 100 s and 150 s

16. A uniform electric field $\vec{E}$ exists between the plates of a charged condenser. A charged particle enters the space between the plates and perpendicular to $\vec{E}$. The path of the particle between the plates is a: [2013]
    (A) straight line
    (B) hyperbola
    (C) parabola
    (D) circle

17. To establish an instantaneous current of 2 A through a 1 μF capacitor; the potential difference across the capacitor plates should be changed at the rate of : [2013]
    (A) 2×10^4 V/s
    (B) 4×10^6 V/s
    (C) 2×10^6 V/s
    (D) 4×10^4 V/s

18. A parallel plate capacitor having a separation between the plates d, plate area A and material with dielectric constant

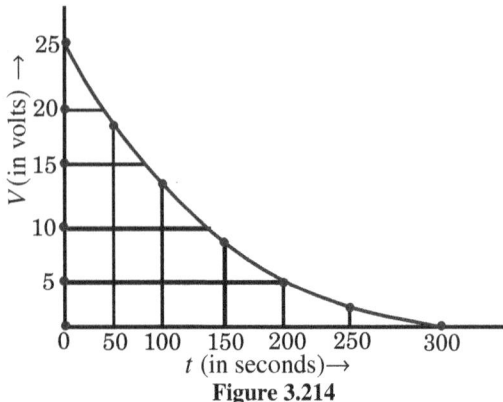

Figure 3.214

κ has capacitance C_0. Now one-third of the material is replaced by another material with dielectric constant 2κ, so that effectively there are two capacitors one with area $A/3$, dielectric constant 2κ and another with area $2A/3$ and dielectric constant κ. If the capacitance of this new capacitor is C then C/C_0 is [2013]
(A) 1 (B) 4/3 (C) 2/3 (D) 1/3

19. Three capacitors, each of 3 μF, are provided. These cannot be combined to provide the resultant capacitance of:
[April 9, 2014]
(A) 1 μF (B) 2 μF (C) 4.5 μF (D) 6 μF

20. A parallel plate capacitor is made of two plates of length l, width w and separated by distance d. A dielectric slab (dielectric constant κ) that fits exactly between the plates is held near the edge of the plates. It is pulled into the capacitor by a force $F = -\partial U/\partial x$ where U is the energy of the capacitor when dielectric is inside the capacitor up to distance x (Fig.3.215). If the charge on the capacitor is Q then the force on the dielectric when it is near the edge is:
[April 11, 2014]
(A) $\dfrac{Q^2 d}{2\omega l^2 \epsilon_o}\kappa$ (B) $\dfrac{Q^2 \omega}{2 d l^2 \epsilon_0}(\kappa - 1)$
(C) $\dfrac{Q^2 d}{2wl^2 \epsilon_o}(\kappa - 1)$ (D) $\dfrac{Q^2 w}{2 d l^2 \epsilon_0}\kappa$

21. The space between the plates of a parallel plate capacitor is filled with a 'dielectric' whose 'dielectric constant' varies with distance as per the relation:
$\kappa(x) = \kappa_o + \lambda x$ (λ = a constant)
The capacitance C, of the capacitor, would be related to its vacuum capacitance C_0 for the relation: [April 12, 2014]
(A) $C = \dfrac{\lambda d}{\ln(1 + \kappa_0 \lambda d)}C_o$ (B) $C = \dfrac{\lambda}{d \cdot \ln(1 + \kappa_0 \lambda d)}C_o$
(C) $C = \dfrac{\lambda d}{\ln(1 + \lambda d/\kappa_o)}C_o$ (D) $C = \dfrac{\lambda}{d \cdot \ln(1 + \kappa_o/\lambda d)}C_o$

22. The gap between the plates of a parallel plate capacitor of area A and distance between plates d, is filled with a dielectric whose permittivity varies linearly from ϵ_1 at one plate to ϵ_2 at the other. The capacitance of capacitor is:
[April 19,2014]
(A) $\dfrac{\epsilon_0(\epsilon_1 + \epsilon_2) A}{d}$ (B) $\dfrac{\epsilon_0(\epsilon_2 + \epsilon_1) A}{2d}$
(C) $\dfrac{\epsilon_0 A}{[d \ln(\epsilon_2/\epsilon_1)]}$ (D) $\dfrac{\epsilon_0(\epsilon_2 - \epsilon_1) A}{[d \ln(\epsilon_2/\epsilon_1)]}$

23. A parallel plate capacitor is made of two circular plates separated by a distance 5 mm and with a dielectric of dielectric constant 2.2 between them. When the electric field in the dielectric is 3×10^4 V/m the charge density of the positive plate will be close to: [2014]

(A) 6×10^{-7} C/m^2 (B) 3×10^{-7} C/m^2
(C) 3×10^4 C/m^2 (D) 6×10^4 C/m^2

24. In Figure 3.216 a system of four capacitors connected across a 10 V battery is shown. Charge that will flow from switch S when it is closed is: [April 11, 2015]
(A) 5 μC from b to a (B) 20 μC from a to b
(C) zero (D) 5 μC from a to b

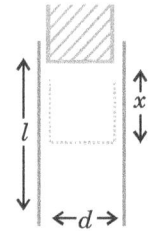
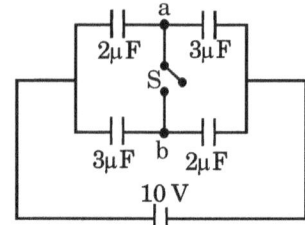

Figure 3.215: Question 20.   Figure 3.216: Question 24.

25. In the given circuit (Fig.3.217), charge Q_2 on the 2 μF capacitor changes as C is varied from 1 μF to 3 μF. Q_2 as a function of 'C' is given properly by which of the sub figures of Fig.3.218: (Sub figures of Fig.3.218, are drawn schematically and are not to scale) [2015]

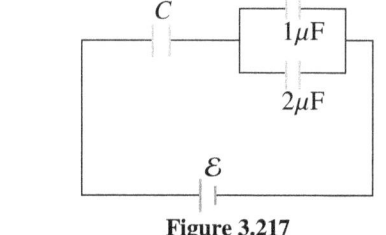

Figure 3.217

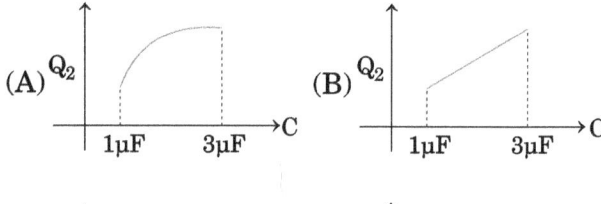
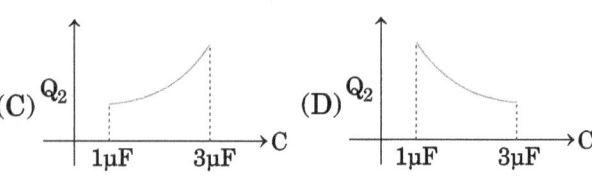

Figure 3.218

26. Three capacitors each of 4 μF are to be connected in such a way that the effective capacitance is 6 μF. This can be done by connecting them: [April 9, 2016]
(A) all in series
(B) all in parallel
(C) two in parallel and one in series
(D) two in series and one in parallel

27. Figure 3.219 shows a network of capacitors where the numbers indicates capacitances in micro farad. The value of capacitance C if the equivalent capacitance between point A and B is to be 1 μF is: [April 10, 2016]
(A) $\frac{32}{23}$ μF (B) $\frac{31}{23}$ μF (C) $\frac{33}{23}$ μF (D) $\frac{34}{23}$ μF

28. A combination of capacitors is set up as shown in the Figure 3.220. The magnitude of the electric field, due to a point charge Q (having a charge equal to the sum of the charges on the 4 μF and 9 μF capacitors), at a point distance 30 m

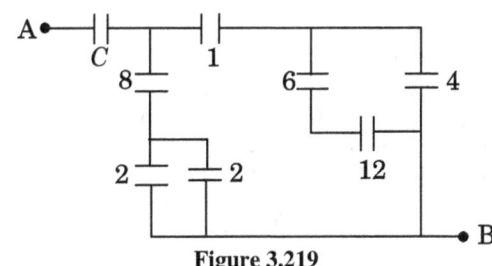

Figure 3.219

from it, would equal: [2016]
(A) 420 N/C (B) 480 N/C
(C) 240 N/C (D) 360 N/C

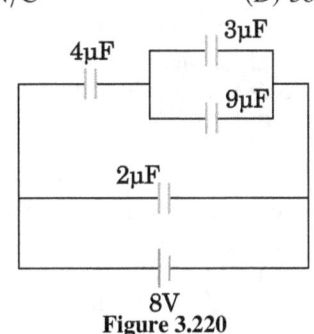

Figure 3.220

29. The energy stored in the electric field produced by a metal sphere is 4.5 J. If the sphere contains 4 μC charge, its radius will be : [Take : $\frac{1}{4\pi\epsilon_0} = 9 \times 10^9$ Nm^2/C^2] [April 8, 2017]
    (A) 20 mm (B) 32 mm (C) 28 mm (D) 16 mm

30. A combination of parallel plate capacitors is maintained at a certain potential difference [Fig.3.221]. When a 3 mm thick slab is introduced between all the plates, in order to maintain the same potential difference, the distance between the plates is increased by 2.4 mm. Find the dielectric constant of the slab. [April 9, 2017]
    (A) 3 (B) 4 (C) 5 (D) 6

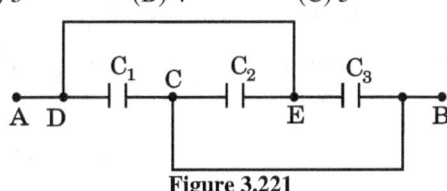

Figure 3.221

31. A capacitance of 2 μF is required in an electrical circuit across a potential difference of 1.0 kV. A large number of 1 μF capacitors are available which can withstand a potential difference of not more than 300 V. The minimum number of capacitors required to achieve this is [2017]
    (A) 24 (B) 32 (C) 2 (D) 16

32. A capacitor C_1 is charged up to a voltage $V = 60$ V by connecting it to battery B through switch (1) as shown in Fig.3.222. Now C_1 is disconnected from battery and connected to a circuit consisting of two uncharged capacitors $C_2 = 3.0$ μF and $C_3 = 6.0$ μF through a switch (2) as shown in the Fig.3.222. The sum of final charges on C_2 and C_3 is: [April 15, 2018]
    (A) 36 μC (B) 20 μC (C) 54 μC (D) 40 μC

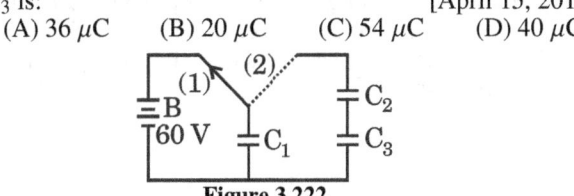

Figure 3.222

33. A parallel plate capacitor with area 200 cm^2 and separation between the plates 1.5 cm, is connected across a battery of emf V. If the force of attraction between the plates is 25×10^{-6} N, the value of V is approximately $\left(\epsilon_0 = 8.85 \times 10^{-12} \frac{C^2}{N \cdot m^2}\right)$. [April 15, 2018]
    (A) 150 V (B) 100 V (C) 250 V (D) 300 V

34. The equivalent capacitance between A and B in the circuit given in Fig.3.223 is: [April 15, 2018]
    (A) 4.9 μF (B) 3.6 μF (C) 5.4 μF (D) 2.4 μF

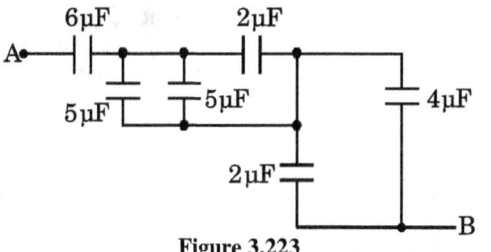

Figure 3.223

35. A parallel plate capacitor of capacitance 90 pF is connected to a battery of emf 20 V. If a dielectric material of dielectric constant $\kappa = 5/3$ is inserted between the plates, the magnitude of the induced charge will be: [April 16, 2018]
    (A) 1.2 nC (B) 0.3 nC (C) 2.4 nC (D) 0.9 nC

36. A parallel plate capacitor with square plates is filled with four dielectrics of dielectric constants $\kappa_1, \kappa_2, \kappa_3, \kappa_4$ arranged as shown in the Figure 3.224. The effective dielectric constant κ will be: [9 Jan. 2019]

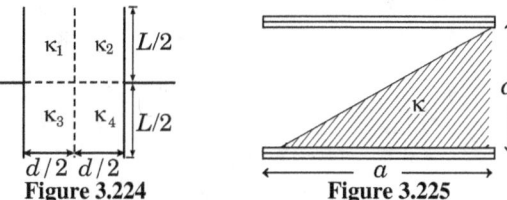

Figure 3.224   **Figure 3.225**

(A) $\kappa = \frac{(\kappa_1 + \kappa_3)(\kappa_2 + \kappa_4)}{\kappa_1 + \kappa_2 + \kappa_3 + \kappa_4}$

(B) $\kappa = \frac{(\kappa_1 + \kappa_2)(\kappa_3 + \kappa_4)}{2(\kappa_1 + \kappa_2 + \kappa_3 + \kappa_4)}$

(C) $\kappa = \frac{(\kappa_1 + \kappa_2)(\kappa_3 + \kappa_4)}{\kappa_1 + \kappa_2 + \kappa_3 + \kappa_4}$

(D) $\kappa = \frac{\kappa_1\kappa_2(\kappa_3 + \kappa_4) + \kappa_3\kappa_4(\kappa_1 + \kappa_2)}{(\kappa_1 + \kappa_2)(\kappa_3 + \kappa_4)}$

37. A parallel plate capacitor is made of two square plates of side 'a', separated by a distance $d(d << a)$. The lower triangular portion is filled with a dielectric of dielectric constant κ, as shown in the Fig.3.225. Capacitance of this capacitor is: [9 Jan. 2019]

(A) $\frac{\kappa\epsilon_0 a^2}{2d(\kappa + 1)}$ (B) $\frac{\kappa\epsilon_0 a^2}{d(\kappa - 1)} \ln \kappa$

(C) $\frac{\kappa\epsilon_0 a^2}{d} \ln \kappa$ (D) $\frac{1}{2}\frac{\kappa\epsilon_0 a^2}{d}$

38. A parallel plate capacitor is of area 6 cm^2 and a separation 3 mm. The gap is filled with three dielectric materials of equal thickness (Fig.3.226) with dielectric constants $\kappa_1 = 10, \kappa_2 = 12$ and $\kappa_3 = 14$. The dielectric constant of a material which when fully inserted in above capacitor, gives same capacitance would be: [10 Jan. 2019]
    (A) 4 (B) 14 (C) 12 (D) 36

39. A parallel plate capacitor having capacitance 12 pF is charged by a battery to a potential difference of 10 V between its plates. The charging battery is now disconnected and a porcelain slab of dielectric constant 6.5 is slipped

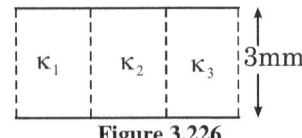

Figure 3.226

between the plates. The work done by the capacitor on the slab is: [10 Jan. 2019]

(A) 692 pJ (B) 508 pJ (C) 560 pJ (D) 600 pJ

40. Seven capacitors, each of capacitance 2 μF, are to be connected in a configuration to obtain an effective capacitance of (6/13) μF. Which of the combinations, shown in Fig.3.227, will achieve the desired value? [Jan. 2019]

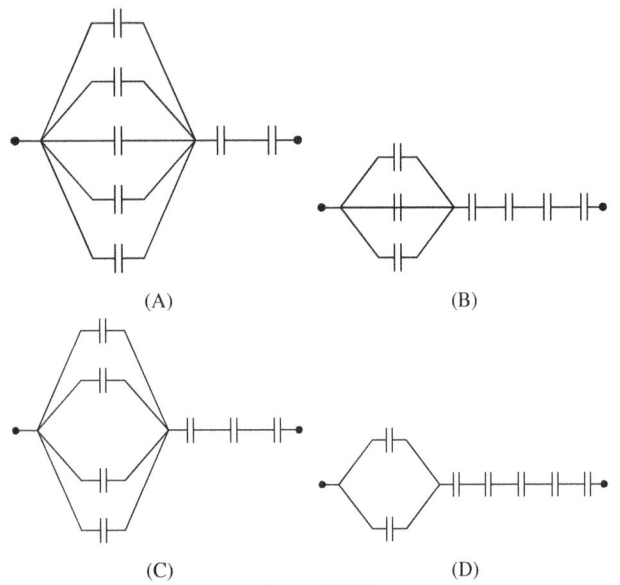

Figure 3.227

41. In the Fig.3.228, the charge on the left plate of the 10 μF capacitor is -30 μC. The charge on the right plate of the 6 μF capacitor is: [11 Jan. 2019]

(A) -12 μC (B) $+12$ μC (C) -18 μC (D) $+18$ μC

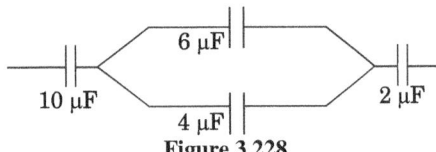

Figure 3.228

42. In the circuit shown in Fig.3.229, find C if the effective capacitance of the whole circuit is to be 0.5 μF. All values in the circuit are in μF. [12 Jan. 2019]

(A) $7/11$ μF (B) $6/5$ μF (C) 4 μF (D) $7/10$ μF

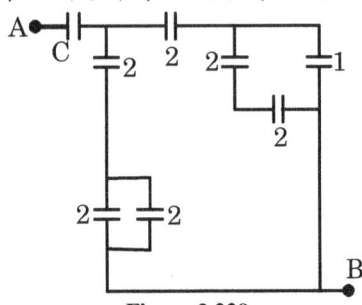

Figure 3.229

43. A parallel plate capacitor with plates of area 1 m^2 each, are at a separation of 0.1 m. If the electric field between the plates is 100 N/C, the magnitude of charge on each plate is:

(Take $\epsilon_0 = 8.85 \times 10^{-12} \frac{C^2}{N.m^2}$) [12 Jan. 2019]

(A) 7.85×10^{-10} C (B) 6.85×10^{-10} C
(C) 8.85×10^{-10} C (D) 9.85×10^{-10} C

44. In Figure 3.230, after the switch 'S' is turned from position 'A' to position 'B', the energy dissipated in the circuit in terms of capacitance 'C' and total charge 'Q' is: [Jan. 2019]

(A) $Q^2/8C$ (B) $3Q^2/8C$ (C) $5Q^2/8C$ (D) $3Q^2/4C$

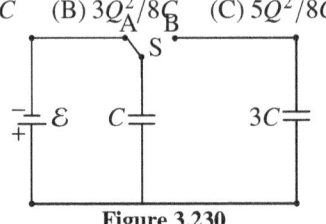

Figure 3.230

45. A capacitor with capacitance 5 μF is charged to 5 μC. If the plates are pulled apart to reduce the capacitance to 2 μF, how much work is done? [9 April 2019]

(A) 6.25×10^{-6} J (B) 3.75×10^{-6} J
(C) 2.16×10^{-6} J (D) 2.55×10^{-6} J

46. Voltage rating of a parallel plate capacitor is 500 V. Its dielectric can withstand a maximum electric field of 10^6 V/m. The plate area is 10^{-4} m^2. What is the dielectric constant if the capacitance is 15 pF? (given $\epsilon_0 = 8.86 \times 10^{-12}$ C^2/Nm^2) [8 April 2019]

(A) 3.8 (B) 8.5 (C) 4.5 (D) 6.2

47. A parallel plate capacitor has 1 μF capacitance. One of its two plates is given $+2$ μC charge and the other plate, $+4$ μC charge. The potential difference developed across the capacitor is: [8 April 2019]

(A) 3 V (B) 1 V (C) 5 V (D) 2 V

48. Figure 3.231 shows charge (q) versus voltage (V) graph for series and parallel combination of two given capacitors. The capacitances are: [10 April 2019]

(A) 40 μF and 10 μF (B) 60 μF and 40 μF
(C) 50 μF and 30 μF (D) 20 μF and 30 μF

49. In the given circuit in Fig.3.232, the charge on 4 μF capacitor will be: [12 April 2019]

(A) 5.4 μC (B) 9.6 μC (C) 13.4 μC (D) 24 μC

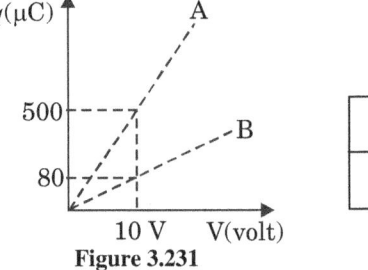

Figure 3.231   Figure 3.232

50. Two identical parallel plate capacitors, of capacitance C each, have plates of area A, separated by a distance d. The space between the plates of the two capacitors, is filled with three dielectrics, of equal thickness and dielectric constants κ_1, κ_2 and κ_3. The first capacitors is filled as shown in Fig. 3.233a, and the second one is filled as shown in Fig. 3.233b. If these two modified capacitors are charged by the same potential V, the ratio of the energy stored in the two, would be (E_1 refers to capacitors shown in Fig.3.233a and E_2 to capacitors in Fig.3.233b: [12 April 2019]

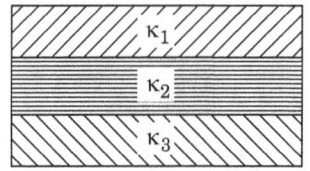
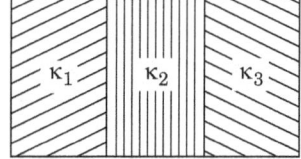

(a)   **(b)**

Figure 3.233

(A) $\dfrac{E_1}{E_2} = \dfrac{\kappa_1\kappa_2\kappa_3}{(\kappa_1+\kappa_2+\kappa_3)(\kappa_2\kappa_3+\kappa_3\kappa_1+\kappa_1\kappa_2)}$

(B) $\dfrac{E_1}{E_2} = \dfrac{(\kappa_1+\kappa_2+\kappa_3)(\kappa_2\kappa_3+\kappa_3\kappa_1+\kappa_1\kappa_2)}{\kappa_1\kappa_2\kappa_3}$

(C) $\dfrac{E_1}{E_2} = \dfrac{9\kappa_1\kappa_2\kappa_3}{(\kappa_1+\kappa_2+\kappa_3)(\kappa_2\kappa_3+\kappa_3\kappa_1+\kappa_1\kappa_2)}$

(D) $\dfrac{E_1}{E_2} = \dfrac{(\kappa_1+\kappa_2+\kappa_3)(\kappa_2\kappa_3+\kappa_3\kappa_1+\kappa_1\kappa_2)}{9\kappa_1\kappa_2\kappa_3}$

51. The parallel combination of two air filled parallel plate capacitors of capacitance C and nC is connected to a battery of voltage, V. When the capacitors are fully charged, the battery is removed and after that a dielectric material of dielectric constant κ is placed between the two plates of the first capacitor. The new potential difference of the combined system is: [9 April 2020]

(A) $\dfrac{nV}{\kappa+n}$ (B) V

(C) $\dfrac{V}{\kappa+n}$ (D) $\dfrac{(n+1)V}{(\kappa+n)}$

52. A 60 pF capacitor is fully charged by a 20 V supply. It is then disconnected from the supply and is connected to another uncharged 60 pF capacitor in parallel. The electrostatic energy that is lost in this process by the time the charge is redistributed between them is (in nJ) [7 Jan. 2020]

53. In Fig.3.234, a parallel plate capacitor has plates of area A separated by distance 'd' between them. It is filled with a dielectric which has a dielectric constant that varies as $\kappa(x) = K(1+\alpha x)$ where 'x' is the distance measured from one of the plates. If $(\alpha d) \ll 1$, the total capacitance of the system is best given by the expression: [7 Jan. 2020]

(A) $\dfrac{AK\epsilon_0}{d}\left(1+\dfrac{\alpha d}{2}\right)$ (B) $\dfrac{A\epsilon_0 K}{d}\left(1+\left(\dfrac{\alpha d}{2}\right)^2\right)$

(C) $\dfrac{A\epsilon_0 K}{d}\left(1+\dfrac{\alpha^2 d^2}{2}\right)$ (D) $\dfrac{AK\epsilon_0}{d}(1+\alpha d)$

54. A capacitor is made of two square plates each of side 'a' making a very small angle θ between them, as shown in Figure 3.235. The capacitance will be close to: [8 Jan. 2020]

(A) $\dfrac{\epsilon_0 a^2}{d}\left(1-\dfrac{\theta a}{2d}\right)$ (B) $\dfrac{\epsilon_0 a^2}{d}\left(1-\dfrac{\theta a}{4d}\right)$

(C) $\dfrac{\epsilon_0 a^2}{d}\left(1+\dfrac{\theta a}{d}\right)$ (D) $\dfrac{\epsilon_0 a^2}{d}\left(1-\dfrac{3\theta a}{2d}\right)$

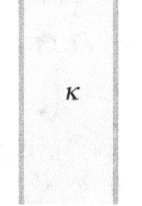
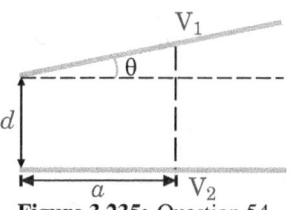

Figure 3.234: Question 53.   **Figure 3.235:** Question 54

55. A 10 μF capacitor is fully charged to a potential difference of 50 V. After removing the source voltage it is connected to an uncharged capacitor in parallel. Now the potential difference across them becomes 20 V. The capacitance of the second capacitor is: [02 Sep. 2020]

(A) 15 μF   (B) 30 μF   (C) 20 μF   (D) 10 μF

56. Effective capacitance of parallel combination of two capacitors C_1 and C_2 is 10 μF. When these capacitors are individually connected to a voltage source of 1 V, the energy stored in the capacitor C_2 is 4 times that of C_1. If these capacitors are connected in series, their effective capacitance will be: [8 Jan. 2020]

(A) 4.2 μF   (B) 3.2 μF   (C) 1.6 μF   (D) 8.4 μF

57. In the circuit shown in the Fig.3.236, the total charge is 750 μC and the voltage across capacitor C_2 is 20 V. Then the charge on capacitor C_2 is : [03 Sep. 2020]

(A) 450 μC   (B) 590 μC
(C) 160 μC   (D) 650 μC

58. A 5 μF capacitor is charged fully by a 220 V supply. It is then disconnected from the supply and is connected in series to another uncharged 2.5 μF capacitor. If the energy change during the charge redistribution is $\dfrac{X}{100}$ J then value of X to the nearest integer is [02 Sep. 2020]

59. A capacitor of capacitance C, is fully charged with voltage V_0. After disconnecting the voltage source, it is connected in parallel with another uncharged capacitor of capacitance $C/2$. The energy loss in the process after the charge is distributed between the two capacitors is: [04 Sep. 2020]

(A) $\dfrac{1}{2}CV_0^2$ (B) $\dfrac{1}{3}CV_0^2$

(C) $\dfrac{1}{4}CV_0^2$ (D) $\dfrac{1}{6}CV_0^2$

60. In the circuit shown in Fig.3.237, the charge on the 5 μF capacitor is: [05 Sep. 2020]

(A) 18.00 μC   (B) 10.90 μC
(C) 16.36 μC   (D) 5.45 μC

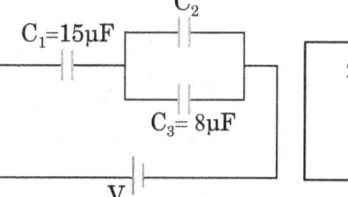
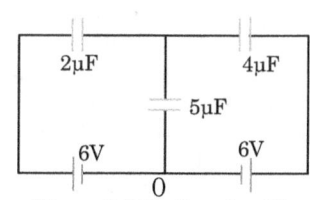

Figure 3.236: Question 57.   **Figure 3.237:** Question 59.

61. Two capacitors of capacitances C and $2C$ are charged to potential differences V and $2V$, respectively. These are then connected in parallel in such a manner that the positive terminal of one is connected to the negative terminal of the other. The final energy of this configuration is : [05 Sep.2020]

(A) $\dfrac{25}{6}CV^2$ (B) $\dfrac{3}{2}CV^2$

(C) zero   (D) $\dfrac{9}{2}CV^2$

62. Consider the combination of two capacitors C_1 and C_2, with $C_2 > C_1$, when connected in parallel, the equivalent capacitance is 15/4 time the equivalent capacitance of the same connected in series. Calculate the ratio of capacitors C_2/C_1. [26 Feb 2021]

(A) 15/11   (B) 111/80
(C) 29/15   (D) imaginary

63. If C and V represent capacity and voltage respectively, then what are the dimensions of λ, where $C/V = \lambda$? [26 Feb 2021]

(A) $[M^{-2}L^{-3}I^2T^6]$ (B) $[M^{-3}L^{-4}I^3T^7]$
(C) $[M^{-1}L^{-3}I^{-2}T^{-7}]$ (D) $[M^{-2}L^{-4}I^3T^7]$

64. For changing the capacitance of a given parallel plate capacitor, a dielectric material of dielectric constant κ is used, which has the same area as the plates of the capacitor. The thickness of the dielectric slab is $3d/4$, where d is the separation between the plates of parallel plate capacitor. The new capacitance (C') in terms of original capaci-

tance (C_0) is given by the following relation [16 March 2021]

(A) $C' = \dfrac{3+\kappa}{4\kappa} C_0$ (B) $C' = \dfrac{4+\kappa}{3} C_0$

(C) $C' = \dfrac{4\kappa}{\kappa+3} C_0$ (D) $C' = \dfrac{4}{3+\kappa} C_0$

65. In a parallel plate capacitor set up, the plate area of capacitor is 2 m² and the plates are separated by 1 m. If the space between the plates is filled with a dielectric material (κ = 3.2) of thickness 0.5 m and area 2 m² (see Fig.3.238) the capacitance of the set-up, to nearest integer, will be ____ ϵ_0. [24 March 2021]

66. Four identical rectangular plates with length, $l = 2$ cm and breadth, $b = 3/2$ cm are arranged as shown in Fig.3.239. The equivalent capacitance between A and C is $x\epsilon_0/d$. The value of x is ____. (Round off to the nearest integer) [24 March 2021]

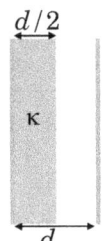
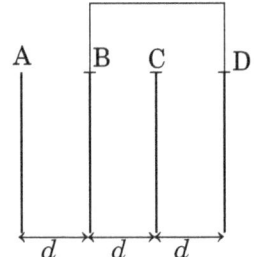

Figure 3.238: Question 65   **Figure 3.239:** Question 66.

67. A parallel plate capacitor whose capacitance C is 14 pF is charged by a battery to a potential difference $V = 12$ V between its plates. The charging battery is now disconnected and a porcelain plate with $\kappa = 7$ is inserted between the plates, then the plate would oscillate back and forth between the plates with a constant mechanical energy of ____ pJ (assume no friction). [24 March 2021]

68. A 2 μF capacitor C_1 is first charged to a potential difference of 10 V using a battery. Then, the battery is removed and the capacitor is connected to an uncharged capacitor C_2 of 8 μF (Fig. 3.240). The charge in C_2 on equilibrium condition is ____ μC. (Round off to the nearest integer) [18 March 2021]

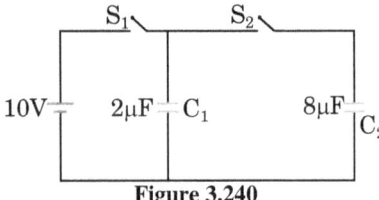

Figure 3.240

69. A parallel plate capacitor has plate area 100 m² and plate separation of 10 m. The space between the plates is filled up to a thickness 5 m with a material of dielectric constant of 10. The resultant capacitance of the system is x pF. The value of $\epsilon_0 = 8.85 \times 10^{-12}$ Fm^{-1}. The value of x to the nearest integer is ____. [18 March 2021]

70. In the reported Fig.3.241, a capacitor is formed by placing a compound dielectric between the plates of parallel plate capacitor. The expression for the capacity of the said capacitor will be (Take, area of plate = A) [July 2021]

(A) $\dfrac{15\,\kappa\epsilon_0 A}{34\,d}$ (B) $\dfrac{15\,\kappa\epsilon_0 A}{6\,d}$

(C) $\dfrac{25\,\kappa\epsilon_0 A}{6\,d}$ (D) $\dfrac{9\,\kappa\epsilon_0 A}{6\,d}$

71. A simple pendulum of mass m, length l and charge $+q$ suspended in the electric field produced by two conducting parallel plates as shown in Fig.3.242. The value of deflection

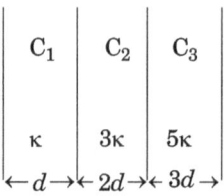

Figure 3.241

of pendulum in equilibrium position will be [July 2021]

(A) $\tan^{-1}\left[\dfrac{q}{mg} \times \dfrac{C_1(V_2-V_1)}{(C_1+C_2)(d-t)}\right]$

(B) $\tan^{-1}\left[\dfrac{q}{mg} \times \dfrac{C_2(V_2-V_1)}{(C_1+C_2)(d-t)}\right]$

(C) $\tan^{-1}\left[\dfrac{q}{mg} \times \dfrac{C_2(V_1+V_2)}{(C_1+C_2)(d-t)}\right]$

(D) $\tan^{-1}\left[\dfrac{q}{mg} \times \dfrac{C_1(V_1+V_2)}{(C_1+C_2)(d-t)}\right]$

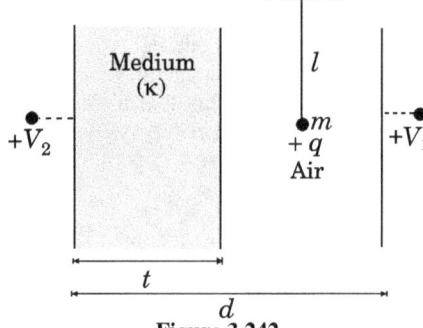

Figure 3.242

72. The material filled between the plates of a parallel plate capacitor has resistivity 200 Ωm. The value of capacitance of the capacitor is 2 pF. If a potential difference of 40 V is applied across the plates of the capacitor, then the value of leakage current flowing out of the capacitor is (Given, the value of relative permittivity of material is 50.) [26 Aug. 2021]

(A) 9.0 μA   (B) 9.0 mA   (C) 0.9 mA   (D) 0.9 μA

73. In Fig.3.243, a parallel-plate capacitor with plate area A has separation d between the plates. Two dielectric slabs of dielectric constant κ_1 and κ_2 of same area $A/2$ and thickness $d/2$ are inserted in the space between the plates. The capacitance of the capacitor will be given by: [26 August 2021]

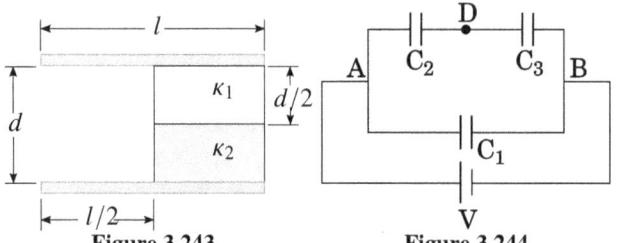

Figure 3.243   **Figure 3.244**

(A) $\dfrac{\epsilon_0 A}{d}\left(\dfrac{1}{2} + \dfrac{K_1 K_2}{K_1+K_2}\right)$ (B) $\dfrac{\epsilon_0 A}{d}\left(\dfrac{1}{2} + \dfrac{K_1 K_2}{2(K_1+K_2)}\right)$

(C) $\dfrac{\epsilon_0 A}{d}\left(\dfrac{1}{2} + \dfrac{K_1+K_2}{K_1 K_2}\right)$ (D) $\dfrac{\epsilon_0 A}{d}\left(\dfrac{1}{2} + \dfrac{2(K_1+K_2)}{K_1 K_2}\right)$

74. Three capacitors $C_1 = 2$ μF, $C_2 = 6$ μF and $C_3 = 12$ μF are connected as shown in Figure 3.244. The ratio of charges on capacitors C_1, C_2 and C_3, respectively- [7 Aug. 2021]

(A) 2 : 1 : 1   (B) 2 : 3 : 3
(C) 1 : 2 : 2   (D) 3 : 4 : 4

75. A parallel plate capacitor is formed by two plates each of area 30π cm² separated by 1 mm. A material of dielectric strength 3.6×10^7 Vm^{-1} is filled between the plates. If the maximum charge that can be stored on the

3.18. PROBLEMS

capacitor without causing any dielectric breakdown is 7×10^{-6} C, the value of dielectric constant of the material is (Use: $1/4\pi\epsilon_0 = 9 \times 10^9$ Nm^{2}C^{-2}): [24 June 22 Shift I]
(A) 1.66   (B) 1.75   (C) 2.25   (D) 2.33

76. Charge on capacitor is increased by 2 C and energy stored becomes 44%. Find initial charge [24 June 22 Shift II]
(A) 20 C   (B) 9 C   (C) 10 C   (D) 11 C

77. The equivalent capacitance between points A and B in shown Fig.3.245, will be ____ μF. [25 June 22 Shift I]

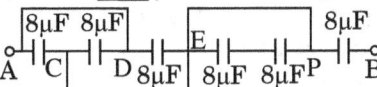

Figure 3.245

78. Two capacitors having capacitance C_1 and C_2 respectively are connected as shown in Fig.3.246. Initially, capacitor C_1 is charged to a potential difference V volt by a battery. The battery is then removed and the charged capacitor C_1 is now connected to uncharged capacitor C_2 by closing the switch S. The amount of charge on the capacitor C_2, after equilibrium is: [26 June 22 Shift I]

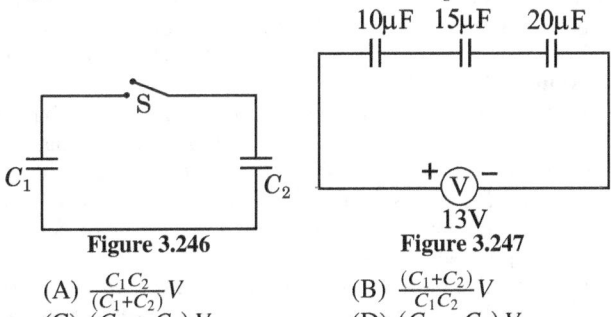

Figure 3.246   **Figure 3.247**

(A) $\frac{C_1 C_2}{(C_1+C_2)} V$ (B) $\frac{(C_1+C_2)}{C_1 C_2} V$
(C) $(C_1 + C_2) V$ (D) $(C_1 - C_2) V$

79. The charge on capacitor of capacitance 15 μF in the Fig.3.247 given below is: [26 June 22 Shift II]
(A) 60 μC   (B) 130 μC   (C) 260 μC   (D) 585 μC

80. A parallel plate capacitor with plate area A and plate separation $d = 2$ m has a capacitance of 4 μF. The new capacitance of the system if half of the space between them is filled with a dielectric material of dielectric constant $\kappa = 3$ (as shown in Fig.3.248) will be: [26 June 22 Shift II]
(A) 2 μF   (B) 32 μF   (C) 6 μF   (D) 8 μF

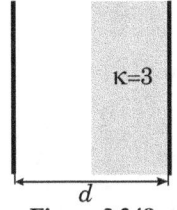

Figure 3.248

81. A parallel plate capacitor is made up of stair like structure with a plate area A of each stair and that is connected with a wire of length b, as shown in the Fig.3.249. The capacitance of the arrangement is $\frac{x}{15}\frac{\epsilon_0 A}{b}$. The value of x is ____.
[27 June 22 Shift II]

82. A capacitor C_1 of capacitance 5 μF is charged to a potential of 30 V using a battery. The battery is then removed and the charged capacitor is connected to an uncharged capacitor C_2 of capacitance 10 μF as shown in Fig.3.250. When the switch is closed charge flows between the capacitors. At equilibrium, the charge on the capacitor C_2 is ____ μC.
[28 June 22 Shift II]

83. A parallel plate capacitor with air between the plate has

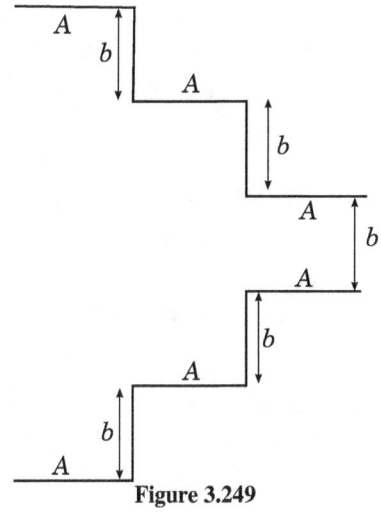

Figure 3.249

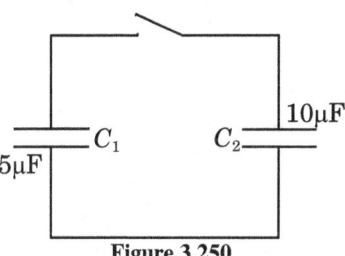

Figure 3.250

a capacitance of 15 pF. The separation between the plate becomes twice and the space between them is filled with a medium of dielectric constant 3.5. Then the capacitance becomes $\frac{x}{4}$ pF. The value of x is ____.
[24 Jan 2023 Shift II]

84. A parallel plate capacitor has plate area 40 cm^2 and plates separation 2 mm. The space between the plates is filled with a dielectric medium of a thickness 1 mm and dielectric constant 5. The capacitance of the system is:
[25 Jan 2023 Shift I]
(A) $24\epsilon_0$ F   (B) $\frac{3}{10}\epsilon_0$ F   (C) $\frac{10}{3}\epsilon_0$ F   (D) $10\epsilon_0$ F

85. A capacitor has capacitance 5 μF when it's parallel plates are separated by air medium of thickness d. A slab of material of dielectric constant 1.5 having area equal to that of plates but thickness $d/2$ is inserted between the plates. Capacitance of the capacitor in the presence of slab will be ____ μF.
[25 Jan 2023 Shift II]

86. A capacitor of capacitance 900 μF is charged by a 100 V battery. The capacitor is disconnected from the battery and connected to another uncharged identical capacitor such that one plate of uncharged capacitor connected to positive plate and another plate of uncharged capacitor connected to negative plate of the charged capacitor. The loss of energy in this process is measured as $x \times 10^{-2}$ J. The value of x is
[30 Jan 2023 Shift I]

87. Two parallel plate capacitors C_1 and C_2 each having capacitance of 10 μF are individually charged by a 100 V D.C. source. Capacitor C_1 is kept connected to the source and a dielectric slab is inserted between it plates. Capacitor C_2 is disconnected from the source and then a dielectric slab is inserted in it. Afterwards the capacitor C_1 is also disconnected from the source and the two capacitors are finally connected in parallel combination. The common potential of the combination will be V. (Assuming Dielectric constant = 10) [31 Jan 2023 Shift II]

88. A parallel plate capacitor with plate area A and plate separa-

tion d is filled with a dielectric material of dielectric constant $\kappa = 4$ as shown in Fig.3.251. The thickness of the dielectric material is x, where $x < d$. Let C_1 and C_2 be the capacitance of the system for $x = d/3$ and $x = 2d/3$, respectively. If $C_1 = 2\,\mu\text{F}$ the value of C_2 is _____ μF.
[06 April 23 Shift I]

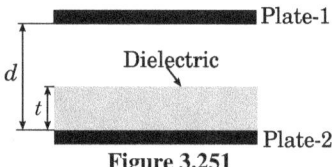

Figure 3.251

89. As shown in the Fig.3.252, two parallel plate capacitors having equal plate area of 200 cm^2 are joined in such a way that $a \ne b$. The equivalent capacitance of the combination is $x\epsilon_0$ F. The value of x is _____. [06 April 23 Shift II]

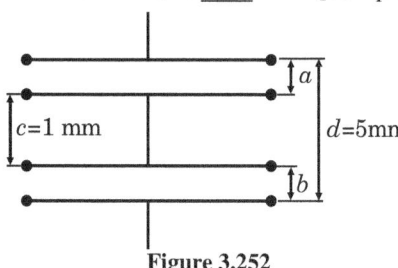

Figure 3.252

90. A 600 pF capacitor is charged by 200 V supply. It is then disconnected from the supply and is connected to another uncharged 600 pF capacitor as shown in Fig.3.253. Electrostatic energy lost in the process is _____ μJ.
[08 April 23 Shift II]

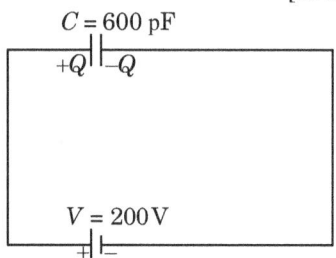

Figure 3.253

91. The equivalent capacitance between terminals A and B in the combination shown in Figure 3.254 is- shown is [10 April 23 Shift I]
(A) $C/2$ (B) $4C$ (C) $2C$ (D) $5C/3$

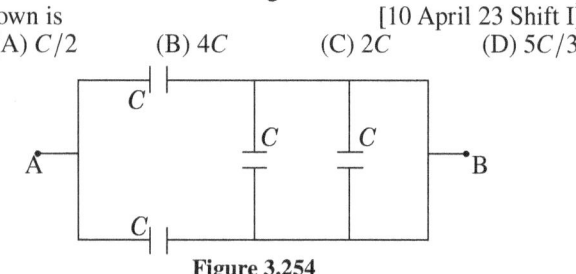

Figure 3.254

92. The distance between two plates of a capacitor is d and its capacitance is C_1, when air is the medium between the plates. If a metal sheet of thickness $2d/3$ and of same area as plate is introduced between the plates, the capacitance of the capacitor becomes C_2. The ratio C_2/C_1 is:
[10 April 23 Shift II]
(A) 2 : 1 (B) 4 : 1 (C) 3 : 1 (D) 1 : 1

93. A parallel plate capacitor of capacitance 2 F is charged to a potential V. The energy stored in the capacitor is E_1. The capacitor is now connected to another uncharged identical capacitor in parallel combination. The energy stored in the combination is E_2. The ratio E_2/E_1 is: [11 April 23 Shift I]
(A) 2 : 1 (B) 1 : 2 (C) 1 : 4 (D) 2 : 3

94. A capacitor of capacitance C is charged to a potential V. The flux of the electric field through a closed surface enclosing the positive plate of the capacitor is:
[11 April 23 Shift II]
(A) $\frac{CV}{2\epsilon_0}$ (B) $\frac{2CV}{\epsilon_0}$ (C) $\frac{CV}{\epsilon_0}$ (D) Zero

95. In the given circuit in Fig.3.255, $C_1 = 2\,\mu$F, $C_2 = 0.2\,\mu$F, $C_3 = 2\,\mu$F, $C_4 = 4\,\mu$F, $C_5 = 2\,\mu$F, $C_6 = 2\,\mu$F, the charge stored on capacitor C_4 is _____ μC. [11 April 23 Shift II]

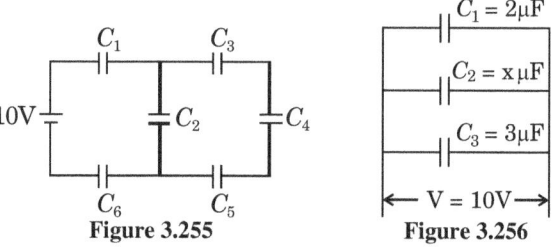

Figure 3.255   Figure 3.256

96. In the given Fig.3.256 the total charge stored in the combination of capacitors is 100 μC. The value of 'x' is _____.
[15 April 23 Shift I]

Section B: JEE Advanced

1. A parallel plate capacitor of plate area A and plate separation d is charged to potential difference V and then the battery is disconnected. A slab of dielectric constant κ is then inserted between the plates of the capacitor so as to fill the space between the plates. If Q, E and W denote respectively, the magnitude of charge on each plate, the electric field between the plates (after the slab is inserted), and work done on the system, in question, in the process of inserting the slab, then- [1991]
(A) $Q = \frac{\epsilon_0 AV}{d}$ (B) $W = \frac{\epsilon_0 AV^2}{2d}\left(\frac{1}{\kappa} - 1\right)$
(C) $E = \frac{V}{\kappa d}$ (D) All of these

2. Two identical metal plates are given positive charges Q_1 and $Q_2 (< Q_1)$ respectively. If they are now brought close together to form a parallel plate capacitor with capacitance C, the potential difference between them is-[1999]
(A) $\frac{Q_1 + Q_2}{2C}$ (B) $\frac{Q_1 + Q_2}{C}$
(C) $\frac{Q_1 - Q_2}{C}$ (D) $\frac{Q_1 - Q_2}{2C}$

3. For the circuit shown in Fig.3.257, which of the following statements is true? [1999]
(A) with S_1 closed, $V_1 = 15$ V, $V_2 = 20$ V
(B) with S_3 closed, $V_1 = V_2 = 25$ V
(C) with S_1 and S_2 closed, $V_1 = V_2 = 0$
(D) with S_1 and S_3 closed, $V_1 = 30$ V, $V_2 = 20$ V

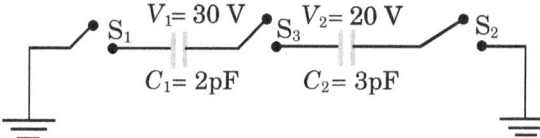

Figure 3.257

4. A parallel plate capacitor of area A, plate separation d and capacitance C is filled with three different dielectric materials having dielectric constants κ_1, κ_2 and κ_3 as shown in Fig.3.258. If a single dielectric material is to be used to have the same capacitance C in this capacitor, then its dielectric constant κ is given by [2000]

(A) $\frac{1}{\kappa} = \frac{1}{\kappa_1} + \frac{1}{\kappa_2} + \frac{1}{2\kappa_3}$ (B) $\frac{1}{\kappa} = \frac{1}{\kappa_1+\kappa_2} + \frac{2\kappa_3}{\kappa_1 k_3}$
(C) $\kappa = \frac{\kappa_1\kappa_2}{\kappa_1+\kappa_2} + 2\kappa_3$ (D) $\kappa = \frac{\kappa_1\kappa_3}{\kappa_1+\kappa_3} + \frac{\kappa_2 k_3}{\kappa_2+\kappa_3}$

5. Consider the situation shown in the Fig.3.259. The capacitor A has a charge q on it whereas B is uncharged. The charge appearing on the capacitor B a long time after the switch is closed, is- [2001]
   (A) Zero (B) $q/2$ (C) q (D) $2q$

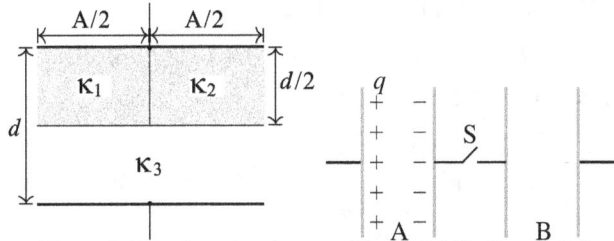

Figure 3.258: Question 4.   **Figure 3.259:** Question 5.

6. Two identical capacitors, have the same capacitance C. One of them is charged to potential V_1 and the other to V_2. The negative ends of the capacitors are connected together. When the positive ends are also connected, the decrease in energy of the combined system is- [2002]
   (A) $\frac{1}{4}C(V_1^2 - V_2^2)$ (B) $\frac{1}{4}C(V_1^2 + V_2^2)$
   (C) $\frac{1}{4}C(V_1 - V_2)^2$ (D) $\frac{1}{4}C(V_1 + V_2)^2$

7. A $2\,\mu F$ capacitor is charged as shown in the Figure 3.260. The percentage of its stored energy dissipated after the switch S is turned to position 2 is [2011]
   (A) 0% (B) 20% (C) 75% (D) 80%

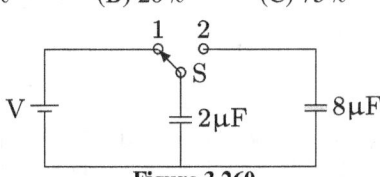

Figure 3.260

8. In the given circuit in Fig.3.39, a charge of $+80\,\mu C$ is given to the upper plate of the $4\,\mu F$ capacitor. Then in the steady state, the charge on the upper plate of the $3\,\mu F$ capacitor is [2012]
   (A) $+32\,\mu C$ (B) $+40\,\mu C$
   (C) $+48\,\mu C$ (D) $+80\,\mu C$

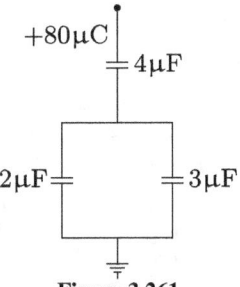

Figure 3.261

9. In the circuit shown in Fig.3.262, there are two parallel plate capacitors each of capacitance C. The switch S_1 is pressed first to fully charge the capacitor C_1 and then released. The switch S_2 is then pressed to charge the capacitor C_2. After some time, S_2 is released and then S_3 is pressed. After some time, [2013]
   (A) the charge on the upper plate of C_1 is $2CV_0$
   (B) the charge on the upper plate of C_1 is CV_0
   (C) the charge on the upper plate of C_2 is 0
   (D) the charge on the upper plate of C_2 is $-CV_0$

10. A parallel plate capacitor has a dielectric slab of dielectric constant κ between its plates that covers 1/3 of the area of its

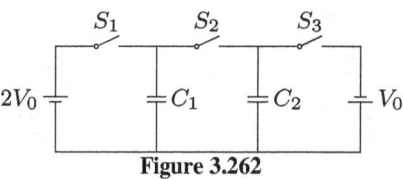

Figure 3.262

plates, as shown in the Figure 3.263. The total capacitance of the capacitor is C while that of the portion with dielectric in between is C_1. When the capacitor is charged, the plate area covered by the dielectric gets charge Q_1 and the rest of the area gets charge Q_2. The electric field in the dielectric is E_1 and that in the other portion is E_2. Choose the correct option/options, ignoring edge effects. [2014]

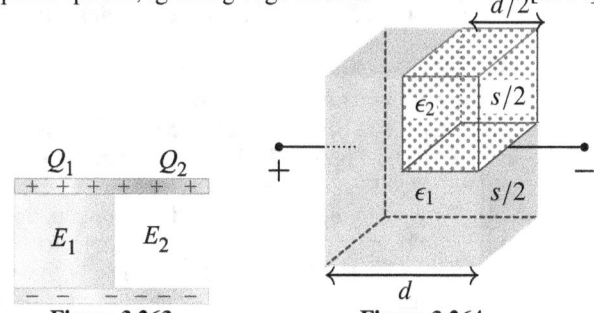

Figure 3.263   **Figure 3.264**

(A) $E_1/E_2 = 1$ (B) $E_1/E_2 = 1/\kappa$
(C) $Q_1^2/Q_2 = 3/\kappa$ (D) $C/C_1 = (2+\kappa)/\kappa$

11. A parallel plate capacitor having plates of area s and plate separation d, has capacitance C_1 in air. When two dielectrics of different relative permittivities ($\epsilon_1 = 2$ and $\epsilon_2 = 4$) are introduced between the two plates as shown in the Figure 3.264, the capacitance becomes C_2. The ratio C_2/C_1 is [2015]
   (A) 6/5 (B) 5/3 (C) 7/5 (D) 7/3

12. Two capacitors with capacitance value $C_1 = (2000 \pm 10)$ pF and $C_2 = (3000 \pm 15)$ pF are connected in series. The voltage applied across this combination is $V = (5.00 \pm 0.02)$ V. The percentage error in the calculation of the energy stored in this combination of capacitors is ___%. [2020]

13. In the circuit shown in Fig.3.265, $C_1 = 12\,\mu F$, $C_2 = C_3 = 4\,\mu F$ and $C_4 = C_5 = 2\,\mu F$. The Charge stored in C_3 is ___ μC. [2022]

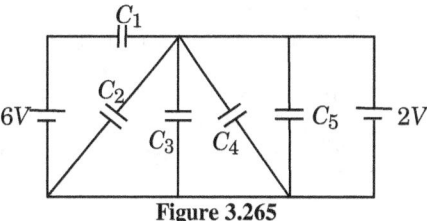

Figure 3.265

14. A medium having dielectric constant $\kappa > 1$ fills the space between the plates of a parallel plate capacitor. The plates have large area, and the distance between them is d. The capacitor is connected to a battery of voltage V. as shown in Figure 3.266a. Now, both the plates are moved by a distance of $d/2$ from their original positions, as shown in Fig.3.266b.
   In the process of going from the configuration depicted in Fig.3.266a to that in Fig.3.266b, which of the following statement(s) is(are) correct? [2022]

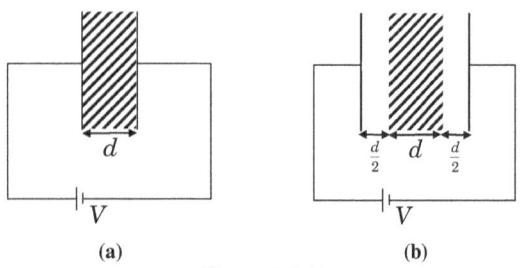

Figure 3.266

(A) The electric field inside the dielectric material is reduced by a factor of 2κ.
(B) The capacitance is decreased by a factor of $\frac{1}{\kappa+1}$.
(C) The voltage between the capacitor plates is increased by a factor of $(\kappa + 1)$.
(D) The work done in the process does not depend on the presence of the dielectric material.

15. A container has a base of $50\,\text{cm} \times 5\,\text{cm}$ and height $50\,\text{cm}$, as shown in the Fig.3.267. It has two parallel electrically conducting walls each of area $50\,\text{cm} \times 50\,\text{cm}$. The remaining walls of the container are thin and non-conducting. The container is being filled with a liquid of dielectric constant 3 at a uniform rate of $250\,\text{cm}^3\text{s}^{-1}$. What is the value of the capacitance of the container after 10 seconds? [Given: Permittivity of free space $\epsilon_0 = 9 \times 10^{-12}\,\text{C}^2\text{N}^{-1}\text{m}^{-2}$, the effects of the non-conducting walls on the capacitance are negligible] [2023]
  (A) 27 pF (B) 63 pF (C) 81 pF (D) 135 pF

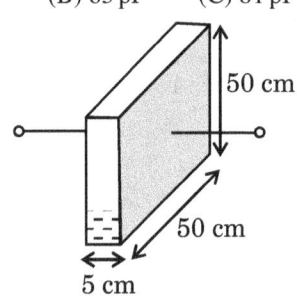

Figure 3.267

3.19 Answer Keys and Solutions
3.19.1 Check Point 1

1. **APPROACH** The charge on the spherical conductor is related to its radius and potential according to $V = kQ/r$ and we can use the definition of capacitance to find the self-capacitance of the sphere.
   SOLUTION (a) Relation between the potential V of the spherical conductor to the charge on it and to its radius is given by-
   $$V = \frac{kQ}{r} \implies Q = \frac{rV}{k}$$
   Substituting numerical values, we get Q:
   $$Q = \frac{(10.0\,\text{cm})(2.00\,\text{kV})}{8.988 \times 10^9\,\text{N} \cdot \text{m}^2/\text{C}^2} = 22.252\,\text{nC}$$
   $$= 22.3\,\text{nC}$$
   (b) The self-capacitance of the sphere, is given by-
   $$C = \frac{Q}{V} = \frac{22.252\,\text{nC}}{2.00\,\text{kV}} = 11.1\,\text{pF}$$
   (c) It doesn't. The self-capacitance of a sphere is a function of its radius.

2. We assume that the capacitor is fully charged, according to relation $Q = CV$, we have-
   $$Q = CV = \left(12.6 \times 10^{-6}\,\text{F}\right)(12.0\,\text{V}) = 1.51 \times 10^{-4}\,\text{C}$$

3. Electrons transferred from one plate to other-
   $$N = 1 \times 10^{12}$$
   Therefore, the magnitude of negative charge transferred
   $$Q = Ne = 1 \times 10^{12} \times 1.6 \times 10^{-19} = 1.6 \times 10^{-7}\,\text{C}$$
   Due to this charge transfer, the potential difference appeared between the plates:
   $$V = 10\,\text{volt (Given)}$$
   Therefore, required capacitance
   $$C = \frac{q}{V} = \frac{1.6 \times 10^{-7}}{10} = 1.6 \times 10^{-8}\,\text{F}.$$

4. **APPROACH** Given that-
   capacity of each smaller drop = C,
   and potential of each smaller drop = V
   So, by applying Eq.3.6: $C = 4\pi\epsilon_0 r$, we can find the radius (r) of smaller drop in terms oc C-
   $$r = \frac{C}{4\pi\epsilon_0} \quad (1)$$
   The electric potential of each smaller drop is-
   $$V = \frac{1}{4\pi\epsilon_0}\frac{q}{r} \quad (2)$$
   If the capacity of larger drop is C', it's radius is R, then from Eq.3.6, we have-
   $$C' = 4\pi\epsilon_0 R \quad (3)$$
   Therefore, to find capacitance, we need radius R.
   Since, collision takes place at constant temperature, therefore volume of liquid remains conserved. So, R can be obtained by using the principle of conservation of volume. Potential (V') of the larger drop can be obtained by using the relation,
   $$V' = \frac{1}{4\pi\epsilon_0}\frac{Q}{R}$$
   Here, we again need radius R and the charge Q of larger drop. Radius is already calculated by using the principle of conservation of volume. The charge Q, can be obtained by applying the principle of conservation of charge.
   SOLUTION If r is radius and q is charge on smaller drop then from Eq.3.6, the capacitance of each smaller drop-
   $$C = 4\pi\epsilon_0 r$$
   and from Eq.3.1, the charge on each drop -
   $$q = CV$$
   By conservation of volume, we have
   $$V_{\text{larger}} = 2 \times V_{\text{smaller}}$$
   $$\implies \frac{4}{3}\pi R^3 = 2 \times \frac{4}{3}\pi r^3 \implies R = 2^{1/3} r$$
   Therefore, capacitance of larger drop:
   $$C' = 4\pi\epsilon_0 R = 2^{1/3} C$$
   Charge on larger drop: $Q = 2q = 2CV$
   Potential of larger drop: $V' = \frac{Q}{C'} = \frac{2CV}{2^{1/3}C} = 2^{2/3} V$

5. **APPROACH** From Eq.3.10, the magnitude of charge on facing surface of either plate:
   $$q = \frac{|Q_1 - Q_2|}{2} = \frac{|+2.0 \times 10^{-8}\text{C} - (-1.0 \times 10^{-8}\text{C})|}{2}$$
   $$= 1.5 \times 10^{-8}\,\text{C}$$
   Now, from Eq.3.7: $V = q/C$. Substitute the given values of C and q and then solve for V.
   SOLUTION Given that-
   $C = 1.5 \times 10^{-3}\,\mu\text{F}$ and calculated charge $q = 1.5 \times 10^{-8}\,\text{C}$, therefore-
   $$V = \frac{q}{C} = \frac{1.5 \times 10^{-8}\,\text{C}}{1.5 \times 10^{-3}\,\mu\text{F}} = 10\,\text{volt}.$$

6. Let Q_1 and V_1 be the initial charge and voltage on the capacitor, and let Q_2 and V_2 be the final charge and voltage on the capacitor, then from $Q = CV$, we can write-

$$Q_1 = CV_1, \quad Q_2 = CV_2$$

On subtracting second relation in first, we get-
$$Q_2 - Q_1 = CV_2 - CV_1 = C(V_2 - V_1)$$
$$\Longrightarrow \quad C = \frac{Q_2 - Q_1}{V_2 - V_1} = \frac{26 \times 10^{-6} \text{ C}}{50 \text{ V}}$$
$$= 5.2 \times 10^{-7} \text{ F} = 0.52 \ \mu\text{F}$$

7. After the first capacitor is disconnected from the battery, the total charge must remain constant. The voltage across each capacitor must be the same when they are connected together, since each capacitor plate is connected to a corresponding plate on the other capacitor by a constant-potential connecting wire. Use the total charge and the final potential difference to find the value of the second capacitor.
$$Q_{\text{Total}} = C_1 V_{1i} \quad Q_{1f} = C_1 V_f \quad Q_{2f} = C_2 V_f$$
Here, subscripts i and f are used for initial and final stage respectively
$$Q_{\text{Total}} = Q_{1f} + Q_{2f} = (C_1 + C_2) V_f$$
$$C_1 V_{1i} = (C_1 + C_2) V_f$$
$$C_2 = C_1 \left(\frac{V_{1i}}{V_f} - 1 \right) = (7.7 \times 10^{-6} \text{ F}) \left(\frac{125 \text{ V}}{15 \text{ V}} - 1 \right)$$
$$= 5.6 \times 10^{-5} \text{ F} = 56 \ \mu\text{F}$$

8. **APPROACH** Let the separation of the spheres be d and their radii be R. Outside the two spheres the electric field is approximately the field due to point charges of $+Q$ and $-Q$, each located at the centres of spheres, separated by distance d. We can derive an expression for the potential at the surface of each sphere and then use the potential difference between the spheres and the definition of capacitance and to estimate the capacitance of the two-sphere system.
SOLUTION The capacitance of the two-sphere system is given by:
$$C = \frac{Q}{V}$$
where V is the potential difference between the spheres. The potential at any point outside the two spheres is:
$$V = \frac{k(+Q)}{r_1} + \frac{k(-Q)}{r_2}$$
where r_1 and r_2 are the distances from the given point to the centres of the spheres.
For a point on the surface of the sphere with charge $+Q$:
$$r_1 = R \text{ and } r_2 \approx d$$
Substituting these values of r_1 and r_2 in above expression, we get:
$$V_{+Q} = \frac{k(+Q)}{R} + \frac{k(-Q)}{d}$$
The potential difference between the spheres is:
$$V = V_Q - V_{-Q} = \frac{kQ}{R} - \frac{kQ}{d} - \left(\frac{-kQ}{R} + \frac{kQ}{d} \right) = 2kQ \left(\frac{1}{R} - \frac{1}{d} \right)$$
Substitute for V in the expression for C to obtain:
$$C = \frac{Q}{2kQ \left(\frac{1}{R} - \frac{1}{d} \right)} = \frac{2\pi\epsilon_0}{\left(\frac{1}{R} - \frac{1}{d} \right)} = \frac{2\pi\epsilon_0 R}{1 - \frac{R}{d}}$$

9. **APPROACH** In given graphs, q is taken along x axis and V is along y axis, therefore, with $q = CV$, the slope of each graph is $1/C$.
SOLUTION The slope for capacitor A:
$$\frac{1}{C_A} = \frac{30 \text{ V}}{3 \text{ nC}} = \frac{30 \text{ V}}{3 \times 10^{-9} \text{ C}} = 1 \times 10^{10} \text{ V/C}$$
Take the reciprocal of the slope to find the capacitance:
$$C_A = \frac{1}{1 \times 10^{10} \text{ V/C}} = 1 \times 10^{-10} \text{ F} = 100 \text{ pF}$$
Repeat this process for capacitor B:
$$\frac{1}{C_B} = \frac{30 \text{ V}}{1.5 \text{ nC}} = \frac{30 \text{ V}}{1.5 \times 10^{-9} \text{ C}} = 2 \times 10^{10} \text{ V/C}$$

$$\Longrightarrow \quad C_B = 5 \times 10^{-11} \text{ F} = 50 \text{ pF}$$

10. The radius of the Earth is 6.4×10^6 m; therefore
$$C = 4\pi\epsilon_0 R = 4\pi \times 8.85 \times 10^{-12} \text{F/m} \times 6.4 \times 10^6 \text{m}$$
$$= 7.1 \times 10^{-4} \text{ F}$$
To alter the potential of the Earth by 1 volt requires a charge of only $Q = CV = 7.1 \times 10^{-4} \text{F} \times 1 \text{ volt} = 7.1 \times 10^{-4}$ C.

11. Capacitance of sphere of radius R is-
$$C = 4\pi\epsilon_0 R = \frac{0.5 \times 10^{-3}}{9 \times 10^9} = \frac{1}{18} \times 10^{-12} \text{ F}$$
Rate to escape of charge from surface:
$$= \frac{80}{100} \times 6.25 \times 10^{10} \times 1.6 \times 10^{-19} = 8 \times 10^{-9} \text{ C/s}$$
Therefore, charge acquired by surface in time t is-
$$q = \left(8 \times 10^{-9} \right) t$$
Now, applying, $q = CV$, we get
$$8 \times 10^{-9} t = \frac{1}{18} \times 10^{-12} \times 1$$
$$\Longrightarrow \quad t = \frac{1}{18} \times 10^{-12} \times 1$$
$$\Longrightarrow \quad t = \frac{10^{-12}}{8 \times 10^{-9} \times 18} = \frac{10^{-3}}{144} \text{ s} = 6.95 \ \mu\text{s}$$

3.19.2 Check Point 2

1. Effective area of each plate:
$$A = \pi r^2 = \pi \times \left(8 \times 10^{-2} \right)^2 = 0.0201 \text{ m}^2$$
Plate separation: $d = 1 \text{ mm} = 1 \times 10^{-3}$ m
Dielectric constant of the medium: $\kappa = 6$
Capacitance of the circular plate capacitor:
$$C = \frac{\epsilon_0 \kappa A}{d} = \frac{8.85 \times 10^{-12} \times 6 \times 0.0201}{1 \times 10^{-3}} = 1.068 \times 10^{-9} \text{ F}$$
Potential difference
$$V = 150 \text{ volt}$$
Therefore, the energy stored in the electric field of capacitor:
$$U = \frac{1}{2} CV^2 = \frac{1}{2} \times \left(1.068 \times 10^{-9} \right) \times (150)^2 = 1.2 \times 10^{-5} \text{ J}$$

2. Using the formula $U = Q^2/2C$ for the energy of a capacitor, we find that, in accordance with Eq.3.84, the required force is
$$F_x = -\left. \frac{\partial U}{\partial x} \right|_Q = \frac{V^2}{2} \frac{\partial C}{\partial x} \quad \text{(i)}$$
Since $d \ll R$, the capacitance of the given capacitor can be calculated by the formula for a parallel-plate capacitor. Therefore, if the dielectric is introduced to a depth x and the capacitor length is l, we have
$$C = \frac{\kappa \epsilon_0 x \cdot 2\pi R}{d} + \frac{\epsilon_0 (l-x) \cdot 2\pi R}{d} = \frac{\epsilon_0 \cdot 2\pi R}{d} (\kappa x + l - x) \quad \text{(ii)}$$
Substituting (ii) into (i), we obtain
$$F_x = \epsilon_0 (\kappa - 1) \pi R V^2 / d$$

3. The work performed by the moment of forces τ upon the rotation of the plate through an angle element $d\alpha$ is equal to the decrease in the electric energy of the system at $Q = $ constant,
$$\tau_z d\alpha = -\left. dU \right|_Q$$
where $U = Q^2/2C$. Hence
$$\tau_z = -\left. \frac{\partial U}{\partial \alpha} \right|_Q = \frac{Q^2}{2} \frac{\partial C / \partial \alpha}{C^2} \quad \text{(i)}$$
In the case under consideration,
$$C = C_1 + C_\kappa$$
where C_1 and C_κ are the capacitances of the parts of the capacitor with and without the dielectric. The area of a sector with an angle α is given by-

$S = \alpha R^2/2$, and hence
$C = \epsilon_0 \alpha R^2/2d + \kappa\epsilon_0(\pi - \alpha)R^2/2d$.
On differentiating with respect to α, we get-
$$\partial C/\partial \alpha = \left(\epsilon_0 R^2/2d\right)(1-\kappa)$$
Substituting this expression into Eq.(i), and considering that $C = Q/V$, we get-
$$\tau_z = \frac{V^2}{2}\frac{\epsilon_0 R^2}{2d}(1-\kappa) = -(\kappa-1)\frac{\epsilon_0 R^2 V^2}{4d} < 0$$
The negative sign of τ_z indicates that the moment of the force is acting clockwise (oppositely to the positive direction of the angle α; see Fig.3.127). This moment tends to pull the dielectric inside the capacitor.

It should be noted that τ_z is independent of the angle α. However, in equilibrium, when $\alpha = 0$, the moment $\tau_z = 0$. This discrepancy is due to the fact that for small values of α we cannot ignore edge effects as was done in the solution of this problem.

4. In this case, if we mentally move the plates apart, keeping voltage V constant, the capacitance ($C = \kappa\epsilon_0 A/h$) decreases, therefore, from the relation $q = CV$ charge also decreases. In spite of this, we shall calculate the force under the assumption that $q = $ const, i.e., with the help of Eq.3.69. The electric potential energy of the capacitor dielectric system is-
$$U = \frac{q^2}{2C} = \frac{q^2}{2\kappa\epsilon_0 A}x$$
where κ is the dielectric constant of the medium, A is the

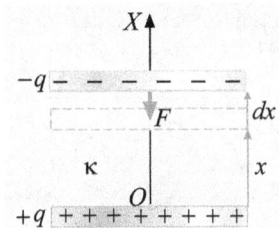

Figure 3.268

facing area of each plate, and x is the distance between them ($x = h$).

Next, let us choose the positive direction of the X-axis as is shown in Fig.3.128. According to Eq.3.69, the force acting on the upper plate of the capacitor is
$$F_x = -\left.\frac{\partial U}{\partial x}\right|_q = -\frac{q^2}{2\kappa\epsilon_0 A} \quad \cdots \text{ (i)}$$
The minus sign in this formula indicates that vector F is directed towards the negative values on the X-axis, i.e. the force is attractive by nature.
Since, $q = \sigma A = (\kappa\epsilon_0 E)A$ and $E = V/h$, therefore Eq.(i) becomes-
$$F_x = -CV^2/2h$$

5. **(1)** The required work is
$$W' = qE_1(x_2 - x_1) = q\frac{\sigma}{2\epsilon_0}(x_2 - x_1) = \frac{q^2}{2\epsilon_0 A}(x_2 - x_1)$$
where E_1 is the intensity of the field created by one plate ($E = \sigma/2\epsilon_0$). It is in this field that the charge located on the other plate moves. This work is completely spent for increasing the electric energy: $\Delta U = W'$.

(2) In this case, the force acting on each capacitor plate will depend on the distance between the plates. Let us write the elementary work of the force acting on a plate during its displacement over a distance dx relative to the other plate:
$$\delta W' = qE_1 dx = \frac{\epsilon_0 A U^2}{2}\frac{dx}{x^2}$$

where we took into account that $q = CU$, $E_1 = V/2x$, and $C = \epsilon_0 A/x$. After integration, we obtain
$$W' = \frac{\epsilon_0 A V^2}{2}\left(\frac{1}{x_1} - \frac{1}{x_2}\right) > 0$$
The increment of the electric energy of the capacitor is
$$\Delta U = \frac{(C_2 - C_1)V^2}{2} = \frac{\epsilon_0 A V^2}{2}\left(\frac{1}{x_2} - \frac{1}{x_1}\right) < 0$$
It should be noted that $\Delta U = -W'$. Thus, by moving the plates apart, we perform a positive work (against the electric forces). The energy of the capacitor decreases in this case. In order to understand this, we must consider a source maintaining the potential difference of the capacitor at a constant value. This source also accomplishes the work W_s. According to the law of conservation of energy, $W_s + W' = \Delta U$, therefore, $W_s = \Delta U - W' = = -2W' < 0$.

6. The resultant force F acting per unit area of each plate can be represented as
$$F = F_0 - F' \quad \ldots \text{(i)}$$
where F_0 is the electric force acting per unit area of a plate from the other plate in the absence of dielectric). In our case, we have
$$F = F_0/\kappa, \quad F_0 = \sigma E = \sigma^2/2\epsilon_0 \quad \ldots\text{(ii)}$$
where E is the field intensity in the region occupied by one plate, created by the charges of the other plate. Considering that $\sigma = \kappa\epsilon_0 V/d$ and substituting (ii) into (i), we obtain
$$F' = F_0(1 - 1/\kappa) = \kappa(\kappa-1)\epsilon_0 U^2/2h^2$$
For example, for $V = 500\,\text{V}$, $d = 1.0\,\text{mm}$ and $\kappa = 81$ (water), we get $F' = 7\,\text{kPa}$ (0.07 atm).

7. As we know that, after elastic collision between two bodies of same masses, their velocities get interchanged. So, if we assume that the velocity of the bullet just before collision is v_0, then velocity of dielectric just after collision will be v_0. Now, the dielectric slab will start moving with the velocity v_0 and when it is coming out of capacitor, the fringing field applies an attractive force on it.
From Eq.3.82, the force of attraction applied by fringing field on the slab is given by-
$$F_x = \frac{1}{2}\frac{\epsilon_0 b V^2}{d}(\kappa - 1)$$
This force is opposite to the velocity of slab, so it produces a negative acceleration which decreases the speed of dielectric slab. According to given problem, the speed of dielectric becomes zero when it just leaves out the capacitor. During this period, the dielectric slab travels a distance a [Fig.3.129].
By work energy theorem,
work done by net force on dielectric
$$= \text{change in the kinetic energy of the slab.}$$
$$\Rightarrow \quad F_x a \cos 180° = K_f - K_i$$
$$\Rightarrow \quad -\frac{1}{2}\frac{\epsilon_0 b V^2}{d}(\kappa - 1)a = 0 - \frac{1}{2}Mv_0^2$$
$$\Rightarrow \quad v_0 = V\left[\frac{\epsilon_0 ab(\kappa - 1)}{Md}\right]^{1/2}$$

8. **(a)** In Fig.3.269, the forces acting on the block kept on the table are:
   (i) Weight Mg of the block, acting in $-y$ direction, i.e., in vertically downward direction.
   (ii) Normal reaction N of the surface, acting in $+y$ direction, i.e., acting in vertically upward direction.
   (iii) Tension (T) of the string acting in $+x$ direction.
   (iv) Force of kinetic friction, $f = \mu N$, acting in $-x$ direction.
   Net force on the block in vertical direction,
   $$\Sigma F_y = N - Mg$$
   Since, the block is in equilibrium in vertical direction,

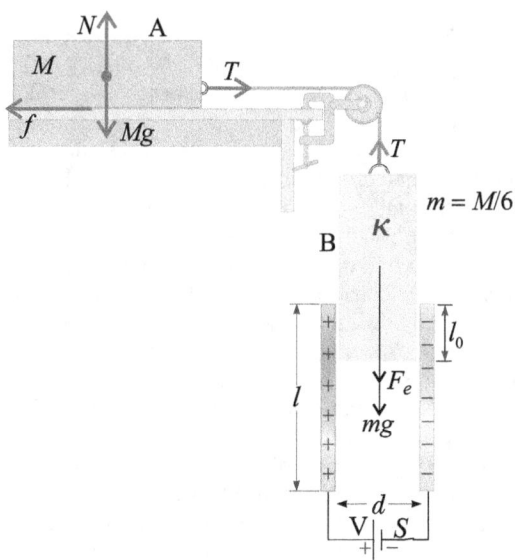

Figure 3.269

therefore-
$$\Sigma F_y = 0 \implies N = Mg$$
Net force on the block in horizontal direction,
$$\Sigma F_x = T - f = T - \mu N$$
Since, $N = Mg$, therefore-
$$\Sigma F_x = T - \mu Mg \qquad \cdots (1)$$
For the non-accelerated motion of the block in horizontal direction:
$$\Sigma F_x = 0 \implies T - \mu Mg = 0$$
$$\implies T = \mu Mg \qquad \cdots (2)$$
Now, Forces acting on the dielectric are-
(i) Weight mg of the dielectric acting in $-y$ direction, i.e., in downward direction (here $m = M/6$)
(ii) Electrostatic attraction $F_e = \frac{1}{2}\frac{\epsilon_0 b V^2}{d}(\kappa - 1)$ [from Eq.3.82] of fringing field of capacitor acting in $-y$ direction, i.e., acting in downward direction.
(iii) Tension of the string T, acting in $+y$ direction, i.e., in upward direction.

Net force on the dielectric in vertically downward direction
$$\Sigma F_y = mg + F_e - T$$
$$= mg + \frac{1}{2}\frac{\epsilon_0 b V^2}{d}(\kappa - 1) - T$$
For motion of the dielectric without acceleration,
$$\Sigma F_y = 0$$
$$mg + \frac{1}{2}\frac{\epsilon_0 b V^2}{d}(\kappa - 1) - T = 0$$
$$\implies T = mg + \frac{1}{2}\frac{\epsilon_0 b V^2}{d}(\kappa - 1) = 0 \qquad \cdots (3)$$
Eliminating T from Eq.(2) and (3), we get-
$$\implies mg + \frac{1}{2}\frac{\epsilon_0 b V^2}{d}(\kappa - 1) = \mu Mg$$
Since, $m = M/6$ and $\mu = 1/4$ therefore we have-
$$\frac{1}{2}\frac{\epsilon_0 b V^2}{d}(\kappa - 1) = \frac{1}{12}Mg$$
$$\implies \frac{\epsilon_0 b V^2}{d}(\kappa - 1) = \frac{1}{6}Mg \qquad \cdots (4)$$
$$\implies V = \sqrt{\frac{Mgd}{6\epsilon_0 b(\kappa - 1)}}$$
This is the voltage corresponding to motion of the block without acceleration.

(b) In case of $V' = 2V$, the dielectric will accelerate the block. If, the acceleration of the system is a, then

For the motion of dielectric:
$$F_e' + mg - T = ma \qquad (5)$$
and for the motion of the block:
$$T - \mu Mg = Ma \qquad (6)$$
To eliminate T, we add (5) and (6). It gives
$$a = \frac{F_e' + (m - \mu M)g}{m + M} \qquad (7)$$
Since, $F_e' = \frac{1}{2}\frac{\epsilon_0 b}{d}(\kappa - 1)V'^2$, and $V' = 2V$, therefore-
$$F_e' = \frac{1}{2}\frac{\epsilon_0 b}{d}(\kappa - 1)V^2 4 = 2\frac{\epsilon_0 b}{d}(\kappa - 1)V^2$$
Using Eq.(4) in above, we get-
$$F_e' = 2 \times \frac{1}{6}Mg = \frac{1}{3}Mg$$
and $\mu = 1/2$, $m = M/6$, therefore, from Eq.(7), we get
$$a = \frac{\frac{1}{3}Mg - \frac{1}{12}Mg}{\frac{7M}{6}} = \frac{g/4}{7/6} = \frac{3}{14}g$$
From equation of motion, $v^2 = v_0^2 + 2as$, with $v_0 = 0$, we get
$$v^2 = 2\left(\frac{3}{14}g\right) \times (l - l_0) \implies v = \sqrt{\frac{3}{7}g(l - l_0)}$$

3.19.3 Conceptual Questions

1. Yes. If two insulators were used the charge would not be able to flow into the insulators and no charge would be stored; thus, conductors must be used.

2. **(a)** The capacitor may be charged and hence there may be potential difference across it's terminals.
   (b) Discharge the capacitor by connecting its terminals together or by grounding it.

3. Underestimate. If the separation between the plates is not very small compared to the plate size, then fringing cannot be ignored and the electric field (for a given charge) will actually be smaller. The capacitance is inversely proportional to potential and, for parallel plates, also inversely proportional to the field, so the capacitance will actually be larger than that given by the formula.

4. To increase the maximum operating voltage of a capacitor, we put a material with higher dielectric strength between the plates, or evacuate the space between the plates. At very high voltages, we have to cool off the plates or choose to make them of a different chemically stable material, because atoms in the plates themselves can ionize, showing thermionic emission under high electric fields.

5. As the temperature of the liquid increases the random motion of the molecules increases and this decreases the alignment of the molecular dipoles along the electric field direction. This reduces the induced electric field within the dielectric and therefore reduces the effect of the dielectric on the net field between the plates. The dielectric constant is the ratio of the net field without the dielectric to the field with the dielectric. Therefore, the dielectric constant becomes closer to 1.0 as the temperature increases.

6. When a capacitor is first connected to a battery, charge flows to one plate. Because the plates are separated by an insulating material, charge cannot cross the gap. An equal amount of charge is therefore repelled from the opposite plate, leaving it with a charge that is equal and opposite to the charge on the first plate. The two conductors of a capacitor will have equal and opposite charges even if they have different sizes or shapes.

7. In order to increase the capacitance from $10.0\,\mu F$ to $18.0\,\mu F$ in a capacitor, you could add a dielectric in the capacitor with a dielectric constant of 1.80. The alternative would

be to change the capacitor geometry by narrowing the plate separation or increasing the plate area.

8. False. The electrostatic energy density is not uniformly distributed the electric field strength is not uniformly distributed.

9. Charge a parallel-plate capacitor using a battery with a known voltage V. Let the capacitor discharge through a resistor with a known resistance R and measure the time constant. This will allow calculation of the capacitance C. Then use $C = \epsilon_0 A/d$ and solve for ϵ_0.

10. Parallel. The equivalent capacitance of the three capacitors in parallel will be greater than that of the same three capacitors in series, and therefore they will store more energy when connected to a given potential difference if they are in parallel.

11. If a large copper sheet of thickness l is inserted between the plates of a parallel-plate capacitor, the charge on the capacitor will appear on the large flat surfaces of the copper sheet, with the negative side of the copper facing the positive side of the capacitor. This arrangement can be considered to be two capacitors in series. From Eq.3.64, the new net capacitance will be $C' = \epsilon_0 A/(d - l)$, so the capacitance of the capacitor will be reduced.

12. The dielectric decreases the electric field between the plates, causing the potential difference to decrease for the same amount of charge. More charge may be placed on the capacitor before the capacitor experiences dielectric breakdown (resulting in charge jumping from one plate to the other, and in a path being burned through the dielectric) because the electric forces between charges on opposite plates are smaller. The capacitor can have a higher maximum operating voltage, allowing it to hold more charge.

13. A force is required to increase the separation of the plates of an isolated capacitor because you are pulling a positive plate away from a negative plate. The work done in increasing the separation goes into increasing the electric potential energy stored between the plates. The capacitance decreases, and the potential between the plates increases since the charge has to remain the same.

14. The dielectric will be pulled into the capacitor by the electrostatic attractive forces between the charges on the capacitor plates and the polarized charges on the dielectric's surface. (Note that the addition of the dielectric decreases the energy of the system.)

15. The work done, $W = QV$, is the work done by an external agent, like a battery, to move a charge through a potential difference, V. To determine the energy in a charged capacitor, we must add the work done to move bits of charge from one plate to the other. Initially, there is no potential difference between the plates of an uncharged capacitor. As more charge is transferred from one plate to the other, the potential difference increases, meaning that more work is needed to transfer each additional bit of charge. The total work is given by $W = \frac{1}{2}QV$. Another explanation is that the charge Q is moved through an average potential difference $\frac{1}{2}V$, requiring total work $W = \frac{1}{2}QV$.

16. If the battery remains connected to the capacitor, the energy stored in the electric field of the capacitor will increase as the dielectric is inserted. Since the energy of the system increases, an external work must be done and the dielectric will have to be pushed into the area between the plates. If it is released, it will be ejected. Note that, we have ignored the fringing field attraction.

17. (a) If the capacitor is isolated, Q remains constant, and $U = \frac{1}{2}\frac{Q^2}{C}$ becomes $U' = \frac{1}{2}\frac{Q^2}{\kappa C}$ and the stored energy decreases.
(b) If the capacitor remains connected to a battery so V does not change, $U = \frac{1}{2}CV^2$ becomes $U' = \frac{1}{2}\kappa CV^2$, and the stored energy increases.

18. For dielectrics consisting of polar molecules, one would expect the dielectric constant to decrease with temperature. As the thermal energy increases, the molecular vibrations will increase in amplitude, and the polar molecules will be less likely to line up with the electric field.

19. When the dielectric is removed, the capacitance decreases. The potential difference across the plates remains the same because the capacitor is still connected to the battery. If the potential difference remains the same and the capacitance decreases, the charge on the plates and the energy stored in the capacitor must also decrease. (Charges return to the battery.) The electric field between the plates will stay the same because the potential difference across the plates and the distance between the plates remain constant.

20. For a given configuration of conductors and dielectrics, C is the proportionality constant between the voltage between the plates and the charge on the plates.

21. The dielectric constant is the ratio of the capacitance of a capacitor with the dielectric between the plates to the capacitance without the dielectric. If a conductor were inserted between the plates of a capacitor such that it filled the gap and touched both plates, both plates get connected to each other and potential difference between them would be zero. So, the capacitance $C = Q/V$ increases to infinite. Now, by definition, $\kappa = C/C_0 = \infty/C_0 = \infty$, i.e., dielectric constant would be infinite.

22. The work you do to pull the plates apart becomes additional electric potential energy stored in the capacitor. The charge is constant and the capacitance decreases but the potential difference increases to drive up the potential energy $\frac{1}{2}QV$. The electric field between the plates is constant in strength but fills more volume as you pull the plates apart.

23. Work has to be done to separate a positively charged plate from a negatively charged plate. When the battery is disconnected, the charge on the plates has nowhere to go and must remain the same. The electric field due to a plane of charge depends only on the charge, not upon the distance from the plane (ignoring edge effects) so the electric field will also remain the same. As you pull the plates apart, the work done by you increases the potential difference between the plates (the voltage difference between the plates will just be the product of the electric field with the separation distance).
When the battery remains connected, the voltage remains the same as the battery voltage. So as the plates are pulled apart, the electric field ($E = V/d$) must decrease to make up for the increase in separation (d), which means the charge must flow off the plates (which it can do, because there's a path to the battery). Thus the force becomes less and less with greater separation. So, in this case, the work done in increasing the separation is less. Therefore, the work done is greater when the capacitor is disconnected from the battery.

24. The electric potential energy stored in capacitor having potential difference of V volt across it's plate and charge q, is given by-
$$U = \frac{CV^2}{2} = \frac{q^2}{2C}$$
Initial electric energy stored in the capacitor-
$$U_i = \frac{q^2}{2C_i}$$
After disconnecting from battery and increasing plate sepa-

ration, electric energy stored in the capacitor-
$$U_f = \frac{q^2}{2C_f}$$
Initial capacitance, $C_i = \frac{\epsilon_0 A}{d}$, $d \to$ initial plate separation.
And after increasing plate separation by d', the final capacitance-
$$C_f = \frac{\epsilon_0 A}{(d+d')} = \frac{C_i d}{(d+d')}.$$
Therefore, final electric potential energy-
$$U_f = \frac{1}{2}\left(\frac{q^2}{C_f}\right) = \frac{1}{2}\frac{q^2}{\left(\frac{C_i d}{d+d'}\right)} = \frac{1}{2}\left(\frac{q^2}{C_i}\right)\left(\frac{d+d'}{d}\right) = U_i\left(1 + \frac{d'}{d}\right)$$
From above expression, it is clear that the energy stored has increased from the work in pulling the charges apart.

3.19.4 Problems

1. **(a)** $Q = CV = (4.00 \times 10^{-6}\text{ F})(12.0\text{ V})$
   $= 4.80 \times 10^{-5}\text{ C} = 48.0\ \mu\text{C}$
   (b) $Q = CV = (4.00 \times 10^{-6}\text{ F})(1.50\text{ V})$
   $= 6.00 \times 10^{-6}\text{ C} = 6.00\ \mu\text{C}$

2. **(a)** $C = \frac{Q}{V} = \frac{10.0 \times 10^{-6}\text{ C}}{10.0\text{ V}}$
   $= 1.00 \times 10^{-6}\text{ F} = 1.00\ \mu\text{F}$
   (b) $V = \frac{Q}{C} = \frac{100 \times 10^{-6}\text{ C}}{1.00 \times 10^{-6}\text{ F}} = 100\text{ V}$

3. $E = \frac{kq}{r^2} \implies q = \frac{(4.90 \times 10^4\text{ N/C})(0.210\text{ m})^2}{(8.99 \times 10^9\text{ N} \cdot \text{m}^2/\text{C}^2)} = 0.240\ \mu\text{C}$
   (a) $\sigma = \frac{q}{A} = \frac{0.240 \times 10^{-6}}{4\pi(0.120)^2} = 1.33\ \mu\text{C/m}^2$
   (b) $C = 4\pi\epsilon_0 r = 4\pi(8.85 \times 10^{-12})(0.120) = 13.3\text{ pF}$

4. **(a)** $C = 4\pi\epsilon_0 R$
   $R = \frac{C}{4\pi\epsilon_0} = kC$
   $= (8.99 \times 10^9\text{ N} \cdot \text{m}^2/\text{C}^2)(1.00 \times 10^{-12}\text{ F})$
   $= 8.99\text{ mm}$
   (b) $C = 4\pi\epsilon_0 R = \frac{4\pi(8.85 \times 10^{-12}\text{ C}^2)(2.00 \times 10^{-3}\text{ m})}{\text{N} \cdot \text{m}^2}$
   $= 0.222\text{ pF}$
   (c) $Q = CV = (2.22 \times 10^{-13}\text{ F})(100\text{ V})$
   $= 2.22 \times 10^{-11}\text{ C}$

5. **(a)** $V = Ed$
   $\implies E = \frac{20.0\text{ V}}{1.80 \times 10^{-3}\text{ m}} = 11.1\text{ kV/m}$
   (b) $E = \frac{\sigma}{\epsilon_0}$
   $\implies \sigma = (1.11 \times 10^4\text{ N/C})(8.85 \times 10^{-12}\text{ C}^2/\text{Nm}^2)$
   $= 98.3\text{ nC/m}^2$
   $C = \frac{\epsilon_0 A}{d}$
   $= \frac{(8.85 \times 10^{-12}\text{ C}^2/\text{Nm}^2)(7.60\text{ cm}^2)(1.00\text{ m}/100\text{ cm}^2)}{1.80 \times 10^{-3}\text{ m}}$
   $= 3.74\text{ pF}$
   (c) $V = \frac{Q}{C} \implies Q = (20.0\text{ V})(3.74 \times 10^{-12}\text{ F}) = 74.7\text{ pC}$

6. $Q = \frac{\epsilon_0 A}{d}(\Delta V) \implies \frac{Q}{A} = \sigma = \frac{\epsilon_0(\Delta V)}{d}$
   $\implies d = \frac{\epsilon_0(\Delta V)}{\sigma} = \frac{(8.85 \times 10^{-12}\text{ C}^2/\text{Nm}^2)(150\text{ V})}{(30.0 \times 10^{-5}\text{ C/m}^2)} = 4.42\ \mu\text{m}$

7. With $\theta = \pi$, the plates are out of mesh and the overlap area is zero. With $\theta = 0$, the overlap area is that of a semi-circle, $\pi R^2/2$. By proportion, the effective area of a single sheet of charge is $(\pi - \theta)R^2/2$

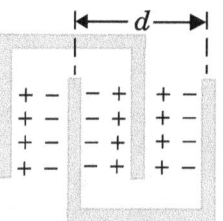

Figure 3.270

When there are two plates in each comb, the number of parallel plate capacitors formed, is 3 as shown in Fig.3.270. Similarly, when there are N plates on each comb, the number of parallel capacitors is $2N - 1$ and the total capacitance is
$$C = (2N-1)\frac{\epsilon_0 A_{\text{effective}}}{\text{distance}} = \frac{(2N-1)\epsilon_0(\pi-\theta)R^2/2}{d/2}$$
$$= \frac{(2N-1)\epsilon_0(\pi-\theta)R^2}{d}$$

8. **(a)** $C = \frac{l}{2k\ln\left(\frac{b}{a}\right)} = \frac{50.0}{2(8.99 \times 10^9)\ln\left(\frac{7.27}{2.58}\right)} = 2.68\text{ nF}$

   (b) Method 1: $V = 2k\lambda\ln\left(\frac{b}{a}\right)$
   $\lambda = \frac{q}{l} = \frac{8.10 \times 10^{-6}\text{ C}}{50.0\text{ m}} = 1.62 \times 10^{-7}\text{ C/m}$
   $V = 2(8.99 \times 10^9)(1.62 \times 10^{-7})\ln\left(\frac{7.27}{2.58}\right) = 3.02\text{ kV}$

   Method 2: $V = \frac{Q}{C} = \frac{8.10 \times 10^{-6}}{2.68 \times 10^{-9}} = 3.02\text{ kV}$

9. Let the radii be b and a with $b = 2a$. Put charge Q on the inner conductor and $-Q$ on the outer. Electric field exists only in the volume between them. The potential of the inner sphere is $V_a = \frac{kQ}{a}$; that of the outer is $V_b = \frac{kQ}{b}$. Then-
   $$V_a - V_b = \frac{kQ}{a} - \frac{kQ}{b} = \frac{Q}{4\pi\epsilon_0}\left(\frac{b-a}{ab}\right)$$
   and
   $$C = \frac{Q}{V_a - V_b} = \frac{4\pi\epsilon_0 ab}{b-a}$$
   Since, $b = 2a$, therefore, $C = \frac{4\pi\epsilon_0 \cdot 2a^2}{a} = 8\pi\epsilon_0 a$
   $\implies a = \frac{C}{8\pi\epsilon_0}$
   The intervening volume is-
   $$\text{Volume} = \frac{4}{3}\pi b^3 - \frac{4}{3}\pi a^3 = 7\left(\frac{4}{3}\pi a^3\right) = 7\left(\frac{4}{3}\pi\right)\frac{C^3}{8^3\pi^3\epsilon_0^3}$$
   $$= \frac{7C^3}{384\pi^2\epsilon_0^3}$$
   $$\text{Volume} = \frac{7(20.0 \times 10^{-6}\text{ C}^2/\text{N} \cdot \text{m})^3}{384\pi^2(8.85 \times 10^{-12}\text{ C}^2/\text{N} \cdot \text{m}^2)^3}$$
   $$= 2.13 \times 10^{16}\text{ m}^3$$
   The outer sphere is 360 km in diameter.

10. **(a)** $C = \frac{ab}{k(b-a)} = \frac{(0.0700)(0.140)}{(8.99 \times 10^9)(0.140 - 0.0700)}$
    $= 15.6\text{ pF}$
    (b) $C = \frac{Q}{\Delta V} \implies \Delta V = \frac{Q}{C} = \frac{4.00 \times 10^{-6}\text{ C}}{15.6 \times 10^{-12}\text{ F}} = 256\text{ kV}$

11. The situation is shown in Fig.3.271a. In equilibrium, we have-
$$\sum F_y = 0 : T\cos\theta - mg = 0$$
$$\sum F_x = 0 : T\sin\theta - Eq = 0$$
Dividing, $\tan\theta = \frac{Eq}{mg}$ so, $E = \frac{mg}{q}\tan\theta$
and, $V = Ed = \frac{mgd\tan\theta}{q}$

12. $C = 4\pi\epsilon_0 R = 4\pi (8.85 \times 10^{-12} \text{C/N} \cdot \text{m}^2)(6.37 \times 10^6 \text{ m})$
$= 7.08 \times 10^{-4}$ F

13. The situation is shown in Fig.3.271b. For an infinitesimal area element of the capacitance a distance y up from the small end, the distance between the plates is $d + x = d + y\tan\theta \approx d + y\theta$.
($\because$ corresponding to a very small value of θ, $\tan\theta \approx \theta$)
Since, the capacitor plates are square, they are of dimension

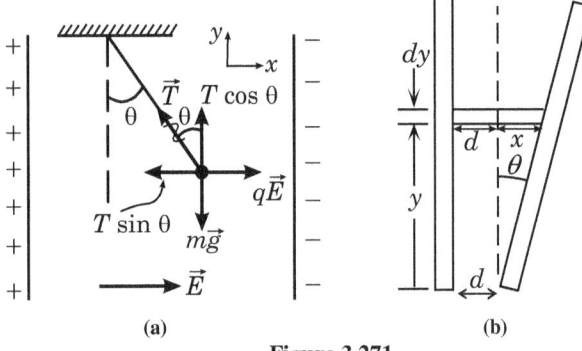

Figure 3.271

$\sqrt{A} \times \sqrt{A}$, and the area of the infinitesimal strip is $dA = \sqrt{A}dy$. The infinitesimal capacitance dC of the strip is calculated, and then the total capacitance is found by adding together all of the infinitesimal capacitances, in parallel with each other.

$$C = \frac{\epsilon_0 A}{d} \implies dC = \frac{\epsilon_0 dA}{d + y\theta} = \frac{\epsilon_0 \sqrt{A} dy}{d + y\theta}$$

$$\implies C = \int dC = \int_0^{\sqrt{A}} \frac{\epsilon_0 \sqrt{A} dy}{d + y\theta} = \frac{\epsilon_0 \sqrt{A}}{\theta} \ln(d + y\theta)\Big|_0^{\sqrt{A}}$$

$$= \frac{\epsilon_0 \sqrt{A}}{\theta}[\ln(d + \theta\sqrt{A}) - \ln d]$$

$$= \frac{\epsilon_0 \sqrt{A}}{\theta} \ln\left(\frac{d + \theta\sqrt{A}}{d}\right) = \frac{\epsilon_0 \sqrt{A}}{\theta} \ln\left(1 + \frac{\theta\sqrt{A}}{d}\right)$$

Using approximation: $\ln(1 + x) \approx x - \frac{1}{2}x^2$, we get-

$$C = \frac{\epsilon_0 \sqrt{A}}{\theta} \ln\left(1 + \frac{\theta\sqrt{A}}{d}\right) = \frac{\epsilon_0 \sqrt{A}}{\theta}\left[\frac{\theta\sqrt{A}}{d} - \frac{1}{2}\left(\frac{\theta\sqrt{A}}{d}\right)^2\right]$$

$$= \left[\frac{\epsilon_0 A}{d}\left(1 - \frac{\theta\sqrt{A}}{2d}\right)\right]$$

14. (a) Capacitors in parallel add. Thus, the equivalent capacitor has a value of
$C_{eq} = C_1 + C_2 = 5.00 \mu F + 12.0 \mu F = 17.0 \mu F$
(b) The potential difference across each branch is the same and equal to the voltage of the battery, i.e., $V = 9.00$ volt
(c) $Q_5 = CV = (5.00 \mu F)(9.00 V) = 45.0 \mu C$
and $Q_{12} = CV = (12.0 \mu F)(9.00 V) = 108 \mu C$

15. (a) In series capacitors add as
$$\frac{1}{C_{eq}} = \frac{1}{C_1} + \frac{1}{C_2} = \frac{1}{5.00 \mu F} + \frac{1}{12.0 \mu F}$$

and $C_{eq} = 3.53 \mu F$
(c) The charge on the equivalent capacitor is
$Q_{eq} = C_{eq} V = (3.53 \mu F)(9.00 V) = 31.8 \mu C$.
Each of the series capacitors has this same charge on it.
So, $Q_1 = Q_2 = 31.8 \mu C$
(b) The potential difference across each is
$$V_1 = \frac{Q_1}{C_1} = \frac{31.8 \mu C}{5.00 \mu F} = 6.35 \text{ V}$$
and $V_2 = \frac{Q_2}{C_2} = \frac{31.8 \mu C}{12.0 \mu F} = 2.65$ V

16. In Fig.3.272a, ①, ② and ③ are three capacitors formed by given arrangement. These are in in parallel order. All plates have same surface area $= a = A/3$. Plates separation for capacitor ①, ② and ③ are d, $d+b$ and $d+2b$ respectively.
If C_1, C_2 and C_3 are respective capacitances of capacitors

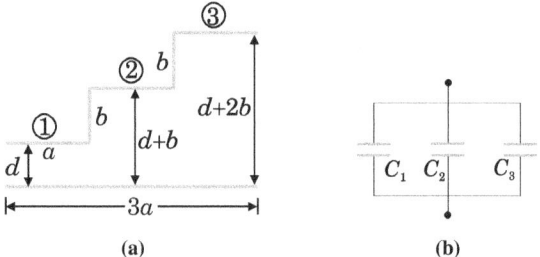

Figure 3.272

①, ② and ③, then-
$$C_1 = \frac{\epsilon_0 A}{3d}, \quad C_2 = \frac{\epsilon_0 A}{3(d+b)} \quad C_3 = \frac{\epsilon_0 A}{3(d+2b)}$$
Their parallel combination is shown in Fig.3.272b.
The equivalent capacitance is given by-
$$C_{eq} = C_1 + C_2 + C_3$$
$$= \frac{\epsilon_0 A}{3d} + \frac{\epsilon_0 A}{3(d+b)} + \frac{\epsilon_0 A}{3(d+2b)}$$
$$= \frac{\epsilon_0 A}{3}\left(\frac{1}{d} + \frac{1}{d+b} + \frac{1}{d+2b}\right)$$
$$= \frac{\epsilon_0 A}{3}\left(\frac{(d+b)(d+2b) + (d+2b)d + (d+b)d}{d(d+b)(d+2b)}\right)$$
$$= \frac{\epsilon_0 A (3d^2 + 6db + 2b^2)}{3d(d+b)(d+2b)}$$

17. (a) Given circuit is redrawn in Fig.3.273a and it's simplified circuit diagram is shown in Fig.3.273b.
From Fig.3.273b, we have-

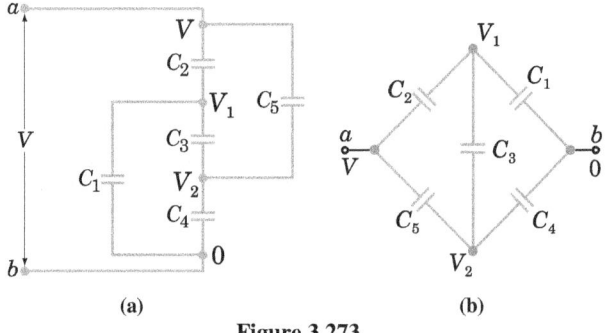

Figure 3.273

$$\frac{C_2}{C_5} = \frac{8.0 \mu F}{4.5 \mu F} \quad \text{and} \quad \frac{C_1}{C_4} = \frac{8.0 \mu F}{4.5 \mu F}$$

From above it is clear that, $\frac{C_2}{C_5} = \frac{C_1}{C_4}$

So, $V_1 = V_2$ and we can remove capacitor C_3. After removing it, C_2 and C_1 will be in series; C_5 and C_4 will also

be in series. Their equivalent are in parallel.
Equivalent of C_2 and C_1 is $C_{12} = \frac{C_2 C_1}{C_1+C_2}$
and equivalent of C_4 and C_5 is $C_{45} = \frac{C_4 C_5}{C_4+C_5}$
The equivalent of C_{12} and C_{45} is-
$$C = C_{12} + C_{54} = \frac{C_2 C_1}{C_1+C_2} + \frac{C_5 C_4}{C_5+C_4}$$
Substituting given values, we get
$$C = \frac{64}{16} + \frac{20.25}{9} = 4 + 2.25 = 6.25 \,\mu F$$
(b) Equivalent capacitance-
$C = \frac{C_1[C_2(C_3+C_4+C_5)+C_5(C_3+C_4)]+C_4(C_2C_3+C_2C_5+C_3C_5)}{C_1(C_3+C_4+C_5)+C_2(C_3+C_4+C_5)+C_3(C_4+C_5)}$
On substituting the given values, we get: $C = 6.0 \,\mu F$

18. Let C be the capacitance of each capacitor in the ladder and let C_{eq} be the equivalent capacitance of the infinite ladder without considering the series capacitor in the first rung. Because the capacitors are infinite therefore, adding one more stage to the ladder will not change the capacitance of the network. The capacitance of the two capacitor combination shown in Fig.3.274 is the equivalent of the infinite ladder, so it has capacitance C_{eq} also.

The equivalent capacitance of the parallel combination of

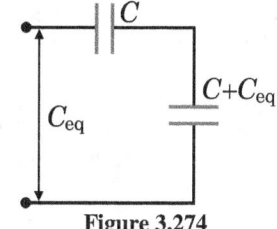

Figure 3.274

C and C_{eq} is $C + C_{eq}$.
It is shown in Fig.3.274.
The equivalent capacitance of the series combination of C and $(C + C_{eq})$ is C_{eq}, so:
$$\frac{1}{C_{eq}} = \frac{1}{C} + \frac{1}{C+C_{eq}}$$
Solve this expression to obtain a quadratic equation in C_{eq}:
$$C_{eq}^2 + CC_{eq} - C^2 = 0$$
On solving for the positive value of C_{eq} gives:
$$C_{eq} = \left(\frac{\sqrt{5}-1}{2}\right) C = 0.618\, C$$

19. **(a)** In Fig.3.153a, points P and Q are interconnected, therefore they are at same potential. As a result of which capacitor C_1 (between terminals A and P) will be in parallel with capacitor C_2 (between terminals A and Q). Similarly, capacitor C_1 (between terminals P and B) is in parallel with capacitor C_2 (between terminals Q and B). There equivalents are in series.
If C_{12} is the equivalent capacitance of C_1 and C_2 connected between AP and AQ, then-
$$C_{12} = C_1 + C_2$$
Now, if C'_{12} is the equivalent capacitance of C_1 and C_2 connected between BP and BQ, then-
$$C'_{12} = C_1 + C_2$$
As, C_{12} and C'_{12} are in series, therefore net capacitance between terminals A and B is
$$\frac{1}{C} = \frac{1}{C_{12}} + \frac{1}{C'_{12}} = \frac{1}{C_1+C_2} + \frac{1}{C_1+C_2}$$
$$\Rightarrow \frac{1}{C} = \frac{1+1}{C_1+C_2}$$
$$\Rightarrow C = (C_1 + C_2)/2 = (4.0\,\mu F + 6.0\,\mu F)/2 = 5\,\mu F$$
(b) In Fig.3.153b, the circuit above dashed line AB is same as that of just below it.

These circuits are equivalent to the circuit of Fig.3.153a. Therefore, above the dashed line, net capacitance between the terminals A and B is $5\,\mu F$ and the capacitance just below the dashed line is also $5\,\mu F$. Now, these are in parallel combination.
Therefore, net capacitance between terminal A and B is
$$C = 5\,\mu F + 5\,\mu F = 10\,\mu F$$

20. $C_p = C_1 + C_2$ and $\frac{1}{C_s} = \frac{1}{C_1} + \frac{1}{C_2}$
Substituting $C_2 = C_p - C_1$, we get-
$$\frac{1}{C_s} = \frac{1}{C_1} + \frac{1}{C_p-C_1} = \frac{C_p-C_1+C_1}{C_1(C_p-C_1)}$$
$$\Rightarrow C_1^2 - C_1 C_p + C_p C_s = 0.$$
$$\Rightarrow C_1 = \frac{C_p \pm \sqrt{C_p^2 - 4C_p C_s}}{2} = \tfrac{1}{2}C_p \pm \sqrt{\tfrac{1}{4}C_p^2 - C_p C_s}$$
We choose arbitrarily the + sign. (This choice can be arbitrary, since with the case of the minus sign, we would get the same two answers with their names interchanged.)
$$C_1 = \tfrac{1}{2}C_p + \sqrt{\tfrac{1}{4}C_p^2 - C_p C_s}$$
$$= \tfrac{1}{2}(9.00\text{ pF}) + \sqrt{\tfrac{1}{4}(9.00\text{ pF})^2 - (9.00\text{ pF})(2.00\text{ pF})}$$
$$= 6.00 \text{ pF}$$
$$C_2 = C_p - C_1 = \tfrac{1}{2}C_p - \sqrt{\tfrac{1}{4}C_p^2 - C_p C_s}$$
$$= \tfrac{1}{2}(9.00\text{ pF}) - 1.50\text{ pF} = 3.00\text{ pF}$$

21. (a) If C_s is the equivalent capacitance of capacitors $15.0\,\mu C$ and $3.00\,\mu C$ attached in series as shown in the circuit in Fig.3.275a, then
$$\frac{1}{C_s} = \frac{1}{15.0} + \frac{1}{3.00} \quad \Rightarrow \quad C_s = 2.50\,\mu F$$
This C_s is in parallel with $6.00\,\mu F$ capacitor. It is shown in Fig.3.275b. If C_p is the parallel resultant of these capacitors, then
$$C_p = 2.50 + 6.00 = 8.50\,\mu F$$

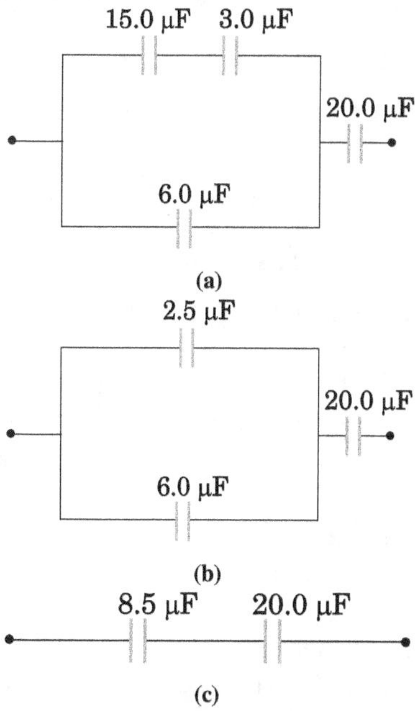

15.0 µF   3.0 µF

20.0 µF

6.0 µF

(a)

2.5 µF

20.0 µF

6.0 µF

(b)

8.5 µF   20.0 µF

(c)

Figure 3.275

This C_p is in series with $20.0\,\mu F$ capacitor as shown in in the circuit in Fig.3.275c. The equivalent of C_p and $20.0\,\mu F$ is-
$$C_{eq} = \left(\frac{1}{8.50\,\mu F} + \frac{1}{20.0\,\mu F}\right)^{-1} = 5.96\,\mu F$$
(b) $Q = CV = (5.96\,\mu F)(15.0\text{ V}) = 89.5\,\mu C$ on $20.0\,\mu F$
Therefore, potential difference across $20\mu F$ capacitor:

$$V_{20\,\mu F} = \frac{Q}{C} = \frac{89.5\ \mu C}{20.0\ \mu F} = 4.47\ V$$
Therefore, potential difference across C_p:
$$V_p = 15.0 - 4.47 = 10.53\ V$$
So, charge stored on $6\ \mu F$ capacitor:
$$Q = CV = (6.00\ \mu F)(10.53\ V) = 63.2\ \mu C \text{ on } 6.00\ \mu F$$
Therefore, charge on $15.0\ \mu F$ and $3.00\ \mu F$ capacitors connected in series:
$$Q_s = 89.5\ \mu C - 63.2\ \mu C = 26.3\ \mu C$$

22. $C = \frac{Q}{V} \implies 6.00 \times 10^{-6} = \frac{Q}{20.0}$
$\implies Q = 120\ \mu C,\ Q_1 = 120\ \mu C - Q_2$
and $V = \frac{Q}{C}: \frac{120 - Q_2}{C_1} = \frac{Q_2}{C_2}$
or $\frac{120 - Q_2}{6.00} = \frac{Q_2}{3.00}$
$(3.00)(120 - Q_2) = (6.00)Q_2$
$\implies Q_2 = \frac{360}{9.00} = 40.0\ \mu C$
and $Q_1 = 120\ \mu C - 40.0\ \mu C = 80.0\ \mu C$

23. $nC = \dfrac{100}{\underbrace{\frac{1}{C} + \frac{1}{C} + \frac{1}{C} + \cdots}_{n\text{ capacitors}}} = \dfrac{100}{n/C} = \dfrac{100C}{n}$

so, $n^2 = 100$ or $n = 10$

24. $C_s = \left(\frac{1}{5.00} + \frac{1}{7.00}\right)^{-1} = 2.92\ \mu F$
$C_p = 2.92 + 4.00 + 6.00 = 12.9\ \mu F$

25. (a) $U = \frac{1}{2}CV^2 = \frac{1}{2}(3.00\ \mu F)(12.0\ V)^2 = 216\ \mu J$
(b) $U = \frac{1}{2}CV^2 = \frac{1}{2}(3.00\ \mu F)(6.00\ V)^2 = 54.0\ \mu J$

26. Use $U = \frac{1}{2}\frac{Q^2}{C}$ and $C = \frac{\epsilon_0 A}{d}$.
If $d_2 = 2d_1$, $C_2 = \frac{1}{2}C_1$, then, the stored energy doubles.

27. Energy density is given by-
$$u = \frac{U}{V} = \frac{1}{2}\epsilon_0 E^2$$
$$\frac{1.00 \times 10^{-7}}{V} = \frac{1}{2}\left(8.85 \times 10^{-12}\right)(3000)^2$$
$$V = 2.51 \times 10^{-3}\ m^3 = \left(2.51 \times 10^{-3}\ m^3\right)\left(\frac{1000\ L}{m^3}\right)$$
$$= 2.51\ L$$

28. $W = U = \int F\,dx$
So, $F = \frac{dU}{dx} = \frac{d}{dx}\left(\frac{Q^2}{2C}\right) = \frac{d}{dx}\left(\frac{Q^2 x}{2\epsilon_0 A}\right) = \frac{Q^2}{2\epsilon_0 A}$

29. With switch closed, distance $d' = 0.500\ d$ and capacitance
$C' = \frac{\epsilon_0 A}{d'} = \frac{2\epsilon_0 A}{d} = 2C$.
(a) $Q = C'V = 2CV = 2(2.00 \times 10^{-6}\ F)(100\ V)$
$= 400\ \mu C$
(b) The force stretching out one spring is-
$$F = \frac{Q^2}{2\epsilon_0 A} = \frac{4C^2 V^2}{2\epsilon_0 A} = \frac{2C^2 V^2}{(\epsilon_0 A/d)d} = \frac{2CV^2}{d}$$
One spring stretches by distance $x = d/4$, so spring constant (k_{sp}),
$$k_{sp} = \frac{F}{x} = \frac{2CV^2}{d}\left(\frac{4}{d}\right) = \frac{8CV^2}{d^2}$$
$$= \frac{8(2.00 \times 10^{-6}\ F)(100\ V)^2}{(8.00 \times 10^{-3}\ m)^2} = 2.50\ kN/m$$

30. The energy transferred is -
$$E_t = \frac{1}{2}QV = \frac{1}{2}(50.0\ C)(1.00 \times 10^8\ V)$$
$$= 2.50 \times 10^9\ J$$
and 1% of this (or $E_{int} = 2.50 \times 10^7\ J$) is absorbed by the tree. If m is the amount of water boiled away, then
$$E_{int} = m(4186\ J/kg \cdot °C)(100°C - 30.0°C)$$
$$+ m(2.26 \times 10^6\ J/kg) = 2.50 \times 10^7\ J$$
Simplifying it, we get $m = 9.79\ kg$.

31. (a) $U = \frac{1}{2}CV^2 + \frac{1}{2}CV^2 = CV^2$
(b) The altered capacitor has capacitance $C' = C/2$. The total charge is the same as before:
$$CV + CV = C(V') + \frac{C}{2}(V') \implies V' = 4V/3$$
(c) $U' = \frac{1}{2}C\left(\frac{4V}{3}\right)^2 + \frac{1}{2}\frac{1}{2}C\left(\frac{4V}{3}\right)^2 = 4C\frac{V^2}{3}$
(d) The extra energy comes from work put into the system by the agent pulling the capacitor plates apart.

32. $U = \frac{1}{2}CV^2$
where $C = 4\pi\epsilon_0 R = \frac{R}{k}$ and $V = \frac{kQ}{R} - 0 = \frac{kQ}{R}$, therefore-
$$U = \frac{1}{2}\left(\frac{R}{k}\right)\left(\frac{kQ}{R}\right)^2 = \frac{kQ^2}{2R}$$

33. $Q_{max} = CV_{max}$, but $V_{max} = E_{max}d$ also $C = \frac{\kappa\epsilon_0 A}{d}$
Therefore, $Q_{max} = \frac{\kappa\epsilon_0 A}{d}(E_{max}d) = \kappa\epsilon_0 A E_{max}$
(a) With air between the plates, $\kappa = 1.00$, $E_{max} = 3.00 \times 10^6\ V/m$ and $E_{max} = 3.00 \times 10^6\ V/m$.
Therefore, $Q_{max} = \kappa\epsilon_0 A E_{max}$
$= (8.85 \times 10^{-12}\ F/m)(5.00 \times 10^{-4}\ m^2)(3.00 \times 10^6\ V/m)$
$= 13.3\ nC$
(b) With polystyrene between the plates, $\kappa = 2.56$ and $E_{max} = 24.0 \times 10^6\ V/m$.
$Q_{max} = \kappa\epsilon_0 A E_{max}$
$= 2.56(8.85 \times 10^{-12}\ F/m)(5.00 \times 10^{-4}\ m^2)$
$(24.0 \times 10^6\ V/m) = 272\ nC$

34. **APPROACH** We can consider this capacitor as two capacitors in in parallel.
SOLUTION In Fig.3.158a, the potential difference is the same on each half of the capacitor, so it can be treated as two capacitors in parallel. Each parallel capacitor has half of the total area of the original capacitor.
$$C = C_1 + C_2 = \kappa_1\epsilon_0\frac{A/2}{d} + \kappa_2\epsilon_0\frac{A/2}{d} = \frac{1}{2}(\kappa_1 + \kappa_2)\epsilon_0\frac{A}{d}$$

35. **APPROACH** We can consider this capacitor as two capacitors in series.
SOLUTION In Fig.3.158b, the intermediate potential at the boundary of the two dielectrics can be treated as the "low" potential plate of one half and the "high" potential plate of the other half, so we treat it as two capacitors in series. Each series capacitor has half of the inter-plate distance of the original capacitor.
$$\frac{1}{C} = \frac{1}{C_1} + \frac{1}{C_2} = \frac{d/2}{\kappa_1\epsilon_0 A} + \frac{d/2}{\kappa_2\epsilon_0 A} = \frac{d}{2\epsilon_0 A}\frac{\kappa_1 + \kappa_2}{\kappa_1\kappa_2}$$
$$\implies C = \frac{2\epsilon_0 A}{d}\frac{\kappa_1\kappa_2}{\kappa_1 + \kappa_2}$$

36. (a) $C = \kappa C_0 = \frac{\kappa\epsilon_0 A}{d}$
$$= \frac{(173)(8.85 \times 10^{-12}\ F/m)(1.00 \times 10^{-4}\ m^2)}{0.100 \times 10^{-3}\ m}$$
$= 1.53\ nF$
(b) The battery delivers the free charge
$Q = C(\Delta V) = (1.53 \times 10^{-9}\ F)(12.0V) = 18.4\ nC$
(c) The surface density of free charge is
$$\sigma = \frac{Q}{A} = \frac{18.4 \times 10^{-9}\ C}{1.00 \times 10^{-4}\ m^2} = 1.84 \times 10^{-4}\ C/m^2$$
The surface density of polarization charge is
$$\sigma_p = \sigma\left(1 - \frac{1}{\kappa}\right) = \sigma\left(1 - \frac{1}{173}\right) = 1.83 \times 10^{-4}\ C/m^2$$
(d) We have $E = \frac{E_0}{\kappa}$ and $E_0 = \frac{V}{d}$; hence,
$$E = \frac{V}{\kappa d} = \frac{12.0\ V}{(173)(1.00 \times 10^{-4}\ m)} = 694\ V/m$$

37. The given combination of capacitors is equivalent to the circuit shown in Fig.3.276.
Put charge Q on point A. Then,
$Q = (40.0\ \mu F)V_{AB} = (10.0\ \mu F)V_{BC}$
$= (40.0\ \mu F)V_{CD}$

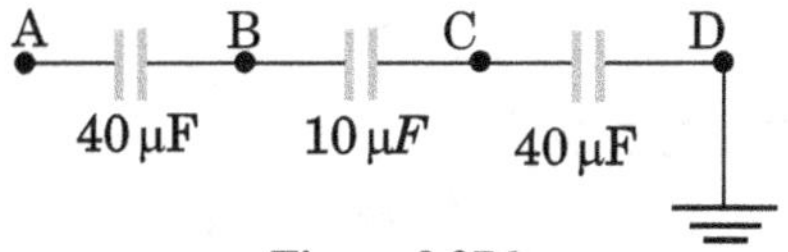

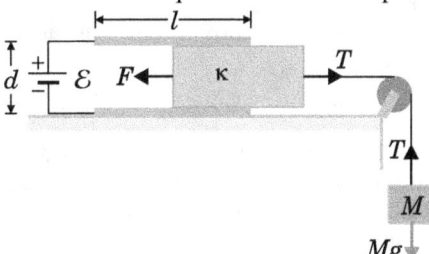

Figure 3.276

So, $V_{BC} = 4V_{AB} = 4V_{CD}$, and the centre capacitor will break down first, at $V_{BC} = 15.0$ V. When this occurs,
$V_{AB} = V_{CD} = \frac{1}{4}(V_{BC}) = 3.75$ V
and $V_{AD} = V_{AB} + V_{BC} + V_{CD} = 3.75$ V $+ 15.0$ V $+ 3.75$ V
$= 22.5$ V

38. The situation is redrawn in Fig.3.277. The forces acting on the dielectric slab are-
   1. Electric force, $F = \frac{\epsilon_0 b \mathcal{E}^2 (\kappa - 1)}{2d}$ (towards left side)
   2. Tension of the string T (towards right side).
   Forces acting on the suspended mass M are-
   1. Gravitational force Mg (downwards)
   2. Tension of the string T (upwards)
   For the translational equilibrium of the suspended mass M,

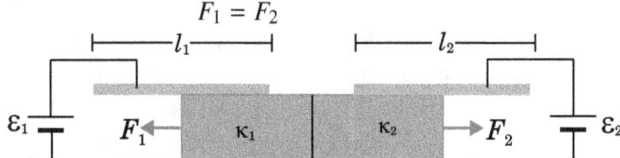

Figure 3.277

we have-
$$T = Mg \qquad \ldots(1)$$
and for the translational equilibrium of the dielectric slab, we have-
$$F = T \qquad \ldots(2)$$
Adding Eq.(1) and (2), we get-
$$F = Mg \implies M = F/g$$
Substituting the value of F in above expression, we get
$$M = \frac{\epsilon_0 b \mathcal{E}^2 (\kappa - 1)}{2dg}$$

39. If F_1 is the electric field force on left dielectric and F_2 on right dielectric, then in translational equilibrium, we have-
$$F_1 = F_2$$

Figure 3.278

$$\frac{\epsilon_0 b \mathcal{E}_1^2 (\kappa_1 - 1)}{2d} = \frac{\epsilon_0 b \mathcal{E}_2^2 (\kappa_2 - 1)}{2d}$$
On simplifying we get-
$$\frac{\mathcal{E}_1}{\mathcal{E}_2} = \frac{\sqrt{\kappa_2 - 1}}{\sqrt{\kappa_1 - 1}}$$

40. There are two uniform electric fields - one in the air, and one in the gap. In each region, the potential difference is the field times the distance in the direction of the field over which the field exists.
$$V = E_{air} d_{air} + E_{glass} d_{glass}$$
$$= E_{air} d_{air} + \frac{E_{air}}{\kappa_{glass}} d_{glass}$$
$$\implies E_{air} = V \frac{\kappa_{glass}}{d_{air}\kappa_{glass} + d_{glass}}$$
$$= (90.0 \text{ V}) \frac{5.80}{(3.00 \times 10^{-3} \text{ m})(5.80) + (2.00 \times 10^{-3} \text{ m})}$$
$$= 2.69 \times 10^4 \text{ V/m}$$
$$E_{glass} = \frac{E_{air}}{\kappa_{glass}} = \frac{2.69 \times 10^4 \text{ V/m}}{5.80} = 4.64 \times 10^3 \text{ V/m}$$
The charge on the plates can be calculated from the field at the plate, using Eq. $E_{air} = \frac{\sigma_{plate}}{\epsilon_0}$. Use Eq.3.54: $Q_{ind} = Q\left(1 - \frac{1}{\kappa}\right)$ to calculate the charge on the dielectric.
$$E_{air} = \frac{\sigma_{plate}}{\epsilon_0} = \frac{Q_{plate}}{\epsilon_0 A}$$
$$Q_{plate} = E_{air} \epsilon_0 A$$
$$= (2.69 \times 10^4 \text{ V/m})(8.85 \times 10^{-12} \text{ C}^2/\text{N} \cdot \text{m}^2)(1.45 \text{ m}^2)$$
$$= 3.45 \times 10^{-7} \text{ C}$$
$$Q_{ind} = Q\left(1 - \frac{1}{\kappa}\right) = (3.45 \times 10^{-7} \text{ C})\left(1 - \frac{1}{5.80}\right)$$
$$= 2.86 \times 10^{-7} \text{ C}$$

41. (a) $C_1 = \frac{\kappa_1 \epsilon_0 A/2}{d}$; $C_2 = \frac{\kappa_2 \epsilon_0 A/2}{d/2}$; $C_3 = \frac{\kappa_3 \epsilon_0 A/2}{d/2}$
$$\left(\frac{1}{C_2} + \frac{1}{C_3}\right)^{-1} = \frac{C_2 C_3}{C_2 + C_3} = \frac{\epsilon_0 A}{d}\left(\frac{\kappa_2 \kappa_3}{\kappa_2 + \kappa_3}\right)$$
$$C = C_1 + \left(\frac{1}{C_2} + \frac{1}{C_3}\right)^{-1} = \frac{\epsilon_0 A}{d}\left(\frac{\kappa_1}{2} + \frac{\kappa_2 \kappa_3}{\kappa_2 + \kappa_3}\right)$$
(b) Using the given values we find:
$C_{total} = 1.76 \times 10^{-12}$ F $= 1.76$ pF

42. (a) Put charge Q on the sphere of radius a and $-Q$ on the other sphere. Relative to $V = 0$ at infinity, the potential at the surface of sphere of radius a is-
$$V_a = \frac{kQ}{a} - \frac{kQ}{d}$$
and the potential of b is
$$V_b = \frac{-kQ}{b} + \frac{kQ}{d}$$
The difference in potential is
$$V_a - V_b = \frac{kQ}{a} + \frac{kQ}{b} - \frac{kQ}{d} - \frac{kQ}{d}$$
and $\qquad C = \frac{Q}{V_a - V_b} = \left(\frac{4\pi\epsilon_0}{(1/a) + (1/b) - (2/d)}\right)$
(b) As $d \to \infty$, $\frac{1}{d}$ becomes negligible compared to $\frac{1}{a}$. Then,
$$C = \frac{4\pi\epsilon_0}{1/a + 1/b}$$
As for two spheres in series, we have -
$$\frac{1}{C} = \frac{1}{4\pi\epsilon_0 a} + \frac{1}{4\pi\epsilon_0 b} \implies C = \frac{4\pi\epsilon_0}{1/a + 1/b}$$
So, the result reduces to that of two spherical capacitors in series.

43. The vertical orientation sets up two capacitors in parallel, with equivalent capacitance
$$C_p = \frac{\epsilon_0 (A/2)}{d} + \frac{\kappa \epsilon_0 (A/2)}{d} = \left(\frac{\kappa + 1}{2}\right)\frac{\epsilon_0 A}{d}$$
Where A is the area of either plate and d is the separation between the plates. The horizontal orientation produces two capacitors in series. If f is the fraction of the horizontal capacitor filled with dielectric, the equivalent capacitance is
$$\frac{1}{C_s} = \frac{fd}{\kappa \epsilon_0 A} + \frac{(1-f)d}{\epsilon_0 A} = \left[\frac{f + \kappa(1-f)}{\kappa}\right]\frac{d}{\epsilon_0 A}$$
$$\implies C_s = \left[\frac{\kappa}{f + \kappa(1-f)}\right]\frac{\epsilon_0 A}{d}$$
Requiring that $C_p = C_s$ gives $\frac{\kappa + 1}{2} = \frac{\kappa}{f + \kappa(1-f)}$,
or $\qquad (\kappa + 1)[f + \kappa(1-f)] = 2\kappa$.
For $\kappa = 2.00$, this yields-
$$3.00[2.00 - (1.00)f] = 4.00 \implies f = 2/3.$$

44. **APPROACH** The force between the plates is equal to the

derivative of the stored electrostatic energy with respect to the plate separation.

To avoid confusion dd (as in $F = -dE/dd$) in relating the force on the electrostatic balance plates to the electric field in the region between them, let l be the variable separation of the plates. We can use the definition of the work done in charging the capacitor to relate the force on the upper plate to the energy stored in the capacitor. Solving this expression for the force and substituting for the energy stored in a parallel-plate capacitor will yield an expression that we can use to decide whether the balance is stable. We can use this same expression and a condition for equilibrium to find the voltage required to balance the object whose mass is M.

SOLUTION (a) Express the work done in charging the capacitor (the energy stored in it) in terms of the force between the plates:

$$dW = dE = -F\, dl \implies F = -\frac{dE}{dl}$$

The energy stored in the capacitor is given by:

$$E = \frac{1}{2}CV_0^2 = \frac{1}{2}\left(\frac{\epsilon_0 A}{l}\right)V_0^2$$

Differentiating E with respect to l, we get -

$$F = -\frac{d}{dl}\left[\frac{1}{2}\left(\frac{\epsilon_0 A}{l}\right)V_0^2\right] = \left(\frac{\epsilon_0 A}{2l^2}\right)V_0^2$$

Changing back to the variable d, we get

$$F = \left(\frac{\epsilon_0 A}{2d^2}\right)V_0^2$$

Because F increases as l decreases, a decrease in plate separation will unbalance the system. Hence, the balance is unstable.

(b) Apply $\sum F = 0$ to the object whose mass is M when the plate separation is d_0 to obtain:

$$Mg - \left(\frac{\epsilon_0 A}{2d_0^2}\right)V^2 = 0 \implies V = d_0\sqrt{\frac{2Mg}{\epsilon_0 A}}$$

45. APPROACH The system is equivalent to two capacitors in parallel. One of the capacitors has plate separation d, plate area $w(L - h)$ and air between the plates. The other has the same plate separation d, plate area wh and dielectric constant κ.

Define κ_{eff} by $C_{\text{eq}} = \frac{K_{\text{eff}}\epsilon_0 A}{d}$, where $A = wL$. For two capacitors in parallel, $C_{\text{eq}} = C_1 + C_2$.

SOLUTION (a) The capacitors are in parallel, so

$$C = \frac{\epsilon_0 w(L-h)}{d} + \frac{K\epsilon_0 wh}{d} = \frac{\epsilon_0 w L}{d}\left(1 + \frac{\kappa h}{L} - \frac{h}{L}\right).$$

This gives $\kappa_{eff} = \left(1 + \frac{Kh}{L} - \frac{h}{L}\right)$.

(b) For gasoline, with $\kappa = 1.95$: $\frac{1}{4}$ full:

$\kappa_{\text{eff}}\left(h = \frac{L}{4}\right) = 1.24$; $\frac{1}{2}$ full: $\kappa_{\text{eff}}\left(h = \frac{L}{2}\right) = 1.48$;

$\frac{3}{4}$ full: $\kappa_{eff}\left(h = \frac{3L}{4}\right) = 1.71$

(c) For methanol, with $\kappa = 33$: $\frac{1}{4}$ full:

$\kappa_{\text{eff}}\left(h = \frac{L}{4}\right) = 9$;

$\frac{1}{2}$ full: $\kappa_{\text{eff}}\left(h = \frac{L}{2}\right) = 17$;

$\frac{3}{4}$ full: $\kappa_{eff}\left(h = \frac{3L}{4}\right) = 25$.

(d) This kind of fuel tank sensor will work best for methanol since it has the greater range of κ_{eff} values.
When $h = 0$, $\kappa_{\text{eff}} = 1$.
When $h = L$, $\kappa_{\text{eff}} = \kappa$.

46. APPROACH The object is equivalent to two identical capacitors in parallel, where each has the same area A, plate separation d and dielectric with dielectric constant κ.

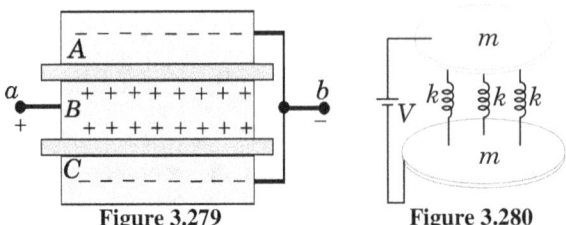

Figure 3.279 **Figure 3.280**

For each capacitor in the parallel combination, $C = \frac{\epsilon_0 A}{d}$.

SOLUTION (a) The charge distribution on the plates is shown in Figure 3.279.

(b) $C = 2\left(\frac{\epsilon_0 A}{d}\right) = \frac{2(4.2)\epsilon_0(0.120\text{ m})^2}{4.5 \times 10^{-4}\text{ m}} = 2.38 \times 10^{-9}$ F.

If two of the plates are separated by both sheets of paper to form a capacitor,

$$C = \frac{\epsilon_0 A}{2d} = \frac{2.38 \times 10^{-9}\text{ F}}{4},$$

smaller by a factor of 4 compared to the capacitor in the problem.

47. SOLUTION (a) For a normal spherical capacitor with air between the plates,

$$C_0 = 4\pi\epsilon_0\left(\frac{r_a r_b}{r_b - r_a}\right)$$

The capacitor shown in Fig.3.168 is equivalent to two parallel capacitors: an upper hemispherical capacitor with capacitance C_U and a lower hemispherical capacitor with capacitance C_L.

$$C_U = \frac{C_0}{2} = 2\pi\epsilon_0\left(\frac{r_a r_b}{r_b - r_a}\right)$$

and $\quad C_L = \frac{\kappa C_0}{2} = 2\pi\kappa\epsilon_0\left(\frac{r_a r_b}{r_b - r_a}\right)$

If C denotes the equivalent capacitance, C is given as-

$$C = C_U + C_L = 2\pi\epsilon_0(1 + \kappa)\left(\frac{r_a r_b}{r_b - r_a}\right).$$

(b) Applying Gauss's theorem — $(\oint \vec{E} \cdot d\vec{A} = Q_{\text{encl}}/\epsilon_0)$ — by considering a hemispherical Gaussian surface for each respective half, we get

$$E_L \frac{4\pi r^2}{2} = \frac{Q_L}{\kappa\epsilon_0} \implies E_L = \frac{Q_L}{2\pi\kappa\epsilon_0 r^2}, \text{ and } E_U \frac{4\pi r^2}{2} = \frac{Q_U}{\epsilon_0}$$

Therefore, $E_U = \frac{Q_U}{2\pi\epsilon_0 r^2}$.

Now, $Q_L = V C_L = V\left(\frac{\kappa C_0}{2}\right) \quad \left[\because C_L = \frac{\kappa C_0}{2}\right]$

$\quad = \kappa\left(\frac{VC_0}{2}\right) = \kappa(VC_U) \quad \left[\because C_U = \frac{C_0}{2}\right]$

$\quad = \kappa Q_U \quad [\because Q_U = VC_U]$

Since, $Q_L + Q_U = Q$, therefore $\kappa Q_U + Q_U = Q$

$\implies Q_U = \frac{Q}{1+\kappa}$, and $Q_L = \kappa Q_U = \frac{\kappa Q}{1+\kappa}$.

This gives, $E_L = \frac{Q_L}{2\pi\kappa\epsilon_0 r^2} = \frac{\kappa Q}{1+\kappa}\left(\frac{1}{2\pi\kappa\epsilon_0 r^2}\right)$

$\quad = \frac{2}{1+\kappa}\left(\frac{Q}{4\pi\epsilon_0 r^2}\right)$

and $E_U = \frac{Q_U}{2\pi\epsilon_0 r^2} = \frac{Q}{1+\kappa}\left(\frac{1}{2\pi\epsilon_0 r^2}\right) = \frac{2}{1+\kappa}\left(\frac{Q}{4\pi\epsilon_0 r^2}\right)$

From above, we find that: $E_U = E_L$.

(c) The free charge density on inner and outer hemispherical plates of upper capacitor:

$$(\sigma_{f,r_a})_U = \frac{Q_U}{2\pi r_a^2} = \frac{Q}{2\pi r_a^2(1+\kappa)}$$

and $(\sigma_{f,r_b})_U = \frac{Q_U}{2\pi r_b^2} = \frac{Q}{2\pi r_b^2(1+\kappa)}$;

Similarly, the free charge density on inner and outer hemispherical plates of lower capacitor:

$(\sigma_{f,r_a})_L = \frac{Q_L}{2\pi r_a^2} = \frac{\kappa Q}{2\pi r_a^2(1+\kappa)}$

and $(\sigma_{f,r_b})_L = \frac{Q_L}{2\pi r_b^2} = \frac{\kappa Q}{2\pi r_b^2(1+\kappa)}$.

(d) $\sigma_{i,r_a} = \sigma_{f,r_a}(1 - 1/\kappa) = \left(\frac{\kappa-1}{\kappa}\right)\frac{Q}{2\pi r_a^2}\left(\frac{\kappa}{\kappa+1}\right)$

$= \left(\frac{\kappa-1}{\kappa+1}\right)\frac{Q}{2\pi r_a^2} \quad \sigma_{i,r_b} = \sigma_{f,r_b}(1-1/\kappa)$

$= \left(\frac{\kappa-1}{\kappa}\right)\frac{Q}{2\pi r_b^2}\left(\frac{\kappa}{\kappa+1}\right) = \left(\frac{\kappa-1}{\kappa+1}\right)\frac{Q}{2\pi r_b^2}$

(e) There is zero bound charge on the flat surface of the dielectric-air interface, or else that would imply a circumferential electric field, or that the electric field changed as we went around the sphere.

The charge is not equally distributed over the surface of each conductor. There must be more charge on the lower half, by a factor of κ, because the polarization of the dielectric means more free charge is needed on the lower half to produce the same electric field.

48. (a) In Fig.3.280, three springs, each with a spring constant of k, are in parallel. Therefore, the effective spring constant (k_e) is

$$k_e = k + k + k = 3k$$

With no voltage source

$$mg = F_{\text{spring}} \quad \cdots (1)$$

With voltage source connected, let the electrostatic force of attraction between the two plates in equilibrium be F_0. The rise in spring force balances this F_0.

$$F_0 = 3\kappa(d_0 - d) \quad \cdots (2)$$

Electric force between plates,

$$F_0 = Q\left(\frac{\sigma}{2\epsilon_0}\right) = Q\left(\frac{Q/A}{2\epsilon_0}\right)$$

$$= \frac{Q^2}{2\epsilon_0 A} = \frac{C^2 V^2}{2\epsilon_0 A} = \frac{\epsilon_0 A V^2}{2d^2}$$

$$\Rightarrow F_0 = \frac{\epsilon_0 A V^2}{2d^2} \quad \cdots (3)$$

From Eq.(2) and (3), we have

$$\frac{\epsilon_0 A V^2}{2d^2} = 3\kappa(d_0 - d)$$

$$V = \sqrt{\frac{6d^2(d_0-d)}{\epsilon_0 A}} \quad \cdots (4)$$

(b) If plate moves by a small distance x the electrostatic force changes by

$$\Delta F_e = \frac{\epsilon_0 A V^2}{2(d+x)^2} - \frac{\epsilon_0 A V^2}{2d^2}$$

$$= \frac{\epsilon_0 A V^2}{2d^2} \cdot \left[\left(1+\frac{x}{d}\right)^{-2} - 1\right]$$

$$= \frac{\epsilon_0 A V^2}{2d^2}\left(-2\frac{x}{d}\right) = -\frac{\epsilon_0 A V^2}{d^3} \cdot x$$

$\left[\because \text{corresponding to } x \ll d, \left(1+\frac{x}{d}\right)^{-2} \approx 1 - \frac{2x}{d}\right]$

Substituting for the value of V from Eq.(4), in above equation, we get

$$|\Delta F_e| = \frac{6\kappa(d_0-d)}{d}\Delta x$$

The spring force changes by
$$|\Delta F_s| = 3\kappa x$$

The two forces ΔF_e and ΔF_s have opposite directions. While approaching the spring pushes the plates away and the electric force causes them to attract more strongly. Therefore,

$$m\frac{d^2x}{dt^2} = -3\kappa x + \frac{6\kappa(d_0-d)}{d}x$$

$$\Rightarrow \frac{d^2x}{dt^2} = -\frac{3\kappa}{m}\left[\frac{3d-2d_0}{d}\right]x$$

Which is of the form of SHM- $\frac{d^2x}{dt^2} = -\omega^2 x$

where, $\omega = \sqrt{\frac{3\kappa}{m}\frac{(3d-2d_0)}{d}}$

Therefore, time period (T) is-

$$T = \frac{2\pi}{\omega} = 2\pi\sqrt{\frac{m \cdot d}{3\kappa(3d-2d_0)}}$$

3.19.5 Multiple Choice Assignments
3.19.5.1 Level 1

Q.No.	1	2	3	4	5	6	7	8	9
Ans.	D	A	D	A	D	C	A	A	A
Q.No.	10	11	12	13	14	15	16	17	18
Ans.	C	C	B	D	C	B	A	C	B
Q.No.	19	20	21	22	23	24	25	26	27
Ans.	A	B	B	C	D	C	A	A	C
Q.No.	28	29	30	31	32	33	34	35	36
Ans.	D	D	C	C	B	C	B	B	B
Q.No.	37	38	39	40	41	42	43	44	45
Ans.	A	B	A	D	B	B	A	C	D
Q.No.	46	47	48	49	50	51	52	53	54
Ans.	D	C	C	B	C	D	A	A	A
Q.No.	55	56	57	58	59	60	61	62	63
Ans.	B	D	A	A	A	A	A	B	C
Q.No.	64	65	66	67	68	69	70	71	72
Ans.	D	C	A	C	C	B	D	D	A
Q.No.	73	74	75	76	77	78	79	80	81
Ans.	C	B	D	B	D	C	C	C	C
Q.No.	82	83	84	85	86	87	88	89	90
Ans.	A	B	D	A	C	A	A	A	B
Q.No.	91	92	93	94	95	96			
Ans.	C	B	C						

3.19.5.2 Level 2

Q.No.	1	2	3	4	5	6	7	8	9
Ans.	B	A	A	C	B	C	A	A	A
Q.No.	10	11	12	13	14	15	16	17	18
Ans.	A	B	C	B	A	D	C	B	C
Q.No.	19	20	21	22	23	24	25	26	27
Ans.	C	C	A	B	C	D	C	A	A
Q.No.	28								
Ans.	B								

3.19.5.3 Level 3

Q.No.	1	2	3	4	5	6	7	8	9
Ans.	A	C	A	B	A	D	C	A	C
Q.No.	10	11	12	13	14	15	16	17	18
Ans.	D	B	C	D	A	A	B	B	D

Q.No.	19	20	21	22	23	24	25		
Ans.	B	A	A	A	D	C	A		

3.19.5.4 Level 4
Section A

Q.No.	1	2	3	4	5	6	7	8	9
Ans.	B	A	A	C	D	A	D	D	B
Q.No.	10	11	12	13	14	15	16	17	18
Ans.	C	C	D	D	C	D	C	C	B
Q.No.	18	20	21	22	23	24	25	26	27
Ans.	D	C	C	D	A	A	D	D	A
Q.No.	28	29	30	31	32	33	34	35	36
Ans.	A	D	C	B	A	C	D	A	D
Q.No.	37	38	39	40	41	42	43	44	45
Ans.	B	C	B	B	D	A	C	B	B
Q.No.	46	47	48	49	50	51	52	53	54
Ans.	B	B	A	D	C	D	(6)	A	A
Q.No.	55	56	57	58	59	60	61	62	63
Ans.	A	C	B	(4)	D	A	B	D	D
Q.No.	64	65	66	67	68	69	70	71	72
Ans.	C	(3)	(2)	864	16	161	A	C	C
Q.No.	73	74	75	76	77	78	79	80	81
Ans.	D	C	D	C	6	A	A	C	23
Q.No.	82	83	84	85	86	87	88	89	90
Ans.	100	105	3	6	225	55	3	5	6
Q.No.	91	92	93	94	95	96			
Ans.	C	C	B	C	4	5			

Section B

Q.No.	1	2	3	4	5	6	7	8	9
Ans.	D	D	D	D	A	C	D	C	BD
Q.No.	10	11	12	13	14	15			
Ans.	AD	D	1.30	8	B	B			

Chapter 4
Current Electricity and DC Circuits

For stationary charge on the surface of conductor, the electric field must be zero inside the conductor (if it weren't, the charges would move). But in case of moving charge in a conductor, there must be an electric field in the conductor. Because, an electric field is needed to set charges into motion, and to keep them in motion in any normal conductor. By regulating electric field or voltage (potential), we can control the flow of charge. In order to have a current in a wire, a potential difference is necessary, which can be provided by a battery.

We first look at electric current from a macroscopic point of view: that is, current as measured in a laboratory. Later in the Chapter we look at currents from a microscopic (theoretical) point of view as a flow of electrons in a wire. We also analyse simple and complicated (symmetric and asymmetric) electric circuits that contain batteries, resistors, and capacitors in various combinations. Most of the circuits analysed are assumed to be in steady state, which means that currents in the circuit are constant in magnitude and direction. A current that is constant in direction is called a direct current (DC). The current that changes direction periodically, is called alternating current (AC). In this chapter we will be studying only DC. Finally, we describe electrical meters for measuring current and potential difference.

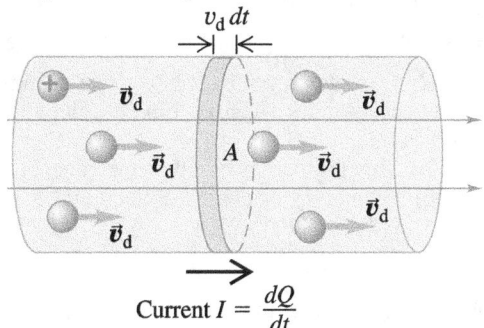

Figure 4.1: The current is the time rate of charge transfer through the cross-sectional area A. The random component of each moving charged particle's motion averages to zero, and the current is in the same direction as $\vec{E}$ whether the moving charges are positive (as shown here) or negative

4.1 Electric Current

Electric current (or just current) is the transfer of charge through a certain surface of a material containing free charges—that is, conductor. The free electric charge always flows from higher potential energy state to lower potential energy state. It is just like a free mass in a gravitational field which moves from higher gravitational potential energy state to lower gravitational potential energy state. If at the position of charge Q, the electric potential is V, then associated potential energy $U = QV$. If Q is positive, then, a higher value of potential V means higher value of potential energy U, and lower value of potential V means lower value of potential energy, U. If Q is negative, then a higher value of potential V means lower value of potential energy U, and lower value of potential V means higher value of potential energy U. So, a free positive charge flows from a higher to lower potential and negative charge flows from lower to higher. Metals such as gold, silver, copper, aluminium etc. are good conductors of electricity. Current can be composed of moving positive and negative charges [such as electrons in metals or positive and negative ions in plasmas (ionized gases) and ionic solutions (electrolytes)]. It may occur within an overall-neutral material such as a conductor, or it may occur as a charged beam such as the electron beam of a television tube.

> Quantitatively, the electric current at any point of a conductor is defined as the rate of flow of charge through the conductor's full cross section at that point.

Figure 4.1 shows a segment of a wire that is carrying a current (charges are moving with drift velocity $\vec{v}_d$* in a particular direction). If ΔQ is the amount of charge that flows through the cross-sectional area A in time Δt, then the average current through A is

$$I_{av} = \frac{\Delta Q}{\Delta t} \qquad (4.1)$$

The instantaneous current through A is, $I = \Delta Q/\Delta t$ in the limit that Δt approaches zero, i.e.,

$$I = \lim_{\Delta t \to 0} \frac{\Delta Q}{\Delta t} = \frac{dQ}{dt} \qquad (4.2)$$

It is a scalar quantity.
If current is steady in a given time interval, then it's instantaneous value at any time within the given time interval will always be equal to it's average value for that time interval, i.e., $I = I_{av}$. Therefore, for steady current Eq.4.2 can also be written as-

$$I = \frac{dQ}{dt} = \frac{\Delta Q}{\Delta t} \qquad (4.3)$$

So, in case of steady current, we can replace I by I_{av}.
Unit of Electric Current: From Eq.4.1, the SI unit of electric current

$$= \frac{\text{unit of charge}}{\text{unit of time}} = \frac{\text{coulomb(C)}}{\text{sec(s)}}$$

It is also called ampere (A)

$$1\text{A} = \frac{1\text{C}}{1\text{s}}$$

The ampere is one of the seven fundamental SI units. In fact, a coulomb is defined in terms of the ampere: *If there is a steady current of one ampere (1 A), then one coulomb (1 C) is the amount of charge that passes a particular cross section in one second (1 s).* One ampere is defined to be one coulomb per second (1A = 1C/s). This unit is named in honour of the French scientist André Marie Ampère (1775-1836).
As 1 C is a huge amount of charge, therefore, the ampere is also a fairly large unit, and it is also convenient to express current in milliamperes (1 mA $= 10^{-3}$ A); microamperes (1 μA $= 10^{-6}$A); nanoamperes (1 nA $= 10^{-9}$A) or even picoamperess (1 pA $= 10^{-12}$A).

4.1.1 Calculating Charge From Electric Current
From Eq.4.2, we have

$$I = dQ/dt \Rightarrow dQ = I dt \qquad (4.4)$$

---
*Drift velocity '$\vec{v}_d$' of free charges is that effective velocity by which they move in the given conductor. It will be discussed later in "subsection 4.2.2".

The amount of charge Q that passes through a particular cross section in some time interval from $t = t_1$ to $t = t_2$, can be found by integration of Eq.4.4 as given below-

$$Q = \int dQ = \int_{t_1}^{t_2} I \, dt \qquad (4.5)$$

Fig.4.2 depicts a typical current versus time graph. In this figure, the shaded area for infinitesimally small time interval from t to $t + dt$ is $I\,dt$ which is the charge dQ that passes through the cross section in time dt. Total amount of charge that passes through the cross section from $t = t_1$ to $t = t_2$ is the area under the graph from point A ($t = t_1$) to point B ($t = t_2$). So, the total charge obtained from Eq.4.5 will be equal to above area.

Similarly, the area between O ($t = 0$) to C ($t = t$) under the I-t curve above t axis, will give the charge that passes through the given cross section of wire in time interval $|t - 0|$ i.e., t.

If current (I) is steady (i.e., time independent), then we can take it out of integral. In this case, Eq.4.5 becomes:

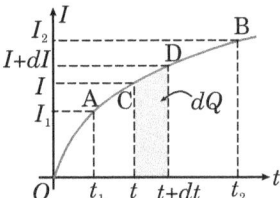

Figure 4.2: Shaded area representing charge dQ which passes through the cross section in time dt.

$$Q = \int dQ = I \int_{t_1}^{t_2} dt = I(t_2 - t_1) = I\Delta t \qquad (4.6)$$

here, $\Delta t = t_2 - t_1$

4.1.2 Rate of Flow of Electrons Through a Given Cross-Section of Conductor

From Eq.4.3, steady current passing through a cross-section of a conductor is given as

$$I = \frac{\Delta Q}{\Delta t} \quad \Rightarrow \quad \frac{\Delta Q}{\Delta t} = I$$

If N is the total number of electrons that pass through a complete cross section in time Δt, then $\Delta Q = Ne$. Now, substituting this value of ΔQ in Eq.4.3: $I = \frac{\Delta Q}{\Delta t}$, we get

$$I = \frac{Ne}{\Delta t}$$

$$\Rightarrow \quad \boxed{\frac{N}{\Delta t} = \frac{I}{e}} \qquad (4.7)$$

Eq.4.7 gives the rate of flow of electrons through a given cross section of a conductor corresponding to current I. If $I = 1$ A, then, $\frac{N}{\Delta t} = \frac{1}{1.6 \times 10^{-19} \text{ C}} = 6.25 \times 10^{18}$

Thus, corresponding to 1 A current in a given conductor, 6.25×10^{18} electrons pass each second through the full cross section of the conductor.

☞ Since electric charge is conserved, therefore, unless charge accumulates within a region, the amount of charge entering a region will always be equal to charge leaving the region.

4.1.3 Current due to Circular Motion of a Point Charge

Consider a point A on the circle of radius r (Fig.**??**). The point charge periodically crosses this point at the intervals of time period T. In other words it crosses the point A once in each revolution. Therefore, in time interval $\Delta t = T$, the charge crossing the point A is $\Delta Q = Q$. Now, from Eq.4.1, the current crossing point A is given by

$$\boxed{I = \frac{\Delta Q}{T} = \frac{Q}{T} = Qf} \qquad (4.8)$$

here, $f = 1/T$ is the frequency of revolution.

If speed of the point charge on circular path is v, then

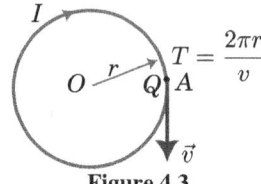

Figure 4.3

$T = \frac{2\pi r}{v}$, therefore from Eq.4.8

$$\boxed{I = \frac{Q}{T} = \frac{Qv}{2\pi r}} \qquad (4.9)$$

EXAMPLE 1. What will be the number of electron passing through a heater wire in one minute, if it carries a current of 8A.
APPROACH From Eq.4.7, the number of electron passing through a cross-section in 1 min is given by:

$$\frac{N}{\Delta t} = \frac{I}{e}$$

Therefore, number of electrons passing through a cross section in time Δt is given by

$$N = \frac{I}{e} \times \Delta t \qquad \cdots (1)$$

SOLUTION Substituting the given values in Eq.(1), we get

$$N = \frac{I}{e} \times \Delta t = \frac{8\text{A}}{1.6 \times 10^{-19}\text{C}} \times 60 = 3 \times 10^{21} \text{ electrons}$$

EXAMPLE 2. An electron moves in a circle of radius 10 cm with a constant speed of 4×10^6 m/s. Find the electric current at a point on the circle.

APPROACH This problem is based on the concept of circular motion of a point charge as discussed in Section 4.1.3, therefore we can directly apply Eq.4.9:

$$I = \frac{Qv}{2\pi r} \qquad \cdots (1)$$

Since, current is defined as the rate of flow of positive charge, therefore, the current will be in opposite sense of the motion of electron on the circular path.

SOLUTION Given that $Q = e = 1.6 \times 10^{-19}$C, $r = 10$ cm $= 0.1$ m, and speed 4×10^6 m/s, therefore from Eq.(1), we have

$$I = \frac{(1.6 \times 10^{-19}\text{C})(4 \times 10^6 \text{ m/s})}{2 \times 3.14 \times 0.1 \text{ m}} \approx 1.02 \times 10^{-12} \text{ A}$$

EXAMPLE 3. The current through a wire depends on time as $I = (2 + 3t)$ A. Calculate the charge crossed through a cross section of the wire in $10s$.

APPROACH From Eq.4.5, the electric charge crossing through a cross-section of a conductor in time t is given by

$$Q = \int_0^t I \, dt \qquad (1)$$

Substitute the values of current and integrate it for $t = 0$ to $t = 10$ s.

SOLUTION Given that, $I = (2 + 3t)$ A, therefore, from Eq.(1), we have-

$$Q = \int_0^t I \, dt = \int_0^{10} (2 + 3t) \, dt$$

$$= \left(2t + \frac{3t^2}{2}\right)_0^{10} = 2(10) + \frac{3}{2} \times 100 = 20 + 150 = 170 \text{ C}$$

4.1.4 The Direction of Electric Current

Current is a scalar quantity, as both charge and time are scalars, but it has a sign associated with it. It is useful to indicate the sign of the current by a directional arrow. Figure4.4b depicts the direction of current flow within a conductor. By historical convention we associate the direction of current flow with the direction of "flow" of positive charges-even though it is actually the negative charges that move in conducting materials such as metals. In a conductor the positive charges-the atomic ions left behind by the electrons - are fixed in an ordered crystal lattice. In an ionized gas or a chemical solution, the charges that actually move and create the current may be positive or negative. This arbitrary convention for current direction causes no real problem, because a flow of positive charge to the right and a flow of the same amount of negative charge to the left represent the same current. By simply measuring the current, it is not possible to determine the sign of the charges that move (the charge carriers). By convention, *the direction of the current is the direction in which positive charge carriers would move, even if the actual charge carriers are negative charges moving in the opposite direction.*

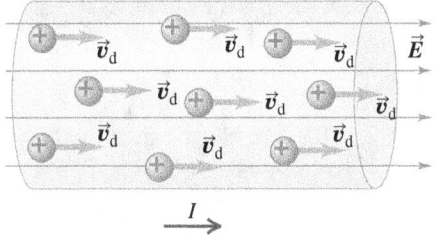

(a) A conventional current is treated as a flow of positive charges, regardless of whether the free charges in the conductor are positive, negative, or both.

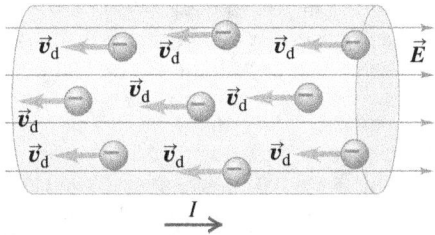

(b) In a metallic conductor, the moving charges are electrons - but the current still points in the direction positive charges would flow.

Figure 4.4: The same current can be produced by (a) positive charges moving in the direction of the electric field $\vec{E}$ or (b) the same number of negative charges moving at the same speed in the direction opposite to $\vec{E}$.

Important Points
1. Since electric charge is conserved, therefore, unless charge accumulates within the conductor, the rate of charge leaving a conductor is exactly the same as the rate of charge entering the conductor. So, conductor remains uncharged when current flows through it.
2. For a given conductor current does not change with change in its cross-section because current is simply rate of flow of charge. In other words, the current must be the same at all points in a current-carrying conductor.
3. If n particles each having a charge q pass per second per unit area, then charge passing through a cross sectional area A of the conductor in time Δt is, $\Delta Q = nqA\Delta t$
   So, current associated with crosssectional area A of the conductor is, $I = \Delta Q/\Delta t = nqA$
4. If a charge q is moving in a circle of radius r with speed v then its time period is $T = 2\pi r/v$. The equivalent current
$$I = q/T = qv/2\pi r.$$

4.1.5 Classification of Materials According to Conductivity

4.1.5.1 Conductors and Insulators

In many materials, such as copper and other metals, some of the electrons are free to move about the entire material. Such materials are called conductors (Examples: Metals like Cu, Ag, Fe, Al, etc.). When such a material is placed in an electric field, the free electrons move in a direction opposite to the field. In other materials, such as wood or glass, all the electrons are bound to nearby atoms and none can move freely. When such a material is placed in an electric field, the electrons may slightly shift opposite to the field but they can't leave their parent atoms or molecules and hence can't move through long distances. These materials are called insulators (Ex. plastic, rubber, wood etc.). Insulators are also called dielectrics.

4.1.5.2 Semiconductor

In semiconductors, the behaviour is like an insulator at low temperature. But at higher temperatures, a small number of electrons are able to free themselves and they respond to the applied electric field. As the number of free electrons in a semiconductor is much smaller than that in a conductor, its behaviour is in between a conductor and an insulator and hence, the name semiconductor. A freed electron in a semiconductor leaves a vacancy in its normal bound position. These vacancies also help in conduction.

4.1.6 Check Point 1

1. •A current of 1.30 A flows in a wire. How many electrons are flowing past any point in the wire per second?
2. •A service station charges a battery using a current of 6.7 A for 5.0 h. How much charge passes through the battery?
3. •What is the current in amperes if 1200Na$^+$ ions flow across a cell membrane in 3.5 μs? The charge on the sodium is the same as on an electron, but positive.
4. ••The current in a wire varies with time according to the relation: $I = (3.0 \text{ A}) + (2.0 \text{ A/s})t$
   (a) How many coulombs of charge pass a cross-section of the wire in the time interval between $t = 0$ and $t = 4.0$ s?
   (b) What constant current would transport the same charge in the same time interval?
5. ••Current passing through a wire decreases linearly from 20 A to 0 in 8 s (Fig.4.5). Find total charge flowing through the wire in the given time interval.

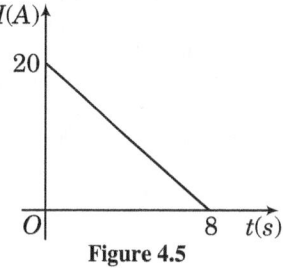

Figure 4.5

6. ••A constant current of 4 A passes through a wire for 8 s. Find total charge flowing through that wire in the given time interval.

4.2 Microscopic Model of Current

4.2.1 In Absence of External Electric Field

We can relate current to the motion of the charge carriers by describing a microscopic model of conduction in a metal. Let us consider an isolated conductor in which the charge carriers are free electrons. These electrons undergo random thermal motion that is analogous to the motion of gas molecules. The speed gained by virtue of temperature is called as thermal speed of an electron.

If at absolute temperature T, the root mean square speed of free electrons is $v_{\rm rms}$, then according to equipartition theorem (a result of classical statistical mechanics), we have

$$\frac{1}{2}mv_{\rm rms}^2 = \frac{3}{2}kT$$

So, thermal speed $v_{\rm rms} = \sqrt{3kT/m}$

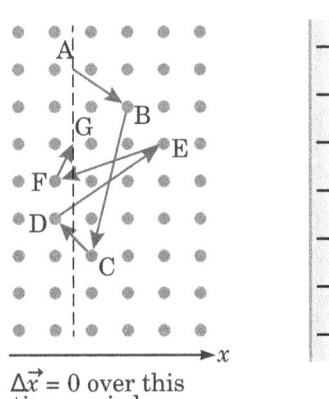
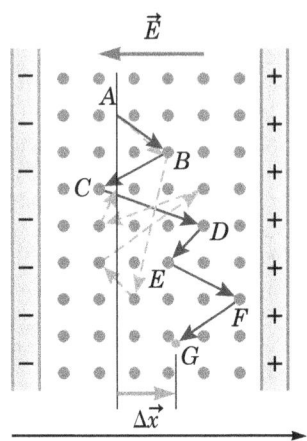

(a) No electric field means no net x displacement.

(b) Electric field causes net x displacement.

Figure 4.6

where m is mass of electron and k is Boltzmann constant and it's value is defined as $k = 1.380649 \times 10^{-23}$ J/K.

At room temperature $T = 300K$, therefore,

$$v_{\rm rms} = \sqrt{\frac{3kT}{m}} = \sqrt{\frac{3 \times 1.380649 \times 10^{-23} {\rm JK}^{-1} \times 300{\rm K}}{9.1 \times 10^{-31} {\rm kg}}}$$
$$\approx 10^5 {\rm m/s}$$

Thus, free electrons move in random directions with relatively large thermal speeds of the order of 10^5 m/s*. In addition, these electrons collide repeatedly with the metal ions†, and their resultant motion is complicated and zigzagged. Imagine watching just one conduction electron for a few moments as shown in Fig.4.6a. Suppose, A is the starting position and G is the ending position of the free electron under consideration. Because the electron's starting position A is directly above its ending position G, therefore, its displacement along x direction is zero during this time. The electron's displacement depends on the time interval during which it is observed. For example, if we had observed the electron for a slightly shorter time, it would have been at position F , and its displacement would have had a component in the negative x direction. However, if we observe a large number of conduction electrons over any length of time, we find that there is no net motion of the electrons in any particular direction and the average velocity is zero.

> ☞ In the absence of potential difference across a conductor, there is no electric field inside the conductor. So, the net electric force on free electrons and hence acceleration of free electrons always remain zero and hence in such case, electrons inside the conductors move randomly.

Mean free path: During any time interval, the ratio of total distance travelled by a free electron with the number of its collision is called it's mean free path. Generally, it is denoted by λ.

Mean free path, $\lambda = \dfrac{\text{total distance travelled}}{\text{number of collisions}}$

At room temperature, the mean free path is of the order of 10^{-10}.

Relaxation time: The time taken by an electron between two successive collisions is called as relaxation time. Generally, it is denoted by τ.

Relaxation time, $\tau = \dfrac{\text{total time taken}}{\text{number of collisions}}$

At room temperature, the relaxation time is of the order of 10^{-14}s.

4.2.2 In Presence of External Electric Field

Suppose, the conductor is placed in an electric field produced between the plates of a charged capacitor. The field exerts a force on each charged particle (negatively charged electrons and positively charged nuclei). But in case of metal, only free electrons (electrons of outer most shell of atoms) are free to move whereas nuclei with inner shell electrons remain fixed at their original positions. So, free electrons of the metal conductor move and rearrange themselves to reach equilibrium with no electric field inside the conductor. During the short time interval before the electrons reach equilibrium, however, the electric field inside the conductor is not zero. Suppose the electric field is in the negative x direction, $\vec{E} = -E_x\hat{i}$, and consider the path of a single conduction electron (Fig.4.6b). The electric field exerts a force on that electron in the positive x direction:

$$\vec{F} = q\vec{E} = (-e)(-E_x\hat{i}) = eE_x\hat{i}$$

When the electron is moving between collisions in this field, it is pulled in the positive x direction. As a result, the electron drifts to the right. If we observe the same conduction electron over the same period of time as in Figure 4.6a, we find the ending position at G is to the right of the starting position at A, so the electron's displacement has a positive x component (Fig.4.6b).

Figure 4.6b shows the path of the electron both with and without an electric field in the metal conductor for comparison. The electric field does not reduce the average number of collisions the conduction electron makes with the positive ions. Instead, the electron's path is shifted to the right. If we looked at all the conduction electrons in the conductor, we would find (1) they undergo many collisions with the ions; (2) their motion between collisions is very fast, with speeds around 10^6m/s and (3) they tend to move toward the right at a very slow velocity known as the ***drift velocity*** (see Example 15) in the direction opposite to the electric field. The typical drift speed (magnitude of drift velocity) $v_{\rm drift}$ is between 10^{-5}m/s and 10^{-4}m/s. Clearly, the magnitude of drift velocity is very small (of the order of 10^{-5} m/s to 10^{-4} m/s) as compared to thermal speed of free electrons at room temperature which is of the order of 10^5 m/s to 10^6 m/s.

If we connect a voltmeter between left and right end of the conductor piece under consideration, and suppose V_L and V_R are potentials at left and right end of the conductor, then the potential difference ΔV between two ends of conductor $\Delta V = V_R - V_L$. In Figure 4.7a, there is no electric field, so there is no net displacement of conduction electrons and no excess charge anywhere in the conductor. The result is that the voltmeter's reading is $\Delta V = V_R - V_L = 0$, which indicates no potential difference across the conductor.

In Figure 4.7b, the conductor is placed in a leftward-pointing electric field so that the conduction electrons drift toward the right. If we wait until the charged particles in the conductor reach equilibrium, the conductor is still neutral overall but the right side of it has excess negative charge and the left side has excess positive charge. In this case, the voltmeter's reading is

---

*The average kinetic energy of the free electrons in a metal is quite large, even at very low temperatures

†Atoms get ionized and become positively charged ions when free electrons get detached from them

4.2. MICROSCOPIC MODEL OF CURRENT

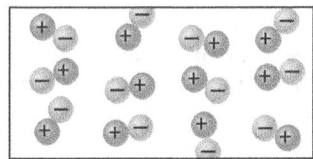

Microscopic model

(a) When $\vec{E} = 0$, then $\Delta V = V_R - V_L = 0$

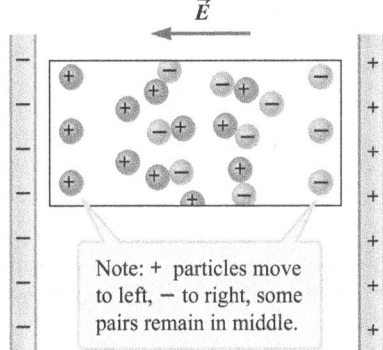

Note: + particles move to left, − to right, some pairs remain in middle.

(b) When $\vec{E} = -E_x \hat{i}$, the conduction electrons drift so that the left side of the conductor becomes positive and the right side becomes negative, and $V_R - V_L < 0$.

Figure 4.7

$\Delta V = V_R - V_L < 0$, indicating that the right side of the conductor is now at a lower potential than the left side.

We cannot directly observe the motion of conduction electrons in a metal, and a voltmeter cannot help in determining whether the free negatively charged electrons really move or perhaps the free positively charged particles move. Without better information, 18th and 19th century experimenters assumed that the positive particles were mobile. Suppose, due to leftward pointing electric field $\vec{E}$, the positive particles move in the direction of electric field, i.e., leftward. The bound negative particles would remain in place, and the net result is that the left side would have excess positive charge and the right side excess negative charge—exactly the way the conductor appears (Fig. 4.7b). The voltmeter (not shown in diagram) measures the potential difference $\Delta V = V_R - V_L < 0$ exactly as before. Note that, in metals, conduction electrons move and positive ions stay relatively motionless, but today we still find it convenient to think about positively charged particles moving in the opposite direction as electrons.

EXAMPLE 4. Find the approximate total distance travelled by an electron in the time-interval in which its displacement is one meter along the wire. ($v_d = 1$ mm/s $= 10^{-3}$ m/s, speed of electron in the wire is 10^6 m/s)

SOLUTION Since, electrons travel along the length of wire with drift (or effective) velocity, therefore

$$\text{time} = \frac{\text{displacement}}{\text{drift velocity}} = \frac{s}{v_d}$$

here, $s = 1$ m, $v_d = 1$ mm/s $= 10^{-3}$ m/s

$\therefore$ time $= \frac{1}{10^{-3}} = 10^3$ sec.

Now distance travelled = speed × time
Given that, speed = 10^6 m/s, time = 10^3 s
$\therefore$ Required distance = $10^6 \times 10^3$ m = 10^9 m

EXAMPLE 5. State whether the given statement is true or false? "Electrons in a conductor have no motion in the absence of a potential difference across it."

DISCUSSION The electrons inside the conductors have random motion. In the absence of potential difference across a conductor, the motion of the electrons is such that their drift velocity is zero.
ANSWER The given statement is false.

4.2.3 Electric Current in Terms of Drift Velocity

Electric current is the apparent motion of positively charged particles. In a metal conductor, the current is a result of the motion of conduction electrons opposite to the applied electric field. So, in a metal conductor, the current is in the same direction as the electric field. Current is not always the result of the flow of electrons. In ionic solutions or semiconductors, it may be due to the motion of both positive and negative types of charge carriers. If a cylindrical conductor is placed between the plates of a charged capacitor, conduction electrons flow in the opposite direction as the electric field, and very soon the particles are in equilibrium again. Such a current does not last very long. In order to study current, we must set up a steady current in the conductor by connecting it to a battery. Battery maintains a constant potential difference across the ends of the conductor. This applied potential difference creates a net electric field inside the conductor that points from the positive terminal toward the negative terminal of the conductor. This is the internal electric field $\vec{E}$ that pushes the electron current through the conductor. The existence of $\vec{E}$ within the cylindrical conductor, does not contradict our result of electrostatics that $\vec{E} = 0$ inside a conductor in the electrostatic case, as we are no longer dealing with the static case. Charges are free to move in a conductor, and hence can move under the action of the electric field. If all the charges are at rest, then $\vec{E}$ must be zero (electrostatics).

☞ An electron current is a nonequilibrium motion of charges sustained by an internal electric field.

Force applied by electric field changes the velocity of each electron opposite to the applied electric field. However, any additional kinetic energy acquired is quickly dissipated by collisions with the lattice ions in the conductor. During the time between collisions with the lattice ions, the free electrons, on average, acquire an additional velocity in the direction opposite to the electric field. The net result of this repeated acceleration and dissipation of energy is that the electrons drift along the cylindrical conductor with drift velocity $\vec{v}_d$. It is directed opposite to the applied electric-field $\vec{E}$. To relate current, a macroscopic quantity, to the microscopic motion of the charges, let's examine a conductor of cross-sectional area A, as shown in Figure 4.8.

If q is the charge of each carrier, and n is the number of charge

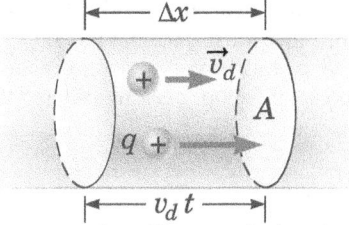

Figure 4.8: A segment of a uniform conductor of cross-sectional area A.

carriers per unit volume, the total amount of charge in this section is then

ΔQ = charge per unit volume × volume = $nq(A\Delta x)$

Suppose, the charge carriers move with drift speed, v_d; then the displacement in a time interval Δt will be $\Delta x = v_d \Delta t$, which implies

$$I_{av} = \frac{\Delta Q}{\Delta t} = \frac{nq(A\Delta x)}{\Delta x/v_d} = nqAv_d$$

$\Rightarrow \qquad \boxed{I_{av} = nqAv_d} \qquad (4.10)$

Here, v_d is drift speed (the magnitude of drift velocity).
Since, current is a scalar quantity and in case of a metallic conductor, the charge carriers are free electrons, therefore, replacing q by $|-e|$ i.e., e, in Eq.4.10, we get-

$$\boxed{I_{av} = neAv_d} \quad (4.11)$$

If v_d is constant, then current will be steady and in this case, we can replace I_{av} by I. So, for steady current,

$$\boxed{I = neAv_d} \quad (4.12)$$

EXAMPLE 6. A steady current passes through a cylindrical conductor. Is there an electric field inside the conductor?

SOLUTION Yes. Under steady state conditions in electrostatics, the charge always remains in electrostatic equilibrium. When a conductor is charged, the whole charge spreads over the surface of the conductor and hence by Gauss's theorem in electrostatics, the net electric field inside the metal becomes zero. In this case, the electric potential at every point inside the metal will be equal to the surface potential and no current flows anywhere inside or at surface of the metal conductor. However, when a potential difference is applied across a conductor and a steady current flows through it, the condition no longer remains static and there exists an electric field inside the conductor.

4.2.3.1 Current Remains Same at Each Point in a Single Branched Closed Loop

Let us consider a typical electric circuit-a battery connected to a light bulb (Fig. 4.9a). Since, there is only a single loop and no other branch is connected to it, therefore it is a single branched electric circuit. If we measure the electric current at any point on this loop we always find the same value. Now a question may arise, why the current is same throughout the loop, when the only driving force is inside the battery? Why not there is a large current in the battery and none in the lamp.

Explanation: To understand it, consider a segment of the loop as shown in Fig.4.9b. Let current is not same throughout the segment and let the current into the bend in Fig.4.9b is greater than the outgoing current. In this case an accumulation of charge takes place at the "knee". This accumulated charge produces it's own electric field which is directed away from the knee. This field opposes the current flowing in (slowing it down) and promote the current flowing out (speeding it up). As incoming current decreases and outgoing current increases, the accumulated charge starts decreasing. Finally, a state reaches when these currents become equal and no further accumulation of charge would be there. So, assuming different currents at different points in a closed loop provide you an accumulated charge which automatically keeps the current same throughout the loop and it does it all so quickly that, we can assume the current is the same through each cross-section of the loop, even in systems that oscillate at radio frequencies.

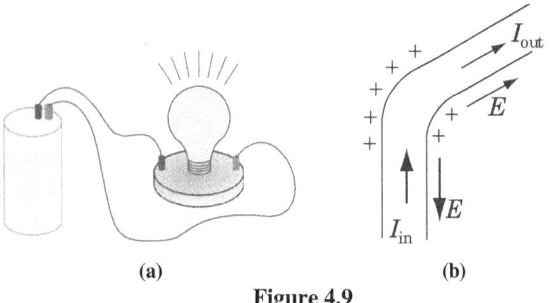

Figure 4.9

4.2.4 Current Density

Electric current is a macroscopic scalar quantity and Eq.4.11 is just a scalar relation between macroscopic quantity I and microscopic quantity v_d. Sometimes we need a vector relation between the macroscopic description to a microscopic model. For it, we introduce a new macroscopic vector quantity, "current density". The magnitude of the current density is the current I per unit cross-sectional area A, i.e.,

$$\boxed{J = \frac{I}{A}} \quad \text{(Definition of current density)} \quad (4.13)$$

where I is uniform over the area A. Current density is a vector that points in the same direction as the electric field.
If we substitute the value of I, from Eq.4.10, in Eq.4.13, we get

$$\boxed{J = \frac{I}{A} = nqv_d} \quad (4.14)$$

In vector form, above relation can also be written as

$$\boxed{\vec{J} = nq\vec{v}_d} \quad (4.15)$$

Thus, from Eq.4.15, we see that $\vec{J}$ and $\vec{v}_d$ point in the same direction for positive value of q, i.e., for positive charge carriers and these are oppositely directed for negative value of q, i.e., for negative charge carriers.

Since, current density is a vector quantity and in the case of a metallic conductor, the charge carriers or free electrons, therefore, we replace q by $-e$, and so, Eq.4.15 gives

$$\boxed{\vec{J} = -ne\vec{v}_d} \quad (4.16)$$

Here, negative sign shows that the direction of current density vector ($\vec{J}$) is opposite to the direction of drift velocity $\vec{v}_d$.
In scalar form Eq.4.16 can also be written as

$$\boxed{J = nev_d} \quad (4.17)$$

Now, current through a surface can be defined as the flux of the current density vector $\vec{J}$ through the surface. That is

$$\boxed{I = \int_S \vec{J} \cdot d\vec{A}} \quad (4.18)$$

here, $d\vec{A}$ is the elementary area vector of the element where the current density is $\vec{J}$.
If $\hat{n}$ is the unit vector normal to the surface area dA and along the current, then $d\vec{A} = \hat{n}\, dA$, therefore above expression becomes

$$\boxed{I = \int_S \vec{J} \cdot d\vec{A} = \int_S \vec{J} \cdot \hat{n}\, dA} \quad (4.19)$$

The SI unit of current density is A/m^2.
If $\vec{J}$ is uniform and the surface is flat, i.e, direction of $\hat{n}$ is fixed, then the flux can be expressed as

$$\boxed{I = \int_S \vec{J} \cdot \overrightarrow{dA} = \int_S \vec{J} \cdot \hat{n}\, dA = \vec{J} \cdot \vec{A} = \vec{J} \cdot \hat{n}A = JA\cos\theta} \quad (4.20)$$

where, θ is the angle between $\vec{J}$ and $\hat{n}$ (Fig.4.10).

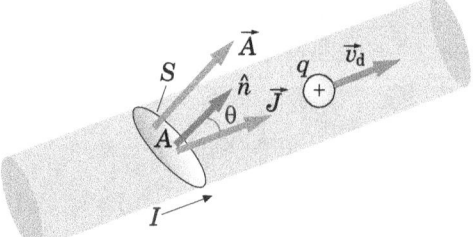

Figure 4.10: Normal to the flat surface S making angle θ with current density vector.

If the current I, is uniform across the surface and $\vec{J}$ is parallel to

4.2. MICROSCOPIC MODEL OF CURRENT

$d\vec{A}$, then Eq.4.20, again gives
$$J = \frac{I}{A}$$
where A is the total area of the surface.

In "Electrostatics", we have seen that we can represent an electric

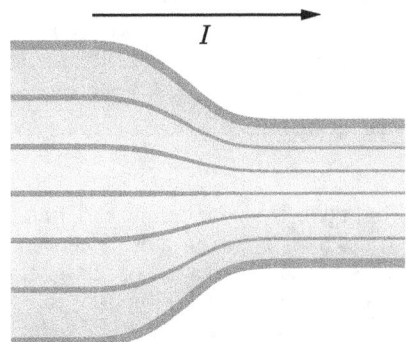

Figure 4.11: Streamlines representing current density in the flow of charge through a constricted conductor.

field with electric field lines. Figure 4.11 shows how current density can be represented with a similar set of lines, which we can call streamlines. The current, which is toward the right in Fig. 4.11, makes a transition from the wider conductor at the left to the narrower conductor at the right. Because charge is conserved during the transition, the amount of charge and thus the amount of current cannot change. However, the current density does change — it is greater in the narrower conductor. The spacing of the streamlines suggests this increase in current density; streamlines that are closer together imply greater current density.

EXAMPLE 7. Uniform Current Density
The current density at a point is $\vec{J} = (2 \times 10^4 \hat{j})$ Am^{-2}. Find the rate of charge flow through a cross sectional area $\vec{A} = (2\hat{i} + 3\hat{j})$ cm^2

APPROACH The rate of flow of charge is current. Since, current density is uniform throughout the cross-section, therefore, to determine current through the given cross-section, we can apply the relation given by Eq.4.20: $I = \int \vec{J} \cdot d\vec{A} = \vec{J} \cdot \vec{A}$. Now, substitute the values of $\vec{J}$ and $\vec{A}$ and simplify for I.

SOLUTION Given that, $\vec{J} = (2 \times 10^4 \hat{j})$ Am^{-2}, and $\vec{A} = (2\hat{i} + 3\hat{j})$ cm^2 = $(2\hat{i} + 3\hat{j}) \times 10^{-4}$ m^2
Therefore, rate of flow of charge, i.e., electric current
$$I = \vec{J} \cdot \vec{A} = (2 \times 10^4) [\hat{j} \cdot (2\hat{i} + 3\hat{j})] \times 10^{-4} \text{ A} = 6 \text{ A}$$
$$[\because \hat{j} \cdot \hat{i} = 0, \hat{j} \cdot \hat{j} = 1]$$

EXAMPLE 8. Non-Uniform Current Density
A potential difference applied to the ends of a wire made up of an alloy drives a current through it. The current density varies as $J = 3 + 2r$, where r is the distance of the point from the axis. If R be the radius of the wire, then find the total current through any cross section of the wire.

APPROACH Since, in this case, the current density is not uniform across the given cross section of the wire, so we use Eq.4.18:
$$I = \int \vec{J} \cdot d\vec{A} \qquad \cdots (1)$$
and to get total current, we integrate the current density over the portion of the wire from $r = 0$ to $r = R$.

SOLUTION The current density vector $\vec{J}$ (along the wire's length) and the differential area vector $d\vec{A}$ (perpendicular to a cross section of the wire) have the same direction. Therefore,
$$\vec{J} \cdot d\vec{A} = JdA\cos 0 = JdA$$
Since, J is given as a function of r, therefore we replace dA with something we can actually integrate between the limits $r = 0$

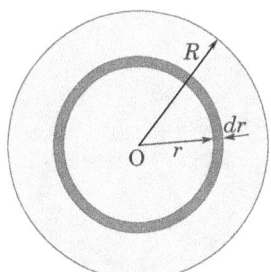

Figure 4.12: If the current is non-uniform, we must first find the current through a thin ring and then sum (via integration) the currents in all such rings from $r = 0$ to $r = R$.

and $r = R$. The simplest replacement is the area $2\pi r dr$ of a thin ring of circumference $2\pi r$ and width dr (Fig.4.12). We can then integrate with respect to r as the variable of integration.
Equation (1): $I = \int \vec{J} \cdot d\vec{A}$ then gives us
$$I = \int \vec{J} \cdot d\vec{A} = \int JdA$$
$$= \int_0^R (3 + 2r) 2\pi r dr = 2\pi \int_0^R \left(3r + 2r^2\right) dr$$
$$= 2\pi \left(\frac{3r^2}{2} + \frac{2}{3}r^3\right)_0^R = 2\pi \left(\frac{3R^2}{2} + \frac{2R^3}{3}\right) \text{ units}$$

EXAMPLE 9. Uniform and Non-uniform Current density
(a) The current density in a cylindrical wire of radius $R = 2.0$ mm is uniform across a cross section of the wire and is $J = 2.0 \times 10^5$ A/m^2. What is the current through the outer portion of the wire between radial distances R/2 and R (Fig.)?
(b) Suppose, instead, that the current density through a cross section varies with radial distance r as $J = ar^2$, in which $a = 3.0 \times 10^{11}$ A/m^4 and r is in meters. What now is the current through the same outer portion of the wire?

(a) APPROACH Since, the current density is uniform across the cross section, the current density J, the current I, and the cross-sectional area A are related by Eq.4.13: $J = I/A$.
We want only the current through a reduced cross-sectional area lying between $r = R/2$ to R. If we denote this area by A' then
$$A' = \pi R^2 - \pi \left(\frac{R}{2}\right)^2 = \pi \left(\frac{3R^2}{4}\right)$$
$$= \frac{3\pi}{4}(0.0020 \text{ m})^2 = 9.424 \times 10^{-6} \text{ m}^2$$
So, we rewrite the relation $I = JA$ as $I = JA'$ and then substitute the data to find I.

SOLUTION $I = JA' = \left(2.0 \times 10^5 \text{ A}^2/\text{m}^2\right)(9.424 \times 10^{-6} \text{ m}^2)$
$= 1.9$ A.

(b) APPROACH Since, in this case, the current density is not uniform across the given cross section of the wire, so we use Eq.4.18:
$$I = \int \vec{J} \cdot d\vec{A} \qquad \cdots (1)$$
and to get total current, we integrate the current density over the portion of the wire from $r = R/2$ to $r = R$.

SOLUTION The current density vector $\vec{J}$ (along the wire's length) and the differential area vector $d\vec{A}$ (perpendicular to a cross section of the wire) have the same direction. Thus,
$$\vec{J} \cdot d\vec{A} = JdA\cos 0 = JdA$$
Since, J is given as a function of r, therefore we have to transform dA in terms of dr. For it, consider an elementary ring of radius r and thickness dr. The elementary area of this ring, $dA = 2\pi r dr$ (Fig.4.13b). Now, substitute this value of dA in Eq. (1) and integrate it from $r = R/2$ to $r = R$, i.e.,

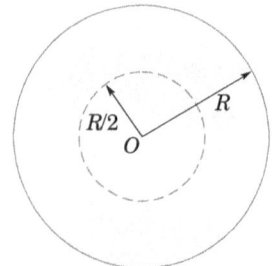
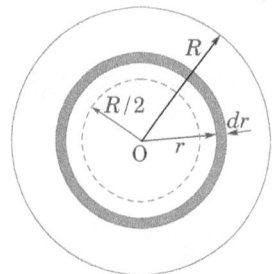

(a) Cross section of a wire of radius R. If the current density is uniform, the current is just the product of the current density and the area.

(b) If the current is non-uniform, we must first find the current through a thin ring and then sum (via integration) the currents in all such rings in the given area

Figure 4.13

$$I = \int \vec{J} \cdot \vec{dA} = \int J dA$$
$$= \int_{R/2}^{R} ar^2 2\pi r \, dr = 2\pi a \int_{R/2}^{R} r^3 \, dr = 2\pi a \left[\frac{r^4}{4}\right]_{R/2}^{R}$$
$$= \frac{\pi a}{2}\left[R^4 - \frac{R^4}{16}\right] = \frac{15}{32}\pi a R^4$$
$$= \frac{15}{32}\pi \left(3.0 \times 10^{11} \text{ A/m}^4\right)(0.0020 \text{ m})^4 = 7.1 \text{ A}$$

EXAMPLE 10. A beam contains 2.0×10^8 doubly charged positive ions per cubic centimeter, all of which are moving north with a speed of 1.0×10^5 m/s. What are the (a) magnitude and (b) direction of the current density $\vec{J}$? (c) What additional quantity do you need to calculate the total current I in this ion beam?

APPROACH The magnitude of the current density is given by Eq.4.14: $J = nqv_d$, where n is the number of particles per unit volume, q is the charge on each particle, and v_d is the drift speed of the particles.

In vector form, we have (see Eq.4.15), $\vec{J} = nq\vec{v}_d$. Current density $\vec{J}$ is related to the current I by (see Eq.4.18):
$$I = \int \vec{J} \cdot d\vec{A}.$$

SOLUTION (a) The particle concentration is
$$n = 2.0 \times 10^8/\text{cm}^3 = 2.0 \times 10^{14} \text{ m}^{-3},$$
the charge is
$$q = 2e = 2\left(1.60 \times 10^{-19}\text{C}\right) = 3.20 \times 10^{-19}\text{C},$$
and the drift speed is 1.0×10^5 m/s. Thus, we find the current density to be
$$J = \left(2.0 \times 10^{14}/\text{m}^3\right)\left(3.2 \times 10^{-19}\text{C}\right)\left(1.0 \times 10^5 \text{ m/s}\right)$$
$$= 6.4 \text{ A/m}^2$$

(b) Since the particles are positively charged the current density is in the same direction as their motion, to the north.
(c) The current cannot be calculated unless the cross-sectional area of the beam is known. Then $I = JA$ can be used.

4.2.5 Calculation of drift velocity

To find the drift velocity of the electrons, we first note that an electron in the conductor experiences an electric force (F_e), given by
$$\vec{F}_e = -e\vec{E}$$
which gives an acceleration
$$\vec{a} = \vec{F}_e/m_e = -e\vec{E}/m_e$$
Let the velocity of a given electron immediate after a collision be $\vec{v}_0$. The velocity of the electron immediately before the next collision is then given by
$$\vec{v} = \vec{v}_0 + \vec{a}t = \vec{v}_0 - \frac{e\vec{E}}{m_e}t$$

where t is the time travelled. The average of $\vec{v}$ over all time intervals is
$$\langle\vec{v}\rangle = \langle\vec{v}_0\rangle - \frac{e\vec{E}}{m_e}\langle t\rangle \quad (4.21)$$
which is equal to the drift velocity $\langle\vec{v}_d\rangle$.
If in presence of an external electric field, the average characteristic time between successive collisions (i.e., the mean free time or relaxation time), is $\vec{\tau}$, i.e., $\langle t\rangle = \tau$, then, Eq.4.21 can also be written as
$$\vec{v}_d = \langle\vec{v}_0\rangle - \frac{e\vec{E}}{m_e}\tau \quad (4.22)$$
As we know that there are large number of free electrons in a conductor and just after collision these electrons can move in random directions in the conductor, so, just after collision, the average initial velocity of electrons opposite to applied electric field is zero. i.e., $\langle\vec{v}_0\rangle = 0$, therefore Eq.4.22 gives

$$\boxed{\vec{v}_d = -\frac{e\vec{E}}{m_e}\tau} \quad (4.23)$$

Here, $-ve$ sign shows that the drift velocity is opposite to applied electric field.
In the absence of electric field (i.e., $\vec{E} = 0$), above equations tells that $\vec{v}_d = 0$ which is an expected result because in absence of an external electric field, the velocity of the electron is completely random, it follows that $\langle\vec{v}_d\rangle = 0$.

☞ In Example 15, we will see that for copper the magnitude of drift velocity is equal to 4.9×10^{-7} m/s, which is only 1.8 mm/h, slower than a sluggish snail.

4.2.6 Check Point 2

1. ●●A certain cylindrical wire carries current. We draw a circle of radius r around its central axis in Fig.4.14a to determine the current I within the circle. Figure 4.14b shows current I as a function of r^2. The vertical scale is set by $I_s = 4.0$ mA, and the horizontal scale is set by $r_s^2 = 4.0$ mm^2. (a) Is the current density uniform? (b) If so, what is its magnitude?

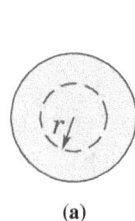
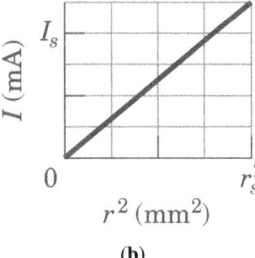

(a) **(b)**
Figure 4.14

2. ●A fuse in an electric circuit is a wire that is designed to melt, and thereby open the circuit, if the current exceeds a predetermined value. Suppose that the material to be used in a fuse melts when the current density rises to 440 A/cm^2. What diameter of cylindrical wire should be used to make a fuse that will limit the current to 0.50 A?

3. ●●A 0.65 mm diameter copper wire carries a tiny current of 2.3 μA. Estimate (a) the electron drift velocity, (b) the current density, and (c) the electric field in the wire. Given: Density of copper is 8.9×10^3 kg/m^3 and molar mass is 63.5×10^{-3} kg.

4. ●●At a point high in the Earth's atmosphere, He^{2+} ions in a concentration of 2.8×10^{12}/m^3 are moving due north at a speed of 2.0×10^6 m/s. Also, a 7.0×10^{11}/m^3 concentration of O$_2^-$ ions is moving due south at a speed of 6.2×10^6 m/s.

Determine the magnitude and direction of the current density $\vec{J}$ at this point.

5. •A small but measurable current of 1.2×10^{-10} A exists in a copper wire whose diameter is 2.5 mm. The number of charge carriers per unit volume is 8.49×10^{28} m^{-3}. Assuming the current is uniform, calculate the (a) current density and (b) electron drift speed.

6. ••The magnitude $J(r)$ of the current density in a certain cylindrical wire is given as a function of radial distance from the centre of the wire's cross section as $J(r) = Br$, where r is in meters, J is in amperes per square meter, and $B = 2.00 \times 10^5$ A/m^3. This function applies out to the wire's radius of 2.00 mm. How much current is contained within the width of a thin ring concentric with the wire if the ring has a radial width of 10.0 μm and is at a radial distance of 1.20 mm?

4.3 Mobility, Conductivity and Resistivity

4.3.1 Mobility

The measurement of how fast an electron can move through a substance (metals or semiconductors) which is under the influence of an external electric field, is known as "mobility". Mathematically, it is the drift speed per unit electric field. Generally, it is denoted by μ.

$$\mu = \frac{v_d}{E} \quad (4.24)$$

In scalar form, Eq.4.23 can be written as

$$v_d = \frac{eE}{m_e}\tau \quad (4.25)$$

Substituting the value of drift speed from Eq.4.25 in Eq.4.24, we get

$$\boxed{\text{Mobility of free electrons} = \frac{e}{m_e}\tau} \quad (4.26)$$

From above equation it is clear that mobility of free electrons, is directly proportional to relaxation time. As the relaxation time depends on the material of the conductor and it's temperature, therefore μ also depends on the material of the conductor and its temperature.

4.3.2 Conductivity

On substituting the value of $\vec{v}_d$ from Eq.4.23 in Eq.4.16, we get-

$$\boxed{\vec{J} = -ne\vec{v}_d = -ne\left(-\frac{e\vec{E}}{m_e}\tau\right) = \frac{ne^2\tau}{m_e}\vec{E}} \quad (4.27)$$

From Eq.4.27, it is clear that current density is directed in the direction of applied electric field $\vec{E}$.
Note that $\vec{J}$ and $\vec{E}$ will be in the same direction for both negative and positive charge carriers.
Now, let us apply same potential difference between the terminals of two different conducting materials. The electric field in each conductor is the same, but from Eq.4.27, it is clear that the current density in each conductor depends on the proportionality constant $ne^2\tau/m_e$. The conductor with the longer average time between collisions (relaxation time) and the greater density of conduction electrons (n) will be the better conductor and have the higher current density. This constant of proportionality $\left(ne^2\tau/m_e\right)$ is the conductivity σ of the material:

$$\boxed{\sigma = \frac{ne^2\tau}{m_e}} \quad (4.28)$$

It is a scalar quantity. On substituting $\left(\frac{ne^2\tau}{m_e}\right) = \sigma$, in Eq.4.27, we get

$$\boxed{\vec{J} = \left(\frac{ne^2\tau}{m_e}\right)\vec{E} = \sigma\vec{E}} \quad (4.29)$$

This expression is shown graphically in Figure 4.15.

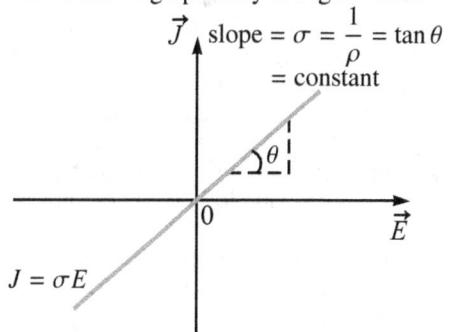

Figure 4.15

In Fig.4.15, ρ is the resistivity of the material of the conductor and it is equal to $1/\sigma$.
Since, conductivity is a scalar, so Equation 4.29 shows that the current density points in the same direction as the electric field. Conductivity depends only on the type of material and it's temperature. So, conductivity is a material property which is temperature dependent.
Eq.4.29 represents microscopic form of "Ohm's Law". In scalar form Eq.4.29, can also be written as

$$\boxed{J = \left(\frac{ne^2\tau}{m_e}\right)E = \sigma E} \quad (4.30)$$

The SI units for conductivity can be found from Equation 4.30:

Unit of $\sigma = \dfrac{\text{unit of } J}{\text{unit of } E} = \dfrac{\text{Am}^{-2}}{\text{Vm}^{-1}} = \dfrac{\text{A/V}}{\text{m}}$

In the SI, the combination of volts per ampere (V/A) is called an ohm:

$$1 \text{ V/A} = 1 \text{ }\Omega$$

where Ω is the uppercase Greek letter omega. The SI units for conductivity are usually given in ohms and meters ($\Omega^{-1} \cdot$ m^{-1}). This is a result of fundamental importance. Equation 4.29 tells us three things:

1. Current is caused by an electric field exerting forces on the charge carriers.
2. The current density, and hence the current $I = JA$, depends linearly on the strength of the electric field. To double the current, you must double the strength of the electric field that pushes the charges along.
3. The current density also depends on the conductivity of the material. Different conducting materials have different conductivities because they have different values of the electron density and, especially, different values of the mean time between electron collisions with the lattice of atoms.

The value of the conductivity is affected by the structure of a metal, by any impurities, and by the temperature. In a conductor, a higher temperature means more vigorously vibrating ions and more collisions for conduction electrons. An increase in the collision frequency means a decrease in the mean free time between collisions (i.e., relaxation time). So, according to Eq.4.28:

$\sigma = ne^2\tau/m_e$, as the temperature of a conductor increases, it's conductivity goes down.

EXAMPLE 11. Write the dimensions of electrical conductivity.

SOLUTION From Eq.4.30, the current density J, electric field E and conductivity σ are related as,
$$J = \sigma E.$$
Thus, $[\sigma] = \left[\dfrac{J}{E}\right] = \left[\dfrac{I/A}{V/d}\right] = \left[\dfrac{Id}{AV}\right] = \left[\dfrac{Id}{A}\dfrac{Q}{W}\right]$ $\left[\because V = \dfrac{W}{Q}\right]$
$= \left[\dfrac{Id}{A}\dfrac{It}{W}\right] = \left[\dfrac{AL}{L^2}\dfrac{AT}{ML^2T^{-2}}\right] = \left[M^{-1}\ L^{-3}\ T^3\ A^2\right]$

EXAMPLE 12. An infinite line charge of uniform electric charge density λ lies along the axis of an electrically conducting infinite cylindrical shell of radius R. At time $t = 0$, the space inside the cylinder is filled with a material of permittivity ε and electrical conductivity σ. The electrical conduction in the material follows Ohm's law. Find an expression of current density as a function of time (t). Also plot a graph of current density as a function of time t.

SOLUTION Since, charged wire of uniform linear charge density λ is placed along the axis of the cylinder, therefore, it produces a radially outward electric field. Due to this field, there is a radially inward motion of free electrons, so, radially outward will be the current. When free electrons reach to the axis, the charge of axial wire decreases. So, there is a continuous decrease of charge of axial wire with time t. As a result, charge density also decreases uniformly, with time. So, charge density will also be a function of t.

If at time t, the linear charge density of the wire on the axis of the cylinder be $\lambda(t)$, then electric field due to this line charge at a radial distance r is given by
$$\vec{E}(r, t) = \dfrac{\lambda(t)}{2\pi\varepsilon r}\hat{r}^* \qquad \cdots (1)$$

The current density at a point is defined as the current flowing

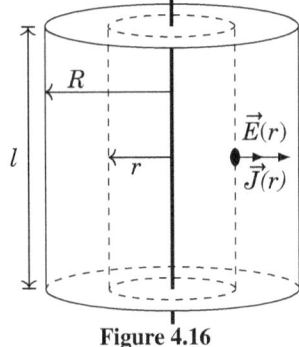

Figure 4.16

across a unit area placed perpendicular to the direction of current flow. From Eq.4.30, the current density in terms of electric field $\vec{E}(r, t)$ at the radial distance r, is given by
$$\vec{J}(r, t) = \sigma\vec{E}(r, t) \qquad \cdots (2)$$
where σ is the electrical conductivity of the conductor.

Substituting the value of $\vec{E}(r, t)$, from Eq.(1) in Eq.(2), we get
$$\vec{J}(r, t) = \sigma\dfrac{\lambda(t)}{2\pi\varepsilon r}\hat{r} \qquad \cdots (3)$$

Now, from Eq.4.19, the current through a cylindrical shell of radius r and length l is given by
$$I = \int_{\text{surface}} \vec{J}(r, t) \cdot d\vec{A} = J(r, t)(2\pi r l) \qquad \cdots (4)$$

Substituting the magnitude of current density from Eq.(3), in Eq.(4), we get

---
*The derivation of this result is given in chapter "Electric Charge and Field"

$$I = \sigma\dfrac{\lambda(t)}{2\pi\varepsilon r}(2\pi r l) = \dfrac{\sigma\lambda(t) l}{\varepsilon} \qquad \cdots (5)$$

Since, electric charge decreases with time t, therefore, by conservation of charge, the current I is equal to the rate of decrease of charge q on axial line segment of length l i.e.,
$$I = -\dfrac{dq}{dt} = -l\dfrac{d\lambda(t)}{dt} \qquad \cdots (6)$$

Substituting this value of I from Eq.(6) in Eq.(5), we get
$$-l\dfrac{d\lambda(t)}{dt} = \dfrac{\sigma\lambda(t) l}{\varepsilon}$$
or
$$\dfrac{d\lambda(t)}{dt} = -\dfrac{\sigma\lambda(t)}{\varepsilon}$$
or
$$\dfrac{d\lambda(t)}{\lambda(t)} = -\dfrac{\sigma}{\varepsilon}dt \qquad \cdots (6)$$

Integrating with initial condition $\lambda(t = 0) = \lambda_0$, we get
$$\int_{\lambda_0}^{\lambda}\dfrac{d\lambda(t)}{\lambda(t)} = -\int_0^t \dfrac{\sigma}{\varepsilon}dt$$
or
$$[\ln \lambda(t)]_{\lambda_0}^{\lambda} = -\dfrac{\sigma}{\varepsilon}[t]_0^t$$
or
$$\ln \lambda(t) - \ln \lambda_0 = -\dfrac{\sigma}{\varepsilon}[t - 0]$$
or
$$\ln\dfrac{\lambda(t)}{\lambda_0} = -\dfrac{\sigma}{\varepsilon}t$$
or
$$\lambda(t) = \lambda_0 e^{-\frac{\sigma t}{\varepsilon}} \qquad \cdots (6)$$

Substituting $\lambda(t)$ in equation (3), we get
$$\vec{J}(r, t) = \dfrac{\sigma\lambda_0}{2\pi\varepsilon r}e^{-\frac{\sigma t}{\varepsilon}}\hat{r} \qquad \cdots (7)$$

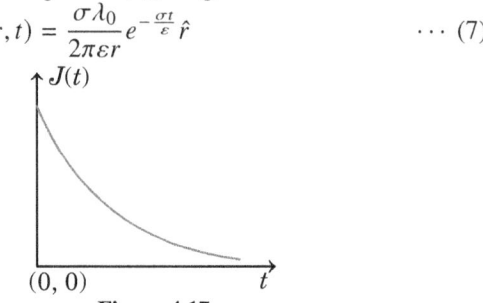

Figure 4.17

Note that current density varies as $1/r$ with radial distance r. It decreases exponentially with time and becomes zero as $t \to \infty$. J vs t graph is shown in Fig.4.17.

EXAMPLE 13. Earth's lower atmosphere contains negative and positive ions that are produced by radioactive elements in the soil and cosmic rays from space. In a certain region, the atmospheric electric field strength is 120 V/m and the field is directed vertically down. This field causes singly charged positive ions, at a density of 620 cm^{-3}, to drift downward and singly charged negative ions, at a density of 550 cm^{-3}, to drift upward (Fig.4.18). The measured conductivity of the air in that region is $2.70 \times 10^{-14}(\Omega.m)^{-1}$. Calculate (a) the magnitude of the current density and (b) the ion drift speed, assumed to be the same for positive and negative ions.

APPROACH For current density, apply Eq.4.30:

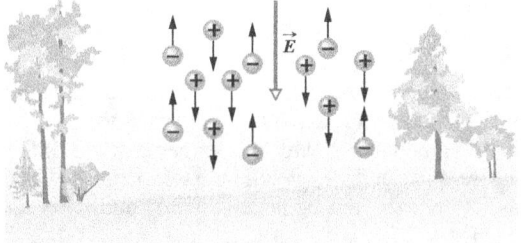

Figure 4.18

$$J = \sigma E \qquad \cdots (1)$$

and for drift speed apply Eq.4.16: $\vec{J} = -ne\vec{v}_d$, i.e.,
$$v_d = \frac{J}{ne} \quad \cdots (2)$$
Since, downward electric field applies downward electric force on positive ions and upward force on negative ions. So, they accelerate in opposite directions. The upward flow of negative ions produces downward electric current whereas downward flow of positive ions also produces downward electric current. So, both ions produce electric current in same direction (here it is in downward direction). Therefore, we consider $n = n_+ + n_-$. On substituting this value of n and $J = \sigma E$ in Eq.(2), we get
$$v_d = \frac{\sigma E}{(n_+ + n_-)e} \quad \cdots (3)$$

SOLUTION (a) From Eq.(1), the magnitude of the current density is
$$J = \sigma E = (2.70 \times 10^{-14} (\Omega.m)^{-1})(120 \text{ V/m})$$
$$= 3.24 \times 10^{-12} \text{ A/m}^2.$$
(b) From Eq.(3), the drift speed is
$$v_d = \frac{\sigma E}{(n_+ + n_-)e}$$
$$= \frac{(2.70 \times 10^{-14}(\Omega.m)^{-1})(120 \text{ V/m})}{[(620+550)\text{cm}^{-3}](1.60 \times 10^{-19}\text{C})} = 1.73 \text{ cm/s}$$

4.3.3 Resistivity (or Specific Resistance)

In many practical applications, it is more convenient to work with the reciprocal of the conductivity, known as the resistivity. Resistivity ρ is a measure of a material's ability to resist conducting electricity:
$$\rho = \frac{1}{\sigma} = \frac{m_e}{ne^2\tau} \quad (4.31)$$
Resistivity is a scalar and it is a material property that depends only on the type of material and it's temperature.
Resistivity is also known as the specific resistance of the material of the conductor.
Substituting $\sigma = 1/\rho$, in Equation 4.29, we get:
$$\vec{J} = (1/\rho)\vec{E}, \text{ i.e.,}$$
$$\boxed{\vec{E} = \rho\vec{J}} \quad (4.32)$$
Conductors have high conductivities and low resistivities whereas insulators have low conductivities and high resistivities. The conductivities and resistivities for semiconductors lie between the values for conductors and insulators. The conductivities and resistivities of semiconductors depend on the amount of impurities in a given sample.

Temperature Dependence

As the temperature of a conductor increases, its resistivity goes up. Mathematically, we express the resistivity at some temperature T as
$$\boxed{\rho(T) = \rho_0[1 + \alpha(T - T_0)]} \quad (4.33)$$
where $\rho(T)$ is the resistivity at some temperature $T°C$, ρ_0 is the resistivity at some reference temperature $T_0°C$ (usually taken to be $20°C$), and α is the temperature coefficient of resistivity. The resistivities listed in Table 4.1 are for substances at $T_0 = 20°C$. The temperature coefficient (α) of resistivity depends on the type of material. It has the dimensions of 1/temperature, i.e., K^{-1}. For most conductors, $\alpha > 0$, so resistivity increases at higher temperatures (Eq. 4.33). For semiconductors, α is negative. The negative value of α indicates that the resistivity of the material decreases with increasing temperature (Fig.4.20a). It is due to an increase in the density of charge carriers at higher temperatures. Some materials, including several metallic alloys and oxides, show a phenomenon called superconductivity. As the temperature decreases, the resistivity at first decreases smoothly, like that of any metal. But then at a certain critical temperature a phase transition occurs and the resistivity suddenly drops to zero, as shown in Fig. 4.20b. Once a current has been established in a superconducting ring, it continues indefinitely without the presence of any driving field.

From Equation 4.33, the temperature coefficient of resistivity can be expressed as
$$\alpha = \frac{(\rho - \rho_0)}{\rho_0(T - T_0)} = \frac{\Delta\rho/\rho_0}{\Delta T}$$
where $\Delta\rho = \rho - \rho_0$ is the change in resistivity in the temperature interval $\Delta T = T - T_0$.

For some metals such as copper, resistivity is nearly propor-

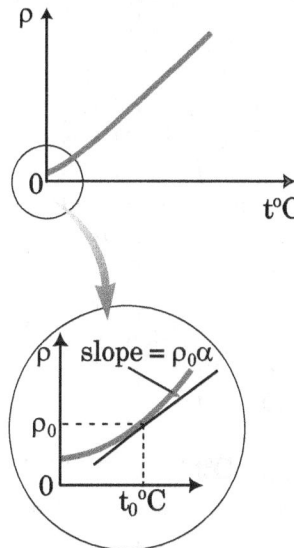

Figure 4.19: Resistivity versus temperature for a metal such as copper. The curve is linear over a wide range of temperatures, and ρ increases with increasing temperature.

tional to temperature as shown in Figure 4.19. At very low temperatures, there exists a non linear region. As the temperature approaches to absolute zero, the resistivity reaches to some finite value. This residual resistivity near absolute zero is caused primarily by the collision of electrons with impurities and imperfections in the metal. In contrast, high-temperature resistivity (the linear region) is predominantly characterized by collisions between electrons and metal atoms.

The resistivity varies over a very wide range. For metals (good conductor) $\rho \approx 10^{-8} \Omega$-m and for insulators $\rho \approx 10^{17} \Omega$ - m
Semiconductors (silicon, germanium) have intermediate value much smaller than insulator but much larger than metals. Temperature coefficient of resistivity is negative for semiconductors and positive for the metals. For superconductors resistivity is zero. Table 4.1 gives the conductivities, resistivities and temperature coefficients for some commonly used metals.

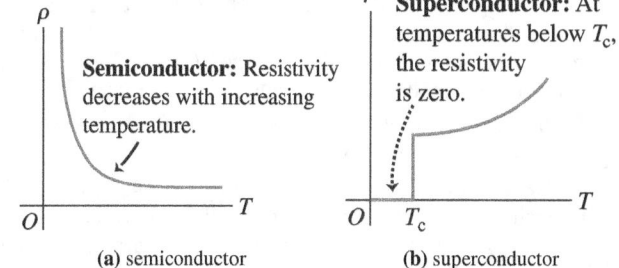

Figure 4.20: Variation of resistivity with absolute temperature for (a) a semiconductor, and (b) a superconductor.

☞ From above discussion we can say that, resistivity depends

Table 4.1: Resistivities, Conductivities, and Temperature Coefficients for Different Materials (at 20°C)

Material	$\rho(\Omega \cdot m)$	$\sigma(\Omega \cdot m)^{-1}$	$\alpha (K^{-1})$
Conductors (or Metals)			
Aluminum	2.65×10^{-8}	3.77×10^7	3.80×10^{-3}
Copper[a]	1.69×10^{-8}	5.96×10^7	6.80×10^{-3}
Copper[b]	1.72×10^{-8}	5.95×10^7	4.30×10^{-3}
Gold	2.24×10^{-8}	4.10×10^7	3.40×10^{-3}
Iron	9.7×10^{-8}	1.00×10^7	5.00×10^{-3}
Lead	2.06×10^{-7}	4.55×10^6	3.90×10^{-3}
Nichrome[c]	1.10×10^{-6}	9.09×10^5	4.00×10^{-3}
Nickel	6.85×10^{-8}	1.43×10^7	6.41×10^{-3}
Platinum	1.06×10^{-7}	9.43×10^6	3.92×10^{-3}
Silver	1.59×10^{-8}	6.30×10^7	3.80×10^{-3}
Tin	1.09×10^{-7}	9.17×10^6	4.50×10^{-3}
Tungsten	5.60×10^{-8}	1.79×10^7	4.50×10^{-3}
Zinc	5.90×10^{-8}	1.69×10^7	3.70×10^{-3}
Semiconductors[d]			
Germanium	$(1 - 500) \times 10^{-3}$		-0.05
Silicon	$0.1 - 60$		-0.07
Insulators			
Glass	$10^9 - 10^{12}$		
Hard rubber	$10^{13} - 10^{15}$		

[a]This copper is widely used for electrical equipment, building wiring, and telecommunication cables.

[b]This is 100% International Annealed copper Standard (IACS). It is used for temper and alloy verification of aluminium

[c]Nichrome (nickel-chromium alloy) used for making heating wires.

[d]Values depend strongly on the presence of even slight amounts of impurities.

on (i) Nature of material (ii) Temperature of material. It does not depend on the size and shape of the material because it is a characteristic of material.

4.3.4 Thermistor

A thermistor is a semiconductor electronic device in which the resistance decreases as its temperature increases. This is used as a thermometer. The temperature coefficient of resistivity is negative for semiconductors, hence thermistors are usually prepared from oxides of various metals such as nickel, iron, cobalt and copper etc. A thermistor is used to detect small changes in temperature of the order of even 10^{-3}°C.

4.3.4.1 Key Points

★ On increasing the temperature of the conductor, there will be increase in thermal agitation of free electrons and hence electrons start colliding quickly so mean free time, 'τ' decreases.

★ Current is flux of current density.

★ Due to principle of conservation of charge: Charge entering at one end of a conductor = charge leaving at the other end, so current does not change with change in cross section and conductor remains uncharged when current flows through it.

Note: For a steady current, we use $I = \vec{J} \cdot \vec{A}$, but for a non-steady current we always use $I = \int \vec{J} \cdot d\vec{A}$.

EXAMPLE 14. Find free electrons per unit volume in a metallic wire of density 10^4 kg/m^3, molar mass of the material of wire is 100×10^{-3} kg/mol and number of free electron per atom is one.

APPROACH Since, there is only one free electron per atom, therefore, the number n of conduction electrons per unit volume is the same as the number of atoms per unit volume.

SOLUTION Number of free charge particle per unit volume

$$n = \begin{pmatrix} \text{atoms} \\ \text{per unit} \\ \text{volume} \end{pmatrix} = \begin{pmatrix} \text{no. of moles} \\ \text{per} \\ \text{unit volume} \end{pmatrix} \begin{pmatrix} \text{no. of atoms} \\ \text{per mole} \end{pmatrix}*$$

$$= \begin{pmatrix} \text{number of moles} \\ \text{per} \\ \text{unit volume} \end{pmatrix} \begin{pmatrix} \text{Avogadro's} \\ \text{number} \end{pmatrix}$$

$$= \frac{\text{mass/molar mass}}{\text{volume}} \times N_A$$

$$= \frac{(\text{density} \times \text{volume})/\text{molar mass}}{\text{volume}} \times N_A$$

$$= \frac{\text{density}}{\text{molar mass}} \times N_A$$

$$= \frac{10^4 \text{ kg/m}^3}{100 \times 10^{-3} \text{ kg/mol}} \times 6.02 \times 10^{23} \text{mol}^{-1}$$

$$= 6.02 \times 10^{28} \text{m}^{-3}$$

$$\Rightarrow n = 6.02 \times 10^{28} \text{m}^{-3}$$

EXAMPLE 15. In a current, the conduction electrons move very slowly

What is the drift speed of the conduction electrons in a copper wire with radius $r = 900$ mm when it has a uniform current $I = 17$ mA? Assume that each copper atom contributes one conduction electron to the current and that the current density is uniform across the wire's cross section (density of copper = 8.96×10^3 kg/m^3 molar mass of copper = 63.54×10^{-3} kg/mol).

APPROACH 1. The drift speed v_d is related to the current density J and the number n of conduction electrons per unit volume according to Eq. $J = nev_d$.

2. Because the current density is uniform, its magnitude J is related to the given current I and wire size by Eq. $J = I/A$, where A is the cross-sectional area of the wire.

3. Because we assume one conduction electron per atom, the number n of conduction electrons per unit volume is the same as the number of atoms per unit volume.

SOLUTION Let us start with the third idea by writing

$$n = \begin{pmatrix} \text{atoms} \\ \text{per unit} \\ \text{volume} \end{pmatrix} = \begin{pmatrix} \text{number of moles} \\ \text{per} \\ \text{unit volume} \end{pmatrix} \begin{pmatrix} \text{Avogadro's} \\ \text{number} \end{pmatrix}$$

$$= \frac{\text{mass/molar mass}}{\text{volume}} \times N_A$$

$$= \frac{(\text{density} \times \text{volume})/\text{molar mass}}{\text{volume}} \times N_A$$

$$= \frac{\text{density}}{\text{molar mass}} \times N_A$$

$$= \frac{8.96 \times 10^3 \text{ kg/m}^3}{63.54 \times 10^{-3} \text{ kg/mol}} \times 6.02 \times 10^{23} \text{mol}^{-1}$$

$$= 8.49 \times 10^{28} \text{m}^{-3}$$

$$\Longrightarrow n = 8.49 \times 10^{28} \text{m}^{-3}$$

Next let us combine the first two points of the approach by writing

$$I = neAv_d$$

Substituting for A with $\pi r^2 (= 2.54 \times 10^6$ m^2) and solving for v_d, we then find

$$v_d = \frac{I}{ne(\pi r^2)}$$

$$= \frac{17 \times 10^{-3} \text{ A}}{(8.49 \times 10^{28} \text{ m}^{-3})(1.6 \times 10^{-19}\text{C})(2.54 \times 10^{-6} \text{ m}^2)}$$

$$= 4.9 \times 10^{-7} \text{ m/s}$$

Lights are fast: Now, a question may arise "If electrons drift so slowly, why do the room lights turn on so quickly when we

---

*The number of atoms per mole is just Avogadro's number $N_A \left(= 6.02 \times 10^{23} \text{ mol}^{-1}\right)$.

4.4 Electrical Resistance, Conductance and Ohm's Law

4.4.1 Electrical Resistance and Conductance

In preceding sections, we used a microscopic model to find the current density $\vec{J}$ when there is an electric field $\vec{E}$. The macroscopic vector quantities $\vec{J}$ and $\vec{E}$ are not very useful when we are building circuits. Instead, it is more convenient to know two scalar quantities: the electric potential difference ΔV and current I.

In order to write $\vec{E} = \rho \vec{J}$ (Eq.4.32) in terms of electric potential ΔV and current I, we consider a segment of straight wire of uniform cross-sectional area A and length l as shown in Figure 4.21. A potential difference $\Delta V = V_b - V_a$ is maintained across the wire, creating in the wire an electric field $\vec{E}$ and a current I. If the field is assumed to be uniform, then a relation between the magnitude of electric field and the magnitude of the potential difference across the wire is given as-
$$E = \frac{\Delta V}{l}$$
Substituting this value of E and $J = I/A$, in Eq.4.30: $J = \sigma E$, we get
$$\frac{I}{A} = \sigma \frac{\Delta V}{l}$$
or
$$\Delta V = \frac{l}{\sigma A} I = \rho \frac{l}{A} I \quad \left[\text{since, } \sigma = \frac{1}{\rho}\right] \quad (4.34)$$
Since, from Eq.4.31, we have
$$\rho = \frac{1}{\sigma} = \frac{m_e}{ne^2\tau}$$
Here, m_e, n, e are already constants for a given material of conductor, but relaxation time τ is a temperature dependent quantity. If temperature of the conductor is maintained constant then τ will also be constant. Therefore, for a given material of conductor at constant temperature, the quantity $\rho l/A$ (or $l/\sigma A$) becomes constant. This constant is denoted by R and it is called the resistance of the conductor, i.e.,
$$R = \frac{l}{\sigma A} = \rho \frac{l}{A} = \frac{m_e}{ne^2\tau} \frac{l}{A} \quad (4.35)$$
Using Eq.4.35, in Eq.4.34, we get
$$\boxed{\Delta V = RI \quad \text{(Ohm's law)}} \quad (4.36)$$

☞ Although, Ohm's law can be stated mathematically by Eq.4.36: $\Delta V = IR$, but this statement is incomplete. In next subsection, we will present three statements of Ohm's law and show why Equation 4.36 alone is not sufficient.

We can also define the resistance in terms of dynamic variables as the ratio of the potential difference across a conductor to the current in the conductor:
$$R = \frac{\Delta V}{I}$$
Therefore, Eq.4.35 can also be written as
$$\boxed{R = \frac{\Delta V}{I} = \frac{l}{\sigma A} = \rho \frac{l}{A} = \frac{m_e}{ne^2\tau} \frac{l}{A}} \quad (4.37)$$

Resistance R is another measure of an object's ability to resist an electric current. It is different from resistivity ρ, which depends on the type of material and temperature. Resistance depends on

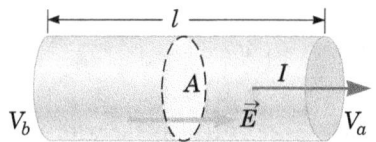

Figure 4.21: A uniform conductor of length l and cross-sectional area A.

the object's resistivity, so resistance also depends on the type of material and temperature. However, resistance R also depends on the geometry of the object.

Again according to Eq.4.35: $R = \rho \frac{l}{A} = \frac{m_e}{ne^2\tau} \frac{l}{A}$, for a given material of conductor at constant temperature $\rho = \frac{m_e}{ne^2\tau} = $ constant, therefore, the resistance of the conductor is directly proportional to it's length (l) and inversely proportional to it's cross sectional area (A), i.e.,

(i) $R \propto l$ and (ii) $R \propto \frac{1}{A}$

So, if two conductors are made from the same material and have the same temperature, they have the same resistivity ρ. Now imagine that one of the conductors is a long, thin wire, and the other is a short, thick one, then, the resistance R of the long, thin wire is higher than the resistance of the short, thick one.

Also from Eq.4.35: $R = \frac{m_e}{ne^2\tau} \frac{l}{A}$, we have

(i) $R \propto \frac{1}{n}$ and (ii) $R \propto \frac{1}{\tau}$

Since, in metals τ decreases as temperature increases, therefore on increasing temperature the resistance R of the conductor increases.

SI Unit of Resistance: From Eq.4.37,
$$\text{SI unit of } R = \frac{\text{SI unit of } \Delta V}{\text{SI unit of } I} = \frac{\text{volt}}{\text{ampere}}.$$
One volt per ampere is defined to be one ohm (Ω):
$$1\Omega = 1 \text{ V/A}$$
Equation 4.37: $R = \Delta V/I$, shows that if a potential difference of 1 V across a conductor causes a current of 1 A, the resistance of the conductor is 1 Ω.

NOTE 1. It is important to distinguish between resistivity and resistance. Resistivity describes just the material, not any particular piece of it. Resistance characterizes a specific piece of the conductor with a specific geometry.

2. Ohm's law is true for a conductor of any shape, but relation
$$R = \frac{l}{\sigma A} = \rho \frac{l}{A} = \frac{m_e}{ne^2\tau} \frac{l}{A}$$
is valid only for a conductor with a constant cross-sectional area.

3. Resistance R is the property of an object (depending on quantities such as l and A), while resistivity is a property of the material itself.

Conductance: Conductance is the measure of how easily electrical current can pass through a material. Generally, it is denoted by an uppercase letter G. It's magnitude is equal to $1/R$, i.e.,
$$\boxed{G = \frac{1}{R}} \quad (4.38)$$

Unit: The siemens (symbolized S) is the SI unit of electrical conductance. The older term for this unit is the mho (ohm spelled backwards).

EXAMPLE 16. What is the resistivity of a wire of 1.0 mm diameter, 2.0 m length, and 50 mΩ resistance?
APPROACH From Eq.4.37: $R = \rho l/A$, the resistivity of the material of the wire, is given by
$$\rho = RA/l \quad \cdots (1)$$
here, R is the resistance, l is the length and A is the cross-sectional

area of the wire.
In this case, the cross-sectional area is
$$A = \pi r^2 = \pi \left(0.50 \times 10^{-3} \text{ m}\right)^2 = 7.85 \times 10^{-7} \text{ m}^2$$
SOLUTION Thus, the resistivity of the wire is
$$\rho = \frac{RA}{l} = \frac{\left(50 \times 10^{-3}\Omega\right)\left(7.85 \times 10^{-7} \text{ m}^2\right)}{2.0 \text{ m}} = 2.0 \times 10^{-8}\Omega \cdot \text{m}$$
EXAMPLE 17. In an aluminium (Al) bar of square cross-section, a square hole is drilled and is filled with iron (Fe) as shown in the Fig.4.22. The electrical resistivities of Al and Fe are $2.7 \times 10^{-8}\Omega$m and $1.0 \times 10^{-7}\Omega$m, respectively. Find the electrical resistance between the two faces P and Q of the composite bar
SOLUTION From Eq.4.37, the resistance of a conductor of

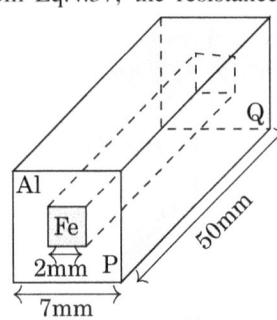

Figure 4.22

length l, cross-sectional area A, and material resistivity ρ is given by
$$R = \rho l/A \qquad \cdots (1)$$
For the given iron bar, we have
$l_{Fe} = 5 \times 10^{-2}$ m, $A_{Fe} = \left(2 \times 10^{-3}\right)^2 = 4 \times 10^{-6}$ m^2, and $\rho_{Fe} = 1.0 \times 10^{-7}\Omega$m.
To get the resistance of iron bar, we substitute these values in Eq.(1)
$$R_{Fe} = \frac{\rho_{Fe} l_{Fe}}{A_{Fe}} = \frac{\left(1 \times 10^{-7}\right)\left(5 \times 10^{-2}\right)}{4 \times 10^{-6}} = 1250\mu\Omega$$
For the aluminium bar, it is given that
$A_{Al} = \left(7 \times 10^{-3}\right)^2 - \left(2 \times 10^{-3}\right)^2 = 4.5 \times 10^{-5}$ m^2, $l_{Al} = 5 \times 10^{-2}$ m, and $\rho_{Al} = 2.7 \times 10^{-8}\Omega$m.
Again to get the resistance of aluminium bar, we substitute these values in Eq.(1)
$$R_{Al} = \frac{\rho_{Al} l_{Al}}{A_{Al}} = \frac{\left(2.7 \times 10^{-8}\right)\left(5 \times 10^{-2}\right)}{4.5 \times 10^{-5}} = 30\mu\Omega$$
These iron and aluminium bars are connected in parallel between P and Q, therefore, the equivalent resistance of the composite bar is
$$R_{eq} = R_{Fe} \| R_{Al} = \frac{R_{Fe} R_{Al}}{R_{Fe} + R_{Al}} = \frac{(1250)(30)}{1250 + 30} = \frac{1875}{64}\mu\Omega$$
EXAMPLE 18. Kiting during a storm. Suppose a kite string of radius 2.00 mm extends directly upward by 0.800 km and is coated with a 0.500 mm layer of water having resistivity 150Ωm. If the potential difference between the two ends of the string is 160 MV, what is the current through the water layer?
SOLUTION Let $r = 2.00$ mm be the radius of the kite string and $t = 0.50$ mm be the thickness of the water layer. The cross-sectional area of the layer of water is
$$A = \pi \left[(r+t)^2 - r^2\right]$$
$$= \pi \left[\left(2.50 \times 10^{-3} \text{ m}\right)^2 - \left(2.00 \times 10^{-3} \text{ m}\right)^2\right]$$
$$= 7.07 \times 10^{-6} \text{ m}^2$$
Using Eq.4.37: $R = \rho l/A$, the resistance of the wet string is
$$R = \frac{\rho l}{A} = \frac{(150 \text{ }\Omega\text{m})(800 \text{ m})}{7.07 \times 10^{-6} \text{ m}^2} = 1.698 \times 10^{10}\Omega$$

The current through the water layer is
$$I = \frac{V}{R} = \frac{1.60 \times 10^8 \text{ V}}{1.698 \times 10^{10}\Omega} = 9.42 \times 10^{-3} \text{ A}$$
☞ The danger is not this current but the chance that the string draws a lightning strike, which can have a current as large as 500000 A (way beyond just being lethal).

4.4.2 Dependence of Resistance on Temperature
From Eq.4.33, the temperature dependence of resistivity of a material of conductor is given by
$$\rho(T) = \rho_0 \left[1 + \alpha\left(T - T_0\right)\right] \qquad (4.39)$$
where $\rho(T)$ is the resistivity at some temperature $T°$C (in degree Celsius), ρ_0 is the resistivity at some reference temperature T_0 (usually taken to be $20°C$), and α is the temperature coefficient of resistivity. Now, substituting this value of ρ, in Eq.4.37: $R = \rho l/A$, the resistance of the conductor,
$$R = \frac{\rho l}{A} = \frac{\rho_0 l}{A}\left[1 + \alpha\left(T - T_0\right)\right]$$
here, $\rho_0 l/A$ is the resistance of the conductor at temperature T_0. If we denote it by R_0, then above equation becomes
$$\boxed{R(T) = R_0 \left[1 + \alpha\left(T - T_0\right)\right]} \qquad (4.40)$$
From Eq.4.40, it is clear that the resistance of a conductor decreases linearly with decrease in temperature and becomes zero at a specific temperature. This temperature is called critical temperature (T_C). At this temperature conductor becomes a superconductor.
The resistance vs temperature graph (Fig.4.23) is similar to the resistivity vs temperature graph shown in Fig.4.19.
EXAMPLE 19. The current-voltage graphs for a given

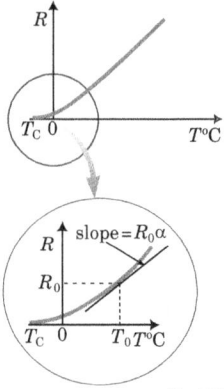

Figure 4.23: Resistance vs temperature (in °C) graph for a conductor

metallic wire at two different temperatures T_1 and T_2 are shown in the Fig.4.24. Which temperature is greater?
SOLUTION Let R_1 and R_2 be the resistances of the metallic

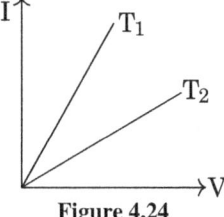

Figure 4.24

wire at temperature T_1 and T_2, respectively. Ohm's law, $V = IR$, gives the slope of the I-V graph as $dI/dV = 1/R$. In the given graph, slope at T_1 is greater than the slope at T_2 i.e., $1/R_1 > 1/R_2$ or $R_1 < R_2$. The resistance of the metallic wire varies with the temperature as
$$R_1 = R_0 \left[1 + \alpha\left(T_1 - T_0\right)\right] \qquad \cdots (1)$$
$$R_2 = R_0 \left[1 + \alpha\left(T_2 - T_0\right)\right] \qquad \cdots (2)$$
where $\alpha > 0$ is the thermal coefficient of resistance and R_0 is the

resistance at temperature T_0. Substitute R_1 and R_2 from equations (1) and (2) into the inequality $R_1 < R_2$ to get $T_1 < T_2$.
So, temperature T_2 is greater than T_1.

EXAMPLE 20. Figure 4.25 shows a person and a cow, each at a radial distance $D = 60.0$ m from the point where lightning of current $I = 100$ kA strikes the ground. The current spreads through the ground uniformly over a hemi-sphere centred on the strike point. The person's feet are separated by radial distance $\Delta L_{per} = 0.50$ m; the cow's front and rear hooves are separated by radial distance $\Delta r_{cow} = 1.50$ m. The resistivity of the ground is $\rho_{gr} = 100 \, \Omega m$. The resistance for both across the person, between left and right feet, and across the cow, between front and rear hooves, is $R = 4.00$ kΩ.
(a) What is the current i_{per} through the person?
(b) What is the current i_{cow} through the cow?

APPROACH The lightning strike sets up an electric field and

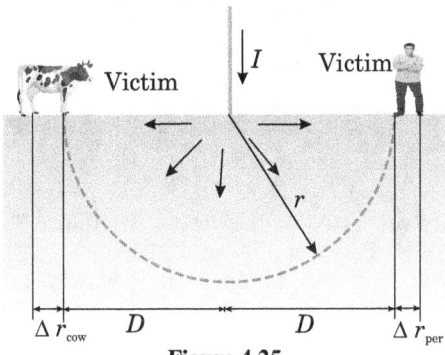

Figure 4.25

an electric potential in the surrounding ground. Because one foot of the person is closer to the strike point than the other foot, a potential difference ΔV_{per} is set up across the person. Similarly, a potential difference ΔV_{cow} also setup across cow's front and rear hooves. These potential differences drive currents i_{per} and i_{cow} through the person and cow respectively.
Therefore, we first calculate the potential and then find respective currents.

SOLUTION Calculating Potential Difference: Because the lightning's current I spreads uniformly over a hemisphere in the ground, therefore, from Eq.4.13, the current density at any given radius r from the strike point is

$$J = \frac{I}{A} = \frac{I}{2\pi r^2} \qquad \ldots (1)$$

where $A = 2\pi r^2$ is the area of the curved surface of a hemisphere. From scalar form of Eq.4.32: $E = \rho J$, the magnitude of the electric field is

$$E = \rho_{gr} J = \frac{\rho_{gr} I}{2\pi r^2} \qquad \ldots (2)$$

From equation $\Delta V = -\int \vec{E} \cdot d\vec{r}$, the potential difference ΔV between a point at radial distance D and a point at radial distance $D + \Delta r$ is

$$\Delta V = -\int_D^{D+\Delta r} E \, dr \qquad \ldots (3)$$

Substituting the value of E, from Eq.(2), into Eq.(3), we get:

$$\Delta V = -\int_D^{D+\Delta r} \frac{\rho_{gr} I}{2\pi r^2} dr = -\frac{\rho_{gr} I}{2\pi} \left[-\frac{1}{r} \right]_D^{D+\Delta r}$$

$$= \frac{\rho_{gr} I}{2\pi} \left(\frac{1}{D+\Delta r} - \frac{1}{D} \right)$$

$$\Rightarrow \quad \Delta V = -\frac{\rho_{gr} I}{2\pi} \frac{\Delta r}{D(D+\Delta r)} \qquad \ldots (4)$$

Calculating Current: This ΔV drives a current through person and cow. Now, to find that currents through person and cow, we use Eq.4.36: $\Delta V = Ri$, i.e.,

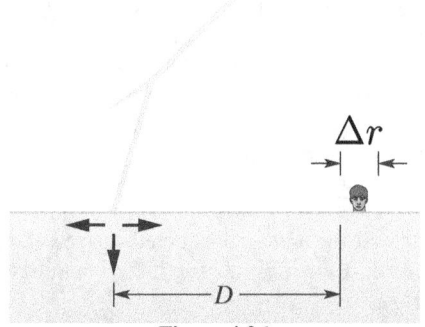

Figure 4.26

$$i = \frac{\Delta V}{R} \qquad \ldots (5)$$

Here, we have used i for current, because we have already used latter I for current due to lightnings.
Substituting the magnitude of ΔV from Eq.(4) in Eq.(5), we get

$$i = \frac{\Delta V}{R} = \frac{\rho_{gr} I}{2\pi} \frac{\Delta r}{D(D+\Delta r)} \frac{1}{R} \qquad \ldots (6)$$

(a) If one of the person's feet is at radial distance D from the strike point and the other foot is at radial distance $D + \Delta r$, the current through the person is given by Eq.(6) as-

$$i_{per} = \frac{\rho_{gr} I}{2\pi} \frac{\Delta r_{per}}{D(D+\Delta r_{per})} \frac{1}{R} \qquad \ldots (7)$$

Substituting known values, including the foot-to-foot separation $\Delta r_{per} = 0.50$ m, gives the current through the person:

$$i_{per} = \frac{(100 \, \Omega.m)(100 \, kA)}{2\pi} \times \frac{0.50 \, m}{(60 \, m)(60.0 \, m + 0.50 \, m)} \frac{1}{4.00 k\Omega}$$

$$= 0.0548 \, A = 54.8 \, mA.$$

This amount of current causes involuntary muscle contraction; the person will collapse but probably soon recover. Note that the person could reduce the current an order of magnitude by standing with feet together so that Δr is only a few centimeters.
(b) Similarly, from Eq.(6), the current through cow

$$i_{cow} = \frac{\rho_{gr} I}{2\pi} \frac{\Delta r_{cow}}{D(D+\Delta r_{cow})} \frac{1}{R} \qquad \ldots (8)$$

Again substituting known values, including the cow's front and rear hooves separation $\Delta r_{cow} = 1.50$ m, gives the current through the cow:
here, $\Delta r_{cow} = 1.50$ m. On substituting it in Eq.(6), we find that the current through the cow is

$$i_{cow} = \frac{(100 \, \Omega.m)(100 \, kA)}{2\pi}$$

$$\times \frac{1.50 \, m}{(60 \, m)(60.0 \, m + 1.50 \, m)} \frac{1}{4.00 \, k\Omega}$$

$$= 0.0162 \, A = 162 \, mA.$$

which is fatal. The cow is in more danger from the ground current because of its greater value of Δr. The cow is, of course, unable to reduce its danger by standing with its hooves together (which would be a bizarre sight).

EXAMPLE 21. Swimming During a Storm Figure 4.26 shows a swimmer at distance $D = 35.0$ m from a lightning strike to the water, with current I = 78 kA. The water has resistivity 30 Ωm, the width of the swimmer along a radial line from the strike is 0.70 m, and his resistance across that width is 4.00 kΩ. Assume that the current spreads through the water over a hemisphere centred on the strike point. What is the current through the swimmer?

APPROACH The lightning strike sets up an electric field and an electric potential in the surrounding water. Because one side of the person is closer to the strike point than the other, a potential difference ΔV is set up across the swimmer which drive currents

i through the him.
Therefore, we first calculate the potential and then find the required current through the swimmer.

SOLUTION Calculating potential difference: Because the lightning's current I spreads uniformly over a hemisphere in the water, therefore, from Eq.4.13, the current density at any given radius r from the strike point is

$$J = \frac{I}{A} = \frac{I}{2\pi r^2} \qquad \ldots (1)$$

where $A = 2\pi r^2$ is the area of the curved surface of a hemisphere. From scalar form of Eq.4.32: $E = \rho J$, the magnitude of the electric field is

$$E = \rho_w J = \frac{\rho_w I}{2\pi r^2} \qquad \cdots (2)$$

From equation $\Delta V = -\int \vec{E} \cdot d\vec{r}$, the potential difference ΔV between a point at radial distance D and a point at radial distance $D + \Delta r$ is

$$\Delta V = -\int_D^{D+\Delta r} E\, dr \qquad \cdots (3)$$

Substituting the value of E, from Eq.(2), into Eq.(3), we get:

$$\Delta V = -\int_D^{D+\Delta r} \frac{\rho_w I}{2\pi r^2} dr = -\frac{\rho_w I}{2\pi}\left[-\frac{1}{r}\right]_D^{D+\Delta r}$$

$$= \frac{\rho_w I}{2\pi}\left(\frac{1}{D+\Delta r} - \frac{1}{D}\right)$$

$$\Rightarrow \quad \Delta V = -\frac{\rho_w I}{2\pi}\frac{\Delta r}{D(D+\Delta r)} \qquad \cdots (4)$$

Calculating Current: This ΔV drives a current through swimmer. Now, to find that current through swimmer, we use Eq.4.36: $\Delta V = Ri$, i.e.,

$$i = \frac{\Delta V}{R} \qquad \cdots (5)$$

Here, we have used i for current, because we have already used latter I for lightnings current.
Substituting the magnitude of ΔV from Eq.(4) in Eq.(5), we get

$$i = \frac{|\Delta V|}{R} = \frac{\rho_w I}{2\pi}\frac{\Delta r}{D(D+\Delta r)}\frac{1}{R} \qquad \cdots (6)$$

If one side of swimmer is at radial distance D from the strike point and the other is at radial distance $D + \Delta r$, then the current through the swimmer is given by Eq.(6).
Substituting the given values in Eq.(6), we obtain

$$i = \frac{(30.0\,\Omega.m)\,(7.80\times 10^4\,A)}{2\pi\,(4.00\times 10^3\,\Omega)\,R}\frac{0.70\,m}{(35.0\,m)(35.0\,m + 0.70\,m)}$$

$$= 5.22 \times 10^{-2}\,A$$

EXAMPLE 22. The dimensions of a conductor of specific resistance ρ are shown in Fig.4.27. Find the resistance of the conductor across AB, CD and EF.

APPROACH To determine R, identify the cross-sectional area

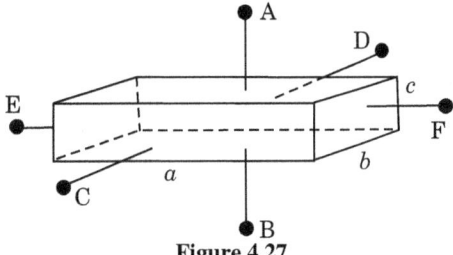

Figure 4.27

and length of conductor in each case and then apply Eq.4.37: $R = \rho l/A$

SOLUTION Resistance across AB: In this case, length of the conductor, $l = c$ and area of cross-section, $A = ab$, therefore

$$R_{AB} = \rho\frac{l}{A} = \frac{\rho c}{ab}$$

Resistance across CD: In this case, length of the conductor, $l = b$ and area of cross-section, $A = ac$, therefore

$$R_{CD} = \rho\frac{l}{A} = \frac{\rho b}{ac}$$

Resistance across EF: In this case, length of the conductor, $l = a$ and area of cross-section, $A = bc$, therefore

$$R_{EF} = \rho\frac{l}{A} = \frac{\rho a}{bc}$$

EXAMPLE 23. Consider a thin square sheet of side L and thickness t, made of material of resistivity ρ. Find the resistance between two opposite faces, shown by the shaded areas in the Fig.4.28.

APPROACH Since, length, thickness and resistivity of the

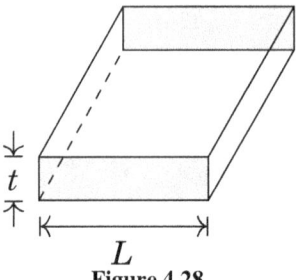

Figure 4.28

square sheet are given, therefore to determine R, we apply Eq.4.37: $R = \rho l/A$

SOLUTION The sheet is of square shape with thickness t, width $w = L$, and length $l = L$. From Eq.4.37, the resistance between the two opposite faces is given by

$$R = \frac{\rho l}{A} = \frac{\rho l}{wt} = \frac{\rho L}{Lt} = \frac{\rho}{t}$$

EXAMPLE 24. Two bars of radius r and $2r$ are kept in contact as shown in Fig.4.29. An electric current I is passed through the bars. Show that-
(A) Electric field in bar AB is 1/4 times of that in bar BC.

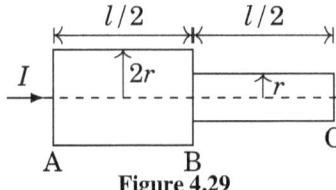

Figure 4.29

(B) Current density across AB is one fourth of that across BC.
(C) Potential difference across AB is 1/4 times that of across BC.

SOLUTION (A) Let ρ be the resistivity of the material. The resistances of the bar AB and BC are given by

$$R_{AB} = \frac{\rho(l/2)}{4\pi r^2} = \frac{\rho l}{8\pi r^2}, \quad R_{BC} = \frac{\rho(l/2)}{\pi r^2} = \frac{\rho l}{2\pi r^2}$$

The currents flowing through both the bars are equal to I as they are connected in series.

The potential drop V, electric field E, and the current density J for the two bars are given by
(A) The potential drop across bar AB,

$$V_{AB} = IR_{AB} = \frac{I\rho l}{8\pi r^2}$$

The potential drop across bar BC,

$$V_{BC} = IR_{BC} = \frac{I\rho l}{2\pi r^2};$$

Clearly, $V_{AB} = \frac{1}{4}V_{BC}$.

(B) Electric field in bar AB,

$$E_{AB} = \frac{V_{AB}}{l/2} = \frac{I\rho}{4\pi r^2},$$

and electric in bar BC,

$$E_{BC} = \frac{V_{BC}}{l/2} = \frac{I\rho}{\pi r^2};$$

Clearly, $E_{AB} = \frac{1}{4}E_{BC}$.

(C) If J_{AB} and J_{BC} are the current densities in bars AB and BC respectively, then-

$$J_{AB} = \frac{I}{4\pi r^2} \quad \text{and} \quad J_{BC} = \frac{I}{\pi r^2}.$$

From above we see that, $J_{AB} = \frac{1}{4}J_{BC}$

4.4.3 Effects of Change in Geometry on Resistance of the Conductor

Let us consider a cylindrical conductor of length l, area of cross-section A and radius r. Now, keeping the temperature constant, if we stretch the conductor along it's length, it's length increases whereas it's cross-sectional area decreases. Since the temperature is constant, so it's resistivity and density of the material of the conductor remain constant.

Since, mass of the conductor is already constant, therefore for constant density it's volume will also be constant. To understand the effect of change of length, we have to write the value of resistance in terms of single variables l only, and to find out the effect of change of cross-sectional area, we have to write the value of resistance in terms of single variable A only.

If we denote the volume of conductor by latter Z^*, then

$$Al = Z = \text{constant} \tag{4.41}$$

Now, from Eq.4.37, we have

$$R = \rho\frac{l}{A} \tag{4.42}$$

Resistance in terms of length and volume: To eliminate A, we substitute the value of A, from Eq.4.41 in Eq.4.42:

$$\boxed{R = \rho\frac{l^2}{Z}} \tag{4.43}$$

So, for constant volume of a conductor, $R \propto l^2$.

If for length l_1 of the conductor, it's resistance is R_1 and for length l_2, it's resistance is R_2, then

$$\boxed{\frac{R_1}{R_2} = \frac{l_1^2}{l_2^2}} \tag{4.44}$$

Resistance in terms of cross-sectional area and volume: To eliminate l, from Eq.4.42, we substitute the value of l, from Eq.4.41 in Eq.4.42:

$$\boxed{R = \rho\frac{Z}{A^2}} \tag{4.45}$$

i.e., $R \propto \frac{1}{A^2}$. So, for constant volume of a conductor, $R \propto 1/A^2$.

If for cross-sectional area A_1 of the conductor, it's resistance is R_1 and for cross-sectional area A_2, it's resistance is R_2, then

$$\boxed{\frac{R_1}{R_2} = \frac{A_2^2}{A_1^2}} \tag{4.46}$$

In terms of radius of cross-section, r, area of cross-section of the conductor, $A = \pi r^2$, therefore from Eq.4.45, we have

$$\boxed{R = \rho\frac{Z}{\pi^2 r^4}} \tag{4.47}$$

i.e., $\quad R \propto \frac{1}{r^4}$

So, if for cross-sectional radii r_1, r_2 of the conductor, it's resistances are R_1 and R_2 respectively, then

$$\boxed{\frac{R_1}{R_2} = \frac{r_2^4}{r_1^4}} \tag{4.48}$$

Note that, if elasticity of the material is taken into consideration, then the variation of area of cross-section is calculated with the help of Young's modulus and Poison's ratio.

Key Points
- On stretching the length of a metal wire to n times of it's original length, the new resistance becomes n^2 times of it's original resistance.
- If a metal wire is stretched such that it's radius is reduced to $\frac{1}{n}$ th of it's original values, then new resistance becomes n^4 times of it's original value. If radius is increased to n times by contraction, then resistance decrease to n^4 time of it's original value.

4.4.4 Effect of Percentage Change in Length of Wire

If initial length of a conductor $l_1 = l$ and it's resistance is $R_1 = R$. Suppose, it's length is increased by $x\%$, then new length of the conductor,

$$l_2 = \left[l + l \times \frac{x}{100}\right] = l\left[1 + \frac{x}{100}\right]$$

On substituting these values in Eq.4.44, we get

$$\frac{R_2}{R_1} = \frac{l^2\left[1 + \frac{x}{100}\right]^2}{l^2}$$

Subtracting 1 from both sides of above equation, we get

$$\frac{R_2}{R_1} - 1 = \frac{l^2\left[1 + \frac{x}{100}\right]^2}{l^2} - 1$$

or $\quad\dfrac{R_2 - R_1}{R_1} = \dfrac{l^2\left[1 + \frac{x}{100}\right]^2 - l^2}{l^2}$

or $\quad\dfrac{R_2 - R_1}{R_1} = \left[1 + \dfrac{x}{100}\right]^2 - 1$

Therefore, relative or fractional change in resistance,

$$\boxed{\left|\frac{R_2 - R_1}{R_1}\right| = \left[1 + \frac{x}{100}\right]^2 - 1} \tag{4.49}$$

and percentage change in resistance,

$$\boxed{\left|\frac{R_2 - R_1}{R_1}\right| \times 100 = \left[\left[1 + \frac{x}{100}\right]^2 - 1\right] \times 100} \tag{4.50}$$

If x is quite small (say $< 5\%$) then $\left[1 + \frac{x}{100}\right]^2 \approx \left[1 + 2\frac{x}{100}\right]$, therefore, for very small values of x, Eq.4.50 becomes

$$\boxed{\left|\frac{R_2 - R_1}{R_1}\right| \times 100 = 2x \, \%} \tag{4.51}$$

Explanation by using Calculus: For small changes in length we can also derive the formula 4.51 by using calculus.
From Eq.4.43, we have

$$R = \rho\frac{l^2}{Z}$$

Taking logarithm of both side of above equation, we get

$$\ln R = \ln \rho + 2\ln l - \ln Z$$

On differentiating both sides with respect to length l, we get

$$\frac{1}{R}\frac{dR}{dl} = \frac{2}{l} \quad \text{(since, } Z \text{ is constant)}$$

---

*we won't use latter V for volume, because, the latter V is reserved for electric potential

or
$$\frac{dR}{R} = 2\frac{dl}{l}$$
So, for small changes in length, the relative change in resistance
$$\left|\frac{dR}{R}\right| = 2\left|\frac{dl}{l}\right|$$
Therefore, for small changes in length, percentage change in resistance
$$\left|\frac{dR}{R}\right| \times 100 = 2\left|\frac{dl}{l}\right| \times 100 \%$$
i.e., for small changes in length, percentage change in resistance of the conductor is equal to two times of percentage change in length of the conductor. Note that the change in length is considered to be small if it is less than 5%.

EXAMPLE 25. A wire has resistance R. If it is stretched to double its length, find the new resistance.
APPROACH In stretching of the wire, it's volume will be conserved. So, we can apply Eq.4.44:
$$\frac{R_1}{R_2} = \frac{l_1^2}{l_2^2} \qquad \cdots (1)$$
SOLUTION Given: $R_1 = R$, $l_2 = 2l_1$. Substituting these values in Eq.(1), we get
$$R_2 = \frac{R_1 l_2^2}{l_1^2} = \frac{R(2l_1)^2}{l_1^2} = 4R$$
EXAMPLE 26. The wire is stretched to increase the length by 1% find the percentage change in the Resistance.
APPROACH Since, the percentage change in length is 1%, which is less that 5%, therefore, we can apply Eq.4.51:
$$\left|\frac{R_2 - R_1}{R_1}\right| \times 100 = 2x\% \qquad \cdots (1)$$
SOLUTION Given: $x = 1$, therefore from Eq.(1), we have
$$\left|\frac{R_2 - R_1}{R_1}\right| \times 100 \approx 2\%$$
Hence percentage increase in the resistance = 2%
EXAMPLE 27. A copper wire is stretched to make it 0.1% longer. What is the percentage change in its resistance?
APPROACH In this case, extension x is 0.1% which is quite small (say < 5%), therefore, we use Eq.4.51:
$$\left|\frac{R_2 - R_1}{R_1}\right| \times 100 = 2x \% \qquad \cdots (1)$$
SOLUTION Given: $x = 0.1$, therefore Eq.(1) gives
$$\left|\frac{R_2 - R_1}{R_1}\right| \times 100 = 2x \% = 2(0.1) \% = 0.2 \%$$
Clearly, the resistance of copper wire increases by 0.2%.

4.4.5 The Radial Resistance of a Hollow Spherical Conductor

Consider a hollow spherical conductor whose inner radius is a and outer radius is b ($a < b$) (see Fig. 4.30a). The resistivity of the material of the conductor is ρ.

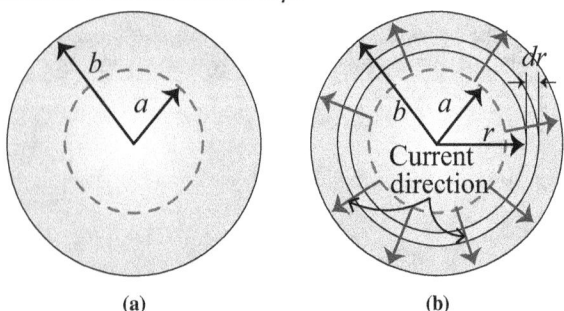

Figure 4.30

To determine its radial resistance, we consider a differential element which is made up of a thin spherical shell of inner radius r and outer radius $r + dr$ as shown in Fig. 4.30b.

In this case, the elementary length of the conductor along the radial current I is dr and the area perpendicular to current flow is the curved surface area of the spherical shell at radius r, i.e.,
$$A = 4\pi r^2.$$
Therefore, the radial resistance of the considered elementary part of the conductor is
$$dR = \rho \frac{dr}{A} = \rho \frac{dr}{4\pi r^2}.$$
To determine the complete radial resistance of the conductor, we integrate it from $r = a$ to $r = b$, i.e.,
$$R = \int_a^b \frac{\rho}{4\pi r^2} dr = \frac{\rho}{4\pi}\left(\frac{1}{a} - \frac{1}{b}\right) \qquad (4.52)$$

Remarks
1. If b/a is kept fixed and both a and b are scaled by a factor s, then
$$R' = \frac{\rho}{4\pi}\left(\frac{1}{sa} - \frac{1}{sb}\right) = \frac{1}{s}R.$$
Hence R scales as $1/s$. In particular, increasing both radii proportionally (with the same material) decreases the radial resistance.
2. If the thickness $b - a = t$ is kept fixed and a is increased (so $b = a + t$), then
$$\frac{1}{a} - \frac{1}{b} = \frac{t}{a(a+t)},$$
which decreases with increasing a. Therefore R decreases when a is increased at constant thickness.
3. A **solid sphere** is a limiting case of a hollow sphere for which the inner radius $a \to 0$.
Thus the radial electric resistance of a solid conducting sphere of radius b is
$$R = \lim_{a \to 0} \frac{\rho}{4\pi}\left(\frac{1}{a} - \frac{1}{b}\right) \to \infty$$

4.4.6 The Radial Resistance of a Coaxial Cable

In Figure 4.31a, we have a cylindrical conductor of length L. The resistivity of the material of the conductor is ρ. It's internal and external radii are a and b respectively.

To determine it's radial resistance, we consider a differential element which is made up of a thin cylinder of inner radius r and outer radius $r + dr$ and length L. In this case, the elementary length of conductor along the radial current I is dr and the area perpendicular to current flow, is the curved area of the cylinder of radius r and length $2\pi r L$, i.e., $A = 2\pi r L$.

Therefore, the radial resistance of considered elementary part of the conductor-
$$dR = \rho \frac{dr}{A}$$
$$\Rightarrow \qquad dR = \rho \frac{dr}{2\pi r L} = \frac{\rho}{2\pi L}\frac{dr}{r}$$
To determine complete radial resistance of the conductor, we have to integrate it from $r = a$ to $r = b$, i.e.,
$$\boxed{R = \int_a^b \frac{\rho}{2\pi L}\frac{dr}{r} = \frac{\rho}{2\pi L}\log_e\left(\frac{b}{a}\right)} \qquad (4.53)$$

Remarks
1. If keeping b/a fixed, we increase the cross-sectional area of the coaxial cable i.e., we increase it's radius, then from Eq.4.53, radial resistance remains unchanged. In this case, if there is a constant potential difference between the ends of the cable, the radial current (also called leakage current) remains unchanged.

4.4. ELECTRICAL RESISTANCE, CONDUCTANCE AND OHM'S LAW

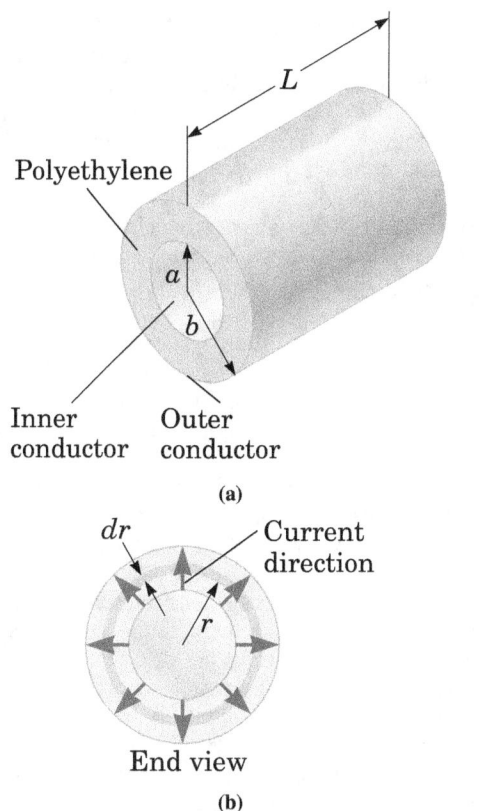

Figure 4.31

2. Now, if $(b - a)$ is fixed i.e., $(b - a) = t$ (constant), then
$$\frac{b}{a} - 1 = \frac{t}{a} \quad \Rightarrow \quad \frac{b}{a} = 1 + \frac{t}{a}$$
From above equation, it is clear that on increasing a (keeping t constant), the value of t/a decreases, as a result b/a also decreases. So, from Eq.4.53, R decreases. Thus, if keeping difference $b - a$ fixed, we increase the cross-sectional area of the cylinder, then radius of cylinder increases and so it's radial resistance decreases. In this case, if there is a constant potential difference between the ends of the cable, the radial current (i.e., leakage current) increases.

4.4.7 Linear Resistance of a Hollow Cylinder

Again consider Figure 4.31a with current along the length of the shown cylindrical conductor. The length of this hollow cylinder is L and it's inner and outer radii are a and b respectively. The resistivity of the material is ρ.

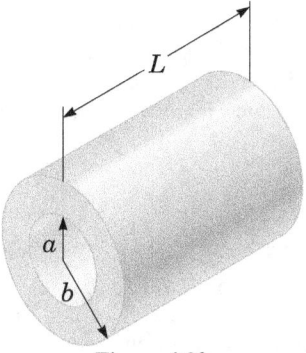

Figure 4.32

When a potential difference is applied between the ends of the cylinder, current flows parallel to the axis. If we denote the effective cross-sectional area as A, then in such a case, $A = \pi(b^2 - a^2)$, and therefore, from Eq.4.37, the net resistance along the length of the cylinder is given by:

$$\boxed{R = \frac{\rho L}{A} = \frac{\rho L}{\pi(b^2 - a^2)}} \quad (4.54)$$

Alternatively, you can also find the required resistance by applying the parallel combination of resistances described in Section 4.11. If we fill the hollow space by inserting the same material as that of the remaining part, then the net resistance of the complete solid cylinder will be given by:
$$\frac{1}{R} = \frac{1}{R_1} + \frac{1}{R_2}$$
Here, $R \rightarrow$ resistance of the solid cylinder of radius b, $R_1 \rightarrow$ resistance of the cylinder of radius a, and $R_2 \rightarrow$ required resistance of the hollow cylinder of inner radius a and outer radius b.
$$\frac{1}{R_2} = \frac{1}{R} - \frac{1}{R_1} = \frac{1}{\rho L/\pi b^2} - \frac{1}{\rho L/\pi a^2} = \frac{\pi}{\rho L}(b^2 - a^2)$$
$$\Rightarrow \quad R_2 = \frac{\rho L}{\pi(b^2 - a^2)}$$

4.4.8 Resistance of a Truncated Cone

Consider a material of resistivity ρ in a shape of a truncated cone of altitude L, and radii a and b ($b > a$) for the right and the left ends respectively, as shown in Figure 4.33.

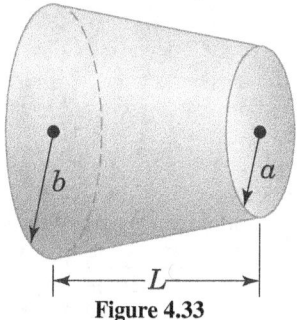

Figure 4.33

To find it's resistance, we consider a thin disc of thickness dx, radius y at a distance x from the left end (Fig.4.34). From the Figure 4.34, in similar triangles $\triangle EGD$ and $\triangle BHD$, we have
$$\frac{GD}{EG} = \frac{HD}{BH}$$
$$\Rightarrow \quad \frac{b - y}{x} = \frac{b - a}{L}$$
or
$$y = (a - b)\frac{x}{L} + b$$

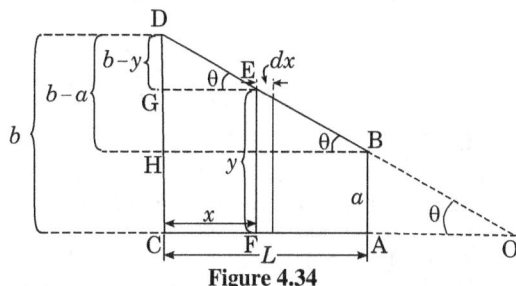

Figure 4.34

Since resistance R is related to resistivity ρ by $R = \rho \frac{l}{A'}$, where l is the length of the conductor and A is the cross section, the contribution to the resistance from the disc having a thickness dx is
$$dR = \frac{\rho dx}{\pi y^2} = \frac{\rho dx}{\pi\left[b + (a - b)\frac{x}{L}\right]^2}$$
Therefore, for complete truncated cone, the resistance

$$R = \int_0^L \frac{\rho dx}{\pi\left[b + (a-b)\frac{x}{L}\right]^2} = \frac{\rho L}{\pi ab}$$

where we have used

$$\int \frac{du}{(\alpha u + \beta)^2} = -\frac{1}{\alpha(\alpha u + \beta)}$$

Note that if $b = a$, then $R = \frac{\rho L}{\pi a^2}$ as expected.

EXAMPLE 28. A long round conductor of cross-sectional area S is made of material whose resistivity depends only on a distance r from the axis of the conductor as $\rho = \frac{\alpha}{r^2}$, where α is a constant. Find:
(a) the resistance per unit length of such a conductor;
(b) the electric field strength in the conductor due to which a current I flows through it.

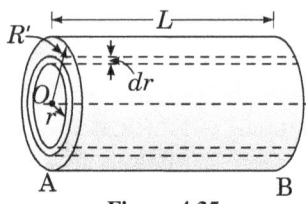

Figure 4.35

SOLUTION (a) Given cylindrical conductor is shown in Fig.4.35. The length of cylindrical conductor is L and it's radius is R'. Now consider an elementary cylindrical shell of radius r and thickness dr. The cross-sectional area of this cylindrical shell-

$$dA = \text{circumference} \times \text{thickness}$$
$$= (2\pi r)dr$$

Therefore, the longitudinal resistance of this cylindrical shell is

$$dR = \rho \frac{L}{dA} = \frac{\alpha}{r^2} \frac{L}{(2\pi r)dr}$$

As the cylinder can be assumed to be made of large number of such parallel shells, therefore, total resistance (R_{total}) of the cylinder can be given as below

$$\frac{1}{R_{\text{total}}} = \int \frac{1}{dR} = \int_0^{R'} \frac{2\pi r^3 dr}{\alpha L} = \frac{2\pi}{\alpha L} \frac{R'^4}{4} = \frac{(\pi R'^2)^2}{2\alpha L \pi}$$

$$= \frac{S^2}{2\alpha \pi L}$$

$$\Rightarrow R_{\text{total}} = \frac{2\alpha \pi L}{S^2}$$

Therefore, resistance per unit length of cylinder

$$= \frac{R_{\text{total}}}{L} = \frac{2\pi \alpha}{S^2}$$

(b) Electric field, $E = \frac{V}{L} = \frac{IR}{L} = \frac{R}{L}I = \frac{2\pi\alpha}{S^2}I$

4.4.9 Ohm's Law

Ohm's law is not a law in the same sense as Newton's laws of motion. It is a pure empirical law. Complete Ohm's law is a collection of following three statements:

First statement: Ohm found experimentally that, the current I in a conductor is directly proportional to the potential difference applied across it, provided all physical conditions and temperatures remain constant. Mathematically, it can be written as

$$\Delta V \propto I \qquad (4.55)$$

This expression of Ohm's law is shown graphically in Figure 4.36a.

If you compare this proportionality relationship $\Delta V \propto I$ (Eq. 4.55) with $\Delta V = RI$ (Eq. 4.36), you can see why Equation 4.36 is called Ohm's law. If the resistance of the conductor is constant- independent of both ΔV and I-then $\Delta V = IR$ is a statement of Ohm's law, where R is the constant of proportionality. Because we have plotted I on the vertical axis and ΔV on the horizontal axis, the slope of the line in Figure 4.36a is $1/R$. For a circuit element that obeys Ohm's law, that slope is constant. The quantity $1/R$ is called the **conductance**, *so the slope of the curve on a graph of I versus ΔV is the conductance.*

Second statement: A second statement of Ohm's law is in terms of the electric field within a conductor and the current density: The electric field $\vec{E}$ in a conductor is directly proportional to the current density $\vec{J}$ in that conductor:

$$\vec{E} \propto \vec{J} \qquad (4.56)$$

According to $\vec{J} = \sigma \vec{E}$ (Eq. 4.29), the slope of the line is the conductivity σ. Because the conductivity is the reciprocal of the resistivity ($\sigma = 1/\rho$), the slope is also $1/\rho$. For a device that obeys Ohm's law, the slope in Figure 4.15 is constant, so the conductivity and resistivity are constants.

Third Statement: A third statement of Ohm's law is that the resistance R, conductance $1/R$, resistivity ρ, and conductivity σ are all constants. If so, the resistance and conductance of a circuit element do not depend on the potential difference across the element or the current in it, and the resistivity and conductivity do not depend on the electric field or current density within the element. No actual circuit element obeys Ohm's law under all conditions, but a circuit that obeys Ohm's law over a wide range of potential differences across it is described as ohmic. Resistors and wires, such as various heating elements, carbon-composition resistors, and film or cermet resistors are examples of ohmic circuit elements.

Figure 4.36a shows graphs of I versus ΔV for three different ohmic circuit elements (such as three different resistors). The curve for each circuit element is a straight line with constant slope equal to the conductance $1/R$. The line with the largest slope represents the circuit element with the lowest resistance, and the line with the smallest slope represents the element with the highest resistance. A circuit element that does not obey Ohm's law over any significant range of ΔV, is said to be nonohmic. Figure 4.36b is a graph of I versus ΔV for a diode, a nonohmic device. The curve for the diode is a flat line for $\Delta V < 0$; for $\Delta V > 0$, the curve is approximately given by $I \propto \Delta V^{3/2}$.

Although $\Delta V = IR$ (Eq. 4.36) holds for a non-ohmic device, the resistance is not constant. Let's use the diode as an example. Although a diode does not obey Ohm's law, any point on the curve in Figure 4.36b is described by $\Delta V = IR$. When $\Delta V < 0$, the current is very nearly zero and we'll assume $I = 0$. According to $\Delta V = IR$, the diode has (nearly) infinite resistance ($R \to \infty$) when $\Delta V < 0$. The resistance is finite but not constant when $\Delta V > 0$. Because $I \propto \Delta V^{3/2}$, i.e., $I = k(\Delta V)^{3/2}$ (here, k is a proportionality constant), so the curve is flatter (has lower slope) for small values of ΔV than for large values, so the resistance is higher for small values of ΔV than for large values and is given by

$$R = \frac{\Delta V}{I} = \frac{\Delta V}{k\Delta V^{3/2}} = \frac{\Delta V^{-1/2}}{k}$$

Another example of a non-ohmic circuit element is the filament of an incandescent light bulb. The resistance of it increases markedly when current heats it enough to produce visible light. The resistance of the filament is relatively low when we turn the light bulb on (it is cold at the beginning), which means a higher amount of current flows through it at the beginning, slowly decreasing as the filament heats up and increases its resistance. It also explains why light bulbs usually burn out just as they are turned on: when you turn the light on, the filament is still cold,

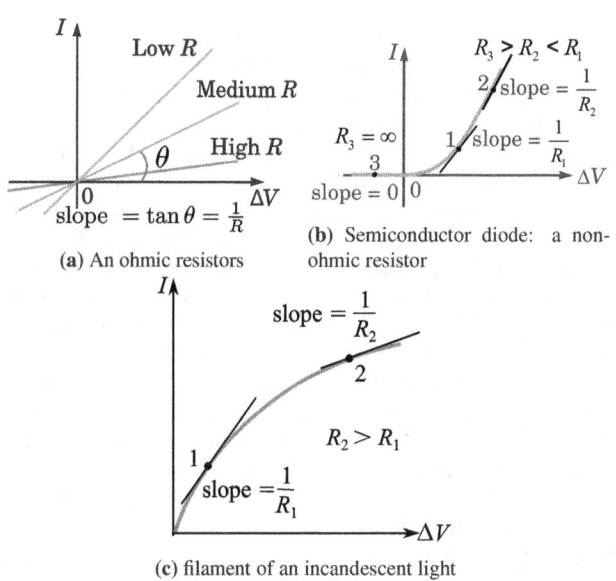

(a) An ohmic resistors

(b) Semiconductor diode: a non-ohmic resistor

(c) filament of an incandescent light bulb: a non-ohmic resistor

Figure 4.36: (a) Current versus potential difference for three ohmic devices. The slope of each line is the conductance $1/R$ of the corresponding circuit element. (b) Current as a function of potential difference for a nonohmic device (junction diode). The resistance is not constant. (c) Current as a function of potential difference for a nonohmic device (filament of incandescent bulb). The resistance is not constant.

so its resistance is low and, as a result, the current is high. This high current can damage the filament and thus burn out the light bulb.

☞ In I-V graph, why we take current on vertical axis, since the slope of this curve is $1/R$. It would be easier to plot current on horizontal axis and ΔV on vertical axis with slope R. We do this because the I-V plot shows how the electric current in the circuit (the dependent variable) depends on the applied voltage (the independent variable). The slope of this I-V curve represents conductance, G (i.e. how well the circuit conducts electric current), which is reciprocal to the resistance, $G = 1/R$.

☞ Ohm's law is not a fundamental law of nature. As it is possible that for an element:
1. V depends on I non linearly (e.g. diode, vacuum tubes, etc.)
2. Relation between V and I depends on the sign of V for the same value [for example, forward and reverse biasing in diode]
3. The relation between V and I is non unique. That is for the same I there is more then one value of V.

☞ Generally,
1. Resistivity of alloys is greater than their metals.
2. Temperature coefficient of alloys is lower than pure metals.
3. Resistance of most of non metals decreases with increase in temperature. (e.g. carbon).
4. The resistivity of an insulator (e.g. amber) is greater then the metal by a factor of 10^{22}

☞ Temperature coefficient (α) of semi conductor including carbon (graphite), insulator and electrolytes is negative.

4.5 Charge at a Junction

EXAMPLE 29. Show that the total amount of charge at the junction of the two materials in Fig.4.37, is $\varepsilon_0 I \left(\sigma_2^{-1} - \sigma_1^{-1}\right)$, where I is the current flowing through the junction, and σ_1 and σ_2 are the conductivities for the two materials.

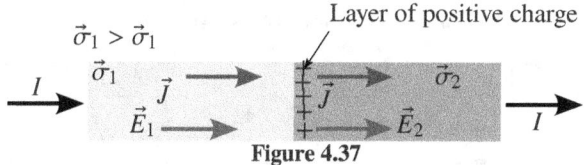

Figure 4.37

SOLUTION In a steady state of current flow, the normal component of the current density $\vec{J}$ must be the same on both sides of the junction. Since $J = \sigma E$, we have $\sigma_1 E_1 = \sigma_2 E_2$

$$E_2 = \left(\frac{\sigma_1}{\sigma_2}\right) E_1$$

Let the charge on the interface be q_{in}, we have, from the Gauss's law:

$$\oint_S \vec{E} \cdot d\vec{A} = (E_2 - E_1) A = \frac{q_{in}}{\epsilon_0}$$

(Since, $\vec{E}_2$ is outward and $\vec{E}_1$ is inward to the junction, that's why we have taken E_2 positive and E_1 negative)

$$\Rightarrow \quad (E_2 - E_1) A = \frac{q_{in}}{\varepsilon_0}$$

$$\Rightarrow \quad E_2 - E_1 = \frac{q_{in}}{A\varepsilon_0}$$

Substituting the expression for E_2 from above, we get

$$q_{in} = A\varepsilon_0 (E_2 - E_1) = A\varepsilon_0 \left(\frac{\sigma_1}{\sigma_2} E_1 - E_1\right) \quad \left[\because E_2 = \left(\frac{\sigma_1}{\sigma_2}\right) E_1\right]$$

$$q_{in} = \varepsilon_0 A E_1 \left(\frac{\sigma_1}{\sigma_2} - 1\right) = \varepsilon_0 A \sigma_1 E_1 \left(\frac{1}{\sigma_2} - \frac{1}{\sigma_1}\right)$$

Since the current $I = JA = (\sigma_1 E_1) A$, the amount of the charge on the interface becomes

$$q_{in} = \varepsilon_0 I \left(\frac{1}{\sigma_2} - \frac{1}{\sigma_1}\right)$$

4.6 Colour Code for Carbon Resistors

To mark the value of the resistance on small circuit elements, we use colour coding (Fig.4.38). The colour coding scheme is given in Table 4.2*. According to this scheme, the first two colour bands (starting with the band nearest an end) represent the first two digits in the value of the resistance, the third colour band represents the power of ten that it must be multiplied by, and the fourth colour band, if present, is the manufactured tolerance (or precision) of the value. No fourth colour band means tolerance of $\pm 20\%$, a silver band $\pm 10\%$, and a gold band $\pm 5\%$.

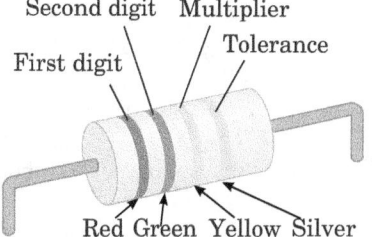

Figure 4.38: A resistor whose four colours are red, green, yellow, and silver has a resistance of $25 \times 10^4 \Omega = 250{,}000 \Omega = 250 k\Omega$, plus or minus 10%.

For a carbon resistor shown in Fig.4.38, the colour codes from nearest end are- red, green, yellow and silver respectively. In these colour bands first two colour bands represent first two digits of resistance (in Ω). From Table 4.2, the first two digits corresponding to first two colour bands are 2 (corresponding to red) and 5 (corresponding to green). Now the multiplier colour

---

*To remember the colour coding, remember the sentence "BB Roy in Great Britain has Very Good Wife". In this sentence each upper case latter is corresponding to first later of colour code given in Table 4.2 in top to bottom order

Table 4.2: Resistance code (in Ω)

Colour	Digit	Multiplier	Tolerance
Black	0	10^0 or 1	
Brown	1	10^1 or 10	1 %
Red	2	10^2	2%
Orange	3	10^3	
Yellow	4	10^4	
Green	5	10^5	
Blue	6	10^6	
Violet	7	10^7	
Gray	8	10^8	
White	9	10^9	
Gold		10^{-1} or 0.1	5%
Silver		10^{-2} or 0.01	10%
No colour			20%

band i.e., third colour band has colour "yellow". From Table 4.2, the multiplier corresponding to yellow colour is 10^4. The tolerance colour band i.e., fourth colour band has colour "silver" so again from Table 4.2, the tolerance is 10 %. Therefore, it's resistance is 25×10^4 Ω ± 10%. Since, 10% of 25×10^4 Ω is 25×10^3 Ω i.e., 25000 Ω, therefore the resistance can also be written as (250000 ± 25000) Ω or (0.25 ± 0.025) MΩ.

EXAMPLE 30. Find the resistance of a carbon resistor if the colour code from left to right indicates green, violet, red and silver [Fig.4.39].

APPROACH Use Table 4.2.

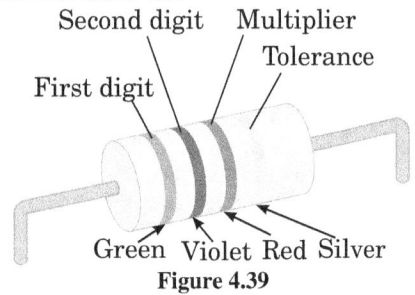

Figure 4.39

SOLUTION
$$\begin{array}{cccc} \text{Green} & \text{Violet} & \text{Red} & \text{Silver} \\ \downarrow & \downarrow & \downarrow & \downarrow \\ 5 & 7 & \times 10^2 & \pm 10\% \end{array}$$

$R = (57 \times 10^2 \pm 10\%)$ Ω

Since, 10 % of 57×10^2 is 57×10^1, therefore
$R = (57 \times 10^2 \pm 57 \times 10^1)$ Ω

or $R = (5.7 \times 10^3 + 0.57 \times 10^3)$ Ω = (5.7 ± 0.57) kΩ

EXAMPLE 31. Find the resistance of a carbon resistor if the colour code from left to right indicates brown, yellow, green and gold.

APPROACH Use Table 4.2.

SOLUTION
$$\begin{array}{cccc} \text{Brown} & \text{Yellow} & \text{Green} & \text{Gold} \\ \downarrow & \downarrow & \downarrow & \downarrow \\ 1 & 4 & \times 10^5 & \pm 5\% \end{array}$$

$R = (14 \times 10^5 \pm 5\%)$ Ω

Since, 5 % of 14×10^5 is 70×10^3, therefore
$R = (14 \times 10^5 \pm 70 \times 10^3)$ Ω

or $R = (1.4 \times 10^6 + 0.07 \times 10^6)$ Ω = (1.4 ± 0.07) MΩ

4.7 Specific Use of Conducting Materials

Some specific uses of conductors and their alloys are given below:

- The heating element of devices like heater, geyser, press etc. are made of nichrome because it has high resistivity and high melting point. It does not react with air and acquires steady state when red heated at 800°C.
- Fuse wire is made of tin lead alloy because it has low melting point and low resistivity. The fuse is used in series, and melts to produce open circuit when current exceeds the safety limit.
- Resistances of resistance box are made of manganin (an alloy of typically 84.2% copper, 12.1% manganese, and 3.7% nickel.) or constantan (copper-nickel alloy) because they have moderate resistivity and very small temperature coefficient of resistance. The resistivity is nearly independent of temperature.
- The filament of bulb is made up of tungsten because it has low resistivity, high melting point of 3300 K and gives light at 2400 K. The bulb is filled with inert gas because at high temperature it reacts with air forming oxide.
- The connection wires are made of copper because it has low resistance and resistivity.

4.7.1 Check Point 3

1. ••A human being can be electrocuted if a current as small as 50 mA passes near the heart. An electrician working with sweaty hands makes good contact with the two conductors he is holding, one in each hand. If his resistance is 2000 Ω, what might the fatal voltage be?
2. ••A coil is formed by winding 250 turns of insulated 16-gauge copper wire (diameter = 1.3 mm) in a single layer on a cylindrical form of radius 12 cm. What is the resistance of the coil? Neglect the thickness of the insulation. (Use Table 4.1)
3. ••Copper and aluminum are being considered for a high-voltage transmission line that must carry a current of 60.0 A. The resistance per unit length is to be 0.150 Ω/km. The densities of copper and aluminum are 8960 and 2600 kg/m^3, respectively. Compute (a) the magnitude J of the current density and (b) the mass per unit length λ for a copper cable and (c) J and (d) λ for an aluminum cable.
4. ••A wire of Nichrome (a nickel-chromium-iron alloy commonly used in heating elements) is 1.0 m long and 1.0 mm^2 in cross-sectional area. It carries a current of 4.0 A when a 2.0 V potential difference is applied between its ends. Calculate the conductivity σ of Nichrome.
5. ••A wire 4.00 m long and 6.00 mm in diameter has a resistance of 15.0 mΩ. A potential difference of 23.0 V is applied between the ends. (a) What is the current in the wire? (b) What is the magnitude of the current density? (c) Calculate the resistivity of the wire material. (d) Using Table 4.1, identify the material.
6. ••A certain wire has a resistance R. What is the resistance of a second wire, made of the same material, that is half as long and has half the diameter?
7. ••A common flashlight bulb is rated at 0.30 A and 2.9 V (the values of the current and voltage under operating conditions). If the resistance of the tungsten bulb filament at room temperature (20°C) is 1.1 Ω, what is the temperature of the filament when the bulb is on?
8. ••When 115 V is applied across a wire that is 10 m long and has a 0.30 mm radius, the magnitude of the current density is 1.4×10^8 A/m^2. Find the resistivity of the wire.
9. ••Figure 4.40a gives the magnitude $E(x)$ of the electric fields that have been set up by a battery along a resistive rod of length 9.00 mm (Fig.4.40b). The vertical scale is set by $E_s = 4.00 \times 10^3$ V/m. The rod consists of three sections of the same material but with different radii. (The schematic diagram of Fig.4.40b does not indicate the different radii.)

The radius of section 3 is 2.00 mm. What is the radius of (a) section 1 and (b) section 2?

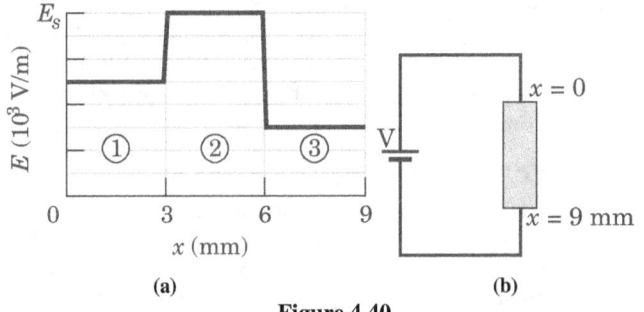

Figure 4.40

10. ●●A wire with a resistance of 6.0 Ω is drawn out through a die so that its new length is three times its original length. Find the resistance of the longer wire, assuming that the resistivity and density of the material are unchanged.

11. ●●In Fig.4.41a, a 9.00 V battery is connected to a resistive strip that consists of three sections with the same cross-sectional areas but different conductivities. Figure 4.41b gives the electric potential $V(x)$ versus position x along the strip. The horizontal scale is set by $x_s = 8.00$ mm. Section 3 has conductivity $3.00 \times 10^7 (\Omega \cdot m)^{-1}$. What is the conductivity of section (a) 1 and (b) 2 ?

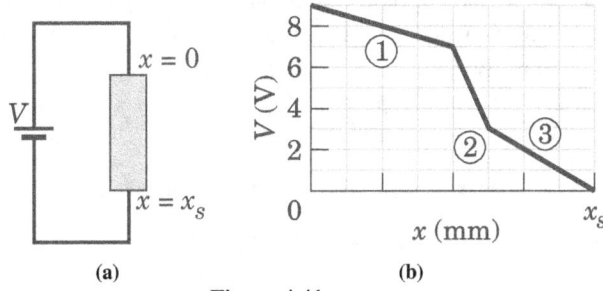

Figure 4.41

12. ●●Two conductors are made of the same material and have the same length. Conductor A is a solid wire of diameter 1.0 mm. Conductor B is a hollow tube of outside diameter 2.0 mm and inside diameter 1.0 mm. What is the resistance ratio R_A/R_B, measured between their ends?

4.8 Power in Electric Circuits and Joule's law

Electric energy is very useful to us because we can easily transform it into other forms of energy. For example, by using motor, we can transform electric energy into mechanical energy. By using a wire resistance known as a "heating element", we can transform electric energy into thermal energy. heating element is used in devices like- electric heaters, stoves, toasters, and hair dryers. In an ordinary lightbulb, the tiny wire filament becomes so hot it glows; only a few percent of the energy is transformed into visible light, and the rest, over 90%, into thermal energy. Lightbulb filaments and heating elements in household appliances have resistances typically of a few ohms to a few hundred ohms. According to microscopic model of current in a metal conductor, when there is an electric current in a conductor, there is electronic motion (motion of electrons) which can be described as a series of accelerations, each of which is terminated by a collision with one of the fixed particles of the conductor. The electrons gain kinetic energy in the free paths between collisions, and give up to the fixed particles, in each collision, the same amount of energy they have gained. The energy acquired by the fixed particles (which are "fixed" only in the sense that their mean position does not change) increases their amplitude of vibration. In other words, it is converted to heat.

Calculation of power transformed by a device To derive the expression for the rate of development of heat in a conductor, we first work out the general expression for the power input to any portion of an electric circuit. The rectangle in Fig.4.42 represents a portion of a circuit in which there is a (conventional) current I from left to right. V_a and V_b are the potentials at terminals a and b. The nature of the circuit between a and b is immaterial-it may be a conductor, motor, generator, battery, or any combination of these. The power input, as we shall show, depends only on the magnitudes and relative directions of the current and the terminal potential difference.

In a time interval dt, a quantity of charge $dq = Idt$ enters the

Figure 4.42

portion of the circuit under consideration at terminal a, and in the same time an equal quantity of charge leaves at terminal b. There has thus been a transfer of charge dq from a potential V_a to a potential V_b. The energy dW given up by the charge is
$$dW = dq(V_a - V_b) = (I\,dt)V_{ab},$$
and the rate at which energy is given up, or the power input P, is

$$\boxed{P = \frac{dW}{dt} = IV_{ab}} \quad (4.57)$$

That is, the power is equal to the product of the current and the potential difference. If the current is in amperes, or coul/sec, and the potential difference in volts, or joules/coul, the power is in joules/sec or watts, since

$$\text{amp} \times \text{volts} = \frac{\text{coul}}{\text{sec}} \times \frac{\text{joules}}{\text{coul}} = \frac{\text{joules}}{\text{sec}} = \text{watts}.$$

Equation 4.57 is a perfectly general relation that holds whatever the nature of the circuit elements between a and b.

In the special case in which the circuit between a and b is a pure resistance R, all of the energy supplied is converted to heat, and in this special case the potential difference V_{ab} is given by
$$V_{ab} = IR$$
Hence, $\boxed{P = IV_{ab} = \frac{V_{ab}}{R}V_{ab} = \frac{V_{ab}^2}{R} = I^2R}$

$$(4.58)$$

If we assume $V_{ab} = V$ (say), then above equation converted to-

$$\boxed{P = IV = \frac{V^2}{R} = I^2R} \quad (4.59)$$

To bring out more explicitly that in this special case the energy appears as heat, we may set $P = dH/dt$, where dH is the heat developed in time dt. Equation 4.58 then becomes

$$\boxed{P = \frac{dH}{dt} = I^2R} \quad (4.60)$$

If the conductor is linear, i.e., if R is a constant independent of I, Eq. 4.60 states that the rate of development of heat is directly proportional to the square of the current. This fact was discovered experimentally by Joule in the course of his measurements of the mechanical equivalent of heat, and it is known as **Joule's law**. Of course, it is a law only in the same sense as is Ohm's law, that is, it expresses a special property of certain materials rather than a general property of all matter. A material which obeys Ohm's law necessarily obeys Joule's law also, and the two are not independent relations.

If in Eq.4.60 the resistance is expressed in ohms and the current in amperes, the rate of development of heat is in joules/sec or watts, since
$$\text{amp}^2 \times \text{ohms} = \text{amp}^2 \times \frac{\text{volts}}{\text{amp}} = \text{amp} \times \text{volts} = \text{watts}.$$
This can readily be converted to cal/sec from the relation
$$1 \text{ calorie} = 4.186 \text{ joules}.$$
Notice that the rate of development of heat in a conductor is not the same thing as the rate of increase of temperature of the conductor. The latter depends on the heat capacity of the conductor and the rate at which heat can escape from the conductor by conduction, convection, and radiation. The rate of loss of heat increases as the temperature of the conductor increases, and the temperature of a current-carrying conductor will rise until the rate of loss of heat equals the rate of development of heat, after which the temperature becomes constant. Thus, when the circuit is closed through an incandescent lamp, the temperature of the filament rises rapidly until the rate of heat loss (mainly by radiation) equals the rate of development of heat, I^2R. On the other hand, a fuse is constructed so that when the current in it exceeds a certain predetermined value, the fuse melts before its final equilibrium temperature can be attained.

EXAMPLE 32. For the situation described in Example 24, show that the heat produced in bar BC is 4 times of the heat produced in bar AB.

SOLUTION From the solution part of Example 24, the resistances of the bar AB and BC are given by
$$R_{AB} = \frac{\rho(l/2)}{4\pi r^2} = \frac{\rho l}{8\pi r^2}, \quad R_{BC} = \frac{\rho(l/2)}{\pi r^2} = \frac{\rho l}{2\pi r^2}$$
The currents flowing through both the bars are equal to I as they are connected in series. The heats produced per unit time in two bars are
$$P_{AB} = I^2 R_{AB} = \frac{I^2 \rho l}{8\pi r^2}, \quad P_{BC} = I^2 R_{BC} = \frac{I^2 \rho l}{2\pi r^2}.$$
From above we can say that $P_{BC} = 4P_{AB}$, i.e., the heat produced in bar BC is 4 times of the heat produced in bar AB.

4.8.1 Check Point 4

1. •What is the maximum power consumption of a 3.0 V portable CD player that draws a maximum of 270 mA of current?
2. •The heating element of an electric oven is designed to produce 3.3 kW of heat when connected to a 240 V source. What must be the resistance of the element?
3. •What is the maximum voltage that can be applied across a 3.3 kΩ resistor rated at $\frac{1}{4}$ watt?
4. •(a) Determine the resistance of, and current through, a 75 W lightbulb connected to its proper source voltage of 110 V. (b) Repeat for a 440 W bulb.
5. ••An electric power plant can produce electricity at a fixed power P, but the plant operator is free to choose the voltage V at which it is produced. This electricity is carried as an electric current I through a transmission line (resistance R) from the plant to the user, where it provides the user with electric power P'. (a) Show that the reduction in power $\Delta P = P - P'$ due to transmission losses is given by $\Delta P = P^2 R/V^2$. (b) In order to reduce power losses during transmission, should the operator choose V to be as large or as small as possible?
6. ••A 120 V hair dryer has two settings of 850 W and 1250 W. (a) At which setting do you expect the resistance to be higher? After making a guess, determine the resistance at (b) the lower setting; and (c) the higher setting.
7. ••A 115 V fish-tank heater is rated at 95 W. Calculate (a) the current through the heater when it is operating, and (b) its resistance.
8. ••You buy a 75 W lightbulb in Europe, where electricity is delivered to homes at 240 V. If you use the lightbulb in the United States at 120 V (assume its resistance does not change), how bright will it be relative to 75 W - 120 V bulbs? [Hint: Assume roughly that brightness is proportional to power consumed.]

4.9 Electric Cells and Batteries

A battery is a device which maintains a constant potential difference across it's two terminals + and −. Dry cells, secondary cells, generator and thermocouple are the devices used for producing potential difference in an electric circuit. The simplest batteries contain two plates or rods made of dissimilar metals (one can be carbon) called electrodes. The electrodes are immersed in a solution or paste, such as a dilute acid, called the electrolyte (Fig.4.43). Such a device is properly called an electric cell, and several cells connected together is a battery, although today even a single cell is called a battery. Figure 4.44a shows a schematic diagram of an open cell. Electrolyte provides continuity for current. Some internal mechanism exerts force $\left(\vec{F}_n\right)$ on the ions (positive and negative) of the solution. This force drives positive ions towards positive terminal and negative ions towards negative terminal. As positive charge accumulates on anode and negative charge on cathode a potential difference and hence an electric field $\vec{E}$ is developed from anode to cathode. This electric field exerts an electrostatic force $\vec{F}_e = q\vec{E}$ on the ions. This force is opposite to that of $\vec{F}_n$. In equilibrium (steady state) $F_n = F_e$ and no further accumulation of charge can takes place.

When the terminals of the battery are connected by a conducting

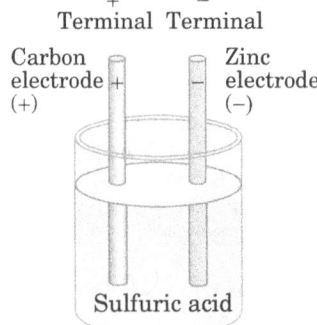

Figure 4.43

wire, an electric field is developed in the wire (Fig.4.44b). The free electrons in the wire move in the opposite direction and enter the battery at positive terminal. Some electrons are withdrawn from the negative terminal. Thus, potential difference and hence, F_e decreases in magnitude while F_n remains the same. Thus, there is a net force on the positive charge towards the positive terminal. Due to this force the positive charge rush towards positive terminal and negative charge rush towards negative terminal. Thus, the potential difference between positive and negative terminals is maintained.

When the emf source is not part of a closed circuit, then $F_n = F_e$ and there is no net motion of charge between the terminals of the battery.

Note: A battery does not create charge; a resistor does not destroy charge

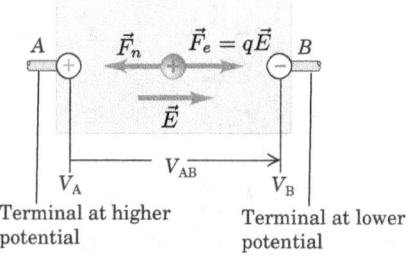

(a) Schematic diagram of a source of emf in an "open-circuit" situation. The electric-field force $\vec{F}_e = q\vec{E}$ and the non electrostatic force $\vec{F}_n$ are shown for a positive charge.

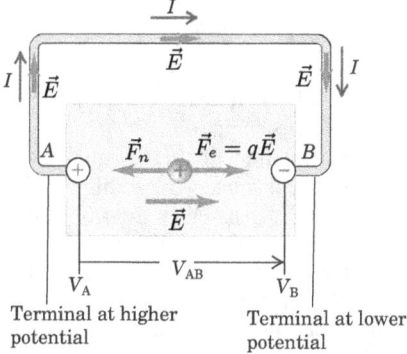

(b) Schematic diagram of an ideal source of emf in A complete circuit. The electric-field force $\vec{F}_e = q\vec{E}$ and the non-electrostatic force $\vec{F}_n$ are shown for a positive charge q The current is in the direction from A to B in the external circuit and from B to A within the source.

Figure 4.44

Frequently Encountered Misconceptions (with Clarifications)

☞ **Point of confusion:** A battery is often thought of as supplying a constant current, no matter what circuit is connected across its terminals.
Clarification: Batteries do not provide a fixed current. Instead, they are designed to maintain a nearly constant potential difference between their terminals. According to Ohm's law ($\Delta V = IR$), the current depends on the external resistance R connected to the battery. The greater the resistance, the smaller the current; the smaller the resistance, the larger the current. It is analogous to pushing an object through a viscous liquid such as oil or molasses: with a steady push (emf), a small object (low R) moves quickly (large I), whereas a large object (high R) moves slowly (small I). Thus, the actual current is determined by the external resistance (or load).
Remark: A battery should therefore be regarded as a *voltage source*, not a current source. The nearly constant voltage it maintains is applied across any wire, circuit, or device connected across its terminals.

☞ **Misconception:** It is often stated that the direction of conventional current is always from the positive terminal of a source toward the negative terminal.
Clarification: This statement is only partially true. If a potential difference is applied across the ends of a conducting wire, the current flows *along the wire*, tangential to its path, no matter how much the wire is curved. In the external resistive part of the circuit, the current indeed flows from the higher-potential end (+) to the lower-potential end (−). However, *inside the source of emf* (such as a cell or battery), the charges are driven by non-electrostatic forces, and the conventional current actually flows from the negative terminal to the positive terminal (from lower to higher potential). Thus, in a complete circuit the current flows $+ \to -$ in the external wire (however curved) and $- \to +$ inside the source.

☞ **Point of confusion:** Current (or charge) decreases or gets "used up" as it flows through a wire or device.
Clarification: Current and charge are not consumed while passing through a conductor or device. The amount of charge that enters one end of a wire comes out from the other end in the same time interval. What actually gets "used up" in a device is energy, not charge. The charges lose energy to the device (for example, as heat or light), but the total flow of charge remains the same.
Remark: Charge is conserved. A device resists the flow of current and dissipates energy, but it does not consume charge or current.

4.10 Symbols for Circuit Components

We make circuit diagrams by using some standard symbols for circuit components and lines to represent wires. These are shown in Figure 4.45. We usually approximate wires as perfect conductors; then all points connected by a wire are at the same potential and are electrically equivalent.

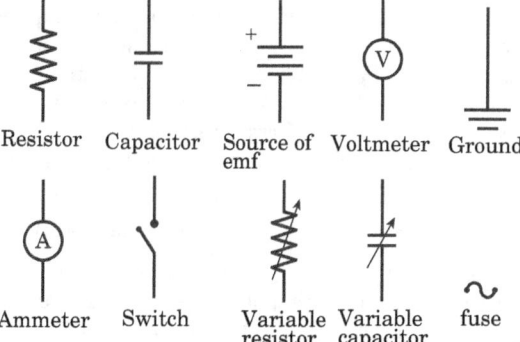

Figure 4.45: Common circuit symbols

4.11 Resistors in Series and Parallel

The function of a resistor in a circuit is to regulate the amount of current. Manufacturers make a wide range of resistances, but often you must combine a number of standard resistors to come up with the resistance required. Here, we will be discussing series and parallel combinations of resistors only. The rest of combinations are the mixed combinations of series and parallel.

4.11.1 Resistors in Series

Fig.4.46a shows three resistors (R_1, R_2 and R_3), placed end to end between points A and B. Resistors that are aligned end to end, with no junctions between them, are called series resistors or, sometimes, "resistors in series". Because there are no junctions, the current I must be the same through each of these resistors. That is, the current out of the last resistor in a series is equal to the current into the first resistor.

If points A, C, D and B are at potentials V_A, V_C, V_D and V_B

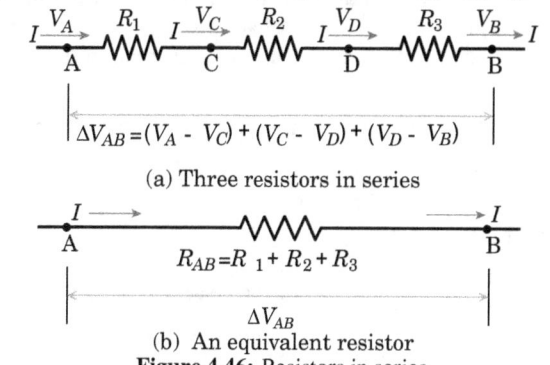

(a) Three resistors in series

(b) An equivalent resistor

Figure 4.46: Resistors in series

respectively, then potential difference across terminals A and B can be written as:

$\Delta V_{AB} = V_A - V_B = (V_A - V_C) + (V_C - V_D) + (V_D - V_B)$ (4.61)

Here, $(V_A - V_C) = IR_1$, the potential difference across R_1,
$(V_C - V_D) = IR_2$, the potential difference across R_2,
and $(V_D - V_B) = IR_3$, the potential difference across R_3.
Substituting these values in Eq.4.61, we get

$\Delta V_{AB} = V_A - V_B = IR_1 + IR_2 + IR_3 = I(R_1 + R_2 + R_3)$

In Fig.4.46b, we replaced the three resistors R_1, R_2 and R_3 with a single resistor R_{AB} having current I and potential difference $\Delta V_{AB} = V_A - V_B$. We can now apply Ohm's law to find that the resistance R_{AB} between points A and B as

$$R_{AB} = \frac{\Delta V_{AB}}{I}$$

here, $\Delta V_{AB} = (V_A - V_C) + (V_C - V_D) + (V_D - V_B)$
$= IR_1 + IR_2 + IR_3 = I(R_1 + R_2 + R_3)$

Therefore,

$$R_{AB} = \frac{I(R_1 + R_2 + R_3)}{I} = R_1 + R_2 + R_3 \quad (4.62)$$

Because the battery has to establish the same potential difference across the load and provide the same current in both cases, the three resistors R_1, R_2 and R_3 act exactly the same as a single resistor of value $R_1 + R_2 + R_3$. We can say that the single resistor R_{AB} is equivalent to the three resistors in series.

There was nothing special about having only three resistors. If we have n resistors in series, their equivalent resistance is

$$\boxed{R_{eq} = R_1 + R_2 + \cdots + R_n} \quad (n \text{ resistors in series}) \quad (4.63)$$

The current and the power output of the battery will be unaltered if the n series resistors are replaced by the single resistor R_{eq}. The key idea in this analysis is that resistors in series all have the same current.

☞ The net resistance after removing some resistors from a group of resistors connected in series involves setting these resistors equal to zero in the Eq.4.63. For example removing R_1, R_4 means set $R_1 = R_4 = 0$.

4.11.2 Resistors in Parallel

Figure 4.47 shows n resistors in parallel. A current I flows into a junction point A, from which n wires sprout, carrying resistors of resistance $R_1, R_2, \ldots, R_n$. These wires come together again at the junction point B, out of which the original (conserved) current I flows. The potential difference between the points A and B is given as V_{AB}, and this is the same as the potential difference across any of the resistors. The current flowing out of the junction point A breaks up into parts that flow through the different resistors. How big these partial currents are is determined by the resistors that they flow through. Thus $I_1 = V_{AB}/R_1$, $I_2 = V_{AB}/R_2$, and so on. The equivalent resistance is given by rewriting the conserved current I as a sum of the partial currents, all of which is written in terms of the potential difference V_{AB} and the individual resistances. In other words, the equivalent resistance R_{eq} is defined by V_{AB}/I, so that $I = V_{AB}/R_{eq}$. Current conservation then implies that

$$I = I_1 + I_2 + \cdots + I_n = \frac{V_{AB}}{R_1} + \frac{V_{AB}}{R_2} + \cdots + \frac{V_{AB}}{R_n}$$

$\Rightarrow \quad \frac{V_{AB}}{R_{eq}} = V_{AB}\left(\frac{1}{R_1} + \frac{1}{R_2} + \cdots + \frac{1}{R_n}\right)$

Dividing both sides by V_{AB}, we get

$$\boxed{\frac{1}{R_{eq}} = \frac{1}{R_1} + \frac{1}{R_2} + \cdots + \frac{1}{R_n}} \quad (n \text{ resistors in parallel}) \quad (4.64)$$

Note: If there are only two resistors, R_1 and R_2, in parallel, then from Eq.4.64, their equivalent resistance is given by

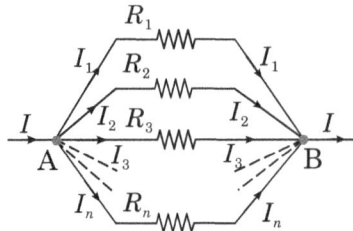

Figure 4.47: An arrangement of n resistors in parallel, carrying current from A to B, with potential difference V_{AB} from A to B.

$$\frac{1}{R_{eq}} = \frac{1}{R_1} + \frac{1}{R_2}$$

or $\boxed{R_{eq} = \frac{R_1 R_2}{R_1 + R_2}}$ (4.65)

Eq.4.65 is very useful for quickly determining the equivalent resistance of two resistors in parallel.

From Eq.4.63 it is clear that, if the resistors are in series (Fig. 4.46), the equivalent resistance R_{eq} is higher than the highest resistance connected in series. From Eq.4.64, we can say that, if the resistors are in parallel (Fig. 4.47), the equivalent resistance R_{eq} is lower than the lowest resistance connected in parallel.

Drinking a milkshake with several straws is analogous to charged particles flowing through several resistors. So, think of drinking a milkshake with just one straw. with only one straw you find it difficult to get a satisfying amount into your mouth with each sip. Now, if you drink it with two straws you will find a better amount in each sip. Think of it this way: Each straw resists the motion of the milkshake, but two straws used in parallel decrease the overall resistance to the milkshake flow. What if you use the two straws in series-by forcing the end of one straw into the end of the other? You would find it even more difficult to drink the milkshake with two straws in series than if you used only one straw, because resistances in series add.

- To get maximum resistance, resistors must be connected in series and in series the resultant is greater than the resistance of the largest resistor in the combination.
- To get minimum resistance, resistance must be connected in parallel and the equivalent resistance of parallel combination is lower than the value of lowest resistance in the combination.

☞ The net resistance after removing some resistors from a group of resistors connected in parallel involves setting these removed resistors equal to infinity in Equation (Eq.4.64). This is because, for an infinite resistance value, current will not pass through that resistance, and we may assume the absence of that resistance. For example, if you want to find the net resistance after removal of resistance R_1, then put $R_1 \to \infty$ in Eq.4.64 and then solve for R_{eq}

EXAMPLE 33. Which electrical elements are connected in series in Fig.4.48?

Answer. Here we see that currents through $S_1, R_1, S_2,$ and R_3 will

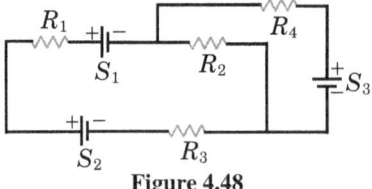

Figure 4.48

always be same so these may be regarded in series. Again current through S_3 and R_4 will also be same, so they are in different series. R_2 is neither in series nor in parallel with any element. So, it will form a mixed combination.

EXAMPLE 34. Find current passing through the battery and each resistor shown in Fig.4.49a.

4.12. VOLTAGE AND POWER RATING

APPROACH Fig.4.49b shows the circuit with currents in each

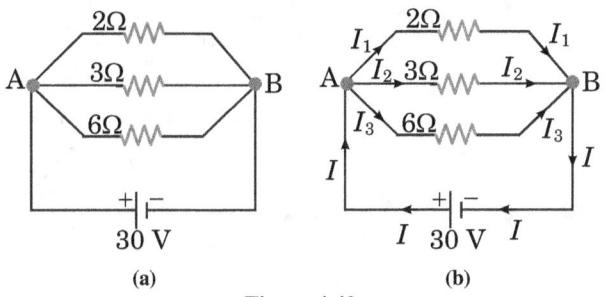

Figure 4.49

branch. Resistors are connected in parallel, therefore, their equivalent resistance R_{eq} can be obtained from Eq.4.64:
$$\frac{1}{R_{eq}} = \frac{1}{R_1} + \frac{1}{R_2} + \frac{1}{R_3} \qquad \cdots (1)$$
Now, by Ohm's law (Eq:4.36), the potential difference across a resistor R is
$$\Delta V = RI \qquad \cdots (2)$$
To get current given by battery, replace R in Eq. (2), by R_{eq} and then solve for I.

Since, resistors are joined in parallel, so, potential difference across each resistor is same. Now, to get current in each resistor, substitute the values of ΔV and R, and then solve for required current.

SOLUTION Given: $R_1 = 2\ \Omega$, $R_2 = 3\ \Omega$ and $R_3 = 6\ \Omega$ and $\Delta V = 30$ V, therefore from Eq.(1), we get
$$\frac{1}{R_{eq}} = \frac{1}{2} + \frac{1}{3} + \frac{1}{6} = \frac{6}{6} = 1$$
or $\qquad R_{eq} = 1\ \Omega$
Therefore, from Eq.(2), current given by battery,
$$I = \frac{\Delta V}{R_{eq}} = \frac{30\ \text{V}}{1} = 30\ \text{A}$$
Again from Eq.(2), current through $2\ \Omega$ resistor
$$I_1 = \frac{\Delta V}{R_1} = \frac{30\ \text{V}}{2} = 15\ \text{A}$$
Current through $3\ \Omega$ resistor
$$I_2 = \frac{\Delta V}{R_2} = \frac{30\ \text{V}}{3} = 10\ \text{A}$$
and current through $6\ \Omega$ resistor
$$I_3 = \frac{\Delta V}{R_3} = \frac{30\ \text{V}}{6} = 5\ \text{A}$$

Checking the Result From above we see that,
$$I_1 + I_2 + I_3 = 15\ \text{A} + 10\ \text{A} + 5\ \text{A} = 30\ \text{A}$$
which is equal to the current given by battery. Hence result is correct.

EXAMPLE 35. All resistances in the Figure 4.50 are in Ω. Find the effective resistance between the points A and B.

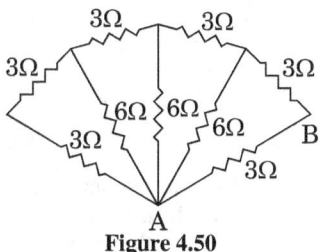

Figure 4.50

SOLUTION In the leftmost loop, two 3Ω resistors are connected in parallel to one 6Ω resistor. Effective resistance of this loop is $R = (3+3)\|6 = 6\|6 = 3\Omega$.
Repeat the same process for the next two loops to get the effective resistance of the three loops (from left) as 3Ω. Thus, the resistance

across AB consists of two 3Ω resistors connected in series and one $3\ \Omega$ resistor connected in parallel, giving the effective resistance
$$R_{AB} = (3+3)\|3 = 6\|3 = 2\ \Omega$$

EXAMPLE 36. Find the equivalent resistance between points A and B in the circuit shown in Fig.4.51. Every resistance shown here has a magnitude of $2\ \Omega$.

APPROACH Points C, O & D are connected by a conducting

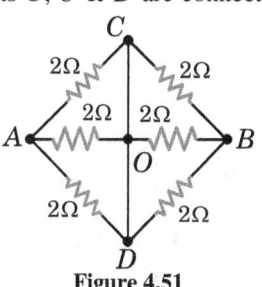

Figure 4.51

wire, therefore, these points are at the same potential. Therefore, resistances AO, AC and AD are in parallel. Similarly BC, BO and BD are also connected in parallel. The resultant of resistances AO, AC and AD is in series with the resultant of BC, BO and BD. The circuit in simplified form is shown in Fig.4.52.

SOLUTION If R_1 is the resultant of parallel resistances between AO, AC and AD, then
$$\frac{1}{R_1} = \frac{1}{2} + \frac{1}{2} + \frac{1}{2} \quad \Longrightarrow \quad R_1 = \frac{2}{3}\Omega$$
Similarly, if R_2 is the resultant of parallel resistances between

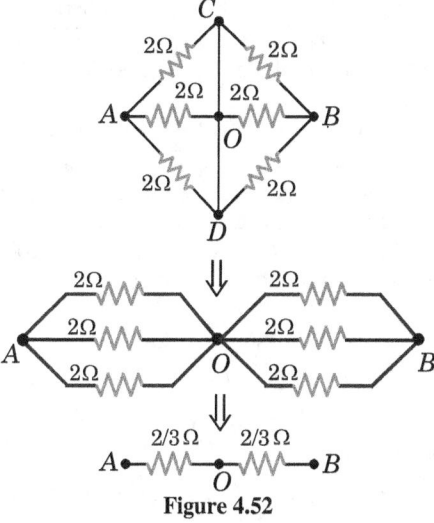

Figure 4.52

BC, BO and BD, then
$$\frac{1}{R_2} = \frac{1}{2} + \frac{1}{2} + \frac{1}{2} \quad \Longrightarrow \quad R_2 = \frac{2}{3}\Omega$$
Now, R_1 is in series with R_2, therefore, their net resistance between AB is
$$R_{AB} = \frac{2}{3}\Omega + \frac{2}{3}\Omega = \frac{4}{3}\Omega$$

4.12 Voltage and Power Rating

Sometimes voltage and power is mentioned with given electrical appliances. The voltage rating simply specifies that a given device can operate safely up to the given voltage and also it consumes the mentioned power at the this voltage. For example, if 220 V and 10 W is written on an electrical bulb, it means that bulb can operate safely at 220 V and at this voltage it consumes 10 W power from the source. From this voltage power rating, we can also determine the resistance of the bulb. For example, corresponding to 220 V and 10 W rating, from Eq.4.58, the resistance

is
$$R = \frac{V^2}{P} = \frac{220^2}{10} = 4840 \, \Omega$$

EXAMPLE 37. In the Fig.4.53 shown B_1, B_2 and B_3 are three bulbs rated as (200 V, 50 W), (200 V, 100 W) and (200 V, 25 W) respectively. Find the current through each bulb and which bulb will give more light?

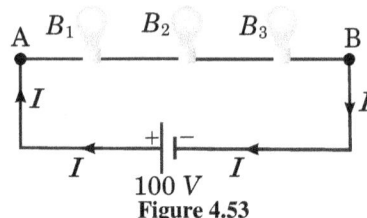

Figure 4.53

APPROACH With the help of voltage-power rating, find the resistance of each bulb. As these bulbs are in series, therefore find their series equivalent resistance and hence current in the circuit. Now, calculate the power consumed by each bulb by using relation $P = I^2 R$. More power means more brightness

SOLUTION Given voltage-power rating of bulb B_1 is (200 V, 50 W), therefore, resistance of bulb B_1 is

$$R_1 = \frac{V_1^2}{P_1} = \frac{(200 \text{ volt})^2}{50 \text{ watt}} = \frac{40000 \text{ volt}^2}{50 \text{ watt}} = 800 \, \Omega$$

Voltage-power rating of bulb B_2 is (200 V, 100 W), therefore resistance of bulb B_2 is

$$R_2 = \frac{V_2^2}{P_2} = \frac{(200 \text{ volt})^2}{100 \text{ watt}} = \frac{40000 \text{ volt}^2}{100 \text{ watt}} = 400 \, \Omega$$

Voltage-power rating of bulb B_3 is (200 V, 25 W), therefore resistance of bulb B_3 is

$$R_3 = \frac{V_3^2}{P_3} = \frac{(200 \text{ volt})^2}{25 \text{ watt}} = \frac{40000 \text{ volt}^2}{25 \text{ watt}} = 1600 \, \Omega$$

All these bulbs are connected in series with a 100 V battery, therefore equivalent resistance of the circuit:
$$R_{eq} = R_1 + R_2 + R_3 = 800 + 400 + 1600 = 2800 \, \Omega$$
The current flowing through each bulb is
$$I = \frac{V}{R_{eq}} = \frac{100 \text{ V}}{2800 \, \Omega} = \frac{1}{28} \text{ A}$$
Power consumed by bulb B_1,
$$P_1' = I^2 R_1 = \frac{1}{28 \times 28} \text{ A}^2 \times 800 \, \Omega = 1.02 \text{ W}$$
Power consumed by bulb B_2,
$$P_2' = I^2 R_2 = \frac{1}{28 \times 28} \text{ A}^2 \times 400 \, \Omega = 0.51 \text{ W}$$
Power consumed by bulb B_3,
$$P_3' = I^2 R_3 = \frac{1}{28 \times 28} \text{ A}^2 \times 1600 \, \Omega = 2.04 \text{ W}$$
Clearly, $P_3' > P_1' > P_2'$, therefore, order of brightness is
$$B_3 > B_1 > B_2$$
Clearly, bulb B_3 will give more light.

Note: In series connection, current through all bulbs remain same, therefore, power consumed by bulb of higher resistance will be more than the power consumed by a bulb having less resistance.
Also, in series, $P_1 : P_2 : P_3 = R_1 : R_2 : R_3$

EXAMPLE 38. Find equivalent resistance between terminals A and B in the circuit shown in Fig.4.54.

APPROACH Let us assign potentials V_1 and V_2 to terminals A and B respectively [Fig.4.55a]. As point D is connected to A by a conducting wire, therefore, potential of point D is also V_1. Similarly, point C is connected to B with a conducting wire, therefore potential of C is also V_2.

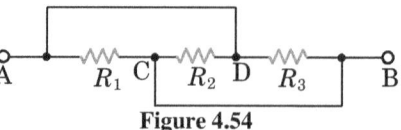

Figure 4.54

Clearly, all three resistors are connected between terminal po-

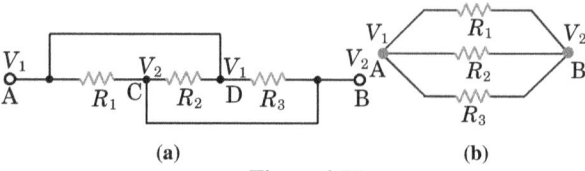

Figure 4.55

tentials V_1 and V_2, so they are in parallel as shown in Fig.4.55b. Now, to calculate equivalent resistance between terminals A and B, apply Eq.4.64.

SOLUTION Applying Eq.4.64, in circuit of Fig.4.55b, we get
$$\frac{1}{R_{eq}} = \frac{1}{R_1} + \frac{1}{R_2} + \frac{1}{R_3}$$
or
$$\frac{1}{R_{eq}} = \frac{R_2 R_3 + R_3 R_1 + R_1 R_2}{R_1 R_2 R_3}$$
or
$$R_{eq} = \frac{R_1 R_2 R_3}{R_2 R_3 + R_3 R_1 + R_1 R_2}$$

Special case: If $R_1 = R_2 = R_3 = R$ (say), then
$$R_{eq} = \frac{R}{3}$$

4.12.1 Check Point 5

1. Find the equivalent resistance of the circuit shown in Fig.4.56 and calculate the current I drawn from the battery.

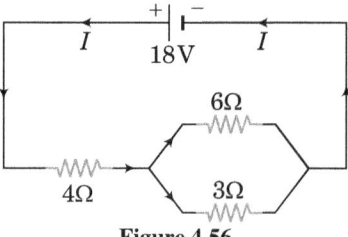

Figure 4.56

2. Six equal resistances are connected between points P, Q and R as shown in the Fig.4.57. Write the name of terminals between which the equivalent resistance of the circuit is maximum.

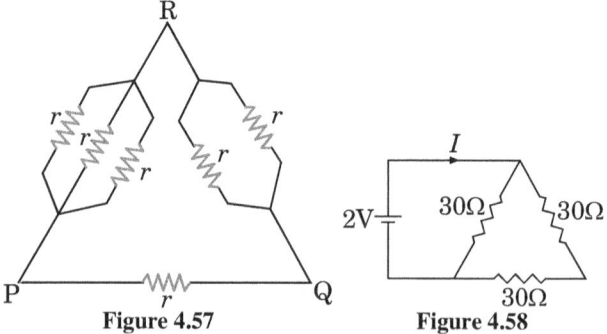

Figure 4.57   Figure 4.58

3. Find the current I in the circuit shown in Fig.4.58.
4. **Bulb brightness in a circuit.** The circuit shown in Fig. 4.59 has three identical lightbulbs, each of resistance R. (a) When switch S is closed, what will be the brightness of bulbs A and B as compared to that of bulb C? (b) What happens when switch S is opened? Use a minimum of mathematics in your answers.
5. Find the equivalent resistance between points A and B of the given circuit shown in Fig.4.60.

4.13. ELECTROMOTIVE FORCE (EMF)

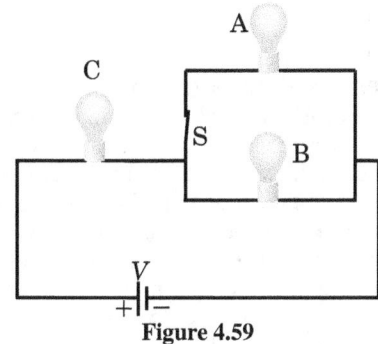

Figure 4.59

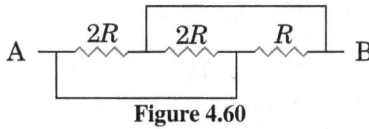

Figure 4.60

6. Each resistor in the network of the Figure 4.61 has a resistance of 10 Ω. Compute the resistance between points A and B.

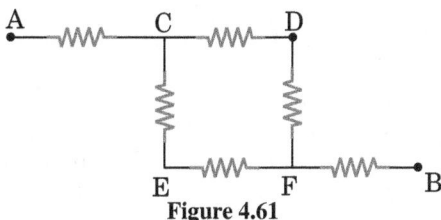

Figure 4.61

7. Find the effective resistance between points A and B of the network shown in Fig.4.62.

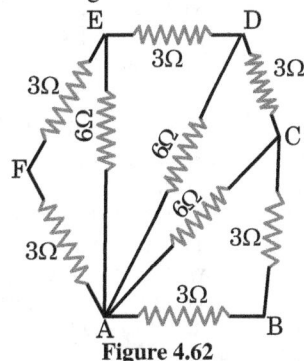

Figure 4.62

Multiple Choice Questions

8. For the circuit shown in the Fig.4.63
   (A) the current I through the battery is 7.5 mA.
   (B) the potential difference across R_L is 18 V.
   (C) ratio of powers dissipated in R_1 and R_2 is 3.
   (D) if R_1 and R_2 are interchanged, magnitude of the power dissipated in R_L will decrease by a factor of 9.

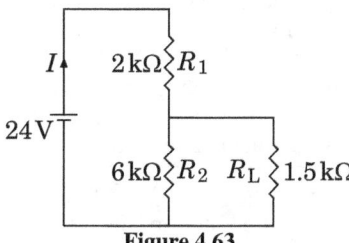

Figure 4.63

9. In Fig.4.64, an electric bulb B_1 rated 220 V and 60 W is connected in series with another electric bulb B_2 rated 220 V and 40 W. The combination is connected across 220 volt source of emf. Which bulb will glow more bright?
   (A) $P'_1 > P'_2$ (B) $P'_1 < P'_2$ (C) $P'_1 = P'_2$ (D) P'_1/P'_2

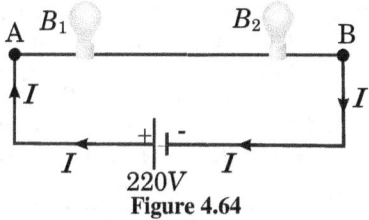

Figure 4.64

10. The three resistors in Fig.4.65 are $R_1 = 25$ Ω, $R_2 = 50$ Ω, and $R_3 = 100$ Ω. What is the total resistance of the circuit?
    (A) 50.3 Ω (B) 58.3 Ω (C) 60.3 Ω (D) 80.3 Ω

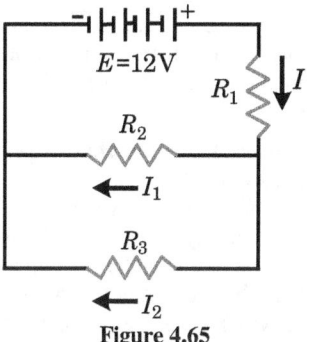

Figure 4.65

11. What are the currents I_1, I_2 and I_3 in the circuit of Fig.4.65 for a 12 V battery?
    (A) 0.206 A, 0.137, 0.0685 A
    (B) 0.506 A, 0.137, 0.0685 A
    (C) 0.606 A, 0.137, 0.0685 A
    (D) 0.706 A, 0.137, 0.0685 A

4.13 Electromotive Force (EMF)

It takes an electric field to drive current through a conductor with non-zero resistance. But unless we actively maintain the field, charge will quickly move to establish electrostatic equilibrium, with no field inside the conductor and no current. To make a constant electric field inside the conductor, we need a device that can maintain a fixed potential difference and therefore an electric field in a current-carrying conductor.

A device that continuously supplies electrical energy to a conductor and maintains a fixed potential difference across it's terminals, is called a source (or seat) of electromotive force. Since, the word force is misleading here, we only use the abbreviation emf. (The term "electromotive force", is something of a misnomer because it is definitely not a force.). An emf source converts some other form of energy to electrical energy by separating positive and negative charge to maintain a fixed potential difference between it's terminals. The most familiar example is a battery, in which chemical reactions drive charge to the two terminals. Others include electric generators, which convert mechanical to electrical energy; photovoltaic cells, which use sunlight to separate charge; and cell membranes, which control ion flow into and out of the cell.

When an emf source is connected to an external circuit, current flows through the circuit from the positive terminal to the negative terminal. In metallic wires and resistors, the electric current is due to the flow of electrons from negative terminal of the emf source to the positive terminal. The electrons leave the negative terminal, travel through the external circuit, and return to the positive terminal. In order for the emf source to maintain the fixed potential difference between the two terminals, it applies a non-electric force on negative charges to bring them (electrons) from the positive terminal to the negative terminal inside it. This increases the potential energy of the charges and, therefore, the electric potential of the charges. It is much similar to lift a weight from lower gravitational potential energy level to higher. Since the source of emf increases the potential energy of charges, therefore it's work done is non-conservative. This work per unit charge is equal to the magnitude of emf $\mathcal{E}$ of the source.

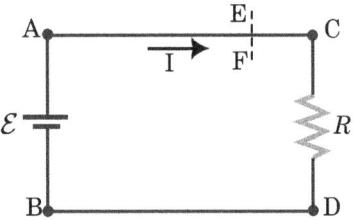

Figure 4.66: A simple circuit consisting of an ideal battery of emf $\mathcal{E}$, a resistance R, and connecting wires that are assumed to be of negligible resistance.

Quantitatively, this work per unit charge is equal to the magnitude of emf $\mathcal{E}$ of the source.

Let us now consider the circuit shown in Fig.4.66 from the point of view of work and energy transfers. Circuit 4.66 consisting of a resistance R connected to an ideal battery of emf $\mathcal{E}$. The resistance is indicated by the symbol —W—. The straight lines indicate connecting wires of negligible resistance. The source of emf ideally maintains a constant potential difference equal to emf $\mathcal{E}$ between terminals A and B, with point A being at the higher potential, i.e.,

$$\boxed{\Delta V_{AB} = V_A - V_B = \mathcal{E}} \quad (4.66)$$

There is negligible potential difference between points A and C and between points D and B, because the connecting wire is assumed to have negligible resistance. The potential drop from point C to D is therefore equal in magnitude to the emf $\mathcal{E}$, and the current I through the resistor is given by $I = \mathcal{E}/R$. The direction of the current in this circuit is clockwise, as shown in the Fig.4.66. If in an infinitesimally small time interval dt, a charge dQ passes through any cross section of this circuit (for example, at intersection with line EF) then the same amount of charge must enter the emf device at its low-potential end and leave at its high-potential end. If inside the emf source, the work done by emf source in bringing the charge dQ from a lower potential terminal to higher, is dW, then the emf $\mathcal{E}$ of the source is defined as

$$\boxed{\mathcal{E} = dW/dQ} \quad (4.67)$$

In words, the emf of an emf device is the work per unit charge that the device does in moving charge from its low-potential terminal to its high-potential terminal. The SI unit of emf is the joule/coulomb, which is the volt (abbreviation V):

1 volt = 1 joule/coulomb.

Flow of charge inside the emf device connected with a circuit means flow of charge at each point in whole circuit. So, emf can also be defined as-

The energy given by the cell in the flow of unit charge in the whole circuit (including the cell) is called the emf of the cell.

Note that, emf depends on:
1. nature of electrolyte
2. metal of electrodes

emf does not depend on:
1. area of plates
2. distance between the electrodes
3. quantity of electrolyte
4. size of cell

4.13.1 Internal Resistance (r)

So far, we have considered only ideal emf sources. An ideal DC emf source, such as an ideal battery, maintains a constant terminal potential $\mathcal{E}$ whether there is current in the emf source or not. Practically, no real emf device can maintain its terminal potential when there is current in the device. If you close the switch while measuring the voltage across the emf device (Fig.4.67b), the terminal voltage decreases slightly. If you open the switch again, the terminal voltage returns to its earlier higher value. We can explain this observation: When you close the switch, there is current in the entire circuit, including the emf source. The real emf source has some ordinary resistivity and hence resistance. So, the current in a real emf source-like the current anywhere else in the circuit-is hindered by it's resistance. We call this the internal resistance of the emf source. It is usually denoted by r.

The real emf device in Figure 4.67 is modelled as two circuit

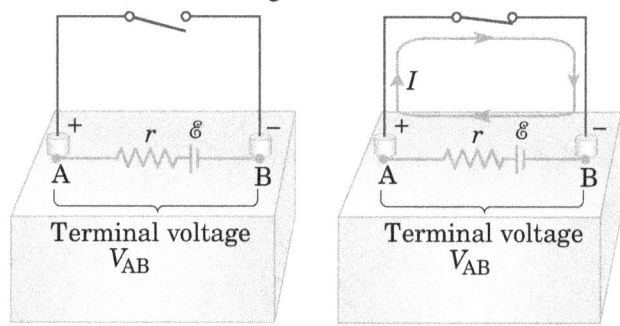

(a) The terminal voltage across a real emf device is higher when there is no current in the device.

(b) The terminal voltage is lower across a real emf device when a current is present.

Figure 4.67: A real emf device is modelled as an ideal emf in series with a resistor.

elements: (1) an ideal emf source with no internal resistance and (2) a resistor with resistance r. Since, we cannot put our voltmeter inside the emf source, so we have to connect our voltmeter to the emf terminals labelled A and B (Fig.4.67) and measure the potential difference between them. In Fig. 4.67a, the emf source is connected to an open switch, so there is no current in the leads or the battery, and therefore no voltage drop across the battery's internal resistor: $\Delta V_r = -Ir = 0$. The terminal potential measured by the voltmeter is

$$\Delta V_{AB} = V_A - V_B = \mathcal{E}$$

Thus, when there is no current in a real emf device, its terminal potential is the same as that of an ideal emf device. So, emf $\mathcal{E}$ of a battery is the maximum possible potential difference that the battery can provide between its terminals, i.e., the voltage at zero current.

When the switch is closed (Fig.4.67b), current I is present in the circuit and therefore in the battery. Now the voltage drop across the internal resistor is given by $\Delta V_r = -Ir$. The terminal potential is the sum of the voltage across the ideal emf device and the voltage drop across the internal resistor. Therefore, the terminal voltage across the real emf source is

$$\boxed{\Delta V_{\text{real emf}} = \Delta V_{AB} = \mathcal{E} - Ir} \quad (4.68)$$

Unless otherwise specified, all emf devices in this textbook are assumed to be ideal, obeying $\Delta V_{AB} = \mathcal{E}$ (Eq.4.66).

The internal resistance of a cell depends on the distance between electrodes ($r \propto d$), effective area of electrodes ($r \propto S$) and nature, concentration ($r \propto c$) and temperature of electrolyte

4.13. ELECTROMOTIVE FORCE (EMF)

$$\left(r \propto \frac{1}{\text{Temperature}}\right)$$

Misconceptions and Clarifications

- **Point of confusion:** The internal resistance of a cell is often thought to be constant and independent of the cell's physical or chemical conditions.

  Clarification: Actually, the internal resistance of a cell varies with several factors: it increases with the distance between electrodes ($r \propto d$), decreases with the electrode area ($r \propto 1/S$), decreases with the concentration of the electrolyte ($r \propto 1/c$), and decreases with the temperature of the electrolyte ($r \propto 1/T$).

  Remark: Thus, the internal resistance decreases with an increase in electrode area, electrolyte concentration, and temperature, but increases if the electrodes are farther apart.

4.13.2 Potential Changes Around a Circuit

Let us consider a DC circuit as shown in Fig.4.68a. It consists of a real battery (internal resistance = r) connected to a light bulb with conducting wires. As a result of the energy stored in the battery, an electric current, I, flows through the circuit. Note that, since there is nowhere else for the current to go, the current through any element of this series circuit must be the same. We assume that the resistance of the wires compared to the resistance of the light bulb, R, is negligible; the light bulb is in its steady state (R = constant).

From Fig.4.68a, the potential difference across the terminals A and B is

$$\Delta V_{AB} = V_A - V_B = \mathcal{E} - Ir$$

This voltage is applied across the resistor R between terminals C and D, therefore from Ohm's law, we can write

$$\Delta V_{AB} = V_{DC}$$

or
$$\mathcal{E} - Ir = IR$$

$$\boxed{I = \frac{\mathcal{E}}{R+r}} \quad \text{(real emf source)} \qquad (4.69)$$

That is, the current equals the source emf divided by the total circuit resistance ($R + r$).

Now, rewriting Eq.4.69 in the form of

$$\mathcal{E} - Ir - IR = 0 \qquad (4.70)$$

A potential gain of $\mathcal{E}$ is associated with the emf, and potential drops of Ir and IR are associated with the internal resistance of the source and the external circuit, respectively. Figure 4.68b is a graph showing how the potential varies as we go around the complete circuit of Fig.4.68a. For convenience we have started at point B and travelled clockwise, in the direction of current, around the circuit. Point B of the graph is at zero potential.

The horizontal axis doesn't necessarily represent actual distances, but rather various points in the loop. If we take the potential to be zero at the negative terminal of the battery, then we have a rise $\mathcal{E}$ and a drop Ir in the battery and an additional drop IR in the external resistor, and as we finish our trip around the loop, the potential is back where it started.

Analysis of potential change in graph of Fig.4.68b, is given in Table 4.3.

4.13.3 Power Supplied by an Ideal emf Source

Note that inside the source of emf, the charge flows from a region where its potential energy is low to a region where its potential is high, so the charge gains electric potential energy*. When charge dQ flows through the ideal source of emf $\mathcal{E}$, its

---

*When a battery is being charged (by a generator or by another battery), within the battery the charge flows from a region where its potential energy is high to a region where its potential energy is low, thus losing electric potential energy. The energy lost is converted to chemical energy and stored in the battery being charged.

Table 4.3: Analysis of potential change in graph of Fig. 4.68b

Part	ΔV	Meaning
BA	$\mathcal{E} - Ir$	A real emf source can regarded as a series combination of an ideal emf source with it's internal resistance r. So, when we cross the emf source from B (at zero potential) to A (at $\mathcal{E} - Ir$ potential), there is increase in potential from 0 to $\mathcal{E}$ and then decrease of Ir. So, along B to A, graph first goes up from 0 to $\mathcal{E}$ and then falls by Ir.
AD	0	AD is a connecting wire with zero resistance. So, there is no potential drop across AD and the graph is horizontal.
DC	IR	Here, we encounter a resistance R. The potential drop across it, is IR and graph again reaches at zero potential
CB	0	CB is also a connecting wire with zero resistance. So, there is no further potential drop across CB and the graph remains horizontal.
Total BADCA	0	After a complete cycle, we again reaches at zero potential, therefore, $\Delta V_{\text{total}} = 0$.

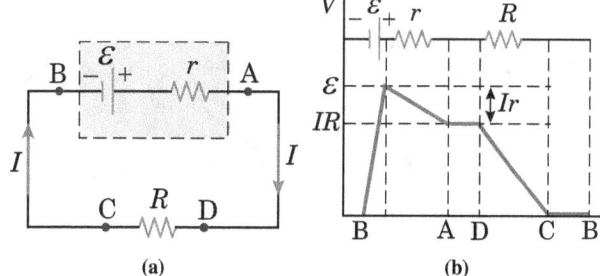

Figure 4.68: Analysis of a simple DC circuit consisting of a nonideal emf source (a battery), a light bulb (in a steady state, so its resistance R is not changing), and connecting wires. (a) A schematic representation of a real-life circuit. (b) Potential rises and drops in a circuit.

potential energy is increased by the amount $(dQ)\mathcal{E}$. The charge then flows through the resistor, where this potential energy is dissipated as thermal energy. The rate at which energy is supplied by the source of emf is the power output of the source:

$$P = \frac{(dQ)\mathcal{E}}{dt} = I\mathcal{E} \qquad (4.71)$$

(Power Supplied by an Ideal emf Source)

In the simple circuit of Figure 4.66, the power output by the ideal source of emf equals the power delivered to the resistor.

The battery in Figure 4.66 can be thought of as a charge pump that pumps the charge from a region where its potential energy is low to a region where its potential energy high. Figure 4.69 shows a mechanical analogue of the simple electric circuit just discussed.

In Fig.4.69a the marbles start at some height h above the bottom and are accelerated between collisions with the nails by the gravitational field. The nails are analogous to the lattice ions in the resistor. During the collisions, the marbles transfer the kinetic energy they obtained between collisions to the nails. Because of the many collisions, the marbles have only a small, approximately constant, drift velocity toward the bottom. Now, when the marbles reach the bottom, a child picks them up (Fig.4.69b), lifts them to their original height h, and starts them again. The child, who does work mgh on each marble of mass m, is analogous to the source of emf. The energy source in this case is the internal chemical energy of the child.

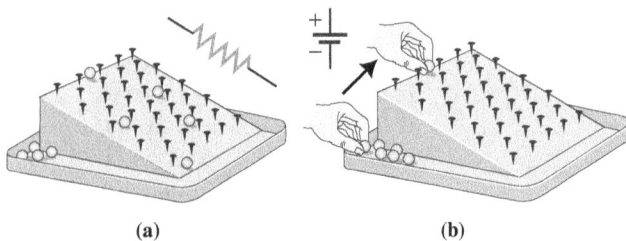

Figure 4.69: A mechanical analogue of a simple circuit consisting of a resistance and source of emf.

4.14 Kirchhoff's Rules for Resistive Circuits

Electric circuits that contain a number of resistors/capacitors can often be analysed by combining individual groups of resistors/capacitors in series and parallel. However, there are many circuits in which no two resistors/capacitors are in series or in parallel. To analyse these networks, we'll use the techniques developed by the German physicist Gustav Robert Kirchhoff.

In Kirchhoff's laws, we often use following two terms-

1. Junction: In any circuit, a junction is a point, where three or more conductors meet. Junctions are also called nodes or branch points.

2. Loop: Any closed conducting path is called a loop.

In Fig. 4.70a points a and b are junctions (because three

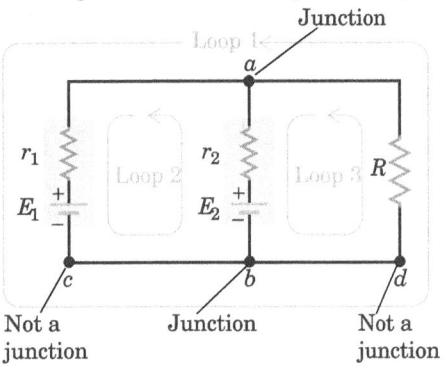

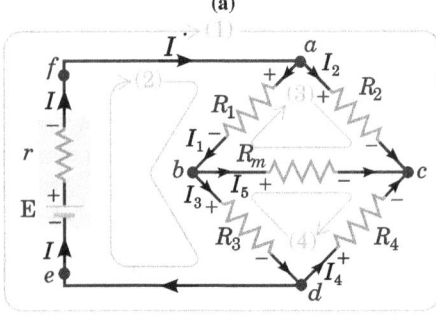

Figure 4.70

conductors are connected at these points), but points c and d are not (because only two conductors are connected at these points). In Fig.4.70b points a, b, c, and d are junctions, but points e and f are not. The inner curved lines with arrow head (the blue lines) in Figs. 4.70a and 4.70b show some possible loops in these circuits.

4.14.1 Kirchhoff's Rules for Circuits having Resistors and Battery

Kirchhoff's rules are the following two statements:

1. Kirchhoff's first rule or junction rule (based on conservation of charge): According to this rule, the algebraic sum of the currents into any junction is zero. That is,

$$\boxed{\sum I = 0 \quad \text{(junction rule)}} \quad (4.72)$$

2. Kirchhoff's second rule or loop rule (based on conservation of energy): According to this rule, the algebraic sum of the potential differences in any loop, including those associated with emfs and those of resistive elements, must equal zero. That is,

$$\boxed{\sum V = 0 \quad \text{(loop rule)}}^* \quad (4.73)$$

Since, loop law (Eq.4.73) is derived for electric field, which is a conservative force field, therefore, it is applicable only for conservative force field. **Explanation of Junction Rule**

The junction rule is based on conservation of electric charge. No charge can accumulate at a junction, so the total charge entering the junction per unit time must equal the total charge leaving per unit time (Fig.4.71a), i.e.,

At any junction: $\sum \dfrac{dQ}{dt} = 0$

∵ Charge per unit time $\dfrac{dQ}{dt}$ is current I,

∴ $\quad \sum I = 0$

Now, if we consider the currents entering a junction to be positive and those leaving to be negative, the algebraic sum of currents into a junction must be zero.

In Figure 4.71a, I_1, I_2 are incoming currents and I_3 is outgoing current, therefore-

$$\sum I = 0 \Rightarrow \quad I_1 + I_2 - I_3 = 0$$
or $\qquad I_1 + I_2 = I_3$

i.e., total incoming current = total outgoing current

It is like a T branch in a water pipe (Fig. 4.71b); if you have 1 litre per minute coming in one pipe, 3 litre per minute coming in other pipe, then outgoing water will be 4 litre per min.

Explanation of loop rule

The loop rule is a statement that the electrostatic force is conservative. Suppose we go around a loop, measuring potential differences across successive circuit elements as we go. When we return to the starting point, we must find that the algebraic sum of these differences is zero; otherwise, we could not say that the potential at this point has a definite value.

For example, considering loop 2 of Fig.4.70b, suppose, V_a, V_b, V_d, V_e, and V_f are electric potentials at points a, b, d, e, and f respectively, then on starting from e along the loop, we have

---

*Since, Kirchhoff's loop law is applicable only for conservative force field, therefore it may not hold when a changing magnetic field is present. A changing magnetic field always produces an induced electric field. If the induced electric field is produced by conservative force fields, such as changing magnetic field due to a stationary charged particle or system of charges, then the induced electric field is conservative. On the other hand, if the changing magnetic field is produced by non-conservative force fields, such as a time varying current in a wire, the induced electric field is non-conservative.

$$\sum V = (V_f - V_e) + (V_a - V_f) + (V_b - V_a)$$
$$+ (V_d - V_b) + (V_e - V_d) = 0$$

Here, in this circuit, there is no circuit element between f and a; and between e and d

Therefore, f and a will be at same potentials and d, e will also be at same potential.

Sign convention:

Charges move from the high-potential end of a resistor toward the low potential end, so assign a $+ve$ sign at the end of resistor from which current enters in it and $-ve$ sign to the end from which current coming out [Fig.4.72c and 4.72d].

So, here it is clear that b, will be at lower potential as compared

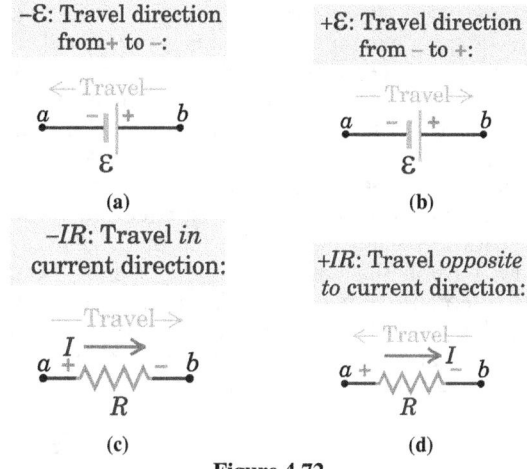

Figure 4.72

to a.

Now, if a resistor is traversed in the direction of the current (Fig.4.72c), i.e., from higher potential end to lower potential end, then there will be a potential drop. In this case, the potential difference ΔV across the resistor is-

$$\Delta V = V_b - V_a = -IR$$

If a resistor is traversed in the direction opposite to current (Fig.4.72d), i.e., from lower potential end to a higher potential end, then there will be a potential gain. In this case, the potential difference ΔV across the resistor is

$$\Delta V = V_a - V_b = +IR$$

- If a source of emf (assumed to have zero internal resistance) is traversed from positive terminal b to negative terminal a [Fig.4.72a], the potential difference ΔV is.

$$\Delta V = V_a - V_b = -\mathcal{E}$$

- If a source of emf (assumed to have zero internal resistance) is traversed from negative terminal a to positive terminal b [Fig.4.72b], the potential difference-

$$\Delta V = V_b - V_a = +\mathcal{E}$$

Problem Solving Strategy:

1. Label the current in each separate branch of the given circuit with a different subscript such as I_1, I_2 and I_3. Each current refers to a segment between two junctions. Choose the direction of each current, using an arrow. The direction can be chosen arbitrarily: if the current is actually in the opposite direction, it will come out with a minus sign in the solution.
2. Identify the unknowns. You will need as many independent equations as there are unknowns. You may write down more equations than this, but you will find that some of the equations will be redundant (that is, not be independent in the sense of providing new information). You may use $\Delta V = IR$ for each resistor, which sometimes will reduce the number of unknowns.
3. Apply Kirchhoff's junction rule at one or more junctions.
4. Apply Kirchhoff's loop rule for one or more loops: follow each loop in one direction only. Pay careful attention to subscripts, and to signs:

    (a) For a resistor, apply Ohm's law; the potential difference is negative (a decrease) if your chosen loop direction is the same as the chosen current direction through that resistor; the potential difference is positive (an increase) if your chosen loop direction is opposite to the chosen current direction.

    (b) For a battery, the potential difference is positive if your chosen loop direction is from the negative terminal toward the positive terminal; the potential difference is negative if the loop direction is from the positive terminal toward the negative terminal.
5. Solve the equations algebraically for the unknowns. Be careful when manipulating equations without any mistake with signs. At the end, check your answers by plugging them into the original equations, or even by using any additional loop or junction rule equations not used previously.

Note: If total number of unknown junctions (junctions whose potentials are not known) is N_j and total number of elements is N_E, then the total number of equations should be $N_j + N_E$.

4.15 Applications of Kirchhoff's Rules

EXAMPLE 39. Using Kirchhoff's rules. Calculate the currents I_1, I_2 and I_3 in the three branches of the circuit in Fig.4.73a.

APPROACH Since, it is your first application of Kirchhoff's

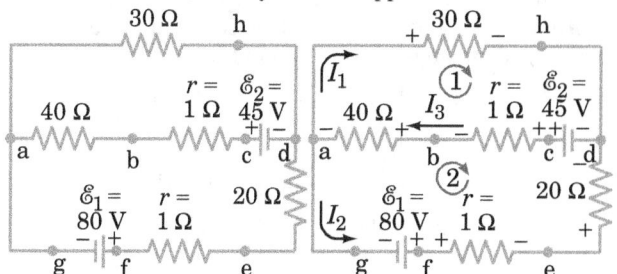

(a) Circuit diagram without polarities of resistors
(b) Circuit diagram with polarities of resistors

Figure 4.73

law, so, we will be simplifying it in steps as follows:

1. **Label the currents and their directions:** Figure 4.73b uses the labels I_1, I_2 and I_3 for the current in the three separate branches. Since (positive) current tends to move away from the positive terminal of a battery, we choose I_2 and I_3 to have the directions shown in Fig.4.73b. The direction of I_1 is not obvious in advance, so we arbitrarily chose the direction indicated. If the current actually flows in the opposite direction, our answer will have a negative sign for that. The polarities of resistors depends on the direction of current. We have assigned $+ve$ sign to the end from which our arbitrary currents enter in resistor and $-ve$ sign to the end from which currents leave it.
2. Identify the unknowns. We have three unknowns (I_1, I_2, and I_3) and therefore we need three equations, which we get by applying Kirchhoff's junction and voltage laws. Now, solve these equations for I_1, I_2 and I_3

SOLUTION We will be applying Kirchhoff's junction law and then Kirchhoff's voltage law in upper loop 1 and lower loop 2.

1. **Junction rule:** We apply Kirchhoff's junction rule to the currents at point a or d. At junction a, I_3 is incoming current whereas I_1 and I_2 are outgoing currents, therefore,

$$I_3 = I_1 + I_2 \quad \cdots (1)$$

This same equation holds at point d, so we get no new infor-

mation by writing an equation for point d.
2. **Loop or voltage law:** We apply Kirchhoff's voltage law to two different closed loops. First we apply it to the upper loop 1, i.e., along ahdcba. We start (and end) at point a. From a to h we have a potential drop $V_{ha} = -(I_1)(30\Omega)$. From h to d there is no change, but from d to c the potential increases by 45 V: that is, $V_{cd} = +45$ V. From c to a the potential decreases through the two resistances by an amount $V_{ac} = -(I_3)(40\Omega + 1\Omega) = -(41\Omega)I_3$. Thus, for closed loop ahdcba, Kirchhoff's loop law gives:

$$\sum V = 0 \quad \text{(Loop law)}$$
$$\Rightarrow V_{ha} + V_{cd} + V_{ac} = 0$$
$$\Rightarrow -30I_1 + 45 - 41I_3 = 0 \qquad \cdots (2)$$

where we have omitted the units (volts and amps) so that we can more easily do the algebra. For our second loop, we take the outer loop ahdefga (We could have chosen the lower loop abcdefga instead.). Again, we start at point a and have $V_{ha} = -(I_1)(30\Omega)$, and $V_{dh} = 0$. But when we take our positive test charge from d to e, it actually is going uphill, against the assumed direction of the current, which is what counts in this calculation. Thus, $V_{cd} = I_2(20\Omega)$ has a positive sign. Similarly, $V_{fc} = I_2(1\Omega)$. From f to g there is a decrease in potential of 80 V since we go from the high potential terminal of the battery to the low. Thus, $V_{gf} = -80$ V. Finally, $V_{ag} = 0$, and the sum of the potential changes around this loop is

$$-30I_1 + (20+1)I_2 - 80 = 0 \qquad \cdots (3)$$

Our major work is done. The rest is algebra.

3. **Solve the equations** We have three equations-labeled (1), (2), and (3) and three unknowns. From Eq. (3), we have

$$I_2 = \frac{80 + 30I_1}{21} = 3.8 + 1.4I_1 \qquad \cdots (4)$$

From Eq. (2) we have

$$I_3 = \frac{45 - 30I_1}{41} = 1.1 - 0.73I_1 \qquad \cdots (5)$$

Substituting Eqs.(4) and (5) into Eq. (1), we get
$$I_1 = I_3 - I_2 = 1.1 - 0.73I_1 - 3.8 - 1.4I_1$$
$$\Rightarrow 3.1I_1 = -2.7$$
$$\Rightarrow I_1 = -0.87 \text{ A}$$

The negative sign indicates that the direction of I_1 is actually opposite to that initially assumed and shown in Fig.4.73b. The answer automatically comes out in amperes because all values were in volts and ohms. From Eq. (4), we have
$$I_2 = 3.8 + 1.4I_1 = 3.8 + 1.4(-0.87) = 2.6 \text{ A}$$
and from Eq. (5), we have
$$I_3 = 1.1 - 0.73I_1 = 1.1 - 0.73(-0.87) = 1.7 \text{ A}$$

Note: The unknowns in different situations are not necessarily currents. It might be that the currents are given and we have to solve for unknown resistance or voltage. The variables are then different, but the technique is the same.

EXAMPLE 40. In the circuit of Fig.4.74, E, F, G, H are cells of emf 2 V, 1 V, 3 V and 1 V respectively, and their internal resistances are 2, 1, 3 and 1 Ω respectively. Calculate:
(i) the potential difference between B and D and
(ii) the potential difference across the terminals of each of the cells G and H.

APPROACH In this problem too, we follow the approach given in Example 39. The circuit is redrawn in Fig.4.75. The step wise approach is as follows:

1. **Label the currents and their directions:** Figure 4.75 uses the labels I_1, I_2 and I_3 for the current in the three separate branches. Since (positive) current tends to move away from the positive terminal of a battery, we choose I_1 and I_2 to have the directions shown in Fig.4.73b. The direction of I_3 is not obvious in advance, so we arbitrarily chose the direc-

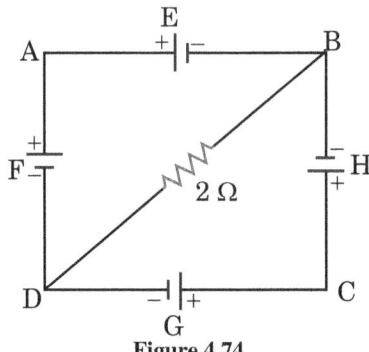

Figure 4.74

tion indicated. If the current actually flows in the opposite direction, our answer will have a negative sign. The polarities of resistors depends on the direction of current. We have assigned $+ve$ sign to the end from which our arbitrary currents enter in resistors and $-ve$ sign to the end from which currents leave.

2. **Identify the unknowns:** We have three unknowns (I_1, I_2, and I_3) and therefore we need three equations, which we get by applying Kirchhoff's junction and voltage laws. Now, solve these equations for I_1, I_2 and I_3. Now, (i) we can determine the potential difference between A and D by applying Ohm's law: $V_{BD} = I_3 R_{BD} = 2I_3$ and (ii) Potential difference across the terminals of the cell G is

$$\Delta V_G = V_D - V_C = \underbrace{-3I_2}_{\substack{\text{voltage drop across} \\ 3\,\Omega \text{ in the direction} \\ \text{of current}}} + \underbrace{3\text{ V}}_{\substack{\text{voltage gain across} \\ 3\text{ V emf in the direction} \\ \text{of current}}}$$

and the potential difference across the terminals of the cell H:

$$\Delta V_H = V_C - V_B = \underbrace{-1\text{ V}}_{\substack{\text{voltage drop across} \\ \text{emf of 1 V in the} \\ \text{direction of current}}} - \underbrace{1 I_2}_{\substack{\text{voltage drop across} \\ 1\,\Omega \text{ resistor}}}$$

By substituting the value of I_2 in above equations, we can find the potential differences across terminals of cells G and H.

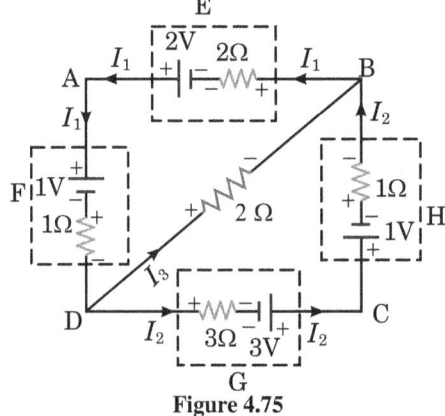

Figure 4.75

SOLUTION We will be applying Kirchhoff's junction law and then Kirchhoff's voltage law in upper loop 1 and lower loop 2.

1. **Junction rule:** By Kirchhoff's junction law either at junction B or D, we have
$$I_1 = I_2 + I_3$$
i.e., $$I_3 = I_1 - I_2 \qquad \cdots (1)$$

2. **Loop or voltage law:** By Kirchhoff's voltage law in upper loop 1, i.e., along ADBA. We start (and end) at point A.
$$-1 - I_1 - 2I_3 - 2I_1 + 2 = 0$$
or $$3I_1 + 2I_3 = 1 \qquad \cdots (2)$$

Substituting the value of I_3 from Eq.(1), in Eq.(2), we get
$$3I_1 + 2(I_1 - I_2) = 1$$
or $\qquad 5I_1 - 2I_2 = 1 \qquad \cdots (3)$

Now, we apply Kirchhoff's voltage law in lower loop 2, along DBCD. We start (and end) at point D.
$$-2I_3 + I_2 + 1 - 3 + 3I_2 = 0$$
or $\qquad 2I_2 - I_3 = 1 \qquad \cdots (4)$

Substituting the value of I_3 from Eq.(1), in Eq.(4), we get
$$2I_2 - (I_1 - I_2) = 1$$
or $\qquad 3I_2 - I_1 = 1 \qquad \cdots (5)$

3. **Solve the equations:** We have three equations-labeled (1), (4), and (5) and three unknowns.

Solving Eq.(3) and (5) for I_1 and I_2, we get
$I_1 = \frac{5}{13}$ A, and $I_2 = \frac{6}{13}$ A

Substituting these values of I_1 and I_2 in Eq.(1), we get
$$I_3 = I_1 - I_2 = \frac{5}{13}\text{A} - \frac{6}{13}\text{A} = -\frac{1}{13}\text{A} \quad \cdots (1)$$

Here, negative sign shows that the actual direction of current is opposite to assumed direction.

(i) Terminal potential across BD is
$$V_{BD} = V_B - V_D = I_3 R_{BD} = \frac{1}{13}\text{A} \times 2\Omega = \frac{2}{13} \text{ V}$$

Note: Since, actual current in branch BD is opposite to shown direction, i.e., from B to D, therefore, terminal B is at higher potential than D.

Potential difference across the terminals of the cell G
$$\Delta V_G = V_D - V_C = \underbrace{-3I_2}_{\substack{\text{voltage drop across} \\ 3\ \Omega \text{ in the direction} \\ \text{of current}}} + \underbrace{3 \text{ V}}_{\substack{\text{voltage gain across} \\ 3\text{ V emf in the direction} \\ \text{of current}}}$$
$$= -3\left(\frac{6}{13}\right)\text{V} + 3 \text{ V} = \frac{21}{13}\text{V}$$

Similarly, the potential difference across the terminals of the cell H
$$\Delta V_H = V_C - V_B = \underbrace{-1 \text{ V}}_{\substack{\text{voltage drop across} \\ \text{emf of 1 V in the direction} \\ \text{of current}}} - \underbrace{1\, I_2}_{\substack{\text{voltage drop across} \\ 1\ \Omega \text{ resistor}}}$$
$$= -1\text{ V} - \frac{6}{13}\text{ V} = -\frac{19}{13}\text{V}$$

Here negative sigh shows that terminal C is at higher potential than B.

4.15.1 Resistors in Series

Although we have already calculated the resultant of resistances in series and in parallel in section 4.11, here we again calculate it by applying Kirchhoff's laws:

In Fig.4.76a, three resistors R_1, R_2 and R_3 are connected in series. This combination is equivalent to one resistor shown in Fig.4.76b, and these two circuits are said to be equivalent because when the same emf device is used, both have the same current.

We use Kirchhoff's loop rule for each circuit in Figure 4.76 to come up with an expression for the equivalent resistance R_{eq} in terms of the three resistances R_1, R_2, and R_3. In both circuits, there is only one emf device, and the current is clockwise in both. Start from the bottom left corner as indicated in Fig.4.76 of each circuit, and apply Kirchhoff's voltage rule. For the circuit in Figure 4.76a,
$$\mathcal{E} - IR_1 - IR_2 - IR_3 = 0$$
or $\qquad \mathcal{E} = I(R_1 + R_2 + R_3) \qquad (4.74)$

Now, for the circuit in Figure 4.76b, by Kirchhoff's voltage law, we have-
$$\mathcal{E} - IR_{eq} = 0$$
or $\qquad \mathcal{E} = IR_{eq} \qquad (4.75)$

Comparing Equation 4.74 to Equation 4.75, we find that
$$R_{eq} = R_1 + R_2 + R_3 \qquad (4.76)$$

We generalize our results for a circuit with any number of resistors in series. The equivalent resistance R_{eq} of N resistors connected in series is

$$\boxed{R_{eq} = R_1 + R_2 + R_3 + \cdots + R_N = \sum_{i=1}^{N} R_i} \qquad (4.77)$$

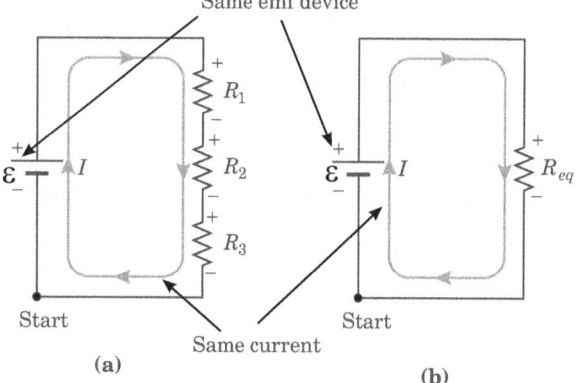

Figure 4.76: (a) Three resistors connected in series. (b) An equivalent circuit.

4.15.2 Resistors in Parallel

The circuit in Figure 4.77a is made up of three loops that involve the emf device. Let's use the loop rule for each.

For the clockwise loop BAA_1B_1B:
$$\mathcal{E} - I_1 R_1 = 0$$
or $\qquad \mathcal{E} = I_1 R_1 \qquad (4.78)$

For the clockwise loop BAA_2B_2B:

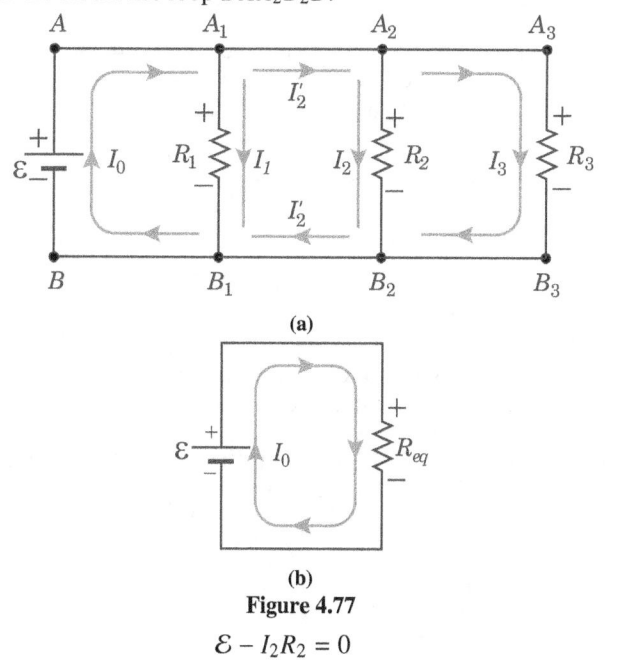

Figure 4.77

$$\mathcal{E} - I_2 R_2 = 0$$
or $\qquad \mathcal{E} = I_2 R_2 \qquad (4.79)$

For the clockwise loop BAA_3B_3B:
$$\mathcal{E} - I_3 R_3 = 0$$
or $\qquad \mathcal{E} = I_3 R_3 \qquad (4.80)$

The current splits twice-once at junction A_1 and once at junction A_2. Let's apply the junction rule at these two points.

The current entering junction A_1 is I_0. The current leaving this

junction is I_1 plus I_2'. (We get the same result if we apply the junction rule at junction B_1.)
$$I_0 = I_1 + I_2' \tag{4.81}$$
The current entering junction A_2 is I_2'. The current leaving this junction is I_2 plus I_3. (We get the same result if we apply the junction rule at junction B_2.)
$$I_2' = I_2 + I_3 \tag{4.82}$$
Substituting this value of I_2' from Eq.4.82 in Eq.4.81, we get-
$$I_0 = I_1 + I_2 + I_3 \tag{4.83}$$
Substitute Equations (4.78) through (4.80) into Equation (4.83).
$$I_0 = \frac{\mathcal{E}}{R_1} + \frac{\mathcal{E}}{R_2} + \frac{\mathcal{E}}{R_3}$$
$$I_0 = \mathcal{E}\left(\frac{1}{R_1} + \frac{1}{R_2} + \frac{1}{R_3}\right) \tag{4.84}$$
Now applying Kirchhoff's loop rule to the equivalent circuit in Fig.4.77b, we get-
$$\mathcal{E} - I_0 R_{eq} = 0$$
Solving for I_0, we get
$$I_0 = \frac{\mathcal{E}}{R_{eq}} \tag{4.85}$$
Compare Equations (4.84) and (4.85) to find an expression for the equivalent resistance R_{eq} in terms of the individual resistances R_1, R_2, and R_3.
$$\frac{1}{R_{eq}} = \frac{1}{R_1} + \frac{1}{R_2} + \frac{1}{R_3} \tag{4.86}$$
We derived Equation (4.86) using three resistors in parallel, and we can extend that equation for any number N of resistors in parallel.
$$\boxed{\frac{1}{R_{eq}} = \sum_{i=1}^{N} \frac{1}{R_i} = \frac{1}{R_1} + \frac{1}{R_2} + \frac{1}{R_3} + \cdots + \frac{1}{R_N}} \tag{4.87}$$

EXAMPLE 41. In the given circuit of Fig.4.78, calculate potential difference between the points P and Q.
APPROACH Redraw the circuit and indicate polarities of dif-

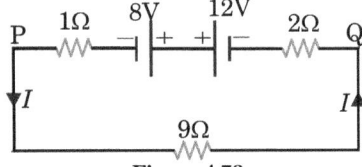

Figure 4.78

ferent resistors and then calculate electric current (I) by applying Kirchhoff's voltage law (KVL) (Figure 4.79). Now, from Fig.4.79, the potential difference between terminals P and Q is equal to potential difference across R.
SOLUTION Starting from negative terminal of 12 V battery

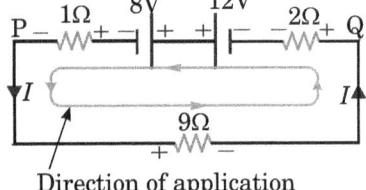

Direction of application
of loop rule
Figure 4.79

and moving along the loop shown in Fig.4.79, Kirchhoff's voltage law (KVL), gives
$$\underbrace{+12\text{ V}}_{\text{Potential gain}} + \underbrace{(-8\text{ V})}_{\text{Potential drop}} + \underbrace{(-1\Omega I)}_{\text{Potential drop}} + \underbrace{(-9\Omega I)}_{\text{Potential drop}}$$

$$+ \underbrace{(-2\Omega I)}_{\text{Potential drop}} = 0$$
or $\quad 4\text{ V} - 12\Omega I = 0$
or $\quad I = \dfrac{4\text{ V}}{12\ \Omega} = \dfrac{1}{3}\text{A}$
Now, potential difference between the points P and Q,
$$V_P - V_Q = 9\ \Omega \times \frac{1}{3}\text{A} = 3\text{ V}$$

EXAMPLE 42. In the given circuit (Fig.4.80) calculate potential difference between A and B.
APPROACH Redraw circuit and indicate the assumed current

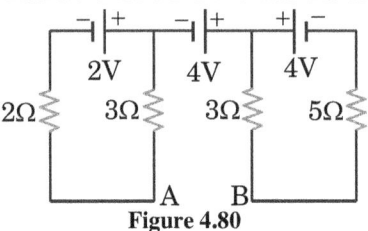

Figure 4.80

with directions. Indicate the closed loops and show the polarities of each circuit element on the basis of direction of current in it (Fig.4.81). To get potential difference between points A and B by using Kirchhoff's voltage law, convert this section into a closed loop ③ by connecting a virtual battery of potential difference $\Delta V = V_A - V_B$. Don't allow any current through this battery
SOLUTION Suppose, I_1 is the currents supplied by 2 V battery

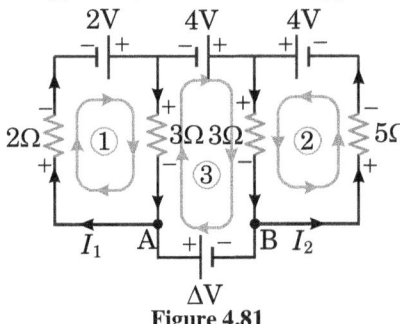

Figure 4.81

connected with loop ① and I_2 is the current supplied by 4 V battery connected with loop ②. Since, for each battery, the current coming out from it's positive terminal is always equal to the current entering in it from it's negative terminal, therefore, current I_1 remains confined in closed loop ① and I_2 remains confined in closed loop ②.There is no current through the connected virtual battery in closed loop ③.
Now, applying Kirchhoff's voltage law (KVL), in indicated direction of closed loop ①, we get
$$\underbrace{+2\text{ V}}_{\text{Potential gain}} + \underbrace{(-3\Omega I_1)}_{\text{Potential drop}} + \underbrace{(-2\Omega I_1)}_{\text{Potential drop}} = 0$$
Omitting units, we get
$$2 - 3I_1 - 2I_1 = 0 \quad \Rightarrow \quad I_1 = \frac{2}{5}\text{A} = 0.4\text{ A}$$
Now applying Kirchhoff's voltage law (KVL) in loop ②, we get
$$\underbrace{+4\text{ V}}_{\text{Potential gain}} + \underbrace{(-3\Omega I_2)}_{\text{Potential drop}} + \underbrace{(-5\Omega I_2)}_{\text{Potential drop}} = 0$$
Omitting units, we get
$$4 - 3I_2 - 5I_2 = 0 \quad \Rightarrow \quad I_2 = \frac{4}{8}\text{A} = 0.5\text{ A}$$
Similarly, in closed loop ③, by KVL, we have
$$\underbrace{+4\text{ V}}_{\text{Potential gain}} + \underbrace{(-3\Omega I_2)}_{\text{Potential drop}} + \underbrace{(+\Delta V)}_{\text{Potential gain}} + \underbrace{(+3\Omega I_1)}_{\text{Potential gain}} = 0$$

(Note that polarities of resistors depend on the direction of current in resistors not on the direction of loop. So, don't confuse in the polarity of 2 Ω resistor in loop ③).
Omitting units, we get
$$4 - 3I_2 + \Delta V + 3I_1 = 0 \quad \Rightarrow \quad \Delta V = 3I_2 - 3I_1 - 4$$
Substituting the values of I_1 and I_2 in above equation, we get
$$\Delta V = 3(0.5) - 3(0.4) - 4 = 1.5 - 1.2 - 4 = -3.7 \text{ V}$$
or $V_a - V_b = -3.7$ V

EXAMPLE 43. A wire of resistance per unit length $\lambda = 10^{-6} \Omega/m$, is turned in the form of a circle of diameter 2 m. A piece of same material is connected in diameter AB. Then find the resistance between points A and B.

APPROACH Resistance per unit length of the wire, $= \lambda$ Diameter of the wire $= 2r$
Therefore, resistance of the diameter, $R_1 = 2r\lambda$
Circumference of the circle formed by wire $= 2\pi r$
Therefore, length of upper semicircular part of the wire
$$= \text{length of lower semicircular part of the wire} = \pi r$$
So, resistance of upper semicircular part of the wire, R_2
$$= \text{resistance of lower semicircular part of the wire}, R_3$$
$$= \pi r \lambda.$$
All these parts are connected between same potential difference, therefore these are in parallel. So, equivalent resistance (R_{eq}) between points A and B is given by
$$\frac{1}{R_{eq}} = \frac{1}{R_1} + \frac{1}{R_2} + \frac{1}{R_3} \quad \cdots (1)$$
Now, put the given values in Eq.(1) and solve for R_{eq}

SOLUTION Given that, $\lambda = 10^{-6} \Omega/m$, therefore

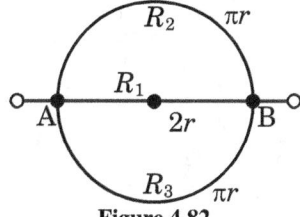

Figure 4.82

$R_1 = 2r\lambda = 2 \text{ m} \times 10^{-6} \Omega/\text{m} = 2 \times 10^{-6} \Omega$
$R_2 = R_3 = \pi r \lambda = \pi \times 1 \text{ m} \times 10^{-6} 2 \text{ m} \times 10^{-6} \Omega/\text{m} = 10^{-6} \pi \Omega$
Substituting these values in Eq.(1), we get
$$\frac{1}{R_{eq}} = \frac{1}{R_1} + \frac{1}{R_2} + \frac{1}{R_3}$$
$$= \frac{1}{2 \times 10^{-6} \Omega} + \frac{1}{10^{-6} \pi \Omega} + \frac{1}{10^{-6} \pi \Omega}$$
$$= 0.88 \times 10^{-6} \Omega = 0.88 \mu\Omega$$

EXAMPLE 44. In circuit diagram of Figure 4.83, the galvanometer reading is zero. If the internal resistance of cells are negligible then find the value of X?

APPROACH The given circuit is redrawn in Figure 4.84. This

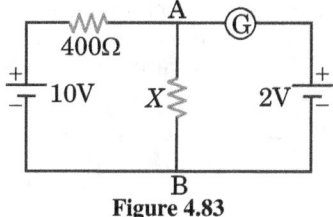

Figure 4.83

circuit shows the assumed current with directions. It also shows the closed loop ① to apply KVL and the polarities of each circuit element on the basis of assumed current in it. To find the resistance X, we apply KVL in loop ①. In this step, we get a relation between two unknowns I and X. So, we need one more equation between I and X to find X which can be obtained by using the fact that galvanometer branch don't has any current. It means the potential drop across galvanometer is zero and hence, potential at point A will be equal to potential at point C and potential at point B is equal to potential at point D. So, potential difference across AB (i.e. across resistor X) is equal to potential difference across CD, i.e., $IX = 2$ V.

SOLUTION Applying KVL in loop ① along the direction as

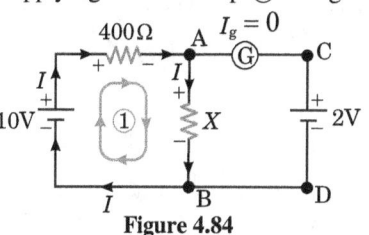

Figure 4.84

shown in Fig.4.84, we get
$$10 - 400I - IX = 0$$
Since, $IX = 2$ V, therefore-
$$10 - 400I - 2 = 0 \quad \text{or} \quad 400I = 8$$
or $I = \dfrac{8}{400}$ A $= \dfrac{1}{50}$ A

Now, $IX = 2 \quad \Rightarrow \quad X = \dfrac{2}{1/50} = 100 \ \Omega$

EXAMPLE 45. Find out the value of current through 2 Ω resistance for the given circuit in Fig.4.85.

APPROACH Given circuit is redrawn in Fig.4.86. We have to

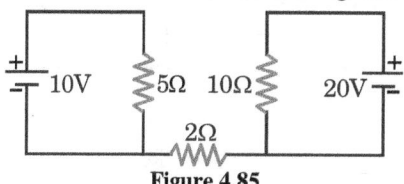

Figure 4.85

focus our attention on the fact that current coming out from the positive terminal of a battery is always equal to the current passing into battery from it's negative terminal. Now, keeping this fact in mind apply Kirchhoff's junction law at junctions C and D of circuit shown in Fig.4.86.

SOLUTION Let current supplied by 10 V battery is I_1, then

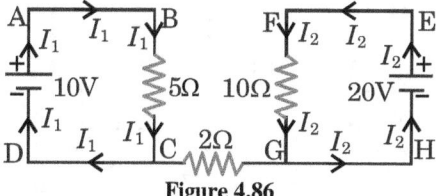

Figure 4.86

current in battery branch DE will be I_1 (Fig.4.86). After passing through 5 Ω resistor, this goes to battery through branch CD. Therefore, by Kirchhoff's junction law at junction C, no part of current I_1 passes through 2 Ω resistor. Now, if we consider that the current supplied by 20 V battery is I_2, then by the same logic, no part of I_2 too passes through 2 Ω resistor. Thus, we can say that no current passes through 2 Ω resistance, i.e., net current through 2 Ω resistance is zero.

EXAMPLE 46. Find the current passing through all resistors in the circuit of Fig.4.87.

APPROACH Given circuit is redrawn in Fig.4.88a. Resistances 2 Ω and 1 Ω are connected in same branch so they are in series. These resistance are enclosed within the dashed box. Their equivalent resistance is 2 Ω + 1 Ω = 3 Ω. It is shown in next simplified circuit diagram shown in Fig.4.88b.

As shown in Fig.4.88b, the potential difference between 3 Ω and 6 Ω resistors is same, therefore they are in parallel. These two

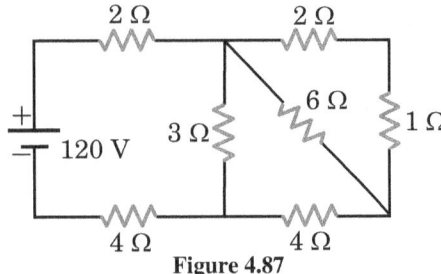

Figure 4.87

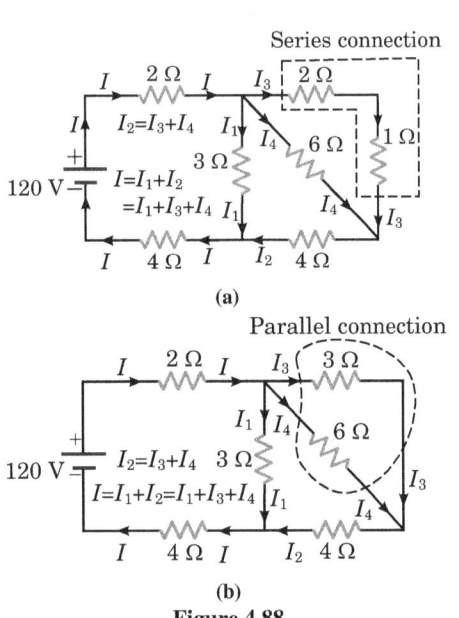

(a)

(b)

Figure 4.88

resistors are enclosed within a dashed curve in Fig.4.88b. Their equivalent resistance is $\frac{6 \times 3}{6+3}\Omega = 2\ \Omega$. It is shown in Fig.4.89a
Now, as shown in Fig.4.89a, 2 Ω and 4 Ω are in series, their

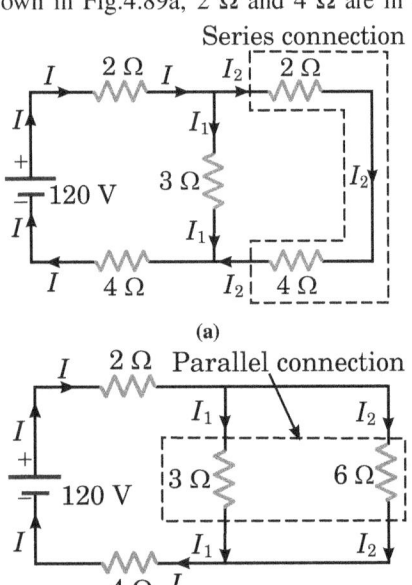

(a)

(b)

Figure 4.89

equivalent resistance is 2 Ω + 4 Ω = 6 Ω. It is shown in next Fig.4.89b. In Fig.4.89b, resistances 3 Ω and 6 Ω are in parallel. Their equivalent resistance is $\frac{6 \times 3}{6+3}\Omega$ i.e., 2 Ω. It is shown in new Fig.4.90a. Now, in Fig.4.90a, all resistors are in series. Their

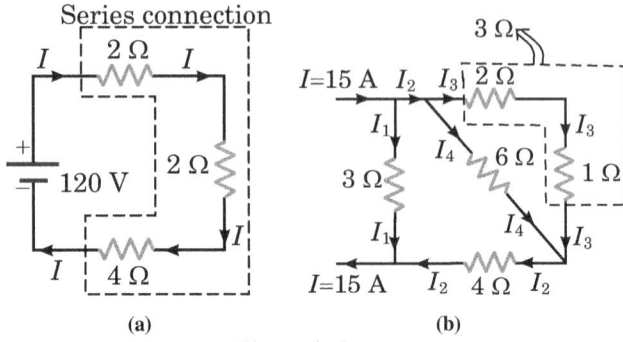

(a)

(b)

Figure 4.90

equivalent resistance is given by
$$R_{eq} = 2\ \Omega + 2\ \Omega + 4\ \Omega = 8\ \Omega$$
Therefore, by Ohm's law, the current supplied by battery is given by
$$I = \frac{\Delta V}{R} = \frac{120\ \text{V}}{8\ \Omega} = 15\ \text{A}$$
At the upper junction of resistors 3 Ω and 6 Ω of Fig.4.89b, by current division rule 4.102, current through 3 Ω resistor is
$$I_1 = \frac{6\ \Omega}{3\ \Omega + 6\ \Omega} \times 15\ \text{A} = \frac{3\ \Omega}{9\ \Omega} \times 15\ \text{A} = 10\ \text{A}$$
Similarly, current through 6 Ω resistor is
$$I_2 = \frac{3\ \Omega}{3\ \Omega + 6\ \Omega} \times 15\ \text{A} = \frac{3\ \Omega}{9\ \Omega} \times 15\ \text{A} = 5\ \text{A}$$
Now, from Fig.4.88b and further simplified diagram 4.90b, we see that I_1 passes through 3 Ω resistor and remaining I_2 divided into two parts I_3 and I_4. Current I_3 passes through 3 Ω (series equivalent of 2 Ω and 1 Ω) resistor. It is enclosed by dashed part. Current I_3 and I_4 again combines at lower junction and give I_2 which passes through lower right 4 Ω resistance. By current division rule, current passes through 3 Ω resistor in dashed part
$$I_3 = \frac{6\ \Omega}{6\ \Omega + 3\ \Omega} \times 5\ \text{A} = \frac{10}{3}\text{A}$$
This current passes through the branch containing series resistors 2 Ω and 1 Ω.
Similarly, current passes through 6 Ω resistor in dashed part
$$I_4 = \frac{3\ \Omega}{6\ \Omega + 3\ \Omega} \times 5\ \text{A} = \frac{5}{3}\text{A}$$
The current through 2 Ω and 4 Ω resistors of main branch of circuit is 15 A.

4.16 Terminal Potential Difference

The potential difference between the two electrodes of a cell in a closed circuit i.e. when current is being drawn from the cell is called terminal potential difference.

4.16.1 When Cell Is Discharging

Figure 4.91 shows a circuit in which a real emf device (cell or battery) of emf $\mathcal{E}$ and internal resistance r, is connected with an external resistance R. Direction of current, in the circuit, is anticlockwise. Due to current in the circuit, the polarities of resistances in closed circuit, are also indicated. Now, apply Kirch-

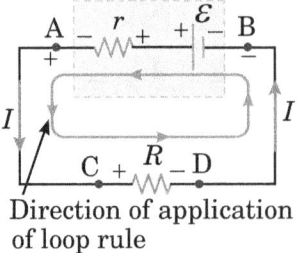

Direction of application of loop rule

Figure 4.91

hoff's loop rule $\Sigma V = 0$ starting from terminal B in anticlockwise direction (Fig.4.91):

$$\underbrace{\mathcal{E}}_{\text{Potential gain across }\mathcal{E}} + \underbrace{(-Ir)}_{\text{Potential drop across }r} + \underbrace{(-IR)}_{\text{Potential drop across }R} = 0$$

or $\mathcal{E} - Ir - IR = 0$ (A)

Therefore, current in the circuit-

$$\boxed{I = \frac{\mathcal{E}}{r+R}} \quad (4.88)$$

Again from Eq.(A),
$$IR = \mathcal{E} - Ir$$

Since, $IR = \Delta V$ = terminal voltage across AB, therefore above equation can also be written as

$$\boxed{\Delta V = \mathcal{E} - Ir} \quad (4.89)$$

So, it ic clear from Eq.4.89, if current is drawn from the cell, it's terminal potential difference is less than emf of cell. Greater is the current drawn from the cell smaller is the terminal voltage.

4.16.1.1 When Cell is in Open circuit

In open circuit $I = 0$, so, from Eq.4.89: $\Delta V = \mathcal{E} - Ir$, we get
$$\Delta V = \mathcal{E} - 0$$
or $\Delta V = \mathcal{E}$ (4.90)

Thus, in an open circuit, the terminal potential difference is equal to emf and is the maximum potential difference which a cell can provide.

4.16.1.2 When Cell is Short Circuited

Short circuited circuit means, left and right terminals of resistor R are connected with a conducting wire. In this case, the current gets an alternate path and the external effective resistance of the circuit becomes zero, i.e., $R = 0$. In short circuited circuit, there is an excessive current flowing through the circuit which may also lead to an electric fire.

From Eq.4.88, we have: $I = \dfrac{\mathcal{E}}{r+R}$

For short circuiting $R = 0$, therefore above equation gives

$$\boxed{I = \frac{\mathcal{E}}{r}} \quad (4.91)$$

and from Eq.4.89, we have

$$\boxed{\Delta V = \mathcal{E} - Ir = \mathcal{E} - \left(\frac{\mathcal{E}}{r}\right)r = 0} \quad (4.92)$$

So, in a short circuited circuit, the current is maximum and terminal potential difference is zero.

EXAMPLE 47. The emf of a primary cell is 2 V, when it is shorted then it gives a current of 4 A. Calculate internal resistance of primary cell.

APPROACH The internal resistance for shorted circuit can be calculated by using Eq.4.91:
$$I = \frac{\mathcal{E}}{r} \quad \cdots (1)$$

SOLUTION From Eq.(1), the internal resistance of the cell
$$r = \frac{\mathcal{E}}{I} = \frac{2}{4} = 0.5 \, \Omega$$

4.16.2 When Cell is Charging

Figure 4.92 shows a circuit in which a real emf device (cell or battery) of emf $\mathcal{E}$ and internal resistance r, is connected with an external voltage source of terminal voltage $\Delta V (> \mathcal{E})$. This voltage source is charging the non-ideal emf source. Direction of current, in the circuit, is clockwise. Due to current in the circuit, the polarities of resistances are also indicated. Although, the current inside emf source (i.e., the cell or battery being charged) is from anode to cathode, the polarity of emf source does not depend on the direction of charging current.

Now, apply Kirchhoff's loop rule $\Sigma V = 0$ starting from terminal

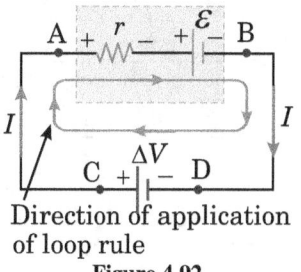

Direction of application of loop rule
Figure 4.92

D in clockwise direction (Fig.4.91):

$$\underbrace{\Delta V}_{\substack{\text{Potential gain across}\\ \text{the voltage source}}} + \underbrace{(-Ir)}_{\substack{\text{Potential drop}\\ \text{across }r}} + \underbrace{(-\mathcal{E})}_{\substack{\text{Potential drop}\\ \text{across }\mathcal{E}}} = 0$$

or $\Delta V - \mathcal{E} - Ir = 0$

Therefore,

$$\boxed{\Delta V = \mathcal{E} + Ir} \quad (4.93)$$

Thus, in case of charging of a cell or battery (Eq.4.93), it's terminal potential difference is greater than the emf of cell.

4.17 Maximum Power Transfer Theorem

In this section, we will find out the value of external resistance R corresponding to maximum output power (i.e., power given to R) provided by a cell of emf '$\mathcal{E}$' and internal resistance 'r' (Fig.4.91). In the circuit of Figure 4.91, power output

$$P = I^2 R = \frac{\mathcal{E}^2}{(r+R)^2} \cdot R \quad (4.94)$$

Differentiating above expression with respect to R, we get

$$\frac{dP}{dR} = \frac{\mathcal{E}^2}{(r+R)^2} - \frac{2\mathcal{E}^2 R}{(r+R)^3} = \frac{\mathcal{E}^2}{(R+r)^3}[R + r - 2R]$$

For maximum power supply
$$\frac{dP}{dR} = 0 \quad \Rightarrow \quad r + R - 2R = 0$$
or $\boxed{r = R}$ (4.95)

So, for maximum power output, the external resistance should be equal to internal resistance of the cell.

Now, to get maximum power ($P_{\max}$) supplied by the cell, we put this value of $R (= r)$, in Eq.4.94:

$$P_{\max} = \frac{\mathcal{E}^2}{(r+r)^2} \cdot r = \frac{\mathcal{E}^2}{4r} \quad (4.96)$$

Again from Eq.4.94 Power output

$$P = \frac{\mathcal{E}^2 R}{(r+R)^2} \quad \Longrightarrow \quad P\left(r^2 + 2rR + R^2\right) = \mathcal{E}^2 R$$

$$\Longrightarrow \quad R^2 + \left(2r - \frac{\mathcal{E}^2}{P}\right)R + r^2 = 0$$

Above equation is quadratic in R. So, for given values of $\mathcal{E}$, P and r, there are two roots of above equation. These are given below

$$R = \frac{-\left(2r - \frac{\mathcal{E}^2}{P}\right) \pm \sqrt{\left(2r - \frac{\mathcal{E}^2}{P}\right)^2 - 4r^2}}{2}$$

If we denote these roots by R_1 and R_2, then their product is given by-

$$R_1 R_2 = r^2 \text{ (product of roots)}$$

Efficiency of the Cell for $R = r$

Total power spent by cell

$$P_{\text{net}} = I^2(r+R) = \frac{\mathcal{E}^2}{(r+r)^2} \cdot 2r = \frac{\mathcal{E}^2}{2r}$$

Output power, i.e., power given to external resistance

$$P_{out} = I^2 R = \left(\frac{\mathcal{E}}{r+r}\right)^2 \times r = \frac{\mathcal{E}^2}{4r}$$

$$\text{Efficiency} = \frac{\text{power output}}{\text{total power spent by cell}}$$

$$= \frac{\frac{\mathcal{E}^2}{4r} \times 100}{\frac{\mathcal{E}^2}{2r}} = \frac{1}{2} \times 100 = 50\%$$

P vs R Graph

Figure 4.93 shows power vs external resistance graph. The power is zero corresponding to $R \to 0$ and $R \to \infty$. The power is maximum corresponding to $R = r$. It also shows that power is same for two values or R_1 and R_2.

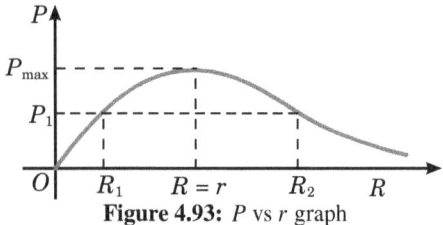

Figure 4.93: P vs r graph

☞ Note that, if r is variable but $\mathcal{E}$ and R are fixed, then I and P both will be maximum when $r = 0$.

☞ If R is variable but $\mathcal{E}$ and r are fixed, then, current in the circuit is maximum when $R = 0$. But power used in R will be maximum, when external resistance is equal to internal resistance (i.e., $R = r$).

4.18 Division of Current in Resistors Joined in Parallel

Fig.4.94 shows two resistors R_1 and R_2 in parallel. At junction A, the incoming current is I which divided to I_1 and I_2 such that I_1 passes through R_1 and I_2 passes through R_2. So, by Kirchhoff's junction law, we have

$$I = I_1 + I_2 \quad (4.97)$$

At junction B, they further combine to give net current I.

By Ohm's law, the potential difference between junctions A and

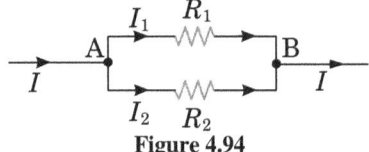

Figure 4.94

B is given by

$$V_A - V_B = I_1 R_1 = I_2 R_2 \quad (4.98)$$

From Eq.4.65, the equivalent resistance between terminals A and B is,

$$R_{eq} = \frac{R_1 R_2}{R_1 + R_2} \quad (4.99)$$

Therefore, by Ohm's law, we have

$$V_A - V_B = I R_{eq} \quad (4.100)$$

Comparing Eq.4.98 and 4.100, we get
$I_1 R_1 = I_2 R_2 = I R_{eq}$, i.e.,

$$\boxed{I_1 = \frac{R_{eq}}{R_1} I \quad \text{and} \quad I_2 = \frac{R_{eq}}{R_2} I} \quad (4.101)$$

If we substitute the value of R_{eq} from Eq.4.99 in Eq.4.101, we get

$$\text{or} \quad \boxed{I_1 = \frac{R_2}{R_1 + R_2} I} \quad (4.102)$$

and

$$\boxed{I_2 = \frac{R_1}{R_1 + R_2} I} \quad (4.103)$$

From Eq.4.103, we see that

$$\frac{I_1}{I_2} = \frac{R_2}{R_1} \quad (4.104)$$

From Eq.4.104, it is clear that that the current is divided in resistors, connected in parallel, in inverse ratio of the resistances.

Generalisation of Result: If there are n resistors in parallel and the current through their equivalent resistor (R_{eq}) is I, then current passing through i^{th} resistor R_i is

$$\boxed{I_i = \frac{R_{eq}}{R_i} I} \quad (4.105)$$

Eq.4.105 can directly be applied to find current quickly, when there are more than two resistors in parallel.

4.19 Division of Voltage in Resistors Joined in Series

In parallel connection, the potential difference between the terminals of resistors always remain same. In this section, we discuss only series connections of resistors.

Let us consider three resistors R_1, R_2 and R_3 in series as shown in Fig.4.95. Their equivalent resistance is

$$R_{eq} = R_1 + R_2 + R_3$$

If V is the voltage applied by battery, then current in circuit

$$I = \frac{V}{R_{eq}}$$

Substituting the value of R_{eq} in above equation, we get

$$I = \frac{V}{R_1 + R_2 + R_3}$$

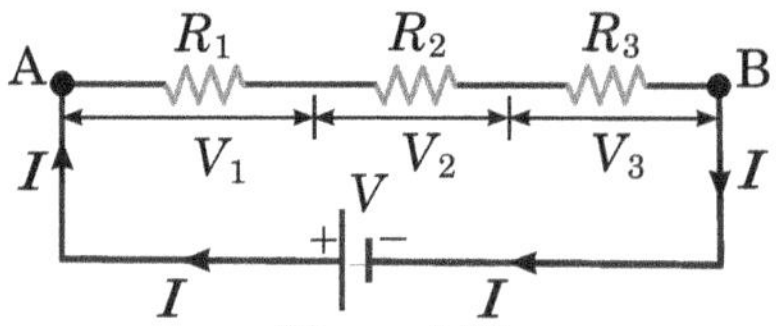

Figure 4.95

Therefore, potential difference across resistor R_1,

$$\boxed{V_1 = IR_1 = \frac{R_1}{R_1 + R_2 + R_3} V} \quad (4.106)$$

Similarly, potential difference across resistor R_2,

$$\boxed{V_2 = IR_2 = \frac{R_2}{R_1 + R_2 + R_3} V} \quad (4.107)$$

and potential difference across resistor R_3,

$$\boxed{V_3 = IR_3 = \frac{R_3}{R_1 + R_2 + R_3} V} \quad (4.108)$$

Generalisation: If there are n resistors in series and their equivalent resistance is R_{eq}, then potential difference between the terminals of ith resistor R_i is,

$$\boxed{V_i = \frac{R_i}{R_{eq}} V} \quad (4.109)$$

4.20 Relative Potential

While solving an electric circuit it is convenient to chose a reference point and assigning its voltage as zero, then all other potentials are measured with respect to this point. This point is also called the common or reference point.

For example, if terminal A of a given circuit is at potential V_A and

terminal B is at potential V_B, then potential of terminal B relative to terminal A will be $V_B - V_A$. In this case, A is considered as a reference potential. Suppose, terminal B is at +10 V and terminal A is at -2 V, then potential of terminal B with respect to terminal A will be
$$V_{B/A} = V_B - V_A = 10\text{ V} - (-2\text{V}) = 12\text{ V}$$
In this case reference terminal A can be regarded as zero potential terminal and it's potential is transfered to terminal B. So, we can say that potential difference between terminals A and B is 12 V with terminal A at 0 V and terminal B at $+12$ V.

4.21 Grouping of Cells
4.21.1 Series Grouping

Suppose n cells having emfs $\mathcal{E}_1, \mathcal{E}_2, \mathcal{E}_3, \ldots, \mathcal{E}_n$ and internal resistances $r_1, r_2, r_3, \ldots, r_n$ respectively, are connected in series as shown in Figure 4.96. The points A and B act as the terminals of the combination. If an external resistance R is connected across the combination, then by Kirchhoff's voltage law (Fig.4.96), we have
$$-IR - Ir_n + \mathcal{E}_n - \cdots - Ir_2 + \mathcal{E}_2 - Ir_1 + \mathcal{E}_1 = 0$$

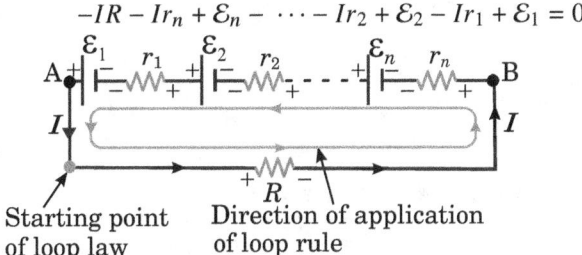

Figure 4.96: Series grouping of cells

Solving for I, we get
$$\boxed{I = \frac{\mathcal{E}_1 + \mathcal{E}_2 + \cdots + \mathcal{E}_n}{R + (r_1 + r_2 + \cdots + r_n)} = \frac{\mathcal{E}_0}{R + r_0}} \quad (4.110)$$

From Eq.4.110, we see that the combination acts as a battery of emf
$$\mathcal{E}_0 = \mathcal{E}_1 + \mathcal{E}_2 + \cdots + \mathcal{E}_n \quad (4.111)$$
having an internal resistance
$$r_0 = r_1 + r_2 + \cdots + r_n \quad (4.112)$$
If all n cell are identical and let-
$$\mathcal{E}_1 = \mathcal{E}_2 = \cdots = \mathcal{E}_n = \mathcal{E} \text{ (say)};$$
$$r_1 = r_2 = \cdots = r_n = r \text{ (say), then}$$
$\mathcal{E}_0 = \mathcal{E}_1 + \mathcal{E}_2 + \cdots + \mathcal{E}_n = n\mathcal{E}$ and
$$r_0 = r_1 + r_2 + \cdots + r_n = nr$$
In this case, from Eq.4.110, the electric current in the circuit
$$\boxed{I = \frac{n\mathcal{E}}{R + nr}} \quad (4.113)$$

☞ If $nr \ll R$, then $R + nr \approx R$ and in this case, Eq.4.113 gives
$$I = \frac{n\mathcal{E}}{R} \simeq n \times \text{current from one cell.}$$
So, in this case, series combination is advantageous.

☞ If $nr \gg R$, then $R + nr \approx nr$ and in this case, Eq.4.113 gives
$$I = \frac{\mathcal{E}}{r} \simeq \text{current from one cell.}$$
So, in this case, series combination is not advantageous.
Now, if the polarities of m ($< n$) cells are reversed, then from Eq.4.113, the equivalent emf will be
$$\mathcal{E}_0 = -\mathcal{E}_1 - \mathcal{E}_2 - \cdots - \mathcal{E}_m + \mathcal{E}_{m+1} + \mathcal{E}_{m+2} + \cdots \mathcal{E}_n \quad (4.114)$$
In this case, net resistance will remain same.
In addition, if cells are identical too, then
$$\mathcal{E}_0 = -m\mathcal{E} + (n-m)\mathcal{E} = (n - 2m)\mathcal{E} \quad (4.115)$$

In this case, from Eq.4.110, net current in the circuit
$$\boxed{I = \frac{\mathcal{E}_0}{R + r_0} = \frac{(n - 2m)\mathcal{E}}{R + nr}} \quad (4.116)$$

• If polarity of m cells is reversed, then equivalent emf is
$$\mathcal{E}_{eq} = (n - m)\mathcal{E} - m\mathcal{E} = (n - 2m)\mathcal{E}$$
while the equivalent resistance is still $nr + R$.

EXAMPLE 48. A battery of six cells each of emf 2 V and internal resistance 0.5 Ω is being charged by DC mains of emf 220 V by using an external resistance of 10 Ω (Fig.4.97). What will be the charging current.

APPROACH Although, we can easily find current in the circuit

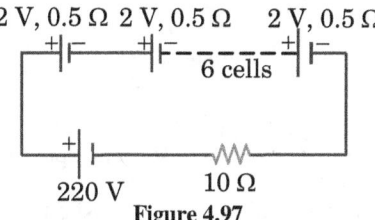

Figure 4.97

by using Kirchhoff's voltage law, we won't do that because this problem is given as an application of the concept discussed in above section.
To find net emf and net internal resistance of battery, which is a series combination of six cells, apply Eq.4.111:
$$\mathcal{E}_0 = \mathcal{E}_1 + \mathcal{E}_2 + \cdots + \mathcal{E}_n \quad \cdots (1)$$
and Eq.4.112:
$$r_0 = r_1 + r_2 + \cdots + r_n \quad \cdots (2)$$
respectively. Finally, to find the circuit current apply Eq.4.93:
$$\Delta V = \mathcal{E}_0 + Ir \quad \cdots (3)$$
SOLUTION Given that all six cells of the battery, are identical such that $\mathcal{E}_1 = \mathcal{E}_2 = \cdots = \mathcal{E}_6 = 2$ V and $r_1 = r_2 = \cdots = r_6 = 0.5$ Ω, therefore, from Eq.(1), we have
$$\mathcal{E}_0 = 2 + 2 + \cdots \text{six times} = 6 \times 2 = 12 \text{ V}$$
and from Eq.(2), net internal resistance of the battery
$$r_0 = 0.5 + 0.5 + \cdots \text{six times} = 6 \times 0.5 = 3.0 \text{ Ω}$$
Since a 10 Ω resistance also connected in series with above combination, therefore, net resistance of the circuit
$$r = 3 + 10 = 13 \text{ Ω}$$
For charging of battery, this resistance can be considered as total internal resistance.
Substituting the given values in Eq.(3), we get
$$\Delta V = \mathcal{E}_0 + Ir$$
or $$I = \frac{\Delta V - \mathcal{E}_0}{r} = \frac{220 - 12}{13} = 16 \text{ A}$$
☞ Students are advised to solve this problem by applying KVL.

EXAMPLE 49. Find the current in the loop shown in Fig.4.98.
APPROACH In this case, we use the series grouping concept of

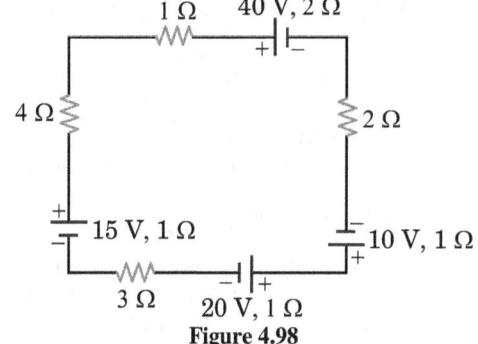

Figure 4.98

cells.
SOLUTION Net emf of the circuit,
$$\mathcal{E}_0 = (40 + 20 - 10 - 15) \text{ V} = 35 \text{ V}$$

Net internal resistance of the circuit,
$$r_0 = (2 + 1 + 1 + 1)\ \Omega = 5\ \Omega$$
Net external resistance of the circuit,
$$R = (1 + 3 + 2 + 4)\ \Omega = 10\ \Omega$$
The given circuit can be simplified as shown in Fig.4.99
Now, from Eq.4.69, the current in circuit,

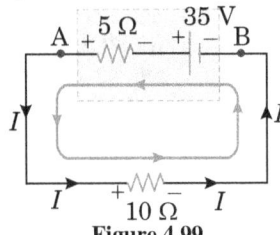

Figure 4.99

$$I = \frac{\mathcal{E}_0}{R+r} = \frac{35}{15}\ \text{A} = \frac{7}{3}\ \text{A}$$

Remark: We suggest students to solve this problem by applying Kirchhoff's voltage law. Take any arbitrary direction of current in the circuit. If actual direction is opposite to it, then you will get a negative answer of I.

4.21.2 Parallel Grouping

Now suppose the cells are connected in parallel as shown in Fig.4.100. The currents are also shown in this figure. Applying Kirchhoff's voltage law anticlockwise along loop ACDBEFA, we get
$$\mathcal{E}_1 - IR - I_1 r_1 = 0$$

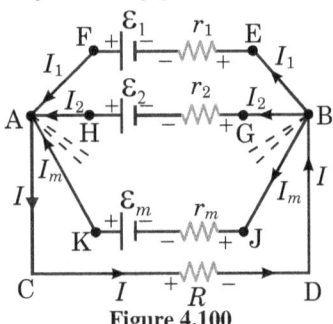

Figure 4.100

or $\quad I_1 = -\dfrac{IR}{r_1} + \dfrac{\mathcal{E}_1}{r_1}$

Similarly, by KVL, in loop ACDBGHA, we have
$$I_2 = -\frac{IR}{r_2} + \frac{\mathcal{E}_2}{r_2}$$
and for m^{th} branch, in loop ACDBJKA, KVL gives
$$I_m = -\frac{IR}{r_m} + \frac{\mathcal{E}_m}{r_m}$$
Now, by Kirchhoff's junction rule at junction A or B, we have
$$I = I_1 + I_2 + \cdots + I_m$$
Substituting the values of $I_1, I_2, \ldots I_m$ in above equation, we get
$$I = -\frac{IR}{r_1} + \frac{\mathcal{E}_1}{r_1} - \frac{IR}{r_2} + \frac{\mathcal{E}_2}{r_2} \cdots - \frac{IR}{r_m} + \frac{\mathcal{E}_m}{r_m}$$
or $\quad I = -IR\sum\left(\dfrac{1}{r}\right) + \sum\left(\dfrac{\mathcal{E}}{r}\right)$

Solving for I and rearranging terms in above equation, we get
$$I = \frac{\sum(\mathcal{E}/r)}{1 + R\sum(1/r)} \quad (4.117)$$

Note: It is very important to observe that R is the external resistance and the net internal resistance of the grouped cells is in series with R. Hence, the denominator must be of the form $(R + r_0)$, where r_0 is the net internal resistance of the grouped cells. To write the denominator in this form, we divide both numerator and denominator on the right-hand side of Eq. 4.117 by $\sum(1/r)$:

$$\boxed{I = \frac{\frac{\sum(\mathcal{E}/r)}{\sum(1/r)}}{R + \frac{1}{\sum(1/r)}} = \frac{\mathcal{E}_0}{R+r_0}} \quad (4.118)$$

We see that the combination acts as a battery of emf $\mathcal{E}_0$ given by
$$\mathcal{E}_0 = \frac{\sum\left(\frac{\mathcal{E}}{r}\right)}{\sum\left(\frac{1}{r}\right)} = \frac{\mathcal{E}_1/r_1 + \mathcal{E}_2/r_2 + \ldots + \mathcal{E}_m/r_m}{1/r_1 + 1/r_2 + \ldots + 1/r_m} \quad (4.119)$$

and net internal resistance (r_0) of grouped cells is
$$\frac{1}{r_0} = \sum \frac{1}{r} \quad (4.120)$$

From Eq.4.118, it is clear that if m identical cells each of emf $\mathcal{E}$ and internal resistance r, are connected in parallel, then
$$\mathcal{E}_0 = \frac{\sum\left(\frac{\mathcal{E}}{r}\right)}{\sum\left(\frac{1}{r}\right)} = \frac{\mathcal{E}/r + \mathcal{E}/r + \cdots + \text{m times}}{1/r + 1/r + \cdots + \text{m times}} = \frac{m\mathcal{E}/r}{m/r} = \mathcal{E}$$
$$\quad (4.121)$$
and $\quad \dfrac{1}{r_0} = \sum \dfrac{1}{r} = \dfrac{1}{r} + \dfrac{1}{r} + \cdots \text{m times} = \dfrac{m}{r}$

or $\quad r_0 = \dfrac{r}{m} \quad (4.122)$

In this case, Eq.4.118, gives
$$I = \frac{\mathcal{E}}{R + \frac{r}{m}} = \frac{m\mathcal{E}}{mR + r} \quad (4.123)$$

☞ If $mR \ll r$; $I = \dfrac{m\mathcal{E}}{r}$ ⇒ Parallel combination is advantageous.

☞ If $mR \gg r$; $I = \dfrac{\mathcal{E}}{r}$ ⇒ Parallel combination is not advantageous.

EXAMPLE 50. Four identical cells each of emf 2 V are joined in parallel providing supply of current to external circuit consisting of two 15 Ω resistors joined in parallel. The terminal voltage of the equivalent cell as read by an ideal voltmeter is 1.6 V. Calculate the internal resistance of each cell.

APPROACH Since, terminal voltage and external resistance (it

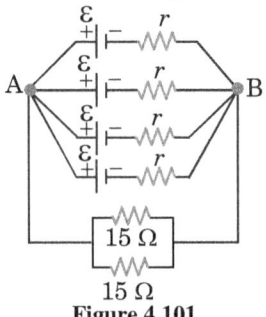

Figure 4.101

is the equivalent of two 15 Ω resistors in parallel) are provided, so, by Ohm's law ($\Delta V = IR$), we can find the current provided by the combination of cells. Now, by applying Eq.4.121 find the net emf of combination and by applying Eq.4.122 find net internal resistance (r_0) of the combination in terms of individual internal resistance (r). Finally, put these values in Eq.4.89: $\Delta V = \mathcal{E}_0 - Ir_0$, and solve for r.

SOLUTION Given that the terminal voltage
$$\Delta V = V_a - V_b = 1.6\ \text{V},$$
Net external resistance, $R = \frac{15 \times 15}{15 + 15} = \frac{15}{2} = 7.5\ \Omega$
Therefore, by Ohm's law, the current provided by equivalent cell
$$I = \frac{\text{terminal potential}}{\text{external resistance}} = \frac{\Delta V}{R} = \frac{1.6}{7.5}\ \text{A}$$
From Eq.4.121, the net emf of the circuit

$\mathcal{E}_0 = \mathcal{E} = 2$ V
and from Eq.4.122, the net internal resistance of the circuit
$$r_0 = \frac{r}{m} = \frac{r}{4}$$
Therefore, Eq.4.89, gives
$$\Delta V = \mathcal{E}_0 - I r_0$$
or
$$r_0 = \frac{\mathcal{E}_0 - \Delta V}{I} = \frac{2 - 1.6}{1.6/7.5} = \frac{7.5}{4} \, \Omega$$
Since, $r_0 = \frac{r}{4}$, therefore above expression gives
$$\frac{r}{4} = \frac{7.5}{4} \, \Omega \quad \Rightarrow \quad r = 7.5 \, \Omega$$

EXAMPLE 51. In the circuit shown in Figure 4.102, $\mathcal{E}_1 = 3$ V, $\mathcal{E}_2 = 2$ V, $\mathcal{E}_3 = 1$ V and $R = r_1 = r_2 = r_3 = 1 \, \Omega$. (a) Find the potential difference between the points A and B and the currents through each branch. (b) If r_2 is short-circuited and the point A is connected to point B, find the currents through $\mathcal{E}_1, \mathcal{E}_2, \mathcal{E}_3$ and the resistor R.

(a) **APPROACH** In Fig.4.102, $\mathcal{E}_1, \mathcal{E}_2$ and $\mathcal{E}_3$ are connected in

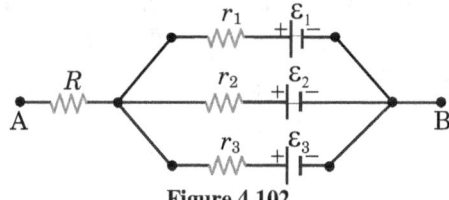

Figure 4.102

parallel. Resistance R is connected in series with this parallel group. Now, find net emf by applying Eq.4.119:
$$\mathcal{E}_0 = \frac{\sum\left(\frac{\mathcal{E}}{r}\right)}{\sum\left(\frac{1}{r}\right)} = \frac{\mathcal{E}_1/r_1 + \mathcal{E}_2/r_2 + \ldots + \mathcal{E}_m/r_m}{1/r_1 + 1/r_2 + \ldots + 1/r_m} \quad \cdots (1)$$
and net internal resistance (r_0) by applying Eq.4.122
$$\frac{1}{r_0} = \sum \frac{1}{r} \quad \cdots (2)$$
Now, the current through any branch having emf $\mathcal{E}$ and internal resistance r, can be calculated by applying Eq.4.89:
$$\Delta V = \mathcal{E} - Ir \quad \cdots (3)$$
or
$$I = \frac{\mathcal{E} - \Delta V}{r} \quad \cdots (4)$$

SOLUTION From Eq.(1), the net internal emf of three batteries would be
$$\mathcal{E}_0 = \frac{3/1 + 2/1 + 1/1}{1/1 + 1/1 + 1/1} = 2 \, \text{V}$$
Given that, $r_1 = r_2 = r_3 = 1 \, \Omega$. Therefore, from Eq.(2), net internal resistance of the equivalent battery
$$\frac{1}{r_0} = \frac{1}{1} + \frac{1}{1} + \frac{1}{1} = 3 \quad \text{or} \quad r_0 = \frac{1}{3}\Omega$$
The equivalent circuit is shown in the Fig.4.103.

Since, equivalent battery is not providing any current, therefore,

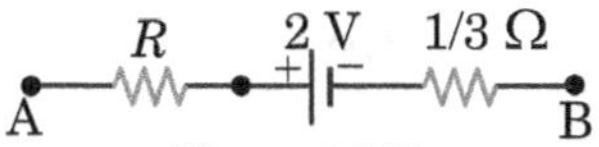

Figure 4.103

this emf would be the terminal voltage across terminals A and B. i.e.,
$$\Delta V = V_A - V_B = 2 \text{ V}$$
Now, from Eq.(4), the current provided by $\mathcal{E}_1$
$$I_1 = \frac{3 \text{ V} - 2 \text{ V}}{1 \, \Omega} = 1 \text{ A}$$
Similarly, the current provided by $\mathcal{E}_2$
$$I_2 = \frac{2 \text{ V} - 2 \text{ V}}{1 \, \Omega} = 0 \text{ A}$$
and the current provided by $\mathcal{E}_3$
$$I_3 = \frac{1 \text{ V} - 2 \text{ V}}{1 \, \Omega} = -1 \text{ A}$$

(b) **APPROACH** Since, r_2 is short circuited therefore the internal resistance of this branch is zero, i.e., $r_2 = 0 \, \Omega$. Now, terminal A is connected to terminal B by a conducting wire. Apply Kirchhoff's voltage law in all three loops and solve for I_1, I_2 and I_3.

SOLUTION Applying KVL in loop containing 3 V battery and

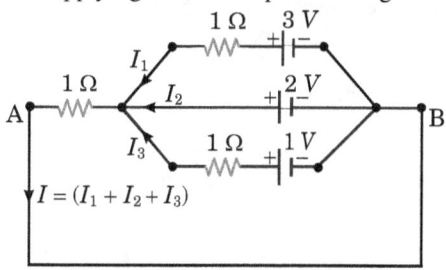

Figure 4.104

1 Ω external resistance.
$$3 - (I_1)(1) - (I_1 + I_2 + I_3)(1) = 0$$
or
$$2I_1 + I_2 + I_3 = 3 \quad \cdots (5)$$
By KVL in loop containing 2 V battery and 1 Ω external resistance.
$$2 - (I_1 + I_2 + I_3)(1) = 0$$
or
$$I_1 + I_2 + I_3 = 2 \text{ A} \quad \cdots (6)$$
and by KVL in loop containing 1 V battery and 1 Ω external resistance.
$$1 - I_3 - (I_1 + I_2 + I_3) = 0$$
or
$$I_1 + I_2 + 2I_3 = 1 \quad \cdots (7)$$
Solving Eq.(6), (7) and (8) for I_1, I_2 and I_3, we get
$$I_1 = 1 \text{ A}, I_2 = 2 \text{ A} \text{ and } I_3 = -1 \text{ A}$$
Here, negative value of I_3 indicates that the actual direction of I_3 is opposite to assumed direction.

EXAMPLE 52. Find the emf and internal resistance of a single battery which is equivalent to a combination of three batteries as shown in Fig.4.105

SOLUTION 10 V and 4 V batteries are connected in parallel

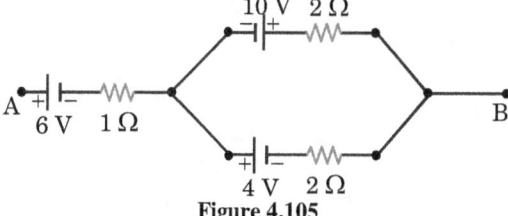

Figure 4.105

with opposite polarities. So, from Eq. 4.119, their equivalent emf,
$$\mathcal{E}_{01} = \frac{\frac{10}{2} + \frac{(-4)}{2}}{\frac{1}{2} + \frac{1}{2}} = \frac{5 - 2}{1} = 3 \text{ V}$$
Net internal resistance of 10 V and 4 V batteries
$$r_{01} = \frac{2 \times 2}{2 + 2} = 1\Omega$$
Now this equivalent is in series with the third battery of 6 V. This circuit is shown in Fig.4.106.

In this case, the emfs are connected in opposite polarities,

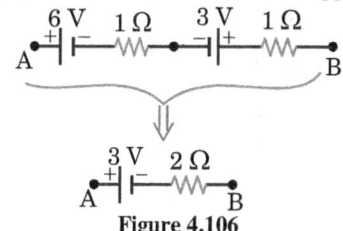

Figure 4.106

therefore their equivalent is given by
$$\mathcal{E}_0 = (6-3) \text{ V} = 3 \text{ V}$$
Series equivalent of internal resistances, is
$$r_0 = (1+1) \, \Omega = 2 \, \Omega$$

4.21.3 Mixed Grouping

Figure 4.107 shows the situation. The equivalent emf and internal resistance of each row are $\mathcal{E}_r = \mathcal{E}_1 + \mathcal{E}_2 + \cdots + \mathcal{E}_n$ and $r_r = r_1 + r_2 + \cdots + r_n$ respectively. There are m such rows in parallel.

Now, from Eq.4.119, we have

$$\mathcal{E}_0 = \frac{\frac{\mathcal{E}_1+\mathcal{E}_2+\cdots+\mathcal{E}_n}{r_1+r_2+\cdots+r_n} + \frac{\mathcal{E}_1+\mathcal{E}_2+\cdots+\mathcal{E}_n}{r_1+r_2+\cdots+r_n} + m \text{ times}}{\frac{1}{r_1+r_2+\cdots+r_n} + \frac{1}{r_1+r_2+\cdots+r_n} + m \text{ times}}$$

or $\quad \mathcal{E}_0 = \dfrac{m\left(\frac{\mathcal{E}_1+\mathcal{E}_2+\cdots+\mathcal{E}_n}{r_1+r_2+\cdots+r_n}\right)}{\frac{m}{r_1+r_2+\cdots+r_n}} = \mathcal{E}_1 + \mathcal{E}_2 + \cdots + \mathcal{E}_n$ (4.124)

and total internal resistance r_0 is given by,
$$\frac{1}{r_0} = \frac{m}{r_1+r_2+\cdots+r_n}$$
$$r_0 = \frac{r_1+r_2+\cdots+r_n}{m} \tag{4.125}$$

Therefore, current I in the external resistance R is given by

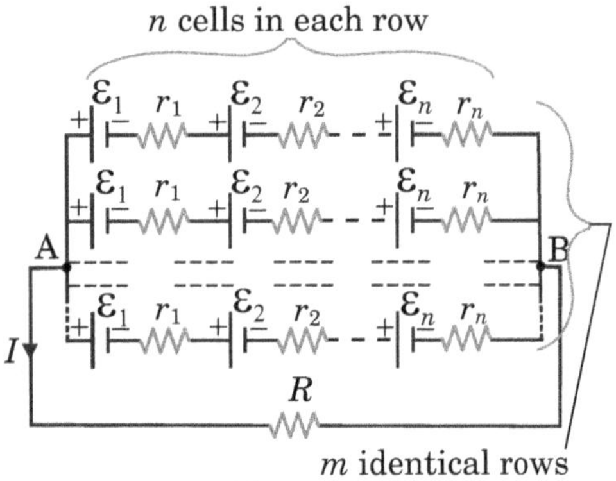

Figure 4.107

$$\boxed{I = \frac{\mathcal{E}_0}{R+r_0} = \frac{\mathcal{E}_1+\mathcal{E}_2+\cdots+\mathcal{E}_n}{R+\frac{r_1+r_2+\cdots+r_n}{m}}} \tag{4.126}$$

If all cells are identical, and suppose,
$$\mathcal{E}_1 = \mathcal{E}_2 = \cdots = \mathcal{E}_n = \mathcal{E} \text{ (say)};$$
$$r_1 = r_2 = \cdots = r_n = r \text{ (say)},$$
then Eq.4.124 gives
$$\mathcal{E}_0 = n\mathcal{E} \tag{4.127}$$
and from Eq.4.125, we get
$$r_0 = \frac{nr}{m} \tag{4.128}$$

In this case, current in external resistance R is

$$\boxed{I = \frac{\mathcal{E}_0}{R+r_0} = \frac{n\mathcal{E}}{R+\frac{nr}{m}} = \frac{mn\mathcal{E}}{mR+nr}} \tag{4.129}$$

This expression can also be written as
$$I = \frac{mn\mathcal{E}}{(\sqrt{mR}-\sqrt{nr})^2 + 2\sqrt{mnrR}}$$

Corresponding to maximum current through R, the denominator of above equation must be minimum. For minimum value of denominator, the square term in it must be zero, i.e.,
$$\sqrt{mR} - \sqrt{nr} = 0 \implies \sqrt{mR} = \sqrt{nr}$$
$$\implies R = \frac{nr}{m} = \text{internal resistance of one row / No.of rows}$$
(4.130)

Therefore,
$$I_{\max} = \frac{m\mathcal{E}}{2r} = \frac{n\mathcal{E}}{2R} \tag{4.131}$$

EXAMPLE 53. m rows each containing n cells in series, are joined in parallel. Maximum current is taken from this combination in a $3 \, \Omega$ resistance. If the total number of cells used is 24 and internal resistance of each cell is $0.5 \, \Omega$, find the value of m and n.

APPROACH From Eq.4.130, the condition of maximum current is
$$R = \frac{nr}{m} \qquad \cdots (1)$$
It is also given that, total number of cell,
$$mn = 24 \qquad \cdots (2)$$
with internal resistance of each cell $r = 0.5 \, \Omega$ and external resistance $R = 3 \, \Omega$
Now, substitute these values in Eq.(1) and then solve Eq. (1) and (2) for m and n.

SOLUTION Substituting the value of R and r in Eq.(1), we get
$$3 = \frac{0.5n}{m} \implies n = \frac{3m}{0.5} = 6m$$
From Eq.(2), $mn = 24$, therefore, $m(6m) = 24$ or $m = 4$
and
$$n = \frac{24}{m} = \frac{24}{2} = 12$$

Important Points
- When cell is charging, the terminal voltage is greater than it's emf $(\mathcal{E})$, i.e., $V = \mathcal{E} + Ir$
- Series combination is useful when internal resistance is less than external resistance of the cell.
- Parallel combination is useful when internal resistance is greater than external resistance of the cell.
- Power in R (external resistance) is maximum, if its value is equal to net resistance of remaining circuit.
- Internal resistance of ideal cell = 0
- if external resistance is zero then current given by circuit is maximum.

4.22 The Wheatstone Bridge

The Wheatstone bridge circuit, shown in Fig. 4.108, is widely used for the rapid and precise measurement of resistance. It was invented in 1843 by the English scientist, Charles Wheatstone. M, N, and R are three adjustable resistors and X represents the unknown resistance. To use the bridge, switches S_1 and S_2 are closed and the resistance of R is adjusted until the galvanometer G shows no deflection. Points B and C must then be at the same potential or, in other words, the potential drop from A to B equals that from A to C. Also, the drop from B to D equals that from C to D. Since the galvanometer current is zero, the current in M equals that in N, say I_1, and the current in R equals that in X, say I_2. Then, since $V_{AB} = V_{AC}$, it follows that
$$I_1 N = I_2 R \tag{4.132}$$
and since $V_{BD} = V_{CD}$,
$$I_1 M = I_2 X \tag{4.133}$$
When Eq.4.133 is divided by Eq.4.132, we find

$$\boxed{\frac{M}{N} = \frac{X}{R}} \quad \text{(When bridge is balanced)} \tag{4.134}$$

or $\boxed{\dfrac{R}{N} = \dfrac{X}{M}}$ (When bridge is balanced) (4.135)

Simplifying Eq.4.134 for X, we get

$$\boxed{X = \frac{M}{N}R} \tag{4.136}$$

4.22. THE WHEATSTONE BRIDGE

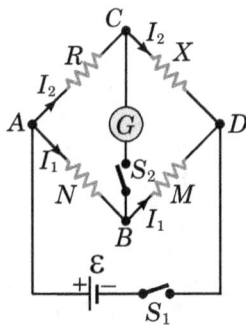

Figure 4.108: Wheatstone bridge circuit.

Hence, if M, N, and R are known, X can be computed. The ratio M/N is usually set at some integral power of 10, such as 0.01, 1, 100, etc., for simplicity in computation.

If any of the resistances are inductive, the potentials V_B and V_C may attain their final values at different rates when S_1 is closed, and the galvanometer, if connected between B and C, would show an initial deflection (due to transient current) even though the bridge were in balance. Hence, S_1 and S_2 are frequently combined in a double key which closes the battery circuit first and the galvanometer circuit a moment later, after the transient currents have died out.

EXAMPLE 54. Determine the value of R in the circuit shown in Fig.4.109, when the current is zero in the branch CD.

APPROACH Since, there is no current in the branch CD of the

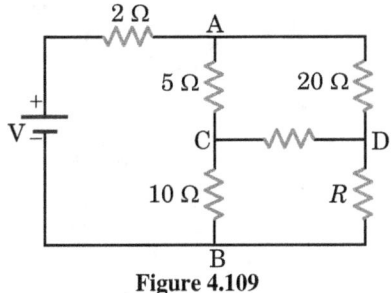

Figure 4.109

circuit, therefore the potential difference across CD is zero. This is what we have discussed in balanced Wheatstone bridge. Apply Eq.4.135 and solve for unknown resistance R.

SOLUTION In this case, from Eq.4.135, we have
$$\frac{20}{5} = \frac{R}{10} \implies R = 40\ \Omega$$

EXAMPLE 55. Find equivalent resistance of the circuit between the terminals A and B.

APPROACH In circuit of Fig.4.110, we see that

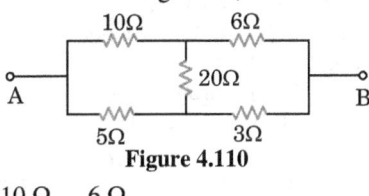

Figure 4.110

$$\frac{10\ \Omega}{5\ \Omega} = \frac{6\ \Omega}{3\ \Omega}$$

Therefore, according to balanced Wheatstone bridge condition (Eq.4.135), we can remove central 20 Ω resistance. The new circuit obtained is shown in Fig.4.111. Here, 10 Ω, 6 Ω resistances are in one series whereas 5 Ω and 3 Ω are in other series. Let series equivalent resistance of 10 Ω, 6 Ω is R_1 and series equivalent of 5 Ω, 3 Ω is R_2, then these equivalents are in parallel. So, first find series equivalents R_1 and R_2 by using Eq.4.63 and then using Eq.4.65, find parallel equivalent of R_1 and R_2.

SOLUTION From Eq.4.63, the series equivalent of 10 Ω and 6 Ω is

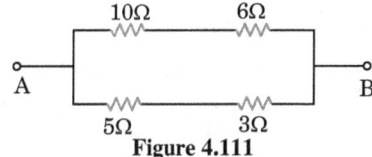

Figure 4.111

$R_1 = 10\ \Omega + 6\ \Omega = 16\ \Omega$

Similarly, the series equivalent of 5 Ω and 3 Ω is
$R_2 = 5\ \Omega + 3\ \Omega = 8\ \Omega$

From Eq.4.65, parallel equivalent resistance of R_1 and R_2 is
$$R_{eq} = \frac{R_1 R_2}{R_1 + R_2} = \frac{16 \times 8}{16 + 8} = \frac{16}{3}\Omega$$

So, equivalent resistance between terminals A and B is $\frac{16}{3}\Omega$.

EXAMPLE 56. Find (a) Equivalent resistance (b) and current in each resistance shown in Fig.4.112.

APPROACH In Fig.4.112, we see that-

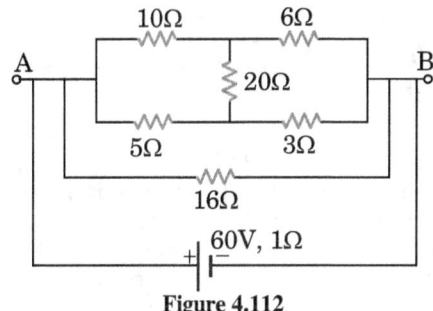

Figure 4.112

$$\frac{10\ \Omega}{5\ \Omega} = \frac{6\ \Omega}{3\ \Omega}$$

Therefore, this part makes a balanced Wheatstone bridge and so terminals of 20 Ω resistor are at same potential, i.e., no current flows through it and hence we can remove it. The next modified circuit diagram is shown in Fig.4.113. This figure also shows arbitrarily assumed currents in all branches. In Example 55, we have already calculated the equivalent resistance of 4 resistance 10 Ω, 6 Ω, 5 Ω and 3Ω. It is $(16/3)\Omega$. Now, this equivalent resistance is in parallel with 16 Ω, and the resultant of these two is in series with internal resistance of battery which is 1 Ω. Now, first find the equivalent resistance and then current provided by battery and then solve for currents in different branches either by using Ohm's law and current division rule or by using Kirchhoff's rules. For time saving, we use, Ohm's law and current division rule.

SOLUTION Equivalent external resistance of the circuit be-

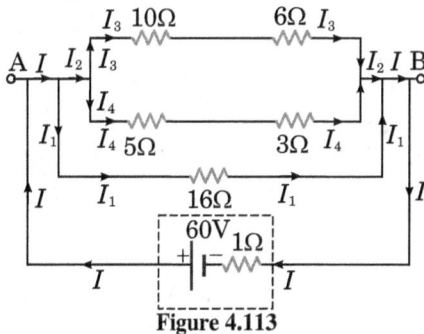

Figure 4.113

tween the terminals A and B, is
$$R_{AB} = \frac{(16/3) \times 16}{16/3 + 16} = 4\ \Omega$$

This resistance is in series with 1 Ω internal resistance of the battery. Therefore, equivalent resistance of the circuit-
$$R_{eq} = 4\ \Omega + 1\ \Omega = 5\ \Omega$$

Therefore by Ohm's law, current supplied by battery

$$I = \frac{60}{5} \text{ A} = 12 \text{ A}$$

Therefore, terminal voltage between A and B, is

$$\Delta V_{AB} = V_A - V_B = \mathcal{E} - Ir = 60 - 12 \times 1 = 48 \text{ V}.$$

By Ohm's law, current passing through 16 Ω resistance,

$$I_1 = \frac{\Delta V_{AB}}{16} = \frac{48}{16} = 3 \text{ A}$$

Remaining current, $I_2 = I - I_1 = 12 \text{ A} - 3 \text{ A} = 9 \text{ A}$

This current is divided into two two parts I_3 and I_4 as shown in Fig.4.113. By current division rule (Eq.4.101), we have

$$I_3 = \frac{(5+3)}{(10+6)+(5+3)} \times 9 \text{ A} = \frac{8}{24} \times 9 \text{ A} = 3 \text{ A}$$

Remaining current, $I_4 = I_2 - I_3 = 9 \text{ A} - 3 \text{ A} = 6 \text{ A}$ passes through series resistances 5 Ω and 3 Ω.

EXAMPLE 57. Find equivalent resistance between terminals A and B in the circuit shown in Fig.4.114.

APPROACH Let us assign potentials V_1 and V_2 to terminals A

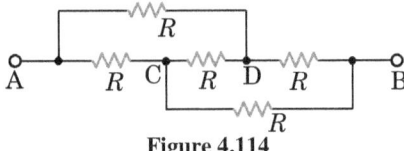

Figure 4.114

and B respectively [Fig.4.115a]. As there is a resistor between points A and D, so there will be a potential difference between A and D and hence we have to assign a new potential value V_3 (say) to point D. Similarly, there is a resistor between points B and C, so there will also be a potential difference between B and C and hence we have to assign a new potential value V_4 (say) to point C. This circuit can be redrawn like Fig.4.115b
Now, from Fig.4.115b, we see that

$$\frac{R_{AD}}{R_{AC}} = \frac{R_{BD}}{R_{BC}}, \quad \text{i.e.,} \quad \frac{R}{R} = \frac{R}{R}$$

So, circuit satisfies the condition of balanced Wheatstone's bridge (see Eq.4.135). In this case, points C and D will be at same potentials and we can remove the resistor connected between points point C and D. So, this circuit can be further simplified like circuit shown in Fig.4.116. Clearly, resistors between points A, D and between D, B are in series. Similarly, resistors between points A, C and between C, B are also in other series. These two series equivalents are in parallel between terminals A and B. Now, first solve for series equivalents and then for final equivalent resistance.

SOLUTION From Eq.4.63, the equivalent resistance of

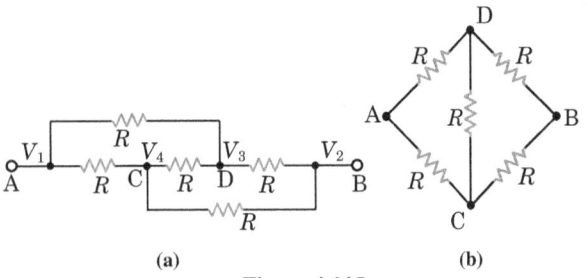

Figure 4.115

resistors connected between points A, D and D, B is

$$R_0 = R_1 + R_2 = R + R = 2R$$

Similarly, the equivalent resistance of resistors connected between points A, C and C, B is

$$R'_0 = R + R = 2R$$

Now, R_0 and R'_0 are in parallel, therefore equivalent resistance between points A and B

$$R_{eq} = \frac{R_0 R'_0}{R_0 + R'_0} = \frac{2R \cdot 2R}{2R + 2R} = R$$

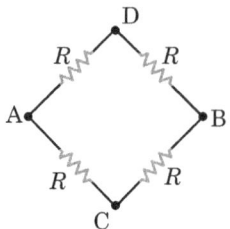

Figure 4.116

EXAMPLE 58. Figure 4.117 shows three resistor configurations (a), (b) and (c) each connected to 3 V battery. If the power dissipated by the configuration (a), (b) and (c) is P_1, P_2 and P_3, respectively, then which of the following option is correct?

(A) $P_1 > P_2 > P_3$ (B) $P_1 > P_3 > P_2$
(C) $P_2 > P_1 > P_3$ (D) $P_3 > P_2 > P_1$

SOLUTION In configuration (a), the resistances form a

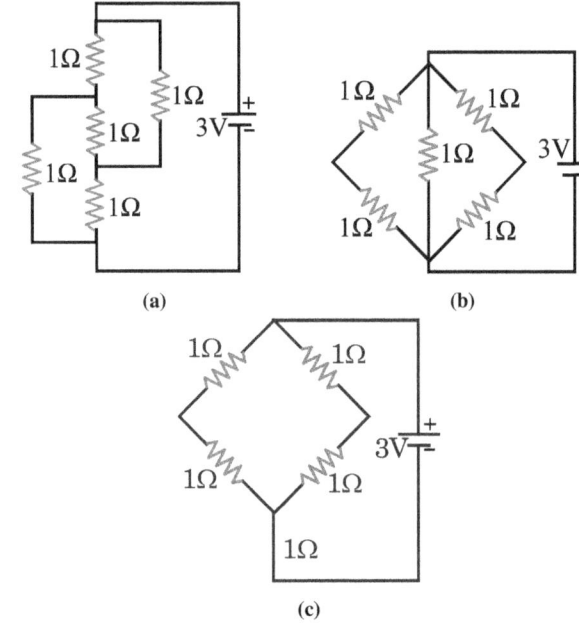

Figure 4.117

balanced Wheatstone bridge. Effective resistance of this configuration is $R_1 = 2\Omega \| 2\Omega = 1\Omega$. The effective resistance of the configuration (b) is $R_2 = (2\Omega \| 2\Omega) \| 1\Omega = 1\Omega \| 1\Omega = 0.5\Omega$. The effective resistance of the configuration (c) is $R_3 = 2\Omega$. Thus, the powers dissipated in the three configurations are

$$P_1 = V^2/R_1 = 9 \text{ W},$$
$$P_2 = V^2/R_2 = 18 \text{ W},$$

and $\quad P_3 = V^2/R_3 = 4.5 \text{ W}$ respectively.

So, the correct option is (C).

EXAMPLE 59. In the circuit shown in Fig.4.118, find the current through the resistor $R(= 2 \text{ Ω})$.

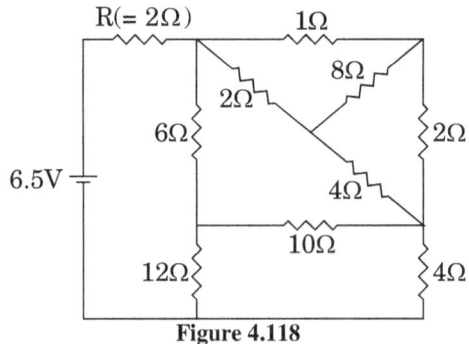

Figure 4.118

SOLUTION The given circuit is redrawn in Fig.4.119a. Con-

4.22. THE WHEATSTONE BRIDGE

sider the resistors that join the nodes B, C, D, and E. These resistors form a balanced Wheatstone bridge between the nodes B and D. Thus, 8Ω resistor in the branch CE can be removed without affecting the circuit (see Fig.4.119b). Effective resistance between the nodes B and D is

$$R_{BD} = (1\Omega + 2\Omega) \| (2\Omega + 4\Omega) = (3\Omega) \| (6\Omega) = 2\Omega$$

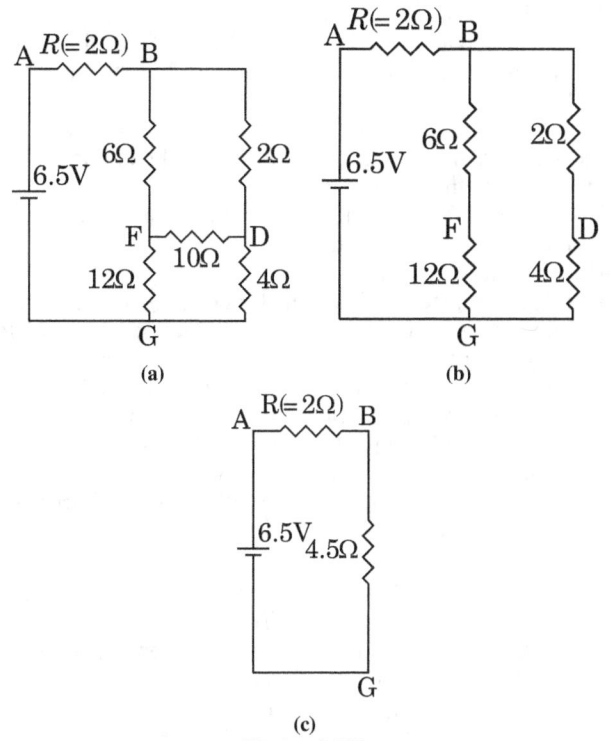

Figure 4.119

The equivalent circuit is shown in Fig.4.120a. In this circuit, consider the resistors which join the nodes B, D, F, and G. These resistors form a balanced Wheatstone bridge between B and G. Thus, 10 Ω resistor between D and F can be removed without affecting the circuit. Therefore, the effective resistance between B and G (see Fig.4.120b), is

$$R_{BG} = (6\Omega + 12\Omega) \| (2\Omega + 4\Omega) = (18\Omega) \| (6\Omega) = 4.5\Omega$$

Thus, the effective resistance of the entire circuit (Figure 4.120c) is

$$R_{\text{eff}} = 2\Omega + 4.5\Omega = 6.5\ \Omega$$

The current through the resistor $R = 2\ \Omega$ is

$$I = V/R_{\text{eff}} = 6.5/6.5 = 1\ \text{A}.$$

EXAMPLE 60. Four resistances of 15 Ω, 12 Ω, 4 Ω and 10 Ω respectively are connected in cyclic order to form Wheatstone's network. What resistance we must connect in parallel with 10 Ω resistance (Fig.4.121) to balance the network.

(A) 10 Ω (B) 5 Ω (C) 15 Ω (D) 20 Ω

SOLUTION Suppose, required resistance in parallel with 10 Ω

Figure 4.121

resistance is R (Fig.4.121).
For a balanced Wheatstone's bridge, we have the relation

$$\frac{15}{10\|R} = \frac{12}{4}$$

where $10\|R$ represents the equivalent Resistance of the parallel combination

$$10\|R = \frac{10R}{10+R}$$

$$\therefore \quad \frac{15}{10\|R} = \frac{15(10+R)}{10R} = \frac{12}{4} = 3$$

$$\Rightarrow \quad 15(10+R) = 30R, \quad \Rightarrow \quad 150 + 15R = 30R$$

$$\Rightarrow \quad 150 = 15R, \quad \Rightarrow \quad R = 10\ \Omega$$

EXAMPLE 61. A battery of internal resistance 4 Ω is connected to the network resistances as shown in the Fig.4.122. For what value of R (in Ω), the battery will deliver the maximum power to the network.

APPROACH Either directly apply the fact given in Eq.4.95 or

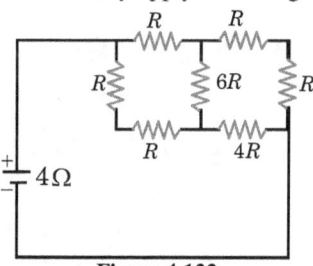

Figure 4.122

apply the concept of maxima and minima for it. In following solution, we will be applying the concept of maxima. However, I advise you to solve it by directly applying Eq.4.95.

SOLUTION Given circuit forms a balanced Wheatstone bridge. Thus, the resistance $6R$ in branch CD can be removed. Equivalent

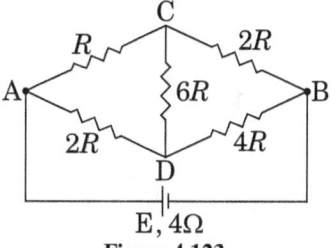

Figure 4.123

resistance between the nodes A and B is

$$R_{eq} = (R + 2R) \| (2R + 4R) = 3R \| 6R = 2R$$

The current supplied by the battery is

$$I = \frac{E}{R_{eq} + r} = \frac{E}{2R + 4}$$

and the power delivered to the network is

$$P = I^2 R_{eq} = \frac{E^2}{2} \frac{R}{(R+2)^2}$$

Corresponding to maximum power transfer, we have

$$\frac{dP}{dR} = \frac{E^2}{2} \frac{(2-R)}{(R+2)^3} = 0,$$

On solving for R, we get

$$R = 2\ \Omega$$

We already studied that the power delivered by the battery is maximum when the load resistance (R_{eq}) is equal to the internal resistance of the battery. Therefore, by applying this fact, you can also find the value of R, directly.

4.23 The meter Bridge

It works on the principle of Wheatstone bridge and is used to find the unknown resistance (X) and its resistivity.

It uses a one metre long wire of uniform cross-sectional area which is made of a material with a high resistivity and low temperature coefficient of resistance (e.g. manganin). The "name meter bridge" is given to it because of the length of wire which is one meter. Fig.4.124a shows a simple form of a Metre Bridge. The wire is stretched along a metre scale between two brass or copper strips. The unknown resistance X and a known resistance R (resistance box) are connected between the gaps CD and CA, respectively. B is the position of Jockey (J) corresponding to zero deflection of galvanometer, i.e., it is the null point of the wire AD. When the sliding contact (jockey J) is at the position B, it divides the wire into two parts of resistance R_{AB} and R_{BD}; these resistances with X and R form a Wheatstone bridge (Fig.4.124b). If $AB = l$ cm, then $BD = (100 - l)$ cm.

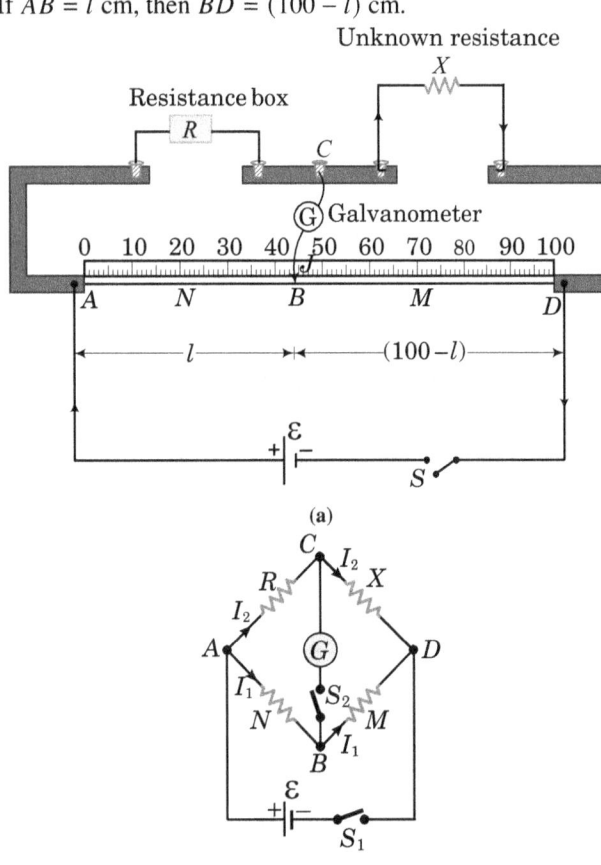

Figure 4.124

Resistance of the wire between A and B, is
$$N = \rho \frac{l}{A}$$
and the resistance of the wire between B and D, is
$$M = \rho \left(\frac{100 - l}{A} \right)$$
here, ρ is the resistivity and A is the cross sectional area of the wire.

Connecting wires and metal strips have negligible resistance.
Applying the condition for balanced Wheatstone bridge (Eq.4.135), we get

$$\frac{R}{N} = \frac{X}{M}$$

or $$X = \frac{M}{N} R = \frac{\left(\rho \frac{(100-l)}{A} \right)}{\rho \frac{l}{A}} R$$

or $$\boxed{X = \left(\frac{100 - l}{l} \right) R} \quad (4.137)$$

Since R and l are known, therefore, the value of X can be calculated from Eq.4.137.

Calculation of Resistivity of X

If L is the length and A_0 is the cross-sectional area of unknown resistive wire of resistance X (calculated from Eq.4.137), then from Eq.4.37, the resistivity ρ_0 of X can be calculated as follows

$$X = \rho_0 \frac{L}{A_0}$$

or $$\rho_0 = \frac{X A_0}{L} \quad (4.138)$$

If r is the radius and D is the diameter of the wire X, then $D = 2r$, and it's cross-sectional area

$$A_0 = \pi r^2 = \frac{\pi D^2}{4}$$

Substituting this value of A_0 in Eq.4.138, we get

$$\rho = \frac{X \pi D^2}{4L} \quad (4.139)$$

So, by knowing X, D and L, we can find resistivity of the given wire by applying Eq.4.139.

Precautions

1. Connections of connecting wires and X should be clean and tight.
2. For batter accuracy, R must be so adjusted that l lies between 40 cm and 60 cm, i.e. the null point must lie between 40 cm and 60 cm.
3. At any position on the wire X, diameter of wire should be measured in two mutually perpendicular directions. D is the average of both readings.
4. The jockey should be moved gently over the bridge wire so that it does not rub the wire.

End Corrections

From Fig.4.124a, we see that in meter bridge, some extra length (over the metallic strips) comes at points A and D. Therefore, we have to include some additional lengths (x_1 and x_2) at the ends A and D respectively. Here, x_1 and x_2 are called the end corrections. So, in this situation, we have to replace length l with $l + x_1$ and also the length $100 - l$ with $100 - l + x_2$.

Now, to calculate x_1 and x_2, take two known resistors R_1 and R_2 in place of R and X and let the new null point length from point A is l_1, then from Eq.4.137, we have

$$\frac{R_1}{R_2} = \frac{l_1 + x_1}{100 - l_1 + x_2} \quad (4.140)$$

Now, interchange the positions of R_1 and R_2 and suppose again the new null point length is l_2. Then,

$$\frac{R_2}{R_1} = \frac{l_2 + x_1}{100 - l_2 + x_2} \quad (4.141)$$

Solving Eqs.4.140 and 4.141 for x_1 and x_2, we get

$$\boxed{x_1 = \frac{R_2 l_1 - R_1 l_2}{R_1 - R_2}} \quad (4.142a)$$

4.23. THE METER BRIDGE

and $$x_2 = \frac{R_1 l_1 - R_2 l_2}{R_1 - R_2} - 100 \quad (4.142b)$$

Key Points

- The balance-point is obtained by trial and error-not by scraping the jockey along the wire.
- The value of R in the resistance box should be chosen so that the balance point comes near to the center of the wire, i.e. from 40 cm to 60 cm from the end A.
- If either length l or $100 - l$ is small, then we cannot neglect the resistance of its end connections at A and D as compared to R_{AB} or R_{BD} respectively. In this case, Eq.4.137 will not be valid.
- The end resistance error can be minimized by interchanging R and X, and balancing again. The average values of l_1 and l_2 are taken to calculate the value of X.
- Since galvanometer is a sensitive instrument, therefore, a high resistance is sometimes connected in series with it until a near balance point is obtained. Then the high resistor is shunted or removed and the final balance point is obtained.
- The lowest resistance that can be measured with this bridge is about 1Ω.

EXAMPLE 62. If resistance R in resistance box is 300 Ω, then the balanced length is found to be 75.0 cm from end A (Fig.4.124a). The diameter of unknown wire is 1 mm and length of the unknown wire is 31.4 cm. Find the specific resistance of the unknown wire.

APPROACH Since, distance of null point from end A is given, therefore, we first determine X by using Eq.4.137:

$$X = \left(\frac{100 - l}{l}\right) R \quad \cdots (1)$$

Now, to find the resistivity of unknown wire, we substitute above value of X and other known values in Eq.4.139:

$$\rho = \frac{X \pi D^2}{4L} \quad \cdots (2)$$

SOLUTION Given: $l = 75.0$ cm, $R = 300$ Ω, $D = 1$ mm, and $L = 31.4$ cm

Substituting the given values in Eq.(1), we get

$$X = \frac{100 - 75}{75}(300) = 100 \text{ Ω}$$

Now, substituting this value of X and other given values in Eq.(2), we get

$$\rho = \frac{X \pi D^2}{4L} = \frac{100 \times 3.14 \times (10^{-3})^2}{4 \times 0.314} = 2.5 \times 10^{-4} \text{Ω.m}$$

EXAMPLE 63. If we use 100 Ω and 200 Ω in place of R and X we get null point deflection, $l = 33$ cm. If we interchange the resistors, the null point length is found to be 67 cm. Find end corrections x_1 and x_2.

SOLUTION End corrections x_1 and x_2 are given by Eq.4.142a and 4.142b respectively. These are

$$x_1 = \frac{R_2 l_1 - R_1 l_2}{R_1 - R_2} = \frac{(200)(33) - (100)(67)}{100 - 200} = 1 \text{ cm}$$

and $$x_2 = \frac{R_1 l_1 - R_2 l_2}{R_1 - R_2} - 100 = \frac{(100)(33) - (200)(67)}{100 - 200} - 100$$
$$= 1 \text{ cm}$$

EXAMPLE 64. In a meter bridge experiment, the value of unknown resistance is 2Ω. To get the balancing point at 40 cm distance from the unknown X side, the resistance in the resistance box will be:

(A) 0.5 Ω   (B) 3 Ω   (C) 20 Ω   (D) 80 Ω

SOLUTION When no current passes through the galvanometer, the bridge is said to be balanced. By changing the known and variable resistances R, this condition can be achieved. Now, from Eq.4.137; we have

$$\frac{X}{R} = \frac{100 - l}{l}$$

Given that $100 - l = 40$ cm, therefore, $l = 60$cm, $X = 2$ Ω, therefore

$$\frac{2}{R} = \frac{40}{60} = \frac{2}{3}$$

or $R = 3$ Ω

Therefore, resistance in the resistance box is 3Ω.
So, the correct choice is B.

4.23.1 Check Point 6

1. ••A battery of six cells each of emf 2 V and internal resistance 0.5Ω is being charged by DC mains of emf 220 V by using an external resistance of 10Ω. What is the potential difference across the battery?
2. ••An electrical circuit is shown in the Fig.4.125. Calculate the potential difference across the resistor of 400 Ω as will be measured by the voltmeter V of resistance 400 Ω either by applying Kirchhoff's rules or otherwise.

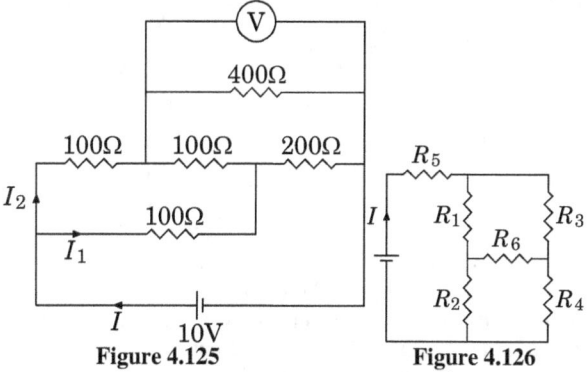

Figure 4.125   Figure 4.126

3. ••In the circuit of Fig.4.126, it is observed that the current I is independent of the value of the resistance R_6. Find the required condition for it.
4. ••If each of the resistances in the network shown in the Fig.4.127 is R, what is the resistance between the terminals A and B?

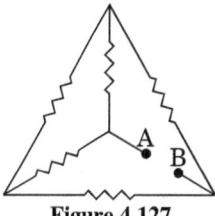

Figure 4.127

5. ••In the circuit shown in Fig.4.128, calculate the heat developed across each resistance in 2 s.

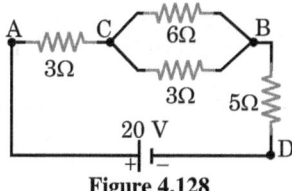

Figure 4.128

6. ••Five equal resistances each of value R are connected to form a network as shown in Fig.4.129. Calculate the equivalent resistance of the network between points B and D.
7. ••Five equal resistances each of value R are again connected to form a network as shown in Figure 4.129. Calculate the equivalent resistance of the network between the points A and C.
8. ••Again in the circuit of Fig.4.129, calculate the equivalent resistance of the network between the points A and B.

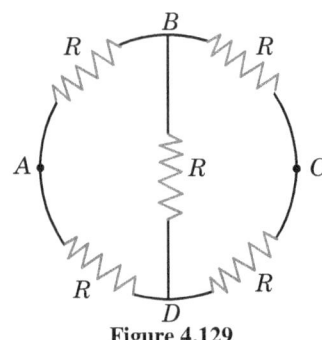

Figure 4.129

9. ●●Calculate currents I_1, I_2, I_3 and I_4 in the circuit of Fig.4.130.

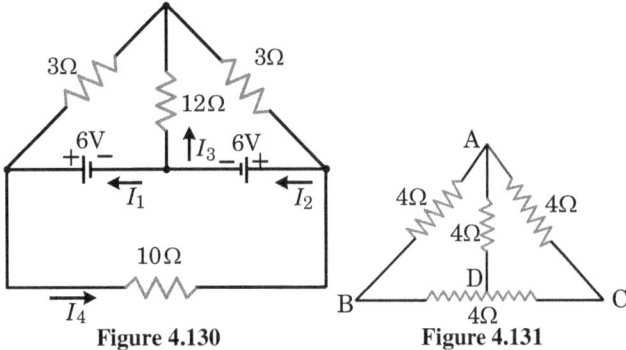

Figure 4.130   Figure 4.131

10. ●●Three resistances of 4 Ω each are connected as shown in Figure 4.131. If the point D divides the resistance into two equal halves, find the resistance between points B and C.

11. ●●In a meter bridge, null point is 20 cm. When the known resistance R is shunted by $10\,\Omega$ resistance, null point is found to be shifted by 10 cm. Find the unknown resistance X.

12. ●●R_1, R_2, R_3 are different values of resistor R and A, B, C are the null points obtained corresponding to R_1, R_2 and R_3, respectively [Fig.4.132]. For which resistor, the value of X will be most accurate and why?

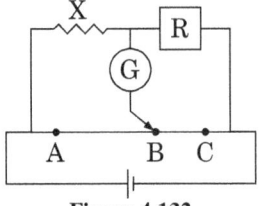

Figure 4.132

13. ●●A metre bridge is set-up as shown in Fig.4.133, to determine an unknown resistance X using a standard 10 Ω resistor. The galvanometer shows null point when tapping-key is at 52 cm mark. The end corrections are 1 cm and 2 cm respectively for the ends A and B. Determine the value of X.

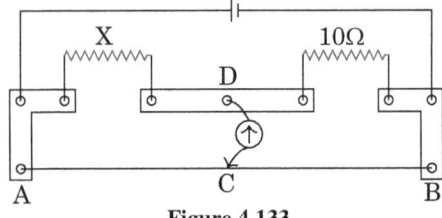

Figure 4.133

14. ●●During an experiment with a metre bridge, the galvanometer shows a null point when the jockey is pressed at 40.0 cm using a standard resistance of 90 Ω, as shown in the Fig.4.134. If the least count of the scale used in the metre bridge is 1 mm, find the unknown resistance R.

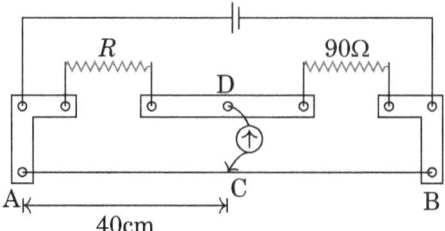

Figure 4.134

15. ●●A resistance of 2 Ω is connected across one gap of a meter-bridge (the length of the wire is 100 cm) and an unknown resistance, greater than 2 Ω, is connected across the other gap. When the resistances are interchanged, the balance point shifts by 20 cm. Neglecting any corrections, find the unknown resistance.

16. ●●In the shown arrangement of the experiment of the meter bridge [Fig.4.135], if AC corresponding to null deflection of galvanometer is x, what should be its value if the radius of the wire AB is doubled?

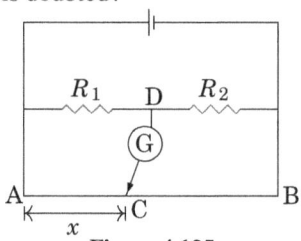

Figure 4.135

Multiple Choice Questions

17. ●●For the resistance network shown in the Fig.4.136, choose the correct option(s)
    (A) The current through PQ is zero.
    (B) $I_1 = 3$ A
    (C) The potential at S is less than that at Q.
    (D) $I_2 = 2$ A

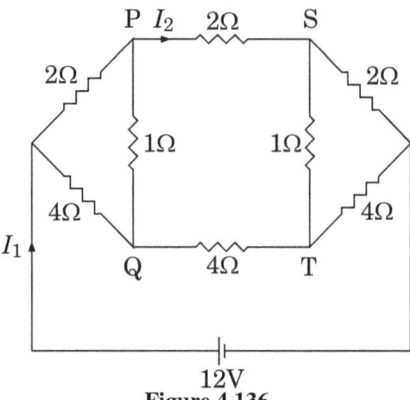

Figure 4.136

18. ●●Two ideal batteries of emf V_1 and V_2 and three resistances R_1, R_2 and R_3 are connected as shown in the Fig.4.137. The current in resistance R_2 would be zero if
    (A) $V_1 = V_2$ and $R_1 = R_2 = R_3$
    (B) $V_1 = V_2$ and $R_1 = 2R_2 = R_3$
    (C) $V_1 = 2V_2$ and $2R_1 = 2R_2 = R_3$
    (D) $2V_1 = V_2$ and $2R_1 = R_2 = R_3$

19. ●●**Statement 1:** In a metre bridge experiment, null point for an unknown resistance is measured. Now, the unknown resistance is put inside an enclosure maintained at a higher temperature. The null point can be obtained at the same point as before by decreasing the value of the standard resistance.

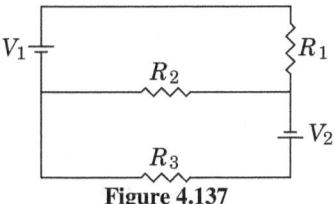

Figure 4.137

Statement 2: Resistance of a metal increase with increase in temperature.
  (A) Statement 1 is true, statement 2 is true; statement 2 is a correct explanation for statement 1.
  (B) Statement 1 is true, statement 2 is true; statement 2 is not a correct explanation for statement 1.
  (C) Statement 1 is true, statement 2 is false.
  (D) Statement 1 is false, statement 2 is true.

20. ••When two resistances X and Y are put in the left hand and right hand gaps in a Wheatstone meter bridge, the null point is at 60 cm. If X is shunted by a resistance equal to half of itself then find the shift in the null point.
   (A) 26.7 cm   (B) 36.7 cm   (C) 46.7 cm   (D) 96.7 cm

4.24 Electrical Meters

4.24.1 The Galvanometer

The galvanometer is the main component in analog meters (like- ammeter and voltmeter) for measuring current and voltage. (Many analog meters are still in use although digital meters, which operate on a different principle, are currently in wide use.) Figure 4.138 illustrates the essential features of a common type called the D'Arsonval galvanometer. It consists of a coil of wire mounted on a cylinder so that it is free to rotate on a pivot in a magnetic field provided by a permanent magnet. The basic operation of the galvanometer uses the fact that a torque acts on a current loop in the presence of a magnetic field*. The torque experienced by the coil is proportional to the current in it: the larger the current, the greater the torque and the more the coil rotates before the spring tightens enough to stop the rotation. Hence, the deflection of a pointer attached to the coil is proportional to the current. Once the instrument is properly calibrated, it can be used in conjunction with other circuit elements to measure either currents or potential differences. The maximum deflection, typically 90° or so, is called full-scale deflection.

If the coil obeys Ohm's law, the current in it is proportional to potential difference. If G is the resistance of galvanometer coil, I_g is the full scale deflection current of the galvanometer, then corresponding potential difference for full scale deflection is given by

$$\boxed{V = I_g G} \qquad (4.143)$$

The best galvanometers can measure currents as small as a pA (10^{-12}A). The deflection of galvanometer coil per unit current is defined as the current sensitivity of galvanometer and is given in units of "division per ampere". Similarly, the deflection of galvanometer coil per unit voltage across it, is defined as the voltage sensitivity of galvanometer and is given in units of "division per volt".

Note: If you try to use the galvanometer to measure a somewhat greater current, the needle will be deflected to its maximum position. Suppose, the galvanometer you are using is designed to operate at currents up to 50 μA and you are trying to measure a current of 60 μA. The needle will deflect so that the meter reads 50 μA. You then know that the current is at least 50μA, but it

---

*Detailed discussion of galvanometer will be given in chapter "Magnetic Effects of Current"

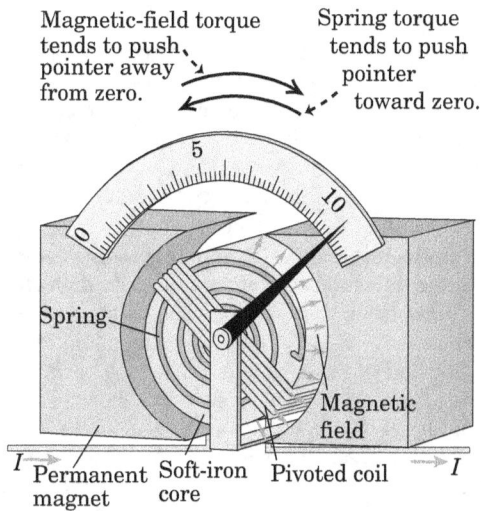

Figure 4.138: Basic parts of a d'Arsonval galvanometer

could be greater.

Ammeters and voltmeters differ in that the ammeter has a low internal resistance and permits the passage of whatever current enters its terminals with little hindrance, whereas the voltmeter has a very large internal resistance and draws only an extremely small current, even when the potential difference applied to its terminals is large.

4.24.2 Ammeter Design

Ammeter is a device that is used to measure an electric current. To directly measure the current passing through a circuit, all the current must go through the ammeter. That is, the ammeter must be put in series with the rest of the circuit (Fig.4.139). Thus we must break the circuit at some point, and take one end of the circuit to one lead of the ammeter, and the other end to the other end of the ammeter. Plus and minus signs are marked near the terminals of the ammeter. The current should enter the ammeter through the terminal marked "plus". When no current passes through the ammeter, the pointer stays at zero which is marked at the left extreme of the scale. To disturb the system's overall resistance as little as possible (here, by minimizing the extra voltage drop due to the ammeter), the overall ammeter resistance R_A must be very small compared to the resistance R of the circuit whose current is to be measured.

To minimize the resistance of ammeter, we connect galvanometer

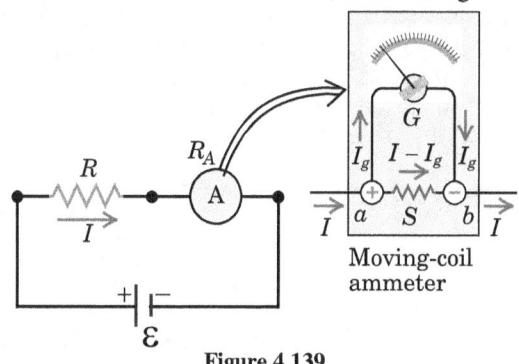

Figure 4.139

in parallel with a small shunt resistor. (Shunt is another word for "parallel.") The shunt resistor is inside the case of the ammeter, so you cannot easily see it. In Figure 4.139, an ammeter is shown in a circuit that consists of a resistor R and a battery. It also shows the schematic detail of the inside of the ammeter. Let current provided by battery to the circuit containing resistor R and ammeter A, is I. The shunt resistor has resistance S. The galvanometer's coil has resistance G. The shunt resistance is generally low compared to the internal resistance of the galvanometer, so only a small current I_g passes through the galvanometer and the pointer is not deflected to its maximum position. By Kirchhoff's junction law at junction a of the moving coil ammeter, the current through the shunt resistor is $I - I_g$. Generally, the galvanometer coil resistance G and I_g are given, and we have to select the shunt resistance for making an ammeter corresponding to given range of the current so that I_g remains unaltered

4.24.2.1 Conversion of Galvanometer to ammeter

Suppose the galvanometer gives full scale deflection when a current I_g flows through it. Now if we want to measure a larger circuit current $I \gg I_g$, we connect a small shunt resistance S in parallel to the coil of the galvanometer, which has a resistance G. The resistance value is so chosen that out of the total current I only I_g flows through the coil and the remaining current flows through S. As potential difference across S = potential difference across G, by Ohm's law, we have
$$(I - I_g)S = I_g G$$

or
$$\boxed{I_g = \left(\frac{S}{G+S}\right)I} \qquad (4.144)$$

From Eq.4.144, we see that $\quad I_g \propto I$

Because, the deflection of the pointer is proportional to the current I_g, so on increasing I the value of I_g and hence the deflection increases. The scale can be graduated to read the value of I directly.

4.24.2.2 Maximum Current Reading of an Ammeter

For it we must know the maximum value of I_g, the maximum current which can pass through the galvanometer. It is also known as full-scale deflection current. The full scale deflection current of galvanometer sets upper limit on the current measured by this ammeter. If we pass current of strength more than I_g through the galvanometer, it may get damaged. This is a standard value for a galvanometer.

As an example, let us suppose that when the current through the ammeter is $I = 2A$, current through the galvanometer is 5 mA and let's suppose that this is the maximum value of I_g for full-scale deflection which can pass through the galvanometer. Therefore, we can say that the maximum current the ammeter can read is 2 A. For any other current less than 2 A, value of current through the galvanometer will be less than 5 mA and reading is possible. But for a current which is greater than 2 A, the current in galvanometer will also be greater than full scale deflection current, so pointer reaches to full scale but can not go beyond it for extra deflection due to extra current. Therefore, for a current larger than than 2 A, deflection is not possible and hence no reading is possible. Now we can say that the range of galvanometer is 0 - 2 A.

☞ From Eq.4.144 we have
$$\boxed{I = I_g\left(\frac{G+S}{S}\right) = I_g\left(\frac{G}{S}+1\right)} \qquad (4.145)$$

From Eq.4.145, we see that, to increase the value of I (i.e., to increase the range of an ammeter for a given G and I_g), we have to decrease the resistance of shunt resistor which is connected in parallel with galvanometer. As shunt resistance decreases, the net resistance of ammeter also decreases.

EXAMPLE 65. A moving coil galvanometer of resistance 100 Ω is used as an ammeter using a resistance 0.1 Ω. The maximum deflection current in the galvanometer is 100 μA. Find the current in the circuit, so that the ammeter shows maximum deflection.

APPROACH A galvanometer of resistance G is converted to an ammeter by connecting a small shunt resistance S in parallel. Between terminals A and B of the circuit (Fig.4.140), Ohm's law

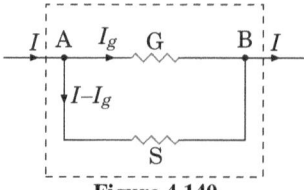

Figure 4.140

gives
$$I_g G = (I - I_g)S$$
$$\implies I = I_g\left(\frac{G+S}{S}\right) \qquad \cdots (1)$$

SOLUTION Given: $G = 100\,\Omega$, $I_g = 100\,\mu A$, $S = 0.1\,\Omega$
Substituting the given values in Eq.(1), we get
$$I = \left(100 \times 10^{-6}\right)\left(\frac{100 + 0.1}{0.1}\right) = 100.1 \text{ mA}$$

4.24.2.3 Modification of Ammeter to Obtain Other Range

As the maximum value of I_g and G are constant for a given galvanometer, by varying the value of shunt resistance S, we can vary the value of I. Therefore by selecting the value of I, we can find the value of shunt resistance. For this shunt value, the range of ammeter can be 0 to I. From above Fig.4.139
$$(I - I_g)S = I_g G$$
Solving for S, we get
$$\boxed{S = \frac{I_g G}{I - I_g}} \qquad (4.146)$$

So, effectively this ammeter will measure current up to I ampere and its effective resistance
$$\boxed{R_A = \left(\frac{GS}{S+G}\right)} \qquad (4.147)$$

In practice G is large as compared to S. Therefore, the effective resistance of the ammeter equals $R_A \approx S$, which is small. An ideal ammeter has zero resistance.

Remarks

1. The reading of an ammeter is always less than the actual current in the circuit. For example in Fig.4.139, in absence of ammeter, the actual current through R is
$$I = \frac{\mathcal{E}}{R} \qquad (4.148)$$
while the current after connecting an ammeter resistance $R_A \left(= \dfrac{GS}{G+S}\right)$ in series with R is,
$$I' = \frac{\mathcal{E}}{R + R_A} \qquad (4.149)$$
On comparing Eq.4.148 and 4.149, we see that $I' < I$. For correct measurement of current, we should have $I' = I$. This is possible when $R_A = 0$. This kind of ammeter is known as an ideal ammeter.

2. **Percentage error** in measuring a current through an ammeter is
$$\left(\frac{I - I'}{I}\right) \times 100 = \left(\frac{\frac{1}{R} - \frac{1}{R+R_A}}{\frac{1}{R}}\right) \times 100 \qquad (4.150)$$

or $\boxed{\text{percentage error} = \left(\frac{R_A}{R + R_A}\right) \times 100\%} \qquad (4.151)$

3. **Current Sensitivity:** We have already seen that on increasing the circuit current, the current through galvanometer and hence the deflection increases proportionally. The current sensitivity (CS) is defined as the ratio of deflection to the galvanometer current producing that deflection i.e. deflec-

tion per unit current.

If corresponding to deflection θ, the galvanometer current is I'_g, then current sensitivity is given by

$$\boxed{CS = \frac{\theta}{I'_g}} \quad (4.152)$$

Here, θ is measured in terms of divisions (for example, 1 division, 2 division, ... etc.). Also note that, if there is a full scale deflection and we denote full scale deflection by ϕ then corresponding galvanometer current I'_g will be equal to I_g. In this case, above equation can also be written as

$$\boxed{CS = \frac{\phi}{I_g}} \quad (4.153)$$

Again, from Eq.4.144, if I is the circuit current, then we have:

$$I_g = \left(\frac{S}{G+S}\right)I$$

Therefore, in terms of circuit current, the current sensitivity can also be written as

$$\boxed{CS = \left(\frac{G+S}{S}\right)\frac{\phi}{I}} \quad (4.154)$$

Now, if we want to find the current sensitivity in absence of shunt, we simply remove shunt from the galvanometer. When shunt is removed, whole current passes through galvanometer, i.e., $i_g = I$. If we denote the current sensitivity, in absence of shunt, by CS_0 then Eq.4.153 gives

$$\boxed{CS_0 = \frac{\phi}{I}} \quad \text{(In absence of shunt)} \quad (4.155)$$

Equation 4.155 can also be obtained by putting $S \to \infty$ in Eq.4.154, because removing shunt is equivalent to put shunt resistor $S \to \infty$. In both cases, whole current passes through the galvanometer and no current passes through shunt. It is given as-

$$\boxed{CS_0 = \lim_{S \to \infty}\left(\frac{G+S}{S}\right)\frac{\phi}{I} = \frac{\phi}{I}} \quad (4.156)$$

From Eq.4.153, 4.155, and 4.154 we see that

$$\boxed{\frac{CS}{CS_0} = \frac{I}{I_g} = \frac{G+S}{S}} \quad (4.157)$$

- In series, removing a resistance means put that resistance equal to zero (because net resistance is the sum of individual resistances).
- In parallel, removing a resistance means put that resistance $\to \infty$.

EXAMPLE 66. A galvanometer having 30 divisions has current sensitivity of 1 division / $20\mu A$. It has a resistance of 25 Ω. How will you convert it into an ammeter measuring upto 1 ampere.

APPROACH From Eq.4.153, the current sensitivity is

$$CS = \frac{\phi}{I_g}$$

So, $\quad I_g = \frac{\phi}{CS} \quad \cdots (1)$

To convert a galvanometer into an ammeter, we need a shunt resistor S, from Eq.4.146, the value of shunt resistor is given by

$$S = \frac{I_g G}{I - I_g} \quad \cdots (2)$$

SOLUTION Given that, $I = 1$ A, $CS = 1$ division / $20\mu A$, $\phi = 30$ divisions and $G = 25$ Ω

Now, from Eq.(1), the current required for full scale deflection, is

$$I_g = \frac{\phi}{CS} = \frac{30}{1 \text{ division}/(20\mu A)} = 600 \mu A = 6 \times 10^{-4} A$$

From Eq.(2), to convert galvanometer into ammeter upto range 1 A, the required shunt resistance S, is given by

$$S = \frac{I_g G}{I - I_g} = \frac{6 \times 10^{-4} A \times 25 \Omega}{\left(1 - (6 \times 10^{-4})\right) A} = 0.015 \Omega$$

EXAMPLE 67. The ammeter shown in Fig.4.141 consists of a 480 Ω coil connected in parallel to a 20 Ω shunt. Find the reading of the ammeter.

APPROACH First of all find the equivalent resistance of the

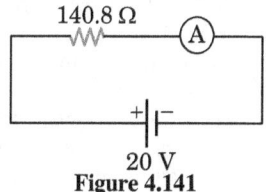

Figure 4.141

ammeter. This ammeter is in series with external resistance, so find the equivalent resistance of the circuit. Now, by applying either kirchhoff's voltage law or Ohm's law, we can find the current in the circuit. As this current passes through the ammeter, therefore this is the reading of the ammeter.

SOLUTION Equivalent resistance of ammeter (R_A) is given by

$$R_A = \frac{(480\Omega)(20\Omega)}{480\Omega + 20\Omega} = 19 \cdot 2 \Omega$$

As R_A is in series with external resistance 140.8 Ω, therefore the equivalent resistance of the circuit is

$$R = 140 \cdot 8 \Omega + 19 \cdot 2 \Omega = 160 \Omega$$

Now, by Ohm's law, the circuit current $I = \frac{20 \text{ V}}{160 \Omega} = 0 \cdot 125 A$.

As this current also passes through the ammeter and hence the reading of the ammeter is $0.125 A$.

EXAMPLE 68. What is the value of shunt which passes 10% of the main current through a galvanometer of 99 Ω ?

APPROACH The shunt resistance can be calculated from Eq.4.146:

$$S = \frac{I_g G}{I - I_g} \quad \cdots (1)$$

Here, letters have their usual meanings.

SOLUTION Given that, $G = 99$ Ω, $I_g = 10\%$ of $I = I/10 = 0.1I$.

Substituting these values in Eq.(1), we get

$$S = \frac{0.1I \times 99}{(I - 0.1I)} = \frac{0.1}{0.9} \times 99 = 11\Omega$$

EXAMPLE 69. The deflection in a moving coil galvanometer falls from 50 divisions to 10 divisions when a shunt of 12 Ω is applied. What is the resistance of the galvanometer?

APPROACH From Eq.4.154, the current sensitivity of a moving coil galvanometer is given by-

$$CS = \left(\frac{G+S}{S}\right)\frac{\phi}{I} \quad \cdots (1)$$

It is given that on putting a shunt $S = 12$ Ω, CS of the galvanometer drops from 50 division/ampere to 10 division/ampere, i.e., corresponding to no shunt $S = \infty$, $CS = 50$ division/ampere and for shunt $S = 12$ Ω, $CS = 10$ division/ampere. Substitute these values in above Eq.(1) and simplify for G.

SOLUTION For $S = \infty$, Eq.(1) gives

$$50 = \lim_{S \to \infty}\left(\frac{G+S}{S}\right)\frac{\phi}{I} = \frac{\phi}{I} \quad \cdots (2)$$

For $S = 12$ Ω, Eq.(1) gives-

$$10 = \left(\frac{G+12}{12}\right)\frac{\phi}{I} \quad \cdots (3)$$

Dividing Eq.(3) by (2), we get

$$\frac{1}{5} = \frac{12}{G+12}$$

or $\quad G + 12 = 5 \times 12 \quad \Rightarrow \quad G = (60-12)\Omega = 48\ \Omega$

4.24.3 Voltmeter Design

A voltmeter (or milli-voltmeter depending on its range) is an electrical device which is used to measures voltage. It always measures voltage (or potential difference) across its terminals. A voltmeter is placed in parallel with a circuit element in order to measure the potential difference across it. For example, in order to measure the potential difference between terminals A and B, i.e., across resistance R (Fig.4.142), we must connect the terminal a of voltmeter to the point A and terminal b of voltmeter to B. Plus and minus signs are marked on the terminals of voltmeter. The terminal marked "plus" should be connected to the point at higher potential. Because we don't want to disturb the circuit by deflecting much current through the voltmeter, so it is always desirable for a voltmeter to have as high resistance as possible. An ideal voltmeter should have infinite resistance so that the current through it is zero.

A voltmeter is constructed by placing a galvanometer in series

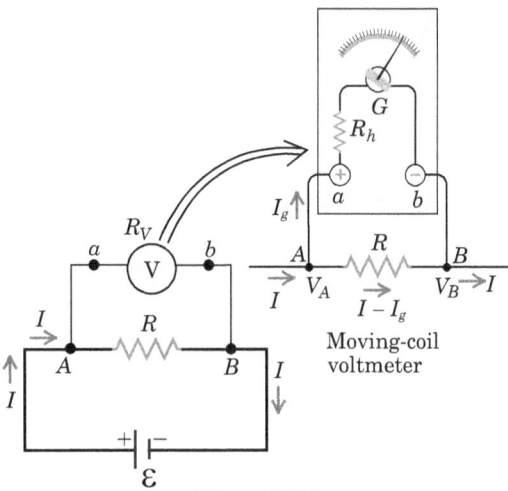

Figure 4.142

with a large resistor R_h (Fig. 4.142). The galvanometer measures the current through R_h and G (G is the resistance of galvanometer coil).
So, the effective resistance of voltmeter R_V, is

$$R_V = (G + R_h) \quad (4.158)$$

In practice R_V is very large as compared to G.
If there is a potential difference between terminals of voltmeter, a current I'_g passes through R_h and G. The voltage drop across it is,

$$V = I'_g (R_h + G) \quad (4.159)$$

This voltage drop is equal to the voltage across the circuit element-in this case, the resistance R (and emf).
Again, from Eq.4.159, the current in the coil is

$$I'_g = \frac{V}{G + R_h} \quad (4.160)$$

The deflection is proportional to the current I'_g and hence to V. The scale on the face of the galvanometer is calibrated to measure voltage directly.
Consider the diagram shown in Fig.4.142. Suppose the galvanometer gives full scale deflection when a current I_g passes through its coil. Then maximum possible potential difference across the voltmeter

$$\boxed{V = I_g(G + R_h) \quad \Rightarrow \quad R_h = \frac{V}{I_g} - G} \quad (4.161)$$

Eq.4.161 gives the magnitude of external resistance to be connected in series to convert the galvanometer into a voltmeter.
- From Eq.4.161, we see that (for a given G and I_g), to increase V (i.e., the range of voltmeter), we have to increase R_h.

Remarks

1. The reading of a voltmeter is always less than the actual voltage in the circuit.
For example, if there is a current I, through a resister R, the actual potential difference across R is

$$V = IR \quad (4.162)$$

Now, if a voltmeter of resistance R_V ($= G + R_h$) is connected in parallel across the resistance R, the new value will be

$$V' = I \times \frac{RR_V}{R + R_V} \quad \text{or} \quad V' = \frac{IR}{1 + \frac{R}{R_V}} \quad (4.163)$$

On comparing Eq.4.162 and 4.163, we get

$$V' = \frac{V}{1 + \frac{R}{R_V}} \quad (4.164)$$

From Eq.4.164, we see that, $V' < V$
If $R_V \to \infty$, then $V' \to V$

2. **Percentage Error in Measurement of Potential Difference**
If actual voltage across resistance R is V and in presence of voltmeter, the measured voltage is V', then percentage error in measuring the potential difference by a voltmeter is,

$$\left(\frac{V - V'}{V}\right) \times 100 = \left(\frac{1}{1 + \frac{R_V}{R}}\right) \times 100\%$$

or $\quad \boxed{\% \text{ error} = \left(\frac{1}{1 + \frac{R_V}{R}}\right) \times 100\%} \quad (4.165)$

3. **Voltage Sensitivity:** The ratio of pointer deflection to the voltage across the galvanometer, i.e. deflection per unit voltage is called voltage sensitivity (VS) of the galvanometer.

$$\boxed{VS = \frac{\theta}{V}} \quad (4.166)$$

If corresponding to deflection θ, the current through galvanometer is I'_g, then $V = I'_g R_V = I'_g(R_h + G)$, so from Eq.4.166, we have

$$\boxed{VS = \frac{\theta}{I'_g R_V} = \frac{CS}{R_V} = \frac{CS}{R_h + G}} \quad (4.167)$$

For full scale deflection ϕ, the current I'_g becomes I_g, in this case Eq.4.167 becomes

$$\boxed{VS = \frac{\phi}{I_g R_V} = \frac{CS}{R_V} = \frac{CS}{R_h + G}} \quad (4.168)$$

EXAMPLE 70. Find potential difference across the resistance 300 Ω in circuits of Fig.4.143a and 4.143b.
SOLUTION In Fig.4.143a, the net resistance of the circuit

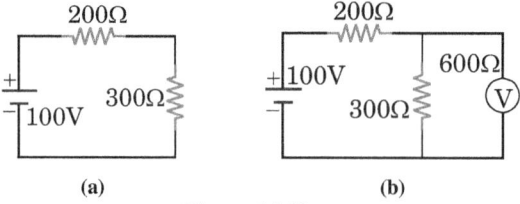

Figure 4.143

$$R_{eq} = 200\ \Omega + 300\ \Omega = 500\ \Omega$$

Therefore, by Ohm's law, current in the circuit-

$$I = \frac{V}{R_{eq}} = \frac{100}{500} = \frac{1}{5}\ A$$

So, potential difference across 300 Ω resistance
$$\Delta V = IR = \frac{1}{5} \times 300 = 60 \text{ V}$$
In circuit of Fig.4.143b, a voltmeter of 600 Ω is also connected in parallel with 300 Ω resistance, therefore net resistance of the circuit
$$R'_{eq} = 200 + \frac{300 \times 600}{300 + 600} = 200 + \frac{180000}{900}$$
$$= 200 + 200 = 400 \text{ Ω}$$
Therefore, current in the circuit
$$I = \frac{V}{R'_{eq}} = \frac{100}{400} = \frac{1}{4} \text{ A}$$
Therefore, potential drop across parallel combination of 300 Ω and voltmeter of resistance 600 Ω is
$$\Delta V = \frac{1}{4} \times \frac{300 \times 600}{300 + 600} = \frac{1}{4} \times 200 = 50 \text{ V}$$
We see that by connecting voltmeter the voltage which was to be measured has changed. Such voltmeters are not good. If its resistance had been very larger than 300 Ω then it would have not affected the voltage by much amount.

EXAMPLE 71. A galvanometer with coil resistance $G = 20$ Ω and full scale deflection current $I_g = 1$ mA has to measure a maximum voltage of 10 V. Find the value of external resistance required to be connected in series. Also, determine the resistance of the voltmeter.

APPROACH To convert a galvanometer into voltmeter, a high resistance in series with it is required. It's magnitude is given by Eq.4.161:
$$R_h = \frac{V}{I_g} - G \qquad \cdots (1)$$
and resistance of voltmeter is given by Eq.4.158:
$$R_V = (G + R_h) \qquad \cdots (2)$$
SOLUTION Given that, $G = 20$ Ω, $V = 10$ volt, $I_g = 1$ mA $= 10^{-3}$ A
Substituting these values in Eq.(1), we get
$$R_h = \frac{V}{I_g} - G = \frac{10}{10^{-3}} - 20 = 9980 \text{ Ω}$$
From Eq.(2), voltmeter resistance is given by
$$R_V = (G + R_h) = 20 + 9980 = 10000 \text{ Ω}$$
Note that, the resistance of the galvanometer is insignificant as compared to the equivalent resistance of the voltmeter.

EXAMPLE 72. A galvanometer has a resistance of 50 Ω and its full scale deflection current is 50 μA. What resistance should be added to it so that it can have a range of 0 - 5V ?

APPROACH In this problem, we want to convert a galvanometer into a voltmeter. For conversion to a voltmeter we add a resistor of high resistance R_h in series to the galvanometer. The required value of R_h is given by Eq.4.161:
$$R = \frac{V}{I_g} - G \qquad \cdots (1)$$
SOLUTION Given: full scale deflection current, i.e.,
$$I_g = 50 \mu A = 5.0 \times 10^{-5} \text{A}$$
The upper limit of voltage, i.e., the maximum voltage to be measured, $V = 5$ volt.
The galvanometer resistance, $G = 50$ Ω.
Substituting these values in Eq.(1), we get
$$R_h = \frac{V}{I_g} - G = \frac{5}{5.0 \times 10^{-5}} - 50 = 99950 \text{ Ω}$$
We see that higher the range of voltmeter, higher is the value of series resistance.

EXAMPLE 73. A galvanometer having 30 divisions has current sensitivity of 1 division / 20μA. It has a resistance of 25 Ω. How will you convert this ammeter into a voltmeter upto 1 volt.
APPROACH From Eq.4.153, the current sensitivity is
$$CS = \frac{\phi}{I_g}$$
So,
$$I_g = \frac{\phi}{CS} \qquad \cdots (1)$$
To convert a galvanometer into voltmeter, a high resistance in series with it is required. It's magnitude is given by Eq.4.161:
$$R_h = \frac{V}{I_g} - G \qquad \cdots (2)$$
SOLUTION Given that, $V = 1$ volt, $CS = 1$ division / 20μA, $\phi = 30$ divisions and $G = 25$ Ω.
Now, from Eq.(1), the current required for full scale deflection, is
$$I_g = \frac{\phi}{CS} = \frac{30}{1 \text{ division}/(20 \mu A)} = 600 \mu A = 6 \times 10^{-4} A$$
Therefore, from Eq.(2), to convert galvanometer into a voltmeter of maximum range of 1 volt, required series resistance,
$$R_h = \frac{V}{I_g} - G = \frac{1 \text{ volt}}{6 \times 10^{-4} \text{ A}} - 25 \text{ Ω}$$
$$= (1666.67 - 25) \text{ Ω} = 1641.67 \text{ Ω}$$

EXAMPLE 74. A voltmeter of resistance 995 Ω and an ammeter of resistance 10 Ω is connected as shown in Fig.4.144 to calculate the unknown resistance R which is connected to the ideal battery. Voltmeter reading is 99.5 volts. The value of resistance R is calculated as $\frac{\text{voltmeter reading}}{\text{ammeter reading}}$ by a student.
(a) Find his answer.
(b) Also find the actual value of resistance.
SOLUTION (a) Voltage across voltmeter = 99.5 V, therefore,

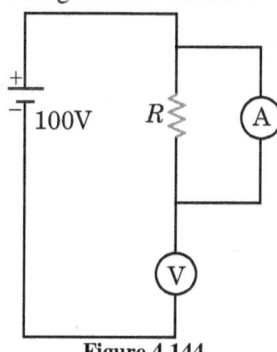

Figure 4.144

Voltage across ammeter = 100 − 99.5 = 0.5 V [Fig.4.144].
Given that, resistance of voltmeter $R_V = 995$ Ω, and ammeter resistance $R_A = 10$ Ω.
Therefore, ammeter reading $= \frac{0.5}{R_A} = \frac{0.5}{10} = 0.05$ A
So, answer of student
$$R = \frac{\text{voltmeter reading}}{\text{ammeter reading}} = \frac{99.5}{0.05} = 1990 \text{ Ω}$$
(b) Current through voltmeter is the circuit current supplied by battery [Fig.4.144]. It is given by
$$I = \frac{\text{voltage across voltmeter}}{\text{resistance of voltmeter}} = \frac{99.5}{995} = 0.1 \text{ A}$$
Current through ammeter $I_A = 0.05$ A
Therefore, current through resistance R is given by
$$I_R = I - I_A = 0.1 - 0.05 = 0.05 \text{ A}$$
and voltage across resistance, R is given by
$$V_R = 0.5 \text{ V}$$
Therefore, $R = \frac{V_R}{I_R} = \frac{0.5}{0.05} = 10 \text{ Ω}$

EXAMPLE 75. A galvanometer has a resistance of 30 Ω and a current of 2 mA is needed for a given full scale deflection. What is the resistance and how is it to be connected to convert the galvanometer (a) into an ammeter of 0.3 A range (b) into a voltmeter of 0.2 V range?

APPROACH (a) Since, full scale current i.e. i_g, galvanometer resistance G and current ranges are given, therefore to get required shunt resistance S, you can apply Eq.4.146:
$$S = \frac{I_g G}{I - I_g} \qquad \cdots (1)$$
Substitute the given values and solve for S.
(b) In this case, apply Eq.4.161:
$$R_h = \frac{V}{I_g} - G \qquad \cdots (2)$$
SOLUTION (a) Given: $G = 30\,\Omega$, $I_g = 2$ mA $= 2.0 \times 10^{-3}$A, $I = 0.3$ A.
Substituting these values in Eq.(1), we get
$$S = \frac{2.0 \times 10^{-3} \times 30}{0.3 - 2.0 \times 10^{-3}} = \frac{0.002 \times 30}{0.3 - 0.002} = 0.2013\,\Omega$$
Thus, to convert a galvanometer into an ammeter of range 0.3 A, a shunt resistance (a resistance in parallel) of 70 Ω should be connected with the galvanometer.
(b) Given: $G = 30\,\Omega$, $I_g = 2$ mA $= 2.0 \times 10^{-3}$A, $V = 0.2$ V.
Substituting these values in Eq.(2), we get
$$R_h = \frac{V}{I_g} - G = \frac{0.2}{0.002} - 30 = \frac{200}{2} - 30 = 70\,\Omega$$
Thus, to convert a galvanometer into a voltmeter of range 0.2 V, a resistance of 70 Ω should be connected in series with the galvanometer.

EXAMPLE 76. The scale of a galvanometer is divided into 150 equal divisions. The galvanometer has current sensitivity of 10 divisions per mA and a voltage sensitivity of 2 divisions per mV. How can the galvanometer be designed to read (a) 6 A/div and (b) 1 V/div?
APPROACH (a) Current sensitivity is given by Eq.4.153:
$$CS = \frac{\phi}{I_g} \qquad \cdots (2)$$
and voltage sensitivity is given by Eq.4.168:
$$VS = \frac{\phi}{I_g G} = \frac{CS}{G} \qquad \cdots (2)$$
or
$$G = \frac{CS}{VS} \qquad \cdots (3)$$
To find I_g apply Eq.(1) and to get the resistance of galvanometer, apply Eq.(3).
To convert a galvanometer into an ammeter of range 0 to I, we need a shunt resistance given by Eq.4.146:
$$S = \frac{I_g G}{I - I_g} \qquad \cdots (4)$$
and to convert a galvanometer to a voltmeter of range 0 to V volt, we need a series resistance R_h given by Eq.4.161
$$R_h = \frac{V}{I_g} - G \qquad \cdots (5)$$
SOLUTION Given that, $CS = 10$ div/mA $= 10^4$div/A, $VS = 2$ div/mV $= 2 \times 10^3$div/volt, $\phi = 150$ div, therefore from Eq.(1), we get
$$I_g = \frac{\phi}{CS} = \frac{150\text{ div}}{10^4\text{div/A}} = 0.015\text{ A}$$
and from (3), we have-
$$G = \frac{CS}{VS} = \frac{10^4\text{div/A}}{2 \times 10^3\text{div/volt}} = \frac{10000\text{ div/A}}{2 \times 10^3\text{div/volt}} = 5.0\,\Omega$$
(a) corresponding to 6A/div, current range is given by
$$I = \text{current per division} \times \text{total number of divisions}$$
$$= 6 \times 150\text{ A}$$
Now, from Eq.(4), shunt resistance is given by
$$S = \frac{0.015 \times 5.0}{6 \times 150 - 0.015} = \frac{0.015 \times 5.0}{900 - 0.015} = 8.3 \times 10^{-5}\,\Omega$$
Thus, to convert a galvanometer into an ammeter of range 6×150 A, a shunt resistance (a resistance in parallel) of 8.3×10^{-5} Ω should be connected with the galvanometer.
(b) In this part, we want to read voltage 1 V/div, i.e., voltage range is $V = 1 \times 150$ volt
So, from Eq.(5), the required series resistance is given by
$$R_h = \frac{V}{I_g} - G = \frac{150}{0.015} - 5.0 = 9995\,\Omega$$
Thus, to convert a galvanometer into a voltmeter of range 150 V, a resistance of 9995 Ω should be connected in series with the galvanometer.

EXAMPLE 77. To verify Ohm's law, a student is provided with a test resister R_T, a high resistance R_1, and a small resistance R_2, two identical galvanometers G_1 and G_2, and a variable voltage source V. Which on of the circuit diagram of Fig.4.145 is a correct circuit to carry out the experiment?
APPROACH To verify Ohm's law, we have to measure the volt-

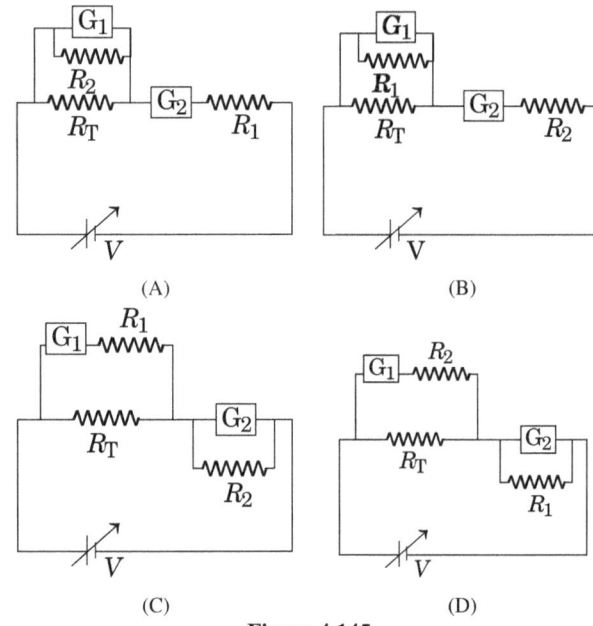

Figure 4.145

age across the test resistance R_T and current passing through it. The voltage can be measured by connecting a high resistance R_1 in series with galvanometer. This combination becomes a voltmeter and should be connected in parallel to R_T. The current can be measured by connecting a low resistance R_2 (shunt) in parallel with galvanometer. This combination becomes an ammeter and should be connected in series to measure the current through R_T.
ANSWER The correct circuit diagram is shown in choice (C).

4.24.4 Check Point 7

1. Calculate the required shunt resistance to convert 1.00 mA, 20 Ω galvanometer into an ammeter with a range of 0 to 50.0 mA ?
2. How can we convert a galvanometer with $G = 20\,\Omega$ and $I_g = 1.0$ mA into a voltmeter with a maximum range of 10 V?
3. Resistance of a milliammeter is R_1, of an ammeter is R_2, of a voltmeter is R_3 and of a kilovoltmeter is R_4. Find the correct order of R_1, R_2, R_3 and R_4.
4. A galvanometer gives full scale deflection with 0.006 A current. By connecting it to a 4990 Ω resistance, it can be converted into a voltmeter of range 0 − 30V. If connected to a $\frac{2n}{249}$ Ω resistance, it becomes an ammeter of range 0 − 1.5 A. Find the value of n.

Multiple Choice Questions

5. On using a shunt, the sensitivity of a galvanometer of resistance 406 ohm, becomes 30 times. The resistance of shunt is
   (A) 88 Ω (B) 14 Ω (C) 6 Ω (D) 16 Ω
6. The sensitivity of a galvanometer of resistance 8722 Ω is increased 90 times. The shunt used is
   (A) 88 Ω (B) 90 Ω (C) 94 Ω (D) 98 Ω

4.25 Stretched-Wire Potentiometer

In Section 4.13.1, we studied that a real battery always has some internal resistance. So, if we want to measure the emf of a real battery, we cannot use a voltmeter. The ordinary moving-coil voltmeter is fairly satisfactory if its resistance is high. An ideal voltmeter which does not change the original potential difference, should have infinite resistance. But in practice, resistance cannot be made infinite. The potentiometer is a device which measures the potential difference between two points of a circuit without drawing any current from it. Thus, it is equivalent to an ideal voltmeter and it can be used to measure the emf of a real battery. Essentially, it balances an unknown potential difference against an adjustable, measurable potential difference.

Construction: Figure 4.146 shows a stretched-wire potentiometer. In it a long manganin or german silver wire AB (usually 5 to 10 metres long), having a uniform cross-sectional area, is stretched between two points A and B on a wooden platform. Usually, separate pieces of wire, each 1 m long, are fixed parallel to each other on the platform. The wires are connected to each other by thick copper strips at left and right sides so that the combination acts as a single wire of desired length. The ends A and B are connected to a driving circuit consisting of a strong battery, a plug key and a rheostat. The driving circuit sends a constant current I through the wire AB. Thus, the potential gradually decreases from A to B. One end of a galvanometer is connected to a metal rod fixed on the wooden platform. A "jockey" can move on this metal rod and may touch the wire AB at any desired point. In this way the galvanometer gets connected to the point of AB which is touched by the jockey. The length of the wire between the end A and this point can be measured with the help of a metre scale fixed on the platform. The other end C of the galvanometer and the high-potential end A of the wire, form the two end points (terminals) of the potentiometer. These points are connected to the points between which the potential difference is to be measured.

Theory: Consider a circuit shown in Fig.4.147. The potential

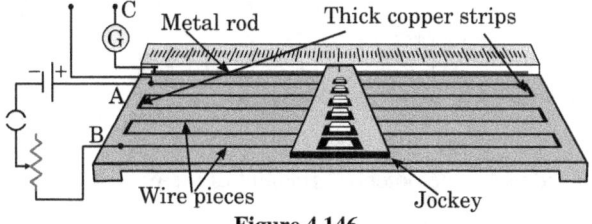

Figure 4.146

of point a is higher than b. Suppose, we want to measure the potential difference between the points a and b. The end A of the wire AB is connected to the point a and the end C of the galvanometer is connected to the point b.

The connecting wire Aa has a negligible resistance and hence potentials of A and a are equal. Suppose, the potential drop across ab is smaller than the potential drop across AB. Then there will be a point P on AB which will have the same potential as b. If the jockey J is moved to touch the wire at this point P, the potential difference across the galvanometer is zero and there will be no current through it. The process of measurement is to search for a point P so that there is no deflection in the galvanometer. If L is the length of wire AB, ρ_0 it's resistivity and A it's cross-sectional area (assumed uniform throughout it's length), then it's resistance R_0 is given by

$$R_0 = \rho_0 \frac{L}{A}$$

If potential difference applied by driving circuit, across AB is V_0, then

$$V_0 = IR_0 = I\rho_0 \frac{L}{A}$$

Therefore, potential gradient (x) across AB, is given by

$$\boxed{x = \frac{V_0}{L} = \frac{\rho_0 I}{A} = \text{constant}} \quad (4.169)$$

So, for any piece of the wire, the potential difference across it is proportional to it's length.

So, if $AP = l$, the potential difference between the points A and P is

$V = $ length $AP \times$ potential gradient, i.e.,

$$\boxed{V = lx = V_0 \frac{l}{L}} \quad (4.170)$$

This is equal to the potential difference between a and b which we had to measure. In order to get the value of the potential difference

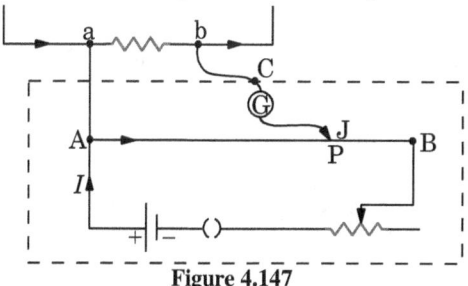

Figure 4.147

V, the total potential drop V_0 across AB must be known. One way to do this is to use a standard cell having known and constant emf in place of ab. If the emf of the standard cell is $\mathcal{E}$ and the potentiometer is balanced (no deflection in the galvanometer) when $AP = l_0$, then from Eq.4.170, we have

$$\mathcal{E} = V_0 \frac{l_0}{L} \quad (4.171)$$

Simplifying for V_0, we get

$$\boxed{V_0 = \frac{L}{l_0}\mathcal{E}} \quad (4.172)$$

Therefore, from Eq.4.170, the potential difference V between a and b is,

$$\boxed{V = \left(\frac{L}{l_0}\mathcal{E}\right)\frac{l}{L} = \frac{l}{l_0}\mathcal{E}} \quad (4.173)$$

This process of finding V_0 is called calibration of the potentiometer. Note that there is no current through the standard cell when the potentiometer is balanced during its calibration. Thus, the emf $\mathcal{E}$ equals the potential difference between its terminals.

EXAMPLE 78. A potentiometer wire of length 100 cm has a total resistance of 10 Ω. It is connected in series with a resistance R and a cell of emf 2 volts and of negligible internal resistance [Fig.4.148]. A cell of emf 10 mV is balanced against a length of 40 cm of potentiometer wire. What is the value of the external resistance R ?

APPROACH As shown in the Fig.4.148, if R is the unknown resistance, then current in the primary circuit

$$I = \frac{V}{(r+R)} = \frac{2}{(10+R)}$$

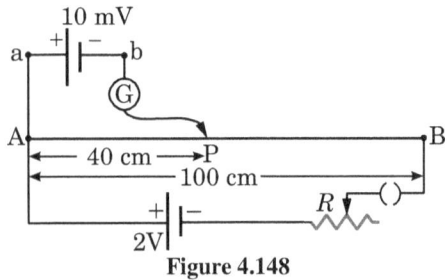

Figure 4.148

If V_0 is the potential difference across the potentiometer wire and R_{AB}, is the total resistance of the potentiometer wire, then

$$V_0 = IR_{AB} = \frac{2}{(10 + R)} \times R_{AB} \quad \cdots (1)$$

Now, if L is the length of potentiometer wire, then from Eq.4.169, the potential gradient along potentiometer wire is given by

$$x = \frac{V_0}{L} \quad \cdots (2)$$

If emf of the cell in secondary circuit is $\mathcal{E}$ and the potentiometer is balanced at $AP = l_0$, then from Eq.4.171, potential difference (V) across AP is given by

$$V = l_0 x = l_0 \left(\frac{V_0}{L}\right) = \mathcal{E}$$

or $\quad V_0 \dfrac{l_0}{L} = \mathcal{E} \quad \cdots (3)$

Substituting the value of V_0 from Eq.(1) in Eq.(3), we get

$$\frac{2}{(10 + R)} \times R_{AB}\frac{l_0}{L} = \mathcal{E}$$

or $\quad R = \dfrac{2 l_0 R_{AB}}{L\mathcal{E}} - 10 \quad \cdots (5)$

Now, substitute the given values in Eq.(5) and solve for R

SOLUTION Given: $l_0 = 40$ cm, $L = 100$ cm, $R_{AB} = 10\ \Omega$, and $\mathcal{E} = 10$ mV $= 10^{-2}$ V

Substituting these values in Eq.(5), we get

$$R = \frac{2 \times 40 \times 10}{100 \times 10^{-2}} - 10 = 790\ \Omega$$

4.25.1 Comparison of Emf's of Two Batteries

The driving circuit of the potentiometer is set up with a strong battery so that the potential difference V_0 across AB is larger than the emf of either battery. One of the batteries is connected between the positive end A and the galvanometer. The jockey is adjusted to touch the wire at a point P_1 so that there is no deflection in the galvanometer. The length $AP_1 = l_1$ is noted. Now, the first battery is replaced by the second and the length $AP_2 = l_2$ for the balance is noted. If the length $AB = L$, then from Eq.4.170, the emf of the first battery is

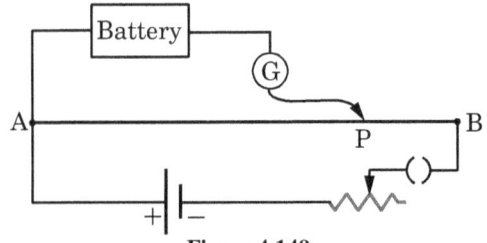

Figure 4.149

$$\mathcal{E}_1 = V_0 \frac{l_1}{L} \quad (4.174)$$

and similarly for the second battery, the emf is

$$\mathcal{E}_2 = V_0 \frac{l_2}{L} \quad (4.175)$$

Therefore, ratio of emfs of above two batteries

$$\boxed{\frac{\mathcal{E}_1}{\mathcal{E}_2} = \frac{l_1}{l_2}} \quad (4.176)$$

Note that no calibration is needed in this case.

One can use a two-way key to connect both the batteries together

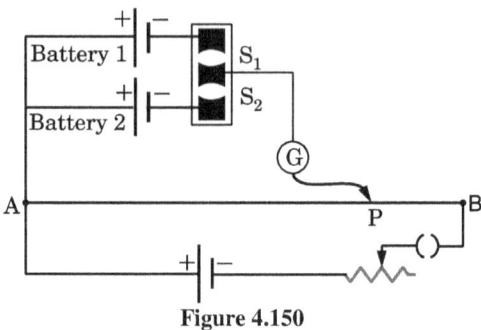

Figure 4.150

as shown in Fig.4.150. When the key is pressed in the plug S_1, the first battery is brought into the circuit. When the key is taken out from S_1 and pressed in the plug S_2, the second battery is brought into the circuit.

The value of emf of a battery can also be obtained by this same method by taking the other battery to be a standard cell. The emf of the standard cell is known and hence the emf of the given battery can be obtained.

EXAMPLE 79. Primary circuit of potentiometer is shown in Fig.4.151. Determine:

(a) current in primary circuit
(b) potential drop across potentiometer wire AB
(c) potential gradient (means potential drop per unit length of potentiometer wire)
(d) maximum potential which we can measure from this potentiometer

SOLUTION (a) By Ohm's law, current in primary circuit is

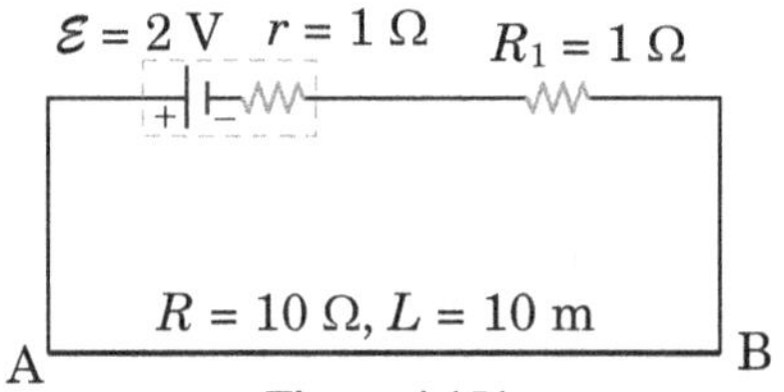

$\mathcal{E} = 2$ V $\quad r = 1\ \Omega \quad R_1 = 1\ \Omega$

$R = 10\ \Omega, L = 10$ m

Figure 4.151

given by

$$I = \frac{\mathcal{E}}{r + R_1 + R} = \frac{2}{1 + 20 + 10} = \frac{2}{31}\ \text{A}$$

(b) Potential drop across potentiometer wire AB, is given by

$$V_{AB} = IR = \frac{2}{31} \times 10 \Rightarrow V_{AB} = \frac{20}{31}\ \text{volt}$$

(c) If x denotes the potential gradient of the potentiometer wire, then

$$x = \frac{V_{AB}}{L} = \frac{20/31}{10} = \frac{2}{31}\ \text{volt/m}.$$

(d) Maximum potential which can be measured by this potentiometer

$$= \text{potential drop across wire AB} = \frac{20}{31}\ \text{volt}$$

EXAMPLE 80. In an experiment to determine the emf of an unknown cell (Fig.4.152), it's emf is compared with a standard cell of known emf $\mathcal{E}_1 = 1.12$ V. The balance point is obtained at 56 cm with standard cell and 80 cm with the unknown cell. Determine the emf of the unknown cell.

APPROACH From Eq.4.176, we have

4.25. STRETCHED-WIRE POTENTIOMETER

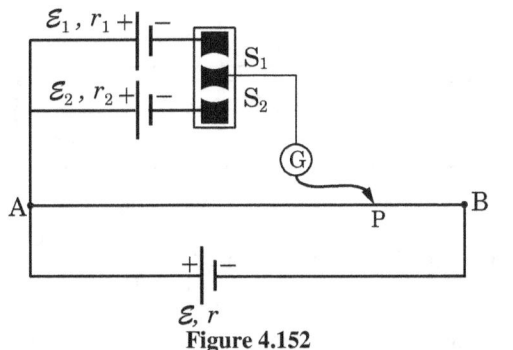

Figure 4.152

$$\frac{\mathcal{E}_1}{\mathcal{E}_2} = \frac{l_1}{l_2} \quad \Rightarrow \quad \mathcal{E}_2 = \frac{l_2}{l_1}\mathcal{E}_1 \quad \cdots (1)$$

SOLUTION Given: $\mathcal{E}_1 = 1.12$ volt, $l_1 = 56$ cm and $l_2 = 80$ cm
Substituting these values in Eq.(1), we get

$$\mathcal{E}_2 = \frac{80}{56} \times 1.12 = 1.6 \text{ volt}$$

EXAMPLE 81. A standard cell of emf $\mathcal{E}_0 = 1.11$ V is balanced against 72 cm length of a potentiometer. The same potentiometer is used to measure the potential difference across the standard resistance $R = 120\ \Omega$. When the ammeter shows a current of 7.8 mA, a balanced length of 60 cm is obtained on the potentiometer.
(a) Determine the current flowing through the resistor.
(b) Estimate the error in measurement of the ammeter.
APPROACH From Eq.4.176, we have

$$\frac{\mathcal{E}_1}{\mathcal{E}_2} = \frac{l_1}{l_2} \quad \cdots (1)$$

Here, $\mathcal{E}_1 = \mathcal{E}_0 = 1.11$ V, $l_1 = 72$ cm, $l_2 = 60$ cm. For $\mathcal{E}_2$ we write $\mathcal{E}_2 = iR$, here i is the current in the given resistor $R = 120\ \Omega$. Current obtained by using Eq.(1) is the actual current and the current given by ammeter is the measured current. The difference in these values gives the error in current measurement. If Δi is the difference in currents obtained by above calculation and by ammeter reading, then % error in measurement in current

$$= \frac{\Delta i}{i} \times 100$$

SOLUTION From Eq.(1), we have

$$\frac{\mathcal{E}_1}{\mathcal{E}_2} = \frac{l_1}{l_2}$$

or

$$\frac{\mathcal{E}_0}{iR} = \frac{l_1}{l_2} \quad \Rightarrow \quad i = \frac{\mathcal{E}_0}{R}\left(\frac{l_2}{l_1}\right)$$

Substituting given values, in above equation, we get

$$i = \frac{1.11\ \text{V}}{120\ \Omega}\left(\frac{60\ \text{cm}}{72\ \text{cm}}\right) = 7.7 \times 10^{-3}\text{A} = 7.7\ \text{mA}$$

So, by potentiometer measurement, the current in the resistor should be 7.7 mA.
(b) According to given problem, the ammeter reading of current in the resistor is 7.8 mA which is greater than reading 7.7 mA which is obtaind by potentiometer in part (a). So, the ammeter has a positive error.
The error in measurement of current is

$$\Delta i = 7.8 - 7.7 = 0.1 \text{ mA}.$$

Therefore, % error produced by ammeter in measurement of current

$$\frac{\Delta i}{i} \times 100 = \frac{0.1}{7.7} \times 100 = 1.3\%$$

4.25.2 Measurement of Internal Resistance of a Battery

Figure 4.153 shows the arrangement for measuring the internal resistance of a battery. The emf of the battery is $\mathcal{E}$ and its internal resistance is r. A known resistance R is connected across the battery together with a plug key K. The potentiometer circuit is set up as usual. The plug key K is opened and the balance point P is searched on the wire AB so that there is no deflection in the galvanometer. As the key is open, there is no current through the resistance R. Hence, there is no current through the battery and the potential difference across the terminals a, b is the same as the emf $\mathcal{E}$ of the battery. If $AP = l$, we have

$$\mathcal{E} = V_0\frac{l}{L} \quad (4.177)$$

with the symbols having their usual meanings.
Now the key K is closed and the new balance point P' is searched.

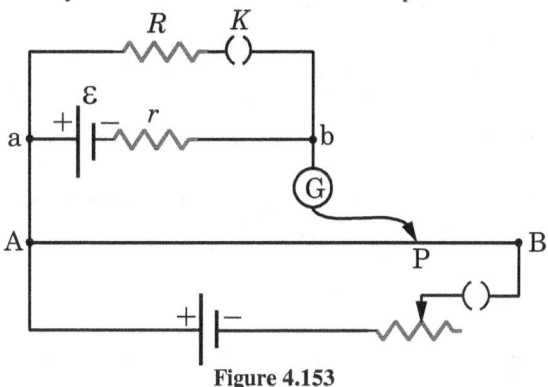

Figure 4.153

In this case, there is a current in the battery. This current is given by

$$i = \frac{\mathcal{E}}{R+r} \quad (4.178)$$

In this case, the potential difference between a and b is

$$V_a - V_b = Ri = \frac{\mathcal{E}R}{R+r} \quad (4.179)$$

If $AP' = l'$, then from Eq.4.170, we have

$$V_a - V_b = V_0\frac{l'}{L} \quad (4.180)$$

Comparing Eq.4.179 and 4.180, we get

$$\frac{\mathcal{E}R}{R+r} = V_0\frac{l'}{L} \quad (4.181)$$

Dividing Eq.4.181 by Eq.4.177, we get

$$\frac{R}{R+r} = \frac{l'}{l}$$

Solving for r, we get

$$\boxed{r = \frac{R(l-l')}{l'}} \quad (4.182)$$

EXAMPLE 82. The internal resistance of a cell is determined by using a potentiometer. In an experiment (Fig.4.153), an external resistance of 60 Ω is used across the given cell. When the key K is closed, the balance length on the potentiometer decreases from 72 cm to 60 cm. calculate the internal resistance of the cell.
APPROACH From Eq.4.182, the internal resistance of a cell is

$$r = \frac{R(l-l')}{l'} \quad \cdots (1)$$

Substitute the given values and solve for r
SOLUTION Given: $R = 60\ \Omega$, $l = 72$ cm, $l' = 60$ cm
Substituting these values in Eq.(1), we get

$$r = \frac{60(72-60)}{60} = 12\ \Omega$$

EXAMPLE 83. The balancing length for a cell is 560 cm in a potentiometer experiment. When an external resistance of 10 Ω is connected in parallel to the cell, the balancing length changes by 60 cm. If the internal resistance of the cell is $N/10\ \Omega$, where N is an integer then find the value of N.

Table 4.4: Difference between potentiometer and voltmeter

Potentiometer	Voltmeter
1. It measures the unknown emf very accurately	It measures the unknown emf approximately.
2. While measuring emf it does not draw any current from the driving source of know emf.	While measuring emf it draws some current from the source of emf.
3. While measuring unknown potential difference the resistance of potentiometer becomes infinite.	While measuring unknown potential difference the resistance of voltmeter is high but finite.
4. It is based on zero deflection method.	It is based on deflection method.
5. It has a high sensitivity.	Its sensitivity is low.
6. it is used for various applications like measurement of internal resistance of cell, calibration of ammeter and voltmeter, measurement of thermo-emf, comparison of emf's etc.	It is only used to measured emf or unknown potential difference.

SOLUTION The formula for the internal resistance of the battery using potentiometer is given by,
$$r = \left(\frac{l-l'}{l'}\right) R$$
where, $l \to$ balancing length when battery is in open circuit $l' \to$ balancing length when battery is shunted by an external resistance R
Given: $l = 560$ cm, $l' = 560 - 60 = 500$ cm
∴ $r = \frac{60}{500} \times 10 = \frac{6}{5}$
Given that, $r = \frac{N}{10}$, therefore
$\frac{N}{10} = \frac{6}{5} \Rightarrow N = 12$

4.25.3 Check Point 8

1. Figure 4.154 shows a potentiometer circuit for determining the internal resistance of a cell. When key K is open, the balance point is found to be at 76.3 cm of the wire. When key K is closed and the value of R is 4.0 Ω, the balance point shifts to 60.0 cm. Find the internal resistance of cell $\mathcal{E}$.
   (A) 1 Ω   (B) 2.1 Ω   (C) 1.1 Ω   (D) 5.1 Ω

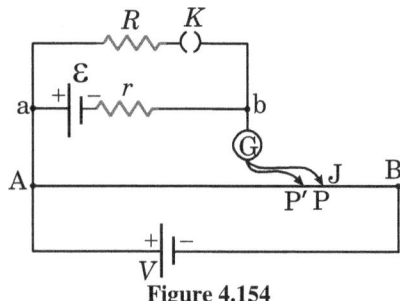

Figure 4.154

Paragraph for Q.2 - 4
A cell having a steady emf of 2 volt is connected across the potentiometer wire of length 10 m. The potentiometer wire is of manganine and having resistance per unit length as 11.5 Ω/m. A new resistance of 5 Ω with negligible length is put in series with the potentiometer wire [Fig.4.155].

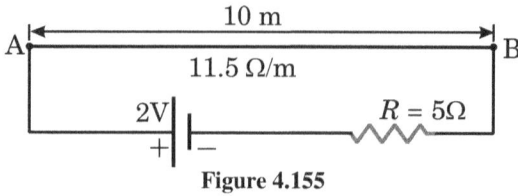

Figure 4.155

2. What is total resistance of wire?
   (A) 11.5 Ω   (B) 115 Ω
   (C) 1150 Ωm (D) 120 Ω
3. What is potential gradient when 5 Ω resistance is not connected
   (A) 0.2 V/m   (B) 0.4 V/m
   (C) 0 V/m   (D) 1.38 V/m
4. What is potential gradient after connecting 5 Ω resistance.
   (A) 0.2 V/m   (B) 0.15 V/m
   (C) 0.104 V/m   (D) 2.0 V/m

4.26 An Infinite Network

In this section, we describe a method to find the equivalent resistance of a circuit that has infinite identical loops. We start by guessing a finite and non zero value (R_{eq}) for the equivalent resistance. For this finite and non-zero equivalent resistance, adding or removing one or more stages to the ladder will not change the resistance of the network. Therefore, to simplify the circuit we can break it in a manner that only one loop is left with us and in place of the remaining portion we connect a resistor of magnitude R_{eq}. Now, we calculate the equivalent resistance of the simplified circuit and put it equal to the guessed value i.e., R_{eq}. This process involves solving a quadratic equation. By solving this equation, we can find the desired value of the equivalent resistance. See following examples for clear understanding of such types of problems.

☞ Note that an infinite ladder should be broken from those points from where the broken part resembles with the original ladder.

EXAMPLE 84. (a) Calculate the equivalent resistance (in terms of R, the resistance of each individual resistor) between points A and B for the infinite ladder of resistors shown in Figure 4.156 assuming the resistors are identical. That is assuming $R = R_1 = R_2$. (b) Repeat Part (a) but do not assume that $R_1 = R_2$ and express your answer in terms of R_1 and R_2. (c) Check your results by showing that your result from Part (b) agrees with your result from part (a) if you substitute R for both R_1 and R_2.

APPROACH Let R be the resistance of each resistor in the ladder

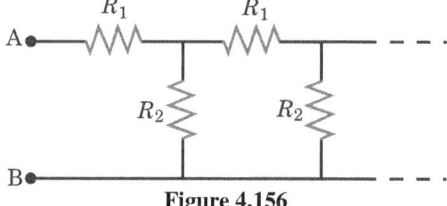

Figure 4.156

and let R_{eq} be the equivalent resistance of the infinite ladder. If the resistance is finite and non-zero, then removing one or more stages to the ladder will not change the resistance of the network. We can apply the rules for resistance combination to the diagram shown in Fig.4.157 to obtain a quadratic equation in R_{eq} that we can solve for the equivalent resistance between points A and B.

SOLUTION (a) The equivalent resistance of the series combination of R and $(R\|R_{eq})$ is R_{eq}, so:

$$R_{eq} = R + R\|R_{eq} = R + \frac{RR_{eq}}{R + R_{eq}}$$

4.26. AN INFINITE NETWORK

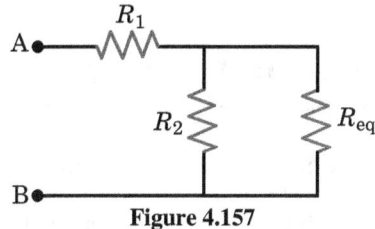

Figure 4.157

or $\quad R_{eq}^2 - RR_{eq} - R^2 = 0$
Solving for R_{eq}, we get

$$R_{eq} = \left(\frac{1 \pm \sqrt{5}}{2}\right)R$$

Since, R_{eq} is a finite and positive value, therefore, we have to ignore negative value. Considering only positive value, we get

$$R_{eq} = \left(\frac{1 + \sqrt{5}}{2}\right)R \approx 1.618R$$

(b) The equivalent resistance of the series combination of R_1 and $(R_2 \| R_{eq})$ is R_{eq}, so:

$$R_{eq} = R_1 + R_2 \| R_{eq} = R_1 + \frac{R_2 R_{eq}}{R_2 + R_{eq}}$$

Simplifying it, we get

$$R_{eq}^2 - R_1 R_{eq} - R_1 R_2 = 0$$

Now, solving for the positive value of R_{eq}, we get:

$$R_{eq} = \frac{R_1 \pm \sqrt{R_1^2 + 4R_1 R_2}}{2}$$

Since, R_{eq} is a finite and positive value, therefore, ignoring negative value, we get

$$R_{eq} = \frac{R_1 + \sqrt{R_1^2 + 4R_1 R_2}}{2}$$

(c) If $R_1 = R_2 = R$, then:

$$R_{eq} = \frac{R + \sqrt{R^2 + 4RR}}{2} = \left(\frac{1 + \sqrt{5}}{2}\right)R$$

It is in agreement with Part (a).

EXAMPLE 85. An infinite ladder network is constructed with 1 Ω and 2 Ω resistors as shown in Fig.4.158. Find the equivalent resistance between terminals A and B.
APPROACH This problem is similar to Example 87 with $R_1 =$

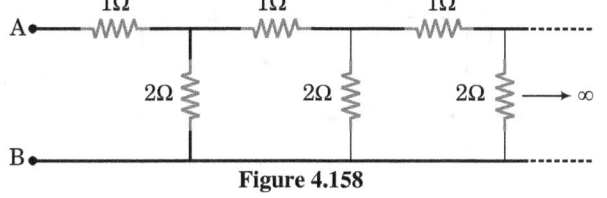

Figure 4.158

1 Ω and $R_2 = 2$ Ω.
SOLUTION Solving like Example 84, we get
$$R_{eq} = 2 \text{ Ω}$$
EXAMPLE 86. Find equivalent resistance between terminals

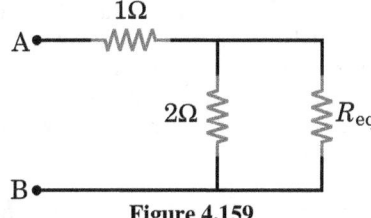

Figure 4.159

A and B of Fig.4.160.
APPROACH Let R_{eq} be the equivalent resistance of the infinite

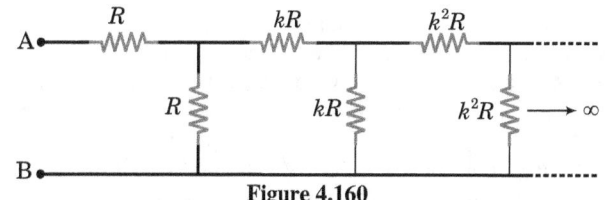

Figure 4.160

ladder. If we break the ladder at dashed line shown in Fig.4.162a and take out it's first stage then remaining ladder will look like the circuit shown in Fig.4.161.

In Fig.4.161, we see that, the circuit is same to circuit shown in

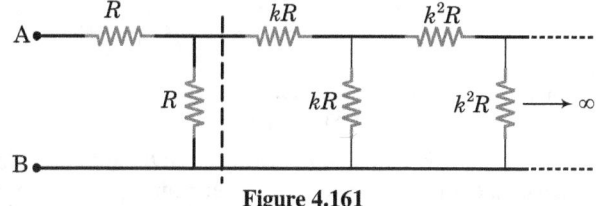

Figure 4.161

Fig.4.160 except that each resistance is k times that of the resistances shown in Fig.4.160. So, it's equivalent resistance will also be k times the resistance of the circuit shown in Fig.4.160, i.e., the resistance of this ladder will be kR_{eq}.

Now, if we connect this resistance kR_{eq} in parallel with the resistor of resistance R of removed first stage of ladder (see Fig.4.162b), the resistance of this circuit will be equal to resistance of original circuit of Fig.4.160.
SOLUTION The equivalent resistance of the series combina-

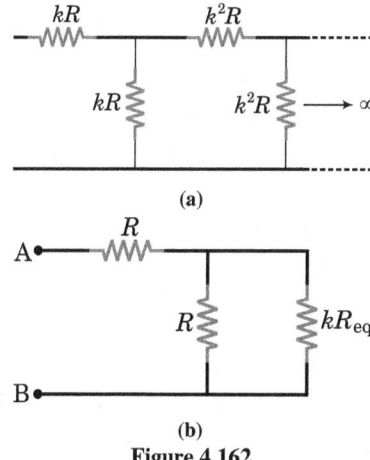

Figure 4.162

tion of R and $(R \| kR_{eq})$ is R_{eq}, so:

$$R_{eq} = R + R\|kR_{eq} = R + \frac{R(kR_{eq})}{R + kR_{eq}}$$

or $\quad R_{eq}(R + R_{eq}) = R(R + R_{eq}) + kRR_{eq}$
or $\quad kR_{eq}^2 + (1 - 2k)RR_{eq} - R^2$
Simplifying it for R_{eq}, we get

$$R_{eq} = \frac{-(1-2k)R \pm \sqrt{(1-2k)^2 R^2 + 4kR^2}}{2k}$$
$$= \frac{(2k-1) \pm \sqrt{(1-2k)^2 + 4k}}{2k}R$$
$$= \frac{(2k-1) \pm \sqrt{1 + 4k^2}}{2k}R$$

Since, R_{eq} is a finite and non-zero value, therefore, ignoring negative value, we get

$$R_{eq} = \left[\frac{(2k-1) + \sqrt{1+4k^2}}{2k}\right]R$$

EXAMPLE 87. (a) Calculate the equivalent resistance (in terms of R, the resistance of each individual resistor) between points A and B for the infinite ladder of resistors shown in Figure 4.163 assuming the resistors are identical. That is assuming $R = R_1 = R_2$. (b) Repeat Part (a) but do not assume that $R_1 = R_2$ and express your answer in terms of R_1 and R_2. (c) Check your results by showing that your result from Part (b) agrees with your result from part (a) if you substitute R for both R_1 and R_2.

APPROACH Let R be the resistance of each resistor in the lad-

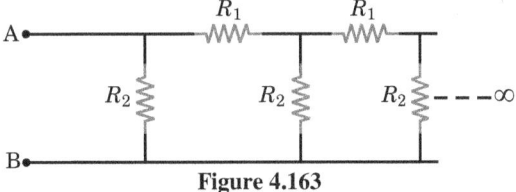

Figure 4.163

der and let R_{eq} be the equivalent resistance of the infinite ladder. If the resistance is finite and non-zero, then removing one or more stages to the ladder will not change the resistance of the network. We can apply the rules for resistance combination to the diagram shown in Fig.4.164 to obtain a quadratic equation in R_{eq} that we can solve for the equivalent resistance between points A and B.

SOLUTION (a) The equivalent resistance of the parallel com-

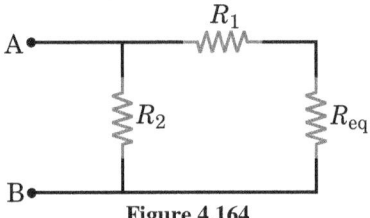

Figure 4.164

bination of R and $(R + R_{eq})$ is R_{eq}, so:
$$R_{eq} = \frac{R(R + R_{eq})}{R + (R + R_{eq})} = \frac{R(R + R_{eq})}{2R + R_{eq}}$$
or $\quad R_{eq}^2 + R R_{eq} - R^2 = 0$
Solving for R_{eq}, we get
$$R_{eq} = \left(\frac{-1 \pm \sqrt{5}}{2}\right) R$$
Since, R_{eq} is a finite and positive value, therefore, we have to ignore negative value. Considering only positive value, we get
$$R_{eq} = \left(\frac{-1 + \sqrt{5}}{2}\right) R \approx \frac{2.24 - 1}{2} = 0.62R$$
(b) The equivalent resistance of the parallel combination of R_2 and $(R_1 + R_{eq})$ is R_{eq}, so:
$$R_{eq} = \frac{R_2(R_1 + R_{eq})}{R_2 + (R_1 + R_{eq})} = \frac{R_2(R_1 + R_{eq})}{R_1 + R_2 + R_{eq}}$$
Simplifying it, we get
$$R_{eq}^2 + R_1 R_{eq} - R_1 R_2 = 0$$
Now, solving for the positive value of R_{eq}, we get:
$$R_{eq} = \frac{-R_1 \pm \sqrt{R_1^2 + 4R_1 R_2}}{2}$$
Since, R_{eq} is a finite and positive value, therefore, ignoring negative value, we get
$$R_{eq} = \frac{-R_1 + \sqrt{R_1^2 + 4R_1 R_2}}{2}$$
(c) If $R_1 = R_2 = R$, then:
$$R_{eq} = \frac{-R + \sqrt{R^2 + 4RR}}{2} = \left(\frac{-1 + \sqrt{5}}{2}\right) R$$

It is in agreement with Part (a).

4.27 Symmetrical Electric Circuits

A symmetrical electric circuit is a circuit that has symmetry with respect to some operation or transformation. This means that if the circuit is subjected to this operation or transformation, the circuit remains unchanged.

For example, a circuit with mirror symmetry has a plane of symmetry, such that if the circuit is reflected across this plane, the circuit looks identical to the original circuit. Another example of a symmetrical electric circuit is a circuit with rotational symmetry, such that if the circuit is rotated by a certain angle around a central point, the circuit looks identical to the original circuit.

Symmetrical electric circuits are important to us because they have properties that are preserved under certain operations or transformations. For example, a circuit with mirror symmetry will have equal currents flowing in opposite directions in the mirrored sections of the circuit. This property can be used to simplify the analysis of the circuit and make it easier to solve for the currents and voltages. Symmetric electrical circuit problems can be solved using the following steps:

1. Draw the circuit diagram with all the components such as resistors, capacitors, and inductors, and mark the points where the symmetry exists.
2. Use the symmetry to simplify the circuit by reducing the number of components. For example, if the circuit is rotationally symmetric, you can assume that the resistance, capacitance, and inductance are the same for all angles.
3. Use symmetry to find the values of the components that are not explicitly given in the circuit. For example, if the circuit has reflection symmetry, the voltages at symmetric points will be equal.
4. Once the circuit is simplified, you can use Kirchhoff's current and voltage laws to solve the circuit. Use Ohm's law to calculate the current through each component and use the junction rule to find the current at each node.

By following these steps, you can solve symmetric electrical circuit problems efficiently and effectively. See following examples for clear understanding of such types of problems.

EXAMPLE 88. Find the effective resistance between points P and Q of the electrical circuit shown in the Figure 4.165.

SOLUTION The given circuit is redrawn in Fig.4.166a. Since,

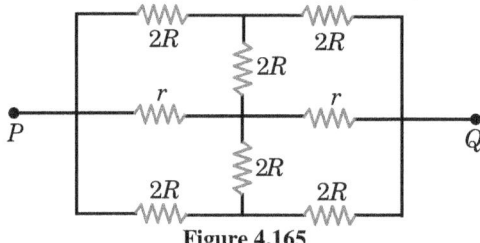

Figure 4.165

the given circuit has mirror symmetry about branch PQ, therefore, same current I_1 will pass through branches PAB and PFE. Also there are same currents, each of magnitude I_3, through branches BO and EO. Current I_2 passes along branch PO.

There is also a mirror symmetry about branch EB. So, current through branch BCQ will be identical to current through branch PAB, i.e., I_1. Current through branch EDQ will be identical to current through branch PFE, i.e., I_1. Current I_3 passes along branch OQ.

By Kirchhoff's voltage law in loop ABOPA, we get
$$2RI_1 + 2RI_3 - I_2 r = 0, \quad \cdots (1)$$
and again by Kirchhoff's voltage law in loop and BCQOB
$$2R(I_1 - I_3) - r(I_2 + 2I_3) - 2RI_3 = 0 \quad \cdots (2)$$

4.27. SYMMETRICAL ELECTRIC CIRCUITS

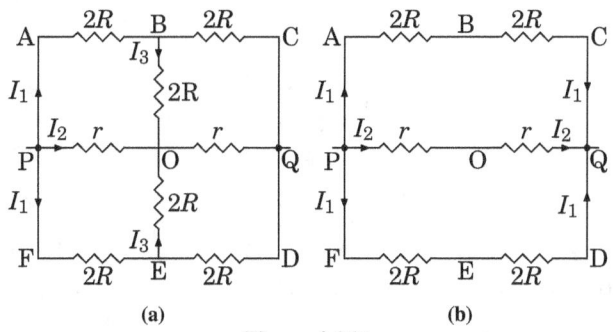

Figure 4.166

On solving Eq.(1) and (2), we get
$$I_3 = 0$$
It means no current flows in the branch BO and EO. Hence, branch BO and EO can be removed from the circuit without affecting the effective resistance. This new circuit diagram is shown in Fig.4.166b.

In the circuit of Fig.4.166b, resistances of branches ABC, FED and POQ are $4R$, $4R$ and $2r$ respectively. These resistances are in parallel connection, therefore, the effective resistance of the circuit between terminals, P and Q is
$$R_e = (4R \| 4R) \| 2r = \frac{2Rr}{R+r}.$$

EXAMPLE 89. Find the equivalent Resistance between terminals A and B of circuit shown in Fig.4.167.

APPROACH 1. The circuit is redrawn in Fig.4.168. Branches

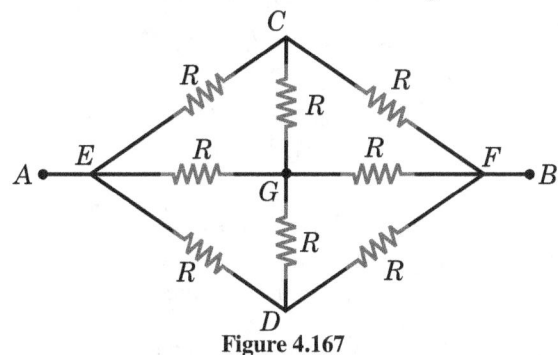

Figure 4.167

EC and ED are symmetrical, therefore, currents through them will be same.

There is also a mirror symmetry about a line passing through

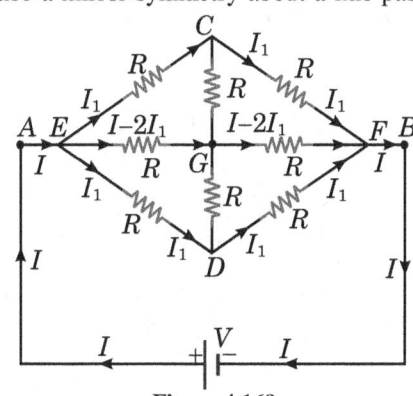

Figure 4.168

CD. Therefore, current distribution while entering through E and an exiting from F will be same. Using all these facts the currents are shown in the Fig.4.168. Since, total current of branch EC passes to branch CF; total current of branch EG passes to GF and also total current of branch ED passes to DF, therefore, no current passes through resistors between C, G and G, D. So, we can remove these resistors. It's equivalent circuit is shown in Fig.4.169. Now, identify series and parallel connections and solve for equivalent resistance.

SOLUTION Method 1. By identifying Series and Parallel

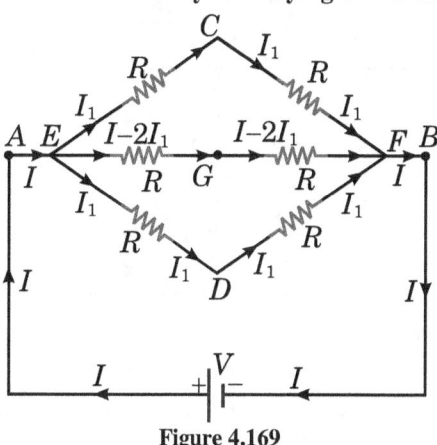

Figure 4.169

Parts: In Fig.4.169, resistors between E, C and C, F are in first series; resistors between E, G and G, F are in second series; and resistors between E, D and D, F are in third series. Equivalents of all these series resistors are in parallel. So, equivalent resistance between terminals A and B is given by
$$\frac{1}{R_{eq}} = \frac{1}{2R} + \frac{1}{2R} + \frac{1}{2R}$$
Therefore, $R_{eq} = \frac{2R}{3}$

Method 2. By using $R_{eq} = \frac{V}{I}$: By Kirchhoff's voltage law either in loop ECFGE or in loop EGFDE, we have
$$-I_1 R - I_1 R + (I - 2I_1)R + (I - 2I_1)R = 0$$
or $\quad -2I_1 R + 2(I - 2I_1)R = 0 \quad$ or $\quad I_1 = \frac{I}{3} \quad \cdots (1)$

If V is the terminal voltage across A and B, then by Ohm's law, in lower branch EDF, we have
$$V = 2I_1 R \quad \cdots (2)$$
Substituting the value of I_1 from Eq.(1) in Eq.(2), we get
$$V = 2\frac{I}{3}R \quad \text{or} \quad \frac{V}{I} = \frac{2R}{3}$$
Therefore equivalent resistance of circuit between terminals A and B is
$$R_{eq} = \frac{V}{I} = \frac{2R}{3}$$

APPROACH 2. In Upper part EGFCE of the circuit of Fig.4.168, we see that
$$\frac{R_{EC}}{R_{EG}} = \frac{R_{CF}}{R_{GF}} \quad \text{i.e.,} \quad \frac{R}{R} = \frac{R}{R}$$
So, it forms a balance, Wheatstone bridge, and hence the resistor between terminals C and G, can be removed.

Similarly, by same argument in lower part EGFDE, we can remove the resistor between terminals G and D. After removing these two resistors, we get the new circuit diagram as shown in Fig.4.169. Now, the solution will be similar to solution of approach 1.

APPROACH 3. In this approach, we observe symmetrical points. These points will be at same potential. So, resistors connected between them will be in parallel.

SOLUTION In Fig.4.168, points C and D are symmetric so they will be at same potential. Keeping it in mind, the circuit can be transformed like shown in Fig.4.170a

Resistors in parallel connection: Between E, C and E, D; Between G, C and G, D; and Between F, C and F, D. Further simplified circuit is shown in Fig.4.170b. It shows that
$$\frac{R_{EC}}{R_{EG}} = \frac{R_{FC}}{R_{FG}}$$

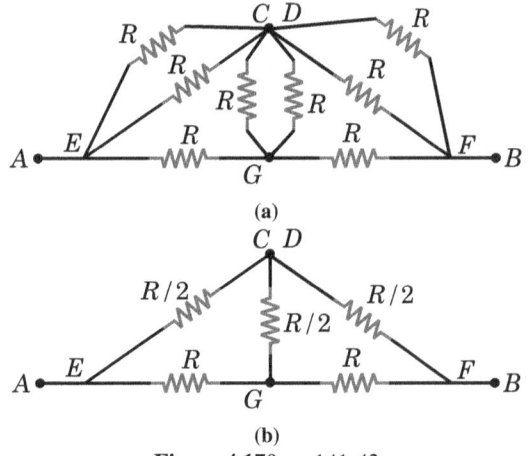

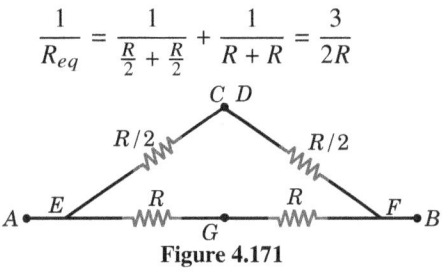

Figure 4.170: ce141-42

So, according to Balanced Wheatstone bridge condition, we can remove the resistor between the terminals G and C. The next simplified circuit is shown in Fig.4.171. From this diagram, we see that

$$\frac{1}{R_{eq}} = \frac{1}{\frac{R}{2} + \frac{R}{2}} + \frac{1}{R+R} = \frac{3}{2R}$$

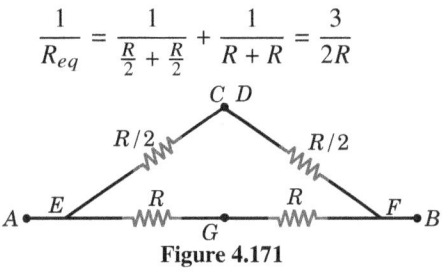

Figure 4.171

or $R_{eq} = \dfrac{2R}{3}$

EXAMPLE 90. In Fig.4.172, resistance of each resistor is R. Find the equivalent resistances of the combinations between terminals A and B.

SOLUTION (a) Given circuit diagram is shown in Fig.4.173.

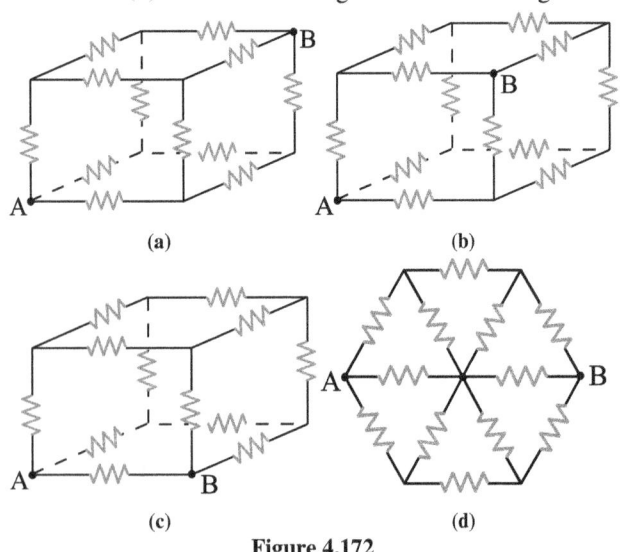

Figure 4.172

Symmetry Observation: All symmetrical points are marked by same letters. These are C and D.

Point C: Between all points marked by letters C and A, there is exactly one resistor of same resistance. These are- 1, 7 and 8. Between points marked by B and C there are two similar resistors from each sides. Theses are- (4, 10), (6, 11), (5, 9), (6, 12), (5, 3), (4, 2).

Point D: Between points marked by D and A, there are exactly

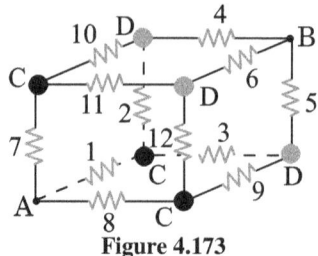

Figure 4.173

two equal resistors from each direction. These are- (1, 3), (8, 9), (1, 2), (7, 10), (8, 12), (7, 11). Between terminal point B and D there is one resistor of same magnitude from each direction. These are- 4, 5 and 6.

So, all points marked by same letters are symmetrical. At all symmetrical points, electric potential will be same. Therefore, all resistors attached between terminals A and B will be in parallel and between terminal points B and D the attached resistors are also in parallel combination. It's new simplified diagram is shown in Fig.4.174.

In Fig.4.174, equivalent resistance between terminals A and C

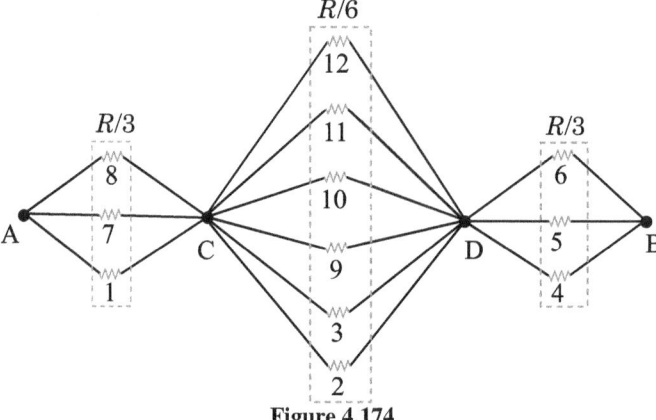

Figure 4.174

is $R/3$; between terminals C and D is $R/6$ and between terminals D and B it is $R/3$. It's next simplified circuit diagram is shown in Fig.4.175.

In Fig.4.175, all resistors are in series, their equivalent resistance

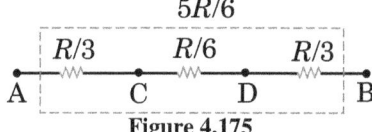

Figure 4.175

is given by-

$$R_{eq} = \frac{R}{3} + \frac{R}{6} + \frac{R}{3} = \frac{5R}{6}$$

or $R_{eq} = 5R/6$

(b) Fig.4.176 shows circuit diagram with two symmetrical points each marked by C and two other symmetrical points each marked by D.

Symmetry observations: Between terminals A and B, there

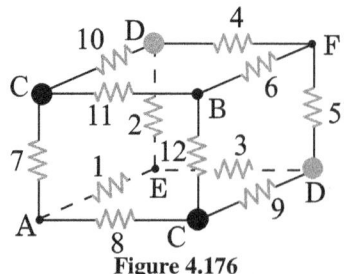

Figure 4.176

are two mirror symmetric points each marked by letter C. Points

4.27. SYMMETRICAL ELECTRIC CIRCUITS

marked by letter D, are also mirror symmetric. All mirror symmetric points are equipotential points. It's simplified circuit diagram is shown in Fig.4.177.

Resistors connected in parallel are shown by dashed boxes.

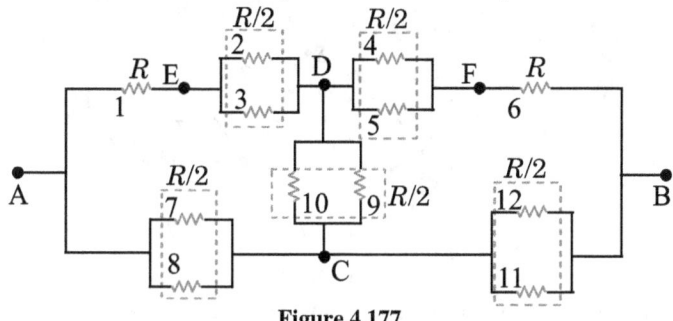

Figure 4.177

Their equivalent resistances are written above respective boxes. Next simplified diagram is shown in Fig.4.178.

In Fig.4.178, the resistors in series are shown in dashed boxes.

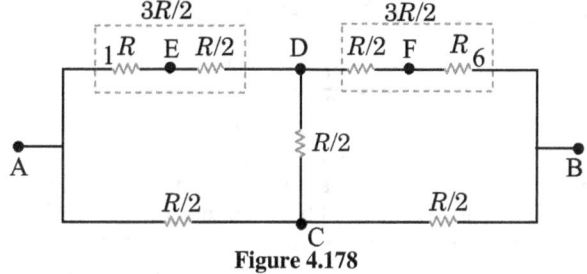

Figure 4.178

The next simplified circuit diagram is shown in Fig.4.179
In Fig.4.179, we observe that the ratio of resistance between

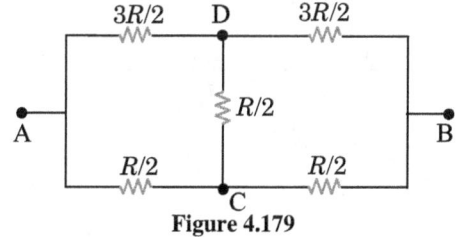

Figure 4.179

terminals A and D with resistance between terminals A and C i.e., $\dfrac{3R/2}{R/2}$ is equal to the ratio of resistance between terminals B and D with resistance between terminals B and C, i.e., $\dfrac{3R/2}{R/2}$. So, it satisfies the Wheatstone's balanced bridge condition (4.135). In this situation, point C and D will be equipotential points and so we can remove the resistor between terminals C and D. It's next simplified circuit diagram is shown in Fig.4.180. Resistors in series with their equivalent resistances, are shown in dashed boxes. Their equivalent resistances are in parallel. The net equivalent resistance between terminals A and B is given by-

$$R_{eq} = \frac{3R \times R}{3R + R} = \frac{3R^2}{4R} = \frac{3R}{4}$$

(c) Symmetry observation: The circuit diagram is shown in Fig.4.181. There are two mirror symmetric points marked by letter C and two other symmetric points marked by D. So, potentials at both C's will be same and at both D's it will also be same. It's first simplified circuit diagram is shown in Fig.4.182
All resistors in parallel with their resultants are shown by dashed boxes. Next simplified diagram is shown in Fig.4.183. Resistors in series with their equivalent resistance are shown by dashed boxes. This series equivalent resistance is in parallel order with

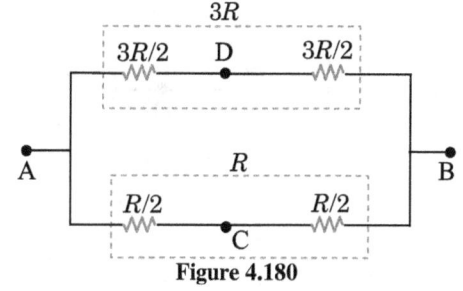

Figure 4.180

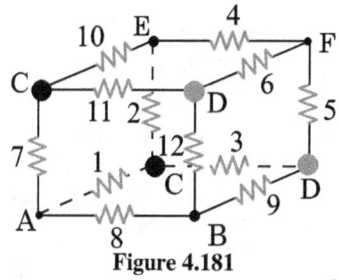

Figure 4.181

the resistor $R/2$. Their equivalent resistance is $\dfrac{2R \times R/2}{2R + R/2} = \dfrac{2R}{5}$

Next simplified circuit diagram is shown in Fig.4.184. Resistors

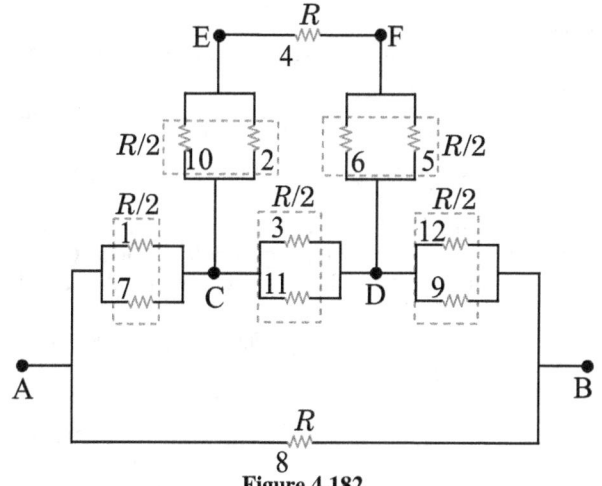

Figure 4.182

connected in branches CE, EF and FD are in series. Their equivalent resistance (R_1), is given by-

$$R_1 = \frac{R}{2} + \frac{2R}{5} + \frac{R}{2} = \frac{7R}{5}$$

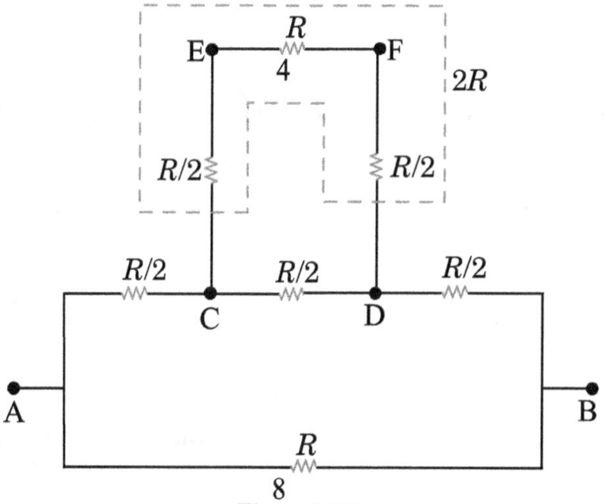

Figure 4.183

This equivalent resistance R_1 is in parallel with resistor number

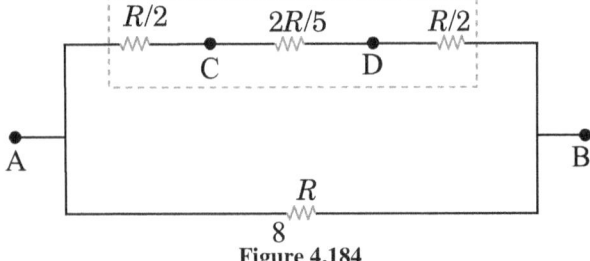

Figure 4.184

8.
Equivalent resistance between terminals A and B:
$$R_{eq} = \frac{R \times 7R/5}{R + 7R/5} = \frac{7R}{12}$$

(d) Given circuit diagram is again redrawn in Fig.4.185a. Each resistor has resistance R. Since, there is a mirror symmetry about a dashed line as shown in Fig.4.185a, therefore, currents passing through these symmetric resistors will also be same except opposite in direction. New diagram with currents is shown in Fig.4.185b. Also note that the points marked by same letter are equipotential points. Here, both C are at same potential and similarly, both D are also at same potential. Therefore currents $i_1 = i_2$.

Note that-
1. At junction A, current passing through resistor 1 goes entirely to B through resistor 3.
2. At junction A, current passing through resistor 7 goes entirely to B through resistor 8.
3. At junction A, current passing through resistor 4 goes entirely to B through resistor 6.

So, no component of either i_1 or i_2 passes to junction E, therefore this junction is unnecessary for registers 9, 11 and 10, 12. On removing this junction from these resistors, we get the new circuit diagram as shown in Fig.4.186a.

Now, in circuit diagram 4.186a, resistors 9 and 11 are in series;

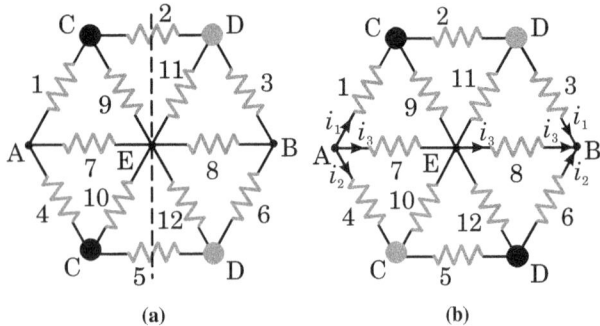

Figure 4.185

similarly, 10 and 12 are also in other series. The equivalent resistances of 9 and 11 is-
$$R_1 = R + R = 2R$$
Similarly, the equivalent resistance of 10 and 12 is also,
$$R_1' = 2R.$$
Now, equivalent R_1 is in parallel with resistor 2 and similarly, R_1' is in parallel with resistor 5 (Fig.4.186b).
In Fig.4.186b, net resistance between upper terminals C and D is given by-
$$R_2 = \frac{R \times 2R}{R + 2R} = \frac{2R}{3}$$
Similarly, net resistance between lower terminals C and D is also $R_2' = \frac{2R}{3}$.

Next simplified circuit diagram is shown in Fig.4.187a.

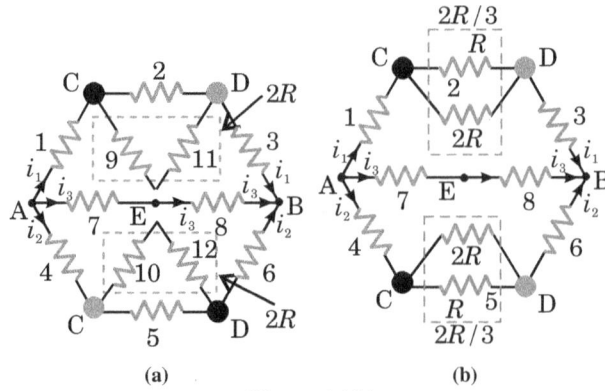

Figure 4.186

The resistor between upper terminals C and D, is in series with resistors 1 and 3. Their series equivalent is given by-
$$R_3 = R + \frac{2R}{3} + R = \frac{8R}{3}$$
Similarly, the resistor between lower terminals C and D, is in series with resistors 4 and 6. Their series equivalent is also given by-
$$R_3' = R + \frac{2R}{3} + R = \frac{8R}{3}$$
Series equivalent of resistors 7 and 8 is-
$$R_4 = R + R = 2R$$
Next simplified circuit diagram is shown in Fig.4.187b. In this figure, all resistors connected between terminals A and B are in parallel. Therefore, equivalent resistance is given by-

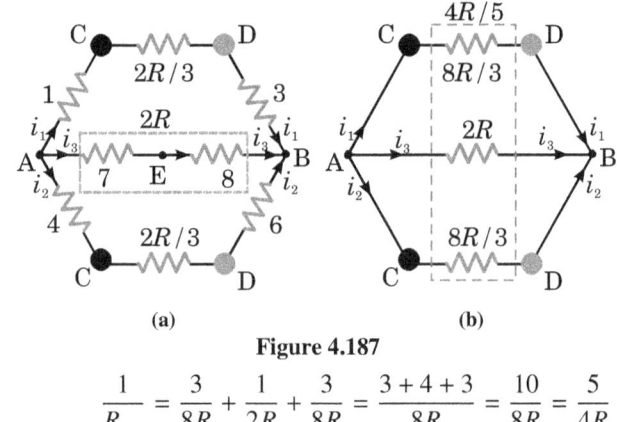

Figure 4.187

$$\frac{1}{R_{eq}} = \frac{3}{8R} + \frac{1}{2R} + \frac{3}{8R} = \frac{3+4+3}{8R} = \frac{10}{8R} = \frac{5}{4R}$$
or $R_{eq} = 4R/5$

4.27.1 Check Point 10
1. Find the equivalent Resistance between terminals A and B of circuit shown in Fig.4.188.

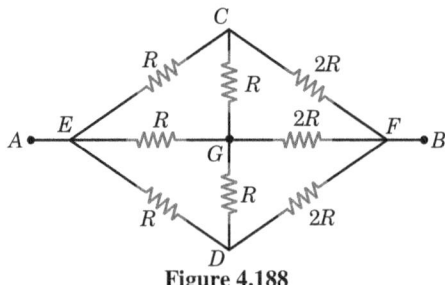

Figure 4.188

Multiple Choice Questions
2. At what value of resistance R_X in the circuit of Fig.4.189 will the total resistance between points A and B is independent of number of sections?

(A) $R(\sqrt{3} - 1)$ (B) $R(\sqrt{3} + 1)$
(C) $R\frac{(\sqrt{3}+1)}{2}$ (D) $R\frac{(\sqrt{3}-1)}{(\sqrt{3}+1)}$

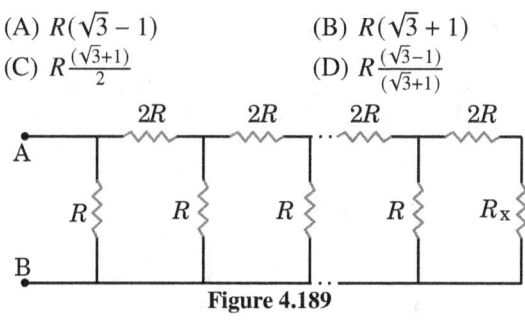

Figure 4.189

4.28 Star and Delta Networks

The combination of resistance in the form of T or Y or star (Fig.4.190a), is called "Star network" whereas combination of resistances in the form of triangle (Fig.4.190b) is called "Delta or π network". These networks have three terminals.

These two networks are electrically equivalent for the resistance

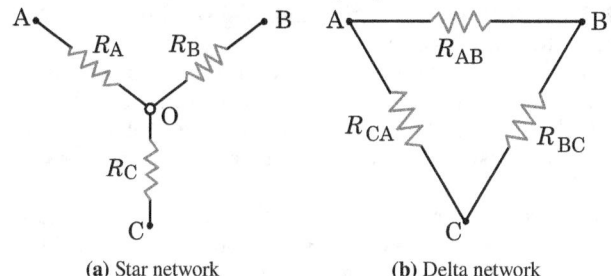

(a) Star network (b) Delta network

Figure 4.190: Star and delta networks

between any two terminals and can be converted from one network to other.

4.28.1 Delta to Star Conversion

Fig.4.191 shows delta and star networks having three terminals A, B, and C. Dashed lines are not the part of any network. They are drawn only to make conversion easy. To convert one network to other, we just attach required resistances along dashed lines and then erase the solid line part from the network. Now, change dashed lines into solid lines. This will be your converted network. In delta network of Fig.4.191, R_{AB}, R_{BC}, and R_{CA} are resistances between terminals A, B; B, C and C, A respectively. In star network of Fig.4.191, R_A, R_B and R_C are the resistances connected between junction O and terminals A, B and C respectively.

The total resistance between terminals A and B in the delta

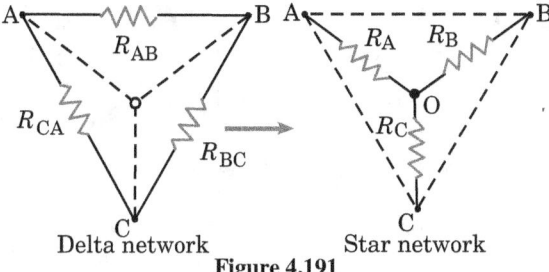

Figure 4.191

network;
$$R_1 = R_{AB} \| (R_{BC} + R_{CA})$$
$$R_1 = \frac{R_{AB}(R_{BC} + R_{CA})}{R_{AB} + R_{BC} + R_{CA}} \quad (4.183)$$

Similarly the resistance between terminals B and C
$$R_2 = \frac{R_{BC}(R_{CA} + R_{AB})}{R_{AB} + R_{BC} + R_{CA}} \quad (4.184)$$

and the resistance between terminals A and C.
$$R_3 = \frac{R_{CA}(R_{AB} + R_{BC})}{R_{AB} + R_{BC} + R_{CA}} \quad (4.185)$$

According to star network;
$$R_1 = R_A + R_B,$$
$$R_2 = R_B + R_C,$$
and $$R_3 = R_C + R_A,$$

Now adding Eq.4.183, 4.184 and 4.185 together
$$R_1 + R_2 + R_3 = \frac{R_{AB}(R_{BC} + R_{CA})}{R_{AB} + R_{BC} + R_{CA}} + \frac{R_{CA}(R_{AB} + R_{BC})}{R_{AB} + R_{BC} + R_{CA}}$$
$$+ \frac{R_{BC}(R_{AB} + R_{CA})}{R_{AB} + R_{BC} + R_{CA}}$$

Substituting the values of R_1, R_2 and R_3, we get
$$(R_A + R_B) + (R_B + R_C) + (R_A + R_C) =$$
$$\frac{R_{AB}(R_{BC} + R_{CA}) + R_{CA}(R_{AB} + R_{BC}) + R_{BC}(R_{AB} + R_{BC})}{R_{AB} + R_{BC} + R_{CA}}$$
$$\Rightarrow 2R_A + 2R_B + 2R_C$$
$$= \frac{R_{AB}R_{BC} + R_{AB}R_{CA} + R_{BC}R_{CA} + R_{AB}R_{BC} + R_{BC}R_{CA}}{R_{AB} + R_{BC} + R_{CA}}$$
$$\Rightarrow 2R_A + 2R_B + 2R_C$$
$$= \frac{2(R_{AB}R_{BC}) + 2(R_{BC}R_{CA}) + 2(R_{AB}R_{CA})}{R_{AB} + R_{BC} + R_{CA}}$$

On simplifying above equation, we get
$$R_A + R_B + R_C = \frac{R_{AB}R_{BC} + R_{BC}R_{CA} + R_{AB}R_{CA}}{R_{AB} + R_{BC} + R_{CA}} \quad (4.186)$$

Now, to find R_A, R_B and R_C, we subtract Eq.4.183, 4.184, & 4.185 one by one from Eq.4.186.
First, Subtracting Eq.4.184 from Eq.4.186
$$R_A + R_B + R_C - (R_B + R_C) = \frac{R_{AB}R_{BC} + R_{BC}R_{CA} + R_{AB}R_{CA}}{R_{AB} + R_{BC} + R_{CA}}$$
$$- \frac{R_{BC}(R_{CA} + R_{AB})}{R_{AB} + R_{BC} + R_{CA}}$$

On simplifying it, we get
$$\boxed{R_A = \frac{R_{AB}R_{AC}}{R_{AB} + R_{BC} + R_{CA}}} \quad (4.187)$$

Note that $R_{AB} = R_{BA}, R_{BC} = R_{CB}$ and $R_{AC} = R_{CA}$
Similarly subtracting Eq.4.183 & 4.185 from 4.186 results in
$$\boxed{R_B = \frac{R_{BA}R_{BC}}{R_{AB} + R_{BC} + R_{CA}}} \quad (4.188)$$

$$\boxed{R_C = \frac{R_{CA}R_{CB}}{R_{AB} + R_{BC} + R_{CA}}} \quad (4.189)$$

From the derived equations for star-equivalent resistances R_A, R_B, & R_C we can conclude the relation between delta-to-star conversions as; the equivalent star resistance is equal to the product of the adjacent delta resistances with a terminal divide by the sum of all three delta resistances.

In case all three resistances are same in a delta network, the equivalent star resistance would become
$$R_{\text{star}} = \frac{RR}{R + R + R} = \frac{R^2}{3R} = \frac{R_{\text{delta}}}{3} \quad (4.190)$$

Since all the resistance throughout the delta network are equal, each three equivalent star resistance would be 1/3 times the delta resistance.

4.28.2 Star to Delta Conversion

Now we are going to convert the star connected resistance into delta connected resistance. Let's derive the equations used for a star to delta conversion.

First figure of Fig.4.192 shows star connected resistance R_A, R_B & R_C. While the required delta equivalent resistance is R_{AB}, R_{BC} & R_{CA} as shown in second figure of Fig.4.192.

In order to find the equivalent delta resistance, multiply the

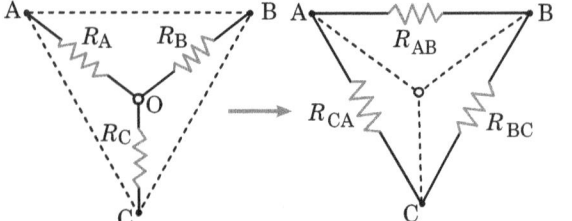

Figure 4.192

previous equation 4.187 & 4.188, as well as 4.188 & 4.189 & 4.187 & 4.189 together.
Multiplying 4.187 & 4.188:

$$R_A R_B = \frac{R_{AB} R_{AC}}{R_{AB} + R_{BC} + R_{CA}} \frac{R_{BC} R_{BA}}{R_{AB} + R_{BC} + R_{CA}}$$

$$R_A R_B = \frac{R_{AB}^2 R_{BC} R_{CA}}{(R_{AB} + R_{BC} + R_{CA})^2} \quad (4.191)$$

Similarly multiplying Eq.4.188 with 4.189 & 4.187 with 4.189

$$R_B R_C = \frac{R_{AB} R_{BC}^2 R_{CA}}{(R_{AB} + R_{BC} + R_{CA})^2} \quad (4.192)$$

$$R_A R_C = \frac{R_{AB} R_{BC} R_{CA}^2}{(R_{AB} + R_{BC} + R_{CA})^2} \quad (4.193)$$

Now adding Eq.4.191, 4.192 & 4.193 together, we get
$R_A R_B + R_B R_C + R_A R_C$
$$= \frac{R_{AB}^2 R_{BC} R_{CA} + R_{AB} R_{BC}^2 R_{CA} + R_{AB} R_{BC} R_{CA}^2}{(R_{AB} + R_{BC} + R_{CA})^2}$$
$$= \frac{R_{AB} R_{BC} R_{CA} (R_{AB} + R_{BC} + R_{CA})}{(R_{AB} + R_{BC} + R_{CA})^2}$$

On simplifying above equation, we get

$$R_A R_B + R_B R_C + R_A R_C = \frac{R_{AB} R_{BC} R_{CA}}{(R_{AB} + R_{BC} + R_{CA})} \quad (4.194)$$

In order to get the individual equivalent delta resistance, we divide Eq.4.194 with 4.187, 4.188 & 4.189 separately. Dividing Eq.4.194 with 4.187, we get

$$\frac{R_A R_B + R_B R_C + R_A R_C}{R_A} = \frac{R_{AB} R_{BC} R_{CA}}{(R_{AB} + R_{BC} + R_{CA})}$$
$$\div \frac{R_{AB} R_{AC}}{(R_{AB} + R_{BC} + R_{CA})}$$

On simplifying for R_{BC}, we get

$$\boxed{R_{BC} = \frac{R_A R_B + R_B R_C + R_A R_C}{R_A}} \quad (4.195)$$

Similarly dividing Eq.4.194 with 4.188 & 4.189 separately results in

$$\boxed{R_{AC} = \frac{R_A R_B + R_B R_C + R_A R_C}{R_B}} \quad (4.196)$$

and

$$\boxed{R_{AB} = \frac{R_A R_B + R_B R_C + R_A R_C}{R_C}} \quad (4.197)$$

Remarks

☞ If the Δ or Y connected network consists of inductances (assumed no mutual coupling forms between the inductors) then the same formula can be used for Y to Δ or Δ to Y conversion.

☞ On the other hand, the Δ or Y connected network consists of capacitances can be converted to an equivalent Y or δ network provided the capacitance value is replaced by its reciprocal in the conversion.

EXAMPLE 91. Find the value of the voltage source (V_S) that delivers 2 A current through the circuit as shown in Fig.4.193.

APPROACH We can make this problem easier by converting all

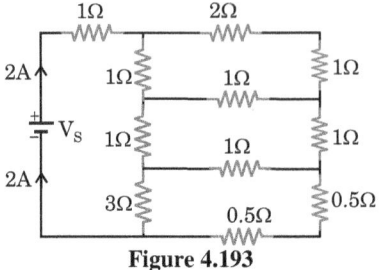

Figure 4.193

delta networks to star networks. First convert all delta networks to star networks and then observe series and parallel combinations. After finding equivalent resistance (R_{eq}) of the circuit, apply $V_S = I R_{eq}$ and then solve for source voltage V_S

SOLUTION In the top right loop of given Fig.4.193, 2Ω and 1Ω resistors are in series. Their equivalent is 3Ω. Similarly, in bottom right loop the both resistors of 0.5Ω are also in series. Their equivalent is 1Ω. Fig.4.194 shows a new circuit diagram in which above mentioned series resistors are replaced with their equivalent resistors. In Fig.4.194, three terminal networks ABC and DEF form delta networks.

Now, we convert the three terminals Δ-networks (ABC & DEF) into their equivalent Y-connected networks along dashed lines. Corresponding to Δ-connected network 'ABC', the equivalent Y-connected resistor values are given as
$R_A = \frac{3 \times 1}{1+1+3} = 0.6\Omega$; $R_B = \frac{1 \times 1}{1+1+3} = 0.2\Omega$; $R_C = \frac{3 \times 1}{1+1+3} = 0.6\Omega$

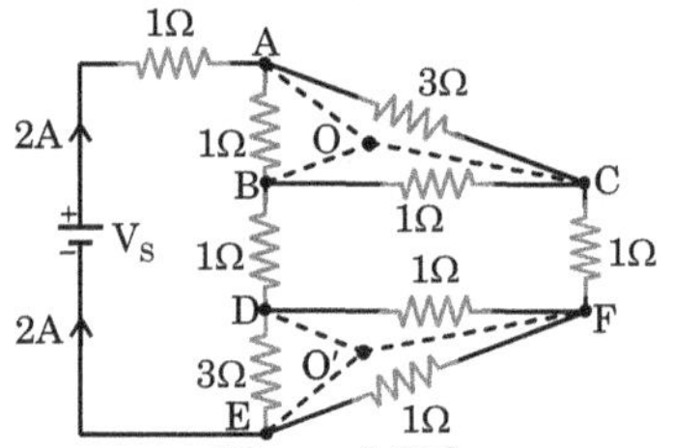

Figure 4.194

Similarly, for the Δ-connected network 'DEF' the equivalent Y-connected resistor values are-
$R_D = \frac{3 \times 1}{1+3+1} = 0.6\Omega$; $R_E = \frac{3 \times 1}{1+3+1} = 0.6\Omega$; $R_F = \frac{1 \times 1}{1+3+1} = 0.2\Omega$

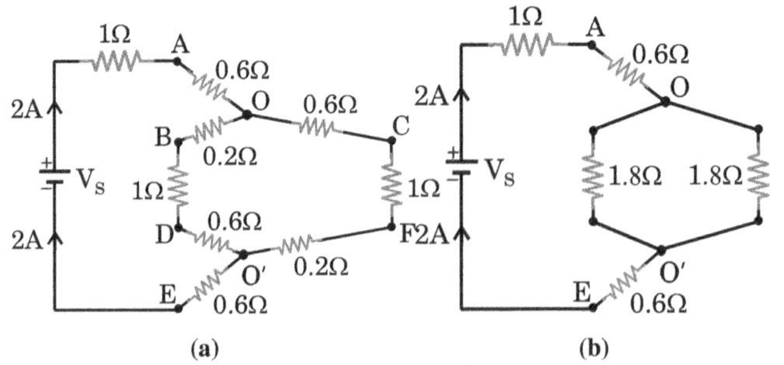

(a)   (b)

Figure 4.195

On replacing both delta networks with their equivalent star networks, we get new circuit diagram as shown in Fig.4.195a.
In this circuit diagram (Fig.4.195a), resistors $R_{OC}(= 0.6\Omega)$, $R_{CF}(= 1\Omega)$ and $R_{FO'}(= 0.2\Omega)$ are in series. Their equivalent resistance is
$$R_1 = 0.6\Omega + 1\Omega + 0.2\Omega = 1.8\Omega$$
Similarly, $R_{BD}(= 0.2\Omega)$, $R_{BD}(= 1\Omega)$ and $R_{DO'}(= 0.6\Omega)$ are in series. Their equivalent resistance is
$$R_2 = 0.2\Omega + 1\Omega + 0.6\Omega = 1.8\Omega$$
On replacing these series resistors with their equivalent resistors,

4.28. STAR AND DELTA NETWORKS

we get the new circuit diagram as shown in next Fig.4.195b. Now, R_1 and R_2 are in parallel. Their equivalent resistance is
$$R_3 = \frac{1.8 \times 1.8}{1.8 + 1.8} = 0.9\Omega.$$
Next simplified circuit diagram is shown in Fig.4.196.
Equivalent resistance of circuit shown in Fig.4.196 is given as

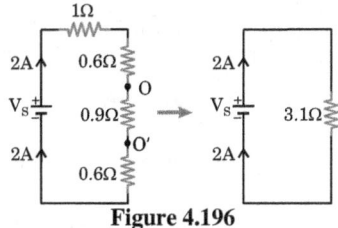

Figure 4.196

$$R_{eq} = 1\Omega + 0.6\Omega + 0.9\Omega + 0.9\Omega = 3.1\Omega$$
The source V_s that delivers 2 A current through the circuit can be obtained as
$$V_S = I \times R_{eq} = 2 \times 3.1 = 6.2 \text{volt}$$

EXAMPLE 92. Determine the equivalent resistance between the terminals A and B of network shown in Fig.4.197.

SOLUTION A 'Δ' is substituted for the 'Y' between points C,

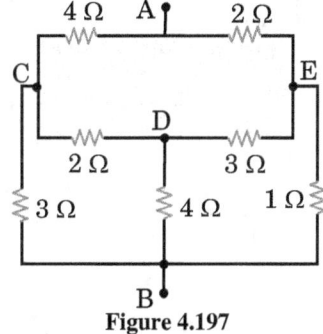

Figure 4.197

D and E as shown in Fig.4.198a; then unknown resistances value for Y to Δ transformation are computed below.

$$R_{CB} = \frac{2 \times 4 + 4 \times 3 + 3 \times 4}{3} = \frac{8 + 12 + 6}{3} = 8.66\Omega;$$

$$R_{EB} = \frac{2 \times 4 + 4 \times 3 + 3 \times 4}{2} = \frac{8 + 12 + 6}{2} = 13\Omega;$$

$$R_{CE} = \frac{2 \times 4 + 4 \times 3 + 3 \times 4}{4} = \frac{8 + 12 + 6}{4} = 6.5\Omega$$

Next we transform 'Δ' connected 3-terminal resistor to an equiv-

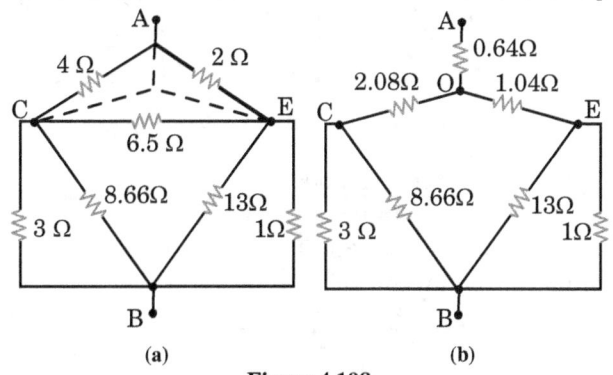

Figure 4.198

alent 'Y' connected network between points 'A'; 'C' and 'E' (see Fig.4.198b) and the corresponding Y connected resistances value are obtained using the following expression.

$$R_A = \frac{4 \times 2}{4 + 2 + 6.5} = 0.64\Omega;$$
$$R_C = \frac{4 \times 6.5}{4 + 2 + 6.5} = 2.08\Omega;$$
$$R_E = \frac{6.5 \times 2}{4 + 2 + 6.5} = 1.04\Omega;$$

In Fig.4.198b, 13Ω, 1Ω are in parallel and 3Ω, 8.66Ω are also in parallel.
Equivalent resistance of 13Ω and 1Ω is
$$R_1 = \frac{13 \times 1}{13 + 1} = \frac{13}{14} \approx 0.93\Omega$$
Similarly, equivalent resistance of 3Ω and 8.66Ω is
$$R_2 = \frac{3 \times 8.66}{3 + 8.66} = \frac{25.98}{11.66} \approx 2.23\Omega$$
Next simplified circuit diagram is shown in Fig.4.199.
Now we can easily find the equivalent resistance between the

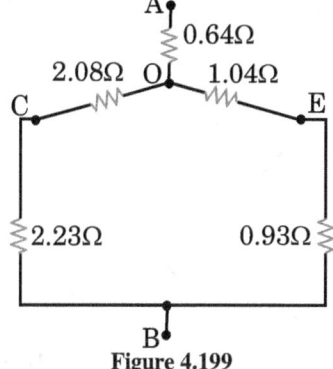

Figure 4.199

terminals 'A' and 'B' as
$$R_{AB} = (2.23 + 2.08) \| (1.04 + 0.93) + 0.64 = 2.21\Omega.$$

EXAMPLE 93. Find the effective resistance between terminals A and B of the circuit shown in Fig.4.200.

SOLUTION Y connected network formed with the terminals A,

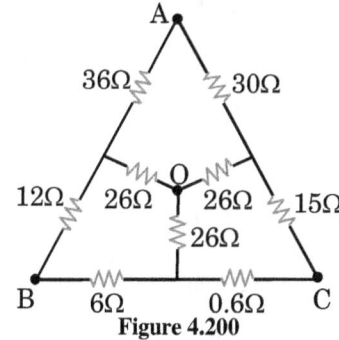

Figure 4.200

B and O is transformed into Δ connected one and its resistance values are given below (Fig.4.201a).

$$R_{AB} = \frac{36 \times 26 + 26 \times 12 + 12 \times 36}{26} \approx 64.61\ \Omega;$$
$$R_{BO} = \frac{36 \times 26 + 26 \times 12 + 12 \times 36}{36} \approx 46.66\ \Omega$$
and $$R_{AO} = \frac{36 \times 26 + 26 \times 12 + 12 \times 36}{12} \approx 140\ \Omega$$

Similarly, Y connected networks formed with the terminals B, C and O (Fig.4.201b); C, A and O (Fig.4.201c) are transformed to Δ connected networks as follows:

$$R_{BC} = \frac{6 \times 0.6 + 0.6 \times 26 + 26 \times 6}{26} \approx 6.74\ \Omega;$$
$$R_{CO} = \frac{6 \times 0.6 + 0.6 \times 26 + 26 \times 6}{6} = 29.2\Omega;$$
$$R_{BO} = \frac{6 \times 0.6 + 0.6 \times 26 + 26 \times 6}{0.6} = 34.60\Omega;$$
and $$R_{CO} = \frac{26 \times 15 + 15 \times 30 + 30 \times 26}{30} = 54.00\Omega;$$
$$R_{AO} = \frac{26 \times 15 + 15 \times 30 + 30 \times 26}{15} = 108\Omega;$$

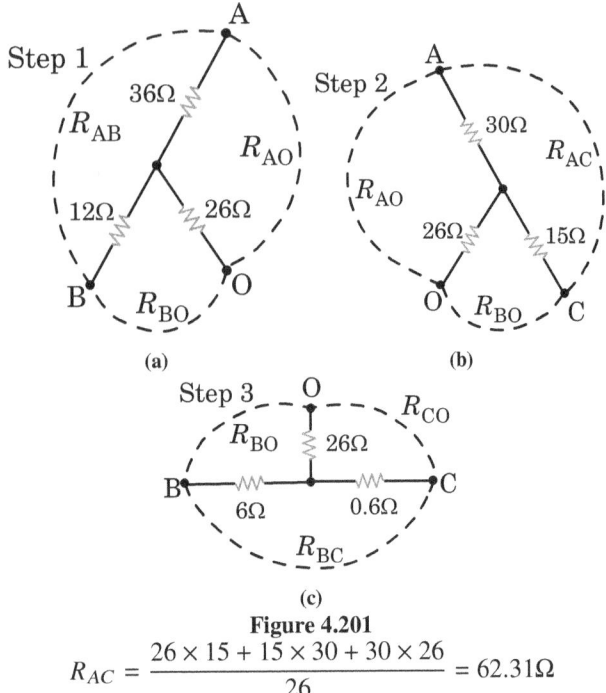

Figure 4.201

$$R_{AC} = \frac{26 \times 15 + 15 \times 30 + 30 \times 26}{26} = 62.31 \Omega$$

Note that the two resistances are connected in parallel (140||108)

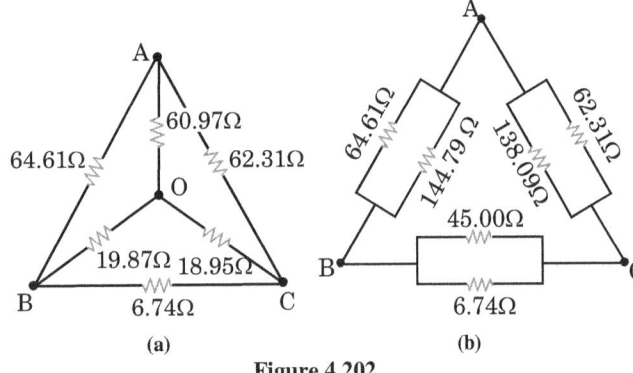

Figure 4.202

between the points 'A' and 'O'. Similarly, between the points 'B' and 'O' two resistances are connected in parallel (46.66||34.6) and resistances 54.0 Ω and 29.2 Ω are connected in parallel between the points 'C' and 'O'.

Therefore, effective resistance along OA,

$$(R_{OA})_{\text{eff}} = \frac{140 \times 108}{140 + 108} \approx 60.97 \, \Omega,$$

effective resistance along OB,

$$(R_{OB})_{\text{eff}} = \frac{46.66 \times 34.6}{46.66 + 34.6} \approx 19.87 \, \Omega,$$

and effective resistance along OC,

$$(R_{OC})_{\text{eff}} = \frac{54.0 \times 29.2}{54.0 + 29.2} \approx 18.95 \, \Omega$$

It is shown in Fig.4.202a

Now, above Y connected network formed with the terminal A, B and C is converted to equivalent Δ connected network as follows (Fig.4.202b).

$$R_{AB} = \frac{60.99 \times 19.87 + 19.87 \times 18.95 + 18.95 \times 60.97}{18.95}$$

$$= \frac{2743.7893}{18.95} = 144.79 \, \Omega$$

Similarly,

$$R_{BC} = \frac{60.99 \times 19.87 + 19.87 \times 18.95 + 18.95 \times 60.97}{60.97}$$

$$= \frac{2743.7893}{60.97} = 45.00 \, \Omega$$

and $R_{CA} = \dfrac{60.99 \times 19.87 + 19.87 \times 18.95 + 18.95 \times 60.97}{19.87}$

$$= \frac{2743.7893}{19.87} = 138.09 \, \Omega$$

In Fig.4.202b, resultant of resistors connected in parallel between terminals A and B is

$$R_1 = \frac{64.61 \times 144.79}{64.61 + 144.79} \approx 44.67 \, \Omega$$

Similarly, resultant of resistors connected in parallel between terminals B and C is

$$R_2 = \frac{6.74 \times 45.00}{6.74 + 45.00} \approx 5.86 \, \Omega$$

and resultant of resistors connected in parallel between terminals C and A is

$$R_3 = \frac{62.31 \times 138.09}{62.31 + 138.09} \approx 42.94 \, \Omega$$

Therefore, equivalent resistance between terminals A and B is

$$(R_{AB})_{\text{eq}} = \frac{(R_2 + R_3) R_1}{R_2 + R_3 + R_1}$$

$$= \frac{(5.86 + 42.94) \times 44.67}{5.86 + 42.94 + 44.67} = 23.32 \, \Omega$$

EXAMPLE 94. Find the equivalent resistance R_{eq} of the network (see Fig.4.203) between the terminals 'A' & 'B' using Y-Δ or Δ-Y transformations.

SOLUTION Convert the three terminals (C-D-E) Δ-connected

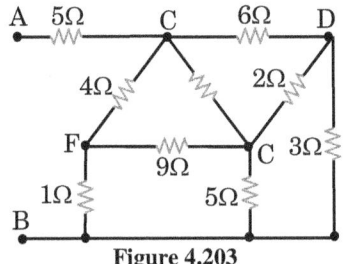

Figure 4.203

network (see Fig.4.203) to an equivalent Y-connected network using the following Δ - Y conversion equations.

$R_{CO} = \frac{6 \times 4}{12} = 2\Omega$; $R_{DO} = \frac{6 \times 2}{12} = 1\Omega$; and $R_{CO} = \frac{2 \times 4}{12} = 0.666\Omega$

Similarly, the Δ-connected network (F-C-B) is converted to an

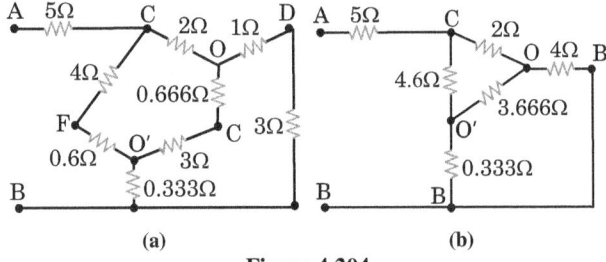

Figure 4.204

equivalent Y-connected network.

$R_{FO'} = \frac{1 \times 9}{15} = 0.6\Omega$; $R_{CO'} = \frac{5 \times 9}{15} = 3\Omega$;

and $R_{BO'} = \frac{1 \times 5}{15} = 0.333\Omega$

After the Δ - Y conversions, the circuit is redrawn and shown in

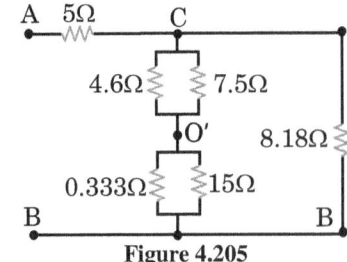

Figure 4.205

Fig.4.204a. Next the series-parallel combinations of resistances reduces the network configuration in more simplified form and it is shown in Fig.4.204b. Fig.4.204b can further be simplified by transforming Y connected network comprising with the three resistors (2Ω, 4Ω and 3.666Ω) to a Δ-connected network and the corresponding network parameters are given below:

$$R_{CO'} = \frac{2 \times 3.666 + 3.666 \times 4 + 4 \times 2}{4} \approx 7.5\Omega;$$

$$R_{CB} = \frac{2 \times 3.666 + 3.666 \times 4 + 4 \times 2}{3.666} \approx 8.18\Omega;$$

and $$R_{BO'} = \frac{2 \times 3.666 + 3.666 \times 4 + 4 \times 2}{2} \approx 15\Omega;$$

Simplified form of the circuit is drawn and shown in Fig.4.205 and one can easily find out the equivalent resistance R_{eq} between the terminals 'A' and 'B' using the series- parallel formula. From Fig.4.205, one can write the expression for the total equivalent resistance R_{eq} between the terminals 'A' and 'B' as

$R_{eq} = 5 + [(4.6| |7.5) + (0.333||15)] ||8.18$
$= 5 + [2.85 + 0.272] ||8.18 = 5 + (3.122| |8.18) = 7.26\ \Omega$

4.28.3 Check Point 11
Multiple Choice Questions
1. What will be the equivalent resistance of the network shown in the Figure 4.206 between the terminals 1 and 2.
   (A) $\frac{2}{5}R$ (B) $\frac{8}{5}R$ (C) $\frac{7}{5}R$ (D) none

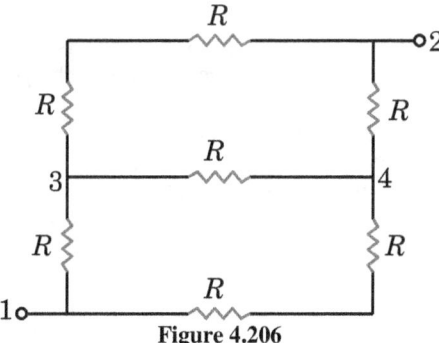
Figure 4.206

2. In network of Fig.4.207, what is total resistance across terminals A and B.
   (A) 6.4Ω (B) 2.4Ω (C) 7.4Ω (D) 5.4Ω

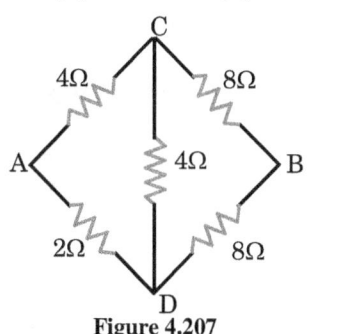
Figure 4.207

4.29 Kirchhoff's Rules for Capacitive Circuits

Kirchhoff's rules can also be used to determine the potential difference and charge on the plates of a capacitor in any electric circuit. In a circuit with capacitors and batteries, two important rules are involved.

1. Junction Rule (Based on Conservation of Charge)
Sum of charges present on the plates of capacitors connected at a junction is equal to zero (If initially all the capacitors are uncharged) (while adding charges of different plates battery can be neglected as net charge on battery is zero). (in brief, electric charge will remain conserved in every isolated region).

In Fig.4.208, the electric charges on plates of isolated system

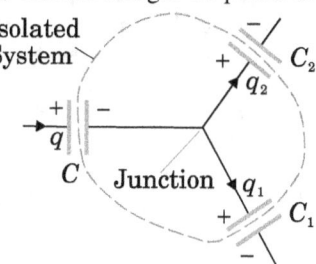
Figure 4.208

are $-q$, q_1 and q_2 respectively, therefore by Kirchhoff's junction rule:
$$\sum q = 0 \Rightarrow -q + q_1 + q_2 = 0$$
or $q = q_1 + q_2 = 0$

2. Voltage or Loop Law (Based on Conservation of Energy)
In a closed circuit, the algebraic sum of the rise up, i.e, voltage gain and voltage drop is zero. i.e., $\sum V = 0$. The direction of the loop is not specified and is chosen in a comfortable manner. There is a voltage gain on moving from a point of lower potential to a point of higher potential, i.e., from $-ve$ plate to $+ve$ plate and a voltage drop when we move from higher potential point to a lower potential point, i.e., from $+ve$ plate to $-ve$ plate.

Sign convention: Voltage gain is taken as positive and voltage drop is taken as negative [Fig.4.209].

Kirchhoff's two rules are all we need to solve a wide variety

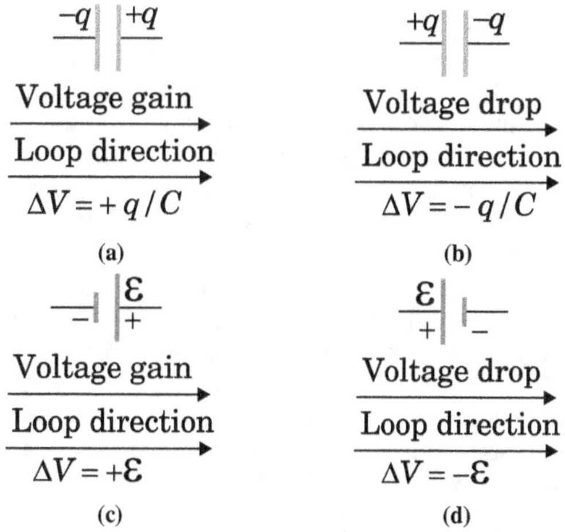

Figure 4.209

of network problems. Usually, some of the emfs, currents, and resistances are known, and others are unknown. We must always obtain from Kirchhoff's rules a number of independent equations equal to the number of unknowns so that we can solve the equations simultaneously.

☞ Resistance offered by a capacitor depends on the type of current source- If current source is a battery of constant emf, then during transient state, capacitor will offer some resistance, but once steady state is reached capacitor will offer infinite resistance. Therefore, in steady state, we can remove all branches containing capacitors.

☞ For simplicity, in transient state of charging of capacitor with a battery or cell, we often neglect it's resistance. An alternating current or a variable current source will result in capacitor offering some resistance which will be discussed in chapter "Alternating Current".

4.30 RC Circuits

A resistor circuit has a steady current. Sometimes a time-varying current is required. A camera's flashbulb requires a short burst of current. A car's turn signal, a patient's electronic pacemaker, and a car's intermittent windshield wipers require a current that repeatedly increases and decreases.

Circuits with resistors and capacitors are called *RC* circuits.

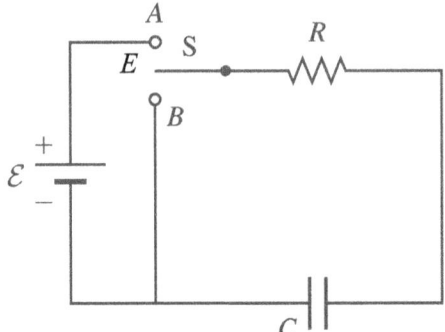

Figure 4.210: An *RC* circuit.

Here, *R* stands for resistance and *C* for capacitance. Fig.4.210 shows an RC circuit with an emf source which can produce such time-varying currents. The switch *S* has three possible positions; in Fig.4.210, it is shown open. When the switch is closed at point *A*, an emf source with terminal potential $\mathcal{E}$ is in the circuit. In this case, the capacitor is charging. When the switch is closed at point *B*, the emf source is not in the circuit and the capacitor is discharging.

4.30.1 Charging

Initially, the capacitor is uncharged and the switch is open (Fig. 4.210). At time $t = 0$, the switch is closed at point *A* so that the circuit consists of a resistor, a capacitor, and the emf source (Fig.4.211). The current in the circuit is initially clockwise as shown. Positive charge builds up on the right plate of the capacitor, and negative charge builds up on the left plate. Charge continues to flow until the voltage across the capacitor $V_C(t)$ equals the terminal potential $\mathcal{E}$ across the emf source. The whole process takes some time and during this time there is an electric current through the connecting wires and the battery. The resistance of the connecting wires and the internal resistance of the battery taken together is shown as the resistance *R*. The capacitor has capacitance *C*.

Suppose, at any instant *t*, the charge on the capacitor and the

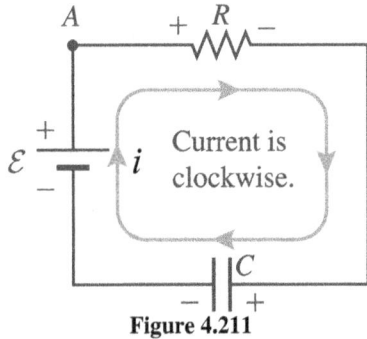

Figure 4.211

current in the circuit are *q* and *i* respectively. The potential drop across the capacitor is q/C and across the resistor it is Ri. Also, the charge deposited on the positive plate in time dt is
$$dq = i\, dt$$
Therefore, current in the circuit
$$i = \frac{dq}{dt}$$

By Kirchhoff's loop law, we have
$$\mathcal{E} - iR - \frac{q}{C} = 0 \qquad (4.198)$$
$$\Rightarrow Ri = \mathcal{E} - \frac{q}{C}$$
$$\Rightarrow R\frac{dq}{dt} = \frac{\mathcal{E}C - q}{C}$$
$$\Rightarrow \int_0^q \frac{dq}{\mathcal{E}C - q} = \int_0^t \frac{1}{CR}dt$$
$$\Rightarrow \ln\frac{\mathcal{E}C - q}{\mathcal{E}C} = \frac{t}{CR}$$
$$\Rightarrow 1 - \frac{q}{\mathcal{E}C} = e^{-t/CR}$$
$$\Rightarrow \boxed{q(t) = \mathcal{E}C\left(1 - e^{-t/CR}\right)} \qquad (4.199)$$

Here, the exponential constant *e*, known as the base for natural logarithms, has the value $e = 2.718\cdots$.

Equation (4.199) gives the charge on the capacitor at time *t*. From Eq.(4.199), it is clear that, *q* increases with increase in time *t*. Figure 4.212a is a graph of *q* as a function of time *t*.

When the switch *S* is closed at $t = 0$: From Eq.(4.199), the charge on the capacitor at time $t = 0$ is-
$$q(0) = C\mathcal{E}\left(1 - e^{-0/RC}\right) = 0$$
as expected.

After a long time, $t \to \infty$: In principle, the charge is maximum, when $t \to \infty$. Let us denote the maximum charge by *Q*, then from Eq.(4.199), we have
$$Q = C\mathcal{E}(1 - e^{-\infty}) = C\mathcal{E}$$
which is also expected.

Instantaneous voltage across capacitor is given by-

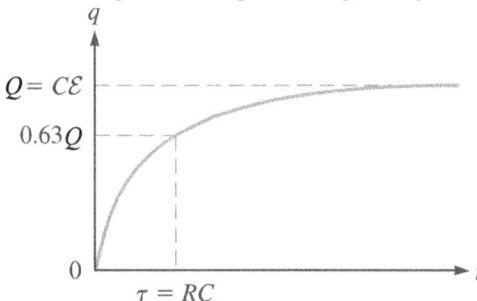

(a) Charge versus time for a charging capacitor.

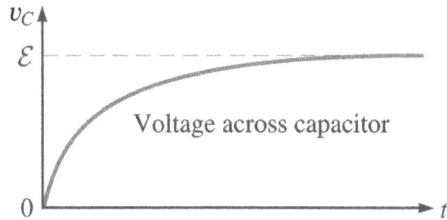

(b) Voltage across capacitor versus time for a charging capacitor.

Figure 4.212

$$v_C = \frac{q}{C}$$
From Eq.(4.199), $q(t) = \mathcal{E}C\left(1 - e^{-t/CR}\right)$, therefore
$$v_C = \mathcal{E}\left(1 - e^{-t/CR}\right) \qquad (4.200)$$

When the switch *S* is closed at $t = 0$: From Eq.(4.200), the voltage across the capacitor at time $t = 0$ is-
$$v_C = \mathcal{E}\left(1 - e^{-0/CR}\right) = \mathcal{E}(1 - 1) = 0$$
as expected.

4.30. RC CIRCUITS

After a long time, $t \to \infty$: When $t \to \infty$, Eq.(4.200) gives-
$$v_C = \mathcal{E}(1 - e^{-\infty}) = \mathcal{E}$$
which is also expected.

Fig.4.212b shows a v_C vs time graph during charging of capacitor. This graph is similar to q-t graph of charging of capacitor (Fig.4.212a)

4.30.2 $i(t)$ for a Charging Capacitor

Now, we find electric current in an RC circuit as a function of time t.

Instantaneous current is defined as- $i = dq/dt$.

Therefore, on differentiating Eq.(4.199), with respect to time t, we get-
$$i(t) = \frac{dq}{dt} = \frac{d}{dt}\left[C\mathcal{E}\left(1 - e^{-t/RC}\right)\right]$$
$$\Rightarrow \quad i(t) = C\mathcal{E}\left[0 - \left(-\frac{1}{RC}\right)e^{-t/RC}\right]$$
$$\Rightarrow \quad i(t) = C\mathcal{E}\left[\left(\frac{1}{RC}\right)e^{-t/RC}\right]$$
$$\Rightarrow \quad i(t) = \left(\frac{\mathcal{E}}{R}\right)e^{-t/RC}$$

From above expression, it is clear that the current in the RC circuit is maximum at $t = 0$, let us denote it by I, then-
$$I = \left(\frac{\mathcal{E}}{R}\right)e^{-0/RC} = \frac{\mathcal{E}}{R}$$
Therefore, the current in an RC circuit can be written as-
$$\boxed{i(t) = Ie^{-t/RC}} \qquad (4.201)$$
here, $I = \mathcal{E}/R$.

When the switch is first closed at $t = 0$: At $t = 0$, the current in the circuit is maximum. From Eq.(4.201), the initial current in RC circuit is-
$$i(0) = Ie^{-0/RC} = I = \frac{\mathcal{E}}{R}$$
as expected for an uncharged capacitor.

After the switch has been closed for a long time, $t \to \infty$, then current in the circuit-
$$i(t \to \infty) = \left(\frac{\mathcal{E}}{R}\right)e^{-\infty/RC} = 0$$
as we expect for a fully charged capacitor.

Fig.4.213a shows the current $i(t)$ time graph for an RC circuit. Since, the voltage drop (v_R) across the resistance R, is given by
$$V_R = i(t)R = Ie^{-t/RC}R = \mathcal{E}e^{-t/RC}$$
Therefore, from Eq. (4.201), the voltage drop across the resistance is given by-
$$v_R(t) = i(t)R = \left(\frac{\mathcal{E}}{R}\right)e^{-t/RC}$$

Fig.4.213b shows the voltage time graph across the resistor R for charging of the capacitor. It is similar to current time graph (Fig.4.213a) for the circuit.

4.30.2.1 Time constant

The constant CR has dimensions of time* and is called time constant of the circuit. It is denoted by τ.
$$\boxed{\tau = CR} \qquad (4.202)$$
In one time constant $\tau(= CR)$, the charge accumulated on the capacitor is
$$q(\tau) = C\mathcal{E}\left(1 - \frac{1}{e}\right) = 0.63\,\mathcal{E}C$$

---
*The following dimensional analysis shows that τ has units of time:
$$[\tau] = [RC] = \left[\left(\frac{V}{I}\right)\left(\frac{Q}{V}\right)\right] = \left[\frac{Q}{Q/\Delta t}\right] = [\Delta t] = [T]$$

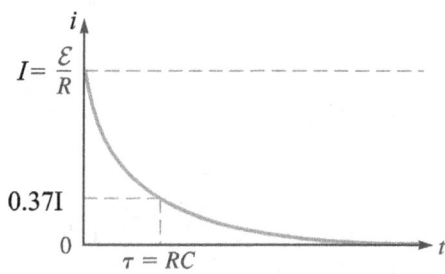

(a) Current versus time in a charging RC circuit.

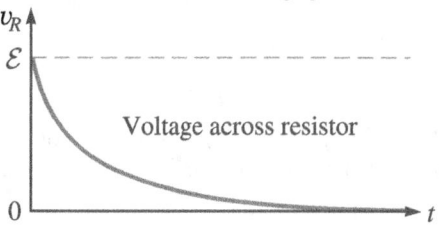

(b) Voltage across resistor versus time in a charging RC circuit.

Figure 4.213

Thus, 63% of the maximum charge is deposited in one time constant(Fig.4.212a).

Now, the current in RC circuit in one time constant drops to-
$$i(\tau) = Ie^{-\tau/\tau} = Ie^{-1} \approx 0.37\,I$$
We see that the time constant τ is the time the current takes to drop to 37% of its initial value (Fig.4.213a).

4.30.3 Discharging

When the switch in Figure 4.210 has been closed at point A for a long time (at least several time constants), the capacitor is fully charged, which means $q = Q$ and its voltage must equal the terminal potential of the emf source, so now the initial voltage across the capacitor must be $V_C(0) = Q/C = \mathcal{E}$. At this time there is no current in the circuit. To discharge the capacitor, the switch is thrown to point B so that there is no emf device in the circuit (Fig.4.214). For convenience, we reset the time to $t = 0$ at the moment the switch is thrown to point B. (This is like resetting a stopwatch.) After the switch is thrown to B, the capacitor discharges through a counterclockwise current in the circuit. When the capacitor is completely discharged, the voltage across it must be zero. So, the voltage across the capacitor goes from $\mathcal{E}$ to zero (Fig.4.214). Similarly, we expect the charge stored by the capacitor to decay from its maximum value.

Suppose a capacitor of capacitance C has a charge Q. At $t = 0$,

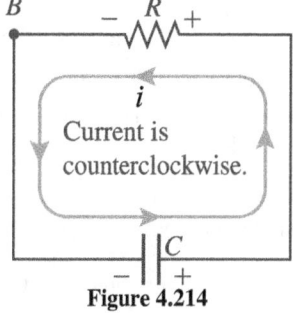

Figure 4.214

the plates are connected through a resistance R. Let the charge on the capacitor be q and the current in the circuit be i at time t. Again by Kirchhoff's loop law, we have
$$\frac{q}{C} - Ri = 0$$
Here, $i = -\dfrac{dq}{dt}$, because the charge q decreases as time passes.

So,
$$\frac{q}{C} - R\left(-\frac{dq}{dt}\right) = 0$$

or
$$\frac{dq}{dt} = -\frac{q}{CR}$$

$$\Rightarrow \frac{dq}{q} = -\frac{1}{CR}dt$$

If during time interval 0 to t, capacitor's charge decreases from Q to q, then integration of above equation within this interval, gives
$$\int_Q^q \frac{dq}{q} = \int_0^t -\frac{1}{CR}dt$$

$$\Rightarrow \ln\frac{q}{Q} = -\frac{t}{CR}$$

$$\boxed{q(t) = Qe^{-t/CR}} \quad (4.203)$$

Figure 4.215a shows a graph of q as a function of t.

When the switch is closed at point B, at $t = 0$, then from

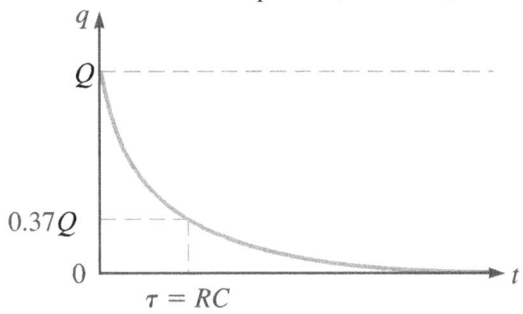

(a) Charge versus time for a discharging RC circuit.

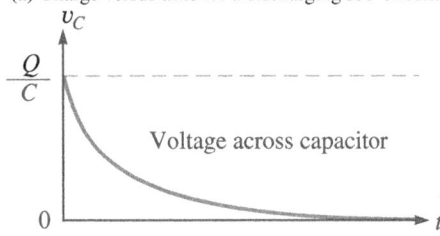

(b) Voltage versus time across capacitor for a discharging RC circuit.

Figure 4.215

Eq.(4.203), we have-
$$q(0) = Qe^{-0/\tau} = Q$$
so the charge is at its maximum.

After a long time, $t \to \infty$, charge $q(t \to \infty) = Q(e^{-\infty}) \to 0$, so the charge stored by the capacitor falls to zero.

Finally, the time constant is the time it takes the capacitor's stored charge to drop to about 37% of its maximum:
$$q(\tau) = Q\left(e^{-1}\right) \approx 0.37Q.$$

The voltage across the capacitor is given by-
$$\boxed{v_C = \frac{q}{C} = \frac{Q}{C}e^{-t/CR} = V_C e^{-t/CR}} \quad (4.204)$$

Here, $V_C = Q/C$ is the maximum voltage across capacitor which is at $t = 0$.

When the switch is closed at point B, at $t = 0$, then from Eq.(4.204), we have-
$$v_C(0) = \frac{Q}{C}e^{-0/\tau} = \frac{Q}{C}$$

Here, Q/C, is the maximum voltage across the capacitor.

After a long time, $t \to \infty$, voltage across the capacitor $v_C(t \to \infty) = \frac{Q}{C}(e^{-\infty}) \to 0$, so the voltage across the capacitor falls to zero.

Fig.4.215b shows voltage time graph across the capacitor in RC discharging circuit. It is similar to charge time graph (Fig.4.215a)

4.30.3.1 Discharging Current

For the charging capacitor, the current is clockwise, which we set to positive (Eq.4.211). In the discharging RC circuit, the current is counterclockwise (Fig.4.214).

From Eq.(4.203), the charge at time t in discharging the capacitor is-
$$q(t) = Qe^{-t/CR}$$

On differentiating above equation with respect to time t, we get
$$\frac{dq}{dt} = Qe^{-t/CR} \times \left(-\frac{1}{CR}\right)$$

or
$$\frac{dq}{dt} = \frac{-Q}{CR}e^{-t/CR}$$

or
$$i = \frac{dq}{dt} = \frac{-Q}{CR}e^{-t/CR}$$

where $\frac{Q}{RC} = \frac{Q/C}{R} = \frac{\mathcal{E}}{R} = I$ is the initial current. The negative sign indicates that as the capacitor discharges, the current direction is opposite its direction when the capacitor was being charged. Therefore,

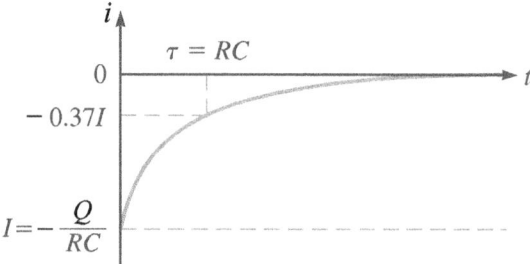

(a) Current versus time for a discharging RC circuit.

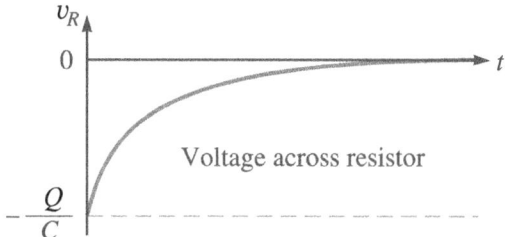

(b) Voltage versus time across resistor for a discharging RC circuit.

Figure 4.216

$$\boxed{i(t) = -Ie^{-t/CR}} \quad (4.205)$$

A graph of i as a function of t is shown in Fig.4.216a. If initially, the capacitor was fully charged, then $i(0) = -I = -\mathcal{E}/R$. After the switch has been closed for a long time, $t \to \infty$, the current goes to zero:
$$i(t \to \infty) = -Ie^{-\infty/\tau} \to 0$$

We see that the time constant is also the time it takes the current in a discharging RC circuit to drop to 37% of its initial value: $I(\tau) \approx 0.37I_0$.

In principle, discharging is complete only at $t = \infty$. The constant CR is the time constant. At $t = CR$, the remaining charge is $q = \frac{Q}{e} = 0.37Q$. Thus, 63% of the discharging is complete in one time constant.

In discharging, the voltage across the resistance as a function of time is given by-
$$v_R = iR = -IRe^{-t/RC}$$

here, $IR = \mathcal{E} = Q/C$ is the maximum voltage across the resistance R in discharging RC circuit.

$$\boxed{v_R = iR = -IRe^{-t/RC}} \quad (4.206)$$

4.30. RC CIRCUITS

Fig.4.216b shows the discharging current as a function of time in RC circuit. It is similar to current time graph shown in Fig.4.216a.

EXAMPLE 95. *RC circuit, with emf* The capacitance in the circuit of Fig.4.211 is $C = 0.30\ \mu F$, the total resistance is $R = 20\ k\Omega$, and the battery emf is 12 V. Determine (a) the time constant, (b) the maximum charge the capacitor could acquire, (c) the time it takes for the charge to reach 99% of this value, (d) the current I when the charge Q on the capacitor is half its maximum value, (e) the maximum current in the circuit, and (f) the charge q on the capacitor when the current i is 0.20 of its maximum value.

SOLUTION (a) The time constant is
$$\tau = RC = \left(2.0\times 10^4\Omega\right)\left(3.0\times 10^{-7}F\right)$$
$$= 6.0\times 10^{-3}s = 6.0\ ms.*$$

(b) The maximum charge on the capacitor would occur when no further current flows, so
$$q_{max} = Q = C\mathcal{E} = \left(3.0\times 10^{-7}F\right)(12V) = 3.6\mu C.$$

(c) In Eq.(4.199), we set $q = 0.99C\mathcal{E}$:
$$0.99C\mathcal{E} = C\mathcal{E}\left(1 - e^{-t/RC}\right)$$
$$\Rightarrow\quad e^{-t/RC} = 1 - 0.99 = 0.01$$

On taking the natural logarithm of both sides, and applying $\ln e^x = x$, we get-
$$\frac{t}{RC} = -\ln(0.01) = 4.6$$
$$\Rightarrow\quad t = 4.6RC = (4.6)\left(6.0\times 10^{-3}s\right)$$
$$= 28\times 10^{-3}s = 28\ ms\ (\text{less than } \tfrac{1}{30}s).$$

(d) From part (b) the maximum charge on the capacitor is 3.6 μC. When the charge is half this value, 1.8 μC, the current i in the circuit can be found using Eq.(4.201):
$$i = \frac{1}{R}\left(\mathcal{E} - \frac{Q}{C}\right)$$
$$= \frac{1}{2.0\times 10^4\Omega}\left(12V - \frac{1.8\times 10^{-6}C}{0.30\times 10^{-6}F}\right) = 300\ \mu A$$

(e) The current is a maximum at $t = 0$ (the moment when the switch is closed) and there is no charge yet on the capacitor ($q = 0$). So the maximum current is
$$I = \frac{\mathcal{E}}{R} = \frac{12V}{2.0\times 10^4\Omega} = 600\mu A.$$

(f) By Kirchhoff's loop law in circuit 4.211, we have-
$$\frac{q}{C} + Ri - \mathcal{E} = 0$$
or $\quad q = C(E - iR)$
or $\quad i = \left(3.0\times 10^{-7}F\right)\left[12V - \left(1.2\times 10^{-4}A\right)\left(2.0\times 10^4\Omega\right)\right]$
$$= 2.9\mu C.$$

EXAMPLE 96. *Discharging RC circuit* In the RC circuit shown in Fig.4.217, the battery has fully charged the capacitor, so $Q = C\mathcal{E}$. Then at $t = 0$ the switch is thrown from position a to b. The battery emf is 20.0 V, and the capacitance $C = 1.02\ \mu F$. The current i is observed to decrease to 0.50 of its initial value in 40 μs. (a) What is the value of q, the charge on the capacitor, at $t = 0$? (b) What is the value of R? (c) What is q at $t = 60\ \mu s$? (d) Draw a voltage across capacitor versus time graph for a discharging RC circuit.

APPROACH At $t = 0$, the capacitor has charge $Q = C\mathcal{E}$, and

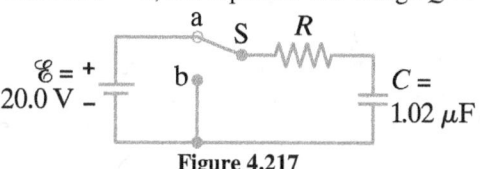

Figure 4.217

then the battery is removed from the circuit and the capacitor begins discharging through the resistor, as in Fig.4.217. At any time t later (Eq. (4.203)) we have
$$q = Qe^{-t/RC} = C\mathcal{E}e^{-t/RC}$$

SOLUTION (a) At $t = 0$, the charge on the capacitor is
$$Q = C\mathcal{E} = \left(1.02\times 10^{-6}F\right)(20.0V) = 2.04\times 10^{-5}C = 20.4\ \mu C$$

(b) To find R, we are given that at $t = 40\ \mu s$, $i = 0.50\ I$. Hence (Eq.4.205):
$$0.50I = Ie^{-t/RC}$$
$$\Rightarrow\quad \frac{1}{2} = e^{-t/RC}$$

Taking natural logarithm of both sides ($\ln 1 = 0$, $\ln 2 = 0.693$), we get-
$$\ln 1 - \ln 2 = -\frac{t}{RC}$$
$$\Rightarrow\quad 0 - 0.693 = -\frac{t}{RC}$$
$$\Rightarrow\quad R = \frac{t}{(0.693)C} = \frac{\left(40\times 10^{-6}s\right)}{(0.693)\left(1.02\times 10^{-6}F\right)} = 57\ \Omega$$

(c) At $t = 60\ \mu s$,
$$q = Qe^{-t/RC} = \left(20.4\times 10^{-6}C\right)e^{-\frac{60\times 10^{-6}s}{(57\Omega)\left(1.02\times 10^{-6}F\right)}} = 7.3\mu C.$$

EXAMPLE 97. *How quickly does voltage drop* A charged capacitor, $C = 35\ \mu F$, is connected to a resistance $R = 120\ \Omega$ as in Fig.4.218. How much time will elapse until the voltage falls to 10% of its original (maximum) value?

APPROACH The voltage across the capacitor decreases accord-

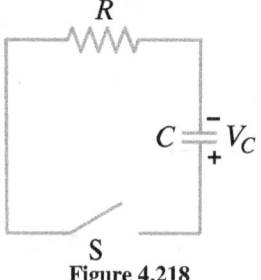

Figure 4.218

ing to Eq.(4.204): $v_C = V_C e^{-t/RC}$

SOLUTION The time constant for this circuit is given by
$$\tau = RC = (120\Omega)\left(35\times 10^{-6}F\right) = 4.2\times 10^{-3}s.$$

After a time t, the voltage across the capacitor will be
$$v_C = V_C e^{-t/RC}$$

We want to know the time t for which $v_C = 0.10V_C$. We substitute into the above equation
$$0.10V = Ve^{-t/RC}\quad \Rightarrow\quad e^{-t/RC} = 0.10$$
or $\quad e^{-t/RC} = \frac{1}{10}$

Taking natural logarithm of both sides, we get-
$$\ln\left(e^{-t/RC}\right) = \ln\frac{1}{10}$$
or $\quad -t/RC = \ln 1 - \ln 10 = 0 - 2.303\log_{10} 10$
or $\quad t/RC = 2.303$

---

*We see that in a typical charging or discharging circuit, the time constant is of the order of a millisecond. Also, four to five time constants are sufficient for 99% of the charging or discharging. Thus, for practical purposes, we can assume that charging or discharging is complete in a fraction of a second.

Solving for t, we find the elapsed time:
$$t = 2.303(RC) = (2.303)\left(4.2 \times 10^{-3} s\right) = 9.6726 \times 10^{-3} s$$
$$\approx 9.7 \text{ ms}$$

NOTE We can find the time for any specified voltage across a capacitor by using $t = RC \ln(v_C/V_C)$.

The graph between voltage across capacitor versus time is shown in Fig.4.219.

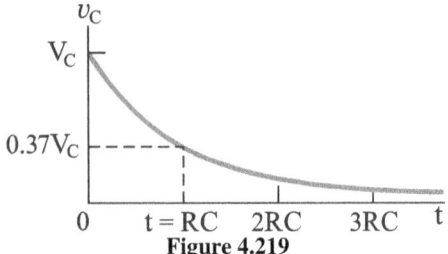
Figure 4.219

Important Points
1. Charge on the capacitor does not change suddenly if there is a resistance in the path (series) of the capacitor.
2. When an uncharged capacitor is connected with battery then its charge is zero initially hence potential difference across it is zero initially. At this time the capacitor can be treated as a conducting wire.
3. The current will become zero finally (that means in steady state) in the branch which contains capacitor (Fig.4.220b). It means a fully charged capacitor offers ∞ resistance.

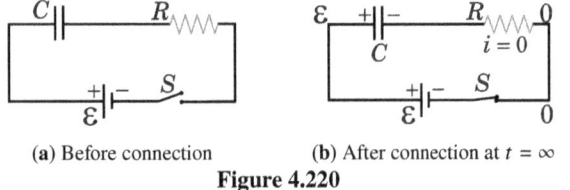
(a) Before connection (b) After connection at $t = \infty$
Figure 4.220

EXAMPLE 98. Find out current in the circuit [Fig.4.221] and charge on capacitor which is initially uncharged in the following situations.
(a) Just after the switch is closed.
(b) After a long time when switch was closed.

SOLUTION (a) Just after closing the switch:

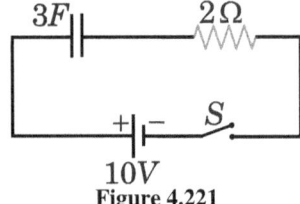
Figure 4.221

At $t = 0$, capacitor behaves like a conducting wire (Fig.4.222a).
So, potential difference across capacitor $= 0$
Therefore, $Q_C = 0$
And potential difference across the resistor,
$$\Delta V_R = 10 \text{ V}$$
Therefore, electric current in the resistor
$$i = \frac{\Delta V_R}{R} = \frac{10 \text{ V}}{2 \text{ }\Omega} = 5 \text{ A}$$

(b) After a long time
After a long time $t = \infty$, the capacitor will get fully charged and then it offers infinite resistance.
So, after a long time (Fig.4.222b), the current in the circuit, becomes zero, and there will not be any potential drop across resistance. So, in this case, the potential difference across capacitor will be equal to potential difference across the battery i.e., $10\ V$,

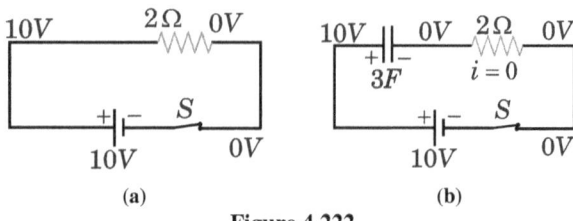
Figure 4.222

Therefore, after long time, magnitude of charge on either plate of capacitor,
$$Q = CV = 3\ F \times 10\ V = 30\ C$$

EXAMPLE 99. The switch in circuit (Fig.4.223a) shifts from 1 to 2 when $v_C > 2V/3$ and goes back to 1 from 2 when $v_C < V/3$. The voltmeter reads voltage as plotted (Fig.4.223b). What is the period T of the wave form in terms of R and C?
(A) $RC \ln 3$ (B) $2RC \ln 2$
(C) $\frac{RC}{2} \ln 3$ (D) $\frac{RC}{3} \ln 3$

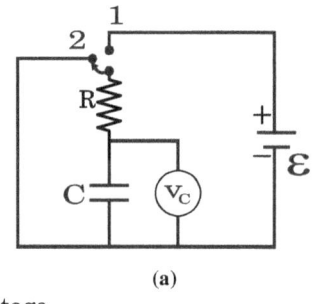
(a)

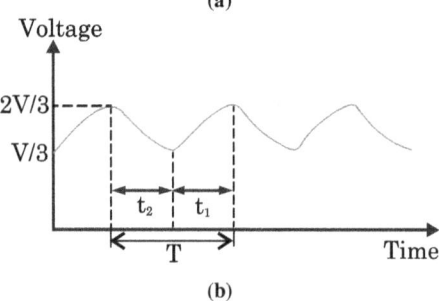
(b)
Figure 4.223

APPROACH When the capacitor is connected to terminal 1, there is a charging through resistor R and when the switch is connected to terminal 2, there is a discharging of capacitor through resistor R. According to given problem, initially, the switch was connected to terminal 1 and the capacitor was charged up to its terminal voltage $2V/3$ and then switch shifted to terminal 2. During the time up to which the switch remains connected to terminal 2, there is a discharging of capacitor. Again according to problem, the capacitor's terminal voltage drops from $2V/3$ to $V/3$ through resistor R, in time t_2 and then switch shifted back to terminal 1. Here, it's terminal voltage rises from $V/3$ to $2V/3$ in time t_1.

Now, from Fig.4.223b, clearly the time period of the wave is -
$$T = t_2 + t_1$$

Time t_2 and t_1 can be calculated by using equations for discharging and charging respectively of capacitor in an RC circuit.

Equations Used

For discharging: During time 't_2' capacitor is discharging with the help of resistor 'R'. Here voltage drops from $2V/3$ to $V/3$, therefore we use-
$$v_C = V_C e^{-t/RC} \quad (1)$$
with $v_C = V/3$, $t = t_2$ and $V_C = $ maximum voltage across capacitor $= 2V/3$ at the place of V.

For Charging: For charging during time interval t_1, we use equation-

4.30. RC CIRCUITS

$$v_C = V_C \left(1 - e^{-t/RC}\right) \quad (2)$$

with v_C = rise in voltage = $\frac{2V}{3} - \frac{V}{3} = \frac{V}{3}$, $V_C = \frac{2V}{3}$, and $t = t_1$.

SOLUTION (B) Calculation of discharging time t_2
From Eq. (1), we have-

$$v_C = V_C e^{-t/RC}$$

or $$\frac{V}{3} = \frac{2V}{3} e^{-t_2/RC}$$

or $$\frac{1}{2} = e^{-t_2/RC}$$

On taking natural log of both sides, we get,

$$\ln \frac{1}{2} = \ln e^{-t_2/RC}$$

$$\Rightarrow \quad \ln 1 - \ln 2 = \frac{-t}{RC}$$

$$\Rightarrow \quad 0 - \ln 2 = \frac{-t_2}{RC}$$

$$\Rightarrow \quad 0 - \ln 2 = \frac{-t_2}{RC}$$

$$\Rightarrow \quad t_2 = RC \ln 2$$

Calculation of charging time t_1
From Eq. (2), we have-

$$v_C = V_C \left(1 - e^{-t/RC}\right)$$

or $$\frac{V}{3} = \frac{2V}{3} \left(1 - e^{-t_1/RC}\right)$$

or $$\frac{1}{2} = \left(1 - e^{-t_1/RC}\right)$$

or $$e^{-t_1/RC} = 1 - \frac{1}{2}$$

or $$e^{-t_1/RC} = \frac{1}{2}$$

On taking natural log of both sides, we get,

$$\frac{-t_1}{RC} = \ln 1 - \ln 2$$

$$\Rightarrow \quad \frac{-t_1}{RC} = 0 - \ln 2$$

$$\Rightarrow \quad \frac{t_1}{RC} = \ln 2$$

$$\Rightarrow \quad t_1 = RC \ln 2$$

Therefore, time period (see Fig.4.223b),

$$T = t_2 + t_1 = RC \ln 2 + RC \ln 2 = 2RC \ln 2$$

EXAMPLE 100. A varying voltage is applied between the terminals A, B so that the voltage across the capacitor varies as shown in the Fig.4.224a. Then-

(A) The voltage between the terminals C and D is constant between $2t_0$ and $3t_0$
(B) The current in the resistor is 0 between $2t_0$ and $3t_0$
(C) The current in the resistor between t_0 and $2t_0$ is twice the current between $3t_0$ and $5t_0$
(D) None of these

APPROACH Charge stored, on the positive plate of a capacitor,

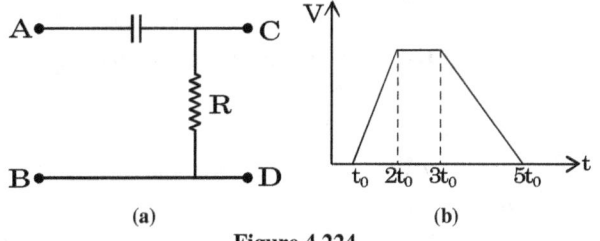

Figure 4.224

is given by-

$$q = CV \quad (1)$$

So, when the voltage across the capacitor is constant, charge stored on it will also be constant. In this case, there will not be any current in capacitor branch. So, in this case, (Fig.4.224b), there will not be any current in resistor R.

On differentiating Eq.(1), with respect to t, we get-

$$\frac{dq}{dt} = C \frac{dV}{dt}$$

or $$i = C \frac{dV}{dt} \quad (2)$$

So, current in circuit is equal to C times of slope of V-t graph. and for constant value of V, current i will be zero.

SOLUTION (ABC) From Fig.4.224a, voltage V across C is constant in time interval 0 to t_0 and again constant in time interval $2t_0$ to $3t_0$, so in these intervals, there will not be any current in R and hence voltage across C and D will also be constant.
Therefore, options (A) and (B) are correct.

From Fig.4.224b, the magnitude of slope of V-t graph is double in time interval t_0 to $2t_0$ as compared to that in interval $3t_0$ to $5t_0$, therefore the current in time interval t_0 to $2t_0$ will also be double that of in time interval $3t_0$ to $5t_0$. So option (C) is also correct.

EXAMPLE 101. Study the circuit diagram shown in Fig.4.225 and mark the correct option(s)

(A) The potential of point a with respect to point b when switch S is open is -6 V.
(B) The points a and b, are at the same potential, when S is opened.
(C) The charge flows through switch S when it is closed is 54 μC
(D) The final potential of b with respect to ground when switch S is closed is 8 V

SOLUTION (A, C) In given circuit diagram Fig. 4.225,

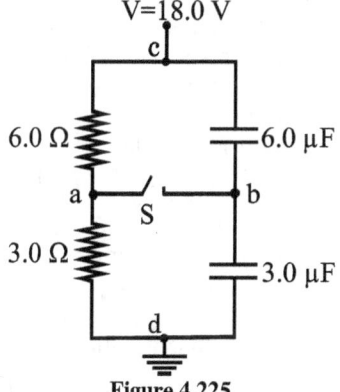

Figure 4.225

terminal d is grounded. So, d can be considered at zero potential. The potential of terminal c is 18.0 V. These potentials are shown in Fig.4.226a. Therefore, potential difference between terminals d and c is

$$V_{cd} = V_c - V_d = 18\ V - 0\ V = 18\ V$$

When switch S is opened: When switch S is opened, capacitors 3.0 μF and 6 μF will be in series. Equivalent capacitance between terminals c and d is-

$$C_{cd} = \frac{3.0\ \mu F \times 6.0\ \mu F}{3.0\ \mu F + 6.0\ \mu F} = 2\ \mu F$$

So, charge on each capacitor (Fig.4.226a),

$$q = C_{cd} V_{cd} = 2\ \mu F \times 18\ V = 36\ \mu C$$

Therefore, net charge on capacitor plates connected with point b

$$q_b = q + (-q) = 36\ \mu C + (-36)\ \mu C = 0\ \mu C$$

If potential at terminal b is V_b, then from Fig.4.226a, potential difference between terminals b and c is

$$V_c - V_b = \frac{q}{C_{bc}} = \frac{36\ \mu C}{6\ \mu F} = 6\ V$$

or $$V_b = V_c - 6\ V = 18\ V - 6\ V = 12\ V$$

Net resistance between terminals c and d is

$$R_{cd} = R_{ca} + R_{ac} = 3.0\ \Omega + 6.0\ \Omega = 9.0\ \Omega$$

Therefore, by Ohm's law, the electric current in branch cad is -

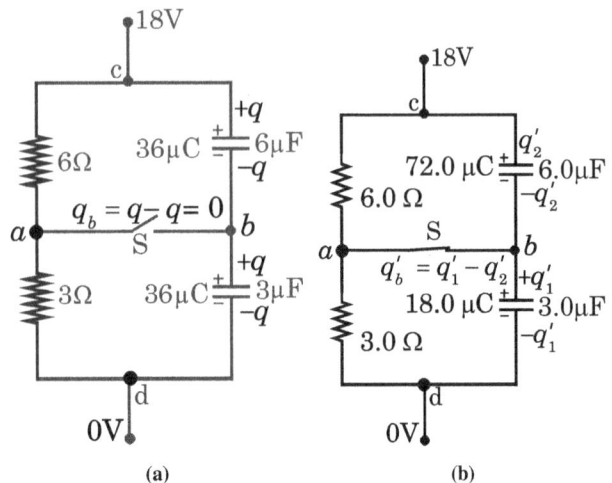

Figure 4.226

$$I = \frac{V_{cd}}{R_{cd}} = \frac{18.0\ V}{9.0\ \Omega} = 2.0\ A$$

If electric potential at point a is V_a, then by Ohm's law between terminals a and c, we can write-

$$V_c - V_a = IR_{ac}$$

or $\qquad V_a = V_c - IR_{ac} = 18\ V - (2.0\ A \times 6.0\ \Omega) = 6\ V$

Therefore, potential difference between terminals a and b is

$$V_{ba} = V_b - V_a = 12\ V - 6\ V = 6\ V$$

☞ So, option (B) is incorrect whereas option (A) is correct, i.e., potential of terminal a with respect to b is $-6\ V$.

When switch S is closed: In this case (Fig.4.226b), the potential of point a will also be the potential of b.

So, the potential of point b is $V_b = 6\ V$

☞ Therefore, option (D) is incorrect.

Now, charge stored on $3\ \mu F$ capacitor-

$$q'_1 = (3\ \mu F)(V_b - V_a) = (3\ \mu F)(6\ V - 0) = 18\ \mu C$$

and charge stored on $6\ \mu F$ capacitor-

$$q'_2 = (6\ \mu F)(V_c - V_b) = (6\ \mu F)(18\ V - 6\ V) = 72\ \mu C$$

Therefore, after switch S is closed (Fig.4.226b), net charge on capacitor plates connected with point b

$$q'_b = 18\ \mu C + (-72)\ \mu C = -54\ \mu C$$

Therefore, charges flown through S after it is closed: $54 - 0 = 54\mu C$.

☞ So, option (C) is also correct.

Examples 102-104

In the circuit shown in Fig.4.227, the battery is an ideal one with emf V. The capacitor is initially uncharged. The switch S is closed at time $t = 0$.

EXAMPLE 102. The charge Q on the capacitor at time t is-

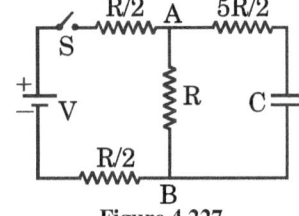

Figure 4.227

(A) $\frac{CV}{2}\left(1 - e^{\frac{t}{RC}}\right)$ \qquad (B) $\frac{CV}{2}\left(1 - e^{-\frac{t}{3RC}}\right)$

(C) $\frac{CV}{2}\left(1 - e^{-\frac{2t}{5RC}}\right)$ \qquad (D) $\frac{CV}{2}\left(1 - e^{-\frac{2t}{9RC}}\right)$

EXAMPLE 103. The current in AB at time t is-

(A) $\frac{V}{2R}\left(1 - e^{-\frac{t}{3RC}}\right)$ \qquad (B) $\frac{2V}{R}\left(1 - e^{-\frac{t}{3RC}}\right)$

(C) $\frac{2V}{R}\left(1 - \frac{e^{-\frac{t}{3RC}}}{6}\right)$ \qquad (D) $\frac{V}{2R}\left(1 - \frac{e^{-\frac{t}{3RC}}}{6}\right)$

EXAMPLE 104. What is its limiting value at $t \to \infty$?

(A) $\frac{V}{2R}$ \qquad (B) $\frac{V}{R}$ \qquad (C) $\frac{2V}{R}$ \qquad (D) $\frac{V}{3R}$

SOLUTION 102(B) Calculation of maximum voltage and maximum charge on capacitor When capacitor is fully charged (Fig.4.228a), there will not be any current in capacitor branch (steady state). If, in steady state, the current in left loop is I, then by Ohm's law, we have-

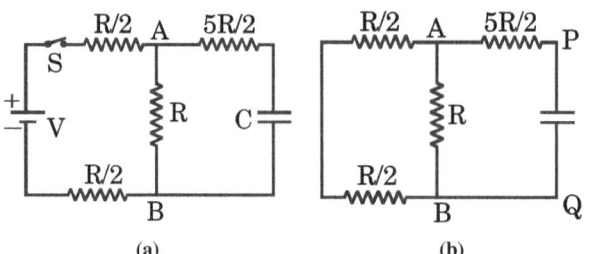

Figure 4.228

$$I = \frac{V}{R_{eq}}$$

Here, R_{eq} is the equivalent resistance of the left loop. It's value is-

$$R_{eq} = \frac{R}{2} + R + \frac{R}{2} = 2R$$

Therefore, electric current in left loop-

$$I = \frac{V}{2R}$$

So, potential difference across AB is

$$V_{AB} = IR = \frac{V}{2R}R = \frac{V}{2}$$

Now, in right loop, the capacitor is connected across AB through a series resistance of $5R/2$. Therefore, in steady state, the maximum voltage across the capacitor is-

$$V_C = V/2$$

So, maximum charge stored in the capacitor-

$$Q = CV_C = CV/2$$

Calculation of time constant To calculate time constant, we need effective resistance across capacitor. For it, we mentally remove the emf source as shown in Fig.4.228b. Now, effective resistance across the capacitor C is-

$$R'_{eq} = \frac{(R/2 + R/2)R}{(R/2 + R/2) + R} + \frac{5R}{2} = \frac{R}{2} + \frac{5R}{2} = 3R$$

Therefore, time constant -

$$\tau_c = R'_{eq}C = 3RC$$

Calculation of instantaneous charge: The charge stored in capacitor plates at time t-

$$q = Q\left(1 - e^{-t/\tau_c}\right) = \frac{CV}{2}\left(1 - e^{-t/3RC}\right)$$

or $\qquad q = \frac{CV}{2}\left(1 - e^{-t/3RC}\right) \qquad (1)$

SOLUTION 103(D) The voltage across AB at time t is-

$$v_{AB} = v_{5R/2} + v_c$$

or $\qquad v_{AB} = \frac{5}{2}Ri + \frac{q}{C} \qquad (2)$

here, $i = \frac{dq}{dt} = \frac{d}{dt}\frac{CV}{2}\left(1 - e^{-t/3RC}\right)$

$$= \frac{CV}{2}\left(0 + \frac{1}{3RC}e^{-t/3RC}\right)$$

$$= \frac{V}{6R}e^{-t/3RC}$$

On substituting this value of i in Eq. (2), we get

$$v_{AB} = \frac{5V}{12}e^{-t/3RC} + \frac{V}{2}\left(1 - e^{-t/3RC}\right)$$

Therefore, the current in AB at time t is-

$$i_{AB} = v_{AB}/R$$
$$\Rightarrow \quad i_{AB} = \frac{5V}{12R}e^{-t/3RC} + \frac{V}{2R}\left(1 - e^{-t/3RC}\right)$$
$$\Rightarrow \quad i_{AB} = \frac{V}{2R}\left(1 - \frac{e^{-t/3RC}}{6}\right) \qquad (3)$$

SOLUTION 104(A) From Eq.(3), we have-
$$\lim_{t \to \infty} i_{AB} = \lim_{t \to \infty} \frac{V}{2R}\left(1 - \frac{e^{-t/3RC}}{6}\right) = \frac{V}{2R}$$

EXAMPLE 105. Find the time constant for the given R-C circuits in Fig.4.229 in correct order (in μs), [Given $R_1 = 1\,\Omega$, $R_2 = 2\,\Omega$, $C_1 = 4\mu F$, $C_2 = 2\mu F$.]

SOLUTION The time constant of a R-C circuit is given by

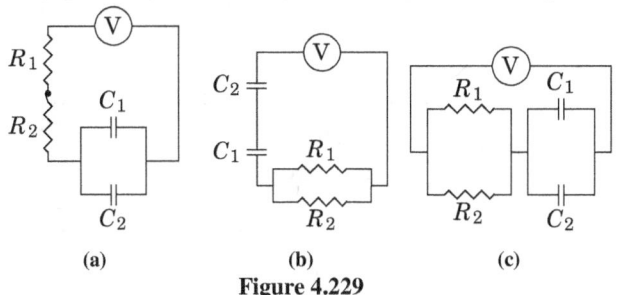

Figure 4.229

$\tau = R_e C_e$, where R_e is the equivalent resistance and C_e is the equivalent capacitance.
(a) For circuit shown in Fig.4.229a, we have
The equivalent resistance of circuit,
$$R_e = R_1 + R_2 = 3\Omega,$$
Equivalent capacitance of the given circuit,
$$C_e = C_1 + C_2 = 6\mu F,$$
Therefore, the time constant
$$\tau_1 = R_e C_e = 18\mu s$$
(b) For circuit shown in Fig.4.229b, we have
The equivalent resistance of circuit,
$$R_e = \frac{R_1 R_2}{R_1 + R_2} = \frac{2}{3}\Omega,$$
Equivalent capacitance of the given circuit,
$$C_e = \frac{C_1 C_2}{C_1 + C_2} = \frac{8}{6}\mu F,$$
Therefore, the time constant
$$\tau_2 = \frac{8}{9}\mu s$$
(c) For circuit shown in Fig.4.229b, we have
The equivalent resistance of circuit,
$$R_e = 2/3\Omega,$$
Equivalent capacitance of the given circuit, $C_e = 6\,\mu F$,
Therefore, the time constant, $\tau_3 = 4\mu s$
Clearly, time constants in given order are 18, 8/9, 4 respectively.

EXAMPLE 106. A $4\,\mu F$ capacitor and a 2.5 MΩ resistance are in series with 12 V battery. Find the time after which the potential difference across the capacitor is 3 times the potential difference across the resistor. [Given $\ln 2 = 0.693$]

SOLUTION The circuit for charging the capacitor is shown in

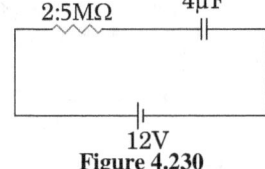

Figure 4.230

the Fig.4.230.
Here, $V = 12$ V is the battery voltage, $R = 2.5$MΩ is the series resistance, and $C = 4\mu F$ is the capacitance. The potential across the capacitor at time t is

$$V_C = V\left[1 - e^{-\frac{t}{RC}}\right] \qquad \cdots (1)$$
and the potential across the resistor is
$$V_R = V - V_C \qquad \cdots (2)$$
Using equations (1) and (2), and the condition $V_C = 3V_R$, we get
$$e^{-\frac{t}{RC}} = 1/4$$
Take logarithm on both sides and simplify to get
$$t = 2RC \ln 2 = 2\left(2.5 \times 10^6\right)\left(4 \times 10^{-6}\right)(0.693)$$
$$= 13.86\text{ s}$$

EXAMPLE 107. Find the charges on the four capacitors of capacitances $1\,\mu F$, $2\,\mu F$, $3\,\mu F$ and $4\,\mu F$ shown in Fig.4.231 in steady state.

APPROACH In this problem, we first find potential difference

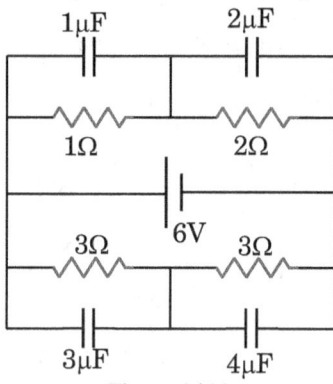

Figure 4.231

across each capacitor, then we calculate charge on each capacitor by applying $Q = CV$.

SOLUTION Given circuit is redrawn in Fig.4.232. Since, in steady state no current flows through capacitors, therefore to determine currents in branches JH and CF, we can remove the branches containing capacitors. Next simplified circuit diagram is shown in Fig.4.233.

Net resistance of branch JH is $1\,\Omega + 2\,\Omega$ i.e., $3\,\Omega$ whereas that

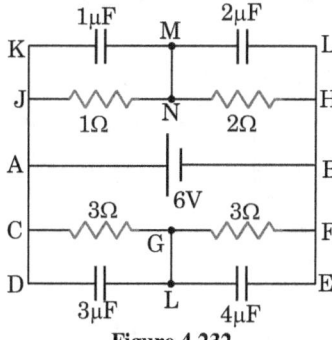

Figure 4.232

of branch CF is $3\,\Omega + 3\,\Omega = 6\,\Omega$. These branches are in parallel, therefore, equivalent resistance across AB is
$$R_{eq} = \frac{3\,\Omega \times 6\,\Omega}{3\,\Omega + 6\,\Omega} = 2\,\Omega$$
So, net current supplied by 6 V battery
$$I = \frac{\Delta V}{R} = \frac{6\text{ V}}{2\,\Omega} = 3\text{ A}$$
Now, by current division rule, current in branch JH
$$I_1 = \left(\frac{6}{3+6}\right)3\text{ A} = 2\text{ A}$$
So, 2 A current passes through resistances $1\,\Omega$ and $2\,\Omega$ connected in series between points JN and NH respectively.
Similarly, the electric current in branch CF
$$I_2 = \left(\frac{3}{3+6}\right)3\text{ A} = 1\text{ A}$$

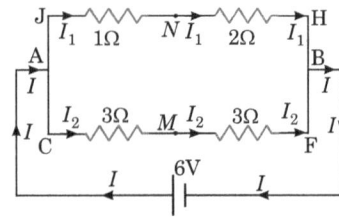

Figure 4.233

So, 1 A current passes through resistances 3 Ω and 3 Ω connected in series between points CM and MF respectively.
Potential difference between points J and N is,
$$V_{JN} = 1 \times 2 = 2 \text{ V}$$
Potential difference between points N and H is,
$$V_{NH} = 2 \times 2 = 4 \text{ V}$$
Potential difference between points C and G is,
$$V_{JN} = 1 \times 3 = 3 \text{ V}$$
Potential difference between points G and F is,
$$V_{GF} = 1 \times 3 = 3 \text{ V}$$
Now, again from Fig.4.232, $V_{KM} = V_{JN} = 2$ V
Therefore, charge on 1 μF capacitor connected between points K and M is
$$Q = CV_{KM} = 1 \times 10^{-6} \times 2 = 2 \text{ }\mu\text{C}$$
Now, $V_{ML} = V_{NH} = 4$ V
Therefore, charge on 2 μF capacitor connected between points M and L is
$$Q = CV_{ML} = 2 \times 10^{-6} \times 4 = 8 \text{ }\mu\text{C},$$
Similarly, $V_{DL} = V_{CG} = 3$ V
Therefore, charge on 3 μF capacitor connected between points D and L is
$$Q = CV_{DL} = 3 \times 10^{-6} \times 3 = 9 \text{ }\mu\text{C}$$
and $V_{LE} = V_{GF} = 3$ V
Therefore, charge on 4 μF capacitor connected between points L and E is
$$Q = CV_{DL} = 4 \times 10^{-6} \times 3 = 12 \text{ }\mu\text{C}$$

EXAMPLE 108. Find the potential difference between the points A and B and between the points B and C of Fig.4.234 in steady state.

SOLUTION Since, in steady state no current flows through

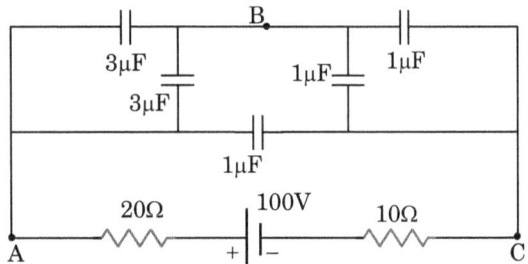

Figure 4.234

capacitors, so no current will pass through any resistor. Therefore, we can remove both resistors and next simplified circuit diagram is shown in Fig.4.235
Between terminals A and B, both 3 μF capacitors are in parallel. Their equivalent capacitance, $C_1 = 3 \text{ }\mu\text{F} + 3 \text{ }\mu\text{F} = 6 \text{ }\mu\text{F}$
Similarly, between terminals B and C, both 1 μF capacitors are also in parallel. Their equivalent capacitance, $C_2 = 1 \text{ }\mu\text{F} + 1 \text{ }\mu\text{F} = 2 \text{ }\mu\text{F}$
On removing these capacitors with their equivalent capacitors, we get the circuit diagram shown in Fig.4.236.
Now, in circuit diagram 4.236, 6 μF and 2 μF capacitors are in series, their equivalent capacitance is

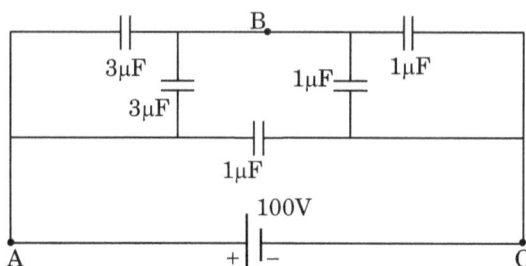

Figure 4.235

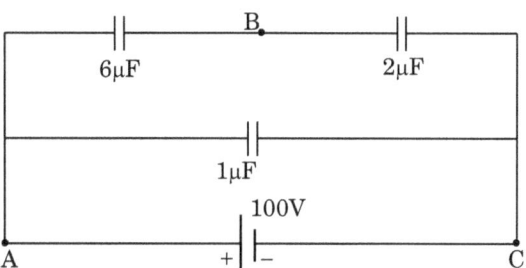

Figure 4.236

$$C_3 = \frac{6\mu\text{F} \times 2\mu\text{F}}{6\mu\text{F} + 2\mu\text{F}} = \frac{3}{2}\mu\text{F}$$

This equivalent capacitance C_3 is in parallel with $1\mu F$ capacitor, therefore, equivalent capacitance between terminals A and C is
$$C_{eq} = C_3 + 1 = \frac{3}{2} + 1 = \frac{5}{2}\mu\text{F}$$
Therefore, charge supplied by 100 V battery
$$Q = C_{eq}V = \frac{5}{2}\mu\text{F} \times 100 \text{ V} = 250 \text{ }\mu\text{C}$$
Charge stored across 1 μF capacitor
$$Q_1 = 1\mu\text{F} \times 100 \text{ V} = 100\mu\text{C}$$
So charge on capacitor of 6 μF
$$Q_2 = Q - Q_1 = 250 - 100 = 150\mu\text{F}$$
C_{eq} between A and B is $6\mu f = C$
Therefore potential drop across AB,
$$V_{AB} = \frac{Q_2}{6\mu\text{F}} = \frac{150\mu\text{C}}{6\mu\text{F}} = 25 \text{ V}$$
Since, potential drop across AC is 100 V, therefore, potential drop across BC*,
$$V_{BC} = 100 \text{ V} - V_{AB} = 100 \text{ V} - 25 \text{ V} = 75 \text{ V}.$$

4.30.4 Check Point 12

1. ••A capacitor of capacitance 100 μF is charged by connecting it to a battery of emf $12V$ and internal resistance 2Ω.
   (a) Find the time constant of the circuit.
   (b) Find the time taken before 99% of the maximum charge is stored on the capacitor.
2. •The plates of a 50 μF capacitor charged to 400 μC are connected through a resistance of 1.0 $k\Omega$. Find the charge remaining on the capacitor 1 s after the connection is made.
3. ••A capacitor of 2.5 μF is charged through a series resistor of 4 $M\Omega$ [Fig.4.237]. In what time the potential drop across the the capacitor will become 3 times that of the resistor. (Given : ln 2 = 0.693)
4. ••Find the time constant for given circuit [Fig. 4.238] if $R_1 = 4\Omega, R_2 = 12\Omega, \quad C_1 = 3\mu F$ and $C_2 = 6\mu F$.

4.31 Solved Problems

EXAMPLE 1. Incandescent bulbs are designed by keeping in mind that the resistance of their filament increases with increase

---
*Students are advised to calculate the potential drop across BC by following the same approach as we used in previous part across AB

4.31. SOLVED PROBLEMS

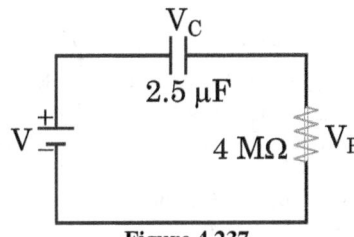

Figure 4.237

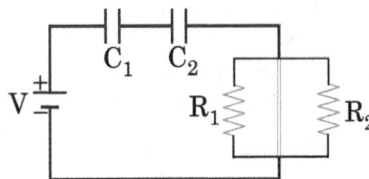

Figure 4.238

in temperature. If at room temperature, 100 W, 60 W and 40 W bulbs have filament resistances R_{100}, R_{60} and R_{40}, respectively, show that
$$\frac{1}{R_{100}} > \frac{1}{R_{60}} > \frac{1}{R_{40}}$$
SOLUTION From Eq.4.58, the power of a bulb having resistance R and operating at voltage V is given by $P = V^2/R$. Suppose, above the room temperature, the temperature of the 100 W, 60 W and 40 W bulbs are T_{100}, T_{60} and T_{40} respectively. The resistances and powers of these three bulbs at operating temperatures are given by
$$R'_{100} = R_{100}(1+\alpha T_{100}), \quad V^2/R'_{100} = 100 \text{ W},$$
$$R'_{60} = R_{60}(1+\alpha T_{60}), \quad V^2/R'_{60} = 60 \text{ W},$$
and $\quad R'_{40} = R_{40}(1+\alpha T_{40}), \quad V^2/R'_{40} = 40 \text{ W}$
where α is the thermal coefficient of resistance. Eliminating R'_{100}, R'_{60}, and R'_{40} we get
$$\frac{1}{R_{100}} = \frac{100}{V^2}(1+\alpha T_{100}),$$
$$\frac{1}{R_{60}} = \frac{60}{V^2}(1+\alpha T_{60}),$$
and $\quad \frac{1}{R_{40}} = \frac{40}{V^2}(1+\alpha T_{40})$
As higher power bulb has a higher temperature i.e., $T_{100} > T_{60} > T_{40}$. Therefore, from above set of equations, we can write-
$$\frac{1}{R_{100}} > \frac{1}{R_{60}} > \frac{1}{R_{40}}$$
EXAMPLE 2. A parallel plate capacitor C with plates of unit area and separation d is filled with a liquid of dielectric constant $\kappa = 2$. The level of liquid is $d/3$ initially [Fig.4.239]. If the liquid level decreases at a constant speed v, express the time constant as a function of time 't'.
SOLUTION Since, liquid level decreases at constant speed v,

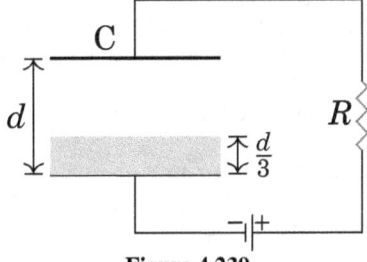

Figure 4.239

therefore in time t, the the fall in liquid level is vt. Now, if at time t, d_u and d_l are the thicknesses of upper part (without dielectric) and lower parts (with dielectric) of the capacitor, then
$$d_u = 2d/3 + vt, \text{ and}$$

$$d_l = d/3 - vt$$
Therefore, the capacitances of the upper part of the capacitor at time t is
$$C_u = \frac{\varepsilon_0 A}{d_u} = \frac{\varepsilon_0}{2d/3 + vt}$$
and the capacitances of the lower part of the capacitor at time t is
$$C_l = \frac{K\varepsilon_0 A}{d_l} = \frac{2\varepsilon_0}{d/3 - vt}$$
Now, capacitors C_u and C_l are connected in series and hence the effective capacitance of the system is
$$C_{\text{eff}} = \frac{C_u C_l}{C_u + C_l} = \frac{6\varepsilon_0}{5d + 3vt},$$
and the time constant (τ) given by
$$\tau = RC_{\text{eff}} = \frac{6\varepsilon_0 R}{5d + 3vt}$$
EXAMPLE 3. A circuit is connected as shown in the Fig.4.240 with the switch S open. When the switch is closed, find the total amount of charge that flows from Y to X
SOLUTION Let q_1 and V_1 be the charge and voltage respectively

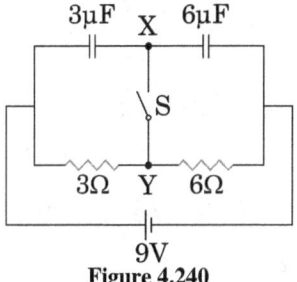

Figure 4.240

for $C_1 = 3\mu F$ capacitor and q_2 and V_2 be the charge and voltage respectively for $C_2 = 6\mu F$ capacitor. When switch S is open, two capacitors are connected in series and hence charge on the two is equal, i.e.,
$$C_1 V_1 = C_2 V_2 \quad \cdots (1)$$
Also, sum of voltages across the two capacitors is equal to the battery voltage, i.e.,
$$V_1 + V_2 = 9 \text{ V} \quad \cdots (2)$$
Solving Eq.(1) and (2), we get
$V_1 = 6$ V, $V_2 = 3$ V, $q_1 = C_1 V_1 = 18\mu C$, and $q_2 = C_2 V_2 = 18\mu C$. Total charge on the plates connected to X is $-q_1 + q_2 = 0$ because they form an isolated circuit.
When the switch S is closed, the voltage V'_1 across C_1 is equal to the voltage across $R_1 = 3$ Ω and voltage V'_2 across C_2 is equal to the voltage across $R_2 = 6$ Ω.
Effective resistance of the circuit is $R_1 + R_2 = 9$ Ω which gives a current $I = V/R_e = 9/(3+6) = 1$ A flowing through R_1 and R_2. Thus, the voltages and charges on the capacitors are
$$V'_1 = IR_1 = 3 \text{ V}, \quad V'_2 = IR_2 = 6 \text{ V},$$
$$q'_1 = C_1 V'_1 = 9\mu C, \quad q'_2 = C_2 V'_2 = 36\mu C$$
Total charge on plates connected to X is
$$-q'_1 + q'_2 = -9 + 36 = 27\mu C,$$
which is flown from Y to X.
EXAMPLE 4. The three resistances of equal value are arranged in the different combination shown in Fig.4.241. Arrange them in increasing order of power dissipation.
SOLUTION The equivalent resistances and power dissipation of the given combination are given by
$$R_I = R + R + R = 3R, \quad P_I = I^2 R_I = 3I^2 R,$$
$$R_{II} = (R+R) \| R = \tfrac{2}{3}R, \quad P_{II} = I^2 R_{II} = \tfrac{2}{3}I^2 R,$$
$$R_{III} = R \| R \| R = \tfrac{1}{3}R, \quad P_{III} = i^2 R_{III} = \tfrac{1}{3}I^2 R,$$
$$R_{IV} = (R \| R) + R = \tfrac{3}{2}R, \quad P_{IV} = I^2 R_{IV} = \tfrac{3}{2}I^2 R$$
Clearly, the correct order is, III < II < IV < I.
EXAMPLE 5. A 100 W bulb B_1 and two 60 W bulbs B_2 and

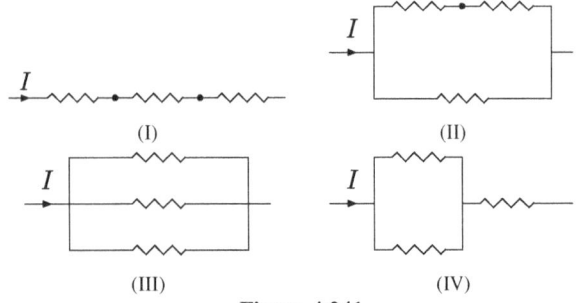

Figure 4.241

B_3, are connected to a 250 V source, as shown in the Fig.4.242. If W_1, W_2 and W_3 are the output powers of the bulb B_1, B_2 and B_3 respectively, then write correct order of powers.

SOLUTION Suppose, the given power ratings are defined at an

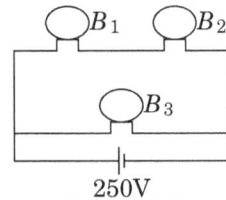

Figure 4.242

operating voltage V. The resistances of the three bulbs are given by
$$R_1 = V^2/100, \quad R_2 = V^2/60, \quad R_3 = V^2/60$$
In the given configuration, the current through B_1 and B_2 is
$$I = \frac{250}{R_1 + R_2} = \frac{250}{V^2}\left(\frac{100 \times 60}{100+60}\right) = \frac{250}{V^2}\left(\frac{75}{2}\right)$$
Output power of bulb B_1
$$W_1 = I^2 R_1 \approx 14(250/V)^2,$$
Output power of bulb B_2
$$W_2 = I^2 R_2 \approx 23(250/V)^2,$$
and Output power of bulb B_3
$$W_3 = (250)^2/R_3 = 60(250/V)^2$$
Clearly, $B_1(100\text{ W})$ is dimmer than $B_2(60\text{ W})$ which in turn is dimmer than $B_3(60\text{ W})$.
Therefore, the correct order is $W_1 < W_2 < W_3$

EXAMPLE 6. In the given circuit, with steady current, find the potential difference across the capacitor.

SOLUTION In the steady state, a capacitor offers infinite resis-

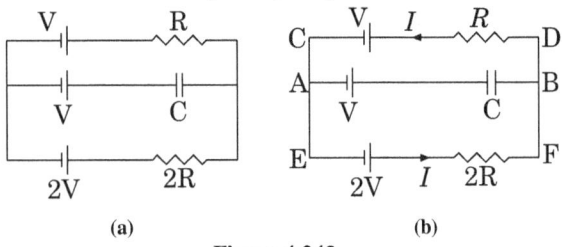

Figure 4.243

tance for directional current and acts like an open circuit element, so there is no current in branch AB of the circuit (see Fig.4.243b). By Kirchhoff's voltage law in closed loop EFDCE (Fig.4.243b), we have
$$2V - 2IR - IR - V = 0 \quad \Rightarrow \quad I = \frac{V}{3R}$$
Now, to find potential difference across the capacitor (V_C (say)), we have to apply Kirchhoff's voltage law either in upper loop ABDCA or in lower loop ABFEA. Let us apply it in lower loop ABFEA:
$$V + V_C + 2IR - 2V = 0$$

$$\Rightarrow \quad V_C = V - 2IR = V - 2\left(\frac{V}{3R}\right)R = \frac{V}{3}$$

EXAMPLE 7. In Fig.4.244, a parallel combination of 0.1 MΩ resistor and a 10 μF capacitor is connected across a 1.5 V source of negligible resistance. Find the time required for the capacitor to get charged upto 0.75 V.

APPROACH From Eq.4.199, at any time t, the charge on a capacitor being charged, is given by
$$q(t) = \mathcal{E}C\left(1 - e^{-t/CR}\right) \quad \cdots (1)$$
Here, $\mathcal{E}$ is the emf of the battery and R is the net resistance connected in series with the capacitor
Put the given values in Eq.(1) and solve for required time (t).

SOLUTION In Fig.4.244, the capacitor plates are directly con-

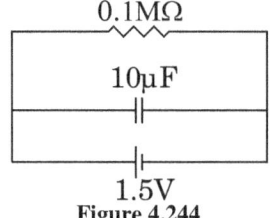

Figure 4.244

nected to a battery of zero internal resistance and no resistance is there in series with it. Therefore, the effective resistance connected in series with the capacitor, $R = 0$. In such case, for any charge on capacitor's plates, Eq.(1) gives
$$q(t) = \mathcal{E}C\left(1 - e^{-t/CR}\right) \quad \Rightarrow \quad e^{-t/CR} = 1 - \frac{q(t)}{\mathcal{E}C}$$
or
$$-\frac{t}{RC} = \ln\left(1 - \frac{q(t)}{\mathcal{E}C}\right)$$
or
$$t = -RC\ln\left(1 - \frac{q(t)}{\mathcal{E}C}\right) \quad \cdots (2)$$
For $R = 0$, Eq.(2) gives, $t = 0$
It shows that when a capacitor is directly connected to an ideal emf, it get fully charged immediately. So, required time is zero.

EXAMPLE 8. In the given circuit shown in Fig.4.245, each battery has emf of 5 V and an internal resistance of 0.2 Ω. Find the reading of the ideal voltmeter (in volts).

SOLUTION Let $\mathcal{E} = 5$ V and $r = 0.2$ Ω.

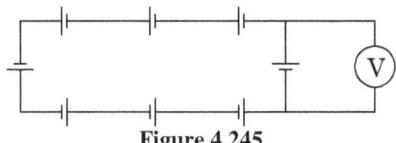

Figure 4.245

Applying Kirchhoff's loop law, $8\mathcal{E} - 8Ir = 0$, we get
$$I = \frac{\mathcal{E}}{r} = 25 \text{ A}.$$
From Eq.4.89, the reading of voltmeter
$$V = \mathcal{E} - Ir = 0 \text{ V}$$
Thus, the voltmeter reading is zero.

EXAMPLE 9. Two batteries of different emfs and different internal resistances are connected as shown in Fig.4.246a. Find the voltage across AB

SOLUTION Circuit is redrawn in Fig.4.246b. Let I be the current in the loop.
By Kirchhoff's voltage law, we have
$$6 - 3 - 2I - 1I = 0 \quad \Rightarrow \quad I = 1 \text{ A}$$
Now, going from B to A along the upper branch, we get,
$$V_B - 1 + 6 = V_A \quad \Rightarrow \quad V_A - V_B = 5 \text{ V}.$$
Thus voltage between terminals A and B is 5 V.

EXAMPLE 10. At time $t = 0$, a battery of 10 V is connected across points A and B in the given circuit (Fig.4.247). If the capacitors have no charge initially, at what time (in seconds) does

4.31. SOLVED PROBLEMS 381

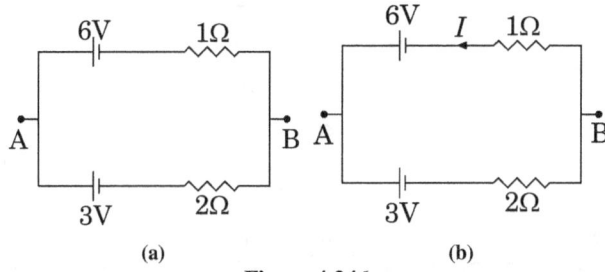
Figure 4.246

the voltage across them become 4 V? [Take $\ln 5 = 1.6, \ln 3 = 1.1$].
SOLUTION The resistors are connected in parallel which gives

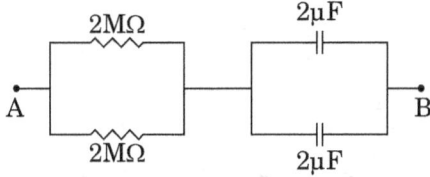
Figure 4.247

the equivalent resistance $R = 2\|2 = 1$ MΩ. The capacitors connected in parallel give the equivalent capacitance $C = 2\|2 = 4\mu$F. The equivalent circuit is shown in the Fig.4.248.

The voltage across the equivalent capacitor is same as the voltage

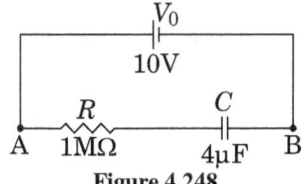
Figure 4.248

across the individual capacitors (parallel combination). Thus, we need to find time t at which the voltage across C become 4 V in the equivalent circuit (charging of a capacitor). The voltage across C at time t is
$$V = V_0 \left[1 - e^{-t/(RC)}\right]$$
which simplifies to
$$t = RC \ln\left(\frac{V_0}{V_0 - V}\right)$$
Substituting $V_0 = 10$ V, $V = 4$ V, $R = 1 \times 10^6 \Omega$ and $C = 4 \times 10^{-6}$ F, in above expression, we get
$$t = 4\ln(5/3) = 4(\ln 5 - \ln 3) = 2 \text{ s}$$

EXAMPLE 11. At $t = 0$, switch S is closed (see Fig.4.249). The charge on the capacitor is varying with time as $q = q_0 \left(1 - e^{-\alpha t}\right)$. Obtain the value of q_0 and α in the given circuit parameters.
SOLUTION Fig.4.250 shows the circuit when switch S is closed.

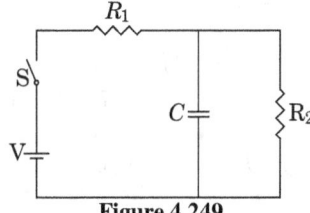
Figure 4.249

Let at any instant t, the charge on the capacitor is q, current through it is $I_c = dq/dt$, and potential across it is $V_c = q/C$.

The capacitor C and resistor R_2 are connected in parallel and hence the potential difference across these two is equal, therefore
$$V_C = (I - I_c) R_2 \qquad \cdots (1)$$
By Kirchhoff's voltage law in the loop containing battery, resistors R_1 and R_2, we have
$$V = IR_1 + (I - I_c) R_2 \qquad \cdots (2)$$

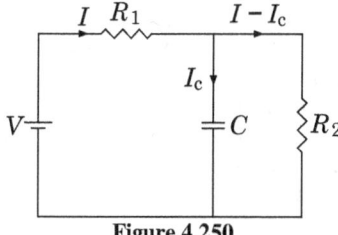
Figure 4.250

Eliminating I from equations (1) and (2) and substituting $V_C = q/C$ and $I_C = dq/dt$, we get
$$\frac{dq}{dt} + \frac{(R_1 + R_2)}{CR_1R_2}q = \frac{V}{R_1} \qquad \cdots (3)$$
On integrating equation (3) with initial condition $q = 0$ at $t = 0$, we get
$$q = \frac{CVR_2}{R_1 + R_2}\left[1 - e^{-\frac{R_1+R_2}{C(R_1R_2)}t}\right] \qquad \cdots (4)$$
Aliter: At any time t, the magnitude of charge on the plates of the capacitor is given by
$$q = q_0 \left(1 - e^{-\alpha t}\right) \qquad \cdots (1)$$
To get current through capacitor, we have to differentiate Eq.(1), with respect t, it is given as
$$I_c = \frac{dq}{dt} = \alpha q_0 e^{-\alpha t} \qquad \cdots (2)$$
At $t = 0$, Eq.(2) gives,
$$I_{c0} = \alpha q_0 e^{-\alpha 0} = \alpha q_0 \qquad \cdots (3)$$
At the moment when switch is just closed, i.e., at $t = 0$, the capacitor behaves like a short circuit (see Fig.4.251a). In this case, no current passes through R_2. Thus, at $t = 0$, the current through the capacitor can also be written as
$$I_{c0} = V/R_1 \qquad \cdots (4)$$
Therefore, from Eq.(3) and (4), we can write
$$V/R_1 = \alpha q_0 \qquad \cdots (5)$$
As $t \to \infty$, Eq.(2) gives

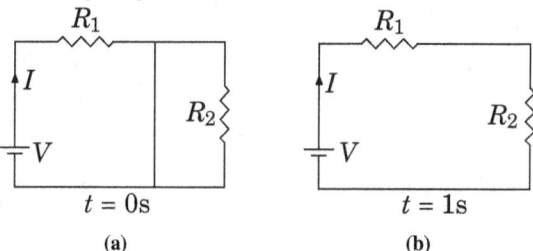
Figure 4.251

$$I_c = \alpha q_0 e^{-\infty} = 0$$
So, no current passes through capacitor when $t \to \infty$. This is called steady state of the capacitor. In steady state, (i.e., when $(t \to \infty)$, the capacitor behaves like an open circuit element.
In steady state, Eq.(1) gives
$$q = q_0 \left(1 - e^{-\infty}\right) = q_0$$
Thus, the charge on capacitor in the steady state is q_0 and the potential across it is
$$V_C = q_0/C$$
The current through R_2 is
$$I = V/(R_1 + R_2)$$
and the potential across R_2 is
$$V_{R_2} = IR_2 = VR_2/(R_1 + R_2)$$
The condition $V_C = V_{R_2}$ gives
$$q_0 = \frac{CVR_2}{R_1 + R_2}$$
On substituting this value of q_0 in Eq.(5), we get
$$\alpha = \frac{R_1 + R_2}{CR_1R_2}.$$

EXAMPLE 12. Draw the circuit for experimental verification

of Ohm's law using a source of variable DC voltage, a main resistance of 100 Ω, two galvanometers and two resistances of values 10^6 Ω and 10^{-3} Ω, respectively. Clearly show the positions of the voltmeter and the ammeter.
SOLUTION Verification of Ohm's law requires the measurement of the voltage and current.

The galvanometer can be converted to a voltmeter by connecting

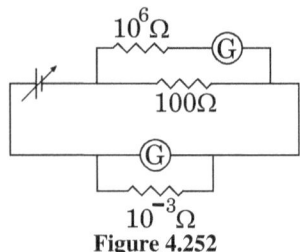
Figure 4.252

a very high resistance $(10^6$ Ω$)$ in series and to an ammeter by connecting a very low resistance $(10^{-3}$ Ω$)$ in parallel, as shown in Fig.4.252.

EXAMPLE 13. Show by diagram, how can we use a rheostat as the potential divider?
SOLUTION The rheostat is a resistance coil with two ends A and B and a slider C. To use rheostat as a potential divider, connect

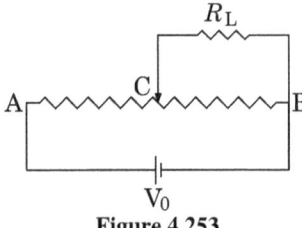
Figure 4.253

A and B across the battery (V_0) and B and C across the load resistance (R_L). As slider C moves from A to B, the potential across R_L changes from V_0 to zero.

EXAMPLE 14. A thin uniform wire AB of length 1 m, an unknown resistance X and a resistance of 12Ω are connected by thick conducting strips, as shown in the Fig.4.254. A battery and galvanometer (with a sliding jockey connected to it) are also available. Connections are to be made to measure the unknown resistance X using the principle of Wheatstone bridge. Answer the following questions.
(a) Are there positive and negative terminals on the galvanometer?

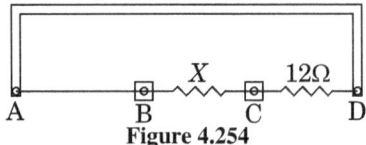
Figure 4.254

ter?
(b) Copy the Fig.4.254 in your answer book and show the battery and the galvanometer (with jockey) connected at appropriate points.
(c) After appropriate connections are made, it is found that no deflection takes place in the galvanometer when the sliding jockey touches the wire at a distance of 60 cm from A. Obtain the value of the resistance X.
SOLUTION The galvanometer does not have positive or negative terminals. The circuit diagram to measure the unknown resistance X is given in the Fig.4.255.
Let R be the total resistance of the potentiometer wire of length 100 cm and N be the null point. The resistance of branch AN is $R_{AN} = 60(R/100) = 0.6R$ and of branch BN is $R_{BN} = 0.4R$. The balancing condition of Wheatstone bridge, $R_{AN}/R_{BN} = 12/X$,

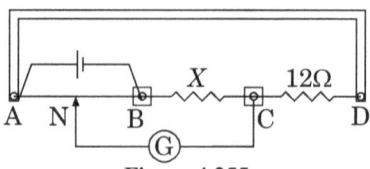
Figure 4.255

gives $X = 8$Ω.
ANSWER (a) No (b) See solution (c) 8 Ω
EXAMPLE 15. In the circuit shown in Fig.4.256, the battery is an ideal one, with emf V. The capacitor is initially uncharged. The switch S is closed at time $t = 0$. Find,
(a) the charge q on the capacitor at time t.

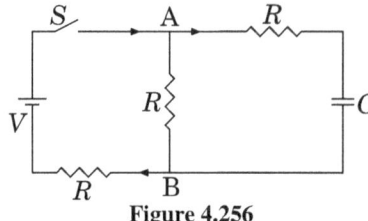
Figure 4.256

(b) the current in AB at time t. What is its limiting value as $t \to \infty$?
SOLUTION (a) Let at time t, the current through the battery is I, q is the charge on the capacitor C, $I_1 = dq/dt$ is the current through C, and $V_C = q/C$ be potential across C (see Fig.4.257 for polarity).

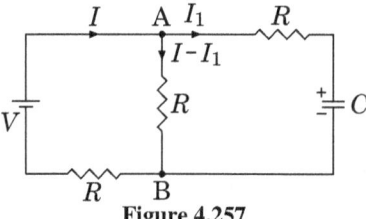
Figure 4.257

The Kirchhoff's loop law gives,
$$V - (I - I_1) R - IR = 0 \quad \cdots (1)$$
$$V - I_1 R - q/C - IR = 0 \quad \cdots (2)$$
To find $I_1 = dq/dt$, we eliminate I from equation (1) and (2). This gives
$$\frac{dq}{dt} + \left(\frac{2}{3RC}\right) q = \frac{V}{3R} \quad \cdots (3)$$
On integrating above equation, we get
$$q = \frac{CV}{2}\left[1 - e^{-\frac{2t}{3RC}}\right] \quad \cdots (4)$$
(b) Current through the capacitor
$$I_1 = \frac{dq}{dt} = \frac{V}{3R} e^{-\frac{2t}{3RC}} \quad \cdots (5)$$
and current through branch AB,
$$I_{AB} = I - I_1 = \frac{V}{2R} - \frac{I_1}{2} = \frac{V}{2R}\left[1 - \frac{1}{3} e^{-\frac{2t}{3RC}}\right] \quad \cdots (5)$$
Taking the limits $t \to \infty$ in equation (5), we get steady state current in AB as $V/(2R)$. It is important to note that capacitor becomes an open circuit element in steady state.
EXAMPLE 16. Find the emf (V) and the internal resistance (r) of a single battery which is equivalent to a parallel combination of two batteries of emfs V_1 and V_2 and internal resistances r_1 and r_2 respectively, with polarities as shown in the Fig.4.258.
SOLUTION The emf of the equivalent battery is $V = V_A - V_B$ and its internal resistance r is the resistance between A and B, when A and B are not connected to external resistance. Let I be the current in the loop when A and B are open.
By Kirchhoff's voltage law, we get

4.31. SOLVED PROBLEMS

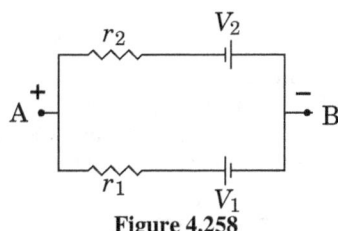

Figure 4.258

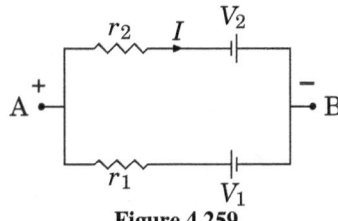

Figure 4.259

$$V_1 - ir_1 - Ir_2 + V_2 = 0,$$

or $\quad I = \dfrac{V_1 + V_2}{r_1 + r_2}.$

Traversing from B to A along the lower branch relates the potential at A and B by

$$V_A = V_B + V_1 - Ir_1 = V_B + V_1 - \dfrac{(V_1 + V_2)\,r_1}{r_1 + r_2}$$

which gives the emf of the equivalent battery as

$$V = V_A - V_B = \dfrac{V_1 r_2 - V_2 r_1}{r_1 + r_2}$$

The internal resistance of the battery is

$$r = r_1 \| r_2 = \dfrac{r_1 r_2}{r_1 + r_2}$$

EXAMPLE 17. A leaky parallel plate capacitor is filled completely with a material having electrical conductivity $\sigma = 7.4 \times 10^{-12} \Omega^{-1}\,\text{m}^{-1}$ and dielectric constant $K = 5$. If the charge on the capacitor at instant $t = 0$ is $q = 8.85\,\mu\text{C}$, then calculate the leakage current at the instant $t = 12$ s.

SOLUTION The charge in a leaky capacitor decreases (leaks) due to presence of in-built resistance. Equivalent circuit of the leaky capacitor is an R-C circuit as shown in the Fig.4.260.

The capacitance of parallel plate capacitor with plate area A,

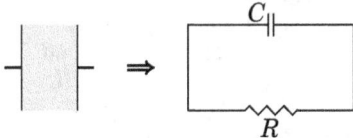

Figure 4.260

separation d, and dielectric constant κ is given as

$$C = \dfrac{\kappa\epsilon_0 A}{d} \qquad \cdots (1)$$

The conductivity is related to the current density J and electric field E by Eq.4.29: $J = \sigma E$. The electric field and the potential difference V are related by $V = Ed$. Thus, the resistance R can be written as

$$R = \dfrac{V}{I} = \dfrac{V}{JA} = \dfrac{Ed}{\sigma E A} = \dfrac{d}{\sigma A} \qquad \cdots (2)$$

The equations (1) and (2) give the time constant of the circuit as

$$\tau = RC = \dfrac{d}{\sigma A}\dfrac{K\epsilon_0 A}{d} = \dfrac{K\epsilon_0}{\sigma} = \dfrac{5\,(8.85 \times 10^{-12})}{7.4 \times 10^{-12}}$$
$$= 5.98\,\text{s}$$

The charge on the capacitor and the current through the circuit at time t are

$$q = q_0 e^{-t/\tau}, \quad \text{and} \quad I = dq/dt = -(q_0/\tau)\,e^{-t/\tau}$$

where $q_0 = 8.85 \times 10^{-6}$ C is the initial charge. The magnitude of the current at $t = 12$ s is

$$I = \dfrac{8.85 \times 10^{-6}}{5.98} e^{-12/5.98} = 0.198 \times 10^{-6}\,\text{A}$$

$$= 0.198\,\mu\text{A}$$

EXAMPLE 18. In the circuit of Fig.4.261a, $E_1 = 3E_2 = 2E_3 = 6$ V and $R_1 = 2R_4 = 6\,\Omega$, $R_3 = 2R_2 = 4\Omega$, $C = 5\mu\text{F}$. Find the current in R_3 and the energy stored in the capacitor.

SOLUTION In the steady state, the capacitor will act as an open

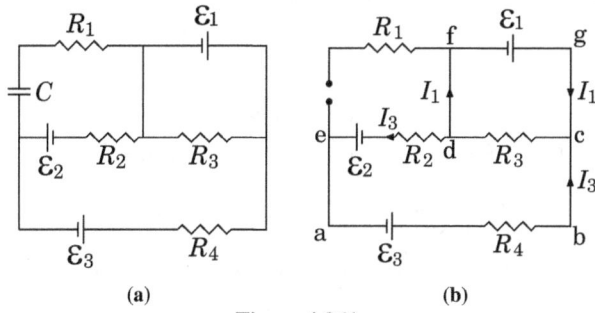

Figure 4.261

circuit element as shown in the Fig.4.261b

Let I_1 be the current flowing through the battery $\mathcal{E}_1$ and I_3 be the current flowing through the battery $\mathcal{E}_3$. Kirchhoff's junction law at junctions c and d gives the current through R_3 as $I_1 + I_3$ and current through R_2 as I_3.

Applying Kirchhoff's voltage law in the loops fgcdf, we get

$$\mathcal{E}_1 - (I_1 + I_3)\,R_3 = 0 \qquad \cdots (1)$$

Again by Kirchhoff's voltage law in loop abcdea, we get

$$\mathcal{E}_3 - I_3 R_4 - (I_1 + I_3)\,R_3 - I_3 R_2 + E_2 = 0 \qquad \cdots (2)$$

From Eq.(1) and (2), we get

$$I_1 + I_3 = \mathcal{E}_1/R_3 = 6/4 = 1.5\,\text{A}$$

Substituting it in Eq.(2), we get

$$I_3 = \dfrac{\mathcal{E}_2 + \mathcal{E}_3 - (I_1 + I_3)\,R_3}{R_2 + R_4} = \dfrac{2 + 3 - (1.5)4}{2 + 3}$$
$$= -1/5\,\text{A}$$

The potential difference across the capacitor C is same as the potential difference across e and f which in turn is equal to the potential difference across e and d i.e.,

$$V = V_e - V_d = \mathcal{E}_2 - I_3 R_2 = 2 - (-0.2)(2) = 2.4\,\text{V}$$

The energy stored in the capacitor is

$$\mathcal{E} = \dfrac{1}{2}CV^2 = \dfrac{1}{2}\left(5 \times 10^{-6}\right)(2.4)^2 = 1.44 \times 10^{-5}\,\text{J}$$

ANSWER 1.5 A, 1.44×10^{-5} J

EXAMPLE 19. An infinite ladder network of resistances is constructed with 1Ω and 2Ω resistances (see Fig.4.262). The 6 V battery between A and B has negligible internal resistance. (a)

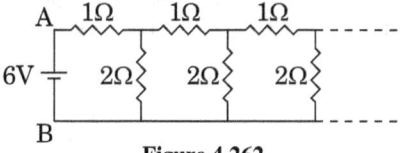

Figure 4.262

Show that the effective resistance between A and B is $2\,\Omega$.
(b) What is the current that passes through the $2\,\Omega$ resistance nearest to the battery?

SOLUTION Let the effective resistance between A and B be R. The ladder network consists of infinite number of units, where each unit consists of two resistances of values $R_1 = 1\,\Omega$ and $R_2 = 2\,\Omega$. The effective resistance of the ladder will not change by removal of one unit, say the unit close to the battery.

Thus, the effective resistance between A and B is equal to the resistance of the circuit shown in the Fig.4.263 i.e.,

$$R = R_1 + (R_2 \| R)$$
$$= R_1 + \dfrac{R_2 R}{R_2 + R} = 1 + \dfrac{2R}{2 + R}$$

or $\quad 2R + R^2 = 2 + 3R \quad$ or $\quad R^2 - R - 2 = 0$

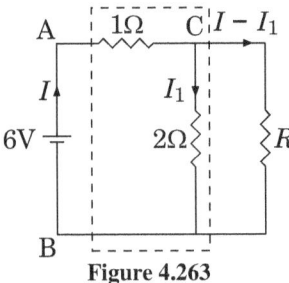

Figure 4.263

or $R^2 - 2R + R - 2 = 0$ or $R(R-2) + 1(R-2) = 0$
or $R = 2\,\Omega, R = -1\,\Omega$

Since, resistance can not be negative, therefore, the acceptable answer is $R = 2\,\Omega$

Thus, the effective resistance between terminals A and B is $R = 2\,\Omega$.

(Answer of part (a))

Now, the current through the battery of emf $\mathcal{E} = 6$ V is $I = \mathcal{E}/R = 6/2 = 3$ A. Since $R_2 = R = 2\,\Omega$, therefore, current I is equally divided at the node C giving $I_1 = I/2 = 1.5$ A.

(Answer of part (b))

EXAMPLE 20. A part of circuit in steady state along with the currents flowing in the branches, the values of resistances etc., is shown in the Fig.4.264. Calculate the energy stored in the capacitor $C(4\,\mu F)$.

SOLUTION Given circuit is redrawn in Fig.4.265. In the steady

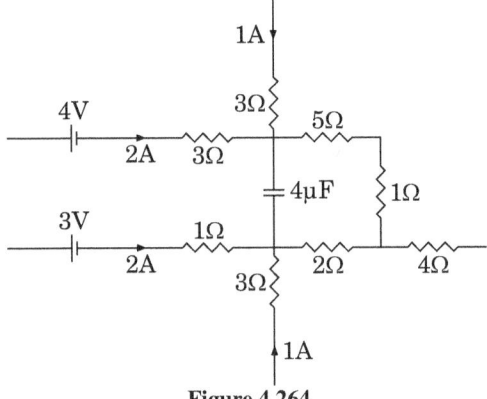

Figure 4.264

state, no current flows through the capacitor as it acts as an open circuit element. By Kirchhoff's junction law at the junctions a and b, the current flowing out of these junctions is 3 A.

Now, traversing from a to b through $5\,\Omega, 1\,\Omega$, and $2\,\Omega$ resistances,

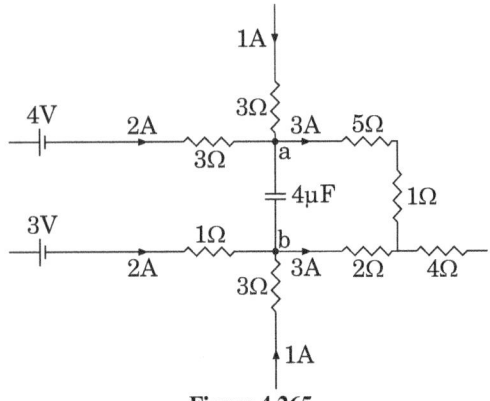

Figure 4.265

we get
$$V_a - (3)(5) - (3)(1) + (3)(2) = V_b$$
Thus, the potential difference across the capacitor is $V_{ab} = V_a - V_b = 12$ V. The energy stored in the capacitor is given by

$$E = \frac{1}{2}CV_{ab}^2 = \frac{1}{2}\left(4 \times 10^{-6}\right)(12)^2 = 2.88 \times 10^{-4}\,\text{J} = 0.288\,\text{mJ}$$

EXAMPLE 21. In the circuit shown in Fig.4.266, E, F, G, H are cells of emf 2, 1, 3 and 1 V, respectively and their internal resistances are 2, 1, 3 and 1$\,\Omega$, respectively. Calculate,
(a) the potential difference between B and D.

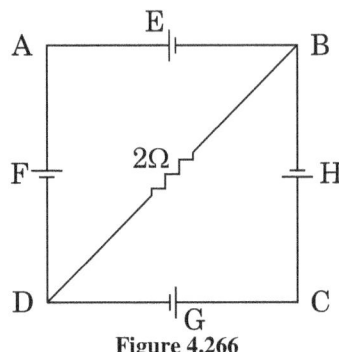

Figure 4.266

(b) the potential difference across the terminals of each of the cells G and H.

SOLUTION The given circuit is redrawn in the Fig.4.267. By

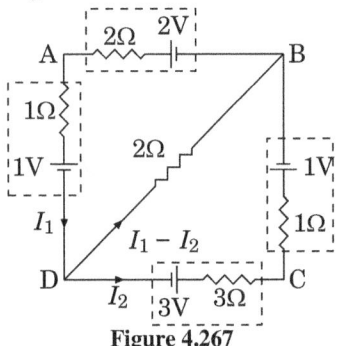

Figure 4.267

Kirchhoff's voltage law in the loop BADB, we have
$$2 - 2I_1 - I_1 - 1 - 2(I_1 - I_2) = 0 \quad \cdots (1)$$
Again by Kirchhoff's voltage law in the loop DCBD, we have
$$3 - 3I_2 - I_2 - 1 + 2(I_1 - I_2) = 0 \quad \cdots (2)$$
On solving Eq.(1) and (2), we get
$$I_1 = 5/13\,\text{A and }I_2 = 6/13\,\text{A}$$

(a) The potential difference between B and D is
$$V_B - V_D = -2(I_1 - I_2) = -2(5/13 - 6/13)$$
$$= 2/13\,\text{V}$$

(b) The potential difference between the cells of batteries G and H are-
$$V_G = 3 - 3I_2 = 3 - 3(6/13) = 21/13\,\text{V}$$
$$V_H = 1 + (1)I_2 = 1 + 6/13 = 19/13\,\text{V}$$

EXAMPLE 22. Two resistors, 400 Ω and 800 Ω are connected in series with a 6 V battery. It is desired to measure the current in the circuit. An ammeter of 10 Ω resistance is used for this purpose. What will be the reading in the ammeter? Similarly, if a voltmeter of 1000 Ω resistance is used to measure the potential difference across the 400 Ω resistor, what will be the reading in the voltmeter?

SOLUTION The ammeter of resistance 10 Ω is connected in series to measure current in the circuit [Fig.4.268].

Effective resistance of the circuit is
$$R_A = 400 + 800 + 10 = 1210\,\Omega$$
The current through the circuit is
$$I = \frac{6}{1210} = \frac{3}{605}\,\text{A}$$

The voltmeter of resistance 1000 Ω is connected in parallel to 400 Ω resistor to measure the voltage across it (Fig.4.269). Effective resistance of the circuit is

4.31. SOLVED PROBLEMS

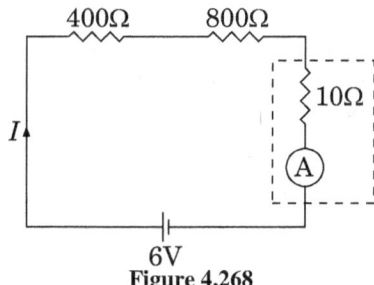

Figure 4.268

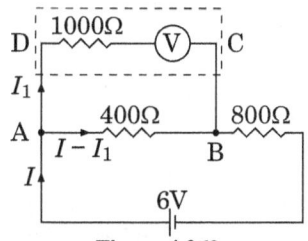

Figure 4.269

$R_V = (400 \| 1000) + 800$
$= \dfrac{400 \times 1000}{400 + 1000} + 800 = 1085.7 \, \Omega$

The current through the battery is
$I = V/R_V = 6/1085.7 = 5.53 \times 10^{-3}$ A.

The current I gets divided into two branches at the node A. By Kirchhoff's voltage law in loop ABCDA, we have
$400(I - I_1) = 1000 I_1$ i.e., $I_1 = 2I/7 = 1.58 \times 10^{-3}$ A

The potential difference across the voltmeter is
$V = I_1(1000) = 1.58$ V

Note that ideal resistances of the ammeter and the voltmeter are zero and infinite, respectively.

EXAMPLE 23. Calculate the steady state current in the 2 Ω resistor shown in the circuit (see Fig 4.270). The internal resistance of the battery is negligible and the capacitance of the condenser is 0.2 μF.

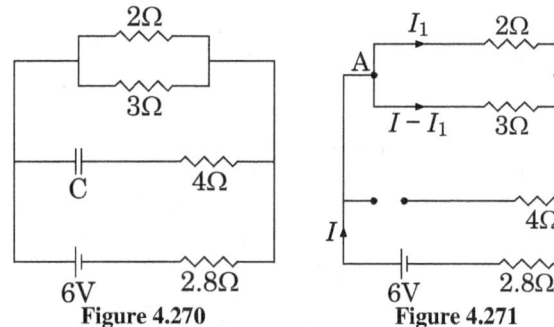

Figure 4.270   Figure 4.271

SOLUTION In the steady state, the capacitor acts as an open circuit element. The equivalent circuit in the steady state is shown in the Fig.4.271.

The effective resistance of the circuit is,
$R = (2 \| 3) + 2.8 = \dfrac{2 \times 3}{2 + 3} + 2.8 = 4 \, \Omega.$

The current I through the circuit is,
$I = V/R = 6/4 = 1.5\, \Omega.$

The current I gets divided into two branches at the node A. Apply Kirchhoff's loop law in the loop containing 2 Ω and 3 Ω resistors to get
$2I_1 = 3(I - I_1) = 3(1.5 - I_1)$
which gives $I_1 = 0.9$ A.

EXAMPLE 24. A steady current passes through a cylindrical conductor. Is there an electric field inside the conductor?

SOLUTION The electric field inside a conductor is zero under the electrostatic equilibrium i.e., when charges are at rest. When a steady current is set through a cylinder, the charges are in motion (not at rest). In this case, electric field inside the conductor is not zero. In fact, it is the electric field inside the conductor that provides the drift velocity for the flow of charges.

EXAMPLE 25. In Fig.4.272, find the potential difference between the points A and B and between the points B and C in the steady state.

SOLUTION In the steady state, the capacitors act as open cir-

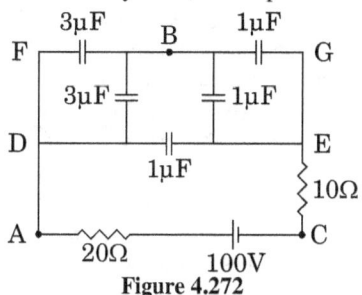

Figure 4.272

cuit elements, breaking all paths for flow of current through the battery. There is no current through 10 Ω and 20 Ω resistors i.e., no potential drop across these resistors. So, these resistors can be replaced by conducting wires. Equivalent circuit for the steady state is shown in the Fig.4.273.

Further, two 3 μF capacitors between B and D are connected

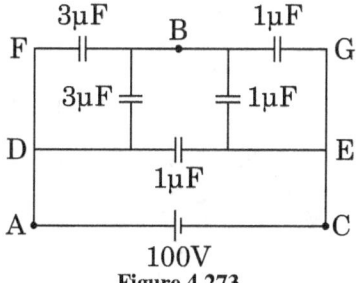

Figure 4.273

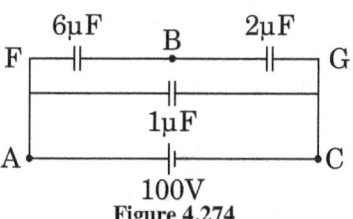

Figure 4.274

in parallel giving their effective capacitance 6μF. Similarly, two 1μF capacitors between B and E are connected in parallel giving their effective capacitance 2 μF. The equivalent circuit is shown in the Fig.4.274. Let $C_{AB} = 6\, \mu$F and V_{AB} be potential across this capacitor. The charge on C_{AB} is $q_{AB} = C_{AB} V_{AB}$. Similarly, let $C_{BC} = 2\mu$F and V_{BC} be the potential difference across this capacitor. The charge on C_{BC} is $q_{BC} = C_{BC} V_{BC}$. The charges on C_{AB} and C_{BC} are equal because they are connected in series i.e., $q_{AB} = q_{BC}$, which gives
$$3 V_{AB} = V_{BC} \qquad \cdots (1)$$
Total potential difference across C_{AB} and C_{BC} is equal to the battery voltage, i.e.,
$$V_{AB} + V_{BC} = 100 \qquad \cdots (2)$$
Solving Eq.(1) and (2) for V_{AB} and V_{BC}, we get
$V_{AB} = 25$ V and $V_{BC} = 75$ V

EXAMPLE 26. In the circuit shown in Fig.4.275, $\mathcal{E}_1 = 3$ V, $\mathcal{E}_2 = 2$ V, $\mathcal{E}_3 = 1$ V and $R = r_1 = r_2 = r_3 = 1\, \Omega$.

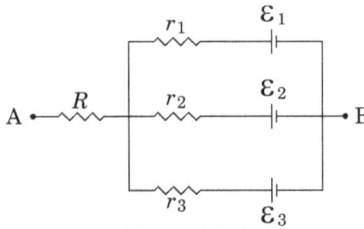

Figure 4.275

(a) Find the potential difference between the points A and B and the currents through each branch.

(b) If r_2 is short circuited and the point A is connected to point B, find the currents through $\mathcal{E}_1, \mathcal{E}_2, \mathcal{E}_3$ and the resistor R.

SOLUTION The circuit is redrawn in Fig.4.276. Suppose $I_1, I_2,$ and I_3 be the currents through the batteries $\mathcal{E}_1 = 3$ V, $\mathcal{E}_2 = 2$ V, $\mathcal{E}_3 = 1$ V, respectively. There is no current through $R = 1$ Ω.

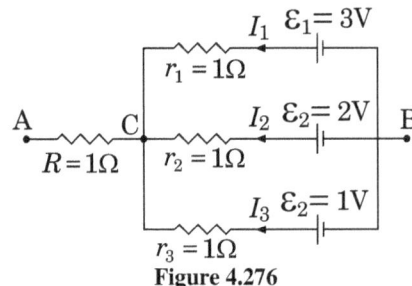

Figure 4.276

By Kirchhoff's junction law at node C, we get
$$I_1 + I_2 + I_3 = 0 \quad \cdots (1)$$
By Kirchhoff's voltage law in the upper loop. we get
$$3 - I_1 + I_2 - 2 = 0 \quad \cdots (2)$$
Again by Kirchhoff's voltage law in lower loop, we get
$$2 - I_2 + I_3 - 1 = 0 \quad \cdots (3)$$
Solving Eq.(1)-(3) for I_1, I_2 and I_3, we get
$$I_1 = 1 \text{ A}, I_2 = 0 \text{ A, and } I_3 = -1 \text{ A}.$$
Since $I_2 = 0$ A and current through R is zero,
$$V_{AB} = V_{CB} = E_2 = 2 \text{ V}.$$
The circuit after shorting $r_2 = 1$ Ω and connecting point A to B is shown in the Fig.4.277.

By Kirchhoff's loop law in loops ABCDA, we have

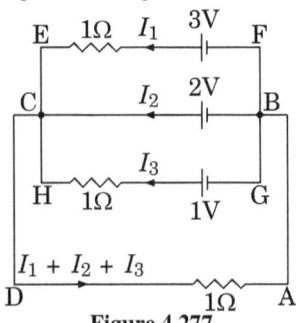

Figure 4.277

$$2 - (I_1 + I_2 + I_3) = 0 \quad \cdots (4)$$
By Kirchhoff's voltage law in loop FECBF, we have
$$3 - I_1 - 2 = 0 \quad \cdots (5)$$
and by Kirchhoff's voltage law, in loop BCHGB, we have
$$2 + I_3 - 1 = 0 \quad \cdots (6)$$
Solving equations (4)-(6) for I_1, I_2 and I_3, we get
$$I_1 = 1 \text{ A}, I_2 = 2 \text{ A, and } I_3 = -1 \text{ A}$$
The current through the resistor R is $I_1 + I_2 + I_3 = 1 + 2 - 1 = 2$ A.
ANSWER (a) 2 V, 1 A, 0, −1 A (b) 1 A, 2 A, −1 A, 2 A

EXAMPLE 27. A 25 W and a 100 W bulb are joined in series and connected to the mains. Which bulb will glow brighter?
SOLUTION Let the powers of the bulbs, $P_1 = 25$ W and $P_2 =$ 100 W, be specified at a voltage V. The resistances of the bulbs are $R_1 = V^2/P_1$ and $R_2 = V^2/P_2$. When connected in series, the currents through the bulbs are equal, say I. The powers consumed by the bulbs, connected in the series, are

$$P_1' = I^2 R_1 = I^2 V^2/P_1 = \frac{1}{25} I^2 V^2 \quad \cdots (1)$$

$$P_2' = I^2 R_2 = I^2 V^2/P_2 = \frac{1}{100} I^2 V^2 \quad \cdots (2)$$

From equations (1) and (2), we have
$$P_1' = 4P_2'$$
Hence, the bulb with 25 W power will glow brighter.

EXAMPLE 28. Ideal gas is contained in a thermally insulated and rigid container and it is heated through a resistance of 100 Ω by passing a current of 1 A. Find the change in internal energy of the gas after 5 min.
SOLUTION The heat generated by the current $I = 1$ A flowing through a resistance $R = 100$ Ω in time interval $t = 5$ min is given by
$$\Delta Q = I^2 Rt = (1)^2(100)(5 \times 60) = 30 \text{ kJ}$$
There is no heat loss as the container is thermally insulated. The work done by the gas, ΔW, is zero because the container is rigid. According to the first law of thermodynamics, we have
$$\Delta Q = \Delta U + \Delta W \quad \cdots (1)$$
Here, $\Delta Q = 30$ kJ, $\Delta W = 0$ and $\Delta U = ?$
On substituting these values in Eq.(1), we get
Change in internal energy of the gas
$$\Delta U = 30 \text{ kJ}$$

EXAMPLE 29. A wire of length L and three identical cells of negligible internal resistances are connected in series. Due to the current, the temperature of the wire is raised by ΔT in time t. A number N of similar cells is now connected in series with a wire of the same material and cross-section but of length $2L$. The temperature of the wire is raised by the same amount ΔT in the same time. Find the value of N.
SOLUTION Let R and m be the resistance and mass of the wire of length L. Let $\mathcal{E}$ be the emf of each cell. The potential difference across the wire when it is connected to three cells in series is $3\mathcal{E}$. The heat produced in time interval t is given by
$$Q_1 = (3\mathcal{E})^2 t/R = 9\mathcal{E}^2 t/R \quad \cdots (1)$$
This heat increases the temperature of the wire. If S is the specific heat of the material of the wire then rise in temperature ΔT is given by
$$Q_1 = 9\mathcal{E}^2 t/R = mS\Delta T$$
In the second case, potential difference across the wire is $N\mathcal{E}$, resistance is $2R$, mass is $2m$, and rise in temperature is same as ΔT. Thus,
$$Q_2 = N^2 \mathcal{E}^2 t/(2R) = (2m)S\Delta T \quad \cdots (2)$$
Solving equations (1) and (2), we get $N = 6$.

EXAMPLE 30. In the circuit shown in Fig.4.278 the heat produced in the 5 Ω resistor due to the current flowing through it is 10 cal/s. Find the heat generated in the 4 Ω resistor.

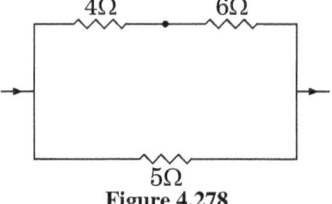

Figure 4.278

SOLUTION Let I_1 and I_2 be the currents through the 5 Ω and 4 Ω resistors, respectively. The heats produced in the 5Ω and 4Ω resistors due to the current flowing through them are
$$P_1 = 5I_1^2 = 10 \text{ cal/s} \quad \cdots (1)$$
$$P_2 = 4I_2^2 \quad \cdots (2)$$

Applying Kirchhoff's voltage rule in loop ABCA (Fig.4.279), we

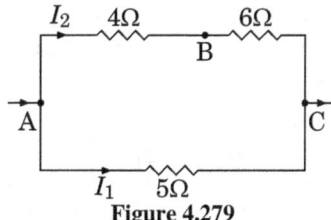

Figure 4.279

get
$$10I_2 - 5I_1 = 0 \quad \text{i.e.,} \quad I_2/I_1 = 1/2$$
Dividing equation (2) by (1) and substituting $I_2/I_1 = 1/2$, we get
$$P_2 = \frac{4}{5}\left(\frac{I_2}{I_1}\right)^2 P_1 = \frac{4}{5}\left(\frac{1}{2}\right)^2 (10) = 2 \text{ cal/s}$$

EXAMPLE 31. An electric bulb rated for 500 W at 100 V is used in a circuit having a 200 V supply (Fig.4.280). Find the value of the resistance R that must be put in series with the bulb, so that the bulb delivers 500 W.

SOLUTION The resistance of the bulb of power rating $P = 500$ W at applied voltage 100 V is
$$R_b = \frac{100^2}{500} = 20 \text{ }\Omega$$
The effective resistance of the circuit when resistance R is

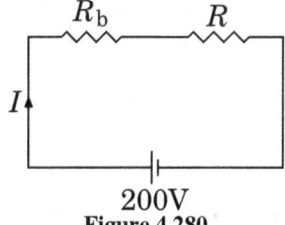

Figure 4.280

connected in series to the bulb is $R_e = R_b + R$. The current through the circuit when connected to 200 V supply is $I = V/R_e = 200/(R_b + R)$. The power delivered by the bulb in this configuration is 500 W if
$$P = I^2 R_b = \left(\frac{200}{R_b + R}\right)^2 R_b = 500$$
Substituting $R_b = 20 \text{ }\Omega$ and solving for R, we get
$$R = 20 \text{ }\Omega$$

EXAMPLE 32. When two identical batteries of internal resistance 1Ω each are connected in series across a resistor R, the rate of heat produced in R is J_1. When the same batteries are connected in parallel across R, the rate is J_2. If $J_1 = 2.25 J_2$ then find the value of R (in Ω).

SOLUTION Let $\mathcal{E}$ be the emf and $r = 1\Omega$ be the internal resistance of each battery. The series and parallel combination of two batteries are shown in the Fig.4.281.

When the two batteries are connected in series (Fig.4.281a),

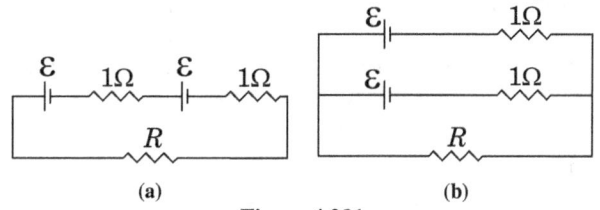

Figure 4.281

effective emf becomes $2\mathcal{E}$ and effective internal resistance is $2r$. Thus, the current through R is $I_1 = 2\mathcal{E}/(2r + R)$ and the heat produced in R is given by
$$J_1 = I_1^2 R = \left(\frac{2\mathcal{E}}{2r + R}\right)^2 R$$

When the batteries are connected in parallel (Fig.4.281b), the effective emf is $\mathcal{E}$ and the effective internal resistance is $r/2$. Thus, the current through R is $I_2 = \mathcal{E}/(r/2 + R)$ and the heat produced in it is given by
$$J_2 = I_2^2 R = \left(\frac{\mathcal{E}}{r/2 + R}\right)^2 R$$
Substituting J_1 and J_2 in the given relation, $J_1 = 2.25 J_2$, and solving for R, we get $R = 4 \text{ }\Omega$.

Note: You can also find the effective emf and the effective internal resistance for series and parallel combination by using Kirchhoff's law.

EXAMPLE 33. A copper wire having cross-sectional area of 0.5 mm^2 and a length of 0.1 m is initially at 25°C and is thermally insulated from the surroundings. If a current of 1.0 A is set up in this wire, (a) find the time in which the wire will start melting. The change of resistance with the temperature of the wire may be neglected. (b) What will this time be, if the length of the wire is doubled? [For Copper, Melting point = 1075°C, Specific resistance = $1.6 \times 10^{-8} \Omega$m, Density $D = 9 \times 10^3$ kg/m^3, Specific heat = 9×10^{-2} cal/(kg°C).]

SOLUTION (a) A current $I = 1.0$ A is flowing through a copper wire of cross-sectional area $A = 0.5 \text{ mm}^2$ and length $l = 0.1$ m. The wire is thermally insulated from the surroundings. Hence, the heat produced in the wire is used to raise its temperature to the melting point $T_m = 1075°$C i.e.,
$$I^2 R t = m S \Delta T \quad \cdots (1)$$
where $R = \rho l/A$ is the resistance ($\rho = 1.6 \times 10^{-8} \Omega$m is resistivity), t is the time required for melting to start, $m = DAl$ is the mass of the wire ($D = 9 \times 10^3$ kg/m^3 is mass density), $S = 9 \times 10^{-2}$ cal/(kg°C) is specific heat, and $\Delta T = T_m - T_0 = 1075 - 25 = 1050°$C is the rise in temperature. Substitute the values in equation (1) to get
$$t = \frac{mS\Delta T}{i^2 R} = \frac{(DAl)S\Delta T}{i^2(\rho l/A)} = \frac{DA^2 S\Delta T}{i^2 \rho},$$
$$= \frac{(9 \times 10^3)(0.5 \times 10^{-6})^2 (9 \times 10^{-2} \times 4.18)(1050)}{(1)^2 (1.6 \times 10^{-8})}$$
$$= 55.55 \text{ s} \quad \cdots (2)$$

(b) From equation (2), the time t is independent of the wire length. Thus, the time required for the melting to start remains equal to 55.55 s when the length of the wire is doubled.

EXAMPLE 34. A heater is designed to operate with a power of 1000 W in a 100 V line. It is connected in combination with a resistance of 10 Ω and a resistance R, to a 100 V mains as shown in the Fig.4.282. What will be the value of R so that the heater operates with a power of 62.5 W?

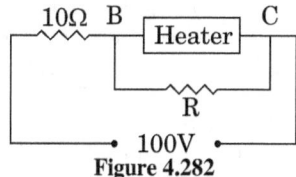

Figure 4.282

SOLUTION The given circuit with branch currents, is shown in Fig.4.283. The heater power is $P = 1000$ W at an applied voltage of $V = 100$ V. The resistance of the heater is $R_h = V^2/P = (100)^2/1000 = 10 \text{ }\Omega$. The potential difference across the heater for it to operate at $P' = 62.5$ W is given by
$$V_h = \sqrt{P' R_h} = \sqrt{(62.5)(10)} = \sqrt{625} = 25 \text{ V}$$
The current through the heater is $I_h = V_h/10 = 2.5$ A.

The potential difference across 10 Ω resistance connected in series to the heater is $V = 100 - V_h = 100 - 25 = 75$ V and current through it is $I = V/10 = 7.5$ A.

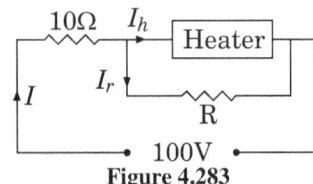

Figure 4.283

The current flowing through the resistance R connected in parallel to the heater is $I_r = I - I_h = 7.5 - 2.5 = 5$ A. The potential difference across this resistor is 25 V. Hence, $R = 25/5 = 5$ Ω.

Paragraph for Problems 35 and 36

Consider a simple RC circuit shown in Fig.4.284a.

Process 1: In the circuit the switch S is closed at $t = 0$ and the capacitor is fully charged to voltage V_0 (i.e., charging continues for time $T \gg RC$). In the process some dissipation (E_D) occurs across the resistance R. The amount of energy finally stored in the fully charged capacitor is E_C.

Process 2: In a different process the voltage is first set to $V_0/3$ and maintained for a charging time $T \gg RC$. Then the voltage is raised to $2V_0/3$ without discharging the capacitor and again maintained for a time $T \gg RC$. The process is repeated one more time by raising the voltage to V_0 and the capacitor is charged to the same final voltage V_0 as in process 1. These two processes are depicted in Fig.4.284b.

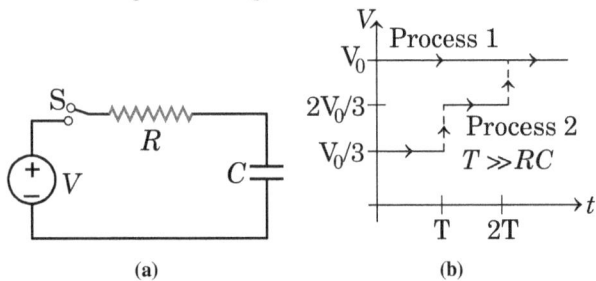

Figure 4.284

EXAMPLE 35. In process 1, Find a relation between the energy stored in the capacitor E_C and heat dissipated across resistance E_D.

SOLUTION In process 1, when the capacitor is fully charged (i.e., $T \gg RC$), charge on the capacitor is $Q = CV_0$ and potential across it is $V_C = V_0$. The electrostatic energy stored in the capacitor is

$$E_C = \frac{1}{2}CV_C^2 = \frac{1}{2}CV_0^2 \qquad \cdots (1)$$

In the charging process, the charge Q flows through the battery at potential $V_B = V_0$ (battery emf). Thus, the work done by the battery is

$$E_B = QV_B = (CV_0)V_0 = CV_0^2 \qquad \cdots (2)$$

The energy supplied (E_B) by the battery is stored in the capacitor (E_C) and dissipated by the resistor (E_D). By conservation of energy

$$E_B = E_C + E_D \qquad \cdots (3)$$

Substituting the values of E_C and E_B from (1) and (2) in Eq.(3), we get

$$CV_0^2 = \frac{1}{2}CV_0^2 + E_D$$

or $\qquad E_D = \frac{1}{2}CV_0^2 \qquad \cdots (4)$

From Eq.(1) and (4), we get $E_C = E_D$.

EXAMPLE 36. In process 2, calculate total energy dissipated across the resistance E_D.

SOLUTION Let us divide the process 2 in following three parts:
(1) battery set to potential $V_{B1} = \frac{1}{3}V_0$
(2) battery set to potential $V_{B2} = \frac{2}{3}V_0$ and
(3) battery set to potential $V_{B3} = V_0$.

In part (1), potential across the capacitor is $V_{C1} = \frac{1}{3}V_0$, charge on the capacitor is $Q_{C1} = CV_{C1} = \frac{1}{3}CV_0$ and the charge flowing through the battery is $Q_{B1} = Q_{C1} = \frac{1}{3}CV_0$. Thus, the work done by the battery is

$$E_{B1} = Q_{B1}V_{B1} = \left(\frac{1}{3}CV_0\right)\left(\frac{V_0}{3}\right) = \frac{1}{9}CV_0^2$$

In part (2), potential across the capacitor is $V_{C2} = \frac{2}{3}V_0$, charge on the capacitor is $Q_{C2} = CV_{C2} = \frac{2}{3}CV_0$ and the charge flowing through the battery is $Q_{B2} = Q_{C2} - Q_{C1} = \frac{1}{3}CV_0$ (additional charge supplied to the capacitor). Thus, the work done by the battery is

$$E_{B2} = Q_{B2}V_{B2} = \left(\frac{1}{3}CV_0\right)\left(\frac{2}{3}V_0\right) = \frac{2}{9}CV_0^2$$

In part (3), potential across the capacitor is $V_{C3} = V_0$, charge on the capacitor is $Q_{C3} = CV_{C3} = CV_0$ and charge flowing through the battery is $Q_{B3} = Q_{C3} - Q_{C2} = \frac{1}{3}CV_0$. Thus, the work done by the battery is

$$E_{B3} = Q_{B3}V_{B3} = \left(\frac{1}{3}CV_0\right)V_0 = \frac{1}{3}CV_0^2$$

Thus, the total work done by the battery is

$$E_B = E_{B1} + E_{B2} + E_{B3} = \frac{2}{3}CV_0^2$$

The total electrostatic energy stored in the capacitor is

$$E_C = \frac{1}{2}CV_0^2$$

The energy conservation gives total energy dissipated in the resistor R as

$$E_D = E_B - E_C = \frac{1}{6}CV_0^2$$

4.32 Solved Objective Questions

1. A steady current flows in a metallic conductor of non-uniform cross-section. The quantity/quantities constant along the length of the conductor is (are)
   (A) current, electric field and drift speed
   (B) drift speed only
   (C) current and drift speed
   (D) current only

   SOLUTION (D) In the steady state, charge cannot accumulate in a conductor. Thus, a constant current $I = dq/dt$ flows along the length of conductor. The current density at a point having cross-section area A is given by $J = I/A$. The current density is related to electric field by $J = \sigma E$, where conductivity σ is a material property. The drift velocity v_d is given by $I = neAv_d$. Thus, both E and v_d are inversely proportional to A.

2. A copper wire of diameter 1.02 mm carries a current of 1.7 A. Find the drift velocity (v_d) of electrons in the wire. Given n, density of free electrons in copper $= 8.5 \times 10^{28}/\text{m}^3$.
   (A) 1.75 mm/sec (B) 2.5 mm/sec
   (C) 1.25 mm/sec (D) 1.5 mm/sec

   APPROACH From Eq.4.11, the current in a conductor in terms of current density is given by

   $$I = neAv_d$$

   or $\qquad v_d = \dfrac{I}{neA} \qquad \cdots (1)$

   SOLUTION (D) Given: $I = 1.7$ A, $n = 8.5 \times 10^{28}/\text{m}^3$, $A = \pi r^2 = 3.14 \times (1.02 \times 10^{-3}/2)^2 \text{ m}^2 = 8.17 \times 10^{-7}\text{m}^2$, and $e = 1.6 \times 10^{-19}$C

   Substituting all these values in Eq.(1), we get

   $$v_d = \frac{1.7}{8.5 \times 10^{28} \times 1.6 \times 10^{-19} \times 8.17 \times 10^{-7}}$$

4.32. SOLVED OBJECTIVE QUESTIONS

$$= 1.5 \times 10^{-3} \text{m/s} = 1.5 \text{ mm/s}$$

3. A cylindrical conductor of length l and inner radius R_1 and outer radius R_2 has specific resistance ρ [Fig.4.285]. A cell of emf $\mathcal{E}$ is connected across the two lateral faces of the conductor. Find the current drown from the cell.

   (A) $I = \dfrac{2\pi l\mathcal{E}}{\rho \ln\left(\frac{R_2}{R_1}\right)}$ (B) $I = \dfrac{2\pi l\mathcal{E}}{\rho \ln\left(\frac{R_1}{R_2}\right)}$

   (C) $I = \dfrac{2\pi l\mathcal{E}}{\rho \ln\left(\frac{R_2}{R_1}\right)^2}$ (D) $I = \dfrac{2\pi}{\rho \ln\left(\frac{R_2}{R_1}\right)}$

APPROACH From Eq.4.53, the radial resistance of a hol-

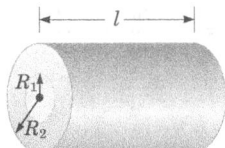

Figure 4.285

low cylinder of length L, inner and outer radii a and b respectively, is given by

$$R = \dfrac{\rho}{2\pi L}\log_e\left(\dfrac{b}{a}\right) \quad \cdots (1)$$

here, ρ is the specific resistance of the material of the cylinder.

Now, by Ohm's law, radial current can be given as

$$I = \dfrac{\mathcal{E}}{R} \quad \cdots (2)$$

Here, $\mathcal{E}$ is the emf applied across the cylindrical faces.
Substituting the value of R from Eq.(1), in Eq.(2), we get

$$I = \dfrac{2\pi L\mathcal{E}}{\rho \log_e\left(\frac{b}{a}\right)} \quad \cdots (3)$$

SOLUTION (A) Given: $a = R_1$, $b = R_2$, $L = l$, $\mathcal{E} = \mathcal{E}$, and $\rho = \rho$
Substituting these values in Eq.(3), we get

$$I = \dfrac{2\pi l\mathcal{E}}{\rho \log_e\left(\frac{R_2}{R_1}\right)} \quad \text{i.e.,} \quad I = \dfrac{2\pi l\mathcal{E}}{\rho \ln\left(\frac{R_2}{R_1}\right)}$$

4. A copper wire of resistance $4\,\Omega$ is melted and redrawn to thrice its original length. Find the resistance of stretched wire.

   (A) $40\,\Omega$ (B) $60\,\Omega$ (C) $45\,\Omega$ (D) $36\,\Omega$

APPROACH From Eq.4.44, we have

$$\dfrac{R_1}{R_2} = \dfrac{l_1^2}{l_2^2} \quad \text{or} \quad R_2 = \left(\dfrac{l_2}{l_1}\right)^2 R_1 \quad \cdots (1)$$

Now, substitute given values in Eq.(1) and solve for R_2.

SOLUTION (D) Given: $R_1 = 4\,\Omega$, $l_2 = 3l_1$
Substituting these values in Eq.(1), we get

$$R_2 = \left(\dfrac{3l_1}{l_1}\right)^2 \times 4\,\Omega = 36\,\Omega$$

Paragraph for Q.5 - 6
A battery of emf 1.4 V and internal resistance 2 Ω is connected to a 100 Ω resistor through an ammeter as shown in Fig.4.286. The resistance of the ammeter is $4/3\,\Omega$. A voltmeter is also connected to find the potential difference across the resistor. The ammeter reading is 0.02 A whereas the voltmeter reading is 1.10 V.

5. What is the resistance of the voltmeter?

   (A) $200\,\Omega$ (B) $300\,\Omega$ (C) $400\,\Omega$ (D) $150\,\Omega$

SOLUTION (A) If R_v is the resistance of voltmeter, then total resistance of the circuit

$$R = \left[2 + \dfrac{4}{3} + \dfrac{100R_v}{100+R_v}\right]$$

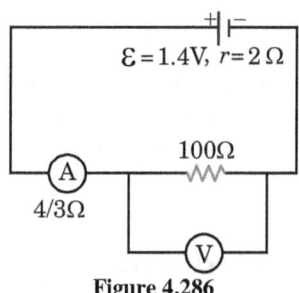

Figure 4.286

Now, current supplied by the battery

$$I = \dfrac{\text{emf}}{\text{total resistance}}$$

$$\implies 0.02 \text{ A} = \dfrac{1.4 \text{ V}}{\left[2 + \frac{4}{3} + \frac{100R_v}{100+R_v}\right]\Omega} \implies R_v = 200\,\Omega$$

6. What is the error in voltmeter reading?

   (A) 0.3 V (B) 0.43 V (C) 0.53 V (D) 0.23 V

SOLUTION (D) In Q.5, we have calculated that, $R_v = 200\,\Omega$ and given circuit current $I = 0.02$ A
Equivalent resistance of voltmeter of resistance $R_v = 200\,\Omega$ and $100\,\Omega$ resistance is

$$R_{eq} = \dfrac{100 \times 200}{100 + 200} = \dfrac{200}{3}$$

Therefore, potential difference across the voltmeter

$$V_v = IR_{eq} = 0.02 \times \left[\dfrac{100 \times 200}{100 + 200}\right] = 1.33 \text{ V}.$$

According to problem, voltmeter reading is 1.10 V
Therefore, error in voltmeter reading

$$\Delta V = 1.33 - 1.10 = 0.23 \text{ V}$$

7. In the circuit shown in Fig.4.287, each cell has emf 5 V and has an internal resistance of 0.2 Ω. What is the reading of ideal voltmeter V.

   (A) 0 V (B) 1 V (C) 2 V (D) 5 V

SOLUTION (A) As internal resistance of an ideal voltmeter

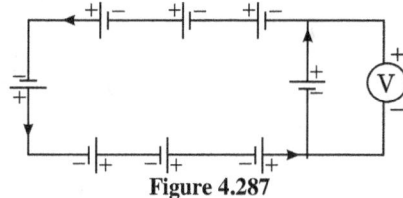

Figure 4.287

is infinite, therefore, the resistance of the battery across which it is connected will not change by its presence as

$$\dfrac{1}{r_1} = \dfrac{1}{r} + \dfrac{1}{\infty} \implies r_1 = r$$

Now, as the given 8 batteries are discharging in series therefore from Eq.4.111 the combination acts as a battery of emf

$$\mathcal{E}_0 = \mathcal{E} + \mathcal{E} + \cdots 8 \text{ times} = 8\mathcal{E} = 8 \times 5.0 = 40.0 \text{ V}$$

and from Eq.4.112, the total internal resistance of the circuit

$$r_0 = r + r + \cdots 8 \text{ times} = 8r = 8 \times 0.2 = 1.6\,\Omega$$

Therefore, current in the circuit

$$I = \dfrac{\mathcal{E}_0}{r_0} = \dfrac{40.0 \text{ V}}{1.6\,\Omega} = 25.0 \text{ A}$$

Hence from Eq.4.68, potential difference across the required battery

$$\Delta V = \mathcal{E} - Ir = 5.0 - 25.0 \times 0.2 = 0 \text{ V}.$$

8. In a gas discharge tube if 3×10^{18} electrons are flowing per second from left to right and 2×10^{18} protons are flowing per second from right to left through a given cross section. The magnitude and direction of current through the cross section

(A) 0.48, left to right (B) 0.80 A, left to right
(C) 0.48 A, right to left (D) 0.80 A, right to left

SOLUTION (D) According to given problem, 3×10^{18} electrons are flowing per second from left to right through a cross-section, so amount of negative charge flowing per second from left to right is
$$\Delta Q_e = N_e e = 3 \times 10^{18} \times 1.6 \times 10^{-19} \text{C} = 0.48 \text{ C}$$
This charge produces a current in opposite direction of it's flow, i.e. in right to left direction. If we denote this current by I_e, then
$$I_e = \frac{\Delta Q_e}{\Delta t} = \frac{0.48 \text{ C}}{1 \text{ s}} = 0.48 \text{ A}$$
Now, 2×10^{18} protons are flowing per second from right to left through the cross-section, so amount of positive charge flowing per second from right to left is
$$\Delta Q_p = N_p e = 2 \times 10^{18} \times 1.6 \times 10^{-19} \text{C} = 0.32 \text{ C}$$
This charge produces a current along the direction of it's flow, i.e. in right to left direction. If we denote this current by I_p, then
$$I_p = \frac{\Delta Q_p}{\Delta t} = \frac{0.32 \text{ C}}{1 \text{ s}} = 0.32 \text{ A}$$
So, both I_e and I_p are in same direction (right to left), therefore, net current passing through the given cross section
$$I = I_e + i_p = 0.48 + 0.32 = 0.80 \text{ A (right to left)}.$$

9. The current in a wire varies with time according to the equation $I = 4 + 2t$, where I is in ampere and t is in second. The quantity of charge which has passed through a cross-section of the wire during the time $t = 2$ s to $t = 6$ s will be
   (A) 60 C (B) 24 C (C) 48 C (D) 30 C

SOLUTION Let dQ be the charge which has passed through the given cross-section in a small time interval dt, then from Eq.4.2, instantaneous current can be written as
$$I = \frac{dQ}{dt} \quad \cdots (1)$$
Therefore, from above equation, charge flowing through the cross-section in time interval $t_1 = 2$ s to $t_2 = 6$ s is
$$Q = \int_{t_1}^{t_2} I dt = \int_2^6 (4+2t) dt = [4t + t^2]_2^6 = 60 - 12 = 48 \text{ C}$$

10. A uniform copper wire of length 1 m and cross-sectional area 5×10^{-7} m² carries a current of 1 A. Assuming that there are 8×10^{28} free electrons per m³ in copper, how long will an electron take to drift from one end of the wire to the other?
    (A) 0.8×10^3 s (B) 1.6×10^3 s
    (C) 3.2×10^3 s (D) 6.4×10^3 s

SOLUTION (D) From Eq.4.11, the drift velocity of electrons is given by
$$I = neAv_d \quad \text{or} \quad v_d = \frac{I}{enA}$$
If l is the length of the wire, the time taken by electrons to cover it, is given by
$$t = \frac{l}{v_d} = \frac{lneA}{I}$$
$$= \frac{1 \times 1.6 \times 10^{-19} \times 8 \times 10^{28} \times 5 \times 10^{-7}}{1} = 6.4 \times 10^3 \text{ s}$$

11. An electric current of 16 A exists in a metal wire of cross section 10^{-6} m² and length 1 m. Assuming one free electrons per atom, the drift speed of the free electrons in the wire will be
    (Density of metal = 5×10^4 kg/m³, atomic weight = 60)
    (A) 5×10^{-3} m/s (B) 2×10^{-3} m/s
    (C) 4×10^{-3} m/s (D) 7.5×10^{-3} m/s

SOLUTION (B) From Eq.4.11, we have
$$I = neAv_d \quad \cdots (1)$$
Here, $n = \dfrac{\text{number of electrons}}{\text{volume}}$
Since, there is only one free electrons per atom of the metal wire, therefore
$$n = \frac{\text{number of atoms}}{\text{volume}} = \frac{\text{number of moles} \times N_A}{\text{volume}}$$
or $\quad n = \dfrac{(\text{mass/molar mass}) \times N_A}{\text{volume}} = \dfrac{m/M}{V} \times N_A$
$$= \frac{m}{V} \frac{N_A}{M} = \left(\frac{\text{density}(\rho)}{\text{molar mass}(M)}\right) N_A$$
here, N_A is Avogadro number. It's value is 6.022×10^{23}. Substituting this value of n in Eq.(1), and simplifying for v_d, we get
$$v_d = \frac{I}{neA} = \frac{IM}{\rho N_A eA}$$
$$= \frac{16 \times 60}{5 \times 10^4 \times 6.022 \times 10^{23} \times 1.6 \times 10^{-19} \times 10^{-6}}$$
$$= 2 \times 10^{-3} \text{ m/s}$$

12. A copper wire is stretched to make it 0.1% longer. The percentage change in its resistance is
    (A) 0.2% increase (B) 0.2% decrease
    (C) 0.1% increase (D) 0.1% decrease

SOLUTION (A) If a wire is stretched by x%, which is less that 5%, then from Eq.4.51, % change in the resistance of wire is given by
$$\left|\frac{R_2 - R_1}{R_1}\right| \times 100 = 2x \%$$
In this problem, the wire is elongated by 0.1% which is less than 5%, therefore putting $x = 0.1$ in above equation, we get
$$\left|\frac{R_2 - R_1}{R_1}\right| \times 100 = 2 \times 0.1 \% = 0.2 \%$$

13. The driver cell of a potentiometer has an emf of 2 V and negligible internal resistance. The potentiometer wire has a resistance of 5Ω and is 1 m long. The resistance which must be connected in series with the wire so as to have a potential difference of 5mV across the whole wire is
    (A) 1985 Ω (B) 1990 Ω
    (C) 1995 Ω (D) 2000 Ω

SOLUTION (C) In order to have a potential drop of 5mV = 5×10^{-3} V across a wire AB of resistance 5 Ω [Fig.4.288], the current flowing in the wire
$$I = \frac{5 \times 10^{-3}}{5} = 1 \times 10^{-3} \text{ A}$$

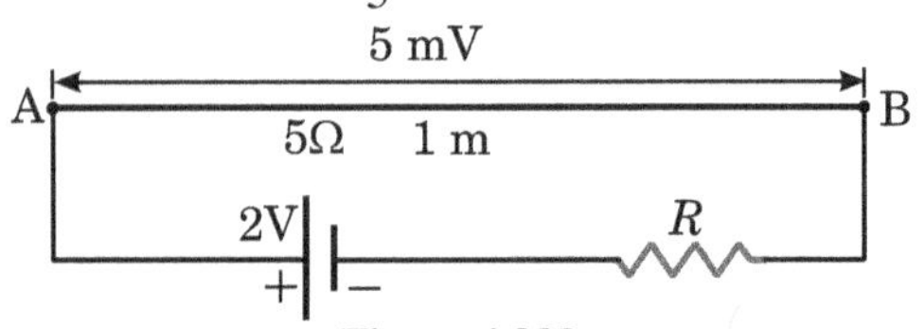

Figure 4.288

If R is the resistance to be connected in series with the wire, then by Ohm's law, we have
$$V = I(R + 5)$$
$$\Rightarrow R + 5 = \frac{V}{I} = \frac{2}{1 \times 10^{-3}} = 2 \times 10^3 \Omega = 2000 \text{ } \Omega$$
$$\Rightarrow R = 1995 \text{ } \Omega$$

14. A battery of 10 volt is connected to a resistance of 20 ohm through a variable resistance R. The amount of charge which has passed in the circuit in 4 minutes, if the variable resistance R is increased at the rate of 5 Ω/min.

(A) 120 coulomb (B) $\frac{120}{\log_e 2}$ coulomb
(C) 120 $\log_e$ 2 coulomb (D) $\frac{60}{\log_e 2}$ coulomb

SOLUTION (B) From Eq.4.36, we have
$$\Delta V = RI \quad \text{(Ohm's law)} \quad \cdots (1)$$
or
$$I = \frac{dq}{dt} = \frac{\Delta V}{R}$$
or
$$\frac{dq}{dR} \cdot \frac{dR}{dt} = \frac{\Delta V}{R}$$

Given: $\Delta V = 10V$, $\frac{dR}{dt} = 5\,\Omega/\text{min} = 5 \times \frac{1}{60}\,\Omega/s = \frac{1}{12}\,\Omega/s$

Substituting these values in above equation, we get
$$\frac{dq}{dR} \cdot \frac{1}{12} = \frac{10}{R}$$
$$\Rightarrow \quad dq = 120 \frac{dR}{R} \quad \cdots (2)$$

Resistance at $t = 0$ is 20 Ω,
Resistance after 4 min is, $R = 5\,\Omega/\text{min} \times 4\text{min} = 20\,\Omega$
Now, to get the amount of charge which has passed in the circuit in 4 minutes, we integrate Eq.(2) from $R = 20\,\Omega$ to $R = 40\,\Omega$, i.e.,
$$q = 120 \int_{20}^{40} \frac{dR}{R} = 120 (\ln 40 - \ln 20) = 120 \ln 2$$

15. In the given circuit (Fig.4.289), $R_1 = 10\,\Omega$, $R_2 = 6\,\Omega$, and applied emf is 10 V. The effective resistance of the circuit is
    (A) 20 Ω (B) 30 Ω (C) 40 Ω (D) 50 Ω

SOLUTION (A) Since, circuit shown in Fig.4.289 has

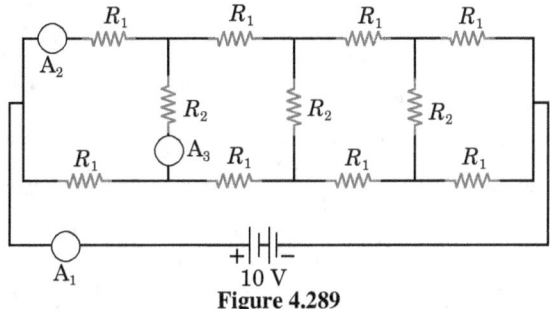

Figure 4.289

mirror symmetry about a horizontal line passing through end terminals. Therefore, across all three resistors R_2 potential is same, i.e., potential difference across all three R_2 is zero and no current passes through any R_2. So, reading of A_3 is zero and we can remove all resistors represented by R_2. The next simplified diagram is shown in Fig.4.290.

Equivalent resistance of the circuit connected with emf

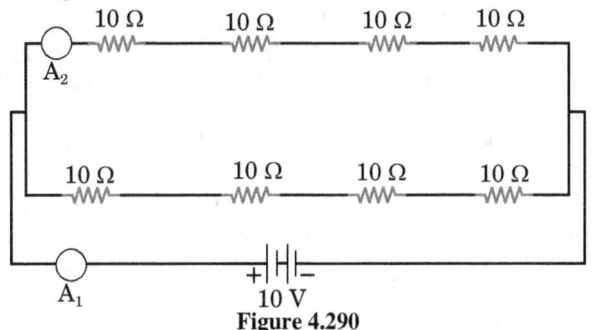

Figure 4.290

source is
$$R_{eq} = \frac{40 \times 40}{40 + 40} = 20\,\Omega$$

16. Again in the circuit of Fig.4.289, $R_1 = 10\,\Omega$, $R_2 = 6\,\Omega$ and applied emf of source is 10 V. The reading of A_1 is
    (A) 1 A (B) 1/2 A (C) 2 A (D) 3 A

SOLUTION From above example 15, the equivalent resistance of the circuit connected with emf source is 20 Ω. Therefore, current supplied by emf source,
$$I = \frac{10}{20} = \frac{1}{2}\,\text{amp}$$
Hence reading of A_1 is 1/2, i.e., 0.5 A.

17. Again in the circuit of Fig.4.289, $R_1 = 10\,\Omega$, $R_2 = 6\,\Omega$ and applied emf of source is 10 V. The reading of A_2 is
    (A) 1 A (B) 2 A (C) 1/4 A (D) 3 A

SOLUTION (C) From above example 17, the reading of A_1 in the circuit of Fig.4.289, is 1/2, i.e., 0.5 amp.
Since, both branches of Fig.4.290 are identical, so current divides into two equal part and hence the reading of A_2, is 1/4 i.e., 0.25 A.

18. One billion electrons pass from A to B in 1 ms. What is the direction and magnitude of current?
    (A) 1.6 A (B) $0.16\mu A$ (C) $0.8\mu A$ (D) $1.6\mu A$

SOLUTION (C) $I = \frac{Ne}{t} = \frac{(10^9)(1.6 \times 10^{-19} C)}{10^{-3}}$
$= 1.6 \times 10^{-7}$ A
$\Rightarrow \quad I = 0.16\mu A$.

The current flows from B to A (opposite to direction of flow of electrons).

19. An electron gun in a TV set shoots out a beam of electrons. The beam current is 10 μA. How many electrons strike the TV screen each second?
    (A) 2.78×10^{14} (B) 6.3×10^{13}
    (C) 6.78×10^4 (D) electrons will not reach

SOLUTION (B) $N = I/e$
$= (1.0 \times 10^{-5} C/s)/(1.6 \times 10^{-19} C)$
$= 6.3 \times 10^{13}$ electrons per second.

20. In above example 19, how much charge strikes the screen in a minute?
    (A) $-600\,\mu C$ (B) $600\,\mu C$
    (C) $1300\,\mu C$ (D) $-1300\,\mu C$

SOLUTION (A) The charge Q striking the screen is given by
$$|Q| = It = (10\mu C/s)(60\,s) = 600\mu C.$$
Since, charge carriers are electrons, therefore, the actual charge is $Q = -600\mu C$.

21. A copper bus bar carrying current 1200 A has a potential drop of 1.2 mV along 24 cm of its length. What is the resistance per meter of the bar?
    (A) $5.2\,\mu\Omega$ (B) $3.2\,\mu\Omega$ (C) $8\,\mu\Omega$ (D) $4.2\,\mu s$

SOLUTION (D) From Ohm's law, we have
$$\Delta V = IR$$
Therefore, the potential difference across 24 cm of length of bar
$$\Delta V_{24} = IR_{24}, \text{ or } (1.2 \times 10^{-3}\,V) = (1200\,A)R,$$
$$\Rightarrow \quad R_{24} = 1.0\mu\Omega.$$
This is the resistance of 24 cm length of the bar. Therefore, resistance per meter i.e., per 100 cm of the bar
$$R_{100} = \frac{R_{24}}{24} \times 100 = 4.2\,\mu\Omega$$

22. A 20 cm long copper tube has an inner diameter of 0.85 cm and an outer diameter of 1.10 cm. Find its electric resistance when used lengthwise. (resistivity of copper, $\rho = 1.7 \times 10^{-8}\,\Omega.m$)
    (A) $51.2\,\mu\Omega$ (B) $89\,\mu\Omega$ (C) $80\,\mu\Omega$ (D) $42\,\mu\Omega$

SOLUTION (B) From Eq.4.37, we have
$$R = \rho \frac{l}{A} \quad \cdots (1)$$
The cross-sectional area, $A = \pi (R_2^2 - R_1^2)$

Here, R_1 and R_2 are the internal and external radii of the copper tube.
Given: $l = 0.20$ m, $\rho = 1.7 \times 10^{-8}\,\Omega.$m,
$R_1 = \dfrac{0.85}{2} \times 10^{-2}$ cm, and $R_2 = \dfrac{1.10}{2} \times 10^{-2}$ cm

Therefore, $A = \dfrac{\pi\left(1.10^2 - 0.85^2\right) \times 10^{-4}}{4}$
$= 3.83 \times 10^{-5}$ m^2;

On substituting these values in Eq.(1), we get
$R = 89\,\mu\Omega$.

Paragraph for 23 - 25 In Fig.4.291, an ammeter and a voltmeter are connected in series to a battery with an emf E = 6 volt. When a certain resistance is connected in parallel with voltmeter, the reading of voltmeter decreases two times, where as the reading of the ammeter increasing the same number of times.

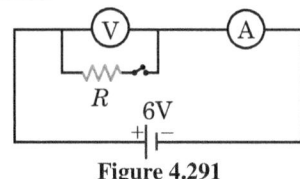

Figure 4.291

23. What is the ratio of resistance of voltmeter to resistance of ammeter.
    (A) 2   (B) 1/2   (C) 1/3   (D) 3
24. What will be voltmeter reading before connecting the resistance.
    (A) 1 V   (B) 2 V   (C) 3 V   (D) 4 V
25. What will be voltmeter reading after connecting the resistance.
    (A) 1 V   (B) 2 V   (C) 3 V   (D) 4 V

SOLUTION 23 to 25: Suppose R_A = Resistance of ammeter, R_V = resistance of Voltmeter. In the first case current is the circuit of Fig.4.291
$$I = \dfrac{6}{(R_A + R_V)} \qquad \cdots (1)$$
If V_A and V_V are voltages across ammeter and voltmeter respectively, then
$$V_V = 6 - V_A = 6 - IR_A \qquad \cdots (2)$$
Substituting the value of I from Eq.(1) in Eq.(2), we get
$$V_V = 6 - \dfrac{6R_A}{(R_A + R_V)} \qquad \cdots (3)$$
After connecting a certain resistance R(say) in parallel with the voltmeter, the reading of ammeter becomes two times i.e. the total resistance become half while the resistance of ammeter remains unchanged. Hence, new voltage across voltmeter
$$V'_V = 6 - I'R_A \qquad \cdots (4)$$
Here, $\quad I' = 2I = \dfrac{12}{(R_A + R_V)}$

Substituting this value of I' in Eq.(4), we get
$$V'_V = 6 - \dfrac{12R_A}{(R_A + R_V)} \qquad \cdots (5)$$
It is also given that, $V'_V = \dfrac{V_V}{2} = \dfrac{1}{2}\left[6 - \dfrac{6R_A}{R_A + R_V}\right]$

Substituting this value in Eq.(5), we get
$$\dfrac{1}{2}\left[6 - \dfrac{6R_A}{R_A + R_V}\right] = 6 - \dfrac{12R_A}{(R_A + R_V)}$$
or $\quad \dfrac{9R_A}{R_A + R_V} = 3$
$\Rightarrow \quad 6R_A = 3R_V \quad$ or $\quad \dfrac{R_V}{R_A} = 2 \qquad \cdots (6)$

It is the answer of Q. 23. So correct choice is (A).
Substituting the value of R_V from Eq.(6) in (3), we get
$$V_V = 6 - \dfrac{6R_A}{(R_A + 2R_A)} = 6 - 2 = 4\text{ volt}$$
It is the answer of Q. 24. So, correct choice is (D).
Substituting the value of R_V from Eq.(6) in (5), we get
$$V'_V = 6 - \dfrac{12R_A}{(R_A + 2R_A)} = 6 - 4 = 2\text{ volt}$$
It is the answer of Q.25. So, correct choice is (B).

26. In Bohr model of hydrogen atom, the electron revolves in a circular orbit of radius 5.3×10^{-11}m with a speed of 2.2×10^6 m/s around the nucleus. Determine the current I in the orbit.
    (A) 1.06 mA   (B) 2.06 mA
    (C) 0.06 mA   (D) 3.06 mA

SOLUTION (A) Consider any point A on the circular orbit of radius r. Electron periodically crosses this point at the intervals of time period T. In other words it crosses the point A once in each revolution.
Therefore, in time interval $\Delta t = T$, the charge crossing the point A is $\Delta Q = e$. Now, from Eq.4.1, the current crossing point A is given by
$$I = \dfrac{\Delta Q}{T} = \dfrac{e}{T} = ef \qquad \cdots (1)$$
here, $f = 1/T$ is the frequency of revolution.
If speed of the electron on circular path is v, then
$T = \dfrac{2\pi r}{v}$, therefore from Eq.(1)
$$I = \dfrac{e}{T} = \dfrac{ev}{2\pi r} \qquad \cdots (2)$$
Given: $r = 5.3 \times 10^{-11}$m, $v = 2.2 \times 10^6$ m/s, $e = 1.6 \times 10^{-19}$ C
Substituting these values in Eq.(2), we get
$$I = \dfrac{1.6 \times 10^{-19} \times 2.2 \times 10^6}{2 \times 3.14 \times 5.3 \times 10^{-11}} \approx 0.106 \times 10^{-2}\text{ A} = 1.06\text{ mA}$$
Note that the current flows in the opposite direction to the electron, which is negatively charged.

27. A wire carries a current of 2.0 A. What is the charge that has flowed through its cross-section in 1.0 s. How many electrons does this correspond to?
    (A) 3.0C, 1.25×10^{19} (B) 2.0 C, 1.25×10^{19}
    (C) 4.0C, 1.25×10^{19} (D) 2.0C, 5.25×10^{19}

SOLUTION (B) $Q = It = (2.0\text{A})(1.0\text{ s}) = 2.0$ C
If N is the number of electrons flowed in 1 s through the given cross-section of the wire, then
$$Q = Ne \quad \Rightarrow \quad N = \dfrac{Q}{e} = \dfrac{2}{1.6 \times 10^{-19}} = 1.25 \times 10^{19}$$

28. A current of 7.5 A is maintained in a wire for 45 s. In this time (a) how much charge and (b) how many electrons flow through the wire?
    (A) 400.5C, 2.1×10^{21} (B) 300.5C, 2.1×10^{21}
    (C) 337.5C, 2.1×10^{21} (D) 700.5C, 2.1×10^{21}

SOLUTION (C) (a) $Q = It = (7.5$ A$)(45$ s$) = 337.5$C
(b) The number of electrons N is given by
$$N = \dfrac{Q}{e} = \dfrac{337.5\text{C}}{1.6 \times 10^{-19}\text{C}} = 2.1 \times 10^{21}$$
where e = 1.6×10^{-19}C is the charge of an electron.

29. What will be the current if charge Q revolves in a circular orbit with the frequency f ?
    (A) f (B) q/f (C) q (D) qf

SOLUTION (D) From Eq.4.8, we have: $\quad I = Qf$

30. Calculate the mean free time between collisions of electrons in copper at room temperature.

4.32. SOLVED OBJECTIVE QUESTIONS

(A) 2.4×10^{-14} s (B) 4×10^{-14} s
(C) 5.4×10^{-14} s (D) $.4 \times 10^{-14}$ s

SOLUTION (A) From Eq.4.31, the mean free time is given as
$$\tau = \frac{1}{\sigma} = \frac{m_e}{ne^2\rho} \quad \cdots (1)$$
Given: $n = 8.5 \times 10^{28}$ m^{-3}, $\rho = 1.72 \times 10^{-8}\Omega$, and we know that $m_e = 9.11 \times 10^{-31}$kg, $e \approx 1.6 \times 10^{-19}$C

Substituting these values in Eq.(1), we get
$$\tau = \frac{1}{\sigma} = \frac{9.11 \times 10^{-31}}{8.5 \times 10^{28} \times (1.6 \times 10^{-19})^2 \times 1.72 \times 10^{-8}}$$
$$= 2.4 \times 10^{-14} \text{ s}$$

☞ The reciprocal of τ gives the average number of collisions per second. In this case it is $1/(2.4 \times 10^{-14})$ s^{-1}, i.e., 4×10^{13} collision per second.

31. A coil of wire has a resistance of 25.00Ω at $20°$C and resistance of 25.17Ω at $35°$C. What is its temperature coefficient of resistance?
(A) 4.5×10^{-40}C^{-1} (B) 5×10^{-40}C^{-1}
(C) 0.5×10^{-40}C^{-1} (D) 4.0×10^{-40}C^{-1}

SOLUTION (A) From Eq.4.40, we have
$$R_t = R_0 [1 + \alpha(t - t_0)]$$
or $$R_t - R_0 = R_0 \alpha (t - t_0)$$
or $$\alpha = \frac{R_t - R_0}{R_0 (t - t_0)} = \frac{\Delta R}{R_0 \Delta t} \quad \cdots (1)$$
Given: $R_0 = 25.00\Omega$, $\Delta R = 25.17 - 25 = 0.17$ Ω, and $\Delta t = 35 - 20 = 15°$C.

Substituting these values in Eq. (1), we get
$$\alpha = \frac{(0.17)}{25.00 \times 15} = 4.5 \times 10^{-4}/°C.$$

32. Kirchhoff's current law obeys conservation of
(A) charge (B) momentum
(C) energy (D) none of these.

SOLUTION (A) Kirchhoff's current law is based on principle of conservation of charge.

33. The slide wire Wheatstone bridge shown in Fig.4.292 is balanced when the uniform slide wire AB is divided as shown. Find the value of the resistance
(A) 3 Ω (B) 4 Ω (C) 2 Ω (D) 7 Ω

SOLUTION (A) $\frac{X}{3\Omega} = \frac{L}{M} = \frac{40 \text{ cm}}{60 \text{ cm}}$ or $X = 2\ \Omega$

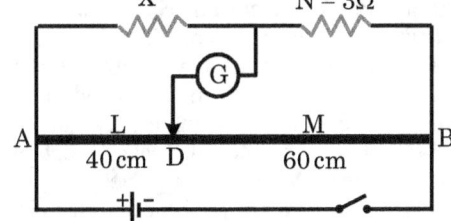

Figure 4.292

34. A dry cell delivering current 2 A has terminal voltage 1.41 V. What is the internal resistance of the cell if its open-circuit voltage is 1.59 V ?
(A) 5.09Ω (B) 6.09Ω (C) 7.09Ω (D) 0.09Ω

SOLUTION (D) The open-circuit voltage is simply the emf of the cell. From Eq.4.68
$$\Delta V = \mathcal{E} - Ir$$
$$\therefore \quad r = \frac{\mathcal{E} - \Delta V}{I} \quad \cdots (1)$$
Given: $I = 2$ A, $\Delta V = 1.41$V and $\mathcal{E} = 1.59$ V
Substituting these values in Eq.(1), we get
$$r = \frac{1.59 - 1.41}{2} = 0.09\ \Omega$$

35. A 8.00 m long wire, of uniform cross-sectional area 8.00 mm^2, has a conductance of $G = 2.45\Omega^{-1}$. What is the resistivity of the material of the wire?
(A) 2.1×10^{-7} S (B) 3.1×10^{-7} S
(C) 4.1×10^{-7} S (D) 5.1×10^{-7} S

SOLUTION (C) From Eq.4.37, we have
$$R = \rho \frac{l}{A}$$
Since, $R = 1/G$, therefore above equation gives
$$\rho = \frac{RA}{l} = \frac{A}{Gl} \quad \cdots (1)$$
Given: $l = 8.00$ m, $A = 8.00$ mm$^2 = 8.00 \times 10^{-6}$ m^2, $G = 2.45\Omega^{-1}$

Substituting these values in Eq.(1), we get
$$\rho = \frac{8.00 \times 10^{-6}}{2.45 \times 8.00} = 4.1 \times 10^{-7} \text{ S}$$

36. A wire 250 cm long and 1 mm^2 in cross-section carries a current of 4 A when connected to a 2 V battery. The resistivity of the wire is
(A) $0.2 \times 10^{-6}\Omega$m (B) $2 \times 10^{-7}\Omega$m
(C) $5 \times 10^{-6}\Omega$m (D) $4 \times 10^{-6}\Omega$m

SOLUTION (A) $\rho = \frac{RA}{l}$
Since, $R = V/I$, therefore, above equation gives
$$\rho = \frac{VA}{Il} \quad \cdots (1)$$
Given: $l = 2.50$ m, $A = 10^{-6}$ m^2, $I = 4$ A and 2 V
Substituting these values in Eq.(1), we get
$$\rho = \frac{VA}{Il} = \frac{2 \times 10^{-6}}{4 \times 2.50} = 0.2 \times 10^{-6}\Omega\text{m}$$

37. An electrical cable of copper has just one wire of radius 9 mm. Its resistance is 5 Ω. This single wire of cable is replaced by 6 different well-insulated copper wires each of radius 3 mm. The total resistance of the cable will now be equal to?
(A) 7.5Ω (B) 45Ω (C) 90Ω (D) 270Ω

SOLUTION (A) If resistance of a single wire of length l, radius r_1, is R_1. Then
$$R_1 = \rho \frac{l}{A_1} = \rho \frac{l}{\pi r_1^2} \quad \cdots (1)$$
Let when the wire is replaced by 6 wires of same length in parallel, the resistance of each wire is R_2. Then
$$R_2 = \rho \frac{l}{A_2} = \rho \frac{l}{\pi r_2^2} \quad \cdots (2)$$
Dividing Eq.(1) by Eq.(2), we get
$$\frac{R_1}{R_2} = \frac{r_2^2}{r_1^2} \quad \text{or} \quad R_2 = \frac{r_1^2}{r_2^2} R_1 \quad \cdots (3)$$
Given: $r_1 = 9$ mm, $R_1 = 5$ Ω, $r_2 = 3$ mm
Substituting these values in Eq.(3), we get
$$R_2 = \frac{9^2}{3^2} \times 5\ \Omega = 45\ \Omega$$
Since, these 6 wires are in parallel, therefore, equivalent resistance
$$R_{eq} = \frac{R_2}{6} = \frac{45}{6} = 7.5\ \Omega$$

38. When cells are arranged in parallel
(A) the current capacity decreases
(B) the current capacity increases
(C) the emf increases
(D) the emf decreases

SOLUTION (B) When cells are connected in parallel, the current capacity increases.

39. An incandescent bulb has a thin filament of tungsten that is heated to high temperature by passing an electric current. The hot filament emits black-body radiation. The filament is observed to break up at random locations after a sufficiently long time of operation due to non-uniform evaporation of tungsten from the filament. If the bulb is powered at constant voltage, which of the following statement(s) is(are) true?
(A) The temperature distribution over the filament is uniform.
(B) The resistance over small sections of the filament decreases with time.
(C) The filament emits more light at higher band of frequencies before it breaks up.
(D) The filament consumes less electrical power towards the end of the life of the bulb.

SOLUTION (C, D) The filament breaks up at random location due to non-uniform evaporation of tungsten from the filament. The non-uniform evaporation is caused by the non-uniform temperature distribution over the filament.
Let l_0 be the length and r_0 be the radius of the filament. Let radius of the filament reduces from r_0 to r in a small segment of length l ($\ll l_0$). The resistance of this small segment increases from $\frac{\rho l}{\pi r_0^2}$ to $\frac{\rho l}{\pi r^2}$ ($\because r < r_0$), where ρ is the resistivity of the tungsten. The heat generated in this

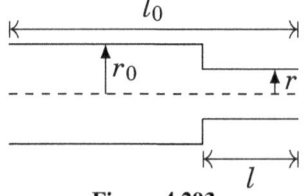

Figure 4.293

small segment in a time interval t, $I^2 t \rho l / (\pi r^2)$, is more than the heat generated in other segment of same length but of radius r_0. This heat increases temperature of this segment which further increases its resistance through (i) temperature dependence of resistance (ii) increase in rate of evaporation. Thus, the temperature T of this segment is higher than the temperature T_0 of other segment. A black body at higher temperature emits more radiation at higher band of frequencies (see Fig.4.294).
Towards the end of life of the bulb, effective resistance R_e

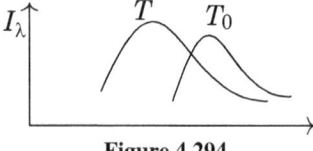

Figure 4.294

of the filament increases (due to decrease in radius caused by evaporation). Thus, the electrical power consumed by the filament, V^2/R_e, decreases.

40. Heater of an electric kettle is made of a wire of length L and diameter d. It takes 4 minutes to raise the temperature of 0.5 kg water by 40 K. This heater is replaced by a new heater having two wires of the same material, each of length L and diameter $2d$. The way these wires are connected is given in options. How much time in minutes will it take to raise the temperature of the same amount of water by 40 K?

(A) 4 if wires are in parallel
(B) 2 if wires are in series
(C) 1 if wires are in series
(D) 0.5 if wires are in parallel

SOLUTION (B, D) The resistance of the wire of length L and diameter d is $R_1 = \frac{4\rho L}{\pi d^2}$ and of length L and diameter $2d$ is $R_2 = \frac{\rho L}{\pi d^2} = \frac{R_1}{4}$. When the two wires of resistances R_2 are connected in series, the effective resistance is
$$R_s = R_2 + R_2 = 2R_2 = R_1/2 \quad \cdots (1)$$
and when they are connected in parallel, the effective resistance is
$$R_p = \frac{R_2 R_2}{R_2 + R_2} = R_2/2 = R_1/8 \quad \cdots (2)$$
Let V be the applied voltage, $m = 0.5$ kg be the mass of the water, and S be the specific heat of the water. Initially, the heat produced by a resistance R_1 in time $t_1 = 4$ min is $V^2 t_1 / R_1$. This heat is used to raise the temperature of the water by $\Delta T = 40$ K. Thus,
$$V^2 t_1 / R_1 = mS\Delta T \quad \cdots (3)$$
Let t_s and t_p be the time taken to raise the temperature of same amount of water by ΔT when the resistances are connected in series and parallel. Thus,
$$V^2 t_s / R_s = mS\Delta T \quad \cdots (4)$$
$$V^2 t_p / R_p = mS\Delta T \quad \cdots (5)$$
Dividing equation (4) by (3) and using the equation (1), we get
$$t_s = \frac{R_s}{R_1} t_1 = \frac{1}{2} \times 4 = 2 \text{ min}$$
Similarly, dividing equation (5) by (3) and using the equation (2), we get
$$t_p = \frac{R_p}{R_1} t_1 = \frac{1}{8} \times 4 = 0.5 \text{ min}.$$

41. Consider two identical galvanometers and two identical resistors with resistance R. If the internal resistance of the galvanometers $R_c < R/2$, which of the following statement(s) about any one of the galvanometers is(are) true?
(A) The maximum voltage range is obtained when all the components are connected in series.
(B) The maximum voltage range is obtained when the two resistors and one galvanometer are connected in series, and the second galvanometer is connected in parallel to the first galvanometer.
(C) The maximum current range is obtained when all the components are connected in parallel.
(D) The maximum current range is obtained when the two galvanometers are connected in series and the combination is connected in parallel with both the resistors.

SOLUTION Let I_g be the full scale deflection current of the galvanometer. When all components are connected in series (Fig.4.295), effective resistance of the circuit is $R_e = 2R + 2R_c$ and maximum current allowed in the circuit is I_g.

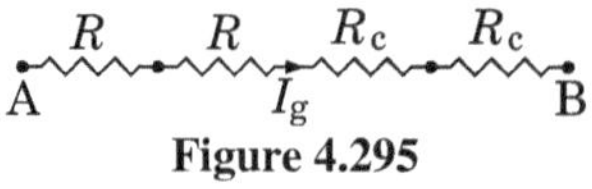

Figure 4.295

Thus, the voltage between A and B is
$$V_{AB} = I_g R_e = 2I_g (R + R_c)$$
Consider the case when two resistors and one galvanometer are connected in series and the second galvanometer is connected in parallel (Fig.4.296).
The maximum current through each galvanometer is I_g and the maximum current through the resistors is $2I_g$. Apply Kirchhoff's voltage law to get the voltage between A and B as

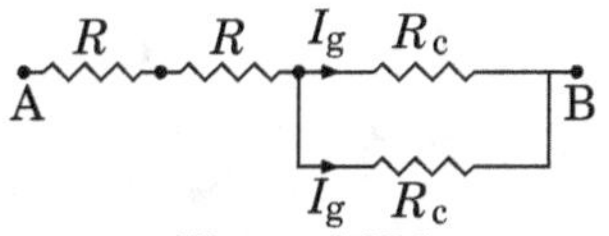

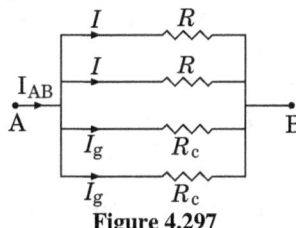

Figure 4.296

$$V'_{AB} = 2I_g R + 2I_g R + I_g R_c$$
$$= 2I_g (R + R_c) + 2I_g (R - R_c/2)$$
$$= V_{AB} + 2I_g (R - R_c/2)$$
$$> V_{AB} \quad (\because R_c < R/2)$$

Consider the case when all four components are connected in parallel (Fig.4.297).

Let I and I_g be the currents through the resistors and the

Figure 4.297

galvanometers. By Ohm's voltage law, $IR = I_g R_c$, which gives $I = I_g R_c/R$. The current between A and B is
$$I_{AB} = 2I + 2I_g = 2I_g (1 + R_c/R)$$

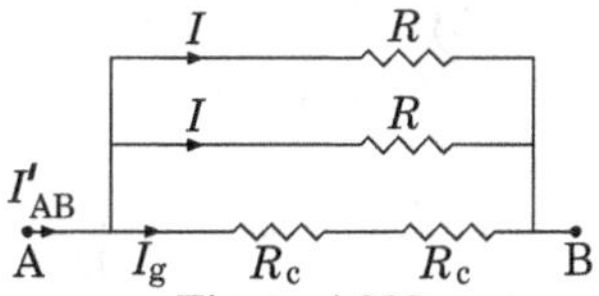

Figure 4.298

At last, consider the case when the two galvanometers are connected in series and the combination is connected in parallel with both the resistors (Fig.4.299).
$$I'_{AB} = I_g + 2I = I_g + 2I_g (2R_c/R)$$

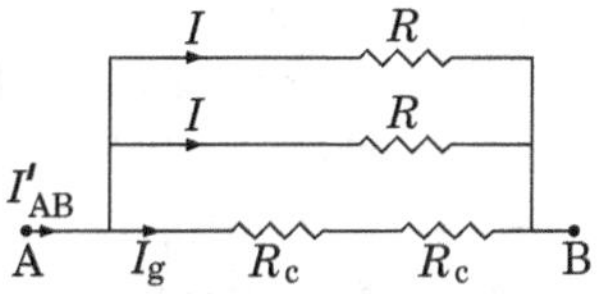

Figure 4.299

$$= 2I_g (1 + R_c/R) - (1 - 2R_c/R) I_g$$
$$= I_{AB} - (1 - 2R_c/R) I_g$$
$$< I_{AB}, \quad \text{since } R_C < R/2$$

42. In the circuit shown in the Fig.4.300, the key is pressed at time $t = 0$. Which of the following statement(s) is(are) true?
    (A) The voltmeter displays -5 V as soon as the key is pressed, and displays $+5$ V after a long time.
    (B) The voltmeter will display 0 V at time $t = \ln 2$ seconds.
    (C) The current in the ammeter becomes $1/e$ of the initial value after 1 second.
    (D) The current in the ammeter becomes zero after a long time.

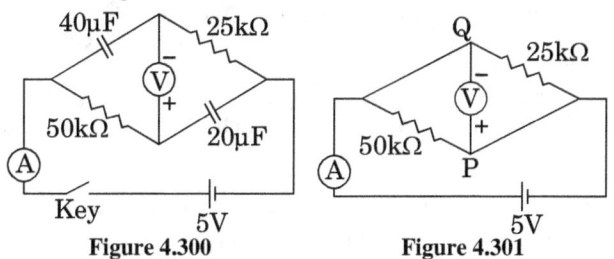

Figure 4.300   Figure 4.301

SOLUTION (A, B, C, D) The impedance (effective resistance) of the capacitors is zero immediately after the key is pressed ($t \to 0^+$). Thus, capacitors will act as short circuit elements (Fig.4.301).

Thus, potential at the node Q is $V_Q = 5$ V and potential at the node P is $V_P = 0$ V. The voltmeter will display $V = V_P - V_Q = 0 - 5 = -5$ V.

As $t \to \infty$, both capacitors get fully charged and reach to steady state. In this case, the impedance of the capacitors become infinite and they act like open circuit elements (Fig.4.302).

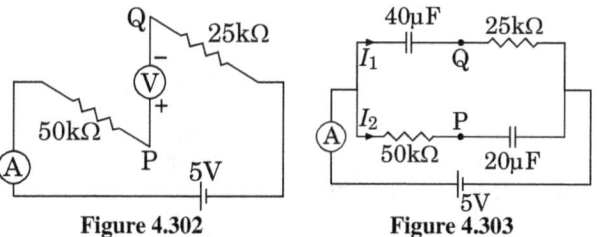

Figure 4.302   Figure 4.303

We assume voltmeter to be ideal with infinite resistance. Thus, no current flows through the circuit and hence ammeter will read zero. Also, potential at the node Q is $V_Q = 0$ V and potential at the node P is $V_P = 5$ V. The voltmeter will display $V = V_P - V_Q = 5 - 0 = 5$ V.

The capacitor $C_1 = 40\,\mu$F is connected across a $V = 5$ V battery by a series resistance $R_1 = 25$ kΩ. Similarly, the capacitor $C_2 = 20\,\mu$F is connected across the same battery by a series resistance $R_2 = 50$ kΩ. We can remove voltmeter from the circuit for current calculation as it has infinite resistance as shown in Fig.4.303. This circuit is used to charge both the capacitors.

From Eq.4.201: $i(t) = Ie^{-t/RC}$ (here, $I = V/R$), the currents through the capacitors at time t (charging of the capacitor) are given by

$$I_1 = (V/R_1)\, e^{-\frac{t}{R_1 C_1}} = 200 e^{-t}\,\mu\text{A} \quad \cdots (1)$$
$$I_2 = (V/R_2)\, e^{-\frac{t}{R_2 C_2}} = 125 e^{-t}\,\mu\text{A} \quad \cdots (2)$$

From equations (1) and (2), the current through the ammeter at time $t = 0$ s is $I_0 = I_1 + I_2 = 325\,\mu$A. The current through the ammeter at time $t = 1$ s is $I = I_1 + I_2 = 200 e^{-1} + 125 e^{-1} = 325/e = I_0/e$.

The potentials at points P and Q at a time t are given by
$$V_P = V - I_2 R_2 = 5\left(1 - e^{-t}\right) \quad \cdots (3)$$
$$V_Q = 0 + I_2 R_1 = 5 e^{-t} \quad \cdots (4)$$

From equations (3) and (4), the voltmeter at time $t = \ln 2$ will display $V = V_p - V_Q = 5 - 10 e^{-\ln 2} = 0$ V.

Therefore all the choices A, B, C and D are correct.

43. A microammeter has a resistance of 100Ω and full scale range of 50μA. It can be used as a voltmeter or as a higher range ammeter provided a resistance is added to it. Pick the correct range and resistance combination(s).
    (A) 50 V range with 10kΩ resistance in series.
    (B) 10 V range with 200kΩ resistance in series.
    (C) 5 mA range with 1Ω resistance in parallel.
    (D) 10 mA range with 1Ω resistance in parallel.

SOLUTION (B, C) The microammeter has a resistance $G = 100\,\Omega$ and the full scale range $I_g = 50\,\mu$A. It can be used to measure the voltage across A and B by connecting a series resistance R (Fig.4.304a). Maximum voltage across A and B that can be measured is
$$V_{\max} = I_g(G + R)$$
For $R = 10$ kΩ, $V_{\max} = 50 \times 10^{-6}(100 + 10000) \approx 0.5$ V and for $R = 200$kΩ, $V_{\max} \approx 10$ V

The microammeter can be used to measure higher cur-

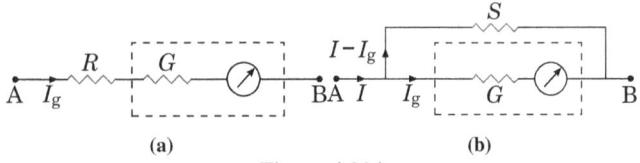

Figure 4.304

rent from A to B by connecting a parallel resistance S (Fig.4.304b). Maximum current from A to B that can be measured is
$$I_{max} = I_g\left(1 + \frac{G}{S}\right)$$
For $S = 1\Omega$,
$$I_{max} = 50 \times 10^{-6}\left(1 + \frac{100}{1}\right) \approx 5 \times 10^{-3} \text{ A} = 5 \text{ mA}$$
Therefore, the correct choices are B and C.

44. Capacitor C_1 of capacitance $1\mu F$ and capacitor C_2 of capacitance $2\mu F$ are separately charged fully by a common battery. The two capacitors are then separately allowed to discharge through equal resistors at time $t = 0$
    (A) The current in each of the two discharging circuits is zero at $t = 0$.
    (B) The currents in the two discharging circuits at $t = 0$ are equal but not zero.
    (C) The currents in the two discharging circuits at $t = 0$ are unequal.
    (D) Capacitor C_1 loses 50% of its initial charge sooner than C_2 loses 50% of its initial charge.

SOLUTION (B, D) Let $C_1 = 1\mu F$, $C_2 = 2\mu F$, and V be the emf of the battery. Initial charges on capacitors C_1 and C_2 are
$$q_{0,1} = C_1 V \text{ and } q_{0,2} = C_2 V$$
respectively.

Consider the discharging of capacitors at time t [Fig.4.305].

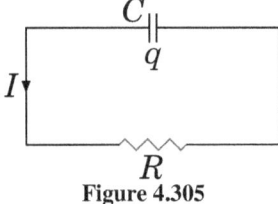

Figure 4.305

Let the charge on the capacitor at any time t is q, its potential is q/C, and current through it is $I = -dq/dt$ (negative sign because the charge q is decreasing with time).
Applying Kirchhoff's voltage law in loop shown in Fig.4.305, we get
$$\frac{q}{C} - IR = 0, \quad \text{i.e.,} \quad \frac{q}{C} - \left(-\frac{dq}{dt}\right)R = 0$$
$$\Rightarrow \quad \frac{dq}{dt} + \frac{q}{RC} = 0 \quad \text{or} \quad \frac{dq}{q} = -\frac{1}{RC}dt$$
Let at time $t = 0$, the charge on capacitor's positive plate is q_0 (maximum charge) and at time t the remaining charge is q, then integration of above equation gives
$$\int_{q_0}^{q} \frac{dq}{q} = -\frac{1}{RC}\int_0^t dt$$
or $\quad [\ln q]_{q_0}^{q} = -\frac{1}{RC}[t]_0^t$
or $\quad \ln \frac{q}{q_0} = -\frac{t}{RC}$
or $\quad q = q_0 e^{-\frac{t}{RC}} \qquad \cdots (1)$
Substituting initial value of charges in equation (1), we get the charges at time t on the two capacitors as

$$q_1 = q_{0,1} e^{-\frac{t}{RC_1}} = VC_1 e^{-\frac{t}{RC_1}} \qquad \cdots (2)$$
$$q_2 = q_{0,2} e^{-\frac{t}{RC_2}} = VC_2 e^{-\frac{t}{RC_2}} \qquad \cdots (3)$$
On differentiating Eq.(2) and (3), we get respective currents ($I = -dq/dt$) through capacitor C_1 and C_2. as
$$I_1 = (V/R)e^{-\frac{t}{RC_1}} \qquad \cdots (3)$$
$$I_2 = (V/R)e^{-\frac{t}{RC_2}} \qquad \cdots (4)$$
At $t = 0$, Eq.(3) and (4) give
$$I_{0,1} = I_{0,2} = V/R$$
Thus, the initial currents through the capacitors C_1 and C_2 are $I_{0,1} = I_{0,2} = V/R$. Note that the initial current depends only on the initial potential of the capacitor and the discharging resistance. From equation (1), if the charge on the capacitor becomes half of its initial value at time T, then
$$\frac{q_0}{2} = q_0 e^{-\frac{T}{RC}} \quad \text{or} \quad T = RC \ln 2$$
Thus, $\quad T_1 = RC_1 \ln 2 \quad$ and $\quad T_2 = RC_2 \ln 2$
Given that $C_1 = 1\ \mu F$ and $C_2 = 2\ \mu F$, therefore,
$$T_1 = T_2/2.$$
Therefore, the correct choices are B and D.

45. A capacitor is charged using an external battery with a resistance x in series. The dashed line, in Fig.4.306, shows the variation of $\ln I$ with respect to time. If the resistance is changed to $2x$, the new graph will be
    (A) P (B) Q (C) R (D) S

SOLUTION (B) The charge q on a capacitor of capaci-

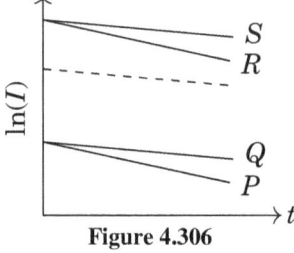

Figure 4.306

tance C while it is being charged by a battery of voltage V and connected in series to a resistor R is given by
$$q = CV\left[1 - e^{-\frac{t}{RC}}\right]$$
To get current, we differentiate it with respect to time t
$$I = \frac{dq}{dt} = \frac{V}{R}e^{-\frac{t}{RC}}$$
Taking logarithm on both sides of above equation, we get
$$\ln I = \ln\left(\frac{V}{R}\right) - \left(\frac{1}{RC}\right)t \qquad \cdots (1)$$
The equation (1) represents a straight line between t and $\ln I$ with an intercept $\ln\left(\frac{V}{R}\right)$ on $\ln I$ axis and slope $-\frac{1}{RC}$. When R is changed from x to $2x$, the intercept decreases from $\ln\left(\frac{V}{x}\right)$ to $\ln\left(\frac{V}{2x}\right)$ and slope increases from $-\frac{1}{xC}$ to $-\frac{1}{2xC}$. So, option B is correct.

46. Which of the set-up shown in Fig.4.307, can be used to verify Ohm's law?

SOLUTION (B) The verification of Ohm's law ($V = IR$) requires the measurements of current through and voltage across the variable resistance. Therefore, correct setup is shown in option (B).

47. In the circuit shown in Fig.4.308a, $P \neq R$, the reading of galvanometer is same with switch S open or closed. Then,
    (A) $I_R = I_G$ (B) $I_P = I_G$
    (C) $I_Q = I_G$ (D) $I_Q = I_R$

SOLUTION (A) Circuit is redrawn in Fig.4.308b. Let G be the resistance of the galvanometer. Consider the case when switch S is open. In this case, resistances R and G are connected in series so, $I_R = I_G$. Also, the resistances P

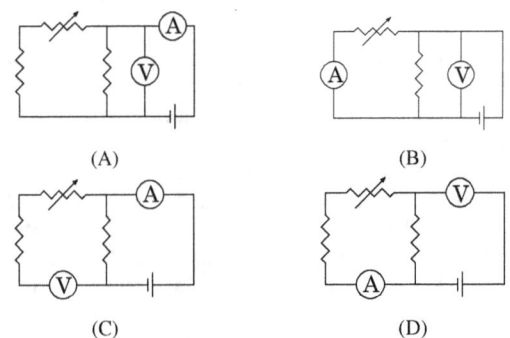

Figure 4.307

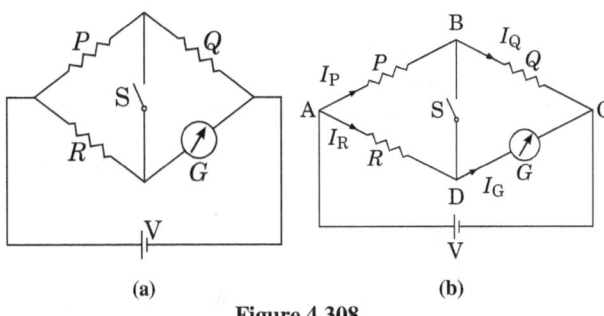

Figure 4.308

and Q are connected in series, therefore, $I_P = I_Q$.
Kirchhoff's voltage law in the loop ABCA (through battery) gives
$$I_P = I_Q = \frac{V}{P+Q} \quad \cdots (1)$$
Similarly, Kirchhoff's voltage law in loop ADCA (through battery) gives
$$I_R = I_G = \frac{V}{R+G} \quad \cdots (2)$$
Now consider the case, when the switch S is closed. Let in this case, the currents through P, Q, R, and G be I'_P, I'_Q, I'_R and I'_G respectively.
Since, it is given that, the current through the galvanometer does not change when S is closed, therefore,
$$I'_G = I_G = \frac{V}{R+G} \quad \cdots (3)$$
Applying Kirchhoff's voltage law in loop ADCA, we get
$$I'_R R + I'_G G = V \quad \cdots (4)$$
Substituting the value of I'_G from Eq.(3) in Eq.(4), we get
$$I'_R R = V - \frac{V}{R+G} G = V\left[1 - \frac{G}{R+G}\right] = V\frac{R}{R+G}$$
$$\Rightarrow \quad I'_R = \frac{V}{R+G} \quad \cdots (5)$$
From Eq.(3) and (5), we see that
$$I'_R = I'_G = I_G \quad \cdots (6)$$
Thus, no current passes through the branch BD and Wheatstone bridge is balanced i.e.,
$$\frac{P}{Q} = \frac{R}{G} \quad \cdots (7)$$
Dividing Eq.(1) by Eq.(2), we get
$$\frac{I_P}{I_G} = \frac{R+G}{P+Q} = \frac{R}{P}\left(\frac{1+(G/R)}{1+(Q/P)}\right) \quad \cdots (8)$$
Using Eq.(7) in Eq.(8), we get
$$\frac{I_P}{I_G} = \frac{R}{P}\left(\frac{1+(Q/P)}{1+(Q/P)}\right) = \frac{R}{P}$$
or $\quad I_P = I_G\left(\frac{R}{P}\right) \quad \cdots (9)$
Similarly, from Eq.(1), (2) and (7), we can also show that
$$I_Q = I_R\left(\frac{R}{P}\right) \quad \cdots (9)$$
Since $P \neq R$, therefore, from Eq.(8) and (9), we get $I_P \neq I_G$, $I_Q \neq I_R$ and $I_Q \neq I_G$.
Therefore, the correct choice is (A).

48. In the circuit shown in the Fig.4.309, the current through
  (A) the 3Ω resistor is 0.50 A
  (B) the 3Ω resistor is 0.25 A
  (C) the 4Ω resistor is 0.50 A
  (D) the 4Ω resistor is 0.25 A

SOLUTION (D) The circuit is redrawn in Fig.4.310. If R_1

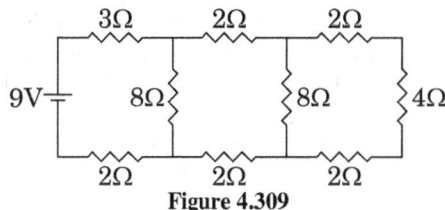

Figure 4.309

and R_2 be the effective resistances of the circuits to the right of CD and AB, respectively, then from Fig.4.310, we have
$R_1 = 2 + 4 + 2 = 8$ Ω and $R_2 = 2 + (8\|R_1) + 2 = 2 + (8\|8) + 2 = 8$ Ω. Now, effective resistance of the

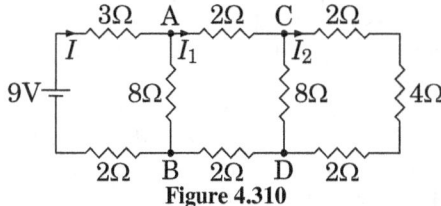

Figure 4.310

complete circuit is
$R = 3 + (8\|R_2) + 2 = 3 + (8\|8) + 2 = 9$ Ω
Thus, the current in 3 Ω resistor is $I = V/R = 9/9 = 1$ A.
Since, the part of circuit, right side to branch AB has the same resistance ($= 8$ Ω) as that of branch AB, therefore, at junction A, the current I is equally divided into two parts. One part $I_1 = I/2 = 0.5$ A passes through branch AC and other same amount of current passes in branch AB having 8 Ω resistance.
Current I_1 is further divided into two equal parts at junction C. Therefore, current through 2 Ω resistance right to C is $I_2/2$, i.e., 0.25 A, thus the current through 4Ω resistor is 0.25 A.

49. A quantity X is given by $\varepsilon_0 L \frac{\Delta V}{\Delta t}$, where ε_0 is the permittivity of free space, L is a length, ΔV is a potential difference and Δt is a time interval. The dimensional formula for X is the same as that of
  (A) resistance   (B) charge
  (C) voltage   (D) current

SOLUTION (D) The electric potential at distance r, due to a point charge q is
$$V = \frac{q}{4\pi\varepsilon_0 r} \quad \cdots (1)$$
From Eq.(1), we can write
$$\varepsilon_0 V = \frac{q}{4\pi r} \quad \cdots (2)$$
Therefore, dimensional formula of X is given by
$$[X] = \left[\varepsilon_0 L \frac{\Delta V}{\Delta t}\right] = [\varepsilon_0 V]\left[\frac{L}{t}\right] \quad \cdots (3)$$
Substituting the value of $\varepsilon_0 V$ from Eq.(2) in Eq.(3), we get
$$[X] = \left[\frac{q}{4\pi r}\right]\left[\frac{L}{t}\right] = \left[\frac{AT}{L}\right]\left[\frac{L}{T}\right] = [AT^{-2}]$$
which are same as the dimensions of current.

4.33 Questions and Problems
4.33.1 Conceptual Questions
1. How would you connect resistors so that the equivalent resistance is larger than the greatest individual resistance? Give an example involving three resistors.
2. How would you connect resistors so that the equivalent resistance is smaller than the least individual resistance? Give an example involving three resistors.
3. What quantity is measured by a battery rating given in ampere-hours $(A \cdot h)$?
4. When an electric cell is connected to a circuit, electrons flow away from the negative terminal in the circuit. But within the cell, electrons flow to the negative terminal. Explain.
5. Is a circuit breaker wired in series or in parallel with the device it is protecting?
6. Under what condition does the potential difference across the terminals of a battery equal its emf? Can the terminal voltage ever exceed the emf? Explain.
7. In an electrolyte, the positive ions move from left to right and the negative ions from right to left. Is there a net current? If yes, in what direction?
8. When a flash-light is operated, what is being used up: battery current, battery voltage, battery energy, battery power, or battery resistance? Explain.
9. The drift speed is defined as $v_d = \Delta l / \Delta t$ where Δl is the distance drifted in a long time Δt. Why don't we define the drift speed as the limit of $\Delta l / \Delta t$ as $\Delta t \to 0$?
10. The equation $P = V^2/R$ indicates that the power dissipated in a resistor decreases if the resistance is increased, whereas the equation $P = I^2 R$ implies the opposite. Is there a contradiction here? Explain.
11. When a current is established in a wire, the free electrons drift in the direction opposite to the current. Does the number of free electrons in the wire continuously decrease?
    or
    Does a conductor become charged when a current is passed through it?
12. In electrostatics we have read that there can be no electric field inside a conductor. And hence there can be no current through it. What is wrong with this argument?
13. If the resistance of a small immersion heater (to heat water for tea or soup) was increased, would it speed up or slow down the heating process? Explain.
14. If a rectangular solid made of carbon has sides of lengths a, $2a$, and $3a$, how would you connect the wires from a battery so as to obtain (a) the least resistance, (b) the greatest resistance?
15. Explain why lightbulbs almost always burn out just as they are turned on and not after they have been on for some time.
16. Which draws more current, a 100 W lightbulb or a 75 W bulb? Which has the higher resistance?
17. Is current used up in a resistor? Explain.
18. Compare the drift velocities and electric currents in two wires that are geometrically identical and the density of atoms is similar, but the number of free electrons per atom in the material of one wire is twice that in the other.
19. A voltage V is connected across a wire of length l and radius r. How is the electron drift velocity affected if (a) l is doubled, (b) r is doubled, (c) V is doubled?
20. Which, if any, of these statements are true? (More than one maybe true.) Explain.
    (a) A battery supplies the energy to a circuit.
    (b) A battery is a source of potential difference; the potential difference between the terminals of the battery is always the same.
    (c) A battery is a source of current; the current leaving the battery is always the same.
21. The thermal energy developed in a current-carrying resistor is given by $U = I^2 Rt$ and also by $U = VIt$. Should we say that U is proportional to I^2 or to I?
22. The current in a wire is doubled. What happens to (a) the current density, (b) the conduction-electron density, (c) the mean time between collisions, and (d) the electron drift speed? Are each of these doubled, halved, or unchanged? Explain.
23. Consider a circuit containing an ideal battery connected to a resistor. Do "work done by the battery" and "the thermal energy developed" represent two names of the same physical quantity?
24. Would you prefer a voltmeter or a potentiometer to measure the emf of a battery?
25. Why is it possible for a bird to sit on a high-voltage wire without being electrocuted?
26. When can the potential difference across a resistor be positive?

4.33.2 Problems
Electric Current
1. ••A wire carries a current of 2.0 A. What is the charge that has flowed through its cross-section in 1.0 s? How many electrons does this correspond to?
2. ••In a given time of 10 s, 40 electrons pass from right to left. In the same interval of time 40 protons also pass from left to right. Is the average current zero? If not, then find the value of average current.
3. ••An insulating belt moves at speed 30 m/s and has a width of 50 cm. It carries charge into an experimental device at a rate corresponding to 100 μA. What is the surface charge density on the belt?

Current Density
4. ••The magnitude J of the current density in a certain lab wire with a circular cross section of radius $R = 2.00$ mm is given by $J = (3.00 \times 10^8) r^2$, with J in amperes per square meter and radial distance r in meters. What is the current through the outer section bounded by $r = 9R/10$ and $r = R$?
5. ••What is the current in a wire of radius $R = 3.40$ mm if the magnitude of the current density is given by (a) $J_a = J_0 r/R$ and (b) $J_b = J_0(1 - r/R)$, in which r is the radial distance and $J_0 = 5.50 \times 10^4$ A/m^2? (c) Which function maximizes the current density near the wire's surface?

Resistance, Ohm's Law
6. •What is the resistance of a toaster if 120 V produces a current of 4.2 A?
7. ••An electric clothes dryer has a heating element with a resistance of 8.6 Ω. (a) What is the current in the element when it is connected to 240 V? (b) How much charge passes through the element in 50 min? (Assume direct current.)
8. ••A hair dryer draws 9.5 A when plugged into a 120 V line. (a) What is its resistance? (b) How much charge passes through it in 15 min? (Assume direct current.)
9. ••A 4.5 V battery is connected to a bulb whose resistance is 1.6 Ω. How many electrons leave the battery per minute?
10. ••An electric device draws 6.50 A at 240 V. (a) If the voltage drops by 15%, what will be the current, assuming nothing else changes? (b) If the resistance of the device were reduced by 15%, what current would be drawn at 240 V?

Resistivity
11. •What is the diameter of a 1.00 m length of tungsten wire

whose resistance is 0.32Ω ?

12. •What is the resistance of a 4.5 m length of copper wire 1.5 mm in diameter?

13. ••A 5.80 m length of 2.0 mm diameter wire carries a 750 mA current when 22.0 mV is applied to its ends. If the drift velocity is 1.7×10^{-5} m/s, determine (a) the resistance R of the wire, (b) the resistivity ρ, (c) the current density j, (d) the electric field inside the wire, and (e) the number n of free electrons per unit volume.

14. ••Calculate the ratio of the resistance of 10.0 m of aluminum wire 2.0 mm in diameter, to 20.0 m of copper wire 1.8 mm in diameter.

15. ••Can a 2.2 mm-diameter copper wire have the same resistance as a tungsten wire of the same length? Give numerical details.

16. ••How much would you have to raise the temperature of a copper wire (originally at 20°C) to increase its resistance by 15% ?

17. ••A certain copper wire has a resistance of 10.0 Ω. At what point along its length must the wire be cut so that the resistance of one piece is 4.0 times the resistance of the other? What is the resistance of each piece?

18. ••Determine at what temperature aluminum will have the same resistivity as tungsten does at 20°C.

19. ••A 100 W lightbulb has a resistance of about 12 Ω when cold (20°C) and 140 Ω when on (hot). Estimate the temperature of the filament when hot assuming an average temperature coefficient of resistivity $\alpha = 0.0045$ (C°)$^{-1}$.

20. ••A length of wire is cut in half and the two lengths are wrapped together side by side to make a thicker wire. How does the resistance of this new combination compare to the resistance of the original wire?

21. • • • For some applications, it is important that the value of a resistance not change with temperature. For example, suppose you made a 3.70 kΩ resistor from a carbon resistor and a Nichrome wire-wound resistor connected together so the total resistance is the sum of their separate resistances. What value should each of these resistors have (at 0° C) so that the combination is temperature independent?

22. • • • Determine a formula for the total resistance of a spherical shell made of material whose conductivity is σ and whose inner and outer radii are r_1 and r_2. Assume the current flows radially outward.

23. • • • The filament of a lightbulb has a resistance of 12Ω at 20°C and 140Ω when hot (as in Problem 19). (a) Calculate the temperature of the filament when it is hot, and take into account the change in length and area of the filament due to thermal expansion (assume tungsten for which the thermal expansion coefficient is $\approx 5.5 \times 10^{-6}$C°$^{-1}$) . (b) In this temperature range, what is the percentage change in resistance due to thermal expansion, and what is the percentage change in resistance due solely to the change in ρ? Use following equation:
$$\rho_T = \rho_0 [1 + \alpha (T - T_0)]$$

24. • • • A 10.0 m length of wire consists of 5.0 m of copper followed by 5.0 m of aluminum, both of diameter 1.4 mm. A voltage difference of 85mV is placed across the composite wire. (a) What is the total resistance (sum) of the two wires? (b) What is the current through the wire? (c) What are the voltages across the aluminum part and across the copper part?

25. • • • A hollow cylindrical resistor with inner radius r_1 and outer radius r_2, and length l, is made of a material whose resistivity is ρ (Fig.4.311). (a) Show that the resistance is given by
$$R = \frac{\rho}{2\pi l} \ln \frac{r_2}{r_1}$$
for current that flows radially outward. [Hint: Divide the resistor into concentric cylindrical shells and integrate.] (b) Evaluate the resistance R for such a resistor made of carbon whose inner and outer radii are 1.0 mm and 1.8 mm and whose length is 2.4 cm.
(Choose $\rho = 15 \times 10^{-5} \Omega \cdot$ m.) (c) What is the resistance

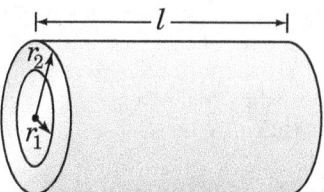

Figure 4.311

in part (b) for current flowing parallel to the axis?

Electric Power

26. ••How many kWh of energy does a 550 W toaster use in the morning if it is in operation for a total of 6.0 min ? At a cost of 9.0 cents/ kWh, estimate how much this would add to your monthly electric energy bill if you made toast four mornings per week.

27. ••At $0.095/kWh, what does it cost to leave a 25 W porch light on day and night for a year?

28. ••What is the total amount of energy stored in a 12 V, 75 Ah car battery when it is fully charged?

29. ••An ordinary flashlight uses two D-cell 1.5 V batteries connected in series. The bulb draws 380 mA when turned on. (a) Calculate the resistance of the bulb and the power dissipated. (b) By what factor would the power increase if four D-cells in series were used with the same bulb? (Neglect heating effects of the filament.)

30. ••A power station delivers 750 kW of power at 12,000 V to a factory through wires with total resistance 3.0 Ω. How much less power is wasted if the electricity is delivered at 50,000 V rather than 12,000 V ?

31. ••A small immersion heater can be used in a car to heat a cup of water for coffee or tea. If the heater can heat 120 mL of water from 25°C to 95°C in 8.0 min, (a) approximately how much current does it draw from the car's 12-V battery, and (b) what is its resistance? Assume the manufacturer's claim of 75% efficiency.

32. • • • The current in an electromagnet connected to a 240 V line is 17.5 A. At what rate must cooling water pass over the coils if the water temperature is to rise by no more than 6.50 C° ?

33. • • • A 1.0 m long round tungsten wire is to reach a temperature of 3100 K when a current of 15.0 A flows through it. What diameter should the wire be? Assume the wire loses energy only by radiation (emissivity $\varepsilon = 1.0$) and the surrounding temperature is 20°C.

General Problems

34. •How many coulombs are there in 1.00 ampere-hour?

35. ••You want to design a portable electric blanket that runs on a 1.5 V battery. If you use copper wire with a 0.50 mm diameter as the heating element, how long should the wire be if you want to generate 15 W of heating power? What happens if you accidentally connect the blanket to a 9.0 V battery?

36. ••What is the average current drawn by a 1.0 hp 120 V motor? (1hp = 746 W.)

37. ••Determine the resistance of the tungsten filament in a

75 W 120 V incandescent lightbulb (a) at its operating temperature of about 3000 K, (b) at room temperature.

38. •• A microwave oven running at 65% efficiency delivers 950 W to the interior. Find (a) the power drawn from the source, and (b) the current drawn. Assume a source voltage of 120 V.

39. ••A 1.00 Ω wire is stretched uniformly to 1.20 times its original length. What is its resistance now?

40. ••A 2800 W oven is connected to a 240 V source. (a) What is the resistance of the oven? (b) How long will it take to bring 120 mL of 15°C water to 100°C assuming 75% efficiency? (c) How much will this cost at 11 cents/kWh?

41. ••A 12.5 Ω resistor is made from a coil of copper wire whose total mass is 15.5 g. What is the diameter of the wire, and how long is it?

42. ••• A fish-tank heater is rated at 95 W when connected to 120 V. The heating element is a coil of Nichrome wire. When uncoiled, the wire has a total length of 3.8 m. What is the diameter of the wire?

43. ••A 100 W, 120 V lightbulb has a resistance of 12 Ω when cold (20°C) and 140 Ω when on (hot). Calculate its power consumption (a) at the instant it is turned on, and (b) after a few moments when it is hot.

44. ••Lightbulb A is rated at 120 V and 40 W for household applications. Lightbulb B is rated at 12 V and 40 W for automotive applications. (a) What is the current through each bulb? (b) What is the resistance of each bulb? (c) In one hour, how much charge passes through each bulb? (d) In one hour, how much energy does each bulb use? (e) Which bulb requires larger diameter wires to connect its power source and the bulb?

45. ••A copper pipe has an inside diameter of 3.00 cm and an outside diameter of 5.00 cm (Fig.4.312). What is the resistance of a 10.0-m length of this pipe?

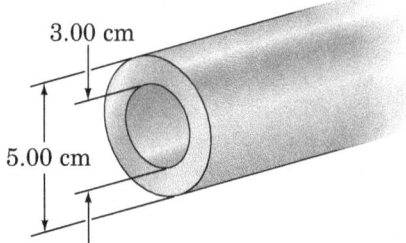

Figure 4.312

46. ••• For the wire in Fig.4.313, whose diameter varies uniformly from a to b as shown, suppose a current $I = 2.0$ A enters at a. If $a = 2.5$ mm and $b = 4.0$ mm, what is the current density (assume uniform) at each end?

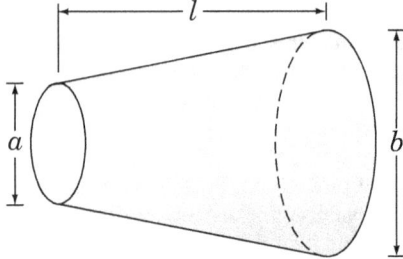

Figure 4.313

47. ••The cross section of a portion of wire increases uniformly as shown in Fig.4.313 so it has the shape of a truncated cone. The diameter at one end is a and at the other it is b, and the total length along the axis is l. If the material has resistivity ρ, determine the resistance R between the two ends in terms of a, b, l, and ρ. Assume that the current flows uniformly through each section, and that the taper is small, i.e., $(b - a) \ll l$.

48. ••• The level of liquid helium (temperature ≤ 4 K) in its storage tank can be monitored using a vertically aligned niobium-titanium (NbTi) wire, whose length l spans the height of the tank. In this level-sensing setup, an electronic circuit maintains a constant electrical current I at all times in the NbTi wire and a voltmeter monitors the voltage difference V across this wire. Since the superconducting transition temperature for NbTi is 10 K, the portion of the wire immersed in the liquid helium is in the superconducting state, while the portion above the liquid (in helium vapor with temperature above 10 K) is in the normal state. Define $f = x/l$ to be the fraction of the tank filled with liquid helium (Fig.4.314) and V_0 to be the value of V when the tank is empty ($f = 0$). Determine the relation between f and V (in terms of V_0).

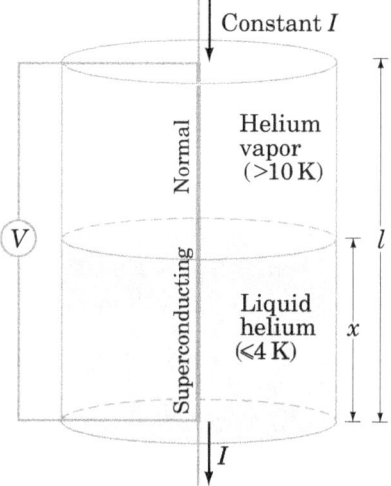

Figure 4.314

49. •••In Fig.4.315, a resistance coil, wired to an external battery, is placed inside a thermally insulated cylinder fitted with a frictionless piston and containing an ideal gas. A current $I = 240$ mA flows through the coil, which has a resistance $R = 550 \Omega$. If the gas temperature remains constant while the 12 kg piston moves upward, what is the limit of the piston's speed?

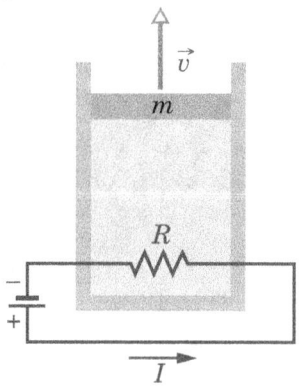

Figure 4.315

50. •••In a hypothetical fusion research lab, high temperature helium gas is completely ionized and each helium atom is separated into two free electrons and the remaining positively charged nucleus, which is called an alpha particle. An

applied electric field causes the alpha particles to drift to the east at 25.0 m/s while the electrons drift to the west at 88.0 m/s. The alpha particle density is 2.80×10^{15} cm^{-3}. What are (a) the net current density and (b) the current direction?

51. • • •A beam of 16 MeV deuterons from a cyclotron strikes a copper block. The beam is equivalent to current of 15 μA. (a) At what rate do deuterons strike the block? (b) At what rate is thermal energy produced in the block?

52. • •A 400 W immersion heater is placed in a pot containing 2.00 L of water at 20°C. (a) How long will the water take to rise to the boiling temperature, assuming that 80% of the available energy is absorbed by the water? (b) How much longer is required to evaporate half of the water?

53. • •A 30 μF capacitor is connected across a programmed power supply. During the interval from $t = 0$ to $t = 3.00$ s the output voltage of the supply is given by $V(t) = 6.00 + 4.00t - 2.00t^2$ volts. At $t = 0.500$ s find (a) the charge on the capacitor, (b) the current into the capacitor, and (c) the power output from the power supply.

54. • • •The switch in Figure 4.316a closes when $V_c > 2V/3$ and opens when $V_c < V/3$. The voltmeter reads a voltage as plotted in Figure 4.316a. What is the period T of the waveform in terms of R_1, R_2, and C?

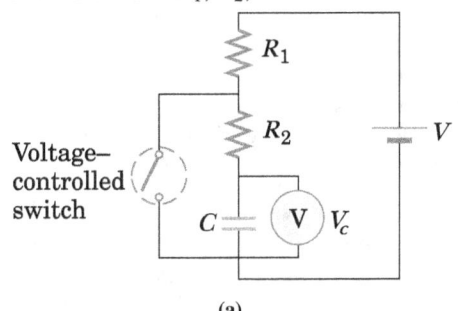

(a)

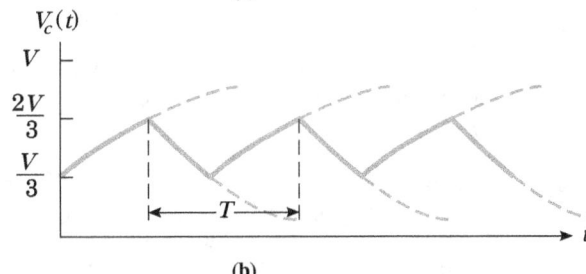

(b)

Figure 4.316

55. • • •Switch S has been closed for a long time, and the electric circuit shown in Figure 4.317 carries a constant current. Take $C_1 = 3.00$ μF, $C_2 = 6.00$ μF, $R_1 = 4.00$ kΩ, and $R_2 = 7.00$ kΩ. The power delivered to R_2 is 2.40 W. (a) Find the charge on C_1. (b) Now the switch is opened. After many milliseconds, by how much has the charge on C_2 changed?

56. • • •The circuit in Figure 4.318 contains two resistors, $R_1 = 2.00$ kΩ and $R_2 = 3.00$ kΩ, and two capacitors, $C_1 = 2.00$ μF and $C_2 = 3.00$ μF, connected to a battery with emf $E = 120$ V. No charge is on either capacitor before switch S is closed. Determine the charges q_1 and q_2 on capacitors C_1 and C_2, respectively, after the switch is closed.

4.34 Multiple Choice Assignments
4.34.1 Level 1
Current carrying conductor

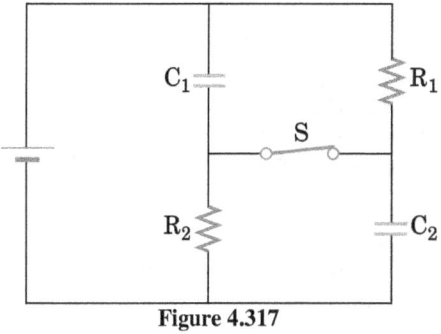

Figure 4.317

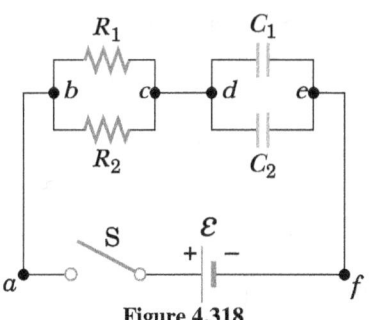

Figure 4.318

1. A current of 5 A exist on a 10 Ω resistance for 4 min. How much charge pass through any cross-section of the resistor in this time?
   (A) 12 C (B) 120 C
   (C) 1200 C (D) 12000 C

2. Current in a conductor is due to
   (A) motion of free electrons in it
   (B) motion of (+) ve ions
   (C) free electrons and holes
   (D) protons

3. The electric current in a liquid is due to the flow of
   (A) electron only
   (B) positive ions only
   (C) negative and positive ions both
   (D) electrons and positive ions both

4. The electric current in a discharge tube containing a gas is due to
   (A) electron only
   (B) positive ions only
   (C) negative ion and positive ions both
   (D) electrons and positive ions both

5. A steady current is passing through a linear conductor of non-uniform cross-section. The net quantity of charge crossing any cross-section per second is
   (A) independent of area of cross-section
   (B) directly proportional to the length of conductor
   (C) directly proportional to the area of cross-section
   (D) inversely proportional to the lengths of conductor

6. A current (I) flows through a uniform wire of diameter (d) when the mean drift velocity is v. The same current will flow through a wire of diameter $d/2$ made of the same material if the mean drift velocity of the electron is
   (A) $v/4$ (B) $v/2$ (C) $4v$ (D) $2v$

7. A wire of non-uniform cross-section is carrying a steady current. Along the wire
   (A) current and current density are constant
   (B) only current is constant
   (C) only current density is constant
   (D) neither current nor current density is a constant

8. When a potential difference (V) is applied across a conductor, the thermal speed of electrons is

(A) zero
(B) proportional to $\sqrt{T}$
(C) proportional to (T)
(D) proportional to V

9. A steady current is passing through a linear conductor of non-uniform cross-section. The current density in the conductor is
   (A) independent of area of cross-section
   (B) directly proportional to area of cross-section
   (C) inversely proportional to area of cross-section
   (D) inversely proportional to the square root of area of cross-section

10. A metallic block has no potential difference applied across it. Then the mean velocity of free electron is
    (A) proportional to T
    (B) proportional to $\sqrt{T}$
    (C) zero
    (D) finite but independent of temperature

Ohm's law & resistance

11. Specific resistance of a wire depends on the
    (A) length of the wire
    (B) area of cross-section of the wire
    (C) resistance of the wire
    (D) material of the wire

12. A cross-sectional area of a copper wire is 3×10^{-6} m^2. The current of 4.2A. is flowing through it. The current density in amp/m^2 through the wire is -
    (A) 1.4×10^3 (B) 1.4×10^4
    (C) 1.4×10^5 (D) 1.4×10^6

13. The resistance of some substances become zero at very low temperature, then these substances are called -
    (A) good conductors   (B) super conductors
    (C) bad conductors   (D) semi conductors

14. The resistance of wire is 20 Ω. The wire is stretched to three times its length. Then the resistance will now be-
    (A) 6.67Ω (B) 60Ω (C) 120Ω (D) 180Ω

15. The dimensions of a mangnin block are 1 cm$\times$1 cm$\times$100 cm. The electrical resistivity of mangnin is 4.4×10^{-7} Ωm. The resistance between the opposite rectangular faces is -
    (A) $4.4 \times 10^{-7}\Omega$ (B) $4.4 \times 10^{-3}\Omega$
    (C) $4.4 \times 10^{-5}\Omega$ (D) $4.4 \times 10^{-1}\Omega$

16. If the temperatures of iron and silicon wires are increased from 30°C to 50°C, the correct statement is-
    (A) resistance of both wires increases
    (B) resistance of both wires decreases
    (C) resistance of iron wire increases and the resistance of silicon wire decreases
    (D) resistance of iron wire decreases and the resistance of silicon wire increases

17. When the resistance of copper wire is 0.1 Ω and the radius is 1 mm, then the length of the wire is (specific resistance of copper is $3.14 \times 10^{-8} \Omega$m)
    (A) 10 cm   (B) 10 m   (C) 100 m   (D) 100 cm

18. When the resistance wire is passed through a die the cross-section area decreases by 1%, the change in resistance of the wire is
    (A) 1% decrease   (B) 1% increase
    (C) 2% decrease   (D) 2% increase

19. In Fig.4.319, two parallelopiped A and B are of the same thickness. The arm of B is double that of A. The value of R_A/R_B is-
    (A) 1   (B) 2   (C) $\frac{1}{2}$ (D) 4

20. When the temperature of a metallic conductor is increased

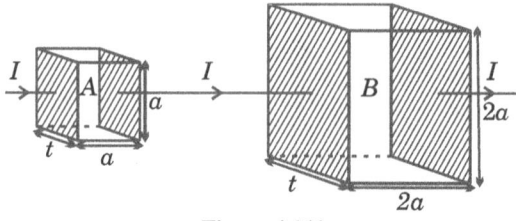

Figure 4.319

its resistance -
(A) always decreases
(B) always increases
(C) may increase or decrease
(D) remains the same

21. In which one of the following substances the resistance decreases with increase in temperature
    (A) carbon   (B) constantan
    (C) copper   (D) silver

22. The resistance of a semi-conductors
    (A) increases with increase in temperature
    (B) decreases with increase in temperature
    (C) does not charge with charge in temperature
    (D) first decreases and then increases with increase in temperature

23. Specific resistance of a wire depends upon
    (A) it's length
    (B) it's cross-sectional area
    (C) it's dimensions
    (D) it's material

24. Ohm's law deals with the relation between
    (A) current and potential difference
    (B) capacity and charge
    (C) capacity and potential
    (D) all are true

25. Ohm's law is valid when the temperature of the conductor is
    (A) constant   (B) very high
    (C) very low   (D) varying

26. A certain piece of copper, of length l and diameter d, is to be shared into a conductor of minimum resistance. It's length and diameter should be respectively
    (A) l, d (B) $2l, d$
    (C) $l/2, 2d$ (D) $2l, d/2$

27. A wire has a resistance of 10 Ω. A second wire of the same material is having length double and radius of cross-section half that of the wire. The resistance of the second wire is
    (A) 20 Ω (B) 40 Ω (C) 80 Ω (D) 10 Ω

28. A cylindrical copper rod is reformed to twice its original length with no change in volume. The resistance between its ends before the change was (R). Now its resistance
    (A) 8R (B) 6R (C) 4R (D) 2R

29. The length of a conductor is halved. Its conductance will be
    (A) halved   (B) unchanged
    (C) doubled   (D) quadrupled

Combination of resistors

30. Net resistance between X and Y, in Fig.4.320, is -
    (A) R (B) $2R$ (C) $\frac{R}{2}$ (D) $4R$

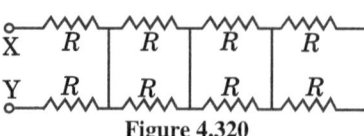

Figure 4.320

31. Net resistance between X and Y in Fig.4.321, is -
    (A) 5 Ω (B) 10 Ω (C) 15 Ω (D) 60 Ω

4.34. MULTIPLE CHOICE ASSIGNMENTS

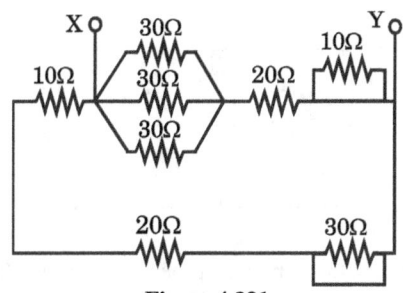

Figure 4.321

32. In Fig.4.322, net resistance between X and Y is -
    (A) 4 Ω (B) 4.55 Ω (C) 2 Ω (D) 20 Ω

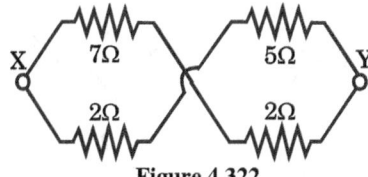

Figure 4.322

33. The equivalent resistance between the terminal points P and Q is 4Ω in the given circuit of Fig.4.323, find out the resistance of R in ohms -
    (A) 7 (B) 4 (C) 2 (D) 5

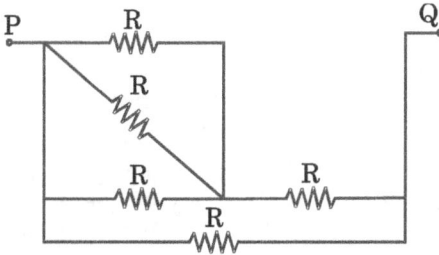

Figure 4.323

34. At a point $\sum I = 0$ in a circuit with one emf source, then
    (A) the resistance of the circuit is zero
    (B) the point is the junction point
    (C) the emf of the source is infinity
    (D) this is not possible

35. For the circuit of Fig.4.324, the potential difference between X and Y, in volts, is -
    (A) 2/3 (B) 4/3 (C) 8/9 (D) 5/3

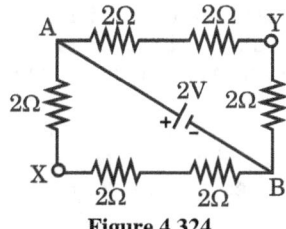

Figure 4.324

36. Reading of an ideal ammeter in ampere for the circuit of Fig.4.325 is-
    (A) 1 (B) 2 (C) 3 (D) 4

37. In a closed circuit the sum of total emf is equal to the sum of the -
    (A) currents
    (B) resistances
    (C) products of current and the resistances
    (D) none of the above

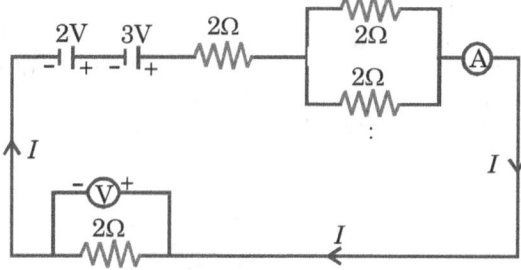

Figure 4.325

38. For the diagram of Fig.4.326, the galvanometer shows zero deflection, the value of R is
    (A) 52 Ω (B) 50 Ω (C) 100 Ω (D) 25 Ω

39. For circuit of Fig.4.327, the value of total resistance between X and Y in ohm is-
    (A) R (B) 4R (C) 5R (D) 6R

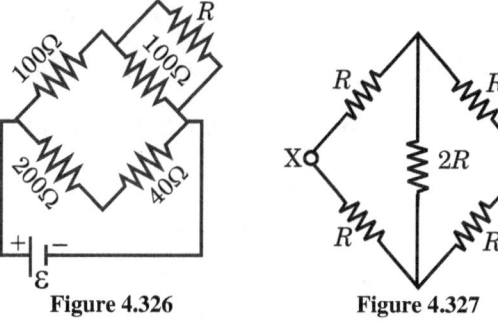

Figure 4.326   Figure 4.327

40. The equivalent resistance in series combination is
    (A) smaller than the largest resistance
    (B) larger than the largest resistance
    (C) smaller than the smallest resistance
    (D) larger than the smallest resistance

41. The equivalent resistance of resistors in parallel is always
    (A) higher than the highest of component resistor
    (B) less than the lowest of component resistors
    (C) in between the lowest and the highest of component resistors
    (D) equal to the sum of the component resistors

42. When n identical resistances of value 'r' each are connected in parallel, the equivalent resistance is x. The resultant resistance when they are connected in series is-
    (A) nx (B) n^2x (C) rnx (D) r^2x/n

43. Two resistance $r_1 \Omega$ and $r_2 \Omega$ are connected in parallel. The equivalent resistance of the combination is equal to-
    (A) $r_1 + r_2$ (B) $[r_1 \cdot r_2/(r_1 + r_2)]$
    (C) $[(r_1 + r_2)/r_1 \cdot r_2]$ (D) $r_1 - r_2$

44. Five identical resistances are connected as shown in Fig.4.328. The equivalent resistance between point (A) and (B) is-
    (A) R (B) 5 R (C) R/5 (D) 2R/5

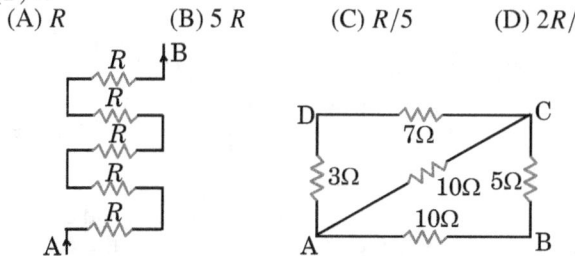

Figure 4.328   Figure 4.329

45. Five resistance are connected as shown in the Figure 4.329. The equivalent resistance between A and B is -

(A) 35 Ω (B) 5 Ω (C) 15/4 Ω (D) 25 Ω

46. The equivalent resistance between points A and B in the Fig.4.330, is one ohm. What is the value of middle resistance-
    (A) 9 Ω (B) 1 Ω (C) 6 Ω (D) 3 Ω

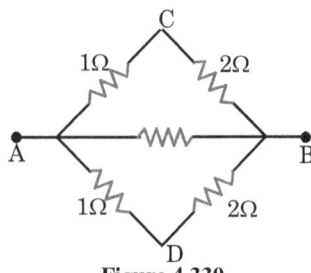

Figure 4.330

47. Four wires of equal length and of resistance 5 Ω each are connected in the form of a square. The equivalent resistance between the diagonally opposite corners of the square is
    (A) 5 Ω (B) 10 Ω (C) 20 Ω (D) 5/4 Ω

48. The effective resistance (in Ω) between B and C in Fig.4.331 is-
    (A) 60 (B) 40 (C) 80/3 (D) 160/9

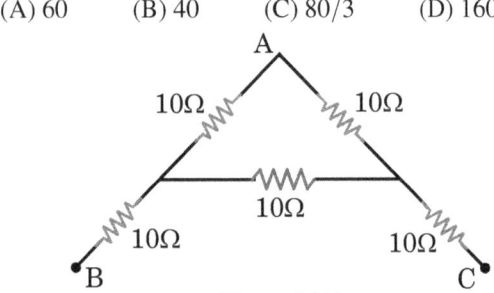

Figure 4.331

49. Four identical resistances are joined as shown in Fig.4.332. The equivalent resistance between points A and B is R_1. The equivalent resistance between points A and C is R_2 then ratio of R_1/R_2 is-
    (A) 1 : 1 (B) 4 : 3 (C) 3 : 4 (D) 1 : 2

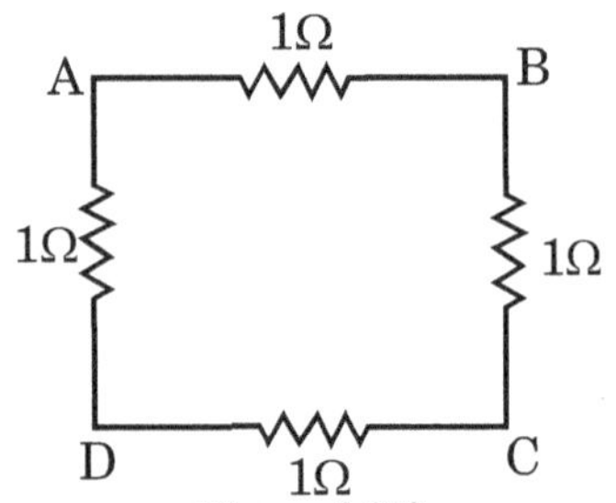

Figure 4.332

50. Kirchhoff's first law i.e., $\Sigma I = 0$ at a junction deals with-
    (A) conservation of charge
    (B) conservation of energy
    (C) conservation of linear momentum
    (D) conservation of angular momentum

51. Kirchhoff's second law is based on the law of conservation of-
    (A) charge (B) energy
    (C) momentum (D) sum of mass and energy

52. In the Fig.4.333, there is no deflection in the galvanometer. The value of R is-
    (A) 2 Ω (B) 30 Ω (C) 6 Ω (D) (2/3) Ω

53. Five resistance are connected as shown in Fig.4.334. The

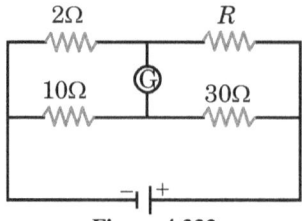

Figure 4.333

effective resistance between the points A and B is
    (A) 10/3 Ω (B) 20/3 Ω (C) 15 Ω (D) 6 Ω

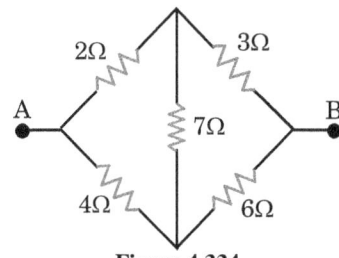

Figure 4.334

54. In Fig.4.335, the reading of ammeter is-
    (A) 1 (B) 2 (C) 2/3 (D) 3

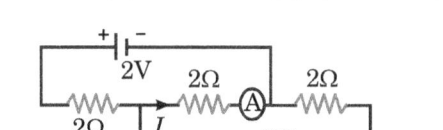

Figure 4.335

55. In the circuit of Fig.4.336, the resultant emf between AB is-
    (A) $\mathcal{E}_1 + \mathcal{E}_2 + \mathcal{E}_3 + \mathcal{E}_4$ (B) $\mathcal{E}_1 + \mathcal{E}_2 + 2\mathcal{E}_3 + \mathcal{E}_4$
    (C) $\mathcal{E}_1 + \mathcal{E}_2 + (\mathcal{E}_3/2) + \mathcal{E}_4$ (D) $\mathcal{E}_1 + \mathcal{E}_2 + (\mathcal{E}_3/4) + \mathcal{E}_4$

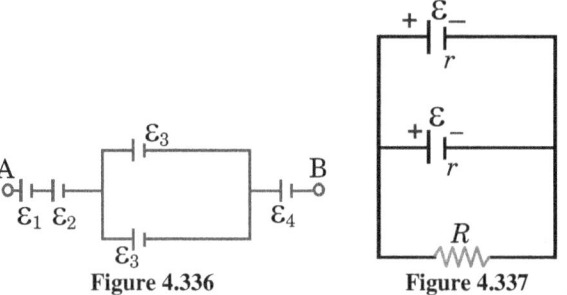

Figure 4.336   Figure 4.337

56. Two cells of same emf $\mathcal{E}$ and internal resistance r are connected in parallel with a resistance of R as shown in Fig.4.337. To get maximum power in the external circuit, the value of R is -
    (A) R = r/2 (B) R = r (C) R = 2r (D) R = 4r

57. A cell of emf $\mathcal{E}$ and internal resistance (r) is connected in series with an external resistance (nr.) then the ratio of the terminal potential difference to emf is
    (A) $1/n$ (B) $1/(n+1)$
    (C) $n/(n+1)$ (D) $(n+1)/n$

58. The terminal potential difference of a cell, when short circuited is
    (A) $\mathcal{E}$ (B) $\mathcal{E}/2$ (C) 0 (D) $\mathcal{E}/3$

59. Five dry cell each of emf 1.5V are connected in parallel. The emf of the combination is
    (A) 7.5 V (B) 0.3 V (C) 3 V (D) 1.5 V

Heating effect of current

60. Two bulbs, one of 50 watt and another of 25 watt are connected in series to the mains, the ratio of the current through

them is -
(A) 2 : 1
(B) 1 : 2
(C) 1 : 1
(D) can't be determined without the p.d. of the main supply

61. Constant voltage is applied between the two ends of a uniform metallic wire. The heat developed is doubled if -
(A) both the length and radius of the wire are halved
(B) both the length and radius of the wire are doubled
(C) the radius of wire is doubled
(D) the length of the wire is doubled

62. Two electric bulbs rated P_1 watt V volt and P_2 watt V volt are connected in parallel across V volt mains then the total power is-
(A) $P_1 + P_2$
(B) $\sqrt{P_1 P_2}$
(C) $\frac{P_1 P_2}{(P_1+P_2)}$
(D) $\frac{(P_1+P_2)}{P_1 P_2}$

63. Lamps used for the house lightening are connected in-
(A) series
(B) parallel
(C) mixed grouping
(D) arbitrary manner

64. Two electric bulbs whose resistances are in the ratio of 1 : 2 are connected in parallel to a constant voltage source. The power's dissipated in them have the ratio
(A) 1 : 2
(B) 1 : 1
(C) 2 : 1
(D) 1 : 4

65. An electric bulb is rated 220 volt and 100 watt. The resistance of the filament of the electric bulb is
(A) 2.2 Ω
(B) 2.2×10^4 Ω
(C) 484 Ω
(D) 100 Ω

66. Three electric bulbs 40 W, 60 W and 100 W are designed to work on a 220 V mains. Which bulb will burn most brightly if they are connected in series across 220 V mains
(A) 100 W bulb
(B) 60 W bulb
(C) 40 W bulb
(D) all bulbs will burn equally brightly

67. If the current in a electric bulb drops by 2% then the power decreases by -
(A) 1%
(B) 2%
(C) 4%
(D) 16%

68. If the current in an electric bulb decreases by 0.5 percent, then the power in the bulb decreases approximately by -
(A) 0.5%
(B) 1%
(C) 2%
(D) 0.25%

Charging and Discharging of Capacitors

69. A capacitor of capacitance 100 μF is charged by connecting it to a battery of emf 12 V and internal resistance 2Ω. The time taken before 99% of the maximum charge is stored on the capacitor-
(A) 0.92 ms
(B) 0.4 ms
(C) 0.8 ms
(D) 0.1 ms

70. A capacitor of capacitance 0.1 μF is charged to certain potential and allow to discharge through a resistance of 10 MΩ. How long will it take for the potential to fall to one half of its original value-
(A) 0.1 s
(B) 0.2346 s
(C) 1.386 s
(D) 0.693 s

71. A 500 μF capacitor is charged at a steady rate of 100 μC/s. The potential difference across the capacitor will be 10 V after an interval of-
(A) 5 s
(B) 20 s
(C) 25 s
(D) 50 s

72. An RC series circuit is connected to a battery of emf $\mathcal{E}$. The time required by the capacitor to acquire maximum charge, depends upon-
(A) R only
(B) C only
(C) RC
(D) applied potential difference

73. Calculate the charge on the plates of the capacitor C in the given circuit 4.338-

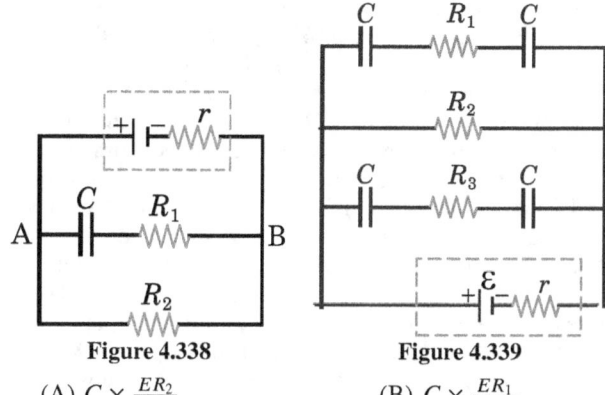

Figure 4.338   Figure 4.339

(A) $C \times \frac{ER_2}{R_2+r}$
(B) $C \times \frac{ER_1}{R_1+r}$
(C) $C \times \frac{ER_1 \cdot R_2}{R_1+R_2}$
(D) $C \times \frac{ER_1}{R_2+r}$

74. In the circuit diagram shown in Fig.4.339 $\mathcal{E} = 5$ volt, $r = 1$ ohm, $R_2 = 4$ Ω, $R_1 = R_3 = 1$ Ω and $C = 3$ μF. Then the numerical value of the charge on each plate of the capacitor is -
(A) 24 μC
(B) 12 μC
(C) 6 μC
(D) 3 μC

4.34.2 Level 2

1. The current (I) and voltage (V) graphs for a given metallic wire at two different temperature (T_1) and (T_2) are shown in Fig.4.340. It is concluded that
(A) $T_1 > T_2$
(B) $T_1 < T_2$
(C) $T_1 = T_2^2$
(D) $T_1 = 2T_2$

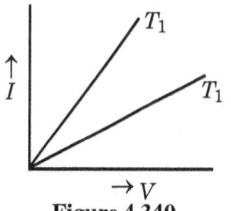 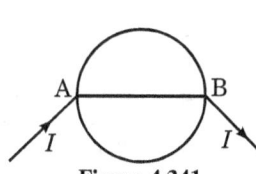

Figure 4.340   Figure 4.341

2. A 3°C rise in temperature is observed in a conductor by passing a certain current. When the current is doubled, the rise in temp -
(A) 15°C
(B) 12°C
(C) 9°C
(D) 3°C

3. A wire of resistance 0.5 Ωm^{-1} is bent into a circle of radius 1 m. The same wire is connected across a diameter AB as shown in Fig.4.341. The equivalent resistance is-
(A) π Ω
(B) $\frac{\pi}{(\pi+2)}$ Ω
(C) $\frac{\pi}{(\pi+4)}$ Ω
(D) $(\pi+1)$ Ω

4. You have three equal resistances. How many different combination can you have with these resistances -
(A) 2
(B) 4
(C) 3
(D) 6

5. (i) The product of a volt and a coulomb is a joule
(ii) The product of a volt and an ampere is joule/s
(iii) the product of a volt and a watt is horse power
(iv) Watt-hour can be measured in terms of electron volt
State if
(A) all the four are correct
(B) (i), (ii) and (iv) are correct
(C) (i) and (iii) are correct
(D) (iii) and (iv) are correct

6. An electron charge e is revolving in a circular orbit of radius r round a nucleus of charge Ze with speed v. The equivalent current is
(A) 0
(B) $e \cdot v/2\pi r$
(C) $Ze \cdot v/2\pi r$
(D) $e \cdot 2\pi r/v$

7. In a Wheat stone's bridge four resistors $P = 9\,\Omega$, $Q = 11\,\Omega$, $R = 4\,\Omega$ and $S = 6\,\Omega$. How much resistance must be put in parallel to the resistance (S) to balance the bridge
   (A) $24\,\Omega$ (B) $(44/9)\,\Omega$
   (C) $26.4\,\Omega$ (D) $18.7\,\Omega$

8. A wire of resistance 2Ω is redrawn so that its length becomes four times. The resistance of the redrawn wire is -
   (A) $2\,\Omega$ (B) $8\,\Omega$ (C) $16\,\Omega$ (D) $32\,\Omega$

9. Two wires of equal lengths and of material (x) and (y) have same resistance. The ratio of the radii of two wires is $1 : 2$. The ratio of the specific resistance of the two materials is -
   (A) $1 : 1$ (B) $1 : 2$ (C) $1 : 4$ (D) $4 : 1$

10. A wire is cut into 4 pieces, which are put together side by side to obtain one conductor. If the original resistance of the wire was (R). The resistance of the bundle will be -
    (A) $R/4$ (B) $R/8$ (C) $R/16$ (D) $R/32$

11. The current-voltage variation for a wire of copper of length L and area A is shown in Fig.4.342. The slope of the line will be
    (A) less if experiment is done at a higher temperature
    (B) more if a wire of silver of same dimensions is used
    (C) will be doubled if the lengths of the wire is doubled
    (D) will be halved if the length is doubled

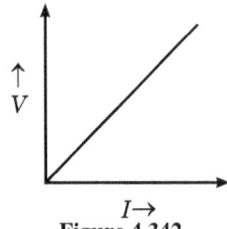

Figure 4.342

12. The internal resistance of cell is $0.1\,\Omega$ and its emf is 2 V. When a current of 2 A is being drawn from it, the potential difference across its terminals will be
    (A) more than 2 V
    (B) 2 V
    (C) 1.8 V
    (D) none of the above

13. A dry cell has an emf of 1.5 V and internal resistance 0.5Ω. If the cell sends a current of 1 A through an external resistance the p.d. of the cell will be
    (A) 1.5 V (B) 1 V (C) 0.5 V (D) 0 V

14. Twelve wires of equal resistance (R) are connected to form a cube. The effective resistance between two diagonal ends will be
    (A) $5/6\,R$ (B) $6/5\,R$ (C) $3\,R$ (D) $12\,R$

15. A wire has resistance $12\,\Omega$. It is bent in the form of a circle. The effective resistance between the two points on any diameter of the circle is
    (A) $12\,\Omega$ (B) $6\,\Omega$ (C) $24\,\Omega$ (D) $3\,\Omega$

16. Five cells each of emf $\mathcal{E}$ and internal resistance (r) are connected in series. If due to oversight one cell is connected wrongly, then the equivalent emf and internal resistance of the combination is
    (A) $5\,\mathcal{E}$ and $5r$ (B) $3\,\mathcal{E}$ and $3r$
    (C) $3\,\mathcal{E}$ and $5r$ (D) $5\,\mathcal{E}$ and $4r$

17. In Fig.4.343, the equivalent resistance between points X and Y
    (A) $16\,\Omega$ (B) $14\,\Omega$ (C) $11\,\Omega$ (D) $18\,\Omega$

18. In the circuit shown in Fig.4.344, the reading of voltmeter is-
    (A) 1.33 V (B) 0.8 V (C) 2.0 V (D) 1.6 V

19. Five identical lamps each resistance $R = 1100\,\Omega$ are con-

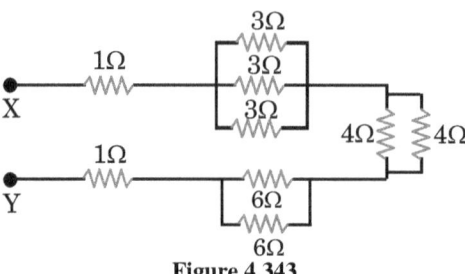

Figure 4.343

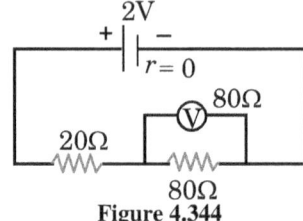

Figure 4.344

nected to 220 V as shown in Fig.4.345. The reading of ideal ammeter (A) is
(A) 1/5 A (B) 2/5 A (C) 3/5 A (D) 1 A

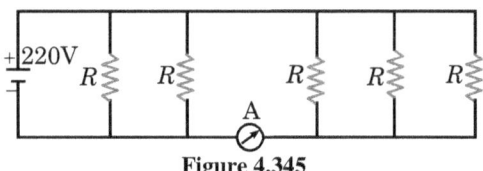

Figure 4.345

20. In Fig.4.346, the potential difference between points B and D, is -
    (A) +0.67 V (B) –0.67 V (C) 2 V (D) 1.33 V

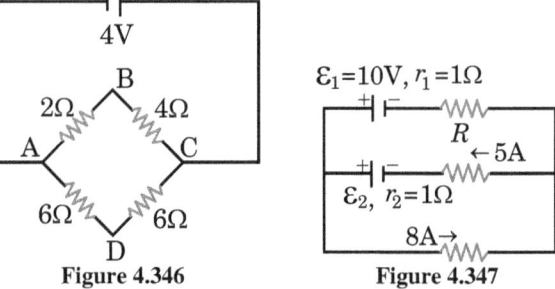

Figure 4.346   Figure 4.347

21. In Fig.4.347, the current through resistance R is
    (A) 3 A (B) 13 A (C) 6.5 A (D) 9 A

22. In the Figure 4.348, the reading of an ideal voltmeter (V) is zero. Then the relation between R, r_1, and r_2 is -
    (A) $R = r_2 - r_1$ (B) $R = r_1 - r_2$
    (C) $R = r_1 + r_2$ (D) $R = \frac{r_1 \cdot r_2}{r_1 + r_2}$

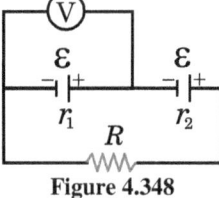

Figure 4.348

23. In Fig.4.349, the ratio of power dissipated in resistors R_1 and R_2 is
    (A) 1 : 4 (B) 4 : 1 (C) 1 : 2 (D) 2 : 1

24. In Fig.4.350, the ratio of current in $3\,\Omega$ and $1\,\Omega$ resistance is
    (A) 1 (B) 1/3 (C) 2/3 (D) 3

25. Fig.4.351 represents a part of a closed circuit. The potential

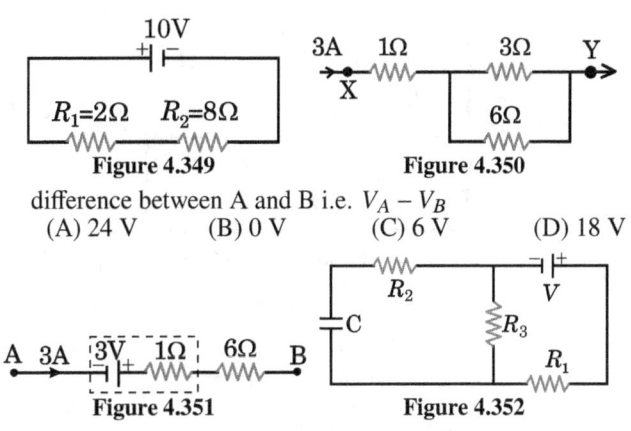

Figure 4.349   **Figure 4.350**

difference between A and B i.e. $V_A - V_B$
(A) 24 V (B) 0 V (C) 6 V (D) 18 V

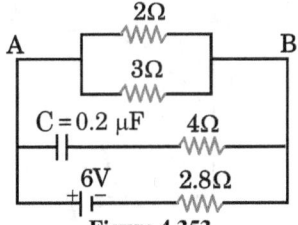

Figure 4.351

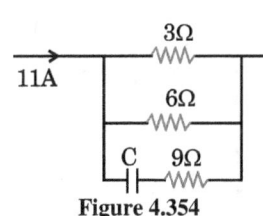

Figure 4.352

26. In Fig.4.352, the steady state voltage drop across capacitor (C) is
    (A) V
    (B) $\dfrac{V_{R_1}}{\left[R_3\left(\dfrac{R_1 \cdot R_3}{R_1+R_3}\right)\right]}$
    (C) $\dfrac{VR_3}{R_1+R_3}$
    (D) $\dfrac{VR_1}{R_1+R_3}$

27. In Fig.4.353, the steady state current in 2Ω resistance is
    (A) 1.5 A (B) 0.9 A (C) 0.6 A (D) 0

[Figure 4.353 and Figure 4.354 diagrams]

Figure 4.353   **Figure 4.354**

28. In Fig.4.354, the current in 3Ω and 6Ω resistance is respectively-
    (A) 7.33 A, 3.067 A (B) 3.67 A, 7.33 A
    (C) 6 A, 3 A (D) 3A, 6A

29. A battery of 20 cells (each having emf 1.8 V and internal resistance 0.1 Ω) is charged by 220 volts and the charging current is 15 A. The resistance to be put in the circuit is
    (A) 10.27 Ωohm (B) 12.27 Ω
    (C) 8.62 Ω (D) 16.24 Ω

30. A battery is connected in series with an external resistance. The current in circuit is 1 A and 0.7 A. When external resistance equals 5 Ω and 8 Ω respectively the internal resistance of the battery is
    (A) 0.2 Ω (B) 0.5 Ω (C) 2 Ω (D) 0.6 Ω

31. In the above question the maximum current is
    (A) 8 A (B) 4 A (C) 3.5 A (D) 1 A

32. A house is served by a 220 V supply line. In a circuit protected by a fuse marked 9 A. The maximum number of 60 W lamps in parallel that can be turned on is
    (A) 44 (B) 22 (C) 20 (D) 33

33. The two head lamps of a car are in parallel and they together consume 48 watts with the help of a 6 V battery. The resistance of each bulb is -
    (A) 0.67 Ω (B) 3.0 Ω (C) 4.0 Ω (D) 1.5 Ω

34. A 25 watt, 220 volt bulb and a 100 watt, 220 volt bulb are connected in series across a 440 volt line
    (A) only 100 watt bulb will fuse
    (B) only 25 watt bulb will fuse
    (C) both bulbs will fuse
    (D) none of the bulb will fuse

35. In the circuit of Fig.4.355, ammeter (A) reads 0.5A. Bulbs L_1 and L_2 are brightly lit, but L_3 is not lit. What is the reason for L_3 not being lit ?

    (A) the ammeter is faulty
    (B) the filament of L_3 is broken
    (C) the resistance of L_3 is much lower than that of L_1 and L_2
    (D) there is a break in the connecting wire between L_2 and L_3

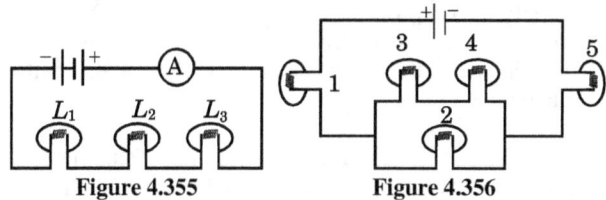

Figure 4.355   **Figure 4.356**

36. All bulbs in Fig.4.356, are identical. Which bulb(s) lights most brightly
    (A) 1 only (B) 2 only
    (C) 3 and 4 only (D) 1 and 5

37. A cell of emf $\mathcal{E}$ volt and internal resistance (r) ohms is connected to an external resistance of (r) ohms. The potential difference across the terminals of the cell will be
    (A) $\mathcal{E}$ volt (B) $\mathcal{E}/2$ volt
    (C) $\mathcal{E}/4$ volt (D) $2\mathcal{E}$ volt

38. When a cell is connected to 1 Ω resistance, 1 A current flows through the circuit. When 3 Ω resistance issued then 0.5 A current flows, then internal resistance of the cell is-
    (A) 1 Ω (B) 1.5 Ω (C) 2 Ω (D) 2.5 Ω

39. An electric kettle has two coils. When one of these is switched on, the water in the kettle boils in 6 minutes. When the other coil is switched on, the water boils in 3 minutes. If the two coils are connected in series, the time taken to boil the water in the kettle is-
    (A) 2 minutes (B) 3 minutes
    (C) 6 minutes (D) 9 minutes

40. In question 39, if the two coils are connected in parallel, then the total time taken to boil the water in kettle is
    (A) 2 minutes (B) 3 minutes
    (C) 6 minutes (D) 9 minutes

41. A resistance coil of 60 Ω is immersed in 42 kg of water. A current of 7 A is passed through it. The rise in temperature of water per minutes is :
    (A) 4°C (B) 8°C
    (C) 1°C (D) 12°C

42. A coil of wire of resistance 50 Ω is embedded in a block of ice and a potential difference of 210 V is applied across it. the amount of ice which melts in 1 second is -
    (A) 0.262 g (B) 2.62 g
    (C) 26.2 g (D) 0.0262 g

43. In the circuit shown in Fig.4.357, the heat produced in 5Ω resistor due to a current flowing in it is 10 cal/s. The heat produced in 4Ω resistor is :
    (A) 4cal/s (B) 1cal/s
    (C) 2cal/s (D) 3cal/s

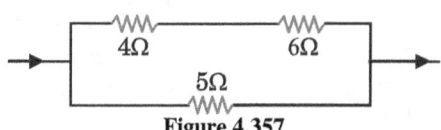

Figure 4.357

44. 'N' equal resistors connected in seres across a source of emf together dissipate 4 watts of power. The power dissipated when the same resistors are connected in parallel across the same source of emf is 64 watts. The number of resistors 'N' is equal to -
    (A) 8 (B) 4 (C) 16 (D) 2

45. The same mass of copper is drawn into two wires 1 mm thick and 2 mm thick. If the two wires are connected in series and the current is passed, the heat produced in the wires will be in the ratio.
    (A) 2 : 1   (B) 4 : 1   (C) 1 : 16   (D) 16 : 1
46. Forty electric bulbs are connected in series across a 220 V supply. After one bulb is fused the remaining 39 are connected again in series across the same supply. The percentage with which the illumination of the bulbs will change will be -
    (A) 10.25%   (B) 7%   (C) 5%   (D) 2.5%
47. A cell of emf $\mathcal{E}$ and internal resistance r supplies currents for the same time through external resistance $R_1 = 100\ \Omega$ and $R_2 = 40\ \Omega$ separately. If the heat developed in both the cases is the same, then the internal resistance of the cell is given by -
    (A) 28.6 Ω   (B) 70 Ω   (C) 63.3 Ω   (D) 140 Ω
48. Two bulbs of 500 watt and 200 watt are manufactured to operate on 220 volt line. The ratio of heat produced in 500 watt and 200 watt, in two cases, when first they are joined in series and secondly in parallel, will be -
    (A) 5/2, 2/5   (B) 5/2, 5/2
    (C) 2/5, 5/2   (D) 2/5, 2/5
49. A capacitor of capacitance 3 μF is first charged by connecting across a 10 V battery, then it is allowed to get discharged through 2Ω and 4Ω resistor by closing the key K (Fig.4.358). The total energy dissipated in 2Ω resistor is equal to -
    (A) 0.15 mJ   (B) 0.5 mJ
    (C) 0.05 mJ   (D) 1.0 mJ

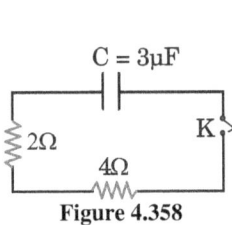

Figure 4.358

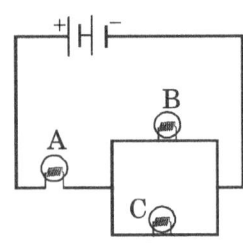

Figure 4.359

50. The bulbs A, B and C are connected as shown in Fig.4.359. The bulbs B and C are identical. If the bulb C is fused,
    (A) both A and B will glow more brightly
    (B) both A and B will glow less brightly
    (C) A will glow less brightly and B will glow more brightly
    (D) A will glow more brightly and B will glow less brightly
51. How much electrical energy in kilo-watt hour is consumed in operating ten 50 watt bulbs for 10 hours per day in a month of 30 days?
    (A) 1500   (B) 15000
    (C) 15   (D) 150
52. A 1 μF capacitor is connected in the circuit shown in Fig.4.360. The emf of the cell is 2 volts and internal resistance is 0.5 Ω. The resistors R_1 and R_2 have values 40 Ω and 1 Ω respectively. The charge on the capacitor must be-
    (A) 2 μC   (B) 1 μC   (C) 1.33 μC   (D) zero
53. In the Figure 4.361, the capacity of the condenser C is 2 μF. The current in 2 Ω resistor is-
    (A) 9 A   (B) 0.9 A   (C) 1/9 A   (D) 1/0.9 A
54. As in Figure 4.362, if a capacitor C is charged by connecting it with resistance R, then energy given by the battery will be

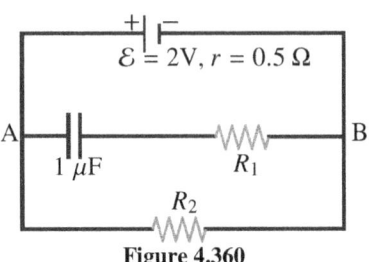

Figure 4.360

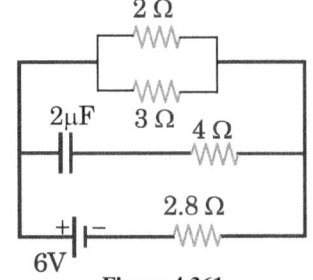

Figure 4.361

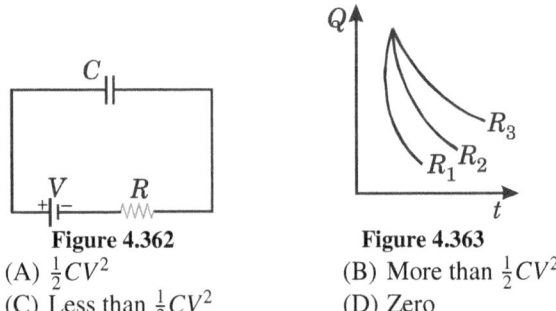

Figure 4.362   Figure 4.363

(A) $\frac{1}{2}CV^2$ (B) More than $\frac{1}{2}CV^2$
(C) Less than $\frac{1}{2}CV^2$ (D) Zero

55. A parallel plate capacitor is charged to a potential difference of 50 V. it is discharged through a resistance. After 1 second, the potential difference between plates becomes 40 V. Then
    (A) Fraction of stored energy after 1 second is 16/25
    (B) Potential difference between the plates after 2 seconds will be 32 V
    (C) Potential difference between the plates after 2 seconds will be 20 V
    (D) Fraction of stored energy after 1 second is 4/5
56. Three identical capacitors are given a charge Q each and they are then allowed to discharge through resistance R_1, R_2 and R_3. Their charges, as a function of time, are shown in Fig.4.363. The smallest of the three resistance is
    (A) R_3 (B) R_2
    (C) R_1 (D) Cannot be predicted

4.34.3 Level 3

1. An electric current is established in a hydrogen gas discharge tube when a high voltage is applied across the two electrodes in the tube. The gas is ionised. Electrons move towards the positive terminal and the positive ions towards the negative terminal. The magnitude of the current in the tube in which 3.1×10^{18} electrons and 1.1×10^{18} protons move past a cross-sectional area of the tube each second will be-
    (A) 1.6 A   (B) 3.2 A   (C) 0.16 A   (D) 0.672 A
2. A charge of 2×10^{-2} C moves at 30 revolution per second in a circle of diameter 0.80 m. The current linked with the circuit will be -
    (A) 0.1 A   (B) 0.2 A   (C) 0.4 A   (D) 0.6 A
3. The current in a copper wire is increased by increasing the potential difference between its end. Which one of the following statements regarding n, the number of charge carriers

4.34. MULTIPLE CHOICE ASSIGNMENTS

per unit volume in the wire and v_d, the drift velocity of the charge carriers, is correct -
(A) n is unaltered but v_d is decreased
(B) n is unaltered but v_d is increased
(C) n is increased but v_d is decreased
(D) n is increased but v_d is unaltered

4. A wire of resistance 32Ω is melted and drawn into a wire of half of its original length. The resistance of new wire and percentage decrease in resistance -
(A) 8Ω, 75% (B) 8Ω, 50%
(C) 16Ω, 75% (D) 16Ω, 50%

5. Consider two conducting wires of same length and material, one wire is solid with radius r. The other is a hollow tube of outer radius $2r$ while inner r. The ratio of resistance of the two wires will be-
(A) 1 : 1 (B) 1 : 2 (C) 1 : 3 (D) 1 : 4

6. A carbon and an aluminium wire connected in series. If the combination has resistance of 30 ohm at 0°C, what is the resistance of carbon and aluminium wire at 0°C so that the resistance of the combination does not change with temperature-
$[\alpha_C = -0.5 \times 10^{-3} (C°)^{-1}$ and $\alpha_{Al} = 4 \times 10^{-3} (C°)^{-1}]$
(A) $\frac{10}{3}\Omega, \frac{80}{3}\Omega$ (B) $\frac{80}{3}\Omega, \frac{10}{3}\Omega$
(C) $10\Omega, 80\Omega$ (D) $80\Omega, 10\Omega$

7. A resistance R_2 is connected in parallel with a resistance R_1 what resistance R_3 must be connected in series with the combination of R_1 and R_2 so that the equivalent resistance is equal to the resistance R_1 -
(A) $\dfrac{R_1^2}{R_1 + R_2}$ (B) $\dfrac{(R_1 + R_2)^2}{R_1}$
(C) $\dfrac{R_2^2}{R_1 + R_2}$ (D) $\dfrac{R_1^2}{R_2}$

8. In Fig.4.364, an infinite ladder network of resistance is constructed with 1Ω and 2Ω resistances. The 6 V battery between A and B has negligible internal resistance. The current that passes through 2Ω resistance nearest to the battery is -
(A) 1 A (B) 1.5 A (C) 2 A (D) 2.5 A

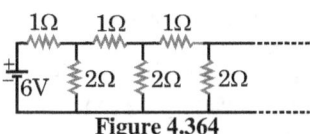

Figure 4.364

9. A potential difference of 200 V is applied to a coil at a temperature of 15°C and the current is 10 A. What will be the mean temperature of the coil when the current has fallen to 5 A, the applied voltage being the same as before (Given $\alpha = \frac{1}{234}C^{\circ-1}$ at 0°C).
(A) 254° (B) 256° (C) 258° (D) 264°

10. In a given electric circuit (Fig.4.365), the potentials at points a, b and c are 30 V, 12 V and 2 V respectively. The respective currents through resistors 10Ω, 20Ω and 30Ω are-
(A) 1, 0.4, 0.6 (B) 2, 0.8, 1.2
(C) 0.6 A, 0.4 A, 1 A (D) None of these

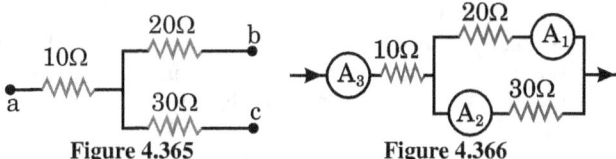

Figure 4.365   Figure 4.366

11. If the reading of ammeter A_1, in Figure 4.366, is 2.4 A, what will the ammeter A_2 and A_3 read? (Neglecting the resistances of ammeters)-
(A) 1.6 A, 2.3 A (B) 1.6 A, 4.0 A
(C) 4.0 A, 1.6 A (D) 2.3 A, 1.6 A

12. The emf of the battery shown in the Figure 4.367, is given by-
(A) 6 V (B) 12 V (C) 18 V (D) 8 V

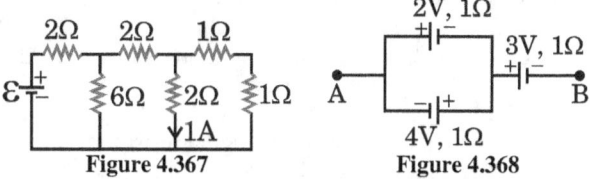

Figure 4.367   Figure 4.368

13. In the circuit of Fig.4.368, the potential difference between points A and B is-
(A) 2 V (B) 6 V (C) 4 V (D) 3 V

14. In the given Fig.4.369, the ratio of current in 8Ω and 3Ω will be-
(A) $\frac{8}{3}$ (B) $\frac{3}{8}$ (C) $\frac{4}{3}$ (D) $\frac{3}{4}$

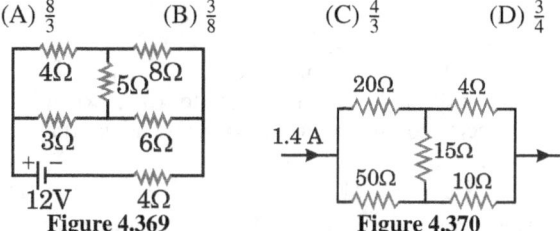

Figure 4.369   Figure 4.370

15. Through an electrolyte, an electric current is due to drift of -
(A) Free electrons
(B) Free electrons and holes
(C) Positive and negative ions
(D) Protons

16. A current flows in a wire of circular cross-section with the free electrons travelling with a mean drift speed $\bar{v}$. If an equal current flows in a wire of twice the radius, new mean drift speed is-
(A) $\bar{v}$ (B) $\bar{v}/2$
(C) $\bar{v}/4$ (D) None of these

17. If a copper wire is stretched to make its radius decrease by 0.1%, then the percentage increase in resistance is approximately-
(A) 0.1% (B) 0.2% (C) 0.4% (D) 0.8%

18. There is a current of 1.344 A in a copper wire whose area of cross-section normal to the length of the wire is 1 mm². If the number of free electrons per cm³ is 8.4×10^{22}, then the drift velocity would be-
(A) 1.0 mm/s (B) 1.0 m/s
(C) 0.1 mm/s (D) 0.01 mm/s

19. In Fig.4.370, the current through 4Ω resistor is-
(A) 1.4 A (B) 0.4 A (C) 1.0 A (D) 0.7 A

20. In Fig.4.371, the reading of the ammeter A when the internal resistance of the battery is zero, is-
(A) $\dfrac{20}{3}$ A (B) $\dfrac{20}{12}$ A
(C) $\dfrac{20}{4}$ A (D) $\left(\dfrac{20}{3} + \dfrac{20}{12}\right)$ A

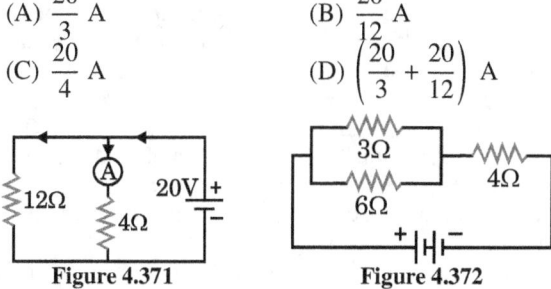

Figure 4.371   Figure 4.372

21. The number of dry cells, each of emf 1.5 volt and internal re-

sistance 0.5 Ω that must be joined in series with a resistance of 20 Ω so as to send a current of 0.6 A through the circuit is-
(A) 2   (B) 8   (C) 10   (D) 12

22. Two batteries of different emf and internal resistance are connected in series with each other and with an external load resistor. The current is 3.0 A. When the polarity of one battery is reversed, the current becomes 1.0 A. The ratio of the emf of the two batteries is-
(A) 2.5   (B) 2.0   (C) 1.5   (D) 1.0

23. In the Fig.4.372, current through 3 Ω resistor is 0.8 A; then the potential drop through 4 Ω resistor is-
(A) 9.6 V   (B) 2.6 V   (C) 4.8 V   (D) 1.2 V

24. A cell supplies a current I_1 through a resistor of resistance R_1 and a current I_2 through a resistor of resistance R_2, then internal resistance of the cell is-
(A) $R_1 - R_2$
(B) $R_1 + R_2$
(C) $\frac{I_1 R_2 + I_1 R_1}{I_1 + I_1}$
(D) $\frac{I_2 R_2 - I_1 R_1}{I_1 - I_2}$

25. The sides of a rectangular block are 2 cm, 3 cm and 4 cm. The ratio of maximum to minimum resistance between its parallel faces is-
(A) 4   (B) 3   (C) 2   (D) 1

26. For a cell the terminal potential difference is 2.2 V when the circuit is open and reduces to 1.8 V when the cell is connected to a resistance $R = 5\Omega$. The internal resistance of the cell (r) is-
(A) $\frac{10}{9} \Omega$ (B) $\frac{9}{10} \Omega$ (C) $\frac{11}{9} \Omega$ (D) $\frac{5}{9} \Omega$

27. The current in a conductor varies with time t is $I = 2t + 3t^2$ where I is in ampere and t in seconds. Electric charge flowing through a section of conductor during $t = 2$ s to $t = 3$ s, is-
(A) 10 C   (B) 24 C   (C) 33 C   (D) 44 C

28. Two wires of resistance R_1 and R_2 have temperature coefficient of resistance α_1 and α_2, respectively. These are joined in series. The effective temperature coefficient of resistance is-
(A) $\frac{\alpha_1 + \alpha_2}{2}$
(B) $\sqrt{\alpha_1 \alpha_2}$
(C) $\frac{\alpha_1 R_1 + \alpha_2 R_2}{R_1 + R_2}$
(D) $\frac{\sqrt{R_1 R_2 \alpha_1 \alpha_2}}{\sqrt{R_1^2 + R_2^2}}$

29. A long resistance wire is divided into $2n$ parts. Then, n parts are connected in series and the other n parts in parallel separately. Both combinations are connected to identical supplies. Then the ratio of heat produced in series to parallel combinations will be-
(A) $1 : 1$ (B) $1 : n^2$ (C) $1 : n^4$ (D) $n^2 : 1$

30. Two bulbs 100 W, 250 V and 200 W, 250 V are connected in parallel across a 500 V line. Then-
(A) 100 W bulb will fuse
(B) 200 W bulb will fuse
(C) Both bulbs will fuse
(D) No bulb will fuse

31. A bulb rated 220 V, 100 W is connected across 160 V line. The power dissipated will be-
(A) 100 W   (B) 75 W   (C) 52 W   (D) 26 W

32. A uniform wire connected across a supply produces heat H per second. If the wire is cut into n equal parts and all the parts are connected in parallel across the same supply, the heat produced per second will be
(A) H/n (B) nH (C) $n^2 H$ (D) H/n^2

33. Two electric bulbs 40 W, 200 V and 100 W, 200 V are connected in series. Then the maximum voltage that can be applied across the combination, without fusing either bulb is-
(A) 280 V   (B) 400 V   (C) 3000 V   (D) 200 V

34. The resistance of 3Ω and 6Ω are joined in series and connected across a battery of emf 10 V and internal resistance 1Ω. The power dissipated by battery is-
(A) 3 W   (B) 8 W   (C) 9 W   (D) 10 W

35. A 24 V battery of internal resistance 4Ω is connected to a variable resistor. The rate of heat production in the resistor is maximum when the current in the circuit is-
(A) 2 A   (B) 3 A   (C) 4 A   (D) 6 A

36. A 4 μF condenser is charged to 400 volts and then its plates are joined through a resistance of 1 kΩ. The heat produced in the resistance is-
(A) 0.16 J   (B) 1.28 J   (C) 0.64 J   (D) 0.32 J

37. If the condenser shown in the circuit (Fig.4.373) is charged to 5 V and left in the circuit, in 12 s the charge on the condenser will become-

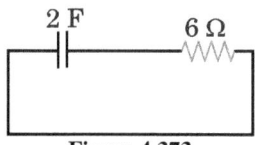

Figure 4.373

(A) $10/e$ C   (B) $e/10$ C   (C) $10/e^2$ C   (D) $e^2/10$ C

38. A capacitor of 2 μF is charged to its maximum emf of 2 V and is discharged through a resistance of 10^4 Ω. Current in the circuit after 0.02 s will be-
(A) 10^{-4} A
(B) 1.4×10^5 A
(C) 7.4×10^{-5} A
(D) 3.7×10^{-5} A

Passage Type Questions

Passage - I Two tungsten lamps with resistance R_1 and R_2 respectively at full incandescence are connected first in parallel and then in series, in a lighting circuit of negligible internal resistance. It is given that $R_1 > R_2$.

39. Which lamp will glow more brightly when they are connected in parallel?
(A) Lamp having lower resistance
(B) Lamp having higher resistance
(C) Both the lamps
(D) None of the two lamps

40. Which lamp will glow more brightly when they are in connected in series?
(A) Lamp having lower resistance
(B) Lamp having higher resistance
(C) Both the lamps
(D) None of the two lamps

41. Would physically bending a supply wire cause any change in the illumination?
(A) Illumination will remain same
(B) Illumination will increase
(C) Illumination will decrease
(D) It is not possible to predict from the given data

Passage - II A set of experiments in the physics lab is designed to develop understanding of simple electrical circuit principles for direct current circuits. The student is given a variety of batteries, resistors, and DC meters; and it directed to wire series and parallel combinations of resistors and batteries making measurements of the currents and voltage drops using the ammeters and voltmeters. The student calculator expected current and voltage values using ohm's law and kirchhoff's circuit rules and then checks the results with the meters.

42. A student connects a 6 volt battery and a 12 V battery in series and then connects this combination across a 10Ω resistor. What is the current in the resistor?
(A) 0.8 A   (B) 0.9 A   (C) 1.8 A   (D) 3.6 A

43. Resistors of 4 Ω and 8 Ω are connected in series. A

battery of 6 V is connected across the series combination. How much power (in watts) is consumed in 8Ω resistor?
(A) 0.67 W (B) 2 W (C) 12 W (D) 24 W

44. A 6 V battery is connected across a 2 Ω resistor. What is the heat energy dissipated in the resistor in 5 minutes?
(A) 430 J (B) 560 J (C) 4300 J (D) 5400 J

Assertion and Reason type questions
Each of the questions given below consist of Statement – I and Statement - II. Use the following keys to choose the appropriate answer.
(A) If both Statement-I and Statement-II are true, and Statement - II is the correct explanation of Statement-I.
(B) If both Statement-I and Statement-II are true but Statement-II is not the correct explanation of Statement-I.
(C) If Statement-I is true but Statement-II is false.
(D) If Statement-I is false but Statement-II is true.

45. **Statement I:** The resistance of a copper wire varies directly as the length and diameter.
Statement II: Because the resistance varies inversely the area of cross-section.

46. **Statement I:** When cells are connected in parallel to the external load, the effective emf increases.
Statement II: Because effective internal resistance of cells decreases.

47. **Statement I:** The total resistance in series combination of resistors increases and in parallel combination of resistors decreases.
Statement II: In series combination of resistors, the effective length of resistors increases and in parallel combination of resistors, the area of cross-section of the resistors increases.

48. **Statement I:** In parallel combination of electrical appliance, total power consumption is equal to the sum of the powers of the individual appliances.
Statement II: In parallel combination, the voltage across each appliance is the same, as required for the proper working of electrical appliance.

49. **Statement I:** In series combination of electrical bulbs of lower power emits more light than that of higher power bulb.
Statement II: The lower power bulb in series gets more current than the higher power bulb.

50. **Statement I:** Each bulb in a frill of 20 bulbs in series when connected to supply voltage will emit more light than each bulb in frill of 19 bulbs in series when connected to same supply voltage.
Statement II: Each bulb in a frill of 20 bulbs in series will get less voltage than that in frill of 19 bulbs.

4.34.4 Level 4 (Previous Years JEE Main & Advanced Questions)
Section-A: JEE Mains
1. If energy consumption of circuit of Fig.4.374, is 150 W then find the value of resistance [2002]
(A) 2 Ω (B) 4 Ω (C) 6 Ω (D) 8 Ω

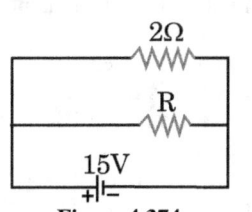
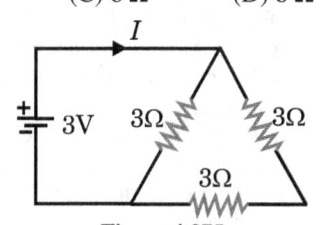

Figure 4.374   Figure 4.375

2. A wire when connected to 220 V mains supply has power dissipation P_1. Now the wire is cut into two equal pieces which are connected in parallel to the same supply. Power dissipation in this case is P_2. The $P_2 : P_1$ is- [2002]
(A) 1 (B) 4 (C) 2 (D) 3

3. The length of a given cylindrical wire is increased by 100%. Due to the consequent decrease in diameter the change in the resistance of the wire will be- [2003]
(A) 100% (B) 50% (C) 300% (D) 200%

4. A 220 volt, 1000 watt bulb is connected across a 110 volt mains supply. The power consumed will be- [2003]
(A) 500 watt (B) 250 watt
(C) 1000 watt (D) 750 watt

5. A 3 volt battery with negligible internal resistance is connected in a circuit as shown in Fig.4.375. The current I in the circuit will be- [2003]
(A) 1.5 A (B) 2 A (C) 1/3 A (D) 1 A

6. The total current supplied to the circuit in Fig.4.376, by the battery is- [2004]
(A) 1 A (B) 2 A (C) 4 A (D) 6 A

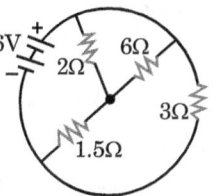

Figure 4.376

7. The resistance of the series combination of two resistance is S. When they are joined in parallel the total resistance is P. If $S = nP$ then the minimum possible value of n is- [2004]
(A) 4 (B) 3 (C) 2 (D) 1

8. An electric current is passed through a circuit containing two wires of the same material, connected in parallel. If the lengths and radii of the wires are in the ratio of 4/3 and 2/3, then the ratio of the currents passing through the wires will be- [2004]
(A) 3 (B) 1/3 (C) 8/9 (D) 2

9. The thermistors are usually made of- [2004]
(A) metals with low temperature coefficient of resistivity
(B) metals with high temperature coefficient of resistivity
(C) metal oxides with high temperature coefficient of resistivity
(D) Semiconducting meterials having low temperature coefficient of resistivity

10. Time taken by a 836 W heater to heat one litre of water from 10°C to 40°C is- [2004]
(A) 50 s (B) 100 s (C) 150 s (D) 200 s

11. A moving coil galvanometer has 150 equal divisions. It's current sensitivity is 10 divisions per milliampere and voltage sensitivity is 2 divisions per millivolt. In order that each division reads 1 volt, the resistance in ohms needed to be connected in series with the coil will be- [2005]
(A) 10^3 (B) 10^5 (C) 99995 (D) 9995

12. In the circuit of Fig.4.377, the galvanometer G shows zero deflection. If the batteries A and B have negligible internal resistance, the value of the resistor R will be- [2005]
(A) 200Ω (B) 100Ω (C) 500Ω (D) 1000Ω

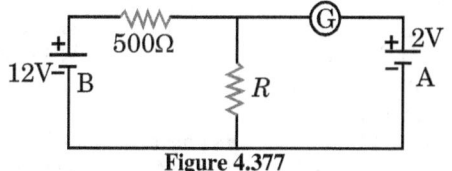

Figure 4.377

13. Two sources of equal emf are connected to an external resistance R. The internal resistances of the two sources are R_1 and R_2 ($R_2 > R_1$). If the potential difference across the source having internal resistance R_2 is zero, then- [2005]
    (A) $R = R_2 \times \frac{(R_1+R_2)}{(R_2-R_1)}$
    (B) $R = R_2 - R_1$
    (C) $R = \frac{R_1 R_2}{(R_1+R_2)}$
    (D) $R = \frac{R_1 R_2}{(R_2-R_1)}$

14. An energy source will supply a constant current into the load if its internal resistance is- [2005]
    (A) equal to the resistance of the load
    (B) very large as compared to the load resistance
    (C) zero
    (D) non-zero but less than the resistance of the load

15. A material 'B' has twice the specific resistance of 'A'. A circular wire made of 'B' has twice the diameter of a wire made of 'A'. then for the two wires to have the same resistance, the ratio l_B/l_A of their respective lengths must be- [2006]
    (A) 1/4   (B) 2   (C) 1   (D) 1/2

16. The Kirchhoff's first law ($\sum I = 0$) and second law ($\sum IR = \sum \mathcal{E}$), where the symbols have usual meanings, are respectively based on- [2006]
    (A) conservation of momentum, conservation of charge
    (B) conservation of charge, conservation of energy
    (C) conservation of charge, conservation of momentum
    (D) conservation of energy, conservation of charge

17. The current I drawn from the 5 volt source in the circuit of Fig.4.378, will be- [2006]
    (A) 0.67 A   (B) 0.17 A   (C) 0.33 A   (D) 0.5 A

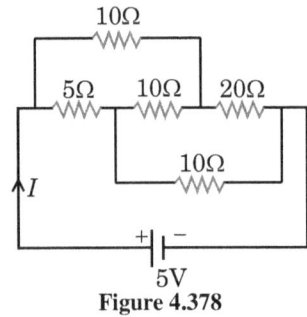

Figure 4.378

18. In a Wheatstone's bridge, three resistances P, Q and R are connected in the three arms and the fourth arm is formed by two resistances S_1 and S_2 connected in parallel. the condition for the bridge to be balance will be- [2006]
    (A) $\frac{P}{Q} = \frac{R(S_1+S_2)}{2 S_1 S_2}$
    (B) $\frac{P}{Q} = \frac{R}{S_1+S_2}$
    (C) $\frac{P}{Q} = \frac{2R}{S_1+S_2}$
    (D) $\frac{P}{Q} = \frac{R(S_1+S_2)}{S_1 S_2}$

19. An electric bulb is rated 220 volt - 100 watt. The power consumed by it when operated on 110 volt will be- [2006]
    (A) 25 W   (B) 50 W   (C) 75 W   (D) 40 W

20. The resistance of wire is 5 Ω at 50°C and 6 Ω at 100°C. The resistance of the wire at 0°C will be [2007]
    (A) 2 Ω   (B) 1 Ω   (C) 4 Ω   (D) 3 Ω

21. A 5 V battery with internal resistance 2 Ω and a 2 V battery with internal resistance 1 Ω are connected to a 10 Ω resistor as shown in the Fig.4.379. The current in the 10 Ω resistor is - [2008]
    (A) 0.03 A P_1 to P_2
    (B) 0.03 A P_2 to P_1
    (C) 0.27 A P_1 to P_2
    (D) 0.27 A P_2 to P_1

    Directions: Questions No.22 and 23 are based on the following paragraph.
    Consider a block of conducting material of resistivity 'ρ' shown in the Fig.4.380. Current 'I' enters at 'A' and leaves

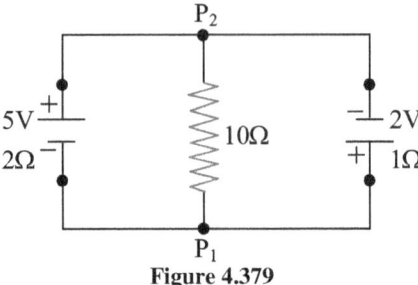

Figure 4.379

from 'D'. We apply superposition principle to find voltage 'ΔV' developed between 'B' and 'C'. The calculation is done in the following steps:
(i) Take current 'I' entering from 'A' and assume it to spread over a hemispherical surface in the block.
(ii) Calculate field $E(r)$ at distance 'r' from A by using Ohm's law $E = \rho J$, where J is the current per unit area at 'r'.
(iii) From the 'r' dependence of $E(r)$, obtain the potential $V(r)$ at r.
(iv) Repeat (i), (ii) and (iii) for current 'I' leaving 'D' and superpose results for 'A' and 'D'.

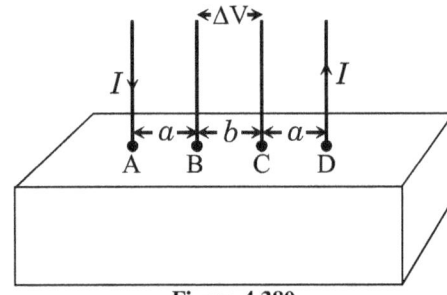

Figure 4.380

22. For current entering at A, in Fig.4.380, the electric field at a distance 'r' from A is- [2008]
    (A) $\rho I/r^2$
    (B) $\rho I/2\pi r^2$
    (C) $\rho I/4\pi r^2$
    (D) $\rho I/8\pi r^2$

23. ΔV measured between B and C is- [2008]
    (A) $\frac{\rho I}{a} - \frac{\rho I}{(a+b)}$
    (B) $\frac{\rho I}{2\pi a} - \frac{\rho I}{2\pi(a+b)}$
    (C) $\frac{\rho I}{2\pi(a-b)}$
    (D) $\frac{\rho I}{\pi a} - \frac{\rho I}{\pi(a+b)}$

24. **Statement-1:** The temperature dependence of resistance is usually given as $R = R_0(1 + \alpha \Delta t)$. The resistance of a wire changes from 100 Ω to 150 Ω when its temperature is increased from 27°C to 227°C. This implies that $\alpha = 2.5 \times 10^{-3}/°C$.
    Statement-2: $R = R_0(1 + \alpha \Delta t)$ is valid only when the change in the temperature ΔT is small and $\Delta R = (R - R_0) \ll R_0$. [2009]
    (A) Statement-1 is true, Statement-2 is true; Statement-2 is a correct explanation for Statement-1
    (B) Statement-1 is true. Statement2 is true; Statement-2 isn't a correct explanation for Statement-1
    (C) Statement-1 is true, Statement-2 is false.
    (D) Statement-1 is false, Statement-2 is true.

25. Two conductors have the same resistance at 0°C but their temperature coefficients of resistance are α_1 and α_2. The respective temperature coefficients of their series and parallel combinations are nearly [2010]
    (A) $\frac{\alpha_1+\alpha_2}{2}$, $\alpha_1 + \alpha_2$
    (B) $\alpha_1 + \alpha_2$, $\frac{\alpha_1+\alpha_2}{2}$
    (C) $\alpha_1 + \alpha_2$, $\frac{\alpha_1 \alpha_2}{\alpha_1+\alpha_2}$
    (D) $\frac{\alpha_1+\alpha_2}{2}$, $\frac{\alpha_1+\alpha_2}{2}$

4.34. MULTIPLE CHOICE ASSIGNMENTS

26. A resistor R and $2\mu F$ capacitor in series is connected through a switch to 200 V direct supply. Across the capacitor is a neon bulb that lights up at 120 V. Calculate the value of R to make the bulb light up $5s$ after the switch has been closed. ($\log_{10} 2.5 = 0.4$) [2011]
    (A) $1.7 \times 10^5 \Omega$
    (B) $2.7 \times 10^6 \Omega$
    (C) $3.3 \times 10^7 \Omega$
    (D) $1.3 \times 10^4 \Omega$

27. If a wire is stretched to make it 0.1% longer, its resistance will: [2011]
    (A) increase by 0.2%
    (B) decrease by 0.2%
    (C) decrease by 0.05%
    (D) increases by 0.05%

28. Two electric bulbs marked 25 W - 220 V and 100 W - 220 V are connected in series to a 440 V supply. Which of the bulbs will fuse? [2012]
    (A) both
    (B) 100W
    (C) 25W
    (D) neither

29. Resistance of a given wire is obtained by measuring the current flowing in it and the voltage difference applied across it. If the percentage errors in the measurement of the current and the voltage difference are 3% each, then error in the value of resistance of the wire is: [2012]
    (A) 6% (B) 0 (C) 1% (D) 3%

30. The supply voltage to room is 120 V. The resistance of the lead wires is 6 Ω. A 60 W bulb is already switched on. What is the decrease of voltage across the bulb, when a 240 W heater is switched on in parallel to the bulb [Fig.4.381]? [2013]

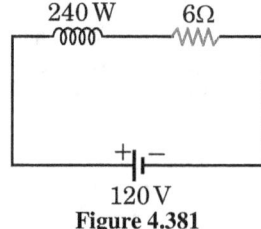

Figure 4.381

    (A) 0 V (B) 2.9 V (C) 13.3 V (D) 10.04 V

31. This questions has Statement I and Statement II. Of the four choices given after the Statements, choose the one that best describes the two Statements. [2013]
    Statement I: Higher the range, greater is the resistance of ammeter.
    Statement II: To increase the range of ammeter, additional shunt needs to be used across it.
    (A) Statement I is true, statement II is true, statement I is the correct explanation of statement II.
    (B) Statement I is true, statement II is true, statement II is not the correct explanation of statement I.
    (C) Statement I is true, statement II is false.
    (D) Statement I is false, statement II is true.

32. The current voltage relation of diode is given by $I = (e^{1000V/T} - 1)$ mA, where the applied voltage V is in volts and the temperature T is in Kelvin. If a student makes an error measuring 0.01 V while measuring the current of 5 mA at 300 K, what will be the error in the value of current in mA? [2014]
    (A) 0.5 mA
    (B) 0.05 mA
    (C) 0.2 mA
    (D) 0.02 mA

33. A uniformly charged solid sphere of radius R has potential V_0 (measured with respect to ∞) on its surface. For this sphere the equipotential surfaces with potentials $\frac{3V_0}{2}$, $\frac{5V_0}{2}$, $\frac{3V_0}{4}$ and $\frac{V_0}{4}$ have radii R_1, R_2, R_3 and R_4 respectively. Then [2015]

    (A) $R_1 = 0$ & $R_2 > (R_4 - R_3)$
    (B) $R_1 \neq 0$ & $(R_2 - R_1) > (R_4 - R_3)$
    (C) $R_1 = 0$ & $R_2 < (R_4 - R_3)$
    (D) $2R < R_4$

34. A galvanometer having a coil resistance of 100 W gives a full scale deflection, when a current of 1 mA is passed through it. The value of the resistance, which can convert this galvanometer into ammeter giving a full scale deflection for a current of 10 A, is: [2016]
    (A) 3 Ω (B) 0.01 Ω (C) 2 Ω (D) 0.1 Ω

35. In the circuit of Fig.4.382, the current in each resistance is: [2017]
    (A) 0 A (B) 1 A (C) 0.25 A (D) 0.5 A

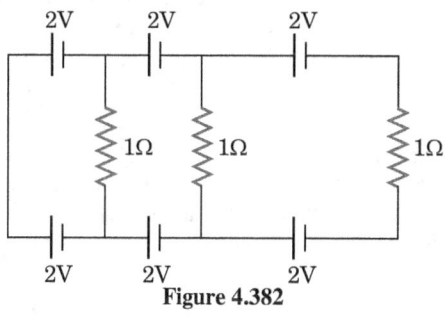

Figure 4.382

36. In the given circuit diagram (Fig.4.383), when the current reaches steady state in the circuit, the charge on the capacitor of capacitance C will be: [2017]

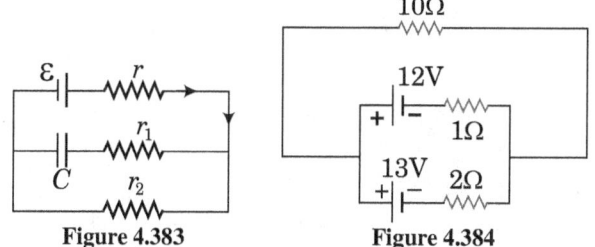

Figure 4.383   Figure 4.384

    (A) $C\mathcal{E}\frac{r_1}{(r_1 + r)}$
    (B) $C\mathcal{E}$
    (C) $C\mathcal{E}\frac{r_1}{(r_2 + r)}$
    (D) $C\mathcal{E}\frac{r_2}{(r + r_2)}$

37. Which of the following statements is false? [2017]
    (A) Kirchhoff's second law represents energy conservation.
    (B) Wheatstone bridge is the most sensitive when all the four resistances are of the same order of magnitude.
    (C) In a balanced Wheatstone bridge, if the cell and the galvanometer are exchanged, the null point is disturbed.
    (D) A rheostat can be used as a potential divider.

38. When a current of 5 mA is passed through a galvanometer having a coil of resistance 15 Ω, it shows full scale deflection. The value of the resistance to be put in series with the galvanometer to convert it into a voltmeter of range 0-10 V is: [2017]
    (A) 4.005×10^3 Ω
    (B) $1.985 \times 10^3 \Omega$
    (C) $2.045 \times 10^3 \Omega$
    (D) 2.535×10^3

39. Two batteries with emf 12 V and 13 V are connected in parallel across a load resistor of 10 Ω as shown in Fig.4.384. The internal resistances of the two batteries are 1 Ω and 2 Ω respectively. The voltage across the load lies between [2018]
    (A) 11.6 V and 11.7 V
    (B) 11.5 V and 11.6 V
    (C) 11.4 V and 11.5 V
    (D) 11.7 V and 11.8 V

40. In a potentiometer experiment, it is found that no current passes through the galvanometer when the terminals of the cell are connected across 52 cm of the potentiometer wire. If the cell is shunted by a resistance of 5 Ω, a balance is found when the cell is connected across 40 cm of the wire. Find the internal resistance of the cell. [2018]
    (A) 1 Ω (B) 1.5 Ω (C) 2 Ω (D) 2.5 Ω

41. On interchanging the resistances, the balance point of a meter bridge shifts to the left by 10 cm. The resistance of their series combination is 1 $k\Omega$. How much was the resistance on the left slot before interchanging the resistances? [2018]
    (A) 990 Ω (B) 505 Ω (C) 550 Ω (D) 910 Ω

42. Space between two concentric conducting spheres of radii a and $b (b > a)$ is filled with a medium of resistivity ρ. The resistance between the two spheres will be [2019]
    (A) $\dfrac{\rho}{2\pi}\left(\dfrac{1}{a} + \dfrac{1}{b}\right)$ (B) $\dfrac{\rho}{4\pi}\left(\dfrac{1}{a} - \dfrac{1}{b}\right)$
    (C) $\dfrac{\rho}{2\pi}\left(\dfrac{1}{a} - \dfrac{1}{b}\right)$ (D) $\dfrac{\rho}{4\pi}\left(\dfrac{1}{a} + \dfrac{1}{b}\right)$

43. In an experiment, the resistance of a material is plotted as a function of temperature (in some range). As shown in the Fig.4.385, it is a straight line. One may conclude that [2019]
    (A) $R(T) = R_0 e^{-T^2/T_0^2}$ (B) $R(T) = R_0 e^{T^2/T_0^2}$
    (C) $R(T) = R_0 e^{-T_0^2/T^2}$ (D) $R(T) = \dfrac{R_0}{T^2}$

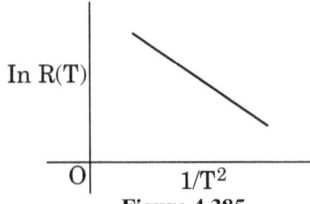

Figure 4.385

44. A metal wire of resistance 3Ω is elongated to make a uniform wire of double its previous length. This new wire is now bent and the ends joined to make a circle. If two points on this circle make an angle 60° at the centre, the equivalent resistance between these two points will be [2019]
    (A) $\dfrac{7}{2}\Omega$ (B) $\dfrac{5}{2}\Omega$ (C) $\dfrac{12}{5}\Omega$ (D) $\dfrac{5}{3}\Omega$

45. In a conductor, if the number of conduction electrons per unit volume is $8.5 \times 10^{28} m^{-3}$ and mean free time is 25 fs (femto second), it's approximate resistivity is (Take, $m_e = 9.1 \times 10^{-31}$ kg) [2019]
    (A) $10^{-7}\Omega.m$ (B) $10^{-5}\Omega.m$
    (C) $10^{-6}\Omega.m$ (D) $10^{-8}\Omega.m$

46. Determine the charge on the capacitor in the circuit shown in Fig.4.386 [2019]
    (A) $2\mu C$ (B) $200\mu C$ (C) $10\mu C$ (D) $60\mu C$

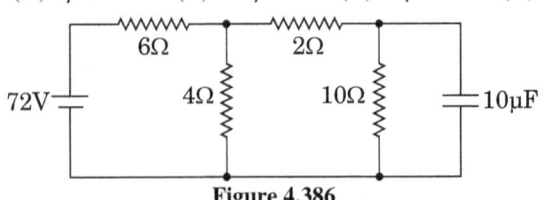

Figure 4.386

47. A wire of resistance R is bent to form a square $ABCD$ as shown in the Fig.4.387. The effective resistance between E and C is [E is mid-point of arm CD] [2019]
    (A) $\dfrac{7}{64}R$ (B) $\dfrac{3}{4}R$ (C) R (D) $\dfrac{1}{16}R$

48. A uniform metallic wire has a resistance of 18Ω and is bent into an equilateral triangle. Then, the resistance between

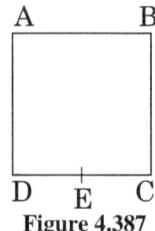

Figure 4.387

any two vertices of the triangle is [2019]
    (A) 12 Ω (B) 8 Ω (C) 2 Ω (D) 4 Ω

49. A copper wire is stretched to make it 0.5% longer. The percentage change in its electrical resistance, if its volume remains unchanged is [2019]
    (A) 2.0% (B) 1.0% (C) 0.5% (D) 2.5%

50. Mobility of electrons in a semiconductor is defined as the ratio of their drift velocity to the applied electric field. If for an n-type semiconductor, the density of electrons is $10^{19} m^{-3}$ and their mobility is $1.6 m^2 (V.s)$, then the resistivity of the semiconductor (since, it is an n-type semiconductor contribution of holes is ignored) is close to [2019]
    (A) 2Ω.m (B) 0.2Ω.m (C) 0.4Ω.m (D) 4Ω.m

51. Drift speed of electrons, when 1.5 A of current flows in a copper wire of cross-section 5 mm^2 is v. If the electron density in copper is $9 \times 10^{28}/m^3$, the value of v in mm/s is close to (Take, charge of electron to be $= 1.6 \times 10^{-19}$C) [2019]
    (A) 0.02 (B) 0.2 (C) 2 (D) 3

52. In the Fig.4.388 shown, what is the current (in ampere) drawn from the battery? You are given: $R_1 = 15$ Ω, $R_2 = 10$ Ω, $R_3 = 20$ Ω, $R_4 = 5$ Ω, $R_5 = 25$ Ω, $R_6 = 30$ Ω, $\mathcal{E} = 15$ V [2019]
    (A) 13/24 (B) 7/18 (C) 20/3 (D) 9/32

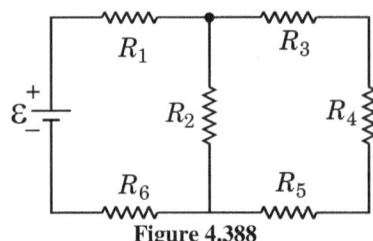

Figure 4.388

53. For the circuit shown with $R_1 = 10$ Ω, $R_2 = 2.0$ Ω, $\mathcal{E}_1 = 2$ V and $\mathcal{E}_2 = \mathcal{E}_3 = 4$ V [Fig.4.389], the potential difference between the points a and b is approximately (in volt) [2019]
    (A) 2.7 (B) 2.3 (C) 3.7 (D) 3.3

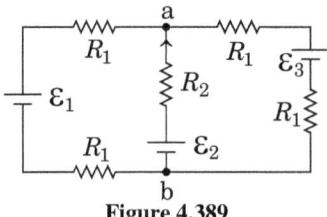

Figure 4.389

54. In the circuit shown (Fig.4.390), the potential difference between A and B is [2019]
    (A) 3 V (B) 1 V (C) 6 V (D) 2 V

55. In a Wheatstone bridge (Fig.4.391), resistances P and Q are approximately equal. When $R = 400$ Ω, the bridge is balanced. On interchanging P and Q, the value of R for balance is 405 Ω. The value of X is close to [2019]
    (A) 404.5Ω (B) 401.5Ω (C) 402.5Ω (D) 403.5Ω

56. In the given circuit diagram (Fig.4.392), the currents

4.34. MULTIPLE CHOICE ASSIGNMENTS

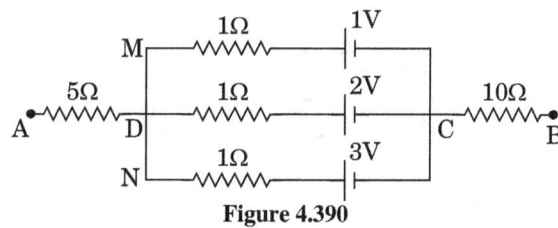

Figure 4.390

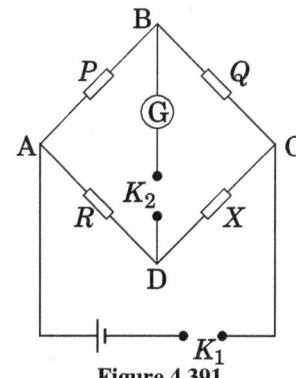

Figure 4.391

$I_1 = -0.3$ A, $I_4 = 0.8$ A and $I_5 = 0.4$ A, are flowing as shown. The currents I_2, I_3 and I_6 respectively, are [2019]
(A) 1.1A, 0.4A, 0.4A
(B) 1.1 A, −0.4 A, 0.4 A
(C) 0.4 A, 1.1 A, 0.4 A
(D) −0.4 A, 0.4 A, 1.1 A

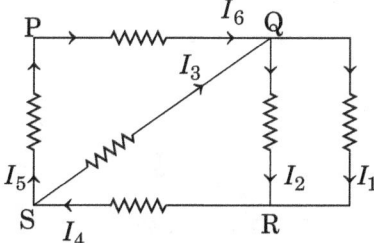

Figure 4.392

57. In the given circuit (Fig.4.393), the cells have zero internal resistance. The currents (in ampere) passing through resistances R_1 and R_2 respectively are [2019]
(A) 0.5, 0
(B) 1, 2
(C) 2, 2
(D) 0, 1

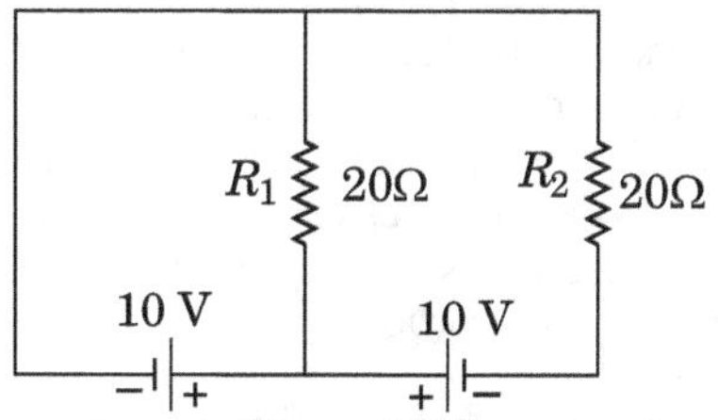

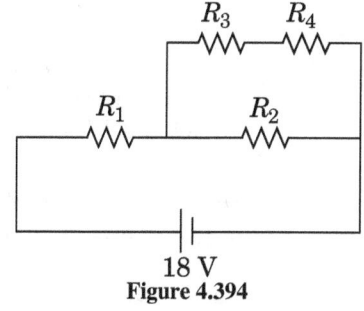

Figure 4.393

58. In the given circuit (Fig.4.394), the internal resistance of the 18 V cell is negligible. If $R_1 = 400$ Ω, $R_3 = 100$ Ω and $R_4 = 500$ Ω and the reading of an ideal voltmeter across R_4 is 5 V, then the value of R_2 will be [2019]
(A) 550 Ω
(B) 230 Ω
(C) 300 Ω
(D) 450 Ω

59. When the switch S in the circuit shown (Fig.4.395) is closed, then the value of current I will be - [2019]
(A) 4 A
(B) 3 A
(C) 2 A
(D) 5 A

60. One kilogram of water at $20°C$ is heated in an electric kettle whose heating element has a mean (temperature averaged) resistance of 20 Ω. The rms voltage in the mains is 200 V. Ignoring heat loss from the kettle, time taken for water to evaporate fully, is close to [Specific heat of water

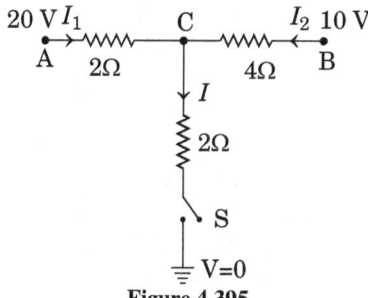

Figure 4.394

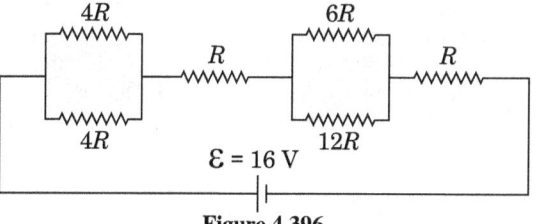

Figure 4.395

$= 4200$ J/(kg°C), Latent heat of water $= 2260$ kJ/kg]
[2019]
(A) 16min
(B) 22min
(C) 3min
(D) 10min

61. The resistive network shown in Fig.4.396, is connected to a DC source of 16 V. The power consumed by the network is 4 W. The value of R is - [2019]
(A) 6 Ω
(B) 8 Ω
(C) 1 Ω
(D) 16 Ω

Figure 4.396

62. A cell of internal resistance r drives current through an external resistance R. The power delivered by the cell to the external resistance will be maximum when [2019]
(A) $R = 2r$
(B) $R = r$
(C) $R = 0.001\,r$
(D) $R = 1000\,r$

63. Two electric bulbs rated at 25 W - 220 V and 100 W - 220 V are connected in series across a 220 V voltage source. If the 25 W and 100W bulbs draw powers P_1 and P_2 respectively, then [2019]
(A) $P_1 = 16$ W, $P_2 = 4W$
(B) $P_1 = 4$ W, $P_2 = 16$ W
(C) $P_1 = 9$ W, $P_2 = 16$ W
(D) $P_1 = 16$ W, $P_2 = 9$ W

64. Two equal resistances when connected in series to a battery consume electric power of 60W. If these resistances are now connected in parallel combination to the same battery, the electric power consumed will be [2019]
(A) 60 W
(B) 30 W
(C) 240 W
(D) 120 W

65. A current of 2 mA was passed through an unknown resistor which dissipated a power of 4.4 W. Dissipated power when an ideal power supply of 11 V is connected across it is [2019]
(A) 11×10^{-4}W
(B) 11×10^{-5}W
(C) 11×10^{5}W
(D) 11×10^{-3}W

66. In a meter bridge experiment, the circuit diagram is shown in Fig.4.397. It's observation table is is given below. Which of the readings is inconsistent? [2019]

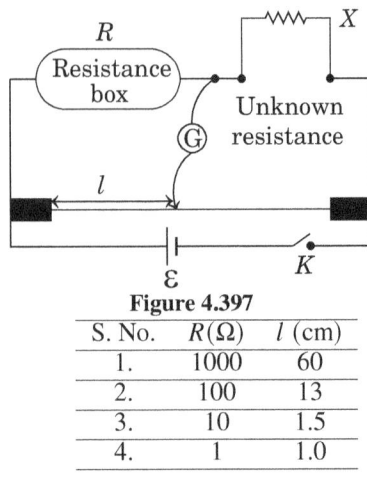

Figure 4.397

S. No.	R (Ω)	l (cm)
1.	1000	60
2.	100	13
3.	10	1.5
4.	1	1.0

(A) 3　　(B) 2　　(C) 1　　(D) 4

67. A galvanometer whose resistance is 50 Ω, has 25 divisions in it. When a current of 4×10^{-4} A passes through it, its needle (pointer) deflects by one division. To use this galvanometer as a voltmeter of range 2.5 V, it should be connected to a resistance of　　　[2019]
(A) 250 Ω　(B) 6200 Ω　(C) 200 Ω　(D) 6250 Ω

68. An ideal battery of 4V and resistance R are connected in series in the primary circuit of a potentiometer of length 1m and resistance 5Ω. The value of R to give a potential difference of 5 mV across 10 cm of potentiometer wire is　[2019]
(A) 395 Ω　(B) 495 Ω　(C) 490 Ω　(D) 480 Ω

69. The galvanometer deflection, when key K_1 is closed but K_2 is open equals θ_0 (Fig.4.398). On closing K_2 also and adjusting R_2 to 5Ω, the deflection in galvanometer becomes $\theta_0/5$. The resistance of the galvanometer is given by (neglect the internal resistance of battery):　[2019]
(A) 22 Ω　(B) 5 Ω　(C) 25 Ω　(D) 12 Ω

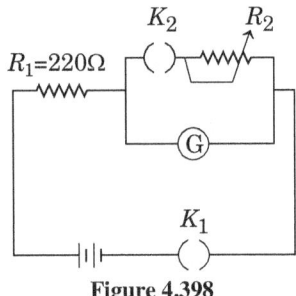

Figure 4.398

70. In the experimental set up of meter bridge shown in the Fig.4.399, the null point is obtained at a distance of 40 cm from A. If a 10 Ω resistor is connected in series with R_1, the null point shifts by 10 cm. The resistance that should be connected in parallel with $(R_1 + 10)$ Ω such that the null point shifts back to its initial position, is [2019]
(A) 60 Ω　(B) 20 Ω　(C) 30 Ω　(D) 40 Ω

71. A galvanometer having a resistance of 20 Ω and 30 divisions on both sides has figure of merit 0.005 ampere/division. The resistance that should be connected in series such that it can be used as a voltmeter up to 15 volt is　[2019]
(A) 100 Ω　(B) 125 Ω　(C) 120 Ω　(D) 80 Ω

72. The resistance of the meter bridge AB in given Fig.4.400 is 4 Ω. With a cell of emf $\mathcal{E} = 0.5V$ and rheostat resistance $R_k = 2$ Ω. The null point is obtained at some point J. When the cell is replaced by another one of emf $\mathcal{E} = \mathcal{E}_2$, the same null point J is found for $R_k = 6$ Ω. The emf ε_2 is　[2019]
(A) 0.6 V　(B) 0.3 V　(C) 0.5 V　(D) 0.4 V

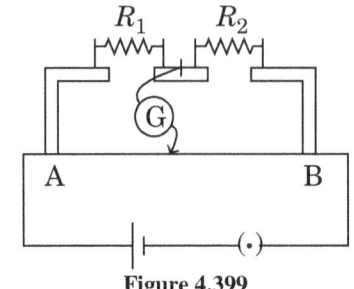

Figure 4.399

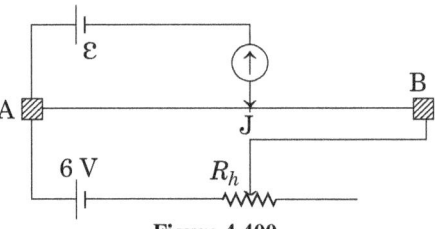

Figure 4.400

73. A potentiometer wire AB having length L and resistance 12 r is joined to a cell D of emf $\mathcal{E}$ and internal resistance r. A cell C having emf $\mathcal{E}/2$ and internal resistance 3r is connected. The length AJ at which the galvanometer as shown in Fig.4.401 shows no deflection is　[2019]
(A) $\frac{5}{12}L$　(B) $\frac{11}{12}L$　(C) $\frac{13}{24}L$　(D) $\frac{11}{24}L$

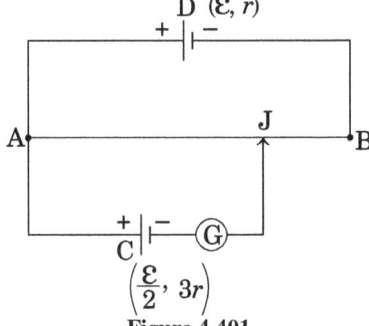

Figure 4.401

74. A 200 Ω resistor has a certain colour code. If one replaces the red colour by green in the code, the new resistance will be　　　[2019]
(A) 100 Ω　(B) 400 Ω　(C) 300 Ω　(D) 500 Ω

75. The Wheatstone bridge shown in Fig.4.402 here, gets balanced when the carbon resistor is used as R_1 has the color code (orange, red, brown). The resistors R_2 and R_4 are 80Ω and 40Ω, respectively. Assuming that the colour code for the carbon resistors gives their accurate values, the colour code for the carbon resistor is used as R_3 would be　　　[2019]
(A) brown, blue, black　　(B) brown, blue, brown
(C) grey, black, brown　　(D) red, green, brown

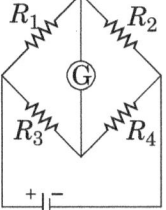

Figure 4.402

76. The actual value of resistance R, shown in the Fig.4.403 is 30 Ω. This is measured in an experiment as shown using the standard formula $R = V/I$, where V and I are the readings

of the voltmeter and ammeter, respectively. If the measured value of R is 5% less, then the internal resistance of the voltmeter is [2019]

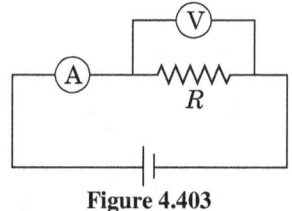

Figure 4.403

(A) 600 Ω (B) 570 Ω (C) 350 Ω (D) 35 Ω

77. A 2 W carbon resistor is colour coded with green, black, red and brown respectively. The maximum current which can be passed through this resistor is [2019]
    (A) 0.4 mA (B) 63 mA (C) 20 mA (D) 100 mA

78. A resistance is shown in the Fig.4.404. Its value and tolerance are given respectively by [2019]
    (A) 270 Ω, 5% (B) 27 kΩ, 20%
    (C) 27 kΩ, 10% (D) 270 kΩ, 10%

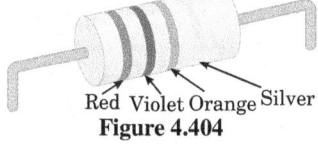

Figure 4.404

79. The resistance of a galvanometer is 50 ohm and the maximum current which can be passed through it is 0.002 A. What resistance must be connected to it in order to convert it into an ammeter of range 0 – 0.5 A? [2019]
    (A) 0.2 Ω (B) 0.5 Ω (C) 0.002 Ω (D) 0.02 Ω

80. A moving coil galvanometer has resistance 50 Ω and it indicates full deflection at 4 mA current. A voltmeter is made using this galvanometer and a 5 kΩ resistance. The maximum voltage, that can be measured using this voltmeter, will be close to [2019]
    (A) 40 V (B) 10 V (C) 15 V (D) 20 V

81. A moving coil galvanometer, having a resistance G, produces full scale deflection when a current I_g flows through it. This galvanometer can be converted into (i) an ammeter of range 0 to I_0 ($I_0 > I_g$) by connecting a shunt resistance R_A to it and (ii) into a voltmeter of range 0 to V ($V = GI_0$) by connecting a series resistance R_V to it. Then, [2019]

    (A) $R_A R_V = G^2 \left(\frac{I_0 - I_g}{I_g}\right)$ and $\frac{R_A}{R_V} = \left(\frac{I_g}{(I_0 - I_g)}\right)^2$

    (B) $R_A R_V = G^2$ and $\frac{R_A}{R_V} = \left(\frac{I_g}{I_0 - I_g}\right)^2$

    (C) $R_A R_V = G^2 \left(\frac{I_g}{I_0 - I_g}\right)$ and $\frac{R_A}{R_V} = \left(\frac{I_0 - I_g}{I_g}\right)^2$

    (D) $R_A R_V = G^2$ and $\frac{R_A}{R_V} = \frac{I_g}{(I_0 - I_g)}$

82. A galvanometer of resistance 100 Ω has 50 divisions on its scale and has sensitivity of 20 µA/ division. It is to be converted to a voltmeter with three ranges of 0 - 2 V, 0 - 10 V and 0 - 20 V. In Fig.4.405, the appropriate circuit to do so is [2019]

83. A torch battery of length l is to be made up of a thin cylindrical bar of radius a and a concentric thin cylindrical shell of radius b is filled in between with an electrolyte of resistivity ρ (Fig.4.406). If the battery is connected to a resistance R, the maximum joule's heating in R will takes place for [2020]

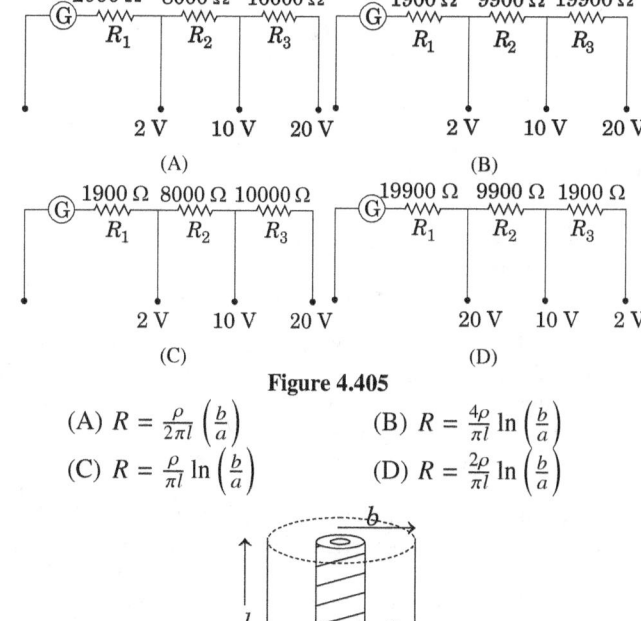

Figure 4.405

(A) $R = \frac{\rho}{2\pi l}\left(\frac{b}{a}\right)$ (B) $R = \frac{4\rho}{\pi l} \ln\left(\frac{b}{a}\right)$

(C) $R = \frac{\rho}{\pi l} \ln\left(\frac{b}{a}\right)$ (D) $R = \frac{2\rho}{\pi l} \ln\left(\frac{b}{a}\right)$

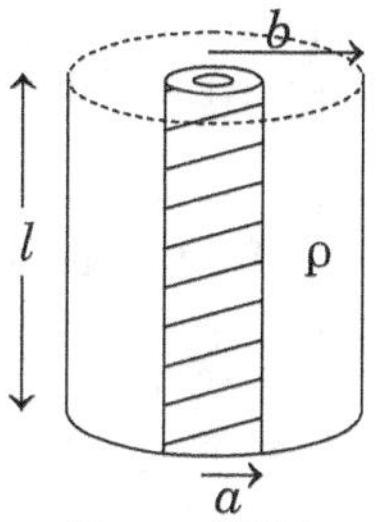

Figure 4.406

84. A current of 5 A passes through a copper conductor (resistivity = 1.7×10^{-8} Ωm) of radius of cross-section 5 mm. Find the mobility of the charges, if their drift velocity is 1.1×10^{-3} m/s. [2019]
    (A) 1.5 m²/V.s (B) 1.3 m²/V.s
    (C) 1.0 m²/V.s (D) 1.8 m²/V.s

85. A circuit to verify Ohm's law uses ammeter and voltmeter in series or parallel connected correctly to the resistor. In the circuit: [Sep. 06, 2020 (II)]
    (A) ammeter is always used in parallel and voltmeter is series
    (B) both ammeter and voltmeter must be connected in parallel
    (C) ammeter is always connected in series and voltmeter in parallel
    (D) both, ammeter and voltmeter must be connected in series

86. Consider four conducting materials copper, tungsten, mercury and aluminium with resistivity ρ_C, ρ_T, ρ_M and ρ_A respectively. Then: [Sep. 02, 2020 (I)]
    (A) $\rho_C > \rho_A > \rho_T$ (B) $\rho_M > \rho_A > \rho_C$
    (C) $\rho_A > \rho_T > \rho_C$ (D) $\rho_A > \rho_M > \rho_C$

87. In the given circuit diagram (Fig.4.407), a wire is joining points B and D. The current in this wire is:
    [||9 Jan. 2020, (I)]
    (A) 0.4 A (B) 2 A (C) 4 A (D) 0

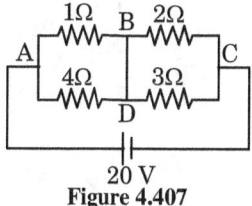

Figure 4.407

88. The current I_1 (in A) flowing through 1 Ω resistor in the circuit shown in Fig.4.408, is: [7 Jan. 2020 (I)]
    (A) 0.4 (B) 0.5 (C) 0.2 (D) 0.25

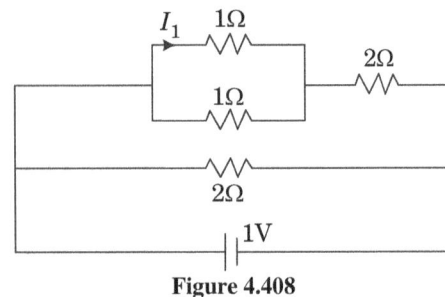

Figure 4.408

89. The series combination of two batteries, both of the same emf 10 V, but different internal resistance of 20 Ω and 5 Ω, is connected to the parallel combination of two resistors 30 Ω and R Ω. The voltage difference across the battery of internal resistance 20 Ω is zero, the value of R (in Ω) is [8 Jan. 2020, (II)]
    (A) 30.00   (B) 40.00   (C) 20.00   (D) 10.00

90. In the Fig.4.409, the current in the 10 V battery is close to: [06 Sep. 2020 (II)]
    (A) 0.71A from positive to negative terminal
    (B) 0.42A from positive to negative terminal
    (C) 0.21A from positive to negative terminal
    (D) 0.36A from negative to positive terminal

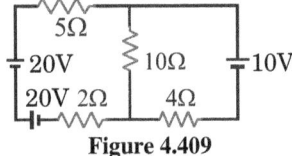

Figure 4.409

91. In the circuit given in the Fig.4.410, currents in different branches and value of one resistor are shown. Then potential at point B with respect to the point A is: [05 Sep. 2020 (II)]
    (A) +2 V   (B) −2 V   (C) −1 V   (D) +1 V

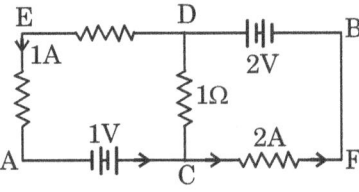

Figure 4.410

92. The value of current I_1 flowing from A to C in the circuit diagram (Fig.4.411), is: [04 Sep. 2020 (II)]
    (A) 2 A   (B) 4 A   (C) 1 A   (D) 5 A

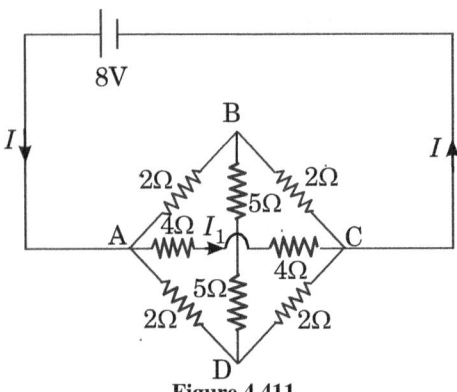

Figure 4.411

93. Four resistances 40 Ω, 60 Ω, 90 Ω and 110 Ω make the arms of a quadrilateral ABCD (Fig.4.412). Across AC is an ideal battery of emf 40 V. The potential difference across BD in V is [04 Sep. 2020 (II)]
    (A) 2   (B) 4   (C) 10   (D) 5

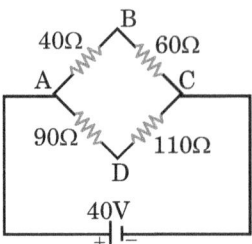

Figure 4.412

94. An ideal cell of emf 10 V is connected in circuit shown in Fig.4.413. Each resistance is 2 Ω. The potential difference (in volts) across the capacitor when it is fully charged, is [02 Sep. 2020 (II)]
    (A) 0.400   (B) 0.800   (C) 0.100   (D) 0.004

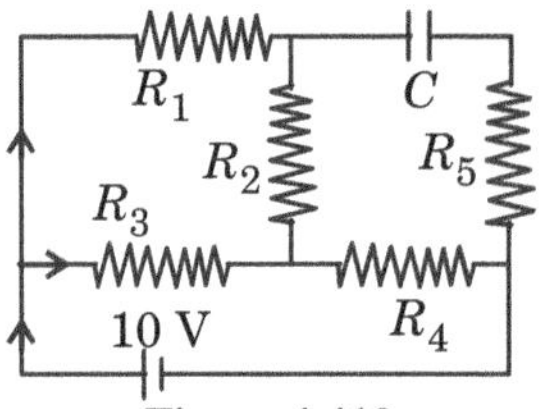

Figure 4.413

95. An electrical power line, having a total resistance of 2Ω, delivers 1 kW at 220 V. The efficiency of the transmission line is approximately: [05 Sep.2020 (I)]
    (A) 72%   (B) 91%   (C) 85%   (D) 96%

96. In a building there are 15 bulbs of 45 W, 15 bulbs of 100 W, 15 small fans of 10 W and 2 heaters of 1 kW. The voltage of electric main is 220 V. The minimum fuse capacity (rated value) of the building will be: [7 Jan. 2020 (II)]
    (A) 10 A   (B) 25 A   (C) 15 A   (D) 20 A

97. Two resistors 400 Ω and 800 Ω are connected in series across a 6 V battery. The potential difference measured by a voltmeter of 10 kΩ across 400 Ω resistor is close to: [03 Sep. 2020 (II)]
    (A) 2V   (B) 1.8V   (C) 2.05V   (D) 1.95V

98. Which of the following will not be observed when a multimeter (operating in resistance measuring mode) probes connected across a component, are just reversed? [03 Sep. 2020 (II)]
    (A) Multimeter shows an equal deflection in both cases i.e. before and after reversing the probes if the chosen component is resistor.
    (B) Multimeter shows NO deflection in both cases i.e. before and after reversing the probes if the chosen component is capacitor.
    (C) Multimeter shows a deflection, accompanied by a splash of light out of connected and NO deflection on reversing the probes if the chosen component is LED.
    (D) Multimeter shows NO deflection in both cases i.e. before and after reversing the probes if the chosen component is metal wire.

99. A potentiometer wire PQ of 1 m length is connected to a standard cell $\mathcal{E}_1$. Another cell $\mathcal{E}_2$ of emf 1.02 V is connected with a resistance 'r' and switch S (as shown in Fig.4.414). With switch S open, the null position is obtained at a distance of 49 cm from Q. The potential gradient in the potentiometer

wire is: [02 Sep.2020 (II)]
(A) 0.02 V/cm (B) 0.01 V/cm
(C) 0.03 V/cm (D) 0.04 V/cm

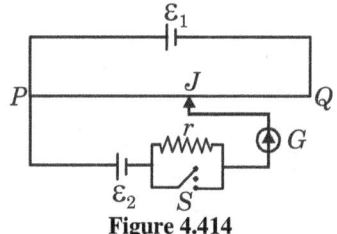

Figure 4.414

100. The length of a potentiometer wire is 1200 cm and it carries a current of 60 mA. For a cell of emf 5 V and internal resistance of 20 Ω, the null point on it is found to be at 1000 cm. The resistance of whole wire is:
[8 Jan. 2020 (I)]
(A) 80 Ω (B) 120 Ω (C) 60 Ω (D) 100 Ω

101. Four resistances of 15 Ω, 12 Ω, 4 Ω and 10 Ω respectively in cyclic order to form Wheatstone's network. The resistance that is to be connected in parallel with the resistance of 10 Ω (Fig.4.415) to balance the network is Ω. [8 Jan.2020 (I)]
(A) 10 Ω (B) 5 Ω (C) 15 Ω (D) 20 Ω

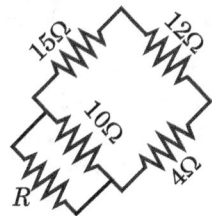

Figure 4.415

102. The balancing length for a cell is 560 cm in a potentiometer experiment. When an external resistance of 10Ω is connected in parallel to the cell, the balancing length changes by 60 cm. If the internal resistance of the cell is $\frac{N}{10}$Ω, where N is an integer then value of N is [7 Jan. 2020 (II)]
(A) 5 (B) 8 (C) 10 (D) 12

103. If you are provided a set of resistances 2 Ω, 4 Ω and 8 Ω. Connect these resistances so as to obtain an equivalent resistance of 46/3 Ω. [28 Aug. 2021 (II)]
(A) 2 Ω and 4 Ω are in parallel with 6 Ω and 8 Ω in series.
(B) 6 Ω and 8 Ω are in parallel with 2 Ω and 4 Ω in series.
(C) 2 Ω and 6 Ω are in parallel with 4 Ω and 8 Ω in.series.
(D) 4 Ω and 6 Ω are in parallel with 2 Ω and 8 Ω in series.

104. Calculate the amount of charge on capacitor of 4 μF (Fig.4.416). The internal resistance of battery is 1 Ω:
[27 Aug. 2021]
(A) 0C (B) 8 μC (C) 16 μC (D) 4 μC

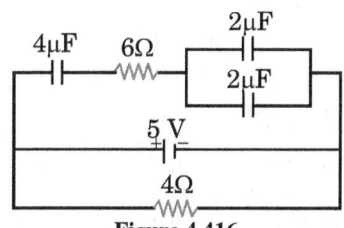

Figure 4.416

105. The colour coding on a carbon resistor is shown in the given Fig.4.417. The resistance value of the given resistor is:
[27 Aug. 2021 (II)]
(A) (7500 ± 375)Ω (B) (5700 ± 285)Ω
(C) (7500 ± 755)Ω (D) (5700 ± 375)Ω

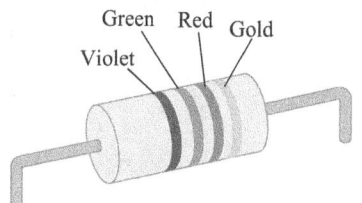

Figure 4.417

106. The ratio of the equivalent resistance of the network (Fig.4.418) between the points A and B when switch is open and switch is closed is x : 8. The value of x is
[27 Aug. 2021 (II)]
(A) 5 (B) 9 (C) 7 (D) 6

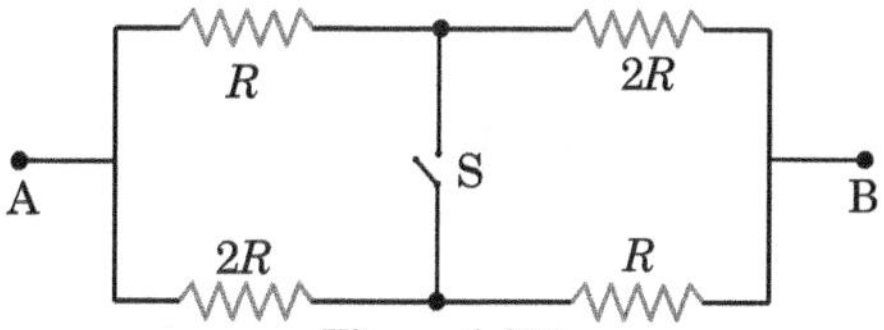

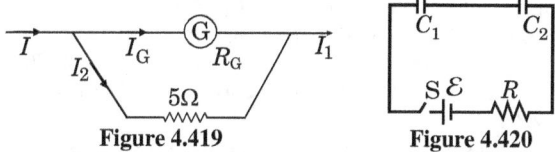

Figure 4.418

107. Consider a galvanometer shunted with 5 Ω resistance and 2% of current passes through it (Fig.4.419). What is the resistance of the given galvanometer? [27 Aug. 2021 (II)]
(A) 226 Ω (B) 245 Ω (C) 344 Ω (D) 300 Ω

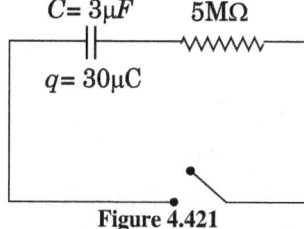

Figure 4.419 Figure 4.420

108. In the circuit of Fig.4.420, the switch S is closed at $t = 0$. The charge on the capacitor C_1 as a function of time will be given by $\left(C_{eq} = \frac{C_1 C_2}{C_1 + C_2}\right)$. [April 16, 2018]
(A) $C_{eq}\mathcal{E}\left[1 - e^{-t/RC_{eq}}\right]$ (B) $C_1\mathcal{E}\left[1 - e^{-tR/C_1}\right]$
(C) $C_2\mathcal{E}\left[1 - e^{-t/RC_2}\right]$ (D) $C_{eq}\mathcal{E}e^{-t/RC_{eq}}$

109. The circuit shown in the Figure 4.421 consists of a charged capacitor of capacity 3 μF and a charge of 30 μC. At time $t = 0$, when the key is closed, the value of current flowing through the 5 MΩ resistor is x μA. The value of x to the nearest integer is [18 March 2021]

C = 3μF   5MΩ

q = 30μC

Figure 4.421

110. A capacitor of capacitance $C = 1$ μF is suddenly connected to a battery of 100 V through a resistance $R = 100$ Ω. The time taken for the capacitor to be charged to get 50 V is [Take, ln 2 = 0.69] [27 July 2021]
(A) 1.44×10^{-4} s (B) 3.33×10^{-4} s
(C) 0.69×10^{-4} s (D) 0.30×10^{-4} s

111. Calculate the amount of charge on capacitor of 4 μF [Fig.4.422]. The internal resistance of battery is 1 Ω.
[26 August 2021]

(A) 8 μC (B) zero (C) 16 μC (D) 4 μC

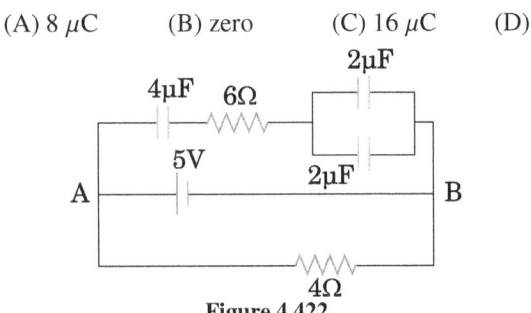

Figure 4.422

112. A capacitor of 50 μF is connected in a circuit as shown in Figure 4.423. The charge on the upper plate of the capacitor is ____ μC. [31 August 2021]

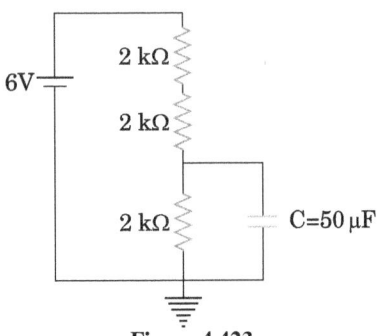

Figure 4.423

113. If equivalent of three resistances $R_1 = 2\,\Omega$, $R_2 = 4\,\Omega$ and $R_3 = 6\,\Omega$ is $22/3\,\Omega$, find their arrangement.
[24 June 2022]
(A) R_1, R_2 in parallel and their equivalent in series with R_3.
(B) R_2, R_3 in parallel and their equivalent in series with R_1.
(C) R_1, R_3 in parallel and their equivalent in series with R_2.
(D) All are connected in parallel

114. What is the current through battery (Fig.4.424)?
[27 July 2022]
(A) 15 A (B) 20 A (C) 10 A (D) 25 A

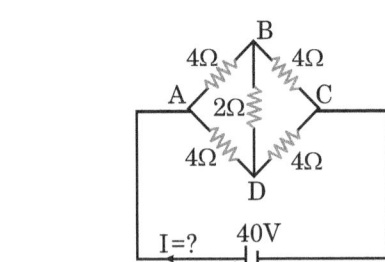

Figure 4.424

115. A meter bridge is shown in Fig.4.425. If a resistance $x\,\Omega$ is connected in series with $4r$, new null point comes at 80 cm, find x. [26 July 2022 (I)]

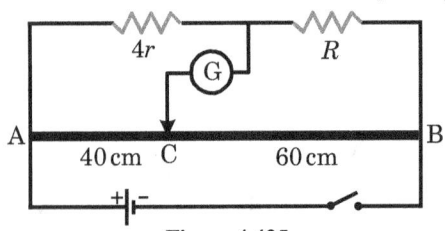

Figure 4.425

(A) 30 (B) 20 (C) 10 (D) 40

116. Find current I in circuit given in Fig.4.426.
[26 July 2022 (II)]

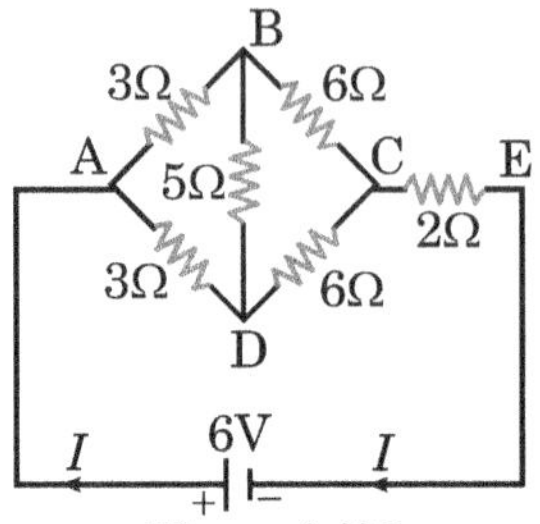

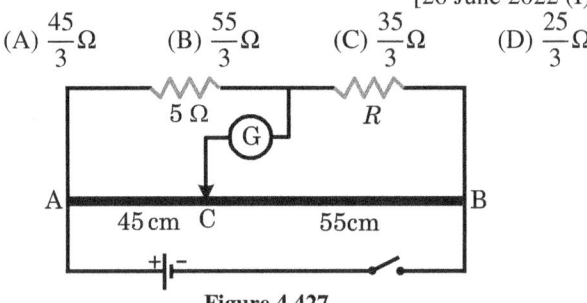

Figure 4.426

(A) 0.723 A (B) 0.523 A (C) 0.923 A (D) 0.623 A

117. In the given meter bridge(Fig.4.427), find the value of R.
[26 June 2022 (I)]
(A) $\dfrac{45}{3}\Omega$ (B) $\dfrac{55}{3}\Omega$ (C) $\dfrac{35}{3}\Omega$ (D) $\dfrac{25}{3}\Omega$

Figure 4.427

118. **Assertion:** Resistance of 80 Ω is cut equally in 4 parts and all resistances are connected in parallel. The net resistance is 5 Ω
Reason: When $2R$ and $3R$ connected in parallel, ratio of heat dissipated in them is 3 : 2 correct statement is/are:
[28 July 2022 (II)]
(A) If both assertion and reason are true and the reason is the correct explanation of the assertion.
(B) If both assertion and reason are true, but the reason is not the correct explanation of the assertion.
(C) If assertion is true, but reason is false.
(D) If both the assertion and reason are true.

119. Two Identical cell give same current across external resistance R, when they are in series combination and when they are in parallel combination. Find internal resistance of cell? [29 June 2022 (II)]
(A) R (B) $3R$ (C) $R/2$ (D) $5R$

120. The value of current I in the circuit of Fig.4.428, is.
[29 July 2022 (I)]

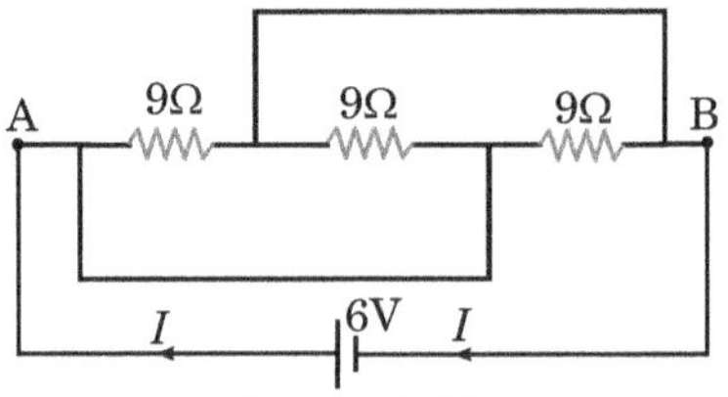

Figure 4.428

(A) 2 A (B) 3 A (C) 4 A (D) 5 A

121. A student is provided with a variable voltage source V, a test resistor $R_T = 10\,\Omega$, two identical galvanometers G_1 and G_2 and two additional resistors, $R_1 = 10\,M\Omega$ and $R_2 = 0.001\,\Omega$. For conducting an experiment to verify ohm's law, the most suitable circuit in Fig.4.429, is: [6 April 2023 (II)]

122. Figure 4.430 shows a part of an electric circuit. The potentials at points A, B and C are 30 V , 12 V and 2 V respectively. The current through the 20 Ω resistor will be,

4.34. MULTIPLE CHOICE ASSIGNMENTS

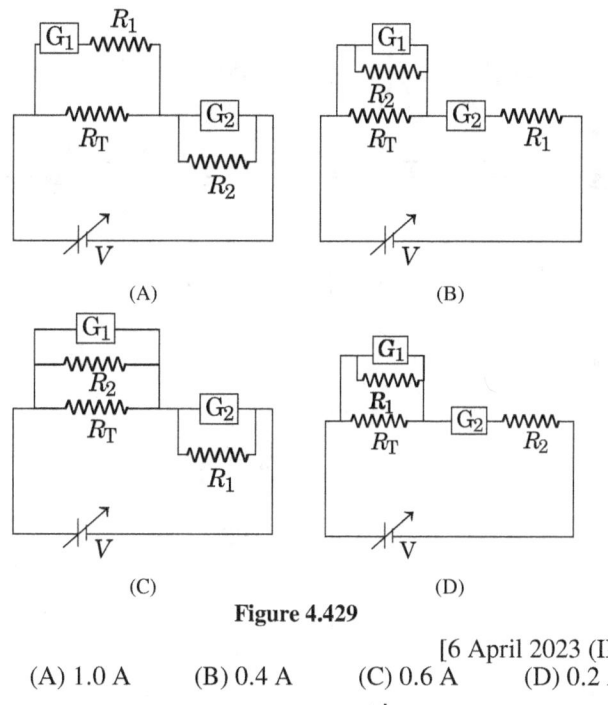

Figure 4.429

[6 April 2023 (II)]
(A) 1.0 A (B) 0.4 A (C) 0.6 A (D) 0.2 A

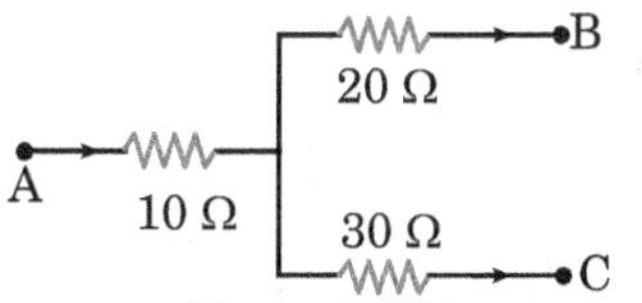

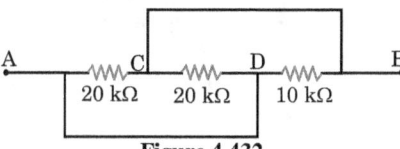

Figure 4.430 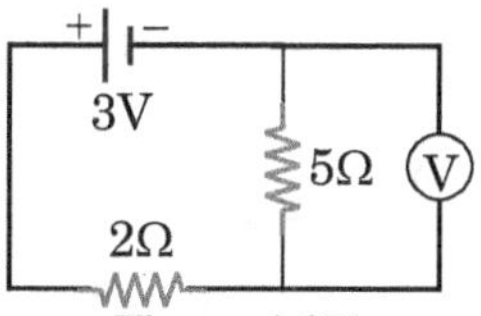 Figure 4.431

123. As shown in the Fig.4.431 the voltmeter reads 2 V across 5 Ω resistor. The resistance of the voltmeter is [6 April 2023 (II)]
(A) 20 Ω (B) 40 Ω (C) 30 Ω (D) 10 Ω

124. Two resistances are given as $R_1 = (10 \pm 0.5)\Omega$ and $R_2 = (15 \pm 0.5)\Omega$. The percentage error in the measurement of equivalent resistance when they are connected in parallel is [6 April 2023 (I)]
(A) 6.33 (B) 2.33 (C) 5.33 (D) 4.33

125. The equivalent resistance between A and B as shown in Fig.4.432 is: [8 April 2023 (II)]
(A) 10 kΩ (B) 5 kΩ (C) 20 kΩ (D) 30 kΩ

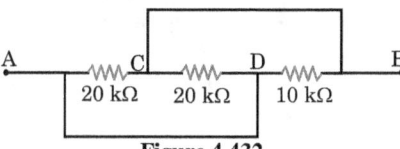

Figure 4.432

126. In Fig.4.433, the resistance of the coil of galvanometer G is 2 Ω. The emf of the cell is 4 V. The ratio of potential difference across C_1 and C_2 is: [8 April 2023 (I)]
(A) 1 (B) 4/5 (C) 5/4 (D) 3/4

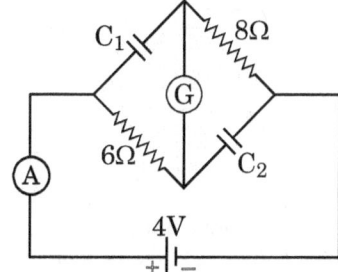

Figure 4.433

127. The equivalent resistance of the circuit shown in Fig.4.434 between points A and B is: [10 April 2023 (I)]
(A) 1.6 Ω (B) 3.2 Ω (C) 2.4 Ω (D) 2.0 Ω

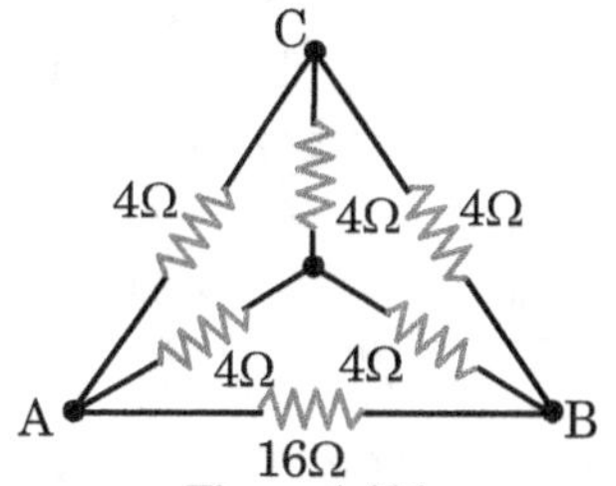

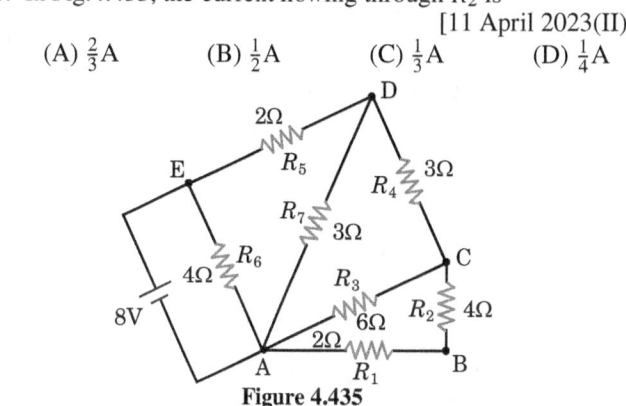

Figure 4.434

128. In Fig.4.435, the current flowing through R_2 is- [11 April 2023(II)]
(A) $\frac{2}{3}$A (B) $\frac{1}{2}$A (C) $\frac{1}{3}$A (D) $\frac{1}{4}$A

Figure 4.435

129. Two identical cells each of emf 1.5 V are connected in series across a 10 Ω resistance (Fig.4.436). An ideal voltmeter connected across 10 Ω resistance reads 1.5 V. The internal resistance of each cell is [11 April 2023 (II)]
(A) 10 Ω (B) 5 Ω (C) 15 Ω (D) 20 Ω

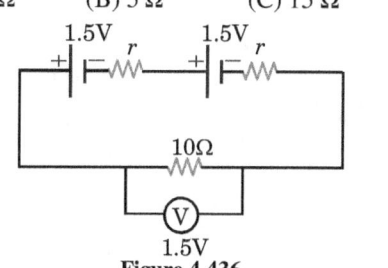

Figure 4.436

130. In the network shown in Fig.4.437, the charge accumulated in the capacitor in steady state will be: [13 April 2023 (II)]
(A) 10.3 μC (B) 4.8 μC (C) 12 μC (D) 7.2 μC

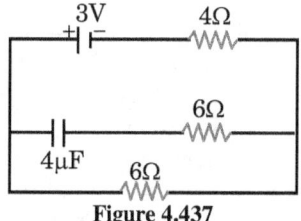

Figure 4.437

131. In the circuit shown in Fig.4.438, the energy stored in the capacitor is n μJ. The value of n is [13 April 2023(II)]
(A) 80 (B) 75 (C) 70 (D) 85

132. A potential V_0 is applied across a uniform wire of resistance R. The power dissipation is P_1. The wire is then cut into two equal halves and a potential of V_0 is applied across the length of each half. The total power dissipation across two wires is P_2. The ratio of $P_2 : P_1$ is $\sqrt{x} : 1$. The value of x is [13 April 2023 (I)]

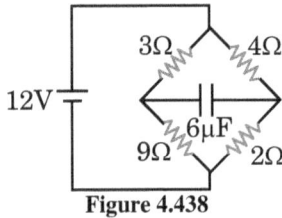

Figure 4.438

(A) 16   (B) 20   (C) 14   (D) 15

133. When a resistance of 5 Ω is shunted with a moving coil galvanometer, it shows a full scale deflection for a current of 250 mA, however when 1050 Ω resistance is connected with it in series, it gives full scale deflection for 25 volt. The resistance of galvanometer is [13 April 2023 (I)]
    (A) 30 Ω   (B) 50 Ω   (C) 40 Ω   (D) 50 Ω

134. For designing a voltmeter of range 50 V and an ammeter of range 10 mA using a galvanometer which has a coil of resistance 54 Ω showing a full scale deflection for 1 mA as in Fig.4.439. [15 April 2023 (I)]

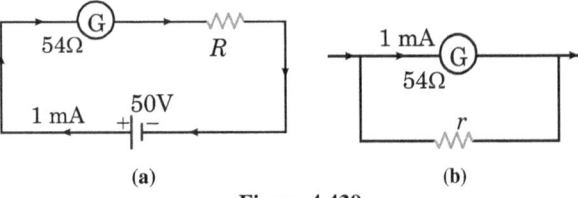

Figure 4.439

(1) for voltmeter $R \approx 50\ k\Omega$
(2) for ammeter $r \approx 0.2\ \Omega$
(3) for ammeter $r \approx 6\Omega$
(4) for voltmeter $R \approx 5\ k\Omega$
(5) for voltmeter $R \approx 500\ \Omega$

Choose the correct answer from the options given below:
(A) (1) and (3)   (B) (3) and (5)
(C) (3) and (4)   (D) (1) and (2)

135. A network of four resistances is connected to 9 V battery, as shown in Fig.4.440. The magnitude of voltage difference between the points A and B is [15 April 2023 (I)]
    (A) 4 V   (B) 6 V   (C) 3 V   (D) 0 V

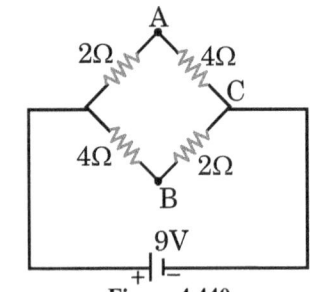

Figure 4.440

136. A meter bridge setup is shown in the Fig.4.441. It is used to determine an unknown resistance R using a given resistor of 15 Ω. The galvanometer (G) shows null deflection when tapping key is at 43 cm mark from end A. If the end correction for end A is 2 cm. then the determined value of R will be ____ Ω. [28 June 22 Shift I]

137. Current measured by the ammeter (A) in the reported circuit when no current flows through 10 Ω resistance. will be ____A. [28 June 22 Shift I]

138. All resistances in Fig.4.443 are 1Ω each. The value of current 'I' is $\frac{a}{5}$ A. The value of a is ____. [28 June 22 Shift II]

139. Two coils require 20 minutes and 60 minutes respectively

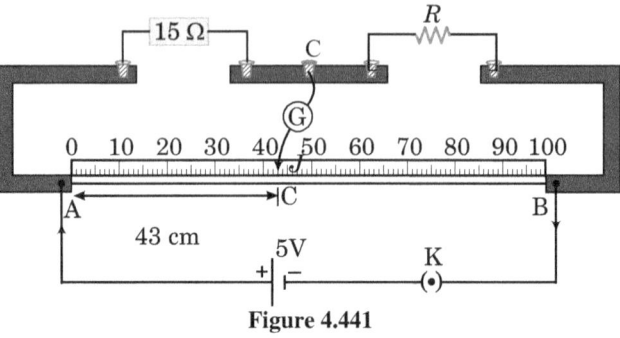

Figure 4.441

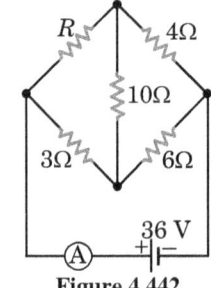

Figure 4.442

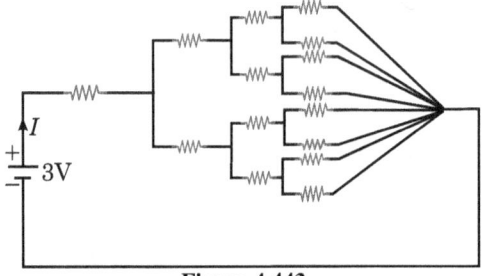

Figure 4.443

to produce same amount of heat energy when connected separately to the same source. If they are connected in parallel arrangement to the same source; the time required to produce same amount of heat by the combination of coils, will be ____ min. [29 June 22 Shift I]

140. The variation of applied potential and current flowing through a given wire is shown in Fig.4.444. The length of wire is 31.4 cm. The diameter of wire is measured as 2.4 cm. The resistivity of the given wire is measured as $x \times 10^{-3}$ Ωcm. The value of x is ____. [Take $\pi = 3.14$] [29 June 22 Shift I]

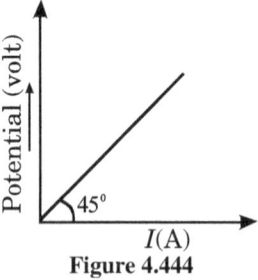

Figure 4.444

141. For the network shown in Fig.4.445, the value $V_B - V_A$ is ____ V. [29 June 22 Shift I]

142. A capacitor is discharging through a resistor R. Consider in time t_1, the energy stored in the capacitor reduces to half of its initial value and in time t_2, the charge stored reduces to one eighth of its initial value. The ratio t_1/t_2 will be: [29 June 22 Shift II]

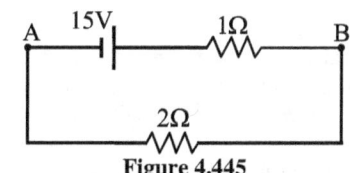
Figure 4.445

(A) 1/2 (B) 1/3 (C) 1/4 (D) 1/6

143. The combination of two identical cells, whether connected in series or parallel combination provides the same current through an external resistance of 2 Ω. The value of internal resistance of each cell is: [29 June 22 Shift II]
    (A) 2 Ω (B) 4 Ω (C) 6 Ω (D) 8 Ω

144. Two resistors are connected in series across a battery as shown in Fig.4.446. If a voltmeter of resistance 2000 Ω is used to measure the potential difference across 500 Ω resister, the reading of the voltmeter will be ____ V.
    [29 June 22 Shift II]

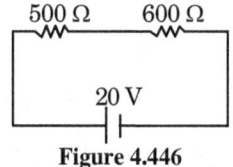
Figure 4.446

145. As shown in the Fig.4.447, a network of resistors is connected to a battery of 24 V with an internal resistance of 3Ω. The currents through the resistors R_4 and R_5 are I_4 and I_5 respectively. The values of I_4 and I_5 are:
    [24 Jan 2023 Shift I]
    (A) $I_4 = \frac{8}{5}A$ & $I_5 = \frac{2}{5}A$ (B) $I_4 = \frac{24}{5}A$ & $I_5 = \frac{6}{5}A$
    (C) $I_4 = \frac{6}{5}A$ & $I_5 = \frac{24}{5}A$ (D) $I_4 = \frac{2}{5}A$ & $I_5 = \frac{8}{5}A$

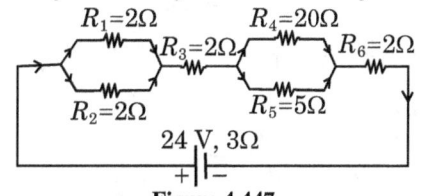
Figure 4.447

146. A hollow cylindrical conductor has length of 3.14 m, while its inner and outer diameters are 4 mm and 8 mm respectively. The resistance of the conductor is $n \times 10^{-3}\Omega$. If the resistivity of the material is $2.4 \times 10^{-8}\Omega$m. The value of n is ____. [24 Jan 2023 Shift I]

147. A cell of emf 90 V is connected across series combination of two resistors each of resistance of 100 Ω as shown in Fig.4.448. A voltmeter of resistance 400 Ω is used to measure the potential difference across each resistor. The reading of the voltmeter will be: [24 Jan 2023 Shift II]
    (A) 40 V (B) 45 V (C) 80 V (D) 90 V

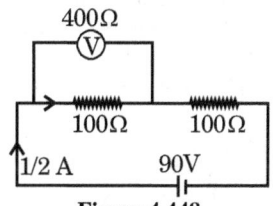
Figure 4.448

148. If a copper wire is stretched to increase its length by 20%. The percentage increase in resistance of the wire is ____ %.
    [24 Jan 2023 Shift II]

149. A uniform metallic wire carries a current 2 A. when 3.4 V battery is connected across it. The mass of uniform metallic wire is 8.92×10^{-3} kg. Density is 8.92×10^3 kg/m³ and resistivity is 1.7×10^{-8} Ω.m. The length of wire is:
    [25 Jan 2023 Shift I]
    (A) 6.8 m (B) 10 m (C) 5 m (D) 100 m

150. In the circuit shown in Fig.4.449, the equivalent resistance between the terminal A and B is ____ Ω.
    [25 Jan 2023 Shift I]

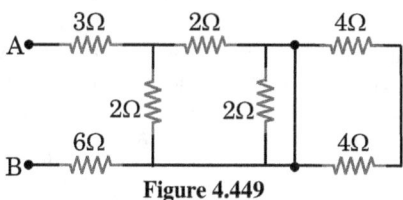
Figure 4.449

151. The resistance of a wire is 5 Ω. It's new resistance in ohm if stretched to 5 times of it's original length will be:
    [25 Jan 2023 Shift II]

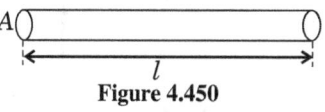
Figure 4.450

(A) 625 (B) 5 (C) 125 (D) 25

152. Two cells are connected between points A and B as shown in Fig.4.451. Cell 1 has emf of 12 V and internal resistance of 3 Ω. Cell 2 has emf of 6 V and internal resistance of 6 Ω. An external resistor R of 4 Ω is connected across A and B. The current flowing through R will be ____ A.
    [25 Jan 2023 Shift II]

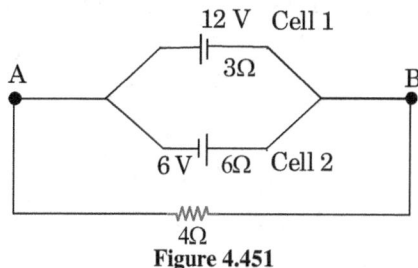
Figure 4.451

153. In a metre bridge experiment the balance point is obtained if the gaps are closed by 2 Ω and 3 Ω. A shunt of $X\Omega$ is added to 3 Ω resistor to shift the balancing point by 22.5 cm. The value of X is- [29 Jan 2023 Shift I]
    (A) 1 Ω (B) 2 Ω (C) 2.5 Ω (D) 5 Ω

154. With the help of potentiometer, we can determine the value of emf of a given cell. The sensitivity of the potentiometer is
    (a) directly proportional to the length of the potentiometer wire
    (b) directly proportional to the potential gradient of the wire
    (c) inversely proportional to the potential gradient of the wire
    (d) inversely proportional to the length of the potentiometer wire
    Choose the correct option for the above statements:
    [29 Jan 2023 Shift II]
    (A) a and d only (B) a and c only
    (C) a only (D) c only

155. When two resistance R_1 and R_2 connected in series and introduced into the left gap of a meter bridge and a resistance of 10 Ω is introduced into the right gap, a null point is found at 60 cm from left side. When R_1 and R_2 are connected in

parallel and introduced into the left gap, a resistance of 3 Ω is introduced into the right-gap to get null point at 40 cm from left end. The product of $R_1 R_2$ is _____ Ω^2
[29 Jan 2023 Shift II]

156. A null point is found at 200 cm in potentiometer when cell in secondary circuit is shunted by 5Ω. When a resistance of 15Ω is used for shunting, null point moves to 300 cm. The internal resistance of the cell is _____ Ω.
[29 Jan 2023 Shift II]

157. The charge flowing in a conductor changes with time as $Q(t) = \alpha t - \beta t^2 + \gamma t^3$. Where α, β and γ are constants. Minimum value of current is: [30 Jan 2023 Shift I]

(A) $\alpha - \frac{3\beta^2}{\gamma}$ (B) $\alpha - \frac{\gamma^2}{3\beta}$
(C) $\beta - \frac{\alpha^2}{3\gamma}$ (D) $\alpha - \frac{\beta^2}{3\gamma}$

158. In the circuit shown in Fig.4.452, the magnitude of current I_1, is _____ A. [30 Jan 2023 Shift I]

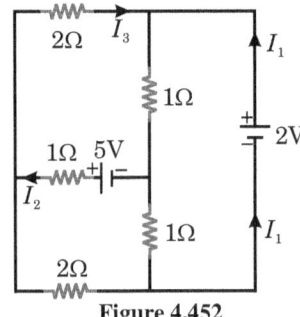

Figure 4.452

159. The equivalent resistance between A and B in the circuit shown in Fig. 4.453 is _____. [30 Jan 2023 Shift II]

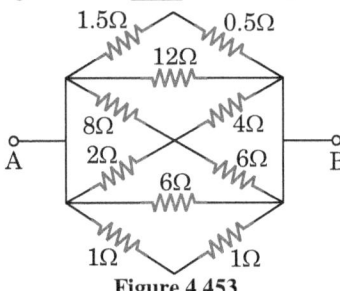

Figure 4.453

(A) $\frac{2}{3}\Omega$ (B) $\frac{1}{2}\Omega$ (C) $\frac{3}{2}\Omega$ (D) $\frac{1}{3}\Omega$

160. If the potential difference between B and D is zero, the value of x is $(1/n)\Omega$. The value of n is _____.
[30 Jan 2023 Shift II]

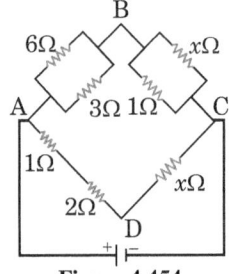

Figure 4.454

161. The drift velocity of electrons for a conductor connected in an electrical circuit is v_d. The conductor in now replaced by another conductor with same material and same length but double the area of cross section. The applied voltage remains same. The new drift velocity of electrons will be
[31 Jan 2023 Shift I]

(A) v_d (B) $\frac{v_d}{2}$ (C) $\frac{v_d}{4}$ (D) $2v_d$

162. Two identical cells, when connected either in parallel or in series gives same current in an external resistance 5 Ω. The internal resistance of each cell will be _____ Ω.
[31 Jan 2023 Shift I]

163. The number of turns of the coil of a moving coil galvanometer is increased in order to increase current sensitivity by 50%. The percentage change in voltage sensitivity of the galvanometer will be: [31 Jan 2023 Shift II]
(A) 100% (B) 50% (C) 75% (D) 0%

164. A water heater of power 2000 W is used to heat water. The specific heat capacity of water is 4200 J kg^{-1}K^{-1}. The efficiency of heater is 70%. Time required to heat 2 kg of water from 10°C to 60°C is _____ s. (Assume that the specific heat capacity of water remains constant over the given temperature range of the water). [31 Jan 2023 Shift II]

165. For the given circuit shown in Fig.4.455, in the steady state, $|V_B - V_D|$ = _____ V. [31 Jan 2023 Shift II]

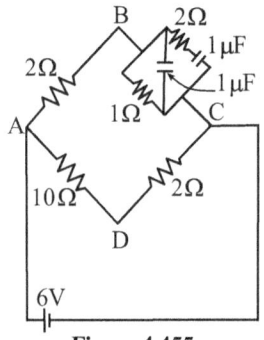

Figure 4.455

166. The equivalent resistance between A and B of the network shown in Fig.4.456, is [01 Feb 23 Shift I]
(A) $2R/3$ (B) $14R$ (C) $21R$ (D) $8R/3$

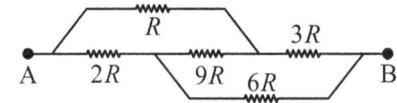

Figure 4.456

167. Given below are two statements: One is labelled as Assertion A and the other is labelled as Reason R

Assertion A : For measuring the potential difference across a resistance of 600 Ω, the voltmeter with resistance 1000 Ω will be preferred over voltmeter with resistance 4000 Ω.

Reason R : Voltmeter with higher resistance will draw smaller current than voltmeter with lower resistance.

In the light of the above statements, choose the most appropriate answer from the options given below.
[01 Feb 23 Shift II]
(A) A is not correct but R is correct
(B) Both A and R are correct and R is the correct explanation of A
(C) Both A and R are correct but R is not the correct explanation of A
(D) A is correct but R is not correct

168. In the given circuit of Fig.4.457, the value of $\left|\frac{I_1+I_3}{I_2}\right|$ is _____ A. [01 Feb 23 Shift II]

169. Two resistances are given as $R_1 = (10 \pm 0.5)$ Ω and

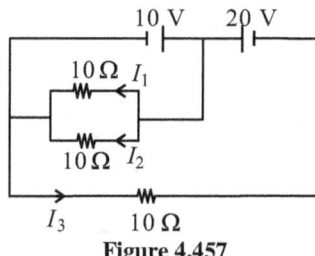

Figure 4.457

$R_2 = (15 \pm 0.5)\ \Omega$. The percentage error in the measurement of equivalent resistance when they are connected in parallel is [06 April 23 Shift I]

(A) 6.33 (B) 2.33 (C) 4.33 (D) 5.33

170. The resistivity (ρ) of semiconductor varies with temperature. Which of the following curve represents the correct behaviour [06 April 23 Shift I]

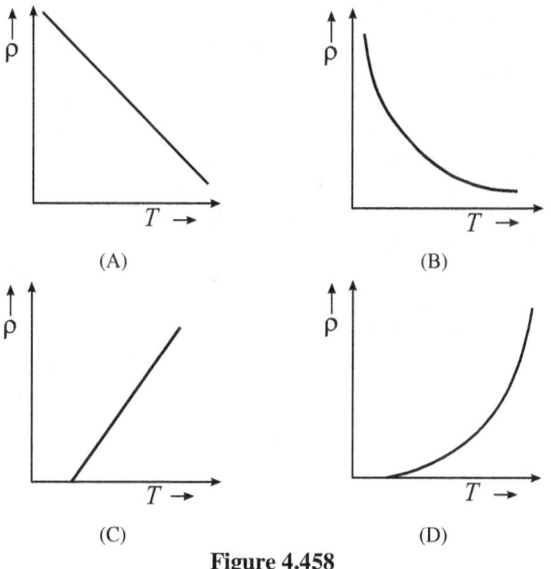

Figure 4.458

171. The length of a metallic wire is increased by 20% and its area of cross section is reduced by 4%. The percentage change in resistance of the metallic wire is ___.
[06 April 23 Shift I]

172. Fig.4.459 shows a part of an electric circuit. The potentials at points a, b and c are 30 V, 12 V and 2 V respectively. The current through the 20 Ω resistor will be.
[06 April 23 Shift II]

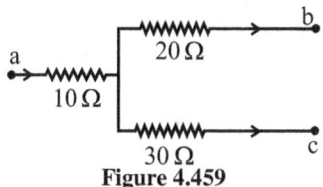

Figure 4.459

(A) 0.4 A (B) 0.2 A (C) 0.6 A (D) 1.0 A

173. A current of 2 A flows through a wire of cross sectional area 25.0 mm^2. The number of free electrons in a cubic meter are 2.0×10^{28}. The drift velocity of the electrons is $\times 10^{-6}$ ms^{-1} (given, charge on electron = 1.6×10^{-19} C)
[08 April 23 Shift II]

174. The equivalent resistance between A and B as shown in Fig.4.460, is: [08 April 23 Shift II]

(A) 5 kΩ (B) 30 kΩ (C) 10 kΩ (D) 20 kΩ

175. The number density of free electrons in copper is nearly 8×10^{28} m^{-3}. A copper wire has its area of cross section

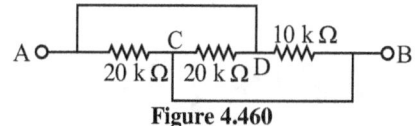

Figure 4.460

$= 2 \times 10^{-6}$ m^2 and is carrying a current of 3.2 A. The drift speed of the electrons is ___ $\times 10^{-6}$ ms^{-1}.
[08 April 23 Shift II]

176. 10 resistors each of resistance 10 Ω can be connected in such as to get maximum and minimum equivalent resistance. The ratio of maximum and minimum equivalent resistance will be ___. [10 April 23 Shift I]

177. In a metallic conductor, under the effect of applied electric field, the free electrons of the conductor
[10 April 23 Shift II]

(A) drift from higher potential to lower potential.
(B) move in the curved paths from lower potential to higher potential
(C) move with the uniform velocity throughout from lower potential to higher potential
(D) move in the straight line paths in the same direction

178. A rectangular parallelepiped is measured as 1 cm × 1 cm × 100 cm. If its specific resistance is 3×10^{-7} Ωm, then the resistance between its two opposite rectangular faces will be ___ × $10^{-7}\Omega$. [10 April 23 Shift II]

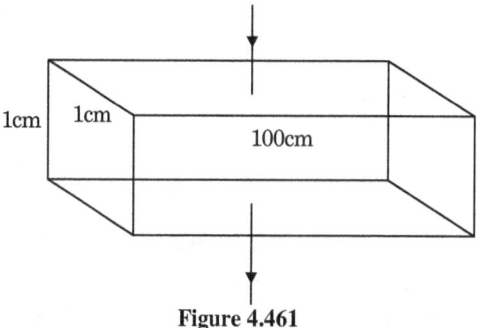

Figure 4.461

179. Two identical heater filaments are connected first in parallel and then in series. At the same applied voltage, the ratio of heat produced in same time for parallel to series will be:
[11 April 23 Shift I]

(A) 4 : 1 (B) 2 : 1 (C) 1 : 2 (D) 1 : 4

180. In the circuit diagram shown in figure given below, the current flowing through resistance 3 Ω is $\frac{x}{3}$ A. The value of x is ___ [11 April 23 Shift I]

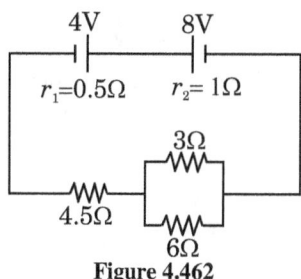

Figure 4.462

181. The current flowing through R_2 in circuit diagram in Fig.4.463 is: [11 April 23 Shift II]

(A) $\frac{2}{3}$ A (B) $\frac{1}{4}$ A (C) $\frac{1}{2}$ A (D) $\frac{1}{3}$ A

182. Two identical cells each of emf 1.5 V are connected in series across a 10 Ω resistance. An ideal voltmeter connected across 10 Ω resistance reads 1.5 V. The internal resistance of each cell is ___ Ω. [11 April 23 Shift II]

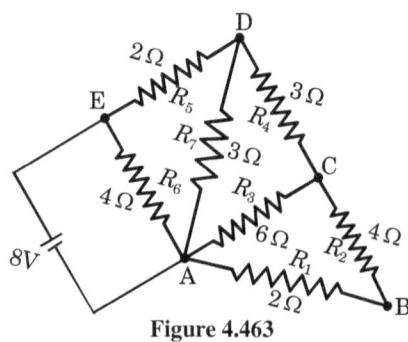

Figure 4.463

183. A wire of resistance 160 Ω is melted and drawn in wire of one-fourth of its length. The new resistance of the wire will be [12 April 23 Shift I]
    (A) 10 Ω (B) 640 Ω (C) 40 Ω (D) 16 Ω

184. The current flowing through a conductor connected across a source is 2A and 1.2 A at 0°C and 100°C respectively. The current flowing through the conductor at 50°C will be ____ ×10² mA. [12 April 23 Shift I]

185. Different combination of 3 resistors of equal resistance R are shown in the Fig.4.464. The increasing order for power dissipation is: [13 April 23 Shift I]

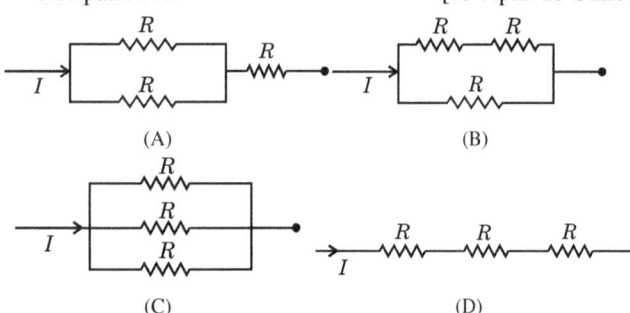

Figure 4.464

(A) $P_A < P_B < P_C < P_D$ (B) $P_C < P_D < P_A < P_B$
(C) $P_B < P_C < P_D < P_A$ (D) $P_C < P_B < P_A < P_D$

186. When a resistance of 5 Ω is shunted with a moving coil galvanometer, it shows a full scale deflection for a current of 250 mA, however when 1050 Ω resistance is connected with it in series, it gives full scale deflection for 25 V. The resistance of galvanometer is ____ Ω. [13 April 23 Shift I]

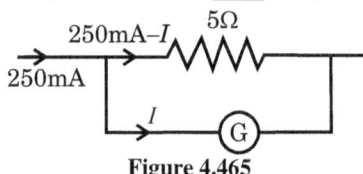

Figure 4.465

187. In the network shown in Fig.4.466, the charge accumulated in the capacitor in steady state will be:
[13 April 23 Shift II]
(A) 7.2 μC (B) 4.8 μC (C) 10.3 μC (D) 12 μC

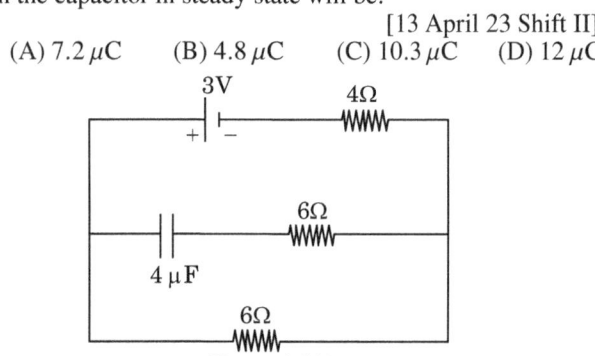

Figure 4.466

188. In the circuit shown, the energy stored in the capacitor is n μJ. The value of n is ____. [13 April 23 Shift II]

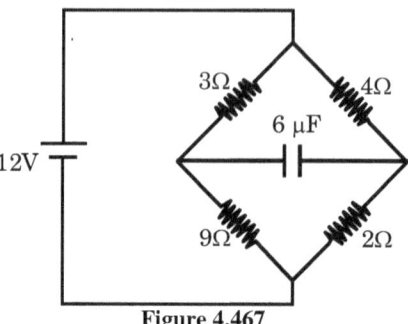

Figure 4.467

189. Given below are two statements:
Statement I: The equivalent resistance of resistors in a series combination is smaller than least resistance used in the combination.
Statement II: The resistivity of the material is independent of temperature.
In the light of the above statements, choose the correct answer from the options given below: [15 April 23 Shift I]
(A) Statement I is false but Statement II is true
(B) Both Statement I and Statement II are false
(C) Statement I is true but Statement II is false
(D) Both Statement I and Statement II are true

190. A network of four resistances is connected to 9 V battery, as shown in Fig.4.468. The magnitude of potential difference between the points A and B is ____ V.
[15 April 23 Shift I]

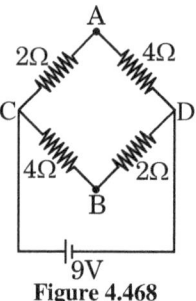

Figure 4.468

Section B: JEE Advanced

1. A battery of internal resistance 4 Ω is connected to the network of resistance as shown in Fig.4.469. In order that maximum power can be delivered to the network, the value of R in ohm should be [1995]
    (A) 4/9 (B) 2 (C) 8/3 (D) 18

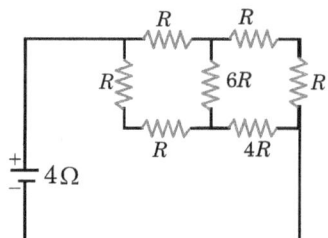

Figure 4.469

2. A series combination of 0.1 MΩ resistor and a 10 μF capacitor is connected across a 1.5 V source of negligible resistance. The time required for the capacitor to get charged up to 0.75 V approximately (in seconds) is [1997]
    (A) ∞ (B) $\log_e 2$ (C) $\log_{10} 2$ (D) zero

3. A capacitor of capacity C is charged in RC circuit. The variation of log I vs time is shown in the Fig.4.470 by dotted line when net resistance of circuit is X. When resistance changes to $2X$ the variation is now shown by   [2004]
   (A) P (B) Q (C) R (D) S

Figure 4.470

4. A capacitor of capacitance 4 μF is charged through a resistor 2.5 MΩ connected in series to a battery of emf 12 volt having negligible internal resistance. Then time in which potential drop across capacitor is 3 times the potential drop across the resistor -   [2005]
   (A) 13.86 s   (B) 6.93 s
   (C) 27.72 s   (D) 3.46 s

5. Time constants for the given circuits in Fig.4.471, are respectively-   [2006]
   (A) 18 μs, $\frac{8}{9}$ μs, 4 μs   (B) 18 μs, 4 μs, $\frac{8}{9}$ μs
   (C) 4 μs, $\frac{8}{9}$ μs, 18 μs   (D) $\frac{8}{9}$ μs, 18 μs, 4 μs

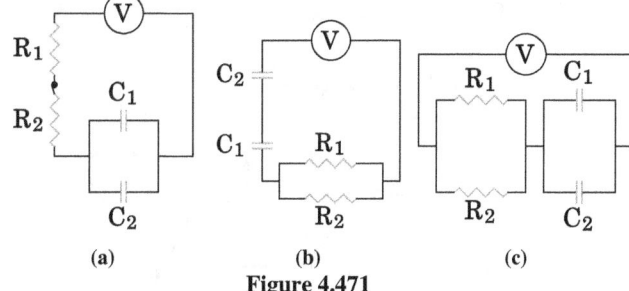

Figure 4.471

6. A circuit is connected as shown in the Fig.4.472 with the switch S open. When the switch is closed, the total amount of charge that flows from Y to X is   [2007]
   (A) 0   (B) 54 μC   (C) 27 μC   (D) 81 μC

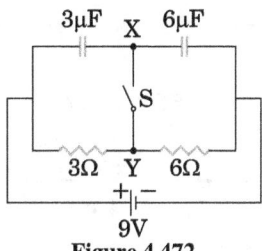

Figure 4.472

7. In Fig.4.473, a parallel plate capacitor C with plates of unit area and separation d is filled with a liquid of dielectric constant $\kappa = 2$. The level of liquid is $d/3$ initially. Suppose the liquid level decreases at a constant speed V, the time constant as a function of time t is -   [2008]
   (A) $\frac{6\epsilon_0 R}{5d+3Vt}$ (B) $\frac{(15d+9Vt)\epsilon_0 R}{2d^2-3dVt-9V^2t^2}$
   (C) $\frac{6\epsilon_0 R}{5d-3Vt}$ (D) $\frac{(15d-9Vt)\epsilon_0 R}{5d^2+3dVt-9V^2t^2}$

8. In the circuit shown in Fig.4.474, the emf of each battery is 5 V and has an internal resistance of 0.2 Ω. The reading in ideal voltmeter V is-   [1997]
   (A) 0 V   (B) 5 V   (C) 40 V   (D) 25 V

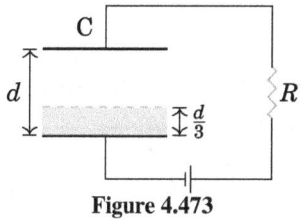

Figure 4.473

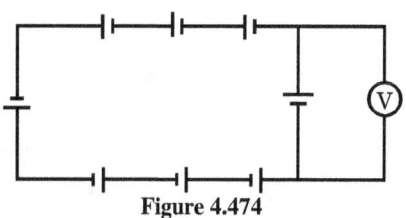

Figure 4.474

9. The equivalent resistance between points A and B of the circuit given in Fig.4.475, is-   [1997]
   (A) 5R   (B) $R/2$ (C) 2R   (D) R

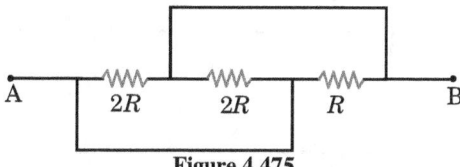

Figure 4.475

10. A steady current flows in a metallic conductor of non-uniform cross-section. The quantity/quantities constant along the length of the conductor is/are   [1997]
    (A) current, electric field and drift speed
    (B) drift speed only
    (C) current and drift speed
    (D) current only

11. In the circuit shown in Fig.4.476, the current through   [1998]
    (A) the 3Ω resistor is 0.50 A
    (B) the 3Ω resistor is 0.25 A
    (C) the 4Ω resistor is 0.50 A
    (D) the 4Ω resistor is 0.25 A

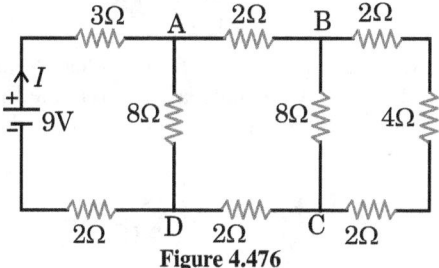

Figure 4.476

12. When a potential difference is applied across, the current passing through   [1999]
    (A) a metal at 0 K is zero
    (B) a semiconductor at 0 K is zero
    (C) a metal at 0 K is finite
    (D) a p-n diode at 300 K is finite, if it is reverse biased

13. Two wires of equal diameters of resistivities ρ_1 and ρ_2 and lengths x_1 and x_2 are joined in series. The equivalent resistivity of the combination is-   [2001]
    (A) $\frac{\rho_1 x_1 + \rho_2 x_2}{x_1 + x_2}$ (B) $\frac{\rho_1 x_2 - \rho_2 x_1}{x_1 - x_2}$
    (C) $\frac{\rho_1 x_2 + \rho_2 x_1}{x_1 + x_2}$ (D) $\frac{\rho_1 x_1 + \rho_2 x_2}{\rho_1 + \rho_2}$

14. In the circuit shown in Fig.4.477, it is observed that the current I is independent of the value of the resistance R_6.

Then the resistance values must satisfy- [2001]
(A) $R_1R_2R_5 = R_3R_4R_6$
(B) $\frac{1}{R_5} + \frac{1}{R_6} = \frac{1}{R_1+R_2} + \frac{1}{R_3+R_4}$
(C) $R_1R_4 = R_2R_3$
(D) $R_1R_3 = R_2R_4 = R_5R_6$

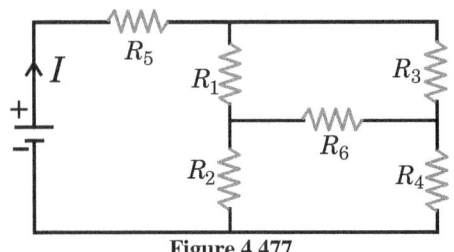

Figure 4.477

15. In the given circuit shown in Fig.4.478, with steady current, the potential drop across the capacitor must be [2001]
(A) V (B) V/2 (C) V/3 (D) 2V/3

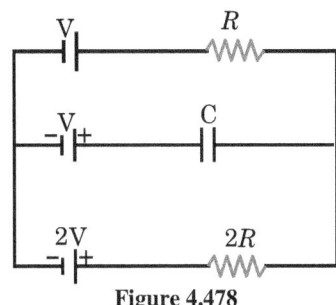

Figure 4.478

16. The effective resistance between points P and Q of the electrical circuit shown in the Fig.4.479, is- [2002]
(A) $\frac{2Rr}{R+r}$ (B) $\frac{8R(Rr)}{3R+r}$ (C) $2r + 4r$ (D) $\frac{5R}{2} + 2r$

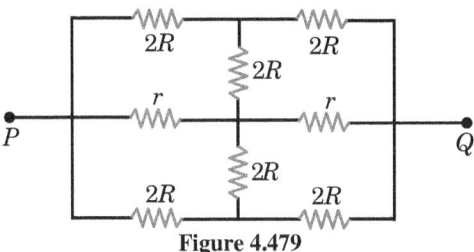

Figure 4.479

17. A 100 W bulb B_1, and two 60 W bulbs B_2 and B_3, are connected to a 250 V source, as shown in the Fig.4.480. If W_1, W_2 and W_3 are the output powers of the bulbs B_1, B_2 and B_3, respectively, then- [2002]

Figure 4.480

(A) $W_1 > W_2 = W_3$ (B) $W_1 > W_2 > W_3$
(C) $W_1 < W_2 = W_3$ (D) $W_1 < W_2 < W_3$

18. Arrange the order of power dissipated in the given circuits shown in Fig.4.481, if the same current is passing through all the circuits. The resistance of each resistor is 'r'.
[13 April 2023 (I)]
(A) $P_2 > P_3 > P_4 > P_1$ (B) $P_1 > P_4 > P_2 > P_3$
(C) $P_3 > P_1 > P_4 > P_2$ (D) $P_2 > P_4 > P_1 > P_3$

(I)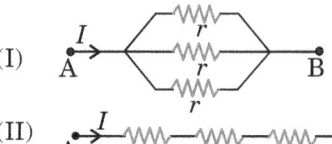

(II)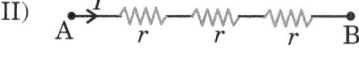

(III)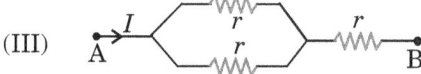

(IV)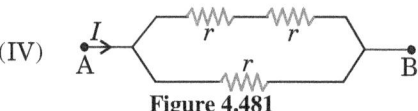

Figure 4.481

19. In the circuit shown in the Fig.4.482, equivalent resistance is maximum- [2004]
(A) between P & Q
(B) between P & R
(C) between R & P
(D) same between all the points

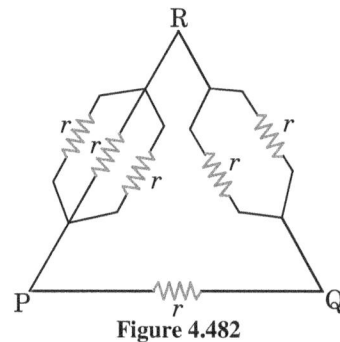

Figure 4.482

20. Find out Current in 2 Ω resistor shown in the circuit of Fig.4.483? [2005]
(A) 0 (B) 2 A (C) 3A (D) 5 A

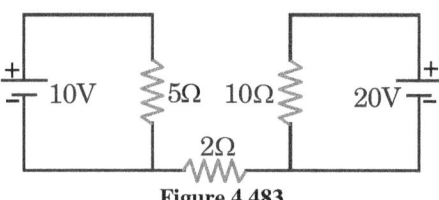

Figure 4.483

21. In the Figure 4.484, length of each wire is $l/2$ and their radii are $2r$ and r. Then- [2006]
(A) current density in both wires is same
(B) power dissipated in QR is 4 times that in the PQ
(C) ratio of potential drops on PQ & QR is 4
(D) resistance of PQ is 4 times that of QR

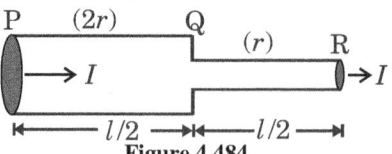

Figure 4.484

22. Figure 4.485 shows three resistor configurations (a), (b) and (c) connected to a 3 V battery. If the power dissipated

by the configuration (a), (b) and (c) are P_1, P_2 and P_3, respectively, then [2008]
(A) $P_1 > P_2 > P_3$
(B) $P_1 > P_3 > P_2$
(C) $P_2 > P_1 > P_3$
(D) $P_3 > P_2 > P_1$

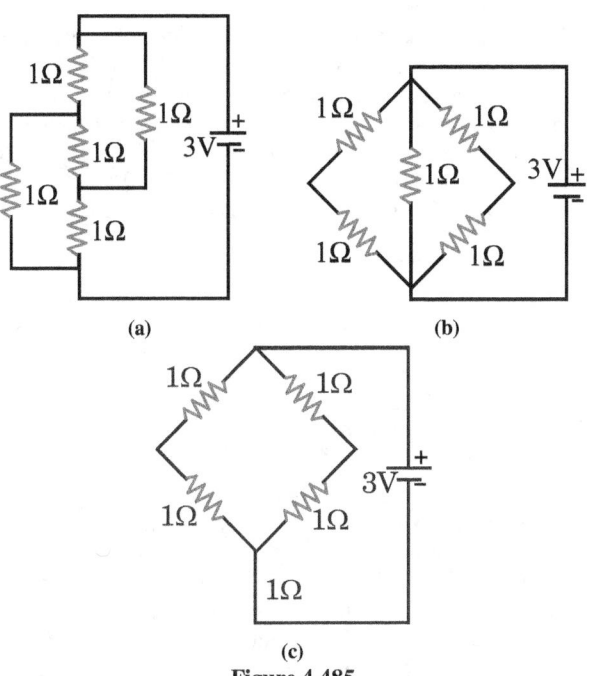

Figure 4.485

23. **STATEMENT 1:** In a Meter Bridge experiment, null point for an unknown resistance is measured. Now, the unknown resistance is put inside an enclosure maintained at a higher temperature. The null point can be obtained at the same point as before by decreasing the value of the standard resistance.
STATEMENT-2: Resistance of a metal increases with increase in temperature. [2008]
(A) Statement-1 is true, statement-2 is true; statement-2 is a correct explanation for statement-1
(B) Statement-1 is true, statement-2 is true; statement-2 is not a correct explanation for statement 1
(C) Statement-1 is true, statement-2 is false
(D) Statement-1 is false, statement-2 is true

24. For the circuit shown in the Fig.4.486 [2009]
(A) the current I through the battery is 7.5 mA
(B) the potential difference across R_L is 18 V
(C) ratio of powers dissipated in R_1 and R_2 is 3
(D) if R_1 and R_2 are interchanged, magnitude of the power dissipated in R_L, will decrease by a factor of 9.

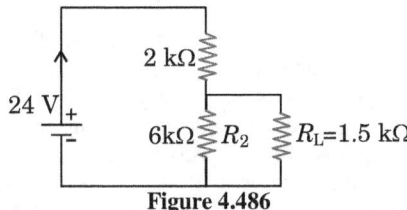

Figure 4.486

25. Consider a thin square sheet of side L and thickness t, made of a material of resistivity ρ. The resistance between two opposite faces, shown by the shaded areas in the Fig.4.487 is [2010]
(A) directly proportional to L
(B) directly proportional to t
(C) independent of L
(D) independent of t

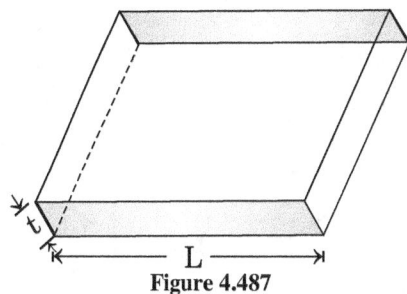

Figure 4.487

26. To Verify Ohm's law, a student is provided with a test resistor R_T, a high resistance R_1, a small resistance R_2, two identical galvanometers G_1 and G_2, and a variable voltage source V. In Fig.4.488, the correct circuit to carry out the experiment is [2010]

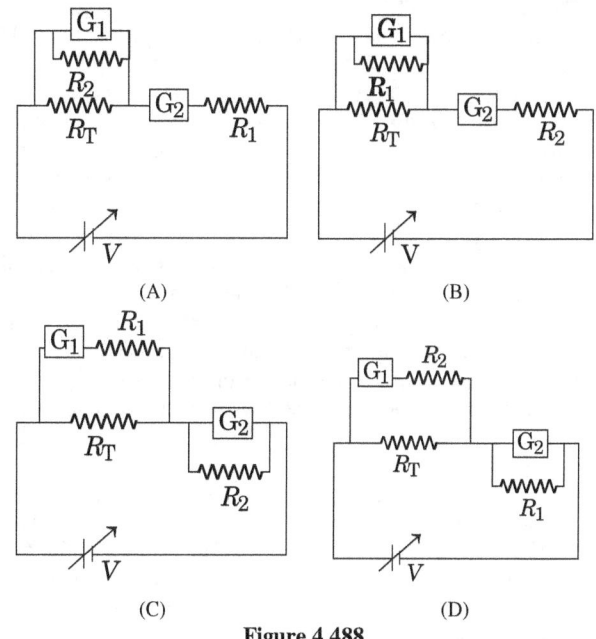

Figure 4.488

27. A meter bridge is set-up as shown in Fig.4.489, to determine an unknown resistance 'X' using a standard 10 Ω resistor. The galvanometer shows null point when tapping key is at 52 cm mark. The end corrections are 1 cm and 2 cm respectively for the ends A and B. The determined value of 'X' is [2011]
(A) 10.2 ohm   (B) 10.6 ohm
(C) 10.8 ohm   (D) 11.1 ohm

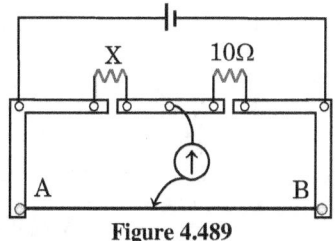

Figure 4.489

28. For the resistance network shown in the Figure 4.490, choose the correct option(s) [2012]

(A) The current through PQ is zero
(B) $I_1 = 3$ A
(C) The potential at S is less than that at Q
(D) $I_2 = 2$ A

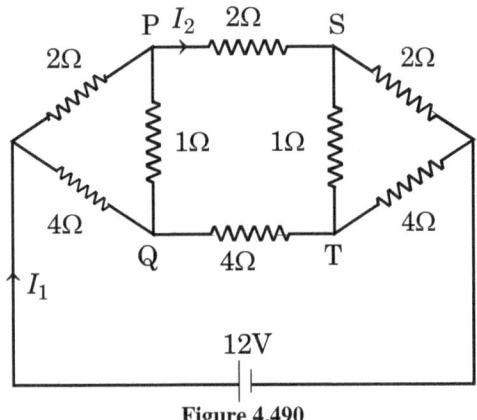

Figure 4.490

Paragraph for Q.29
A thermal power plant produces electric power of 600 kW at 4000 V, which is to be transported to a place 20 km away from the power plant for consumers' usage. It can be transported either directly with a cable of large current carrying capacity or by using a combination of step-up and step-down transformers at the two ends. The drawback of the direct transmission is the large energy dissipation. In the method using transformers, the dissipation is much smaller. In this method, a step-up transformer is used at the plant side so that the current is reduced to a smaller value. At the consumers' end, a step-down transformer is used to supply power to the consumers at the specified lower voltage. It is reasonable to assume that the power cable is purely resistive and the transformers are ideal with the power factor unity. All the currents and voltage mentioned are rms values.

29. If the direct transmission method with a cable of resistance $0.4 \ \Omega\text{km}^{-1}$ is used, the power dissipation (in %) during transmission is [2013]
    (A) 20   (B) 30   (C) 40   (D) 50

30. Two ideal batteries of emf V_1 and V_2 and three resistances R_1, R_2 and R_3 are connected as shown in the Fig.4.491. The current in resistance R_2 would be zero if [2014]
    (A) $V_1 = V_2$
    (B) $V_1 = V_2$ and $R_1 = R_2 = R_3$
    (C) $V_1 = 2V_2$ and $2R_1 = 2R_2 = R_3$
    (D) $2V_1 = V_2$ and $2R_1 = R_2 = R_3$

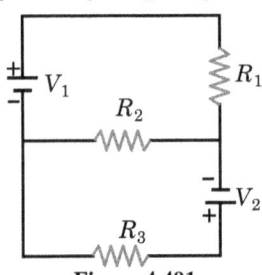

Figure 4.491

31. Heater of an electric kettle is made of a wire of length L and diameter d. It takes 4 minutes to raise the temperature of 0.5 kg water by 40 K. This heater is replaced by a new heater having two wires of the same material, each of length L and diameter $2d$. The way these wires are connected is given in the options. How much time in minutes will it take to raise the temperature of the same amount of water by 40 K?

[2014]
(A) 4 if wires are in parallel
(B) 2 if wires are in series
(C) 1 if wires are in series
(D) 0.5 if wires are in parallel.

32. During an experiment with a metre bridge, the galvanometer shows a null point when the jockey is pressed at 40.0 cm using a standard resistance of 90 Ω, as shown in the Fig.4.492. The least count of the scale used in the metre bridge is 1 mm. The unknown resistance is [2014]
    (A) $60 \pm 0.15 \Omega$ (B) $135 \pm 0.56 \Omega$
    (C) $60 \pm 0.25 \Omega$ (D) $135 \pm 0.23 \Omega$

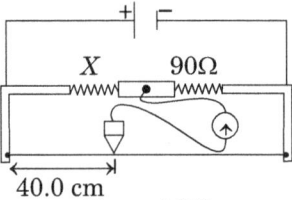

Figure 4.492

33. An infinite line charge of uniform electric charge density λ lies along the axis of an electrically conducting infinite cylindrical shell of radius R. At time $t = 0$, the space inside the cylinder is filled with a material of permittivity ϵ and electrical conductivity σ. The electrical conduction in the material follows Ohm's law. Which one of the graphs, in Fig.4.493, best describes the subsequent variation of the magnitude of current density $j(t)$ at any point in the material? [2016]

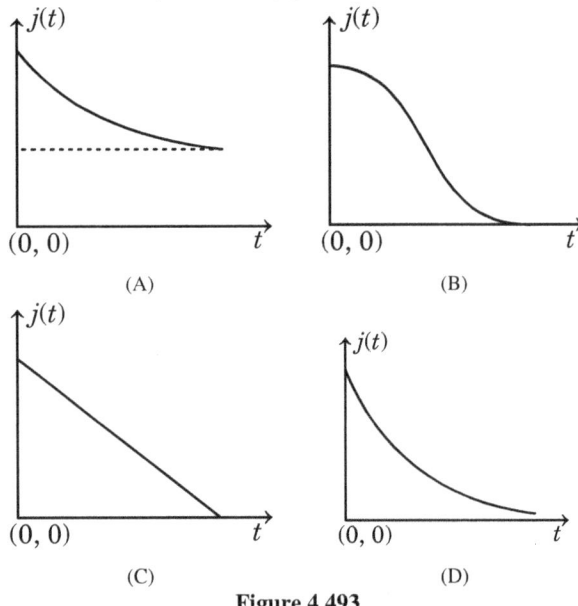

Figure 4.493

34. An incandescent bulb has a thin filament of tungsten that is heated to high temperature by passing an electric current. The hot filament emits black-body radiation. The filament is observed to break up at random locations after a sufficiently long time of operation due to non-uniform evaporation of tungsten from the filament. If the bulb is powered at constant voltage, which of the following statement(s) is(are) true? [2016]

(A) The temperature distribution over the filament is uniform
(B) The resistance over small sections of the filament decreases with time
(C) The filament emits more light at higher band of frequencies before it breaks up
(D) The filament consumes less electrical power towards the end of the life of the bulb.

35. Consider two identical galvanometers and two identical resistors with resistance R. If the internal resistance of the galvanometers $R_C < R/2$, which of the following statement(s) about any one of the galvanometers is(are) true? [2016]
    (A) The maximum voltage range is obtained when all the components are connected in series
    (B) The maximum voltage range is obtained when the two resistors and one galvanometer are connected in series, and the second galvanometer is connected in parallel to the first galvanometer
    (C) The maximum current range is obtained when all the components are connected in parallel
    (D) The maximum current range is obtained when the two galvanometers are connected in series and the combination is connected in parallel with both the resistors.

36. In the circuit shown in Fig.4.494, the key is pressed at time $t = 0$. Which of the following statement(s) is(are) true? [2016]
    (A) The voltmeter displays $-5V$ as soon as the key is pressed, and displays $+5V$ after a long time
    (B) The voltmeter will display 0V at time $t = \ln 2$ seconds
    (C) The current in the ammeter becomes 1/e of the initial value after 1 second
    (D) The current in the ammeter becomes zero after a long time

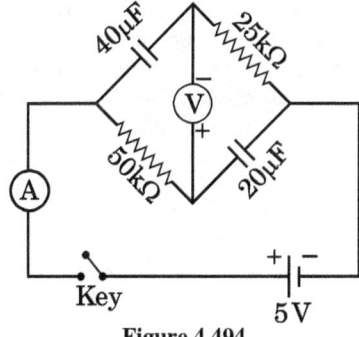

Figure 4.494

PARAGRAPH-1 [2017]
Consider a simple RC circuit as shown in Fig.4.495A.
Process 1: In the circuit, the switch S is closed at $t = 0$ and the capacitor is fully charged to voltage V_0 (i.e., charging continues for time $T \gg RC$). In the process, some dissipation (E_D) occurs across the resistance R. The amount of energy finally stored in the fully charged capacitor is E_C.
Process 2: In a different process, the voltage is first set to $V_0/3$ and maintained for a charging time $T \gg RC$. Then the voltage is raised to $2V_0/3$ without discharging the capacitor and again maintained for a time $T \gg RC$. The process is repeated one more time by raising the voltage to V_0 and the capacitor is charged to the same final voltage V_0 as in Process 1.
These two processes are depicted in Figure 4.495B.

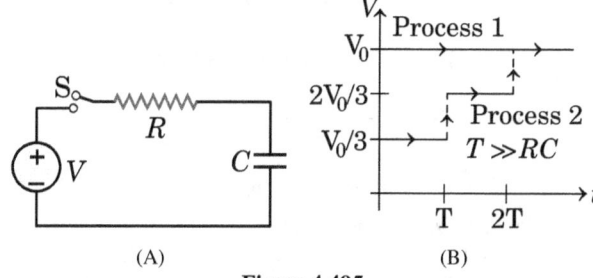

Figure 4.495

37. In Process 1, the energy stored in the capacitor E_C and heat dissipated across resistance E_D are related by:
    (A) $E_C = E_D$ (B) $E_C = 2E_D$
    (C) $E_C = \frac{1}{2}E_D$ (D) $E_C = E_D \ln 2$

38. In Process 2, total energy dissipated across the resistance, E_D is:
    (A) $E_D = \frac{1}{3}\left(\frac{1}{2}CV_0^2\right)$ (B) $E_D = 3\left(\frac{1}{2}CV_0^2\right)$
    (C) $E_D = \frac{1}{2}CV_0^2$ (D) $E_D = 3CV_0^2$

39. In the balanced condition, the values of the resistances of the four arms of a Wheatstone bridge are shown in the Fig.4.496. The resistance R_3 has temperature coefficient $0.0004°C^{-1}$. If the temperature of R_3 is increased by $100°C$, the voltage developed between S and T will be ___ volt. [2020]
    (A) 0.270 V (B) 0.370 V
    (C) 0.250 V (D) 0.280 V

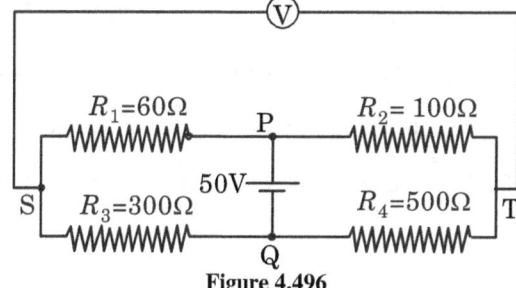

Figure 4.496

Question Stem for Question No. 40 and 41
Question Stem
In the circuit shown below, the switch S is connected to position P for a long time so that the charge on the capacitor becomes q_1 μC. Then S is switched to position Q. After a long time, the charge on the capacitor is q_2 μC. [2021]

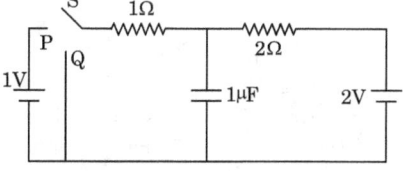

Figure 4.497

40. The magnitude of q_1 is _____.
41. The magnitude of q_2 is _____.
42. In order to measure the internal resistance r_1 of a cell of emf $\mathcal{E}$, a meter bridge of wire resistance $R_0 = 50$ Ω, a resistance $R_0/2$, another cell of emf $\mathcal{E}/2$ (internal resistance r) and a galvanometer G are used in a circuit, as shown in the Fig.4.498. If the null point is found at $l = 72$ cm, then the value of $r_1 = \cdots$ Ω [2021]

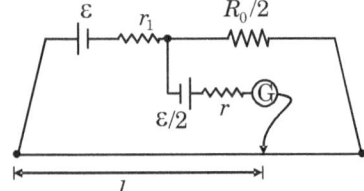

Figure 4.498

Paragraph Questions 43–44

In the circuit shown in Fig. 4.499, the switch S is connected to position P for a long time so that the charge on the capacitor becomes q_1 μC. Then S is switched to position Q. After a long time, the charge on the capacitor is q_2 μC.

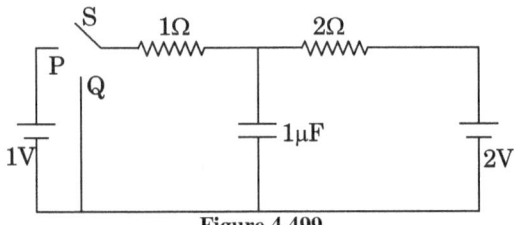

Figure 4.499

43. The magnitude of q_1 is... [2021]
44. The magnitude of q_2 is... [2021]
45. The Fig.4.500 shows a circuit having eight resistances of 1 Ω each, labelled R_1 to R_2, and two ideal batteries with voltages $\mathcal{E}_1 = 12$ V and $\mathcal{E}_2 = 6$ V. Which of the following statement(s) is(are) correct? [2022]
    (A) The magnitude of current flowing through R_1 is 7.2 A.
    (B) The magnitude of current flowing through R_2 is 1.2 A.
    (C) The magnitude of current flowing through R_3 is 4.8 A.
    (D) The magnitude of current flowing through R_4 is 2.4 A.

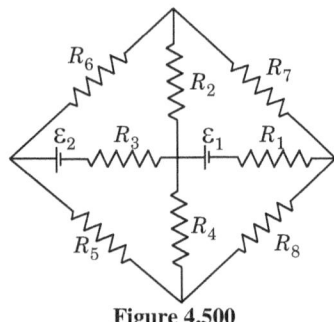

Figure 4.500

46. Two resistances $R_1 = X\Omega$ and $R_2 = 1\Omega$ are connected to a wire AB of uniform resistivity, as shown in the Fig.4.501. The radius of the wire varies linearly along its axis from 0.2 mm at A to 1 mm at B. A galvanometer (G) connected to the centre of the wire, 50 cm from each end along its axis, shows zero deflection when A and B are connected to a battery. The value of X is ____. [2022]
47. In Circuit-1 and Circuit-2 shown in the figures, $R_1 = 1\Omega, R_2 = 2\Omega$ and $R_3 = 3\Omega$. P_1 and P_2 are the power dissipations in Circuit-1 and Circuit-2 when the switches S_1 and S_2 are in open conditions, respectively. Q_1 and Q_2 are the power dissipations in Circuit-1 and Circuit-2 when the switches S_1 and S_2 are in closed conditions, respectively. Which of the following statement(s) is(are) correct?

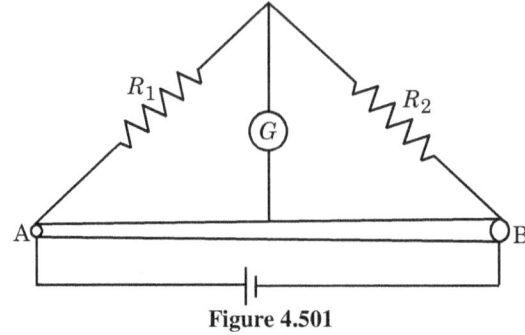

Figure 4.501

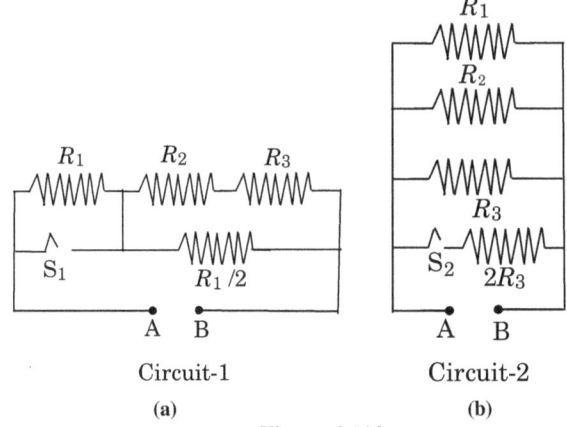

Figure 4.502

(A) When a voltage source of 6 V is connected across A and B in both circuits, $P_1 < P_2$
(B) When a constant current source of 2 A is connected across A and B in both circuits, $P_1 > P_2$
(C) When a voltage source 6 V is connected across A and B in Circuit-1, $Q_1 > P_1$
(D) When a constant current source of 2 A is connected across A and B in both circuits, $Q_2 < Q_1$

48. Fig.4.503 shows a circuit having eight resistances of 1 Ω each, labelled R_1 to R_8, and two ideal batteries with voltages $\mathcal{E}_1 = 12$ V and $\mathcal{E}_2 = 6$ V. Which of the following statement(s) is(are) correct? [2022]

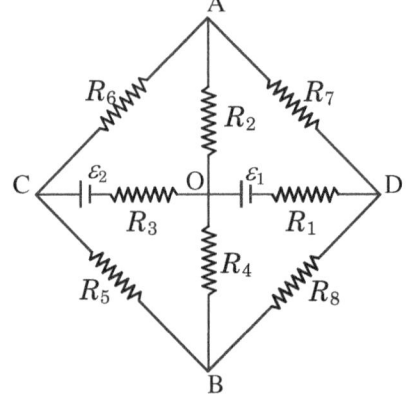

Figure 4.503

(A) The magnitude of current flowing through R_1 is 7.2 A.
(B) The magnitude of current flowing through R_2 is 1.2 A.
(C) The magnitude of current flowing through R_3 is 4.8 A.
(D) The magnitude of current flowing through R_5 is 2.4 A.

4.35. ANSWER KEYS AND SOLUTIONS

49. In a circuit shown in the Fig.4.504, the capacitor C is initially uncharged and the key K is open. In this condition, a current of 1 A flows through the 1 Ω resistor. The key is closed at time $t = t_0$. Which of the following statement(s) is(are) correct? [Given: $e^{-1} = 0.36$] [2023]
   (A) The value of the resistance R is 3 Ω.
   (B) For $t < t_0$, the value of current I_1 is 2 A.
   (C) At $t = t_0 + 7.2\,\mu s$, the current in the capacitor is 0.6 A.
   (D) For $t \to \infty$, the charge on the capacitor is 12 μC.

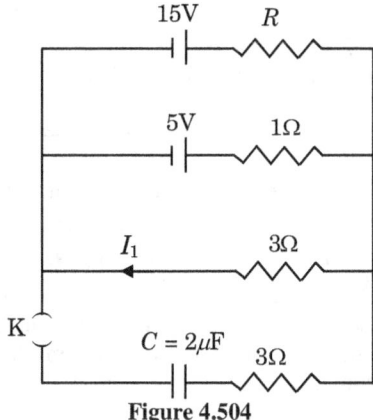

Figure 4.504

Integer Type
1. When two identical batteries of internal resistance 1Ω each are connected in series across a resistor R_1 the rate of heat produced in R is J_1. When the same batteries are connected in parallel across R_2 the rate is J_2. If $J_1 = 2.25 J_2$ then the value of R in Ω is [2010]
2. At time $t = 0$, a battery of 10V is connected across points A and B in the circuit given in Fig.4.505. If the capacitors have no charge initially, at what time (in seconds) does the voltage across them becomes 4 volt? [take $\ln 5 = 1.6$, $\ln 3 = 1.1$] [2010]

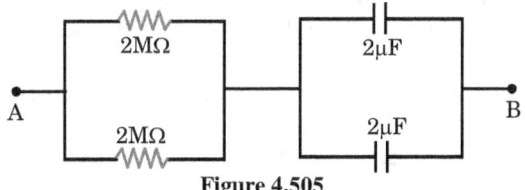

Figure 4.505

3. Two batteries of different emfs and different internal resistances are connected as shown in Fig.4.506. The voltage across AB in volts is [2011]

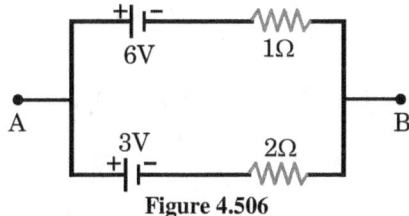

Figure 4.506

4. A galvanometer gives full scale deflection with 0.006 A current. By connecting it to a 4990 Ω resistance, it can be converted into a voltmeter of range 0 – 30 V. If connected to a $2n/249$ Ω resistance, it becomes an ammeter of range 0 – 1.5 A. The value of n is [2014]
5. A moving coil galvanometer has 50 turns and each turn has an area $2 \times 10^{-4}\,m^2$. The magnetic field produced by the magnet inside the galvanometer is 0.02 T. The torsional constant of the suspension wire is 10^{-4} Nm.rad^{-1}. When a current flows through the galvanometer, a full scale deflection occurs if the coil rotates by 0.2 rad. The resistance of the coil of the galvanometer is 50 Ω. This galvanometer is to be converted into an ammeter capable of measuring current in the range 0 - 1.0 A. For this purpose, a shunt resistance is to be added in parallel to the galvanometer. The value of this shunt resistance, in ohms, is ⋯ [2018]

4.35 Answer Keys and Solutions
4.35.1 Check Point 1
1. **APPROACH** Use the definition of current, Eq.4.1:
$$I_{av} = \Delta Q/\Delta t$$
SOLUTION $I_{av} = \dfrac{\Delta Q}{\Delta t}$
If N electrons are flowing past any point in the wire in time Δt, then number of electrons per second will be $N/\Delta t$
Since, $Q = Ne$, therefore,
$$I_{av} = \frac{\Delta Q}{\Delta t} = \frac{Ne}{\Delta t}$$
Therefore, $\dfrac{N}{\Delta t} = \dfrac{I_{av}}{e} = \dfrac{1.30}{1.60 \times 10^{-19}}$
$= 8.13 \times 10^{18}$ electrons /s

2. **APPROACH** Apply Eq.4.2: $I = \Delta Q/\Delta t$, and solve for electric charge ΔQ
SOLUTION $I_{av} = \dfrac{\Delta Q}{\Delta t} \Rightarrow \Delta Q = I_{av} \Delta t$
$= (6.7\,A)(5.0\,h)(3600\,s/h) = 1.2 \times 10^5 C$

3. **APPROACH** Use the definition of current, Eq.4.1:
$I_{av} = \Delta Q/\Delta t$ with ΔQ = (no. of ions) × (charge on an ion)
SOLUTION Substituting the given values in above expression, we get
$$I = \frac{\Delta Q}{\Delta t} = \frac{(1200\,\text{ions})(1.60 \times 10^{-19} C/\text{ion})}{3.5 \times 10^{-6}\,s}$$
$= 5.5 \times 10^{-11}$ A

4. **APPROACH** (a) To determine the charge passing through given cross-section, we apply Eq.4.5:
$$Q = \int dQ = \int_0^t I\,dt \quad \cdots (1)$$
(b) To find the constant current corresponding to same time interval and same charge as obtained in part (a), we apply Eq.4.1:
$$I_{av} = \frac{\Delta Q}{\Delta t} \quad \cdots (2)$$
SOLUTION (a) From above Eq.(1), we can write
$$I = \frac{dQ}{dt} \Rightarrow \int_0^Q dQ = \int_0^4 I\,dt$$
$$\Rightarrow Q = \int_0^4 (3 + 2t)dt$$
$$= [3t + t^2]_0^4 = [12 + 16] = 28\,C$$
(b) From Eq.(2), we have
$$I = \frac{\Delta Q}{\Delta t} = \frac{28}{4} = 7\,A$$

5. **APPROACH** Area under I-t graph gives the net charge that flows through any given cross-section of the conductor.
SOLUTION Current versus time graph is as shown in Fig.4.5. If ΔQ is the net charge that passes through any given cross section of the conductor in 8 seconds, then
ΔQ = Area under $I - t$ graph.
$= \dfrac{1}{2} \times$ base $\times$ height $= \dfrac{1}{2} \times 8 \times 20 = 80\,C$

6. **APPROACH** From Eq.4.5, the charge passes through any cross-section of a conductor, having current I, is

$$Q = \int dQ = \int_0^t I\,dt \qquad \cdots (1)$$

SOLUTION Since, $I = 4$ A (constant), and $t = 8$ s therefore above Eq.(1) gives-

$$Q = \int dQ = 4\int_0^8 dt = 4(8-0)$$
$$= 4 \times 8 = 32 \text{ C}$$

4.35.2 Check Point 2

1. (a) Circular area depends, on r^2, so the horizontal axis of the graph in Fig.4.14b is effectively the same as the area (enclosed at variable radius values), except for a factor of π. The fact that the current increases linearly in the graph means that $I/A = J = $ constant. Thus, the answer is "yes, the current density is uniform."
   (b) We find $I/(\pi r^2) = (0.005 \text{ A})/(\pi \times 4 \times 10^{-6} \text{ m}^2)$
   $= 398 \approx 4.0 \times 10^2 \text{ A/m}^2$.

2. The cross-sectional area of wire is given by $A = \pi r^2$, where r is its radius (half its thickness). The magnitude of the current density vector is, $J = I/A = I/\pi r^2$, so

$$r = \sqrt{\frac{I}{\pi J}} = \sqrt{\frac{0.50 \text{ A}}{\pi (440 \times 10^4 \text{ A/m}^2)}} = 1.9 \times 10^{-4} \text{ m}$$

The diameter of the wire is therefore
$d = 2r = 2(1.9 \times 10^{-4} \text{ m}) = 3.8 \times 10^{-4}$ m.

3. (a) From Eq.4.17: $J = neAv_d$, we have
$$J = nev_d$$
$$\Rightarrow v_d = \frac{J}{ne} = \frac{I}{neA} \qquad \cdots (1)$$

Since, each Cu atom is giving only one electron, therefore number of electrons per unit volume will be equal to number of Cu atoms per unit volume. If there are N atoms of Cu in mass m of of the material, N_A is the Avogadro number and M is the molar mass of Cu, then

Number of moles $= \dfrac{N}{N_A} = \dfrac{m}{M} \Rightarrow N = N_A \dfrac{m}{M}$

$\therefore n = \dfrac{N}{\text{Volume}} = \dfrac{N_A m/M}{\text{Volume}} = \dfrac{N_A m/M}{\text{Volume}} = \dfrac{N_A}{M} \times \rho_D$

Putting these values in Eq.(1), we get

$$v_d = \frac{I}{\left(\frac{N_A}{M}\rho_D\right)e\left[\pi\left(\frac{1}{2}d\right)^2\right]} = \frac{4Im}{N\rho_D e\pi d^2}$$

$$\Rightarrow v_d = \frac{4(2.3\times 10^{-6}\text{ A})(63.5\times 10^{-3}\text{ kg})}{(6.02\times 10^{23})(8.9\times 10^3 \text{ kg/m}^3)(1.60\times 10^{-19}\text{C})\pi(0.65\times 10^{-3}\text{ m})^2}$$
$= 5.1 \times 10^{-10}$ m/s

(b) Calculate the current density from Eq.4.13:
$$J = \frac{I}{A} = \frac{I}{\pi r^2} = \frac{4I}{\pi d^2}$$
$$= \frac{4(2.3 \times 10^{-6} \text{ A})}{\pi(6.5 \times 10^{-4}\text{ m})^2} = 6.931 \text{ A/m}^2 \approx 6.9 \text{ A/m}^2$$

(c) The electric field is calculated from Eq.4.32: $\vec{E} = \rho\vec{J}$
$$J = \frac{1}{\rho}E \Rightarrow E = \rho J$$
$$\Rightarrow E = \left(1.68 \times 10^{-8} \Omega \cdot \text{m}\right)\left(6.931 \text{ A/m}^2\right)$$
$$= 1.2 \times 10^{-7} \text{ V/m}$$

4. We are given a charge density and a speed (like the drift speed) for both types of ions. From that we can use Eq.4.11: $I = neAv_d$ to determine the current per unit area. Both currents are in the same direction in terms of conventional current - positive charge moving north has the same effect as negative charge moving south - and so they can be added.

$$I = neAv_d$$
$$\Rightarrow \frac{I}{A} = (nev_d)_{\text{He}} + (nev_d)_{\text{O}}$$
$$= \left[\left(2.8\times 10^{12} \text{ ions /m}^3\right) 2\left(1.60 \times 10^{-19} \text{C/ion}\right)\right.$$
$$\left.\left(2.0 \times 10^6 \text{ m/s}\right)\right]$$
$$+ \left[\left(7.0 \times 10^{11} \text{ ions /m}^3\right)\left(1.60 \times 10^{-19} \text{C/ion}\right)\right.$$
$$\left.\left(6.2 \times 10^6 \text{ m/s}\right)\right]$$
$$= 2.486 \text{ A/m}^2 \approx 2.5 \text{ A/m}^2, \text{ North}$$

5. (a) The magnitude of the current density vector is
$$J = \frac{I}{A} = \frac{I}{\pi d^2/4} = \frac{4(1.2 \times 10^{-10}\text{ A})}{\pi(2.5 \times 10^{-3}\text{ m})^2} = 2.4 \times 10^{-5} \text{ A/m}^2$$

(b) The drift speed of the current-carrying electrons is
$$v_d = \frac{J}{ne} = \frac{2.4 \times 10^{-5} \text{ A/m}^2}{(8.47 \times 10^{28}/\text{m}^3)(1.60 \times 10^{-19}\text{C})}$$
$$= 1.8 \times 10^{-15} \text{ m/s}$$

6. In this case, we note that the radial width $\Delta r = 10$ μm is small enough (compared to $r = 1.20$ mm) that we can make the approximation
$$I = \int J\,dA = \int Br2\pi r\,dr \approx Br2\pi r\Delta r$$

Thus, the enclosed current is $2\pi Br^2\Delta r = 18.1$ μA. Performing the integral will also give you the same answer.

4.35.3 Check Point 3

1. Since the potential difference V and current I are related by $V = IR$, where R is the resistance of the electrician, the fatal voltage is $V = (50 \times 10^{-3} \text{ A})(2000\Omega) = 100$ V.

2. **APPROACH** The resistance of the coil is given by $R = \rho l/A$, where l is the length of the wire, ρ is the resistivity of copper, and A is the cross-sectional area of the wire.
   SOLUTION Since each turn of wire has length $2\pi r$, where r is the radius of the coil, therefore,
   $l = (250)2\pi r = (250)(2\pi)(0.12 \text{ m}) = 188.5$ m.
   If r_w is the radius of the wire itself, then its cross-sectional area is
   $$A = \pi r_w^2 = \pi\left(0.65 \times 10^{-3} \text{ m}\right)^2 = 1.33 \times 10^{-6} \text{ m}^2.$$
   According to Table4.1, the resistivity of copper is $\rho = 1.69 \times 10^{-8} \Omega \cdot$ m.
   Therefore, the resistance of the copper coil is
   $$R = \frac{\rho l}{A} = \frac{(1.69 \times 10^{-8} \Omega \cdot \text{m})(188.5 \text{ m})}{1.33 \times 10^{-6} \text{ m}^2} = 2.4 \text{ }\Omega.$$

3. We use $R/L = \rho/A = 0.150 \Omega/$km.
   (a) For copper
   $J = I/A = (60.0 \text{ A})(0.150\Omega/\text{km})/(1.69 \times 10^{-8}\Omega \cdot \text{m})$
   $= 5.32 \times 10^5$ A/m^2.
   (b) We denote the mass densities as ρ_m. For copper,
   $$(m/L)_c = \left(\frac{mA}{AL}\right)_c = \left(\frac{m}{\text{volume}}A\right)_c = (\rho_m A)_c$$
   $$= \left(8960 \text{ kg/m}^3\right)\left(1.69 \times 10^{-8}\Omega \cdot \text{m}\right)/(0.150\Omega/\text{km})$$
   $= 1.01$ kg/m
   (c) For aluminum
   $J = (60.0 \text{ A})(0.150\Omega/\text{km})/(2.75 \times 10^{-8}\Omega \cdot \text{m})$
   $= 3.27 \times 10^5$ A/m^2.
   (d) The mass density of aluminum is

$(m/L)_a = (\rho_m A)_a$
$= (2700 \text{ kg/m}^3)(2.75 \times 10^{-8} \Omega \cdot \text{m})/(0.150 \Omega/\text{km})$
$= 0.495 \text{ kg/m}$

4. We find the conductivity of Nichrome (the reciprocal of its resistivity) as follows:
$$\sigma = \frac{1}{\rho} = \frac{l}{RA} = \frac{l}{(V/I)A} = \frac{lI}{VA}$$
$$= \frac{(1.0 \text{ m})(4.0 \text{ A})}{(2.0 \text{ V})(1.0 \times 10^{-6} \text{ m}^2)} = 2.0 \times 10^6/\Omega \cdot \text{m}$$

5. (a) $I = V/R = 23.0 \text{ V}/15.0 \times 10^{-3} \Omega = 1.53 \times 10^3$ A.
   (b) The cross-sectional area is $A = \pi r^2 = \frac{1}{4}\pi D^2$. Thus, the magnitude of the current density vector is
   $$J = \frac{I}{A} = \frac{4I}{\pi D^2}$$
   $$= \frac{4(1.53 \times 10^{-3} \text{ A})}{\pi (6.00 \times 10^{-3} \text{ m})^2} = 5.41 \times 10^7 \text{ A/m}^2$$
   (c) The resistivity is
   $$\rho = \frac{RA}{l} = \frac{(15.0 \times 10^{-3} \Omega) \pi (6.00 \times 10^{-3} \text{ m})^2}{4(4.00 \text{ m})}$$
   $$= 10.6 \times 10^{-8} \Omega \cdot \text{m}$$
   (d) The material is platinum.

6. The thickness (diameter) of the wire is denoted by D. We use $R \propto l/A$ and note that $A = \frac{1}{4}\pi D^2 \propto D^2$. The resistance of the second wire is given by
   $$R_2 = R\left(\frac{A_1}{A_2}\right)\left(\frac{l_2}{l_1}\right) = R\left(\frac{D_1}{D_2}\right)^2\left(\frac{l_2}{l_1}\right)$$
   $$= R(2)^2\left(\frac{1}{2}\right) = 2R$$

7. The resistance at operating temperature T is $R = V/I = 2.9 \text{ V}/0.30 \text{ A} = 9.67 \Omega$. From Table 4.1, $\alpha = 4.50 \times 10^{-3}$ K^{-1} Thus, from $R = R_0[1 + \alpha(T - T_0)]$, we find that
   $$T = T_0 + \frac{1}{\alpha}\left(\frac{R}{R_0} - 1\right)$$
   $$= 20°C + \left(\frac{1}{4.5 \times 10^{-3} \text{K}^{-1}}\right)\left(\frac{9.67 \Omega}{1.1 \Omega} - 1\right)$$
   $$= 1.8 \times 10^3 \text{°C}$$
   Since a change in Celsius is equivalent to a change on the Kelvin temperature scale, the value of α used in this calculation is not inconsistent with the other units involved.

8. We use $J = E/\rho$, where E is the magnitude of the (uniform) electric field in the wire, J is the magnitude of the current density, and ρ is the resistivity of the material. The electric field is given by $E = V/l$, where V is the potential difference along the wire and l is the length of the wire. Thus $J = V/l\rho$ and
   $$\rho = \frac{V}{lJ} = \frac{115 \text{ V}}{(10 \text{ m})(1.4 \times 10^8 \text{ A/m}^2)} = 8.2 \times 10^{-8} \Omega \cdot \text{m}.$$

9. (a) Since the material is the same, the resistivity ρ is the same, which implies (by Eq.4.32: $\vec{E} = \rho\vec{J}$) that the electric fields (in the various rods) are directly proportional to their current-densities. Thus, $J_1 : J_2 : J_3$ are in the ratio 2.5 : 4 : 1.5 (see Fig.4.40). Now, the currents in the rods must be the same (they are "in series") so
   $$J_1 A_1 = J_3 A_3, \quad J_2 A_2 = J_3 A_3.$$
   Since $A = \pi r^2$, this leads (in view of the aforementioned ratios) to
   $$4r_2^2 = 1.5 r_3^2, \quad 2.5 r_1^2 = 1.5 r_3^2.$$
   Thus, with $r_3 = 2$ mm, the latter relation leads to $r_1 = 1.55$ mm.

   (b) The $4r_2^2 = 1.5 r_3^2$ relation leads to $r_2 = 1.22$ mm.

10. **APPROACH** Since the wire is stretched without changing it's volume, therefore, we apply Eq.4.44:
    $$\frac{R_1}{R_2} = \frac{l_1^2}{l_2^2} \quad \cdots (1)$$
    SOLUTION Given: $l_2 = 3l_1$, $R_1 = 6.0 \Omega$ and $R_2 =?$
    Substituting the given values in Eq.(1), we get dshjkd jsdfh sdfjh fjhsd fjsdhf jsdhfd jfh sdfjkhsd fjhsdf jhf jfh jfkhs jfhksd fjsdkhfsd
    $$R_2 = \frac{l_2^2}{l_1^2} R_1 = \frac{(3l_1)^2}{l_1^2} \times 6 \Omega = 54 \Omega$$

11. The absolute values of the slopes (for the straight-line segments shown in the graph of Fig. 4.41b) are equal to the respective electric field magnitudes. Thus, applying Eq. $J = I/A$ and Eq. $J = \sigma E$ to the three sections of the resistive strip, we get
    $$J_1 = \frac{I}{A} = \sigma_1 E_1 = \sigma_1 \left(0.50 \times 10^3 \text{ V/m}\right)$$
    $$J_2 = \frac{I}{A} = \sigma_2 E_2 = \sigma_2 \left(4.0 \times 10^3 \text{ V/m}\right)$$
    $$J_3 = \frac{I}{A} = \sigma_3 E_3 = \sigma_3 \left(1.0 \times 10^3 \text{ V/m}\right)$$
    We note that the current densities are the same since the values of I and A are the same (see the problem statement) in the three sections, so $J_1 = J_2 = J_3$.
    (a) Thus we see that $\sigma_1 = 2\sigma_3 = 2(3.00 \times 10^7 (\Omega \cdot \text{m})^{-1}) = 6.00 \times 10^7 (\Omega \cdot \text{m})^{-1}$.
    (b) Similarly, $\sigma_2 = \sigma_3/4 = (3.00 \times 10^7 (\Omega \cdot \text{m})^{-1})/4 = 7.50 \times 10^6 (\Omega \cdot \text{m})^{-1}$.

12. In this problem we compare the resistances of two conductors that are made of the same materials. The resistance of conductor A is given by
    $$R_A = \frac{\rho l}{\pi r_A^2},$$
    where r_A is the radius of the conductor. If r_o is the outside radius of conductor B and r_i is its inside radius, then its cross-sectional area is $\pi(r_o^2 - r_i^2)$, and its resistance is
    $$R_B = \frac{\rho L}{\pi(r_o^2 - r_i^2)}.$$
    Therefore, the ratio of the resistances is
    $$\frac{R_A}{R_B} = \frac{r_o^2 - r_i^2}{r_A^2} = \frac{(1.0 \text{ mm})^2 - (0.50 \text{ mm})^2}{(0.50 \text{ mm})^2} = 3.0.$$

    ☞ The resistance R of an object depends on how the electric potential is applied to the object. Also, R depends on the ratio l/A, according to $R = \rho l/A$.

4.35.4 Check Point 4

1. Use Eq.4.57 to find the power from the voltage and the current.
   $$P = IV = (0.27 \text{ A})(3.0 \text{ V}) = 0.81 \text{ W}$$

2. Use Eq.4.58 to find the resistance from the voltage and the power.
   $$P = \frac{V^2}{R} \to R = \frac{V^2}{P} = \frac{(240 \text{ V})^2}{3300 \text{ W}} = 17 \Omega$$

3. Use Eq.4.58 to find the voltage from the power and the resistance.
   $$P = \frac{V^2}{R} \Rightarrow V = \sqrt{RP} = \sqrt{(3300 \Omega)(0.25 \text{ W})} = 29 \text{ V}$$

4. Use Eq.4.58 to find the resistance, and the current.

(a) $P = \dfrac{V^2}{R} \Rightarrow R = \dfrac{V^2}{P} = \dfrac{(110\text{ V})^2}{75\text{ W}} = 161.3\,\Omega \approx 160\,\Omega$

$P = IV \Rightarrow I = \dfrac{P}{V} = \dfrac{75\text{ W}}{110\text{ V}} = 0.6818\text{ A} \approx 0.68\text{ A}$

(b) $P = \dfrac{V^2}{R} \Rightarrow R = \dfrac{V^2}{P} = \dfrac{(110\text{ V})^2}{440\text{ W}} = 27.5\,\Omega \approx 28\,\Omega$

$P = IV \Rightarrow I = \dfrac{P}{V} = \dfrac{440\text{ W}}{110\text{ V}} = 4.0\text{ A}$

5. (a) From Eq.4.58, if power P is delivered to the transmission line at voltage V, there must be a current $I = P/V$. As this current is carried by the transmission line, there will be power losses of I^2R due to the resistance of the wire. This power loss can be expressed as $\Delta P = I^2 R = P^2 R/V^2$. Equivalently, there is a voltage drop across the transmission lines of $V' = IR$. Thus, the voltage available to the users is $V - V'$, and so the power available to the users is
$P' = (V - V')I = VI - V'I = VI - I^2 R = P - I^2 R.$
The power loss is
$\Delta P = P - P' = P - (P - I^2 R) = I^2 R = P^2 R/V^2.$

(b) Since $\Delta P \propto \dfrac{1}{V^2}$, therefore, V should be as large as possible to minimize ΔP.

6. (a) Since $P = V^2/R \Rightarrow R = V^2/P$ says that the resistance is inversely proportional to the power for a constant voltage, we predict that the 850 W setting has the higher resistance.

(b) $R = \dfrac{V^2}{P} = \dfrac{(120\text{ V})^2}{850\text{ W}} = 17\,\Omega$

(c) $R = \dfrac{V^2}{P} = \dfrac{(120\text{ V})^2}{1250\text{ W}} = 12\,\Omega$

7. (a) Use Eq.4.58 to find the current:
$P = IV \Rightarrow I = \dfrac{P}{V} = \dfrac{95\text{ W}}{115\text{ V}} = 0.83\text{ A}$

(b) Use Eq.4.58 to find the resistance:
$P = \dfrac{V^2}{R} \Rightarrow R = \dfrac{V^2}{P} = \dfrac{(115\text{ V})^2}{95\text{ W}} \approx 140\,\Omega$

8. The power (and thus the brightness) of the bulb is proportional to the square of the voltage, according to Eq.4.58, $P = \dfrac{V^2}{R}$. Since the resistance is assumed to be constant, if the voltage is cut in half from 240 V to 120 V, the power will be reduced by a factor of 4. Thus the bulb will appear only about 1/4 as bright in the United States as in Europe.

4.35.5 Check Point 5

1. In Fig.4.56, the 6 Ω and 3 Ω resistances are in parallel. Their equivalent resistance R_1, is given as
$R_1 = \dfrac{6 \times 3}{6 + 3} = 2\,\Omega$

Now, this resistance, $R_1 (= 2\,\Omega)$ is in series with 4 Ω resistance (Fig.4.507). Therefore, equivalent resistance of the circuit between the terminals of battery is
$R_{eq} = 4 + R_1 = 4 + 2 = 6\,\Omega$

So, the equivalent resistance of the network connected with battery is 6 Ω.

Now, current drawn from the battery is

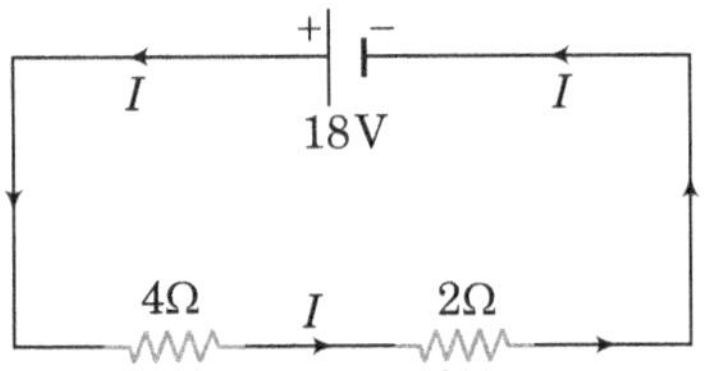

Figure 4.507

$I = \dfrac{\Delta V}{R} = \dfrac{18\text{ V}}{6\,\Omega} = 3\text{ A}$

2. **APPROACH** Between terminals P and R (Fig.4.57), three resistances each of magnitude r are connected in parallel order. Their net equivalent is $r_1 = r/3$.

Also, between terminals Q and R (Fig.4.57), two resistances each of magnitude r are connected in parallel order. Their net equivalent is $r_2 = r/2$.

Equivalent resistance between terminals P and Q: In Fig.4.57, we see that the resistance r connected between terminals P and Q is in parallel with series combinations of r_1 and r_2.

Equivalent resistance between terminals Q and R: In this case, the resistance r_2 is in parallel with series combinations of r and r_1.

Equivalent resistance between terminals P and R: In this case, the resistance r_1 is in parallel with series combinations of r and r_2.

SOLUTION The equivalent resistances between P and Q:
$R_{PQ} = r\|(r/2 + r/3) = r\|5r/6$
$= \dfrac{r(5r/6)}{r + 5r/6} = \dfrac{5}{11}r,$

The equivalent resistances between Q and R:
$R_{QR} = (r/2)\|(r + r/3) = \dfrac{4}{11}r,$

and the equivalent resistances between P and R:
$R_{PR} = (r/3)\|(r + r/2) = \dfrac{3}{11}r.$

Clearly, between terminals P and Q, the equivalent resistance of the circuit is maximum.

3. The effective resistance of the given circuit between the terminals of battery
$R_e = (30 + 30)\|30 = 60\|30$
$= \dfrac{30 \times 60}{30 + 60} = 20\,\Omega$

Thus, current $I = V/R_e = 2/20 = 1/10$ A.

4. **APPROACH** Note that brightness of a bulb is directly proportional to the electric power consumed by bulb. So, our task is to determine power consumed by bulbs in different situations. To determine, power in each case, we use Eq.4.60: $P = I^2 R$. From this equation we see that $P \propto I^2$.

When switch S is closed, A and B will be in parallel and their equivalent resistance will be in series with C.

Calculation of Power

1. When switch S is closed
$R_1 = R + \dfrac{R \times R}{R + R} = \dfrac{3}{2}R$

In this case, current provided by battery,
$I = \dfrac{V}{3R/2} = \dfrac{2V}{3R}$

This current passes through bulb C but at junction switch S, by current division rule, it splits into two equal parts because the resistance of bulb A equals that of B. The magnitude of each part is $I/2$, i.e., $V/3R$. One part passes through bulb A and other through bulb B.

In this case, power consumed by bulb C is
$P_C = I^2 R = \left(\dfrac{2V}{3R}\right)^2 R = \dfrac{4V^2}{9R}$

and power consumed by bulb A, $P_A = \left(\dfrac{V}{3R}\right)^2 R = \dfrac{V^2}{9R}$

similarly, power consumed by bulb B, $P_B = \dfrac{V^2}{9R}$

Clearly, $P_A = P_B = \dfrac{P_C}{4}$

2. When switch S is open

In this case, bulbs C and B will be in series and no current passes through A. The equivalent resistance of bulbs C and B is
$$R_2 = R + R = 2R$$
In this case, current provided by battery,
$$I' = \frac{V}{2R}$$
Now, power consumed by bulb C,
$$P'_C = I'^2 R = \left(\frac{V}{2R}\right)^2 R = \frac{V^2}{4R}$$
and power consumed by bulb B,
$$P'_B = I'^2 R = \left(\frac{V}{2R}\right)^2 R = \frac{V^2}{4R}$$
Clearly, $P'_C = P'_B = \frac{V^2}{4R}$ and $P'_A = 0$.
Also note that, $P'_C = 9P_C/16$, i.e., $P'_C < P_C$; $P'_B = 9P_B/4$, i.e., $P'_B > P_B$.
SOLUTION (a) Since, $P_A = P_B = P_C/4$, therefore bulbs A and B will be equally bright, but they will be less bright than bulb C
(b) When the switch S is open, no current can flow through bulb A, so, power consumed by it is zero and hence it will be dark. We now have a simple one-loop series circuit, and we expect bulbs B and C to be equally bright. However, the equivalent resistance of this circuit is greater than that of the circuit with the switch closed. When we open the switch, we increase the resistance and reduce the current leaving the battery. Thus, bulb C will be dimmer when we open the switch ($P'_C < P_C$). Also, bulb B gets more current when the switch is open, and so it will be brighter than with the switch closed (Mathematically, we have already shown in approach that $P'_B > P_B$); and B will be as bright as C.

5. **APPROACH** This problem is similar to EXAMPLE 38 with $R_1 = 2R$, $R_2 = 2R$ and $R_3 = R$, connected in parallel as shown in Fig.4.55.
SOLUTION The equivalent resistance is
$$R_{eq} = (2R\|2R)\|R = R\|R = R/2$$

6. From Fig.4.61, we see that resistors in arms CD and EF are in series which add up to $10 + 10 = 20\,\Omega$.
Similarly, resistors in arms CE and EF are also in series which add up to 20Ω
Now, these two equivalents are in parallel, so resistance between points C and F, is given by
$$R_{CF} = \frac{20 \times 20}{20 + 20} = 10\,\Omega$$
This R_{AB} is in series with remaining resistors between A and B. Therefore, net resistance between A and B is
$$R_{AB} = 10\Omega + 10\Omega + 10\Omega = 30\,\Omega$$

7. In Fig.4.62, resistors AF and FE of 3Ω each are in series with each other. Therefore, the network AEF is a parallel combination of two 6Ω resistors. Thus the resistance between points A and E is given by
$$R_{AE} = \frac{6 \times 6}{6 + 6} = 3\Omega$$
Now, R_{AE} is in series with resistor ED and their equivalent is in parallel with resistor AD. Solving as before, we get $R_{AD} = 3\,\Omega$. On repeating the process, we get
$$R_{AB} = 3\,\Omega$$

8. (A, D) In Fig.4.63, R_2 and R_L are in parallel, therefore, the effective resistance of R_2 and R_L, is
$$R_{R_2\|R_L} = R_2 R_L/(R_2 + R_L)$$
$$= 6 \times \frac{1.5}{6 + 1.5} = 1.2\text{ k}\Omega$$

Since, this equivalent resistance, $R_{R_2\|R_L}$, is connected in series with R_1. Therefore, the total effective resistance of the circuit connected across the terminals of battery is
$$R_{eff} = R_1 + R_{R_2\|R_L} = 2 + 1.2 = 3.2 \text{ k}\Omega$$
The current through the battery is
$$I = V/R_{eff} = 24/3.2 = 7.5 \text{ mA}$$
The respective potentials across R_1, R_2, and R_L and the corresponding power dissipations are
$$V_{R_1} = IR_1 = 7.5 \times 2 = 15 \text{ V},$$
$$P_{R_1} = V_{R_1}^2/R_1 = 15^2/\left(2 \times 10^3\right),$$
$$V_{R_2} = V - V_{R_1} = 24 - 15 = 9 \text{ V},$$
$$P_{R_2} = V_{R_2}^2/R_2 = 9^2/\left(6 \times 10^3\right),$$
$$V_{R_L} = V_{R_2} = 9 \text{ V},$$
$$P_{R_L} = V_{R_L}^2/R_L = 9^2/\left(1.5 \times 10^3\right)$$
Thus, the ratio, $P_{R_1}/P_{R_2} = 25/3$.
When R_1 and R_2 are interchanged, the effective resistance becomes
$$R'_{eff} = R_2 + R_{R_1\|R_L}$$
$$= 6 + (2 \times 1.5)/(2 + 1.5) = 48/7 \text{ k}\Omega$$
The current through the battery is
$$I' = 24/(48/7) = 3.5 \text{ mA}.$$
The potentials across R_1, R_2, and R_L and the corresponding power dissipations are
$$V'_{R_2} = I'R_2 = 3.5 \times 6 = 21 \text{ V},$$
$$P'_{R_2} = V'^2_{R_2}/R_2 = 21^2/\left(6 \times 10^3\right),$$
$$V'_{R_1} = V - V'_{R_2} = 24 - 21 = 3 \text{ V},$$
$$P'_{R_1} = V'^2_{R_1}/R_1 = 3^2/\left(2 \times 10^3\right),$$
$$V'_{R_L} = V'_{R_1} = 3 \text{ V},$$
$$P'_{R_L} = V'^2_{R_L}/R_L = 3^2/\left(1.5 \times 10^3\right),$$
Thus, the ratio, $P'_{R_L}/P_{R_L} = 3^2/9^2 = 1/9$.
Therefore, the correct choices are A and D.

9. **APPROACH** With the help of voltage-power rating, find the resistance of each bulb. As these bulbs are in series, therefore find their series equivalent resistance and hence current in the circuit. Now, calculate the power consumed by each bulb by using relation $P = I^2 R$. More power means more brightness
SOLUTION (B) Given voltage-power rating of bulb B_1 is (220 V, 60 W), therefore, resistance of bulb B_1 is
$$R_1 = \frac{V_1^2}{P_1} = \frac{(220)^2}{60}\Omega$$
Voltage-power rating of bulb B_2 is (220 V, 40 W), therefore resistance of bulb B_2 is
$$R_2 = \frac{V_2^2}{P_2} = \frac{(220)^2}{40}\Omega$$
Both B_1 and B_2 are connected in series with a 100 V battery, therefore equivalent resistance of the circuit:
$$R_{eq} = R_1 + R_2 = \left[\frac{(220)^2}{60} + \frac{(220)^2}{40}\right]\Omega$$
$$= \frac{(220)^2}{24}\Omega$$
Since, bulbs are in series, therefore same current passes through each bulb. The current flowing in this circuit is
$$I = \frac{V}{R_{eq}} = \frac{220 \text{ V}}{(220)^2/24\,\Omega} = \frac{24}{220} \text{ A}$$
Power consumed by bulb B_1,

$$P'_1 = I^2 R_1 = \left(\frac{24}{220}\right)^2 \times \frac{(220)^2}{60} W = \frac{24^2}{60} W$$

Power consumed by bulb B_2,

$$P'_2 = I^2 R_2 = \left(\frac{24}{220}\right)^2 \times \frac{(220)^2}{40} W = \frac{24^2}{40} W$$

Clearly, $P'_2 > P'_1$, therefore, bulb B_2 is brighter than bulb B_1.

10. In Fig.4.65, R_2 and R_3 are in parallel order. Their equivalent is in series with R_1.
   Now, equivalent of R_2 and R_3 is given by

   $$R_{23} = \frac{R_2 R_3}{R_2 + R_3} = \frac{50 \times 100}{50 + 100} = \frac{100}{3} \Omega$$

   This equivalent is in series with R_1, therefore in given circuit, the equivalent resistance connected with battery is

   $$R_{eq} = R_1 + R_{23} = 25 + \frac{100}{3} = \frac{175}{3} = 58.3 \, \Omega$$

11. (A) Current supplied by battery,

    $$I = \frac{\mathcal{E}}{R_{eq}} = \frac{12 \, V}{58.3 \Omega} = 0.206 \, A$$

    The potential V' across R_2 and R_3 is

    $$V' = \mathcal{E} - R_1 I = 12 \, V - (25\Omega)(0.206 A) = 6.85 \, V$$

    Therefore, $I_2 = \frac{V'}{R_2} = \frac{6.85 \, V}{50 \Omega} = 0.137 \, A$

    $$I_3 = \frac{V'}{R_3} = \frac{6.85 \, V}{100 \Omega} = 0.0685 \, A$$

4.35.6 Check Point 6

1. **APPROACH** Follow the approach given in Example 48.
   SOLUTION In case of charging of battery, the terminal potential difference
   $$\Delta V = \mathcal{E}_0 + Ir = 12 + 16 \times 3 = 60 \text{ volt.}$$

2. The resistance of voltmeter is 400 Ω as shown in the Fig.4.508a. The equivalent resistance of the two 400 Ω

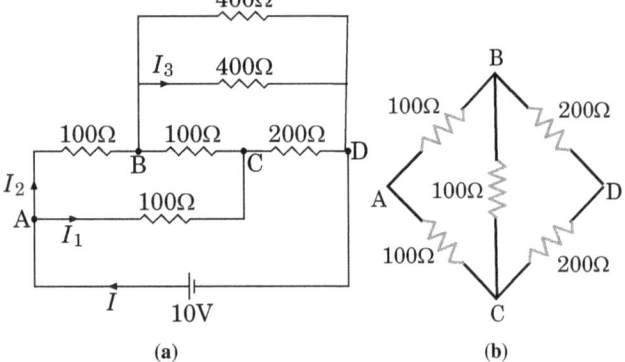

Figure 4.508

resistors connected in parallel is 200 Ω.
Fig.4.508b shows that the given circuit is a balanced Wheatstone bridge:

$$\frac{R_{AB}}{R_{BD}} = \frac{R_{AC}}{R_{CD}}$$

Therefore, no current passes through 100Ω resistor connecting B and C.
In this case, the resistance of upper branch connecting A and D is

$$R_1 = 100 + (400 \| 400) = 100 + 200 = 300 \Omega$$

and that of lower branch connecting A and D is

$$R_2 = 100 + 200 = 300 \Omega$$

Thus, the equivalent resistance between A and D is

$$R = R_1 \| R_2 = 150 \Omega$$

Thus, the currents I, I_1, and I_2 are given by

$$I = V/R = 10/150 = 1/15 \, A,$$
$$I_1 = I_2 = I/2 = 1/30 \, A$$

The current through 400 Ω resistor is

$$I_3 = I_2/2 = 1/60 \, A,$$

and voltage across it is

$$V_{400} = 400 I_3 = 400(1/60) = 20/3 \, V$$

3. **APPROACH** Effective resistance of the circuit depends on R_6 unless Wheatstone bridge formed by R_1, R_2, R_3, R_4 and R_6 is balanced. Thus, the current becomes independent of R_6 only when the bridge is balanced i.e.,

   $$\frac{R_1}{R_2} = \frac{R_3}{R_4} \qquad \cdots (1)$$

   SOLUTION From Eq.(1), the required condition is:
   $$R_1 R_4 = R_2 R_3$$

4. **SOLUTION** Given circuit is redrawn in the Figure 4.509 after moving out the terminals A and B from inside the triangle. It is a balanced Wheatstone bridge. Hence, resistance

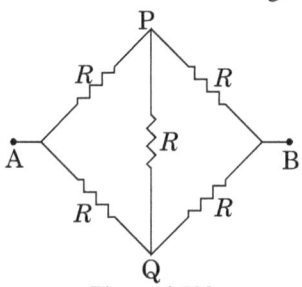

Figure 4.509

in the branch PQ can be removed without affecting the effective resistance of the circuit. Hence, the circuit has two branches, APB and AQB, each of resistance $2R$, connected in parallel. Thus, the effective resistance between A and B is

$$R_{AB} = (2R) \| (2R) = R$$

5. In Fig.4.128, resistances 6 Ω and 3 Ω between junctions C and B are in parallel. So, their equivalent resistance is

   $$R_1 = \frac{6 \times 3}{6 + 3} = 2 \, \Omega$$

   This resistance (R_1) is in series with the resistance of 3 Ω between node A and junction C; and resistance of 5 Ω connected between junction B and node D. Therefore, equivalent resistance of the circuit between the terminals of battery is

   $$R_{eq} = 3 + 2 + 5 = 10 \, \Omega$$

   Now, current given by battery,

   $$I = \frac{\Delta V}{R_{eq}} = \frac{20 \, V}{10 \, \Omega} = 2 \, A$$

   So, 2 A current passes through 3 Ω resistance connected between node A and junction C. At junction C it get divided into two parts. One part I_1 (say) passes through 6 Ω resistance and remaining part I_2 (say) passes through 3 Ω resistance. At junction B, these two parts again get combined and give the current $I = I_1 + I_2 = 2 \, A$. So, there will be 2 A current through resistance 5 Ω connected between junction B and node D.
   By current division rule, we have
   current through 6 Ω resistance, $I_1 = \frac{3}{6 + 3} \times 2 = \frac{2}{3} \, A,$
   current through 3 Ω resistance, $I_2 = \frac{6}{6 + 3} \times 2 = \frac{4}{3} \, A$
   Now, heat developed across 3 Ω resistance connected between node A and junction C is,

   $$H_1 = I^2 Rt = 2^2 \times 3 \times 2 = 24 \, J$$

   Similarly, heat developed across 5 Ω resistance connected between junction B and node D is,

$$H_2 = I^2 Rt = 2^2 \times 5 \times 2 = 40 \text{ J}$$

Heat developed across 6 Ω resistance connected between junctions C and B is,

$$H_3 = I_1^2 (6 \, \Omega)(2 \text{ s}) = \left(\frac{2}{3}\right)^2 \times 6 \times 2 = \frac{16}{3} \text{ J}$$

Similarly, heat developed across 3 Ω resistance connected between junctions C and B is,

$$H_4 = I_2^2 (3 \, \Omega)(2 \text{ s}) = \left(\frac{4}{3}\right)^2 \times 3 \times 2 = \frac{32}{3} \text{ J}$$

6. **APPROACH** In Fig.4.510a, we have to find the equivalent resistance between terminals B and D. In this figure, we see that resistors between points D, A and between A, B are in series. Similarly, resistors between points D, C and between C, B are also in other series. When you replace these series resistors with their equivalents, you see that all resistors between terminals B and D are in parallel. Find their parallel equivalent.

SOLUTION From Eq.4.63, the series equivalent resistance

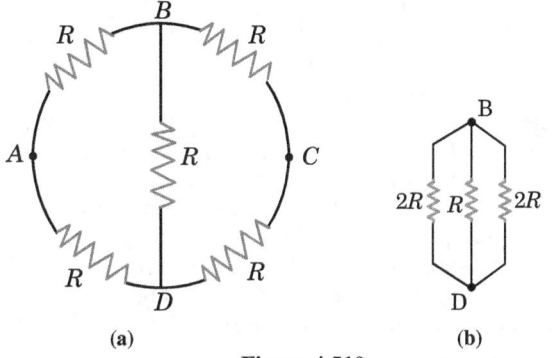

Figure 4.510

of resistors connected between points D, A and A, B is

$$R_0 = R_1 + R_2 = R + R = 2R$$

Similarly, the series equivalent resistance of resistors connected between points D, C and C, B is

$$R_0' = R + R = 2R$$

Now, on replacing the resistors in series with their equivalents, we get new circuit diagram as shown in Fig.4.510b. In Fig.4.510b, all resistors between points B and D are in parallel, therefore equivalent resistance between points B and D is given by

$$\frac{1}{R_{eq}} = \frac{1}{2R} + \frac{1}{R} + \frac{1}{2R} = \frac{2}{R}$$

or $R_{eq} = R/2$.

7. **APPROACH** In Fig.4.510a, we see that
$$\frac{R_{AB}}{R_{AD}} = \frac{R_{CB}}{R_{CD}}, \text{ i.e., } \frac{R}{R} = \frac{R}{R}$$

So, circuit satisfies the condition of balanced Wheatstone's bridge (Eq.4.135). In this case, points B and D will be at same potentials and we can remove the resistor connected between points point B and D. So, this circuit can be further simplified like circuit shown in Fig.4.511. Clearly, resistors between points A, B and between B, C are in series. Similarly, resistors between points A, D and between D, C are also in other series. These two series equivalents are in parallel between terminals A and C. Now, first solve for series equivalents and then for final equivalent resistance.

SOLUTION From Eq.4.63, the equivalent resistance of resistors connected between points A, B and B, C is
$$R_0 = R_1 + R_2 = R + R = 2R$$

Similarly, the equivalent resistance of resistors connected between points A, D and D, C is

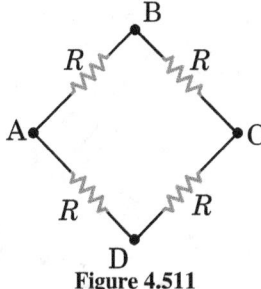

Figure 4.511

$$R_0' = R + R = 2R$$

Now, R_0 and R_0' are in parallel, therefore equivalent resistance between points A and C

$$R_{eq} = \frac{R_0 R_0'}{R_0 + R_0'} = \frac{2R \cdot 2R}{2R + 2R} = R$$

8. **APPROACH** Carefully observe series and parallel combinations and each time replace these combinations with their equivalents.

SOLUTION The Given circuit is redrawn in Fig.4.512a.

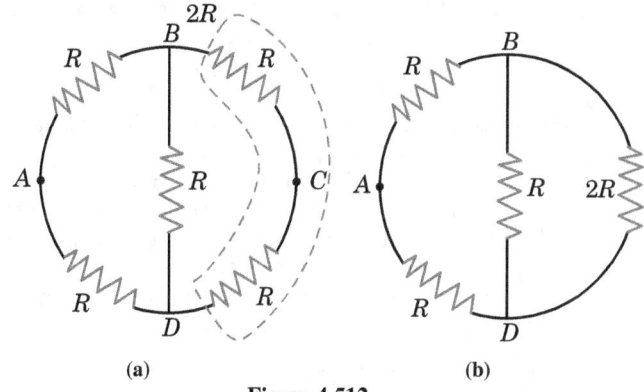

Figure 4.512

Resistors connected between points B, C and between points C, D are in series. Their equivalent resistance is $R + R = 2R$. This equivalent resistance $2R$ is in parallel connection with resistance R connected between terminals B and D [Fig.4.512b]. The parallel equivalent of these two resistances is $\frac{2R \times R}{2R + R} = \frac{2R}{3}$. Now, this equivalent $\frac{2R}{3}$ is in series connection with resistance R connected between terminals D and A [Fig.4.513a]. Their series equivalent is $R + \frac{2R}{3} = \frac{5R}{3}$. Now, this equivalent resistance of $\frac{5R}{3}$ is in parallel with the resistance R connected between terminals A and B [Fig.4.513b]. So, their equivalent resistance

$$R_{eq} = \frac{R \times \frac{5R}{3}}{R + \frac{5R}{3}} = \frac{5R}{8}.$$

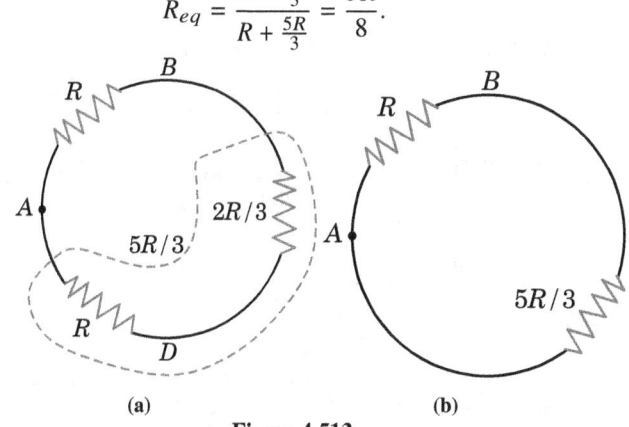

Figure 4.513

9. The given circuit is redrawn in Fig.4.514. In loop BCDEB, by Kirchhoff's voltage law, we have
$$-6 + 6 - 10I_4 = 0$$

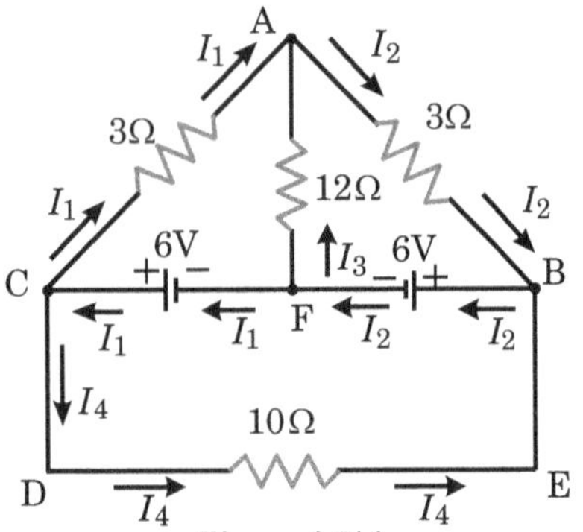

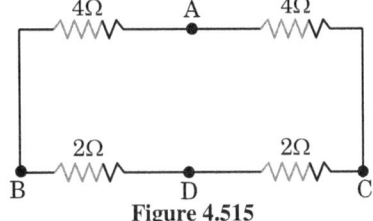

Figure 4.514

$$\Rightarrow \quad I_4 = 0 \quad \cdots (1)$$

This result tells us that I_1 and I_2 flow through their respective 3Ω resistances [Fig.4.514].

Now, in loop FCAF, by Kirchhoff's voltage law, we have
$$+6 - 3I_1 + 12I_3 = 0 \quad \cdots (2)$$

Similarly, by Kirchhoff's voltage law, in loop BFAB, we have
$$-6 - 12I_3 - 3I_2 = 0 \quad \cdots (3)$$

Now, at junction F, by Kirchhoff's junction law, we have
$$I_2 = I_1 + I_3 \quad \cdots (4)$$

Solving simulateously Eq.(1) to (4), we get
$I_1 = 0.222$ A, $I_2 = -0.222$ A, $I_3 = -0.444$ A and $I_4 = 0$ A.

10. In Fig.4.131, point D divides the lower 4 Ω resistance in two equal parts i.e., in two parts of 2 Ω each.
If we denote the resistances of branches BA, BD, AC and DC by R_{BA}, R_{BD}, R_{AC} and R_{DC}, then clearly
$$\frac{R_{BA}}{R_{BD}} = \frac{R_{AC}}{R_{DC}} = \frac{1}{2}$$
So, Wheatstone's balanced bridge condition is satisfied and hence we can remove the resistor connected between terminals A and D. The new circuit obtained is given in Fig.4.515.
Now, from Fig.4.515, the resistance between points B and

Figure 4.515

C is given by
$$R = \frac{(4+4)(2+2)}{(4+4)+(2+2)} = \frac{32}{12}\Omega = \frac{8}{3}\Omega$$

11. **APPROACH** Apply Eq.4.137:
$$X = \left(\frac{100-l}{l}\right)R \quad \cdots (1)$$
in both cases and then solve for X.
SOLUTION In first case, $l = 20$ cm, therefore Eq.(1) gives
$$X = \left(\frac{100-20}{20}\right)R = 4R$$
Now, when known resistance R is shunted (i.e., connected in parallel) with 10 Ω, it's net resistance
$$R' = \frac{10R}{10+R}$$
Clearly, $R' < R$, i.e., known resistance decreases and hence it's new null point length also decreases. So, new null point distance from left end will be $l' = 20$ cm $- 10$ cm $= 10$ cm.

Again from Eq.(1), we get
$$X = \left(\frac{100-10}{10}\right)R' = 9R'$$
Equating above two values of X, we get
$$4R = 9R' = 9 \times \frac{10R}{10+R}$$
or $\quad 40 + 4R = 90 \quad \Rightarrow \quad R = 50/4$
Therefore, $X = 4R = 50$ Ω

12. Let l is the the total length of the wire and x is the null point distance from left arm of the meter bridge. Wheatstone bridge is balanced when
$$X = Rx/(l-x) \quad \cdots (1)$$
To find the the relative error in measurement of X, we have to differentiate it with respect to x.
Taking logarithm of Eq.(1), we get
$$\ln X = \ln R + \ln\left(\frac{x}{(l-x)}\right)$$
On differentiating equation (1) with respect to x, we get
$$\frac{\Delta X}{X} = \frac{l-x}{x}\left(\frac{(l-x)\Delta x + x\Delta x}{(l-x)^2}\right)$$
or $\quad \frac{\Delta X}{X} = \frac{l\Delta x}{x(l-x)} \quad \cdots (2)$
where Δx is the measurement error in x. The value of ΔX in equation (2) attains its minimum value when the denominator is maximum i.e., $x = l/2$.
Now, from Fig.4.132, we see that corresponding to resistance R_2, the null point is at B for which x is very close to $l/2$. Thus, corresponding to R_2, the relative error in X, i.e., $\Delta X/X$ will be minimum and hence reading will be most accurate.

13. Let C be the null point on the wire. Given AC = 52 cm and NB = 100 - 52 = 48 cm.
Let A' and B' represent the two end points with end

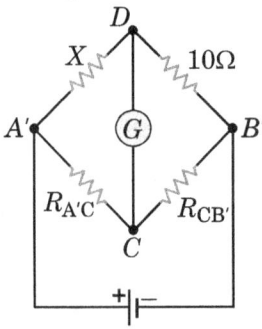

Figure 4.516

corrections i.e., $A'C = x = 52 + 1 = 53$ cm and $CB' = 48 + 2 = 50$ cm. The resistance of branch $A'C$ is $R_{A'C} = 53\lambda$ and that of branch CB' is $R_{CB'} = 50\lambda$, where λ is the resistance per unit length (in Ω/cm). Since, bridge is balanced at null point C, therefore, we can write
$$\frac{X}{R_{A'C}} = \frac{10\;\Omega}{R_{CB'}}$$
or $\quad X = \frac{10}{R_{B'C}}R_{A'C} = \frac{10}{50\lambda} \times 53\lambda$
or $\quad X = \frac{530}{50} = 10.6$ Ω

14. Let λ be the linear resistance density (in Ω/cm) of the meter bridge wire. Total length of the wire AB is 100 cm and null point is obtained at C, such that $AC = x = 40$ cm.
The resistance of length AC of the meter bridge wire, $R_{AC} = \lambda x$ and resistance of length BC of potentiometer is $R_{BC} = \lambda(100-x)$. It's equivalent Wheatstone bridge is shown in Fig.4.517.

Since, bridge is balanced, therefore, we can write

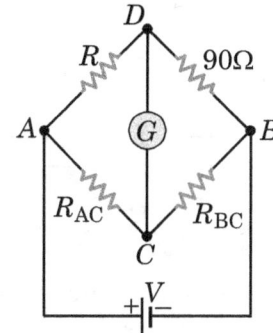

Figure 4.517

$$\frac{R}{R_{AC}} = \frac{90\,\Omega}{R_{BC}}$$

or $\quad R = \dfrac{90}{R_{BC}} R_{AC} = \dfrac{90}{\lambda(100-x)} \lambda x$

or $\quad R = \dfrac{90x}{(100-x)} \quad \cdots (1)$

Substituting, $x = 40$ in Eq.(1), we get

$$R = \frac{90 \times 40}{(100 - 40)} = 60\,\Omega$$

The least count of scale gives error in measurement of x, i.e., $\Delta x = 0.1$ cm. To find error in measurement of R, we have to differentiate equation (1) with respect to x:

$$\frac{dR}{dx} = \frac{(100-x)90 - 90x(-1)}{(100-x)^2} = \frac{9000}{(100-x)^2}$$

or $\quad dR = \dfrac{9000}{(100-x)^2} dx \quad \cdots (2)$

To find the relative error in R, divide Eq.(2) by Eq.(1):

$$\frac{dR}{R} = \frac{100}{x(100-x)} dx = \frac{dx}{x} + \frac{dx}{100-x}$$

or $\quad \dfrac{\Delta R}{R} = \dfrac{\Delta x}{x} + \dfrac{\Delta x}{100-x} \quad \cdots (3)$

Substituting $x = 40$ and $\Delta x = 0.1$, in above Eq.(3), we get

$$\frac{\Delta R}{R} = \frac{0.1}{40} + \frac{0.1}{100-40} = \frac{1}{400} + \frac{1}{600} = \frac{1}{4}$$

or $\quad \Delta R = \dfrac{1}{4} \times 60\,\Omega = 0.25\,\Omega$

Clearly, the error in the measurement of resistance is $0.25\,\Omega$. Therefore, the value of the unknown resistance R, including its error, is given by

$$R = (60 \pm 0.25)\,\Omega$$

15. Let X be the unknown resistance. The null point is obtained when Wheatstone bridge is balanced i.e.,

$$\frac{X}{2} = \frac{\rho l_1 / A}{\rho l_2 / A} = \frac{l_1}{l_2},$$

where l_1 and l_2 are as shown in the Fig.4.518a.
Since $X > 2\,\Omega$, we get $l_1 > l_2$. Also, $l_1 + l_2 = 100$ cm.
When the resistances are interchanged, the null point shifts by 20 cm (Fig.4.518b). As $X > 2\,\Omega$, the null point will shift towards the left i.e.,

$$l_1' = l_1 - 20 \text{ and } l_2' = l_2 + 20$$

The balance condition gives

$$\frac{2}{X} = \frac{l_1'}{l_2'} = \frac{l_1 - 20}{l_2 + 20}$$

On solving for x, we get

$$X = 3\,\Omega$$

16. The resistance of a wire (resistivity $= \rho$) of length l, cross-sectional area A is

$$R = \frac{\rho l}{A} = \frac{\rho l}{\pi r^2}$$

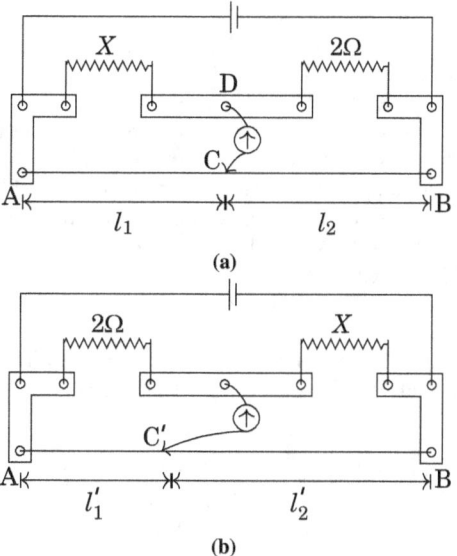

Figure 4.518

here r is the radius of the wire.

Clearly, for constant l, $R \propto \dfrac{1}{r^2}$.

Therefore, if the radius of the wire AB is doubled keeping length unchanged, then its resistance becomes one fourth and hence, current $(I = V/R)$ becomes four times.
If R' and I' are new resistance and new current respectively, then

$$R' = R/4, \text{ and } I' = 4I$$

In this case, new potential gradient along the length of wire

$$\frac{V}{l} = \frac{I'R'}{l} = \frac{(4I)(R/4)}{l} = \frac{IR}{l} = \frac{V}{l}$$

We see that, the potential drop per unit length, V/l, remains the same.
Therefore, new null point is also at $AC = x$.

17. (A, B, C, D) Given circuit is redrawn in Fig.4.519. Let I be the current leaving node A through branch AP and I' be the current entering node B through branch SB. Kirchhoff's junction law gives the currents in other branches. Since,

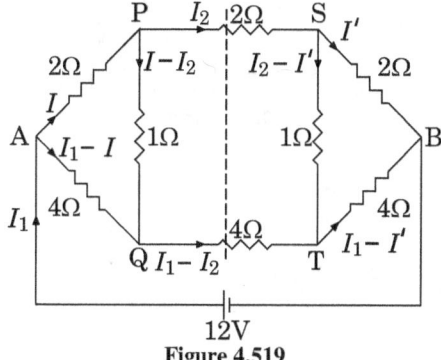

Figure 4.519

circuit has mirror symmetry about the dashed line as shown in Fig.4.519, therefore by symmetry, we can write,

$$I = I'$$

This can also be shown by applying Kirchhoff's voltage law in loops APQA and SBTS
By Kirchhoff's voltage law in loop APQA, we have

$$2I + (I - I_2) - 4(I_1 - I) = 0 \quad \cdots (1)$$

Similarly, by Kirchhoff's voltage law in loop SBTS, we have

$$2I' - 4(I_1 - I') - (I_2 - I') = 0 \quad \cdots (2)$$

Solving Eq.(1), for I, we get

$$I = (4I_1 + I_2)/7 \quad \cdots (3)$$

and solving Eq.(2) for I', we get
$$I' = (4I_1 + I_2)/7 \quad \cdots (4)$$
From Eq.(3) and (4), we have
$$I = I' = (4I_1 + I_2)/7 \quad \cdots (5)$$
Again, by symmetry, point P is symmetric to pint Q therefore they are at same potential. So, current in branch PQ will be zero. Point S is symmetric to point T therefore they will also be at same potential. So, current in branch ST will be zero. This can also be shown by applying Kirchhoff's loop law in PSTQP which gives
$$2I_2 + (I_2 - I') - 4(I_1 - I_2) - (I - I_2) = 0 \quad \cdots (6)$$
Substituting the values of I and I' from equation (3) and (4) into equation (6) and solving for I_1, we get
$$I_1 = \frac{3}{2} I_2 \quad \cdots (7)$$
Substituting this value of I_1 in Eq.(5), we get
$$I = I' = \frac{1}{7}\left(4\left(\frac{3}{2}I_2\right) + I_2\right) = I_2 \quad \cdots (8)$$
Now, applying Kirchhoff's voltage law in loop APSBA (through battery), we get,
$$6I_2 = 12 \quad \text{or} \quad I_2 = 2\text{ A}$$
Substituting this value in Eq.(8), we get
$$I = I' = I_2 = 2\text{ A} \quad \cdots (9)$$
and substituting the value of I_2 in Eq.(7), we get
$$I_1 = \frac{3}{2} I_2 = \frac{3}{2}(2\text{ A}) = 3\text{ A} \quad \cdots (10)$$
Taking the path QAPS, the potential at S is
$$V_S = V_Q + 4(I_1 - I) - 2I - 2I_2 = V_Q - 4\text{ V} \quad \cdots (10)$$
Clearly, all choices A, B, C and D are correct.

☞ It may be noted that the problem can be solved by using Kirchhoff's laws only without utilizing symmetry arguments.

18. (A, B, D) Let I_1 and I_2 are the currents as shown in the Fig.4.520. On applying Kirchhoff's voltage law in the loop

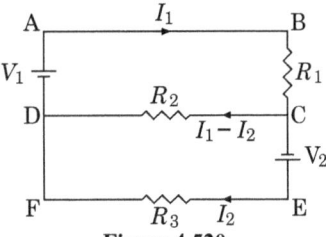

Figure 4.520

ABCDA, we get
$$I_1 R_1 + (I_1 - I_2) R_2 = V_1 \quad \cdots (1)$$
Applying Kirchhoff's voltage law in loop CEFDC, we get
$$I_2 R_3 - (I_1 - I_2) R_2 = V_2 \quad \cdots (2)$$
To find the current $(I_1 - I_2)$ through R_2, we multiply equation (1) by R_3 and (2) by R_1 and then subtract Eq.(2) in Eq.(1)
$$I_1 - I_2 = \frac{V_1 R_3 - V_2 R_1}{R_1 R_3 + R_2 R_3 - R_1 R_2}$$
The current through R_2 becomes zero when $V_1 R_3 = V_2 R_1$. Therefore, the correct choices are A, B and D.

19. (D) Let the unknown resistance be X and standard resistance be R. The null point is obtained when Wheatstone bridge is balanced (Fig.4.521). Thus,
$$\frac{X}{R} = \frac{\rho l_1 / A}{\rho l_2 / A} = \frac{l_1}{l_2} \quad \cdots (1)$$
The resistance of a metal increases with temperature. Thus X increases when its temperature is increased.
From Eq.(1), we can write

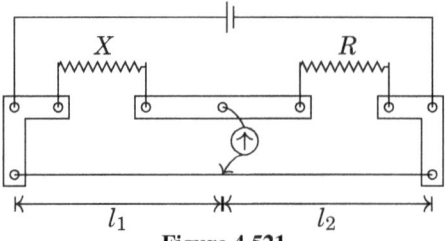

Figure 4.521
$$R = \frac{l_2}{l_1} X \quad \cdots (2)$$
From Eq.(2), we can say that corresponding to constant values of l_1 and l_2, any change in change in X changes R as given below
$$\Delta R = \frac{l_2}{l_1} \Delta X \quad \cdots (3)$$
From Eq.(3), it is clear that to obtain the null point at the same location (same l_1 and l_2), R should be increased as
$$\Delta R = \frac{l_2}{l_1} \Delta X.$$
Thus, the correct choice is D.

20. (A) Arrangement is shown in the Fig. 4.522.
$$\frac{X}{Y} = \frac{60}{40} = \frac{3}{2}$$
When X is shunted then resistance in the left gap becomes
$$X' = \frac{X \times \frac{X}{2}}{X + \frac{X}{2}} = \frac{X}{3}$$
Now,

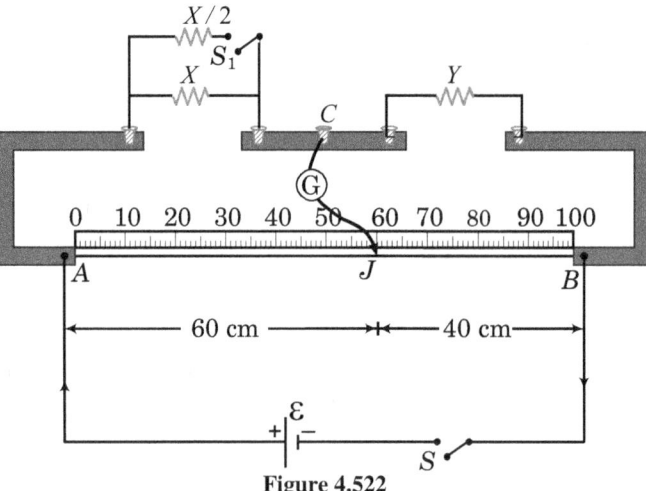

Figure 4.522

$$\frac{X/3}{Y} = \frac{l}{(100-l)} \Rightarrow \frac{1}{3} \times \frac{3}{2} = \frac{l}{(100-l)} \Rightarrow l = 33.3\text{ cm}$$
∴ Shift in null point = 60 − 33.3 = 26.7 cm.

4.35.7 Check Point 7

1. **APPROACH** Given: $I_g = 1.00$ mA $= 10^{-3}$ A, $G = 20$ Ω, $I = 50.0 \times 10^{-3}$ A and $S = ?$
Required shunt resistance and the given quantities are related by Eq.4.146:
$$S = \frac{I_g G}{I - I_g} \quad \cdots (1)$$
Substitute the given values in Eq.(1) and solve for S.
From Eq.4.147, ammeter resistance
$$R_A = \frac{GS}{S+G} \quad \cdots (2)$$
Now, substitute the given values in Eq.(2) and solve for required ammeter resistance R_A.
SOLUTION Substituting given values in Eq.(1), we get

$$S = \frac{(10^{-3})(20)}{(50.0 \times 10^{-3}) - (10^{-3})} = 0.408 \, \Omega$$

Now, substituting given values and the value of S calculated from Eq.(1), in Eq.(2), we get

$$R_A = \frac{GS}{G+S} = \frac{20 \times 0.408}{20 + 0.408} = 0.4 \, \Omega$$

Thus, we see that the shunt resistance is too small as compared to the galvanometer resistance that the ammeter resistance is approximately equal to the shunt resistance. This shunt resistance converts the given galvanometer into a low resistance ammeter with the desired range of 0 to 50.0 mA. At full scale deflection $I = 50.0$ mA, the current through the galvanometer is 1.0 mA while the current through the shunt is 49.0 mA. If the current I is less than 50.0 mA, the coil current and the deflection are proportionally less, but the ammeter resistance is still 0.4Ω

2. **APPROACH** To convert a galvanometer into a voltmeter, we connect a heavy resistance R_h in series with it. According to Eq.4.161, it's value is

$$R_h = \frac{V}{I_g} - G \quad \cdots (1)$$

Substitute the given values in above Eq.(1) and solve for R_h
SOLUTION Given: $G = 20\Omega$, Using $R = \frac{V}{I_g} - G$, $I_g = 1.0$ mA, 10 V and $R_h = ?$
Substituting these values in Eq.(1), we get

$$R = \frac{10}{10^{-3}} - 20 = 9980 \, \Omega$$

Thus, a resistance of 9980 Ω is to be connected in series with the galvanometer to convert it into the voltmeter of desired range.
Note: At full scale deflection current through the galvanometer, the voltage drop across the galvanometer is

$$V_g = I_g G = 20 \times 10^{-3} \text{ volt} = 0.02 \text{ volt}$$

and the voltage drop across the series resistance R_h is

$$V_{R_h} = I_g R_h = 9980 \times 10^{-3} \text{ volt} = 9.98 \text{ volt}$$

Thus, we see that most of the voltage appears across the series resistor R_h.

3. We have already studied that to increase the range of an ammeter we have to decrease the resistance of shunt resistor which is connected in parallel with galvanometer. As shunt resistance decreases, the net ammeter resistance decreases. To increase the range of voltmeter, we increase the resistance of the high resistance connected in series with the galvanometer. In this case, net resistance of voltmeter increases. Therefore, the correct order is

$$R_4 > R_3 > R_1 > R_2$$

4. Fig.4.523, shows a galvanometer's conversion into a voltmeter.
From eq.4.161, we have

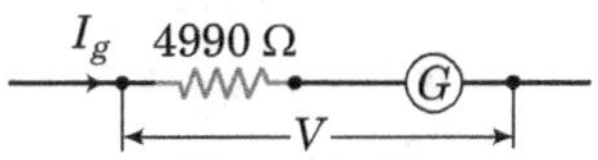

![Figure 4.523]
I_g — 4990 Ω — G
←— V —→
Figure 4.523

$$V = I_g(G + R_h) \quad \cdots (1)$$

Substituting given values, we get

$$30 = \frac{6}{1000}(G + 4990) \Rightarrow G = 10 \, \Omega$$

Now, in Fig.4.524, the above galvanometer is converted to an ammeter.
From Eq.4.146, we have

$$S = \frac{I_g G}{I - I_g} \quad \cdots (2)$$

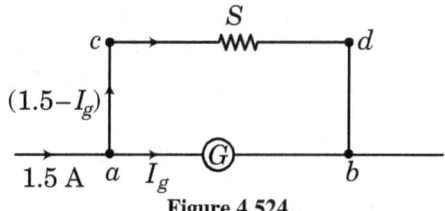

Figure 4.524

Given: $I = 1.5$ A, $I_g = 0.006$ A, $S = \frac{2n}{249}$ and $G = 10 \, \Omega$ (as calculated above).
Substituting these values in Eq.(2), we get

$$\frac{2n}{249} = \frac{0.006 \times 10}{1.5 - 0.006}$$

On solving for n, we get

$$n = 5$$

5. (B) From Eq.4.154, the current sensitivity in presence of shunt of resistance S is

$$CS = \left(\frac{G+S}{S}\right)\frac{\phi}{I} \quad \cdots (1)$$

In absence of shunt, the current sensitivity is

$$CS_0 = \lim_{S \to \infty}\left(\frac{G+S}{S}\right)\frac{\phi}{I} = \frac{\phi}{I} \quad \cdots (2)$$

Dividing Eq.(1) by (2), we get

$$\frac{CS}{CS_0} = \frac{G+S}{S} \quad \cdots (3)$$

Given: $CS = 30 \, CS_0$, $G = 406 \, \Omega$ therefore Eq.(3) gives

$$30 = \frac{406+S}{S} \Rightarrow 29S = 406$$

or

$$S = \frac{406}{29} = 14 \, \Omega$$

6. (D) Proceed as Q.5, you get

$$\frac{CS}{CS_0} = \frac{G+S}{S} \quad \cdots (1)$$

Given: $CS = 90CS_0$, $G = 8722$ ohm therefore Eq.(1) gives

$$90 = \frac{8722+S}{S} \Rightarrow 90S = 8722 + S$$

or

$$S = \frac{8722}{89} = 98 \, \Omega$$

4.35.8 Check Point 8

1. **APPROACH** From Eq.4.182, the internal resistance of a cell is given by

$$r = \frac{R(l-l')}{l'} \quad \cdots (1)$$

SOLUTION (C) Given: $l = 76.3$ cm, $l' = 60.0$ cm, and $R = 4.0 \, \Omega$
Substituting these values in Eq.(1), we get

$$r = \frac{4.0(76.3 - 60.0)}{60.0} = 1.1 \, \Omega$$

2. (D) Resistance per unit length of the potentiometer wire, $\lambda = 11.5 \, \Omega/m$
Length of potentiometer wire $L = 10$ m
∴ Resistance of wire $= \lambda L = 11.5 \times 10 = 115$
Since, a resistance of 5 Ω with negligible length is also put in series with the potentiometer wire,
So, total resistance $= 115 + 5 = 120 \, \Omega$

3. (A) Fall in potential per meter i.e., potential gradient of the wire

$$x = \frac{\text{applied voltage}}{\text{length of wire of potential}} = \frac{2}{10} \text{V/m} = 0.2 \text{ V/m}$$

4. (A) In Fig.4.155, current flowing in the wire,

$$I = \frac{2}{115+5} = \frac{2}{120} = \frac{1}{60} \text{A}$$

So, potential difference across the potentiometer wire AB is

$$\Delta V = IR_{AB} = \tfrac{1}{60} \times 115 = \tfrac{23}{12} \text{ V} \approx 1.92 \text{ V}$$

Therefore, required potential gradient across the wire AB,

$$= \frac{\text{voltage across potentiometer wire AB}}{\text{length of the wire AB}}$$

$$= \frac{1.92}{10} \text{ V/m} = 0.192 \text{ V/m} \approx 0.2 \text{ V/m}$$

4.35.9 Check Point 9

1. (B) **2.** (A) **3.** (A)

4.35.10 Check Point 10

1. APPROACH This problem is similar to Example 89, so you can apply any one approach of Example 89. Here, we solve it by observing symmetric points for same potential.

SOLUTION From Fig.4.525, we see that points C and D

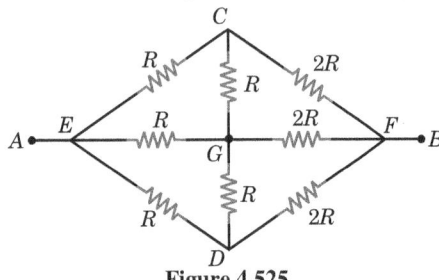

Figure 4.525

are symmetric points so they are at same potential. Next simplified circuit diagram is shown in Fig.4.526a.

Resistors in parallel connection: Resistors are in parallel between E, C and E, D; between G, C and G, D; and between F, C and F, D. Further simplified circuit is shown in Fig.4.526b. It shows that

$$\frac{R_{EC}}{R_{EG}} = \frac{R_{FC}}{R_{FG}}$$

So, according to Balanced Wheatstone bridge condition, we

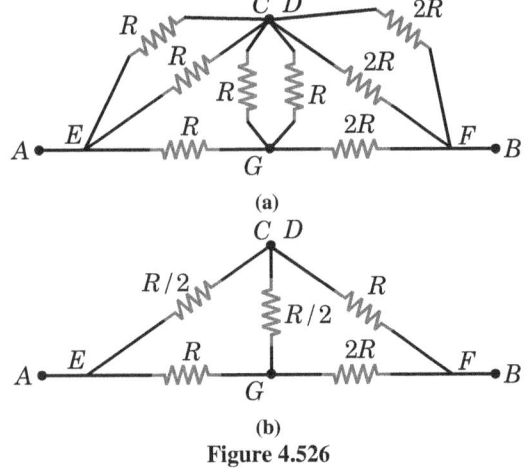

Figure 4.526

can remove the resistor between the terminals G and C. The new circuit diagram so obtained is shown in Fig.4.527. After removing the resistor between terminal G and C, the resistor connected between terminals E and C comes in series with the resistor connected between terminals C and F.

Similarly, the resistor connected between terminals E and G

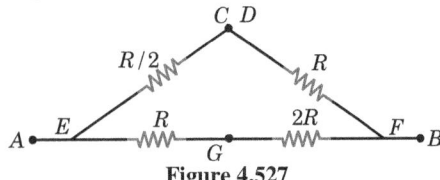

Figure 4.527

comes in series with resistor connected between terminals G and F. Their equivalents are in parallel. So, final equivalent resistance R_{eq} is given by

$$\frac{1}{R_{eq}} = \frac{1}{\frac{R}{2} + R} + \frac{1}{R + 2R} = \frac{2}{3R} + \frac{1}{3R} = \frac{3}{3R} = \frac{1}{R}$$

or $R_{eq} = R$

2. APPROACH The given circuit of Fig.4.189 is again redrawn in Fig.4.528a. Now, according to given problem we have to find the value of R_x, for which total resistance between points A and B is independent of number of sections. It means, if we cut the given circuit at dashed line CD (Fig.4.528a) and remove the last shown section after CD and then connect R_x between CD, we get the same equivalent resistance between terminals A and B as before.

So, between terminals C and D, the equivalent resistance of removed part shown in Fig.4.528b, must also be R_x. i.e.,

$$R \parallel (2R + R_x) = R_x$$

or $$\frac{R(2R + R_x)}{R + (2R + R_x)} = R_x \quad \cdots (1)$$

We can again reduce the number of sections by replacing the obtained new last section with R_x. If we continue this procedure, we get final circuit as shown in Fig.4.529.

Fig.4.529 shows that the equivalent resistance between terminals A and B is also be R_x. To get the value of R_x, solve above Eq.(1) for R_x.

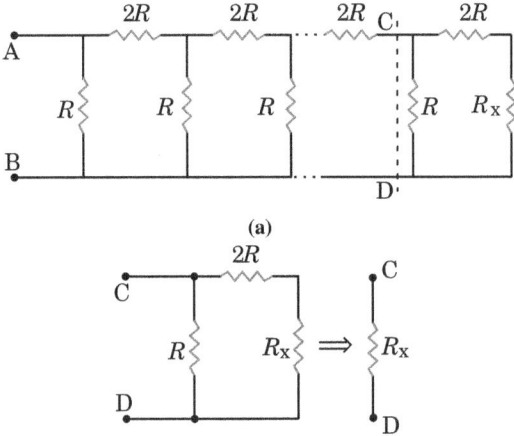

Figure 4.528

SOLUTION (A) From Eq.(1), we have

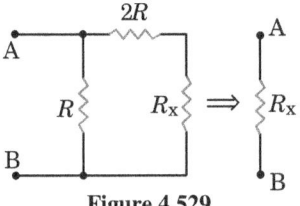

Figure 4.529

$$\frac{R(2R + R_x)}{R + (2R + R_x)} = R_x$$

or $$R_x^2 + 2RR_x - 2R^2 = 0$$

Solving for R, we get

$$R = \frac{-2R \pm \sqrt{4R^2 + 8R^2}}{2} = \frac{-2R \pm 2\sqrt{3}R}{2} = R(-1 \pm \sqrt{3})$$

On rejecting the negative root, we get

$$R_x = R(\sqrt{3} - 1)$$

4.35.11 Check Point 11

1. (C) The given network (Fig.4.206) can be reduced in the form as shown in Fig.4.530a using series and parallel grouping.

Using Delta-Star Transformation, the Figure 4.530a can be reduced to Figure 4.530b. Now the reduced network is very simple and using series and parallel grouping, the equivalent resistance between terminal 1 and 2 becomes $(7/5)R$.

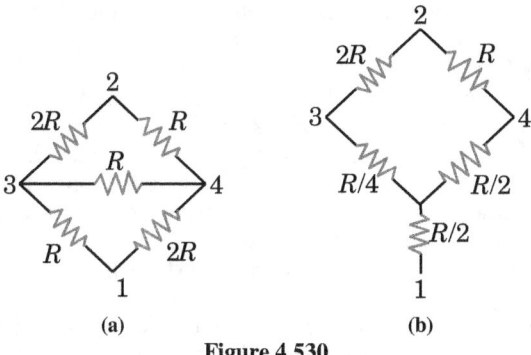

Figure 4.530

2. In network of Fig.4.531, we cannot remove the resistor connected between terminals C and D because Wheatstone's balanced bridge condition does not hold: $4/8 \neq 2/8$
Therefore, we break delta $\triangle ACD$ into star Y connection (Fig.4.532a).
$$R_A = \frac{R_{AB} \times R_{AC}}{R_{AB} + R_{AC} + R_{BC}} = \frac{4 \times 2}{4 + 2 + 4} = \frac{8}{10} = 0.8 \Omega$$

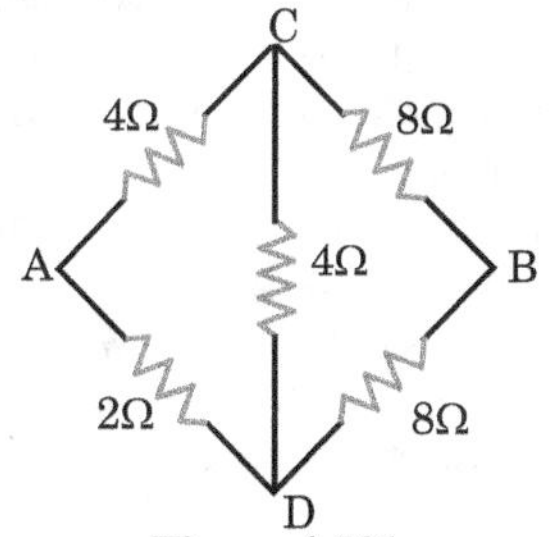

Figure 4.531

$$R_B = \frac{R_{AB} \times R_{BC}}{R_{AB} + R_{BC} + R_{AC}} = \frac{4 \times 4}{4 + 2 + 4} = \frac{16}{10} = 1.6 \Omega$$
$$R_C = \frac{R_{BC} \times R_{AC}}{R_{AB} + R_{BC} + R_{AC}} = \frac{4 \times 2}{4 + 2 + 4} = \frac{8}{10} = 0.8 \Omega$$
Connect R_A, R_B and R_C like Fig.4.532a and then remove

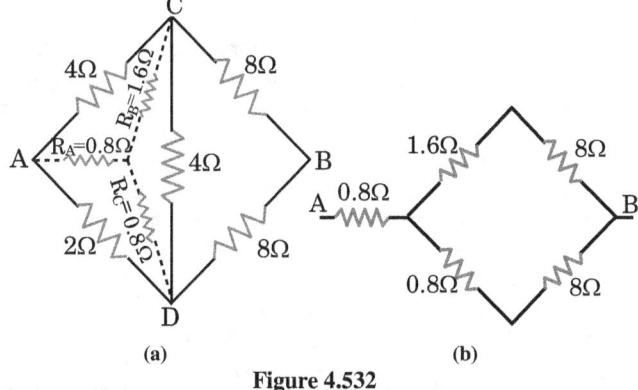

Figure 4.532

delta circuit from above it, you will get a circuit as shown in Fig.4.532b. In this circuit, net resistance across terminals A and B is
$$R = 0.8 + [(8 + 1.6) \| (0.8 + 8)] = 0.8 + \left(\frac{9.6 \times 8.8}{9.6 + 8.8}\right) = 5.4 \Omega$$

4.35.12 Check Point 12
1. **(a)** The time constant is given by-
$$\tau = CR = (100 \, \mu F)(2 \, \Omega) = 200 \, \mu s$$
(b) The charge stored at time t is

$$q = \mathcal{E}C\left(1 - e^{-t/CR}\right)$$
For 99% of maximum charge, i.e., $q = 0.99 \mathcal{E}C$, Eq.4.199, gives-
$$9.99 \mathcal{E}C = \mathcal{E}C\left(1 - e^{-t/(200\mu s)}\right)$$
$$\Rightarrow 9.99 = \left(1 - e^{-t/(200\mu s)}\right) \Rightarrow -\frac{t}{200\mu s} = \ln(0.01)$$
$$\Rightarrow t = 920 \, \mu s = 0.92 \text{ ms}.$$

2. The time constant '(τ)', is
$$\tau = CR = (50 \, \mu F)(1 \cdot 0 \, k\Omega) = 50 \text{ ms}$$
At $t = 1$ s, $t/CR = 1$ s/50 ms $= 20$.
The charge remaining on the capacitor is
$$q = Qe^{-t/CR} = (400 \, \mu C)e^{-20} = 8 \cdot 2 \times 10^{-7} \, \mu C$$

3. By Kirchhoff's loop law in circuit 4.237, we have
$$V - v_C - v_R = 0$$
When, $v_C = 3v_R$, then above equation gives-
$$V = v_C + \frac{v_C}{3} \implies v_C = \frac{3}{4}V$$
Now, from Eq.(4.200), the voltage gain across capacitor during it's charging is given by-
$$v_C = \mathcal{E}\left(1 - e^{-t/RC}\right)$$
here, for given circuit, $\mathcal{E} = V$, therefore, above equation gives
$$v_C = V\left(1 - e^{-t/RC}\right)$$
Substituting, $v_C = \frac{3}{4}V$, we get
$$\Rightarrow \frac{3}{4}V = V\left(1 - e^{-t/RC}\right)$$
$$\Rightarrow \frac{3}{4} = 1 - e^{-t/RC} \Rightarrow \frac{1}{4} = e^{-t/RC} \Rightarrow 4 = e^{t/RC}$$
$$\Rightarrow \frac{t}{RC} = \ln 4 \Rightarrow t = RC \ln 4 = 2RC \ln 2$$
$$= 2 \times 4 \times 10^6 \times 2.5 \times 10^{-6} \times 0.693 = 13.86 \text{ s}$$

4. In Fig.4.238 capacitors C_1 and C_2 are in series. Their effective capacitance
$$C = \frac{C_1 C_2}{C_1 + C_2} = \frac{3 \times 6}{3 + 6} = 2 \, \mu F,$$
Resistances R_1, R_2 are in parallel, therefore their effective resistance-
$$R = \frac{R_1 R_2}{R_1 + R_2} = \frac{4}{4} + 12 = 3 \Omega$$
The equivalent circuit diagram is shown in Fig.4.533. In

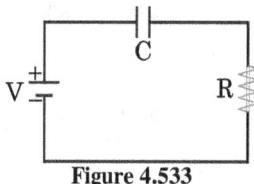

Figure 4.533

this diagram equivalent resistor R and equivalen capacitor C are in series, therefore time constant-
$$\tau = RC = (3)\left(2 \times 10^{-6}\right) = 6 \, \mu s$$

4.35.13 Conceptual Questions
1. Connect the resistors in series. For example, if resistors of $5.0 \, \Omega$, $10.0 \, \Omega$ and $20.0 \, \Omega$ are connected in series, then their equivalent resistance R_{eq} is given by
$$R_{eq} = 5.0 + 10.0 + 20.0 = 35.0 \, \Omega.$$
Clearly $R_{eq} = 35.0 \, \Omega$ is larger than the greatest individual resistance $20.0 \, \Omega$ connected in series with other smaller resistances.

2. Connect the resistors in parallel. For example, if resistors of $5.0 \, \Omega$, $10.0 \, \Omega$ and $20.0 \, \Omega$, are connected in parallel, their

equivalent resistance R_{eq} is given by
$$\frac{1}{R_{eq}} = \frac{1}{5.0} + \frac{1}{10.0} + \frac{1}{20.0} = \frac{4.0 + 2.0 + 1.0}{20.0} = \frac{7.0}{20.0}$$
or $R_{eq} = \frac{20.0}{7.0} \approx 2.9 \, \Omega$

Clearly, R_{eq} is smaller than the smallest resistor 5.0 Ω connected in parallel.

3. A battery rating in ampere-hours gives the total amount of charge available in the battery.

4. Series, because the circuit breaker trips and opens the circuit when the current in that circuit loop exceeds a certain preset value. The circuit breaker must be in series to sense the appropriate current.

5. The potential difference between the terminals of a battery will equal the emf of the battery when there is no current in the battery. At this time, the current though, and hence the potential drop across the internal resistance is zero. This only happens when there is no load placed on the battery-that includes measuring the potential difference with a voltmeter! The terminal voltage will exceed the emf of the battery when current is driven backward through the battery, in at its positive terminal and out at its negative terminal.

6. Since, in an electrolyte, both positive and negative charge can flow, therefore, both types of charges produce electric current

    Flow of positive charge produces the electric current in it's direction whereas the flow of negative charge produces current opposite to it.

    In the given electrolyte, the positive ions move from left to right, therefore they produce electric current from left to right. Suppose, this current is I_1. Negative ions are moving from right to left, so they also produce an electric current opposite to the direction of their motion, i.e., from left to right. Suppose, this current is I_2.

    So, the net current produced by the motion of both ions,
    $$I = I_1 + I_2$$
    It is from left to right.

7. Yes. The direction of current will be along the direction of flow of positive charge, i.e., in the direction of flow of protons which is from east to west.

8. When a flash-light is operated, the battery energy is being used up.

9. Note that $\lim_{\Delta t \to 0} \frac{\Delta l}{\Delta t}$ represents instantaneous speed. Under the influence of an electric field produced by applied voltage, electrons are always accelerated opposite to it. During their motion, they may collide with stationary atoms or other moving electrons. After each collision, their direction of motion and magnitude of speed get changed but they are still accelerated opposite to the applied field. If we observe for a very long time, we find that electrons are little drifted opposite to applied field. This is the reason why we define the drift speed by, $\frac{\Delta l}{\Delta t}$, where Δl is the distance drifted in a long time Δt. The direction of drift of electrons is always oppsite to the current i.e., opposite to the applied field. Now, the instantaneous speed of free electrons defined as $\lim_{\Delta t \to 0} \frac{\Delta l}{\Delta t}$, gives instantaneous shift of electrons which may be in any direction. Drift is not defined as the random shift of electrons.

10. If the emf in a circuit remains constant and the resistance in the circuit is increased, less current will flow, and the power dissipated in the circuit will decrease. Both power equations ($P = V^2/R$ and $P = I^2 R$) support this result. If the current in a circuit remains constant and the resistance is increased, then the emf ($V = IR$) must increase and the power dissipated in the circuit will increase. Both equations also support this result. There is no contradiction, because the voltage, current, and resistance are related to each other by $V = IR$.

11. No, because rate of flow of electrons from wire to positive terminal of voltage source is always equal to the rate of flow of electrons from negative terminal to the wire. So, always a charge neutrality remains maintained throughout the wire.

12. In absence of a potential difference across a metallic conductor, the electric field inside it remains zero. In this case, charge remains stationary on the surface of the metal. But if we apply a potential difference across a metallic conductor, an electric field produces inside it which applies an electric force on free electrons opposite to direction of electric field. As a result, an electric current produces in it in the direction of applied field. So, there is a misconception that electric field always remains zero inside a metal. It is true only when there is no potential difference across the metallic conductor.

13. If the resistance of a small immersion heater were increased, it would slow down the heating process. The emf in the circuit made up of the heater and the wires that connect it to the wall socket is maintained at a constant rms* value. If the resistance in the circuit is increased, less current will flow, and the power dissipated in the circuit will decrease, slowing the heating process.

14. Resistance is proportional to length and inversely proportional to cross-sectional area.
    (a) For the least resistance, you want to connect the wires to maximize area and minimize length. Therefore, connect them opposite to each other on the faces that are $2a$ by $3a$.
    (b) For the greatest resistance, you want to minimize area and maximize length. Therefore, connect the wires to the faces that are $1a$ by $2a$.

15. When a light is turned on, the filament is cool, and has a lower resistance than when it is hot. The current through the filament will be larger, due to the lower resistance. This momentary high current will heat the wire rapidly, possibly causing the filament to break due to thermal stress or vaporize. After the light has been on for some time, the filament is at a constant high temperature, with a higher resistance and a lower current. Since the temperature is constant, there is less thermal stress on the filament than when the light is first turned on.

16. When connected to the same potential difference, the 100 W bulb will draw more current ($P = IV$). The 75 W bulb has the higher resistance ($V = IR$ or $P = V^2/R$).

17. No. Energy is dissipated in a resistor but current, the rate of flow of charge, is not used up.

18. In the two wires described, the drift velocities of the electrons will be about the same, but the current density, and therefore the current, in the wire with twice as many free electrons per atom will be twice as large as in the other wire.

19. (a) If the length of the wire doubles, its resistance also doubles, and so the current in the wire will be reduced by a factor of two. Drift velocity is proportional to current, so the drift velocity will be halved.
    (b) If the wire's radius is doubled, the drift velocity remains

---

*The term "rms" means "root mean squared". In current electricity, it is used for alternating current (ac) or alternating voltage. The rms ac value is equivalent to that dc value which produces the same heating effect as ac.

the same. (Although, since there are more charge carriers, the current will quadruple.)

(c) If the potential difference doubles while the resistance remains constant, the drift velocity and current will also double.

20. (a) True. The chemical reactions in the electrolytes separate the positive and negative charges. This creates a potential difference. The charges flowing in the circuit have energy due to this potential difference.

(b) False. It is true that a battery is a source of potential difference. But the potential difference is always the same only for an ideal battery. In a real battery there are energy losses in the battery itself so the terminal voltage is not always the same.

(c) False. The current leaving the battery depends upon the resistance in the circuit.

21. In expression, $U = VIt$, both V and I are time dependent quantities, therefore we cannot say that U is proportional to I. However, in expression, $U = I^2 Rt$, only I is time dependent, therefore, we can say that U is proportional to I^2.

22. (a) Since $J = \frac{I}{A}$, when I is doubled, J is doubled.

(b) The conduction electron density is a property of the material, so it is unchanged.

(c) The mean time between collisions 'τ' is a property of the material and is unchanged for a constant temperature.

(d) Since $J = nev_d$ and J is doubled but n and e are constants, v_d is also doubled.

23. When we connect a voltage source across the terminals of a resistor, a current passes through it. When a current I passes through the resistor R the rate at which electric energy converted to thermal energy, is $I^2 R$. This power is provided by voltage source. Therefore for a circuit containing only an ideal battery and a resistor, "work done by the battery" and "the thermal energy developed" represent two names of the same physical quantity

24. Potentiometer is used to measure emf.

25. The whole wire is very nearly at one uniform potential. There is essentially zero difference in potential between the bird's feet. Then negligible current goes through the bird. The resistance through the bird's body between its feet is much larger than the resistance through the wire between the same two points.

26. The potential difference across a resistor is positive when it is measured against the direction of the current in the resistor.

4.35.14 Problems

1. **APPROACH** Use the definition of current, Eq.4.1:
$I = \Delta Q / \Delta t$ with $\Delta Q = Ne$
SOLUTION $I = \frac{\Delta Q}{\Delta t}$
$\therefore \quad \Delta Q = I \Delta t = (2.0 \text{ A})(1.0 \text{ s}) = 2.0 \text{ C}$
Now, $\Delta Q = Ne \Rightarrow N = \frac{\Delta Q}{e} = \frac{2.0}{1.6 \times 10^{-19}} = 1.25 \times 10^{19}$

2. **SOLUTION** No, the average current is not zero. Direction of current is the direction of motion of positive charge or opposite to the direction of motion of negative charge. So, both electrons and protons produce currents from left to right and net current will be their sum.
$I_{av} = I_{electron} + I_{proton}$
$= \frac{\Delta Q_1}{\Delta t_1} + \frac{\Delta Q_2}{\Delta t_2} = \frac{40e}{10} + \frac{40e}{10} = 8e$
$= 8 \times 1.6 \times 10^{-19} \text{ A} = 1.28 \times 10^{-18} \text{ A}$

3. Let σ is the charge per unit area of the belt, w is the belt width and v is the speed of the belt. If belt moves a distance Δl in time Δt, then $\Delta l = v \Delta t$. In this time interval, the area of belt which reaches to experimental device is, $\Delta A = w \Delta l$. If belt carries ΔQ charge to device in time Δt, then from Eq.4.3, we have
$\Delta Q = I \Delta t$ $\quad \cdots (1)$
Since, $\Delta Q = \sigma \Delta A = \sigma w \Delta l$, therefore Eq.(1), gives
$\sigma w \Delta l = I \Delta t$
or $\quad \sigma = \frac{I \Delta t}{w \Delta l} = \frac{I}{w \Delta l / \Delta t} = \frac{I}{wv}$
Now, substituting the given values in above expression, we get
$\sigma = \frac{I}{wv} = \frac{100 \times 10^{-6} \text{ A}}{(50 \times 10^{-2} \text{ m})(30 \text{ m/s})} = 6.7 \times 10^{-6} \text{C/m}^2$

4. Assuming $\vec{J}$ is directed along the wire (with no radial flow) we integrate, starting with Eq.4.18,
$I = \int J dA = \int_{9R/10}^{R} \left(kr^2\right) 2\pi r dr = \frac{1}{2} k\pi \left(R^4 - 0.656 R^4\right)$
where $k = 3.0 \times 10^8$ and SI units are understood. Therefore, if $R = 0.002$ m, we obtain $I = 2.59 \times 10^{-3}$ A.

5. (a) The current resulting from this non-uniform current density is
$I = \int_{\text{cylinder}} J_a dA = \frac{J_0}{R} \int_0^R r \cdot 2\pi r dr = \frac{2}{3} \pi R^2 J_0$
$= \frac{2}{3} \pi \left(3.40 \times 10^{-3} \text{ m}\right)^2 \left(5.50 \times 10^4 \text{ A/m}^2\right) = 1.33 \text{ A}$

(b) In this case,
$I = \int_{\text{cylinder}} J_b dA = \int_0^R J_0 \left(1 - \frac{r}{R}\right) 2\pi r dr = \frac{1}{3} \pi R^2 J_0$
$= \frac{1}{3} \pi \left(3.40 \times 10^{-3} \text{ m}\right)^2 \left(5.50 \times 10^4 \text{ A/m}^2\right) = 0.666 \text{ A}$

(c) The result is different from that in part (a) because J_b is higher near the centre of the cylinder (where the area is smaller for the same radial interval) and lower outward, resulting in a lower average current density over the cross section and consequently a lower current than that in part (a). So, J_a has its maximum value near the surface of the wire.

6. From Eq.4.37, we have
$R = \frac{V}{I} = \frac{120 \text{ V}}{4.2 \text{ A}} = 29 \Omega$

7. (a) Use Eq.4.37, to find the current.
$V = IR \Rightarrow I = \frac{V}{R} = \frac{240 \text{ V}}{8.6 \Omega} = 27.91 \text{ A} \approx 28 \text{ A}$

(b) From Eq.4.2, electric current, $I = \frac{\Delta Q}{\Delta t}$
$\Rightarrow \Delta Q = I \Delta t = (27.91 \text{ A})(50 \text{ min})(60 \text{ s/min}) = 8.4 \times 10^4 \text{C}$

8. (a) Solve Eq.4.37 for resistance.
$R = \frac{V}{I} = \frac{120 \text{ V}}{9.5 \text{ A}} = 12.63 \Omega \approx 13 \Omega$

(b) Applying the definition of average current, Eq.4.2: $I = \frac{\Delta Q}{\Delta t}$, we get
$\Delta Q = I \Delta t = (9.5 \text{ A})(15 \text{ min})(60 \text{ s/min}) = 8600 \text{C}$

9. To find current, we use Ohm's Law, Eq.4.37:
$I = \frac{\Delta V}{R}$ $\quad \cdots (1)$
Then to find the number of electrons flowing in time Δt, we apply Eq.4.2:
$I = \frac{\Delta Q}{\Delta t}$, with $\Delta Q = Ne$,

i.e., $\quad I = \dfrac{Ne}{\Delta t} \quad \cdots$ (2)

Here, N is the number of electrons leaving the battery in time Δt.
Now, comparing Eq.(1) and (2), we get
$$\dfrac{Ne}{\Delta t} = \dfrac{\Delta V}{R}$$
or $\quad N = \dfrac{\Delta V}{eR}\Delta t \quad \cdots$ (3)

Given: $\Delta V = 4.5$ V, $R = 1.6$ Ω and $\Delta t = 60$ s
Substituting these values in Eq.(3), we get
$$N = \dfrac{\Delta V}{eR}\Delta t = \dfrac{4.5 \text{ V}}{1.6 \times 10^{-19} \times 1.6\, \Omega} \times 60\text{ s}$$
$$= 1.1 \times 10^{21} \text{ electrons}$$
Thus, 1.1×10^{21} electrons leave the battery per minute.

10. **(a)** If the voltage drops by 15%, and the resistance stays the same, then by Eq.4.37, $V = IR$, the current will also drop by 15%.
$$I_{\text{final}} = 0.85 I_{\text{initial}} = 0.85(6.50 \text{ A}) = 5.525 \text{ A} \approx 5.5 \text{ A}$$
(b) If the resistance drops by 15% (the same as being multiplied by 0.85), and the voltage stays the same, then by Eq.4.37, the current must be divided by 0.85.
$$I_{\text{final}} = \dfrac{I_{\text{initial}}}{0.85} = \dfrac{6.50 \text{ A}}{0.85} = 7.647 \text{ A} \approx 7.6 \text{ A}$$

11. To find the diameter, use Eq. 4.37: $R = \rho l/A$, with the area as $A = \pi r^2 = \pi d^2/4$.
$$R = \rho \dfrac{l}{A} = \rho \dfrac{4l}{\pi d^2}$$
$$\Rightarrow d = \sqrt{\dfrac{4l\rho}{\pi R}} = \sqrt{\dfrac{4(1.00 \text{ m})(5.6 \times 10^{-8}\,\Omega\cdot\text{m})}{\pi(0.32\,\Omega)}}$$
$$= 4.7 \times 10^{-4} \text{ m}$$

12. Use Eq.4.37: $R = \rho l/A$ to calculate the resistance, with the area as $A = \pi r^2 = \pi d^2/4$.
$$R = \rho\dfrac{l}{A} = \rho \dfrac{4l}{\pi d^2}$$
$$= \left(1.68 \times 10^{-8}\,\Omega\cdot\text{m}\right)\dfrac{4(4.5 \text{ m})}{\pi(1.5 \times 10^{-3}\text{ m})^2}$$
$$= 4.3 \times 10^{-2}\,\Omega$$

13. (a) Use Ohm's law to find the resistance.
$$V = IR \Rightarrow R = \dfrac{V}{I} = \dfrac{0.0220 \text{ V}}{0.75 \text{ A}} = 0.02933\,\Omega \approx 0.029\,\Omega$$
(b) Find the resistivity from Eq.4.37: $R = \rho l/A$.
$$R = \dfrac{\rho l}{A} \Rightarrow \rho = \dfrac{RA}{l} = \dfrac{R\pi r^2}{l}$$
$$\Rightarrow \rho = \dfrac{(0.02933\,\Omega)\pi(1.0 \times 10^{-3}\text{ m})^2}{(5.80 \text{ m})}$$
$$= 1.589 \times 10^{-8}\,\Omega\cdot\text{m} \approx 1.6 \times 10^{-8}\,\Omega\cdot\text{m}$$
(c) Use Eq.4.13: $J = I/A$, to find the current density.
$$J = \dfrac{I}{A} = \dfrac{I}{\pi r^2} = \dfrac{0.75}{\pi(0.0010 \text{ m})^2}$$
$$= 2.387 \times 10^5 \text{ A/m}^2 \approx 2.4 \times 10^5 \text{ A/m}^2$$
(d) Use Eq.4.32: $E = \rho J$ to find the electric field.
$$E = \rho J = \left(1.589 \times 10^{-8}\,\Omega\cdot\text{m}\right)\left(2.387 \times 10^5 \text{ A/m}^2\right)$$
$$= 3.793 \times 10^{-3} \text{ V/m} \approx 3.8 \times 10^{-3} \text{ V/m}$$
(e) Find the number of electrons per unit volume from the absolute value of Eq.4.17: $J = nev_d$.
$$J = nev_d$$
$$\Rightarrow n = \dfrac{J}{v_d e} = \dfrac{2.387 \times 10^5 \text{ A/m}^2}{(1.7 \times 10^{-5}\text{ m/s})(1.60 \times 10^{-19}\text{C})}$$
$$= 8.8 \times 10^{28}\, e^-/\text{m}^3$$

14. Use Eq.4.37: $R = \rho l/A$ to calculate the resistances, with the area as $A = \pi r^2 = \pi d^2/4$.
$$R = \rho\dfrac{l}{A} = \rho\dfrac{4l}{\pi d^2}$$
$$\therefore \dfrac{R_{\text{Al}}}{R_{\text{Cu}}} = \dfrac{\rho_{\text{Al}}\dfrac{4l_{\text{Al}}}{\pi d_{\text{Al}}^2}}{\rho_{\text{Cu}}\dfrac{4l_{\text{Cu}}}{\pi d_{\text{Cu}}^2}} = \dfrac{\rho_{\text{Al}} l_{\text{Al}} d_{\text{Cu}}^2}{\rho_{\text{Cu}} l_{\text{Cu}} d_{\text{Al}}^2}$$
$$= \dfrac{(2.65 \times 10^{-8}\,\Omega\cdot\text{m})(10.0 \text{ m})(1.8 \text{ mm})^2}{(1.68 \times 10^{-8}\,\Omega\cdot\text{m})(20.0 \text{ m})(2.0 \text{ mm})^2} = 0.64$$

15. Use Eq.4.37: $R = \rho l/A$ to express the resistances, with the area as $A = \pi r^2 = \pi d^2/4$, and so $R = \rho\dfrac{l}{A} = \rho\dfrac{4l}{\pi d^2}$.
$$R_w = R_{\text{Cu}} \Rightarrow \rho_w \dfrac{4l}{\pi d_w^2} = \rho_{\text{Cu}}\dfrac{4l}{\pi d_{\text{Cu}}^2}$$
$$\Rightarrow R_w = R_{\text{Cu}} \Rightarrow \rho_w \dfrac{4l}{\pi d_w^2} = \rho_{\text{Cu}}\dfrac{4l}{\pi d_{\text{Cu}}^2}$$
$$\Rightarrow d_w = d_{\text{Cu}}\sqrt{\dfrac{\rho_w}{\rho_{\text{Cu}}}} = (2.2 \text{ mm})\sqrt{\dfrac{5.6 \times 10^{-8}\,\Omega\cdot\text{m}}{1.68 \times 10^{-8}\,\Omega\cdot\text{m}}}$$
$$= 4.0 \text{ mm}$$
The diameter of the tungsten should be 4.0 mm.

16. From Eq.4.40, we have
$$R = R_0\left[1 + \alpha(T - T_0)\right] = 1.15 R_0$$
$$\Rightarrow 1 + \alpha(T - T_0) = 1.15$$
$$\Rightarrow T - T_0 = \dfrac{0.15}{\alpha} = \dfrac{0.15}{0.0068\,(\text{C}°)^{-1}} = 22\text{C}°$$
So raise the temperature by 22C° to a final temperature of 42°C.

17. Since the resistance is directly proportional to the length, the length of the long piece must be 4.0 times the length of the short piece.
$$l = l_{\text{short}} + l_{\text{long}} = l_{\text{short}} + 4.0 l_{\text{short}} = 5.0 l_{\text{short}}$$
$$\Rightarrow l_{\text{short}} = 0.20 l, \quad l_{\text{long}} = 0.80 l$$
Make the cut at 20% of the length of the wire.
$$l_{\text{short}} = 0.20 l, \quad l_{\text{long}} = 0.80 l$$
$$\Rightarrow R_{\text{short}} = 0.2 R = 2.0\,\Omega, \quad R_{\text{long}} = 0.8 R = 8.0\,\Omega$$

18. Use Eq.4.33 for the resistivity.
$$\rho_{T\text{Al}} = \rho_{0\text{Al}}\left[1 + \alpha_{\text{Al}}(T - T_0)\right] = \rho_{0W}$$
$$\Rightarrow T = T_0 + \dfrac{1}{\alpha_{\text{Al}}}\left(\dfrac{\rho_{0W}}{\rho_{0\text{Al}}} - 1\right)$$
$$= 20°\text{C} + \dfrac{1}{0.00429\,(\text{C}°)^{-1}}\left(\dfrac{5.6 \times 10^{-8}\,\Omega\cdot\text{m}}{2.65 \times 10^{-8}\,\Omega\cdot\text{m}} - 1\right)$$
$$= 279.49°\text{C} \approx 280°\text{C}$$

19. From Eq.4.40, we have
$$R = R_0\left[1 + \alpha(T - T_0)\right]$$
$$\Rightarrow T = T_0 + \dfrac{1}{\alpha}\left(\dfrac{R}{R_0} - 1\right)$$
$$= 20°\text{C} + \dfrac{1}{0.0045\,(\text{C}°)^{-1}}\left(\dfrac{140\,\Omega}{12\,\Omega} - 1\right)$$
$$= 2390°\text{C} \approx 2400°\text{C}$$

20. The resistance depends on the length and area as $R = \rho l/A$. Cutting the wire and running the wires side by side will halve the length and double the area.
$$R_2 = \dfrac{\rho\left(\tfrac{1}{2}l\right)}{2A} = \dfrac{1}{4}\dfrac{\rho l}{A} = \dfrac{1}{4} R_1$$

21. The total resistance is to be 3700 ohms (R_{total}) at all temperatures. Write each resistance in terms of Eq.4.40:
$$R_T = R_0\left[1 + \alpha(T - T_0)\right] \quad \text{(with } T_0 = 0°\text{C)}$$
$$R_{\text{total}} = R_{0C}\left[1 + \alpha_C T\right] + R_{0\text{N}}\left[1 + \alpha_{\text{N}} T\right]$$
$$= R_{0C} + R_{0C}\alpha_C T + R_{0\text{N}} + R_{0\text{N}}\alpha_{\text{N}} T$$

$$= R_{0C} + R_{0N} + (R_{0C}\alpha_C + R_{0N}\alpha_N)T$$

For the above to be true, the terms with a temperature dependence must cancel, and the terms without a temperature dependence must add to R_{total}. Thus we have two equations in two unknowns.

$$0 = (R_{0C}\alpha_C + R_{0N}\alpha_N)T \Rightarrow R_{0N} = -\frac{R_{0C}\alpha_C}{\alpha_N}$$

$$R_{\text{total}} = R_{0C} + R_{0N} = R_{0C} - \frac{R_{0C}\alpha_C}{\alpha_N} = \frac{R_{0C}(\alpha_N - \alpha_C)}{\alpha_N}$$

$$\Rightarrow R_{0C} = R_{\text{total}} \frac{\alpha_N}{(\alpha_N - \alpha_C)}$$

$$= (3700\Omega)\frac{0.0004\,(C^\circ)^{-1}}{0.0004\,(C^\circ)^{-1} + 0.0005\,(C^\circ)^{-1}}$$

$$= 1644\Omega \approx 1600\Omega$$

$$\therefore R_{0N} = R_{\text{total}} - R_{0C} = 3700\Omega - 1644\Omega$$

$$= 2056\Omega \approx 2100\Omega$$

22. We choose a spherical shell of radius r and thickness dr as a differential element. The area of this element is $4\pi r^2$. Use Eq. 4.37, but for an infinitesimal resistance. Then integrate over the radius of the sphere.

$$R = \rho\frac{l}{A} \Rightarrow dR = \rho\frac{dl}{A} = \frac{dr}{4\pi\sigma r^2}$$

$$\Rightarrow R = \int dR = \int_{r_1}^{r_2} \frac{dr}{4\pi\sigma r^2} = \frac{1}{4\pi\sigma}\left(-\frac{1}{r}\right)_{r_1}^{r_2}$$

$$= \frac{1}{4\pi\sigma}\left(\frac{1}{r_1} - \frac{1}{r_2}\right)$$

23. (a) Let the values at the lower temperature be indicated by a subscript "0". Thus $R_0 = \rho_0\frac{l_0}{A_0} = \rho_0\frac{4l_0}{\pi d_0^2}$. The change in temperature results in new values for the resistivity, the length, and the diameter. Let α represent the temperature coefficient for the resistivity, and α_T represent the thermal coefficient of expansion, which will affect the length and diameter.

$$R = \rho\frac{l}{A} = \rho\frac{4l}{\pi d^2}$$

$$= \rho_0[1 + \alpha(T - T_0)]\frac{4l_0[1 + \alpha_T(T - T_0)]}{\pi\{d_0[1 + \alpha_T(T - T_0)]\}^2}$$

$$= \rho_0\frac{4l_0}{\pi d_0^2}\frac{[1 + \alpha(T - T_0)]}{[1 + \alpha_T(T - T_0)]}$$

$$= R_0\frac{[1 + \alpha(T - T_0)]}{[1 + \alpha_T(T - T_0)]}$$

$$\Rightarrow R[1 + \alpha_T(T - T_0)] = R_0[1 + \alpha(T - T_0)]$$

$$\Rightarrow T = T_0 + \frac{(R - R_0)}{(R_0\alpha - R\alpha_T)}$$

$$= 20^\circ\text{C} + \frac{(140\Omega - 12\Omega)}{[(12\Omega)(0.0045\text{C}^{0-1}) - (140\Omega)(5.5 \times 10^{-6}\text{C}^{\circ-1})]}$$

$$= 20^\circ\text{C} + 2405^\circ\text{C} = 2425^\circ\text{C} \approx 2400^\circ\text{C}$$

(b) The net effect of thermal expansion is that both the length and diameter increase, which lowers the resistance.

$$\frac{R}{R_0} = \frac{\rho_0\frac{4l}{\pi d^2}}{\rho_0\frac{4l_0}{\pi d_0^2}} = \frac{ld_0^2}{l_0 d^2}$$

$$= \frac{l_0[1 + \alpha_T(T - T_0)]}{l_0}\frac{d_0^2}{\{d_0[1 + \alpha_T(T - T_0)]\}^2}$$

$$= \frac{1}{[1 + \alpha_T(T - T_0)]}$$

$$= \frac{1}{[1 + (5.5 \times 10^{-6}\text{C}^{-1})(2405^\circ\text{C})]} = 0.9869$$

$$\%\text{ change} = \left(\frac{R - R_0}{R_0}\right)100 = \left(\frac{R}{R_0} - 1\right)100$$

$$= -1.31 \approx -1.3\%$$

The net effect of resistivity change is that the resistance increases.

$$\frac{R}{R_0} = \frac{\rho\frac{4l_0}{\pi d_0^2}}{\rho_0\frac{4l_0}{\pi d_0^2}}$$

$$= \frac{\rho}{\rho_0} = \frac{\rho_0[1 + \alpha(T - T_0)]}{\rho_0} = [1 + \alpha(T - T_0)]$$

$$= [1 + (0.0045\text{C}^{0-1})(2405^\circ\text{C})] = 11.82$$

$$\%\text{ change} = \left(\frac{R - R_0}{R_0}\right)100 = \left(\frac{R}{R_0} - 1\right)100$$

$$= 1082 \approx 1100\%$$

24. (a) Calculate each resistance separately using Eq.4.37, and then add the resistances together to find the total resistance.

$$R_{\text{Cu}} = \frac{\rho_{\text{Cu}}l}{A} = \frac{4\rho_{\text{Cu}}l}{\pi d^2}$$

$$= \frac{4(1.68 \times 10^{-8}\Omega \cdot \text{m})(5.0\text{ m})}{\pi(1.4 \times 10^{-3}\text{ m})^2} = 0.054567\Omega$$

Similarly, $R_{\text{Al}} = \frac{\rho_{\text{Al}}l}{A} = \frac{4\rho_{\text{Al}}l}{\pi d^2}$

$$= \frac{4(2.65 \times 10^{-8}\Omega \cdot \text{m})(5.0\text{ m})}{\pi(1.4 \times 10^{-3}\text{ m})^2} = 0.086074\Omega$$

$$\therefore R_{\text{total}} = R_{\text{Cu}} + R_{\text{Al}}$$
$$= 0.054567\Omega + 0.086074\Omega = 0.140641\Omega$$
$$\approx 0.14\,\Omega$$

(b) The current through the wire is the voltage divided by the total resistance.

$$I = \frac{V}{R_{\text{total}}} = \frac{85 \times 10^{-3}\text{ V}}{0.140641\Omega} = 0.60438\text{ A} \approx 0.60\text{ A}$$

(c) For each segment of wire, Ohm's law is true. Both wires have the current found in (b) above.

$$V_{\text{Cu}} = IR_{\text{Cu}} = (0.60438\text{ A})(0.054567\Omega) \approx 0.033\text{ V}$$
$$V_{\text{Al}} = IR_{\text{Al}} = (0.60438\text{ A})(0.086074\Omega) \approx 0.052\text{ V}$$

Notice that the total voltage is 85 mV.

25. (a) Divide the cylinder up into concentric cylindrical shells of radius r, thickness dr, and length l. See Fig.4.534. The resistance of one of those shells, from Eq.4.37, is found. Note that the "length" in Eq.4.37 is in the direction of the current flow, so we must substitute in dr for the "length" in Eq.4.37. The area is the surface area of the thin cylindrical shell. Then integrate over the range of radii to find the total resistance.

$$R = \rho\frac{l}{A} \Rightarrow dR = \rho\frac{dr}{2\pi rl};$$

$$R = \int dR = \int_{r_1}^{r_2} \rho\frac{dr}{2\pi rl} = \frac{\rho}{2\pi l}\ln\frac{r_2}{r_1}$$

(b) Substitute the given values to calculate the resistance from the above formula.

$$R = \frac{\rho}{2\pi l}\ln\frac{r_2}{r_1} = \frac{15 \times 10^{-5}\Omega \cdot \text{m}}{2\pi(0.024\text{ m})}\ln\left(\frac{1.8\text{ mm}}{1.0\text{ mm}}\right)$$

$$= 5.8 \times 10^{-4}\Omega$$

(c) For resistance along the axis, we again use Eq.4.37, but the current is flowing in the direction of length l. The area is the cross-sectional area of the face of the hollow cylinder.

$$R = \frac{\rho l}{A} = \frac{\rho l}{\pi(r_2^2 - r_1^2)}$$

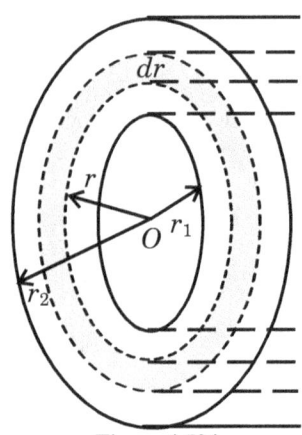

Figure 4.534

$$= \frac{(15 \times 10^{-5} \Omega \cdot m)(0.024 \text{ m})}{\pi \left[(1.8 \times 10^{-3} \text{ m})^2 - (1.0 \times 10^{-3} \text{ m})^2\right]}$$
$$= 0.51 \Omega$$

26. To find the kWh of energy, multiply the kilowatts of power consumption by the number of hours in operation.
Energy $= P$ (in kW) t (in h)
$$= (550 \times 10^{-3} \text{ kW}) \left(\frac{6.0}{60} \text{ h}\right) = 0.055 \text{ kWh}$$
To find the cost of the energy used in a month, multiply times 4 days per week of usage, times 4 weeks per month, times the cost per kWh.
$$\text{Cost} = \left(0.055 \frac{\text{kWh}}{\text{d}}\right)\left(\frac{4 \text{ d}}{1 \text{ week}}\right)\left(\frac{4 \text{ week}}{1 \text{ month}}\right)\left(\frac{9.0 \text{ cents}}{\text{kWh}}\right)$$
$$= 7.9 \text{ cents / month}$$

27. To find the cost of the energy, multiply the kilowatts of power consumption by the number of hours in operation times the cost per kWh.
$$\text{Cost} = (25 \text{ W})\left(\frac{1 \text{ kW}}{1000 \text{ W}}\right)(365 \text{ day})\left(\frac{24 \text{ h}}{1 \text{ day}}\right)\left(\frac{\$0.095}{\text{kWh}}\right) \approx 21$$

28. The $A \cdot h$ rating is the amount of charge that the battery can deliver. The potential energy of the charge is the charge times the voltage.
$$U = QV = (75 \text{ A} \cdot \text{h})\left(\frac{3600 \text{ s}}{1 \text{ h}}\right)(12 \text{ V})$$
$$= 3.2 \times 10^6 \text{ J} = 0.90 \text{ kWh}$$

29. (a) Calculate the resistance from Eq.4.36: $R = V/I$ and the power from Eq.4.58: $P = IV$.
$$R = \frac{V}{I} = \frac{3.0 \text{ V}}{0.38 \text{ A}} = 7.895 \Omega \approx 7.9 \text{ } \Omega$$
$$P = IV = (0.38 \text{ A})(3.0 \text{ V}) = 1.14 \text{ W} \approx 1.1 \text{ W}$$
(b) If four D-cells are used, the voltage will be doubled to 6.0 V. Assuming that the resistance of the bulb stays the same (by ignoring heating effects in the filament), the power that the bulb would need to dissipate is given by Eq.4.58: $P = V^2/R$. A doubling of the voltage means the power is increased by a factor of 4. This should not be tried because the bulb is probably not rated for such a high wattage. The filament in the bulb would probably burn out, and the glass bulb might even explode if the filament burns violently.

30. Find the current used to deliver the power in each case, and then find the power dissipated in the resistance at the given current.
$$P = IV \Rightarrow I = \frac{P}{V} \quad \therefore \quad P_{\text{dissipated}} = I^2 R = \frac{P^2}{V^2} R$$
$$P_{\text{dissipated} \atop 12,000 \text{ V}} = \frac{(7.5 \times 10^5 \text{ W})^2}{(1.2 \times 10^4 \text{ V})^2}(3.0 \Omega) = 11719 \text{ W}$$

$$P_{\text{dissipated} \atop 50,00 \text{ V}} = \frac{(7.5 \times 10^5 \text{ W})^2}{(5 \times 10^4 \text{ V})^2}(3.0 \Omega) = 675 \text{ W}$$
difference = $11719 \text{ W} - 675 \text{ W} = 1.1 \times 10^4 \text{ W}$

31. (a) By conservation of energy and the efficiency claim, 75% of the electrical power dissipated by the heater must be the rate at which energy is absorbed by the water.
$$I = \frac{mc\Delta T}{0.75 Vt}$$
$$= \frac{(0.120 \text{ kg})(4186 \text{ J/kg})(95°\text{C} - 25°\text{C})}{(0.75)(12 \text{ V})(480 \text{ s})}$$
$$= 8.139 \text{ A} \approx 8.1 \text{ A}$$
(b) Use Ohm's law to find the resistance of the heater.
$$V = IR \Rightarrow R = \frac{V}{I} = \frac{12 \text{ V}}{8.139 \text{ A}} = 1.5 \Omega$$

32. The water temperature rises by absorbing the heat energy that the electromagnet dissipates. Express both energies in terms of power, which is energy per unit time.
$$P_{\text{electric}} = P_{\text{to heat} \atop \text{water}} \Rightarrow IV = \frac{Q_{\text{heat water}}}{t} = \frac{mc\Delta T}{t}$$
$$\Rightarrow \frac{m}{t} = \frac{IV}{c\Delta T} = \frac{(17.5 \text{ A})(240 \text{ V})}{(4186 \text{ J/kg} \cdot °\text{C})(6.50 \text{C}°)}$$
$$= 0.154 \text{ kg/s} \approx 0.15 \text{ kg/s}$$
Since, density of water is 10^3kg/m^3, therefore volume of water passing per second
$$= \frac{0.154 \text{ kg/s}}{10^3 \text{kg/m}^3} = 0.154 \times 10^{-3} \text{ m}^3/\text{s}$$
$$= 0.154 \text{ L/s} = 154 \text{ mL/s}$$
Thus, required rate is 154 mL/s.

33. For the wire to stay at a constant temperature, the power generated in the resistor is to be dissipated by radiation. Use Eq.4.58: $P = I^2 R$ and Stefan-Boltzmann's radiation law: $P = \varepsilon \sigma A \left(T_{\text{high}}^4 - T_{\text{low}}^4\right)$, both expressions of power (energy per unit time). We assume that the dimensions requested and dimensions given are those at the higher temperature, and do not take any thermal expansion effects into account. We also use Eq.4.37: $R = \rho l / A$, for resistance.
$$I^2 R = \varepsilon \sigma A \left(T_{\text{high}}^4 - T_{\text{low}}^4\right)$$
$$\Rightarrow I^2 \frac{4\rho l}{\pi d^2} = \varepsilon \sigma \pi d l \left(T_{\text{high}}^4 - T_{\text{low}}^4\right)$$
$$\Rightarrow d = \left(\frac{4 I^2 \rho}{\pi^2 \varepsilon \sigma \left(T_{\text{high}}^4 - T_{\text{low}}^4\right)}\right)^{1/3}$$
$$= \left(\frac{4(15.0 \text{ A})^2 (5.6 \times 10^{-8} \Omega \cdot m)}{\pi^2 (1.0)(5.67 \times 10^{-8} \text{ W/m}^2 \cdot \text{K}^4)\left[(3100 \text{ K})^4 - (293 \text{ K})^4\right]}\right)^{1/3}$$
$$= 9.92 \times 10^{-5} \text{ m} \approx 0.099 \text{ mm}$$

34. The ampere-hour is a unit of charge.
$$(1.00 \text{ A} \cdot \text{h})\left(\frac{1 \text{ C/s}}{1 \text{ A}}\right)\left(\frac{3600 \text{ s}}{1 \text{ h}}\right) = 3600 \text{ C}$$

35. Use Eqs.4.37: $R = \rho l /A$ and 4.58: $P = V^2/R$
$$R = \rho \frac{l}{A} = \rho \frac{l}{\pi r^2} = \frac{4\rho l}{\pi d^2}; \quad P = \frac{V^2}{R} = \frac{V^2}{\frac{4\rho l}{\pi d^2}}$$
$$\Rightarrow l = \frac{V^2 \pi d^2}{4 \rho P} = \frac{(1.5 \text{ V})^2 \pi (5.0 \times 10^{-4} \text{ m})^2}{4(1.68 \times 10^{-8} \Omega \cdot m)(15 \text{ W})}$$
$$= 1.753 \text{ m} \approx 1.8 \text{ m}$$
If the voltage increases by a factor of 6 without the resistance changing, the power will increase by a factor of 36. The blanket would theoretically be able to deliver 540 W of power, which might make the material catch on fire or burn

36. Use Eq.4.58: $P = IV$ to calculate the current.
$$P = IV \Rightarrow I = \frac{P}{V} = \frac{746 \text{ W}}{120 \text{ V}} = 6.22 \text{ A}$$

37. (a) The resistance at the operating temperature can be calculated directly from Eq.4.58: $P = V^2/R$
$$P = \frac{V^2}{R} \Rightarrow R = \frac{V^2}{P} = \frac{(120 \text{ V})^2}{75 \text{ W}} = 190 \text{ }\Omega$$

(b) The resistance at room temperature is found by applying Eq.4.40: $R = R_0 [1 + \alpha (T - T_0)]$ and solving for R_0.
$$R = R_0 [1 + \alpha (T - T_0)]$$
$$R_0 = \frac{R}{[1 + \alpha (T - T_0)]}$$
$$= \frac{192 \Omega}{[1 + (0.0045 \text{ K}^{-1})(3000 \text{ K} - 293 \text{ K})]} = 15 \Omega$$

38. (a) The power delivered to the interior is 65% of the power drawn from the source.
$$P_{\text{interior}} = 0.65 P_{\text{source}}$$
$$\Rightarrow P_{\text{source}} = \frac{P_{\text{interior}}}{0.65} = \frac{950 \text{ W}}{0.65} = 1462 \text{ W}$$
$$\approx 1500 \text{ W}$$

(b) The current drawn is current from the source, and so the source power is used to calculate the current.
$$P_{\text{source}} = IV_{\text{source}}$$
$$\Rightarrow I = \frac{P_{\text{source}}}{V_{\text{source}}} = \frac{1462 \text{ W}}{120 \text{ V}} = 12.18 \text{ A} \approx 12 \text{ A}$$

39. The volume of wire is unchanged by the stretching. The volume is equal to the length of the wire times its cross-sectional area, and since the length was increased by a factor of 1.20, the area was decreased by a factor of 1.20. Use Eq.4.37: $R = \rho l/A$.
$$R_0 = \rho \frac{l_0}{A_0}$$
Given: $l = 1.20 l_0$, $A = \frac{A_0}{1.20}$, therefore, new resistance
$$R = \rho \frac{l}{A} = \rho \frac{1.20 l_0}{\frac{A_0}{1.20}} = (1.20)^2 \rho \frac{l_0}{A_0}$$
$$= 1.44 R_0 = 1.44 \Omega$$

40. (a) Use Eq.4.58: $P = V^2/R$
$$P = \frac{V^2}{R} \Rightarrow R = \frac{V^2}{P} = \frac{(240 \text{ V})^2}{2800 \text{ W}} = 20.57 \Omega \approx 21 \Omega$$

(b) Only 75% of the heat from the oven is used to heat the water. Use equation $Q = mc \Delta T$.
$0.75 (P_{\text{oven}}) t =$ Heat absorbed by water $= mc \Delta T$
$$\Rightarrow t = \frac{mc \Delta T}{0.75 (P_{\text{oven}})}$$
$$= \frac{(0.120 \text{ L}) \left(\frac{1 \text{ kg}}{1 \text{ L}}\right) (4186 \text{ J/kg} \cdot \text{C}^\circ)(85 \text{C}^\circ)}{0.75(2800 \text{ W})}$$
$$= 20.33 \text{ s} \approx 20 \text{ s (2 significant figures)}$$

(c) $\frac{11 \text{ cents}}{\text{kWh}} (2.8 \text{ kW})(20.33 \text{ s}) \frac{1 \text{ h}}{3600 \text{ s}} = 0.17 \text{ cents}$

41. The mass of the wire is the density of copper times the volume of the wire, and the resistance of the wire is given by Eq.4.37: $R = \rho \frac{l}{A}$. We represent the mass density by ρ_m and the resistivity by ρ.
$$R = \rho \frac{l}{A} \Rightarrow A = \frac{\rho l}{R}, \quad m = \rho_m l A = \rho_m l \frac{\rho l}{R}$$
$$\Rightarrow l = \sqrt{\frac{mR}{\rho_m \rho}} = \sqrt{\frac{(0.0155 \text{ kg})(12.5 \Omega)}{(8.9 \times 10^3 \text{ kg/m}^3)(1.68 \times 10^{-8} \Omega \cdot \text{m})}}$$
$$= 35.997 \text{ m} \approx 36.0 \text{ m}$$

Now, $A = \frac{\rho l}{R} = \pi \left(\frac{1}{2} d\right)^2$
$$\Rightarrow d = \sqrt{\frac{4 \rho l}{\pi R}} = \sqrt{\frac{4(1.68 \times 10^{-8} \Omega \cdot \text{m})(35.997 \text{ m})}{\pi(12.5 \Omega)}}$$
$$= 2.48 \times 10^{-4} \text{ m}$$

42. The resistance can be calculated from the power and voltage, and then the diameter of the wire can be calculated from the resistance.
$$P = \frac{V^2}{R} \Rightarrow R = \frac{V^2}{P}, \quad R = \frac{\rho L}{A} = \frac{\rho L}{\pi \left(\frac{1}{2} d\right)^2}$$
$$\Rightarrow \frac{V^2}{P} = \frac{\rho L}{\pi \left(\frac{1}{2} d\right)^2}$$
$$\Rightarrow d = \sqrt{\frac{4 \rho L P}{\pi V^2}} = \sqrt{\frac{4(100 \times 10^{-8} \Omega \cdot \text{m})(3.8 \text{ m})(95 \text{ W})}{\pi (120 \text{ V})^2}}$$
$$= 1.787 \times 10^{-4} \text{ m} \approx 1.8 \times 10^{-4} \text{ m}$$

43. Use Eq.4.58: $P = V^2/R$
(a) $P = \frac{V^2}{R} = \frac{(120 \text{ V})^2}{12 \Omega} = 1200 \text{ W}$
(b) $P = \frac{V^2}{R} = \frac{(120 \text{ V})^2}{140 \Omega} = 103 \text{ W}$

44. (a) The current can be found from Eq.4.58: $P = IV$
$I_A = P_A/V_A = 40 \text{ W}/120 \text{ V} = 0.33 \text{ A}$,
$I_B = P_B/V_B = 40 \text{ W}/12 \text{ V} = 3.3 \text{ A}$
(b) The resistance can be found from Eq.4.58: $P = V^2/R$:
$$R = \frac{V^2}{P} \Rightarrow R_A = \frac{V_A^2}{P_A} = \frac{(120 \text{ V})^2}{40 \text{ W}} = 360 \text{ }\Omega$$
and $R_B = \frac{V_B^2}{P_B} = \frac{(12 \text{ V})^2}{40 \text{ W}} = 3.6 \text{ }\Omega$
(c) The charge is the current times the time, i.e.,
$Q = It \Rightarrow Q_A = I_A t = (0.33 \text{ A})(3600 \text{ s}) = 1200 \text{ C}$
and $Q_B = I_B t = (3.3 \text{ A})(3600 \text{ s}) = 12,000 \text{ C}$
(d) The energy is the power times the time, and the power is the same for both bulbs.
$E = Pt \Rightarrow E_A = E_B = (40 \text{ W})(3600 \text{ s}) = 1.4 \times 10^5 \text{ J}$
(e) Bulb B requires a larger current, and so should have larger diameter connecting wires to avoid overheating the connecting wires.

45. Eq. 4.37: $R = \rho l/A$ can be used. The area to be used is the cross-sectional area of the pipe.
$$R = \frac{\rho l}{A} = \frac{\rho l}{\pi \left(r_{\text{outside}}^2 - r_{\text{inside}}^2\right)}$$
$$= \frac{(1.68 \times 10^{-8} \Omega \cdot \text{m})(10.0 \text{ m})}{\pi \left[(2.50 \times 10^{-2} \text{ m})^2 - (1.50 \times 10^{-2} \text{ m})^2\right]}$$
$$= 1.34 \times 10^{-4} \Omega$$

46. We assume that all of the current that enters at a leaves at b, so that the current is the same at each end. The current density is given by Eq.4.13
$$J_a = \frac{I}{A_a} = \frac{I}{\pi \left(\frac{1}{2} a\right)^2} = \frac{4I}{\pi a^2}$$
$$= \frac{4(2.0 \text{ A})}{\pi (2.5 \times 10^{-3} \text{ m})^2} = 4.1 \times 10^5 \text{ A/m}^2$$
$$J_b = \frac{I}{A_b} = \frac{I}{\pi \left(\frac{1}{2} b\right)^2} = \frac{4I}{\pi b^2}$$

$$= \frac{4(2.0 \text{ A})}{\pi (4.0 \times 10^{-3} \text{ m})^2} = 1.6 \times 10^5 \text{ A/m}^2$$

47. Using Eq.4.37: $R = \rho l/A$, we find the infinitesimal resistance first of a thin vertical slice at a horizontal distance x from the centre of the left side towards the centre of the right side (Fig.4.535). Let the radius of this slice is y and it's thickness is dx. This thickness corresponds to the variable l in Eq.4.37.

In Fig.4.535, $\triangle ATP$ and $\triangle ACB$ are similar, therefore
$$\frac{TP}{AT} = \frac{CB}{AC} \Rightarrow \frac{y - \frac{a}{2}}{x} = \frac{b-a}{2l}$$
$$\Rightarrow \quad y = \frac{1}{2}\left[a + \left(\frac{b-a}{l}\right)x\right]$$

Therefore, area of this slice is,
$$A = \pi y^2 = \frac{\pi}{4}\left[a + \left(\frac{b-a}{l}\right)x\right]^2$$

Integrate over all the slices to find the total resistance.
$$R = \rho \frac{l}{A} \Rightarrow dR = \rho \frac{dx}{\frac{\pi}{4}\left[a + \frac{x}{l}(b-a)\right]^2}$$

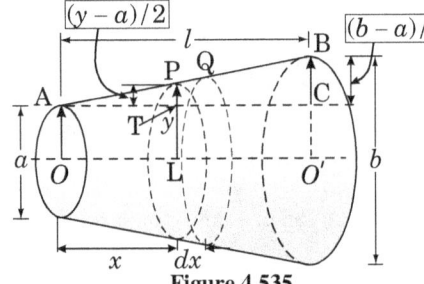

Figure 4.535

$$\Rightarrow R = \int dR = \int_0^l \rho \frac{dx}{\frac{\pi}{4}\left[a + \frac{x}{l}(b-a)\right]^2}$$
$$= -\frac{4\rho}{\pi} \frac{l}{b-a} \frac{1}{\left[a + \frac{x}{l}(b-a)\right]} = \frac{4\rho}{\pi} \frac{l}{ab}$$

48. When the tank is empty, the entire length of the wire is in a non-superconducting state, and so has a non-zero resistivity, which we call ρ. Then the resistance of the wire when the tank is empty is given by $R_0 = \rho \frac{l}{A} = \frac{V_0}{I}$. When a length x of the wire is superconducting, that portion of the wire has 0 resistance. Then the resistance of the wire is only due to the length $l - x$, and so
$$R = \rho \frac{l-x}{A} = \rho \frac{l}{A} \frac{l-x}{\ell} = R_0 \frac{l-x}{l}.$$
This resistance, combined with the constant current, gives $V = IR$
$$V = IR = \left(\frac{V_0}{R_0}\right) R_0 \frac{l-x}{l} = V_0\left(1 - \frac{x}{l}\right) = V_0(1-f)$$
$$\Rightarrow \quad f = 1 - \frac{V}{V_0}$$

Thus a measurement of the voltage can give the fraction of the tank that is filled with liquid helium.

49. For the temperature of the gas to remain unchanged, the rate of the thermal energy dissipated through the resistor, $P_R = I^2R$, must be equal to the rate of increase of mechanical energy of the piston, $P_m = mg(dh/dt) = mgv$. Thus,
$$I^2R = mgv \Rightarrow v = \frac{I^2R}{mg} = \frac{(0.240 \text{ A})^2(550\Omega)}{(12 \text{ kg})(9.8 \text{ m/s}^2)} = 0.27 \text{ m/s}$$

50. (a) The total current density is equal to the sum of the contributions from the alpha particles and the electron. Using the general expression $J = nqv$, and noting that $n_e = 2n_\alpha$ (two electrons for each α particle), we have
$$J_{\text{total}} = n_\alpha q_\alpha v_\alpha + n_e q_e v_e = n_\alpha(2e)v_\alpha + (2n_\alpha)(e)v_e$$
$$= 2n_\alpha e(v_\alpha + v_e)$$
$$= 2(2.80 \times 10^{21}/\text{m}^3)(1.6 \times 10^{-19}\text{C})(88 \text{ m/s} + 25 \text{ m/s})$$
$$= 1.01 \times 10^5 \text{ A/m}^2 = 10.1 \text{ A/cm}^2$$

(b) The direction of the current is eastward (same as the motion of the alpha particles).

51. (a) Using $I = dq/dt = e(dN/dt)$, we obtain
$$\frac{dN}{dt} = \frac{I}{e} = \frac{15 \times 10^{-6} \text{ A}}{1.6 \times 10^{-19}\text{C}} = 9.4 \times 10^{13}/\text{s}$$

(b) The rate of thermal energy production is
$$P = \frac{dU}{dt} = \left(\frac{dN}{dt}\right) U_1$$
$$= (9.4 \times 10^{13}/\text{s})(16\text{MeV})\left(\frac{1.6 \times 10^{-13} \text{ J}}{1\text{MeV}}\right) = 240 \text{ W}$$

52. (a) The mass of the water is
$$m = \rho V = (1000 \text{ kg/m}^3)(2.0 \text{ L})(10^{-3} \text{ m}^3/\text{L}) = 2.00 \text{ kg}$$
The energy required to raise the water temperature to the boiling point is
$$Q_1 = mc\Delta T = (2.00 \text{ kg})(4187 \text{ J/kg} \cdot \text{C}°)(100°\text{C} - 20°\text{C})$$
$$= 6.70 \times 10^5 \text{ J}$$
With $P = 400$ W at 80% efficiency, we find the time needed to be
$$\Delta t_1 = \frac{Q_1}{P_{\text{eff}}} = \frac{6.70 \times 10^5 \text{ J}}{(0.80)(400 \text{ W})} = 2.09 \times 10^3 \text{ s} \approx 35 \text{ min}$$

(b) The energy required to vaporize half of the water is
$$Q_2 = L_V(m/2)$$
$$= (2.256 \times 10^6 \text{ J/kg})(2.00 \text{ kg}/2) = 2.256 \times 10^6 \text{ J}$$
Thus, the additional time elapsed is
$$\Delta t_2 = \frac{Q_2}{P_{\text{eff}}} = \frac{2.256 \times 10^6 \text{ J}}{(0.80)(400 \text{ W})} = 7.05 \times 10^3 \text{ s} \approx 118 \text{ min}$$
or about 1.96 h.

53. (a) At $t = 0.500$ s, the charge on the capacitor is
$$q = CV = C\left(6.00 + 4.00t - 2.00t^2\right)$$
$$= (30 \times 10^{-6} \text{ F})\left[6.00 + 4.00(0.500) - 2.00(0.500)^2\right]$$
$$= 225 \times 10^{-6}\text{C} = 225\mu\text{C}$$

(b) The current flowing into the capacitor is
$$I = \frac{dq}{dt} = C\frac{dV}{dt} = C\frac{d}{dt}\left(6.00 + 4.00t - 2.00t^2\right)$$
$$= C(4.00 - 4.00t)$$
$$= (30 \times 10^{-6} \text{ F})[4.00 - 4.00(0.500)]$$
$$= 60.0 \times 10^{-6} \text{ A} = 60.0\mu\text{A}$$

(c) The corresponding power output is
$$P = IV$$
$$= (60.0 \times 10^{-6} \text{ A})\left[6.00 + 4.00(0.500) - 2.00(0.500)^2\right]$$
$$= 4.50 \times 10^{-4} \text{ W}$$

54. Start at the point when the voltage has just reached $\frac{2}{3}V$ and the switch has just closed. The voltage is $\frac{2}{3}V$ and is decaying towards $0V$ with a time constant R_2C
$$V_C(t) = \left[\frac{2}{3}V\right] e^{-t/R_2C}$$
After the switch opens, the voltage is $\frac{1}{3}V$, increasing toward V with time constant $(R_1 + R_2)C$:
$$V_C(t) = V - \left[\frac{2}{3}V\right] e^{-t/(R_1+R_2)C}$$
When $V_C(t) = \frac{2}{3}V$, then
$$\frac{2}{3}V = V - \frac{2}{3}Ve^{-t/(R_1+R_2)c}$$
$$\Rightarrow \quad e^{-t/(R_1+R_2)C} = \frac{1}{2}$$
$$\Rightarrow \quad t_2 = (R_1 + R_2)C \ln 2$$

Therefore, $T = t_1 + t_2 = (R_1 + 2R_2) C \ln 2$

55. **(a)** With the switch closed, current exists in a simple series circuit as shown. The capacitors carry no current. For R_2 we have

$$P = I^2 R_2 \implies I = \sqrt{\frac{P}{R_2}} = \sqrt{\frac{2.40 \text{ V·A}}{7000 \text{ V/A}}} = 18.5 \text{ mA}$$

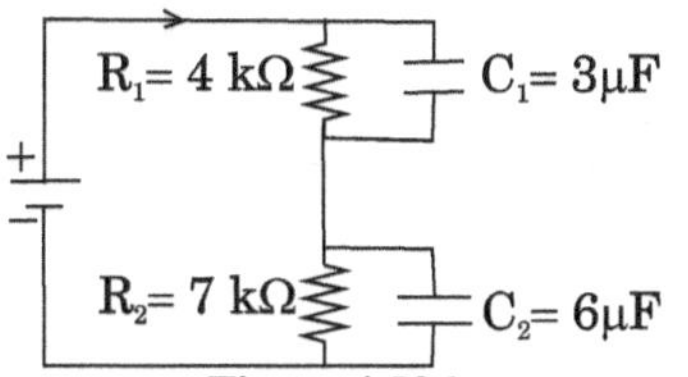

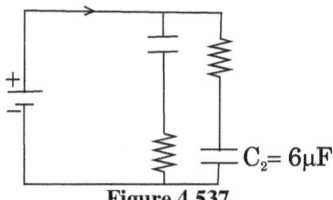

Figure 4.536

The potential difference across R_1 and C_1 is
$V_1 = IR_1 = (1.85 \times 10^{-2} \text{ A})(4000 \text{ V/A}) = 74.1$ V
The charge on C_1
$Q = C_1 V_1 = (3.00 \times 10^{-6} \text{ C/V})(74.1 \text{ V}) = 222 \ \mu$C.
The potential difference across R_2 and C_2 is
$V_2 = IR_2 = (1.85 \times 10^{-2} \text{ A})(7000 \Omega) = 130$ V.
The charge on C_2
$Q = C_2 V = (6.00 \times 10^{-6} \text{ C/V})(130 \text{ V}) = 778 \ \mu$C.
The battery emf is
$IR_{eq} = I(R_1 + R_2)$
$= (1.85 \times 10^{-2} \text{ A})(4000 + 7000) \text{ V/A} = 204$ V.

(b) In equilibrium after the switch has been opened, no current exists. The potential difference across each resistor is zero. The full 204 V appears across both capacitors. The new charge on C_2
$Q = C_2 V = (6.00 \times 10^{-6} \text{ C/V})(204 \text{ V}) = 1222 \ \mu$C

Figure 4.537

for a change of $1222 \ \mu\text{C} - 778 \ \mu\text{C} = 444 \ \mu$C.

56. **APPROACH** First reconstruct the circuit so that it becomes a simple RC circuit containing a single resistor and single capacitor in series, connected to the battery, and then determine the total charge q stored in the equivalent circuit.
The total resistance between points b and c is:
$$R = \frac{(2.00 \text{ k}\Omega)(3.00 \text{ k}\Omega)}{2.00 \text{ k}\Omega + 3.00 \text{ k}\Omega} = 1.20 \text{ k}\Omega.$$
The total capacitance between points d and e is:
$C = 2.00 \ \mu\text{F} + 3.00 \ \mu\text{F} = 5.00 \ \mu$F.
The potential difference between point d and e in this series RC circuit at any time is:
$V = \mathcal{E}\left[1 - e^{-t/RC}\right] = (120.0 \text{ V})\left[1 - e^{-1000t/6}\right]$.
Therefore, the charge on each capacitor between points d and e is:
$q_1 = C_1 V = (2.00 \ \mu\text{F})(120.0 \text{ V})\left[1 - e^{-1000t/6}\right]$
$= (240 \ \mu\text{C})\left[1 - e^{-1000t/6}\right]$
and $q_2 = C_2 V = (3.00 \ \mu\text{F})(120.0 \text{ V})\left[1 - e^{-1000t/6}\right]$
$= (360 \ \mu\text{C})\left[1 - e^{-1000t/6}\right]$

4.35.15 Multiple Choice Assignments
4.35.16 Level 1

Q.No.	1	2	3	4	5	6	7	8	9
Ans.	C	A	C	D	A	C	B	B	C
Q.No.	10	11	12	13	14	15	16	17	18
Ans.	C	D	D	B	D	A	C	B	D
Q.No.	19	20	21	22	23	24	25	26	27
Ans.	A	B	A	B	D	A	A	C	C
Q.No.	28	29	30	31	32	33	34	35	36
Ans.	C	C	B	C	B	A	B	A	A
Q.No.	37	38	39	40	41	42	43	44	45
Ans.	C	D	A	B	B	B	B	B	B
Q.No.	46	47	48	49	50	51	52	53	54
Ans.	D	A	C	C	A	B	C	A	C
Q.No.	55	56	57	58	59	60	61	62	63
Ans.	A	A	C	C	D	C	B	A	B
Q.No.	64	65	66	67	68	69	70	71	72
Ans.	C	C	C	C	B	A	D	D	C
Q.No.	73	74	75	76	77	78	79	80	81
Ans.	A	C							

4.35.17 Level 2

Q.No.	1	2	3	4	5	6	7	8	9
Ans.	B	B	C	C	B	B	C	D	C
Q.No.	10	11	12	13	14	15	16	17	18
Ans.	C	C	C	B	A	D	C	C	A
Q.No.	19	20	21	22	23	24	25	26	27
Ans.	C	A	A	B	A	C	D	C	B
Q.No.	28	29	30	31	32	33	34	35	36
Ans.	A	A	C	C	D	D	B	C	D
Q.No.	37	38	39	40	41	42	43	44	45
Ans.	B	A	D	A	C	B	C	B	D
Q.No.	46	47	48	49	50	51	52	53	54
Ans.	D	C	C	C	C	D	C	B	B
Q.No.	55	56	57	58	59	60	61	62	63
Ans.	AB	C							

4.35.18 Level 3

Q.No.	1	2	3	4	5	6	7	8	9
Ans.	D	D	B	A	C	B	A	B	D
Q.No.	10	11	12	13	14	15	16	17	18
Ans.	A	B	B	A	D	C	C	C	C
Q.No.	19	20	21	22	23	24	25	26	27
Ans.	C	C	C	B	C	D	A	A	B
Q.No.	28	29	30	31	32	33	34	35	36
Ans.	C	B	C	C	C	A	D	B	D
Q.No.	37	38	39	40	41	42	43	44	45
Ans.	A	C	A	B	A	C	B	D	D
Q.No.	46	47	48	49	50	51	52	53	54
Ans.	D	A	A	C	D				

4.35.19 Level 4
Section A

Q.No.	1	2	3	4	5	6	7	8	9
Ans.	C	B	C	B	A	C	A	B	C
Q.No.	10	11	12	13	14	15	16	17	18
Ans.	C	D	B	B	B	B	B	D	A
Q.No.	19	20	21	22	23	24	25	26	27
Ans.	A	C	B	B	B	D	D	B	A
Q.No.	28	29	30	31	32	33	34	35	36
Ans.	C	A	D	D	C	CD	B	A	D
Q.No.	37	38	39	40	41	42	43	44	45

Ans.	C	B	B	B	C	A	C	D	D
Q.No.	46	47	48	49	50	51	52	53	54
Ans.	B	A	D	B	C	A	D	D	D
Q.No.	55	56	57	58	59	60	61	62	63
Ans.	C	C	A	C	D	B	B	B	A
Q.No.	64	65	66	67	68	69	70	71	72
Ans.	C	B	D	C	A	A	A	D	B
Q.No.	73	74	75	76	77	78	79	80	81
Ans.	C	D	B	B	C	C	A	D	B
Q.No.	82	83	84	85	86	87	88	89	90
Ans.	C	A	C	C	B	B	C	A	C
Q.No.	91	92	93	94	95	96	97	98	99
Ans.	D	C	A	B	D	D	D	B	A
Q.No.	100	101	102	103	104	105	106	107	108
Ans.	D	A	D	A	B	A	B	B	A
Q.No.	109	110	111	112	113	114	115	116	117
Ans.	2	C	A	100	A	C	B	C	B
Q.No.	118	119	120	121	122	123	124	125	126
Ans.	D	A	A	A	B	A	D	B	B
Q.No.	127	128	129	130	131	132	133	134	135
Ans.	B	C	B	D	B	16	D	A	C
Q.No.	136	137	138	139	140	141	142	143	144
Ans.	19	10	8	15	144	10	D	A	8
Q.No.	145	145	147	148	149	150	151	152	153
Ans.	D	2	A	44	B	10	3	1	B
Q.No.	154	155	156	157	158	159	160	161	162
Ans.	B	30	5	D	1.5	A	2	A	5
Q.No.	163	164	165	166	167	168	169	170	171
Ans.	4	300	1	D	A	2	C	B	25
Q.No.	172	173	174	175	176	177	178	179	180
Ans.	A	25	A	125	100	B	3	A	1
Q.No.	181	182	183	184	185	186	187	188	189
Ans.	D	5	A	15	D	50	A	75	B
Q.No.	190	191	192	193	194	195	196	197	198
Ans.	31								

Section B

Q.No.	1	2	3	4	5	6
Ans.	B	B	B	A	A	C
Q.No.	7	8	9	10	11	12
Ans.	A	A	B	D	D	B
Q.No.	13	14	15	16	17	18
Ans.	A	C	C	A	D	A
Q.No.	19	20	21	22	23	24
Ans.	A	A	B	C	D	AD
Q.No.	25	26	27	28	29	30
Ans.	C	C	B	ABCD	B	ABD
Q.No.	31	32	33	34	35	36
Ans.	BD	C	D	CD	BC	ABCD
Q.No.	37	38	39	40	41	42
Ans.	A	A	A	1.33	0.67	(3)
Q.No.	43	44	45	46	47	48
Ans.	1.33	0.67	ABCD	(5)	ABC	ABCD
Q.No.	49	50	51	52	53	54
Ans.	ABCD					

Integer Answer Type

Q.No.	1	2	3	4	5	6	7	8	9
Ans.	4	2	5	5	5.55				

www.ingramcontent.com/pod-product-compliance
Lightning Source LLC
Chambersburg PA
CBHW081057290526
45795CB00006B/1892